Cold Pressed Oils

Cold Pressed Oils

Green Technology, Bioactive Compounds, Functionality, and Applications

Edited by

Mohamed Fawzy Ramadan
Professor of Biochemistry,
Zagazig University,
Zagazig, Egypt

Academic Press is an imprint of Elsevier
125 London Wall, London EC2Y 5AS, United Kingdom
525 B Street, Suite 1650, San Diego, CA 92101, United States
50 Hampshire Street, 5th Floor, Cambridge, MA 02139, United States
The Boulevard, Langford Lane, Kidlington, Oxford OX5 1GB, United Kingdom

Copyright © 2020 Elsevier Inc. All rights reserved.

No part of this publication may be reproduced or transmitted in any form or by any means, electronic or mechanical, including photocopying, recording, or any information storage and retrieval system, without permission in writing from the publisher. Details on how to seek permission, further information about the Publisher's permissions policies and our arrangements with organizations such as the Copyright Clearance Center and the Copyright Licensing Agency, can be found at our website: www.elsevier.com/permissions.

This book and the individual contributions contained in it are protected under copyright by the Publisher (other than as may be noted herein).

Notices
Knowledge and best practice in this field are constantly changing. As new research and experience broaden our understanding, changes in research methods, professional practices, or medical treatment may become necessary.

Practitioners and researchers must always rely on their own experience and knowledge in evaluating and using any information, methods, compounds, or experiments described herein. In using such information or methods they should be mindful of their own safety and the safety of others, including parties for whom they have a professional responsibility.

To the fullest extent of the law, neither the Publisher nor the authors, contributors, or editors, assume any liability for any injury and/or damage to persons or property as a matter of products liability, negligence or otherwise, or from any use or operation of any methods, products, instructions, or ideas contained in the material herein.

Library of Congress Cataloging-in-Publication Data
A catalog record for this book is available from the Library of Congress

British Library Cataloguing-in-Publication Data
A catalogue record for this book is available from the British Library

ISBN: 978-0-12-818188-1

For information on all Academic Press publications
visit our website at https://www.elsevier.com/books-and-journals

Publisher: Charlotte Cockle
Acquisitions Editor: Nina Rosa de Araujo Bandeira
Editorial Project Manager: Lena Sparks
Production Project Manager: Bharatwaj Varatharajan
Cover Designer: Matthew Limbert

Typeset by SPi Global, India

Dedication

Dedicated to my beloved family.

Contents

Contributors

Numbers in parenthesis indicate the pages on which the authors' contributions begin.

Ali Abbas (491), Department of Chemistry, Government Postgraduate Taleem-ul-Islam College, Chiniot, Punjab, Pakistan

Mariam Abdur-Rahman (267), Chemistry Department, Faculty of Science, Cairo University, Giza, Egypt

Hussein H. Abulreesh (683), Department of Biology, Faculty of Applied Science, Umm Al-Qura University, Makkah, Saudi Arabia

Anjana Adhikari-Devkota (273), Graduate School of Pharmaceutical Sciences, Kumamoto University, Kumamoto, Japan

Muhammad Haseeb Ahmad (105), Institute of Home and Food Sciences, Faculty of Life Sciences, Government College University, Faisalabad, Pakistan

Naveed Ahmad (309), Department of Chemistry, University of Education, Lahore, Faisalabad Campus, Faisalabad, Pakistan

Nazir Ahmad (105,191), Institute of Home and Food Sciences, Faculty of Life Sciences, Government College University, Faisalabad, Pakistan

Akeem Omolaja Akinfenwa (181), Chemistry Department, Faculty of Applied Sciences, Cape Peninsula University of Technology, Bellville, South Africa

Sumia Akram (345), Division of Science and Technology, University of Education Lahore, Lahore, Pakistan

Milica Fotirić Akšić (637), University of Belgrade—Faculty of Agriculture, Belgrade-Zemun, Serbia

Mehmet Aksu (335,525), Agriculture and Rural Development Support Institution, Isparta Provincial Coordination Unit, Isparta, Turkey

Ambrogina Albergamo (159), Department of Biomedical, Dental, Morphological and Functional Imaging Sciences (BIOMORF), University of Messina; Science4Life s.r.l., spin-off of the University of Messina, Messina, Italy

Muhammad Ali (105), Institute of Home and Food Sciences, Faculty of Life Sciences, Government College University, Faisalabad, Pakistan

Farooq Anwar (309,315,491,725), Department of Chemistry, University of Sargodha, Sargodha, Pakistan

Zeliha Ustun Argon (39,241,449,597), Department of Biosystems Engineering, Eregli Faculty of Engineering and Natural Sciences; Medical and Cosmetic Plants Application and Research Center, Necmettin Erbakan University, Konya, Turkey

Muhammad Sajid Arshad (105), Institute of Home and Food Sciences, Faculty of Life Sciences, Government College University, Faisalabad, Pakistan

Nosheen Asghar (191), Department of Food Science, Nutrition and Home Economics, Government College University, Faisalabad, Pakistan

Rizwan Ashraf (345), Department of Chemistry, University of Agriculture, Faisalabad, Pakistan

Abdelrahaman Assaeedi (683), Department of Biology, Faculty of Applied Science, Umm Al-Qura University, Makkah, Saudi Arabia

Adel M.A. Assiri (683), Biochemistry Department, Faculty of Medicine, Umm Al-Qura University, Makkah, Saudi Arabia

Buket Aydeniz-Guneser (129,497), Department of Food Engineering, Engineering Faculty, Usak University, Usak, Turkey

Maria Francesca Belcaro (459), DiSIA-PHYTOLAB (Pharmaceutical, Cosmetic, Food supplement Technology and Analysis), University of Florence, Florence, Italy

Roberta Bernini (97), Department of Agriculture and Forest Science (DAFNE), University of Tuscia, Viterbo, Italy

Muhammad Iqbal Bhanger (725), H.E.J. Research Institute of Chemistry, International Centre for Chemical and Biological Sciences, University of Karachi, Karachi, Pakistan

Ronnie Böck (467), Department of Biological Sciences, University of Namibia, Windhoek, Namibia

Sandra Bulut (15), Faculty of Technology Novi Sad, University of Novi Sad, Novi Sad, Serbia

Veysel Umut Celenk (39,241,449,597), Drug Research and Pharmacokinetic Development and Applied Center, ARGEFAR; Central Research Testing and Analysis Laboratory Research and Application Center, EGE-MATAL, Ege University, İzmir, Turkey

Tossaton Charoonratana (391), Department of Pharmacognosy, College of Pharmacy, Rangsit University, Muang Pathum Thani, Thailand

Ahmad Cheikhyoussef (181,277,289,467,477), Science and Technology Division, Multidisciplinary Research Centre, University of Namibia; Ministry of Higher Education, Training and Innovation, Windhoek, Namibia

Natascha Cheikhyoussef (181,277,289,467,477), Ministry of Higher Education, Training and Innovation, Windhoek, Namibia

Sook Chin Chew (65), School of Foundation Studies, Xiamen University Malaysia Campus, Sepang, Selangor, Malaysia

Moncef Chouaibi (373,625,611), Food Preservation Laboratory, High Institute of Food Industry; Research Unit: 'Bio-Preservation and Valorization of Agricultural Products UR13-AGR 02', University of Carthage, High Institute of Food Industries of Tunisia, Tunis, Tunisia

Monika Choudhary (659), Punjab Agricultural University, Ludhiana, India

Ivanka Ćirić (637), Innovation Centre of Faculty of Chemistry Ltd, Belgrade, Serbia

Rosaria Costa (159), Department of Biomedical, Dental, Morphological and Functional Imaging Sciences (BIOMORF), University of Messina, Messina, Italy

Valeria da Silva Santos (405), Independent Consultant on Fats and Oils Technology, Pelotas, Rio Grande do Sul, Brazil

Khaled Ben Daoued (373), Food Preservation Laboratory, High Institute of Food Industry, Tunis, Tunisia

Hari Prasad Devkota (273), Graduate School of Pharmaceutical Sciences, Kumamoto University, Kumamoto, Japan

Giacomo Dugo (159), Department of Biomedical, Dental, Morphological and Functional Imaging Sciences (BIOMORF), University of Messina; Science4Life s.r.l., spin-off of the University of Messina, Messina, Italy

Alessandra Durazzo (97,459), CREA-Research Centre for Food and Nutrition, Rome, Italy

Peter Eck (81), Department of Food & Human Nutritional Sciences, University of Manitoba, Winnipeg, MB, Canada

Khaled Elbanna (683), Department of Biology, Faculty of Applied Science, Umm Al-Qura University, Makkah, Saudi Arabia; Department of Agricultural Microbiology, Faculty of Agriculture, Fayoum University, Fayoum, Egypt

Rafaat M. Elsanhoty (683), Department of Industrial Biotechnology, Institute of Genetic Engineering and Biotechnology, Sadat City University, Sadat City, Egypt

Sayed A. El-Toumy (711), Chemistry of Tannins Department, National Research Centre, Cairo, Egypt

Pelin Günç Ergönül (291,231,255,323,575), Department of Food Engineering, Faculty of Engineering, Manisa Celal Bayar University, Manisa, Turkey

N.A. Michael Eskin (81), Department of Food & Human Nutritional Sciences, University of Manitoba, Winnipeg, MB, Canada

Gabriel Deschamps Fernandes (405), Independent Consultant on Fats and Oils Technology, Pelotas, Rio Grande do Sul, Brazil

Nesrine Gaout (373), Food Preservation Laboratory, High Institute of Food Industry, Tunis, Tunisia

Saba Ghufran (345), Department of Chemistry, University of Agriculture, Faisalabad, Pakistan

Zinar Pinar Gumus (39,241,449,597), Central Research Testing and Analysis Laboratory Research and Application Center, EGE-MATAL, Ege University, İzmir, Turkey

Onur Guneser (497), Department of Food Engineering, Engineering Faculty, Usak University, Usak, Turkey

Salem Hamdi (373,611,625), Food Preservation Laboratory, High Institute of Food Industry; Food Conservation and Valorization Laboratory, High Institute of Food Industries of Tunisia, Tunis, Tunisia

Alan-Javier Hernández-Álvarez (113,665), School of Food Science & Nutrition, University of Leeds, Leeds, United Kingdom.

Nevena Hromiš (15), Faculty of Technology Novi Sad University of Novi Sad, Novi Sad, Serbia

Ahmed A. Hussein (181,703,711), Chemistry Department, Faculty of Applied Sciences, Cape Peninsula University of Technology, Bellville, South Africa

Muhammad Imran (105,191), Institute of Home and Food Sciences, Faculty of Life Sciences, Government College University, Faisalabad, Pakistan

Yasemin Incegul (335,525), Department of Food Engineering, Faculty of Engineering, Suleyman Demirel University, Isparta, Turkey

Nazish Jahan (439), Department of Chemistry, University of Agriculture, Faisalabad, Pakistan

Martha Kandawa-Schulz (467), Department of Chemistry and Biochemistry, University of Namibia, Windhoek, Namibia

Aftab Ahmed Kandhro (31), Dr. M. A. Kazi Institute of Chemistry, University of Sindh, Jamshoro, Pakistan

Onur Ketenoglu (53,365), Department of Food Engineering, Faculty of Engineering, Cankiri Karatekin University, Cankiri, Turkey

Khalil-ur-Rahman (439), Department of Biochemistry, University of Agriculture, Faisalabad, Pakistan

Muhammad Kamran Khan (105,191), Institute of Home and Food Sciences, Faculty of Life Sciences, Government College University, Faisalabad, Pakistan

Manal Khider (683), Department of Dairy Science, Faculty of Agriculture, Fayoum University, Fayoum, Egypt

Johannes Kiefer (97), Technische Thermodynamik, Universität Bremen, Bremen, Germany

Mustafa Kiralan (53,335,365,515,525), Department of Food Engineering, Faculty of Engineering, Balıkesir University, Balikesir, Turkey

Sündüz Sezer Kiralan (53,335,515,525), Department of Food Engineering, Faculty of Engineering, Balıkesir University, Balikesir, Turkey

Dilşat Bozdoğan Konuşkan (7), Food Engineering Department, Faculty of Agriculture, Hatay Mustafa Kemal University, Hatay, Turkey

Abdul Hafeez Laghari (31), Pakistan Council of Scientific and Industrial Research, Laboratories Complex, Karachi, Pakistan

Anja I. Lampe (97), Technische Thermodynamik, Universität Bremen, Bremen, Germany

Vera Lazić (15), Faculty of Technology Novi Sad, University of Novi Sad, Novi Sad, Serbia

Massimo Lucarini (97,459), CREA-Research Centre for Food and Nutrition, Rome, Italy

Alfred Maroyi (147,277,477), Medicinal Plants and Economic Development (MPED) Research Centre, Department of Botany, University of Fort Hare, Alice, South Africa

Aspasia Mastralexi (547), Laboratory of Food Chemistry and Technology, School of Chemistry, Aristotle University of Thessaloniki, Thessaloniki, Greece

Mekjell Meland (637), Norwegian Institute of Bioeconomy Research, Ås, Norway

Guiomar Melgar-Lalanne (113,665), Institute for Basic Sciences, Veracruzana University, Veracruz, Mexico

Ayaz Ali Memon (31), National Center of Excellence in Analytical Chemistry, University of Sindh, Jamshoro, Pakistan

Najma Memon (725), National Centre of Excellence in Analytical Chemistry, University of Sindh, Jamshoro, Pakistan

Martin Mondor (113,665), Saint-Hyacinthe Research and Development Centre, Agriculture and Agri-Food Canada, Saint-Hyacinthe, QC, Canada

Kamel Msaada (373,611,625), Laboratory of Aromatic and Medicinal Plants, Biotechnology Center in Borj Cedria Technopole, Hammam-Lif, Tunisia

Collen Musara (147,477), Medicinal Plants and Economic Development (MPED) Research Centre, Department of Botany, University of Fort Hare, Alice, South Africa

Muhammad Mushtaq (345), Department of Chemistry, GC University, Lahore, Pakistan

Zarina Mushtaq (105,191), Institute of Home and Food Sciences, Faculty of Life Sciences; Department of Food Science, Nutrition and Home Economics, Government College University, Faisalabad, Pakistan

Muhammad Nadeem (105), Department of Dairy Technology, University of Veterinary and Animal Sciences, Lahore, Pakistan

Ruchira Nandasiri (81), Department of Food & Human Nutritional Sciences, University of Manitoba; Richardson Centre for Functional Foods & Nutraceuticals, Winnipeg, MB, Canada

Maja Natić (637), University of Belgrade—Faculty of Chemistry, Belgrade, Serbia

Stefano Ferrari Nicoli (459), CREA—Research Centre for Food and Nutrition, Rome, Italy

Ettore Novellino (459), Department of Pharmacy, University of Napoli Federico II, Napoli, Italy

Kar Lin Nyam (295), Faculty of Applied Sciences, UCSI University, Kuala Lumpur, Malaysia

Ali Osman (385,537,541), Agricultural Biochemistry Department, Faculty of Agriculture, Zagazig University, Zagazig, Egypt

Zeynep Aksoylu Özbek (219,231,255,323,575), Department of Food Engineering, Faculty of Engineering, Manisa Celal Bayar University, Manisa, Turkey

Onur Özdikicierler (547), Laboratory of Food Chemistry and Technology, School of Chemistry, Aristotle

University of Thessaloniki, Thessaloniki, Greece; Food Engineering Department, Faculty of Engineering, Ege University, Izmir, Turkey

Gulcan Ozkan (53,335,515,525), Department of Food Engineering, Faculty of Engineering, Suleyman Demirel University, Isparta, Turkey

Zahra Piravi-Vanak (429), Food Technology and Agricultural Products Research Center, Standard Research Institute of Iran, Karaj, Iran

Senka Popović (15), Faculty of Technology Novi Sad, University of Novi Sad, Novi Sad, Serbia

Rahman Qadir (309,315,491,725), Department of Chemistry, University of Sargodha, Sargodha, Pakistan

Biljana Rabrenović (637), University of Belgrade—Faculty of Agriculture, Belgrade-Zemun, Serbia

Antonio Raffo (97,459), CREA-Research Centre for Food and Nutrition, Rome, Italy

Muhammad Abdul Rahim (105), Institute of Home and Food Sciences, Faculty of Life Sciences, Government College University, Faisalabad, Pakistan

Ateeq Rahman (277), Department of Chemistry and Biochemistry, Faculty of Science, University of Namibia, Windhoek, Namibia

Mohamed Fawzy Ramadan (1,53,289,365,677,683,695, 719), Agricultural Biochemistry Department, Faculty of Agriculture, Zagazig University, Zagazig, Egypt; Deanship of Scientific Research, Umm Al-Qura University, Makkah, Saudi Arabia

Leila Rezig (373,625,611), High Institute of Food Industries, El Khadra City; University of Carthage, National Institute of Applied Sciences and Technology, LR11ES26, 'Laboratory of Protein Engineering and Bioactive Molecules'; Food Preservation Laboratory, High Institute of Food Industry, Tunis, Tunisia

Annalisa Romani (97,459), DiSIA-PHYTOLAB (Pharmaceutical, Cosmetic, Food supplement Technology and Analysis), University of Florence, Florence, Italy

Ranko Romanić (15,197), Faculty of Technology Novi Sad, University of Novi Sad, Novi Sad, Serbia

Antonello Santini (97,459), Department of Pharmacy, University of Napoli Federico II, Napoli, Italy

Engy Shams-Eldin (267), Special Food and Nutrition Department, Food Technology Research Institute, Agriculture Research Center, Giza, Egypt

Yan Yi Sim (295), Faculty of Applied Sciences, UCSI University, Kuala Lumpur, Malaysia

Eliana B. Souto (459), Department of Pharmaceutical Technology, Faculty of Pharmacy, University of Coimbra, Coimbra; CEB—Centre of Biological Engineering, University of Minho, Campus de Gualtar, Braga, Portugal

Bushra Sultana (345), Department of Chemistry, University of Agriculture, Faisalabad, Pakistan

Danijela Šuput (15), Faculty of Technology Novi Sad, University of Novi Sad, Novi Sad, Serbia

Chin Xuan Tan (357,587), Department of Allied Health Sciences, Faculty of Science, Universiti Tunku Abdul Rahman, Kampar Perak, Malaysia

Seok Shin Tan (357,587), Department of Nutrition and Dietetics, School of Health Sciences, International Medical University, Kuala Lumpur, Malaysia

Seok Tyug Tan (357,587), Department of Healthcare Professional, Faculty of Health and Life Sciences, Management and Science University, Shah Alam, Selangor, Malaysia

Usha Thiyam-Höllander (81), Department of Food & Human Nutritional Sciences, University of Manitoba; Richardson Centre for Functional Foods & Nutraceuticals, Winnipeg, MB, Canada

Suna Timur (449), Department of Biochemistry, Faculty of Science, Ege University, İzmir, Turkey

Maria Z. Tsimidou (547), Laboratory of Food Chemistry and Technology, School of Chemistry, Aristotle University of Thessaloniki, Thessaloniki, Greece

Silvia Urciuoli (97), DiSIA-PHYTOLAB (Pharmaceutical, Cosmetic, Food supplement Technology and Analysis), University of Florence, Florence, Italy

Chiara Vita (459), DiSIA-PHYTOLAB (Pharmaceutical, Cosmetic, Food supplement Technology and Analysis), University of Florence, Florence, Italy

Dragana Dabić Zagorac (637), Innovation Centre of Faculty of Chemistry Ltd, Belgrade, Serbia

Foreword

Oils and fats obtained from seeds, herbs, spices, medicinal plants, vegetables, fruits, and agricultural by-products are of great economic importance, and the need for widely usable bioactive lipids and natural antioxidants continues to grow. However, conventional methods to extract and process oils, such as solvent extraction, refining, bleaching, and deodorization, can alter the functional properties and stability of their phytonutrients, produce hazardous waste products, and cause major health concerns regarding human diets. Cold pressing is a technique that offers a safe, nonhazardous method for oil extraction and processing in which the main bioactive constituents are preserved through the omission of heat, chemical treatments, and refining processes.

Due to the vast application of cold pressed oils in the food industry and pharmaceuticals, there was a need for a comprehensive book on cold pressed oils. *Cold Pressed Oils: Green Technology, Bioactive Compounds, Functionality, and Applications* meets that need and presents recent advances in the chemistry and functionality of bioactive phytochemicals in cold pressed oils.

This volume creates a multidisciplinary forum of discussion on recent advances in the chemistry and functionality of the bioactive phytochemicals in lipids found in cold pressed oils. Each chapter explores a different cold pressed oil, with a focus on cold press extraction and processing, composition, physicochemical characteristics, organoleptic attributes, nutritional quality, oxidative stability, food applications, and functional and health-promoting traits. Each chapter contains background information and concluding remarks for better understanding of the subject matter.

Edited by a team of experts, *Cold Pressed Oils: Green Technology, Bioactive Compounds, Functionality, and Applications* brings a diversity of developments in food science to scientists, chemists, nutritionists, and students in nutrition, lipids chemistry and technology, agricultural science, pharmaceuticals, cosmetics, nutraceuticals, and many other fields.

Mohamed Fawzy Ramadan
Makkah, Saudi Arabia

Preface

Cold pressing is a technique that offers a safe, nonhazardous method for edible oil extraction and processing in which the bioactive constituents are preserved through the omission of thermal and chemical treatments as well as refining processes. On the other hand, it is acknowledged that the contribution of specialty oils to our health and well-being is recognized by their chemical composition. Fatty acid profiles and a wide range of specific bioactive lipids (i.e., sterols, polar lipids, tocols, phenolics) have been shown to affect the biological functions of our bodies.

This book aims at building a multidisciplinary forum of discussion on advances in cold pressed oil technology, chemistry, and functionality of lipid bioactive phytochemicals found in cold pressed oils, focusing on oil processing, composition, physicochemical properties, nutritional quality, organoleptic attributes, oxidative stability, food and nonfood uses, as well as health-promoting traits.

This book contains several chapters that describe different cold pressed oils. With the aim of providing a major reference work for those involved with the oils industry as well as undergraduate and graduate students, this volume presents a comprehensive review of the results that have led to advancements in cold pressed oil production, processing, functionality, and applications. As far as possible, the chapters have followed a similar outline, describing the properties and processing of cold pressed oils, with a focus on the extraction, chemical composition, quality of different cold pressed oils, and applications of cold pressed oils in food and nonfood applications as well as nutraceutical products. I hope that the book will be a valuable source for people involved in cold pressed oils.

I sincerely thank all authors for their valuable contributions and for their cooperation during the book's preparation. The help and support given to me by the Elsevier staff, especially Lena Sparks and Nina Bandeira, were essential for the completion of my task, and are appreciated.

Mohamed Fawzy Ramadan
Makkah, Saudi Arabia

Chapter 1

Introduction to cold pressed oils: Green technology, bioactive compounds, functionality, and applications

Mohamed Fawzy Ramadan
Agricultural Biochemistry Department, Faculty of Agriculture, Zagazig University, Zagazig, Egypt

1 Introduction

In 2015, the United Nations Sustainable Development Goals (UNSDGs) were announced (https://sustainabledevelopment.un.org). These seventeen goals offer a vision of a fairer, peaceful, more prosperous, and sustainable world. They imagine a future that will be free of hunger and poverty, and safe from the worst impact of environmental hazard and climate change. In food—the way it is grown, processed, transported, stored, marketed, and consumed—lies the fundamental connection between people and the path to sustainable economic development. The third UNSDG, "Good Health and Well-Being," aims to promote a healthy life and human well-being which is closely related to the use of environmentally friendly processing techniques in food systems as well as the functionality of foodstuffs.

We live in an era where rapid innovations are being made, and these unique technologies could be applied to enhance our edible system. Scientists are searching for new foodstuffs that have properties that can be manipulated and designed at the molecular level to improve their safety, quality, and healthfulness. The scientific research being performed now will have a great effect on the way we eat in the future (McClements, 2019).

Traditional methods of oil extraction use excessive amounts of organic solvents and need high-energy input. Current environmental issues associated with organic solvent disposal demand alternative methods for the extraction of edible oils that are environmentally friendly and energy-efficient. Green technologies identified some alternative methods suitable for edible oil extraction. This has led to the improvement of more energy-efficient and eco-friendly green techniques that reduced the utilization of toxic organic solvents and enabled high-quality products to be developed.

The green extraction methods, including cold pressing extraction (CPE), ultrasonic-aided extraction (UAE), microwave-aided extraction (MAE), subcritical extraction (SWE), and supercritical extraction (SFE), have received attention due to their eco-friendliness and energy efficiency. These green extraction techniques are applied to minimize the utilization of toxic organic solvents, and to extract bioactive lipid-soluble compounds, with the focus on developing a better-quality final product. These benefits have been welcomed by the oilseed industry and manufacturers.

Edible unrefined oils include cold pressed oils (CPO) and virgin oils. Unrefined oils are a category covered by technical regulations in the field of edible oils (Codex, 1999): "Cold pressed unrefined vegetable oil is produced without heating, precleaning, dehulling and milling mechanically. Cold pressed unrefined oil can only be purified by washing with water, precipitating, filtrating and centrifuging."

2 Green technologies and processing of vegetable oils

All technologies to eliminate the use of hazardous toxic solvents and chemicals refer to green technology. One definition of green technology is as follows: "Green Extraction is based on the discovery and design of extraction processes which will reduce energy consumption, allows the use of alternative solvents and renewable natural products, and ensure a safe and high-quality extract." The "Six Principles of Green Extraction of Natural Products" are directions to build an innovative and green label and standard.

Cold Pressed Oils. https://doi.org/10.1016/B978-0-12-818188-1.00001-3
Copyright © 2020 Elsevier Inc. All rights reserved.

- Principle 1: Variety selection and use of renewable plant resources.
- Principle 2: Reduce energy consumption using innovative technologies and energy recovery.
- Principle 3: Use alternative solvents (water or agro-solvents).
- Principle 4: Reduce unit operation and favor a safe and controlled process.
- Principle 5: Produce coproducts to include the agro-refining industry.
- Principle 6: Aim for biodegradable and nondenatured extracts without contaminants.

Extraction, according to these principles, is a new concept to protect the consumer and environment, and enhance the competitiveness of industries to be more economic, innovative, and ecologic (Chemat, Vian, & Cravotto, 2012). Many studies have been conducted to find and apply research techniques related to green technology in different applications.

The use of alternative nontraditional techniques of oil extraction has gained attention during the last years. These novel techniques have been applied in the oilseed industry to minimize detrimental changes in the nutritional quality and physicochemical and sensory traits of the extracted oils while reducing the carbon footprint from solvents (Matthäus & Brühl, 2003).

Conventional oil extraction methods are replaced by modern ones, usually called green or clean techniques, because of the long time requirement and high solvent consumption of the former (Chemat et al., 2012; Parker, Adams, Zhou, Harris, & Yu, 2003; Rodríguez-pérez, Quirantes-piné, & Fernández-gutiérrez, 2015; Tiwari, 2015). Pressurized liquid (PLE), pulsed electric field (PEF), high hydrostatic pressure (HHP), high voltage electrical discharges (HVED), SFE, UAE, and MAE are green technologies considered as alternatives to conventional methods (Soquetta, Terra, & Bastos, 2018). According to the Codex Alimentarius, cold pressing is performed only by mechanical processes without thermal application, and the product is produced without destroying the oil nature (Matthäus & Spener, 2008).

3 Advantages versus disadvantages of cold pressing technology

Methods utilized for oil extraction might alter minor compounds that have functional traits and contribute to oil oxidative stability. Recently, CPO has increasingly been considered as these oils have high nutritional values. Vold pressing techniques are becoming an interesting substitute for traditional methods because of consumers' desire for safe and natural edible products (El Makawya, Ibrahimb, Mabrouka, Ahmedc, & Ramadan, 2019; Kiralan, Çalik, Kiralan, & Ramadan, 2018; Ramadan, 2013).

The advantages of this technology at an industrial level include lower energy consumption and lower investment cost. This extraction does not use toxic solvents or thermal conditioning of the seeds, and does not generate wastewater. It ensures a safe working environment for employees, has a lower environmental impact in comparison with solvent extraction, and shows higher flexibility because processing diverse types of seeds is fast and easy. CPO are preferred to refined oils as they contain more antioxidants and bioactive substances like sterols, carotenoids, and phenolics. More natural biologically active substances such as phenolic compounds and tocols are present in CPO, which could improve oxidative stability (Bhatnagar & Krishna, 2014; Prescha, Grajzer, Dedyk, & Grajeta, 2014).

The main disadvantage of cold pressing techniques is the high capital or investment required compared to conventional methods. In addition, CPO have low efficiency and are not always of the same quality. Most CPO contain high amounts of polyunsaturated fatty acids (PUFA), which might be disadvantageous in terms of oxidative stability. CPO could also contain higher amounts of pro-oxidative compounds, so their shelf life might be shorter compared to refined oils (Brühl, 1996; Rotkiewicz, Konopka, & Żylik, 1999).

4 Cold pressing process

Vegetable oils can be obtained from oilseeds using different systems of the press, solvent extraction, or a combination of both methods. Seeds contain high amounts of oil are prepressed then solvent extracted, or direct solvent extraction could be performed on seeds with lower oil content. The extraction technology can be selected depending on the production cost, material traits, availability, usage goal of the cake, and environmental factors (Ghazani, Garcia Llatas, & Marangoni, 2014; Matthäus & Brühl, 2003; Sloan, 2000).

Based on the oilseed structure and composition, some fractions of oil might remain in the meal or cake. This should be considered when comparing the press and cold press for the oil yield and meal composition. To increase the oil yield from cold pressing, some pretreatments could be applied to seeds before pressing, such as enzyme application, microwave treatment, steaming, and roasting. Cold pressed virgin oils do not require expensive refining. Only centrifugation or filtration is necessary to obtain high-quality CPO. Minor bioactive lipids that are commonly lost during refining are retained in CPO. A producer could select the production type based upon the aim of production, uses of oil and cake, and the amount of processed seeds.

The cold press machine has a simple working scheme wherein oilseeds are fed into one inlet, and two exits provide oil and a nonoiled cake. Oil yield depends on pretreatment (i.e., peeling, drying, and enzymatic treatment) and process parameters applied to the oilseeds or raw materials. Cold pressing could be investigated under three main systems: expellers, expanders, and twin-cold systems (for pilot- or laboratory-scale production) (Çakaloğlu, Özyurt, & Ötleş, 2018).

5 Features and specific phytochemicals of cold pressed oils (CPO)

The contribution of lipid-soluble bioactives to human health is determined by their composition. Fatty acids profile (especially omega-9, omega-6, and omega-3) and high-value minor lipid compounds (i.e., tocols, sterols, glycolipids, phospholipids, aroma compounds, and phenolics) exhibit health-promoting traits and positively influence the biological functions of our body (Ibrahim, Attia, Maklad, Ahmed, & Ramadan, 2017; Kiralan et al., 2017).

The techniques used to extract oils as well as the processing steps such as bleaching, refining, and deodorization influence their bioactive constituents. CPO usually contain unique phytochemicals with health-promoting traits. For example, cold pressed pomegranate seed oil contains punicic acid (C18:3-9*cis*, 11*trans*, 13*cis*), and α-eleostearic. Punicic acid, also known as tricosanic acid, is an omega-5 long-chain PUFA and a conjugated α-linolenic acid isomer with structural similarities to conjugated linoleic acid and α-linolenic acid (Costa, Silva, & Torres, 2019; Lansky & Newman, 2007; Viladomiu, Hontecillas, Yuan, Lu, & Bassaganya-Riera, 2013). The potential health benefits of these conjugated fatty acids have made them increasingly interesting for scientists and consumers (Carvalho, Melo, & Mancini-Filho, 2010; Grossmann, Mizuno, Schuster, & Cleary, 2010).

Another example is pinolenic acid (PNLA; all *cis*-5,-9,-12-18:3), which is found in cold pressed pine nuts oil. This acid, constituting 14%–19% of fatty acids in pine nuts, is the basic polymethylene interrupted fatty acid (Δ5-UPIFA) (Destaillats, Cruz-Hernandez, Giuffrida, & Dionisi, 2010; Ryan, Galvin, O'Connor, Maguire, & O'Brien, 2006; Wolff & Bayard, 1995). Pinolenic acid has antiinflammatory traits that protect and strengthen the stomach and stomach lining (Chen, Zhang, Wang, & Zu, 2011; Xie, Miles, & Calder, 2016). Cold pressed Korean pine nuts oils are used as nutritional supplements thanks to pinolenic acid and antioxidants.

6 Cold pressed oils (CPO) in the literature

A careful search on "cold pressed oils" (as keywords) in the titles, abstracts, and keywords of publications in the Scopus database (www.scopus.com) revealed that the total number of scholarly outputs published is high (c.975 as of December 2019). Apart from the total published scholarly outputs, c.600 were research articles and c.200 reviews. Fig. 1 shows the scholarly output on CPO since 2000. It is clear that the scholarly output published annually on CPO has increased

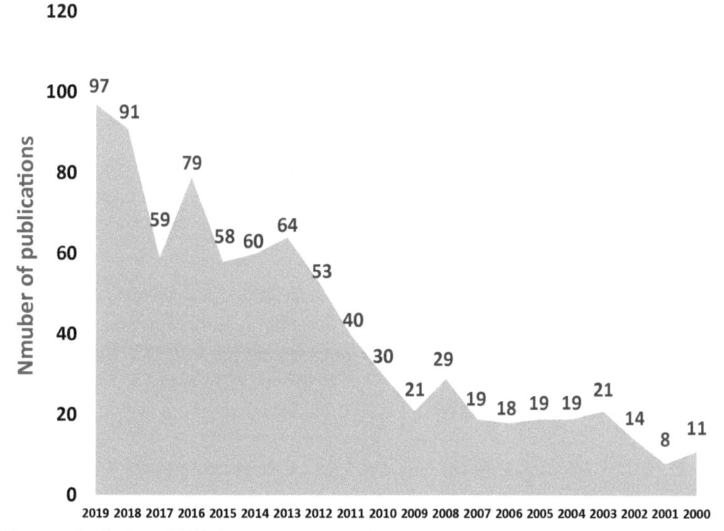

FIG. 1 Scholarly output on cold pressed oil since 2000 (www.scopus.com).

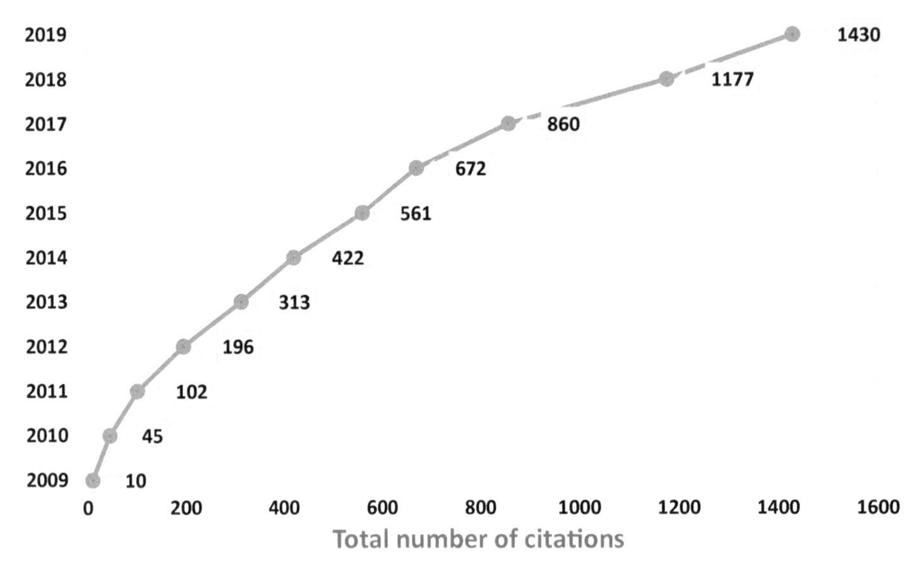

FIG. 2 Total number of annual citations (2009–19) from articles published on cold pressed oil (www.scopus.com).

dramatically over the last two decades, from 11 articles in 2000 to 97 article in 2019. The publications were mainly related to the subject areas of agricultural and biological sciences, biochemistry, genetics and molecular biology, chemical engineering, chemistry, and engineering. The United States, Poland, China, Italy, Turkey, Japan, Germany, Egypt, and Canada have emerged as main contributors. The most prolific journals were the *Journal of Agricultural and Food Chemistry*, *European Journal of Lipid Science and Technology*, *Journal of Food Science*, *Food Chemistry*, and *Journal of the American Oil Chemists Society*. On the other side, the sum of annual citations (2009–19) from articles published on CPO have significantly increased during the last 10 years (Fig. 2). These measurable indicators reflect the importance of cold pressed oils as a topic in the international scientific community.

Several books have already been published on the composition and functional properties of oils and fats from herbs, oilseeds, medicinal plants, animal sources, and marine sources. However, it is hard to find a book focused on the composition and functionality of CPO. This book contains comprehensive chapters focusing on CPO, which contain unique bioactive components that have led to their being considered health-promoting oils. The following chapters describe several CPO. Each chapter is dedicated to a particular cold pressed oil. Most CPO have unique composition and characteristics that make them valuable sources for novel foods, nutraceuticals, and pharmaceuticals. The chapters have the following topics:

- Cold press extraction and processing of oils
- CPO recovery, content, use of oil cake, and economy
- Fatty acids composition and acyl lipids profile of CPO
- Minor bioactive lipids (carotenoids, phenolics, sterols, tocols, hydrocarbons, aroma compounds) in CPO
- The contribution of bioactive constituents in CPO to organoleptic traits
- Applications of CPO
- Health-promoting properties of CPO and oil constituents

References

Bhatnagar, A. S., & Krishna, G. A. G. (2014). Lipid classes and subclasses of cold-pressed and solvent-extracted oils from commercial Indian Niger (*Guizotia abyssinica* (L.) Cass.) seed. *Journal of American Oil Chemists' Society*, *91*(7), 1205–1216.

Brühl, L. (1996). Trans fatty acids in cold pressed oils and in dried seeds. *Lipid/Fett*, *98*(11), 380–383.

Çakaloğlu, B., Özyurt, V. H., & Ötleş, S. (2018). Cold press in oil extraction. A review. *Ukrainian Food Journal*, *7*, 640–654.

Carvalho, E. B. T., Melo, I. L. P., & Mancini-Filho, J. (2010). Chemical and physiological aspects of isomers of conjugated fatty acids. *Ciência e Tecnologia de Alimentos*, *30*, 295–307.

Chemat, F., Vian, M. A., & Cravotto, G. (2012). Green extraction of natural products: Concept and principles. *International Journal of Molecular Sciences*, *13*, 8615–8627.

Chen, X., Zhang, Y., Wang, Z., & Zu, Y. (2011). In vivo antioxidant activity of *Pinus koraiensis* nut oil obtained by optimised supercritical carbon dioxide extraction optimised supercritical carbon dioxide extraction. *Natural Product Research*, *25*, 1807–1816.

Codex. (1999). *Codex standards for named vegetable oils, 1999. Codex-Stan 210, Amended 2005*. Joint FAO/WHO Food Standards Programme.

Costa, A. M. M., Silva, L. O., & Torres, A. G. (2019). Chemical composition of commercial cold-pressed pomegranate (*Punica granatum*) seed oil from Turkey and Israel, and the use of bioactive compounds for samples' origin preliminary discrimination. *Journal of Food Composition and Analysis, 2019*(75), 8–16.

Destaillats, F., Cruz-Hernandez, C., Giuffrida, F., & Dionisi, F. (2010). Identification of the botanical origin of pine nuts found in food products by gas-liquid chromatography analysis of fatty acid profile. *Journal of Agricultural and Food Chemistry*, *58*(4), 2082–2087.

El Makawya, A. I., Ibrahimb, F. M., Mabrouka, D. M., Ahmedc, K. A., & Ramadan, M. F. (2019). Effect of antiepileptic drug (Topiramate) and cold pressed ginger oil on testicular genes expression, sexual hormones and histopathological alterations in mice. *Biomedicine & Pharmacotherapy*, *110*, 409–419.

Ghazani, S. M., Garcia Llatas, G., & Marangoni, A. G. (2014). Micronutrient content of cold-pressed, hot-pressed, solvent extracted and RBD canola oil: Implications for nutrition and quality. *European Journal of Lipid Science and Technology*, *116*, 380–387.

Grossmann, M. E., Mizuno, N. K., Schuster, T., & Cleary, M. P. (2010). Punicic acid is an omega-5 fatty acid capable of inhibiting breast cancer proliferation. *International Journal of Oncology*, *36*, 421–426.

Ibrahim, F. M., Attia, H. N., Maklad, Y. A. A., Ahmed, K. A., & Ramadan, M. F. (2017). Biochemical characterization, anti-inflammatory properties and ulcerogenic traits of some cold pressed oils in experimental animals. *Pharmaceutical Biology*, *55*, 740–748.

Kiralan, M., Çalik, G., Kiralan, S., & Ramadan, M. F. (2018). Monitoring stability and volatile oxidation compounds of cold pressed flax seed, grape seed and black cumin seed oils upon photo-oxidation. *Journal of Food Measurement and Characterization*, *12*, 616–621.

Kiralan, M., Ulaş, M., Özaydin, A. G., Özdemir, N., Özkan, G., Bayrak, A., et al. (2017). Blends of cold pressed black cumin oil and sunflower oil with improved stability: A study based on changes in the levels of volatiles, tocopherols and thymoquinone during accelerated oxidation conditions. *Journal of Food Biochemistry*, *41*, e12272.

Lansky, E. P., & Newman, R. A. (2007). *Punica granatum* (pomegranate) and its potential for prevention and treatment of inflammation and cancer. *Journal of Ethnopharmacology*, *109*, 177–206.

Matthäus, B., & Brühl, L. (2003). Cold-pressed edible rapeseed oil production in Germany. *Nahrung/Food*, *47*(6), 413–419.

Matthäus, B., & Spener, F. (2008). What we know and what we should know about virgin oils—A general introduction. *European Journal of Lipid Science and Technology*, *110*, 597–601.

McClements, D. J. (2019). The science of foods: Designing our edible future. In D. J. McClements (Ed.), *Future foods: How modern science is transforming the way we eat*. Cham: Springer International Publishing.

Parker, T. D., Adams, D. A., Zhou, K., Harris, M., & Yu, L. (2003). Fatty acid composition and oxidative stability of cold-pressed edible seed oils. *Food Chemistry and Toxicology*, *68*(4), 1240–1243.

Prescha, A., Grajzer, M., Dedyk, M., & Grajeta, H. (2014). The antioxidant activity and oxidative stability of cold-pressed oils. *Journal of American Oil Chemists' Society*, *91*, 1291–1301.

Ramadan, M. F. (2013). Healthy blends of high linoleic sunflower oil with selected cold pressed oils: Functionality, stability and antioxidative characteristics. *Industrial Crops and Products*, *43*, 65–72.

Rodríguez-pérez, C., Quirantes-piné, R., & Fernández-gutiérrez, A. (2015). Optimization of extraction method to obtain a phenolic compounds-rich extract from *Moringa oleifera* Lam leaves. *Industrial Crops and Products*, *66*, 246–254.

Rotkiewicz, D., Konopka, I., & Żylik, S. (1999). State of works on the rapeseed oil processing optimalization. I. Oil obtaining. *Rosliny Oleiste/Oilseed Crops*, 151–168 [in Polish].

Ryan, E., Galvin, K., O'Connor, T. P., Maguire, A. R., & O'Brien, N. M. (2006). Fatty acid profile, tocopherol, squalene and phytosterol content of Brazil, pecan, pine, pistachio and cashew nuts. *International Journal of Food Sciences and Nutrition*, *57*(3–4), 219–228.

Sloan, A. E. (2000). The top ten functional food trends. *Food Technology*, *54*(4), 33–62.

Soquetta, M. B., Terra, L. M., & Bastos, C. P. (2018). Green technologies for the extraction of bioactive compounds in fruits and vegetables. *CYTA-Journal of Food*, *16*(1), 400–412.

Tiwari, B. K. (2015). Ultrasound: A clean, green extraction technology. *Trends in Analytical Chemistry*, *71*, 100–109.

Viladomiu, M., Hontecillas, R., Yuan, L., Lu, P., & Bassaganya-Riera, J. (2013). Nutritional protective mechanisms against gut inflammation. *Journal of Nutritional Biochemistry*, *24*, 929–939.

Wolff, R. L., & Bayard, C. C. (1995). Fatty acid composition of some pine seed oils. *Journal of American Oil Chemist' Society*, *72*(9), 1043–1046.

Xie, K., Miles, E. A., & Calder, P. C. (2016). A review of the potential health benefits of pine nut oil and its characteristic fatty acid pinolenic acid. *Journal of Functional Foods*, *23*, 464–473.

Chapter 2

Minor bioactive lipids in cold pressed oils

Dilşat Bozdoğan Konuşkan
Food Engineering Department, Faculty of Agriculture, Hatay Mustafa Kemal University, Hatay, Turkey

1 Introduction

Nowadays consumers prefer natural and healthy, beneficial food products (Ananth, Deviram, Mahalakshmi, Sivasudha, & Tietel, 2019). Cold pressed oil is one of these products. In recent years, cold pressing has become the preferred method for oil extraction from oilseeds and fruits. This method is also user-friendly and inexpensive compared to other extraction methods (Ananth et al., 2019; Ramadan, 2013). Cold pressed oils refer to oils that are extracted by oil extraction without heat and chemical treatment, which may contain a higher amount of minor bioactive compounds, including natural antioxidants, than other refined oils (Bozdogan Konuskan, Arslan, & Oksuz, 2019; Teh & Birch, 2013). These minor bioactive lipids that are naturally found in cold pressed oils are phytosterols, phospholipids, tocopherols, phenolic compounds, hydrocarbons (squalene), pigments (carotenoids and chlorophyll), and flavor and aroma compounds (Gao, Liu, Jin, & Wang, 2019; Gornas, Siger, & Seglina, 2013; Mildner-Szkudlarz, Rozanska, Siger, Kowalczewski, & Rudzinska, 2019). Minor fractions play an essential role in determining the nutritional and health impact of edible oils (Teh & Birch, 2013). Cold pressed oils are more resistant against lipid oxidation and they have a higher initial oxidation level due to their natural antioxidants (Grosshagauer, Steinschaden, & Pignitter, 2019).

2 Phytosterols

Phytosterols, also called plant sterols, constitute the major proportion of the unsaponifiable fraction of lipids (Wasowicz, 2003; Yang, Oyeyinka, Xu, Ma, & Zhou, 2018), which can occur in vegetable oils either in free form or esterified with fatty acids (Manai-Djebali & Queslati, 2017). Phytosterols are by-products of the isoprenoid biosynthetic pathway via squalene from acetyl coenzyme-A. The phytosterols are cell membrane structural compounds found in the regulation of membrane fluidity, permeability, and metabolism (Shahzad et al., 2017). Based on the chemical structure, phytosterols can be classified into three groups: 4-desmethylsterols (cholestane series, i.e., normal phytosterols), 4-monomethylsterols (4 α-methylcholestane series), and 4,4-dimethylsterols (lanostane series, also known as triterpene alcohols) (Manai-Djebali & Queslati, 2017; Wasowicz, 2003). Generally, plant sterols are 4-desmethylsterols because they do not contain any methyl groups at the fourth position of the sterol ring structure (Moreau et al., 2018). Phytosterols are a family of triterpenes comprising of a 27–30 carbon ring-based structure with hydroxyl groups, and they are highly similar to cholesterol in both structure and biological function (Figueiredo et al., 2018; Shi et al., 2019). Phytosterols consist of a tetracyclic structure and a side chain in position C-17 (Manai-Djebali & Queslati, 2017; Moreau et al., 2018).

Free phytosterols contain a double bond in the B-ring between C-5 and C-6, or C-7 and C-8, also called Δ^5- and Δ^7-sterols, while the place of the double bond in the ring is characteristic to specific plant types (Moreau et al., 2018). Phytosterols with a saturated ring structure, called stanols, occur in nature in minor concentrations (Wasowicz, 2003).

The oxidation products of phytosterols in vegetable oils were identified and quantitated by Dutta and Appelqvist (1997). These are 7α and 7β hydroxy-sito-and campesterol, 7-ketosito and 7- ketocampesterol, 5α, 6α-epoxy-sito- and campesterol, 5β, 6β-epoxy-sito and campesterol, and dihydroxysitosterol and dihydroxycampesterol. A study investigating the contents of sterol oxidation products of cold pressed oils and refined oils from Polish markets showed that the content of oxyphytosterols in refined oils was 2–2.5 times higher than in cold pressed oil. Epimers of 7-hydroxy-phytosterols

Cold Pressed Oils. https://doi.org/10.1016/B978-0-12-818188-1.00002-5
Copyright © 2020 Elsevier Inc. All rights reserved.

and 7-keto-phytosterols were the dominant compounds in cold pressed oil, while epoxy derivatives were dominant compounds in refined oil (Rudsinzka, Kazus, & Wasquwicz, 2001).

Vegetable oils, oilseeds, and nuts are the richest sources of phytosterols. Beta-sitosterol (29 carbons), campesterol (28 carbons), and stigmasterol (29 carbons) are the three most common sterols (Manai-Djebali & Queslati, 2017; Moreau et al., 2018; Shahzad et al., 2017). The total phytosterol content and profile can vary according to variety, agronomic and climatic conditions, maturity, extraction and refining methods, and preextraction and storage conditions (Bozdogan Konuskan & Mungan, 2016; Fernandez-Cuesta, Leon, Velasco, & De La Rosa, 2013; Lukic, Lukic, Krapac, Sladonja, & Pilizota, 2013; Manai-Djebali & Queslati, 2017; Temime et al., 2008).

Several studies have demonstrated that phytosterols protect against many chronic ailments such as cardiovascular diseases (Ros, 2010; Shahzad et al., 2017), cancer (Jones & AbuMweis, 2009; Rubis et al., 2010), ulcers (Plat et al., 2014), diabetes (Misawa et al., 2012), and inflammation (Grattan, 2013; Gylling & Simonen, 2015). Phytosterols have the capacity to reduce dietary cholesterol absorption in the intestine and serum low-density lipoprotein-cholesterol concentrations. Some research has shown that the consumption of 2 g/day of sterol or stanols may reduce the risk of heart disease by 25% (Hicks & Moreau, 2001; Shahzad et al., 2017; Wasowicz, 2003).

Phytosterols have been reported to have a protective effect against various forms of cancer such as breast (Grattan, 2013), prostate (Awad, Fink, Williams, & Kim, 2001), lung (Manai-Djebali & Queslati, 2017), liver and stomach (Ramprasath & Awad, 2015), and ovary and colon cancers (Baskar, Ignacimuthu, Paulraj, & Al Numair, 2010). In vivo studies have shown that diets enriched with phytosterols (2%, w/w) contributed to improve lipid profiles and decreased atherosclerotic lesions in apolipoprotein E-knockout (apo E-KO) mice (Manai-Djebali & Queslati, 2017; Moghadasian, 2006). Raicht, Cohen, Fazzini, Sarwal, and Takashi (1980) reported that the development of methyl nitrosourea induced tumors in mice when fed with 0.2% β-sitosterol in their diet for 28 weeks. The results showed a 39% reduction in overall tumor numbers and a 60% reduction in tumors per rat.

The phytosterol composition is a very useful parameter for detecting adulterations or to check authenticity since vegetable oils have a specific sterol profile known as a "fingerprint" (Piravi-Vanak, Ghasemi, Ghavami, Ezzatpanah, & Zolfonoun, 2012; Yorulmaz & Bozdogan Konuskan, 2016). Standardized methods for total phytosterol analysis have been developed by the American Oil Chemists' Society, the Association of Official Analytical Chemists, and the International Organization for Standardization (Moreau et al., 2018). Generally, the determination of phytosterols including extraction of lipid fraction followed by alkaline hydrolysis (saponification), extraction of unsaponifiable matter, derivatization of phytosterols, and separation/quantification are done by chromatography (Figueiredo et al., 2018). Saponification is practically a universal step prior to phytosterol analysis by gas chromatography (GC), and is included in most of the above references for total phytosterol analysis (Moreau et al., 2018). Thin-layer chromatography (TLC), solid-phase extractions (SPE), and high-performance liquid chromatography are the three main methods used to separate phytosterols classes and total sterols from unsaponifiables (Azadmard-Damirchi & Dutta, 2010). The qualitative and quantitative analysis of phytosterols in vegetable oils is analyzed by gas chromatography. Both flame ionization (FID) and MS (electron impact, or EI) detection are commonly used (Moreau et al., 2018).

3 Phospholipids

Phospholipids (PL), one of the minor lipid components in seed oils, are highly abundant structural and functional lipids found in cell membranes (Gao & Wu, 2019; Herchi et al., 2012). Major phospholipids are derivatives of glycerol, where the 1- and 2- positions are acylated by fatty acids while the 3-positions are esterified with phosphoric acid (Pokorny, 2003). Major PL are lecithin (phosphatidylcholine), cephalin (phosphatidylethanolamine, phosphatidylserine), phosphatidylinositol, sphingomyelin, and phosphatidic acid. Lecithin, cephalin, and phosphatidylinositol, which are the most common phosphatides, can be regarded as triglycerides in which one fatty acid root switches with phosphoric acid (Gunstone & Norris, 1983). The major functions of PL in foods are their surface-active properties, and they act as emulsifiers and stabilizers of emulsions. Phospholipids are completely removed from the oil during degumming or mucilage removal processes, since they cause productivity loss in the oil extraction as they have emulsifier properties (Gunstone & Norris, 1983; Kayahan, 2003). PL usually occur to the content of 1%–2% in freshly extracted soybean or corn oils. They are present in lesser amounts in other seed oils, for example, 0.7%–0.9% in crude cottonseed oil and 0.3%–0.4% in crude peanut oil (Bailey, 1951). PL increase the oxidative stability of fats and oils by acting as synergists with tocols and phenolics (Pokorny, 2003). Phospholipids have attracted remarkable interest because of their health benefits (Herchi et al., 2012). Recent research has shown that PL could be a good source of arachidonic acid, which is important in metabolism, especially in the synthesis of prostaglandins and leukotrienes (Gao & Wu, 2019).

4 Tocols (tocopherols and tocotrienols)

Tocopherols and tocotrienols, together abbreviated as tocols, are natural lipophilic antioxidants that protect oxidation in vegetable oils (Ozcan, Al-Juhaimib, Ahmed, Osman, & Gassem, 2019; Schwartz, Ollilainen, Piironen, & Lampi, 2008). Tocols (vitamin E) comprise a chromanol ring with a C16 phytol side chain and are reclassified in two types according to which the side chain is either saturated (tocopherols) or contains three double bonds at carbons 3, 7, and 11 (tocotrienols) (Lachman, Hejtmankova, Orsak, Popov, & Martinek, 2018). Both tocopherols and tocotrienols exist in four different isomers called alpha, beta, gamma and delta; these differ in the methylation pattern of the benzopyran ring with three methyl groups (at C-5, C-7, and C-8) (Boschin & Arnoldi, 2011). Among the tocopherols, alpha and gamma tocopherols are the most effective lipid-soluble antioxidants in vegetable oils. Alpha-tocopherol shows the highest vitamin E activity while gamma-tocopherol has the highest antioxidant activity (Böhmdorfer, Patel, Netscher, Gille, & Rosenau, 2011; Boschin & Arnoldi, 2011).

Vegetable oils contain not only alpha-tocopherol but also other tocopherols, especially gamma and delta-tocopherol (Saldeen & Saldeen, 2005). The soybean and corn oils are usually dominated by gamma-tocopherol, while in olive oil the more abundant form is alpha-tocopherol (Szymanska & Kruk, 2008). The tocols have quench singlet oxygen and free radicals scavenge, particularly lipid peroxy radicals of polyunsaturated fatty acids (PUFA), therefore terminating lipid peroxidation chain reactions (Azzi, 2019; Maeda & Dellapenna, 2007). Tocols also play an important role as an antioxidant in oil stability. Due to their potent antioxidant properties and biological effects at the molecular level, they have reduced the risk of many diseases such as cardiovascular diseases, antiinflammatory and antidiabetic effects, osteoporosis, hyperlipidemia, neurodegenerative diseases, and cancer (Bartosinska, Buszewska-Forajta, & Siluk, 2016; Saldeen & Saldeen, 2005; Schwartz et al., 2008; Zhang et al., 2019). Vitamin E is also indispensable for immune defense. It has been suggested that tocopherols, acting as hormones or as secondary donors of genetic information, control the expression of some genes (Nogala Kalucka, 2003).

Vitamin E deficiency causes the damage of cellular membranes resulting from oxidation of the unsaturated fatty acids in lipids, and vitamin E deficiency can also display itself as muscular pain and progressing muscular disorder (Nogala Kalucka, 2003). The tocols in oilseeds could be determined by analytical techniques such as GC, liquid chromatography with diode array detection, thin-layer chromatography equipped with evaporative light scattering detection, and liquid chromatography-mass spectrometry (Zhang et al., 2019). For tocopherols, the richest dietary sources are vegetable oils and the products made from these oils. The tocopherol contents in seed oils range from 2 to 8 mg/100 g of coconut oil to 113 to 183 mg/100 g of corn oil (Nogala Kalucka, 2003). The amounts of tocopherols in vegetable oils vary according to variety, extraction method, and refining (Flakelar, Luckett, Howitt, Doran, & Prenzler, 2015).

5 Phenolic compounds

Phenolic compounds (PC), the most abundant secondary metabolites in plants, are compounds containing phenol function (Xu, Wang, Pu, Tao, & Zhang, 2017). The simplest phenolic compound is benzene, which contains one hydroxyl group (Cemeroğlu, Yemenicioğlu, & Özkan, 2001). PC have a common chemical structure comprising an aromatic ring with one or more hydroxyl groups that can be divided into several classes. These are flavonoids, phenolic acids, tannins, stilbenes, and lignans (Xu et al., 2017). PC are important in terms of their health functions, effects on taste, odor, pigment formation, antioxidant and antimicrobial effect, enzyme inhibition, and control criteria in different foods (Acar & Gokmen, 2016; Xu et al., 2017). Many studies have demonstrated that PC have various effects such as antioxidant, antimicrobial, anticarcinogenic, antiinflammatory, and estrogen-related prevention of cardiovascular diseases, cancers, diabetes, and diseases associated with oxidative stress (Lin et al., 2016; Louis, Thandapilly, & Kalt, 2014; Siger, Nogala Kalıcka, & Lampart Szczapa, 2008; Xu et al., 2017; Zhang, Lin, & Li, 2016).

A phenolic compound can interrupt the radical chain reaction by donating a hydrogen atom to the free radicals and therefore converting itself to a radical. Phenolics can also act as metal chelators and oxygen scavenger. Phenolics are removed to a certain portion during the refining process, thus PC are found in higher amounts in cold pressed oil than in refined oil (Grosshagauer et al., 2019). Natural olive oil is known to be the most stable oil because of its high amount of PC (Siger et al., 2008).

It has been stated that the most important PC derivatives are in the rapeseed oil including 2,6 dimethoxy-4-vinylphenol (257 μg/100 g) and ferulic acid (5.6 μg/100 g), while vanillic acid (11.4 μg/100 g) is in pumpkin seed oil and ferulic acid (5.8 μg/100 g) is in corn oil. The total amount of PC was determined as 79 mg gallic acid/kg oil in soy oil, 124 mg gallic acid/kg oil in canola oil, 8397 mg gallic acid/100 g oil in palm fruit, and 20–43 mg synapic acid/100 g oil in rapeseed oil (Yemiscioğlu, Özdikicierler, & Gümüskesen, 2016). PC content and profile in plant oil generally depends on the variety, environmental conditions, extraction methods, and storage conditions (Boskou, 1996).

6 Squalene

Squalene, a molecular formula of $C_{30}H_{50}$, is formed by the combination of six isoprene radicals to form symmetry at the midpoint of the squalene molecule (Kayahan, 2016). Squalene is a triterpene hydrocarbon, often found in nature, and acts as a precursor to cholesterol and other sterols (Lu, Jiang, & Chen, 2004). It forms the majority of hydrocarbons in the unsaponified fraction of olive oil (Minguez-Mosquera, Rejano-Navarro, Gandul-Rojas, Sanchez-Gomez, & Garrido-Fernandez, 1991). Olive oil contains 0.2%–07% squalene, while other oils contain 0.02%–0.03% (Bayram & Özçelik, 2012). Squalene is usually stored under the skin after ingestion and is therefore used as a moisturizer and emollient in cosmetics. It also has protective cell properties against free radicals as a potential oxidation inhibitor (Bozdogan & Altan, 2008). Squalene has been reported to inhibit the activity of beta-hydroxy-beta-methylglutaryl-CoA (HMG-CoA) reductase, which plays a key role in cholesterol synthesis. Thus, the cell membrane is displaced, and the risk of larynx, colon, and pancreatic cancers is reduced (Bayaz & Mehenktaş, 2004). According to one study, consuming a certain amount of olive oil every day reduces the risk of breast cancer in women by 25% (Trichopoulou, Katsouyanni, & Stuver, 1995). In a study investigating the effects of squalene on skin, colon, and lung cancer in mice, it was reported that squalene, when consumed daily, has antitumor, antibacterial, and anticarcinogenic effects. Researchers have reported that squalene also plays an important role in eye health, especially for rod photoreceptor cells of the retina (Asman & Wahrburg, 2007).

7 Pigments

The content of pigments in cold pressed oils is higher than in refined oils (Aachary, Liang, Hydamaka, Eskin, & Thiyam-Hollander, 2016). The major pigments in cold pressed oils are carotenoids and chlorophyll (Kayahan, 2003).

7.1 Carotenoids

The characteristic yellow-red color of most vegetable oils is due to the presence of various carotenoid pigments (Bailey, 1951). Carotenoids are isoprenoid components comprising of two C_{20} (geranylgeranyl diphosphate) molecules' coming together and usually containing 40 carbon (C_{40}) atoms (Bailey, 1951; Minguez-Mosquera, Hornero-Méndez, & Pérez-Gálvez, 2002). Carotenoids are lipophilic compounds that exist in acyclic, monocyclic, and bicyclic forms (Bailey, 1951; Minguez-Mosquera et al., 2002). Carbon atoms in the carbon chain forming the molecular skeleton of carotenoids are connected with alternately single or dual conjugated dual bonds (Altan & Kola, 2009). In general, carotenoids can be defined as terpenic alcohols, hydrocarbons, and organic acids. While a large part of carotenoids is present in the structure of the hydrocarbons, the rest exists in the structure of the alcohol (Kayahan, 2003).

The most prominent feature of hydrocarbon carotenoids is that no oxygen is included in their form. In these carotenoids, the most notable one is carotene ($C_{40}H_{56}$) (Kayahan, 2003; Rodriguez Amaya, 1997). There are three isomers of carotene in nature: α, β, and γ. α and β-carotenes are usually together in nature. All three carotenes are physiologically provitamins of vitamin A (Divya, Puthusseri, & Neelwarne, 2012). Another carotenoid in the structure of hydrocarbon is lycopene, giving the tomato its red color (Kayahan, 2003; Minguez-Mosquera et al., 2002). Major carotenoid alcohols are lutein (xanthophylls), cryptoxanthin, and rubicxanthin. Lutein is a dioxi derivative of α-carotene showing optical activity (Kayahan, 2003; Rodriguez Amaya, 1997).

Colors of carotenoids are the reason for the polienic conjugated bonds named chromophore structure (Kayahan, 2003). Carotenoids must include at least seven conjugated double bonds to have a noticeable light yellow color. Because of this, poliens including less conjugated double bonds are not technically regarded as carotenoids (Altan & Kola, 2009).

Carotenoids, which are natural color substances, usually give the foods in which they are present yellow, red, and orange colors. Of these, α-carotene, β-carotene, lutein, and zeaxanthin give a yellow color, lutein gives a green color with yellow, and lycopene gives a red color (Zeb & Mehmood, 2004). It is known that the oil with the highest carotenoid content is crude palm oil (500–700 mg/kg oil). Carotenoid content of the other crude vegetable oils is below 100 mg/kg oil (Yemiscioğlu et al., 2016).

Some important characteristics of carotenoids include absorbing light, being isomerized and oxidized easily, being able to connect hydrophobic surfaces, and being a singlet oxygen quencher (Rodriguez Amaya, 1997). Carotenoids in oils play an important role in the stability of the oil as a singlet oxygen quencher in addition to their coloring properties (Cert, Moreda, & Pérez-Camino, 2000; Psomiadou & Tsimidou, 2002). Carotenoids, together with PC and tocols, are involved in the oxidative stability of oils and have synergist antioxidant effects (Luaces, Perez, Garcia, & Sanz, 2005;

Mba, Dumont, & Ngadi, 2018). Carotenoids play a major role in human nutrition and health due to the fact that they take part in provitamin A and display anticancer properties. Studies have shown that antioxidant carotenoids have protective effects against skin disorders, eye disorders, cancer, and cardiovascular diseases (Campestrini et al., 2019; Rodriguez Concepcion et al., 2018).

7.2 Chlorophyll

Chlorophyll is in porifyrin structure since it is classified into porifyrin group compounds. Porifyrin consists of four pyrrole units connected by methyin bridges. The most important feature of porifyrins is that they tend to chelate with metal ions. The color of porifyrins depends on the compounds in the ring structure and the structure of the central atom. Magnesium is the central atom that chelates metal in chlorophyll (Minguez-Mosquera et al., 2002). Chlorophyll pigment, containing one atom of magnesium at the center of the molecule, consists of chlorophyll *a* and chlorophyll *b* (Cemeroğlu et al., 2001; Psomiadou & Tsimidou, 2001). The structural difference between two chlorophylls stems from the fact that chlorophyll *a* contains one more oxygen atom and two fewer hydrogen atoms than chlorophyll *b* (Kayahan, 2003). The green color of all plant tissues originates from chlorophyll pigment (Minguez-Mosquera et al., 2002). Some oils, such as early harvest olive oil, frost-exposed seed oils, rapeseed, and soybean oil, have a green color from chlorophyll and similar compounds (Kayahan, 2003). Since chlorophyll is temperature-sensitive, it loses its magnesium during heat treatment, which breaks into pheophytin and turns brown-yellow (Kayahan, 2003). With the stimulating effect of oxygen, heat, light, and enzymes, the chlorophyll molecule is easily broken down (Del Giovine & Fabietti, 2005). The breakdown of chlorophylls is much faster than that of carotenoids (Boskou, 1996).

Chlorophyll a and b (1–10 ppm) are responsible for the greenish color of olive oil, and chlorophyll breakdown products a and b (0.2–24 ppm) are responsible for the brown color of the oil (Boskou, 1996; Kayahan, 2003). Among chlorophyll, the one having the highest amount is pheophytin (70%–80%), which is found in olive oil and other vegetable oils (Boskou, 1996; Minguez-Mosquera et al., 1991). Chlorophyll and pheophytin act as pro-oxidants in the photooxidation of oils (Aparicio, Roda, Albi, & Gutierrez, 1999; Tan, Chong, & Low, 1997). Chlorophyll transforms into pheophytin using O_2 in a light environment. It shows antioxidant activity together with phenolic antioxidants in the dark and plays an important role in the stability of olive oil (Psomiadou & Tsimidou, 2001; Velasco & Dobarganes, 2002).

The amount of carotenoid and chlorophyll in the vegetable oils depends on variety, maturity, environmental factors, extraction systems, and storage conditions (Minguez-Mosquera et al., 1991; Sibbett, Connell, Luh, & Ferguson, 1994).

8 Aroma and flavor compounds

Cold pressed oils have a characteristic taste, smell, and flavor (Mounts, 1985). Flavor is tightly correlated with the qualitative and quantitative composition of volatile components in oil (Dun et al., 2019). Taste and odor substances in oils can be examined in two groups, as natural taste and odor substances as well as taste and odor substances formed as a result of technological processes or oxidative degradation. The natural taste and odor substances are hydrocarbons and have a polyethylene structure. As a result of technological processes or oxidative deterioration, taste and odor substances such as aldehyde, ketone, and alcohol are formed in the structure of the oils (Sidar, 2011). Tridecadien, pentadecen, hexadecadien, nonadecen, triacosahexen, and octacosatrien were determined in fresh olive oil and peanut oil (Kayahan, 2003). It is also stated that the characteristic smell of palm kernel oil originates from methyl-nonyl and other ketones. In addition, oils can absorb any foreign taste and odor easily (Kayahan, 2003). The flavor of oils depends on the variety, ripeness degree, environmental condition, growing region, storage, and extraction methods. Many studies have investigated the volatile compounds attributed to plant edible oils. Hexanol, *trans*-2-butenal, and acetic acid were the main volatile compounds in linseed oil, while *trans*-2-butenal and acetic acid, accompanied by *trans*, *trans*-3,5-octadiene-2-one and trans, *trans*-2,4-heptadienal dominated the flavor components of camelina oil (Dun et al., 2019).

Wei et al. (2012) compared the volatile components in rapeseed oils obtained by cold pressing without any treatment, cold pressing with thermal treatment, and microwave radiation. The contents of the degradation products of glucosinolates obtained by cold pressing with thermal treatment and microwave radiation decreased, while the contents of oxidized volatiles and heterocyclic compounds increased. Can-Cauich, Sauri-Duch, Moo-Huchin, Betancur-Ancona, and Cuevas-Glory (2019) reported that the total phenolic compounds, total carotenoids, and squalene of pumpkin oils obtained by mechanical pressing were higher than those obtained by solvent extraction.

References

Aachary, A. A., Liang, J., Hydamaka, A., Eskin, N. A. M., & Thiyam-Hollander, U. (2016). A new ultrasound-assisted bleaching technique for impacting chlorophyll content of cold-pressed hempseed oil. *LWT-Food Science and Technology*, *72*, 439–446.

Acar, J., & Gokmen, V. (2016). In I. Saldamlı (Ed.), *Vol. 5. Phenolic compounds and natural pigments: Food chemistry* (p. 557). Ankara: Hacettepe University Publishers.

Altan, A., & Kola, O. (2009). *Oil processing technology* (p. 227s). Ankara: Bizim Büro Publisher.

Ananth, D. A., Deviram, G., Mahalakshmi, V., Sivasudha, T., & Tietel, Z. (2019). Phytochemical composition and antioxidant characteristics of traditional cold pressed oils in South India. *Biocatalysis and Agricultural Biotechnology*, *17*, 416–421.

Aparicio, R., Roda, L., Albi, M. A., & Gutierrez, F. (1999). Effects of various compounds on virgin olive oil stability measured by rancimat. *Journal of Agricultural and Food Chemistry*, *47*, 4150–4155.

Asman, G., & Wahrburg, U. (2007). *Health effects of the minor components of olive oil (Part I)*. Germany: The Institute of Arteriosclerosis Research, University of Münster.

Awad, A. B., Fink, C. S., Williams, H., & Kim, U. (2001). In vitro and in vivo effects of phytosterols on the growth and dissemination of human prostate cancer PC-3 cells. *European Journal of Cancer Prevention*, *10*(6), 507–513.

Azadmard-Damirchi, S., & Dutta, P. C. (2010). Phytosterol classes in olive oils and their analysis by common chromatographic methods. In V. R. Preedy, & R. R. Watson (Eds.), *Olives and olive oil in health and disease prevention* (pp. 249–259). Oxford: Elsevier.

Azzi, A. (2019). Tocopherols, tocotrienols and tocomonoenols: Many similar molecules but only one vitamin E. *Redox Biology*, *26*, 101259.

Bailey, A. E. (1951). *Industrial oil and fat products. Second completely revised and augmented edition* (p. 967s). New York: Interscience Publishers, Inc.

Bartosinska, E., Buszewska-Forajta, M., & Siluk, D. (2016). GC-MS and LC-MS approaches for determination of tocopherols and tocotrienols in biological and food matrices. *Journal of Pharmaceutical and Biomedical Analysis*, *127*, 156–169.

Baskar, A. A., Ignacimuthu, S., Paulraj, G. M., & Al Numair, K. S. (2010). Chemopreventive potential of β-sitosterol in experimental colon cancer model—An in vitro and in vivo study. *BMC Complementary and Alternative Medicine*, *10*(24), 10.

Bayaz, M., & Mehenktaş, C. (2004). *Lipid based bioactive compounds. Turkey 8*. Bursa: Food Congress. 26–28 May 2004.

Bayram, B., & Özçelik, B. (2012). Bioactive compounds in olive oil and their health benefits. *Food*, *11*, 77–84.

Böhmdorfer, S., Patel, A., Netscher, T., Gille, L., & Rosenau, T. (2011). On the dimers of β-tocopherol. *Tetrahedron*, *67*, 4858–4861.

Boschin, G., & Arnoldi, A. (2011). Legumes are valuable sources of tocopherols. *Food Chemistry*, *127*, 1199–1203.

Boskou, D. (1996). *Olive oil chemistry and technology* (p. 160). Thessaloniki, Greece: Department of Chemistry Aristotle Üniversity of Thessaloniki.

Bozdogan, K. D., & Altan, A. (2008). Olive and olive oil bioactive compounds and physiological effects. *Food*, *33*(6), 297–302.

Bozdogan Konuskan, D., Arslan, M., & Oksuz, M. (2019). Physicochemical properties of cold pressed sunflower, peanut, rapeseed, mustard and olive oils grown in the Eastern Mediterranean region. *Saudi Journal of Biological Sciences*, *26*, 340–344.

Bozdogan Konuskan, D., & Mungan, B. (2016). Effects of variety, maturation and growing region on chemical properties, fatty acid and sterol compositions of virgin olive oils. *Journal of the American Oil Chemists' Society*, *93*, 1499–1508.

Campestrini, L. H., Melo, P. S., Peres, L. E. P., Calhelha, R. C., Ferreira, I. C. F. R., & Alencar, S. M. (2019). A new variety of purple tomato as a rich source of bioactive carotenoids and its potential health benefits. *Heliyon*, *5*, e02831.

Can-Cauich, C., Sauri-Duch, E., Moo-Huchin, V. M., Betancur-Ancona, D., & Cuevas-Glory, L. F. (2019). Effect of extraction method and specie on the content of bioactive compounds and antioxidant activity of pumpkin oil from Yucatan, Mexico. *Food Chemistry*, *285*, 186–193.

Cemeroğlu, B., Yemenicioğlu, A., & Özkan, M. (2001). *Fruit and vegetable processing technology 1. Composition of fruits and vegetables, cold storage* (p. 328s). Food Technology Association Publications. No: 24.

Cert, A., Moreda, W., & Pérez-Camino, M. C. (2000). Chromatographic analysis of minor constituents in vegetable oils. *Journal of Chromatography A*, *881*, 131–148.

Del Giovine, L., & Fabietti, F. (2005). Copper chlorophyll in olive oils: Identification and determination by LIF capillary electrophoresis. *Food Control*, *16*, 267–272.

Divya, P., Puthusseri, B., & Neelwarne, B. (2012). Carotenoid content, its stability during drying and the antioxidant activity of commercial coriander (*Coriandrum sativum* L.) varieties. *Food Research International*, *45*, 342–350.

Dun, Q., Yao, L., Deng, Z., Li, H., Li, J., Fan, Y., et al. (2019). Effects of hot and cold-pressed processes on volatile compounds of peanut oil and corresponding analysis of characteristic flavor components. *LWT-Food Science and Technology*, *112*, 107648.

Dutta, P. C., & Appelqvist, L. A. (1997). Studies on phytosterol oxides. I. Effect of storage on the content in potato chips prepared in different vegetable oils. *JAOCS*, *74*, 647–657.

Fernandez-Cuesta, A., Leon, L., Velasco, L., & De La Rosa, R. (2013). Changes in squalene and sterols associated with olive maturation. *Food Research International*, *54*(2), 1885–1889.

Figueiredo, L. C., Bonafe, E. G., Martins, J. G., Martins, A. F., Marruyama, S. A., Junior, O. O. S., et al. (2018). Development of an ultrasound assisted method for determination of phytosterols in vegetable oil. *Food Chemistry*, *240*, 441–447.

Flakelar, C. L., Luckett, D. J., Howitt, J. A., Doran, G., & Prenzler, P. D. (2015). Canola (*Brassica napus*) oil from Australian cultivars shows promising levels of tocopherols and carotenoids, along with good oxidative stability. *Journal of Food Composition and Analysis*, *42*, 179–186.

Gao, P., Liu, R., Jin, Q., & Wang, X. (2019). Comparison of solvents for extraction of walnut oils: Lipid yield, lipid compositions, minor-component content and antioxidant capacity. *LWT-Food Science and Technology*, *110*, 346–352.

Gao, Y., & Wu, S. (2019). Comprehensive analysis of the phospholipids and phytosterols in *Schisandra chinensis* oil by UPLC-Q/TOF-MS. *Chemistry and Physics of Lipids*, *221*, 15–23.

Gornas, P., Siger, A., & Seglina, D. (2013). Physicochemical characteristics of the cold-pressed Japanese quince seed oil: New promising unconventional bio-oil from by-products for the pharmaceutical and cosmetic industry. *Industrial Crops and Products*, *48*, 178–182.

Grattan, B. J. (2013). Plant sterols as anticancer nutrients: Evidence for their role in breast cancer. *Nutrients*, *5*, 359–387.

Grosshagauer, S., Steinschaden, R., & Pignitter, M. (2019). Strategies to increase the oxidative stability of cold pressed oils. *LWT-Food Science and Technology*, *106*, 72–77.

Gunstone, F. D., & Norris, F. A. (1983). *Lipids in food: Chemistry, biochemistry and technology* (p. 164s). Pergamon Press.

Gylling, H., & Simonen, P. (2015). Phytosterols, phytostanols, and lipoprotein metabolism. *Nutrients*, *7*, 7965–7977.

Herchi, W., Bouali, I., Bahashwan, S., Rochut, S., Boukhchina, S., Kallel, H., et al. (2012). Changes in phospholipid composition, protein content and chemical properties of flaxseed oil during development. *Plant Physiology and Biochemistry*, *54*, 1–5.

Hicks, K. B., & Moreau, R. A. (2001). Phytosterols and phytostanols: Functional food and cholesterol buster. *Food Technology*, *55*, 63–67.

Jones, P. J., & AbuMweis, S. S. (2009). Phytosterols as functional food ingredients: Linkages to cardiovascular disease and cancer. *Current Opinion in Clinical Nutrition and Metabolic Care*, *12*(2), 147–151.

Kayahan, M. (2003). *Oil chemistry* (p. 220). Ankara: ODTÜ Publishers.

Kayahan, M. (2016). In I. Saldamli (Ed.), *Vol. 5. Lipids. p: 197. Food chemistry* (p. 557). Ankara: Hacettepe University Publishers.

Lachman, J., Hejtmankova, A., Orsak, M., Popov, M., & Martinek, P. (2018). Tocotrienols and tocopherols in colored-grain wheat, tritordeum and barley. *Food Chemistry*, *240*, 725–735.

Lin, D., Xiao, M., Zhao, J., Zhuohau, L., Xing, B., Li, X., et al. (2016). An overview of plant phenolic compounds and their importance in human nutrition and management of type 2 diabetes. *Molecules*, *21*(10), 1374.

Louis, X. L., Thandapilly, S. J., & Kalt, W. (2014). Blueberry polyphenols prevent cardiomyocyte death by preventing calpain activation and oxidative stress. *Food & Function*, *5*(8), 1785–1794.

Lu, H. T., Jiang, Y., & Chen, F. (2004). Determination of squalene using high-performance liquid chromatography with diode array detection. *Chromatographia*, *59*, 367–371.

Luaces, P., Perez, A. G., Garcia, J. M., & Sanz, C. (2005). Effects of heat-treatments of olive fruit on pigment composition of virgin olive oil. *Food Chemistry*, *90*, 169–174.

Lukic, M., Lukic, I., Krapac, M., Sladonja, B., & Pilizota, V. (2013). Sterols and triterpene diols in olive oil as indicators of variety and degree of ripening. *Food Chemistry*, *136*, 251–258.

Maeda, H., & Dellapenna, D. (2007). Tocopherol functions in photosynthetic organisms. *Current Opinion in Plant Biology*, *10*, 260–265.

Manai-Djebali, H., & Queslati, I. (2017). Olive oil phytosterols and human health. In T. Fritjof, & B. Henning (Eds.), *Olive oil* (p. 40). Nova Science Publishers, Inc. ISBN: 978-1-53612-563-4.

Mba, O. I., Dumont, M. J., & Ngadi, M. (2018). Characterization of tocopherols, tocotrienols and total carotenoids in deep-fat fried French fries. *Journal of Food Composition and Analysis*, *69*, 78–86.

Mildner-Szkudlarz, S., Rozanska, M., Siger, A., Kowalczewski, P. L., & Rudzinska, M. (2019). Changes in chemical composition and oxidative stability of cold-pressed oils obtained from by-product roasted berry seeds. *LWT-Food Science and Technology*, *111*, 541–547.

Minguez-Mosquera, M. I., Hornero-Méndez, D., & Pérez-Gálvez, A. (2002). Carotenoids and Provitamin A in functional foods. In M. I. Minguez-Mosquera (Ed.), *Methods of analysis for Functional foods and nutraceuticals* (p. 57s). CRC Press LLC.

Minguez-Mosquera, M. I., Rejano-Navarro, L., Gandul-Rojas, B., Sanchez-Gomez, A. H., & Garrido-Fernandez, J. (1991). Colorpigment correlation in virgin olive oil. *JAOC*, *68*, 332–336.

Misawa, E., Tanaka, M., Nomaguchi, K., Nabeshima, K., Yamada, M., Toida, T., et al. (2012). Oral ıngestion of aloe vera phytosterols alters hepatic gene expression profiles and ameliorates obesity-associated metabolic disorders in zucker diabetic fatty rats. *Journal of Agricultural and Food Chemistry*, *60*(11), 2799–2806.

Moghadasian, M. H. (2006). Dietary phytosterols reduce cyclosporine induced hypercholesterolemia in apolipoprotein E-knockout mice. *Transplantation*, *81*, 207–213.

Moreau, R. A., Nyström, L., Whiteaker, B. D., Winkler-Moser, J. K., Baer, D. J., Gebauer, S. K., et al. (2018). Phytosterols and their derivatives: Structural diversity, distribution, metabolism, analysis, and health-promoting uses. *Progress in Lipid Research*, *70*, 35–61.

Mounts, T. L. (1985). Effects of oil processing conditions on flavor stability-degumming, refining, hydrogenation and deodorization. In D. B. Min, & T. H. Smouse (Eds.), *Flavor chemistry of fats and oils* (pp. 79–83). Illinois: The American Oil Chemists' Society Champaign.

Nogala Kalucka, M. (2003). Fat soluble vitamins. In Z. Sikorski, & A. Kolakowska (Eds.), *Chemical and functional properties of food lipids* (p. 118). CRC Press. ISBN: 1-58716-105-2.

Ozcan, M., Al-Juhaimib, F. Y., Ahmed, I. A. M., Osman, M. A., & Gassem, M. A. (2019). Effect of different microwave power setting on quality of chia seed oil obtained in a cold press. *Food Chemistry*, *278*, 190–196.

Piravi-Vanak, Z., Ghasemi, J. B., Ghavami, M., Ezzatpanah, H., & Zolfonoun, R. (2012). The influence of growing region on fatty acids and sterol composition of Iranian olive oils by unsupervised clustering methods. *Journal of the American Oil Chemists' Society*, *89*, 371–378.

Plat, J., Hendrikx, T., Bieghs, V., Jeurissen, M. L. J., Walenbergh, S. M. A., Van Gorp, P. J., et al. (2014). Protective role of plant sterol and stanol esters in liver ınflammation: Insights from mice and humans. *PLoS ONE*, *9*(10), e110758.

Pokorny, J. (2003). Phospholipids. In Z. Sikorski, & A. Kolakowska (Eds.), *Chemical and functional properties of food lipids, p:100* (pp. 79–92). CRC Press. ISBN: 1-58716-105-2.

Psomiadou, E., & Tsimidou, M. (2001). Pigments in greek virgin olive oils: Occurrence and levels. *Journal of the Science of Food and Agriculture*, *81*, 640–647.

Psomiadou, E., & Tsimidou, M. (2002). Stability of virgin olive oil. 1. Autoxidation studies. *Journal of Agricultural and Food Chemistry*, *50*, 716–721.

Raicht, R. F., Cohen, B. I., Fazzini, E. P., Sarwal, A. N., & Takashi, M. (1980). Protective effect of plant sterols against chemically induced colon tumors in rats. *Cancer Research*, *40*(2), 403–405.

Ramadan, M. F. (2013). Healthy blends of high linoleic sunflower oil with selected cold pressed oils: Functionality, stability and antioxidative characteristics. *Industrial Crops and Products*, *43*, 65–72.

Ramprasath, V. R., & Awad, A. B. (2015). Role of phytosterols in cancer prevention and treatment. *Journal of AOAC International*, *98*, 679–684.

Rodriguez Amaya, D. B. (1997). *Carotenoids and food preparation: the retention of provitamin a carotenoidsin prepared, processed, and stored foods.* Ph.D. Campinas, SP, Brazil: Departamento de Ciências de Alimentos Faculdade de Engenharia de Alimentos Universidade Estadual de Campinas C.P. 6121.

Rodriguez Concepcion, M., Avalos, J., Bonet, M. L., Boronat, A., Gomez-Gomez, L., Hornero-Mendez, D., et al. (2018). A global perspective on carotenoids: Metabolism, biotechnology, and benefits for nutrition and health. *Progress in Lipid Research*, *70*, 62–93.

Ros, E. (2010). Health benefits of nut consumption. *Nutrients*, *2010*(2), 652–682.

Rubis, B., Polrolniczak, A., Knula, H., Potapinska, O., Kaczmarek, M., & Rybczynska, M. (2010). Phytosterols in physiological concentrations target multidrug resistant cancer cells. *Medicinal Chemistry*, *6*(4), 184–190.

Rudsinzka, M., Kazus, T., & Wasquwicz, E. (2001). Sterols and their oxidized derivatives in refined and cold pressed plant oils. *Rosliny Oleiste*, *22*, 477.

Saldeen, K., & Saldeen, T. (2005). Importance of tocopherols beyond a-tocopherol: Evidence from animal and human studies. *Nutrition Research*, *25*, 877–889.

Schwartz, H., Ollilainen, V., Piironen, V., & Lampi, A. M. (2008). Tocopherol, tocotrienol and plant sterol contents of vegetable oils and industrial fats. *Journal of Food Composition and Analysis*, *21*, 152–161.

Shahzad, N., Khan, W., Shadab, M. D., Asgar, A., Sundeep, S. S., Sharma, S., et al. (2017). Phytosterols as a natural anticancer agent: Current status and future perspective. *Biomedicine & Pharmacotherapy*, *88*, 786–794.

Shi, T., Zhu, M., Huo, X., Long, Y., Zeng, X. Z., & Chen, Y. (2019). H NMR combined with PLS for the rapid determination of squalene and sterols in vegetable oils. *Food Chemistry*, *287*, 46–54.

Sibbett, G. S., Connell, J. H., Luh, B. S., & Ferguson, L. (1994). *Producing olive oil. Olive production manual publication 3353.* University of California Division of Agriculture and Naturel Resources.

Sidar, H. (2011). *Oil extraction from terebinth seeds: Effect of enzyme and surfactant on the aqueous extraction.* Istanbul Technical University, Science Institute. Master thesis, 83 p.

Siger, A., Nogala Kalıcka, M., & Lampart Szczapa, E. (2008). The content and antioxidant activity of phenolic compounds in cold-pressed plant oils. *Journal of Food Lipids*, *15*, 137–149.

Szymanska, R., & Kruk, J. (2008). Tocopherol content and isomers' composition in selected plant species. *Plant Physiology and Biochemistry*, *46*, 29–33.

Tan, Y. A., Chong, C. L., & Low, K. S. (1997). Crude palm oil characteristics and chlorophyll content. *Journal of the Science of Food and Agriculture*, *75*, 281–288.

Teh, S. S., & Birch, J. (2013). Physicochemical and quality characteristics of cold-pressed hemp, flax and canola seed oils. *Journal of Food Composition and Analysis*, *30*, 26–31.

Temime, S. B., Manai, H., Methenni, K., Baccouri, B., Abaza, L., Daoud, D., et al. (2008). Sterolic composition of Chetoui virgin olive oil: Influence of geographical origin. *Food Chemistry*, *110*, 368–374.

Trichopoulou, A., Katsouyanni, K., & Stuver, S. (1995). Consumption of olive oil and specific food groups in relation to breast cancer risk in Greece. *Journal of the National Cancer Institute*, *87*, 110–116.

Velasco, J., & Dobarganes, C. (2002). Oxidative stability of virgin olive oil. *European Journal of Lipid Science and Technology*, *104*, 661–676.

Wasowicz, E. (2003). Cholesterol and phytosterols. In Z. Sikorski, & A. Kolakowska (Eds.), *Chemical and functional properties of food lipids* (p. 100). CRC Press. ISBN: 1-58716-105-2.93-107.

Wei, F., Yang, M., Zhou, Q., Zheng, C., Peng, J., Liu, C. S., et al. (2012). Varietal and processing effects on the volatile profile of rapeseed oils. *LWT-Food Science and Technology*, *48*, 323–329.

Xu, C. C., Wang, B., Pu, Y. Q., Tao, J. S., & Zhang, T. (2017). Advances in extraction and analysis of phenolic compounds from plant materials. *Chinese Journal of Natural Medicines*, *15*(10), 0721–0731.

Yang, F., Oyeyinka, A., Xu, W., Ma, Y., & Zhou, S. (2018). In vitro bioaccessibility and physicochemical properties of phytosterol linoleic ester synthesized from soybean sterol and linoleic acid. *LWT-Food Science and Technology*, *92*, 265–271.

Yemiscioğlu, F., Özdikicierler, O., & Gümüskesen, A. S. (2016). A new approach in vegetable oil refining: Minimal refining. *Academic Food*, *14*(2), 172–179.

Yorulmaz, H. O., & Bozdogan Konuskan, D. (2016). Antioxidant activity, sterol and fatty acid compositions of Turkish olive oils as an indicator of variety and ripening degree. *Journal of Food Science*, *54*(12), 4067–4077.

Zeb, A., & Mehmood, S. (2004). Carotenoids contents from various sources and their potential health applications. *Pakistan Journal of Nutrition*, *3*(3), 199–204.

Zhang, M., Lin, J. M., & Li, X. S. (2016). Quercetin ameliorates LPS-induced inflammation in human peripheral blood mononuclear cells by inhibition of the TLR2-NF-kB pathway. *Genetics and Molecular Research*, *15*(2), 15028297.

Zhang, L., Wang, S., Yang, R., Mao, J., Wang, X., Zhang, Q., et al. (2019). Simultaneous determination of tocopherols, carotenoids and phytosterols in edible vegetable oil by ultrasound-assisted saponification, LLE and LC-MS/MS. *Food Chemistry*, *289*, 313–319.

Chapter 3

Valorization of by-products from the production of pressed edible oils to produce biopolymer films

Senka Popović, Nevena Hromiš, Danijela Šuput, Sandra Bulut, Ranko Romanić, and Vera Lazić
Faculty of Technology Novi Sad, University of Novi Sad, Novi Sad, Serbia

1 Introduction

One of the most important challenges for agriculture and food industry in the present and the upcoming future is to feed the growing world population (FAO, 2018). The production of a huge amount of agricultural products and their usage for food production lead to significant amount of residues, called agro-waste. A certain amount of agro-waste is considered as by-products, but even these by-products are currently underutilized. Trends and potential solutions are mainly aimed at the production of biofuel and animal feed, as well as extraction of valuable components (i.e., proteins, polysaccharides, phenols, etc.). In this context, production of biodegradable natural polymers—biopolymers—can make a significant contribution to recovery and utilization of obtained wastes and/or by-products (Popović, Lazić, Hromiš, Šuput, & Bulut, 2018). Oilseeds and oilseed products play an important role in providing a nutritionally balanced diet. After production of oil from oilseeds, valuable by-product, cake, or meal remains. Considering their composition (high content of proteins and polysaccharides), these materials present promising substrates for eco-friendly biopolymer packaging materials.

2 By-products of oilseeds processing

Depending on modern concepts of sustainable development and environmental protection, industrial production as the ultimate goal has the maximum utilization of all resources and practical production without any waste. Sustainable development concerns the safe present and leads to a secure future since, at the same time, the needs for raw materials and energy sources have been reduced.

The oilseed industry creates significant quantities of by-products that are currently underused. Depending on the type and quality of oilseeds, the oil extraction process, and the quality of the obtained by-products, their application could be not only for animal feed but also for a wide range of food products. Most research on the subject showed that in addition to an adequate hygienic approach of complete technological processing and health safety, the by-products of a particular process can have favorable functional characteristics and are suitable for human consumption. Of particular importance is the use of various oilseed cakes obtained in the processing of pressing; therefore, without the use of organic solvents, oilseed cake maintains the favorable nutritional properties of the starting material.

3 Oilseeds processing technology

Usage of by-products deriving from the processing of seeds and oilseed has become a necessity in terms of creating new products instead of creating waste products. The use of by-products has become an integral part of the usual production process, by increasing the efficiency of processing and using energy. A scheme of the technological process of production of edible oils, followed by cakes and meals, as by-products, is shown in Fig. 1.

Edible oils can be produced as nonrefined (cold pressed and virgin) or as refined edible oils. In the production of edible nonrefined oils, cake is obtained as a by-product. From the cake after the production of crude oils, the remaining oil is extracted by solvent extraction and remains as a secondary product. Crude pressed oil, crude extracted oil, or a mixture

Cold Pressed Oils. https://doi.org/10.1016/B978-0-12-818188-1.00003-7
Copyright © 2020 Elsevier Inc. All rights reserved.

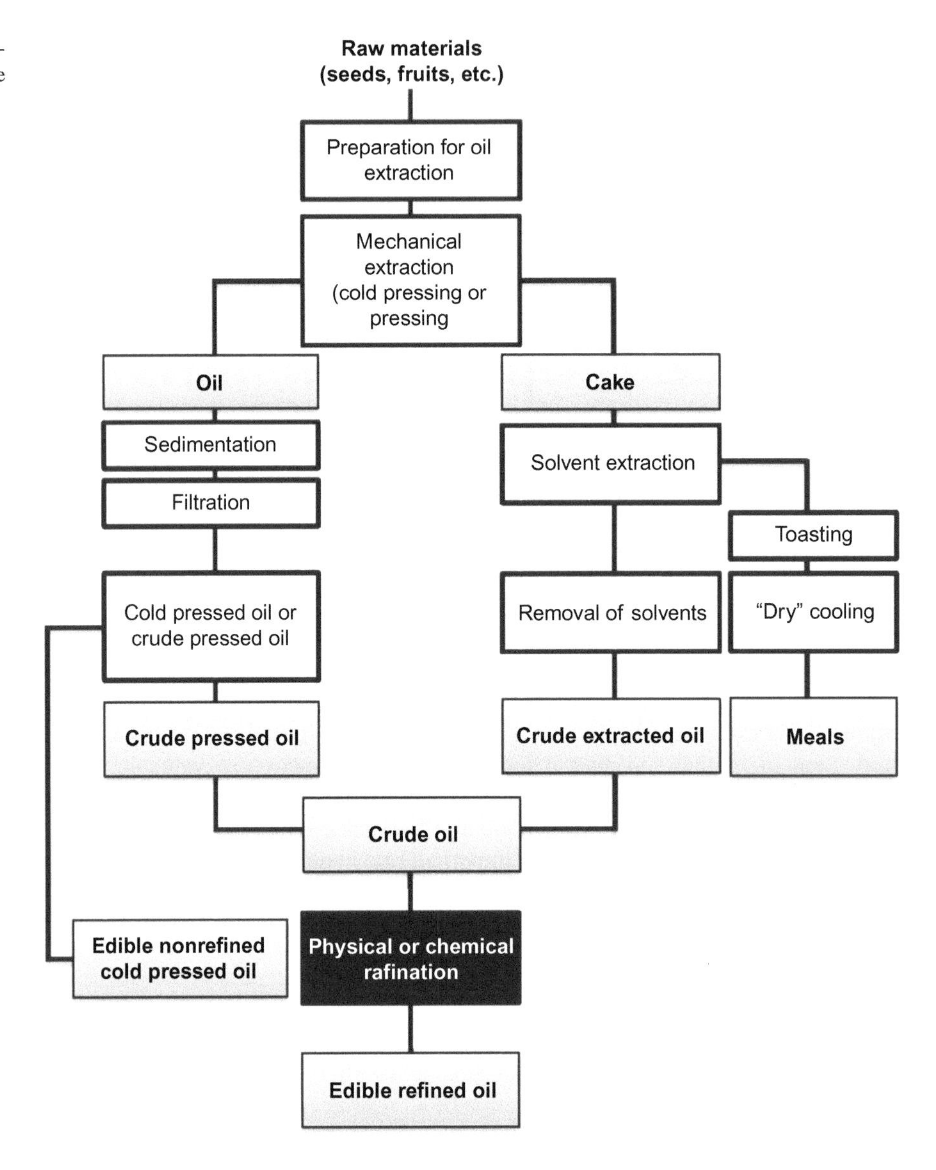

FIG. 1 Technological scheme of production of nonrefined and refined edible oils, cakes, and meals.

is passed through a whole series of steps of the chemical or physical refining process, to obtain edible refined oil (Dimić, 2005; Romanić, 2015; Shim, Gui, Wang, & Reaney, 2015).

4 Cakes composition and characteristics of some oilseeds

The cake is a by-product (Fig. 2) that remains after the extraction of oil from oilseeds by the process of pressing, primarily in the production of edible nonrefined or cold pressed oils. Cakes can be edible or inedible. Some edible oilseed cakes have a high nutritional value. Their composition and nutritional value depend on the type, the conditions of growing the raw material, and oilseeds processing route (Dimić, Romanić, Peričin, & Panić, 2006; Peričin, Radulović, Mađarev, & Dimić, 2007).

Cakes and flour, obtained after the oil extraction from oilseeds by pressing, contain significant amounts of residual oil and are rich in proteins, nutrients (vitamins and carotenoids), and minerals. Table 1 shows the chemical composition of some oilseed cakes.

However, the cake of most oilseeds also contains undesirable antinutritive ingredients (i.e., chlorogenic acid in sunflower seed, glucosinolates, tannins, synapin, phytic acid in rapeseed, etc.), which must be removed so that the cake becomes nutritionally acceptable. Some cakes obtained after the extraction of oil do not contain antinutritive ingredients,

FIG. 2 Cakes of sunflower seeds obtained from the production of cold pressed oil.

TABLE 1 Chemical composition of some oilseed cakes (Peričin et al., 2007; Radočaj, 2011).

Oil cake	Dry matter (%)	Crude proteins (%)	Crude fiber (%)	Ash (%)	Oil (%)
Sunflower seed	90.54	30.07	14.73	4.95	15.62
Soybean	92.69	41.77	3.41	5.09	10.90
Rapeseed	90.46	29.24	9.86	–	18.25
Hull-less pumpkin seed	94.83	61.02	5.11	8.42	8.37
Hempseed	93.65	22.21	13.09	5.24	9.53

such as seeds of hull-less pumpkin oilseed, which have strong potential to be used as a feedstock, for the production of fortified foods, functional products, food ingredients, or for pharmaceutical purposes (Dimić et al., 2006; Radočaj, 2011).

There are many studies regarding the usage and functional properties of soybean (Escamilla-Silva, Guzman-Maldonado, Cano-Medinal, & Gonzalez-Alatorre, 2003; Singh, Kumar, Sabapathy, & Bawa, 2008) and rapeseed proteins (Khattab & Arntfield, 2009). The use of the cake after pressing of peanuts was also examined, confirming its use for the production of extruded product snacks (Prinyawiwatkul, Beuchat, Phillips, & Resurreccion, 1995). Adapted products obtained from the extraction of oil of various oilseeds, such as soybeans, rapeseed, sunflower, coconut, cotton, palm, sesame, and flax, have already been widely used in both the food and pharmaceutical industries (Sunil, Prakruthi Appaiah, Prasanth Kumar, & Gopala Krishna, 2015; Teh & Bekhit, 2015).

Oil cakes have not been used sufficiently. They could have a much wider application, including biopolymer-based materials production, with potential application for food and pharmaceutical packaging.

5 Oilseed by-products application for biopolymer packaging materials production

Biopolymer-based packaging material is usually produced in the form of film or coating. Films and coatings can have the same chemical composition, but the main difference is their physical appearance: films are self-standing materials, while coatings are formed on the substrate surface (Nandane & Jain, 2015). For film production, two different processes can be applied; the first is wet process or casting, and the other is dry process, which implies thermo-mechanical processing such as compression molding or extrusion is applied in order to obtain a film (Garrido, Etxabide, Peñalba, Guerrero, & de la Caba, 2013; Garrido, Leceta, Cabezudo, de la Caba, & Guerrero, 2016).

Films are primarily developed using the solvent casting method (Fig. 3), where the film-forming macromolecules are dissolved/dispersed in an appropriate medium, whose evaporation leads to film-forming.

The application of biopolymer films as packaging materials is influenced by many film characteristics: mechanical, barrier, structural, optical, thermal, biological, etc.

Mechanical properties are of great importance due to the preservation of packaged product integrity (Prakash Maran, Sivakumar, Thirugnanasambandham, & Kandasamy, 2013). The mechanical properties of biopolymer films depend on the

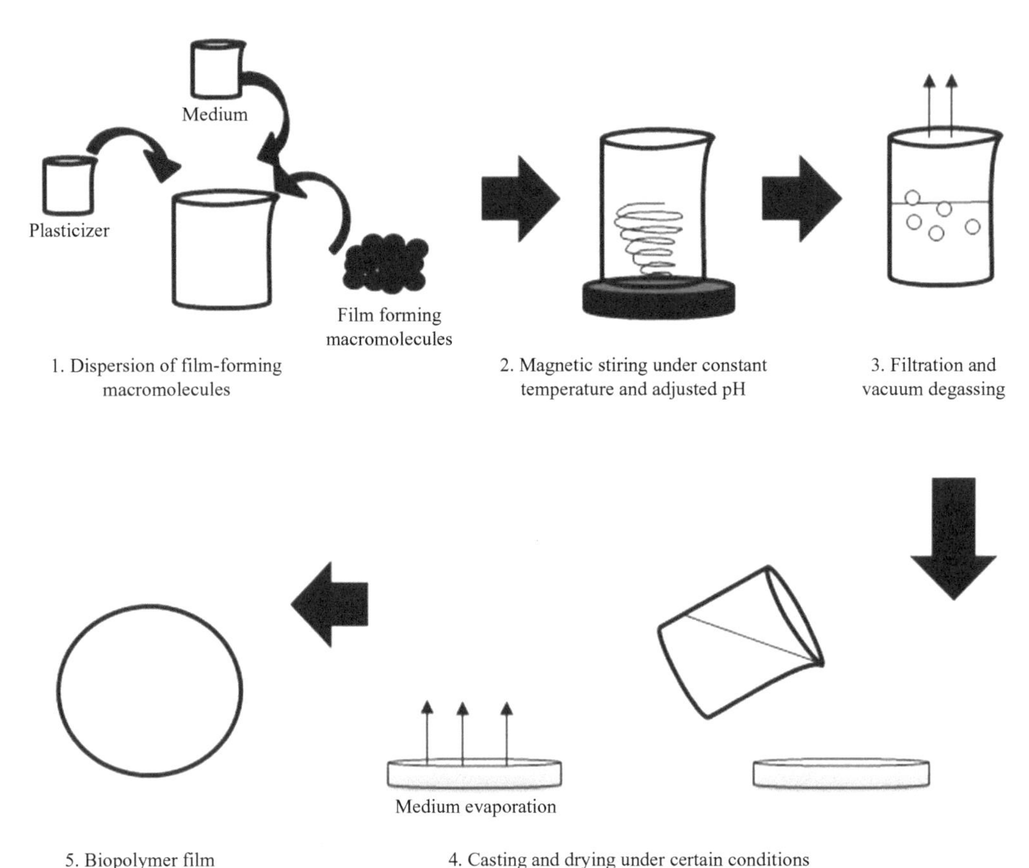

FIG. 3 Solvent casting method.

structure of the polymer chains, the coherence of the polymer matrix, the interaction of the additive and the matrix, the mode of film production and the presence of plasticizers, additives used to improve film elasticity, with glycerol as representative and most commonly used plasticizer. The basic mechanical properties of packaging films are their elongation at break, tensile strength, and module of elasticity.

Barrier properties depend on the chemical composition of the film, the process of obtaining films, and the method of applying the film to the surface of the product (Rodrıguez, Oses, Ziani, & Mate, 2006). The most important barrier properties are gas, water vapor, light, and aroma permeability. Gas permeability is determined by measuring the gas diffusion through the film, which is quantified by a gas chromatograph (García, Martino, & Zaritzky, 2000). Although hydrocolloid films are a good barrier to gases, they represent a very poor barrier to water vapor. Their hydrophilic nature is one of the limiting factors for widespread usage (Falguera, Quinterob, Jimenez, Munoz, & Ibarz, 2011). Water vapor permeability can be reduced by adding lipid components to a film solution, such as waxes or essential oils, which increases the hydrophobicity of the film (Anker, Berntsen, Hermansson, & Stading, 2001; Ayranci & Tunc, 2003; Sanchez-Gonzalez, Vargas, Gonzalez-Martinez, Chiralt, & Chafer, 2011), or mixing with various other biopolymers (Xu, Kim, Hanna, & Nag, 2005).

Water sensitivity and hydrophilicity of biopolymer films is commonly assessed through film moisture content in specific ambient conditions, water swelling, water uptake, solubility in water (or total soluble matter) and water contact angle.

The structural and thermal characteristics of biopolymer films depend on factors such as the ratio of crystalline and amorphous zones, the mobility of polymer chains, and specific interactions between functional groups of polymers and amorphous zones. There are a number of techniques based on which the structure of films can be studied, including scanning electron microscopy, X-ray diffraction, Fourier transform infrared (FTIR) spectroscopy, differential scanning calorimetry, and thermo-mechanical analysis.

Sensory characteristics are significant in the production of packaging. Preferably, materials should be transparent, odorless, tasteless, and clean. The hydrocolloid materials are substantially more neutral than lipid materials and wax-based materials. The optical properties of biopolymer materials depend on the composition and method of production of these materials.

The active function of biopolymer films is reflected in the protection of food products from oxidation and microbiological failure, which leads to improved quality and improvement of the safety of the packaged product (Kim, Yang, Noh, Chung, & Min, 2012; Lee, Noh, & Min, 2012). It is possible to implement additives, such as antimicrobial agents, antioxidants, and pigments in biopolymer films, and then control the rate of release of these additives into foods (Chillo et al., 2008).

Biopolymer films are completely biodegradable, unless previously chemically modified. The biodegradation kinetics depends on the type of polymer (molecular weight, structure, and amorphism) and the substances used to improve the properties (plasticizers, fillers, etc.). The degree of biodegradability can be adjusted depending on the application of biopolymer packaging materials (Guilbert & Gontard, 2005).

5.1 Oil cakes valorization for biopolymer films production

Pumpkin oil cake (PuOC), a by-product obtained after extracting oil from pumpkin (*Cucurbita pepo*) seed by cold pressing, represents a valuable source of natural macromolecules. The first evaluation of PuOC application as a substrate for biopolymer films occurred in 2010, when Popović, Lazić, Lj, Vaštag, and Peričin (2010) examined the possibility of replacing a part of gelatin with PuOC to form new, cheaper biopolymer materials, with adequate characteristics to act as packaging material. The films with the best tensile strength were obtained when 40% of gelatin was substituted with PuOC. Elongation at break increased with an increase of PuOC percentage in films, and finally a film with 95% of PuOC had 20% higher elongation at break than a film made of pure gelatin. In addition, films prepared with increased content of plasticizer glycerol had up to three times higher elongation at break than films prepared with a lower glycerol/protein ratio. Other examined characteristics (swelling properties, solubility, color, and antioxidant activity) were similar or improved when PuOC was added to gelatin. Obtained results showed that PuOC possesses film-forming ability.

Further experiments evaluated the feasibility of solo PuOC to form biodegradable films (Popović, Peričin, Vaštag, Popović, & Lazić, 2011). PuOC was used as a substrate, and the effect of heating treatment and pH of film solutions on films properties was determined. The film with the highest tensile strength and elongation at break, as well as with the lowest gas permeability, was produced at pH 12 and temperature 90°C. A film prepared at pH 12 and 50°C showed the highest value of total soluble matter and soluble protein. Another author's observation was that obtained films showed antioxidant activity. A film produced at pH 10 and 60°C had the best radical scavenging activity. The influence of process parameters was also observed using scanning electron microscopy. A film prepared at 90°C had a very compact structure surface, with only nanoholes observed. On the other hand, a film prepared at 50°C had pores that could be described as micro-holes, which explained why this film possesses worse gas barrier properties.

Further investigation of the potential application of PuOC films was to examine the possibility to form pouches. PuOC films are not heat-sealable so their lamination with zein (material with heat sealing ability) was done (Bulut, Lazić, Popović, Hromiš, & Šuput, 2017a, 2017b, 2017c). Visual, mechanical, physicochemical, and structural properties of the films were evaluated. The obtained results showed that the PuOC film exhibited higher elongation at break, compared to the bilayer PuOC/Zein film. The zein monolayer film showed the lowest elongation at break. However, the zein film showed the highest tensile strength, followed by the PuOC film, while the PuOC/zein film showed by far the lowest value for tensile strength. Examined physicochemical properties indicated that the zein film was the least sensitive to moisture, due to its hydrophobic nature. On the other side, the PuOC film showed moisture sensitivity. The combination of a hydrophilic PuOC and a hydrophobic zein layer led to reduced moisture content, swelling, and total soluble matter of the bilayer PuOC/Zein film, compared to the PuOC film. FTIR analysis was carried out and there were no significant differences between PuOC and zein spectra, due to the predominantly protein nature of PuOC and the pure protein nature of zein.

Bulut et al. (2016) compared barrier properties of laminated PuOC/zein biopolymer pouches and commercial high-barrier pouches made of polyamide-polyethylene along with the possibility to maintain modified atmosphere in new biopolymer pouches. As a model food product, semi-hard mozzarella cheese was used. Comparing O_2 permeability of biopolymer and commercial materials, it was shown that both materials have excellent O_2 barrier properties. Considering CO_2, PuOC/zein materials have higher permeability, which can be an advantage for packaging food products with high respiration rates, such as cheese.

In order to improve mechanical and barrier characteristics, Bulut et al. (2017a, 2017b, 2017c) investigated the influence of glycerol and guar-xanthan in different concentrations on a bilayer film based on PuOC/zein. The results obtained showed that the concentration of glycerol and guar-xanthan has influence on the mechanical and gas barrier properties of the PuOC layer of duplex PuOC/zein films. Optimal mechanical properties had a film with the lowest concentration of glycerol and the highest concentration of guar-xanthan. All films produced showed a poor barrier for water vapor and a very good barrier to O_2. With increase of additive concentration, transmission rate of CO_2 increased.

The aging process of biopolymer materials is one of the key factors for their application in food packing. Duplex PuOC/zein films, as well as pouches made from them, were prepared, and the changes in their properties were analyzed during 4 weeks of storage (Bulut et al., 2017a, 2017b, 2017c). The heat seal quality of formed pouches and composition of the gas atmosphere in the pouches were also monitored. The results showed that thickness and tensile strength of the duplex films stayed unchanged during 3 weeks of storage, afterwards a slight increase was observed. On the other side, elongation at break, moisture content, total soluble matter, and swelling of the obtained film showed a decreasing trend during the whole storage period. After 1 month of storage, the transmission rate of O_2 and CO_2 through films increased. Analyzing the quality of pouches formed during that time, a decrease in heat seal strength of the PuOC/zein pouches was observed, followed by a change of modified atmosphere in the formed pouches. This was followed by an increase in percentage of O_2 in the pouches, up to seven times during the storage period. However, the percentage of CO_2 decreased up to 18 times after 1 week and stayed stable for the next 3 weeks. The results obtained for gas content in pouches are probably the consequence of decreases in heat seal strength and gas transmission rate through material.

Earlier work (Popović, 2013; Popović et al., 2010) pointed at the antioxidant activity of PuOC films without incorporating any active components. In order to increase the activity of PuOC films, Bulut et al. (2018a, 2018b) produced an active film PuOC with the addition of caraway essential oil, as well as caraway essential oil and chitosan, in different proportions. The authors analyzed the antioxidant activity of obtained films after productions and after 1 month of storage period. By the addition of chitosan in the PuOC film, activity was significantly increased. However, addition of chitosan only impacted the film with the lowest value of caraway essential oil, and its activity was similar to that of the film containing caraway essential oil. There was no significant increase in the antioxidant activity of films with chitosan and higher values of caraway essential oil, compared to PuOC film activity. A storage period of 1 month has caused a mild decline in the antioxidant activity of the produced films.

Bulut et al. (2018a, 2018b) produced active films based on PuOC with winter savory (*Satureja montana*) and basil (*Ocimum basilicum*) essential oil, with different concentrations of Tween 20. The film based on PuOC without essential oils and Tween 20 was used as the control. Analysis of antioxidant activity indicates that incorporation of essential oils and Tween 20 increased activity of the PuOC films, compared to the control film. The addition of different concentrations of Tween 20 affected the activity of the PuOC film with winter savory essential oil more than the activity of the film with basil essential oil. With an increase of Tween 20 concentration, antioxidant activity of the PuOC film with winter savory increased regarding all measured parameters. The effect of Tween 20 on the antioxidant activity of the film with basil essential oil only was observed in a hydroxyl radical scavenging activity assay, where the antioxidant activity decreased with the increasing concentration of Tween 20.

Sunflower oil cake (SFOC) represents an inexpensive source of proteins which can be isolated (Ramachandran, Singh, Larroche, Soccol, & Pandey, 2007) and used to form films with good characteristics, including barrier and adhesive characteristics as well as resistance to oily and organic solvents (De Graaf, Harmgen, Vereijken, & Monikes, 2001; Nesterenko, Alric, Violleau, Silvestre, & Durrieu, 2013). However, there are insufficient data on the properties of biopolymer films synthesized from the solo SFOC. Lazić et al. (2017) obtained biopolymer films from solo SFOC by the casting method. The film-forming solution was prepared by pH adjustment and heating, after which it was fine- and coarse-filtrated. The films obtained were dark brownish-green, with mildly expressed scent on the sunflower cake. Films were firm and flexible. Compared to many synthetic polymers, the values of tensile strength and elongation at break for biopolymer films obtained from whole SFOC are fairly low, but sufficient for these films to find application in the food industry (Park, Byun, Kim, Whiteside, & Bae, 2014; Souza, Goto, Mainardi, Coelho, & Tadini, 2013). The film samples obtained represent a good barrier to UV light at wavelengths less than about 500 nm. Tested film samples can be used for packaging products that are susceptible to oxidative changes, which are catalyzed by UV light.

Šuput et al. (2018) further investigated the effects of processing parameters (pH and temperature) during the casting procedure on biopolymer film properties. The obtained films were dark and shiny, with a hint of sunflower smell and good tactile characteristics (smooth, flexible). Tensile strength values increased along with temperature and pH values increments. Water vapor permeability values were uniform, and lower values were observed in samples synthesized at higher temperatures. Maximal light transmission value was reached at 800 nm. The values obtained for swelling and solubility are in a characteristic range for most biopolymer films. The authors proposed films obtained at 90°C and a pH value of 12 as optimal. In a further step, Lazić et al. (2018) activated biopolymer films obtained from SFOC by the addition of essential oils of parsley and rosemary. Antioxidant activity was more pronounced with the addition of parsley essential oil. The addition of essential oils influenced all tested biopolymer film properties. Regarding mechanical properties, tensile strength values were reduced, and the elongation at break was increased as essential oils were added. Water vapor permeability values decreased on increasing the concentration of essential oil of parsley and increased with the addition of higher concentrations of rosemary oil. The addition of lipophilic component (parsley essential oil) reduced the swelling degree and solubility. All examined films, control and active, regardless of the added type of oil, present good protection in the UV wavelength range, similar to the basic SFOC film.

5.2 Application of protein isolates and concentrates derived from edible oil industry by-products for biopolymer films production

The potential application of PuOC protein isolates (PuOC PI) as a substrate for biopolymer film production was evaluated by Popović, Peričin, Vaštag, Lazić, and Popović (2012). Preparations of film were examined at different pH values (2–12) with different plasticizer content. New biodegradable films were formed in a wide range of pH. Films could not be synthesized at pH = 4–8. The pH of film-forming solutions influenced all examined film characteristics (mechanical, physicochemical, barrier, and antioxidant). Further analysis was to evaluate the influence of the plasticizer concentration. Too-low concentration led to brittle films, while too-high concentration left films wet and sticky. Only films with intermediate concentration of plasticizer were suitable. The amount of glycerol added significantly affected the mechanical characteristics and solubility of the films. PuOC PI films were shown to represent an excellent barrier for O_2, N_2, CO_2, and air. For comparison, the gas barrier properties of these films are 150–250 times better than the commercially used plastic films polyethylene and biaxially oriented polypropylene.

Sunflower protein isolates (SFPI) and concentrates (SFPC) have been shown to have the ability to form biodegradable and/or edible films (Ayhllon-Meixueiro, Vaca-Garcia, & Silvestre, 2000; Orliac, Silvestre, Rouilly, & Rigal, 2003; Rouilly, Meriaux, Geneau, Silvestre, & Rigal, 2006). While these proteins can be extracted from seeds, they can also be extracted from the residual cake produced in the oil industry, currently used for animal feeding only. During oil extraction, the proteins are exposed to treatment with organic solvents, at high pressures or temperatures, which modify their structure and, consequently, their functionality. Plastic properties of SFPI have been demonstrated and SFPI-based materials could be formed by casting (Ayhllon-Meixueiro et al., 2000), thermo-molding (Orliac, Rouilly, Silvestre, & Rigal, 2002), injection-molding (Orliac et al., 2003), and extrusion (Rouilly et al., 2006). Flexible biodegradable films, obtained from sunflower protein isolates and concentrates, generally have mechanical and barrier properties in the same range of other protein films.

Ayhllon-Meixueiro et al. (2000) investigated biodegradable films from sunflower protein isolates. Films were obtained by dissolution of SFPI in the alkaline range, addition of a plasticizer, casting, and drying. Films were excessively brittle when the glycerol content in the film was $<16\%$, and they became sticky when it was $>50\%$. Therefore, the optimal glycerol/SFPI ratio was 1:2. Elongation at break increased with increasing protein content in the film. On the other hand, the tensile strength values of films remained constant when the protein content of the film was varied. The mechanical properties of the SFPI film depend not only on the protein content but also on the protein unfolding extent. SFPI films were more elastic and resistant than films obtained from isolate of soy proteins at comparable contents of plasticizer.

Orliac and Silvestre (2003) focused on aging and the physical properties of films based on SFPI obtained by thermo-molding. Regarding mechanical properties, it was reported that an increase in the content of the plasticizer led to a decrease in mechanical resistance and an increase in elongation at break. The authors concluded that the protein network is almost insoluble and that film dry matter solubilized in water is mostly consisted of plasticizer and low molecular weight polypeptides. Film water vapor permeability was similar to the water permeability of low-density polyethylene films and lower than for most other protein films. Films' water vapor permeability increases along with the addition of plasticizer content. During long-term storage, a decrease in the mechanical properties of films was observed, which was probably due to plasticizer volatilization.

Orliac et al. (2003) described the thermoplastic behavior of mixtures of sunflower protein isolate, glycerol, and water to demonstrate their suitability for the manufacture of injection-molded objects. They managed to make injected objects containing only small amounts of plasticizer, entirely biodegradable in a liquid medium, which allowed degradation prediction. The objects obtained had good mechanical properties, but their water-resistance was low. These materials were less elastic and less water-resistant than films obtained by thermo-molding (Orliac & Silvestre, 2003).

Rouilly et al. (2006) obtained biodegradable protein isolate-based films by film extrusion. It was demonstrated that the highest temperature, well above the SFPI denaturation temperature, the highest glycerol content, and medium water content gave the most regular and smoothest film. Soaked in water, the film was insoluble.

Salgado, Molina Ortiz, Petruccelli, and Mauri (2010) conducted research on SFPI films naturally activated with antioxidant compounds with a soy isolate protein film as a reference. All the films had thickness, water vapor permeability, densities, and moisture contents with no significant differences regardless of the protein isolate used. The soy isolate films had similar resistance but higher values of deformation at break and elastic modulus than those obtained from sunflower protein isolates. Although films prepared from sunflower isolates with different content of phenolic compounds did not differ in terms of thermal, mechanical, and barrier properties, they presented a very distinct aspect due to differences in color. All the sunflower protein isolate films had antioxidant properties, which is absent in soy films. In further research, Salgado, Molina Ortiz, Petruccelli, and Mauri (2011) added bovine plasma hydrolysates (BPH) in order to improve the antioxidant properties of soybean and sunflower protein-based films. BPH implied decrease in tensile strength, elastic modulus, and glass transition temperature along with increase in elongation at break and water vapor permeability. BPH considerably affected films' antioxidant properties. The increase in the antioxidant capacity of the films with the addition of BHP is the same regardless of the origin of the film, although higher antioxidant activity was also proven in SFPI samples compared to SPI samples. The antioxidant capacity proved to be the result of the sum of the characteristics of the protein matrix and characteristics of the hydrolysate. No synergistic or antagonistic effects were observed.

Salgado et al. (2012) investigated the antioxidant and antimicrobial capacity of two sunflower protein concentrate films including sunflower protein concentrate film (SFPC) and sunflower protein concentrate obtained with isoelectric precipitation (SFPCIP). Film-forming dispersions were prepared by dispersing sunflower protein concentrates and glycerol in distilled water. Films were prepared by casting. The resulting sunflower protein films (SFPC and SFPCIP) showed very high water solubility. In agreement with the high water solubility of both films, the percentages of release of proteins and phenolic compounds to the aqueous phase were also very high. Both SFPC filmogenic dispersions and film water-soluble fractions exhibited significantly higher values than their SFPCIP counterparts, indicating a greater antioxidant capacity. The phenolic compounds naturally present in the sunflower protein concentrate, considering either the filmogenic dispersions or the resulting protein films, showed no antimicrobial activity.

Literature data regarding biopolymer films based on proteins favor soy proteins for formulating biodegradable films, due to their good film-forming ability (Monedero, Fabra, Talens, & Chiralt, 2009; Sun et al., 2011). Soy protein isolate (SPI) films' properties were shown to be influenced by numerous factors, such as molecular distribution, extraction process, SPI concentration, plasticizer type, and concentration, as well as pH in film-forming solution, changing drying conditions, and application of high temperature and pressure in dry process of film formation (Chiralt, González-Martínez, Vargas, & Atarés, 2018; Denavi et al., 2009; Garrido et al., 2013; Garrido et al., 2016; Han, Li, Chen, & Li, 2017; Nandane & Jain, 2015; Park et al., 2014; Tang, Jiang, Wen, & Yang, 2005; Wu, Cai, Wang, Mei, & Shi, 2018).

Variation of these factors leads to changes in the functional properties of SPI films, mechanical properties, water sensitivity and hydrophylicity of films (moisture content, water swelling, water uptake, solubility, and contact angle), and water vapor and oxygen permeability.

Different studies were conducted in order to improve SPI-based films' properties without compromising the low O_2 permeability of these films in dry conditions. These studies included the formation of composite films with different hydrocolloids, like guar gum (Sui, Zhang, Ye, Liu, & Yu, 2016) or gelatin (Cao, Fu, & He, 2007), with lipids (Galus, 2018) or bilayer films with PLA (González & Igarzabal, 2013). Further attempts to improve the properties of SPI-based films include application of cellulose and starch nanocrystals in order to form nanocomposite films (González & Igarzabal, 2015; Han, Yu, & Wang, 2018; Kumar, 2014), as well as enzyme-mediated or chemical crosslinking with dialdehyde carboxymethyl cellulose (Zheng, Yu, & Pilla, 2017), methyl methacrylate (González & Igarzabal, 2017), rutin and epicatechin (Friesen, Chang, & Nickerson, 2015), and genipin (González, Strumia, & Igarzabal, 2011). Chemical modifications of soy proteins, such as acetylation and graft polymerization of acrylic acid (Zhao, Xu, Mu, Xu, & Yang, 2016), were shown to decrease water vapor permeability (Foulk & Bunn, 2001). In addition, application of high pressure and high temperature was shown to increase tensile strength and decrease water sensitivity and permeability of films. In this sense, film production processes: compression molding using acetylated soy proteins; extrusion in combination with other thermoplastic compounds and high-pressure homogenization were proposed (Cao et al., 2007; Foulk & Bunn, 2001; Wu et al., 2018).

In order to add more value and widen the application range of soy protein-based edible films, different bioactive compounds of synthetic or natural origin were added into SPI-based films, such as tutin, epicatechin (Friesen et al., 2015), red grape extract (Ciannamea, Stefani, & Ruseckaite, 2016), mango kernel extract (Maryam Adilah, Jamilah, & Nur Hanani, 2018), natamycin and thymol (González & Igarzabal, 2013), mandelic acid (Kumar, Anandjiwala, & Kumar, 2016), silver Ag^+ ions and Ag nanoparticles (Sun et al., 2011), cortex phellodendron extract (Liang & Wang, 2018), pine needle extract (Yu et al., 2018), citronella essential oil (Arancibia, López-Caballero, Gómez-Guillén, & Montero, 2014), and nisin in combination with citric, lactic, malic, or tartaric acids (Eswaranandam, Hettiarachchy, & Johnson, 2006). Added active ingredients influence most of the film properties including mechanical, barrier, and water sensitivity. In this way, together with transforming an edible film to an antioxidant or antimicrobial edible film, or both, film functional properties can be optimized.

When soy protein biopolymer film production is reviewed, it can be noted that by far the most commonly used source of soy protein is SPI. Soy protein concentrate (SPC) has limited data for usage in biopolymer film production, due to the high content of nonprotein components, which may significantly influence film characteristics (Ciannamea, Stefani, & Ruseckaite, 2014). Cho, Park, Batt, and Thomas (2007) produced films based on membrane-processed SPC and compared their characteristics to the films based on commercial SPI. The mechanical and barrier characteristics examined were comparable, while protein solubility was higher for SPC-based films, and, in line with this result, the opacity of SPC-based films was lower than that of SPI-based films for pH under 9. In the work reported by Ciannamea, Sagüés, Saumell, Stefani, and Ruseckaite (2013), for compression-molded SPC-based films plasticized with glycerol, tensile strength, and elongation at break were similar to the films produced by casting, and also in the range of values reported for SPI-based films (Ciannamea, Espinosa, Stefani, & Ruseckaite, 2018). In molded films, film water content, tensile strength (slightly), elongation at break, and oxygen permeability were higher, while film solubility and water vapor permeability were lower (Ciannamea et al., 2014). The composite casted film based on SPC and corn starch exhibited tensile strength that exceeded tensile strength values of both individual films. With the addition of Cloisite 30B nanoclay as a filler into the composite film, tensile strength was additionally increased (Kishore Pradhan & Das, 2017). An increase in the tensile strength of the SPC-Cloisite 30B nanocomposite was also observed when a chitosan layer was coated on these films (Kishore Pradhan, Das, & Nayak, 2018). Antioxidant films based on SPC were formed with the addition of red grape extract into films produced by casting and compression molding. Both films showed antioxidant activity, but the release of antioxidants from casted films was lower, so the activity of these films was considerably reduced. The addition of a hydrophobic active compound into the casted films led to a certain moisture sensitivity decrease, as well as a water vapor permeability decrease (Ciannamea et al., 2016).

It was shown for soy protein-based films, similar to other protein-based films, that they suffer aging during long-term storage. Physical changes involved in the film aging process occur because films are in a nonequilibrium condition at an ambient temperature and tend to approach the thermodynamic equilibrium through time-dependent spontaneous relaxation and molecular rearrangements. In addition to physical changes, the aging process imparts also chemical transformations including aggregation and thiol oxidation (Ciannamea et al., 2018; Ciannamea, Stefani, & Ruseckaite, 2015).

Jang, Lim, and Song (2011) and Jang, Shin, and Song (2011) developed films based on rapeseed protein (RP) derivate and their physical properties were examined. Various concentrations of different plasticizers and emulsifier were incorporated into the film-forming solution. The authors suggested that the RP film could not be formed without plasticizers and emulsifier, and optimal conditions for the preparation of the RP film were given. The influence of plasticizer and emulsifier content was observed on mechanical properties of film only by improving them. The water vapor permeability of the film was not affected by plasticizer content; however, it decreased with the addition of an emulsifier. To improve the properties of the RP films, films were blended with polysaccharide agarose (*Gelidium corneum*), or gelatin, which influenced the mechanical properties of the film by increasing its tensile strength and decreasing its elongation at break.

The authors suggested an appropriate composition of blend film containing RP and gelatin, with optimal properties as packaging materials. This film was chosen for further research, to develop an active packaging film based on RP with different amounts of grapefruit seed extract, for packaging of strawberries. Incorporation of grapefruit seed extract into the RP blend, apart from introducing properties of an active film, significantly decreased tensile strength, while elongation at break and water vapor permeability were increased. The pure RP film did not show any antimicrobial activity against *Escherichia coli* and *Listeria monocytogenes*. However, by increasing the concentration of grapefruit seed extract up to a certain level, antimicrobial activity was significantly increased, but above this point, the microbial growth inhibition effect decreased. The results obtained showed that active RP films with grapefruit extract have higher antimicrobial activity against *L. monocytogenes* than *E. coli*. In strawberries packed in those films, the initial populations of total aerobic bacteria and yeast and molds significantly decreased during storage, and sensory evaluation of the film-packaged strawberries obtained better sensory scores. These results suggested that RP film containing grapefruit seed extract could be used to extend the shelf life of strawberries. Zhang et al. (2019) prepared composite films by mixing rapeseed protein hydrolysate with chitosan. The effects of different hydrolysis degrees of the protein, as well as different ratios of RP hydrolysate and chitosan, on the physical properties of obtained films and antimicrobial properties against *E. coli*, *Staphylococcus aureus*, and *Bacillus subtilis* were evaluated. Hydrolysis degree and chitosan content significantly influenced the mechanical properties of RP films. The pure RP film had the worst mechanical properties among the films; however, with the increase of proteolytic hydrolysis, the properties of composite films were improved. An increase of chitosan ratio in the composite film gradually increased tensile strength and elongation at break, but only to a certain ratio, after which elongation at break values decreased. According to the results obtained for antimicrobial properties, pure RP films did not show antimicrobial activity, while the pure chitosan film showed activity against different tested bacteria. The composite films had higher activity against *S. aureus* and *B. subtilis* than *E. coli*.

Based on a literature review, Jangchud and Chinnan (1999a, 1999b) were the first authors to produce and characterize biopolymer films based on peanut protein concentrate. Other authors used peanut protein isolate (PPI) and analyzed the effect of different concentrations of peanut protein concentrate and drying temperature on film formation. The results obtained showed that a film with acceptable characteristics could be formed at 70°C and 16h drying (Jangchud & Chinnan, 1999b). They also reported that the film formed without plasticizer was very brittle, so they determined the effect of different plasticizers at various concentrations (Jangchud & Chinnan, 1999a). Glycerol was found to be the most appropriate plasticizer, giving, in defined concentration, optimal mechanical properties of film, without compromising barrier properties. The influences of different pH values and drying temperatures in film production were also tested (Jangchud & Chinnan, 1999b). The results showed that pH increment influenced the darkness of the film. The temperatures only affected color parameters so that the film appeared more yellowish. Films obtained at higher pH and higher temperatures showed improved appearance and color. On increasing the film drying temperature, film water sensitivity decreased, as did water vapor and oxygen permeability, giving a film with improved mechanical properties. The authors suggested a film composition and production process that leads to films with optimal properties (Jangchud & Chinnan, 1999b).

Liu, Tellez-Garay, and Castell-Perez (2004) developed a film based on PPI and reported much lower values for film properties (tensile strength, elongation at break, total soluble matter, water vapor permeability, and oxygen permeability) compared to peanut protein concentrate films. In addition, much higher parameters of color, L^*, a^*, b^*, and for opacity, were reported for PPI films compared to peanut protein concentrate films (Sun, Sun, & Xiong, 2013). The authors tried to modify the physical and mechanical properties of PPI films, with different physical and chemical treatments. The physical treatments applied included heat denaturation, and ultrasound processes on film-forming solution, as well as ultraviolet irradiation on obtained films. The chemical treatments used were the addition of aldehydes and anhydrides (acetic anhydride, succinic anhydride, formaldehyde solution, and glutaraldehyde solution) into a film-forming solution. Heat denaturation was the most effective treatment, making the films stronger, more resistant to water, and less permeable to water vapor and oxygen. Only the heat denaturation treatment showed a positive effect on almost all tested properties. However, none of the used treatments improved the flexibility of the film. Meanwhile, with an increase of the heat denaturation temperature, the barrier properties decreased. When chemical treatments were used, only the formaldehyde and glutaraldehyde solution showed significant influences on PPI films. Tensile strength increased and water vapor permeability and total soluble matter decreased, compared to pure PPI films, but did not show an influence on elongation at break and oxygen permeability. The group of authors tried to improve the properties of PPI films with addition of different chemicals, by crosslinking PPI with citric acid (Reddy, Jiang, & Yang, 2012), blending with pea starch (Sun et al., 2013), and glycating with gum arabic conjugates (Li et al., 2015) or xylose (Lin et al., 2015). Crosslinking of PPI with different concentrations of citric acid improved the tensile strength but decreased the elongation at break. The highest tensile strength was obtained at the lowest citric acid concentration. The strength of films obtained in this research is several folds higher than those reported earlier, but much lower amounts of glycerol as plasticizer was used in this research compared to previous studies (Jangchud & Chinnan, 1999a, 1999b; Liu et al., 2004). Crosslinking slightly decreased the water vapor permeability, but there was no significant difference between the samples crosslinked with various concentration of citric acid (Reddy et al., 2012). Sun et al. (2013) produced biopolymer films based on mixtures of pea starch and PPI at different ratios. With an increase of PPI ratio in films, a slight increase of composite film opacity, as well as an increase of the color parameters, was observed. This indicated that composite blend films were darker and yellowish in color. The addition of PPI in blend films significantly decreased tensile strength, water vapor permeability, and total soluble matter, but increased elongation at break. Glycating PPI with gum arabic conjugates for 3days significantly decreased water vapor permeability and elongation at break, but increased the strength of the film. However, as the glycation proceeded from 3 to 9days, the tensile strength of films decreased, and the water vapor permeability and the elongation at break increased gradually (Li et al., 2015). Crosslinking PPI films with xylose improved the film properties, suggesting that the tensile strength of the modified peanut protein films developed in this study was much higher than the tensile strength of peanut protein films modified with other chemicals reported in the literature (Lin et al., 2015). Zhong et al. (2017) developed an active biopolymer film based on PPI with a different concentration of thymol. The pure PPI film did not show any antimicrobial activity against *E. coli*, *S. aureus*, *Pseudomonas aeruginosa*, and *L. plantarum*. Adding thymol into films led to the inhibition of all test bacteria even at the minimum concentration applied into the film formulation. As the concentration increased, the zones of inhibition increased significantly. At all concentrations assayed, *S. aureus* showed the largest inhibition zones, and *P. aeruginosa* was the most resistant bacterium.

Previously, all films have been made from peanut proteins by the solution casting method. Compared to solution casting, compression molding is more convenient, provides better properties, and is more environmentally friendly. However, plant proteins have poor thermo-plasticity, and are therefore not ideally suited to form thermoplastic products (Reddy, Jiang, & Yang, 2013). Reddy et al. (2013) suggested that peanut protein extracted from peanut meal could be compression molded to form films with good tensile properties.

Based on the reviewed literature, in the first report on sesame protein films, Lee, Song, Jo, and Song (2014) examined the effect of nanoclay type and content (Cloisite Na^+ and Cloisite 10A) on the physical properties of these films. Their results suggested that the addition of nanoclays into a film-forming solution improved the properties of the SP film. The tensile strength of the SP film increased; however, elongation at break and water vapor permeability of the SP nanocomposite films decreased with the addition of the nanoclays. Sharma and Singh (2016) prepared sesame protein isolate-based biopolymer films, and the effects on film properties of different protein concentration, alkaline pH range, temperatures of heating film-forming solution, and concentration of glycerol were evaluated. The optimized values obtained with the response surface methodology indicated that the films prepared at the lowest protein and plasticizer concentration, and the highest pH and temperature of heating, resulted in the lowest water vapor permeability, the maximum tensile strength, and the lowest total soluble matter. Films were less transparent, and dark yellowish, which might be due to the darker color of the protein and the alkaline pH of the film. The same authors tried to modify the properties of SP-based films, crosslinking SP with three different carboxylic acids (malic, citric, and succinic acid) in different concentrations (Sharma, Kumar Sharma, & Singh, 2018; Sharma, Sing Saini, & Kumar Sharma, 2018), or blending with gum rosin at different ratios (Sharma & Singh, 2018). Their results suggested that those chemicals had a positive influence on almost all films' properties. With increasing concentrations of carboxylic acids, the tensile strength of films increased, but elongation, water vapor, solubility, transparency, and moisture content decreased. Among the three organic acids, crosslinked films prepared with succinic acid showed the lowest water sensitivity and the highest strength. Composite films based on SP with lower amounts of gum rosin decreased water vapor permeability. Different UV light types were applied to film-forming solutions or to preformed films, and the properties of obtained films were analyzed (Fathi, Almasi, & Pirouzifard, 2018). Moisture content, total soluble matter, and water vapor permeability decreased with UV exposure. Results showed that moisture content values were lower when UV was applied on film-forming solutions than on preformed films. UV radiation implied a tensile strength increase, and elongation at break decreased. Sharma, Kumar Sharma, et al. (2018), Sharma, Sing Saini, et al. (2018), and Sharma et al. (2019) examined the influence of coatings based on sesame protein, crosslinked with organic acids (citric, malic, and succinic acid), and guar gum and calcium chloride, on improving the quality and increasing the shelf life of fresh fruits (pineapple and mango). The results indicated that all coating formulations have the potential to enhance the shelf life and maintain the quality of fresh-cut pineapple and mango for 15 days. Coatings from crosslinked proteins were found to be more effective compared to pure SP-based coatings.

5.3 Related by-products valorization for biopolymer films production (flours and meals)

In addition to the application of protein isolates and concentrates for biopolymer films synthesis, researchers explored the potential of two less expensive and more abundant raw materials, defatted meal and defatted flour, for composite biopolymer-based film production. The disadvantage of defatted soy meal (DSM) as a source for film production is that composition of DSM varies from season to season, among species and locations. Lee and Min (2013) tried to deal with this issue by making model films composed of soy polysaccharide, soy protein, and soy oil in different ratios, but the results for the DSM-based films and model films were different. For films produced by compression molding, DSM-based films showed tensile strength and water vapor permeability that fall into the ranges of values for SPI-based films, while elongation at break was lower for these films (Lee & Min, 2014). In the paper reported by Lee, Paek, and Min (2011), films based on DSM showed higher tensile strength and elongation at break compared to SPI-based films produced under the same conditions. Also in this work, it was shown that the water vapor permeability of the films increased with an increase in the concentration of plasticizer (glycerol) and decreased with high-pressure homogenization application in film production. Cold plasma treatment was proposed as a potential method for improvement of different properties of DSM-based films in order to improve their applicability. Film properties, such as printability surface roughness, ink adhesion, water contact angle, glass transition temperature, oxygen availability, and biodegradability could also be tailored using different cold plasma treatments (Oh, Roh, & Min, 2016). The DSM-based film produced by casting or compression molding, with an incorporated lactoperoxidase system, was proven to be an effective active antimicrobial film, inhibiting the growth of *Salmonella* (Lee & Min, 2013, 2014). DSM was tested as an antimicrobial packaging solution after fermentation with different strains producing bacteriocin-like substances. Among different strains, *B. subtilis* was selected because DSM fermented with this strain showed the widest inhibition zone against different foodborne microorganisms. Fermented DSM was used in film production and the film obtained showed increased antimicrobial activity toward *E. coli* and slowed growth of spoilage and pathogen microorganisms in a variety of food. A minor decrease of tensile strength and increase in elongation at break was reported for the films obtained from fermented DSM compared to the films from native DSM (Kim et al., 2004; Kim et al., 2005). In addition to antimicrobial film formulations, a film with antioxidant activity was formed based on DSM with the addition of tea polyphenols. This film was shown to have antioxidant activity in vitro, as well as the ability to slow browning in apple cuts (Sun et al., 2018).

Defatted soy flour (DSF) was used for production of composite films with the addition of pectin, with or without added transglutaminase. The films obtained with transglutaminase had higher tensile strength, while elongation at break was lower, compared to the film without added transglutaminase (Mariniello et al., 2003). In another study, the addition of DSF (10%–20%) into pectin-based films (80%–90%) resulted in increased stiffness and strength, and decreased water solubility and water vapor transmission rate of pectin films. When pectin-DSF composite film was crosslinked with glutaraldehyde, films with improved mechanical properties as well as water resistance were obtained (Liu, Liu, Fishman, & Hicks, 2007). In order to combine the advantageous properties of soy flour-based films and biobased LLDPE films, two different approaches were tested: I LLDPE film with the addition of up to 20% of DSF was obtained by an extrusion process and II multilayers coextruded films with soy flour loaded layer in the middle of LLDPE layers in order to protect the DSF-based layer from moisture. Every material contributed to the properties of the resulting material: DSF increased the strength and oxygen barrier properties of the composite films, while LLDPE improved the water resistance and water vapor barrier properties (Thellen, Hauver, & Ratto, 2014). When SPI and DSF extruded and compression-molded materials were compared for their mechanical and water sensitivity properties, the tensile strength of SPI-based material was higher, while the impact strength was higher in DSF-based material. In addition, DSF-based material showed a lower percentage of water absorption. Further improvement of these materials was achieved by forming composite material with a different share of PLA and/or sugarcane bagasse fiber for reinforcement (Boontima, Noomhorm, Puttanlek, Uttapap, & Rungsardthong, 2015).

Riveros, Martin, Aguirre, and Grosso (2018) produced a biopolymer film based on whole defatted peanut flour, characterized its properties, and analyzed the possible application as a packaging material for sunflower oil, determining its efficacy to preserve the chemical quality of the product during storage. Biopolymer film based on defatted peanut flour was developed by the casting method. The film produced was transparent and light yellow. Compared to results obtained for films based on peanut protein, peanut flour film presented an excellent water vapor barrier, with a lower value than for PPI films. Those films also exhibited good flexibility, but weak strength. With respect to the mechanical properties of the films, values of the peanut flour films were comparable to the results obtained for peanut protein-based films (Li et al., 2015; Liu et al., 2004; Sun et al., 2013; Zhong et al., 2017). Riveros et al. (2018) also suggested that biopolymer films based on defatted peanut flour could prolong sunflower oil's shelf life and that of possibly other food products with similar physicochemical characteristics. The films showed a protective effect against lipid oxidation, without the incorporation of any additive in the film formulation.

6 Conclusion

Utilization of different oil industry by-products (cake, meal, and flour), as well as products derived from them (protein concentrates and isolates), for production of new eco-friendly materials have been reviewed in this chapter. The main advantages of these materials are, in addition to biodegradability, the wide accessibility and renewability of their sources. Biopolymer materials obtained from oil industry by-products possess good gas barrier characteristics in low moisture conditions, while their mechanical, physicochemical, thermal, water vapor barrier, and water sensitivity characteristics, which are important for their application as packaging materials, can be improved by applying different optimization processes. Currently, research is focused on the widening area of biopolymer film applications. The share of biopolymer packaging production and application is growing and the goal is to move the boundaries further, step by step. However, the complete replacement of synthetic polymer packaging with biopolymer-based packaging seems unrealistic at this time.

References

Anker, M., Berntsen, J., Hermansson, A. M., & Stading, M. (2001). Improved water vapour barrier of whey protein films by addition of an acetylated monoglyceride. *Innovative Food Science and Emerging Technologies*, *3*(1), 81–92.

Arancibia, M. Y., López-Caballero, M. E., Gómez-Guillén, M. C., & Montero, P. (2014). Release of volatile compounds and biodegradability of active soy protein lignin blend films with added citronella essential oil. *Food Control*, *44*, 7–15.

Ayhllon-Meixueiro, F., Vaca-Garcia, C., & Silvestre, F. (2000). Biodegradable films from isolate of sunflower (*Helianthus annuus*) proteins. *Journal of Agricultural and Food Chemistry*, *48*(7), 3032–3036.

Ayranci, E., & Tunc, S. (2003). A method for the measurement of the oxygen permeability and the development of edible films to reduce the rate of oxidative reactions in fresh foods. *Food Chemistry*, *80*(3), 423–431.

Boontima, B., Noomhorm, A., Puttanlek, C., Uttapap, D., & Rungsardthong, V. (2015). Mechanical properties of sugarcane bagasse fiber-reinforced soy based biocomposite. *Journal of Polymers and the Environment*, *23*(1), 97–106.

Bulut, S., Lazić, V., Popović, S., Hromiš, N., & Šuput, D. (2017a). Influence of storage period on properties of biopolymer packaging materials and pouches. *Acta Periodica Technologica*, *48*, 53–62.

Bulut, S., Lazić, V., Popović, S., Hromiš, N., & Šuput, D. (2017b). Influence of different concentrations of glycerol and guar-xanthan on properties of PuOC-Zein bi-layer film. *Field and Vegetable Crops Research*, *54*(1), 19–24.

Bulut, S., Lazić, V., Popović, S., Hromiš, N., & Šuput, D. (2017c). Mono- and bilayer biopolymer films: Synthesis and characterization. *Journal on Processing and Energy in Agriculture*, *21*(4), 214–218.

Bulut, S., Lazić, V., Popović, S., Hromiš, N., Šuput, D., Malbaša, R., et al. (2018a). Influence of surfactant Tween 20 on antioxidant activity of biopolymer films. In O. Đuragić (Ed.), *Proceedings of the 4th international congress, food technology, quality and safety, and 18th international symposium feed technology (FoodTech 2018), Novi Sad, Serbia, October 25–28, 2018* (pp. 122–127).

Bulut, S., Lazić, V., Popović, S., Hromiš, N., Šuput, D., Malbaša, R., et al. (2018b). Antioxidant activity of biopolymer films with addition of caraway essential oil and chitosan. In I. Pavkov, & L. Babić (Eds.), *Proceedings of 30. jubilee national conference with international attending processing energy in agriculture-PTEP 2018, Brzeće, Serbia, April 15–April 18, 2018* (pp. 16–17).

Bulut, S., Lazić, V., Popović, S., Šuput, D., Hromiš, N., & Popović, L. (2016). Possibility to maintain modified atmosphere in pouches made from biopolymer materials. In O. Đuragić (Ed.), *Proceedings of 3rd international congress "Food Technology, Quality and Safety" & 17th international symposium "Feed Technology" (FoodTech 2016), Novi Sad, Serbia, October 23–25, 2016* (pp. 122–127).

Cao, N., Fu, Y., & He, J. (2007). Preparation and physical properties of soy protein isolate and gelatin composite films. *Food Hydrocolloids*, *21*(7), 1153–1162.

Chillo, S., Flores, S., Mastromatteo, M., Conte, A., Gerschenson, L., & Del Nobile, M. A. (2008). Influence of glycerol and chitosan on tapioca starch-based edible film properties. *Journal of Food Engineering*, *88*(1), 159–168.

Chiralt, A., González-Martínez, C., Vargas, M., & Atarés, L. (2018). Edible films and coatings from proteins. In R. Y. Yada (Ed.), *Proteins in food processing* (2nd ed., pp. 477–500). Cambridge: Woodhead Publishing.

Cho, S. Y., Park, J.-W., Batt, H., & Thomas, R. (2007). Edible films made from membrane processed soy protein concentrates. *LWT—Food Science and Technology*, *40*(3), 418–423.

Ciannamea, E. M., Espinosa, J. C., Stefani, P. M., & Ruseckaite, R. A. (2018). Long-term stability of compression-moulded soybean protein concentrate films stored under specific conditions. *Food Chemistry*, *243*, 448–452.

Ciannamea, E. M., Sagüés, M. F., Saumell, C., Stefani, P. M., & Ruseckaite, R. A. (2013). Soybean protein films. Characterization and potential as novel delivery devices of *Duddingtonia flagrans* chlamydospores. *Biological Control*, *66*(2), 92–101.

Ciannamea, E. M., Stefani, P. M., & Ruseckaite, R. A. (2014). Physical and mechanical properties of compression moulded and solution casting soybean protein concentrate based films. *Food Hydrocolloids*, *38*(part 1), 193–204.

Ciannamea, E. M., Stefani, P. M., & Ruseckaite, R. A. (2015). Storage-induced changes in functional properties of glycerol plasticized-soybean protein concentrate films produced by casting. *Food Hydrocolloids*, *45*, 247–255.

Ciannamea, E. M., Stefani, P. M., & Ruseckaite, R. A. (2016). Properties and antioxidant activity of soy protein concentrate films incorporated with red grape extract processed by casting and compression moulding. *LWT—Food Science and Technology*, *74*(2), 353–362.

De Graaf, L. A., Harmgen, P. F. H., Vereijken, J. M., & Monikes, M. (2001). Requirements for nonfood applications for pea proteins—A review. *Nahrung/Food*, *45*(6), 408–411.

Denavi, G. A., Pérez-Mateos, M., Añón, M. C., Montero, P., Mauri, A. N., & Gomez-Guillen, M. C. (2009). Structural and functional properties of soy protein isolate and cod gelatin blend films. *Food Hydrocolloids*, *23*(8), 2094–2101.

Dimić, E. (Ed.). (2005). *Cold pressed oils*. Novi Sad: University of Novi Sad, Faculty of Technology Novi Sad.

Dimić, E., Romanić, R., Peričin, D., & Panić, B. (2006). Use of by products of naked pumpkin seed processing. *Journal of Edible Oil Industry*, *37*(3–4), 29–35.

Escamilla-Silva, E. M., Guzman-Maldonado, S. H., Cano-Medinal, A., & Gonzalez-Alatorre, G. (2003). Simplified process for the production of sesame protein concentrate. Differential scanning calorimetry and nutritional, physicochemical and functional properties. *Journal of the Science of Food and Agriculture*, *83*(9), 972–979.

Eswaranandam, S., Hettiarachchy, N. S., & Johnson, M. G. (2006). Antimicrobial activity of citric, lactic, malic, or tartaric acids and nisin-incorporated soy protein film against *Listeria monocytogenes*, *Escherichia coli* O157:H7, and *Salmonella gaminara*. *Journal of Food Science*, *69*(3), FMS79–FMS84.

Falguera, V., Quinterob, J. P., Jimenez, A., Munoz, J. A., & Ibarz, A. (2011). Edible films and coatings: Structures, active functions and trends in their use. *Trends in Food Science & Technology*, *22*(6), 292–303.

Fathi, N., Almasi, H., & Pirouzifard, M. K. (2018). Effect of ultraviolet radiation on morphological and physicochemical properties of sesame protein isolate based edible films. *Food Hydrocolloids*, *85*, 136–143.

Food and Agriculture Organization of the United Nations (FAO). (2018). *The State of food security and nutrition in the world 2018. Building climate resilience for food security and nutrition*. Rome: Food and Agriculture Organization of the United Nations. Retrieved from: http://www.fao.org/3/i9553en/i9553en.pdf.

Foulk, J., & Bunn, J. (2001). Properties of compression-moulded, acetylated soy protein films. *Industrial Crops and Products*, *14*(1), 11–22.

Friesen, K., Chang, C., & Nickerson, M. (2015). Incorporation of phenolic compounds, rutin and epicatechin, into soy protein isolate films: Mechanical, barrier and cross-linking properties. *Food Chemistry*, *172*(1), 18–23.

Galus, S. (2018). Functional properties of soy protein isolate edible films as affected by rapeseed oil concentration. *Food Hydrocolloids*, *85*, 233–241.

García, M. A., Martino, M. N., & Zaritzky, N. E. (2000). Microstructural characterization of plasticized starch-based films. *Starch/Staerke*, *52*(1), 118–124.

Garrido, T., Etxabide, A., Peñalba, M., Guerrero, P., & de la Caba, K. (2013). Preparation and characterization of soy protein thin films: Processing-properties correlation. *Materials Letters*, *105*, 110–112.

Garrido, T., Leceta, I., Cabezudo, S., de la Caba, K., & Guerrero, P. (2016). Tailoring soy protein film properties by selecting casting or compression as processing methods. *European Polymer Journal*, *85*, 499–507.

González, A., & Igarzabal, C. A. (2013). Soy protein-poly (lactic acid) bilayer films as biodegradable material for active food packaging. *Food Hydrocolloids*, *33*(2), 289–296.

González, A., & Igarzabal, C. A. (2015). Nanocrystal-reinforced soy protein films and their application as active packaging. *Food Hydrocolloids*, *43*, 777–784.

González, A., & Igarzabal, C. A. (2017). Study of graft copolymerization of soy protein-methyl methacrylate: Preparation and characterization of grafted films. *Journal of Polymers and the Environment*, *25*(2), 214–220.

González, A., Strumia, M. C., & Igarzabal, C. A. (2011). Cross-linked soy protein as material for biodegradable films: Synthesis, characterization and biodegradation. *Journal of Food Engineering*, *106*(4), 331–338.

Guilbert, S., & Gontard, N. (2005). Agro-polymers for edible and biodegradable films. Review of agricultural polymeric materials, physical and mechanical characteristics. In J. H. Han (Ed.), *Innovations in food packaging* (pp. 237–259). San Diego: Elsevier Academic Press.

Han, Y., Li, K., Chen, H., & Li, J. (2017). Properties of soy protein isolate biopolymer film modified by graphene. *Polymers*, *9*(8), 312.

Han, Y., Yu, M., & Wang, L. (2018). Soy protein isolate nanocomposites reinforced with nanocellulose isolated from licorice residue: Water sensitivity and mechanical strength. *Industrial Crops and Products*, *117*, 252–259.

Jang, S. A., Lim, G. O., & Song, K. B. (2011). Preparation and mechanical properties of edible rapeseed protein films. *Journal of Food Science*, *76*(2), 218–223.

Jang, S. A., Shin, Y. J., & Song, K. B. (2011). Effect of rapeseed protein–gelatin film containing grapefruit seed extract on 'Maehyang' strawberry quality. *International Journal of Food Science and Technology*, *46*, 620–625.

Jangchud, A., & Chinnan, M. S. (1999a). Peanut protein film as affected by drying temperature and pH of film forming solution. *Journal of Food Science*, *64*, 153–157.

Jangchud, A., & Chinnan, M. S. (1999b). Properties of peanut protein film: Sorption isotherm and plasticizer effect. *LWT—Food Science and Technology*, *32*(1), 89–94.

Khattab, R. Y., & Arntfield, S. D. (2009). Functional properties of raw and processed canola meal. *LWT—Food Science and Technology*, *42*(6), 1119–1124.

Kim, H.-W., Kim, K.-M., Ko, E.-J., Lee, S.-K., Ha, S.-D., Song, K.-B., et al. (2004). Development of antimicrobial edible film from defatted soybean meal fermented by *Bacillus subtilis*. *Journal of Microbiology and Biotechnology*, *14*(6), 1303–1309.

Kim, H.-W., Ko, E.-J., Ha, S.-D., Song, K.-B., Park, S.-K., Chung, D.-H., et al. (2005). Physical, mechanical, and antimicrobial properties of edible film produced from defatted soybean meal fermented by *Bacillus subtilis*. *Journal of Microbiology and Biotechnology*, *15*(4), 815–822.

Kim, I. H., Yang, H. J., Noh, B. S., Chung, S. J., & Min, S. C. (2012). Development of a defatted mustard meal-based composite film and its application to smoked salmon to retard lipid oxidation. *Food Chemistry*, *133*(4), 1501–1509.

Kishore Pradhan, N., & Das, M. (2017). Biodegradability determination of soy protein concentrate-corn starch-C30B nanocomposites. *Journal of Advance Nanobiotechnology*, *2*(1), 15–28.

Kishore Pradhan, N., Das, M., & Nayak, P. L. (2018). Chitosan-coated soya protein concentrate-closite 30B nanocomposites films for morphological, mechanical, thermal and biodegradability properties. *Asian Journal of Microbiology, Biotechnology & Environmental Sciences*, *20*, 224–230.

Kumar, R. (2014). Arylated and nonarylated soy protein isolate based nanocomposites. In R. Kumar (Ed.), *Polymer-matrix composites* (pp. 83–107). N.Y: Nova Science Publishers Inc.

Kumar, R., Anandjiwala, R., & Kumar, A. (2016). Thermal and mechanical properties of mandelic acid-incorporated soy protein films. *Journal of Thermal Analysis and Calorimetry*, *123*(2), 1273–1279.

Lazić, V., Šuput, D., Popović, S., Hromiš, N., Bulut, S., & Romanić, R. (2017). Synthesis and characterization of protein biopolymer films obtained from sunflower oil cake. In R. Romanić, & I. Lončarević (Eds.), *Proceedings of the 58th oil industry conference production and processing of oilseeds, Herceg Novi, Montenegro, June 18–23, 2017* (pp. 175–182).

Lazić, V., Šuput, D., Popović, S., Hromiš, N., Bulut, S., & Vitas, J. (2018). Active biopolymer films based on sunflower oil cake. *Journal of Edible Oil Industry*, *49*(1), 11–16.

Lee, H., & Min, S. C. (2013). Antimicrobial edible defatted soybean meal-based films incorporating the lactoperoxidase system. *LWT—Food Science and Technology*, *54*(1), 42–50.

Lee, H. N., & Min, S. C. (2014). Development of antimicrobial defatted soybean meal-based edible films incorporating the lactoperoxidase system by heat pressing. *Journal of Food Engineering*, *120*, 183–190.

Lee, H. B., Noh, B. S., & Min, S. C. (2012). *Listeria monocytogenes* inhibition by defatted mustard meal-based edible films. *International Journal of Food Microbiology*, *153*(1), 99–105.

Lee, H. N., Paek, H. J., & Min, S. C. (2011). Defatted soybean meal-based edible film development. *Food Engineering Progress*, *15*, 305–310.

Lee, J. H., Song, N. B., Jo, W. S., & Song, K. B. (2014). Effects of nano-clay type and content on the physical properties of sesame seed meal protein composite films. *International Journal of Food Science and Technology*, *49*, 1869–1875.

Li, C., Zhu, W., Xue, H., Chen, Z., Chen, Y., & Wang, X. (2015). Physical and structural properties of peanut protein isolate-gum Arabic films prepared by various glycation time. *Food Hydrocolloids*, *43*, 322–328.

Liang, S., & Wang, L. (2018). A natural antibacterial-antioxidant film from soy protein isolate incorporated with cortex phellodendron extract. *Polymers*, *10*(1), 71.

Lin, W. J., Liu, H. Z., Shi, A. M., Liu, L., Wang, Q., & Adhikari, B. (2015). Effect of glycosylation with xylose on the mechanical properties and water solubility of peanut protein films. *Journal of Food Science and Technology*, *52*(10), 6242–6253.

Liu, L., Liu, C.-K., Fishman, M. L., & Hicks, K. B. (2007). Composite films from pectin and fish skin gelatin or soybean flour protein. *Journal of Agricultural and Food Chemistry*, *55*(6), 2349–2355.

Liu, C. C., Tellez-Garay, A. M., & Castell-Perez, M. E. (2004). Physical and mechanical properties of peanut protein films. *LWT—Food Science and Technology*, *37*(7), 731–738.

Mariniello, L., Di Pierro, P., Esposito, C., Sorrentino, A., Masi, P., & Porta, R. (2003). Preparation and mechanical properties of edible pectin/soyflour films obtained in the absence or presence of transglutaminase. *Journal of Biotechnology*, *102*, 91–198.

Maryam Adilah, Z. A., Jamilah, B., & Nur Hanani, Z. A. (2018). Functional and antioxidant properties of protein-based films incorporated with mango kernel extract for active packaging. *Food Hydrocolloids*, *74*, 207–218.

Monedero, F. M., Fabra, M. J., Talens, P., & Chiralt, A. (2009). Effect of oleic acid-beeswax mixtures on mechanical, optical and water barrier properties of soy protein isolate based films. *Journal of Food Engineering*, *91*(4), 509–515.

Nandane, A., & Jain, R. (2015). Study of mechanical properties of soy protein based edible film as affected by its composition and process parameters by using RSM. *Journal of Food Science and Technology*, *52*(6), 3645–3650.

Nesterenko, A., Alric, I., Violleau, F., Silvestre, F., & Durrieu, V. (2013). A new way of valorizing biomaterials: The use of sunflower proteins for alpha-tocopherol microencapsulation. *Food Research International*, *53*(1), 115–124.

Oh, Y. A., Roh, S. H., & Min, S. C. (2016). Cold plasma treatments for improvement of the applicability of defatted soybean meal-based edible film in food packaging. *Food Hydrocolloids*, *58*, 150–159.

Orliac, O., Rouilly, A., Silvestre, F., & Rigal, L. (2002). Effects of additives on the mechanical properties, hydrophobicity and water uptake of thermo-moulded films produced from sunflower protein isolate. *Polymer*, *43*(20), 5417–5425.

Orliac, O., & Silvestre, F. (2003). New thermo-moulded biodegradable films based on sunflower protein isolate: Aging and physical properties. *Macromolecular Symposia*, *197*(1), 193–206.

Orliac, O., Silvestre, F., Rouilly, A., & Rigal, L. (2003). Rheological studies, production, and characterization of injection-moulded plastics from sunflower protein isolate. *Industrial & Engineering Chemistry Research*, *42*(8), 1674–1680.

Park, H. J., Byun, Y. J., Kim, Y. T., Whiteside, W. S., & Bae, H. J. (2014). Processes and applications for edible coating and film materials from agropolymers. In J. Han (Ed.), *Innovations in food packaging* (2nd ed., pp. 258–276). London: Academic Press-Elsevier.

Peričin, D., Radulović, L., Mađarev, S., & Dimić, E. (2007). Bioprocessing for value added products from oil cakes. *Journal of Edible Oil Industry*, *38*(1–2), 35–40.

Popović, S. (2013). *The study of production and characterization of biodegradable, composite films based on plant proteins*. Novi Sad: University of Novi Sad, Faculty of Technology.

Popović, S., Lazić, V., Hromiš, N., Šuput, D., & Bulut, S. (2018). Biopolymer packaging materials for food shelf-life prolongation. In A. M. Grumezescu, & A. M. Holban (Eds.), *Biopolymers for food design* (pp. 223–277). Elsevier: Academic Press.

Popović, S., Lazić, V., Lj, P., Vaštag, Ž., & Peričin, D. (2010). Effect of the addition of pumpkin oil cake to gelatine to produce biodegradable composite films. *International Journal of Food Science and Technology*, *45*(6), 1184–1190.

Popović, S., Peričin, D., Vaštag, Ž., Lazić, V., & Popović, L. (2012). Pumpkin oil cake protein isolate films as potential gas barrier coating. *Journal of Food Engineering*, *110*, 374–379.

Popović, S., Peričin, D., Vaštag, Ž., Popović, L., & Lazić, V. (2011). Evaluation of edible film-forming ability of pumpkin oil cake; effect of pH and temperature. *Food Hydrocolloids*, *25*(3), 470–476.

Prakash Maran, J., Sivakumar, V., Thirugnanasambandham, K., & Kandasamy, S. (2013). Modeling and analysis of film composition on mechanical properties of maize starch based edible films. *International Journal of Biological Macromolecules*, *62*(1), 565–573.

Prinyawiwatkul, W., Beuchat, L. R., Phillips, R. D., & Resurreccion, A. V. A. (1995). Modelling the effects of peanut flour, feed moisture content, and extrusion temperature on physical properties of an extruded snack product. *International Journal of Food Science and Technology*, *30*(1), 37–44.

Radočaj, O. (2011). *Optimization of technology of hull-less pumpkin press-cake spread rich in omega fatty acids*. Novi Sad: University of Novi Sad, Faculty of Technology.

Ramachandran, S., Singh, S. K., Larroche, C., Soccol, C. R., & Pandey, A. (2007). Oil cakes and their biotechnological applications—A review. *Bioresource Technology*, *98*(10), 2000–2009.

Reddy, N., Jiang, Q., & Yang, Y. (2012). Preparation and properties of peanut protein films crosslinked with citric acid. *Industrial Crops and Products*, *39*, 26–30.

Reddy, N., Jiang, Q., & Yang, Y. (2013). Thermoplastic films from peanut proteins extracted from peanut meal. *Industrial Crops and Products*, *43*, 159–164.

Riveros, C. G., Martin, M. P., Aguirre, A., & Grosso, N. R. (2018). Film preparation with high protein defatted peanut flour: Characterization and potential use as food packaging. *International Journal of Food Science and Technology*, *53*, 969–975.

Rodrıguez, M., Oses, J., Ziani, K., & Mate, J. I. (2006). Combined effect of plasticizers and surfactants on the physical properties of starch based edible films. *Food Research International*, *39*(8), 840–846.

Romanić, R. (2015). *Chemometric approach to the optimization of the production of cold pressed oil from high-oleic sunflower seeds*. Novi Sad: The University of Novi Sad, Faculty of Technology.

Rouilly, A., Meriaux, A., Geneau, C., Silvestre, F., & Rigal, L. (2006). Film extrusion of sunflower protein isolate. *Polymer Engineering and Science*, *46*(11), 1635–1640.

Salgado, P. R., Drago, S. R., Molina Ortiz, S. E., Petruccelli, S., Andrich, O., González, R. J., et al. (2012). Production and characterization of sunflower (*Helianthus annuus* L.) protein-enriched products obtained at pilot plant scale. *LWT—Food Science and Technology*, *45*(1), 65–72.

Salgado, P. R., Molina Ortiz, S. E., Petruccelli, S., & Mauri, A. N. (2010). Biodegradable sunflower protein films naturally activated with antioxidant compounds. *Food Hydrocolloids*, *24*(5), 525–533.
Salgado, P. R., Molina Ortiz, S. E., Petruccelli, S., & Mauri, A. N. (2011). Sunflower protein concentrates and isolates prepared from oil cakes have high water solubility and antioxidant capacity. *Journal of the American Oil Chemists' Society*, *88*(3), 351–360.
Sanchez-Gonzalez, L., Vargas, M., Gonzalez-Martinez, C., Chiralt, A., & Chafer, M. (2011). Use of essential oils in bioactive edible coatings. *Food Engineering Reviews*, *3*(1), 1–16.
Sharma, L., Kumar Sharma, H., & Singh, C. (2018). Edible films developed from carboxylic acid cross-linked sesame protein isolate: Barrier, mechanical, thermal, crystalline and morphological properties. *Journal of Food Science and Technology*, *55*(2), 532–539.
Sharma, L., Sing Saini, C., & Kumar Sharma, H. (2018). Development of crosslinked sesame protein and pineapple extract-based bilayer coatings for shelf-life extension of fresh-cut pineapple. *Journal of Food Processing and Preservation*, *42*(2), e13527.
Sharma, L., Sing Saini, C., Kumar Sharma, H., & Singh, S. K. (2019). Biocomposite edible coatings based on cross linked-sesame protein and mango puree for the shelf life stability of fresh-cut mango fruit. *Journal of Food Process Engineering*, *42*(1), e12938.
Sharma, L., & Singh, C. (2016). Sesame protein based edible films: Development and characterization. *Food Hydrocolloids*, *61*, 139–147.
Sharma, L., & Singh, C. (2018). Composite film developed from the blends of sesame protein isolate and gum rosin and their properties thereof. *Polymer Composites*, *39*, 1480–1487.
Shim, Y. Y., Gui, B., Wang, Y., & Reaney, M. J. T. (2015). Flaxseed (*Linum usitatissimum* L.) oil processing and selected products. *Trends in Food Science & Technology*, *43*(2), 162–177.
Singh, P., Kumar, R., Sabapathy, S. N., & Bawa, A. S. (2008). Functional and edible uses of soy protein products. *Comprehensive Reviews in Food Science and Food Safety*, *7*(1), 14–28.
Souza, A. C., Goto, G. E. O., Mainardi, J. A., Coelho, A. C. V., & Tadini, C. C. (2013). Cassava starch composite films incorporated with cinnamon essential oil: Antimicrobial activity, microstructure, mechanical and barrier properties. *Food Science and Technology*, *54*(2), 346–352.
Sui, C., Zhang, W., Ye, F., Liu, X., & Yu, G. (2016). Preparation, physical, and mechanical properties of soy protein isolate/guar gum composite films prepared by solution casting. *Journal of Applied Polymer Science*, *133*(18), 43382.
Sun, Q., Li, X., Wang, P., Du, Y., Han, D., Wang, F., et al. (2011). Characterization and evaluation of the Ag^+ loaded soy protein isolate-based bactericidal film-forming dispersion and films. *Journal of Food Science*, *76*(6), E438–E443.
Sun, Q., Sun, C., & Xiong, L. (2013). Mechanical, barrier and morphological properties of pea starch and peanut protein isolate blend films. *Carbohydrate Polymers*, *98*, 630–637.
Sun, L.-L., Tian, B. Y., Wang, Y.-X., Guan, Y.-J., Chen, X.-I., & Wang, T. (2018). Effect of tea polyphenols on the properties and antioxidant activity of defatted soybean meal films. *Science and Technology of Food Industry*, *02*, TS206.4.
Sunil, L., Prakruthi Appaiah, P. K., Prasanth Kumar, P. K., & Gopala Krishna, A. G. (2015). Preparation of food supplements from oilseed cakes. *Journal of Food Science and Technology*, *52*(5), 2998–3005.
Šuput, D., Lazić, V., Popović, S., Hromiš, N., Bulut, S., Pezo, L., et al. (2018). Process parameters influence on biopolymer films based on the sunflower oil cake. *Journal on Processing and Energy in Agriculture*, *22*(3), 125–128.
Tang, C.-H., Jiang, Y., Wen, Q.-B., & Yang, X.-Q. (2005). Effect of transglutaminase treatment on the properties of cast films of soy protein isolates. *Journal of Biotechnology*, *120*(3), 296–307.
Teh, S.-S., & Bekhit, A. E.-D. A. (2015). Utilization of oilseed cakes for human nutrition and health benefits. In K. R. Hakeem, M. Jawaid, & O. Alothman (Eds.), *Agricultural biomass based potential materials* (pp. 191–229). Switzerland: Springer Int. Pub.
Thellen, C., Hauver, C., & Ratto, J. A. (2014). Development of a melt-extrudable biobased soy flour/polyethylene blend for multilayer film applications. *Journal of Applied Polymer Science*, *131*(17), 40707.
Wu, Y., Cai, L., Wang, C., Mei, C., & Shi, S. (2018). Sodium hydroxide-free soy protein isolate-based films crosslinked by pentaerythritol glycidyl ether. *Polymers*, *10*(12), 1300.
Xu, Y., Kim, K., Hanna, M., & Nag, D. (2005). Chitosan-starch composite film: Preparation and characterization. *Industrial Crops and Products*, *21*(1), 185–192.
Yu, Z., Sun, L., Wang, W., Zeng, W., Mustapha, A., & Lin, M. (2018). Soy protein-based films incorporated with cellulose nanocrystals and pine needle extract for active packaging. *Industrial Crops and Products*, *112*, 412–419.
Zhang, C., Wang, Z., Li, Y., Yang, Y., Ju, X., & He, R. (2019). The preparation and physiochemical characterization of rapeseed protein hydrolysate-chitosan composite films. *Food Chemistry*, *272*, 694–701.
Zhao, Y., Xu, H., Mu, B., Xu, L., & Yang, Y. (2016). Biodegradable soy protein films with controllable water solubility and enhanced mechanical properties via graft polymerization. *Polymer Degradation and Stability*, *133*, 75–84.
Zheng, T., Yu, X., & Pilla, S. (2017). Mechanical and moisture sensitivity of fully bio-based dialdehyde carboxymethyl cellulose cross-linked soy protein isolate films. *Carbohydrate Polymers*, *157*, 1333–1340.
Zhong, T., Liang, Y., Jiang, S., Yang, L., Shi, Y., Guo, S., et al. (2017). Physical, antioxidant and antimicrobial properties of modified peanut protein isolate based films incorporating thymol. *RSC Advances*, *7*, 41610–41618.

Chapter 4

Cold pressed *Torreya grandis* kernel oil

Abdul Hafeez Laghari[a], Aftab Ahmed Kandhro[b], and Ayaz Ali Memon[c]

[a]*Pakistan Council of Scientific and Industrial Research, Laboratories Complex, Karachi, Pakistan,* [b]*Dr. M. A. Kazi Institute of Chemistry, University of Sindh, Jamshoro, Pakistan,* [c]*National Center of Excellence in Analytical Chemistry, University of Sindh, Jamshoro, Pakistan*

List of abbreviations

ABTS	2,2′-azino-bis(3-ethylbenzothiazoline-6-sulfonic acid)
CO_2	carbon dioxide
DPPH	2,2-diphenyl-1-picrylhydrazyl
GC-MS	gas chromatography mass spectrometry
LDL	low-density lipoprotein
MUFA	monounsaturated
ORAC	oxygen radical absorbance capacity
PUFA	polyunsaturated fatty acids
RMB	renminbi
T. grandis	*Torreya grandis*
TC	total cholesterol
TG	triglycerides

1 Introduction

Torreya grandis (*T. grandis*) is a large, evergreen, ornamental coniferous tree belonging to the Cephalotaxaceae Taxaceae family. *Torreya genus* is an ancient nut tree native to Asia and North America. It has been traditionally cultivated in subtropical hilly areas of China near the city of Zhuji in Zhejiang province, and is also found in Japan (He et al., 2016; Shi, Zheng, Liu, et al., 2018; Shi, Zheng, Mao, et al., 2018). *T. grandis* has also been listed as a national key protected wild plant (second group) in China (Saeed et al., 2007; Shen et al., 2014).

There are *T. grandis* varieties including *Torreya grandis*, *Torreya jiulongshanensis*, *Torreya jackii*, *Torreya fargesii*, *Torreya yunnanensis*, *Torreya grandis* cv. Merrillii all over the world. Of these, *T. grandis* Fort. ex. Lindl. and *T. grandis* Fort. var. Merrillii are the two representative varieties widely planted in China (Saeed et al., 2010; Shi, Zheng, Liu, et al., 2018; Shi, Zheng, Mao, et al., 2018). This is a large, evergreen coniferous tree and a primitive member of the gymnosperms yew family (Taxaceae) having the ability to survive for 400–500 years It has dioecious flowers and drupe-like fruits with nut-seeds which take 3 years from blossoming to maturation and develop an attractive flavor and crispy taste with enormous biological functions. *T. grandis* is one of the world's rare dry fruits because of its high quality and assortment functions. Li, Luo, Cheng, Feng, and Yu (2005) stated the influence of the environment and cultivation management resulting in variety difference (morphology) of *T. grandis*, i.e., varying seed shapes and qualities, including Yuanfei, Xifei, Xiangyafei, Zhimafei, and others. Two important cultivars with significant differences in external appearance are Yuanfei (Mufei) and Xifei; the former is a seedling, while the latter is a grafted variety that is widely cultivated for nut production. The *T. grandis* kernel is the edible part of the *T. grandis* seed, which possesses about 76.1%–94.3% oil (He et al., 2016; Ni & Shi, 2014), with an abundance of unsaturated fatty acids (76.1%–94.3%) of the total fatty acids, and like edible oils the majority are oleic and linoleic acids. Thus, the *T. grandis* kernel is considered a good source of precious oil, which also contains biologically important nonmethylene-interrupted unsaturated fatty acid-sciadonic acid (5c, 11c, 14c-eicosatrienoic acid) at 8.3%–16.4% (He et al., 2016; Ni & Shi, 2014). Extraction of *T. grandis* kernel oil through cold pressing has been reported for different varieties/cultivars (Feng et al., 2011; He et al., 2016; Shi, Zheng, Liu, et al., 2018; Shi, Zheng, Mao, et al., 2018). He et al. (2016) have reported chemical components of seven different *T. grandis* cultivar kernel oils extracted through cold pressing, with yields in the range of 23.3%–39.7%.

Cold Pressed Oils. https://doi.org/10.1016/B978-0-12-818188-1.00004-9
Copyright © 2020 Elsevier Inc. All rights reserved.

Shi, Zheng, Liu, et al. (2018) and Shi, Zheng, Mao, et al. (2018) stated that variety and the technique for the oil extraction process may result in a great diversity of the chemical profiles and biological properties. Previous reports (Dong et al., 2014; He et al., 2016; Ni et al., 2015; Ni & Shi, 2014; Yu, Ni, Wu, & Sang, 2016) on cold press extraction of *T. grandis* claimed that no correlation between chemical profile and biological properties. He et al. (2016) have extracted oil of *T. grandis* (*T. grandis* Fort. ex. Lindl. and *T. grandis* Fort. var. Merrillii) via three techniques—hot pressing, cold pressing, and solvent extraction—to evaluate the effects of the variety and oil extraction method on the quality, fatty acid composition, and antioxidant capacity. There were significant differences in physicochemical parameters and bioactivity in two species of oil; however, there was no considerable effect from the extraction technique used to obtain the oil.

Feng et al. (2011) reported the essential oil composition of peels of *T. grandis* which were discarded and causing environmental pollution. Cold press and hydrodistillation techniques were employed to extract essential oils. The overall yield of oil by two methods is not compared and the yield of hydrodistillation only is reported to be 1.25% on dry basis, and 0.5% on wet basis. However, cold pressing was found to be better for extracting 65 essential oil components quantitatively.

In a report (Feng et al., 2011), the yield of Torreya seeds in Zhejiang province was 1000–1500 tons per year, which was three to four times higher compared to that of 1990, when it was around 350 tons per year. The price also increased from 40 to 80 RMB to 100–150 RMB per kilogram. These figures show encouraging market trends for Torreya seeds.

In addition to the many benefits of cold press techniques, there are some disadvantages that may be eliminated through CO_2 supercritical extraction, such as reducing the temperature from 40°C to obtain the most valuable content which is friable at this temperature. In addition, refraining extraction of unwanted compounds like proteins, carbohydrates, inorganic salts, or metals is an advantage of CO_2 extraction. However, the expenses of later techniques are two to three times higher than cold pressing (https://restingniceface414.wordpress.com/2017/07/27/why-cold-pressed-oils-arent-the-best-cold-pressed-vs-c02-supercritical-extraction). Thus, it is still reasonable to apply cold pressing.

2 Extraction and processing of cold pressed *Torreya grandis* oil

Few reports were published on the extraction of *T. grandis* seed oil, its physicochemical properties, and potential functions. Ni et al. (2015) reported on two varieties: *T. grandis* var. Merrillii and *T. grandis* var. jiulongshanensis. Yang et al. (2019) optimized solvent extractions to extract valuable bioactive constituents in which antinociceptive and antiinflammatory potential were evaluated (Saeed et al., 2007). Later on, supercritical carbon dioxide extraction of oils from two varieties, *T. grandis* var. Merrillii and *T. grandis* var. jiulongshanensis, was carried out to identify the physicochemical and antioxidant properties.

The first comprehensive study taking seven cultivars from different regions of China to extract oil through cold press technique and comparison of chemical components was reported by He et al. (2016). The yield of oil and its components were comparable to solvent and supercritical extraction, with the advantage of obtaining some abundant bioactive substances such as tocopherols (α-, β-, γ-, and δ-isomers), sterols including β-sitosterol, campesterol, and stigmasterol, in addition to phenolics.

A recent report (Shi, Zheng, Liu, et al., 2018; Shi, Zheng, Mao, et al., 2018) carried out a comparison of hot pressing, cold pressing, and solvent extraction of kernel oil to investigate lipid characterization and free radical scavenging. Two varieties, *T. grandis* Fort. ex. Lindl. and *T. grandis* Fort. var. Merrillii, were extracted with three processing methods to evaluate the effect on yield as well as physicochemical parameters. Acid value and peroxide value were two to three times higher in solvent and hot pressing compared to cold pressing. There were no significant differences in iodine value and saponification values for both species. The fatty acid profile was also the same for all extraction techniques.

3 Fatty acids of cold pressed *Torreya grandis* oil

It is a fact that the major saturated fatty acids of plant seed oil are lauric (dodecanoic), myristic (tetradecanoic), palmitic (hexadecanoic), and stearic (octadecanoic), while the unsaturated fatty acids are oleic (*cis*-9-octadecenoic), linoleic (*cis*-9, *cis*-12-octadecadienoic), and linolenic (all *cis*-9,12,15-octadecatrienoic) (Harwood, 1980). Following the abundance and trend, the same has been observed in *T. grandis*. The major fatty acids reported were hexadecanoic, 9,12-octadecadienoic, 9-octadecenoic, octadecanoic, 8,11-octadecadienoic, 8-octadecenoic, 9,12,15-octadecatrienoic, 11,13-ecosadienoic, and 11-eicosenoic acid (He et al., 2016; Shi, Zheng, Liu, et al., 2018; Shi, Zheng, Mao, et al., 2018).

Fatty compositions for oils extracted by solvent (hexane) and supercritical CO_2 were compared. Supercritical CO_2 (at 50°C for 4 h) extraction exhibited several advantages including nonexplosive, nontoxic, and availability in high purity with nonsolvent residues (Ni et al., 2015). As expected, major fatty acids were oleic and linoleic, while almost all fatty acids were quite similar in both varieties.

He et al. (2016) reported the fatty acid composition of cold pressed kernel oils from different *T. grandis* cultivars (Table 1). Ten major fatty acids with very similar concentrations for the seven *T. grandis* were found in cold pressed oil, which was comparable to soybean oil fatty acid content while being much higher than peanut oil. Eicosatrienoic acid was found in the range of 6.29%–8.37%, which was lower than that of other cultivars (9.8% and 11.2%) extracted with the supercritical extractor as well as conventional techniques (Dong et al., 2014; Wolff, Pédrono, Marpeau, Christie, & Gunstone, 1998). The fatty profile of Zhuji oil obtained through the supercritical technique was reported (Feng et al., 2011), and almost all fatty acids were similar as later revealed in oil obtained through cold pressing (He et al., 2016), but some additional fatty acids were present in Zhugi which were not reported in cold pressed oil such as C4:0, C6:0, C14:0, C20:4, C20:5, and C24:0.

Shi, Zheng, Liu, et al. (2018) and Shi, Zheng, Mao, et al. (2018) reported the effect of extraction method on the fatty acid profiles for two varieties of *T. grandis*: *T. grandis* Fort. ex. Lindl., and *T. grandis* Fort. var. Merrillii. The fatty acid profiles are listed in Table 2. Apart from oleic acid (C18:1) in *T. grandis* Fort. ex. Lindl., all fatty acids were either comparable or higher than those of seven cultivars reported by He et al. (2016). Results were also claimed to be consistent with previous reports (Ni & Shi, 2014). It was observed that short-chain C4 fatty acids are mostly found in conventional oil, while higher C16 onwards fatty acids are present in cold pressed oil.

4 Minor bioactive lipids of cold pressed *Torreya grandis* oil

Some reports are available about cold pressed *T. grandis* minor lipids (Yang et al., 2018). However, few works have been carried out on the individual chemical contents of oil obtained from the cold press technique. Only three sterols have been determined through GC-MS in seven cultivars: β-sitosterol (0.90–1.29 mg/g), campesterol (0.06–0.32 mg/g), and stigmasterol (0.04–0.18 mg/g).

5 Contribution of bioactive compounds in cold pressed *Torreya grandis* oil

Feng et al. (2011) have also focused on the peel of *T. grandis* Fort, which was left to rot for a long time due to lack of awareness, thereby generating environmental problems and hampering the development of the *Torreya* industry. Essential oils of peel were obtained through warm (steam distillation) and cold press techniques. Their chemical composition, as well as the quantities of obtained essential oil, were interestingly similar and even some valuable volatiles, i.e., limonene, δ-cadinene, and 3-carene, were found with higher recoveries through cold pressing. The authors did not recommend a technique, but cold pressing is arguably better, with low cost/time but with the same yield as that obtained by warm extraction. This type of analysis is yet to be performed for kernel oils.

6 Health-promoting traits of cold pressed *Torreya grandis* oil and oil constituents

The use of *T. grandis* parts in medication has been reported. Some researchers have compared bioactivities of this species with others. Shi, Zheng, Liu, et al. (2018) and Shi, Zheng, Mao, et al. (2018), among others, investigated the antioxidant capacities of the *Torreya grandis* kernel, sea buckthorn, and sesame oils, and the results demonstrated that the antioxidant capacities of *T. grandis* kernel, sea buckthorn, and roasted sesame oils were high (Kandhro et al., 2014; Memon, Memon, Luthria, Pitafi, & Bhanger, 2012; Shi, Zheng, Liu, et al., 2018; Shi, Zheng, Mao, et al., 2018).

He et al. (2016) concluded that on the basis of β- and γ-isomers, β-sitosterol, and phenolics, kernel oil obtained through cold pressing has a relatively better oxidative stability compared to other sources such as sunflower seed oil and olive oil.

Shi, Zheng, Liu, et al. (2018) and Shi, Zheng, Mao, et al. (2018) reported the effect of extraction techniques on the yield of bioactive molecules and their role in antioxidant activity. The authors have correlated the antioxidant activity, tested by DPPH·, ORAC, and ABTS, with β-sitosterol and campesterol. Both of these components, as well as antioxidant activity, were found to be better for *T. grandis* Fort. ex. Lindl. when compared to *T. grandis* Fort. var. Merrillii. However, the species difference in yield of these components on the basis of extraction techniques was not significant, but antioxidant activity in hot pressing was better than that in solvent and cold pressing. This might be due to the production of more phenolics upon strong heating, which resulted in the enhancement of the activity. Nevertheless, it was proved that heating may result in change or damage to the chemical profile, which may contribute to other health benefits. Antioxidant activity for oil obtained through supercritical extraction of *T. grandis* var. Merrillii and *T. grandis* var. jiulongshanensis was low compared to the abovementioned results of cold pressing.

Phillips, Ruggio, Toivo, Swank, and Simpkins (2002) and de Jong, Plat, and Mensink (2003) reported that β-sitosterol, campesterol, and stigmasterol can effectively decrease serum LDL cholesterol and atherosclerotic risk, suggesting that cold

TABLE 1 Fatty acid profile of cold pressed *T. grandis* oil.

No.	Specie	Palmitic acid (C16:0)	Heptadecanoic acid (C17:0)	Stearic acid (C18:0)	Oleic acid (C18:1)	Linoleic (C18:2)	Linolenic acid (C18:3)	Arachidic acid (C20:0)	Eicosenoic acid (C20:1)	Eicosadienoic acid (C20:2)	Eicosatrienoic acid (C20:3)	Ref.
1	Seven cultivars											He et al. (2016)
	Zhuji	8.21	ND	3.21	37.5	39.3	0.32	0.14	0.49	2.16	8.37	
	Panan	7.86	ND	4.00	31.7	45.6	0.36	0.18	0.48	2.22	7.55	
	Shengzhou 1	8.44	ND	2.70	32.5	44.6	0.40	0.13	0.50	2.42	8.16	
	Shengzhou 2	8.42	0.11	2.34	30.4	47.0	0.39	0.15	0.46	2.30	8.25	
	Dazinai	7.87	0.08	2.52	30.7	47.7	0.38	0.12	0.48	2.38	7.67	
	Tedanai	7.68	0.08	2.68	35.4	44.1	0.37	0.12	0.49	2.26	6.78	
	Huangshan	9.00	0.12	3.16	31.4	43.7	0.51	0.17	0.65	3.03	8.29	
2	*T. grandis* Fort. ex. Lindl.	8.32	ND	3.06	22.3	46.0	0.74		0.85	0.83 (*cis* 5,11) 3.71 (*cis* 5,14)	12.9	Shi, Zheng, Mao, et al. (2018)
3	*T. grandis* Fort. var. Merrillii	7.37	ND	3.15	35.1	41.8	0.49		0.71	0.81 (*cis* 5,11) 2.33 (*cis* 5,14)	9.13	

TABLE 2 Physical parameters of cold pressed *T. grandis* oil.

	T. grandis cultivar (He et al., 2018)							*T. grandis* Fort. ex. Lindl. (Shi, Zheng, Mao, et al., 2018)	*T. grandis* Fort. var. Merrillii (Shi, Zheng, Mao, et al., 2018)
Parameter	**Zhuji**	**Panan**	**Shengzhou 1**	**Shengzhou 2**	**Dazinai**	**Tedanai**	**Huangshan**		
Refractive index (25°C)	1.472	1.472	1.473	1.474	1.474	1.474	1.473	Not reported	Not reported
Acid value	0.54	0.33	0.70	0.42	0.80	0.61	0.31	0.66	0.95
Iodine value	124.8	119.3	121.8	126.8	119.4	119.3	121.0	141.4	124.9
Peroxide value	1.05	0.64	1.15	1.19	1.77	1.72	0.87	0.33	0.19
Saponification value	187.8	181.0	182.6	186.0	184.0	190.5	195.7	191.5	194.3

pressed *T. grandis* kernel oils, containing sitosterol, campesterol, and stigmasterol in considerable quantity, could be better for human health.

The quality of fats plays a very important role in food processing technology. Fat oxidation is the main reason for deterioration in the quality of foods and can directly affect many quality characteristics such as the flavor, color, texture, nutritive value, and safety of the food. The quality and dietary character of the edible oils are topics of concern among food scientists, nutritionists, and consumers (Sherazi, Kandhro, Mahesar, Talpur, & Latif, 2010). Most edible oils contain monounsaturated acids (MUFAs) and polyunsaturated fatty acids (PUFAs). Among the MUFAs, oleic acid (C18:1 *cis*-9) was the major fatty acid present. This acid is considered to be responsible for lowering the LDL (bad) cholesterol levels, while PUFAs have beneficial effects on both normal health and chronic diseases, such as regulation of lipid levels (Mori et al., 2000), cardiovascular functions (Kris-Etherton, Harris, & Appel, 2002), and immune functions (Hao, Ma, Hwang, Kim, & Zhang, 2000; Kandhro et al., 2008). Mostly in the industrial catalytic hydrogenation process, some natural fatty acids are destroyed and new artificial *trans* isomers are produced that behave similarly to saturated fats. These isomers lack the essential metabolic activity of the parent compounds and inhibit the enzymatic desaturation of essential fatty acids. The fatty acid groups and ratio represented between the saturated/unsaturated fatty acid show the relation between two major fatty acid groups. This clearly indicates a proportion of saturated fatty acids and unsaturated fatty acids. The prevalence of unsaturated over saturated fatty acids (smaller ratio) is considered to be positive from a nutritional point of view (Kandhro et al., 2008).

Oil of *T. grandis* cultivars obtained through cold pressing (He et al., 2016) was found to contain a high unsaturated fatty acid/saturated fatty acid ratio, which is highly favorable for the reduction of serum cholesterol and deposition of fatty material on inner walls of arteries, and thus is useful in the prevention of heart disease (Reena & Lokesh, 2007). The unsaturated and saturated fatty acid ratio of the *T. grandis* kernel oil was 7.05–8.52, which was higher than those of soybean oil (3.69) and olive oil (3.02–5.82) (Moghaddam et al., 2012; Nehdi, 2011). Some cultivars of *T. grandis* kernel oils had higher levels of unsaturated fatty acid and lower levels of saturated fatty acid than other common edible oils such as soybean, peanut, and corn (Ni et al., 2015). Considering these facts, *T. grandis* cold pressed oil is proven as nutritious and beneficial to human health and could be used in cooking as well as other applications in the food industry.

7 Edible and nonedible applications of cold pressed *Torreya grandis* oil

Oil obtained at low temperatures by pressing the seed below 49°C possesses more nutritional properties. Cold pressed oil is also important if the oil is used directly as engine fuel because this oil carries lower levels of phosphorous. High levels of phosphorous in the oil can be harmful to a diesel engine and are one of the compounds with a maximum limit set in the standard for any oil for use as engine fuel. Zhao, Wei, and Julson (2014) described the application of cold pressed oil. They investigated the best properties of cold pressed oils from nonedible oilseeds for future bio-jet fuel production. Bio-jet fuel produced from nonedible oilseeds can be an alternative to fossil fuels with the benefits of increasing safety of national energy, and economic growth for rural areas. Efficient oil extraction from oilseeds is critical for economic production of bio-jet fuels. Oilseeds produce oil that may be subsequently upgraded into saturated, unbranched, and long-chain hydrocarbon fuels, which are suitable for bio-jet fuel production (Maher & Bressler, 2007). Using this oil for bio-jet fuel production has potential advantages, such as high energy density, low moisture content, and high relative stability (Atabani et al., 2013; Cheng, Li, Huang, Zhou, & Cen, 2014). Converting crude oils to bio-jet fuels involves three main steps. First, the oil is extracted from oilseeds. It is then upgraded into hydrocarbon fuel. Finally, this fuel is delivered to a petroleum refinery for production of bio-jet fuel. Yetim, Sagdic, and Ozturk (2008) utilized seven different plant seeds to produce cold pressed oil. They demonstrated that the cold pressed seed oils from different plant seeds may serve as potential dietary sources of ω-3 and MUFAs in improving human nutrition and may have potential in new functional foods. This cold pressed oil from plant seeds containing no added synthetic antioxidants may exhibit very good oxidative stability and be suitable for therapeutic food preparations. Therefore, *T. grandis* seeds can be considered important sources for the high-value food and nutraceutical supplement industries.

He et al. (2016) have reported significant health-promoting bioactive chemical components such as different isomers (α, β, γ, δ) of tocopherols, sterols (β-sitosterol, campesterol, and stigmasterol), and phenolics. They also determined a high percentage of total oil (>50%) extracted from kernels of *Torreya grandis* with greater quantities of protein. This study clearly indicated that the cold pressed oil of *T. grandis* kernel contains potentially important nutrition and health benefits and could be used as oil in the human diet. According to its available functional ingredients, the *T. grandis* kernel is a good candidate for obtaining edible oil and could be used for preparing different food products in the food industry.

8 Other issues

No adulteration has been reported so far in *T. grandis* kernel oil obtained through cold pressing techniques. However, many chemicals, as well as biological studies, are reported in the oil obtained through techniques other than cold pressing, but limited work has been carried out on cold pressed oil. There is a need to perform in vivo studies on cold press *T. grandis* oil. Different cultivars are reported and their oil is extracted using cold pressing, but no environmental effects have been studied. In addition, no field-type study has been performed.

References

Atabani, A., Silitonga, A., Ong, H., Mahlia, T., Masjuki, H., Badruddin, I. A., et al. (2013). Non-edible vegetable oils: A critical evaluation of oil extraction, fatty acid compositions, biodiesel production, characteristics, engine performance, and emissions production. *Renewable and Sustainable Energy Reviews*, *18*, 211–245.

Cheng, J., Li, T., Huang, R., Zhou, J., & Cen, K. (2014). Optimizing catalysis conditions to decrease aromatic hydrocarbons and increase alkanes for improving jet biofuel quality. *Bioresource Technology*, *158*, 378–382.

de Jong, A., Plat, J., & Mensink, R. P. (2003). Metabolic effects of plant sterols and stanols. *The Journal of Nutritional Biochemistry*, *14*(7), 362–369.

Dong, D., Wang, H., Xu, F., Xu, C., Shao, X., & Li, H. (2014). Supercritical carbon dioxide extraction, fatty acid composition, oxidative stability, and antioxidant effect of *Torreya grandis* seed oil. *Journal of the American Oil Chemists' Society*, *91*(5), 817–825.

Feng, T., Cui, J.-j., Xiao, Z.-b., Tian, H.-x., Yi, F.-p., & Ma, X. (2011). Chemical composition of essential oil from the peel of Chinese *Torreya grandis* Fort. *Organic Chemistry International*, *2011*, 1–5.

Hao, Y., Ma, D. H.-K., Hwang, D. G., Kim, W.-S., & Zhang, F. (2000). Identification of antiangiogenic and anti-inflammatory proteins in the human amniotic membrane. *Cornea*, *19*(3), 348–352.

Harwood, J. (1980). Plant acyl lipids: Structure, distribution, and analysis. In *Lipids: Structure and function* (pp. 1–55). Elsevier.

He, Z., Zhu, H., Li, W., Zeng, M., Wu, S., Chen, S., et al. (2016). Chemical components of cold pressed kernel oils from different *Torreya grandis* cultivars. *Food Chemistry*, *209*, 196–202.

Kandhro, A. A., Saleem, R., Laghari, A. H., Sultana, R., Ahmed, K. M., & Memon, D. H. (2014). Quantification of pure refined olive oil adulterant in extra virgin olive oil using diamond cell ATR-FTIR spectroscopy. *Journal of the Chemical Society of Pakistan*, *36*(4), 699–706.

Kandhro, A., Sherazi, S., Mahesar, S., Bhanger, M., Talpur, M. Y., & Rauf, A. (2008). GC-MS quantification of fatty acid profile including trans FA in the locally manufactured margarines of Pakistan. *Food Chemistry*, *109*(1), 207–211.

Kris-Etherton, P. M., Harris, W. S., & Appel, L. J. (2002). Fish consumption, fish oil, omega-3 fatty acids, and cardiovascular disease. *Circulation*, *106*(21), 2747–2757.

Li, Z., Luo, C., Cheng, X., Feng, X., & Yu, W. (2005). Component analysis and nutrition evaluation of seeds of *Torreya grandis* 'Merrillii'. *Journal of Zhejiang Forestry College*, *22*(5), 540–544.

Maher, K., & Bressler, D. (2007). Pyrolysis of triglyceride materials for the production of renewable fuels and chemicals. *Bioresource Technology*, *98*(12), 2351–2368.

Memon, A., Memon, N., Luthria, D., Pitafi, A., & Bhanger, M. (2012). Phenolic compounds and seed oil composition of *Ziziphus mauritiana* L. fruit. *Polish Journal of Food and Nutrition Sciences*, *62*(1), 15–21.

Moghaddam, G., Heyden, Y. V., Rabiei, Z., Sadeghi, N., Oveisi, M. R., Jannat, B., et al. (2012). Characterization of different olive pulp and kernel oils. *Journal of Food Composition and Analysis*, *28*(1), 54–60.

Mori, T. A., Burke, V., Puddey, I. B., Watts, G. F., O'Neal, D. N., Best, J. D., et al. (2000). Purified eicosapentaenoic and docosahexaenoic acids have differential effects on serum lipids and lipoproteins, LDL particle size, glucose, and insulin in mildly hyperlipidemic men. *The American Journal of Clinical Nutrition*, *71*(5), 1085–1094.

Nehdi, I. (2011). Characteristics, chemical composition and utilisation of *Albizia julibrissin* seed oil. *Industrial Crops and Products*, *33*(1), 30–34.

Ni, Q., Gao, Q., Yu, W., Liu, X., Xu, G., & Zhang, Y. (2015). Supercritical carbon dioxide extraction of oils from two *Torreya grandis* varieties seeds and their physicochemical and antioxidant properties. *LWT: Food Science and Technology*, *60*(2), 1226–1234.

Ni, L., & Shi, W.-Y. (2014). Composition and free radical scavenging activity of kernel oil from *Torreya grandis*, *Carya cathayensis*, and *Myrica rubra*. *Iranian Journal of Pharmaceutical Research: IJPR*, *13*(1), 221.

Phillips, K. M., Ruggio, D. M., Toivo, J. I., Swank, M. A., & Simpkins, A. H. (2002). Free and esterified sterol composition of edible oils and fats. *Journal of Food Composition and Analysis*, *15*(2), 123–142.

Reena, M. B., & Lokesh, B. R. (2007). Hypolipidemic effect of oils with balanced amounts of fatty acids obtained by blending and interesterification of coconut oil with rice bran oil or sesame oil. *Journal of Agricultural and Food Chemistry*, *55*(25), 10461–10469.

Saeed, M. K., Deng, Y., Dai, R., Li, W., Yu, Y., & Iqbal, Z. (2010). Appraisal of antinociceptive and anti-inflammatory potential of extract and fractions from the leaves of *Torreya grandis* Fort Ex. Lindl. *Journal of Ethnopharmacology*, *127*(2), 414–418.

Saeed, M. K., Deng, Y., Parveen, Z., Dai, R., Ahmad, W., & Yu, Y. (2007). Studies on the chemical constituents of *Torreya grandis* Fort. Ex Lindl. *Journal of Applied Sciences*, *7*(2), 269–273.

Shen, C., Hu, Y., Du, X., Li, T., Tang, H., & Wu, J. (2014). Salicylic acid induces physiological and biochemical changes in *Torreya grandis* cv. Merrillii seedlings under drought stress. *Trees*, *28*(4), 961–970.

Sherazi, S., Kandhro, A. A., Mahesar, S., Talpur, M. Y., & Latif, Y. (2010). Variation in fatty acids composition including trans fat in different brands of potato chips by GC-MS. *Pakistan Journal of Analytical & Environmental Chemistry*, *11*(1), 6.

Shi, L., Zheng, L., Liu, R., Chang, M., Huang, J., Zhao, C., et al. (2018). Physicochemical property, chemical composition and free radical scavenging capacity of cold pressed kernel oils obtained from different *Eucommia ulmoides* Oliver cultivars. *Industrial Crops and Products*, *124*, 912–918.

Shi, L.-K., Zheng, L., Mao, J.-H., Zhao, C.-W., Huang, J.-h., Liu, R.-J., et al. (2018). Effects of the variety and oil extraction method on the quality, fatty acid composition and antioxidant capacity of *Torreya grandis* kernel oils. *LWT: Food Science and Technology*, *91*, 398–405.

Wolff, R. L., Pédrono, F., Marpeau, A. M., Christie, W. W., & Gunstone, F. D. (1998). The seed fatty acid composition and the distribution of Δ5-olefinic acids in the triacylglycerols of some taxaceae (Taxus and Torreya). *Journal of the American Oil Chemists' Society*, *75*(11), 1637–1641.

Yang, J. Y., Peng, B., Wang, M., Zou, X.-G., Yin, Y.-L., & Deng, Z.-Y. (2019). Characteristics and emulsifying properties of two protein fractions derived from the emulsion formed during aqueous extraction of Camellia oil. *Food Hydrocolloids*, *87*, 644–652.

Yang, R., Zhang, L., Li, P., Yu, L., Mao, J., Wang, X., et al. (2018). A review of chemical composition and nutritional properties of minor vegetable oils in China. *Trends in Food Science & Technology*, *74*, 26–32.

Yetim, H., Sagdic, O., & Ozturk, I. (2008). Fatty acid compositions of cold press oils of seven edible plant seeds grown in Turkey. *Chemistry of Natural Compounds*, *44*(5), 634–636.

Yu, Y.-J., Ni, S., Wu, F., & Sang, W.-G. (2016). Chemical composition and antioxidant activity of essential oil from *Torreya grandis* cv. *merrillii* Arils. *Journal of Essential Oil-Bearing Plants*, *19*(5), 1170–1180.

Zhao, X., Wei, L., & Julson, J. (2014). First stage of bio-jet fuel production: Non-food sunflower oil extraction using cold press method. *AIMS Energy*, *2*(2), 193–209.

Chapter 5

Cold pressed grape (*Vitis vinifera*) seed oil

Zeliha Ustun Argon[a,b], Veysel Umut Celenk[c], and Zinar Pinar Gumus[d]

[a]*Department of Biosystems Engineering, Eregli Faculty of Engineering and Natural Sciences, Necmettin Erbakan University, Konya, Turkey,* [b]*Medical and Cosmetic Plants Application and Research Center, Necmettin Erbakan University, Konya, Turkey,* [c]*Drug Research and Pharmacokinetic Development and Applied Center, ARGEFAR, Ege University, İzmir, Turkey,* [d]*Central Research Testing and Analysis Laboratory Research and Application Center, EGE-MATAL, Ege University, İzmir, Turkey*

Abbreviations

°C degrees Celsius
CO_2 carbon dioxide
kN kilo Newtons
min minutes
Mton Megatonne
W Watt

1 Introduction

Grapes (*Vitis vinifera*) are one of the world's most grown fruit crops, spread over approximately 7.5 million hectares around the world, with 78 million tons of production per year. The fruits are consumed both as fresh and as processed products such as wine, juice, jam, grape seed extract, jelly, vinegar, dried grapes, and grape seed oil (FAO and OIV, 2016). The total world grape production is distributed among Europe (39%), Asia (34%), and America (18%), and the main grape producers are China, Italy, the USA, France, Spain, and Turkey, according to their production volumes (OIV, 2017).

The processes for grape juice and wine production leave a great amount of skins, seeds, and pulp. Grape seeds are approximately 20%–26% of the pomace, which is mostly a by-product of wineries. Recovery of agricultural and food processing by-products and their usage in different fields can decrease waste problems and extend alternative applications for limited sources (Kamel & Dawson, 1985). Therefore, the oil extracted from the pomace, which formerly was discarded, is classified as a value-added and environmentally friendly product in the industry and contributes to lessening the waste disposals (American Chemical Society News, 1919). Grape seed oil is commercially available and it is the major by-product of the processing industry. The pomace has been used as an organic fertilizer and as a source for energy or fiber in the US and Canada. In Europe, this material has been preferred mostly for oil extraction as a valuable by-product or a raw material for animal feed and tannin (Kamel & Dawson, 1985).

Grape seeds were listed by FAO (FAO, 1992) within the minor oil crops in the early 1990s. Since then, grape seed oil has been one of the most commonly preferred gourmet culinary oils with its unique nutty and light flavor in salad dressings, baking, and marinades, and with a high smoke point it is used in frying. In addition, grape seed oils are preferred for supplements, cosmeceutical, personal care, and pharmaceutical products with fatty acids and high levels of nutrients, bioactive components, and antioxidants (Callaway & Pate, 2009; Gomez, Lopez, & De La Ossa, 1996; Lutterodt, Slavin, Whent, Turner, & Yu, 2011; Passos, Silva, Da Silva, Coimbra, & Silva, 2009; Stafne, Sleezer, & Clark, 2015).

2 Extraction of cold pressed oil, highly valued phytochemicals and lipid compounds

Grape seeds are the main by-products of wine and grape juice industries. An individual grape generally contains two seeds. These seeds count for approximately 3%–6% of the grape's weight, which can be affected by the type and size of the grape. The annual amount of discarded grape seeds is over 3 Mton around the world, and 20% of this is considered to be waste during production. Grape seeds have proteins, fiber, carbohydrates, minerals, lipids, and polyphenolic compounds (5%–8%). The oil content may vary between 10% and 20% and the percentages of the oil components may vary all

Cold Pressed Oils. https://doi.org/10.1016/B978-0-12-818188-1.00005-0
Copyright © 2020 Elsevier Inc. All rights reserved.

depending upon origin, variety, soil characteristics, climate, ripeness level, and manufacturing processes. These wastes can be used for different purposes such as antioxidants extraction, biotechnological production of fine chemicals, biomethane generation, composting, animal feeding, and incineration owing to their chemical structure (Choi & Lee, 2009; Dávila, Robles, Egüés, Labidi, & Gullón, 2017; Spigno, Marinoni, & Garrido, 2017).

The chemical properties and the amount of valuable substances within grape seed waste are important factors to determine the intended processes and their production. The high value-added substances can be classified as unsaturated fatty acids and antioxidants (phenolics and tocopherols). These substances might be used to obtain edible vegetable oil and natural antioxidants, which may be preferred as food ingredients and dietary supplements to promote overall wellness and to prevent disease. These secondary metabolites showed antioxidant, antimicrobial, antiinflammatory, anticancer, and cardiovascular-protective effects, which support human health and decrease the risk of different and mostly chronic diseases (Dávila et al., 2017).

The extraction process of valuable compounds could be defined in the following steps: extraction of intended bioactive materials from a complex plant sample; removal of potential impurities; increasing the concentration of the intended compounds so that sensitivity of bioassays will increase; preparing the analyte for detection and separation; and determining a valid method which is not affected by the variations in different samples. After these steps, an appropriate extraction method can be applied to the samples in order to separate, identify, and characterize the bioactive components (Hogervorst, Miljic, & Pu, 2017).

Various methods can be used to extract the oil from different parts of the plants such as cold pressing, supercritical CO_2, aqueous enzymatic, and solvent extraction. Oil is extracted from the seeds, which are separated from the pomace, dried and screened. For industrial production, cold pressing (screw pressing or continuous mechanical expression) and/or solvent extraction processes are mostly used to obtain grape seed oil. Cold pressing, which applies a lower process temperature and no solvent, is preferred to ensure a higher-quality oil extraction, although the yield is lower (55%–95%).

Solvent and cold press extractions are used to avoid the oxidation of fatty acids and bioactive substances, and to decrease the losses. If *n*-hexane solvent extraction or expeller pressing methods are used, a desolventization process should be considered to decrease gums (phosphatides) in the process of degumming similar to the refining process in the edible oil industry (Lampi & Heinonen, 2009). Traditional refining processes are completed with neutralization, degumming, bleaching, deodorization, and, if necessary, dewaxing steps to remove the undesired impurities from the crude oil. Although a regular refining process must be applied for the solvent-extracted minor crude vegetable oils to obtain the required quality, if necessary, only low temperature refining with degumming, dehydration, and filtration processes can be applied for cold pressed minor vegetable oils (Yang et al., 2018).

Over recent years, cold pressed plant oils have become more popular due to a more natural, safer, and better nutritional profile. Cold pressing, as a seed oil extraction procedure, does not apply either heat or chemicals during or prior to the process. Therefore, the amount of beneficial phytochemicals in the oil increases. In addition, this process is ecological and simple, and does not require high energy. Cold pressed edible seed oils may contain considerable levels of phytochemical components such as carotenoids and tocopherols, which are well-known for their potential health benefits, and their characterization and consumption may provide a basis for improving the human diet. Cold pressed oils are also known to have a remarkable amount of phenolic components, which have powerful antioxidative effects and may decrease and prevent cellular damage of free radicals. Therefore, the oil can be used as an ingredient of functional foods. Antioxidant activity is affected by the number and position of hydroxyl groups, solubility polarity, and phenolic compounds' stability during processing. In addition to the phytochemical components, the peroxide value and oxidative stability index of cold pressed oils, which are known as stability properties, are important for indicating the shelf life. In addition, cold pressed oils contain acids, alcoholic compounds, and aroma active esters (Bail, Stuebiger, Krist, Unterweger, & Buchbauer, 2008; Fiori, 2007; Maier, Schieber, Kammerer, & Carle, 2009; Nunes, Rodrigues, & Oliveira, 2017; Parry et al., 2005, 2006; Rombaut et al., 2015; Siger, Nogala-Kalucka, & Lampart-Szczapa, 2008). In comparison to traditional high-temperature solvent extraction, cold pressing extraction is a good alternative, but the yield is usually lower and it is difficult to obtain a product of constant quality due to differences in geographical location, processing techniques, and species (Bail et al., 2008; Lutterodt et al., 2011; Siger et al., 2008; Yu, Zhou, & Parry, 2005). The addition of enzymes could increase the oil extraction yield for grape seeds (Soto, Chamy, & Zúñiga, 2007).

Studies demonstrated that oil yield is positively related with pressure and temperature. A smaller diameter of the die at the meal discharge section and lower speed of screw rotation also cause higher oil yields. Suitable pretreatments of the seeds such as flaking, moisture conditioning, and cooking can also enhance oil yields. However, the application parameters of pretreatments should be determined according to seed bioactive components and the aimed results otherwise the pretreatment processes could cause oil degradation, loss of quality and sensory properties, and reduced process effectiveness (Rombaut et al., 2015).

The valuable compounds of food by-products are recovered in five stages: macroscopic matrix pretreatment, separation of molecules, extraction of molecules, purification, and formation of product. The extract quality and the composition are affected by the extraction technique, the type of solvent, the raw material's origin, the part of the plant used (seeds, skin, and leaves), the storage conditions of plant material, and the applied pretreatments. Moreover, the quality of the extract is determined according to the chemical and biological properties, such as its biological activities (Hogervorst et al., 2017).

Many different emerging technologies can apply to by-products of grape processing to recapture the valuable compounds. Solid–liquid extraction is used conventionally. Alternative extraction methods are ultrasound-assisted extraction (UAS), pulsed ohmic heating (POH), microwave-assisted extraction (MAE), pressurized liquid extraction (PLE), accelerated solvent extraction (ASE), high voltage electrical discharges (HVED), supercritical fluid extraction (SFE), infrared-assisted extraction (IRAE), and pulsed electric fields (PEF) (Maroun, Rajha, Vorobiev, & Louka, 2017; Nunes et al., 2017).

The chemical composition of grape seeds is rich in fatty acids (13%–19%), proteins (11%), carbohydrates (60%–70%), and antioxidants. Fatty acids can be extracted with ultrasound-assisted extraction (UAE) (150 W, 90 min) and supercritical CO_2 extraction (400 bar and 41°C) other than the Soxhlet method. Tocopherols and tocotrienols belong to the vitamin E family, well-known for its antioxidant, antiinflammatory, and antithrombotic effects. Polyphenols can be recovered with UAE, Soxhlet, screw pressing, supercritical CO_2 extraction (53.8 MPa, 104°C, and 17 kg/h), and gas-assisted mechanical expression (GAME) (combination of supercritical CO_2 and pressing) methods (Maroun et al., 2017).

3 Fatty acids and acyl lipids profile of cold pressed oil

Fatty acids (FA) have important roles in the human body and other organisms. Fatty acids construct the main components of biological matter with proteins and carbohydrates (Yang et al., 2018). Grape seed oil consists mainly of neutral lipids, especially triacylglycerols (TAG). Depending on the method used for extraction and the refining extent, the oil may also contain some free fatty acids and phospholipids. Most of the physical and chemical properties of oils are determined by the fatty acid composition. The nutritional quality of the oil is also affected by the fatty acid profile. Thus, oils are generally characterized by their fatty acid composition and classified according to their different functions as polyunsaturated fatty acid (PUFA), monounsaturated fatty acid (MUFA), or saturated fatty acid (SFA).

Triacylglycerol (TAG) is defined as an important storage lipid and plays a significant role in carbon storage and energy homeostasis (Wase, Black, & Dirusso, 2018). In a study, Lisa and Holcapek (2011) found the detailed characterization of vine white and vine red grapes triacylglycerol profile to be LLL, OLL, LLP, OLP, OLO, and SLL. Also in this study, average equivalent carbon numbers were 14.81 and 14.77, average carbon numbers were 17.81 and 17.78, average double bonds were 1.50 and 1.51, essential fatty acids were 63.8% and 65.6%, saturated fatty acids were 13.6% and 14.6%, MUFA were 22.5% and 19.6%, and PUFA were 63.84% and 65.73%, respectively.

Acyl lipids are accepted with the negative effects to oil quality except triacylglycerol groups. To remove these components, generally a refining process is preferred. In addition to the acyl lipids, some minor lipid forms can also be found in seed oils. Grape seed oil contains other acyl lipids to some extent, which are coextracted during processing. Seed membranes naturally contain phospholipids, and free fatty acids are mainly products of hydrolysis or glycerolipids' metabolic intermediates. Some studies determined 18.6 μg/g of phosphorus in commercial crude grape seed oil, which indicates the presence of phospholipids. Other results related with different seeds also show that there is a great difference in the phospholipid contents of the oils related with the extraction efficiency and the refining extent. The presence of free fatty acids indicates that the deterioration products of TAG and phospholipids in the hexane-extracted grape seed oils can be removed with physical refining and molecular distillation (Lampi & Heinonen, 2009).

Grape seed (*Vitis vinifera* L.) contains approximately 8%–15% (w/w) of quality oil, which has a high level of unsaturated fatty acids, such as oleic and linoleic acids, mono and diglycerides. (Cao & Ito, 2003; Fiori, 2007; Lampi & Heinonen, 2009; Passos et al., 2009; Talbot, 2015). Various studies about cold pressed grape seed oil fatty acid composition revealed that it is rich especially in unsaturated fatty acids (UFAs) such as linoleic acid (72%–76%, w/w). Analysis results showed that grape seed oils contain seven abundant TAG species, which are mainly linoleic (18:2), oleic (18:1), palmitic (16:0), and stearic (18:0) acid. These fatty acids compose more than 90% of the oils. Generally, grape seed fatty acid composition is reported as palmitic acid (16:0; 6.3–11.6), stearic acid (18:0; 3.6–5.4), oleic acid (18:1 n-9; 12.7–20.9), linoleic acid (18:2 n-6; 61.3–74.6), and linolenic acid (18:3 n-3; 0.3–1.8) (Lampi & Heinonen, 2009).

The fatty acid components also showed differences between various types of grapes. The main fatty acids of the Ruby Red grape variety were linoleic (66.0%), oleic (21.9%), palmitic (7.10%), and stearic (4.24%), while the Muscadine, Concord, and Chardonnay grape varieties have the same major fatty acids with different amounts for linoleic (70.2%, 75.3%, 68.8%), oleic (15.4%, 13.9%, 19.3%), palmitic (7.70%, 7.05%, 7.75%), and stearic acids (4.72%, 2.52%, 3.63%)

(Lutterodt et al., 2011). In another study for the Monastrell, Garnacha Tintorerea, Petit Verdot, and Syrah varieties, the main fatty acids levels ranged between 60.94% and 69.16% for linoleic acid, 16.07% and 24.88% for oleic acid, 7.86% and 9.19% for palmitic acid, and 4.42% and 5.87% for stearic acid (Pardo et al., 2009). The difference of oleic acid content from Egypt and Italy, France, and Spain shows that the origin is another effective factor in fatty acid composition (Nunes et al., 2017). In addition, different solvent extraction techniques applied for grape seeds give a similar arrangement for the main fatty acids (Freitas, Dariva, Jacques, & Caramão, 2013).

In summary, linoleic (58%–78%) and oleic (10%–20%) acids are the two major fatty acids of grape seed oil. MUFA are 14%–22% and PUFA are 66%–75% in grape seed oil, which are very important from a health perspective. SFA are generally 9%–12% of total fatty acids, causing an unexpectedly high smoking point (190–230°C). This fatty acid profile provides various applications for edible and pharmaceutical purposes. The differences between the samples may result from the differences of vine varieties, vegetation, fruit maturity, extraction parameters, soil composition, tannin removal processing, drying, and storage conditions. In addition, climatic conditions effect the oleic-linoleic acid balance of plant oils (Bail et al., 2008; Matthäus, 2008; Nunes et al., 2017; Rombaut et al., 2015).

The antioxidant capacity, which shows the ability of the oil to protect the DNA, membrane lipids, and proteins against free radical activities, of grape seed oil has been studied for different samples. Bail et al. (2008) found the TEAC (trolox equivalent antioxidant capacity) value for nine different grape seed oils between 0.14 and 1.16 μg/g. Muscadine, Concord, Ruby Red, and Chardonnay varieties grape seed oils' relative DPPH· scavenging activity was determined as 1.11, 2.22, 0.66, and 0.07 mmol TE/g, respectively (Lutterodt et al., 2011). Commercial grape seed oils from Turkey showed the DPPH· radical scavenging activity between 31.0% and 45.3% (Ustun-Argon, 2019b). The results show that the antioxidant level of grape seed oil is very high, therefore the oil can be considered as an active ingredient of the products where the antioxidant activities are important, such as in nutraceuticals, cosmetics, and foods. The oxidative stability of grape seed oil was determined for Monastrell, Garnacha Tintorera, Petit Verdot, and Syrah varieties at 98°C as 6.38, 8.14, 8.07, and 7.90 h, respectively (Pardo et al., 2009). Oxidative stability values for Muscadine, Concord, Ruby Red, and Chardonnay varieties were 19.7, 26.9, 40.0, 23.4, and 46.5 h at 80°C with a 7 L/h airflow rate (Lutterodt et al., 2011).

The iodine value of grape seed oil range in the samples from France was 136–143, samples from Spain ranged between 124 and 138, and samples from Italy ranged from 129 to 136, which was considered to be narrower than the other studies (Crews et al., 2006). Kamel and Dawson (1985) found the iodine value of grape seed oil to be 132, which is compliant with the other results. In the same study, the refractive index, specific gravity, saponification value, unsaponifiable matter, hydroxyl value, acid value, free fatty acids, and ester number were 1.4741, 0.904 (25/25°C), 194, 0.93%, 16, 0.78 (oleic acid %), and 192, respectively. Other physicochemical quality parameters of the grape seed oils were determined for four different varieties, and the results were free acidity 0.37%–1.47%, peroxide index 5.99–13.50 meq/kg, K_{270} 0.35–0.46, and wax 30.6–59.3 mg/kg (Pardo et al., 2009).

4 Minor bioactive lipids in cold pressed oil

Grape seed oil contains minor bioactive lipids such as tocopherols (tocols), phytosterols, phenolic components, and other fat-soluble compounds. Tocopherols are the most powerful natural fat-soluble antioxidants (Göktürk-Baydar & Akkurt, 2001). Carotenoids are responsible the strong yellow color of oils and have antioxidant effects. Some carotenoids, such as β-carotene, β-cryptoxanthin, and α-carotene, have provitamin A activity. Phytosterols are considered to be another important class of nonsaponifiable lipids, which may help to reduce the serum cholesterol level if it has been consumed at 2 g regularly on a daily basis, and may aid in treatment of different diseases such as cardiovascular problems and cancer (Duba & Fiori, 2015; Garavaglia, Markoski, Oliveira, & Marcadenti, 2016; Lampi & Heinonen, 2009).

4.1 Tocopherols and tocotrienols

Tocols are one of the most important natural fat-soluble antioxidants in the human diet. Tocopherols exist in seed oils in four forms: α (5,7,8-trimethyltocol), β (5,8-dimethyltocol), γ (7,8-dimethyltocol), and δ (8-methyltocol), which may show vitamin E activity or act as the most potent antioxidants. Considering the oils that are rich in vitamin E, wheat germ oil with 223 mg of vitamin E is found at the highest level, followed by grape seed oil, sunflower seed oil, hazelnut oil, and rice bran oil with 100 mg, 54 mg, 47 mg, and 32 mg vitamin E, respectively. It can be seen from these values that among the edible oils, grape seed oil is one of the richest natural sources of vitamin E. Since lipid oxidation is considered to be a major problem in the food and health industry, the oils with high tocopherol content can be applied where a high level of antioxidant protection is necessary (Göktürk-Baydar & Akkurt, 2001). In a study from China, which evaluated the minor vegetable oils, the tocopherol content of grape seed oil was as follows: α-tocopherol 249.5 mg/kg, β-tocopherol 12.1 mg/kg, γ-tocopherol 140.0 mg/kg, and δ-tocopherol 21.9 mg/kg. The total tocopherol content was 423.5 mg/kg (Yang et al., 2018). In another study from Turkey, tocopherol contents of 18 different wine and table-type grapes were evaluated. The total tocopherol contents were found to be between 328 and 578 mg/kg. The highest rate belonged to the Kalecik Karası type, while Razakı cultivars had the lowest rate of total tocopherols (Göktürk-Baydar & Akkurt, 2001). Garavaglia et al. (2016) found that the Marufo variety has the highest concentration of both tocopherol and tocotrienol compounds among the oils from 10 traditional Portuguese grape varieties. Different studies investigating total tocopherol content from three countries with 30 grape seed oil samples (Crews et al., 2006) and eight grape seeds (Lampi & Heinonen, 2009) reported <0.05–229 mg tocols/kg oil and 0.6–1.0 mg tocols/g of oil, respectively. The studies showed that tocol contents depends on variation, origin of grapes, varieties, process, and environmental cultivation conditions. For example, the total tocopherols content of crude grape seed oil was 1.01 mg/g, and it decreased by 10% after physical refining including degumming, dewaxing, and bleaching. Comparing the supercritical fluid extraction and hexane extraction under optimal conditions, the α-tocopherol content of supercritical CO_2 extracted grape seed oil was six times higher than that of hexane-extracted samples (Anonymous, 2019; Garavaglia et al., 2016; Göktürk-Baydar & Akkurt, 2001; Lampi & Heinonen, 2009).

Grape seed oils are also rich in unsaturated forms of vitamin E, which are known as tocotrienols. The forms of γ-tocotrienol and α-tocotrienol are the most abundant and these two components showed the highest variations between grape varieties. A study that evaluated the grape seed oils from France, Italy, and Spain identified γ-tocotrienol as the major component, ranging between <0.05 and 785 mg/kg among other α-, β-, and Δ-tocotrienols (Crews et al., 2006). Compared to oils of cereal grains and palm, grape seed oil was one of the rare examples of α-tocotrienol-rich oils (Lampi & Heinonen, 2009).

4.2 Carotenoids

Carotenoids are the most important source of vitamin A, and are the components responsible from the antioxidant activity of the oil and its stability (Rekik et al., 2016). Studies about the carotenoid content of grape seed oil are very limited. A study that was conducted about cold pressed grape seed oils in the Brazilian market found the β-carotene amount was between 33.9 and 59.8 for four different samples (Shinagawa, de Santana, Araujo, Purgatto, & Mancini-Filho, 2018). A study about refined oils and cold pressed Polish commercial oils was not able to detect any carotene compounds (Rafalowski, Zegarska, Kuncewicz, & Borejszo, 2008). Another study, which compared sesame seed oil and fenugreek oil with grape seed oil, found carotene values to be 3.98, 56.7, and 15.2 ppm, respectively (Rekik et al., 2016).

4.3 Phytosterols

Phytosterols are the largest part of nonsaponifiable lipids in oils. They may exist in the form of esters of fatty acids or free alcohols. Depending on the varieties, the content of phytosterols in grape seed oil differs widely. The phytosterol profile of grape seed oil from China found β-sitosterol (145.5 mg/100 g), campesterol (39.0 mg/kg), and stigmasterol (24.9 mg/100 g), with the dominance of β-sitosterol (Yang et al., 2018). Crews et al. (2006) have conducted a study about grape seed oil sterol compositions in three different countries with 30 different samples and found β-sitosterol (172.3–823.5 mg/100 g), stigmasterol (28.5–96.1 mg/100 g), campesterol (26.3–94.4 mg/100 g) with minor amounts of cholesterol, 24-methylene-cholesterol, campestenol, Δ7-campesterol, Δ5,23-stigmastadienol, Δ7-stigmastenol, and Δ-avenasterols. Research based on commercial grape seed oils from Brazil showed similar results for four different samples as β-sitosterol (83.5–91.9 mg/100 g), stigmasterol (30.5–32.6 mg/100 g), and campesterol (12.7–13.7 mg/100 g) (Shinagawa et al., 2018). In another study, the total phytosterol value of eight grape varieties was between 3.16 and 18.61 mg/g oil (Lampi & Heinonen, 2009). Studies showed that sitosterols constitute 60%–80% of total sterols and are accepted as the major sterol of grape seed oil, followed by stigmasterol and campesterol. The results also indicate that industrial extraction can decrease the phytosterol levels.

Grape seed oil can contain steradienes in minor amounts, which are formed with the dehydration of sterols. These compounds can form during solvent extraction, filtration, acid-earth bleaching, or deodorization. One study reported campestadiene and stigmastadiene at the rate of 0.05–6.7 mg/kg, where the acceptable level for virgin olive oils is ≤0.15 mg/kg. The detected levels were not related to the variety, type, or origin of the grape (Crews et al., 2006; Lampi & Heinonen, 2009).

4.4 Phenolic compounds

Polyphenols are important components for the biological activities with their antimicrobial and antioxidant effects. These compounds help to prevent oxidation and to increase the oil stability. At the same time, nutritional value and quality of the edible oils are affected by the amounts and compositions of the phenolics (Liang, Appukuttan Aachary, & Hollader, 2015; Rekik et al., 2016; Shinagawa et al., 2018; Siger et al., 2008; Slavin, Cheng, Luther, Kenworthy, & Yu, 2009).

Since the cold pressing process does not apply heat to the seeds, antioxidant phenolics are tend to be present in higher amounts in the product (Bail et al., 2008). The total phenolic content (TPC) of grape seed oil found in different studies was 75.8 mg/kg (Yang et al., 2018), and 48–360 mg GAE/kg (Rombaut et al., 2015). A study conducted by Lutterodt et al. (2011) compared the Muscadine, Concord, Ruby Red, and Chardonnay varieties grape seed oils and found TPCs to be 0.44, 0.80, 0.16, 0.23 mg GAE/g oil, respectively. The lower level of TPC can be explained by binding of some phenolic acids to the seed material and the lower solubility rates of low molecular weight phenolics in the oil. Commercial grape seed oil samples from the Brazilian market showed 13.92–27.87 mg/100 g TPC in their content (Shinagawa et al., 2018). TPC values of cold pressed grape seed oils from Austria were 59–108 μg GAE/g oil (Bail et al., 2008), while for Spain, samples TPC values were 16.82–34.43 mg/kg for Monastrell, Garnacha Tintorera, Petit Verdot, and Syrah varieties (Pardo et al., 2009). Although most of the studies focused on the TPC, some studies also identified the phenolic components of grape seed oils. Ustun-Argon (2019b) found 26–77 different phenolic components in the cold pressed grape seed oils, which are sampled from the Turkish market (Table 1).

TABLE 1 Phenolic contents of cold pressed grape seed oil.

				Sample		
Name	Formula	A	B	C	D	E
Metlin negative						
α-Linolenic acid	$C_{18}H_{30}O_2$	+		+	+	+
Δ2-cis-hexadecenoic acid	$C_{16}H_{30}O_2$		+	+		
Traumatic acid	$C_{12}H_{20}O_4$	+		+		
Sucrose	$C_{12}H_{22}O_{11}$	+	+	+		
Stearidonic acid	$C_{18}H_{28}O_2$	+		+	+	+
Stearic acid	$C_{18}H_{36}O_2$	+		+	+	+
Urocanic acid	$C_6H_6N_2O_2$					+
trans-EKODE-(*E*)-Ib	$C_{18}H_{30}O_4$					+
Ricinoleic acid	$C_{18}H_{34}O_3$		+			
Methyl N-(a-methylbutyryl)glycine	$C_9H_{16}O_4$		+			
Elaidic acid	$C_{18}H_{34}O_2$		+	+	+	
Quercetin	$C_{15}H_{10}O_7$	+			+	
Pyrocatechol	$C_6H_6O_2$	+				
Pinocembrin	$C_{15}H_{12}O_4$	+		+	+	+
Petroselinic acid	$C_{18}H_{34}O_2$	+				+
Methyl N-(a-methylbutyryl)glycine	$C_9H_{16}O_4$	+		+	+	+
m-Coumaric acid	$C_9H_8O_3$	+			+	+
Matairesinol	$C_{20}H_{22}O_6$	+	+	+		
Luteolin	$C_{15}H_{10}O_6$	+	+	+	+	
Kaempferol	$C_{15}H_{10}O_6$	+		+	+	+
Genkwanin	$C_{16}H_{12}O_5$					+

TABLE 1 Phenolic contents of cold pressed grape seed oil—cont'd

		Sample				
Name	**Formula**	**A**	**B**	**C**	**D**	**E**
Isosteviol	$C_{20}H_{30}O_3$				+	+
Hieracin	$C_{15}H_{10}O_7$	+		+	+	+
Hexadecanedioic acid	$C_{16}H_{30}O_4$			+		
Gallic acid	$C_7H_6O_5$	+			+	
Ethyl-p-coumarate	$C_{11}H_{12}O_3$	+		+	+	+
Dihydroxyphenylacetic acid	$C_8H_8O_4$				+	
DL-b-hydroxycaprylic acid	$C_8H_{16}O_3$	+		+	+	
Diosmetin	$C_{16}H_{12}O_6$	+		+	+	
D-(+)-3-phenyllactic acid	$C_9H_{10}O_3$	+		+		
Apigenin	$C_{15}H_{10}O_5$					+
Abscisic acid (*cis,trans*)	$C_{15}H_{20}O_4$					+
Corosolic acid	$C_{30}H_{48}O_4$	+		+	+	+
cis-9,10-Epoxystearic acid	$C_{18}H_{34}O_3$	+		+	+	+
Betulonic acid	$C_{30}H_{46}O_3$	+		+		
Betulinic acid	$C_{30}H_{48}O_3$	+	+	+	+	+
Abscisic acid (*cis,trans*)	$C_{15}H_{20}O_4$	+		+		
9-Thiastearic acid	$C_{17}H_{34}O_2S$					+
9-OxoOTrE	$C_{18}H_{28}O_3$	+		+	+	+
9-OxoODE	$C_{18}H_{30}O_3$	+	+	+	+	+
(+)9-HODE	$C_{18}H_{32}O_3$					+
9,10-DiHOME	$C_{18}H_{34}O_4$		+			
9(E),11(E)-Conjugated linoleic acid	$C_{18}H_{32}O_2$	+	+	+		+
4-Nitrophenol	$C_6H_5NO_3$	+		+		
4-Hydroxybenzaldehyde	$C_7H_6O_2$	+		+	+	+
4-Hydroxy-3-methoxy cinnamaldehyde	$C_{10}H_{10}O_3$			+		
4-Formyl indole	C_9H_7NO	+	+	+	+	+
3-Hydroxycapric acid	$C_{10}H_{20}O_3$	+		+		
3-Hydroxy-tetradecanoic acid	$C_{14}H_{28}O_3$			+		
3,4-Dihydroxybenzoic acid	$C_7H_6O_4$	+				
3,4-Dihydroxy-benzaldehyde	$C_7H_6O_3$	+	+	+		
2,2′-3-MethylCyclohexane-1,1diacetic acid	$C_{11}H_{18}O_4$				+	
2-Pyrimidine acetic acid	$C_6H_6N_2O_2$	+		+	+	+
2-Hydroxy-hexadecanoic acid	$C_{16}H_{32}O_3$	+	+	+		+
2-Hexyldecanoic acid	$C_{16}H_{32}O_2$	+	+	+	+	+
2-Hydroxycinnamic acid	$C_9H_8O_3$			+		
2-Hydroxymyristic acid	$C_{14}H_{28}O_3$				+	

Continued

TABLE 1 Phenolic contents of cold pressed grape seed oil—cont'd

Name	Formula	Sample				
		A	B	C	D	E
1-Palmitoyl lysophosphatidic acid	$C_{19}H_{39}O_7P$	+				
1-Oleoyl lysophosphatidic acid	$C_{21}H_{41}O_7P$	+				
18α-Glycyrrhetinic acid	$C_{30}H_{46}O_4$	+		+		
16-Hydroxy hexadecanoic acid	$C_{16}H_{32}O_3$	+		+	+	+
13,14-Dihydro prostaglandin F1α	$C_{20}H_{38}O_5$	+		+		
13(*S*)-HODE	$C_{18}H_{32}O_3$			+		
13(*S*)-HOTrE	$C_{18}H_{30}O_3$	+			+	+
13(*R*)-HODE	$C_{18}H_{32}O_3$				+	+
(+)-Naringenin	$C_{15}H_{12}O_5$	+		+	+	+
(+)12,13-DiHOME	$C_{18}H_{34}O_4$	+		+		+
Metlin positive						
Phthalic acid mono-2-ethylhexyl ester	$C_{16}H_{22}O_4$		+			
Tetramethylpyrazine	$C_8H_{12}N_2$			+		
Sucrose	$C_{12}H_{22}O_{11}$				+	+
U-74389G	$C_{37}H_{50}N_6O_2$	+		+		
Stearidonic acid	$C_{18}H_{28}O_2$	+		+	+	+
Stearamide	$C_{18}H_{37}NO$	+	+	+	+	+
Quercetin	$C_{15}H_{10}O_7$			+	+	
Quercetin 3-methyl ether	$C_{16}H_{12}O_7$	+		+	+	
Prostaglandin D2 methyl ester	$C_{21}H_{34}O_5$				+	
PGF2α methyl ester	$C_{21}H_{36}O_5$	+		+	+	+
PGF1α alcohol	$C_{20}H_{38}O_4$	+		+	+	+
Omega-3 arachidonic acid	$C_{20}H_{32}O_2$	+		+		
O-2545	$C_{26}H_{36}N_2O_2$	+	+	+	+	+
MG(18:2(9Z,12Z)/0:0/0:0)	$C_{21}H_{38}O_4$	+	+	+	+	+
MG(18:1(9Z)/0:0/0:0)	$C_{21}H_{40}O_4$		+		+	+
Luteolin	$C_{15}H_{10}O_6$	+		+	+	
Linoleoyl ethanolamide	$C_{20}H_{37}NO_2$	+		+	+	
Linoelaidic acid	$C_{18}H_{32}O_2$				+	
Linoleic acid ethyl ester	$C_{20}H_{36}O_2$	+		+		
Niacin (nicotinic acid)	$C_6H_5NO_2$					+
Hepoxilin A3	$C_{20}H_{32}O_4$			+		
Herbacetin	$C_{15}H_{10}O_7$	+				
Ethyl-p-coumarate	$C_{11}H_{12}O_3$	+		+		
Ethyl syringate	$C_{11}H_{14}O_5$	+	+	+		+
Enoxolone	$C_{30}H_{46}O_4$	+		+	+	+

TABLE 1 Phenolic contents of cold pressed grape seed oil—cont'd

Name	Formula	Sample				
		A	B	C	D	E
Dioctyl phthalate	$C_{24}H_{38}O_4$				+	
Dehydrocholic acid	$C_{24}H_{34}O_5$					+
Methyl linolenate	$C_{19}H_{32}O_2$				+	
N,N′-Dicyclohexylurea	$C_{13}H_{24}N_2O$		+			
Dehydrocholic acid	$C_{24}H_{34}O_5$		+			
1-Monopalmitin	$C_{19}H_{38}O_4$		+			
2-Linoleoyl glycerol	$C_{21}H_{38}O_4$			+		
2,3-Dinor prostaglandin E1	$C_{18}H_{30}O_5$			+	+	
9(Z),11(E),13(E)-octadecatrienoic acid ethyl ester	$C_{20}H_{34}O_2$	+		+	+	+
8-IsoPGF2β	$C_{20}H_{34}O_5$	+				
8-Iso prostaglandin F1β	$C_{20}H_{36}O_5$	+		+	+	
8-Iso misoprostol	$C_{22}H_{38}O_5$	+				
4-Hydroxy-3-methoxycinnamaldehyde	$C_{10}H_{10}O_3$	+	+	+		+
4-Formyl indole	C_9H_7NO	+				
3-Methoxy prostaglandin F1α	$C_{21}H_{38}O_6$	+		+	+	+
1-Monopalmitin	$C_{19}H_{38}O_4$	+		+	+	+
19(*R*)-Hydroxy-PGF1α	$C_{20}H_{36}O_6$	+		+	+	
19(*R*)-Hydroxy-PGE1	$C_{20}H_{34}O_6$	+				
15(*R*)-PGE1	$C_{20}H_{34}O_5$	+				
11β-Prostaglandin F1β	$C_{20}H_{36}O_5$	+		+	+	+
15(*R*),19(*R*)-hydroxy prostaglandin E1	$C_{20}H_{34}O_6$			+		
11-deoxy-PGF1β	$C_{20}H_{36}O_4$	+		+	+	+
11-deoxy-PGE1	$C_{20}H_{34}O_4$	+		+		
13-OxoODE	$C_{18}H_{30}O_3$			+	+	+

5 Contribution of bioactive compounds in cold pressed oil to organoleptic traits and functions in food or nonfood products

The main bioactive compounds of grapes are linoleic acid, flavonoids, vitamin E and procyanidins, oligostilbenes, and stilbenes resveratrols. Although these components can be extracted from some specific parts of the fruit, linoleic acid is found in high quantities in grape seed oil, followed by oleic acid and palmitic acids.

The unsaturated fatty acid profile of grape seed oil is very important since it has the fatty acids essential for human health. Therefore, it is highly recommended to include grape seed oil in one's daily diet or to take its supplements to support the body against coronary and chronical diseases (Cao & Ito, 2003; Fiori, 2007; Lampi & Heinonen, 2009; Maroun et al., 2017; Nassiri-Asl & Hosseinzadeh, 2016; Soto, Falqué, & Domínguez, 2015). Unsaturated fatty acids are preferred in cosmetics due to their skin moisturizing effects. In addition, a low level of linolenic acid helps to reduce unfavorable odor and increase the stability duration compared with higher amounts of this fatty acid. However, it should be noted that levels of fatty acids are affected by oil extraction methods, seed origin, vine location, harvesting time, composition of soil, and processes such as drying and tannin-removal (Nunes et al., 2017).

Since the cold press method includes no heat or chemical treatments, the last product has a relatively higher amount of antioxidants, an important level of aroma compounds and phytochemicals. These properties cause grape seed oil to be preferred for supplements and a healthy diet. Due to its antioxidant effects, grape seed oil is popular not only in the cosmetic market but also in pharmaceutical, culinary, and medical applications such as intracellular antioxidant activities and anticancer studies, dietary supplements, food additives, and nutraceutical formulations (Dávila et al., 2017).

Additionally, tocopherols and tocotrienols with α-tocopherol, α-tocotrienol, γ-tocopherol, and γ-tocotrienol were identified as the most prevalent tocol components. The total tocol amount of grape seed oil can reach up to 1208 mg/kg of oil and β and δ isomers of tocopherols and tocotrienols found in minor amounts. α-tocopherol is the most abundant biological isoform among all vitamers. These components play important roles as antioxidants and prevent degenerative diseases. In cosmetic products, vitamin E is preferred in day/night creams due to their antioxidant and oxidative stress preventive effects. Tocopherols also have an important role for protection of PUFA against free radical oxidation. Phytosterols are an additional important bioactive component of grape seed oil (Dávila et al., 2017; Lavelli, Kerr, García-Lomillo, & González-SanJosé, 2017; Lisa & Holcapek, 2011; Maroun et al., 2017; Nassiri-Asl & Hosseinzadeh, 2016; Nunes et al., 2017).

Phenolics in grape seed oil also show antioxidant properties and supporting effects for treatment of diseases including cancer. These molecules have also been used as effective agents in stand-alone products or additive for sun protection products with the effect of modulating the UV radiation's pro-carcinogenic and immunosuppressive effects on the skin (Nunes et al., 2017).

The stability properties of cold pressed oils, such as oxidative stability index and peroxide value, support a longer shelf life for the oil (Bail et al., 2008; Çelenk, Gümüş, Ustun-Argon, Büyükhelvacıgil, & Karasulu, 2018; Maier et al., 2009; Nunes et al., 2017; Parry et al., 2005; Ustun-Argon, 2019a; Ustun-Argon, Gökyer, Gümüş, & Büyükhelvacıgil, 2017). Additionally, grape seed oil contains a high content of tannins such as oligomeric proanthocyanosides (OPCs), which is a thousand times higher than some other oils. Therefore, it shows higher stability for oxidation (Cao & Ito, 2003).

6 Edible and nonedible applications of cold pressed oil

Grape seed oil is primarily used as a gourmet oil and preferred by chefs in salad dressings, baking, homemade mayonnaise, and marinades, due to its nutty and light flavor. In addition, it is suitable for frying and stir-frying as it has a relatively high smoke point. Grape seed oil has been in the nutraceutical market with a focus on health and wellness due to its high levels of essential fatty acids, phytochemicals, and natural antioxidants (Kochhar, 2011; Lampi & Heinonen, 2009).

The unsaturated fatty acid profile and antioxidant components in grapes and grape seed oil make the product very significant and beneficial for human health. Therefore, grape seed oil is very important as the main ingredient and intermediate for formulations of dietary supplements, nutraceuticals, and food additives. These products are mostly preferred as antioxidants and supportive products for cardiovascular and other chronic illnesses (Maroun et al., 2017).

Grape seed oil also has been examined in different medical applications as a protective agent for some diseases such as acute liver injury, which is caused by environmental pollutants. Antioxidants' components also have an important role in health applications due to their high commercial values. Some studies have shown that antioxidants of grape seed oil can be applied for the protection or treatment of ulcers, cancer, inflammation, allergies, cardiovascular and toxical diseases, prevention of oxidation of low-density lipoproteins and thrombosis, lowering cholesterol serum levels, and regulation of autonomic nerve. Specifically, grape seed flavanols were found to be effective for curing high blood pressure, and a protector against lung oxidative stress induced by bleomycin in lung fibrosis, and pancreatic dysfunction caused by oxidative stress (Dávila et al., 2017; Garavaglia et al., 2016).

Grape seed oil is also very popular for skin products and cosmeceuticals due to its antioxidant, emollient, moisturizing, anti-wrinkle, and skin moisturizing effects. *Vitis vinifera* seed oil is a sustainable and natural ingredient for technical and functional applications, and it can be a good raw material alternative for "eco cosmetic" products (Lampi & Heinonen, 2009; Nunes et al., 2017). Since *Vitis vinifera* cold pressed and hydrogenated oils are accepted to be safe to use in cosmeceuticals, red grape seed oil is also used in wine therapy for massaging the whole body in an exfoliating step (Soto et al., 2015). The saturated stearic and palmitic fatty acids in grape seed oil can be used in the manufacturing of soap, detergents, and other cosmetic products (Domínguez, Sanchez-Hernandez, & Lores, 2017).

Another component of grape seed is squalene, a specific bioactive lipid. Squalene extraction is one of the new and value-added products, which could be obtained from the wine lees with several processes by 2% (w/w) of the pomace. Even though the amount of extracted material amount is low, at 13 mg/100 g grape seed oil, the squalene is subject to use in different applications such as cosmetic, food, and pharmaceutical industries (Dávila et al., 2017; Yang et al., 2018). Similarly, the oils rich in unsaturated fatty acids can also be evaluated for synthesis of value-added biomaterials such as thermoset resin, polyols, and lubricants by transforming unsaturated fatty acids' double bonds into more reactive oxirane rings (Dávila et al., 2017). In a wide evaluation range, bioactive compounds in the grape and its byproducts also have been used as coloring agents, antioxidant, texturizing agents, antioxidant dietary fiber, and antimicrobials (Lavelli et al., 2017).

7 Health-promoting traits of cold pressed oil and oil constituents

Cold pressed plant oils have a rich nutritional profile and contain more natural components than oils processed by other methods, as the manufacturing process does not include chemical or heat applications. Grape seed oil is rich in unsaturated fatty acids, phenolics, and vitamin E with other phytochemicals. Due to its high polyphenol content, grape seed oil shows a high level of antioxidant capacity, which can help to decrease free radicals damage (Bail et al., 2008; Fiori, 2007; Maier et al., 2009; Nunes et al., 2017; Parry et al., 2005, 2006; Rombaut et al., 2015; Siger et al., 2008). The PUFA and MUFA content of the oils show antithrombotic and antiatherogenic properties, and affect the fluidity of the membrane, concentration of lipoproteins, enzyme function of membranes, and other compounds' modulations (Simopoulos, 1999; Yang et al., 2018).

The consumption of cold pressed oils may help to prevent diseases caused by free radicals and to treat them by improving human health, particularly for infants, elderly people, and aircrews with high-quality nutritional profiles. Free radicals damage in biological molecules (DNA, membrane lipids, and proteins) can be found in two ways, reversible and irreversible, which can induce coronary diseases, arthritis, cancer, cholesterol, risk of chronic diseases, ulcers, and age-related health problems (Fiori, 2007; Parry et al., 2006; Siger et al., 2008). Oleic acid (omega-9) can support the body to decrease the levels of total and LDL-cholesterol and the glycemic index. It has also been determined that grape seed oil has many pharmaceutical effects, such as regulation of autonomic nerves, dilation of blood vessels, prevention of thrombosis, pancreatic problems, and lung fibrosis, and it shows antimutagenic, antitoxic, antiallergic, antitumor, antiinflammatory, and hepatoprotective effects. Additionally conjugated linoleic acid (CLA) can be produced from grape seed oil. Many studies have shown that CLA could be preferred due to its macrophage activities and modulation of lymphocyte for the treatments of forestomach, colon, skin, and mammary carcinogenesis. CLA also shows anti-hyperinsulinemic and anti-atherogenic effects, which make it an important ingredient for nutritional supplements, along with food antioxidants (Cao & Ito, 2003; Dávila et al., 2017; Domínguez et al., 2017; Lisa & Holcapek, 2011).

Grape seed oil is important for skin applications with its antioxidant, smoothing, softening, moisturizing, and UV protective properties. It is also effective in topical applications for itching and flaking feet (Garavaglia et al., 2016; Kochhar, 2011; Nunes et al., 2017). Linoleic acid is accepted as one of the least comedogenic fatty acids and helps to prevent eczema. Linoleic acid can also corporate with linolenic acid in cell membranes lipid components, regenerate the lipid barrier of epidermis if it is damaged, help limit the loss of water, and decrease inflammation in the skin (Babu, Chen, & Kanikkannan, 2015; Nunes et al., 2017). Additionally, from the perspective of toxicology, polyaromatic hydrocarbon (PAH) levels of some grape seed oil samples are found to be higher than desired. Therefore, the amounts of these components in the last product should be carefully determined (Kochhar, 2011; Lampi & Heinonen, 2009).

8 Authenticity and adulteration of cold pressed grape seed oil

Adulteration is an important issue for the safety, efficacy, quality, and economic sustainability of nutraceuticals and cosmeceuticals, whose usages and the commercial products containing grape seed are increasing. Adulteration can be defined in different ways, such as lack of bioactive components, contamination, and lack of standardization of components, which is vital for consistency of batches (Villani et al., 2015). Therefore, many studies have been conducted to find the most convenient method to prove the authentication of plant seed extracts and oil. Near-infrared (NIR), mid-infrared (MIR), and Raman spectroscopy with chemometrics have been used to determine intrinsic quality parameters authentication, and adulteration of edible fats and oils. The methods developed are preferred due to being rapid, simple, and generally nondestructive. Some methods are focused on the triacylglycerol composition of the edible oils, and to obtain quantitative results for grape seed, HPLC/APCI-MS/UV–vis, HPLC/APCI-MS, and HPLC/APCI-MS–MS analytical techniques are generally used (Indelicato et al., 2017; Nunes, 2014). In recent studies, grape seed oil adulteration with refined soybean and other lower-quality refined edible oils was determined using a combination of multivariate data analysis, ATR-MIR, and synchronous fluorescence data. In principal component analysis (PCA), soft independent modeling of class analogies (SIMCA) results were found successful at determining adulteration below 5%, and with the partial least-squares regression (PLSR) method, oil adulterants that are lower than 0.55% in an unknown mixture can be detected (Akin, Karuk Elmas, Arslan, Yılmaz, & Kenar, 2019; Karuk-Elmas et al., 2019).

9 Conclusion

The potential of grape seed oil, which is a by-product of the wine-making and fruit processing industries, has been recognized for different applications, from food to cosmetics, nutraceuticals, and the pharmaceuticals sector due to its antioxidant and nutritious properties. New developments and the new applications show that industrial interest is focused on preserving the biologically active components of grape seed oil while trying to find cost-effective methods. These studies will lead to the creation of new value-added products based on grape seed oil in different industries and more earnings from a more economical utilization of these by-products.

References

Akin, G., Karuk Elmas, S. N., Arslan, F. N., Yılmaz, İ., & Kenar, A. (2019). Chemometric classification and quantification of cold pressed grape seed oil in blends with refined soybean oils using attenuated total reflectance-mid infrared (ATR-MIR) spectroscopy. *LWT- Food Science and Technology*, *100*, 126–137. https://doi.org/10.1016/j.lwt.2018.10.046.

American Chemical Society News (1919). New food oil from grape seed. *Journal of the Franklin Institute*, *188*(6), 852–853.

Anonymous (2019). *Grape seed oil*. Available at: https://grapeseedoil.com/. Accessed 25 March 2019.

Babu, R. J., Chen, L., & Kanikkannan, N. (2015). Fatty alcohols, fatty acids, and fatty acid esters as penetration enhancers. In N. Dragicevic, & H. I. Maibach (Eds.), *Percutaneous penetration enhancers chemical methods in penetration enhancement: Modification of the stratum corneum* (pp. 133–150). Berlin: Springer.

Bail, S., Stuebiger, G., Krist, S., Unterweger, H., & Buchbauer, G. (2008). Characterisation of various grape seed oils by volatile compounds, triacylglycerol composition, total phenols, and antioxidant capacity. *Food Chemistry*, *108*, 1122–1132. https://doi.org/10.1016/j.foodchem.2007.11.063.

Callaway, J. C., & Pate, D. W. (2009). In R. A. Moreau, & A. Kamal-Eldin (Eds.), *Gourmet and health-promoting specialty oils, gourmet and health-promoting specialty oils*. Urbana, IL: AOCS Press. https://doi.org/10.1016/B978-1-893997-97-4.50011-5.

Cao, X., & Ito, Y. (2003). Supercritical fluid extraction of grape seed oil and subsequent separation of free fatty acids by high-speed counter-current chromatography. *Journal of Chromatography A*, *1021*, 117–124. https://doi.org/10.1016/j.chroma.2003.09.001.

Çelenk, V. U., Gümüş, Z. P., Ustun-Argon, Z., Büyükhelvacıgil, M., & Karasulu, E. (2018). Analysis of chemical compositions of 15 different cold-pressed oils produced in Turkey: A case study of tocopherol and fatty acid analysis. *Journal of the Turkish Chemical Society, Section A: Chemistry*, *5*(1), 1–18.

Choi, Y., & Lee, J. (2009). Antioxidant and antiproliferative properties of a tocotrienol-rich fraction from grape seeds. *Food Chemistry*, *114*(4), 1386–1390. https://doi.org/10.1016/j.foodchem.2008.11.018.

Crews, C., et al. (2006). Quantitation of the main constituents of some authentic grape-seed oils of different origin. *Journal of Agricultural and Food Chemistry*, *54*, 6261–6265.

Dávila, I., Robles, E., Egüés, I., Labidi, J., & Gullón, P. (2017). The biorefinery concept for the industrial valorization of grape processing by-products. In M. Galanakis Charis (Ed.), *Handbook of grape processing by-products* (pp. 29–53). UK: Elsevier.

Domínguez, J., Sanchez-Hernandez, J. C., & Lores, M. (2017). Vermicomposting of winemaking by-products. In M. Galanakis Charis (Ed.), *Handbook of grape processing by-products: Sustainable solutions* (pp. 55–78). UK: Elsevier Inc. https://doi.org/10.1016/B978-0-12-809870-7.00003-X.

Duba, K. S., & Fiori, L. (2015). Supercritical CO2 extraction of grape seed oil: Effect of process parameters on the extraction kinetics. *The Journal of Supercritical Fluids*, *98*, 33–43. https://doi.org/10.1016/j.supflu.2014.12.021.

FAO (1992). *Minor oil crops*. Intermediate Technology Development Group. Available at: http://www.fao.org/3/X5043E/X5043E00.htm#Contents. Accessed 1 December 2018.

FAO and OIV (2016). *Table and dried grapes*. FAO-OIV Focus 2016.

Fiori, L. (2007). Grape seed oil supercritical extraction kinetic and solubility data: Critical approach and modeling. *The Journal of Supercritical Fluids*, *43*, 43–54. https://doi.org/10.1016/j.supflu.2007.04.009.

Freitas, L. D. S., Dariva, C., Jacques, R. A., & Caramão, E. B. (2013). Effect of experimental parameters in the pressurized liquid extraction of Brazilian grape seed oil. *Separation and Purification Technology*, *116*, 313–318. https://doi.org/10.1016/j.seppur.2013.06.002 Elsevier B.V.

Garavaglia, J., Markoski, M. M., Oliveira, A., & Marcadenti, A. (2016). Grape seed oil compounds: Biological and chemical actions for health. *Nutrition and Metabolic Insights*, *9*, 59–64. https://doi.org/10.4137/NMI.S32910.

Göktürk-Baydar, N., & Akkurt, M. (2001). Oil content and oil quality properties of some grape seeds. *Turkish Journal of Agriculture and Forestry*, *25*, 163–168.

Gomez, A. M., Lopez, C. P., & De La Ossa, E. M. (1996). Recovery of grape seed oil by liquid and supercritical carbon dioxide extraction: A comparison with conventional solvent extraction. *Chemical Engineering Journal*, *61*, 227–231.

Hogervorst, J. C., Miljic, U., & Pu, V. (2017). Extraction of bioactive compounds from grape processing by-products. In M. Galanakis Charis (Ed.), *Handbook of grape processing by-products* (pp. 105–135). London, UK: Academic Press, Elsevier.

Indelicato, S., Bongiornob, D., Pitonzoc, R., Di Stefano, V., Calabreseb, V., Indelicatoe, S., et al. (2017). Triacylglycerols in edible oils: Determination, characterization, quantitation, chemometric approach and evaluation of adulterations. *Journal of Chromatography A*, *1515*, 1–16. https://doi.org/10.1016/j.chroma.2017.08.002 Elsevier B.V.

Kamel, B. S., & Dawson, H. (1985). Characteristics and composition of melon and grape seed oils and cakes. *Journal of the American Oil Chemists' Society*, *62*(5), 881–883.

Karuk-Elmas, Ş. N., et al. (2019). Synchronous fluorescence spectroscopy combined with chemometrics for rapid assessment of cold-pressed grape seed oil adulteration: Qualitative and quantitative study. *Talanta*, *196*, 22–31. https://doi.org/10.1016/j.talanta.2018.12.026 Elsevier B.V.

Kochhar, S. P. (2011). Minor and speciality oils. In F. D. Gunstone (Ed.), *Vegetable oils in food technology composition, properties and uses* (pp. 291–341). UK: Wiley-Blackwell Second.

Lampi, A.-M., & Heinonen, M. (2009). Berry seed and grapeseed oils. In R. A. Moreau, & A. Kamal-Eldin (Eds.), *Gourmet and health-promoting specialty oils* (pp. 215–235). Urbana, Illinois: AOCS Press.

Lavelli, V., Kerr, W. L., García-Lomillo, J., & González-SanJosé, M. L. (2017). Applications of recovered bioactive compounds in food products. In M. Galanakis Charis (Ed.), *Handbook of grape processing by-products* (pp. 233–266). UK: Elsevier Inc.

Liang, J., Appukuttan Aachary, A., & Hollader, U. T. (2015). Hemp seed oil: Minor components and oil quality. *Lipid Technology*, *27*(10), 231–233. https://doi.org/10.1002/lite.201500050.

Lisa, M., & Holcapek, M. (2011). Triacylglycerols in nut and seed oils. In V. R. Preedy, R. R. Watson, & V. B. Patel (Eds.), *Nuts & seeds in health and disease prevention* (pp. 43–54). London, UK: Academic Press, Elsevier.

Lutterodt, H., Slavin, M., Whent, M., Turner, E., & Yu, L. (2011). Fatty acid composition, oxidative stability, antioxidant and antiproliferative properties of selected cold-pressed grape seed oils and flours. *Food Chemistry*, *128*(2), 391–399. https://doi.org/10.1016/j.foodchem.2011.03.040 Elsevier.

Maier, T., Schieber, A., Kammerer, D. R., & Carle, R. (2009). Residues of grape (*Vitis vinifera* L.) seed oil production as a valuable source of phenolic antioxidants. *Food Chemistry*, *112*, 551–559. https://doi.org/10.1016/j.foodchem.2008.06.005.

Maroun, R. G., Rajha, H. N., Vorobiev, E., & Louka, N. (2017). Emerging technologies for the recovery of valuable compounds from grape processing by-products. In M. Galanakis Charis (Ed.), *Handbook of grape processing by-products* (pp. 155–181). UK: Elsevier Inc.

Matthäus, B. (2008). Review article virgin grape seed oil: Is it really a nutritional highlight? *European Journal of Lipid Science and Technology*, *110*, 645–650. https://doi.org/10.1002/ejlt.200700276.

Nassiri-Asl, M., & Hosseinzadeh, H. (2016). Review of the pharmacological effects of *Vitis vinifera* (grape) and its bioactive constituents: An update. *Phytotherapy Research*, *1403*(April), 1392–1403. https://doi.org/10.1002/ptr.5644.

Nunes, C. A. (2014). Vibrational spectroscopy and chemometrics to assess authenticity, adulteration and intrinsic quality parameters of edible oils and fats. *Food Research International*, *60*, 255–261. https://doi.org/10.1016/j.foodres.2013.08.041 Elsevier Ltd.

Nunes, M. A., Rodrigues, F., & Oliveira, M. B. P. P. (2017). Grape processing by-products as active ingredients for cosmetic proposes. In M. Galanakis Charis (Ed.), *Handbook of grape processing by-products* (pp. 267–292). UK: Elsevier Inc.

OIV (2017). *OIV statistical report on world vitiviniculture*. OIV, International Organisation of Vine and Wine, Intergovernmental Organisation.

Pardo, J. E., et al. (2009). Characterization of grape seed oil from different grape varieties (*Vitis vinifera*). *European Journal of Lipid Science and Technology*, *111*, 188–193. https://doi.org/10.1002/ejlt.200800052.

Parry, J., et al. (2005). Fatty acid composition and antioxidant properties of cold-pressed marionberry, boysenberry, red raspberry, and blueberry seed oils. *Journal of Agricultural and Food Chemistry*, *53*(3), 566–573. https://doi.org/10.1021/jf048615t.

Parry, J., et al. (2006). Characterization of cold-pressed onion, roasted pumpkin, and Milk thistle seed oils. *Journal of the American Oil Chemists' Society*, *83*(10), 847–854.

Passos, C. P., Silva, R. M., Da Silva, F. A., Coimbra, M. A., & Silva, C. M. (2009). Enhancement of the supercritical fluid extraction of grape seed oil by using enzymatically pre-treated seed. *The Journal of Supercritical Fluids*, *48*, 225–229. https://doi.org/10.1016/j.supflu.2008.11.001.

Rafalowski, R., Zegarska, Z., Kuncewicz, A., & Borejszo, Z. (2008). Fatty acid composition, tocopherols and Beta-carotene content in polish commercial vegetable oils. *Pakistan Journal of Nutrition*, *7*(2), 278–282.

Rekik, D. M., et al. (2016). Evaluation of wound healing properties of grape seed, sesame, and fenugreek oils. *Evidence-based Complementary and Alternative Medicine*, *2016*, 1–12. https://doi.org/10.1155/2016/7965689 Hindawi Publishing Corporation.

Rombaut, N., et al. (2015). Optimization of oil yield and oil total phenolic content during grape seed cold screw pressing. *Industrial Crops and Products*, *63* (July), 26–33. https://doi.org/10.1016/j.indcrop.2014.10.001 Elsevier B.V.

Shinagawa, F. B., de Santana, F. C., Araujo, E. S., Purgatto, E., & Mancini-Filho, J. (2018). Chemical composition of cold pressed Brazilian grape seed oil. *Food Science and Technology*, *38*(1), 164–171.

Siger, A., Nogala-Kalucka, M., & Lampart-Szczapa, E. (2008). The content and antioxidant activity of phenolic compounds in cold-pressed plant oils. *Journal of Food Lipids*, *15*(2), 137–149. https://doi.org/10.1111/j.1745-4522.2007.00107.x.

Simopoulos, A. P. (1999). Essential fatty acids in health and chronic disease. *The American Journal of Clinical Nutrition*, *70*, 560–569.

Slavin, M., Cheng, Z., Luther, M., Kenworthy, W., & Yu, L. (2009). Antioxidant properties and phenolic, isoflavone, tocopherol and carotenoid composition of Maryland-grown soybean lines with altered fatty acid profiles. *Food Chemistry*, *114*(1), 20–27. https://doi.org/10.1016/j.foodchem.2008.09.007.

Soto, C., Chamy, R., & Zúñiga, M. E. (2007). Enzymatic hydrolysis and pressing conditions effect on borage oil extraction by cold pressing. *Food Chemistry*, *102*(3), 834–840. https://doi.org/10.1016/j.foodchem.2006.06.014.

Soto, M. L., Falqué, E., & Domínguez, H. (2015). Relevance of natural phenolics from grape and derivative products in the formulation of cosmetics. *Cosmetics*, *2*(3), 259–276. https://doi.org/10.3390/cosmetics2030259.

Spigno, G., Marinoni, L., & Garrido, G. D. (2017). State of the art in grape processing by-products. In M. Galanakis Charis (Ed.), *Handbook of grape processing by-products* (pp. 1–27). UK: Elsevier Inc.

Stafne, E., Sleezer, S. M., & Clark, J. R. (2015). In A. G. Reynolds (Ed.), *Grapevine breeding programs for the wine industry*. (1st ed.). Cambridge, UK: Woodhead Publishing in Food.

Talbot, G. (2015). In G. Talbot (Ed.), *Specialty oils and fats in food and nutrition: Properties, processing and applications, specialty oils and fats in food and nutrition: Properties, processing, and applications*. Amsterdam: Woodhead Publishing Limitedhttps://doi.org/10.1016/C2014-0-01770-4.

Ustun-Argon, Z., Gökyer, A., Gümüş, Z. P., & Büyükhelvacıgil, M. (2017). Evaluation of some medicinal herbs cold pressed oils according to their physicochemical properties with Chemometry. *International Journal of Secondary Metabolite*, *4*(2), 473–481. https://doi.org/10.21448/ijsm.377319.

Ustun-Argon, Z. (2019a). Phenolic compounds, antioxidant activity and fatty acid compositions of commercial cold-pressed hemp seed (*Cannabis Sativa* L) oils from Turkey. *International Journal of Scientific and Engineering Research*, *10*(4), 166–171.

Ustun-Argon, Z. (2019b). Phenolic compounds, fatty acid compositions and antioxidant activity of commercial cold-pressed grape seed (*Vitis vinifera*) oils from Turkey. *International Journal of Scientific and Engineering Research*, *10*(4), 1211–1218.

Villani, T. S., Reichert, W., Ferruzzi, M. G., Pasinetti, G. M., Simon, J. E., & Wu, Q. (2015). Chemical investigation of commercial grape seed derived products to assess quality and detect adulteration. *Food Chemistry*, *170*, 271–280. https://doi.org/10.1016/j.foodchem.2014.08.084 Elsevier.

Wase, N., Black, P., & Dirusso, C. (2018). Progress in lipid research innovations in improving lipid production: Algal chemical genetics. *Progress in Lipid Research*, *71*, 101–123. https://doi.org/10.1016/j.plipres.2018.07.001 Elsevier.

Yang, R., et al. (2018). A review of chemical composition and nutritional properties of minor vegetable oils in China. *Trends in Food Science & Technology*, *74*, 26–32. https://doi.org/10.1016/j.tifs.2018.01.013.

Yu, L. L., Zhou, K. K., & Parry, J. (2005). Antioxidant properties of cold-pressed black caraway, carrot, cranberry, and hemp seed oils. *Food Chemistry*, *91* (4), 723–729. https://doi.org/10.1016/j.foodchem.2004.06.044.

Chapter 6

Cold pressed black cumin (*Nigella sativa* L.) seed oil

Onur Ketenoglu[a], Sündüz Sezer Kiralan[b], Mustafa Kiralan[b], Gulcan Ozkan[c], and Mohamed Fawzy Ramadan[d,e]
[a]*Department of Food Engineering, Faculty of Engineering, Cankiri Karatekin University, Cankiri, Turkey,* [b]*Department of Food Engineering, Faculty of Engineering, Balıkesir University, Balikesir, Turkey,* [c]*Department of Food Engineering, Faculty of Engineering, Suleyman Demirel University, Isparta, Turkey,* [d]*Agricultural Biochemistry Department, Faculty of Agriculture, Zagazig University, Zagazig, Egypt,* [e]*Deanship of Scientific Research, Umm Al-Qura University, Makkah, Saudi Arabia*

Abbreviations

BCO	black cumin oil
CD	conjugated diene value
CPBCO	cold pressed black cumin oil
CT	conjugated triene value
DPPH	2,2-diphenyl-1-picrylhydrazyl
MAE	microwave-assisted extraction
MFC	minimal fungicidal concentration
MIC	minimal inhibitory concentration
MUFA	monounsaturated fatty acids
PUFA	polyunsaturated fatty acids
PV	peroxide value
SFA	saturated fatty acids
SFE	supercritical fluid extraction
TQ	thymoquinone
UAE	ultrasound-assisted extraction

1 Introduction

Black cumin seed, also known as nigella (*Nigella sativa* L., family Ranunculaceae), is an annual flowering plant native to southwest Asia (Sharma, Ahirwar, Jhade, & Gupta, 2009). The seeds are used in folk medicine for the treatment and prevention of several diseases and ailments all over the world. Usage of the seeds is popular in certain traditional medicinal systems such as Unani and Tibb, Ayurveda, and Siddha (Ahmad et al., 2013). In addition to their medicinal uses, these seeds are used in food applications. The seeds have a characteristic bitter taste and smell, and are therefore especially used in traditional bread and cheeses to enhance the flavor of these foods. The seeds are used in the preparation of a traditional sweet dish composed of *N. sativa* paste (Cheikh-Rouhou et al., 2007; Hassanien, Assiri, Alzohairy, & Oraby, 2015; Ramadan, 2007).

The seeds of black cumin are of particular interest because they contain phytochemicals with significant antioxidant properties and health benefits. The seeds are rich in oil, which contribute to human health and also nutrition due to major (essential fatty acids) and minor compounds (phenolic compounds, tocopherols, and sterols) (Cheikh-Rouhou et al., 2007; Sultan et al., 2009; Takruri & Dameh, 1998).

Black cumin oil (BCO) has been extracted from seeds using various methods. One of these extraction methods is cold pressing. The cold pressing method is simpler, less expensive, and less labor-intensive than the solvent extraction technique. In other respects, the cold pressing method involves neither heat nor chemical treatments, which enables this oil to meet consumers' demand for natural and health food products. According to reports, these healthy oils might enhance

Cold Pressed Oils. https://doi.org/10.1016/B978-0-12-818188-1.00006-2
Copyright © 2020 Elsevier Inc. All rights reserved.

human health and also prevent various diseases (Ananth, Deviram, Mahalakshmi, Sivasudha, & Tietel, 2019; Siger, Nogala-Kalucka, & Lampart-Szczapa, 2008; Thanonkaew, Wongyai, McClements, & Decker, 2012).

BCO contains essential fatty acids in human nutrition. It is also rich in polyunsaturated fatty acids (PUFA). The major PUFA in the BCO is linoleic acid ($C_{18:2}$), at a level of 58.8–61.2 g/100 g total fatty acids. The primary monounsaturated fatty acid (MUFA) in the BCO is oleic acid with the range of 22.6–24.5 g/100 g (Lutterodt et al., 2010).

Tocols (tocopherols and tocotrienols) are one of the minor compounds in BCO and they protect the oil from lipid oxidation and improve the oxidative stability of the oil (Ramadan, 2013). BCO contains high levels of total tocols (1731 μg/g). Tocols in BCO consist of c. 61% of tocopherols and c. 39% of tocotrienols. Among tocopherols, γ-tocopherol is the main isomer found in BCO, at 938 μg/g. As for tocotrienols, the primary isomer present in BCO was γ-tocotrienol (376 μg/g) (Rudzińska, Hassanein, Abdel-Razek, Ratusz, & Siger, 2016). The other minor compounds found in BCO are sterols, which play important roles for human health such as cholesterol-lowering activity (Plat et al., 2019). The total sterol content of BCO was 1746 mg/kg of oil. The major sterols identified in BCO were β-sitosterol, stigmasterol, Δ5-Avenasterol, and campesterol. β-Sitosterol was the dominant sterol in BCO, at 828 mg/kg (Kostadinović Veličkovska, Brühl, Mitrev, Mirhosseini, & Matthäus, 2015). Among minor compounds, the well-known characteristic compound for BCO is thymoquinone (TQ), which is the most active component of the seeds, essential oil, and fixed oil with many healthful properties such as antioxidant and antiinflammatory (Ahmad et al., 2019; Danaei, Memar, Ataee, & Karami, 2019; Majdalawieh, Fayyad, & Nasrallah, 2017). Although thymoquinone is considered as an active component of essential oils, it is also one of the minor components determined in fixed oils (Lutterodt et al., 2010). BCO is reported to contain high levels of thymoquinone (460–873 mg/100 g oil) (Kiralan et al., 2017; Lutterodt et al., 2010).

2 Cold press extraction and processing of oils

Many extraction systems and methods have been used for obtaining oil from such oilseeds as black cumin. Among these techniques, the most common application is the cold pressing method, also known as mechanical pressing. This method is well-known for its capability of producing a high-quality product, in addition to such advantages as lower energy requirements, a more ecological (nontoxic) approach, fewer investment costs, and being simpler than traditional solvent extraction (Soxhlet) or novel extraction methods. In addition, Singh and Bargale (2000) reported that a protein-rich cake could be obtained by cold pressing without any residual toxic solvent.

In a cold press system, a screw press is operated under mechanical pressure, which enables the solid-liquid phase separation. Depending on the nature of the extraction step, the temperature can be applied during the cold pressing to make the oil less viscous. Cold press equipment produces two main streams from the feed: the oil stream and the cake. The major processing parameters such as temperature, screw rotation speed, and initial proximate composition of the feed determine the quality and the yield of the extract. Pretreatment of the oilseed is also effective in the process.

Although cold pressed oils are ready to consume without any refining process and have a characteristic seed flavor and smell along with high-quality composition, the method has some disadvantages. The extraction yield in cold pressing is comparably lower than traditional solvent extraction. In addition, the majority of raw seeds used in the traditional production of cooking oils with the Soxhlet method are not appropriate for cold pressing.

From these aspects, the utilization of novel extraction technologies has gained interest in the extraction of black cumin oil. In the literature, a great number of studies regarding novel extraction techniques deal mostly with microwave-assisted extraction (MAE), ultrasound-assisted extraction (UAE), supercritical fluid extraction (SFE), and some other specialized extraction methods. The applications of these novel techniques in the extraction of black cumin oil have been studied by many researchers.

3 Ultrasound-assisted extraction (UAE) of black cumin oil

In a recent study by Moghimi, Farzaneh, and Bakhshabadi (2018), the ultrasound pretreatment was investigated in terms of different physicochemical properties of black cumin oil. The oil yield, acidity, color index, refractive index, peroxide value, and total phenolic compounds were examined under varying processing conditions such as ultrasonic power and extraction time. Their results revealed that all tested parameters except the refractive index of the extracted black cumin oil were increased by either increasing ultrasonic power or ultrasound time, individually. Their highest oil extraction efficiency was approximately 40%, when the highest levels of the factors were applied.

Abdullah and Koc (2013) studied the oil extraction kinetics of black seed using ultrasound technology. The researchers tested such parameters as temperature, solvent-to-solid ratio, and time on the oil yield using response surface methodology. They concluded that the temperature difference and the gradient of solvent concentration positively affected oil extraction during the washing stage and diffusion stage, respectively. They also stated that the fatty acid composition of the extracted oil did not change during ultrasound processing.

In another study (Kausar, Abidin, & Mujeeb, 2017), different extraction methods with varying solvents were compared for the isolation of thymoquinone from black seed. According to their results, methanol was the best solvent for the isolation of thymoquinone. In addition, researchers stated that ultrasound technology was the most efficient way for their purpose. Their results stated that thymoquinone content was found to be approximately 15% (*w*/w) with the utilization of UAE when methanol was used as the extraction medium.

4 Microwave-assisted extraction (MAE) of black cumin oil

The use of microwave technology is yet another novel technology for the extraction of black cumin oil. Bakhshabadi et al. (2017) have recently studied the microwave pretreatment for the extraction of black cumin seed oil under different process conditions such as time and microwave power. Their findings revealed that most of the properties of the extracted oil were increased with the enhanced microwave parameters. However, the researchers also noted that a decrease in the oxidative stability of the oil was observed. In addition, it was stated that microwave utilization did not significantly affect the refractive index values.

In another study, researchers investigated the effects of the pulsed electric field (PEF) and microwave usage in the extraction of BCO (Bakhshabadi, Mirzaei, Ghodsvali, Jafari, & Ziaiifar, 2018). Their results concluded that both PEF and microwave applications had a positive impact on the extraction yield and oxidative stability. The authors also concluded that the refractive indexes of the extracted oils did not significantly change with the pretreatments.

Kiralan, Özkan, Bayrak, and Ramadan (2014) compared different extraction systems such as cold pressing, Soxhlet, and MAE to determine their effects on the BCO. According to their results, the overall fatty acid composition did not significantly change according to the extraction technique. The authors stated that the black cumin oil extracted with MAE was a good source of thymoquinone. In another study regarding the microwave pretreatment of the black cumin seeds (Kiralan & Kiralan, 2015), the authors stated that remarkable losses occurred when microwave heating was utilized in addition to the formation of oxidized products of volatile compounds.

5 Supercritical fluid extraction (SFE) of black cumin oil

The extraction of black cumin oil with supercritical fluid extraction using carbon dioxide has also been studied as an alternative extraction method. Salea, Widjojokusumo, Hartanti, Veriansyah, and Tjandrawinata (2013) optimized the process conditions for the extraction of black cumin oil and thymoquinone in two design experiments: Taguchi design and full factorial design. The researchers investigated different SFE parameters such as extraction pressure, temperature, CO_2 flow rate, and extraction time. The authors noticed that there was a difference between the calculated optimum conditions of Taguchi and full factorial designs; however, the difference was reported as insignificant.

In another study by Venkatachallam, Pattekhan, Divakar, and Kadimi (2010), the chemical composition of black cumin seed extracts was studied by supercritical CO_2. The authors clearly stated the advantages of supercritical CO_2 technology in terms of obtaining more phenolic compounds from the black cumin seeds.

6 Cold pressed oil recovery, uses of oil cake, and economy

Oilseed cakes occur as by-products at the end of the pressing process and they are rich in some major and minor components such as carbohydrates, proteins, minerals, fiber, and some lipids (Table 1; Acar et al., 2016; Thilakarathna, Madhusankha, & Navaratne, 2018). They are widely used as fodder in animal nutrition as well as biodiesel production (Sielicka & Małecka, 2017). Krimer Malešević, Vaštag, Popović, Popović, and Peričin-Starčevič (2014) determined phenolic acids in oil cake. The predominant phenolic acids in seed cakes are *p*-hydroxybenzoic, syringic, and *p*-coumaric acids. Black cumin seed cakes are rich in proteins, and particularly some essential amino acids. Therefore, they are considered as important sources for animal feeding. Egg production increases in chickens fed by oilseed cakes (10%) (Attia et al., 2008).

Kachel-Jakubowska, Kraszkiewicz, and Krajewska (2016) investigated the usabilities of waste oilseed cakes for obtaining energy. They tested many seedcakes using a calorimeter in order to determine their calorific values. According to their results, black cumin was found to have a calorific value of approximately 20 MJ/kg. Dursun (2018) investigated the production of bio-hydrogen from waste black seeds. The author stated that this residual black seed could be successfully used as a potential source for bio-hydrogen. It was also pointed out that the production capacity of this reaction could be increased by a better decomposition of the seed using some chemical or biological pretreatments. In another study, the

TABLE 1 Some physicochemical properties of oil cake.

Parameter	Concentration
Moisture (%)	4.37–6.97
Crude protein (%)	18.4–23.9
Ash (%)	4.98–9.45
Oil (%)	16.1–20.1
Fiber (%)	6.03–13.8
Total carbohydrate (%)	23.1–28.1
Total phenolics (mg GAE/100g)	1458
Fatty acid	
C16:0	12.2
C18:0	3.3
C18:1	24.7
C18:2	58.6

Adapted from Acar, R., Geçgel, Ü., Dursun, N., Hamurcu, M., Nizamlıoğlu, A., Özcan, M. M., et al. (2016). Fatty acids, minerals contents, total phenol, antioxidant activity and proximate analyses of *Nigella sativa* seed cake and seed cake oil. Journal of Agroalimentary *Processes and Technologies, 22 (1)*, 35–38 and Thilakarathna, R. C. N., Madhusankha, G. D. M. P., & Navaratne, S. B. (2018). Comparison of physico-chemical properties of Indian and Ethiopian origin black cumin (*Nigella sativa*) seed cake. *International Journal of Food Science and Nutrition, 3 (4)*, 30–31.

usability of the seed cake was reported in the production of bio-oil due to its high heating value (38.48), which was nearly equal to diesel fuel (Şen & Kar, 2011).

In addition to the general use of oil cake, the extracts obtained from oilseed cakes had higher values in total phenolic content (39.5 mg GAE/g extract) as well as exhibiting antiradical activity in terms of 2,2-diphenyl-1-picrylhydrazyl (DPPH $^{\bullet}$) antiradical activity and reducing power (FRAP) tests. Furthermore, these extracts from oil cake were used to improve the oxidative stability of cold pressed flaxseed oil, which revealed strong antioxidant properties in model assays (Sielicka & Małecka, 2017). The antioxidant activity of the black cumin seedcakes was also studied by Mariod, Ibrahim, Ismail, and Ismail (2009). The researchers pointed out the possible use of phenolic-rich fractions of black cumin seedcake as an enhancer for the oxidative stability of corn oil. According to the authors, some fractions even showed more antioxidative properties than BHA. The residual black seeds have also been proven to be effective on a kind of nematode in chamomile plants (Ismail, Mohamed, & Mahfouz, 2009). The researchers stated that the use of black cumin seed residue had positive effects on the chamomile oil quality, and the level of this effect depended on the dose of the seed residue used.

7 Fatty acid composition and acyl lipids profile of cold pressed oil

Lipid classes of CPBCO are given in Table 2. Triacylglycerol represented the main lipid fraction in CPBCO contributed to 57.5% of total neutral lipids (NL), while the NL profile was characterized by the high level of free fatty acids (14.2% of total NL) (Atta, 2003). The fatty acid profile of the CPBCO is presented in Table 3 (Gharby et al., 2015). CPBCO contains high levels of PUFA followed by MUFA then SFA. The major fatty acids in CPBCO are linoleic ($C_{18:2}$) and oleic ($C_{18:1}$) for unsaturated fatty acids, while palmitic acid ($C_{16:0}$) is the major SFA. Linoleic acid as PUFA is the most dominant (47.5%–61.2% of total acids) in BCO. Oleic acid is the major fatty acid in MUFA, with the range of 18.9%–24.5% of total fatty acids. The major SFA is palmitic acid, with the range of 12.0%–13.2% of total fatty acids.

8 Minor bioactive lipids in cold pressed oil

Tocopherols and tocotrienols, collectively known as tocols, are important minor active constituents in vegetable oils. These compounds exhibit strong antioxidant activity and also have potential health benefits including preventing various cancer types, heart disease, and other chronic ailments (Ko, Lee, & Kim, 2008; Shahidi & De Camargo, 2016). The composition of tocols in CPBCO from different countries is shown in Table 4. Although the tocopherol composition varies depending on

TABLE 2 Lipid classes of BCO (%).

Polar lipids (PL)	3.7
Monoacylglycerol (MG)	4.8
Diacylglycerol (DG)	5.1
Free sterols (FS)	3.0
Free fatty acids (FFA)	14.2
Triacylglycerol (TG)	57.5
Sterol esters (SE)	2.5

Source: Atta M. B. (2003). Some characteristics of nigella (*Nigella sativa* L.) seed cultivated in Egypt and its lipid profile. *Food Chemistry, 83,* 63–68.

TABLE 3 Fatty acid composition of CPBCO (%).

Fatty acid	USA[a]	Egypt[b]	Turkey[c]	Morocco[d]
$C_{14:0}$	–	11.1	0.13	1.0
$C_{16:0}$	12.9–13.2	12.1	12.01	13.1
$C_{16:1}$	–	0.5	0.25	0.2
$C_{18:0}$	2.56–2.80	3.7	2.77	2.3
$C_{18:1}$	22.6–24.5	18.9	23.9	23.8
$C_{18:2}$	58.8–61.2	47.5	57.4	58.5
$C_{18:3}$	0.21–0.28	2.1	0.25	0.4
$C_{20:0}$	0.13–0.15	–	0.15	0.5
$C_{20:1}$	0.31–0.35			
$C_{20:2}$	–	–	2.33	–
$C_{22:0}$	–	0.9	–	–
$C_{22:1}$	–	0.7	–	–
$C_{24:0}$	–	0.2	0.31	–

[a]*Adapted from Lutterodt, H., Luther, M., Slavin, M., Yin, J.-J., Parry, J., Gao, J.-M., et al. (2010). Fatty acid profile, thymoquinone content, oxidative stability, and antioxidant properties of cold-pressed black cumin seed oils.* LWT-Food Science and Technology, 43 (9), *1409–1413.*
[b]*Adapted from Atta M. B. (2003). Some characteristics of nigella (*Nigella sativa *L.) seed cultivated in Egypt and its lipid profile.* Food Chemistry, 83, *63–68.*
[c]*Adapted from Kiralan, M., Özkan, G., Bayrak, A., & Ramadan, M. F. (2014). Physicochemical properties and stability of black cumin (*Nigella sativa*) seed oil as affected by different extraction methods.* Industrial Crops and Products, 57, *52–58.*
[d]*Adapted from Gharby, S., Harhar, H., Guillaume, D., Roudani, A., Boulbaroud, S., Ibrahimi, M., et al. (2015). Chemical investigation of*Nigella sativa *L. seed oil produced in Morocco.* Journal of the Saudi Society of Agricultural Sciences, 14 (2), *172–177.*

the location, the major tocopherol isomer is γ-tocopherol, with values ranging from 21 to 208 mg/kg oil. Beside to tocopherols, Ramadan (2013) determined that tocotrienols of CPBCO had higher levels of β-tocotrienol (1195 mg/kg) in CPBCO, followed by α-tocotrienol with the amount of 47.6 mg/kg oil.

Phytosterols have some remarkably beneficial effects on human health, such as cholesterol-lowering activity, anticancer, antioxidative, antiinflammatory, and antiatherosclerotic activities (Blanco-Vaca, Cedó, & Julve, 2018; Moreau et al., 2018). The main sterols in CPBCO are β-sitosterol, stigmasterol, △5-avenasterol, and campesterol. β-Sitosterol is predominant; this component contributes approximately 50% of the total phytosterol contents, with stigmasterol, △5-avenasterol, and campesterol together accounting for more than 90% of the total phytosterols. The sterol contents in CPBCO are given in Table 5 (Gharby et al., 2015).

Phenolics are a diverse group of plant secondary metabolites that have many health properties such as anticancer, antiinflammatory, antimicrobial, and antioxidant activities (Mark, Lyu, Lee, Parra-Saldívar, & Chen, 2019). One of the

TABLE 4 The contents of tocopherols and tocotrienols (mg/kg) in CPBCO from different countries.

Tocol	Egypt[a]	Turkey[b]	Macedonia[c]
α-Tocopherol	64.22	7.30	10
β-Tocopherol	53.88	15.47	23
γ-Tocopherol	208.9	34.23	21
δ-Tocopherol	14.48	8.37	–
α-Tocotrienol	47.60	–	–
β-Tocotrienol	1195	–	202
γ-Tocotrienol	–	–	–
δ-Tocotrienol	14.30	–	–

[a]*Adapted from Ramadan, M. F. (2013). Healthy blends of high linoleic sunflower oil with selected cold pressed oils: Functionality, stability and antioxidative characteristics.* Industrial Crops and Products, 43, *65–72.*
[b]*Adapted from Kiralan, M., Özkan, G., Bayrak, A., & Ramadan, M. F. (2014). Physicochemical properties and stability of black cumin (*Nigella sativa*) seed oil as affected by different extraction methods.* Industrial Crops and Products, 57, *52–58.*
[c]*Adapted from Kostadinović Veličkovska, S., Brühl, L., Mitrev, S., Mirhosseini, H., & Matthäus, B. (2015). Quality evaluation of cold-pressed edible oils from Macedonia.* European Journal of Lipid Science and Technology, 117 (12), *2023–2035.*

TABLE 5 Sterol content in CPBCO.

	Source (country)			
Sterol	Turkey[a] (%)[b]	Morocco[c] (%)	Egypt[d] (g/kg)	Macedonia[e] (mg/kg)
Cholesterol	–	0.9	–	12.1
Brassicasterol	–	–	–	3.48
24-Metylencholesterol	–	–	–	20.8
Campesterol	14.88	13.1	0.23	182.8
Campestanol	–	–	–	37.0
Stigmasterol	17.48	17.8	0.32	252
Δ^7-Campesterol	–	–	–	10.45
$\Delta^{5,23}$-Stigmastadienol	–	–	–	1.74
Chlerosterol	–	–	–	15.6
β-Sitosterol	58.05	49.4	1.19	828
Sitostanol	–	–	–	68.4
Δ^5-Avenasterol	7.27	12.4	1.06	202
Δ^7-Stigmasterol	1.24	0.6	–	–
$\Delta^{5,24}$-Stigmastadienol	–	–	–	31.34
Δ^7-Avenasterol	1.62	2.1	0.84	33.60
Δ^7-Stigmastenol	–	–	–	44.74
Lanosterol	–	–	0.12	–

[a]*Adapted from Kiralan, M., Özkan, G., Bayrak, A., & Ramadan, M. F. (2014). Physicochemical properties and stability of black cumin (*Nigella sativa*) seed oil as affected by different extraction methods.* Industrial Crops and Products, 57, *52–58.*
[b]*The units are given in brackets under the countries.*
[c]*Adapted from Gharby, S., Harhar, H., Guillaume, D., Roudani, A., Boulbaroud, S., Ibrahimi, M., et al. (2015). Chemical investigation of* Nigella sativa *L. seed oil produced in Morocco.* Journal of the Saudi Society of Agricultural Sciences, 14 (2), *172–177.*
[d]*Adapted from Ramadan, M. F., Asker, M. M. S., & Tadros, M. (2012). Antiradical and antimicrobial properties of cold-pressed black cumin and cumin oils.* European Food Research and Technology, 234 (5), *833–844.*
[e]*Adapted from Kostadinović Veličkovska, S., Brühl, L., Mitrev, S., Mirhosseini, H., & Matthäus, B. (2015). Quality evaluation of cold-pressed edible oils from Macedonia.* European Journal of Lipid Science and Technology, 117 (12), *2023–2035.*

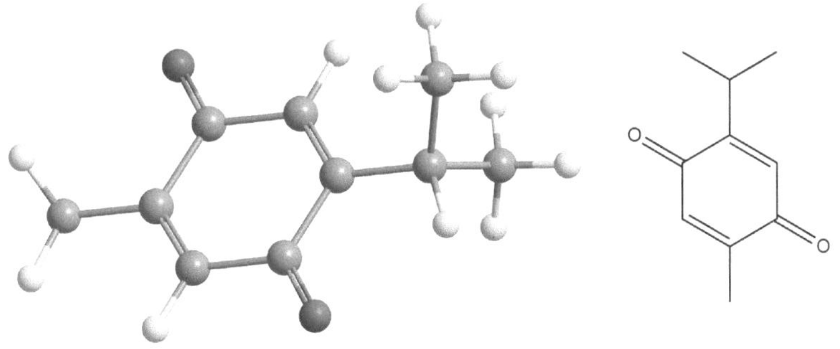

FIG. 1 Chemical structure for thymoquinone.

phenolics, thymoquinone (TQ, Fig. 1), is an active compound of black seed and is responsible for most of its biological activity. This compound is the major component of the essential oil of black cumin and is also present in the fixed oil (Ali & Blunden, 2003). CPBCOs differed in their TQ values, which varied from 3.48 to 8.73 mg per gram of oil (Lutterodt et al., 2010). Depending on the oil extraction techniques, higher levels of TQ in CPBCO were reported (14 μg/g), while Soxhlet-extracted oil (6.2 μg/g) and MAE extracted oil (5.6 μg/g) had lower amounts of TQ (Kiralan et al., 2014).

Chlorophyll and carotenoid pigments present in vegetable oils could negatively influence oxidative stability (Kamal-Eldin, 2006; Yanishlieva, Aitzetmüller, & Raneva, 1998). Small amounts of chlorophyll and carotenoid were reported (0.30 and 0.18 mg/kg, respectively) in CPBCO (Kiralan et al., 2014).

9 Composition of cold pressed oil in comparison to other cold pressed oils

CPBCO contains significant levels of unsaturated fatty acids. Lutterodt et al. (2010) reported that the maim fatty acid in CPBCO was linoleic acid ($C_{18:2}$) accounting for 58.8%–61.2% of the total fatty acids, followed by oleic acid ($C_{18:1}$) at a level of 22.6%–24.5% and palmitic acid ($C_{16:0}$) at 13.0–13.3 g/100 g total fatty acids, respectively. CPBCO is comparable to the cold pressed mullein and milk thistle oils, which have high PUFA contents with linoleic acid ($C_{18:2}$) as the predominant fatty acid (Table 6; Parry et al., 2006). In addition, the amounts of oleic acid in the parsley seed oils were similar to the levels in cold pressed milk thistle oil.

CPBCO is rich in tocols especially tocotrienols. Similarly, cold pressed coriander, cumin, and clove oils are characterized by high amounts of total tocopherols, especially tocotrienols (Table 7; Parry et al., 2006). CPBCO contains a high level of β-tocotrienol, while γ-tocopherol, α-tocotrienol, and δ-tocotrienol were found to be higher in cold pressed coriander, cumin, and clove oils. Among tocopherols, γ-tocopherol is the main tocopherol of CPBCO and its amount is reported as 208.9 mg/kg. Similar to CPBCO, γ-tocopherol was reported as the dominant tocol in cold pressed mullein and pumpkin oils.

β-Sitosterol is the main phytosterol of CPBCO, at 828.7 mg/kg (Table 8). The reported amounts of β-sitosterol in some seed oils are as follows: apricot kernel oil (3394 mg/kg), sunflower oil (1634 mg/kg), flaxseed oil (1451 mg/kg), sesame oil (3376 mg/kg), hemp seed oil (2311 mg/kg), and rapeseed oil (3656 mg/kg).

10 Edible and nonedible applications of cold pressed oil

CPBCO can be used in different food applications. Due to the satisfactory oxidative stability of CPBCO, the oil can be blended with vegetable oils having poor oxidative stabilities in order to enhance their resistance to oxidation. In addition, these vegetable oils are also enriched in minor compounds such as tocols and phenolics, which contribute to human health and organoleptic characteristics. It was reported that blending CPBCO with rapeseed oil (addition of 10% and 20% of BCO) resulted in lower ratios of PUFA/SFA, thus enabling an increase in the oxidative stability of rapeseed oil. The addition of 5%, 10%, and 20% of CPBCO to rapeseed oil also increased the level of α- and γ-tocopherols. On the other side, β-and γ-tocotrienols were found in reasonable concentrations that were not found in rapeseed oil (Rudzińska et al., 2016).

Another study by Kiralan et al. (2017) emphasized that CPBCO improved the oxidative stability of refined sunflower oil (SO) at concentrations of 5%, 10%, and 20% (w/w) when such chemical analyses as conjugated diene (CD) and conjugated triene (CT) values, as well as peroxide value (PV), were evaluated under thermal storage (60°C). In addition, the blends were rich in γ-tocopherol and TQ wherein an SO: BCO (4:1) blend showed the highest oxidative stability within oil blends.

In addition to mixing with other edible oils, CPBCO is also used in some other food applications. It was used in Domiati cheese with probiotic cultures at levels of 0.1% and 0.2% (w/w). The use of CPBCO inhibited the growth of foodborne

TABLE 6 Fatty acid composition of cold pressed oils (%).

Fatty acid	Cardamom[a]	Mullein[a]	Pumpkin[a]	Milk thistle[a]	Black cumin[b]
$C_{14:0}$	1.5	TR[c]	0.1	0.1	–
$C_{16:0}$	26.4	6.0	8.9	8.9	12.90–13.25
$C_{16:1}$	1.6	0.1	0.1	0.1	–
$C_{18:0}$	2.3	2.7	6.4	4.8	2.56–2.80
$C_{18:1}$	49.2	16.1	36.3	23.8	22.6–24.5
$C_{18:2}$	15.2	73.1	47.2	60.8	58.8–61.2
$C_{18:3}$	2.7	1.0	0.2	0.2	0.21–0.28
$C_{20:0}$	0.4	0.7	0.5	–	0.13–0.15
$C_{20:1}$	0.5	0.2	0.4	1.2	0.31–0.35
$C_{20:2}$	–				–
$C_{22:0}$	–				–
$C_{22:1}$	–				–
$C_{24:0}$	–				–
ΣSFA	30.8	9.4	15.9	13.8	15.6–16.2
ΣMUFA	51.3	16.4	36.7	25.2	23.0–24.9
ΣPUFA	17.9	74.2	47.4	61.1	58.9–61.3

[a]*Adapted from Parry, J., Hao, Z., Luther, M., Su, L., Zhou, K., & Yu, L. L. (2006). Characterization of cold-pressed onion, parsley, cardamom, mullein, roasted pumpkin, and milk thistle seed oils.* Journal of the American Oil Chemists' Society, 83 (10), *847–854.*
[b]*Adapted from Lutterodt, H., Luther, M., Slavin, M., Yin, J.-J., Parry, J., Gao, J.-M., et al. (2010). Fatty acid profile, thymoquinone content, oxidative stability, and antioxidant properties of cold-pressed black cumin seed oils.* LWT-Food Science and Technology, 43 (9), *1409–1413.*
[c]*TR, trace.*

TABLE 7 Tocol content in cold pressed oils (mg/kg).

Tocols	Black cumin[a]	Cumin[a]	Coriander[a]	Clove[b]	Mullein[b]	Pumpkin[b]
α-Tocopherol	64.22	831.2	45.95	1489	27.1	26.8
β-Tocopherol	53.88	38.68	38.76	55.96	–	–
γ-Tocopherol	208.9	4012	3944	4184	213.3	216.3
δ-Tocopherol	14.48	82.19	110.2	186.4	76.2	19.2
α-Tocotrienol	47.60	1147	1327	1110	–	–
β-Tocotrienol	1195	42.71	40.10	55.45	–	–
γ-Tocotrienol	–	62.09	507.2	84.96	–	–
δ-Tocotrienol	14.30	9205	10,056	9498	–	–

[a]*Adapted from Ramadan, M. F. (2013). Healthy blends of high linoleic sunflower oil with selected cold pressed oils: Functionality, stability and antioxidative characteristics.* Industrial Crops and Products, 43, *65–72.*
[b]*Adapted from Parry, J., Hao, Z., Luther, M., Su, L., Zhou, K., & Yu, L. L. (2006). Characterization of cold-pressed onion, parsley, cardamom, mullein, roasted pumpkin, and milk thistle seed oils.* Journal of the American Oil Chemists' Society, 83 (10), *847–854.*

TABLE 8 Some sterols in cold pressed oils (mg/kg).

Sterol	Apricot kernel oil	Sunflower oil	Flaxseed oil	Sesame seed oil	Hemp seed oil	Rapeseed oil	Black cumin oil
24-Metylencholesterol	7.98	3.24	72.58	92.24	29.57	148.8	20.89
Campesterol	199.1	227.1	763.9	973.3	504.5	2979	182.8
Campestanol	29.41	48.67	43.29	71.87	52.96	66.08	37.08
Stigmasterol	24.37	273.5	282.2	329.9	90.10	21.49	252.4
Δ^7-Campesterol	4.20	74.63	41.34	29.30	33.36	39.80	10.45
Chlerosterol	28.15	292.0	26.04	52.63	21.32	34.23	15.67
β-Sitosterol	3394	1634	1451	3376	2311	3656	828.7
Sitostanol	123.9	18.45	28.32	76.40	52.96	53.34	68.42
Δ^5-Avenasterol	336.1	69.12	399.3	507.6	219.0	262.2	202.4
$\Delta^{5,24}$-Stigmastadienol	44.12	37.97	58.59	79.23	11.35	39.80	31.34
Δ^7-Avenasterol	25.63	471.4	22.78	35.65	53.30	81.99	33.60
Δ^7-Stigmastenol	17.22	121.0	28.32	50.93	28.89	13.53	44.74

Adapted from Kostadinović Veličkovska, S., Brühl, L., Mitrev, S., Mirhosseini, H., & Matthäus, B. (2015). Quality evaluation of cold-pressed edible oils from Macedonia. *European Journal of Lipid Science and Technology, 117 (12)*, 2023–2035.

pathogens (*Staphylococcus aureus*, *Listeria monocytogenes*, *Escherichia coli*, and *Salmonella enteritidis*) inoculated in cheese during storage at 4°C for 42 days (Mohamed, Elsanhoty, & Hassanien, 2014). BCO was also used in soft cheese with 0.1% or 0.2% (w/w), and presented antibacterial activity against pathogens (*E. coli* ATCC 8739, *S. aureus* ATCC 6538, *L. monocytogenes* Scott A, and *S. enteritidis* PT4) (Hassanien, Mahgoub, & El-Zahar, 2014).

CPBCO was used in marine products for feed for rainbow trout. It was added to the fish feed using a spraying technique and the trout were harvested and stored at 2°C in ice for 23 days. No negative sensory properties were observed in the tested groups according to sensory analysis. In addition, CPBCO improved the oxidative stability of trout depending on the concentration. At the end of storage, thiobarbituric acid values were found to be lower in the sample with 0.7% and 1.00% CPBCO containing feed, while the highest values were observed in the samples with 0.4% CPBCO and the control group (Öz, Dikel, Durmuş, & Özoğul, 2017). Another application of CPBCO in marine products was fish (*Barbus grypus*) fillets. The effects of CPBCO on the microbiological, chemical, and sensory properties of fish (*B. grypus*) fillets during storage at 2°C were examined by Ozpolat and Duman (2017). The results of microbiological analyses showed that CPBCO inhibited the growth of some pathogen microorganisms such as *Enterobacteriaceae*, yeasts, and mold. In addition, sensory analysis exhibited that fish fillets with CPBCO were preferred by panelists.

11 Health-promoting traits of cold pressed oil and oil constituents

In a study about the effect of CPBCO on hypertension, 70 healthy volunteers (aged between 34 and 63 years) with systolic blood pressure (BP) from 110 to 140 (mmHg) and diastolic BP from 60 to 90 (mmHg) were selected, and they received 2.5 mL CPBCO or placebo two times a day for 8 weeks. It was reported that the systolic and diastolic BPs of the BCO-treated group were significantly lower than those of the baseline and placebo groups (Fallah Huseini et al., 2013).

CPBCO could also be used to treat infertility. A study by Kolahdooz, Nasri, Modarres, Kianbakht, and Huseini (2014) demonstrated that usage of 5 mL of BCO for 2 months improved semen quality in infertile men having abnormal semen parameters without any adverse effects. A honey-based formulation from CPBCO has been used in the symptomatic improvement of patients with functional dyspepsia who received the antisecretory therapy. In addition, the rate of *Helicobacter pylori* infection was low in the CPBCO-treated group (Mohtashami et al., 2015).

The bioactive constituent of the volatile oil and fixed oil, TQ, has been reported to have antiinflammatory, antioxidant, and antineoplastic impacts both in vivo and in vitro (Gali-Muhtasib, Roessner, & Schneider-Stock, 2006). Antioxidant enzymes are important in defending the cells against free radical-mediated oxidative injury (Ji, 1993). A recent study demonstrated that an increase in antioxidant enzymes and low molecular weight antioxidants was observed in TQ-treated diabetic rats (Sankaranarayanan & Pari, 2011).

12 Oxidative stability and antiradical and antimicrobial activity

The oxidative stability and antiradical activity of CPBCO have been reported as higher than those of most vegetable oils, mainly because they have high phenolics contents. Kiralan et al. (2014) demonstrated that CPBCO exhibited high resistance to oxidation during thermal conditions (60°C and 100°C) according to oxidation monitoring parameters (PV, CD, and CT). Besides these parameters, volatile compounds were also determined by headspace solid-phase micro-extraction with the GC-MS technique. The major compounds of the oil were *p*-cymene (44.7%), followed by TQ (28.6%). In thermal conditions, some volatiles disappeared suddenly; however, TQ, α-longipinene, 4-terpineol, carvacrol, and isolongifolene reduced slowly and remained stable. Most of these compounds presented strong antioxidant activity, resulting in their contribution to the oxidative stability of BCO (Edris, 2011). Symoniuk, Ratusz, Ostrowska-Ligęza, and Krygier (2018) evaluated the oxidative stability of 10 different cold pressed oils using the Rancimat test. According to the Rancimat results at 110°C, CPBCO had higher stability value (38.2 h) among oil samples.

Ramadan, Asker, and Tadros (2012) evaluated the antiradical activity of CPBCO and compared it with olive oil using stable DPPH$^{\bullet}$ and galvinoxyl radicals. Two radical systems showed that CPBCO had higher radical scavenging activity than olive oil. After 1 h of incubation with DPPH$^{\bullet}$ free radicals, 57% of DPPH$^{\bullet}$ free radicals were deactivated by CPBCO, while the inhibition rate was lower (quench only 45% of DPPH$^{\bullet}$). The other radical system exhibited the same pattern, with CPBCO quenching 45% of galvinoxyl radicals; however, olive oil was able to quench only about 38% after 1 h of reaction. In addition to antiradical activity, CPBCO had strong antimicrobial activity. Ramadan et al. (2012) determined the possible antimicrobial activity of CPBCO against *Pseudomonas aeruginosa*, *S. aureus*, *B. subtilis*, *E. coli*, *Aspergillus flavus*, *Aspergillus niger*, *Saccharomyces cerevisiae*, and *Candida albicans*. CPBCO inhibited the growth of all tested microorganisms except *A. flavus* and *A. niger*. In addition, the results for minimal inhibitory concentration (MIC) revealed that CPBCO was effective against Gram-positive bacteria, Gram-negative bacteria, and yeasts, with MIC ranging between 1 and 2 mL/L. Asdadi et al. (2014) examined the antifungal activity of CPBCO against some Candida strains. CPBCO showed strong antifungal activity when compared to conventional fungicides. The minimal fungicidal concentration (MFC) of CPBCO was higher than conventional fungicides (Fluconazole and Amphotericine B) against *Candida glabrata*, *Candida dubliniensis*, *C. albicans*, and *Candida krusei*.

References

Abdullah, M., & Koc, A. B. (2013). Kinetics of ultrasound-assisted oil extraction from black seed (*Nigella sativa*). *Journal of Food Processing and Preservation*, *37*(5), 814–823.

Acar, R., Geçgel, Ü, Dursun, N., Hamurcu, M., Nizamlıoğlu, A., Özcan, M. M., et al. (2016). Fatty acids, minerals contents, total phenol, antioxidant activity and proximate analyses of *Nigella sativa* seed cake and seed cake oil. *Journal of Agroalimentary Processes and Technologies*, *22*(1), 35–38.

Ahmad, A., Husain, A., Mujeeb, M., Khan, S. A., Najmi, A. K., Siddique, N. A., et al. (2013). A review on therapeutic potential of *Nigella sativa*: A miracle herb. *Asian Pacific Journal of Tropical Biomedicine*, *3*(5), 337–352.

Ahmad, A., Mishra, R. K., Vyawahare, A., Kumar, A., Rehman, M. U., Qamar, W., et al. (2019). Thymoquinone (2-isopropyl-5-methyl-1, 4-benzoquinone) as a chemopreventive/anticancer agent: Chemistry and biological effects. *Saudi Pharmaceutical Journal*, *27*(8), 1113–1126. https://doi.org/10.1016/j.jsps.2019.09.008.

Ali, B. H., & Blunden, G. (2003). Pharmacological and toxicological properties of *Nigella sativa*. *Phytotherapy Research*, *17*(4), 299–305.

Ananth, D. A., Deviram, G., Mahalakshmi, V., Sivasudha, T., & Tietel, Z. (2019). Phytochemical composition and antioxidant characteristics of traditional cold pressed seed oils in South India. *Biocatalysis and Agricultural Biotechnology*, *17*, 416–421.

Asdadi, A., Harhar, H., Gharby, S., Bouzoubaâ, Z., Yadini, A. E., Moutaj, R., et al. (2014). Chemical composition and antifungal activity of *Nigella sativa* L. oil seed cultivated in Morocco. *International Journal of Pharmaceutical Science Invention*, *3*(11), 9–15.

Atta, M. B. (2003). Some characteristics of nigella (*Nigella sativa* L.) seed cultivated in Egypt and its lipid profile. *Food Chemistry*, *83*, 63–68.

Attia, Y. A., El-Din, A. E.-R. E. T., Zeweil, H. S., Hussein, A. S., Qota, E. S. M., & Arafat, M. A. (2008). The effect of supplementation of enzyme on laying and reproductive performance in Japanese quail hens fed nigella seed meal. *Journal of Poultry Science*, *45*(2), 110–115.

Bakhshabadi, H., Mirzaei, H., Ghodsvali, A., Jafari, S. M., & Ziaiifar, A. M. (2018). The influence of pulsed electric fields and microwave pretreatments on some selected physicochemical properties of oil extracted from black cumin seed. *Food Science & Nutrition*, *6*(1), 111–118.

Bakhshabadi, H., Mirzaei, H., Ghodsvali, A., Jafari, S. M., Ziaiifar, A. M., & Farzaneh, V. (2017). The effect of microwave pretreatment on some physico-chemical properties and bioactivity of Black cumin seeds' oil. *Industrial Crops and Products*, *97*, 1–9.

Blanco-Vaca, F., Cedó, L., & Julve, J. (2018). Phytosterols in cancer: From molecular mechanisms to preventive and therapeutic potentials. *Current Medicinal Chemistry*, *25*, 1.

Cheikh-Rouhou, S., Besbes, S., Hentati, B., Blecker, C., Deroanne, C., & Attia, H. (2007). *Nigella sativa* L.: Chemical composition and physicochemical characteristics of lipid fraction. *Food Chemistry*, *101*(2), 673–681.

Danaei, G. H., Memar, B., Ataee, R., & Karami, M. (2019). Protective effect of thymoquinone, the main component of *Nigella sativa*, against diazinon cardio-toxicity in rats. *Drug and Chemical Toxicology*, *42*(6), 585–591.

Dursun, N. (2018). *Research of the efficiency of biohydrogen gain from waste black cumin (*Nigella sativa*) in anaerobic bioreactors*. Ph.D. thesis Şanlıurfa/Turkey: Harran University Graduate School of Natural and Applied Sciences Department of Environmental Engineering.

Edris, A. E. (2011). The chemical composition and the content of volatile oil: Potential factors that can contribute to the oxidative stability of *Nigella sativa* L. crude oil. *Journal of Dietary Supplements*, *8*(1), 34–42.

Fallah Huseini, H., Amini, M., Mohtashami, R., Ghamarchehre, M. E., Sadeqhi, Z., Kianbakht, S., et al. (2013). Blood pressure lowering effect of *Nigella sativa* L. seed oil in healthy volunteers: A randomized, double-blind, placebo-controlled clinical trial. *Phytotherapy Research*, *27*(12), 1849–1853.

Gali-Muhtasib, H., Roessner, A., & Schneider-Stock, R. (2006). Thymoquinone: A promising anti-cancer drug from natural sources. *The International Journal of Biochemistry & Cell Biology*, *38*(8), 1249–1253.

Gharby, S., Harhar, H., Guillaume, D., Roudani, A., Boulbaroud, S., Ibrahimi, M., et al. (2015). Chemical investigation of *Nigella sativa* L. seed oil produced in Morocco. *Journal of the Saudi Society of Agricultural Sciences*, ***14***(2), 172–177.

Hassanien, M. F. R., Assiri, A. M. A., Alzohairy, A. M., & Oraby, H. F. (2015). Health-promoting value and food applications of black cumin essential oil: An overview. *Journal of Food Science and Technology*, *52*(10), 6136–6142.

Hassanien, M. F. R., Mahgoub, S. A., & El-Zahar, K. M. (2014). Soft cheese supplemented with black cumin oil: Impact on food borne pathogens and quality during storage. *Saudi Journal of Biological Sciences*, *21*(3), 280–288.

Ismail, A. E., Mohamed, M. M., & Mahfouz, S. A. (2009). Effect of waste residues from black seed and jojoba seed oil extraction as organic amendments on *Meloidogyne incognita*, growth and oil of chamomile. *Pakistan Journal of Nematology*, *27*(2), 297–307.

Ji, L. L. (1993). Antioxidant enzyme response to exercise and aging. *Medicine and Science in Sports and Exercise*, *25*(2), 225–231.

Kachel-Jakubowska, M., Kraszkiewicz, A., & Krajewska, M. (2016). Possibilities of using waste after pressing oil from oilseeds for energy purposes. *Agricultural Engineering*, *20*(1), 45–54.

Kamal-Eldin, A. (2006). Effect of fatty acids and tocopherols on the oxidative stability of vegetable oils. *European Journal of Lipid Science and Technology*, *108*(12), 1051–1061.

Kausar, H., Abidin, L., & Mujeeb, M. (2017). Comparative assessment of extraction methods and quantitative estimation of thymoquinone in the seeds of *Nigella sativa* L by HPLC. *International Journal of Pharmacognosy and Phytochemical Research*, *9*(12), 1425–1428.

Kiralan, M., & Kiralan, S. S. (2015). Changes in volatile compounds of black cumin oil and hazelnut oil by microwave heating process. *Journal of the American Oil Chemists' Society*, *92*(10), 1445–1450.

Kiralan, M., Özkan, G., Bayrak, A., & Ramadan, M. F. (2014). Physicochemical properties and stability of black cumin (*Nigella sativa*) seed oil as affected by different extraction methods. *Industrial Crops and Products*, *57*, 52–58.

Kiralan, M., Ulaş, M., Özaydin, A., Özdemir, N., Özkan, G., Bayrak, A., et al. (2017). Blends of cold pressed black cumin oil and sunflower oil with improved stability: A study based on changes in the levels of volatiles, tocopherols and thymoquinone during accelerated oxidation conditions. *Journal of Food Biochemistry*, *41*(1)e12272.

Ko, S.-N., Lee, S.-M., & Kim, I.-H. (2008). The concentration of tocols from rice bran oil deodorizer distillate using solvent. *European Journal of Lipid Science and Technology*, *110*(10), 914–919.

Kolahdooz, M., Nasri, S., Modarres, S. Z., Kianbakht, S., & Huseini, H. F. (2014). Effects of *Nigella sativa* L. seed oil on abnormal semen quality in infertile men: A randomized, double-blind, placebo-controlled clinical trial. *Phytomedicine*, *21*(6), 901–905.

Kostadinović Veličkovska, S., Brühl, L., Mitrev, S., Mirhosseini, H., & Matthäus, B. (2015). Quality evaluation of cold-pressed edible oils from Macedonia. *European Journal of Lipid Science and Technology*, *117*(12), 2023–2035.

Krimer Malešević, V., Vaštag, Ž., Popović, L., Popović, S., & Peričin-Starčevič, I. (2014). Characterisation of black cumin, pomegranate and flaxseed meals as sources of phenolic acids. *International Journal of Food Science & Technology*, *49*(1), 210–216.

Lutterodt, H., Luther, M., Slavin, M., Yin, J.-J., Parry, J., Gao, J.-M., et al. (2010). Fatty acid profile, thymoquinone content, oxidative stability, and antioxidant properties of cold-pressed black cumin seed oils. *LWT- Food Science and Technology*, *43*(9), 1409–1413.

Majdalawieh, A. F., Fayyad, M. W., & Nasrallah, G. K. (2017). Anti-cancer properties and mechanisms of action of thymoquinone, the major active ingredient of *Nigella sativa*. *Critical Reviews in Food Science and Nutrition*, *57*(18), 3911–3928.

Mariod, A. A., Ibrahim, R. M., Ismail, M., & Ismail, N. (2009). Antioxidant activity and phenolic content of phenolic rich fractions obtained from black cumin (*Nigella sativa*) seedcake. *Food Chemistry*, *116*(1), 306–312.

Mark, R., Lyu, X., Lee, J. J., Parra-Saldívar, R., & Chen, W. N. (2019). Sustainable production of natural phenolics for functional food applications. *Journal of Functional Foods*, *57*, 233–254.

Moghimi, M., Farzaneh, V., & Bakhshabadi, H. (2018). The effect of ultrasound pretreatment on some selected physicochemical properties of black cumin (*Nigella sativa*). *Nutrire*, *43*(1), 18.

Mohamed, K. M., Elsanhoty, R. M., & Hassanien, M. F. R. (2014). Improving thermal stability of high linoleic corn oil by blending with black cumin and coriander oils. *International Journal of Food Properties*, *17*(3), 500–510.

Mohtashami, R., Huseini, H. F., Heydari, M., Amini, M., Sadeqhi, Z., Ghaznavi, H., et al. (2015). Efficacy and safety of honey based formulation of *Nigella sativa* seed oil in functional dyspepsia: A double blind randomized controlled clinical trial. *Journal of Ethnopharmacology*, *175*, 147–152.

Moreau, R. A., Nyström, L., Whitaker, B. D., Winkler-Moser, J. K., Baer, D. J., Gebauer, S. K., et al. (2018). Phytosterols and their derivatives: Structural diversity, distribution, metabolism, analysis, and health-promoting uses. *Progress in Lipid Research*, *70*, 35–61.

Öz, M., Dikel, S., Durmuş, M., & Özoğul, Y. (2017). Effects of black cumin oil (*Nigella sativa*) on sensory, chemical and microbiological properties of rainbow trout during 23 days of storage at 2 ± 1°C. *Journal of Aquatic Food Product Technology*, *26*(6), 665–674.

Ozpolat, E., & Duman, M. (2017). Effect of black cumin oil (*Nigella sativa* L.) on fresh fish (*Barbus grypus*) fillets during storage at 2 ± 1°C. *Food Science and Technology*, *37*(1), 148–152.

Parry, J., Hao, Z., Luther, M., Su, L., Zhou, K., & Yu, L. L. (2006). Characterization of cold-pressed onion, parsley, cardamom, mullein, roasted pumpkin, and milk thistle seed oils. *Journal of the American Oil Chemists' Society*, *83*(10), 847–854.

Plat, J., Baumgartner, S., Vanmierlo, T., Lütjohann, D., Calkins, K. L., Burrin, D. G., et al. (2019). Plant-based sterols and stanols in health & disease: "Consequences of human development in a plant-based environment?" *Progress in Lipid Research*, *74*, 87–102.

Ramadan, M. F. (2007). Nutritional value, functional properties and nutraceutical applications of black cumin (*Nigella sativa* L.): An overview. *International Journal of Food Science & Technology*, *42*(10), 1208–1218.

Ramadan, M. F. (2013). Healthy blends of high linoleic sunflower oil with selected cold pressed oils: Functionality, stability and antioxidative characteristics. *Industrial Crops and Products*, *43*, 65–72.

Ramadan, M. F., Asker, M. M. S., & Tadros, M. (2012). Antiradical and antimicrobial properties of cold-pressed black cumin and cumin oils. *European Food Research and Technology*, *234*(5), 833–844.

Rudzińska, M., Hassanein, M. M. M., Abdel-Razek, A. G., Ratusz, K., & Siger, A. (2016). Blends of rapeseed oil with black cumin and rice bran oils for increasing the oxidative stability. *Journal of Food Science and Technology*, *53*(2), 1055–1062.

Salea, R., Widjojokusumo, E., Hartanti, A. W., Veriansyah, B., & Tjandrawinata, R. R. (2013). Supercritical fluid carbon dioxide extraction of *Nigella sativa* (black cumin) seeds using taguchi method and full factorial design. *Biochemical Compounds*, *1*, 1. https://doi.org/10.7243/2052-9341-1-1.

Sankaranarayanan, C., & Pari, L. (2011). Thymoquinone ameliorates chemical induced oxidative stress and β-cell damage in experimental hyperglycemic rats. *Chemico-Biological Interactions*, *190*(2–3), 148–154.

Şen, N., & Kar, Y. (2011). Pyrolysis of black cumin seed cake in a fixed-bed reactor. *Biomass and Bioenergy*, *35*(10), 4297–4304.

Shahidi, F., & De Camargo, A. C. (2016). Tocopherols and tocotrienols in common and emerging dietary sources: Occurrence, applications, and health benefits. *International Journal of Molecular Sciences*, *17*(10), 1745.

Sharma, N. K., Ahirwar, D., Jhade, D., & Gupta, S. (2009). Medicinal and phamacological potential of *Nigella sativa*: A review. *Ethnobotanical Review*, *13*, 946–955.

Sielicka, M., & Małecka, M. (2017). Enhancement of oxidative stability of flaxseed oil through flaxseed, evening primrose and black cumin cake extracts. *Journal of Food Processing and Preservation*, *41*(4)e13070.

Siger, A., Nogala-Kalucka, M., & Lampart-Szczapa, E. (2008). The content and antioxidant activity of phenolic compounds in cold-pressed plant oils. *Journal of Food Lipids*, *15*(2), 137–149.

Singh, J., & Bargale, P. C. (2000). Development of a small capacity double stage compression screw press for oil expression. *Journal of Food Engineering*, *43*(2), 75–82.

Sultan, M. T., Butt, M. S., Anjum, F. M., Jamil, A., Akhtar, S., & Nasir, M. (2009). Nutritional profile of indigenous cultivar of black cumin seeds and antioxidant potential of its fixed and essential oil. *Pakistan Journal of Botany*, *41*(3), 1321–1330.

Symoniuk, E., Ratusz, K., Ostrowska-Ligęza, E., & Krygier, K. (2018). Impact of selected chemical characteristics of cold-pressed oils on their oxidative stability determined using the Rancimat and pressure differential scanning calorimetry method. *Food Analytical Methods*, *11*(4), 1095–1104.

Takruri, H. R., & Dameh, M. A. (1998). Study of the nutritional value of black cumin seeds (*Nigella sativa* L.). *Journal of the Science of Food and Agriculture*, *76*(3), 404–410.

Thanonkaew, A., Wongyai, S., McClements, D. J., & Decker, E. A. (2012). Effect of stabilization of rice bran by domestic heating on mechanical extraction yield, quality, and antioxidant properties of cold-pressed rice bran oil (*Oryza sativa* L.). *LWT- Food Science and Technology*, *48*(2), 231–236.

Thilakarathna, R. C. N., Madhusankha, G. D. M. P., & Navaratne, S. B. (2018). Comparison of physico-chemical properties of Indian and Ethiopian origin black cumin (*Nigella sativa*) seed cake. *International Journal of Food Science and Nutrition*, *3*(4), 30–31.

Venkatachallam, S. K. T., Pattekhan, H., Divakar, S., & Kadimi, U. S. (2010). Chemical composition of *Nigella sativa* L. seed extracts obtained by supercritical carbon dioxide. *Journal of Food Science and Technology*, *47*(6), 598–605.

Yanishlieva, N. V., Aitzetmüller, K., & Raneva, V. (1998). β-Carotene and lipid oxidation. *Lipid/Fett*, *100*(10), 444–462.

Further reading

Kıralan, M. (2014). Changes in volatile compounds of black cumin (*Nigella sativa* L.) seed oil during thermal oxidation. *International Journal of Food Properties*, *17*(7), 1482–1489.

Uddin, M. S., Ferdosh, S., Haque Akanda, M. J., Ghafoor, K., Rukshana, A. H., Ali, M. E., et al. (2018). Techniques for the extraction of phytosterols and their benefits in human health: A review. *Separation Science and Technology*, *53*(14), 2206–2223.

Chapter 7

Cold pressed rapeseed (*Brassica napus*) oil

Sook Chin Chew
School of Foundation Studies, Xiamen University Malaysia Campus, Sepang, Selangor, Malaysia

Abbreviations

$^1O_2$	singlet oxygen
ABTS$^{•+}$	2,2′-azino-bis (3-ethylbenzothiazoline-6-sulphonic acid)
DPPH	2,2-diphenyl-1-picrylhydrazyl
LDL	low-density lipoprotein
MUFA	monounsaturated fatty acid
PC-8	plastochromanol-8
PET	polyethylene terephthalate
PUFA	polyunsaturated fatty acid
PV	peroxide value
SFA	saturated fatty acid
TAG	triacylglycerol

1 Introduction

Rapeseeds (Brassica napus, family Brassicaceae) are broadleaf plants with four petals arrayed in a cross, and are cultivated mainly for their oil-rich seeds. Rapeseed is the second major oilseed crop globally with the increasing world demand and production, followed by soybean oil. The production area of rapeseed increased to 1 million hectares, and rapeseed production reached 2.8 million tons in Poland, while worldwide production of rapeseed oil reached 26.98 million tons in 2013/2014 (FAO, 2019; Wroniak, Rękas, Siger, & Janowicz, 2016). The cultivation and production of rapeseed are 31 million hectares and 60 million tons, respectively per year. There are about 6.7 million hectares of the rapeseed planting area and the total yield accounts for 20% of the world's supply, which is about 12 million tons. Rapeseed acreage has reached 7.3 million hectares in China, which accounts for one-third of the world's total. The quality of rapeseed varieties has been improving continuously in China during the past 20 years. Rapeseed with low erucic acid and low glucosinolates accounts for about 70% of all rapeseed plantation area (Yang et al., 2013). The oil extracted from the rapeseed varieties with low erucic acid (<2%) and glucosinolates (<30 μmol/g) contents is also known as canola oil. In Europe, rapeseed is mainly planted for the production of edible oil and secondarily for animal feed, bio-oil, and biodiesel.

Most of the seeds reserve (oil and protein) is situated inside the cells. The cells have a double cell wall constituted of cellulose, hemicellulose, lignin, and pectin. There is a high oil content in the seeds, which is approximately 38%–42% of the mass of the whole seed and 30% of the protein fraction of a nondecorticated whole seed. Protein accounts for 16%–19% of the total mass of the whole seeds including the film surrounding the seed. Polysaccharides and fibers are the major components in the film surrounding the seeds, which makes up 20%–28% of the total mass of the seed (Perrier et al., 2017). The properties of rapeseed stalk fibers have been identified as similar to those of hardwood and nonwood that are used in pulp and paper manufacturing, by analyzing holocellulose, cellulose, and lignin ratios (Tofanica, Cappelletto, Gavrilescu, & Mueller, 2011).

Rapeseeds (barrel variety) with an insignificant amount of erucic acid and glucosinolates contain 8.1% moisture, 18.7% protein, 45.2% oil, 20.3% crude fiber, and 3.7% ash (Laoretani, Fernández, Crapiste, & Nolasco, 2014). According to the Codex Alimentarius Standard for Named Vegetable Oils, virgin oil is the oil obtained by mechanical processes like expelling or pressing, as well as heat processing without changing the nature of the oil. However, any heat application does not allow for cold pressed oils (Xie et al., 2019). Rapeseed oil is recognized as a healthy oil due to its attractive fatty

Cold Pressed Oils. https://doi.org/10.1016/B978-0-12-818188-1.00007-4
Copyright © 2020 Elsevier Inc. All rights reserved.

acid profile, which is high in monounsaturated fatty acids (MUFAs) and polyunsaturated fatty acids (PUFAs), as well as low in saturated fatty acids (SFAs). Cold pressed rapeseed oils have a distinct aroma and taste with a hint of nuts.

2 Extraction and processing

2.1 Seeds processing

Rapeseeds are required to reduce the moisture content for storage purposes. The rapeseeds are normally stored at a moisture content of 5.5%–8.5% wet basis once the rapeseed is harvested, which also depends on the storage temperature and the oil content of the seeds (Laoretani et al., 2014). Low-quality rapeseeds that are too moist, unripe, contaminated, damaged, and undergoing hydrolytic and oxidative changes may result in producing low-quality oil with low oxidative stability.

Rapeseed is still processed by the traditional method, which involves prepressing and extraction, which requires a long time and high temperature. Double-low rapeseed improves its nutritional value of oil and protein by decreasing the glucosinolates and erucic acid contents. The obtained rapeseed meal had a lower protein content, deeper color, and unpleasant taste using the traditional method. Thus, the method requires improvement to utilize the source of double-low rapeseed. As a result, three key types of equipment have been developed: dehulling, cold pressing, and expansion machines.

Dehulling is a process to remove the fibrous envelopes (hulls) of the seeds to acquire kernels, which is the fraction high in oil and protein contents. Dehulling could hence obtain the meal of high protein content that is highly sought after for animal feed. Hulls can be used as energy fuel for the supply of oil mills. The increasing global demand for proteins and the recent success of dehulled sunflower meals have led to the potential applications of dehulled rapeseeds, which contain 10%–18% hulls, being reconsidered. Rapeseeds being dehulled before pressing has the advantage of maintaining a screw press temperature lower than 40°C, thus restricting the seed enzyme activities, and decreasing the transfer of undesired compounds from seeds to the oil. Moreover, oil pressing from dehulled seeds improves the sensory characteristics and results in a lower amount of waxes compared to whole seed pressed oil. The seeds were ground to a particle size <630 mm, followed by cold pressing, which could help to improve the expression yield of oil. This can increase the oil yield by about 15%–20% as the reduced particle size leads to cell damage and thus facilitates the oil expression (Koubaa, Mhemdi, & Vorobiev, 2016).

Dehulling rapeseed can bring advantages like increased protein content, increased effective value of feed, improved meal taste for feed, improved oil quality, and decreased refining cost. Cold pressing can bring advantages like obtaining of green natural quality rapeseed oil, avoiding loss of available amino acids, increased nutritional effective value of the meal, and decreased energy expenditure on rapeseed processing. The expansion helps to improve the material's extraction property, speed up extraction of oil, increase the yield of oil, and decrease production cost. In 2003, the process of dehulling-expansion for double-low rapeseed had been practiced. It enabled high protein rapeseed meal to be produced, but the high oil content of the expansion meal increased the burden of the extraction system and high residual oil in the meal. Therefore, it is still difficult to conduct this practice.

Subsequently, a new idea was developed to improve the dehulling, cold pressing, and expansion. Dehulling is conducted by separating the hull and kernel by adjusting the winding quantity of the fan. Cold pressing is a process to obtain oil and cake. Cold pressed oil was purified through the plate filter and then canned to get the oil. Next, the cold pressed cake is sent to the expansion process. The expansion is conducted by heating the cold pressed cake to 90°C in a cooker and then delivering it to an expander. After expansion, the crude oil could be extracted and sent to refining plant for routine refining process. This new process increased the oil contained in the kernel to about 45% when the rapeseeds were dehulled, and crude fiber content decreased to 3%–5%. This is because the dehulled rapeseeds contain high oil and low crude fiber. The twin-screw press gives a high compression ratio compared to the traditional single-screw press (Huang, Li, Huang, Niu, & Wan, 2005).

2.2 Cold pressing

Solvent extraction is usually used for seeds with low oil content (<20%), like soybean. A pressing method can be used for seeds with high oil content, like rapeseeds. However, the pressing method is relatively inefficient and leads to residual oil in the meal. This residual oil can be extracted subsequently by solvent from the meal. Solvent extraction will leave a very low amount of oil (about 1%) in the defatted meal. Therefore, solvent extraction gives an accurate oil content of the samples, by comparing the oil extraction yield from treated and untreated pressed rapeseed. Solvent extraction is the most efficient method with high extraction yield, but this method may cause plant security problems, emissions of volatile organic compounds into the atmosphere, high operation costs, poor-quality products caused by high processing temperature, and a relatively high number of processing steps. Mechanical pressing is simpler, safer, and involves fewer steps than solvent extraction (Azadmard-Damirchi, Habibi-Nodeh, Hesari, Nemati, & Achachlouei, 2010).

Seed variety, growing conditions, harvesting time, postharvest storage, seed drying temperature, dehulling, extraction techniques (solvent extraction, cold or hot pressing), and processing methods are factors that might affect the oil quality and minor components of rapeseed oil. Impurities present in the rapeseeds may negatively affect the oil extraction process in terms of decreased oil yield, efficiency, and oil quality. This will result in a decrease in seed quality and incur the higher costs during the industrial refining process. Adverse weather conditions and improper cultivation techniques can contribute to the contamination level caused by mold grain, weeds, and pests. High-quality oilseeds, optimized parameters in the oil pressing process, immediate cleaning of the crude oil, storage conditions, and the type of packaging material are factors that affect the production of high-quality cold pressed rapeseed oil.

There are two types of oil pressing methods: cold press and hot press. No heat treatment is applied in the cold press method, whereas seeds are pretreated by heat in the hot press method. Cold press oil has generally better preserved native properties compared to other types of oil that involve heat treatment. Thus, there is an increasing demand in the market for cold pressed oil (CPO) (Azadmard-Damirchi et al., 2010).

Cold pressing is normally performed using hydraulic or expeller (screw-type) press equipped with a cooling system. There are three primary stages of cold pressing: seed defatting, seed fragmentation, and seed conditioning. Seed defatting improves the oil quality with brighter color and high resistance to oxidation by reducing the chlorophylls content and other substances like metals and pesticides permeating through the hulls or shells into the oil. However, this process is not normally used in industry due to some technical difficulties. Seed fragmentation (crushing) is carried out to break the seed structure and hull partially to open some cells and enlarge the oil spillage area, as well as to lower the cell resistance to allow the oil extraction. Crushed seeds are needed to process directly as the quality and stability of CPO are time-dependent. Lastly, the seeds are conditioned with heat pretreatment in an oven by exposing the pulp to temperatures of up to 100°C to the optimal humidity. This pretreatment positively affects the oil extraction yield (Makata, 2015).

There is an increasing trend in the establishment of small- and medium-sized extraction plants to focus on the production of CPO with a capacity between 0.5 and 25 tons per day. These plants normally produce cold press rapeseed oil through the use of only the screw press. The CPO is then purified with physical processes such as filtration, sedimentation, or centrifugation (Ghazani, García-Llatas, & Marangoni, 2014).

The temperature of the press head, speed of the screw press, diameter of the nozzle and type of the press are factors in cold pressing that can affect the quality of oil to different degrees. The cold pressing parameters can greatly affect the chlorophyll, phosphorus, and free fatty acids contents in the cold pressed rapeseed oil. An increase in the press head temperature will increase the chlorophyll content in the oil. However, the nozzle diameter and screw press speed have no influence on the free fatty acids value. The use of a small nozzle (6mm) results in increasing chlorophyll content if the speed of the screw press decreases from 95 to 20rpm. This is because the seed remains for a longer time in the screw press due to it being used at a lower speed, and therefore more components can be extracted from the seeds with the oil. The intensity of the "seedlike" sensory attribute will be increased by increasing the speed of the screw press using a small nozzle of 6mm, as the temperature stress of the seed material increases. Greater diameter of the nozzle and increase of the press head temperature from 60°C to 90°C would decrease the development of the seedlike taste of the CPO (Matthäus & Brühl, 2004). The screw and hydraulic presses are widely used for cold pressing extrusion with maximum outlet temperatures of 50°C and 20–25°C (ambient temperature), respectively (Makata, 2015).

Cold pressing has a lower oil extraction yield than hot pressing. However, low oil extraction yield prevents the development of cold pressing to become commercially viable regardless of the benefits of cold pressing. It is suggested to find a new pretreatment method for cold pressing to overcome the low extraction yield problem. The new pretreatment is required to preserve the desirable bioactive compounds, such as phytosterols, tocopherols, and phenolics in the extracted oil (Azadmard-Damirchi et al., 2010). This can be improved by heat pretreatment on the oilseeds via roasting or microwave irradiation.

Supercritical carbon dioxide extraction (SCE) can be applied in the extraction of rapeseed oil. The main advantages of this method are the absence of toxicity and explosion risks, low processing temperature, low cost, and the renewable property of carbon dioxide. SCE can be applied directly after rapeseed crushing or flacking. The energy consumption of SCE is about 10.2kWh/kg oil. It is recommended to extract the significant amount of rapeseed oil mechanically with low energy consumption and extract the residual oil using SCE.

The benefits of CPO are free from chemical from the refining process and free from microbiological contamination including mycotoxin and metals (Fe and Cu) that may act as pro-oxidants. Chlorophylls are usually removed during the refining process. The nutritive bioactive compounds like tocopherols, sterols, and carotenoids would be removed from the refining process.

2.3 Microwave pretreatment

Microwave pretreatment of seeds is gaining attention from researchers. Microwave radiation can reduce the processing time and save energy as the energy is delivered directly throughout the volume of the material. Microwave radiation ruptures the cell membrane of the seeds to generate permanent pores. Thus, the oil moves through the permeable cell walls to give a high extraction oil yield using the microwave radiation method. Microwave treatment will vaporize the water from the vegetable substrate microstructure, thus rupturing the cell membrane. The low moisture content that resulted in the microwave-treated seeds makes them more brittle and easier to rupture their tissues, which facilitates the oil extractability during mechanical pressing. However, the low moisture content resulted in poor elasticity in the seed. The seeds will be pressed into powder, which disallow to maintain the operating pressure in the pressing chamber and enhance the pressing of the oil (Wroniak et al., 2016). Besides that, microwave radiation can give a rapid and uniform treatment to the seeds. Microwave pretreatment can be conducted at a frequency of 2450 MHz for two periods of radiation (120 s and 240 s) for the seeds.

Rapeseed is a dicotyledonous plant in which the embryo composes 80% of the mature seed. The embryo is formed by the cotyledon storage parenchyma cells, containing protein bodies and lipid bodies embedded in the cytoplasmic protein network. Wroniak et al. (2016) showed that the microscopic studies of microwave-treated rapeseed demonstrated a structural change in protein bodies because of thermally induced denaturation. The lipoprotein membrane surrounding the individual lipid bodies also undergo damage. Thsee kernel microstructural changes might facilitate the passage of oil from lipid bodies and release from the cell membrane. Thus, the extraction oil yield would be increased by the microwave pretreatment on the rapeseed. Microwave pretreatment enhances the recovery of oil yield with an improved amount of nutraceuticals, which helps to produce oil with a longer shelf life and enhanced nutritional value.

Seed roasting may cause the formation of volatile compounds such as heterocyclic compounds like pyrazines in the Maillard reaction, which contributed to a pleasant roasting flavor. Moreover, seed roasting can help to reduce the undesirable pungent aroma in the seed oil, which is caused by the presence of glucosinolates like 4-isothiocyanato-1-butene (Gracka, Jeleń, Majcher, Siger, & Kaczmarek, 2012). Previous research showed high retention of bioactive compounds like phytosterols, tocopherols, and phenolics, and improvement of oxidative stability of oil after the microwave pretreatment (Xie et al., 2019). Wroniak et al. (2016) reported that microwave pretreatment did not alter the fatty acids composition of cold pressed rapeseed oil.

Microwave pretreatment decreased the moisture content of rapeseeds, which affected the oil extraction yield from the seeds. The microwave pretreatment of rapeseeds for 3 min to a moisture content of 9% could help to increase the oil extraction yield by up to 64.5% (Wroniak et al., 2016). The moisture content of seeds and microwave exposure time play significant roles in the oil extraction yield. Azadmard-Damirchi et al. (2010) reported that the cold pressed rapeseed oil increased the oil yield by 10% by microwave pretreatment on the seeds for 2 and 4 min.

Kraljić et al. (2013) found that conditioning rapeseeds at 80°C for 30 min increased the amount of tocopherols, phytosterols, and phenolics in the oil. Similar results were reported by Rękas, Wroniak, and Rusinek (2015), who identified a significant increase in the total tocopherols content and a slight increase in the total sterol content in CPO from seeds that had undergone roasting at 80–140°C for 60 min. Siger, Kaczmarek, and Rudzińska (2015) found higher content of tocopherols, plastochromanol-8, phenolic compounds, canolol, and phytosterols in CPO from seeds that had undergone roasting at 140°C, 160°C, and 180°C.

3 Chemical composition

3.1 Chemical properties

The peroxide values (PV) of all the cold pressed rapeseed oils tested from different varieties ranged from 0.38 to 0.84 meq O_2/kg oil. Generally, the PV of fresh vegetable oils is less than 10 meq O_2/kg oil. High temperature, visible light, and oxygen will promote primary oxidation, which increases the PV. Oils with higher PV may bring health adverse effects like stimulating cardiovascular and inflammatory diseases by increasing reactive oxygen species and secondary oxidation products (Lobo, Patil, Phatak, & Chandra, 2010). In addition, oils with higher PV will have a shorter shelf life and be unsuitable for consumption. According to Esuoso and Odetokun (1995), the free fatty acid value of edible oils should not exceed 5%. Thus, the low free fatty acid value in cold pressed rapeseed oil (0.65% oleic) shows their high quality and suitability for consumption (Konuskan, Arslan, & Oksuz, 2019). Unsaponifiable matter represents the substance dissolved in oils. They cannot be saponified by the caustic alkali but are soluble in nonpolar solvents. Konuskan et al. (2019) reported 0.97% of unsaponifiable matter presented in the cold pressed rapeseed oil.

The *p*-anisidine values of all cold pressed rapeseed oils (0.14–0.90) tested from different varieties did not exceed 1.0, which indicates the insignificant effect of the cold pressing process on the secondary oxidation status of the oil. In addition, the cold pressed rapeseed oil presented low conjugated diene (1.32%–1.75%) and conjugated triene (0.07%–0.20%) values, which showed the negligible impact of cold pressing on the formation of oxidation by-products such as unsaturated α- and β-diketones and β-ketones (Rękas, Wroniak, & Szterk, 2016). The acid values of the cold pressed rapeseed oils ranged from 0.1 to 3.9 mg KOH/g oil (McDowell, Elliott, & Koidis, 2017) and 0.42 to 1.47 mg KOH/g oil (Rękas et al., 2016). The acid values reported are followed to the recommended threshold value that set at 4 mg KOH/g oil for cold pressed oils (McDowell et al., 2017).

Rapeseeds are required to dry until they reach 7% moisture content after harvesting under European weather conditions. Incorrect drying temperatures may influence the quality of the seeds. Too high a temperature or too long a drying time may result in overdrying of the seeds, which may damage the seeds and reduce their mechanical resistance. An increase in the drying temperature to more than 93°C results in an elevation of free fatty acid value in the oil. On the other hand, the oil obtained from seeds with 7% moisture content is suitable for consumption for 9 months, while the oil from seeds with 9% moisture content is suitable for consumption for 6 months only. Thus, the moisture content of seeds may affect the shelf life and sensory properties of oils (Siger, Józefiak, & Górnaś, 2017).

3.2 Fatty acid composition

Triacylglycerols (TAG) are the major constituents in cold pressed rapeseed oils (97%–99%), while accompanying components like phenolics, phytosterols, tocopherols, carotenoids, chlorophylls, as well as phospholipids, mono- and diglycerides, and free fatty acids, are present in oil in minor quantities (1%–3%) (Wroniak & Rękas, 2016). Xie et al. (2019) identified 67 individual TAG species in 15 types of cold pressed rapeseed oil samples using LC-MS. OOO, OOL, and OOLn (>50 mg/g) represent the most abundant TAG species in all samples. Microwave pretreatment had an insignificant effect on the TAG species in cold pressed rapeseed oils. The fatty acid profile of cold pressed rapeseed oil has been reported in several previous studies in the literature, as shown in Table 1. Rapeseed oil is recognized as one of the healthiest vegetable oils due to its unique fatty acid profile, which is rich in MUFA (50%–70%) and PUFA, mainly α-linolenic acid (7%–10%) and linolenic acid (17%–21%), as well as low in saturated fatty acids (5%–7%).

Oleic acid, linoleic acid, γ-linolenic acid, and palmitic acid represent the predominant fatty acids in cold pressed rapeseed oil. However, stearic acid, eicosenoic acid, erucic acid, arachidic acid, behenic acid, and palmitoleic acid were exhibited in minor composition in cold pressed rapeseed oil. Oleic acid is associated with low incidence of coronary heart diseases and colon, breast, and skin cancer. Oleic acid is more stable than linoleic acid to withstand high heat processing.

TABLE 1 Fatty acids profile of cold pressed rapeseed oil.

Fatty acid	Rękas et al. (2016)	McDowell et al. (2017)	Konuskan et al. (2019)
Palmitic acid (C16:0)	3.53–4.62	5.3	3.97
Palmitoleic acid (C16:1)	–	0.1	0.17
Stearic acid (C18:0)	1.52–2.02	1.9	2.12
Oleic acid (C18:1)	61.02–76.64	56.2	63.68
Linoleic acid (C18:2)	7.74–21.04	21.0	17.43
γ-linolenic acid (C18:3n6)	7.42–11.45	10.0	6.75
Arachidic acid (C20:0)	0.48–0.64	0.7	0.70
Eicosenoic acid (C20:1)	1.17–1.32	1.5	2.82
Behenic acid (C22:0)	–	0.8	0.37
Erucic acid (C22:1)	ND		1.93
SFA	5.54–7.03	8.3	7.16
MUFA	62.21–77.95	60.1	68.60
PUFA	15.12–30.91	31.0	24.18

Thus, a high level of oleic acid is desirable in the aspects of nutrition and stability. Linoleic acid is an essential fatty acid that plays a role in the maintenance of physiological functions of the human body. Linoleic acid helps to maintain the skin's integrity, cell membranes, the immune system, and eicosanoids production. The omega-6 fatty acid family consists of linoleic acid and its longer chain derivatives like γ-linolenic acid and arachidonic acid.

Stearic acid is a neutral saturated fatty acid that does not increase the LDL-C level in humans. High levels of erucic acid present in the oil are harmful to human health as erucic acid may cause myocardial conductance, lipidosis in children, and elevated blood cholesterol. Therefore, regulations have been imposed to restrict the maximum level of erucic acid of 2%–5% of the total fatty acids owing to the potential adverse health effects at excessive levels. Behenic acid is present in a minor amount in the rapeseed oil (0.37%). Bioavailability of behenic acid is low due to its long-chain structure, thus it would cause less effect on cholesterol content in humans (Konuskan et al., 2019).

A nutritionally desirable ratio of omega-6 to omega-3 fatty acids is in the range of 1.6–2.5 is present in rapeseed oil, accompanied by a ratio of linoleic to α-linolenic essential fatty acids of 2:1 (Wroniak et al., 2016; Xie et al., 2019). According to Makata (2015), the desirable ratio of omega-6 to omega-3 fatty acids is 3–5 to 1. An inappropriate dietary ratio of these fatty acids has been involved in havocking disorders in the body functions, leading to antagonistic reactions and eventually illness (Makata, 2015). The saturated fatty acids profile in rapeseed oil is relatively low compared to commonly consumed vegetable oils like sunflower oil (11%) and soybean oil (15%) (Perrier et al., 2017).

Canola seeds are rapeseeds with low erucic acid and glucosinolate levels that contain high amounts of oil (38%–50%) and proteins (36%–44%) (Xie et al., 2019). The refining process slightly increases the amount of *trans* fatty acids of linoleic acid and γ-linolenic acid, which are absent in cold pressed rapeseed oil (McDowell et al., 2017). Thus, cold pressing preserves the quality of fatty acids in the rapeseed oil. On the other hand, microwave pretreatment of rapeseed does not change the fatty acid composition of the cold pressed rapeseed oil (Wroniak et al., 2016).

3.3 Phospholipids composition

Rapeseed oil is made up of a different combination of TAG, fatty acids, phospholipids, and unsaponifiable components. During the processing of rapeseeds and oil pressing, degradation of membranes due to mechanical and/or thermal destruction causes the phospholipids to migrate freely into the extracted oil. The previous study reported a low phosphorus content (<20 mg/kg) presented in the cold pressed rapeseed oil that was preserved at temperatures <45°C, while high phosphorus content (125–476 mg/kg) presented in the pressed oil that was obtained at a higher pressing temperature and from seeds conditioned at different temperatures (Ambrosewicz-Walacik, Tańska, & Rotkiewicz, 2015).

The extraction method plays a significant role in the phosphorus content and phospholipids profile of oils. The phosphorus content of CPO was reported as 6.2–8.9 mg/kg, 152.0–195.4 mg/kg for CPO from heat-treated seeds, and 209.9 mg/kg for hot-pressed oil. It was found that heat treatment of whole and milled seeds elevated the phosphorus content in the oil, especially phosphatidic acid content in CPO and solvent extracted oil with petroleum ether. Generally, low phosphorus content in CPO caused a higher phosphorus content in the cold pressed meal (1136.6–1445.2 mg/kg). The majority of the phospholipids remained in cold pressed meal. The phospholipid profile in the cold pressed rapeseed oil was not identified due to the low phosphorus content present. On the other hand, phosphatidylcholine (48.5%–62.5%) and phosphatidyl ethanolamine (37.5%–51.5%) were detected in cold pressed rapeseed oil obtained from heated seeds, while phosphatidic acid and phosphatidyl inositol were not detected (Ambrosewicz-Walacik et al., 2015).

Phospholipids can act as antioxidants to protect against lipid oxidation. However, phospholipids may precipitate in the oil during the storage and cause a dark color. The precipitated phospholipids may carry trace metal ions to act as pro-oxidants to affect the oxidative stability of the oil (Chew, Tan, & Nyam, 2017a). Thus, low phospholipid content in cold pressed rapeseed oil makes it ideal to use as an edible oil to preserve the quality. Phosphatidylcholine helps in the synthesis of the lipid bilayer and liposome formation in the cell membrane, as well as being able to function as an emulsifier in industry.

4 Minor bioactive compounds

4.1 Phenolics

Rapeseed contains the highest amount of phenolics among oilseeds. Rapeseed oil contains high phenolic acid levels as rapeseeds contain 10 times more phenolic acids than other oilseed crops. The phenolic compounds of rapeseed present as phenolic acids and polyphenolic tannins. Phenolic acids may present in the forms of free and bound, like esters and glycosides. Phenolic acids occur mainly in sinapic acid and its low-molecule derivative in rapeseeds (Siger,

Nogala-Kalucka, Lampart-Szczapa, & Hoffman, 2005). Sinapic acid is the most common phenolic compound present in cold pressed rapeseed oil and also its derivatives, which contributes 80% of phenolic compounds in rapeseed. Sinapic acid can exist as glucopyranosyl sinapate, while only a small amount of sinapic acid (less than 16%) is present as a free form. These compounds are transferred to the oil to a small degree (Kraljić et al., 2013). Defatted rapeseed meal may present up to 2% of phenolic acids in free, esterified, and insoluble-bound forms. Table 2 shows the phenolic acids present in cold pressed rapeseed oil as determined by high-performance liquid chromatography (HPLC) with reversed phase column.

Yang et al. (2013) analyzed the total phenolic content of CPO from rapeseed cultivars in China, and their phenolic contents ranged from 9 to 104 mg/100 g oil. The total phenolic contents of cold pressed rapeseed oils in China mostly fell in the range of 15–65 mg/100 g. Cold pressed rapeseed oil from Jiangxi had the lowest phenolics content (9 mg/100 g), while cold pressed rapeseed oil from Hubei had the highest phenolics content (104 mg/100 g). The mean total phenolics content in the CPO from the upper, middle, and lower Yangtze River were 43, 35, and 29 mg/100 g, respectively. This result showed that the total phenolic content decreased with increasing the longitude of the rapeseed cultivars' growing location. The factors of genetics, agronomics, growing conditions, and the degree of maturation can explain this variability of the level of antioxidants (Yang et al., 2013).

The extraction method may affect the phenolics content in the rapeseed oil. Heat application on the seeds prior to extraction causes the decarboxylation of sinapic acid into 4-vinylsyringol, which is also known as canolol. Rapeseed oil extracted from heated seeds contains up to 120 times more canolol content than oil extracted from the seeds without heating (Spielmeyer, Wagner, & Jahreis, 2009). Canolol is more oil-soluble than sinapic acid, as well as being able to act as an antioxidant with antimutagenic and anticarcinogenic activity. Refining of rapeseed oils may destroy the canolol content to nearly undetectable levels. More than 90% of phenolic compounds are lost during the refining process (Kraljić et al., 2015). Siger et al. (2005) reported 236.1 μg/100 g of sinapic acid in cold pressed rapeseed oil. However, there no sinapic acid was detected in refined rapeseed oil. The antioxidant activity of phenolic compounds depends on the number and position of hydroxyl groups in their structure, polarity, solubility, and stability during processing. Canolol possesses greater antioxidant activity than other bioactive compounds like tocopherols, ascorbic acid, β-carotene, rutin, and quercetin. Terpinc et al. (2011) reported that canolol's antioxidant activity exceeded that of sinapic acid by 15%. Therefore, the reactivity of canolol content in rapeseed oil during the processing is required to understand how to maximize the phenolics content in the final oil product.

Spielmeyer et al. (2009) showed that microwave roasting of rapeseeds for 4.5 min at 135°C results in the increase of canolol content from 5.8 to 340.1 μg/g. However, microwave roasting at the temperature higher than 160°C can result in a decrease of canolol content. Canolol has been determined generally in cold pressed rapeseed oil at a level of 9–81 μg/g. On the other hand, Shrestha and De Meulenaer (2014) reported 20 min to be the optimum time duration for seed roasting at 180°C to achieve the highest canolol content in the CPO produced. Siger et al. (2015) investigated seeds roasting at 140°C, 160°C, and 180°C for 5, 10, and 15 min on the canolol content in cold pressed rapeseed oil produced. The highest canolol content occurred in the oil produced from the seeds roasted at 180°C for 15 min. Wroniak et al. (2016) reported that microwave pretreatment for 7 min helped to increase the canolol content in cold pressed rapeseed oil to 926.4 μg/g when the seeds were

TABLE 2 Phenolic acids of cold pressed rapeseed oil.

Phenolic acid (μg/100 g)	McDowell et al. (2017)	Tańska, Mikolajczak, and Konopka (2018)
4-Hydroxybenzoic acid	0.4–3.6	–
Cinnamic acid	0.8–13.9	–
p-Coumaric acid	0–1.6	3.3
Vanillic acid	0–2.3	ND
Canolol	0–19.0	–
Ferulic acid	0–6.9	5.4
Syringic acid	0–13.1	–
Sinapic acid	8.9–189.2	12.3
Total	10.6–209.6	21.1

ND, not detected.

adjusted to a moisture content of 7%. The canolol content present in the cold pressed rapeseed oil from the seeds of the Bakara and Kana varieties ranged from 10.0 to 16.97 mg/g oil, respectively.

4.2 Tocopherols

The major tocopherols present in rapeseed oil are γ-tocopherol (60%–74%) and α-tocopherol (26%–35%), which are found among different rapeseed cultivars. The average α- to γ-tocopherol ratio was 0.61 with 144 mg/kg of α-tocopherol and 235 mg/kg of γ-tocopherol being reported. The total tocopherol content in the cold pressed rapeseed oil was in the range of 113–824 mg/kg with a mean of 379 mg/kg. The mean content of α-tocopherol, γ-tocopherol, and total tocopherol contents were 181, 140, and 110 mg/kg, 289, 230, and 186 mg/kg, and 470, 370, and 297 mg/kg in cold pressed rapeseed oils from the upper, middle, and lower Yangtze River, respectively. These results showed that the tocopherol contents increased as the longitude of the cultivars' growing location increased. Planting location, cultivar, and degree of maturity may explain the variability of tocopherol contents from different varieties (Yang et al., 2013). Tocotrienol was not detected in the cold pressed rapeseed oil (Azadmard-Damirchi et al., 2010). Interactions between genotype and environment are a major factor that affects the variability of tocopherol content in rapeseed oil (Marwede, Schierholt, Möllers, & Becker, 2004).

Plastochromanol-8 (PC-8) is a homolog of γ-tocotrienol, which was present in the cold pressed rapeseed oil. The only difference between PC-8 and γ-tocotrienol is that the former has a longer side chain in the structure. The antioxidant activity of PC-8 is 1.5 times greater than that of α-tocopherol (Siger et al., 2017). The major tocochromanols detected in the rapeseed oil samples were α-, γ-, δ-tocopherol, and PC-8. Tocopherol proportions of 60%–74% γ-tocopherol and 26%–35% α-tocopherol are commonly found in the various rapeseed varieties. γ-tocopherol was the predominant component (41.41 and 38.79 mg/100 g, respectively), followed by α-tocopherol (23.40 and 30.28 mg/100 g, respectively), and δ-tocopherol contents were 0.61 and 0.50 mg/100 g, respectively (Wroniak et al., 2016).

Siger et al. (2017) reported the total tocopherol contents in cold pressed rapeseed oils to be 91.4–103.7 mg/100 g produced from unroasted seeds and 92.0–97.8 mg/100 g produced from roasted seeds. Hot pressed oil presented a 20% higher total tocopherol content than CPO, especially 30% more α-tocopherol and 16% more γ-tocopherol (Wroniak, Krygier, & Kaczmarczyk, 2008). Maillard reactions occur during the seed roasting process and fat-soluble melanoidins may be extracted together with the oil. These components may contribute to antioxidant activity in the oil. The increase in the γ-tocopherol content in the oil produced from the seeds roasted at 160°C for 5 min is probably due to its coelution in HPLC with another component generated after roasting (Wijesundera, Ceccato, Fagan, & Shen, 2008).

Solvent-extracted rapeseed oil presented the highest content of tocopherol (596 μg/g) compared to cold pressed rapeseed oil. Microwave pretreatment helps to increase the tocopherol content in the CPO. CPO from untreated rapeseeds presented the lowest amount of tocopherols (510 μg/g). These results suggested that damage to the oilseed cell membrane by microwave pretreatment allowed increased release extracted from untreated rapeseeds to 45% in oil extracted from pretreated rapeseeds (Azadmard-Damirchi et al., 2010). However, α-tocopherol was found to decrease with increasing microwave pretreatment time (Azadmard-Damirchi et al., 2010; Wroniak et al., 2016). α-tocopherol is susceptible to high temperature application, and presented lower stability than other tocopherol homologs during accelerated storage (Chew, Tan, & Nyam, 2017b). Table 3 summarizes the tocopherol contents of cold pressed rapeseed oil from several studies.

The Harris coefficient is expressed as a ratio of α-tocopherol equivalent (mg) to the mass (g) of PUFA in 100 g of the oil, which is used to determine the nutritional value of the oil. The Harris coefficient of the analyzed cold pressed rapeseed oils from different varieties ranged from 0.84 to 1.69, and the cold pressed high oleic rapeseed oil presented the highest Harris coefficient. All cold pressed rapeseed oils contain proper physiological Harris coefficient values of the ratio of α-tocopherol to PUFA (mg/g) of 0.6:1, which is a minimum value to protect against PUFA peroxidation (Rękas et al., 2016).

4.3 Phytosterols

Sterols exhibit in a free or esterified form with fatty acids in plant oils. The combined determination of free and esterified phytosterols provides an informative approach to identify the authenticity of plant oils. Sterols are unsaponifiable components in oils. Phytosterols can improve the nutritional value of oil as they help to lower serum cholesterol levels, containing antiinflammatory, antibacterial, antiulcerative, and antitumor properties in humans, as well as improving the oxidative stability of vegetable oil. Azadmard-Damirchi et al. (2010) reported six phytosterols in cold pressed rapeseed oil: cholesterol, brassicasterol, stigmasterol, campesterol, sitosterol, and Δ^5-avenasterol (Table 4). Sitosterol was the predominant (50%–56%) phytosterol in cold pressed rapeseed oil, followed by campesterol (32%–36%) and brassicasterol (8%–9%). Solvent-extracted rapeseed oil contained the highest phytosterols content (8120 μg/g), which showed that solvent extraction is more efficient in extracting phytosterols in oil.

TABLE 3 Tocopherol contents (mg/100 g) of cold pressed rapeseed oil.

Tocopherol	Azadmard-Damirchi et al. (2010)	Yang et al. (2013)	Rękas et al. (2016)	Siger et al. (2017)	Tańska et al. (2018)
α-Tocopherol	18.4	4.2–32.2	21.3–28.4	38.2–45.4	26.6
γ-Tocopherol	32.6	7.1–50.2	31.0–42.4	52.1–57.1	39.8
δ-Tocopherol	ND	ND	0.6–1.3	1.0–1.1	2.3
β-Tocopherol	ND	ND	ND	0.1–0.2	ND
PC-8	ND	ND	ND	2.4–3.4	3.1
Total	51.0	11.3–82.4	25.2–32.2	91.4–103.7	71.8

ND, not detected.

TABLE 4 Phytosterols content (mg/100 g) of rapeseed oil extracted by different methods.

Phytosterol	Cold pressed	Solvent-extracted	Extracted by press after microwave pretreatment for 4 min
Cholesterol	1.5	2.5	2.3
Brassicasterol	51.0	75.0	71.0
Campesterol	211.0	290.0	275.0
Stigmasterol	2.6	3.5	3.2
Sitosterol	372.0	410.0	400.8
Δ^5-Avenasterol	22.0	31.0	29.0
Total	660.1	812.0	781.3

Source: Azadmard-Damirchi, S, Habibi-Nodeh, F, Hesari, J, Nemati, M & Achachlouei, BF 2010, 'Effect of pretreatment with microwaves on oxidative stability and nutraceuticals content of oil from rapeseed', *Food Chemistry*, 121, 1211–1215.

Yang et al. (2013) reported the phytosterols content in the cold pressed rapeseed oil from different cultivars in China were within the range reported by Codex standards for the phytosterols content in low erucic acid rapeseed oil, which ranged from 432 to 1190 mg/100 g with a mean of 826 mg/100 g. β-sitosterol, campesterol, and brassicasterol contributed 53.6%, 30.7%, and 12.2%, respectively, to the total phytosterols content in cold pressed rapeseed oil. The presence of brassicasterol is a unique feature identifying rapeseed oil. Phytosterols content in the CPO increased with increasing microwave treatment time. The phytosterols content of CPO from untreated seeds were lowest (6601 μg/g). Microwave pretreatment helps to enrich the phytosterols content of extracted oil (Azadmard-Damirchi et al., 2010).

4.4 Carotenoids

Carotenoids act as an antioxidant in the oil by trapping free radicals or scavenging singlet oxygen physically. Carotenoids contribute to the activity of vitamin A. Yang et al. (2013) reported up to 95 mg/kg of total carotenoids content in the crude rapeseed oil, which was composed of carotene (7%–10%) and xanthophylls (85%–90%), including lutein (50%), 13-*cis*-lutein (15%), and 9-*cis*-lutein (20%) (Ratnayake & Daun, 2004). The carotenoids content present in the cold pressed rapeseed oils from different varieties ranged from 6.66 to 17.39 mg/kg (Rękas et al., 2016). Table 5 shows the total carotenoids content of cold pressed rapeseed oils from different studies.

β-Carotene can react with peroxyl radicals to inhibit the propagation stage and facilitate the termination stage of the lipid oxidation mechanism. β-carotene can also be quenching singlet oxygen effectively during the inhibition of

TABLE 5 Total carotenoids content (mg/kg) of cold pressed rapeseed oil.

Carotenoid	Yang et al. (2013)	Tańska et al. (2018)
Lutein	28–352	32.6
Zeaxanthin	ND	1.2
β-Carotene	1.4–6.7	18.8
α-Carotene	ND	Tr
Total	29.4–358.7	52.6

ND, not detected; Tr, traces.

photo-oxidation. It is the most effective pro-vitamin A, with an average of 2.4 mg/kg presented in the cold pressed rapeseed oil. The cold pressed rapeseed oils from Hubei and Sichuan presented the highest content of β-carotene (6.0–6.7 mg/kg) from the analyzed cultivars in China (Yang et al., 2013).

Lutein is an oxygenated carotenoid that can filter out visible blue light, which can cause free radical damage to the eyes (Subagio & Morita, 2003). There was 40–120 mg/kg of lutein content present in the cold pressed rapeseed oils in China. However, carotenoids can also cause lipid oxidation in a dried food model because of their heat instability. β-carotene was reported as acting as a pro-oxidant during lipid oxidation in the presence of light, while lutein can act as a pro-oxidant in dark or light conditions (Yang et al., 2013). Cold pressed rapeseed oil contains higher carotenoid content than refined rapeseed oil. Kreps, Vrbiková, and Schmidt (2014) showed that the β-carotene content of crude rapeseed oil was 63.6 mg/kg oil, and reduced to 10.2 mg/kg in refined rapeseed oil due to thermal treatment in the refining process.

4.5 Chlorophyll

Chlorophyll adds a green color to the oil, and is important for photosynthesis in the plant. However, chlorophyll can act as a pro-oxidant and inhibit the function of a catalyst in the hydrogenation process. Chlorophyll promotes lipid oxidation in the presence of light, but it acts as an antioxidant in dark conditions. The chlorophyll content in the cold pressed rapeseed oils from different cultivars in China ranged from 0.9 to 51.0 mg/kg. This deviation could be due to the differences of maturation stage at the time of harvest and climatic variations in the Yangtze River Valley in China. The fully ripe seeds contain lower amounts of chlorophyll due to the metabolism of chlorophyll when the seed matures, while immature rapeseeds contain high amounts of chlorophyll (Yang et al., 2013). The composition and amount of chlorophyll in the rapeseeds depend on seed maturity. The chlorophyll would degrade gradually as the seeds ripened, as shown by the chlorophyll content in the physiologically mature seeds (35 days before maturity) is 1239 mg/kg, while in fully matured seeds it is 4 mg/kg (Möllers, 2002). The average chlorophyll content in cold pressed rapeseed oil was 2.92 mg/kg (Rękas et al., 2016).

5 Oxidative stability

Rapeseed contains more than 40% lipids, which gives it lower tolerance to humidity. Improper storage of rapeseed may lead to the degradation of bioactive compounds and loss of nutrition value of the oil produced. According to the Australian Oilseed Federation (AOF) Standards Committee, a maximum moisture concentration at seed delivery is 8%. The harvested seeds are generally dried to a moisture content of 7%, which is appropriate for storage purposes in Poland (Rybacki, Skawiński, & Lampkowski, 2001). During storage, the seeds in the top layer have more moisture than those in deeper layers. This is partly due to the diffusion and adsorption of water from the air by the seeds, and partly due to the water absorption secreted during seed respiration. Gopalakrishnan, Cherian, and Sim (1996) reported a 50% decrease in the tocopherol levels of oil produced from rapeseeds stored at 20°C for 10 days and at 5°C for 30 days. On the other hand, Goffman and Mollers (2000) reported losses of tocopherol content in cold pressed rapeseed oil produced from seeds stored at 40°C.

Cold pressed rapeseed oil is less resistant to oxidation compared to refined rapeseed oil. This is because of higher hydroperoxide and free fatty acid values present in the CPO. Refining process helps to remove the undesirable components in the crude oil to increase shelf stability, but it also simultaneously destroys some of the bioactive compounds. Availability of oxygen, presence of light, and temperature are the major factors that affect the rate of oxidation. On the other hand, the presence of free fatty acids, mono- and diacylglycerols, metal, and thermally oxidized compounds will facilitate auto-oxidation in oils even in the absence of light. The formation of lipid peroxy radicals and hydroperoxides in the auto-oxidation of oils depends only on oxygen availability and temperature. Thus, the container used for vegetable oils packaging is important to protect the oils against oxidation. Oxygen permeability is an important characteristic of plastic packaging, while glass containers are able to prevent oxygen permeation fully, and polyethylene terephthalate (PET) is able to reduce the rate of oxygen exchange (Robertson, 2012; Wroniak & Rękas, 2016).

Wroniak and Rękas (2016) investigated the effect of storage test under light and dark on cold pressed rapeseed oil using amber glass and amber PET bottles at 4 and 20°C. There was no change in the fatty acid profile of cold pressed rapeseed oil stored at 4°C for 12 months. However, there was a slightly change (less than 0.3%) in the fatty acid profile of cold pressed rapeseed oil stored at room temperature for 12 months compared to the control oil sample. Therefore, storage of oil at 4°C was found to be the most appropriate to preserve the quality and extend the shelf life of cold pressed rapeseed oil, but storage at room temperature is common in households.

The substantial loss of tocopherols during storage might be due to the tocopherols acting as hydrogen donors or oxygen quenchers to terminate the oxidation mechanism. Tocopherols protect lipids against oxidation by acting as hydrogen donors to lipid peroxy radicals, which leads to the formation of lipid hydroperoxide and tocopheroxy radicals. These radicals are more stable than lipid peroxy radicals and help to decrease the rate of auto-oxidation in oils. In addition, tocopherols protect lipids against photo-oxidation by singlet oxygen ($^1O_2$) quenching when the oil is exposed to the light. α-tocopherol exhibited the highest reaction rate with $^1O_2$, followed by β-, γ-, and δ-tocopherols. Thus, α-tocopherol will always undergo the greatest degradation during storage among all the tocopherols. The total phytosterols content of cold pressed rapeseed oil decreased by 15.1% (amber glass) and 16.3% (amber PET) when stored at room temperature with access of light and oxygen. On the other hand, the total phytosterols content of cold pressed rapeseed oil decreased by 1.7% and 2.9% for glass and PET containers, respectively. Phytosterols showed higher storage stability over the entire storage period. This might be due to the tocopherols being consumed first when protecting PUFAs against oxidation (Wroniak & Rękas, 2016).

Wroniak and Rękas (2016) reported a strong correlation between the hydroperoxides formation and oil exposure to oxygen, which showed that the oxygen concentration is the most critical factor that affects the rate of oxidation in the cold pressed rapeseed oil. It can be concluded that the retention of rapeseed oil nutritional value in the storage conditions can be ordered as follows: storage at 4°C > 20°C in the dark > storage at 20°C in the presence of light. Amber glass and amber PET containers were able to ensure comparable retention of cold pressed rapeseed oil quality during 12 months of storage. Amber PET is suitable as a cheaper alternative to amber glass for CPO packaging. Thus, it is suggested to label cold pressed rapeseed oils with information on storage conditions recommended after opening.

Azadmard-Damirchi et al. (2010) reported higher stability of the oils as follows: oil pressed from 4 min microwave pretreated seeds > oil pressed from 2 min microwave pretreated seeds > solvent extracted oil > CPO. Rapeseeds with microwave pretreatment increase the antioxidant components and thus contribute to higher oxidative stability. This is because microwave pretreatment could enhance the extractability of components such as canolol, tocopherols, and phytosterols to improve the antioxidant capacity. Lipophilic Maillard-type browning reaction products and phospholipids present in the cold pressed rapeseed oil from microwave-treated rapeseeds may contribute to the antioxidant activity (Wroniak et al., 2016). McDowell et al. (2017) showed that the oxidation stability of cold pressed rapeseed oils had significant variation, whereby the peak PV values after 600 h in a 60°C dark oven ranged from 44.9 to 111.6 meq O_2/kg oil.

The fatty acid composition mainly affected the induction period of the cold pressed rapeseed oil. Previous studies reported the values of induction period of 3.51–6.54 h for cold pressed rapeseed oil from different varieties in Poland (Rękas et al., 2016), and 2.1–4.5 h for cold pressed rapeseed oil (Koski et al., 2002). The lowest amount of PUFA (15.1%) present in high oleic cold pressed rapeseed oil contributed to the highest oxidative stability while the highest amount of PUFA (30.9%) present in yellow seeded cold pressed rapeseed oil contributed to the lowest induction period, according to Rękas et al. (2016).

Sensory assessment is one of the most important parameters to assess the quality of CPO, particularly the intensity of sensory attributes. Seed-like, nutty, woody, and astringent are typical attributes of cold pressed rapeseed oil, while off-flavors can be described as rancid, fusty, musty, and bitter (Rękas et al., 2016). Matthäus and Brühl (2004) showed that the "seedlike" attribute is not detectable during the storage of seed with 5% broken seeds at 40°C after 6 days, while attributes like "fusty," "woody," or "musty" had increased during the storage. In addition, free fatty acids increased and composition of the volatile components changed during the storage of seeds.

6 Health-promoting traits

6.1 Antioxidant activity

High levels of free radicals in the body may cause oxidative stress, and subsequently influence the functions of proteins, lipids, DNA, and other biomolecules, which interrupt the metabolism and cause mutagenesis. Antioxidants help to scavenge the free radicals in the human body to regulate metabolism and keep the body healthy. Synthetic antioxidants have a limited application due to many associated health reasons. Hence, natural antioxidants are emphasized by the public for their importance for health benefits. Plant oils rich in omega-3 PUFA and bioactive compounds may be introduced to diets to have a therapeutic effect on the human body due to their ability to scavenge free radicals. Siger et al. (2005) showed that the antioxidant activity exerted by cold pressed rapeseed oil is higher than that of refined rapeseed oil. Moreover, the anti-radical properties of rapeseed and sunflower extracts are able to inhibit the formation of conjugated dienes (Matthaus, 2002).

Different assays can be used to examine the antioxidant activity of plant oils, which are 2,2′-azino-bis (3-ethylbenzothiazoline-6-sulphonic acid) ($ABTS^{\bullet+}$), 2,2-diphenyl-1-picrylhydrazyl (DPPH·), β-carotene bleaching, and ferric reducing antioxidant power. DPPH· assay is suitable to examine the hydrophobic system, which contains lipophilic antioxidants including carotenoids, chlorophyll, tocopherols, and phytosterols. Moreover, ABTS assay is suitable to examine the hydrophilic and lipophilic systems containing lipophilic bioactives and phenolics, as $ABTS^{\bullet+}$ can dissolve in both aqueous and organic solvents (Chew et al., 2017b).

The antioxidant activity of phenolic extracts of rapeseed oil showed 60% of free radical scavenging activity by DPPH· assay and 80% of inhibition of hydroperoxide formation (Vuorela, Meyer, & Heinonen, 2004). DPPH· assay showed that refined rapeseed oil contains the highest antioxidant activity, followed by cold pressed rapeseed oil and extra virgin olive oil. On the other hand, $ABTS^{\bullet+}$ showed that there was no significant difference between cold pressed rapeseed oil and extra virgin olive oil, while refined rapeseed oil contains the highest antioxidant activity. The higher phenolic content in cold pressed rapeseed oil is one of the reasons for the higher stability of cold pressed rapeseed oil. However, the higher antioxidant activity in refined oil than CPO showed by $DPPH^{\bullet}$ assay might be due to the insensitivity of the assay caused by other reducing substances present in the oil. When individual phenolic acids were identified using liquid chromatography, cold pressed rapeseed oil was shown to have significantly higher phenolic acids than refined rapeseed oil. Both the cold pressed and refined rapeseed oils contain similar fatty acid, tocopherol, and phytosterol profiles, while antioxidant assay showed that refined rapeseed oil had higher levels of antioxidant potential (McDowell et al., 2017). High antioxidants in CPO help in the inhibition or prevention of diet-dependent lifestyle diseases, such as obesity, ischemic heart disease, and hypertension.

6.2 Antihypercholesterolemic activity

The Mediterranean diet rich in MUFAs has been proved to decrease cardiovascular morbidity and mortality (Trichopoulou, Costacou, Bamia, & Trichopoulos, 2003). MUFAs help to reduce the susceptibility of low-density lipoprotein (LDL) to oxidation. Accumulation of lipoproteins, especially LDL, into the intimae of arteries results in atherosclerosis. LDL particles undergo oxidative changes in the arterial wall, which leads to the atherosclerotic process. Oxidized LDL increases the risk of coronary heart disease and occurrence of atherosclerotic plaque. Circulation of oxidized LDL increases the level of oxidative stress and correlates with the risk factors of metabolic syndrome (Palomäki et al., 2010).

The study by Palomäki et al. (2010) showed that the intake of cold pressed rapeseed oil significantly reduced the level of oxidized LDL compared to the intake of butter. Total cholesterol and LDL cholesterol were significantly lower after the intake of cold pressed rapeseed oil compared to those levels after the intake of butter. However, there were no significant differences in high-density lipoprotein cholesterol, TAG, glucose, and HbA1C between these two intake periods. The use of rapeseed oil did not have a significant effect on HDL cholesterol levels (Gulesserian & Widhalm, 2002; Valsta, Jauhiainen, Aro, Katan, & Mutanen, 1992), but caused a significant reduction in total and LDL cholesterol levels (Valsta et al., 1992). MUFAs are major fatty acids in rapeseed oil, thus they may contribute to the protective effect of the decreased level of oxidized LDL. Previous studies also showed that a diet with a high MUFA content reduced the oxidized LDL levels among healthy nonobese men as well as among subjects at high cardiovascular risk (Egert, Kratz, Kannenberg, Fobker, & Wahrburg, 2010; Fitó et al., 2007). Thus, cold pressed rapeseed oil can be used as a supplement to offer a cholesterol-lowering effect.

Intake of cold pressed rapeseed oil also proved to improve arterial elasticity, although there was no significant difference between the intake of cold pressed rapeseed oil and butter on arterial elasticity. A short-term modification of the diet with cold pressed rapeseed oil reduced the level of total cholesterol (8%) circulating LDL cholesterol (11%), and oxidized LDL (16%). In addition, modification of the diet with cold pressed rapeseed oil helped in treating hyperlipidemia, diminishing oxidative stress, and delaying the progression of atherosclerosis in subjects with metabolic syndrome (Palomäki et al., 2010).

Oil that is high in PUFAs helps to regulate blood plasma triglyceride levels in patients with dyslipidemia, lowering blood pressure and protecting against coronary heart disease (Grosshagauer, Steinschaden, & Pignitter, 2019). Linolenic acid and γ-linolenic acid help to regulate the lipid levels in the human body by reducing cholesterol and triglyceride level in the blood, which can help to prevent cardiovascular diseases. In addition, linolenic acid and γ-linolenic acid possess anti-inflammatory and antiallergic properties to improve the immune system and the brain function of the nervous system. Furthermore, they serve as a precursor of the prostaglandin and hormones, as well as regulate the blood pressure (Makata, 2015). High phenolics in cold pressed rapeseed oil presented high antioxidant activity, which lowers the risk of getting hypercholesterolemia. In addition, phytosterols and tocopherols present in the oil help to decrease the absorption of dietary cholesterol and reduce the risk of getting hyperlipidemia and cardiovascular diseases.

6.3 Anticancer activity

Canolol has been shown to reduce apoptosis in human cancer SW480 cells and is one of the most potent antimutagenic compounds in the Salmonella typhimurium TA102 modified Ames test (Kuwahara et al., 2004). Phytosterols are effective in regulating the membrane functions and signal transduction pathways that control the growth of a tumor and apoptosis. Phytosterols help to restrain the growth and lower the metastatic capability of human breast, colon, leukemia, and prostate cancers (Awad, Chan, Downie, & Fink, 2000; Awad & Fink, 2000). Moreover, PUFAs, especially linoleic acid, help to restrain the growth of human breast cancer, colon cancer, skin cancer, stomach cancer, and leukemia in vitro and in vivo (Hubbard, Lim, Summers, & Erickson, 2000; Kritchevsky, 2000; Phoon, Desbordes, Howe, & Chow, 2001). Thus, high phytosterols, phenolics, and linoleic acids in cold pressed rapeseed oil are responsible for the oil's cytotoxic activity.

6.4 Insulin sensitivity and glucose tolerance

A diet with high SFAs is associated with insulin resistance, obesity, and metabolic syndrome. Substitution of SFAs with MUFAs helps to modulate insulin sensitivity and glycemic control. However, there are other reasons than fatty acid composition that may affect glucose tolerance and insulin sensitivity. Previous research reported steeper glucose disappearance rates (Uusitupa et al., 1994) and lowered fasting plasma glucose levels (Södergren et al., 2001) in participants who consumed a canola oil-based diet compared with a high SFA diet. In general, canola oil that is high in oleic acid shows a positive effect in regulating glucose and insulin levels compared to a diet with high SFAs. Research is continuing to study the effects of rapeseed oil on glycemic control, fasting blood glucose, and insulin sensitivity to introduce healthy oil (Lin et al., 2013).

7 Edible and nonedible applications

Rapeseeds and their derivatives can be used entirely for various applications, such as human food (oil), animal feed (cake), and nonfood areas (energy, lubricants, and surfactants). Rapeseed oil is the most consumed vegetable oil in the European Union and makes up 42% of total consumed oils. It is used in the production of salad dressings, mayonnaise, and other emulsified products like biscuits and canning (Perrier et al., 2017). Cold pressed rapeseed oil can be used as an additive to fresh food and as an ingredient enhancing various products like bread and mayonnaise with specific bioactive nutrients. Functional food acts as an integral part of a healthy diet and has characteristics of conventional foods and gives health benefits to humans. The health benefits of functional foods should be proven with scientific research.

Cold pressed rapeseed oil can be used as a renewable fuel for adapted diesel engines, while press cake can be used as a valuable protein feed to protect energy resources. Purification is a solid-liquid separation of solid particles, especially seed peel and seed pulp from the oil. Purification is subdivided into sedimentation and filtration. The oil quality is strongly affected by the purification process as the purification is at the end of the oil production in decentral plants. The use of a safety filter at the end of the purification is to hold back remaining particles in the oil and help to detect problems during the purification process by a disproportionate rise of the pressure difference at the filter unit. The purification process is needed to adjust specifically to the properties of the muddy oil because of the wide range of the contamination and the particle size distribution of the muddy oil.

The safety filter has to take up more particles than after filtration with a chamber filter press if a sedimentation system is used for purification. A candle filter with a candle made of cotton fiber is suitable to use as a safety filter. The contamination of the muddy oil was 44 mg/kg and could be decreased to an average of 11 mg/kg by filtration with a candle filter. This can help to meet the limitation value for contamination according to the "Quality Standard for Rapeseed Oil as a Fuel," which requires contamination to be lower than 25 mg/kg. The level of contamination and the particle size distribution in the oil can affect the lifespan of a filter and the cost of filtration (Remmele & Widmann, 2002).

The residual cold pressed rapeseed meal can be used to produce rapeseed protein concentrates that possess techno-functional characteristics like emulsification and foaming. The meal can also be used as a material for the chemical industry in nonfood applications such as adhesives, detergents, paints, varnishes, and biodegradable polymers. Fetzer, Herfellner, and Eisner (2019) showed that the protein was isolated at pH 5.7–7.0 after being cold pressed through ultrafiltration or a combination of acidic precipitation followed by ultrafiltration. The protein isolation yield of cold pressed rapeseed meal was 36.5%–40.6%, which is higher than the yield of prepressed rapeseed meal. This result shows the positive effect of low temperature processing during defatting.

8 Other issues (allergenic seed protein)

According to Puumalainen et al. (2015), the Finnish and French atopic children showed clinical allergies to rapeseeds in skin prick tests, but the mechanism of sensitization is unknown. The possible allergenic seed proteins were extracted from commercial cold pressed and refined rapeseed oils, and examined them by gel-based tandem nanoflow liquid chromatography with mass spectrometry. Napin was identified as the main allergen in rapeseeds, while cruciferin was detected in cold pressed rapeseed oil, but not in refined oils, which may act as a new potential allergen. Napin is the seed storage protein in rapeseed and contributes approximately 20% of the total protein. Napin is comprised of a small and large chain connected by disulfide bonds; it is resistant to pepsin digestion and will undergo denaturation by heat and low pH. Hence napin would not be destroyed during conventional food processing. Cruciferin (60%) is the main seed storage protein in rapeseeds, comprised of two subunits by a cysteine S—S bridge. Cruciferin represents one of the main allergens in white mustard and hazelnut (Puumalainen et al., 2015).

The study showed the reaction of five children to these proteins when examined with immunoglobulin E immunoblotting (Puumalainen et al., 2015). Therefore, cold pressed rapeseed oil might be one possible source of sensitization for these allergens. Cold pressed rapeseed oil usually contains traces of seed proteins. During the processing of cold pressed rapeseed oil, low temperatures are involved, and sediment from the seeds is precipitated. The recommended daily consumption of cold pressed rapeseed oils is about two tablespoons per day, or 30 mL, and this contains 360–1380 ng of protein (Puumalainen et al., 2015). The amount of allergen to cause the allergic reaction should be focused in the future research. Refined rapeseed oil is suitable for individuals allergic to rapeseed as refined oil is more purified and contains less protein.

The allergy mechanism of cold pressed rapeseed oil is unknown as there is not enough evidence to show the oil's effects on atopic children. Currently, there is no empirical evidence of allergic reactions after consumption of rapeseed oil. However, further research is needed to evaluate the clinical importance of exposure to low concentration allergens in rapeseed oil.

References

Ambrosewicz-Walacik, M., Tańska, M., & Rotkiewicz, D. (2015). Effect of heat treatment of rapeseed and methods of oil extraction on the content of phosphorus and profile of phospholipids. *Polish Journal of Natural Science*, *30*(2), 123–136.

Awad, A. B., Chan, K. C., Downie, A. C., & Fink, C. S. (2000). Peanuts as a source of beta-sitosterol, a sterol with anticancer properties. *Nutrition and Cancer*, *36*, 238–241.

Awad, A. B., & Fink, C. S. (2000). Phytosterols as anticancer dietary components: Evidence and mechanism of action. *Journal of Nutrition*, *130*, 2127–2130.

Azadmard-Damirchi, S., Habibi-Nodeh, F., Hesari, J., Nemati, M., & Achachlouei, B. F. (2010). Effect of pretreatment with microwaves on oxidative stability and nutraceuticals content of oil from rapeseed. *Food Chemistry*, *121*, 1211–1215.

Chew, S. C., Tan, C. P., & Nyam, K. L. (2017a). Optimization of degumming parameters in chemical refining process to reduce phosphorus contents in kenaf seed oil. *Separation and Purification Technology*, *188*, 379–385.

Chew, S. C., Tan, C. P., & Nyam, K. L. (2017b). Comparative storage of crude and refined kenaf (*Hibiscus cannabinus* L.) seed oil during accelerated storage. *Food Science and Biotechnology*, *26*, 63–69.

Egert, S., Kratz, M., Kannenberg, F., Fobker, M., & Wahrburg, U. (2010). Effects of high-fat and low-fat diets rich in monounsaturated fatty acids on serum lipids, LDL size and indices of lipid peroxidation in healthy non-obese men and women when consumed under controlled conditions. *European Journal of Nutrition*, *50*, 71–79.

Esuoso, K. O., & Odetokun, S. M. (1995). Proximate chemical composition and possible industrial utilization of Blighiasapida seed and seed oils. *Rivista Italiana Sostanze Grasse (Italy)*, *72*, 311–313.

Fetzer, A., Herfellner, T., & Eisner, P. (2019). Rapeseed protein concentrates for non-food applications prepared from pre-pressed and cold-pressed press cake via acidic precipitation and ultrafiltration. *Industrial Crops and Products*, *132*, 396–406.

Fitó, M., Guxens, M., Corella, D., Sáez, G., Estruch, R., de la Torre, R., et al. (2007). Effect of traditional Mediterranean diet on lipoprotein oxidation: A randomized controlled trial. *Archives of Internal Medicine*, *167*, 1195–1203.

Food and Agriculture Organization (FAO) (2019). Statistics division. Available from: http://faostat.fao.org [2 February 2019].

Ghazani, S. M., García-Llatas, G., & Marangoni, A. G. (2014). Micronutrient content of cold-pressed, hot-pressed, solvent extracted and RBD canola oil: Implications for nutrition and quality. *European Journal of Lipid Science and Technology*, *116*, 1–8.

Goffman, F. D., & Mollers, C. (2000). Changes in tocopherol and plastochromanol-8 contents in seeds and oil of oilseed rape (*Brassica napus* L.) during storage as influenced by temperature and air oxygen. *Journal of Agricultural and Food Chemistry*, *48*, 1605–1609.

Gopalakrishnan, N., Cherian, G., & Sim, J. S. (1996). Chemical changes in the lipids of canola and flax seed during storage. *Fett-Lipid*, *98*, 168–171.

Gracka, A., Jeleń, H. H., Majcher, M., Siger, A., & Kaczmarek, A. (2012). Flavoromics approach in monitoring changes in volatile compounds of virgin rapeseed oil caused by seed roasting. *Journal of Chromatography A*, *1428*, 292–304.

Grosshagauer, S., Steinschaden, R., & Pignitter, M. (2019). Strategies to increase the oxidative stability of cold pressed oils. *LWT- Food Science and Technology*, *106*, 72–77.

Gulesserian, T., & Widhalm, K. (2002). Effect of a rapeseed oil substituting diet on serum lipids and lipoproteins in children and adolescents with familial hypercholesterolemia. *Journal of the American College of Nutrition*, *21*, 103–108.

Huang, F., Li, W., Huang, Q., Niu, Y., & Wan, C. (2005). New process of dehulling-cold pressing-expansion for double-low rapeseed. *Quality, Nutrition, and Processing: Processing Technology*, *38*, 126–130.

Hubbard, N. E., Lim, D., Summers, L., & Erickson, K. L. (2000). Reduction of murine mammary tumor metastasis by conjugated linoleic acid. *Cancer Letters*, *150*, 93–100.

Konuskan, D. B., Arslan, M., & Oksuz, A. (2019). Physicochemical properties of cold pressed sunflower, peanut, rapeseed, mustard and olive oils grown in the Eastern Mediterranean region. *Saudi Journal of Biological Sciences*, *26*, 340–344.

Koski, A., Psomiadou, E., Tsimidou, M., Hopia, A., Kefalas, P., Wähälä, K., et al. (2002). Oxidaitve stability and minor constituents of virgin olive oil and cold-pressed rapeseed oil. *European Food Research and Technology*, *214*, 292–298.

Koubaa, M., Mhemdi, H., & Vorobiev, E. (2016). Influence of canola seed dehulling on the oil recovery by cold pressing and supercritical CO_2 extraction. *Journal of Food Engineering*, *182*, 18–25.

Kraljić, K., Škevin, D., Barišić, L., Kovačević, M., Obranović, M., & Jurčević, I. (2015). Changes in 4-vinylsyringol and other phenolics during rapeseed oil refining. *Food Chemistry*, *187*, 236–242.

Kraljić, K., Škevin, D., Pospivšil, M., Obranović, M., Nederal, S., & Bosolt, T. (2013). Quality of rapeseed oil produced by conditioning seeds at modest temperatures. *Journal of the American Oil Chemists' Society*, *90*, 589–599.

Kreps, F., Vrbiková, L., & Schmidt, Š. (2014). Influence of industrial physical refining on tocopherol, chlorophyll and beta-carotene content in sunflower and rapeseed oil. *European Journal of Lipid Science and Technology*, *116*, 1572–1582.

Kritchevsky, D. (2000). Antimutagenic and some other effects of conjugated linoleic acid. *British Journal of Nutrition*, *83*, 459–465.

Kuwahara, H., Kanazawa, A., Wakamatu, D., Morimura, S., Kida, K., Akaike, T., et al. (2004). Antioxidative and antimutagenic activities of 4-vinyl-2,6-dimethoxyphenol (Canolol) isolated from canola oil. *Journal of Agricultural and Food Chemistry*, *52*, 4380–4387.

Laoretani, D., Fernández, M., Crapiste, G., & Nolasco, S. (2014). Effect of drying operating conditions on canola oil tocopherol content. *Antioxidants*, *3*, 190–199.

Lin, L., Allemekinders, H., Dansby, A., Campbell, L., Durance-Tod, S., Berger, A., et al. (2013). Evidence of health benefits of canola oil. *Nutrition Reviews*, *71*(6), 370–385.

Lobo, V., Patil, A., Phatak, A., & Chandra, N. (2010). Free radicals, antioxidants and functional foods: Impact on human health. *Pharmacognosy Reviews*, *4*, 118–126.

Makata, H. (2015). Cold-pressed oils as functional food. In *Plant lipids science, technology, nutritional value and benefits to human health* (pp. 185–200). Kerala, India: Research Signpost.

Marwede, V., Schierholt, A., Möllers, C., & Becker, H. C. (2004). Genotype x environment interactions and heritability of tocopherol contents in canola. *Crop Science*, *44*, 728–731.

Matthaus, B. (2002). Antioxidant activity of extracts isolated from residues of oilseeds, such as rapeseed or sunflower. *Agro Food Industry Hi-Tech*, *13*(4), 22–25.

Matthäus, B., & Brühl, L. (2004). Cold-pressed edible rapeseed oil production in Germany. *Inform*, *15*(4), 266–269.

McDowell, D., Elliott, C. T., & Koidis, A. (2017). Characterization and comparison of UK, Irish, and French cold pressed rapeseed oils with refined rapeseed oils and extra virgin olive oils. *European Journal of Lipid Science and Technology*, *119*, 1600327–1600339.

Möllers, C. H. (2002). Potential and future prospects for rapeseed oil. In F. D. Gunstone (Ed.), *Rapeseed and canola oil. Production, processing, properties and uses* (pp. 186–212). Oxford: Blackwell Publishing Ltd.

Palomäki, A., Pohjantähti-Maaroos, H., Wallenius, M., Kankkunen, P., Aro, H., Husgafvel, S., et al. (2010). Effects of dietary cold-pressed turnip rapeseed oil and butter on serum lipids, oxidized LDL and arterial elasticity in men with metabolic syndrome. *Lipids in Health and Disease*, *9*, 137–144.

Perrier, A., Delsart, C., Boussetta, N., Grimi, N., Citeau, M., & Vorobiev, E. (2017). Effect of ultrasound and green solvents addition on the oil extraction efficiency from rapeseed flakes. *Ultrasonics Sonochemistry*, *39*, 58–65.

Phoon, M. C., Desbordes, C., Howe, J., & Chow, V. T. K. (2001). Linoleic and linolelaidic acids differentially influence proliferation and apoptosis of MOLT-4 leukaemia cells. *Cell Biology International*, *25*, 777–784.

Puumalainen, T. J., Puustinen, A., Poikonen, S., Turjanmaa, K., Palosuo, T., & Vaali, K. (2015). Proteomic identification of allergenic seed proteins, napin and cruciferin, from cold-pressed rapeseed oils. *Food Chemistry*, *175*, 381–385.

Ratnayake, W. M. N., & Daun, J. K. (2004). Chemical composition of canola and rapeseed oils. In F. D. Gunstone (Ed.), *Rapeseed and canola oil. Production, processing, properties and uses* (pp. 37–78). Oxford: Blackwell Publishing.

Rękas, A., Wroniak, M., & Rusinek, R. (2015). Influence of roasting pretreatment on high-oleic rapeseed oil quality evaluated by analytical and sensory approaches. *International Journal of Food Science and Technology*, *50*, 2208–2214.

Rękas, A., Wroniak, M., & Szterk, A. (2016). Chatacterization of some quality properties and chemical composition of cold-pressed oils obtained from different rapeseed varieties cultivated in Poland. *Polish Journal of Natural Science*, *31*, 249–261.

Remmele, E., & Widmann, B. (2002). Purification of cold pressed rapeseed oil to use as a fuel for adapted diesel engines. In: *12th European Conference on Biomass for Energy, Industry and Climate Protection, Amsterdam, The Netherlands*, pp. 1142–1144.

Robertson, G. L. (2012). Vegetable oils. In G. L. Robertson (Ed.), *Food packaging: Principles and practice* (pp. 503–505). New York: CRC Press.

Rybacki, R., Skawiński, P., & Lampkowski, M. (2001). The status of rapeseed dehydration in the raw material base area of ZT Kruszwica S.A. *Oilseed Crops*, *22*, 539–549.

Shrestha, K., & De Meulenaer, B. (2014). Effect of seed roasting on canolol, tocopherol, and phospholipids contents, Maillard type reactions, and oxidative stability of mustard and rapeseed oils. *Journal of Agricultural and Food Chemistry*, *62*, 5412–5419.

Siger, A., Józefiak, M., & Górnaś, P. (2017). Cold-pressed and hot-pressed rapeseed oil: The effects of roasting and seed moisture on the antioxidant activity, canolol, and tocopherol level. *Acta Scientarium Polonorum, Technologia Alimentaria*, *16*(1), 69–81.

Siger, A., Kaczmarek, A., & Rudzińska, M. (2015). Antioxidant activity and phytochemicals content in cold-pressed rapeseed oil obtained from the roasting seeds. *European Journal of Lipid Science and Technology*, *177*, 1225–1237.

Siger, A., Nogala-Kalucka, M., Lampart-Szczapa, E., & Hoffman, A. (2005). Antioxidant activity of phenolic compounds of selected cold-pressed and refined plant oils. *Rośliny Oleiste—Oilseed Crops*, *26*(2), 549–559.

Södergren, E., Gustafsson, I. B., Basu, S., Nourooz-Zadeh, J., Nälsén, C., Turpeinen, A., et al. (2001). A diet containing rapeseed oil-based fats does not increase lipid peroxidation in humans when compared to a diet rich in saturated fatty acids. *European Journal of Clinical Nutrition*, *55*, 922–931.

Spielmeyer, A., Wagner, A., & Jahreis, G. (2009). Influence of thermal treatment of rapeseed on the canolol content. *Food Chemistry*, *112*, 944–948.

Subagio, A., & Morita, N. (2003). Prooxidant activity of lutein and its dimyristate esters in corn triacylglyceride. *Food Chemistry*, *81*, 97–102.

Tańska, M., Mikolajczak, N., & Konopka, I. (2018). Comparison of the effect of sinapic and ferulic acids derivatives (4-vinylsyringol vs. 4-vinylguaiacol) as antioxidants of rapeseed, flaxseed, and extra virgin olive oils. *Food Chemistry*, *240*, 679–685.

Terpinc, P., Polak, T., Šegatin, N., Hanzlowsky, A., Poklar, N., & Abramovič, H. (2011). Antioxidant properties of 4-vinyl derivatives of hydroxycinnamic acids. *Food Chemistry*, *128*, 62–69.

Tofanica, B. M., Cappelletto, E., Gavrilescu, D., & Mueller, K. (2011). Properties of rapeseed (*Brassica napus*) stalks fibers. *Journal of Natural Fibers*, *8*, 241–262.

Trichopoulou, A., Costacou, T., Bamia, C., & Trichopoulos, D. (2003). Adherence to a Mediterranean diet and survival in a Greek population. *The New England Journal of Medicine*, *348*, 2599–2608.

Uusitupa, M., Schwab, U., Mäkimattila, S., Karhapää, P., Sarkkinen, E., Maliranta, H., et al. (1994). Effects of two high-fat diets with different fatty acid compositions on glucose and lipid metabolism in healthy young women. *American Journal of Clinical Nutrition*, *59*, 1310–1316.

Valsta, L. M., Jauhiainen, M., Aro, A., Katan, M. B., & Mutanen, M. (1992). Effects of a monounsaturated rapeseed oil and a polyunsaturated sunflower oil diet on lipoprotein levels in humans. *Arteriosclerosis, Thrombosis, and Vascular Biology*, *12*, 50–57.

Vuorela, S., Meyer, A. S., & Heinonen, M. (2004). Impact of isolation method on the antioxidant activity of rapeseed meal phenolics. *Journal of Agricultural and Food Chemistry*, *52*(26), 8202–8207.

Wijesundera, C., Ceccato, C., Fagan, P., & Shen, Z. (2008). Seed roasting improves the oxidative stability of canola (*B. napus*) and mustard (*B. juncea*) seed oils. *European Journal of Lipid Science and Technology*, *110*, 360–367.

Wroniak, M., Krygier, K., & Kaczmarczyk, M. (2008). Comparison of the quality of cold pressed and virgin rapeseed oils with industrially obtained oils. *Polish Journal of Food and Nutrition Sciences*, *58*, 85–89.

Wroniak, M., & Rękas, A. (2016). Nutritional value of cold-pressed rapeseed oil during long term storage as influenced by the type of packaging material, exposure to light & oxygen and storage temperature. *Journal of Food Science and Technology*, *53*, 1338–1347.

Wroniak, M., Rękas, A., Siger, A., & Janowicz, M. (2016). Microwave pretreatment effects on the changes in seeds microstructure, chemical composition and oxidative stability of rapeseed oil. *LWT- Food Science and Technology*, *68*, 634–641.

Xie, Y., Wei, F., Xu, S., Wu, B., Zheng, C., Lv, X., et al. (2019). Profiling and quantification of lipids in cold-pressed rapeseed oils based on direct infusion electrospray ionization tandem mass spectrometry. *Food Chemistry*, *285*, 194–203.

Yang, M., Zheng, C., Zhou, Q., Huang, F., Liu, C., & Wang, H. (2013). Minor components and oxidative stability of cold-pressed oil from rapeseed cultivars in China. *Journal of Food Composition and Analysis*, *29*, 1–9.

Chapter 8

Application of green technology on extraction of phenolic compounds in oilseeds (Canola)

Ruchira Nandasiri[a,b], N.A. Michael Eskin[a], Peter Eck[a], and Usha Thiyam-Höllander[a,b]
[a]*Department of Food & Human Nutritional Sciences, University of Manitoba, Winnipeg, MB, Canada,* [b]*Richardson Centre for Functional Foods & Nutraceuticals, Winnipeg, MB, Canada*

Abbreviations

ASE	accelerated solvent extraction
CC	column chromatography
DCCC	droplet countercurrent chromatography
FFA	free fatty acids
GHz	giga hertz
Hz	hertz
kHz	kilohertz
MAE	microwave aided extraction
MHz	mega hertz
MSAE	mega sonic assisted aqueous extraction
nm	nanometer
PAH	polycyclic aromatic hydrocarbons
psi	pounds per square inch
SFE	supercritical fluid extraction
SPE	solid-phase extraction
SWE	subcritical water extraction
UAE	ultrasonic aided extraction
UV	ultraviolet

1 Introduction

The production of oil from oilseeds has its origins in ancient times. Recent studies, however, have demonstrated that oilseeds could be a rich source of bioactive compounds with potential for use in pharmaceutical and cosmetic industries. Canola (*Brassica napus*), a relatively new crop, is the major oil crop of Canada, accounting for more than 20% of global production (Azargohar, Nanda, Rao, & Dalai, 2013). Canola oil consists mainly of triglycerides and other minor compounds such as free phenolic acids (vanillic, ferulic, *p*-coumaric, chlorogenic, caffeic, etc.), esterified phenolic acids, free fatty acids, and proteins (Abuzaytoun & Shahidi, 2006; Alam et al., 2016; Alkan, Tokatli, & Ozen, 2012). During the past few years, researchers have revealed that esterified phenolic acids are the predominant phenolic compounds in canola oil, with sinapine being the most abundant (Cai, Arntfield, & Charlton, 1999).

Each year thousands of metric tons of defatted canola meal, the by-product of oil production, are underutilized and wasted by the industry (Casseus, 2009). In fact, defatted oilseed meal could be a valuable source of antioxidants, and a rich source for phenolic compounds (Alam et al., 2016; Chen, Thiyam-Hollander, Barthet, & Aachary, 2014). Oilseeds are rich in plant secondary metabolites including hydroxycinnamic acids derivatives, glycosides, flavonoids, and lignans (Morley et al., 2013). Furthermore, with reference to canola meal, phenolic compounds such as sinapic acid esters

Cold Pressed Oils. https://doi.org/10.1016/B978-0-12-818188-1.00008-6
Copyright © 2020 Elsevier Inc. All rights reserved.

(sinapine and glucopyransol sinapate) account for 90%–95% of the total phenolic compounds, while free sinapic acid accounts for less than 10% (Chen et al., 2014; Morley et al., 2013). Consequently, the rich phenolic content of oilseed meal, especially canola, has garnered considerable interest by researchers as a natural source of antioxidants and nutraceuticals (Chen et al., 2014).

Extraction of phenolic compounds from canola oil is carried out over an extended period using different extraction systems (Chen et al., 2014). Among them, the solvent extraction technique was extensively used, due to its simple design, relatively higher yield, and the availability of economically affordable apparatus (Luthria, 2006; Wanasundara & Shahidi, 1994). Yet this technique is associated with many drawbacks, including the high cost of unit operation and the excessive use of solvents (Luthria, 2006).

The above limitations encouraged food engineers and technologists to investigate alternative methods that were eco-friendly and green. High selectivity, short extraction time, lower solvent usage, energy competency, and environmental friendliness are all commonly attributed with green extraction which is highly applicable to the oilseed industry (Carabias-Martínez, Rodríguez-Gonzalo, Revilla-Ruiz, & Hernández-Méndez, 2005). Among the green extraction technologies, ultrasonic aided extraction (UAE), microwave-aided extraction (MAE), subcritical (SWE) and supercritical (SFE) extraction, as well as accelerated solvent extraction (ASE), have gained considerable attention from researchers due to their energy competency, eco-friendliness, and relatively higher yield. These common green extraction techniques can be readily applied in the oilseed industry to minimize the utilization of toxic solvents, extract phenolic compounds, and meet the current energy and solvent demanding conventional extraction methods with the focus on developing a better-quality final product.

2 Canola: A source of phenolic compounds

"Canola" belongs to the *Brassica* genus, which includes rutabaga, mustard, turnips, cauliflowers, cabbages, and broccoli (Casseus, 2009). The name canola originated from Canadian oil, low acid, used to describe a modified rapeseed variety containing a glucosinolate level of $<18\,\mu$mol/g whole seeds, and <1% erucic acid in oil (Obied et al., 2013). An ancient oilseed, rapeseed was first introduced in Asia and Europe as a source of lamp oil and, later, as cooking oil. Afterward, its physical characteristics made it an important lubricating oil for steam engines for use in merchant ships and warships (Casseus, 2009). Rapeseed is not indigenous to Canada, but was introduced by a Polish farmer who immigrated to Western Canada in the 1920s. Its subsequent conversion to canola involved two Canadian plant breeders, Baldur Stefannson and Keith Downey (Daun, Eskin, & Hickling, 2011). Together they changed the agricultural landscape in Canada, making canola the second major crop in Canada after wheat. Global Production now ranks canola varieties third among the world's oilseed crops, after soybean and palm (Carré & Pouzet, 2014). Canola has a high oil yield (40%) and is one of the richest sources of phenolic compounds compared to other oilseeds (Table 1) (Naczk et al., 1998).

2.1 Canola oil

Canola oil is composed of 6%–14% α-linolenic acid, 50%–65% oleic acid, and <7% of saturated fatty acids (Ghazani & Marangoni, 2013; Gunstone, 2011). However, cold pressed canola oil contains a comparatively higher content of tocopherols (60–70mg/100g) and phytosterols. Generally, canola oil contains a ratio of 1:2 between α-:γ-tocopherols, and a ratio of 1:1 between free and esterified forms of phytosterols (Ghazani & Marangoni, 2013; Gunstone, 2011). It usually has relatively lower levels of saturated fats, and a ratio of two to one (2,1) for linoleic to linolenic acid (Xu & Diosady, 2012).

TABLE 1 Phenolic composition of most common oilseed products (Naczk, Amarowicz, Sullivan, & Shahidi, 1998).

Oilseed product	g/kg dry basis
Peanut flour	0.63
Soybean meal	4.60
Rapeseed/canola flour	6.4–12.8
Canola meal	15.4–18.4

Overall, the oil quality and fatty acid composition in canola make it ideal for human consumption. The replacement of *trans* rich hydrogenated canola oil with high oleic canola oil for industrial/commercial frying also provides a much healthier and stable oil for the consumer market (DeBonte, Iassonova, Liu, & Loh, 2012; Ghazani & Marangoni, 2013).

2.2 Canola meal

Canola meal is the second largest protein meal produced in the world after soybean meal (Aachary, Thiyam-Hollander, & Eskin, 2015). It has a high fiber content, which can interfere with its digestibility (Azargohar et al., 2013). However, this problem has been mitigated with the production of genetically modified varieties with less than 2% erucic acid in its oil, and less than 30 μmol/g of one or any mix of the four recognized aliphatic glucosinolates (gluconapin, progoitrin, glucobrassicanapin, and napoleiferin) (Cai et al., 1999). Canola meal contains the common essential amino acids, with the levels of lysine, methionine, cysteine, threonine, and tryptophan being higher than those of the other cereal varieties (Azargohar et al., 2013; Cai et al., 1999).

2.3 Natural phenolic compounds of canola

Sinapic acid and its derivatives are dominant phenolic antioxidants in *Brassicaceae* species such as rapeseed, canola (*B. napus* L.), and mustard (*Brassica juncea/Sinapis alba*), and are well-known to be effective as synthetic antioxidants (Jun, Wiesenborn, & Kim, 2014; Nićiforović & Abramovič, 2014). Sinapic acid (3,5-dimethoxy-4-hydroxycinnamic acid) can be found in both the free and esterified forms. Sinapine, the choline ester of sinapic acid, and sinapoyl glucose (1-*O*-β-D-glucopyranosyl sinapate), the sugar ester of sinapic acid, are two common sinapoyl esters found in oilseeds (Nićiforović & Abramovič, 2014). Sinapic acid itself comprises more than 73% of the free phenolic acids whilst sinapine (choline ester of sinapic acid), the key phenolic ester, constitutes around 80% of overall phenolic compounds (Fig. 1) (Chen et al., 2014).

Furthermore, decarboxylation of sinapic acid could occur throughout oilseed processing at relatively higher temperatures. It was subsequently found that "canolol," a novel antioxidant compound (2,6-dimethoxy-4-vinyl phenol/vinylsyringol), could be formed by decarboxylation at high operating temperatures and pressures of oil extraction (Galano, Francisco-Márquez, & Alvarez-Idaboy, 2011). Moreover, *p*-hydroxybenzoic, vanillic, gentisic, protocatechuic, syringic, *p*-coumaric, ferulic, caffeic, and chlorogenic acids are some other common phenolic compounds which are present in rapeseed, particularly in canola seed/meal (Fig. 2) (Kozlowska, Rotkiewicz, Zadernowski, & Sosulski, 1983).

The amount of phenolic compounds in canola extract depends on the extraction method used, suggesting that the extraction solvent and isolation steps should be examined to maximize the phenolic compounds in canola seed extracts (Jun et al., 2014). Thus, the solvent selection is the primary factor affecting the amount and the rate of antioxidants extracted by the conventional method of extraction (Hassas-Roudsari, Chang, Pegg, & Tyler, 2009). However, most of

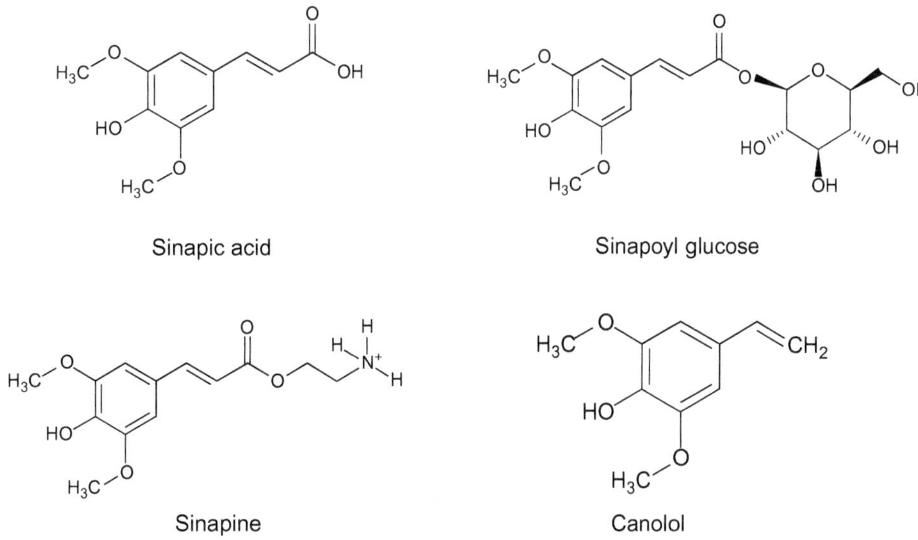

Sinapic acid

Sinapoyl glucose

Sinapine

Canolol

FIG. 1 Chemical structures of sinapic acid derivatives.

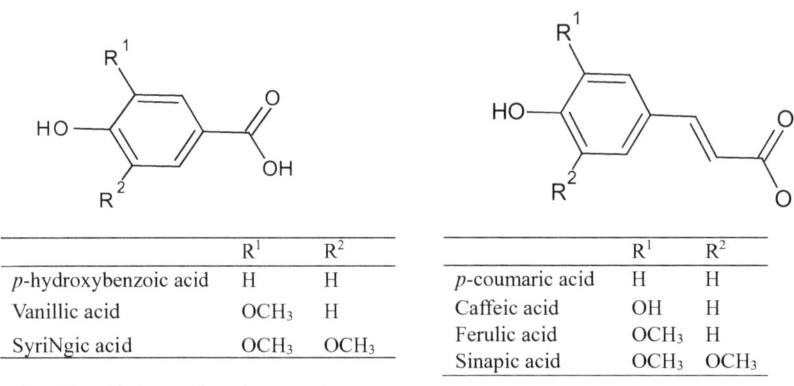

	R^1	R^2
p-hydroxybenzoic acid	H	H
Vanillic acid	OCH_3	H
SyriNgic acid	OCH_3	OCH_3

	R^1	R^2
p-coumaric acid	H	H
Caffeic acid	OH	H
Ferulic acid	OCH_3	H
Sinapic acid	OCH_3	OCH_3

FIG. 2 Structure of major phenolic acids in canola and rapeseed.

the phenolic compounds extracted into the oil during the extraction process depreciate during the chemical refining, while a high proportion of unsaturated fatty acids makes them highly susceptible to oxidation (Chen et al., 2014).

3 Industrial approach toward mechanical and solvent extraction

As mentioned earlier, rapeseed and canola undergo two extraction processes: mechanical extraction followed by solvent extraction. During mechanical extraction, the oil is separated from the seed by mechanical pressure. There are two main techniques of extraction, batch hydraulic pressing and continuous mechanical pressing, which can be further subdivided into cold and hot pressing (Akpan, 2012). The major difference between cold and hot pressing is the application of temperature. In general, it is believed that cold pressed oils have a better ability to preserve the organoleptic properties compare to the hot pressed oils. However, the recovery yield of the cold pressed oils is much lower than for the hot pressed oils. In addition, hot pressed oils have a rich aroma and flavor and might contain some novel phenolic compounds, including canolol and its derivatives, which may be potent antioxidants (Galano et al., 2011). However, the major economic drawback of mechanical extraction is that the majority of the oil gets trapped during the process. This bottleneck can be mitigated by solvent extraction (Carré & Pouzet, 2014). Solvent extraction depends mainly on the dispersion extraction procedure of oilseeds, where the expression "extraction" describes the ability of a solvent for the aggregation of fluid via liquid-solid components (Akpan, 2012). Solvent extraction has been carried out using many different organic solvents, although the primary one is hexane (Carré & Pouzet, 2014). Most industries used a combination of mechanical extraction and solvent extraction to obtain higher yields, followed by additional purification or refining steps to remove problematic minor components (Fig. 3).

Due to the complexity of the extraction process, industries are currently facing many challenges including plant and ecological protection issues, organic compound volatility (0.2–2.0 gal/ton), elevated production costs (shipping, cargo, and dumping), and a massive quantity of organic solvent used and additional phases in processing. The pressed cake also requires additional processing steps including de-solemnization and toasting to remove excess solvents. However, these conditions may alter the nutritional quality of the meal due to the loss of thermo-labile components (Akpan, 2012).

3.1 Extraction of phenolic compounds from canola

Conventional solvent extraction was commonly used as a convenient method for the extraction of phenolic compounds from rapeseed meal or hulls in the early days (Amarowicz, Naczk, & Shahidi, 2000; Wanasundara & Shahidi, 1994). Various solvents, including methanol, ethanol, acetone, and ethyl acetate, with different proportions of water, have been extensively studied for extracting oilseeds including canola products (Wanasundara & Shahidi, 1994). Among the factors affecting the yield of phenolics were the oilseed cultivar, solvent type, extraction time, temperature, sample-to-solvent ratio, and the properties of the extracted materials (Stalikas, 2007). In addition, extraction efficiency can also be affected by the chemical nature of target phytochemicals, extraction technique, particle size, and the existence of tampering materials such as hulls, seeds, and other impurities (Stalikas, 2007).

The concentration of some phenolic compounds was shown to increase when extracted at high temperatures (Azargohar et al., 2013). This was especially true for canolol, which was formed from sinapic acid by decarboxylation (Galano et al., 2011). This phenomenon was confirmed by alterations in the UV spectrum, with the two main peaks for sinapic acid changing from 230 and 320 nm to three main peaks 210, 225, and 310 nm. Approximately 40% of sinapic acid was reported

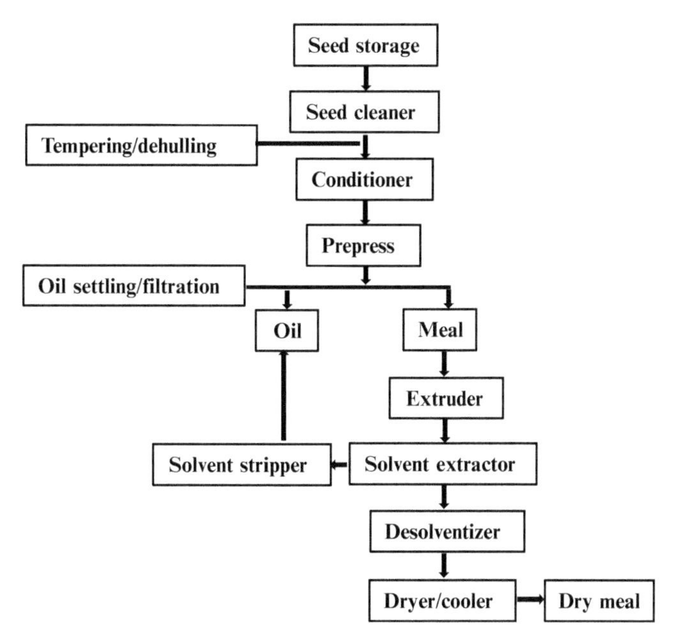

FIG. 3 Oil manufacturing procedure from rapeseed/canola *(Adapted from Gunstone, F. D. (2011). Vegetable oils in food technology: Composition, properties and uses (2nd ed.) (pp. 107–136).)*

to be transformed into novel elements during heat treatment (Cai et al., 1999). In addition, during the heating process, some new phenolic compounds were formed (Fig. 4) (2,6-dimethoxyphenol, 4-ethylphenol, 2-methylphenol, phenol, etc.), fatty acids (e.g., oleic acid), phenolic acids (e.g., 4-hydroxybenzenesulfonic acid), ketonic compounds, and benzyl alcohol (Azargohar et al., 2013; Cai et al., 1999).

It is evident from the results that rapeseed and canola are rich sources of phenolic content and could be applied in the pharmaceutical and nutraceutical industry. However, the extraction of these phenolic compounds has been mainly focused on solvent extraction using different solvents. The classical solvent extraction method is Soxhlet. This has been extensively employed to extract many compounds including phenolics from a solid matrix using a variety of different solvents. The preferential solvent extractants used alongside with Soxhlet are hexane, acetone, ethyl acetate, methanol, ethanol, propanol, and their combinations (Garcia-Salas, Morales-Soto, Segura-Carretero, & Fernández-Gutiérrez, 2010). The Soxhlet/reflux extraction methods normally operate near the boiling point of the solvent employed. While this method is relatively simple,

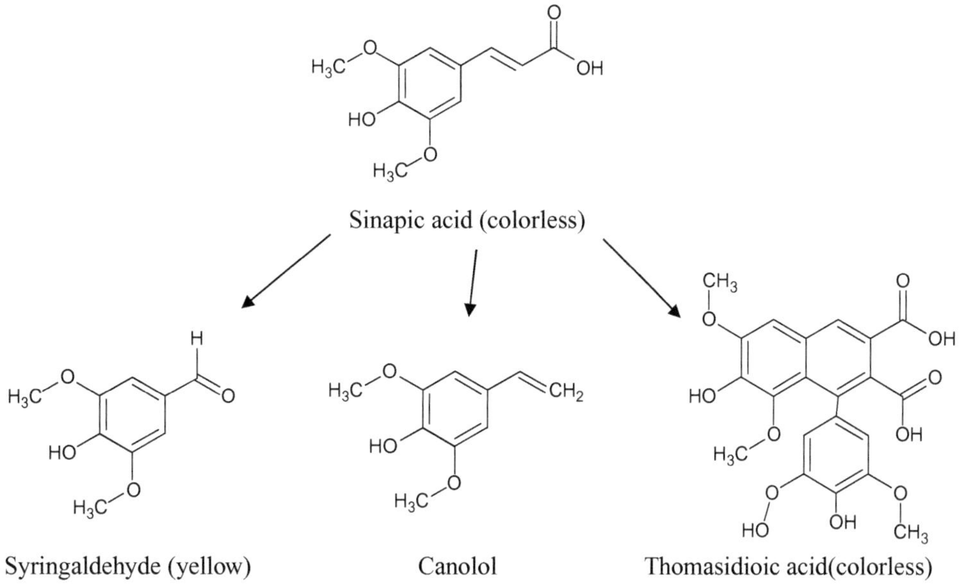

Sinapic acid (colorless)

Syringaldehyde (yellow) Canolol Thomasidioic acid(colorless)

FIG. 4 Heat-related novel phenolic compounds associated with rapeseed/canola.

it is energy-demanding and uses a large amount of solvents (Luthria, 2006). Current environmental concerns demand a reduction in use of solvents and more green extraction methods. This has led to the development of more energy-efficient and eco-friendly green techniques that still produce high yields. Such processes also reduce the utilization of toxic solvents and develop better-quality products.

These advantages have been welcomed by the oilseed industry and encouraged by manufacturers and associates in food science and technology. Green techniques have thus gained attention from the food industry and been adopted by many industries due to its great applicability and environmental benefits. Green extraction technologies such as ultrasonic-aided extraction (UAE), microwave-aided extraction (MAE), supercritical fluid extraction (SFE), and subcritical water extraction (SWE) are receiving closer attention from the oilseeds industry.

4 Green technology

The use of new alternative nonconventional methodologies has gained considerable attention during the last decade. These novel techniques have been applied in the oilseed industry to minimize any detrimental changes in the nutritional quality and physicochemical and sensory properties of the extracted oils, while at the same time reducing the carbon footprint from fossil-derived solvents (Koubaa et al., 2016). A general definition of green extraction is one in which there is a reduction in energy consumption and solvent use while obtaining sustainable natural products, with the guarantee of a safe and pure extract (Chemat, Vian, & Cravotto, 2012). Furthermore, this technique could be applied on a lab or industrial scale by improving its current operation, by the use of nondevoted equipment, and application of innovative operations using alternative solvents (Chemat et al., 2012). The basic concept of green technology is to develop and utilize procedures that minimize/eliminate the use of environmentally hazardous materials (Mustafa & Turner, 2011).

4.1 Principles of green extraction

The concept and principles of green extraction of natural products were presented by Chemat et al. (2012). This was in response to the urgent need to meet the competitiveness of the world market while at the same time protecting the environment. Rather than continuing to use the previous environmentally and energy-demanding methods, new and sustainable processes were desperately needed. The new green extraction methods required innovation that optimized the consumption of raw materials, solvents, and energy. The following six principles for the green extraction of natural products were presented and were designed to innovate not only processes but also all aspects of solid-liquid extraction.

Principle 1: Innovation for the selection of varieties and the use of renewable plant resources.
Principle 2. Use of alternative solvents and principally water and agro-solvents.
Principle 3: Reduce energy consumption by energy recovery and using innovative technologies.
Principle 4: Production of coproducts instead of waste to include the bio- and agro-refining industry.
Principle 5: Reduce unit operations and favor safe, robust, and controlled processes.
Principle 6: Aim for a nondenatured and biodegradable extract.

These six principles are considered essential to meet the demands for extracting natural products in the 21st century. Such guidelines presented by Chemat et al. (2012) were needed to safeguard both the environment and consumers, while increasing competition between industries with more eco-friendly, cost-effective, and innovative processes.

4.2 Green extraction techniques

4.2.1 Ultrasonic/ultrasound aided extraction (UAE)

Ultrasound/ultrasonic aided extraction (UAE) is one of the most common and widely used green extraction methods by industry. This technique usually uses higher frequencies over 20 kHz under mechanic vibrations (Tiwari, 2015). The principle underlying this technique is acoustic cavitation, which involves a series of high-pressure (compression) and low-pressure (rarefaction) cycles (favoring penetration and transport) following each other at high frequency. This results in the formation of small bubbles which collapse rapidly, creating physical and chemical changes in the media that assist the transport phenomena displacing separation stability (Bogdanov, 2014). In addition, mechanic vibrations enable disruption and thinning of the cell membranes (Fig. 5), creating a consistent mass diffusion of analytes from the solid matrix into the solvent (Takeuchi et al., 2009). Consequently, during the implosion of bubbles, temperature and pressure rise up to

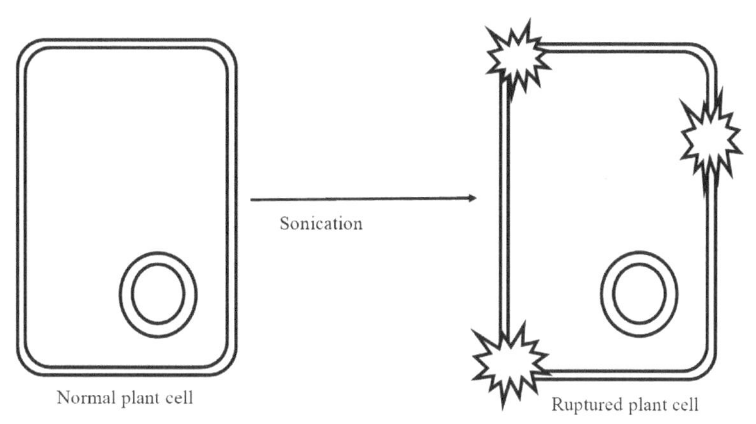

FIG. 5 Schematic diagram of illustrating the effect of ultrasonic treatment on cells.

5000 K (4727°C) and 2000 atm (29,392 psi), respectively, creating a momentum high enough to disrupt the cellular matrix (Bogdanov, 2014).

The UAE technique can be further subdivided into two types depending on its intensity: low-intensity sonication ($<1\,W\,cm^{-2}$) and high-intensity sonication (10–$1000\,W\,cm^{-2}$). Low-intensity sonication ($<1\,W\,cm^{-2}$) is usually applied in process control as well as quality assurance whereas high-intensity sonication (10–$1000\,W\,cm^{-2}$) is usually used in the extraction process (Tiwari, 2015). This UAE technique is again subdivided into two types based on the mode of application: direct and indirect. Direct approaches use ultrasonic horns/probes to sonicate the sample. The probe usually gets immersed in the sample, reducing the barrier between the sample and the source in the direct method of application. Indirect applications involve an ultrasonic bath (Santos, Veggi, & Meireles, 2012). The major distinction between the two techniques is the intensity of the ultrasonic waves produced, as the probe is able to generate 100 times higher intensity than the sonication bath (Santos et al., 2012).

Optimization of ultrasonic extraction conditions normally depends on the type of compounds and the properties of the matrix. The most common extraction conditions for the ultrasonic extraction of phenolic compounds include a sonication power of 90–150 W, a frequency of 20–60 Hz, a time of 2–30 min and 1–5 cycles (Santos-Buelga, Gonzalez-Manzano, Dueñas, & Gonzalez-Paramas, 2012). An increase in extraction time, cycles, power, or frequency could aid the auto-generation of heat, which could affect the degradation of thermally labile phenolic compounds (Santos-Buelga et al., 2012). The major advantages of the application of UAE include its ability to operate at ambient temperature and pressure. This method particularly favors the extraction of the more thermo-labile compounds. Thus, a reduction in extraction time and in use of solvents combined with its simple equipment design are all additional advantages associated with UAE (Capote & De Castro, 2006). However, obligatory filtration and rinsing, and inability to refurbish the solvents, are the main shortcomings of UAE. Other drawbacks associated with UAE include that the intensity of the ultrasonic waves generated may not be uniform (Kiani, Zhang, Delgado, & Sun, 2011). This is evident by regions of high and low intensities identified using the foil test (Fig. 6). High cavitation areas could be located by regions of the foil with holes (Kiani et al., 2011). The use of this technique is encouraged by the food industry, although it still has a number of limitations. Nevertheless, this technique has demonstrated promising results in seed oils by shortening the extraction time, reducing the solvent usage, and producing high value-added compounds in pilot- and lab-scale experiments.

However, to date there are still a very limited number of studies conducted on a commercial scale to estimate the efficacy and effectiveness of the UAE technique. Nevertheless, a wide range of foods applications have been tested using this method including emulsification and homogenization (Kentish et al., 2008), crystallization (Gogate & Kabadi, 2009),

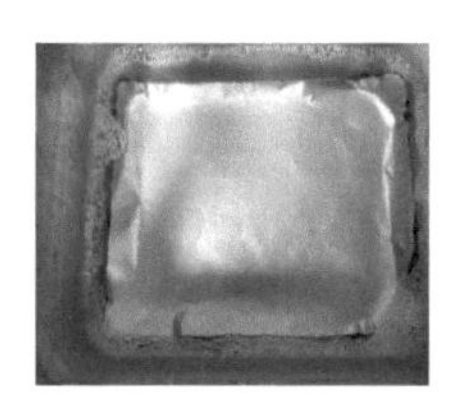

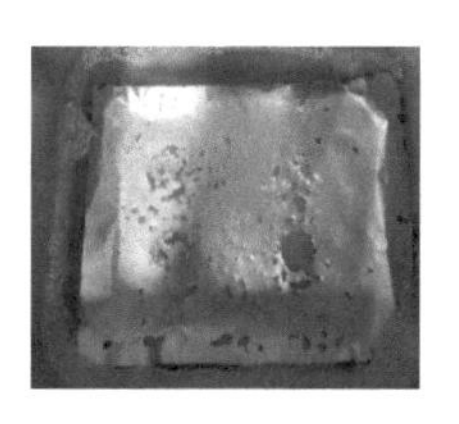

FIG. 6 Foil test in ultrasonic water bath (VWR 250HT).

and alteration of viscosity (Gallego-Juárez, Rodriguez, Acosta, & Riera, 2010; Jambrak et al., 2010). Furthermore, a very limited amount of work has been carried out to assess the economic benefit of UAE and the energy requirement per kg/ton of treated seeds compared to the more traditional techniques. A recent study by Perrier et al. (2017) found that combining ultrasound application with isopropanol appeared to be an alternative solution to hexane for obtaining higher oil recovery from rapeseed flakes.

4.2.2 Megasonic-assisted aqueous extraction (MSAE)

In recent years, ultrasonic/ultrasound aided extraction (USE) has shown promising results in industrial oil extractions, while the application of MSAE has been very limited. Megasonic separation is similar to USE, but uses much higher frequencies (0.4–2 MHz) (Gaber, Trujillo, Mansour, Taylor, & Juliano, 2019). During the application of MSAE, the acoustic field exerts Bjerknes radiation forces, which facilitate the movement of particles and droplets toward pressure nodal or antinodal planes based on the density and compressibility of the particles (Gaber et al., 2019). This phenomenon results in the amalgamation of the droplets into a larger mass with lower density. In contrast, bubble microstreaming with the application of MSAE also contributes to the oil removal by aiding the action of cell-wall breaking enzyme activity or by facilitating the removal of oil from solid matter (Juliano et al., 2017). Other parameters including the level of frequency, ultrasound power, liquid to solid material ratio, duration of sonication, temperature of sonication, and solvent temperature also influence the effectiveness of MSAE (Gaber et al., 2019).

Application of the MSAE technique includes the industrial extraction of oil, the separation of fat from milk and whey, creaming enhancement (Juliano et al., 2011), and removal of residual oils from by-products such as canola meal (Gaber et al., 2019). A recent study by Gaber et al. (2019) found that MSAE improved oil quality by having very low free fatty acid (FFA) levels (0.36%) compared to regular industrial oil extractions (between 0.5% and 0.8%) (Table 2). Free fatty acids (FFA) are considered as one of the quality parameters of frying. Lower FFA levels are generally associated with deodorization at lower temperatures and inhibit the smoking of oils at lower temperature levels (Cmolik & Pokorny, 2000). Furthermore, MSAE produces oils with a lower chlorophyll content, which reduces the amount of bleaching clay needed for de-pigmentation. In addition, the MSAE oils are much lower in peroxides with less off-flavors and off-odors. These findings further support MSAE as a viable alternative method for oil extraction.

4.2.3 Microwave-assisted extraction (MAE)

Microwave-assisted extraction (MAE) is a distinct technique commonly applied for the extraction of phenolic compounds as well as to improve the extraction efficiencies. Microwaves are characterized as nonionizing electromagnetic radiations in the range of frequency from 300 MHZ to 300 GHz (Carré & Pouzet, 2014). MAE consists of two perpendicular oscillating electric and magnetic fields, and its heating principle is based on immediate consequences of the microwaves on polar molecules through an ionic transmission and dipole revolution mechanism (Krishnaswamy, Orsat, Gariépy, & Thangavel, 2013; Santos-Buelga et al., 2012) (Fig. 7).

During the process of extracting, chemical compounds absorb microwaves approximately related to the solvent's dielectric constant. The dielectric constant and the level of microwave absorption have a positive correlation, providing relatively higher microwave energy with the increase in its frequency (Santos-Buelga et al., 2012). In addition, solvents with a higher dielectric constant appear to contain higher microwave-absorbing properties, which enhance the extraction yields (Santos-Buelga et al., 2012) (Table 3).

TABLE 2 Canola oil quality parameters in megasonic trial and industrial specifications.

Oil quality parameter	Units	Megasonic assisted extraction (MSAE)	Industrial specification
Free fatty acid	g acid oleic, %	0.36	0.5–0.8
Chlorophyll	ppm	1.0	10–30
Peroxide	ppm	0.90	10.0
Phosphorous	ppm	32.0	70–100

Adapted from Gaber, M. A. F. M., Trujillo, F. J., Mansour, M. P., Taylor, C. & Juliano, P. (2019). Megasonic-assisted aqueous extraction of canola oil from canola cake. *Journal of Food Engineering, 252*, 60–68.

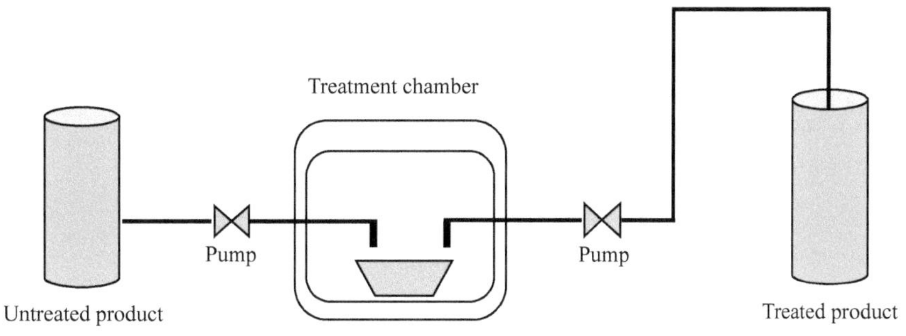

FIG. 7 Schematic representation of continuous laboratory-scale microwave equipment.

TABLE 3 Dielectric constant of common extractants at 20°C.

Solvent	Dielectric constant
Hexane	1.89
Toluene	2.4
Dichloromethane	8.9
Acetone	20.7
Ethanol	24.3
Water	78.5
Methanol	32.6

Thus, blending solvents extractants together could alter the extracting selectivity and the ability to interrelate with microwaves. With relatively higher microwave frequencies, solvent molecules fail to realign themselves and commence vibrating. For example, this phenomenon could lead to molecules vibrating at a rate of 4.9×10^9 times per second at a frequency of 2.45 GHz. This would further direct the development of thermal energy through friction and vibrations (Manadal, Mohan, & Siva, 2007). Thus, once the plant materials are exposed to microwaves, the water tends to evaporate and build up pressure in the cell matrix. The exertion of such forces ruptures the cell wall, thereby enabling better and more efficient extraction by the solvent (Veggi, Martinez, & Meireles, 2012).

Microwaves have been used for the extraction of anthocyanins from grapes (Liazid, Guerrero, Cantos, Palma, & Barroso, 2011), phenolics from tomatoes (Li et al., 2012), flaxseed (Beejmohun et al., 2007), and grape seeds (Krishnaswamy et al., 2013). Reductions in the extraction time and energy-related costs are key advantages of this non-conventional technology. Furthermore, in comparison to the untreated samples, oil-aided extraction via microwave processing demonstrated comparable or elevated oil quality indicating the potential benefits to oilseed industry. In addition, this technique has proven to improve the shelf-life as well as the oxidative stability of the oils with promising results.

The key leverage of MAE is its fast extraction of analytes, together with the extraction of thermolabile compounds (Santos-Buelga et al., 2012). Furthermore, the above is a rapid method of extraction that produces higher yields while lowering solvent consumption compared to conventional extraction techniques. However, few studies have been conducted so far to examine the performance of this technique on a small scale in uninterrupted mode.

4.2.4 *Accelerated solvent extraction (ASE)*

Most traditional extractions are based on solvent reflux techniques which require extensive time, a large amount of samples, and an excessive amount of solvents (Mustafa & Turner, 2011; Szydłowska-Czerniak & Tułodziecka, 2014). Another major drawback of traditional techniques is that they operate at atmospheric pressure, which prevents the operation temperature from exceeding the boiling point. This limits the mass transfer of the analytes across the matrix. This is markedly different for accelerated solvent extraction (ASE), which employs high pressure and high temperature to increase extraction efficiencies (Barros, Dykes, Awika, & Rooney, 2013).

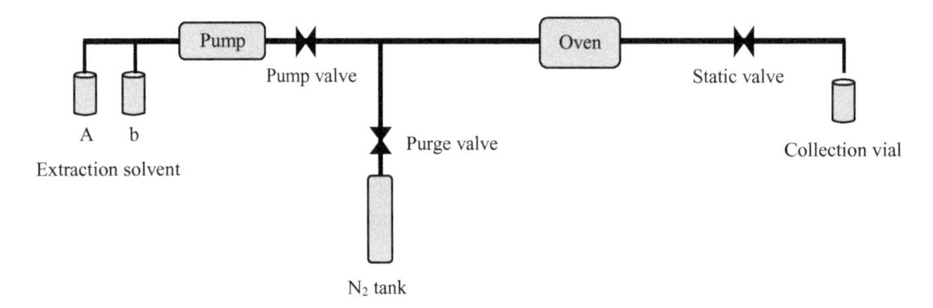

FIG. 8 Schematic diagram of an ASE system.

Dionex Corporation first introduced pressurized liquid extraction in 1995 and named this technique accelerated solvent extraction (ASE). It is also known as pressurized liquid extraction, pressurized solvent extraction, and enhanced solvent extraction (Mustafa & Turner, 2011). It generally operates at higher pressure conditions (~2000 psi), which allows the temperature of the extraction to exceed the flashpoint of the solvent, thereby enhancing the extraction kinetics of the matrix. During this process, stainless steel extraction cells filled with 1–50 g of sample (solid/semisolid) are then extracted with organic solvents. Afterward, the extracted sample is flushed with clean solvent, followed by N_2 flush, and then collected in a glass bottle under nitrogen (Fig. 8) (Ajila et al., 2011; Luthria, Vinjamoorib, Noelc, & Ezzelld, 2004).

The purging step with N_2 prevents any losses of extractants and facilitates the removal of solvents from the extraction cells and solvent tubes (Mustafa & Turner, 2011). The elevated temperature and pressure increase the mass transfer and reduce the viscosity of the solvent, facilitating easy diffusion into the sample matrix (Barros et al., 2013). The presence of moisture in the sample matrix often affects the extraction efficiency. Therefore, a hydro-matrix is often used to absorb water and moisture from the sample matrix to aid the extraction process (Mustafa & Turner, 2011). During the extraction process, dispersing agents are also often used to reduce the cell volume and the usage of solvents. This method uses small amounts of samples and substantially fewer solvents, and produces higher yields in a relatively short period of time (Table 4) (Luthria, 2006).

ASE 350 itself is fully automated and the extractor could be used with different solvents (Luthria, 2006). This could be applied for the selective extraction of the compound of interest with specific polarity. Furthermore, the use of extraction cells eliminates the additional filtration steps associated with extraction, since the matrix residue is preserved inside the extraction cell (Carabias-Martínez et al., 2005). In addition, the application of an inert atmosphere during the extraction facilitates the yield of readily oxidizable phenolic compounds (Santos-Buelga et al., 2012). In addition, the new model ASE 350 is equipped with both static and dynamic extraction techniques. This further allows the system to introduce fresh solvents during the extraction process (Mustafa & Turner, 2011). Furthermore, the static mode consists of one or more extraction cycles with replacement of solvents between the cycles, whereas in the dynamic mode the pump executes solvents at a constant flow rate to achieve maximum extraction efficiency (Mustafa & Turner, 2011). With static extraction, however, complete extraction might not occur due to the limited volume of extraction fluids, therefore, multiple static

TABLE 4 Assessment of different extraction techniques.

Technique	Average time[a](h)	Solvent usage[a](mL)	Cost per sample (USD)
Soxhlet	4–48	150–500	20–25
Automated Soxhlet	1–4	50–100	15–20
Sonication	0.5–1	150–200	17–23
Supercritical fluid extraction	0.5–2	5–50	15–20
Microwave	0.5–1	25–50	12–18
Accelerated solvent extraction	0.2–0.3	5–50	10–15

[a]*Based on 2000 samples/year costs (USD).*

extraction cycles are required to achieve the complete extraction (Luthria, 2006). The common solvents used with ASE are methanol, ethanol, and acetone with different proportions of water during the extraction process. Research has indicated that extractions through ASE could be carried out with a multitude of solvents at various pH and temperatures (Barros et al., 2013). This technique could also use water as an extractant, and is referred to as pressurized hot water extraction or superheated water extraction (Mustafa & Turner, 2011).

Thus, its use of higher temperatures could produce polymeric phenolic compounds with antioxidant properties. However, elevated temperatures associated with ASE also appears to improve the organoleptic properties of the oils extracted as well as aid in the formation of novel phenolic compounds with higher antioxidant properties such as canolol (Li & Guo, 2016a, 2016b). Additionally, the extraction efficiency of ASE could be improved using smaller particle size samples (grinding), drying (reduction of moisture), or incorporation of a desiccant. In addition, ASE could be operated with one or combinations of two organic solvents during the extraction process depending on the polarity of the compound of interest (Santos-Buelga et al., 2012). Other researchers found that acidification could yield better results in terms of flavonoids specifically anthocyanins in comparison with methanol alone (Ponmozhi, Geetha, Saravana Kumar, & Suganya Devi, 2011). Furthermore, ASE has been employed in the extraction process of phenolic mixtures from grape pomace, rosemary, marjoram, oregano, essential oil from lamiaceae plant, and edible oil from different oilseeds (Hossain, Barry-Ryan, Martin-Diana, & Brunton, 2011; Matthäus & Brühl, 2001; Rajha et al., 2014; Rodríguez-Solana, Salgado, Domínguez, & Cortés-Diéguez, 2015). Low solvent use, speed and convenience, high yields, and the low extraction costs per sample makes ASE one of the best methods for green extraction (Table 4).

4.2.5 Subcritical water extraction (SWE)

Subcritical water extraction is one of the eco-friendliest techniques by using water as the extraction solvent to extract bioactive compounds, essential oils, and organic pollutants (Yu, Zhu, Zhong, Li, & Ma, 2015). This technique uses water at its subcritical stage where water is forced to gain the liquid state beyond its boiling point and below its supercritical point (374°C, 3205 psi) (Plaza & Turner, 2015). This process is also known as hot water extraction, pressurized (hot) water extraction, pressurized low polarity water extraction, high-temperature water extraction, superheated water extraction, or hot-liquid water extraction. It is a very promising "green" technique using water as the only extraction solvent (Li & Guo, 2016a, 2016b). The unique solvent properties of water make it ideal for extracting both polar and nonpolar compounds, depending on extraction temperature.

Generally, temperatures between 100°C and 374°C (as the critical point of water is 374°C at 3200 psi) are used together with pressure $<$ 3200 psi, to maintain water in the liquid state (Li & Guo, 2016a, 2016b). Room temperature (25°C), however, has also been used for SWE. The major drawback with water as an extractant is its high polarity. Increasing the temperature of water (over 100°C) significantly changes the physical and chemical properties as well as the dielectric constant of water (Teo, Tan, Yong, Hew, & Ong, 2010). In addition, higher temperatures would facilitate the disruption of hydrogen bonding thereby, making dispersion forces more dominant (Plaza & Turner, 2015). Hence, the ability of water to dissolve less polar compounds including flavanols at an elevated temperature makes it a favorable solvent for extraction compared to other conventional methods (Santos-Buelga et al., 2012).

Further, with the increase in temperature permittivity, the thickness and surface tension of the water are reduced, leading to an increase in its diffusivity characteristics (Teo et al., 2010). This changes the dielectric constant of water from 78.5 at 25°C to 34.8 at 200°C, which is in close proximity to the dielectric constant of methanol ($\varepsilon = 33$) (Table 5). By varying the temperature, water could be used to mimic a methanol–water mixture or ethanol-water mixtures or acetone ($\varepsilon = 21$)

TABLE 5 Chemical and physical properties of water at different temperatures and saturation pressures.

Property	25°C 0.1 MPa	100°C 0.1 MPa	200°C 1.5 MPa	350°C 17 MPa
Dissociation constant, K_w	1.0×10^{-14}	5.6×10^{-13}	4.9×10^{-12}	1.2×10^{-12}
pK_w	13.99	12.25	11.31	11.92
Dipole moment	1.85	1.85	1.85	1.85
Density (g/cm^3)	0.997	0.958	0.865	0.579
Relative static permittivity, ε_r	78.5	55.4	34.8	14.1

(Plaza & Turner, 2015). This is a major advantage of using SWE, where water could be applied to impersonate different solvent systems. Additionally, it could help to reduce or replace different environmentally harmful solvents.

This method has been applied for environment protection to extract organic pollutant polycyclic aromatic hydrocarbons (PAH), to extract volatile compounds from plant materials, to identify the pesticide content from soil, to recover polyphenol compounds including catechins and proanthocyanidins in wine industry (García-Marino, Rivas-Gonzalo, Ibáñez, & García-Moreno, 2006), to extract lignans and associated compounds from flaxseed meal (Kanmaz, 2014), and to extract canola meal phenolic compounds (Hassas-Roudsari et al., 2009).

A study conducted by Ibañez et al. (2003) found that SWE extracted most active antioxidant compounds including carnosol, rosmanol, carnosic acid, methyl carnosate, and flavonoids such as cirsimaritin and genkwanin. Kim and Mazza (2006) later reported that the extraction efficiency of SWE was improved at high temperatures and relatively higher NaOH concentrations. They further reported that extraction temperatures ranging from 140°C to 160°C extracted higher yields of lignans and other bioactive peptides (Cacace & Mazza, 2006). Moreover, García-Marino et al. (2006) demonstrated that SWE extracts had greater antioxidant activity compared to both ethanol extracts and ultrasound-assisted extracts (UAE), and were similar to Soxhlet extracts.

The higher capital required for operating/pressurized system, however, could be the major drawback of this application (Liang & Fan, 2013). Furthermore, the operation costs of SWE are relatively high compared to those for other extraction techniques. Nevertheless, the use of SWE for extracting antioxidants is a feasible option as the purity and extraction efficiency of the phenolic antioxidants will compensate for the higher extraction costs (Ramos, Kristenson, & Brinkman, 2002). Since water as an extraction material is cheap, renewable, and nontoxic, SWE has a huge potential for scaling up in the future.

Thus, the use of SWE in the extraction of antioxidants seems to be a feasible option since price will not be affected due to its purity and the extraction efficiency of the phenolic antioxidants (Ramos et al., 2002). Extraction of thermo-labile compounds is another problem associated with this technique. Nevertheless, the use of water as the extraction material is cheap, renewable, and nontoxic, and thus has a huge potential for scaling up in near future.

4.2.6 Supercritical fluid extraction (SFE)

Supercritical fluid extraction (SFE) is a distinct technique, which is quite similar to subcritical water extraction. This extraction technique has been extensively applied, including natural phenolic compounds. Interest in SFE has increased over time due to the high quality of the final product, the absence of toxic residues, and the preservation of the active ingredients (Pereira & Meireles, 2010). Thus, supercritical fluids are very useful for both the extraction of bio-actives and the removal of undesirable compounds. Furthermore, this method can be programmed (at a specific pressure and temperature) to extract specific compounds (Pereira & Meireles, 2010). In reality, any type of solvent can become a supercritical solvent. Thus, cost, toxicity, technical feasibility, and salvation power could determine the most appropriate solvent for exact function (Pereira & Meireles, 2010).

CO_2, however, is the most commonly used in SFE due to its relatively lower cost, availability, and nontoxic nature. In addition, propane, ethane, hexane, pentane, and butane are some other types of solvents that can also be applied in SFE as cosolvents. It has been shown that cosolvents increase the solubility in oil, making them very important in improving the efficacy of this process (Hussain, 2014). These cosolvents form hydrogen bonds with solutes of oil and increase the density of supercritical carbon dioxide, creating elevated solubility of the solvent in solutes of crude oil. Furthermore, the effect of cosolvents solubility followed the order of ethanol > methanol > propanol > acetone (Hussain, 2014). Of these, however, methanol and ethanol are the most commonly used cosolvents in SFE. Moreover, it was found that using cosolvents such as 10% ethanol improved the SFE extraction of the phenolic compounds (HadiNezhad, Rowland, & Hosseinian, 2015). A comparative study between SFE and conventional extraction showed that while there was no difference between the total extraction of phenolic compounds, the concentrations of individual phenolic compounds were between four and 10 times greater using SFE (Pinelo et al., 2007). Other studies have found that prehydrolyzation of the seeds improved the solubility of the matrix which resulted in higher extraction yields (Comin, Temelli, & Saldaña, 2011). These researchers also found that combined mode of static/dynamic extraction (running and soaking in CO_2) further increased the overall extraction yield and shortened the total extraction time. These studies indicated that there are many important factors affecting extracting efficiency of the SFE process including the design, position of the solute, particle size, shape, and porosity of the solid material, moisture content of the solid material, solvent flow rate, temperature, and pressure (Pereira & Meireles, 2010).

4.3 Application of green technology for the canola industry

Plant secondary metabolites, including phenolic compounds, are a mixture of different classes. The complexity of the phenolics can affect their extraction efficiency and yield requiring additional steps to exclude unwanted substances when isolating specific phenolic compounds (Santos-Buelga et al., 2012). The most common methods of isolation include solid-phase extraction (SPE), column chromatography (CC), and droplet countercurrent chromatography (DCCC) (Santos-Buelga et al., 2012). Coupling an isolation technique with green extraction could facilitate better extraction efficiencies for the oilseed industry. Thus, capital investment and a restricted range of applications make green extraction limiting. Yet many small-scale pilot versions have been applied by the oilseed industry for measuring the feasibility and the extraction efficiencies of green and eco-friendly extraction techniques. Canola/rapeseed, being major oilseeds, have been widely studied with respect to their applicability to green technology, particularly their phenolic compounds. For example, UAE extraction of rapeseed resulted in better extraction efficiency and higher total phenolic content compared to MAE (Szydłowska-Czerniak & Tułodziecka, 2014; Yang et al., 2013). These studies nevertheless confirmed that ability of both ultrasound or microwave technology to produce a phenolic antioxidant as a potential nutraceutical from canola by-products (Yang et al., 2013).

The application of SWE has also proven to be an effective technology for producing greener extracts compared to conventional methods of hydro-distillation (Khajenoori, Asl, Hormozi, Eikani, & Bidgoli, 2009). Temperature, pressure, and solvent modifier intensities are some factors that affect the solubility of the phenolic compounds during the subcritical extraction process (Hussain, 2014). Application of SFE and SWE to rapeseed has both demonstrated that controlling the extraction parameters could facilitate extraction of individual phenolic compounds, including sinapic acid (HadiNezhad et al., 2015).

Of the abovementioned techniques discussed so far, ASE is the most economical and eco-friendly green method for extraction of phenolic compounds from oilseeds. For instance, 10g of oilseed sample can be extracted in about 10–12min using the ASE system, compared to 24h using Soxhlet extraction (Luthria et al., 2004). Furthermore, elevated temperatures facilitate the release of targeted analytes by providing thermal energy to overcome interactions (hydrogen bonds, dipole attraction, etc.) (Galano et al., 2011). The ability of the phenolic compounds to withstand the higher temperatures makes it an effective method for their extraction from oilseeds.

5 Summary and conclusion

The above illustrations indicate the importance of green technology toward the oilseed industry as well as its applications. The advantages of extraction of phenolic compounds over conventional methods make green extraction one of the most applicable methods of extraction of phenolic compounds on oilseeds. Furthermore, current interest in environmental conservation has gained attention in many fields including the research field. Many studies have been conducted to find and apply research techniques related to green technology in different applications. At the time of writing, the oilseed industry elutes a lot of waste chemicals as well as harmful substances in the process of oil refining. This has been a major issue until now and industry is still searching for an alternative method that could reduce the carbon footprint and aid in reducing production costs. The literature data presented in this review demonstrated that alternative methods including MAW, ASE, USE, SWE, and SFE could be of interest. However, the major disadvantage of all these alternative techniques is the high capital or investments associated with them compared to the conventional methods. Many instances the higher capital and operational costs have been mistaken with the novel techniques that cannot contend with traditional methods. Adding this to facts that the above techniques involve clean technology, we could expect green extraction to prosper in many fields, including the oilseed industry, in the near future.

References

Aachary, A., Thiyam-Hollander, U., & Eskin, N. A. M. (2015). Applied food protein chemistry. In Z. Ustanol (Ed.), *Canola/rapeseed proteins and pepsides* (pp. 193–218): Wiley-Blackwell.

Abuzaytoun, R., & Shahidi, F. (2006). Oxidative stability of flax and hemp oils. *Journal of the American Oil Chemists' Society*, *83*, 855–861.

Ajila, C. M., Brar, S. K., Verma, M., Tyagi, R. D., Godbout, S., & Valéro, J. R. (2011). Extraction and analysis of polyphenols: Recent trends. *Critical Reviews in Biotechnology*, *31*, 227–249.

Akpan, N. (2012). From agriculture to petroleum oil production: What has changed about Nigeria's rural development? *International Journal of Developing Societies*, *1*, 97–106.

Alam, M. A., Subhan, N., Hossain, H., Hossain, M., Reza, H. M., Rahman, M. M., et al. (2016). Hydroxycinnamic acid derivatives: A potential class of natural compounds for the management of lipid metabolism and obesity. *Nutrition & Metabolism*, *13*, 27.

Alkan, D., Tokatli, F., & Ozen, B. (2012). Phenolic characterization and geographical classification of commercial extra virgin olive oils produced in Turkey. *Journal of the American Oil Chemists' Society*, *89*, 261–268.

Amarowicz, R., Naczk, M., & Shahidi, F. (2000). Antioxidant activity of various fractions of non-tannin phenolics of canola hulls. *Journal of Agricultural and Food Chemistry*, *48*(7), 2755–2759. https://doi.org/10.1021/jf991160.

Azargohar, R., Nanda, S., Rao, B. V. S. K., & Dalai, A. K. (2013). Slow pyrolysis of deoiled canola meal: Product yields and characterization. *Energy & Fuels*, *27*, 5268–5279.

Barros, F., Dykes, L., Awika, J. M., & Rooney, L. W. (2013). Accelerated solvent extraction of phenolic compounds from sorghum brans. *Journal of Cereal Science*, *58*, 305–312.

Beejmohun, V., Fliniaux, O., Grand, É., Lamblin, F., Bensaddek, L., Christen, P., et al. (2007). Microwave-assisted extraction of the main phenolic compounds in flaxseed. *Phytochemical Analysis*, *18*, 275–282.

Bogdanov, M. G. (2014). *Ionic liquids as alternative solvents for extraction of natural products* (pp. 127–166). Berlin Heidelberg: Springer.

Cacace, J. E., & Mazza, G. (2006). Pressurized low polarity water extraction of lignans from whole flaxseed. *Journal of Food Engineering*, *77*, 1087–1095.

Cai, R., Arntfield, S. D., & Charlton, J. L. (1999). Structural changes of sinapic acid and sinapine bisulfate during autoclaving with respect to the development of colored substances. *Journal of the American Oil Chemists' Society*, *76*, 433–441.

Capote, F. P., & De Castro, M. D. L. (2006). Ultrasound in analytical chemistry. *Analytical and Bioanalytical Chemistry*, *387*, 249–257.

Carabias-Martínez, R., Rodríguez-Gonzalo, E., Revilla-Ruiz, P., & Hernández-Méndez, J. (2005). Pressurized liquid extraction in the analysis of food and biological samples. *Journal of Chromatography A*, *1089*, 1–17.

Carré, P., & Pouzet, A. (2014). Rapeseed market, worldwide and in Europe. *OCL*, *21*, D102.

Casseus, L. (2009). *Canadian agriculture at a glance: Canola: A Canadian success story* [Internet document]. https://www150.statcan.gc.ca/n1/pub/96-325-x/2007000/article/10778-eng.htm. (Accessed 29 March 2019).

Chemat, F., Vian, M. A., & Cravotto, G. (2012). Green extraction of natural products: Concept and principles. *International Journal of Molecular Sciences*, *13*, 8615–8627.

Chen, Y., Thiyam-Hollander, U., Barthet, V. J., & Aachary, A. A. (2014). Value-added potential of expeller-pressed canola oil refining: Characterization of sinapic acid derivatives and tocopherols from byproducts. *Journal of Agricultural and Food Chemistry*, *62*, 9800–9807.

Cmolik, J., & Pokorny, J. (2000). Physical refining of edible oils. *European Journal of Lipid Science and Technology*, *102*, 472–486.

Comin, L. M., Temelli, F., & Saldaña, M. A. (2011). Supercritical CO_2 extraction of flax lignans. *Journal of the American Oil Chemists' Society*, *88*, 707–715.

Daun, J. K., Eskin, N. A. M., & Hickling, D. (2011). *Canola: Chemistry, production, processing, and utilization*. AOCS Press.

DeBonte, L., Iassonova, D., Liu, L., & Loh, W. (2012). Commercialization of high oleic canola oils. *Lipid Technology*, *24*, 175–177.

Gaber, M. A. F. M., Trujillo, F. J., Mansour, M. P., Taylor, C., & Juliano, P. (2019). Megasonic-assisted aqueous extraction of canola oil from canola cake. *Journal of Food Engineering*, *252*, 60–68.

Galano, A., Francisco-Márquez, M., & Alvarez-Idaboy, J. R. (2011). Canolol: A promising chemical agent against oxidative stress. *Journal of Physical Chemistry B*, *115*, 8590–8596.

Gallego-Juárez, J. A., Rodriguez, G., Acosta, V., & Riera, E. (2010). Power ultrasonic transducers with extensive radiators for industrial processing. *Ultrasonics Sonochemistry*, *17*, 953–964.

García-Marino, M., Rivas-Gonzalo, J. C., Ibáñez, E., & García-Moreno, C. (2006). Recovery of catechins and proanthocyanidins from winery by-products using subcritical water extraction. *Analytica Chimica Acta*, *563*, 44–50.

Garcia-Salas, P., Morales-Soto, A., Segura-Carretero, A., & Fernández-Gutiérrez, A. (2010). Phenolic-compound-extraction systems for fruit and vegetable samples. *Molecules*, *15*, 8813–8826.

Ghazani, S. M., & Marangoni, A. G. (2013). Minor components in canola oil and effects of refining on these constituents: A review. *Journal of the American Oil Chemists' Society*, *90*, 923–932.

Gogate, P. R., & Kabadi, A. M. (2009). A review of applications of cavitation in biochemical engineering/biotechnology. *Biochemical Engineering Journal*, *44*, 60–72.

Gunstone, F. D. (2011). *Vegetable oils in food technology: Composition, properties and uses* (2nd ed., pp. 107–136).

HadiNezhad, M., Rowland, O., & Hosseinian, F. (2015). The fatty acid profile and phenolic composition of *Descurainia sophia* seeds extracted by supercritical CO_2. *Journal of the American Oil Chemists' Society*, *92*, 1379–1390.

Hassas-Roudsari, M., Chang, P. R., Pegg, R. B., & Tyler, R. T. (2009). Antioxidant capacity of bioactives extracted from canola meal by subcritical water, ethanolic and hot water extraction. *Food Chemistry*, *114*, 717–726.

Hossain, M. B., Barry-Ryan, C., Martin-Diana, A. B., & Brunton, N. P. (2011). Optimisation of accelerated solvent extraction of antioxidant compounds from rosemary (*Rosmarinus officinalis* L.), marjoram (*Origanum majorana* L.) and oregano (*Origanum vulgare* L.) using response surface methodology. *Food Chemistry*, *126*, 339–346.

Hussain, S. (2014). Solubility of crude oil in pure and modified supercritical carbon dioxide using supercritical fluid extractor. *Academia Journal of Scientific Research*, *2*, 38–49.

Ibañez, E., Kubátová, A., Señoráns, F. J., Cavero, S., Reglero, G., & Hawthorne, S. B. (2003). Subcritical water extraction of antioxidant compounds from rosemary plants. *Journal of Agricultural and Food Chemistry*, *51*, 375–382.

Jambrak, A. R., Herceg, Z., Šubarić, D., Babić, J., Brnčić, M., Brnčić, S. R., et al. (2010). Ultrasound effect on physical properties of corn starch. *Carbohydrate Polymers*, *79*, 91–100.

Juliano, P., Bainczyk, F., Swiergon, P., Supriyatna, M. I. M., Guillaume, C., Ravetti, L., et al. (2017). Extraction of olive oil assisted by high-frequency ultrasound standing waves. *Ultrasonics Sonochemistry*, *38*, 104–114.

Juliano, P., Kutter, A., Cheng, L. J., Swiergon, P., Mawson, R., & Augustin, M. A. (2011). Enhanced creaming of milk fat globules in milk emulsions by the application of ultrasound and detection by means of optical methods. *Ultrasonics Sonochemistry*, *18*, 963–973.

Jun, H.-I., Wiesenborn, D. P., & Kim, Y.-S. (2014). Antioxidant activity of phenolic compounds from canola (*Brassica napus*) seed. *Food Science and Biotechnology*, *23*, 1753–1760.

Kanmaz, E. Ö. (2014). Subcritical water extraction of phenolic compounds from flaxseed meal sticks using accelerated solvent extractor (ASE). *European Food Research and Technology*, *238*, 85–91.

Kentish, S., Wooster, T. J., Ashokkumar, M., Balachandran, S., Mawson, R., & Simons, L. (2008). The use of ultrasonics for nanoemulsion preparation. *Innovative Food Science & Emerging Technologies*, *9*, 170–175.

Khajenoori, M., Asl, A. H., Hormozi, F., Eikani, M. H., & Bidgoli, H. N. (2009). Subcritical water extraction of essential oils from Zataria multiflora bioss. *Journal of Food Process Engineering*, *32*, 804–816.

Kiani, H., Zhang, Z., Delgado, A., & Sun, D.-W. (2011). Ultrasound assisted nucleation of some liquid and solid model foods during freezing. *Food Research International*, *44*, 2915–2921.

Kim, J.-W., & Mazza, G. (2006). Optimization of extraction of phenolic compounds from flax Shives by pressurized low-polarity water. *Journal of Agricultural and Food Chemistry*, *54*, 7575–7584.

Koubaa, M., Mhemdi, H., Barba, F. J., Roohinejad, S., Greiner, R., & Vorobiev, E. (2016). Oilseed treatment by ultrasounds and microwaves to improve oil yield and quality: An overview. *Food Research International*, *85*, 59–66.

Kozlowska, H., Rotkiewicz, D. A., Zadernowski, R., & Sosulski, F. W. (1983). Phenolic acids in rapeseed and mustard. *Journal of the American Oil Chemists' Society*, *60*, 1119–1123.

Krishnaswamy, K., Orsat, V., Gariépy, Y., & Thangavel, K. (2013). Optimization of microwave-assisted extraction of phenolic antioxidants from grape seeds (*Vitis vinifera*). *Food and Bioprocess Technology*, *6*, 441–455.

Li, H., Deng, Z., Wu, T., Liu, R., Loewen, S., & Tsao, R. (2012). Microwave-assisted extraction of phenolics with maximal antioxidant activities in tomatoes. *Food Chemistry*, *130*, 928–936.

Li, J., & Guo, Z. (2016a). Concurrent extraction and transformation of bioactive phenolic compounds from rapeseed meal using pressurized solvent extraction system. *Industrial Crops and Products*, *94*, 152–159.

Li, J., & Guo, Z. (2016b). Identification and quantification of phenolic compounds in rapeseed originated lecithin and antioxidant activity evaluation. *LWT—Food Science and Technology*, *73*, 397–405.

Liang, X., & Fan, Q. (2013). Application of sub-critical water extraction in pharmaceutical industry. *Journal of Materials Science and Chemical Engineering*, *01*, 1–6.

Liazid, A., Guerrero, R. F., Cantos, E., Palma, M., & Barroso, C. G. (2011). Microwave assisted extraction of anthocyanins from grape skins. *Food Chemistry*, *124*, 1238–1243.

Luthria, D. L. (2006). Application of green chemistry principles for extraction of phytolipids and phenolic compounds. *Indian Journal of Chemistry*, *45*, 2291–2296.

Luthria, D., Vinjamoorib, D., Noelc, K., & Ezzelld, J. (2004). *Accelerated solvent extraction* [Internet document]. USDA-Agricultral Research Service. URL. https://www.ars.usda.gov/research/publications/publication/?seqNo115=160306. (Accessed 29 March 2019).

Manadal, V., Mohan, Y., & Siva, H. (2007). Microwave assisted extraction—An innovative and promising extraction tool for medicinal plant research. *Pharmacognosy Reviews*, *1*, 7–18.

Matthäus, B., & Brühl, L. (2001). Comparison of different methods for the determination of the oil content in oilseeds. *Journal of the American Oil Chemists' Society*, *78*, 95–102.

Morley, K. L., Grosse, S., Leisch, H., Lau, P. C. K., Boudet, A.-M., Rosazza, J. P. N., et al. (2013). Antioxidant canolol production from a renewable feedstock via an engineered decarboxylase. *Green Chemistry*, *15*, 3312.

Mustafa, A., & Turner, C. (2011). Pressurized liquid extraction as a green approach in food and herbal plants extraction: A review. *Analytica Chimica Acta*, *703*, 8–18.

Naczk, M., Amarowicz, R., Sullivan, A., & Shahidi, F. (1998). Current research developments on polyphenolics of rapeseed/canola: A review. *Food Chemistry*, *62*, 489–502.

Nićiforović, N., & Abramovič, H. (2014). Sinapic acid and its derivatives: Natural sources and bioactivity. *Comprehensive Reviews in Food Science and Food Safety*, *13*, 34–51.

Obied, H. K., Song, Y., Foley, S., Loughlin, M., Rehman, A., Mailer, R., et al. (2013). Biophenols and antioxidant properties of Australian canola meal. *Journal of Agricultural and Food Chemistry*, *61*, 9176–9184.

Pereira, C. G., & Meireles, M. A. A. (2010). Supercritical fluid extraction of bioactive compounds: Fundamentals, applications and economic perspectives. *Food and Bioprocess Technology*, *3*, 340–372.

Perrier, A., Delsart, C., Boussetta, N., Grimi, N., Citeau, M., & Vorobiev, E. (2017). Effect of ultrasound and green solvents addition on the oil extraction efficiency from rapeseed flakes. *Ultrasonics Sonochemistry*, *39*, 58–65.

Pinelo, M., Ruiz-Rodríguez, A., Sineiro, J., Señoráns, F. J., Reglero, G., & Núñez, M. J. (2007). Supercritical fluid and solid–liquid extraction of phenolic antioxidants from grape pomace: A comparative study. *European Food Research and Technology*, *226*, 199–205.

Plaza, M., & Turner, C. (2015). Pressurized hot water extraction of bioactives. *TrAC Trends in Analytical Chemistry*, *71*, 39–54.

Ponmozhi, P., Geetha, M., Saravana Kumar, M., & Suganya Devi, P. (2011). Extraction of anthocyanin and analysing its antioxidant properties from *Pithecellobium dulce* fruit pericarp. *Asian Journal of Pharmaceutical and Clinical Research*, *4*, 41–45.

Rajha, H. N., Ziegler, W., Louka, N., Hobaika, Z., Vorobiev, E., Boechzelt, H. G., et al. (2014). Effect of the drying process on the intensification of phenolic compounds recovery from grape pomace using accelerated solvent extraction. *International Journal of Molecular Sciences*, *15*, 18640–18658.

Ramos, L., Kristenson, E., & Brinkman, U. A. T. (2002). Current use of pressurised liquid extraction and subcritical water extraction in environmental analysis. *Journal of Chromatography A*, *975*, 3–29.

Rodríguez-Solana, R., Salgado, J. M., Domínguez, J. M., & Cortés-Diéguez, S. (2015). Comparison of Soxhlet, accelerated solvent and supercritical fluid extraction techniques for volatile (GC-MS and GC/FID) and phenolic compounds (HPLC-ESI/MS/MS) from *Lamiaceae* species. *Phytochemical Analysis*, *26*, 61–71.

Santos, D. T., Veggi, P. C., & Meireles, M. A. A. (2012). Optimization and economic evaluation of pressurized liquid extraction of phenolic compounds from jabuticaba skins. *Journal of Food Engineering*, *108*, 444–452.

Santos-Buelga, C., Gonzalez-Manzano, S., Dueñas, M., & Gonzalez-Paramas, A. M. (2012). Extraction and isolation of phenolic compounds. *Methods in Molecular Biology (Clifton, NJ)*, *864*, 427–464. https://doi.org/10.1007/978-1-61779-624-1_17.

Stalikas, C. D. (2007). Extraction, separation, and detection methods for phenolic acids and flavonoids. *Journal of Separation Science*, *30*, 3268–3295.

Szydłowska-Czerniak, A., & Tułodziecka, A. (2014). Antioxidant capacity of rapeseed extracts obtained by conventional and ultrasound-assisted extraction. *Journal of the American Oil Chemists' Society*, *91*, 2011–2019.

Takeuchi, T. M., Pereira, C. G., Braga, M. E. M., Maróstica, M. R., Leal, P. F., Angela, M., et al. (2009). Low-pressure solvent extraction (solid–liquid extraction, microwave assisted, and ultrasound assisted) from Condimentary plants. In A. Meireles (Ed.), *Exctracting bioactive compounds for food products*. Boca Raton, FL: CRC Press.

Teo, C. C., Tan, S. N., Yong, J. W. H., Hew, C. S., & Ong, E. S. (2010). Pressurized hot water extraction (PHWE). *Journal of Chromatography A*, *1217*, 2484–2494.

Tiwari, B. K. (2015). Ultrasound: A clean, green extraction technology. *TrAC Trends in Analytical Chemistry*, *71*, 100–109.

Veggi, P. C., Martinez, J., & Meireles, M. A. A. (2012). *Fundamentals of microwave extraction* (pp. 15–52): Boston, MA: Springer:

Wanasundara, U. N., & Shahidi, F. (1994). Canola extract as an alternative natural antioxidant for canola oil. *Journal of the American Oil Chemists' Society*, *71*, 817–822.

Xu, L., & Diosady, L. (2012). Processing of canola proteins. In *Canola and rapeseed production, processing, food quality, and nutrition* (pp. 59–78): CRC Press.

Yang, M., Huang, F., Liu, C., Zheng, C., Zhou, Q., & Wang, H. (2013). Influence of microwave treatment of rapeseed on minor components content and oxidative stability of oil. *Food and Bioprocess Technology*, *6*, 3206–3216.

Yu, X., Zhu, P., Zhong, Q., Li, M., & Ma, H. (2015). Subcritical water extraction of antioxidant phenolic compounds from XiLan olive fruit dreg. *Journal of Food Science and Technology*, *52*, 5012–5020.

Chapter 9

Cold pressed paprika (*Capsicum annuum*) seed oil

Alessandra Durazzo[a], Massimo Lucarini[a], Antonio Raffo[a], Annalisa Romani[b], Silvia Urciuoli[b], Roberta Bernini[c], Antonello Santini[d], Anja I. Lampe[e], and Johannes Kiefer[e]

[a]*CREA-Research Centre for Food and Nutrition, Rome, Italy*, [b]*DiSIA-PHYTOLAB (Pharmaceutical, Cosmetic, Food supplement Technology and Analysis), University of Florence, Florence, Italy*, [c]*Department of Agriculture and Forest Science (DAFNE), University of Tuscia, Viterbo, Italy* [d]*Department of Pharmacy, University of Napoli Federico II, Napoli, Italy*, [e]*Technische Thermodynamik, Universität Bremen, Bremen, Germany*

1 Introduction

In recent years, the demand of consumers for cold pressed oil as an alternative to conventional products has been increasing. This is probably due to the perception of these oils as natural, nutritional, and safe food products with better nutritive/healthy properties (Fine et al., 2015). Cold pressed oils are easy to obtain. The process does not involve any heat treatment or solvent extraction. Therefore, the cold pressing technique results in a product that is free of organic solvents. The mechanical pressing does not require much energy and it well meets the needs of small and medium-sized companies. Celenk, Gumus, Ustun Argon, Buyukhelvacigil, and Karasulu (2018) underlined how this technology is cost-effective, less labor-intensive, safer, simpler, and more environmentally friendly compared to the other commonly used oil extraction procedures. Moreover, cold pressed seed oils, being not refined, contain more lipophilic phytochemicals. A conventional oil processing procedure might remove them during refining. Therefore, the benefits of the seed oils, as well as their nutritive properties, are documented (Keskin Çavdar, 2019; Thanonkaew, Wongyai, McClements, & Decker, 2012; Yang et al., 2018). For instance, Siger, Nogala-Kalucka, and Lampart-Szczapa (2008) studied the content and antioxidant activity of phenolic compounds in cold pressed plant oils. They concluded that cold pressed oils may retain higher levels of natural antioxidants. In addition, they exhibit acceptable shelf stability and improved safety without the need to add any synthetic antioxidants to increase shelf life and time usage. Yang et al. (2018) underlined how different types of vegetable oils have their own specific advantages and biological activities, and that appropriate vegetable oils could be accordingly selected to meet individual needs. Rotkiewicz, Konopka, and Zylik (1999) evidenced that low productivity and difficulties in obtaining a product of constant quality represent disadvantages of this process.

2 Extraction and processing of cold pressed oil

Paprika is defined in the world market as a ground, bright red, usually nonpungent powder derived from dried fruits of several varieties of *Capsicum annuum* L., having dark red color and a characteristic flavor. In many instances, the term paprika is used to designate also pepper fruit of the varieties of *Capsicum annuum* L. The paprika seeds contain 20–25% (*w*/w) of fatty oils. Besides, within the seeds broken fragments of pods can usually be found together with ribs, out of which carotenoids are dissolved into the oil during pressing increasing its value. Based on physical, chemical, and organoleptic investigations of the oil obtained by pressing, a study by Domokos, Bernáth, and Perédi (1993) has shown that the paprika seed oil is acceptable also for diathetic, medicinal, and cosmetic purposes.

There are only a few studies in the literature concerning the cold extraction of paprika seed oil, but more information regarding the solvent extraction technique. The extraction solvent most commonly employed in the industry is *n*-hexane. Generally, for complete extraction of dehydrated tissue, it is necessary to repeat the process several times in order to obtain an acceptable yield. The discontinuous extraction with recycling of the extraction solvent is the most widely used approach. The main disadvantage of this method is that the solvent needs to be evaporated after the extraction. This causes the thermal degradation of the carotenoids. Moreover, at the end of the process, the solvent residue concentration has to be reduced to below legal limits. This sometimes requires even more drastic process conditions.

Cold Pressed Oils. https://doi.org/10.1016/B978-0-12-818188-1.00009-8
Copyright © 2020 Elsevier Inc. All rights reserved.

Further studies have shown the applicability of supercritical carbon dioxide as an extraction medium to increase the yield in paprika seed oil production (Jarén-Galàn, Nienaber, & Schwartz, 1999). A recent work evaluated and compared the methods used to extract red pepper seed oils, namely: Soxhlet, cold pressing, supercritical carbon dioxide, and microwave-assisted extraction. The red pepper seeds composition (on a dry-weight basis) was determined to be ash 3.05%, water 6.63%, oil 18.39%, protein 28.33%, and total carbohydrate 43.60%. The predominant fatty acids were linoleic, palmitic, and oleic acids. The highest and lowest contents of linoleic acid (76.54%) were found in microwave-assisted (76.54%) and Soxhlet (73.65%) extracted pepper seed oils, respectively. γ-tocopherol was the main tocopherol at 278.65 mg/100 g seed oil, followed by α-tocopherol and δ-tocopherol (Chouaibi, Rezig, Hamdi, & Ferrari, 2019). There is a discrepancy with previously reported data referred to the cold pressing technique. The chemical composition of the seeds is reported as 26.10% oil, 6.25% moisture, and 76.65% dried extracted meal. The dried meal was found to contain 28.92% protein, 29.10% fiber, 5.61% ash, and 36.37% N-free extract (Bosland & Votava, 2000). In a recent study, the seed composition has been reported as 70.95% moisture and 29.05% dry matter. In detail, the dry matter consisted of 19.28% crude protein, 19.57% crude lipids, 4.88% ash, and 56.28% carbohydrates. In addition, the mineral and amino acid composition of the seeds have been reported in the same study. They contained 2.88 g/100 g phytic acid, 2.89 g/100 g tannins, and 2.20 mg/g trypsin inhibitors (Embaby & Mokhtar, 2011). A study conducted in 2015 describes in detail the cold extraction procedure as reproduced on a laboratory scale for the extraction process, for which a cold press machine (12 kg seed/h capacity) was employed. Constant pressing parameters were selected as a total of 10 mm die exit, 40 rpm screw rotation speed, and maximum exit temperature of 40°C. After collection and weighing of the oil (liquid phase) and meal-oily cake (solid phase), the suspended materials in the oil have to be removed by filtration through a 40 mm screen. In the next step, a refrigerated centrifuge separates the oil phases at 6797 g for 10 min at 10°C. The separated clear upper oil phase was stored into amber-colored and capped glasses. The weighed and nitrogen flushed samples were stored in a fridge until analysis. The research group (Yılmaz, Sevgi Aarsunar, Aydeniz, & Güneşer, 2015) calculated the oil yield in percentage of kg oil per kg seeds. It was in the range 5.12%–6.65% respect to 13.57% obtained by solvent extraction and 18.4% obtained using the supercritical carbon dioxide technique (Li, Song, You, & Sun, 2011).

3 Compounds of nutritional and nutraceutical interest in cold pressed paprika seed oil

Cold pressed spice paprika seed oil is rich in unsaturated fatty acids, in particular, linoleic acid (70.79%–74.31%, *w*/w), while saturated acids are present at lower concentration. Palmitic acid and stearic acid, the most abundant saturated acids, are present only at amounts of 11.08%–12.20% and 3.10%–3.75% *w*/w, respectively (Koncsek, Helyes, & Daood, 2017; Kostadinović Veličkovska et al., 2018). Because of the high content of linoleic acid, paprika seed oil can play an important role in human health. It is known to help to prevent the onset of serious diseases like diabetes, cancer, and cardiovascular diseases (Jandacek, 2017; Teixeira dos Santos et al., 2018; Zhao & Schooling, 2019). The chemical structures of the main fatty acids present in cold pressed paprika seed oil are reported in Fig. 1.

Koncsek et al. (2017) reported the industrial production of cold pressed spice paprika seed oil by evaluating the effect of raw material variety and growing season factors on the fatty acid profile, carotenoid, and tocopherol contents. As evidenced by its shiny deep red color, a considerable amount of carotenoids is present (from 629.35 to 848.39 μg/g) and, in particular, of β-carotene, lutein, and capsantin present in greater amounts. Their chemical structures are shown in Fig. 2.

Among the tocopherols, γ-tocopherol was the main compound (57.85–83.57 mg/100 g) and α-tocopherol was present in low amounts (4.50–16.41 mg/100 g). On the other hand, β-tocopherol and δ-tocopherol were not present (Koncsek et al., 2017). Fig. 3 reports the chemical structures of α-and γ-tocopherol. The high level of γ-tocopherol is responsible for the

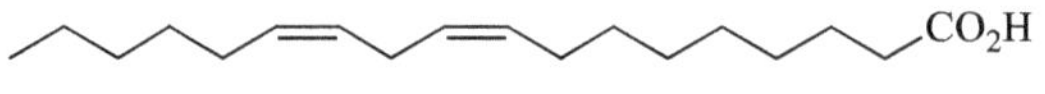

CO_2H

Linoleic acid 18:2 (9c,12c)

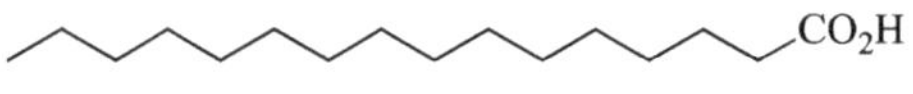

CO_2H

Palmitic acid 16:0

CO_2H

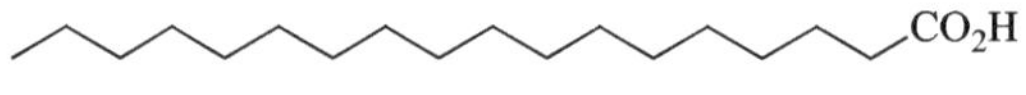

Stearic acid 18:0

FIG. 1 Chemical structures of the main fatty acids in cold-pressed paprika seed oil.

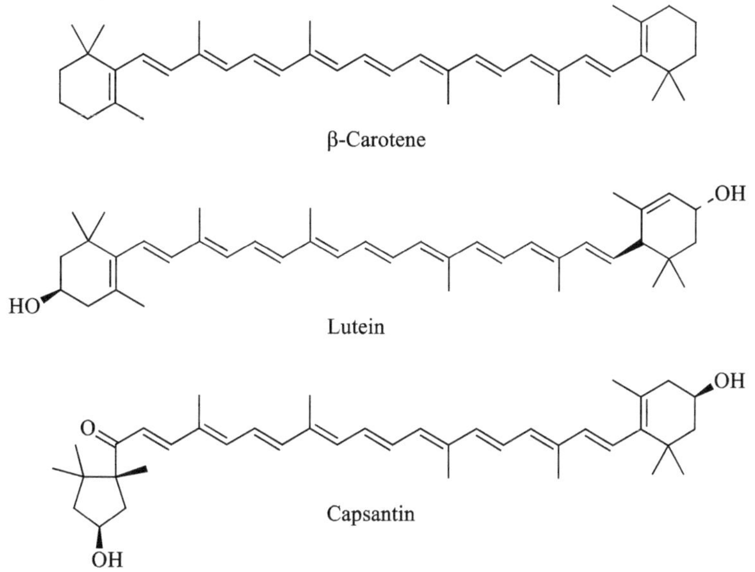

FIG. 2 Chemical structure of β-carotene, lutein, and capsantin.

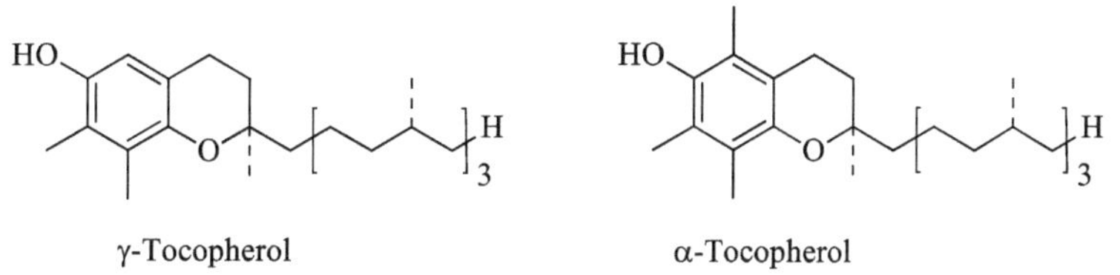

FIG. 3 Chemical structures of γ-and α-tocopherol.

good stability of paprika seed oils and their efficiency against oxidative damage, inducing a low incidence of cardiovascular diseases and prostate cancer. In their conclusion, Koncsek et al. (2017) underlined how cold pressing followed by gravitational sedimentation without other cleaning or refining processes seemed to be suitable to preserve the valuable components of spice paprika seeds oil.

Generally, a bioactive component can be defined as a compound that occurs in nature, is part of the food chain, and has the ability to interact with one or more compounds in living tissue showing an effect on human health (Biesalski et al., 2009). Antioxidant properties are an expression of the interactions between bioactive molecules and other components of a food matrix (Durazzo & Lucarini, 2018). Regarding antioxidants, the review of Yeung et al. (2019), which was based on scientific literature landscape analysis of published studies since 1991, reported that a transition of scientific interest has been observed from research focused on antioxidant vitamins and minerals into more research attention dedicated to phytochemicals having antioxidant properties.

Kostadinović Veličkovska et al. (2018) evaluated the in vitro antioxidant activity of some traditional and nontraditional cold pressed edible oils from Macedonia such as "sweet" and "bitter" apricot kernel, paprika seed, and sesame seed oils by three assays (DPPH, β-carotene, and TEAC). This study evidenced that sesame seed oil and paprika seed oils are the richest sources of phenolic compounds (total phenolic content: 117.4 mg gallic acid/L oil) with linoleic acid and γ-tocopherol as the main bioactive compounds (69.6 g and 25.7 mg per 100 g of oil, respectively). Based on this study, the TEAC assay is the best method for the determination of the antioxidant activity of paprika seed oil (97.9 mg Trolox/L oil).

The assessment of the interaction of biologically active compounds in terms of antioxidant properties represents the first step for evaluation of health properties and nutraceutical applications (Durazzo et al., 2018). As reported by Santini, Tenore, and Novellino (2017), Santini et al. (2018), and Santini and Novellino (2018), the term "nutraceutical," which is derived from the words "nutrient" and "pharmaceutical," was originally defined by Stephen De Felice (De Felice, 1995). It indicates "a food or part of a food that provides medical or health benefits, including the prevention and/or treatment of a disease." Recently this definition was updated to reflect the health beneficial properties required from a nutraceutical to distinguish it clearly from food supplements. In particular, they can be defined as follows: (i) for food of vegetal origin, a nutraceutical is the phytocomplex; and (ii) for food of animal origin, a nutraceutical is the pool of secondary metabolites. Both are concentrated and administered in the proper pharmaceutical form. They are capable of providing beneficial health effects, including the prevention and/or the treatment of a disease substantiated by clinical data (Santini et al., 2018).

In general, however, studies regarding these properties directly linked to cold pressed paprika seed oil are relatively scarce. A recent study evaluated the hypolipidemic effect and mechanism of paprika seed oil (PSO) on Sprague-Dawley rats, observing how this effect may be attributed to the inhibition of lipid synthesis via suppressing the expression of HMG-CoAR, CYP7A1, and FAS, and meanwhile promoting the metabolism and excretion of lipids via upregulating the expression of LDLR, HSL, TRPV1, and PPARα (Chen, Ding, Song, & Kan, 2017).

4 Flavor volatile compounds

While the volatile profiles of paprika and paprika oleoresins have been investigated (Fernández-García & Pérez-Gálvez, 2010), little information is available on the volatile compounds present in cold pressed oil extracted from pepper fruit seeds or in the seed itself. The only paper reporting on the volatile profile of cold pressed oil obtained from pepper seeds was that by Yılmaz et al. (2015). They used the seeds of capia-type red pepper, a pepper (*Capsicum annuum* L.) type commonly used in Turkey, as an oil source. In this study, cold pressed oil was extracted from cleaned and dried pepper seeds, but also from the same seeds after roasting or an enzyme treatment. Forty volatile compounds were identified and quantified in the three types of oils. Interestingly, the volatile profile of cold pressed pepper seed oil was quite similar to that one observed for green pepper fruit (Luning, de Rijk, Wichers, & Roozen, 1994). The monoterpene (*E*)-β-ocimene was by far the major compound of the volatile fraction. Other monoterpenes, such as (*E,Z*)- and (*Z,Z*)-2,6-dimethyl-2,4,6-octatriene, limonene, and (*Z*)-β-ocimene, were present at lower levels. Other compounds present at intermediate levels in green pepper fruits, such as 1-hexanol, 2-heptanone, heptanal, and nonanal, were also found in this oil. The main difference with respect to the green pepper fruit profile was the higher level of hexanal. This compound is likely formed during the drying process of the seeds. The absence of cyclic olefins, such as toluene and xylenes, and of the ester 6-methylheptyl-2-propenoate, were further differences observed. The cold pressed oil extracted from roasted seeds was characterized by the marked increase of some pyrazines (2-methylpyrazine and 2,5-dimethylpyrazine) and phenols (guaiacol and 4-vinylguaiacol), resulting from Maillard reactions promoted by the roasting treatment. On the contrary, the oil obtained from seeds previously subjected to an enzyme treatment with hemicellulase and protease had a volatile profile quite similar to that one of the untreated seeds.

The above oils were evaluated through sensory analysis and determination of consumer acceptability. While the sensory profile of the three oils did not show significant differences for the nine evaluated sensory attributes, it was dominated by undesirable aroma notes, described as vegetable, soil, and woody aroma. This resulted in low acceptance of these oils in the consumer study. It is plausible that the presence of off-odorants at a low level in the oil did cause this negative effect on sensory quality with respect to the pleasant smell of fresh green pepper fruit. This result, according to the authors, highlighted the need for additional treatment of physical refining or deodorization to obtain an oil with adequate consumer acceptance. The authors suggested that cold pressing was not an appropriate technique for producing pepper seed oil due to the low sensory quality of the oil and the low extraction yield. A more suitable strategy would be based on solvent extraction to accomplish a higher yield in combination with a refining treatment to improve the oil sensory quality.

In relation to this last remark, another study obtained a pepper seed oil by subcritical butane extraction applied to roasted seeds purchased from the local market in China (Gu et al., 2017). They reported a quite different volatile profile. The major compound was limonene, instead of (*E*)-β-ocimene, whereas relatively high levels of cyclic olefins (1,4-xylene and toluene) and products of fatty acids degradation (decadienal) denoted marked differences from the profile reported above for the cold pressed oil. However, in this study a different separation technique was used for the analysis of the volatile compounds, involving a relatively strong thermal treatment on the oil sample. Consequently, it is not possible to attribute unambiguously the observed differences in the reported volatile profile to differences in the oil extraction technique, rather than to differences in the composition of the dried seeds used, the conditions of the roasting process, or the analytical technique for volatile determination. In any case, in this study no evaluation of the sensory quality of the obtained oil was performed. The volatile profiles reported in these two studies may be compared to that one determined directly on powdered seeds of two pepper varieties of *Capsicum annuum* L., named "Reus long pairal" and "Sweet Italian" (Silva, Azevedo, Pereira, Valentão, & Andrade, 2013). In this case, the main compound was the ester methyl benzoate, not reported among the main volatiles in the fruit, but structurally related to toluene, one of the main pepper fruit volatiles (Luning et al., 1994). Other chemical groups, in order of concentration, are aldehydes (hexanal, benzaldehyde), alcohols (1-coten-3-ol), and monoterpenes ((*Z*)-β-ocimene). Interestingly, 2-isobutyl-3-methoxypyrazine, one of the most important odorant in the fruit, was also found in the seeds of the variety "Sweet Italian." Part of the differences with respect to the profiles observed in the oils may be due to the different release of volatiles from an oily matrix (Gu et al., 2017; Yılmaz et al., 2015) or an aqueous solution, as carried out in the study of Silva et al. (2013). In addition, it is worth mentioning that also the extent of cleaning of the seeds used for oil extraction is an important factor for the sensory quality of the final oil.

In the abovementioned study by Koncsek et al. (2017), cold pressed pepper seed oils were obtained by applying industrial processing procedures on a starting material represented by pepper dried pods. The authors described that in the applied industrial process, the seeds, stems, and pericarp were separated from the previously crushed dried pods with the help of an industrial sorting sieving machine. Particles of the pericarp and vein remained attached to the seeds. As a result, the obtained oils cold pressed from these partially cleaned seeds were transparent with a shiny deep red color, due to the considerable amounts of carotenoids present in the pericarp and vein particles. The authors also reported that these cold pressed oils were characterized by a pleasant smell and flavor representing the mild aroma of the dried spice paprika, which was presumably due to the contribution of compounds extracted from the pericarp and vein particles to the flavor of the oil. Thus, even if the volatile profile was not characterized, it is plausible that the analyzed oil contained certain amounts of the odorants generally contributing to the aroma of the dried spice paprika. They included 2-isobutyl-3-methoxypyrazine, 2-methylpropanal, 2- and 3-methylbutanal, 2,3-butanedione, 1-penten-3-one, hexanal, 1-octen-3-one, dimethyl trisulfide, 1-octen-3-o, 2,4-heptadienal, and 2,4-nonadienal (Van Ruth, Roozen, Cozijnsen, & Posthumus, 1995).

5 Edible and nonedible applications of cold-pressed oil

Recently, cold pressed vegetable oils as ecological and natural products have started to be used in the cosmetic market. For instance, Ligęza, Wyglądacz, Tobiasz, Jaworecka, and Reich (2016) suggested that natural cold pressed oils can be applied to the skin as cosmetics; they do not have an irritating effect and only rarely cause allergic reactions.

6 Infrared spectroscopy combined with chemometrics applied to cold pressed oil

The analysis of oils is usually carried out using chromatographic techniques such as gas chromatography (GC) and high-performance liquid chromatography (HPLC). In the context of cold pressed paprika seed oil, Koncsek et al. (2017) used GC for the fatty acids and HPLC for the carotenoids. Silva et al. (2013) determined sterols and triterpenes using HPLC with a diode array detector, organic acids using HPLC with a UV detector, fatty acids using GC coupled with mass spectrometry (GC/MS), and volatile compounds using GC/MS after headspace solid-phase micro-extraction. Chromatography methods, however, share the disadvantage of being time-consuming and expensive.

The application of spectroscopic techniques with chemometric data analysis can be a solution to this problem. In particular, FTIR spectroscopy is a promising tool for the analysis of food products and entire food matrices such as full meals; see, for example, Durazzo et al. (2018). From an experimental viewpoint, attenuated total reflection FTIR (ATR-FTIR) is usually the method of choice to analyze food samples. It requires virtually no sample preparation if individual ingredients such as oil are to be analyzed. For the measurement, a droplet of the oil is placed on the ATR crystal and the data can be recorded immediately. Recording a spectrum with reasonable signal-to-noise ratio and sufficient spectral resolution and range takes only a few seconds. The resulting spectra represent a molecular fingerprint of the sample. Owing to the molecular complexity of oils, however, sophisticated data analysis tools must be employed to extract the desired information. Common methods include principal component analysis (PCA), hierarchical cluster analysis, support vector machines (SVM), artificial neural networks (ANNs), and partial least square regression (PLSR). FTIR spectroscopy coupled with these multivariate statistical data analysis tools has enormous potential to facilitate the measurement of quality attributes, but also for the rapid authentication and detection of adulteration of the oil.

Regarding paprika seeds, spectroscopic analysis has been reported by Simonovska et al. (2019). They analyzed samples from the pericarp, the placenta, the seeds, and the stalk of red hot pepper (*Capsicum annuum* L.) using ATR-FTIR and NMR spectroscopy. For this purpose, they performed an extraction with *n*-hexane and supercritical carbon dioxide before the analysis. It was shown that both NMR and FTIR spectroscopy provide useful information on the triacylglycerols composition, the degree of saturation, and the capsaicin content. FTIR spectroscopy was found particularly useful for identification of capsaicinoids in the extracts. In order to highlight the potential of FTIR spectroscopy, e.g., to distinguish between compounds typical for paprika seed oil, Fig. 4 illustrates the spectra (acquired using a Nicolet iS10 FT-IR spectrometer equipped with a diamond crystal ATR cell in the range of 4000–500 cm^{-1} at resolution of 4 cm^{-1}, in our laboratory) of the two C18 fatty acids that were found in the oil at reasonably high amounts: linoleic and stearic acids. They exhibit very similar chemical structures and only distinguish themselves by two CC double bonds, as can be seen in Fig. 1. This small difference results in a characteristic signature marked by an arrow in the diagram. This peak is a result

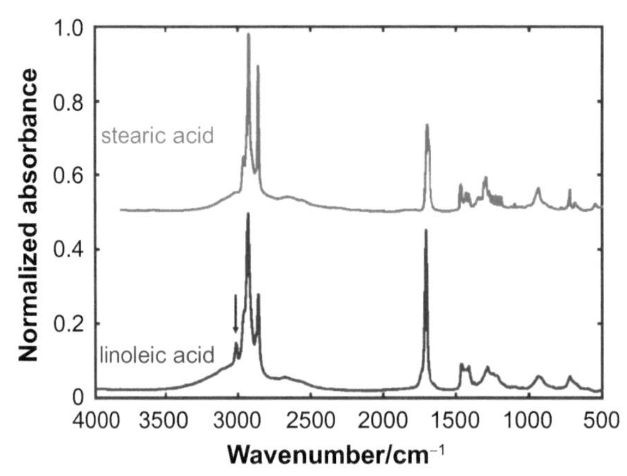

FIG. 4 FTIR spectra (acquired using a Nicolet iS10 FT-IR spectrometer equipped with a diamond crystal ATR cell in the range of 4000–500 cm^{-1} at resolution of 4 cm^{-1}) of unsaturated and saturated C18 fatty acids, linoleic and stearic acids, respectively. The spectra plotted are based on parameters from the NIST Standard Reference Database 69: NIST Chemistry WebBook.

of =C—H stretching vibrations. Published work has even shown that unsaturated fatty acids, in which the position of the C=C double bonds are different can be distinguished with the naked eye (Kiefer et al., 2010). This highlights the potential of using FTIR, in particular when the data are analyzed with the help of chemometric tools.

References

Biesalski, H. K., Dragsted, L. O., Elmadfa, I., Grossklaus, R., Müller, M., Schrenk, D., et al. (2009). Bioactive compounds: Definition and assessment of activity. *Nutrition, 25*, 1202–1205.

Bosland, P. W., & Votava, E. J. (2000). *Peppers: Vegetable and spice capsicums*. New York (USA): CABI Publishing.

Celenk, V., Gumus, Z. P., Ustun Argon, Z., Buyukhelvacigil, M., & Karasulu, E. (2018). Analysis of chemical compositions of 15 different cold pressed oils produced in Turkey: A case study of tocopherol and fatty acid analysis. *Journal of the Turkish Chemical Society Section A: Chemistry, 5*(1), 1–18.

Chen, X., Ding, Y., Song, J., & Kan, J. (2017). Hypolipidaemic effect and mechanism of paprika seed oil on Sprague-Dawley rats. *Journal of the Science of Food and Agriculture, 97*(12), 4242–4249.

Chouaibi, M., Rezig, L., Hamdi, S., & Ferrari, G. (2019). Chemical characteristics and compositions of red pepper seed oils extracted by different methods. *Industrial Crops and Products, 128*, 363–370.

De Felice, S. L. (1995). The nutraceutical revolution: Its impact on food industry R&D. *Trends in Food Science and Technology, 6*, 59–61.

Domokos, J., Bernáth, J., & Perédi, J. (1993). Examination of Hungarian paprika (*Capsicum annuum*) seed oils. *Acta Horticulturae, 331*, 49–52.

Durazzo, A., D'Addezio, L., Camilli, E., Piccinelli, R., Turrini, A., Marletta, L., et al. (2018). From plant compounds to botanicals and Back: A current snapshot. *Molecules, 23*(8). pii: E1844.

Durazzo, A., & Lucarini, M. (2018). A Current shot and re-thinking of antioxidant research strategy. *Brazilian Journal of Analytical Chemistry, 5*(20), 9–11. https://doi.org/10.30744/brjac.2179-3425.2018.5.20.9-11.

Embaby, H. E., & Mokhtar, S. M. (2011). Chemical composition and nutritive value of lantana and sweet pepper seeds and nabak seed kernels. *Journal of Food Science, 76*, 736–741.

Fernández-García, E., & Pérez-Gálvez, A. (2010). Flavouring compounds in red pepper fruits (*Capsicum genus*) and processed products. In Y. H. Hui (Ed.), *Handbook of fruit and vegetable flavours* (pp. 909–934). Hoboken, New Jersey: John Wiley and Sons, Inc.

Fine, F., Brochet, C., Gaud, M., Carre, P., Simon, N., Ramli, F., et al. (2015). Micronutrients in vegetable oils: The impact of crushing and refining processes on vitamins and antioxidants in sunflower, rapeseed, and soybean oils. *European Journal of Lipid Science and Technology, 118*(5), 680–697.

Gu, L. B., Pang, H. L., Lu, K. K., Liu, H. M., Wang, X. D., & Qin, G. Y. (2017). Process optimization and characterization of fragrant oil from red pepper (*Capsicum annuum* L.) seed extracted by subcritical butane extraction. *Journal of the Science of Food and Agriculture, 97*(6), 1894–1903.

Jandacek, R. J. (2017). Linoleic acid: A nutritional quandary. *Healthcare (Basel), 5*(2), 25.

Jarén-Galàn, M., Nienaber, U., & Schwartz, S. J. (1999). Paprika (*Capsicum annuum*) oleoresin extraction with supercritical carbon dioxide. *Journal of Agricultural and Food Chemistry, 47*, 3558–3564.

Keskin Çavdar, H. (2019). Active compounds, health effects, and extraction of unconventional plant seed oils. In M. Ozturk, & K. Hakeem (Eds.), *Plant and human health, volume 2*. Cham: Springer.

Kiefer, J., Noack, K., Bartelmess, J., Walter, C., Dörnenburg, H., & Leipertz, A. (2010). Vibrational structure of the polyunsaturated fatty acids eicosapentaenoic acid and arachidonic acid studied by infrared spectroscopy. *Journal of Molecular Structure, 965*, 121–124.

Koncsek, A., Helyes, L., & Daood, H. G. (2017). Bioactive compounds of cold pressed spice paprika seeds oils. *Journal of Food Processing & Preservation, 42*(3), e13403.

Kostadinović Veličkovska, S., Catalin Moţ, A., Mitrev, S., Gulaboski, R., Brühl, L., Mirhosseini, H., et al. (2018). Bioactive compounds and "in vitro" antioxidant activity of some traditional and non-traditional cold-pressed edible oils from Macedonia. *Journal of Food Science and Technology*, *55*(5), 1614–1623.

Li, G., Song, C., You, J., & Sun, Z. (2011). Optimisation of red pepper seed oil extraction using supercritical CO_2 and analysis of the composition by reversed-phase HPLC-FLD-MS/MS. *International Journal of Food Science and Technology*, *2011*(46), 44–51.

Ligęza, M., Wygłądacz, D., Tobiasz, A., Jaworecka, K., & Reich, A. (2016). Natural cold pressed oils as cosmetic products. *Family Medicine and Primary Care Review*, *18*(4), 443–447.

Luning, P. A., de Rijk, T., Wichers, H. J., & Roozen, J. P. (1994). Gas chromatography, mass spectrometry, and sniffing port analyses of volatile compounds of fresh bell peppers (*Capsicum annuum*) at different ripening stages. *Journal of Agricultural and Food Chemistry*, *42*, 977–983.

Rotkiewicz, D., Konopka, I., & Zylik, S. (1999). *State of works on the rapeseed oil processing optimalization. I. Oil obtaining. Rośliny Oleiste/oilseed crops XX*: (pp. 151–168). [in Polish].

Santini, A., Cammarata, S. M., Capone, G., Ianaro, A., Tenore, G. C., Pani, L., et al. (2018). Nutraceuticals: Opening the debate for a regulatory framework. *British Journal of Clinical Pharmacology*, *84*(4), 659–672.

Santini, A., & Novellino, E. (2018). Nutraceuticals: Shedding light on the grey area between pharmaceuticals and food. *Expert Review of Clinical Pharmacology*, *11*(6), 545–547.

Santini, A., Tenore, G. C., & Novellino, E. (2017). Nutraceuticals: A paradigm of proactive medicine. *European Journal of Pharmaceutical Sciences*, *96*, 53–61.

Siger, A., Nogala-Kalucka, M., & Lampart-Szczapa, E. (2008). The content and antioxidant activity of phenolic compounds in cold pressed plant oils. *Journal of Food Lipids*, *15*, 137–149.

Silva, L. R., Azevedo, J., Pereira, M. J., Valentão, P., & Andrade, P. B. (2013). Chemical assessment and antioxidant capacity of pepper (*Capsicum annuum* L.) seeds. *Food and Chemical Toxicology*, *53*, 240–248.

Simonovska, J. M., Yancheva, D. Y., Mikhova, B. P., Momchilova, S. M., Knez, Ž. F., Primožić, M. J., et al. (2019). Characterization of extracts from red hot pepper (*Capsicum annuum* L.). *Bulgarian Chemical Communications*, *51*, 103–112.

Teixeira dos Santos, A. L., Duarte, C. K., Santos, M., Zoldan, M., Almeida, J. C., Gross, J. L., et al. (2018). Low linolenic and linoleic acid consumption are associated with chronic kidney disease in patients with type 2 diabetes. *PLoS One*. https://doi.org/10.1371/journal.pone.0195249.

Thanonkaew, A., Wongyai, S., McClements, D. J., & Decker, E. A. (2012). Effect of stabilization of rice bran by domestic heating on mechanical extraction yield, quality, and antioxidant properties of cold pressed rice bran oil (*Oryza sativa* L.). *LWT- Food Science and Technology*, *48*(2), 231–236.

Van Ruth, S. M., Roozen, J. P., Cozijnsen, J. L., & Posthumus, M. A. (1995). Volatile compounds of rehydrated French beans, bell peppers and leeks. Part II. Gas chromatography/sniffing port analysis and sensory evaluation. *Food Chemistry*, *54*(1), 1–7.

Yang, R., Zhang, L., Li, P., Yu, L., Mao, J., Wang, X., et al. (2018). A review of chemical composition and nutritional properties of minor vegetable oils in China. *Trends in Food Science & Technology*, *74*, 26–32.

Yeung, A. W. K., Tzvetkov, N. T., El-Tawil, O. S., Bungău, S. G., Abdel-Daim, M. M., & Atanasov, A. G. (2019). Antioxidants: Scientific literature landscape analysis. *Oxidative Medicine and Cellular Longevity*, *2019*, 8278454.

Yılmaz, E., Sevgi Aarsunar, E., Aydeniz, B., & Güneşer, O. (2015). Cold pressed capia pepper seed (*Capsicum annuum* L.) oils: Composition, aroma and sensory properties. *European Journal of Lipid Science and Technology*, *117*(7), 1016–1026.

Zhao, J. V., & Schooling, M. C. (2019). Effect of linoleic acid on ischemic heart disease and its risk factors: A Mendelian randomization study. *BMC Medicine*, *17*, 61.

Chapter 10

Cold pressed sesame (*Sesamum indicum*) oil

Muhammad Imran[a], Muhammad Kamran Khan[a], Muhammad Ali[a], Muhammad Nadeem[b], Zarina Mushtaq[a], Muhammad Haseeb Ahmad[a], Muhammad Sajid Arshad[a], Nazir Ahmad[a], and Muhammad Abdul Rahim[a]

[a]*Institute of Home and Food Sciences, Faculty of Life Sciences, Government College University, Faisalabad, Pakistan,* [b]*Department of Dairy Technology, University of Veterinary and Animal Sciences, Lahore, Pakistan*

Abbreviations

ALA Alpha linolenic acid
CP Cold pressing
CPO Cold pressed oils
CPSSO Cold pressed sesame seed oil
DPPH• Diphenyl-β-picrylhydrazyl
EFAs Essential fatty acids
HDL High-density lipoprotein
HP Hot-press
LDL Low-density lipoprotein
MUFAs Monounsaturated fatty acids
OS Oilseed
PAHs Polycyclic aromatic hydrocarbons
PBDEs Polybrominated biphenyl-ethers
PCBs Polychlorinated-biphenyls
PUFAs Polyunsaturated fatty acids
SS Sesame seeds
SSO Sesame seed oil
UPLC Ultra-performance liquid chromatography

1 Introduction

Sesame seed oil (SSO) is an edible vegetable oil derived from sesame seeds (SS) (*Sesamum indicum L.*) through cold pressing and filtering. SSO can be produced from black-hulled and white dehulled SS. SSO has excessive content of oil (approximately 50%) with superior oxidative stability in comparison with other vegetable oils, both at storage temperatures and during frying (Salunkhe, Chavan, Adsule, & Kadam, 1991). Sesame is one of the most commonly eaten foods either in fresh or processed. It is also an important component of many processed food products with its excellent flavor, attractive color, and high levels of many macro- and micro-nutrients. SSO is extensively used in food processing and nutraceutical industries in many different countries due to its high content of oil, protein, and resistance against oxidation. Besides being used as cooking oil in South India, it is used as a flavor enhancer in Middle Eastern, African, and Southeast Asian cuisines due to its nutty taste and aroma (Goldberg, Novotny, Kieffer, Mor, & Thiele, 1995). SSO is a rich dietary source of vitamin E and essential fatty acids (EFAs) (omega-3 and omega-6) sterols, tocopherols, and tocotrienols concentration (Xu, Chen, & Hu, 2005). Vitamin E levels in foods are of great importance to determine the ingestion of nutrients within a community. Due to the deficiency of subject within the field of hot-press (HP) method for the extraction of SSO on an industrial scale, the principle of cold pressing (CP) is the optimization of SSO extraction factors such as temperature and moisture contents, to achieve oil yield and meal with high quality and SSO rich in sesamin, sesamolin, and sesaminol lignan fractions, which are known to play a vital role in its oxidative stability and antioxidative activity (Rostami, Farzaneh, Boujmehrani, Mohammadi, & Bakhshabadi, 2014).

Cold Pressed Oils. https://doi.org/10.1016/B978-0-12-818188-1.00010-4
Copyright © 2020 Elsevier Inc. All rights reserved.

2 History of sesame oil

India, Burma, China, and Sudan are vital manufacturers of sesame, accounting for approximately 60% of its worldwide cultivation (Elleuch, Besbes, Roiseux, Blecker, & Attia, 2007). For thousands of years, the seeds of sesame have been used to treat several diseases in Asian kingdoms. The intake of SS reported several health-promoting effects. For example, γ-tocopherol plasma considerably rises and improves vitamin E action, which is supposed to inhibit human aging-linked syndromes like heart disease and cancer. Literature reports indicated that the addition of sesame in food could enhance antioxidant capacity (Elleuch, Bedigian, & Zitoun, 2011). SS is one of the world's most essential and earliest crops, used for edible oil extraction (Abou-Gharbia, Shehata, & Shahidi, 2000). The sesamum is a genus belonging to the Pedaliaceae family, which consists of 16 genera plus 60 species (Hassan, 2012). SS have also high protein content (Anilakumar, Pal, Khanum, & Bawa, 2010). The main manufacturers of SSO globally are South Africa, Mediterranean countries, Ethiopia, Mexico, Sudan, India, Guatemala, Nigeria, Japan, and Myanmar (Salunkhe et al., 1991). Cold pressing methods or normal oil-refining methods are being used for SO production. North America and Europe presented cold pressed SSO (CPSSO) in health and food shops. CPSSO is pale yellow, which differs from Indian SO or East Asia dark brown oil obtained from roasted SS. Conventional refined SO is light yellow in color, odorless, and clear (Biswas, Sana, Badal, & Huque, 2001).

3 Extraction and purification of cold pressed oil

The plant material is purified from broken and damaged seeds through purification or separation techniques. It is mandatory to remove solid impurities from fresh cloudy oil that was extracted by pressing process so that its shelf life and further processing may be fulfilled to obtain a pure end product. Sedimentation is the key process to remove solid impurities after the pressing technique. Before pressing for superior quality of extracted oil and useable cake production, the preconditioning of seeds is done at less than 50°C. Cold pressed (CP) extraction is performed through a Komet single screw oil presser. For 17 days chlorophyll, free fatty acid (FFA) and phosphorous contents are increased, which causes oxidation of oil at higher rates. To avoid these degrading processes, the cold pressing process temperature should be maintained at less than 45°C (Kostadinović Veličkovska, Brühl, Mitrev, Mirhosseini, & Matthäus, 2015).

4 Optimization of the extraction process

Oils could be recovered by natural ways, and sometimes with additives or chemical materials (not commonly used). The CP technique has become the most suitable approach for extraction of seed oils. This technique is reasonably user-friendly and cheaper than commercial techniques for oil extraction. The supercritical CO_2 technique for oil extraction is the most environmentally friendly, but the major drawback of this process compared to CP is its significant capital investment required (Gecgel et al., 2015). To obtain a superior quality of extracted SSO, mechanical CP (less than 45°C) and a filtration process are used for quality improvement. Oil quality can be checked by two common parameters: peroxide value and FFAs content (El Khier, Ishag, & Yagoub, 2008). According to Fukuda and Namiki (1988), the traditional procedure for SO extraction is cleaning, dehulling, roasting, grinding, and oil extraction. Yen and Shyu (1989) reported that roasting is a critical point which affects the composition, organoleptic properties, color, and rate of oxidation of extracted oil.

Solvent extraction is used by the seed oil pressing industry. The solvents mainly used are petroleum ether, *n*-hexane, and petroleum benzene. In experimental conditions, propane and CO_2 act as solvents for extraction of SSO. The extracted yield of oil is calculated from the relation between the essential oil mass obtained and the raw material mass used in the extraction. The extracted yield of oil was determined by keeping pressurized solvents fixed for comparison among different experimental conditions. The total extraction time differed due to different experimental conditions, which are set for complete extraction with compressed liquid solvents. Extraction time and temperature were used for determining the extraction percentage in comparison with the *n*-hexane Soxhlet extraction yield, and the solubility was determined by the linear part of the extraction curve (mass of oil and mass of solvent). The extraction with CO_2 solvent is slower than propane solvent extraction (Corso et al., 2010). The extractions were performed in a laboratory scale unit in a temperature and pressure range of 313–333 K and 19–25 MPa for carbon dioxide and 303–333 K and 8–12 MPa for propane extractions, respectively. The result data indicated that solvent and density were principally functional variables for the extraction of oil using supercritical CO_2 while the temperature variable was noted important for the maximum yield efficiency for oil extraction using propane solvent. Propane was a much quicker solvent for vegetable oils extraction by the solvent extraction method, and CO_2 has a poor performance compared to propane. Characteristics including oxidative stability and chemical profile of present FAs were studied by DSC and gas chromatography. Essential phytonutrients present in seed oils can be extracted using an efficient method like CP. Phytocompounds like EFAs, alpha tocopherols, and phytosterols showed poor quality when other techniques like chemical extraction and HP extraction were applied, because of their process's deodorization and refining (Ananth, Deviram, Mahalakshmi, Sivasudha, & Tietel, 2019). Extraction of oilseed (OS) can be performed by cold/lower temperatures through oil pressing applications. The final product yield depends solely on the raw material source, type, and the application/technique of extraction. During hot or higher temperature oil extraction, fundamental nutrients and biologically active nutrients are denatured (García, Ruiz-Méndez, Romero, & Brenes, 2006). Valuable FAs and compounds like tocopherols are present in high degrees in CP edible oils (Butinar, Bučar-Miklavčič, Mariani, & Raspor, 2011).

5 Fatty acid and acyl lipids profile of cold pressed oil

SS contains about 40%–60% oil, which is mainly composed of monounsaturated fatty acids (MUFAs) and polyunsaturated fatty acids (PUFAs). SSO contain monosaturated fat (oleic acid) and polyunsaturated fat (linoleic acid); the remaining oil content is primarily the saturated fat, palmitic acid, and stearic acid, accounting for almost 96% of total FAs (Elleuch et al., 2011). The physical properties of oil showed the state to be liquid at room temperature. SSO contain high levels of unsaturated fatty acids, especially oleic and linoleic acid, at up to 38.8% and 46.2%, respectively. SSO can be classified into oleic acid and linoleic acid oil groups. Physicochemical analysis was carried out on SSO and the results showed that seed oil contains the dominant saturated fatty acid (SFAs), usually palmitic (8.58%) and stearic (5.44%), wherein the oil can be useful as edible oils and for industrial applications (Nzikou et al., 2009). The major FAs composition of the cold pressed sesame seed oil (CPSSO) was determined using gas chromatography. The FAs generally found are monosaturated in all samples, such as myristic acid (C14:0), palmitic acid (C16:0), stearic acid (C18:0), oleic acid (C18:1), and linoleic acid (C18:2). Ranges of oleic acid and palmitic acid in peanut, neem, and SSO samples were between 31.8% and 49.6%, whereas maximum palmitic acid was present in neem (Ananth et al., 2019). Edible SSO contains SFAs, MUFAs, and PUFAs. Levels of bioactive compounds in SSO such as phytosterol, polyphenol, lignan, triterpene, carotenoid, chlorophyll, tocopherols, and tocotrienols were also reported (Ghanbari, Anwar, Alkharfy, Gilani, & Saari, 2012). The PUFAs are classified into ω-(3, 6 and 9). ω-(3 and 6) and are known as EFAs, "because they're attained only through foods consumption and cannot be synthesized by means of mammals on their very own" (Assies et al., 2004). PUFAs are used for prevention and treatment of predominant depressive illness and heart diseases (Assies et al., 2004). PUFAs such as linoleic and linolenic acids could increase high-density lipoprotein (HDL) cholesterol (good cholesterol), while decreasing low-density lipoprotein (LDL) cholesterol (bad cholesterol). In addition, these FAs also reduce bad cholesterol without affecting good cholesterol levels (Lichtenstein et al., 2006). In a report by the American Heart Association, LDL cholesterol was reduced by the consumption of nutrient-rich oil.

6 Roasted and unroasted cold pressed sesame oil

1. The FAs structures of unroasted (without any treatment) and roasted SSO are composed of similar quantities of oleic and linoleic acids and a smaller quantity of alpha linolenic acid (ALA).
2. Roasted SSO showed higher resistance toward instability than unroasted SSO. It seems that inactivation of reactive biological enzymes (lipase plus lipoxygenase) through heat treatment does not distinctively improve the resistance toward oxidation of SSO.
3. Antioxidant capacity toward rancidity of unroasted SSO was 157.8 mg Trolox/kg oil, and for roasted SSO was 114.9 mg Trolox/kg oil.
4. Total phytosterols quantities in unroasted and roasted SSO were similar.
5. Roasted SSO gave the highest trends of oxidative stability, which were prominently higher than those of unroasted SSO (Kostadinović Veličkovska et al., 2015) (Table 1).

7 Antioxidant activity

SO are naturally enriched with different bioactive substances. Other than precious unsaturated fatty acids, these oils are richer with more natural antioxidants, which are tocopherols and phenolic components, than their refined complements (Van Hoed et al., 2006). To resist auto-oxidation in cold pressed sesame oils, antioxidants and bioactive molecules contribute toward a long and stable shelf life. The shelf life of cold pressed sesame oils is generally limited and dependent due to PUFAs content, principally α-linolenic acid and the amount of antioxidants, wherein the normal range of stability is 6–12 months (Choe & Min, 2006). The high amount of PUFAs and some substances enhance the oil decomposition at very high temperature and reduced the use of oils for frying purpose. Frying also harmfully affects acceptability to users of final products in term of color and taste (Warner, 2008). Antioxidant ability, FAs profile, and tocopherols of six cold pressed oils (CPO) (sesame, peanut, castor, neem, iluppai, and coconut) were determined. Lipophilic extracts revealed good α,α-

TABLE 1 The main bioactive compounds present in sesame seed oil.

Bioactive component	Category of component	Quantity
Plant sterol	Campestanol/$C_{28}H_{50}O$	0.01–0.02 mg/g
	Sitostanol	0.02–0.04 mg/g
	Avenasterol/$C_{29}H_{48}O$	0.6–0.8 mg/g
	Stigmasterol/$C_{29}H_{48}O$	0.3–0.4 mg/g
	Beta-sitosterol/$C_{29}H_{50}O$	1.9–2.6 mg/g
Polyunsaturated fatty acids	Palmitic acid/$C_{16}H_{32}O_2$	12.3%–14.4%
	Oleic acid/$C_{18}H_{34}O_2$	40%–50%
	Linoleic acid/$C_{18}H_{32}O_2$	0.3%–45%
	Linolenic acid/$C_{18}H_{30}O_2$	0.6%–0.8%
Tocopherol	γ-tocopherol/$C_{28}H_{48}O_2$	0.4–0.6 mg/g
Lignans	Sesamin/$C_{20}H_{18}O_6$	5–6 mg/g
	Sesamolin/$C_{20}H_{18}O_7$	1.9–2.4 mg/g

diphenyl-β-picrylhydrazyl (DPPH·) free radical scavenging potential compared to hydrophilic extract. Lipophilic extract of SSO showed free radical scavenging activity and high antioxidant ability. The existence of different tocopherols (α, γ, and δ) was studied by ultra-performance liquid chromatography (UPLC) analysis and counted in the six CPO samples. γ-tocopherol was mostly present in peanut oil (104.1 mg/kg), then in SSO (278.7 mg/kg), and castor oil (395.3 mg/kg). The total content of vitamin E active lipophilic antioxidants occurring in alpha and gamma tocopherols contributed to the oxidation-resistant capacity of CPSSO (Lerma-García, Herrero-Martínez, Simó-Alfonso, Mendonça, & Ramis-Ramos, 2009; Tuberoso, Kowalczyk, Sarritzu, & Cabras, 2007). CPSSO target the immune system to prevent chronic disease (Lin, Zhong, & Santiago, 2017).

The OS samples analysis by GCMS showed FAs composition such as myristic, linoleic, stearic, palmitic, and oleic acids. The maximum level of SFAs (85.5%) present in copra oil, MUFA (64.0%) in castor seed oil, and PUFA (31.7%) in SSO were determined. The phytoconstituents from CPO play an encouraging role to improve human health without any side effects (Ananth et al., 2019). Currently, consumers prefer nutritionally safe and naturally rich foods with phytonutrients, which prevent disease and maintain the immune system. Free radicals are formed and accumulated due to the environment, changes in food consumption, physical activities, aging, and emotional stress. Many degenerative diseases are caused by free radicals because they promote cell and tissue destruction by deactivating biomolecules, like DNA proteins and lipids. Antioxidants save cells from oxidative damage by neutralizing the free radicals. CPO extracts contain large number of antioxidant compounds, which have the ability to defend the human immune system (Topkafa, 2016).

8 Oxidative stability

Oxidation is one of the major cause of quality deterioration in lipids. Oxidation changes the properties of edible oils such as their nutritional, chemical, and sensory characteristics (Vujasinovic, Djilas, Dimic, Romanic, & Takaci, 2010; Zine, Gharby, & El Hadek, 2013). The stability of oils may be affected in domestic conditions during storage because they are exposed to light, may have contact with air when open, or be kept at ambient temperatures (Martínez, Barrionuevo, Nepote, Grosso, & Maestri, 2011). Peroxide value, acid value, *p*-anisidine value, and conjugated dienoic and trienoic acids are mainly used to check the oxidative stability of CPO. DPPH· assay may be used to check the oxidative changes resistance in oils. The single electron transfer mechanism is used in DPPH· assay, which measures the antioxidant ability of oil to decrease DPPH· radicals (Espín, Soler-Rivas, & Wichers, 2000). The total antioxidant activity of nonfractionated oil and the antiradical activity of lipophilic and hydrophilic antioxidants of oil can be determined by using DPPH· assay (Prior, Wu, & Schaich, 2005). There is a very strong relation between oxidative stability and MUFA levels in oils (Kostadinović Veličkovska et al., 2015).

9 Adulteration of cold pressed sesame oil

SO contains various bioactive compounds such as tocopherols, sesamol, sesamin, sesamolin, inorganic elements, and PUFAs, which provide many health benefits. SSO is utilized as "flavor enhancers" in several food products in different Asian countries due to its good taste and pleasant flavor (Zhang et al., 2016). SSO is also used in production of different products such as cosmetics, pharmaceuticals, shortenings, and margarines (Warra, 2011). Adulteration of SSO is an issue that should be solved. Different oils such as hazelnut, maize, soybean, sunflower, and canola are used for adulteration with SSO (Seo et al., 2010). Techniques such as the capillary-electrophoresis single-strand conformation polymorphism technique based on PCR (polymerase chain-reaction) and ion mobility have been tested (Zhang et al., 2012). Different approaches are being used to detect adulteration in SSO by the addition of other edible oils such as PCR-single-strand conformation polymorphism analysis (Zhang et al., 2012), ion-mobility spectrometry (IMS) for fingerprinting (Zhang et al., 2016), GC/MS (Peng et al., 2015), HPLC (Park, Chang, & Lee, 2010), electronic-nose (Hai & Wang, 2006), and nuclear magnetic resonance spectrometry (Nam et al., 2014). SSO adulteration is also examined by Fourier transform infrared spectroscopy (FTIR), which is a nondestructive, very time-efficient, effective, and less expensive alternative for detecting adulteration of SSO (Ozulku, Yildirim, Toker, Karasu, & Durak, 2017).

10 Chemical contaminants

Agricultural use of pesticides such as polycyclic aromatic hydrocarbons (PAHs), polychlorinated-biphenyls (PCBs), and polybrominated biphenyl-ethers (PBDEs) may lead to food contamination (Meeker, Johnson, Camann, & Hauser, 2009). PAHs, PCBs, and PBDEs bioaccumulate in the food chain and affect both the environment and human health (ATSDR, 2000). Chemical contaminants with PAHs, PCBs, and PBDEs are reported in CPSSO at various levels of occurrence (Roszko, Szterk, Szymczyk, & Waszkiewicz-Robak, 2012).

11 Applications and health-promoting traits of cold pressed sesame oil

By the definition provided internationally, all virgin oils are CP oils but not all CP oils are virgin due to restriction of roasting of seeds prior to CP when virgin oils are produced. The cold pressed oils enriched with tocols, carotenoids (vitamin A), phenolics, and antioxidants induce positive impact on human health and are considered fit for human consumption (Matthäus & Spener, 2008).

SS exhibited important biological characteristics that contribute to human nutrition. For example:

1. SSO loaded with PUFA is used in the production of margarine and cooking oils.
2. SS has dominant quantities of lignans, sesamin, and sesamolin, which have a positive effect on lipid levels present in serum and proper functioning of the liver (Crews et al., 2006).
3. SSO has a cholesterol-decreasing property in humans (Hirata et al., 1996).
4. SSO avoids hypertension and boosts vitamin E delivery in animals (Kamal-Eldin, Pettersson, & Appelqvist, 1995).
5. SSO may be considered as a laxative (Anilakumar et al., 2010).
6. SSO has been observed to stop the formation of malignant melanoma in vitro and the propagation of cancer cells in the human gut (Smith & Salerno, 1992).
7. SSO is beneficial for protection against UV rays.
8. SSO has a lethal effect on lice present in children's hair.
9. Sesamin possesses bactericide plus insect-killing activities.
10. Sesamolin is an insect killer and is useful as a synergist against pyrethrum insecticides (Morris, 2002).
11. CPSSO is popular in Asian foods due to its nutty flavor.
12. CPSSO provides high amounts of PUFA.
13. CPSSO retards the oxidative impact in vivo, decreases cholesterol levels in the blood, and gives the liver protection against damage caused by oxidation (Boskou, 2017).

References

Abou-Gharbia, H. A., Shehata, A. A. Y., & Shahidi, F. (2000). Effect of processing on oxidative stability and lipid classes of sesame oil. *Food Research International*, *33*(5), 331–340.

Ananth, D. A., Deviram, G., Mahalakshmi, V., Sivasudha, T., & Tietel, Z. (2019). Phytochemical composition and antioxidant characteristics of traditional cold pressed seed oils in South India. *Biocatalysis and Agricultural Biotechnology*, *17*, 416–421.

Anilakumar, K. R., Pal, A., Khanum, F., & Bawa, A. S. (2010). Nutritional, medicinal and industrial uses of sesame (*Sesamum indicum* L.) seeds-an overview. *Agriculturae Conspectus Scientificus*, *75*(4), 159–168.

Assies, J., Lok, A., Bockting, C. L., Weverling, G. J., Lieverse, R., Visser, I., et al. (2004). Fatty acids and homocysteine levels in patients with recurrent depression: An explorative pilot study. *Prostaglandins, Leukotrienes and Essential Fatty Acids*, *70*(4), 349–356.

Biswas, T. K., Sana, N. K., Badal, R. K., & Huque, E. M. (2001). Biochemical study of some oil seeds (Brassica, sesame and linseed). *Pakistan Journal of Biological Sciences*, *4*(8), 1002–1005.

Boskou, D. (2017). Edible cold pressed oils and their biologically active components. *Journal of Experimental Food Chemistry*, *3*, e108.

Butinar, B., Bučar-Miklavčič, M., Mariani, C., & Raspor, P. (2011). New vitamin E isomers (gamma-tocomonoenol and alpha-tocomonoenol) in seeds, roasted seeds and roasted seed oil from the Slovenian pumpkin variety 'Slovenska golica'. *Food Chemistry*, *128*(2), 505–512.

Centers for Disease Control and Prevention (2000). Agency for toxic substances and disease registry. In *Report of the CDC/ATSDR Working Group on a shared vision for environmental public health at CDC/ATSDR*. Atlanta, GA: National Center for Environmental Health.

Choe, E., & Min, D. B. (2006). Mechanisms and factors for edible oil oxidation. *Comprehensive Reviews in Food Science and Food Safety*, *5*(4), 169–186.

Corso, M. P., Fagundes-Klen, M. R., Silva, E. A., Cardozo Filho, L., Santos, J. N., Freitas, L. S., et al. (2010). Extraction of sesame seed (*Sesamun indicum* L.) oil using compressed propane and supercritical carbon dioxide. *The Journal of Supercritical Fluids*, *52*(1), 56–61.

Crews, C., Hough, P., Brereton, P., Godward, J., Lees, M., Guiet, S., et al. (2006). Quantitation of the main constituents of some authentic sesame seed oils of different origin. *Journal of Agricultural and Food Chemistry*, *54*(17), 6266–6270.

El Khier, M. K. S., Ishag, K. E. A., & Yagoub, A. E. A. (2008). Chemical composition and oil characteristics of sesame seed cultivars grown in Sudan. *Research Journal of Agriculture and Biological Sciences*, *4*(6), 761–766.

Elleuch, M., Bedigian, D., & Zitoun, A. (2011). Sesame (*Sesamum indicum* L.) seeds in food, nutrition, and health. In *Nuts and seeds in health and disease prevention* (pp. 1029–1036). Academic Press.

Elleuch, M., Besbes, S., Roiseux, O., Blecker, C., & Attia, H. (2007). Quality characteristics of sesame seeds and by-products. *Food Chemistry*, *103*(2), 641–650.

Espín, J. C., Soler-Rivas, C., & Wichers, H. J. (2000). Characterization of the total free radical scavenger capacity of vegetable oils and oil fractions using 2, 2-diphenyl-1-picrylhydrazyl radical. *Journal of Agricultural and Food Chemistry*, *48*(3), 648–656.

Fukuda, Y., & Namiki, M. (1988). Recent studies on sesame seed and oil. *Nippon Shokuhin Kogyo Gakkaishi*, *35*(8), 552–562.

García, A., Ruiz-Méndez, M. V., Romero, C., & Brenes, M. (2006). Effect of refining on the phenolic composition of crude olive oils. *Journal of the American Oil Chemists' Society*, *83*(2), 159–164.

Gecgel, U., Demirci, A. S., Dulger, G. C., Gecgel, U., Tasan, M., Arici, M., et al. (2015). Some physicochemical properties, fatty acid composition and antimicrobial characteristics of different cold-pressed oils. *La Rivista Italiana Delle Sostanze Grasse*, *92*, 187–200.

Ghanbari, R., Anwar, F., Alkharfy, K. M., Gilani, A. H., & Saari, N. (2012). Valuable nutrients and functional bioactives in different parts of olive (*Olea europaea* L.)-a review. *International Journal of Molecular Sciences*, *13*(3), 3291–3340.

Goldberg, D. L., Novotny, R., Kieffer, E., Mor, J., & Thiele, M. (1995). Complementary feeding and ethnicity of infants in Hawaii. *Journal of the American Dietetic Association*, *95*(9), 1029–1031.

Hai, Z., & Wang, J. (2006). Electronic nose and data analysis for detection of maize oil adulteration in sesame oil. *Sensors and Actuators B: Chemical*, *119*(2), 449–455.

Hassan, M. A. (2012). Studies on Egyptian sesame seeds (*Sesamum indicum* L.) and its products 1-physicochemical analysis and phenolic acids of roasted Egyptian sesame seeds (*Sesamum indicum* L.). *World Journal of Dairy & Food Sciences*, *7*(2), 195–201.

Hirata, F., Fujita, K., Ishikura, Y., Hosoda, K., Ishikawa, T., & Nakamura, H. (1996). Hypocholesterolemic effect of sesame lignan in humans. *Atherosclerosis*, *122*(1), 135–136.

Kamal-Eldin, A., Pettersson, D., & Appelqvist, L. Å. (1995). Sesamin (a compound from sesame oil) increases tocopherol levels in rats fedad libitum. *Lipids*, *30*(6), 499–505.

Kostadinović Veličkovska, S., Brühl, L., Mitrev, S., Mirhosseini, H., & Matthäus, B. (2015). Quality evaluation of cold-pressed edible oils from Macedonia. *European Journal of Lipid Science and Technology*, *117*(12), 2023–2035.

Lerma-García, M. J., Herrero-Martínez, J. M., Simó-Alfonso, E. F., Mendonça, C. R., & Ramis-Ramos, G. (2009). Composition, industrial processing and applications of rice bran γ-oryzanol. *Food Chemistry*, *115*(2), 389–404.

Lichtenstein, A. H., Appel, L. J., Brands, M., Carnethon, M., Daniels, S., Franch, H. A., et al. (2006). Diet and lifestyle recommendations revision 2006: A scientific statement from the American Heart Association Nutrition Committee. *Circulation*, *114*(1), 82–96.

Lin, T. K., Zhong, L., & Santiago, J. (2017). Anti-inflammatory and skin barrier repair effects of topical application of some plant oils. *International Journal of Molecular Sciences*, *19*(1), 70.

Martínez, M., Barrionuevo, G., Nepote, V., Grosso, N., & Maestri, D. (2011). Sensory characterisation and oxidative stability of walnut oil. *International Journal of Food Science & Technology*, *46*(6), 1276–1281.

Matthäus, B., & Spener, F. (2008). What we know and what we should know about virgin oils–a general introduction. *European Journal of Lipid Science and Technology*, *110*(7), 597–601.

Meeker, J. D., Johnson, P. I., Camann, D., & Hauser, R. (2009). Polybrominated diphenyl ether (PBDE) concentrations in house dust are related to hormone levels in men. *Science of the Total Environment*, *407*(10), 3425–3429.

Morris, J. B. (2002). Food, industrial, nutraceutical, and pharmaceutical uses of sesame genetic resources. In J. Janick, & A. Whipkey (Eds.), *Trends in new crops and new uses* (pp. 153–156). ASHS Press.

Nam, Y. S., Noh, K. C., Roh, E. J., Keum, G., Lee, Y., & Lee, K. B. (2014). Determination of edible vegetable oil adulterants in sesame oil using 1*H* nuclear magnetic resonance spectroscopy. *Analytical Letters*, *47*(7), 1190–1200.

Nzikou, J. M., Matos, L., Bouanga-Kalou, G., Ndangui, C. B., Pambou-Tobi, N. P. G., Kimbonguila, A., et al. (2009). Chemical composition on the seeds and oil of sesame (*Sesamum indicum* L.) grown in Congo-Brazzaville. *Advance Journal of Food Science and Technology*, *1*(1), 6–11.

Ozulku, G., Yildirim, R. M., Toker, O. S., Karasu, S., & Durak, M. Z. (2017). Rapid detection of adulteration of cold pressed sesame oil adultered with hazelnut, canola, and sunflower oils using ATR-FTIR spectroscopy combined with chemometric. *Food Control*, *82*, 212–216.

Park, Y. W., Chang, P. S., & Lee, J. (2010). Application of triacylglycerol and fatty acid analyses to discriminate blended sesame oil with soybean oil. *Food Chemistry*, *123*(2), 377–383.

Peng, D., Bi, Y., Ren, X., Yang, G., Sun, S., & Wang, X. (2015). Detection and quantification of adulteration of sesame oils with vegetable oils using gas chromatography and multivariate data analysis. *Food Chemistry*, *188*, 415–421.

Prior, R. L., Wu, X., & Schaich, K. (2005). Standardized methods for the determination of antioxidant capacity and phenolics in foods and dietary supplements. *Journal of Agricultural and Food Chemistry*, *53*(10), 4290–4302.

Rostami, M., Farzaneh, V., Boujmehrani, A., Mohammadi, M., & Bakhshabadi, H. (2014). Optimizing the extraction process of sesame seed's oil using response surface method on the industrial scale. *Industrial Crops and Products*, *58*, 160–165.

Roszko, M., Szterk, A., Szymczyk, K., & Waszkiewicz-Robak, B. (2012). PAHs, PCBs, PBDEs and pesticides in cold-pressed vegetable oils. *Journal of the American Oil Chemists' Society*, *89*(3), 389–400.

Salunkhe, D. K., Chavan, J. K., Adsule, R. N., & Kadam, S. S. (1991). *World oilseeds: Chemistry, technology and utilization*: (pp. 371–402). Sesame.

Seo, H. Y., Ha, J., Shin, D. B., Shim, S. L., No, K. M., Kim, K. S., et al. (2010). Detection of corn oil in adulterated sesame oil by chromatography and carbon isotope analysis. *Journal of the American Oil Chemists' Society*, *87*(6), 621–626.

Smith, D. E., & Salerno, J. W. (1992). Selective growth inhibition of a human malignant melanoma cell line by sesame oil in vitro. *Prostaglandins, Leukotrienes and Essential Fatty Acids*, *46*(2), 145–150.

Topkafa, M. (2016). Evaluation of chemical properties of cold pressed onion, okra, rosehip, safflower and carrot seed oils: Triglyceride, fatty acid and tocol compositions. *Analytical Methods*, *8*(21), 4220–4225.

Tuberoso, C. I., Kowalczyk, A., Sarritzu, E., & Cabras, P. (2007). Determination of antioxidant compounds and antioxidant activity in commercial oilseeds for food use. *Food Chemistry*, *103*(4), 1494–1501.

Van Hoed, V., Depaemelaere, G., Ayala, J. V., Santiwattana, P., Verhé, R., & De Greyt, W. (2006). Influence of chemical refining on the major and minor components of rice brain oil. *Journal of the American Oil Chemists' Society*, *83*(4), 315–321.

Vujasinovic, V., Djilas, S., Dimic, E., Romanic, R., & Takaci, A. (2010). Shelf life of cold-pressed pumpkin (*Cucurbita pepo* L.) seed oil obtained with a screw press. *Journal of the American Oil Chemists' Society*, *87*(12), 1497–1505.

Warner, K. (2008). Chemistry of frying oils. In C. C. Akoh, & D. B. Min (Eds.), *Food lipids: Chemistry, nutrition, and biotechnology* (pp. 189–202). Boca Raton (FL): CRC Press Taylor & Francis Group.

Warra, A. A. (2011). Sesame (sesamum indicum l.) seed oil methods of extraction and its prospects in cosmetic industry: A review. *Bayero Journal of Pure and Applied Sciences*, *4*(2), 164–168.

Xu, J., Chen, S., & Hu, Q. (2005). Antioxidant activity of brown pigment and extracts from black sesame seed (*Sesamum indicum* L.). *Food Chemistry*, *91*(1), 79–83.

Yen, G. C., & Shyu, S. L. (1989). Oxidative stability of sesame oil prepared from sesame seed with different roasting temperatures. *Food Chemistry*, *31*(3), 215–224.

Zhang, Q., Liu, C., Sun, Z., Hu, X., Shen, Q., & Wu, J. (2012). Authentication of edible vegetable oils adulterated with used frying oil by Fourier Transform Infrared Spectroscopy. *Food Chemistry*, *132*(3), 1607–1613.

Zhang, L., Shuai, Q., Li, P., Zhang, Q., Ma, F., Zhang, W., et al. (2016). Ion mobility spectrometry fingerprints: A rapid detection technology for adulteration of sesame oil. *Food Chemistry*, *192*, 60–66.

Zine, S., Gharby, S., & El Hadek, M. (2013). Physicochemical characterization of *opuntia ficus-indica* seed oil from Morocco. *Biosciences, Biotechnology Research Asia*, *10*(1), 99–105.

Chapter 11

Cold pressed amaranth (*Amaranthus tricolor*) oil

Martin Mondor[a], Guiomar Melgar-Lalanne[b], and Alan-Javier Hernández-Álvarez[c]
[a]*Saint-Hyacinthe Research and Development Centre, Agriculture and Agri-Food Canada, Saint-Hyacinthe, QC, Canada,* [b]*Institute for Basic Sciences, Veracruzana University, Veracruz, Mexico,* [c]*School of Food Science & Nutrition, University of Leeds, Leeds, United Kingdom*

List of abbreviations

μm	microgram
ABTS	2,2′-azino-bis-3-ethylbenzothiazoline-6-sulfonic acid
AC	atherogenic coefficient value
Amo	amaranth oil
atm	atmospheres
CO_2	carbon dioxide
CVD	cardiovascular diseases
DNA	deoxyribonucleic acid
g	gram
HDL	high-density lipoprotein cholesterol
kg	kilogram
LDL	low-density lipoprotein cholesterol
mg	milligram
ml	milliliter
mm	millimeter
NLCs	nanostructured lipid carriers
ppm	parts per million
rpm	revolutions per minute
SFE	supercritical fluid extraction
spp.	species
UV	ultraviolet
UVA	ultraviolet A radiations
UVB	ultraviolet B radiations
VLDL	very-low-density lipoprotein cholesterol
***w*/w**	weight/weight

1 Introduction

The *Amaranthus* genus includes more than 60 distinct species, most of which are not cultivated (Wolosik & Markowska, 2019). They belong to the order *Caryophyllales*, the family *Amaranthaceae*, and the subfamily Amaranthoideae (Wolosik & Markowska, 2019), and are used for their grains, leaves, and even as an ornamental plant. Amaranth is native to Mesoamerica, currently Mexico and Central America, and is widely cultivated and consumed in Asia, Central America, Mexico, and Southeastern and Eastern Africa.

Amaranthus spp. are annual herbaceous plants with a short growing period of about 4–6 weeks. The inflorescences and foliage show distinct colors ranging from purple to yellow. The grain is small, around 0.9–1.7 mm in diameter, with a weight around 0.6–1 g/1000 seed, and covered in a hard hull (Fig. 1). The grains are lenticular with a color ranging from white to pink and even brown or black (Wolosik & Markowska, 2019). This crop is generally cultivated in arid zones, and

Cold Pressed Oils. https://doi.org/10.1016/B978-0-12-818188-1.00011-6
Crown Copyright © 2020 Published by Elsevier Inc. All rights reserved.

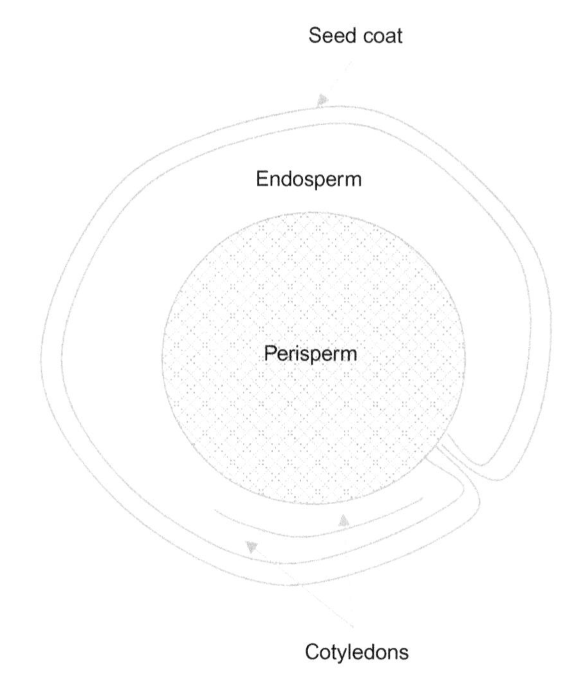

FIG. 1 Amaranth seed structure.

has a great capacity to produce biomass. The species most domesticated for their grains are *A. cruentus*, *A. caudatus*, and *A. hypocondriacus* (Das, 2012).

Amaranthus spp. is defined as a pseudocereal, meaning that it is like true cereals (wheat, barley, sorghum, millet, oat, rye, and corn) in its physical appearance. It is edible with a high starch content and can be milled into flour. However, amaranth has one of the highest contents of high-quality proteins of all the vegetables. This characteristic has made this crop one of the most promising foods of the millennium (Rollán, Gerez, & LeBlanc, 2019).

The biological value of amaranth protein is estimated as 90.4% raw, and 85.5% after heat treatment is applied, which is comparable to egg proteins, and shows an elevated level of lysine (0.83 g/100 g) (Palombini et al., 2013; Písaříková, Kráčmar, & Herzig, 2005). Furthermore, as it is gluten-free, amaranth could be safely consumed by people presenting gluten intolerance and celiac disease. The protein fraction is mainly composed of albumins and globulins, which have interesting techno-functional properties, that are similar to and even superior to other industrial sources such as soy and casein, making amaranth a promising alternative in many food applications (Janssen et al., 2017).

The saccharides content in amaranth is slightly lower than that in cereals. Starch is the most common, ranging from 48% in A. cruentus to 69% in *A. hypocondriacus*, and is mainly stored in small granules in the perisperm, not in the endosperm like in cereals (Pavlik, 2012). The starch is mainly composed of two major biomolecules: amylose and amylopectin. However, the amylose content is lower in comparison with other cereal starches, with values ranging from 0.1% to 11.1%. This peculiarity is responsible for many of the physicochemical properties of the starch (Kong, Bao, & Corke, 2009). Monosaccharides such as glucose and fructose are present only in trace amounts, and there is no presence of free monosaccharides. However, sucrose is the predominant free sugar with close to 2 g/100 g, which is higher than other cereals like wheat (less than 1 g/100 g) (Li et al., 2019).

The lipid content of amaranth seeds is lower than that of true cereals, about 4%–8%, but its biological value is extremely high due to its high squalene content. Moreover, it is a rich source of unsaturated fatty acids (mostly ω-6) with antioxidant properties. Linolenic acid is the most common unsaturated fatty acid (43%), followed by oleic acid (22%). The most common saturated fatty acid present is palmitic acid. Another peculiarity of amaranth oil is its high content in phospholipids (up to 10%), tocopherols (2%), and phytosterols (82%) (Bozorov et al., 2018). Amaranth oil extracts have traditionally been used to treat urinary infections, gynecological conditions, diarrhea, respiratory pain, and diabetes (Peter & Gandhi, 2017).

The lipid fraction of amaranth is usually isolated from the seeds by Soxhlet extraction with nonpolar solvents. The oil yield depends on the particle size, the conditions of extraction (time and temperature) and the solvent used (Kraujalis, Venskutonis, Pukalskas, & Kazernavičiūtė, 2013). The most environmentally friendly extraction technology used is supercritical fluid extraction (SFE) with CO_2 (Czaplicki, Ogrodowska, Zadernowski, & Derewiaka, 2012). The use of cold pressing has remained almost unexplored due to its low extraction rate (Czaplicki et al., 2012).

Thus, the aim of this chapter is to explore the solvent-free technologies used to extract the oil fraction in *Amaranthus* spp., with emphasis on cold pressing to extract compounds of interest such as squalene.

2 Extraction and processing of cold pressed amaranth oil

As previously mentioned, the oil content in the amaranth seed is relatively low compared with other cereal grains (between 4% and 8%). However, the fatty acid profile of amaranth oil is similar to that of corn oil, containing mainly oleic, linoleic, and palmitic acids with a high presence of squalene, which is a precursor of all steroids, making amaranth oil an interesting source of this bioactive compound, better than fish oils (Ariza Ortega, Martínez Zavala, Cano Hernández, & Díaz Reyes, 2012).

The physical characteristics of the seeds, covered with a rigid hull, complicate oil extraction, requiring the implementation of some pretreatments that optimize the process (Fig. 2). The most explored pretreatment has been the previous moisturization of the seed to obtain a humidity of around 15% before milling (Sakhare, Inamdar, Preetham Kumar, & Dharmaraj, 2017). Unfortunately, to the best of our knowledge, enzymatic and roasting pretreatments commonly used in other seeds have not yet been explored.

At the laboratory level, the method of choice is solvent extraction with nonpolar organic solvents using Soxhlet. In this case, the oil extraction yield depends on the amaranth particle size, the equipment used, and the conditions (temperature and time of the exposure and extraction cycles used) (Kraujalis et al., 2013). Some modifications to this method have been used, such as the previous pressurization, which accelerates and optimizes the extraction, increasing pressure (100–400 atm) and temperature (50°C–200°C) (Dunford & Zhang, 2003). The results obtained with pressurized extraction increased the extraction yield by up to five times when compared to conventional Soxhlet extraction, while the composition of fatty acids did not change (Kraujalis et al., 2013). However, from a health point of view, consumption of green extracted oils (with no use of organic solvents) should be preferred to refined oils, since these present a higher content of antioxidants and other active metabolites and less potential toxicity (Grosshagauer, Steinschaden, & Pignitter, 2019). The most used green extraction technology is SFE using carbon dioxide as an extraction fluid (Czaplicki et al., 2012; Kraujalis & Venskutonis, 2013; Westerman, Santos, Bosley, Rogers, & Al-Duri, 2006). This technique reduces extraction time, avoids the use of organic solvents, and requires less energy consumption than conventional solvent extraction techniques. Moreover, it also allows a better quantity of the compounds of interest, and even fractionation of these compounds by adjusting the pressure and temperature (Kraujalis & Venskutonis, 2013).

Few reports have explored the use of cold pressing technologies for amaranth oil. In the studies that were found comparing this technology with others (mainly Soxhlet extraction and SFE), the technical conditions for the cold extraction (temperature, time, and equipment) were not described. Czaplicki et al. (2012) compared the content of squalene, tocopherols, and phytosterols in amaranth oils obtained by different technologies (SFE, extraction chloroform/methanol mixture, and cold pressing), and found that cold pressing extraction appeared to be the worst method for extracting bioactive compounds. To obtain the cold pressed oil, seeds were pressed using an expeller with a 4-mm nozzle and a pressing temperature lower than 40°C. After pressing, the oil was purified by centrifugation at 10,000 rpm (no time or temperature were indicated). The results showed that the extraction of squalene and tocopherols was lower with cold pressing than with supercritical fluid extraction and solvent extraction; only the extraction of α-tocopherol was slightly higher with cold pressing. Finally, the content of sterols was also lower with cold pressing than with SFE and solvent extraction.

Raiciu et al. (2016) studied cold pressed oil from *Amaranthus caudatus* (Greek amaranth) seeds and compared the results with six other oleaginous plants essential for human health. The specific methodology for the extraction employed was not detailed. The results showed that the content in polyunsaturated fatty acids (PUFAs) was around 8%, with linoleic acid being the main one (47.4%), showing a content higher than camelina, flax, and rosehip, and lower than safflower and hemp; the oleic content was found to be around 25.5%. The authors also found elevated levels of squalene (10.8%), vitamin E, and β-sitosterol.

Bozorov et al. (2018) determined the triacylglycerides of different varieties of *Amaranthus hypocondriacus* and *Amaranthus cruentus* oil produced by cold pressing, and compared the results obtained with traditional extraction with Soxhlet

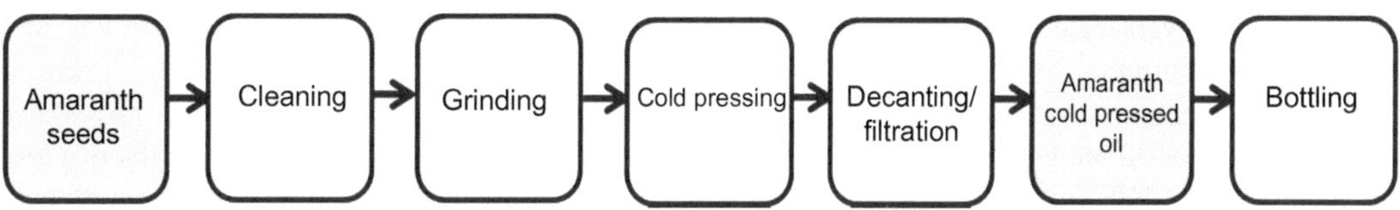

FIG. 2 Amaranth oil cold pressing extraction process.

and the subsequent solvent removal and drying. Unfortunately, the conditions used in the cold pressed procedure and the solvent used in the Soxhlet extraction were not mentioned. The oil extraction yield with the cold pressing method was significantly lower than with the Soxhlet extraction in all cases. The squalene content from cold pressing varied from 5.29% to 6.79%. Moreover, nonsignificant differences in the compositions of triacylglycerides between cold pressed and traditional Soxhlet extraction were found.

All the studies with amaranth oil obtained using cold pressing resulted in a poor oil extraction yield with less bioactive compounds than in other techniques. However, the lack of information about the technological conditions used and the fact that no pretreatments for the seed have been described for cold pressing may be responsible for the poor results obtained. Hence, comparisons between different cold pressing techniques and seed pretreatments are strongly recommended to improve understanding of the bioactive compound extraction rates.

2.1 Fatty acid and acyl lipid (neutral lipids, glycolipids, and phospholipids) profile of cold pressed amaranth oil

The literature on the fatty acid and acyl lipid composition of cold pressed amaranth oil is scarce. Most of the studies on the fatty acid and acyl lipid composition of amaranth oil were carried out for solvent-extracted oil. Moreover, besides the extraction method, the lipid content and profile in the amaranth seeds depend on many other factors including species, cultivar, and environmental growth conditions.

Bozorov et al. (2018) reported the chemical compositions of fatty acids of cold pressed amaranth oil for four varieties (*Kharkov*, *Lera*, *Andijan*, and *Helios*). The oil content ranged from 6.39% to 7.81%, although the authors did not specify whether this rate was obtained with cold pressing or Soxhlet extraction with hexane. For the four varieties tested, the main fatty acids obtained were linoleic acid with molar percentage varying between 40.1% for the *Kharkov* variety and 47.3% for the *Andijan* variety, followed by oleic acid with molar percentage varying between 24.5% for the *Andijan* variety and 31.3% for the *Kharkov* variety, and palmitic acid with molar percentage varying between 22.2% for the *Andijan* variety and 23.19% for the *Kharkov* variety. Stearic, linolenic, arachidic, and myristic acids were also detected, but in lesser concentrations. The sum of saturated fatty acids made up 26.7%–27.2%, while unsaturated fatty acids made up 72.7%–73.2%. The molar ratio of saturated and unsaturated fatty acids was 0.37. This was found to be higher than the ratio reported by Venskutonis and Kraujalis (2013) for solvent-extracted amaranth oil, which was in the range of 0.26 to 0.32.

Among the acyl lipids present in amaranth oil, triacylglycerides represent the major nonpolar compounds of the free lipids, while glycolipids and phospholipids represent the main fraction of the bound lipids (Lorenz & Hwang, 1985). The lipid profile of *Amaranthus* spp. depends on the extraction method used, and also on the species, cultivar, and year. In general, *Amaranthus* spp. has more than 75% unsaturated fatty acids (Kraujalis & Venskutonis, 2013). Bozorov et al. (2018) compared the triacylglyceride composition of cold pressed and solvent-extracted amaranth oils (Table 1). For most triacylglycerides, no significant differences were found between the cold pressed and the solvent-extracted oils.

The most abundant fatty acid found in all species tested was linoleic acid, followed by oleic acid, while the most abundant saturated fatty acid was palmitic (Aguilar et al., 2013). In general, colder years are associated with a higher ratio of unsaturated fatty acids, although in amaranth, only changes in the palmitic acid content were significant (Hlinková, Bednárová, Havrlentová, Šupová, & Čičová, 2013). Differences in the content and composition of unsaturated fatty acids were found when comparing distinct species and cultivars using Soxhlet extraction. The content of unsaturated fatty acids ranged from 77.8% in *A. cruentus* G6 to 82.00% in *A. hypocondriacus* var. Dorado. Kraujalis and Venskutonis (2013) observed similar results with 10 different cultivars of Lithuanian *Amaranthus* spp. In that study, the content of linoleic acid varied from 37% to 50%, that of oleic from 19% to 34%, and that of palmitic from 18% to 21%. In *A. caudatus*, a high α-linoleic acid content (683 g/kg of fatty acids) was observed, but the content reduced with the maturity of the plant, contrary to linoleic acid (516 g/kg fatty acids), which increased during the plant growth with the vegetative stage (Peiretti, Meineri, Gai, Longato, & Amarowicz, 2017). Amaranth also resulted in a rich source of eicosapentaenoic acid and had a similar composition to wheat germ oil (Bialek, Bialek, Jelinska, & Tokarz, 2016). The oleic/linoleic acid ratio ranged from 0.31 to 1.01 depending on the species, cultivar, and year (Hlinkova et al., 2013). Differences between the fatty acid profiles of amaranth and quinoa were also investigated. *A. cruentus* was compared with *Chenopodium quinoa*, and amaranth showed a higher concentration of saturated fatty acids and a higher n-6/n-3 ratio than quinoa oil, which results in a lower oil quality in the case of amaranth (Palombini et al., 2013).

To our knowledge, no data are available for the composition of cold pressed amaranth oil in terms of glycolipids and phospholipids composition. For solvent-extracted amaranth oil, Lorenz and Hwang (1985) reported that glycolipids represented 6.4% of the total lipids, mainly in the form of monogalactosyl and digalactosyl compounds, while phospholipids

TABLE 1 Triacyl-glycerides composition of amaranth oils extracted by cold pressing and by solvent extraction.

	Amaranthus hypochondriacus				Amaranthus cruentus			
	Kharkov		Lera		Andijan		Helios	
	Cold pressed	Solvent-extracted	Cold pressed	Solvent-extracted	Cold pressed	Solvent-extracted	Cold pressed	Solvent-extracted
Triacyl-glycerides	Quantity (molar %)							
L2P	15.68	13.87	16.00	15.61	15.26	15.39	13.34	15.15
L2C+LOP	15.58	19.09	16.60	19.54	18.12	17.72	18.12	16.10
L2O	12.98	11.85	13.60	12.90	12.69	12.15	11.94	13.06
LO2	11.63	12.88	10.79	11.93	11.62	11.36	12.09	11.77
LOS	11.00	10.86	12.66	9.49	11.20	9.74	11.62	13.17
O3	8.09	8.98	7.47	6.05	6.91	7.50	8.16	9.95
L3	4.96	4.37	5.90	5.47	7.22	7.68	5.42	5.88
LP2	4.60	7.61	5.10	8.45	7.91	7.96	7.35	4.72
O2P	3.99	4.07	5.30	4.12	4.48	4.45	5.08	5.57
O2S	3.99	1.70	1.83	2.03	1.85	2.21	1.71	1.64
OP2	2.76	3.28	3.77	3.08	2.07	2.73	2.90	2.16
OPS	0.38	1.34	1.37	1.23	0.50	0.95	2.15	0.74
LPS	0.35	0.09	0.13	0.11	0.17	0.15	0.12	0.09

L, linoleic acid; O, oleic acid; P, palmitic acid; S, stearic acid.
Adapted from Bozorov, S.S., Berdiev, N.Sh., Ishimov, U.J., Olimjonov, S.S., Ziyavitdinov, J.F., Asrorov, A.M., & Salikhov, S.I. (2018). Chemical composition and biological activity of seed oil of amaranth varieties. *Nova Biotechnologica et Chimica, 17*(1), 66–73.

represented 3.6%, mainly in the form of phosphatidylcholine, phosphatidylethanolamine, and phosphatidylinositol. In other work, Gamel, Mesallam, Damir, Shekib, and Linssen (2007) reported that glycolipids represented 9.5% of the lipid fraction, while phospholipids represented 10.2%. Gamel et al. (2007) studied the impact of cooking, popping, and germination on the glycolipids and phospholipids fraction of amaranth seeds for two amaranth varieties (*A. caudatus* and *A. cruentus*). Cooking and popping of *A. cruentus* did not affect the glycolipid and phospholipid content. However, a 10.0% decrease in the phospholipids content of popped *A. caudatus* was observed compared to that of the raw form. The researchers also observed that germination increased the phospholipid fraction for both varieties. Moreover, using higher temperatures for drying the germinated seeds increased the diacylglycerol contents and decreased the phospholipid contents of both varieties.

3 Minor bioactive lipids

3.1 Sterols

Sterols are commonly present in oils and fats and could be found in the form of free sterols and sterol esters of fatty acids. For amaranth oil, about 20% of sterols are present in their free form while 80% of sterols are esterified (León-Camacho, García-González, & Aparicio, 2001). Fifteen sterols were identified by León-Camacho et al. (2001) in amaranth (*Amaranthus cruentus* L.) oil with 24-methylen-cholesterol, campesterol, stigmasterol, Δ^7-campesterol, clerosterol, β-sitosterol, Δ^5-avenasterol, Δ^7-stigmastenol, and Δ^5-avenasterol being the most important ones with concentration (% of total sterols) of 0.3%, 1.6%, 0.9%, 24.8%, 42.0%, 1.3%, 2.0%, 15.2%, and 11.9%, respectively. Terpenic alcohols and methyl sterols were also found in amaranth: taraxerol (348.7 ppm), dammaradienol (189.0 ppm), β-amyrin (213.8 ppm), gramisterol (499.0 ppm), cycloartenol (401.8 ppm), 24-methylene-cycloartanol (446.7 ppm), citrostadienol (320.5 ppm), and four unidentified terpenic alcohols and methyl sterols with concentrations of 34.9, 35.6, 49.2, and 24.6 ppm

(León-Camacho et al., 2001). The content of total sterols was found to be 24.6×10^3 ppm. Czaplicki et al. (2012) compared the sterol content of amaranth oil extracted by supercritical CO_2 extraction, solvent extraction, and cold pressing. They obtained values of 2.49 g/100 g of oil, 2.14 g/100 g of oil, and 1.93 g/100 g of oil, respectively. Those levels are similar to the content of 2.73 g/100 g of oil reported by Berger, Gremaud, et al. (2003) and Berger, Monnard, et al. (2003) for crude oil extracted with *n*-hexane and for refined oil (2.59 g/100 g of oil). Czaplicki et al. (2012) identified and quantified the following 10 sterols for each amaranth oil (values are for cold pressed oil in mg/100 g of oil): campesterol (37.9), 5,24-stigmastadienol (39.4), 24-metylene-cholesterol (29.7), Δ7-stigmastenol (279.6), α-spinasterol + sitosterol (1028), stigmasterol (20.9), sitostanol (26.1), Δ5-avenasterol (14.3), Δ7-ergosterol (315.2), and Δ7-avenasterol (140.3), while cycloartenol and citrostadienol were identified only in amaranth oil extracted by supercritical CO_2. The sterols present in the highest concentrations were the sum of α-spinasterol and sitosterol, which for the cold pressed oil accounted for 53% of total analyzed sterols. Those results are in agreement with those of Fernando and Bean (1985) and of Grajeta (1997), who reported a spinasterol content ranging between 48% and 53%, and 46% and 54% of the total amaranth sterols, respectively. However, they are in contradiction with the results of León-Camacho et al. (2001), who did not observe any presence of α-spinasterol, and with those of Marcone, Kakadu, and Yada (2004), who reported only the presence of β-sitosterol, stigmasterol, and campesterol in different varieties of amaranth seeds. This raises the observation that the identification of sterols in some of these studies is questionable. To summarize, Czaplicki et al. (2012) determined that the oil with the highest amount of sterols was the one obtained by supercritical fluid extraction, followed by those obtained by solvent extraction and by cold pressing. While the content of sterols was the lowest in the cold pressed oil, this content was still significant for the diet of a potential consumer.

3.2 Tocols

Tocopherols and tocotrienols are a class of organic chemical compounds possessing vitamin E and antioxidant activities. They can be found in oilseeds at various levels, and they are collectively known as tocols. The tocols found in amaranth seeds are α-tocopherol, β-tocopherol, δ-tocopherol, γ-tocopherol, β-tocotrienol, δ-tocotrienol, and γ-tocotrienol (Venskutonis & Kraujalis, 2013), with α-tocopherol and β-tocopherol being present in the highest amounts. Compared to other grains, the content of β-tocopherol in amaranth seeds is high, with values of 0.71–6.74 mg/100 g (Budin, Breene, & Putnam, 1996), 546 ppm (León-Camacho et al., 2001), 1.96–4.39 mg/100 g (Bruni, Guerrini, Scalia, Romagnoli, & Sacchetti, 2002), and 8.24–21.18 mg/100 g (Kraujalis & Venskutonis, 2013). Czaplicki et al. (2012) compared the content of α-tocopherol, β-tocopherol, δ-tocopherol, and γ-tocopherol in amaranth oil extracted by supercritical CO_2 extraction, solvent extraction, and cold pressing. They obtained α-tocopherol values of 22.8, 23.9, and 23.0 mg/100 g of oil, β-tocopherol values of 53.0, 34.0, and 49.0 mg/100 g of oil, δ-tocopherol values of 41.1, 31.8, and 40.3 mg/100 g of oil, and γ-tocopherol values of 14.8, 11.2, and 15.6 mg/100 g of oil for oil extracted by supercritical CO_2, solvent, and cold pressing, respectively. The oil obtained by supercritical CO_2 extraction contained the highest content of tocopherols, followed by the oil extracted by solvent and the oil obtained from cold pressing. In another work, Ogrodowska et al. (2017) measured the tocopherols content of a commercial 100% cold pressed amaranth oil, and found contents of 21.26, 20.85, and 9.34 mg/100 g of oil for α-tocopherol, β-tocopherol + γ-tocopherol, and δ-tocopherol, respectively. Differences in terms of tocol content between the different aforementioned studies could be attributed to amaranth species, application of various extraction methods, and even different growing locations (Skwaryło-Bednarz, 2012).

3.3 Carotenoids

Amaranth seeds contain carotenoids at various levels. Tang et al. (2016) reported the presence of lutein (0.36–0.44 mg/100 g) and zeaxanthin (0.014–0.030 mg/100 g) but did not find any β-carotene, while Su, Rowley, Itsiopoulos, and O'Dea (2002) found lutein (13 mg/100 g) and β-carotene (4.0 mg/100 g). Despite the fact that amaranth seeds contain carotenoids, to our knowledge, no specific work has been carried out to quantify the level of carotenoids in amaranth oil. However, considering that extra-virgin cold pressed olive oil contains lutein and β-carotene, while refined extra-light olive oil does not, one may expect that cold pressed amaranth oil would contain some carotenoids (Su et al., 2002).

3.4 Phenolic compounds (other than tocols)

Phenolic compounds are relatively hydrophilic compounds that include phenolic acids, flavonoids, and tannins (Tang & Tsao, 2017). Several phenolic acids, flavonoids, and their glycosides have been identified in amaranth species. Gallic acid, 3,4-dihydroxybenzoic acid, and *p*-hydroxybenzoic acid are the three primary free phenolics found in amaranth seeds at 11.0–440, 4.7–136, and 8.5–20.9 mg/kg dry seed, respectively (Klimczak, Malecka, & Pacholek, 2002; Pasko,

Sajewicz, Gorinstein, & Zachwieja, 2008; Steffensen et al., 2011). Bound ferulic acids were also found in amaranth seeds after alkaline and enzyme hydrolyses [*trans*-ferulic acid (620 mg/kg), and *cis*-ferulic acid (203 mg/kg)] (Bunzel, Ralph, & Steinhart, 2005). Vanillic, caffeic, sinapic, and cinnamic acids were found but in smaller amounts (Venskutonis & Kraujalis, 2013). Flavonoids are found in amaranth seeds but at relatively low concentrations, with rutin being the most important one (80 mg/kg) (Kalinova & Dadakova, 2009). Orientin, morin, vitexin, isovitexin, and traces of neohesperidin and hesperidin were also detected (Pasko et al., 2008). Finally, the tannins content reported in various *Amaranthus* species varied from 0.4 to 5.2 mg/g (Venskutonis & Kraujalis, 2013). To our knowledge, no specific work has been carried out to quantify the level of phenolic acids, flavonoids, and tannins in cold pressed oil.

3.5 Hydrocarbons

Squalene, a triterpene, is found in significant amount (6%–8%) in amaranth oil (Ariza Ortega et al., 2012; Bozorov et al., 2018; Lozano-Grande, Gorinstein, Espitia-Rangel, Dávila-Ortiz, & Martinez-Ayala, 2018; Schnetzler & Breene, 2018; Sun, Wiesenborn, Rayas-Duarte, Mohamed, & Hagen, 1995), and food products enriched with squalene from amaranth oil can be considered as good squalene carriers in the human diet (Piotrowska, Żebrowska, Waszkiewicz-Robak, & Małecka, 2007). Czaplicki et al. (2012) reported a squalene concentration of 5.74% in cold pressed amaranth oil, which was slightly lower than for oil extracted by solvent (6.00%) and by supercritical CO_2 extraction (6.95%). In addition to the presence of squalene, León-Camacho et al. (2001) also reported a number of saturated (13) and unsaturated (10) hydrocarbons with one double bond (C21-C33) in *Amaranthus cruentus* crude oil at concentrations ranging from 1.91 to 64.9 ppm. In another study, Ciecierska and Obiedzinski (2013) quantified the total content of polycyclic aromatic hydrocarbons in cold pressed amaranth oil and reported a value of 101.6 mg/kg, with phenanthrene, anthracene, fluoranthene, and pyrene being the most important ones.

3.6 Flavor and aroma compounds

To our knowledge, no data are available on the flavor and aroma compounds of cold pressed amaranth oil. However, it is known that the main volatile compounds of raw amaranth seeds contributing to the seeds' flavor and aroma are 2,4-dimethyl-1-heptene, 4-methylheptane, branched C11·H24 alkane, and dodecene C12·H24 isomer (Gamel & Linssen, 2008). Those compounds represent about 70% of the total volatile compounds of the raw seeds, and so we may expect that they also contribute to the flavor and aroma of cold pressed amaranth oil.

3.7 Contribution of bioactive compounds in cold pressed amaranth oil to organoleptic traits and functions in food or nonfood products

To date, there have not yet been any studies analyzing the bioactive compounds in cold pressed amaranth oil and their contribution to organoleptic traits. However, as is the case with other vegetable oils, we may expect that the fatty acid profile of cold pressed amaranth oil will contribute significantly to its organoleptic traits, while the oxidation of the fatty acid chains will result in the production of off-flavors. Fortunately, tocopherols present in the oil are powerful antioxidants, which will protect unsaturated fatty acid chains, resulting in the retardation of oxidative rancidity. Similarly to olive oil, it is conceivable that phenolic compounds present in cold pressed amaranth oil will contribute to its positive attributes of bitterness and pungency (Inarejos-García, Santacatterina, Salvador, Fregapane, & Gómez-Alonso, 2010). The functions of bioactive compounds found in cold pressed amaranth oil in food and nonfood products are numerous and include good antioxidant properties, cholesterol-lowering properties, and rejuvenating and regenerative skin action. The main edible and nonedible applications of cold pressed amaranth oil will be discussed in the following sections.

4 Health-promoting traits of cold pressed amaranth oil and oil constituents

4.1 Reduction of the manifestations of chronic oxidative stress

Persistent oxidative stress in the body is responsible for the development of chronic diseases such as duodenal peptic ulcer, cancer, and other diseases (Bartosz, 2009; Cherkas et al., 2018; Negre-Salvayre et al., 2010; Valko et al., 2007). These are the results of antioxidant depletion, tissue damage with the accumulation of oxidative destruction products, and hypoxia. A number of studies performed on animal models, as well as some clinical trials on humans, demonstrated the potential of amaranth oil in the modulation of oxidative stress (Lelisieieva et al., 2006; Tang & Tsao, 2017; Yelisyeyeva et al., 2012, 2014; Yelisyeyeva, Cherkas, Semen, Kaminskyy, & Lutsyk, 2009). Potentially active substances include PUFAs, carotenoids, tocopherols, tocotrienols, and squalene (Martirosyan, Miroshnichenko, Kulakova, Pogojeva, & Zoloedov, 2007). However, the effects of other compounds present in lower concentrations cannot be excluded (Tang & Tsao, 2017). Lelisieieva et al. (2006) showed that modification of antioxidant activity induced by amaranth oil could maintain oxygen homeostasis and morphofunctional state, and inhibit the proliferation of cancer cells. A clinical trial performed by Cherkas et al. (2018) on 75 patients suffering from duodenal peptic ulcer demonstrated that amaranth oil supplementation in addition to standard antihelicobacter treatment reduced the accumulation of 4-hydroxynonenal (lipid peroxidation product) in the gastric mucosa, which reduced the persistence of oxidative stress and resulted in duodenal peptic ulcer healing. In another work, Yelisyeyeva et al. (2012) studied the effects of amaranth oil supplementation (1 mL of Amo 28 days) on heart rate variability in 36 patients suffering from type 2 diabetes. Heart rate variability is a measure indicating the amount of variation in heartbeats within a specific time frame. Decreased heart rate variability is associated with oxidative stress-related diseases (Cherkas et al., 2015; Cherkas & Yatskevych, 2014). The results indicated an improvement in heart rate variability and increased production of endogenous oxygen and enhancement of the cardiorespiratory function. This provides some evidence that the reduction of oxidative stress could lead to improvement in heart rate variability. Cherkas et al. (2009, 2018) also observed improvement of heart rate variability in patients suffering from duodenal peptic ulcer and supplemented with amaranth oil. However, Kanikowska et al. (2019) investigated whether amaranth oil supplementation (20 mL/day over 14 days) impacted oxidative metabolism in patients with obesity during calorie restriction, and no significant effect of oil supplementation was observed.

4.2 Prevention of cardiovascular disease

Cardiovascular disease is the most common cause of mortality. High levels of low-density cholesterol (LDL) represent a significant risk factor for the onset of cardiovascular disease, as with the presence of oxidized LDL (Shahidi & Costa de Camargo, 2016). Research works on amaranth oil and its constituents indicate that regular consumption of amaranth oil could be beneficial for patients with cardiovascular disease (or to prevent the onset of cardiovascular disease) by reducing the LDL level in blood. Works on animals demonstrated that feeding chickens (Qureshi, Lehmann, & Peterson, 1996), hamsters (Berger, Gremaud, et al., 2003), and hypercholesterolemic rabbits (Plate & Arêas, 2002) amaranth oil decreases blood cholesterol levels. However, in other trials with hamsters, the cholesterol-lowering properties of amaranth oil were not evident (Andrade de Castro et al., 2013; Berger, Monnard, et al., 2003). Some of the above discrepancies concerning the cholesterol-lowering properties of amaranth oil are probably related to the experimental design, including the diet, the quantity and type of saturated fat, the polyunsaturated:saturated fat ratio, and the quantity of cholesterol. In their work, Preobrazhenskaya, Pokrovskij, Berezhnova, and Levchenko (2016) used an animal model with male albino rats to study the effect of amaranth oil supplementation (0.25 mL Amo/kg) on doxorubicin-induced experimental cardiomyopathy. Experimental doxorubicin-induced cardiomyopathy results in hypercholesterolemia and disturbance in lipoprotein metabolism. Levels of LDL, HDL, and total cholesterol in the plasma of normal rats were on average 33.1, 20.0, and 46.5, respectively, while they were 68.2, 10.8, and 120.0, respectively, for rats injected with doxorubicin. Rats supplemented with amaranth oil had levels of LDL, HDL, and total cholesterol of 42.8, 16.5, and 84.0, respectively. The results indicated that low doses of amaranth oil did not return plasma lipid profile parameters in doxorubicin-induced cardiomyopathy groups to the normal levels but tended to decrease them. In another work, Martirosyan et al. (2007) investigated the effects of amaranth oil in patients suffering from CVD and hypertension with obesity. They hypothesized that amaranth oil consumption could be of significant benefit for patients with CVD. A randomized placebo-controlled clinical trial of 125 patients (men and women, aged 32–68) was carried out. The patients were randomly given between 3 and 18 mL of amaranth oil daily and otherwise were on similar low-salt diets. The clinical trial was carried out over a 3-week period. After the treatment, the blood serum of the patients was characterized for its level of total cholesterol, triglycerides, HDL, LDL, very low-density lipoprotein cholesterol (VLDL), and atherogenic coefficient (AC) value. The fatty acid composition of patients' erythrocyte membranes was also characterized. The results indicated that amaranth oil decreases the level of total cholesterol, triglycerides, LDL, and VLDL in the blood serum significantly, and that the lowering effect was concentration-dependent. An increase in the concentration of PUFAs of erythrocyte membranes of patients was also observed. These results suggest that amaranth oil could be used for the prevention and treatment of CVD. Potentially active substances include squalene and tocols (Lozano-Grande et al., 2018), and sterols (AbuMweiss & Jones, 2007; EFSA Report, 2008).

5 Edible applications of cold pressed amaranth oil

The main application of amaranth in the food industry comes from amaranth flour and its fraction, which is used in the production of cereal-based products with the main purpose of obtaining gluten-reduced products (Venskutonis &

Kraujalis, 2013). Production of edible films and coating from amaranth flour (Avila-Sosa et al., 2010, 2012; Tapia-Blacido, do Amaral Sobral, & Menegalli, 2013) or amaranth starch were also tested (Chandla, Saxena, & Singh, 2017). However, due to their low oil content, amaranth seeds cannot compete with most other commercial oilseeds, and so the market for edible amaranth oil is very limited (Schnetzler & Breene, 2018).

5.1 Amaranth oil in diets for rainbow trout

The supply of wild fishmeal has shown a downward trend for many years due to overfishing and has resulted in the need for raising fish commercially in tanks or in fishponds (pisciculture). One of the challenges in fish farming is to maximize fish growth rates while at the same time improving or at least maintaining the chemical body composition and fatty acids composition. Traditionally, fish oil has been used in feeds for rainbow trout. In their work, Poczyczyński, Gomułka, Woźniak, and Szostak (2014) studied the potential of amaranth oil to substitute fish oil partially in diets for farmed rainbow trout. Rainbow trout were fed with feeds containing 5.0% amaranth oil/5.0% fish oil, 7.2% amaranth oil/2.8% fish oil, or 10% fish oil (control) for 30days. Following this 30-day period, growth rate, chemical body composition, and fatty acids composition were determined. The results indicated that chemical body composition and fatty acids composition were not affected by the substitution of fish oil by amaranth oil. However, trout fed with diets containing amaranth oil showed higher specific growth rate than those fed with fish oil. The highest specific growth rate (3.75%/day) was found in trout fed with feed containing the highest level of amaranth oil. This work illustrated the potential use of amaranth oil in diets for rainbow trout farming. However, as mentioned by the authors, their results should be considered as a preliminary study. Long-term studies on the impact of substitution of fish oil by amaranth oil in feeds for rainbow trout or other fish are needed.

5.2 Infant formula

Infant formulas are designed to replace or supplement breast milk, and their composition should be as similar as possible to that of breast milk. The fat portion of infant formulas is generally achieved with blends of vegetable oils to balance the fatty acids profile (Innis, 1991). Hence, composition of infant formulas depends on the type and portion of oils used. Another approach consists of modifying the oil by *trans*-esterification and interesterification reactions to change its melting behavior, digestion, absorption, and nutrition value to resemble breast milk fat (Akoh & Kim, 2008). Pina-Rodriguez and Akoh (2009, 2010) used enzymatic interesterification techniques to modify amaranth oil for use in milk-based infant formulas to deliver a lipid component similar to that in breast milk. Original and modified amaranth oils were both characterized for their fatty acid composition, melting profile, chemical characteristics, and oxidative stability, as well as for phytosterol, tocopherol, and squalene contents. The results indicated that modified amaranth oil has strong potential as a partial fat substitute or complement for milk-based infant formula (Pina-Rodriguez & Akoh, 2009).

5.3 Yoghurt

In their work, Piotrowska et al. (2007) supplemented fruit yoghurts with amaranth oil obtained by pressing to develop a good squalene carrier. Squalene is recognized for its beneficial health effects. The amaranth oil used in their study contained 6.3% squalene. Fruit yoghurts were supplemented with levels of amaranth oil varying between 0% and 0.7%. A sensory assessment of the supplemented yoghurts was conducted by trained experts. It was found that up to 0.3% oil could be added to the fruit yoghurts without negatively affecting the organoleptic properties of the yoghurts. The authors concluded that fruit yoghurts would be a good squalene carrier, with a 250g portion supplementing about 47.5mg of squalene.

5.4 Amaranth oil oleogel for use in food products

Amaranth oil, as with other vegetable oils, is susceptible to oxidation, which might result in rancidity and limit the potential use of the oil in the formulation of food products. However, oleogelation is a novel approach to provide structure to liquid oils that are based on the use of organogelators. These molecules provide a solid-like structure to oils (Manzocco et al., 2017). In their work, Kamali, Sahari, Barzegar, and Gavlighi (2019) were the first to investigate the characteristics of oleogels produced from amaranth oil and using monoglycerides from palmitic and stearic acid as organogelators. Oleogels were prepared by adding organogelators to amaranth oil at concentrations of 7%, 9%, 10%, and 12% (*w*/w) before stirring and heating the mixtures to about 65°C to ensure the monoglycerides melted completely. The mixtures were cooled and allowed to crystallize at refrigerator temperature for 24h. Characteristics of the amaranth oil oleogels were compared to those of the liquid oil. The results indicated that the oleogels had higher oxidative stability than the oil when stored for

2 months at 25°C. Furthermore, oleogelation created a favorable crystalline network in oil and did not allow it to be released without augmenting the solid fat content of the oil. However, evaluation of fatty acids by gas chromatography showed a 5%–12% and 2%–6% reduction in oleic and linoleic acids, respectively. As a conclusion, oleogels made from amaranth oil and monoglycerides can be proposed as a solid-like oil without *trans*-fatty acids that could be used in food product formulation.

6 Nonedible applications of cold pressed amaranth oil

6.1 Cosmetics

Amaranth oil is well-known for its high content of ω-fatty acids, especially linoleic and oleic acids, phytosterols, and tocopherols (Korac & Khambholja, 2011), and squalene. Squalene can be found in various plant sources such as olive oil, peanut oil, grape seed oil, and soybean oil, but amaranth oil has the highest content. Squalene is also secreted by the sebaceous glands in the human body and forms part of lipids on the skin surface (300–500 μg/g) and on internal organs such as the small intestine and liver (500 μg/g) (Kostyuk, Potapovich, Stancato, et al., 2012; Popa, Babeanu, Popa, Nita, & Dinu-Parvu, 2015). The presence of those compounds gives amaranth oil an emollient, rejuvenating, and regenerative skin action (Korac & Khambholja, 2011; Venskutonis & Kraujalis, 2013), and makes it recognized in the cosmetic industry for its beneficial effects. In their study, Lacatusu et al. (2018) wanted to develop a plant cosmetic formulation with strong protection against both UVA and UVB radiations based on nanostructured lipid carriers. The carriers were prepared by homogenization of a mixture of an aqueous phase consisting of 2.5% surfactant and 0.3% hesperidin (natural antioxidant), with an oil phase consisting of 10% lipids mixture and 1.4% UV filters. Amaranth oil with 34.5% and 82% squalene was considered in this study for preparation of the lipid phase. An in vitro study of the release of hesperidin and UV-filters from the carriers and in vitro determination of antioxidant activity by the chemiluminescence method were carried out. Antioxidant activity was also measured by the ABTS method (Badea et al., 2017). The results indicated that the nano-delivery system provided a sustained release of the hesperidin while maintaining significant amounts of UV filters in the formulation. The carrier prepared with amaranth oil containing 82% squalene was found to be the most effective scavenging system of oxygen free radicals, while a better ability to inhibit ABTS cation radicals was observed for the carrier prepared with amaranth oil containing 34% squalene. Incorporation of these carriers in Corbopol hydrogels resulted in an average absorption of 83% and 99% UVA and UVB radiations, respectively. The results of this study demonstrated the great potential of amaranth oil for innovative sunscreen formulations with superior photoprotection and enhanced antioxidant properties. In another work, Wolosik, Zareba, Surazynski, and Markowska (2017) studied the potential of amaranth oil for the prevention and treatment of ultraviolet (UV) irradiation-related pathologies such as sunburn, photoaging, photoimmunosuppression, and photocarcinogenesis. Experiments were performed on human skin fibroblasts incubated for 24 h with varying concentrations of amaranth oil (0.1%, 0.25%, and 0.5%) and irradiated for 15 min under a UVA-emitting lamp. The effect of amaranth oil concentration before and after UVA exposure was then characterized by studying the collagen, and DNA biosynthesis. The results indicated that exposure of human skin fibroblast to UVA is less harmful to collagen formation when 0.1% amaranth oil is used. Collagen helps in the maintenance of the skin dermis structure (Roy et al., 2013). The assessment results of pre- and postUVA effects on DNA biosynthesis in human skin fibroblasts showed that 0.1% and 0.5% amaranth oil abolished postUVA effect. The authors concluded that amaranth oil provides pre- and postUVA protection in low concentrations and could be considered in combination with other compounds in cosmetic formulation.

6.2 Amaranth oil for the development of squalene-based nanocarriers for dual release of drug and bioactive compounds

Nanostructured lipid carriers (NLCs) are used as a transport system for drugs. They can be easily formulated in water-based systems and are one of the most convenient colloidal carriers systems (Barbinta-Patrascu et al., 2014; Kraft, Freeling, Wang, & Ho, 2015). Advantages of NLCs compared to other alternatives are increased drug loading, better encapsulation efficiency, chemical stability, bioavailability, and controlled release of lipophilic compounds (Liu et al., 2014; Wun How, Rasedee, Manickam, & Rosli, 2013). However, one limitation is that NLCs generally have low encapsulation efficiency for hydrophilic drugs (Severino et al., 2014; Vrignaud, Anton, Gayet, Benoit, & Saulnier, 2011; Vrignaud, Hureaux, Wack, Benoit, & Saulnier, 2012). In their work, Ott et al. (2015) used amaranth oil for the development of squalene-based nanocarriers for the dual release of pemetrexed disodium, an antitumor drug, and hesperidin, a bio-flavonoid. Vegetable fractions were obtained from cold pressed amaranth seeds and were further processed by supercritical and Soxhlet extractions to obtain fractions with different levels of squalene (e.g., 5.8%, 34.7%, and 83.4%). Lacatusu et al. (2014) described a method used for the production of NLCs using homogenization. For each formulation, 3% of oil fraction (with different squalene levels) was used, while the amount of pemetrexed disodium and hesperidin was varied. The entrapment efficiency of both pemetrexed disodium and hesperidin was assessed, as well as antioxidant activity, and in vitro release tests were carried out (Ott et al., 2015). The results indicated an excellent entrapment efficiency, with values of 89% for pemetrexed disodium and 94% for hesperidin. Antioxidant activity as high as 98.2% for nanocarriers that coencapsulate 0.8% pemetrexed disodium and 0.6% hesperidin was obtained, while in vitro corelease tests demonstrated that pemetrexed disodium and hesperidin were gradually released despite their different lipophilicity. The 83.4% squalene fraction showed the slowest release of both molecules. Overall, the results indicated the great potential of amaranth oil fractions for the development of squalene-based NLCs to improve the performance of medical treatments used to treat cancer or other diseases.

7 Other issues

Vegetable oils are important from both an economic and a nutritional point of view. Some edible cold pressed oils are expensive, which makes it tempting to adulterate them to achieve more profit. Adulterations can be done by mixing cold pressed oil with refined ones, or by replacing expensive oils with cheaper ones (Azadmard-Damirchi & Torbati, 2015; Jee, 2002). Vegetable oils differ significantly in their triacylglycerol profiles, which can be considered as good fingerprints to determine authenticity, origin, and adulteration (Indelicato et al., 2017). The most common analytical methods used to assess triacylglycerol profile are gas chromatography, gas chromatography coupled with mass spectrometry, liquid chromatography, and liquid chromatography coupled with mass spectrometry (Indelicato et al., 2017). However, to our knowledge, contrary to the adulteration of cold pressed olive oil, the adulteration of cold pressed amaranth oil is not a widespread problem.

8 Conclusion

Oils are essential to the food industry for many uses, and oil from amaranth is valuable due to its composition of fatty acids and its high content in oleic and linolenic acids, and squalene, which is found in relatively high amounts. However, there is an important information gap in terms of the technical conditions (use of different equipment, physical or chemical extraction, temperature, solvent/sample ratios, etc.) and pretreatments (roasting, enzymatic assisted, drying technology, thermal conditioning, among others). Therefore, comparisons between different cold pressing techniques, seed pretreatments, and extraction conditions are strongly recommended in order to understand the behavior of the bioactive compounds during extraction through the wide variety of methods available to preserve these molecules. Thus, to improve the final quality of cold pressed amaranth oil, some technological strategies have to be developed. This aspect needs to be investigated further and the lipid detailed characterization extended to several species to verify if similar behavior occurs.

References

AbuMweiss, S. S., & Jones, P. J. H. (2007). Plant sterols: Natural plant components with potential beneficial health effects. *Inform*, *18*(12), 825–827.

Aguilar, E. G., Peiretti, E. G., Uñates, M. A., Marchevsky, E. J., Escudero, N. L., & Camiña, J. M. (2013). Amaranth seed varieties, A chemometric approach. *Food Measure*, *7*, 199–206.

Akoh, C. C., & Kim, B. H. (2008). Structured lipids. In C. C. Akoh, & D. B. Min (Eds.), *Food lipids* (3rd ed., pp. 841–872). Boca Raton, FL: CRC Press.

Andrade de Castro, L. I. A., Manolio Soares, R. A., Saldiva, P. H. N., Ferrari, R. A., Migue, A. M. R. O., Almeida, C. A. S., et al. (2013). Amaranth oil increased fecal excretion of bile acid but had no effect in reducing plasma cholesterol in hamsters. *Lipids*, *48*, 609–618.

Ariza Ortega, J. A., Martínez Zavala, A., Cano Hernández, M., & Díaz Reyes, J. (2012). Analysis of trans fatty acids production and squalene variation during amaranth oil extraction. *Central European Journal of Chemistry*, *10*(6), 1773–1778.

Avila-Sosa, R., Hernandez-Zamoran, E., Lopez-Mendoza, E., Palou, E., Jimenez Munguia, M. T., Nevarez-Moorillon, G. V., et al. (2010). Fungal inactivation by Mexican oregano (*Lippia berlandieri Schauer*) essential oil added to amaranth, chitosan, or starch edible films. *Journal of Food Science*, *75*(3), M127–M133.

Avila-Sosa, R., Palou, E., Jimenez Munguia, M. T., Nevarez-Moorillon, G. V., Navarro Cruz, A. R., & Lopez-Malo, A. (2012). Antifungal activity by vapor contact of essential oils added to amaranth, chitosan, or starch edible films. *International Journal of Food Microbiology*, *153*, 66–72.

Azadmard-Damirchi, S., & Torbati, M. (2015). Adulterations in some edible oils and fats and their detection methods. *Journal of Food Quality and Hazards Control*, *2*, 38–44.

Badea, G., Badea, N., Brasoveanu, L., Mihaila, M., Stan, R., Istrati, D., et al. (2017). Naringenin improves the sunscreen performance of vegetable nanocarriers. *New Journal of Chemistry*, *41*, 480–492.

Barbinta-Patrascu, M. E., Ungureanu, C., Iordache, S. M., Iordache, A. M., Bunghez, I. R., Ghiurea, M., et al. (2014). Eco-designed biohybrids based on liposomes, mint-nanosilver and carbon nanotubes for antioxidant and antimicrobial coating. *Materials Science & Engineering C, Materials for Biological Applications*, *39*, 177–185.

Bartosz, G. (2009). Reactive oxygen species: Destroyers or messengers? *Biochemical Pharmacology*, *77*, 1303–1315.

Berger, A., Gremaud, G., Baumgartner, M., Rein, D., Monnard, I., Kratky, E., et al. (2003). Cholesterol-lowering properties of amaranth grain and oil in hamsters. *International Journal for Vitamin and Nutrition Research*, *73*(1), 39–47.

Berger, A., Monnard, I., Dionisi, F., Gumy, D., Hayes, K. C., & Lambelet, P. (2003). Cholesterol-lowering properties of amaranth flakes, crude and refined oils in hamsters. *Food Chemistry*, *81*, 119–124.

Bialek, A., Bialek, M., Jelinska, M., & Tokarz, A. (2016). Fatty acid profile of new promising unconventional plant oils for cosmetic use. *International Journal of Cosmetic Science*, *38*(4), 382–388. https://doi.org/10.1111/ics.12301.

Bozorov, S. S., Berdiev, N. S., Ishimov, U. J., Olimjonov, S. S., Ziyavitdinov, J. F., Asrorov, A. M., et al. (2018). Chemical composition and biological activity of seed oil of amaranth varieties. *Nova Biotechnologica et Chimica*, *17*(1), 66–73.

Bruni, R., Guerrini, A., Scalia, S., Romagnoli, C., & Sacchetti, G. (2002). Rapid techniques for the extraction of vitamin E isomers from *Amaranthus caudatus* seeds: Ultrasonic and supercritical fluid extraction. *Phytochemical Analysis*, *13*, 257–261.

Budin, J. T., Breene, W. M., & Putnam, D. H. (1996). Some compositional properties of seeds and oils of eight *Amaranthus* species. *Journal of the American Oil Chemists' Society*, *73*, 475–481.

Bunzel, M., Ralph, J., & Steinhart, H. (2005). Association of non-starch polysaccharides and ferulic acid in grain amaranth (*Amaranthus caudatus* L.) dietary fiber. *Molecular Nutrition & Food Research*, *49*, 551–559.

Chandla, N. K., Saxena, D. C., & Singh, S. (2017). Amaranth (*Amaranthus* spp.) starch isolation, characterization, and utilization in development of clear edible films. *Journal of Food Processing & Preservation*, *41*, e13217.

Cherkas, A., Abrahamovych, O., Golota, S., Nersesyan, A., Pichler, C., Serhiyenko, V., et al. (2015). The correlations of glycated hemoglobin and carbohydrate metabolism parameters with heart rate variability in apparently healthy sedentary young male subjects. *Redox Biology*, *5*, 301–307.

Cherkas, A., & Yatskevych, O. (2014). The amplitude of heart rate oscillations is dependent on metabolic status of sinoatrial node cells. *OA Medical Hypothesis*, *2*, 1–8.

Cherkas, A., Yelisyeyeva, O., Semen, K., Zarkovic, K., Kaminskyy, D., Gasparovic, A. C., et al. (2009). Persistent accumulation of 4-hydroxynonenal-protein adducts in gastric mucosa after *Helicobacter pylori* eradication. *Collegium Antropologicum*, *33*, 815–821.

Cherkas, A., Zarkovic, K., Cipak Gasparovic, A., Jaganjac, M., Milkovic, L., Abrahamovych, O., et al. (2018). Amaranth oil reduces accumulation of 4-hydroxynonenal-histidine adducts in gastric mucosa and improves heart rate variability in duodenal peptic ulcer patients undergoing *Helicobacter pylori* eradication. *Free Radical Research*, *52*(2), 135–149.

Ciecierska, M., & Obiedzinski, M. W. (2013). Polycyclic aromatic hydrocarbons in vegetable oils from unconventional sources. *Food Control*, *30*, 556–562.

Czaplicki, S., Ogrodowska, D., Zadernowski, R., & Derewiaka, D. (2012). Characteristics of biologically-active substances of amaranth oil obtained by various techniques. *Polish Journal of Food and Nutrition Sciences*, *62*(4), 235–239.

Das, S. (2012). Domestication, phylogeny and taxonomic delimitation in underutilized grain Amaranthus (Amaranthaceae)-a status review. *Feddes Repertorium*, *123*, 273–282. https://doi.org/10.1002/fedr.201200017.

Dunford, N. T., & Zhang, M. (2003). Pressurized solvent extraction of wheat germ oil. *Food Research International*, *36*(9), 905–909. https://doi.org/10.1016/S0963-9969(03)00099-1.

EFSA Report. (2008). Consumption of food and beverages with added plant sterols in the European Union. *EFSA Journal*, *133*, 1–21.

Fernando, T., & Bean, G. (1985). A comparison of the fatty acids and sterols of seeds of weedy and vegetable species of *Amaranthus* spp. *Journal of the American Oil Chemists' Society*, *62*(1), 89–91.

Gamel, T. H., & Linssen, J. P. H. (2008). Flavor compounds of popped amaranth seeds. *Journal of Food Processing and Preservation*, *32*, 656–668.

Gamel, T. H., Mesallam, A. S., Damir, A. A., Shekib, L. A., & Linssen, J. P. (2007). Characterization of amaranth seed oils. *Journal of Food Lipids*, *14*, 323–334.

Grajeta, H. (1997). Nutritive value and utilization of amaranth. *Bromatologia i Chemia Toksykologiczna*, *30*, 17–23 [in Polish].

Grosshagauer, S., Steinschaden, R., & Pignitter, M. (2019). Strategies to increase the oxidative stability of cold pressed oils. *LWT - Food Science and Technology*, *106*, 72–77. https://doi.org/10.1016/j.lwt.2019.02.046.

Hlinková, A., Bednárová, A., Havrlentová, M., Šupová, J., & Čičová, I. (2013). Evaluation of fatty acid composition among selected amaranth grains grown in two consecutive years. *Biologia*, *68*, 641–650.

Inarejos-García, A. M., Santacatterina, M., Salvador, M. D., Fregapane, G., & Gómez-Alonso, S. (2010). PDO virgin olive oil quality-minor components and organoleptic evaluation. *Food Research International*, *43*, 2138–2146.

Indelicato, S., Bongiorno, D., Pitonzo, R., Di Stefano, V., Calabrese, V., Indelicato, S., et al. (2017). Triacylglycerols in edible oils: Determination, characterization, quantitation, chemometric approach and evaluation of adulterations. *Journal of Chromatography A*, *1515*, 1–16.

Innis, S. M. (1991). Essential fatty acids in growth and development. *Progress in Lipid Research*, *30*, 39–103.

Janssen, F., Pauly, A., Rombouts, I., Jansens, K. J. A., Deleu, L. J., & Delcour, J. A. (2017). Proteins of Amaranth (*Amaranthus* spp.), buckwheat (*Fagopyrum* spp.), and quinoa (*Chenopodium* spp.): A food science and technology perspective. *Comprehensive Reviews in Food Science and Food Safety*, *16*(1), 39–58. https://doi.org/10.1111/1541-4337.12240.

Jee, M. (2002). Adulteration and authentication of oils and fats: An overview. In M. Jee (Ed.), *Oils and fats authentication*. CRC Press.

Kalinova, J., & Dadakova, E. (2009). Rutin and total quercetin content in amaranth (*Amaranthus spp.*). *Plant Foods for Human Nutrition*, *64*, 68–74.

Kamali, E., Sahari, M. A., Barzegar, M., & Gavlighi, M. A. (2019). Novel oleogel formulation based on amaranth oil: Physicochemical characterization. *Food Science & Nutrition*, *7*, 1986–1996.

Kanikowska, D., Kanikowska, A., Rutkowski, R., Włochal, M., Orzechowska, Z., Juchacz, A., et al. (2019). Amaranth (*Amaranthus cruentus* L.) and canola (*Brassica napus* L.) oil impact on the oxidative metabolism of neutrophils in the obese patients. *Pharmaceutical Biology*, *57*(1), 140–144.

Klimczak, I., Malecka, M., & Pacholek, B. (2002). Antioxidant activity of ethanolic extracts of amaranth seeds. *Nahrung/Food*, *46*, 184–186.

Kong, X., Bao, J., & Corke, H. (2009). Physical properties of Amaranthus starch. *Food Chemistry*, *113*(2), 371–376. https://doi.org/10.1016/j.foodchem.2008.06.028.

Korac, R. R., & Khambholja, K. M. (2011). Potential of herbs in skin protection from ultraviolet radiation. *Pharmacognosy Reviews*, *5*(10), 164–173.

Kostyuk, V., Potapovich, A., Stancato, A., et al. (2012). Photooxidation products of skin surface squalene mediate metabolic and inflammatory responses to solar UV in human keratinocytes. *PLoS One*, *7*(8), e44472.

Kraft, J. C., Freeling, J. P., Wang, Z., & Ho, R. J. Y. (2015). Emerging research and clinical development trends of liposome and lipid nanoparticle drug delivery systems. *Journal of Pharmaceutical Sciences*, *103*, 29–52.

Kraujalis, P., & Venskutonis, P. R. (2013). Supercritical carbon dioxide extraction of squalene and tocopherols from amaranth and assessment of extracts antioxidant activity. *Journal of Supercritical Fluids*, *80*, 78–85.

Kraujalis, P., Venskutonis, P. R., Pukalskas, A., & Kazernavičiūtė, R. (2013). Accelerated solvent extraction of lipids from *Amaranthus* spp. seeds and characterization of their composition. *LWT- Food Science and Technology*, *54*(2), 528–534. https://doi.org/10.1016/j.lwt.2013.06.014.

Lacatusu, I., Niculae, G., Badea, N., Stan, R., Popa, O., Oprea, O., et al. (2014). Design of soft lipid nanocarriers based on bioactive vegetable oils with multiple health benefits. *Chemical Engineering Journal*, *246*, 311–321.

Lacatusu, I., Vasilica Arsenie, L., Badea, G., Popa, O., Oprea, O., & Badea, N. (2018). New cosmetic formulations with broad photoprotective and antioxidative activities designed by amaranth and pumpkin seed oils nanocarriers. *Industrial Crops and Products*, *123*, 424–433.

Lelisieieva, O. P., Kamins'kyi, D. V., Cherkas, A. P., Ambarova, L. I., Vyshemyrs'ka, L. D., Dzhura, O. R., et al. (2006). Characteristics of amaranth oil effect on the antioxidant system of the liver and blood in mice with malignant lymphoma growth. *Ukrainskĭ Biokhimicheskĭ Zhurnal*, *78*, 117–123.

León-Camacho, M., García-González, D. L., & Aparicio, R. (2001). A detailed and comprehensive study of amaranth (*Amaranthus cruentus* L.) oil fatty profile. *European Food Research and Technology*, *213*, 349–355.

Li, W., Liu, Y., Liu, M., Zheng, Q., Li, B., Li, Z., et al. (2019). Sugar accumulation is associated with leaf senescence induced by long-term high light in wheat. *Plant Science*, *287*, 110169. https://doi.org/10.1016/j.plantsci.2019.110169.

Liu, Y. J. T., Wang, L., Zhao, Y., He, M., Zhang, X., Niu, M., et al. (2014). Nanostructured lipid carriers versus microemulsions for delivery of the poorly water-soluble drug luteolin. *International Journal of Pharmaceutics*, *476*, 169–177.

Lorenz, L., & Hwang, Y. S. (1985). Lipids in amaranth. *Nutrition Reports International*, *31*, 83–89.

Lozano-Grande, M. A., Gorinstein, S., Espitia-Rangel, E., Dávila-Ortiz, G., & Martinez-Ayala, A. L. (2018). Plant sources, extraction methods, and uses of squalene. *International Journal of Agronomy*, *2018*, 1829160. 13 pages.

Manzocco, L., Valoppi, F., Calligaris, S., Andreatta, F., Spilimbergo, S., & Nicoli, M. C. (2017). Exploitation of κ-carrageenan aerogels as template for edible oleogel preparation. *Food Hydrocolloids*, *71*, 68–75.

Marcone, M. F., Kakadu, Y., & Yada, R. Y. (2004). Amaranth as a rich dietary source of β-sitosterol and other phytosterols. *Plant Foods for Human Nutrition*, *58*, 207–211.

Martirosyan, D. M., Miroshnichenko, L. A., Kulakova, S. N., Pogojeva, A. V., & Zoloedov, V. I. (2007). Amaranth oil application for coronary heart disease and hypertension. *Lipids in Health and Disease*, *6*, 1–12.

Negre-Salvayre, A., Auge, N., Ayala, V., Basaga, H., Boada, J., Brenke, R., et al. (2010). Pathological aspects of lipid peroxidation. *Free Radical Research*, *44*(10), 1125–1171.

Ogrodowska, D., Tanska, M., Brandt, W., Czaplicki, C., Skrajda, M., & Dabrowski, G. (2017). Effect of the drying process on the physicochemical characteristics and oxidative stability of microencapsulated amaranth. *Rivista Italiana Delle Sostanze Grasse*, *94*(4), 257–264.

Ott, C., Lacatusu, I., Badea, G., Adriana Grafu, I., Istrati, D., Babeanu, N., et al. (2015). Exploitation of amaranth oil fractions enriched in squalene for dual delivery of hydrophilic and lipophilic actives. *Industrial Crops and Products*, *77*, 342–352.

Palombini, S. V., Claus, T., Maruyama, S. A., Gohara, A. K., Souza, A. H. P., Souza, N. E. D., et al. (2013). Evaluation of nutritional compounds in new amaranth and quinoa cultivars. *Food Science and Technology*, *33*(2), 339–344.

Pasko, P., Sajewicz, M., Gorinstein, S., & Zachwieja, Z. (2008). Analysis of selected phenolic acids and flavonoids in *Amaranthus cruentus* and *Chenopodium quinoa* seeds and sprouts by HPLC. *Acta Chromatographica*, *20*, 661–672.

Pavlik, V. (2012). The revival of amaranth as a third-millennium food. *Neuroendocrinology Letters*, *33*(37), 3.

Peiretti, P. G., Meineri, G., Gai, F., Longato, E., & Amarowicz, R. (2017). Antioxidative activities and phenolic compounds of pumpkin (*Cucurbita pepo*) seeds and amaranth (*Amaranthus caudatus*) grain extracts. *Natural Product Research*, *31*(18), 2178–2182. https://doi.org/10.1080/14786419.2017.1278597.

Peter, K., & Gandhi, P. (2017). Rediscovering the therapeutic potential of Amaranthus species: A review. *Egyptian Journal of Basic and Applied Sciences*, *4*(3), 196–205. https://doi.org/10.1016/j.ejbas.2017.05.001.

Pina-Rodriguez, A. M., & Akoh, C. C. (2009). Synthesis and characterization of a structured lipid from amaranth oil as a partial fat substitute in milk-based infant formula. *Journal of Agricultural and Food Chemistry*, *57*, 6748–6756.

Pina-Rodriguez, A. M., & Akoh, C. C. (2010). Composition and oxidative stability of a structured lipid from amaranth oil in a milk-based infant formula. *Journal of Food Science*, *75*(2), C140–C146.

Piotrowska, A., Żebrowska, M., Waszkiewicz-Robak, B., & Małecka, K. (2007). Sensory quality of yoghurts enriched with squalene from amaranthus oil. *Polish Journal of Food and Nutrition Sciences*, *57*(4B), 437–444.

Písaříková, B., Kráčmar, S., & Herzig, I. (2005). Amino acid contents and biological value of protein in various amaranth species. *Czech Journal of Animal Science*, *50*(4), 169–174.

Plate, A. Y. A., & Arêas, J. A. G. (2002). Cholesterol-lowering effect of extruded amaranth (*Amaranthus caudatus* L.) in hypercholesterolemic rabbits. *Food Chemistry*, *76*, 1–6.

Poczyczyński, P., Gomułka, P., Woźniak, M., & Szostak, I. (2014). Preliminary study on the partial substitution of fish oil with amaranth oil in diets for rainbow trout (*Oncorhynchus mykiss*) fingerlings: Effects on body/ composition and fatty acids contents. *Turkish Journal of Fisheries and Aquatic Sciences*, *14*, 457–462.

Popa, O., Babeanu, N. E., Popa, I., Nita, S., & Dinu-Parvu, C. E. (2015). Methods for obtaining and determination of squalene from natural sources. *BioMed Research International*, *2015*, 367202. 16 pages.

Preobrazhenskaya, N. S., Pokrovskij, M. V., Berezhnova, T. A., & Levchenko, Y. A. (2016). Hypolipidemic effects of amaranth oil in experimental doxorubicine cardiomyopathy. *Research Journal of Pharmaceutical, Biological and Chemical Sciences*, *7*(3), 1017–1021.

Qureshi, A. A., Lehmann, J. W., & Peterson, D. M. (1996). Amaranth and its oil inhibit cholesterol biosynthesis in 6-week-old female chickens. *Journal of Nutrition*, *126*, 1972–1978.

Raiciu, A. D., Popescu, M., Ivopol, G. C., Bordei, N., Alexandru, G., Crisan, I., et al. (2016). Therapeutic applications of vegetable oils and GC-MS evaluation of ω-3, ω-6 and ω-9 amounts in six oleaginous plants. *Revista de Chimie*, *67*, 2449–2453.

Rollán, G. C., Gerez, C. L., & LeBlanc, J. G. (2019). Lactic fermentation as a strategy to improve the nutritional and functional values of pseudocereals. *Frontiers in Nutrition*. Retrieved from https://www.frontiersin.org/article/10.3389/fnut.2019.00098.

Roy, A., Sahu, R. K., Matlam, M., Deshmukh, V. K., Dwivedi, J., & Jha, A. K. (2013). In vitro techniques to assess the proficiency of skin care cosmetic formulations. *Pharmacognosy Reviews*, *7*, 97–106.

Sakhare, S. D., Inamdar, A. A., Preetham Kumar, K. V., & Dharmaraj, U. (2017). Evaluation of roller milling potential of amaranth grains. *Journal of Cereal Science*, *73*, 55–61. https://doi.org/10.1016/j.jcs.2016.11.006.

Schnetzler, K. A., & Breene, W. M. (2018). Food uses and amaranth product research: A comprehensive review. In O. Paredes-Lopez (Ed.), *Amaranth biology, chemistry, and technology* (pp. 155–184). Boca Raton, FL: Taylor and Francis group.

Severino, P., Andreani, T., Jäger, A., Chaud, M. V., Santana, M. H. A., Silva, A. M., et al. (2014). Solid lipid nanoparticles for hydrophilic biotech drugs: Optimization and cell viability studies (Caco-2 & HEPG-2 cell lines). *European Journal of Medicinal Chemistry*, *81*, 28–34.

Shahidi, F., & Costa de Camargo, A. (2016). Tocopherols and tocotrienols in common and emerging dietary sources: Occurrence, applications, and health benefits. *International Journal of Molecular Sciences*, *17*, 1745.

Skwaryło-Bednarz, B. (2012). Assessment of content of fat and tocopherols in seeds of Amaranthus in relation to diversified fertilization with macroelements. *Ecological Chemistry and Engineering*, *19*, 273–279.

Steffensen, S. K., Rinnan, A., Mortensen, A. G., Laursen, B., de Troiani, R. M., Noellemeyer, E. J., et al. (2011). Variations in the polyphenol content of seeds of field grown Amaranthus genotypes. *Food Chemistry*, *129*, 131–138.

Su, Q., Rowley, K. G., Itsiopoulos, C., & O'Dea, K. (2002). Identification and quantification of major carotenoids in selected components of the Mediterranean diet: Green leafy vegetables, figs and olive oil. *European Journal of Clinical Nutrition*, *56*(11), 1149–1154.

Sun, H., Wiesenborn, D., Rayas-Duarte, P., Mohamed, A., & Hagen, K. (1995). Bench-scale processing of amaranth seed for oil. *Journal of the American Oil Chemists' Society*, *72*, 1551–1555.

Tang, Y., Li, X., Chen, P. X., Zhang, B., Liu, R., Hernandez, M., et al. (2016). Assessing the fatty acid, carotenoid, and tocopherol compositions of amaranth and quinoa seeds grown in Ontario and their overall contribution to nutritional quality. *Journal of Agricultural and Food Chemistry*, *64*, 1103–1110.

Tang, Y., & Tsao, R. (2017). Phytochemicals in quinoa and amaranth grains and their antioxidant, anti-inflammatory, and potential health beneficial effects: A review. *Molecular Nutrition & Food Research*, *61*(7), 1600767.

Tapia-Blacido, D. R., do Amaral Sobral, P. J., & Menegalli, F. C. (2013). Effect of drying conditions and plasticizer type on some physical and mechanical properties of amaranth flour films. *LWT- Food Science and Technology*, *50*, 392–400.

Valko, M., Leibfrit, D., Moncol, J., Cronin, M. T., Mazur, M., & Telser, J. (2007). Free radicals and antioxidants in normal physiological functions and human disease. *International Journal of Biochemistry & Cell Biology*, *39*(1), 44–84.

Venskutonis, P. R., & Kraujalis, P. (2013). Nutritional components of amaranth seeds and vegetables: A review on composition, properties, and uses. *Comprehensive Reviews in Food Science and Food Safety*, *12*, 381–412.

Vrignaud, S., Anton, N., Gayet, P., Benoit, J.-P., & Saulnier, P. (2011). Reverse micelle-loaded lipid nanocarriers: A novel drug delivery system for the sustained release of doxorubicin hydrochloride. *European Journal of Pharmaceutics and Biopharmaceutics*, *79*, 197–204.

Vrignaud, S., Hureaux, J., Wack, S., Benoit, J. P., & Saulnier, P. (2012). Design, optimization and in vitro evaluation of reverse micelle-loaded lipid nanocarriers containing erlotinib hydrochloride. *International Journal of Pharmaceutics*, *436*, 194–200.

Westerman, D., Santos, R. C. D., Bosley, J. A., Rogers, J. S., & Al-Duri, B. (2006). Extraction of Amaranth seed oil by supercritical carbon dioxide. *The Journal of Supercritical Fluids*, *37*(1), 38–52. https://doi.org/10.1016/j.supflu.2005.06.012.

Wolosik, K., & Markowska, A. (2019). *Amaranthus Cruentus* taxonomy, botanical description, and review of its seed chemical composition. *Natural Product Communications*. https://doi.org/10.1177/1934578X19844141.

Wolosik, K., Zareba, I., Surazynski, A., & Markowska, A. (2017). The possible pre- and post-UVA radiation protective effect of amaranth oil on human skin fibroblast cells. *Pharmacognosy Magazine*, *2017*(13), S339–S343.

Wun How, C., Rasedee, A., Manickam, S., & Rosli, R. (2013). Tamoxifen-loaded nanostructured lipid carrier as a drug delivery system: Characterization, stability assessment and cytotoxicity. *Colloids and Surfaces, B: Biointerfaces*, *112*, 393–399.

Yelisyeyeva, O., Cherkas, A., Semen, K., Kaminskyy, D., & Lutsyk, A. (2009). Study of aerobic metabolism parameters and heart rate variability and their correlations in elite athletes: A modulatory effect of amaranth oil. *Clinical and Experimental Medical Journal*, *3*, 293–307.

Yelisyeyeva, O. P., Semen, K. O., Ostrovska, G. V., Kaminskyy, D. V., Sirota, T. V., Zarkovic, N., et al. (2014). The effect of Amaranth oil on monolayers of artificial lipids and hepatocyte plasma membranes with adrenalin-induced stress. *Food Chemistry*, *147*, 152–159.

Yelisyeyeva, O., Semen, K., Zarkovic, N., Kaminskyy, D., Lutsyk, O., & Rybalchenko, V. (2012). Activation of aerobic metabolism by Amaranth oil improves heart rate variability both in athletes and patients with type 2 diabetes mellitus. *Archives of Physiology and Biochemistry*, *118*(2), 47–57.

Chapter 12

Cold pressed orange (*Citrus sinensis*) oil

Buket Aydeniz-Guneser
Department of Food Engineering, Engineering Faculty, Usak University, Usak, Turkey

Abbreviations list

B. cereus	*Bacillus cereus*
BAH	German Medicines Manufacturers' Association
DSC	differential scanning calorimeters
FAO	food and agriculture organization
FMC	food machinery corporation
JBT	John Bean technologies
MICs	minimum inhibitory concentration
P. aeruginosa	*Pseudomonas aeruginosa*
PUFA	polyunsaturated fatty acid
S. Typhi	*Salmonella Typhi*
SPME-GC/MS	solid-phase microextraction gas chromatography–mass spectrometry
tr	*trans*
USDA	the United States department of agriculture

1 Introduction

Varieties of citrus species are widely grown and cultivated in both tropical and subtropical regions of the world. The citrus genus originated in India more than 30 million years ago and is distributed throughout the world (Matheyambath, Padmanabhan, & Paliyath, 2016; Nayak et al., 2015; Sawamura, 2010). The most popular members of the citrus belonging to the Rutaceae family and Aurantioideae subfamily are sweet orange (*Citrus sinensis*), sour orange (*Citrus aurantium*), lemon (*Citrus limon*), key lime (*Citrus aurantifolia*), grapefruit (*Citrus paradisi*), mandarin (*Citrus reticulata*), and bergamot (*Citrus bergawia*) (Anwar et al., 2008; Calabrase, 2003).

According to data from 2016, nearly 675 million metric tons of fruit are cultivated worldwide every year. Based on the production quantities (million metric tonnes), watermelons (117), bananas (113), and apples (89) are located in the first three ranks, followed by grapes (77) and oranges (73) (FAO, 2019). Citrus varieties are among the favorite types of fruits produced and consumed extensively. In the FAO's forecast for 2017, citrus production around the world rose to 146.5 million tonnes, a 0.7% increment compared to the previous year's records (145.5 million tonnes to 2016) (FAO, 2019).

The orange is one of the top-rated citrus fruits, and orange production accounts for more than 50% of global citrus production. World production of oranges was forecast at 73 million metric tons for 2017/18. Fresh orange consumption is 29.5 million metrics tonnes, while orange account using by industrial processing are 17.7 million metrics tons (USDA, 2018). Orange production (in thousand metric tonnes) in the main producing countries from the 2013/14 season to the 2017/18 one is shown in Table 1 (USDA, 2018).

The cold pressing technique is a unique process applied to the extraction of edible oils from various oily seeds, kernels, peels, and fruits, etc. This technique plays an important role in producing specialty oils with typical characteristic aroma compounds and functional and nutritional compositions. Moreover, cold pressed oils are accepted as rich in polyunsaturated fatty acids (PUFAs) and preferred for their desirable flavor characteristics, with bioactive components having therapeutic effects (Aydeniz, Güneşer, & Yılmaz, 2014; Yilmaz, Aydeniz, Güneşer, & Arsunar, 2015).

The cold pressing technique has important advantages compared to other extraction techniques, such as lower energy cost, no needed solvent, and advanced equipment. Cold pressed oils do not contain any chemical contaminants or additives and require physical or chemical refining processes. Lower oil yield compared to solvent extraction is considered the most

Cold Pressed Oils. https://doi.org/10.1016/B978-0-12-818188-1.00012-8
Copyright © 2020 Elsevier Inc. All rights reserved.

TABLE 1 World orange production.

	Years (1000 metric tonnes)					
Production	**2013/14**	**2014/15**	**2015/16**	**2016/17**	**2017/18**	**2018/19 (Feb)**
Brazil	17.870	16.714	14.414	20.890	15.708	17.750
China	7.600	6.600	6.900	7.000	7.300	7.200
European Union	6.550	5.954	6.038	6.739	6.269	6.512
United States	6.140	5.763	5.523	4.616	3.555	5.022
Mexico	4.533	4.515	4.603	4.630	4.530	4.630
Egypt	2.570	2.635	2.930	3.000	3.120	3.420
Turkey	1.700	1.650	1.800	1.850	1.905	1.885
South Africa	1.723	1.645	1.275	1.363	1.550	1.620
Morocco	1.001	868	925	1.037	1.021	1.200
Vietnam	590	566	637	768	770	770
Argentina	800	800	800	700	600	500
Australia	430	430	455	480	515	500
Costa Rica	315	220	335	322	315	310
Guatemala	154	161	177	179	180	180
Israel	69	86	86	81	76	90
Other	209	166	179	183	182	183
Total	52.254	48.773	47.077	53.838	47.596	51.772
Fresh dom. consumption						
China	6.865	6.043	6.446	6.717	7.058	6.950
European Union	5.549	5.333	5.407	5.950	5.735	5.874
Brazil	6.036	5.196	4.940	4.761	4.933	4.976
Mexico	3.312	2.947	2.929	2.473	2.573	2.470
Egypt	1.385	1.350	1.380	1.380	1.480	1.690
Turkey	1.284	1.310	1.366	1.402	1.386	1.400
United States	1.357	1.263	1.346	1.184	1.253	1.277
Morocco	820	688	811	822	826	950
Vietnam	661	602	695	811	832	835
Russia	467	438	470	425	458	475
Saudi Arabia	274	384	371	357	362	370
Iraq	305	247	262	258	335	345
Australia	206	175	235	250	245	245
Bangladesh	113	118	176	169	221	241
Argentina	524	450	469	350	280	230
Other	1.708	1.650	1.757	1.643	1.634	1.674
Total	30.866	28.194	29.060	28.952	29.611	30.002

TABLE 1 World orange production—cont'd

	Years (1000 metric tonnes)					
Production	**2013/14**	**2014/15**	**2015/16**	**2016/17**	**2017/18**	**2018/19 (Feb)**
For processing						
Brazil	11.832	11.506	9.466	16.116	10.771	12.770
United States	4.420	4.133	3.684	3.001	2.014	3.350
Mexico	1.200	1.550	1.650	2.100	1.900	2.100
European Union	1.474	1.251	1.286	1.491	1.253	1.363
China	715	650	600	580	570	590
Costa Rica	208	125	230	238	232	227
Argentina	200	278	270	273	257	223
South Africa	471	403	142	123	201	220
Egypt	85	85	100	100	100	130
Turkey	100	80	100	100	98	95
Other	200	200	129	141	141	160
Total	20.905	20.261	17.657	24.263	17.537	21.228

Adapted from USDA (2019).

significant disadvantage of the cold pressing technique. Studies reported that some pretreatments such as heating, roasting, steaming microwave irradiation, and enzyme addition applied on oily material could enhance the oil yield (Aydeniz et al., 2014; Dündar Emir, Aydeniz, & Yılmaz, 2015; Khoddami, Man, & Roberts, 2014).

When both desired aroma compounds and a unique composition were considered together, it can be seen that cold pressed oils have a wide range of applications as a flavor or basic natural perfume agent in cosmetic, pharmacology, aromatherapy, personal care, manufacturing, and the food industry (Aydeniz-Güneşer & Yilmaz, 2017; Malacrida, Kimura, & Jorge, 2012; Matthaus & Özcan, 2012).

Cold pressed orange oil is added at a 0.01% level (*v*/v) to juices and concentrated juices in order to give a desirable unique flavor and aroma. Beverages and bakery and confectionery products such as cookies, chocolate, caramels, licorice, jelly, and chewing gum containing cold pressed orange oil as a flavoring agent are popular with many consumers (Braddock, 1999; Ringblom, 2004).

Recently, the cold pressing technique is located between the eco-friendly techniques to extraction oil from food waste and by-products (Aydeniz et al., 2014; Aydeniz-Güneşer & Yilmaz, 2017). Citrus seeds are accepted as valuable and cheap oil sources rich in omega and PUFAs, and phenolics such as naringin and hesperidin which have antiinflammatory and antihypertensive effects (Benavente-García & Castillo, 2008; Manners, 2007).

This chapter focuses on orange seed and peel oils obtained from the cold pressing technique and reviews the compositional properties, aromatic and sensory profile, major and minor bioactive compounds, health-promoting traits, and food and industrial applications of cold pressed orange oils.

2 Some physicochemical properties of cold pressed orange oils

Citrus seeds and citrus peel are preferred in the obtaining of aromatic essential oils. High oil contents of popular citrus types play an important role in consumer preference. Various factors such as geographical origin, soil type, climatic conditions, and harvest time can cause changes in the oil content of citrus seeds. The oil contents of lemon (*Citrus lemon*), orange (*Citrus sinensis*), and grapefruit (*Citrus paradisi*) seeds were determined as 34%–45%, 27%–52%, and 36%–49%, respectively, by the solvent extraction technique (Anwar et al., 2008; Saidani, Dhifi, & Marzouk, 2004). Furthermore, different studies exist in the literature indicating oranges' oil content, oil characteristics, and seed properties (Malacrida et al., 2012; Matthaus & Özcan, 2012; Saidani et al., 2004). Saidani et al. (2004) indicated that sweet oranges (51.8%) have a higher oil

TABLE 2 Some physicochemical properties of grapefruit and orange seeds and seed oils (El-Adawy, El-Bedawy, Rahma, & Gafar, 1999; Habib, Hammam, Sakr, & Ashoush, 1986; Malacrida et al., 2012).

Property	Orange seed
Oil yield (%)	51.80–41.50
Protein content (%, N*6.25)	17.40
Dietary fiber content (%)	22.53
Ash content (%)	2.95–3.17
	Orange seed oils
Refractive Index (25°C)	1.4681
Specific gravity (g/ml) (25°C)	0.914–0.933
Acid value (mg KOH/g oil)	0.21–0.673
Iodine number (g I/100 g oil)	99.2–102.57
Peroxide value (meq O_2/kg oil)	6.37
Saponification number (mg KOH/g oil)	190.2–196.8

content than bitter oranges (34%). In another study, the oil contents in bitter and sour orange seeds determined by a Twisselmann-type extractor were 56.5% and 57.4%, respectively (Matthaus & Özcan, 2012). Table 2 shows the physicochemical properties of seed oils obtained as a by-product during grapefruit and orange processing.

Similarly, fatty acid and tocopherol compositions having impact on oils characterization were reported by several researchers (Table 3). The major fatty acids were oleic, linoleic, and palmitic acids in *Citrus sinensis* (bitter) and *Citrus sinensis* (sour) seed oils according to Matthaus and Özcan (2012). Same researchers also reported 178 and 175 ppm as total

TABLE 3 Fatty acid and tocopherol compositions of citrus seed oils (Anwar et al., 2008; Malacrida et al., 2012; Saidani et al., 2004).

Fatty acid (%)	Grapefruit seed oils	Lemon seed oils	Orange seed oils
C16:0	32.17	17.17–21.03	24.73–26.42
C16:1	0.20	0.65	0.40
C18:0	3.64	2.30–3.67	3.40–5.27
C18:1	21.93	20.80–26.20	14.90–26.00
C18:2	36.10	24.70–44.31	38.44–40.30
C18:3	4.36	6.90–8.96	3.92–4.58
C20:0	0.29	0.31	0.38
Totally saturated	36.10	25.28	30.98–32.45
Totally unsaturated	64.10	74.72	59.85–69.02
Totally essential fatty acids	40.46		43.02
Tocopherol (mg/kg oil)			
α-tocopherol	380	102.49	300.19
γ-tocopherol	43.41	1.33	Not detected
δ-tocopherol	9.08	18.98	18.60

TABLE 4 The physicochemical properties of orange seed oils obtained by cold pressing (Aydeniz-Güneşer & Yilmaz, 2017).

Property	Cold pressed	Microwave + cold pressed
Oil yield[a] (%)	52.93 ± 0.38^B	62.99 ± 2.36^A
Specific gravity (g/mL (25°C)	0.92 ± 0.01^A	0.92 ± 0.01^A
Refractive Index (25°C)	1.47 ± 0.01^A	1.47 ± 0.01^A
Viscosity (25°C, cP)	59.55 ± 0.52^A	59.75 ± 0.15^A
Turbidity (NTU)	13.50 ± 4.05^B	34.25 ± 8.67^A
Color L	33.59 ± 3.22^A	31.67 ± 2.19^A
*a**	0.39 ± 0.18^B	1.77 ± 0.189^A
*b**	15.69 ± 6.20^A	15.05 ± 3.85^A
Sediment content (%)	6.65 ± 0.97^A	7.57 ± 2.82^A
Free fatty acid (%, linoleic acid)	0.27 ± 0.01^B	0.34 ± 0.04^A
Acid value (mg KOH/g oil)	0.54 ± 0.01^A	0.69 ± 0.08^A
Peroxide value (meq O_2/kg oil)	13.19 ± 3.23^A	15.49 ± 1.02^A
p-anisidine value	0.87 ± 0.19^A	0.34 ± 0.04^B
Iodine number (g I/100g oil)	117.95 ± 0.89^A	119.33 ± 4.03^A
Saponification number (mg KOH/g oil)	204.08 ± 2.11^A	201.06 ± 0.82^A
Unsaponifiable matter (%)	0.70 ± 0.15^A	0.95 ± 0.01^A
Total phenolics (μg GA/100g)	4.071 ± 179^A	5.607 ± 291^A
TEAC (μmol Trolox/100g oil)	12.431 ± 1164^B	16.512 ± 1168^A

[A-B] means in the same rows followed by different superscript letters were significantly different ($P < .05$).
[a]*Oil yield values are calculated over the total oil content of the orange seeds.*

tocopherol contents (the majority consists of α-tocopherol), and 2038 and 2506ppm as total sterols (the majority consists of β-sitosterol) in bitter and sour orange seed oils, respectively. Most of the reported values in Tables 2 and 3 are recognized as extracted seed oils by laboratory-scale solvent extraction. Citrus oil production by cold pressing is attracting increasing attention for people who do not prefer refined foods and recent studies have focused on the cold pressed technique.

In another study, Aydeniz-Güneşer and Yilmaz (2017) evaluated the effects of microwave and conventional-heat treatment on all quality parameters of orange seed oils produced via cold pressing (Table 4). It is apparent that microwave treatment on orange seeds before cold pressing could enhance oil yield to 63% by 15.8% increments. Turbidity measurements and a^* values indicating red-green colors in the control group (conventional heat treatment) were lower than those of the microwave-treated group.

Although there were no statistically significant differences for acid, peroxide, and iodine values, it was possible to say that these three values of microwave-treated samples have slightly higher values compared to another group. All orange seed oils produced by cold pressing could be fairly suitable with upper limits (4.0mg KOH/g oil acid value, and 15meq active O_2/kg oil peroxide value for cold pressed and virgin oils) appeared in the Codex Standard for Named Vegetable Oils (Alimentarius, 1999).

Acid and peroxide values were also reported for orange seed oils obtained from solvent extraction. Habib et al. (1986) determined the free fatty acid values of orange and mandarin seed oils as 0.21 and 0.65mg KOH/g oil, respectively. Similarly, El-Adawy et al. (1999) reported the acid value as 0.67mg KOH/g oil and peroxide value as 6.37meq active O_2/kg oil for orange seed oil extracted with *n*-hexane. Saloua, Eddine, and Hedi (2009) observed the 2.33meq active O_2/kg oil peroxide value and 1.86 *p*-anisidine value in orange seed oils.

Press meal, in other words the oily cake, is another important by-product obtained during the cold pressing technique, and it has been used in possible purposes due to its contained proteins, oil, and dietary fiber. Aydeniz-Güneşer and

TABLE 5 Proximate composition of the oily meal released during cold pressing (Aydeniz-Güneşer & Yilmaz, 2017).

			Orange seed meal	
Property	Control orange seed	Property	Control	Microwave + cold pressed
Seed size (mm)		Moisture (%)	11.11 ± 0.09^A	9.72 ± 1.46^A
Length	11.69 ± 0.49	Water activity (25°C)	0.71 ± 0.01^A	0.65 ± 0.06^B
Width	5.06 ± 0.13	Ash (%)	4.19 ± 0.09^A	4.41 ± 0.03^A
Height	4.01 ± 0.12	Oil[a] (%)	17.97 ± 0.58^A	13.57 ± 0.74^B
1000-seed weight (g)	197.21 ± 2.90	Protein[a] (%)	20.68 ± 0.29^B	25.61 ± 1.08^A
Skin: flesh ratio	0.29 ± 0.02	Color L	54.14 ± 0.99^A	51.82 ± 0.24^A
Color L	49.33 ± 3.17	Color *a**	6.20 ± 0.47^B	8.60 ± 0.28^A
*a**	2.42 ± 0.45	Color *b**	22.27 ± 0.73^A	22.48 ± 0.46^A
*b**	17.47 ± 1.71			
Moisture (%)	43.99 ± 0.13			
Water activity (25°C)	0.96 ± 0.01			
Oil[†] (%)	40.75 ± 2.17			
Protein[a] (%)	19.22 ± 0.09			
Ash (%)	1.58 ± 0.01			

[A-B] means in the same rows followed by different superscript letters were significantly different ($P < .05$).
[a] *Values are on dry weight basis.*

Yilmaz (2017) analyzed the pressed meals exiting after cold pressing of the orange seeds. Researchers observed that the oil and water contents in orange seed press meal could remain at up to 11% and 18% after the cold pressing, respectively. Moreover, press meal was accepted as a rich source of protein (21%–25%) and dietary fiber (82%–84%). Proximate composition of press meal of microwave roasted seed was statistically different from that of regular roasted seed in terms of water activity oil, protein, and color a^* values (more red). Even though press meal obtained from microwave-roasted seeds has higher protein and color a^* values, the oil and water activity values in this group were lower than those of the control group (press meal obtained from regular roasted orange seeds). Table 5 sets out the selected properties.

3 Extraction and processing of cold pressed orange oil

Essential oils have been known probably since ancient societies which have fragrance culture. From Mayans and Aztecs to Arabs, Egyptians, and Romans, essential oils were used for different purposes in perfumery, sacred ceremony, religious rituals, and also mummification (Chemat & Sawamura, 2010). Various and complex volatile substances have important effects on the essential oils aromatic composition and their chemical profile. The selection technique of extraction of essential oils from plant materials is closely related to crucial parameters such as the extraction temperature, extraction time, solvent type, extraction pressure, and the presence of oxygen (d'Acampora Zellner, Dugo, Dugo, & Mondello, 2015).

Conventional and innovative essential oil extraction techniques having the different working principles were reported by several researchers (Eikani, Golmohammad, & Rowshanzamir, 2007; Kaufmann & Christen, 2002; Lucchesi, Chemat, & Smadja, 2004; Vian, Fernandez, Visinoni, & Chemat, 2008). The most popular extraction procedures are as follows:

- **i.** Steam and hydro-distillation (SD)
- **ii.** Solvent extraction (SE)
- **iii.** Supercritical fluid extraction (SFE)
- **iv.** Pressurized-fluid extraction

v. Simultaneous distillation-extraction (SDE)
vi. Soxhlet extraction
vii. Solvent-free microwave-assisted hydrodistillation (SFMAHD)
viii. Water-free microwave-assisted hydrodistillation (WFMAHD)
ix. Dynamic headspace techniques (DHS)
x. Static headspace techniques (SHS)
xi. Solvent-assisted flavor evaporation (SAFE)
xii. Ultrasound accelerated solvent extraction
xiii. Solid-phase microextraction (SPME)
xiv. Direct thermal desorption (DTD)
xv. Cold pressing (Baser & Buchbauer, 2015; Chemat & Sawamura, 2010; d'Acampora Zellner et al., 2015)

It was reported that essential oils and aroma compounds that have characteristic and desirable odorants could be extracted from the different parts such as the flowers, leaves, twigs, peel, seeds, fruit, and pulp of citrus species by the cold pressing technique (Mukhopadhyay, 2000). Cold pressed essential oils have a yellowish-brown color, a fresh floral odor and are rich in limonene, β-myrcene, and α-pinene (Anwar, Ahmed, Speciale, Cimino, & Saija, 2015).

Cold pressed citrus oils including lemon, lime, bergamot, mandarin, sweet and bitter orange, and grapefruit oils contain more than 200 volatile components (mostly terpenes and their oxygenated derivatives, aldehydes, aliphatic and aromatic hydrocarbons, alcohols, and esters) and nonvolatile components (mainly oxygen heterocyclic compounds such as coumarins, psoralens, flavones, fatty acids, sterols, fat-soluble color pigments, and lipid waxes) (d'Acampora Zellner et al., 2015; Russo et al., 2015).

One of the most popular extraction systems is the JBT (John Bean Technologies) oil recovery system (formerly named the Food Machinery Corporation or FMC) to obtain cold pressed orange oil effectively. The working principle of the JBT system (Fig. 1) is based on the pressure and shredding mechanisms, another means of crushing the fruit together with the fruit peel. Four different products—peel, core, juice, and oil/water emulsion—are obtained at the end of extraction.

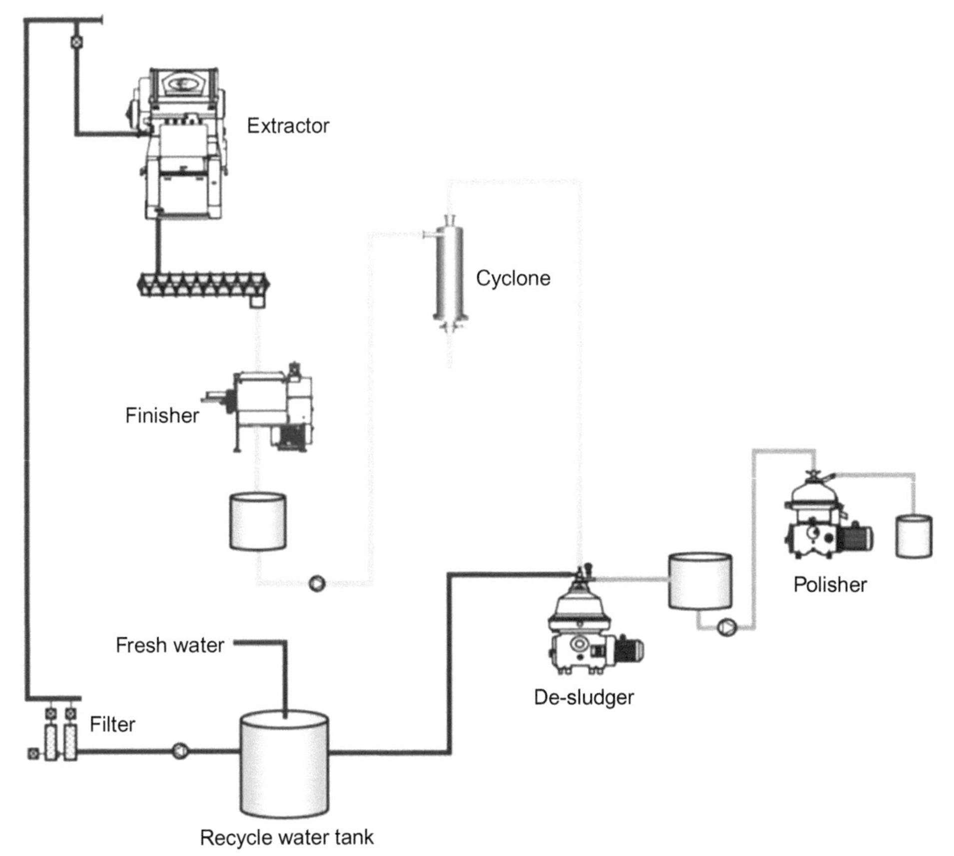

FIG. 1 Typical JBT oil recovery system.

The obtained juicy part (oil-in-water emulsion) contains not only fruit juice but also oil glands in peel. The oily phase was separated from the emulsion medium with the aid of a centrifugation process, then cold pressed orange oil was recovered (Di Giacomo & Di Giacomo, 2003; Ringblom, 2004). The JBT process has important advantages such as lower energy requirement and being able to extract the fruit juice and peel oil simultaneously (JBT, 2018).

Oil yield may vary significantly according to varieties of the fruit, harvest time, soil-climate interactions, ripening degree, extraction procedure applied to oily material conditions, and citrus types. The different citrus peel oils yield were determined as 0.2%–0.4% for grapefruit peel oil, 0.35%–0.5% for lime peel oil, and 0.5%–1.15% for lemon peel oil. Moreover, 3–7.5 g essential oil could be extracted from 1 kg of orange peel (0.3%–0.75% yield) (Kesterson & Braddock, 1975).

Another one of the most widely used machines is the Brown Oil Extractor. This system allows oil to be recovered from citrus peel before the citrus juice extraction. In the first step in this extraction unit, citrus fruits are exposed to pairs of rotating rollers with sharp needles, and the oil in the citrus peel is washed with water to remove it as an emulsion. The emulsion containing peel oil is obtained from a centrifugation process applied in the final stage. Although the Brown Oil Extractor and JBT systems have similar technical points, the peel oils recovered from both systems may exhibit different characteristics. The Brown Oil Extractor achieved the maximum oil yield and has a greater fruit processing capacity per unit time than the JBT system (Kesterson, Braddock, & Crandall, 1979; Somogyi, Barrett, & Hui, 1996).

3.1 Quality control of cold pressed orange oils

Cold pressed oils have contanied various impurities including phospholipids, pigments, free fatty acids, gummies, and waxes. because of unrefined oils that could exhibit sensitivity to heat, oxygen, and light. All of these factors that cause degradation of the oil quality play a role in consumer preferences and purchasing intentions for cold pressed oils. Unlike refined vegetable oils, cold pressed oils should be stored in dark glass bottles sealed tightly away from oxygen and sunlight (JBT, 2018). Another important point on cold pressed orange oils is to avoid metal contamination. Stainless steel should be preferred to iron and copper as a packaging material, to inhibit oil degradation and protect flavor components (JBT, 2018).

4 Fatty acids profile of cold pressed orange oil

It has been noted that different part of citrus varieties could be evaluated as a valuable oil source, containing oil in a wide range from 27% to 52%. It is important to mention that the fat-soluble compounds such as carotenoids, phenolics, tocopherols, and phytosterols also have important roles in the oil content of the citrus (Anwar et al., 2008; Malacrida et al., 2012). Jorge, da Silva, and Aranha (2016) confirmed total carotenoid, α-tocopherol, and phytosterol contents as 19, 136, and 1304 mg/kg oil in orange seed oils, respectively. The fatty acid composition also has major importance in determining oil characterization and oil quality. In a study (Aydeniz-Güneşer & Yilmaz, 2017), fatty acids composition were quantified in cold pressed orange seed oils and it was observed that 65% of the composition occurred as unsaturated fatty acids, mainly oleic (25%) and linoleic (37%). A balanced saturated/unsaturated fatty acids ratio, which is used as an indicator of oil nutritional quality, was approximately 0.47 in orange seed oils (Table 6).

TABLE 6 Fatty acid compositions (%) of orange seed oils obtained by cold pressing (Aydeniz-Güneşer & Yilmaz, 2017).

	Cold pressed	Microwave + cold pressed
Fatty acid (%)		
Palmitic (C16:0)	25.71 ± 0.23	26.10 ± 0.08
Palmitoleic (C16:1)	0.52 ± 0.01	0.52 ± 0.01
Stearic (C18:0)	5.75 ± 0.11	5.87 ± 0.03
Oleic (C18:1 n-9)	25.33 ± 0.12	24.81 ± 0.06
Linoleic (C18:2 n-6)	36.59 ± 0.17	36.61 ± 0.05
Linolenic (C18:3 n-3)	4.14 ± 0.02	4.19 ± 0.01

The fatty acid compositions of the cold pressed orange seed oil are summarized in Table 6. There was no effect of microwave treatment before cold pressing on the fatty acid composition. It is possible to find different results and discussions about fatty acid compositions in the literature. Saidani et al. (2004) analyzed the major fatty acids in sweet orange seed oils as 34% palmitic, 15% oleic, and 40% linoleic acid. Likewise, Matthaus and Özcan (2012) determined major four fatty acids as palmitic (27%), oleic (22%), linoleic (39%), and linolenic (4%) in orange seed oils grown in Turkey. In another study, fatty acid composition with a majority of linoleic acid (76%) was quantified in the Osage orange (*Maclura pomifera* (Rafin)) (Saloua et al., 2009).

A recent study by Matsuo, Miura, Araki, and Yoshie-Stark (2019) investigated the peel oil characteristics of *C. Natsudaidai* cultivars, a hybrid orange citrus. The results of this study indicated that most (70%) of the fatty acids present comprised unsaturated fatty acids such as linoleic acids (19–53 mg/100 g oil). Al Juhaimi et al. (2018) studied the effect of different drying temperatures (60°C, 70°C, and 80°C) on the antioxidant capacity and fatty acid composition of Kinnow mandarin, Orlando orange, and Eureka lemon seed oils. Although the quality parameters analyzed in all the citrus seed oils were significantly affected by drying conditions, the Orlando orange seed had the highest total phenolics content and antioxidant capacity values. In the fatty acid compositions of three citrus seed oils, linoleic and palmitic acids were the most abundant unsaturated and saturated fatty acids. Researchers observed that the Orlando orange has more palmitic acid (27%) and stearic acid (4%) than the Kinnow mandarin and Eureka lemon, and also approximately 23% oleic and 39% linoleic acids, respectively.

5 Minor bioactive compounds profile of cold pressed orange oils

The distilled products of concentrated cold pressed orange essential oil have characteristic organoleptic and volatile aroma compounds (Licandro & Odio, 2002). Hydrocarbons and alcohols (monoterpenic, sesquiterpenic, and aliphatic), aldehydes (monoterpenic and aliphatic), ketones (monoterpenic and sesquiterpenic), esters (monoterpenic and aliphatic), and their oxides are most abundant volatile components in cold pressed and also hydro distilled bitter orange oils (Anwar et al., 2015). It is accepted that a significant part of cold pressed citrus oils comprise volatile fractions (85%–99%) and also contain a minor amount of nonvolatile fractions (1%–15%) such as citropten, meranzin, isomeranzin, bergapten, cnicidin, oxypeucedanin, sinensetin, cnidilin, and phellopterin, classified in coumarins, psoralens, and polymethoxflavones (Anwar et al., 2015; Russo et al., 2015).

Russo et al. (2015) developed high-resolution HPLC methods to investigate the oxygen heterocyclic compounds in various cold pressed citrus essential oils. Researchers detected 38 different nonvolatile components in lemon, bergamot, grapefruit, orange, and mandarin essential oils. Ten compounds (mostly meranzim, isomeranzim as coumarins, and epoxybergamottin as polymethoxflavones) and six oxygen heterocyclic compounds (mostly nobiletin heptamethoxyflavone, and tangeretin as polymethoxflavones) were determined in bitter orange and sweet orange, respectively (Russo et al., 2015).

Physicochemical properties and the fatty acid, sterol, and tocopherol compositions of cold pressed orange seed oil were characterized by Aydeniz-Güneşer and Yilmaz (2017). In addition, the effects of microwave pretreatment (360 W, total 30 min) and conventional roasting (30 min at 150°C) before cold pressing on all analyzed values were compared. Fifteen different sterols were quantified in cold pressed orange seed oil. β-sitosterol (78%), campesterol (10%), and stigmasterol (3%) were determined as the major sterols in composition. When compared to tocopherol contents of both oil samples, cold pressed orange seed oils have a higher α-tocopherol content (283 ppm) than that (256 ppm) of the microwave pretreated group. It was shown that microwave pretreatment caused significant increases in oil yield, turbidity, total phenolics, and antioxidant capacities, although it has no statistically distinct differences on the sterol and tocopherol compositions (Table 7). Researchers claimed the cold pressed orange seed oils could be used in the food and beverage industries and also in nonfood applications.

In a recent study (Aydeniz-Güneşer & Yilmaz, 2017), the thermal behavior and the oxidative induction time of cold pressed orange oils were examined by DSC. Cold pressed orange oils exhibited crystallization onset temperatures range between 1.53°C and 1.90°C, and orange oils started to melt trending at around −25°C. It was noteworthy that microwave roasting (at 600 w power) of orange seeds prior to cold pressing improved the oxidative stability of orange oil (Table 8), in addition to having no effect on temperatures and enthalpies of crystallization and melting.

6 Flavor and sensory characterization of cold pressed orange oils

The volatile components are one of the major important factors in determining oil sensory quality. It is thought to be a linear relationship between the functional properties of thymol, limonene, α-pinene, terpinene, and oil stability and quality. Monoterpenic-, sesquiterpenic-, and aliphatic aldehydes, ketones, alcohols, esters, and oxides are accepted as the most popular volatile fractions found in citrus oils. Limonene, a monoterpene, is an aromatic volatile compound obtained from

TABLE 7 Sterol and tocopherol compositions (%) of the orange seed oils produced by cold pressing (Aydeniz-Güneşer & Yilmaz, 2017).

	Orange seed oil	
	Cold pressed	Microwave + cold pressed
Sterol (%)		
Cholesterol	0.66 ± 0.03^{A}	0.77 ± 0.06^{A}
Brassicasterol	0.07 ± 0.02^{B}	0.49 ± 0.32^{A}
24-Methylen cholesterol	0.05 ± 0.01^{B}	0.18 ± 0.07^{A}
Campesterol	9.43 ± 0.12^{A}	9.97 ± 0.17^{A}
Campestanol	0.21 ± 0.04^{A}	0.26 ± 0.08^{A}
Stigmasterol	3.26 ± 0.15^{A}	3.47 ± 0.06^{A}
Delta-7 campesterol	0.29 ± 0.02^{B}	0.65 ± 0.02^{A}
Delta-5,23 stigmastadienol	nd	0.19 ± 0.07
Chlerosterol	1.16 ± 0.05^{A}	0.83 ± 0.15^{B}
Beta-sitosterol	78.72 ± 0.91^{A}	77.65 ± 1.49^{A}
Sitostanol	0.62 ± 0.10^{B}	2.32 ± 0.99^{A}
Delta-5 avenasterol	4.59 ± 0.05^{A}	1.52 ± 0.93^{B}
Delta-5,24 stigmastadienol	0.17 ± 0.06^{B}	0.74 ± 0.37^{A}
Delta-7 stigmastenol	0.51 ± 0.11^{B}	0.77 ± 0.23^{A}
Delta-7 avenasterol	0.26 ± 0.12^{A}	0.18 ± 0.07^{B}
Tocopherol (mg/kg oil)		
α-tocopherol	283.40 ± 24.60^{A}	256.65 ± 8.21^{A}

A-B means in the same rows followed by different superscript letters were significantly different ($P < .05$).
nd: not detected.

TABLE 8 Thermal properties of the orange seed oils (Aydeniz-Güneşer & Yilmaz, 2017).

	Cold pressed	Microwave + cold pressed
Crystallization		
Onset_c (°C)	−1.90 ± 0.36	−1.53 ± 0.14
T_c (°C)	−5.12 ± 0.18	−4.63 ± 0.53
ΔH_c (J/g)	−30.56 ± 0.54	−31.20 ± 2.73
Melting		
Onset_m (°C)	−25.79 ± 0.28	−26.16 ± 0.08
T_{m1} (°C)	−23.04 ± 0.41	−23.78 ± 0.18
T_{m2} (°C)	−2.35 ± 0.06	−2.36 ± 0.17
T_{m3} (°C)	−0.63 ± 0.04	−0.47 ± 0.03
T_{m4} (°C)	5.97 ± 0.31	6.42 ± 0.28
ΔH_m (J/g)	56.81 ± 2.45	59.41 ± 6.05
OIT (min)	29.51 ± 5.08	44.28 ± 6.94

the rind of orange by cold pressing and has a critical role in the characteristic citrus smell (Anwar et al., 2015). It was reported that D-limonene can be obtained as a distilled by-product in the course of vacuum distillation of cold pressed orange essential oil (Licandro & Odio, 2002).

Aydeniz-Güneşer and Yilmaz (2019) investigated bioactive and volatile aromatic compounds of orange seed oil obtained by the cold pressing technique under mild operation conditions. According to the SPME-GC/MS method, 33 different volatiles were quantified in cold pressed oils obtained from oven preroasted and microwave preroasted orange seeds. The volatile composition comprise mainly monoterpene hydrocarbons and monoterpene alcohols, wherein D-limonene has the highest content (4500–5900 μg/kg oil) in all cold pressed oil samples followed by b-myrcene (87–124 μg/kg oil), a-terpineol (44–49 μg/kg oil), and b-pinene (31–47 μg/kg oil), respectively (Table 9).

The volatile profile of treated orange seeds with the microwave under 60 w power before cold pressing contained the main pyrazines including methyl pyrazine, 2,5-dimethlypyrazine, 3-methoxy-1-butanol, and octanal. Microwave treatment led to an increase in green, sweet, bitter almond, burnt, and roasted aroma levels and a decrease in creamy, woody, spicy, fresh citrus, and floral aroma levels compared to cold pressed oils obtained from nontreatment seeds. Similarly, Moufida and Marzouk (2003) observed that D-limonene levels changed from 63% to 90% in various orange species such as blood, sweet, and bitter orange wherein D-limonene was detected as a principal aromatic compound.

7 Health-promoting traits of cold pressed orange oil and oil constituents

Citrus species contain various bioactive compounds that provide demonstrated pharmacological and physiological benefits, and protective and reducing effects against chronic and degenerative diseases. In recent years, plant-originated phenolic compounds have drawn much attention as not only nutraceuticals but also potential cancer-preventing agents.

Minor bioactive lipids (sterols, tocopherols, carotenoids, hydrocarbons, and flavor and aroma compounds) in citrus provide important contributions via their health-promoting effects such as antioxidant, antimicrobial, antimutagenic, anticholesterol, antidiabetes, antiinflammatory, and antihypertensive activities (Aydeniz et al., 2014; Choi, Sawamura, & Song, 2010; Luther et al., 2007). Moreover, different parts of citrus such as lemon, orange, and mandarin are accepted as rich sources of flavonoids including naringin, hesperidin, and limonoid glucosides such as limonin, nomilin, and obacunone which have a free radical scavenger effect, antihypertensive effect, and inhibitory effect of stomach and lung carcinogenesis (Benavente-García & Castillo, 2008; Lam, Zhang, & Hasegawa, 1994; Manners, 2007; Miller et al., 1994). Both functional and clinical properties of essential oils rich in biologically active compounds were confirmed by several researchers in the literature.

In a recent study (Aydeniz-Güneşer & Yilmaz, 2019), the phenolic composition of cold pressed orange oils was determined and it was showed that the majority of phenolic composition (78%) occurred from flavonoids such as naringin and hesperidin (Table 10). In addition to flavonoids, phenolic acids, especially tr-ferulic acid (222 ppm), were also detected. Before the cold pressing process, microwave treatment at 360 w for a total of 30 min has a positive effect on the orange seeds and improved the flavonoid contents such as eriocitrin, rutin, naringin, naringenin, and neohesperidin. Researchers concluded that flavonoids contained in orange seeds could be transferred into the seed oil during cold pressing, and orange oil consumption provides positive impact on health.

Carotenoids are known as the main color pigments in citrus varieties, and carotenoid derivatives such as β-carotene, lutein, violaxanthin, anteraxanthin, and cryptoxanthin were commonly detected in orange fruit (Cebadera-Miranda et al., 2019). Total carotenoid content was determined as 7.7 ppm, and β-carotene and lutein contents were calculated as 7.37 and 7.35 ppm in cold pressed orange oils, respectively (Aydeniz-Güneşer & Yilmaz, 2019).

Aromatherapy, defined as "the art based on the use of essential oils for a medicinal purpose," has been subdivided into clinical, home-care, and aesthetic aromatherapies, and used to ease aches, spasms in influenza, asthma symptoms, and muscle pains, and reduce psychological symptoms such as stress, anxiety, depression, insomnia, and panic attacks (Kim, Nam, & Paik, 2005; Kumagai, Sawamura, & Son, 2010; Soden, Vincent, Craske, Lucas, & Ashley, 2004).

Seed and peel essential oils obtained from various citrus types have been used widely as massage oils and spray oils in aromatherapy and aroma baths as well as cosmetics and medicines (Kumagai et al., 2010).

Characteristic constituents of essential oil have various therapeutical and pharmacological properties. It has been reported that myrcene and caryophyllene belong to monoterpenes have sedative, analgesic, antibacterial, diuretic, and immune stimulant effectiveness. Terpineol located in alcohols group exhibited a mental stimulant effect as well as antivirus and antibacterial activities. Citronellol (CT) and monoterpene alcohol were naturally present in the citrus essential oil. The oil's health benefits include antifungal efficacy *in vitro* and ataraxic, antiinflammatory, anticonvulsant, antihyperalgesic, and antidiabetic efficacy in vivo (Brito et al., 2015; de Sousa et al., 2006; Imanishi, 2006; Srinivasan & Muruganathan, 2016).

TABLE 9 Volatile aromatics composition of the orange seed oil samples (Aydeniz-Güneşer & Yilmaz, 2019).

No	RI[a]	Volatile compound	Aroma/flavor description[b]	Concentration (μg/kg oil)	
				Cold pressed	Microwave + cold pressed
1	–	3-Methylbutanal	Fruity, sweet	nd	39.05 ± 6.97
2	716	Acetoin	Creamy, buttery	24.23 ± 0.87	18.11 ± 2.83
3	799	Hexanal	Green, grass	8.90 ± 0.03	16.30 ± 0.95
4	813	Furfural	Sweet, caramel, baked bread	24.07 ± 5.97	45.36 ± 2.10
5	825	Methyl pyrazine	Nutty, roasted cacao	nd	20
6	–	2-Furan menthol	Roasted cacao, burnt	4.07 ± 0.77	7.96 ± 0.33
7	877	Isoamyl acetate	Banana, fruity	2.80 ± 0.15	2.52 ± 0.53
8	891	Butrylactone	Creamy, fatty, caramel	0.38 ± 0.28	nd
9	911	2,5-Dimethlypyrazine	Nutty, roasted	nd	4.14 ± 0.25
10	914	Butyl isobutyrate	Pear, pineapple	0.89 ± 0.27	nd
11	924	α-Thujene	Woody, green herb	2.86 ± 0.26	2.38 ± 0.01
12	930	α-Pinene	Herbal, turpentine, woody	24.96 ± 0.22	17.68 ± 1.31
13	942	Isopropyl pentanoate	Fruity, pear	11.52 ± 0.08	11.59 ± 0.15
14	954	Benzaldehyde	Bitter almond	5.24 ± 1.12	6.50 ± 0.13
15	972	β-Pinene	Woody, green pine	47.59 ± 2.58	31.34 ± 0.72
16	991	β-Myrecene	Spicy, terpenic, herbal	124.89 ± 13.55	87.33 ± 6.50
17	1000	α-Phellandrene	Terpenic, citrus	3.99 ± 0.22	3.57 ± 0.18
18	1004	Octanal	Waxy, fatty	nd	2.03 ± 0.53
19	1014	3-Carene	Sweet	2.89 ± 0.49	2.33 ± 0.19
20	–	3-Methoxy-1-butanol	Alcohol	nd	2.70 ± 0.47
21	1015	Hexyl acetate	Fruity, green apple	4.15 ± 0.63	6.51 ± 4.27
22	1021	β-Cymene	Terpenic	22.27 ± 1.21	14.38 ± 0.73
23	1027	D-Limonene	Fresh citrus	5902.07 ± 308.59	4568.84 ± 287.69
24	1052	α-Ocimene	Fruity, floral	1.70 ± 0.07	1.19 ± 0.04
25	1058	γ-Terpinene	Terpenic	32.11 ± 1.02	27.97 ± 1.97
26	–	1-Octenol	Green herb, fatty, herbal	1.04 ± 0.18	0.55 ± 0.13
27	1088	(*Z*)-Linalooloxide	Earthy, floral	1.88 ± 2.67	nd
28	1089	α-Terpinolene	Herbal, woody, citrus	12.19 ± 2.51	10.14 ± 0.08
29	1107	Phenylethyl alcohol	Floral, sweet, rose	0.52 ± 0.38	nd
30	1139	(*E*)-Limonene oxide	Citrus	0.93 ± 0.16	1.32 ± 0.27
31	1179	4-Carvomenthol	Woody, spicy	0.85 ± 0.17	0.77 ± 0.01
32	1190	α-Terpineol	Floral, lilac	49.32 ± 8.79	44.17 ± 0.24
33	–	Decyl acetate	Waxy, fatty, creamy	0.87 ± 0.29	0.70 ± 0.02

[a]RI (Kovat Index) on HP 5 MS column. nd, not detected.
[b]Aromatic definitions of the volatile compounds are found from the web pages of http://www.thegoodscentscompany.com/index.html#, http://www.flavornet.org/flavornet.html.

TABLE 10 Flavonoid, phenolic acid and pigment composition of orange seed oils (Aydeniz-Güneşer & Yilmaz, 2019).

	Cold pressed	Microwave + cold pressed
Flavonoids (mg/kg)		
Catechin	14.87 ± 1.22^A	15.25 ± 0.35^A
Eriocitrin	31.01 ± 1.22^B	85.78 ± 3.80^A
Rutin	52.59 ± 0.39^B	76.48 ± 0.29^A
Naringin	234.28 ± 0.56^A	299.80 ± 27.20^A
Naringenin	10.38 ± 0.15^A	13.23 ± 1.66^A
Hesperidin	909.67 ± 5.63^A	903.40 ± 36.20^A
Neohesperidin	100.99 ± 0.81^A	125.90 ± 19.20^A
Kaempherol	8.64 ± 0.06^A	9.56 ± 0.83^A
Phenolic acids (mg/kg)		
Gallic acid	42.43 ± 3.57^A	29.41 ± 3.40^A
Syringic acid	6.93 ± 0.01^B	7.13 ± 0.01^A
tr-Ferulic acid	222.97 ± 2.76^B	364.30 ± 55.10^A
Rosmaniric acid	58.08 ± 0.48^A	77.90 ± 14.40^A
tr-2-Hydrocinnamic acid	41.65 ± 0.40^A	47.22 ± 5.41^A
Pigments (mg/kg oil)		
Total carotenoid content	7.64 ± 0.48^A	5.49 ± 0.14^B
β-Carotene content	7.37 ± 0.46^A	5.30 ± 0.14^B
Lutein content	7.35 ± 0.46^A	5.29 ± 0.14^B
Total chlorophyll (pheophytin a)	0.20 ± 0.02^B	0.34 ± 0.05^A

All values are average of three determinations ± standard error.
$^{A-B}$ Different lower case letters within the same row are significantly different ($P \leq .05$).

D-limonene is the main terpene and characteristic aromatic component found in the orange peel oils. Clinical studies related to D-limonene have shown that its benefits include antitumor, hepatoprotective, and chemopreventive effects in hepatocellular carcinoma and skin cancer (Giri, Parija, & Das, 1999; Imanishi, 2006; Stratton, Dorr, & Alberts, 2000). Additionally, therapeutic applications of aromatic essential oils can be found in different studies in the literature. Orange essential oils are accepted as very suitable in therapies for anxiety and nausea. Lehrner, Marwinski, Lehr, Johren, and Deecke (2005) assessed the impact of orange and lavender essential oils inhalation on patients' emotional moods in dental clinics. Researchers showed that the ambient odor of diffused orange oil has a distinct effect in reducing anxiety as well as increasing positive mood and calmness during the waiting time in dental clinics. Similar results were demonstrated for pediatric patients (aged 6–9 years), as reported by Jafarzadeh, Arman, and Pour (2013). In this study, orange essential oils were diffused in a dental operating room air (approximately 10 m^2) with the aid of an aroma diffuser for 2 min each 10 min. Following inhalation, it was observed to decrease anxiety symptoms as indicated by the salivary cortisol levels and pulse rates measured in the children.

Goes, Antunes, Alves, and Teixeira-Silva (2012) evaluated the effectivity of sweet orange essential oil (with the aid of inhalation in a surgical mask for 5 min) on male patients having anxiety symptoms. Sweet orange essential oil exhibited potential anxiolytic activity in humans. Extracted oils from fruit peel or flowers of Citrus aurantium (orange bigarade), Citrus sinensis (sweet orange), and *Citrus paradisi* (grapefruit) are found to be beneficial to alternative and complementary therapies in ailments such as agitation, fatigue, challenging behaviors, stress, and insomnia.

Several studies have considered that cold pressed oils rich in bioactive molecules can pose some antibacterial, antifungal, and antimutagenic activities. Crowell (1997) and Choi et al. (2010) investigated the antibacterial and antifungal activities of essential oils and identified that these activities were exhibited not only in the food matrix but also in the human metabolism.

Muthaiyan et al. (2012) reported the inhibitory effect of cold pressed orange oils against *Staphylococcus aureus* strains by means of oil-induced cell lysis and gene expression in the cell wall. In different studies (Nannapaneni et al., 2008, 2009; O'Bryan, Crandall, Chalova, & Ricke, 2008), cold pressed oils obtained from the Valencia orange inhibit the growth of pathogenic bacteria such as *Salmonella* spp., *Escherichia coli* O157: H7, *Campylobacter jejuni*, *Campylobacter coli*, *Campylobacter lari*, *Arcobacter butzleri*, and *Arcobacter cryaerophilus*. Van Hung, Chi, and Phi (2013) evaluated the antifungal activity of cold pressed oils produced by citrus varieties grown in Vietnam against *Mucor*, *Penicillium*, and *Fusarium*. The orange essential oils had inhibition percentages of 35%, 36.5%, and 59.5% on the growth of *Penicillium expansum*, *Mucor hiemalis*, and *Fusarium proliferatum*, respectively.

In another study, Phi, Van Hung, Chi, and Tuan (2015) determined the antimicrobial efficiency of different orange, lime, and pomelo varieties' essential oils. Limonene and myrcene-rich orange oils showed a significant inhibitory effect against *B. cereus*, *S. Typhi*, and *P. aeruginosa*.

Aydeniz-Güneşer, Demirel Zorba, and Yılmaz (2018) investigated the antimicrobial activity of cold pressed orange seed oils by disc diffusion and MICs methods, and compared their activities with the antibiotic discs. Cold pressed orange seed oils had inhibition minimum (6.6 mm) and maximum (9.5 mm) inhibition zones against *Bacillus cereus Holl.* No. 8 and *Staphylococcus aureus* ATCC 6538P, respectively. Moreover, microwave application on orange seeds before the cold pressing process had a remarkable impact on *Klebsiella pneumonia* ATCC 700603 and *E. coli* ATCC 2592 (10.21 mm and 11.00 mm inhibition zones, respectively (Table 11). No orange seed oils showed an inhibition effect on yeast species such as *Saccharomyces cerevisiae* ATCC 9763 and *Candida albicans* ATCC 10231. Values recorded from the MICs

TABLE 11 Antimicrobial activity of cold pressed citrus seed oils against bacteria and yeast species tested by disc diffusion assay (Aydeniz-Güneşer et al., 2018).

	Orange seed oil	
Microorganism	Cold pressed	Microwave + cold pressed
Inhibition zone diameter (mm)		
Staphylococcus aureus ATCC 29213	8.00 ± 0.76	9.50±0.76
Staphylococcus aureus ATCC 25923	8.38 ± 1.76	8.75±0.71
Staphylococcus aureus RSKK1009	9.25 ± 1.75	9.25±0.70
Staphylococcus aureus ATCC 6538P	9.50 ± 0.92	8.75±1.30
Micrococcus luteus ATCC 4698	–[a]	–
Bacillus cereus NCIMB 7464	–	7.87±1.46
Bacillus cereus Holl.	6.62 ± 0.52	6.50 ± 0.84
E.coli 0157:H7 ATCC 43895	7.62 ± 0.52	7.12 ± 0.35
Escherichia coli ATCC 25922	8.37 ± 1.30	11.00 ± 2.88
Escherichia coli ATCC 8739	7.12 ± 0.35	7.50 ± 0.76
Salmonella typhimurium ATCC 51812	–	7.50 ± 0.53
Salmonella typhimurium ATCC 14028	7.50 ± 0.53	7.50 ± 0.53
Salmonella enteritidis ATCC 13076	7.94 ± 0.68	7.75 ± 1.00
Pseudomonas aeroginosa ATCC 27853	8.00 ± 1.31	9.62 ± 2.85
*Klebsiella pneumoniae*ATCC700603	8.75 ± 1.75	10.12 ± 2.59
Saccharomyces cerevisiae ATCC 9763	–	–
Candida albicans ATCC 10231	–	–
Candida utilis	–	9.75 ± 2.26

All values were expressed as the average of four determinations ± standard deviation.
[a]*The zone of inhibition was smaller than the standard size (6 mm).*

TABLE 12 Antimicrobial activity of cold pressed citrus seed oils against bacteria and yeast species tested by minimum inhibition concentration (MIC) assay (Aydeniz-Güneşer et al., 2018).

	Orange seed oil	
	Cold pressed	Microwave + cold pressed
Microorganism	Concentration (%)	
Staphylococcus aureus ATCC 29213	–[a]	–
Staphylococcus aureus ATCC 25923	–	–
Staphylococcus aureus RSKK1009	–	–
Staphylococcus aureus ATCC 6538P	–	–
Micrococcus luteus ATCC 4698	–	–
Bacillus cereus NCIMB 7464	–	–
Bacillus cereus Holl.	–	–
E.coli 0157:H7 ATCC 43895	–	–
Escherichia coli ATCC 25922	100	100
Escherichia coli ATCC 8739	–	–
Salmonella typhimurium ATCC 51812	–	–
Salmonella typhimurium ATCC 14028	–	–
Salmonella enteritidis ATCC 13076	100	100
Pseudomonas aeroginosa ATCC 27853	–	–
Klebsiella pneumoniae ATCC700603	16	50
Saccharomyces cerevisiae ATCC 9763	–	–
Candida albicans ATCC 10231	–	–
Candida utilis		

All values were expressed as the average of six determinations ± standard deviation.
[a]*No inhibition at the maximum concentration (100% oil) used.*

method showed that *Klebsiella pneumonia* ATCC 700603 was inhibited at 16% concentration of cold pressed orange seed oil (Table 12). The researchers concluded that cold pressed orange seed oils may be used as natural antimicrobial agent in oily-based formulations.

8 Adulteration and authenticity of cold pressed oil

As mentioned before, oils can be extracted by both steam distillation and cold pressing techniques. A working group on contaminants by BAH (German Medicines Manufacturers' Association) reported a remarkable finding on the pesticide content of citrus essential oils. The researchers claimed that citrus essential oils obtained from cold pressing were more likely to contain pesticides than the steam-distilled essential oils were. It was observed that high temperatures at steam distillation could cause the degradation of thermolabile pesticides, although no heat treatment was applied during the cold pressing process. In the analyzed 600 essential oil samples, pesticide residues were not determined in distilled essential oils, in contrast to most (more than 50%) of the analyzed cold pressed oils having positive results for pesticide residues (Klier, Knödler, Peschke, Riegert, & Steinhoff, 2015). Similarly, Saitta, Di Bella, Salvo, Lo Curto, and Dugo (2000) explained the presence of organochlorines in lemon, orange, mandarin, and bergamot essential oils produced by cold pressing grown in Italy.

9 Conclusion

The sweet (*Citrus sinensis*) and sour orange (*Citrus aurantium*) are popular and commercial citrus varieties all over the world, and they account for 50% of global citrus production with 73 million metric tonnes with production quantity in 2017/2018. Cold pressing is the most suitable technique to obtain citrus seeds, peel, and fruit oils rich in aroma, taste, and natural bioactive components. Cold pressed orange oils are preferred mainly for food applications as well as cosmetic, pharmaceutical, and aromatherapy because of their flavor and functional properties such as antimicrobial, antioxidant, and antiinflammatory effects.

Cold pressed orange oil has high unsaturated fatty acid (linoleic and linolenic) levels and a balanced saturated/unsaturated fatty acids ratio (0.47); this oil is therefore accepted as a healthy oil source. Moreover, its carotenoid, flavonoid, and tocopherol contents have critical importance for consumers' healthy lifestyle. Clinical tests on D-limonene, a major volatile compound detected in cold pressed orange oils, drew attention due to orange oil's antitumor, hepatoprotective, and chemopreventive activity. This chapter discussed the compositional, characteristics, sensorial properties, bioactive and volatile compounds, and health-related claims of cold pressed orange oils. In addition, the usage and basic composition of oil-free meal gained after cold pressing, possible edible applications, and nonedible applications of cold pressed orange oils were briefly reviewed.

References

Al Juhaimi, F., et al. (2018). The effect of drying temperatures on antioxidant activity, phenolic compounds, fatty acid composition and tocopherol contents in citrus seed and oils. *Journal of Food Science and Technology*, *55*(1), 190–197. https://doi.org/10.1007/s13197-017-2895-y 2017/11/04. Springer India.

Alimentarius, C. (1999). Codex standard for named vegetable oils. *Codex Stan*, *210*, 1–13.

Anwar, F., Naseer, R., Bhanger, M. I., Ashraf, S., Talpur, F. N., & Aladedunye, F. A. (2008). Physico-chemical characteristics of citrus seeds and seed oils from Pakistan. *Journal of the American Oil Chemists' Society*, *85*(4), 321–330. https://doi.org/10.1007/s11746-008-1204-3.

Anwar, S., Ahmed, N., Speciale, A., Cimino, F., & Saija, A. (2015). Bitter orange (*Citrus Aurantium* L.) oils. In *Essential oils in food preservation, flavor and safety* (pp. 259–268). Academic Press.

Aydeniz, B., Güneşer, O., & Yılmaz, E. (2014). Physico-chemical, sensory and aromatic properties of cold press produced safflower oil. *Journal of the American Oil Chemists' Society*. https://doi.org/10.1007/s11746-013-2355-4.

Aydeniz-Güneşer, B., Demirel Zorba, N. N., & Yılmaz, E. (2018). Antimicrobial activity of cold pressed citrus seeds oils, some citrus flavonoids and phenolic acids. *Rivista Italiana Delle Sostanze Grasse*, *95*(2), 119–131.

Aydeniz-Güneşer, B., & Yilmaz, E. (2017). Effects of microwave roasting on the yield and composition of cold pressed orange seed oils. *Grasas y Aceites*, *68*(1), 175. https://doi.org/10.3989/gya.0800162.

Aydeniz-Güneşer, B., & Yilmaz, E. (2019). Comparing the effects of conventional and microwave roasting methods for bioactive composition and the sensory quality of cold-pressed orange seed oil. *Journal of Food Science and Technology*, *56*(2), 634–642. https://doi.org/10.1007/s13197-018-3518-y.

Baser, K. H. C., & Buchbauer, G. (2015). *Handbook of essential oils: Science, technology and applications*. CRC Press 1112 pages.

Benavente-García, O., & Castillo, J. (2008). Update on uses and properties of Citrus flavonoids: New findings in anticancer, cardiovascular, and anti-inflammatory activity. *Journal of Agricultural and Food Chemistry*, *56*(15), 6185–6205. https://doi.org/10.1021/jf8006568. American Chemical Society.

Braddock, R. J. (1999). *Handbook of citrus by-products and processing technology*. John Wiley & Sons, 247 p.

Brito, R. G., Dos Santos, P. L., Quintans, J. S., de Lucca Júnior, W., Araújo, A. A., Saravanan, S., et al. (2015). Citronellol, a natural acyclic monoterpene, attenuates mechanical hyperalgesia response in mice: Evidence of the spinal cord lamina I inhibition. *Chemico-Biological Interactions*, *239*, 111–117. https://doi.org/10.1016/j.cbi.2015.06.039.

Calabrase, F. (2003). Origin and history. In *Citrus: The genus citrus* (pp. 1–16). Dugo: CRC Press.

Cebadera-Miranda, L., Domínguez, L., Dias, M. I., Barros, L., Ferreira, I. C., Igual, M., et al. (2019). Sanguinello and Tarocco (*Citrus sinensis* [L.] Osbeck): Bioactive compounds and colour appearance of blood oranges. *Food Chemistry*, *270*, 395–402. https://doi.org/10.1016/j.foodchem.2018.07.094.

Chemat, F., & Sawamura, M. (2010). Techniques for oil extraction. In *Citrus essential oils* (pp. 9–36). Wiley Online Books. https://doi.org/10.1002/9780470613160.ch2.

Choi, H.-S., Sawamura, M., & Song, H.-S. (2010). Functional properties. In M. Sawamura (Ed.), *Citrus essential oils* (pp. 229–296). Wiley Online Books. https://doi.org/10.1002/9780470613160.ch6.

Crowell, L. P. (1997). *Monoterpenes in breast cancer chemoprevention*: (pp. 191–197). Kluwer Academic Publiser.

d'Acampora Zellner, B., Dugo, P., Dugo, G., & Mondello, L. (2015). Analysis of essential oils. In *Handbook of essential oils: Science, technology, and applications* (pp. 151–185). CRC Press.

Di Giacomo, A., & Di Giacomo, A. (2003). Essential oils production. In *Citrus: The genus citrus* (pp. 114–147). CRC Press.

Dündar Emir, D., Aydeniz, B., & Yılmaz, E. (2015). Effects of roasting and enzyme pretreatments on yield and quality of cold-pressed poppy seed oils. *Turkish Journal of Agriculture and Forestry*, *39*, 260–271. https://doi.org/10.3906/tar-1409-34.

Eikani, M. H., Golmohammad, F., & Rowshanzamir, S. (2007). Subcritical water extraction of essential oils from coriander seeds (*Coriandrum sativum* L.). *Journal of Food Engineering*, *80*(2), 735–740. https://doi.org/10.1016/j.jfoodeng.2006.05.015.

El-Adawy, T. A., El-Bedawy, A. A., Rahma, E. H., & Gafar, A. M. (1999). Properties of some citrus seeds. Part 3. Evaluation as a new source of protein and oil. *Food/Nahrung*, *43*(6), 385–391. https://doi.org/10.1002/(SICI)1521-3803(19991201)43:6<385::AID-FOOD385>3.0.CO;2-V John Wiley & Sons, Ltd.

FAO (2019). (2019). http://www.fao.org/faostat/en/#data/QC available at 22/01/2019.

Giri, R. K., Parija, T., & Das, B. R. (1999). D-limonene chemoprevention of hepatocarcinogenesis in AKR mice: Inhibition of c-jun and c-myc. *Oncology Reports*, *6*(5), 1123–1127.

Goes, T. C., Antunes, F. D., Alves, P. B., & Teixeira-Silva, F. (2012). Effect of sweet orange aroma on experimental anxiety in humans. *The Journal of Alternative and Complementary Medicine*, *18*(8), 798–804. https://doi.org/10.1089/acm.2011.0551 Mary Ann Liebert, Inc., publishers.

Habib, M. A., Hammam, M. A., Sakr, A. A., & Ashoush, Y. A. (1986). Chemical evaluation of Egyptian citrus seeds as potential sources of vegetable oils. *Journal of the American Oil Chemists' Society*, *63*(9), 1192–1196. https://doi.org/10.1007/BF02663951.

Imanishi, J. (2006). Medical aromatherapy. In *Citrus essential oils: Flavor and fragrance* (pp. 297–343). John Wiley & Sons.

Jafarzadeh, M., Arman, S., & Pour, F. F. (2013). Effect of aromatherapy with orange essential oil on salivary cortisol and pulse rate in children during dental treatment: A randomized controlled clinical trial. *Advanced Biomedical Research*, *2*, 10. https://doi.org/10.4103/2277-9175.107968. Medknow Publications & Media Pvt Ltd.

JBT (2018). *Citrus peel oil recovery systems operations manual. (2018). https://www.jbtc.com/-/media/files/foodtech/solutions/citrus-processing/citrus-peel-oil-recovery-manual.ashx available at 22/01/2019.*

Jorge, N., da Silva, A. C., & Aranha, C. P. M. (2016). Antioxidant activity of oils extracted from orange (*Citrus sinensis*) seeds. *Anais da Academia Brasileira de Ciências*, *88*, 951–958. Available at: http://www.scielo.br/scielo.php?script=sci_arttext&pid=S0001-37652016000300951&nrm=iso.

Kaufmann, B., & Christen, P. (2002). Recent extraction techniques for natural products: Microwave-assisted extraction and pressurised solvent extraction. *Phytochemical Analysis*, *13*(2), 105–113. https://doi.org/10.1002/pca.631. John Wiley & Sons, Ltd.

Kesterson, J. W., & Braddock, R. J. (1975). Total pell oil content of the major Florida citrus cultivars. *Journal of Food Science*, *40*(5), 931–933. https://doi.org/10.1111/j.1365-2621.1975.tb02236.x.

Kesterson, J. W., Braddock, R. J., & Crandall, P. G. (1979). Brown oil extractor: A new commercial method for the production of citrus essential oils in Florida. *Perfumer & Flavorist*, *4*(4), 9.

Khoddami, A., Man, Y. B. C., & Roberts, T. H. (2014). Physico-chemical properties and fatty acid profile of seed oils from pomegranate (*Punica granatum* L.) extracted by cold pressing. *European Journal of Lipid Science and Technology*, *116*(5), 553–562. https://doi.org/10.1002/ejlt.201300416. John Wiley & Sons, Ltd.

Kim, M. J., Nam, E. S., & Paik, S. I. (2005). The effects of aromatherapy on pain, depression, and life satisfaction of arthritis patients. *Journal of Korean Academy of Nursing*, *35*(1), 186–194. Available at: https://doi.org/10.4040/jkan.2005.35.1.186.

Klier, B., Knödler, M., Peschke, J., Riegert, U., & Steinhoff, B. (2015). Pesticide residues in essential oils: Evaluation of a database. *Pharmeuropa Bio & Scientific Notes*, *10*, 131–149.

Kumagai, C., Sawamura, M., & Son, U. S. (2010). Aromatherapy. In *Citrus essential oils* (pp. 297–341). Wiley Online Books. https://doi.org/10.1002/9780470613160.ch7.

Lam, L. T., Zhang, J., & Hasegawa, S. (1994). Citrus limonoid reduction of chemically induced tumorigenesis. *Food Technology*, *48*(11), 104–108.

Lehrner, J., Marwinski, G., Lehr, S., Johren, P., & Deecke, L. (2005). Ambient odors of orange and lavender reduce anxiety and improve mood in a dental office. *Physiology & Behavior*, *86*(1), 92–95. https://doi.org/10.1016/j.physbeh.2005.06.031.

Licandro, G., & Odio, C. E. (2002). Citrus by-products. In *Citrus: The genus citrus* (pp. 159–178). London: Taylor & Francis.

Lucchesi, M. E., Chemat, F., & Smadja, J. (2004). Solvent-free microwave extraction of essential oil from aromatic herbs: Comparison with conventional hydro-distillation. *Journal of Chromatography A*, *1043*(2), 323–327. https://doi.org/10.1016/j.chroma.2004.05.083.

Luther, M., Parry, J., Moore, J., Meng, J., Zhang, Y., Cheng, Z., et al. (2007). Inhibitory effect of chardonnay and black raspberry seed extracts on lipid oxidation in fish oil and their radical scavenging and antimicrobial properties. *Food Chemistry*, *104*(3), 1065–1073. https://doi.org/10.1016/j.foodchem.2007.01.034.

Malacrida, C. R., Kimura, M., & Jorge, N. (2012). Phytochemicals and antioxidant activity of citrus seed oils. *Food Science and Technology Research*, *18*(3), 399–404. https://doi.org/10.3136/fstr.18.399.

Manners, G. D. (2007). *Citrus limonoids*: Analysis, bioactivity, and biomedical prospects. *Journal of Agricultural and Food Chemistry*, *55*(21), 8285–8294. https://doi.org/10.1021/jf071797h. American Chemical Society.

Matheyambath, A. C., Padmanabhan, P., & Paliyath, G. (2016). Citrus fruits. In B. Caballero, P. M. Finglas, & F. B. T. Toldrá (Eds.), *Encyclopedia of food and health* (pp. 136–140). Oxford: Academic Press. https://doi.org/10.1016/B978-0-12-384947-2.00165-3.

Matsuo, Y., Miura, L. A., Araki, T., & Yoshie-Stark, Y. (2019). Proximate composition and profiles of free amino acids, fatty acids, minerals and aroma compounds in citrus natsudaidai peel. *Food Chemistry*, *279*, 356–363. https://doi.org/10.1016/j.foodchem.2018.11.146.

Matthaus, B., & Özcan, M. M. (2012). Chemical evaluation of citrus seeds, an agro-industrial waste, as a new potential source of vegetable oils. *Grasas y Aceites*, *63*(3), 313–320. https://doi.org/10.3989/gya.118411.

Miller, E. G., Sanders, A. P. G., Couvillon, A. M., Binnie, W. H., Hasegawa, S., & Lam, L. K. T. (1994). *Citrus limonoids* as inhibitors of oral carcinogenesis. *Food Technology*, 110–114.

Moufida, S., & Marzouk, B. (2003). Biochemical characterization of blood orange, sweet orange, lemon, bergamot, and bitter orange. *Phytochemistry*, *62*(8), 1283–1289. https://doi.org/10.1016/S0031-9422(02)00631-3.

Mukhopadhyay, M. (2000). *Natural extracts using supercritical carbon dioxide*. CRC Press. 360 pages.

Muthaiyan, A., Martin, E. M., Natesan, S., Crandall, P. G., Wilkinson, B. J., & Ricke, S. C. (2012). Antimicrobial effect and mode of action of terpeneless cold-pressed Valencia orange essential oil on methicillin-resistant *Staphylococcus aureus*. *Journal of Applied Microbiology*, *112*(5), 1020–1033. https://doi.org/10.1111/j.1365-2672.2012.05270.x. John Wiley & Sons, Ltd (10.1111).

Nannapaneni, R., Muthaiyan, A., Crandall, P. G., Johnson, M. G., O'Bryan, C. A., Chalova, V. I., et al. (2008). Antimicrobial activity of commercial citrus-based natural extracts against *Escherichia coli* O157:H7 isolates and mutant strains. *Foodborne Pathogens and Disease*, *5*(5), 695–699. https://doi.org/10.1089/fpd.2008.0124. Mary Ann Liebert, Inc., publishers.

Nannapaneni, R., Chalova, V. I., Crandall, P. G., Ricke, S. C., Johnson, M. G., & O'Bryan, C. A. (2009). Campylobacter and Arcobacter species sensitivity to commercial orange oil fractions. *International Journal of Food Microbiology*, *129*(1), 43–49. https://doi.org/10.1016/j.ijfoodmicro.2008.11.008.

Nayak, B., Dahmoune, F., Moussi, K., Remini, H., Dairi, S., Aoun, O., et al. (2015). Comparison of microwave, ultrasound and accelerated-assisted solvent extraction for recovery of polyphenols from *Citrus sinensis* peels. *Food Chemistry*, *187*, 507–516.

O'Bryan, C. A., Crandall, P. G., Chalova, V. I., & Ricke, S. C. (2008). Orange essential oils antimicrobial activities against Salmonella spp. *Journal of Food Science*, *73*(6), M264–M267. https://doi.org/10.1111/j.1750-3841.2008.00790.x. John Wiley & Sons, Ltd (10.1111).

Phi, N. T. L., Van Hung, P., Chi, P. T. L., & Tuan, P. D. (2015). Impact of growth locations and genotypes on antioxidant and antimicrobial activities of citrus essential oils in Vietnam. *Journal of Essential Oil-Bearing Plants*, *18*(6), 1421–1432. https://doi.org/10.1080/0972060X.2015.1004124. Taylor & Francis.

Ringblom, U. (2004). *The orange book*. Sweden: Lund University Library. Tetra Pak Processing Systems AB.

Russo, M., Bonaccorsi, I., Costa, R., Trozzi, A., Dugo, P., & Mondello, L. (2015). Reduced time HPLC analyses for fast quality control of citrus essential oils. *Journal of Essential Oil Research*, *27*(4), 307–315. https://doi.org/10.1080/10412905.2015.1027419. Taylor & Francis.

Saidani, M., Dhifi, W., & Marzouk, B. (2004). Lipid evaluation of some Tunisian citrus seeds. *Journal of Food Lipids*, *11*(3), 242–250. https://doi.org/10.1111/j.1745-4522.2004.01136.x.

Saitta, M., Di Bella, G., Salvo, F., Lo, Curto S., & Dugo, G. (2000). Organochlorine pesticide residues in Italian citrus essential oils, 1991 – 1996. *Journal of Agricultural and Food Chemistry*, *48*(3), 797–801. https://doi.org/10.1021/jf990331z. American Chemical Society.

Saloua, F., Eddine, N. I., & Hedi, Z. (2009). Chemical composition and profile characteristics of Osage orange *Maclura pomifera* (Rafin.) Schneider seed and seed oil. *Industrial Crops and Products*, *29*(1), 1–8. https://doi.org/10.1016/j.indcrop.2008.04.013.

Sawamura, M. (2010). Introduction and overview. In *Citrus essential oils* (pp. 1–8). John Wiley & Sons, Ltd. https://doi.org/10.1002/9780470613160.ch1.

Soden, K., Vincent, K., Craske, S., Lucas, C., & Ashley, S. (2004). A randomized controlled trial of aromatherapy massage in a hospice setting. *Palliative Medicine*, *18*(2), 87–92. https://doi.org/10.1191/0269216304pm874oa. SAGE Publications Ltd STM.

Somogyi, L., Barrett, D. M., & Hui, Y. H. (1996). *Processing fruits*. (Vol. 2). CRC Press. 570 pages.

de Sousa, D. P., Gonçalves, J. C. R., Quintans-Júnior, L., Cruz, J. S., Araújo, D. A. M., & de Almeida, R. N. (2006). Study of anticonvulsant effect of citronellol, a monoterpene alcohol, in rodents. *Neuroscience Letters*, *401*(3), 231–235.

Srinivasan, S., & Muruganathan, U. (2016). Antidiabetic efficacy of citronellol, a citrus monoterpene by ameliorating the hepatic key enzymes of carbohydrate metabolism in streptozotocin-induced diabetic rats. *Chemico-Biological Interactions*, *250*, 38–46. https://doi.org/10.1016/j.cbi.2016.02.020.

Stratton, S. P., Dorr, R. T., & Alberts, D. S. (2000). The state-of-the-art in chemoprevention of skin cancer. *European Journal of Cancer*, *36*(10), 1292–1297. https://doi.org/10.1016/S0959-8049(00)00108-8.

USDA (2018). https://apps.fas.usda.gov/psdonline/circulars/citrus.pdf available at 23.01.2019.

Van Hung, P., Chi, P. T. L., & Phi, N. T. L. (2013). Comparison of antifungal activities of Vietnamese citrus essential oils. *Natural Product Research*, *27*(4–5), 506–508.

Vian, M. A., Fernandez, X., Visinoni, F., & Chemat, F. (2008). Microwave hydrodiffusion and gravity, a new technique for extraction of essential oils. *Journal of Chromatography A*, *1190*(1), 14–17. https://doi.org/10.1016/j.chroma.2008.02.086.

Yilmaz, E., Aydeniz, B., Güneşer, O., & Arsunar, E. S. (2015). Sensory and physico-chemical properties of cold press-produced tomato (*Lycopersicon esculentum* L.) seed oils. *Journal of the American Oil Chemists' Society*. *92*(6). https://doi.org/10.1007/s11746-015-2648-x.

Chapter 13

Cold pressed *Fagus sylvatica* L. seed oil

Collen Musara and Alfred Maroyi

Medicinal Plants and Economic Development (MPED) Research Centre, Department of Botany, University of Fort Hare, Alice, South Africa

1 Introduction

The beech (*Fagus sylvatica* L.) is a common forest tree that belongs to the Fagaceae family (Wühlisch, 2008) and is considered as a climax species in most parts of Central and South Europe, including the British Isles and extending to Crimea (Geßler et al., 2007a, 2007b; Packham, Thomas, Atkinson, & Degen, 2012). *F. sylvatica* is the scientific name of the European beech or common beech, which is a strikingly beautiful tree and one of the most essential large deciduous trees and widespread broadleaved trees native to Europe (Durrant Houston, de Rigo, & Caudullo, 2016). The species is anemophilous and mostly allogamic (Schaffalitzky de Muckadell, 1955). An interesting fact is that the *F. sylvatica* tree can survive for hundreds of years with coppiced stands living for more than 1000 years, hence it can maintain its high growth rate until late maturity (Durrant Houston et al., 2016). It is a plant whose fruits (beechnuts) have been used since ancient times for the production of oil, both edible and technical, and this practice in the second half of the 19th century ceased (Siger, Dwiecki, et al., 2017; Siger, Józefiak, & Górnaś, 2017). These trees grow up to 30–45 m and yield nuts with a size of 12–18 mm (Krüssmann, 1977; Prasad & Gülz, 1989; Tutin, 1984). The fruits are ellipsoid of 2.5–3.5 cm in length and not considered edible, albeit they are eaten by monkeys (Packham et al., 2012).

The beech is usually single-stemmed with smooth, thin silver-gray bark, often with slight horizontal etchings and a 3 m trunk diameter, which resembles an elephant's foot. Its low-slung branches often droop to the ground (Durrant Houston et al., 2016). It has a terrestrial habitat and in cultivated forest stands, trees are normally harvested at 80–120 years of age (Wühlisch, 2008). The leaves are lime green with silky hairs, alternate, simple, and entire or with a slightly crenate margin, 10×7 cm, 6–7 lateral veins on each side of the leaf, 50–100 mm $\times$ 63.5 mm round-toothed leaf blade; the leaves have leaf stalks (Durrant Houston et al., 2016). As they mature, the leaves become darker shiny green and lose their hairs; they are stalked, with an oval to elliptic shape and a wavy edge, and they are often not abscissed in the autumn and instead remain on the tree until the spring (marcescence) (Johnson & More, 2006; Mitchell, 1974). The reddish-brown, torpedo-shaped leaf buds are long and slender, 30 mm $\times$ 3 mm thick, and they form short stalks, and have a distinctive criss-cross pattern (Durrant Houston et al., 2016). The beech is wind-pollinated, monoecious, with the tassel-like male catkins hanging from long stalks at the end of twigs, while female flowers grow in pairs, surrounded by a cup. Hot, sunny, and dry summers are ideal for abundant flower and seed production. The *F. sylvatica* male flowers are borne in the small catkins, which are a hallmark of the Fagales order (beeches, chestnuts, oaks, walnuts, hickories, birches, and hornbeams) while the female flowers produce beechnuts (Durrant Houston et al., 2016). The bitter edible nuts are sharply tri-angled in prickly four-lobed seed cases and are an important source of food for several animals; they play a major part in seed dispersal by hiding the seeds so that they cannot all be retrieved (Packham et al., 2012). Fruiting normally occurs every 5–8 years with small quantities of seeds being produced, but the fruit is dry and does not split open when ripe (Durrant Houston et al., 2016).

An analysis of pollen records indicates that the species has spread across Europe from small scattered populations left after the last glaciation (Magri, 2008). The distribution of beech in Europe is highly limited by the high summer temperatures, drought, and moisture availability probably because it is likely to become less competitive (Fang & Lechowicz, 2006; Hultén & Fries, 1986; Kramer et al., 2010). The common beech requires a growing season of at least 140 days (Magri, 2008), and is usually found at altitudes of more than 1000 m (Horgan et al., 2003; Packham et al., 2012). *F. sylvatica* is a hardy, most shade-tolerant broadleaved tree species (Praciak et al., 2013). The beech seedlings survive and grow below the canopy of established trees, hence natural regeneration is possible in silvicultural systems (Durrant Houston et al., 2016). The predominance of beech means a reduction of light level in the understorey vegetation level, and this will prevent most woodland plants from growing with only specialist shade-tolerant plants surviving beneath a beech canopy (Durrant Houston et al., 2016). *F. sylvatica* requires a wide range of soils with a pH range from 3.5 to 8.5 and a well-drained, fertile drier soil such as chalk, limestone, and light loams (Augusto, Ranger, Binkley, & Rothe, 2002). It tolerates rigorous winter

Cold Pressed Oils. https://doi.org/10.1016/B978-0-12-818188-1.00013-X
Copyright © 2020 Elsevier Inc. All rights reserved.

cold but is sensitive to winter/spring frost (Paule, 1995). The optimal growth of *F. sylvatica* is in humid soils situated on calcareous or volcanic parent rocks and humid atmosphere with precipitation well distributed throughout the year (Durrant Houston et al., 2016). Conversely, the beech will not thrive in waterlogged or compacted soils or regularly flooded or stagnant waters, since it requires good drainage (Geßler et al., 2007a, 2007b; Packham et al., 2012). Beech woodland is shady and is characterized by a dense carpet of fallen litter of around 900 g/m^2 per year that gives a significant amount of potash in its leaves, which may improve water retentions, thus contributing to soil conservation (Durrant Houston et al., 2016). The beech trees are characterized by a superficial, shallow root system, making them vulnerable to drought when compared to coniferous forests (Granier et al., 2007). Despite the trees' shallow roots, the natural cascading branch habit creates an amazing sight of foliage from ground to sky. Most varieties are native, have attractive purple leaves that attract beneficially, a fall color, and are drought tolerant and deer resistant (Durrant Houston et al., 2016).

Previous studies have shown that *F. sylvatica*, which grows in the wild, has seeds whose oil are high content of oils rich in nutritional and industrial values and hence are suitable as raw materials for seed oil (Prasad & Gülz, 1989). Natural oils have become more popular due to the increasing concern of consumers with regard to the use of synthetic chemical oils in food production. The worldwide market for healthy natural oil has grown rapidly because processing of natural products is safe, economical, and environmentally friendly. Natural oils of *F. sylvatica*, which are considered to be more nutritious, healthy, less polluted, and of better quality than other oils by many consumers are of interest (Przybylski & Mag, 2002; Vavpot, Williams, & Williams, 2013). This chapter will explore the functional properties of the *F. sylvatica* seeds, and usable products developed using *F. sylvatica* seed oil. It will also report the physical and chemical characteristics of *F. sylvatica* seed oil (Cornea & Rudenco, 1969; Rankoff & Popow, 1941), the fatty acid composition of its triacylglycerols (Neubeller & Buchloh, 1971), the lipid composition of the seed oil (Prasad & Gülz, 1989), and the total composition of the individual lipid classes namely triacylglycerols, diacylglycerols, monoacylglycerols, free fatty acids, phospholipids, and sterols including the fatty acid composition of acyl lipids. Extensive studies on the chemical composition and physicochemical properties of the oil obtained from *F. sylvatica* seeds were reported by Siger, Dwiecki, et al. (2017) and Siger, Józefiak, et al. (2017).

2 Properties of *F. sylvatica*

F. sylvatica has potential as an important source of seed oil, which is characterized by good taste and stability, even though the nuts are reported to contain low molecular weight toxic components (Kristensen, Larsen, & Sorensen, 1974). *F. sylvatica* seeds are key raw materials and since they contain the richest sources of natural microconstituents, like tocopherols, carotenoids, phytosterols, and phenolic compounds; they are utilized in the pharmaceutical and cosmetics industry (Luzia & Jorge, 2013; Oomah, Ladet, Godfrey, Liang, & Girard, 2000). *F. sylvatica* is considered an unconventional oilseed crop characterized by high lipids (27.3%) content. The fatty acid content in seed oil has linoleic (42.3%), and oleic (37.5%) acids as the major constituents (Prasad & Gülz, 1989). Other phytochemical compounds included triacylglycerols (94.8%), β-sitosterol (89.3%), stigmasterol (10.7%), γ-linolenic acid (4.2%), sterols (0.9%), diacylglycerols (0.8%), phospholipids (0.7%), γ-tocopherol (75.4 mg/100 g), and δ-tocopherol (34.1 mg/100 g) (Siger, Dwiecki, et al., 2017; Siger, Józefiak, et al., 2017).

The seeds of *F. sylvatica* exhibited storage behavior intermediate (Bonner, 1990; Gosling, 1991) between those of orthodox and recalcitrant seeds (Roberts, 1973) and characterized by a higher sensitivity to drying and storage conditions compared to seeds of other species (León-Lobos & Ellis, 2002; Pukacka, Hoffmann, Goslar, Pukacki, & Wójkiewicz, 2003). The optimal storage conditions of *F. sylvatica* seeds are proposed in the range from 255 K for 15% H_2O to 280 K for 9% H_2O, hence the *F. sylvatica* L. seeds indicated intermediate storage behavior (Pukacka & Wójkiewicz, 2003). The oil was reported to be a good source for edible applications because of its good taste and stability (Prasad & Gülz, 1989). *F. sylvatica* reproduction is predominantly sexual and is remarkable for the strength of its masting pattern, which seems to be based on an intrinsic biennial production of seeds (Pidek et al., 2010). Good masting years are also synchronous with years of high pollen production (Pidek et al., 2010). Seed production in a mast year can range up to 1500–4000 seeds per m^2 (Harmer, 1994). *F. sylvatica* has no persistent seed bank, and the seeds germinate in the spring following their dispersal (Grime, Hodgson, & Hunt, 2007).

3 Economic importance (uses of oil cake, economy)

The production of oils in the many countries has permitted and promoted the manufacturing and marketing of blended oils, thus boosting the economy (Ramadan, Amer, & Awad, 2008). Consumer interest in so-called "green processing" is on the increase because cold pressed oils are considered ecologically friendly and natural by consumers (Vavpot, Williams, & Williams, 2014). In the cold pressed *F. sylvatica* products, the minor constituents can have either pro-oxidative (free fatty acids, hydroperoxides, chlorophylls, and carotenoids) or antioxidative (tocopherols, phenols, and phospholipids) effects (Pekkarinen, Hopia, & Heinonen, 1998). The tocopherols along with other phenolic compounds protect the oil from autoxidation (Gutfinger, 1981; Salvador, Aranda, Gómez-Alonso, & Fregapane, 1999), which is the major cause of induction of toxic compounds and deterioration of vegetable oils.

F. sylvatica is characterized by the high content of γ- and δ-tocopherols, which are used in the food and pharmaceutical industries and in the production of cosmetics. γ-tocopherol is an effective scavenger of reactive nitrogen species and prevents DNA bases nitration, which makes *F. sylvatica* seeds oil an interesting raw material in the production of cosmetics (Siger, Dwiecki, et al., 2017; Siger, Józefiak, et al., 2017). This homologue of vitamin E specifically slows down the development of atherosclerosis (Christen et al., 1997). Siger, Dwiecki, et al. (2017) and Siger, Józefiak, et al. (2017) reported that oilseed crops through fat content above 15% of dry mass are considered as economically viable raw materials in oil production, thus *F. sylvatica* seed oil is more important. *F. sylvatica* seed oil may contain a higher level of lipophilic compounds including natural antioxidants and may bring nutraceutical and functional benefits to food systems compared to conventionally available oils, which may cause serious health disorders due to the generation of hazardous oxidation products (Ramadan, 2013).

Cold pressed oils are a good source of beneficial components, such as antioxidative phenolic compounds and other health-beneficial phytochemicals. *F. sylvatica* seed oil comprises also of γ-linolenic acid (GLA) through the formation of dihomo-gamma linolenic acid, whose intake is associated with many health benefits. Nourishments rich in oleic acid have been demonstrated to lower total cholesterol (Kok & Kromhout, 2004). GLA is able to reduce inflammation and may be effective in treating many diseases like rheumatoid arthritis, diabetic neuropathy, atopic eczema, cyclical mastalgia, aging-related disorders, and hyperactivity disorders (Siger, Dwiecki, et al., 2017; Siger, Józefiak, et al., 2017). It could be also considered as a novel antitumor agent (Materac, Marczyński, & Bodek, 2013). Previous clinical studies proved that essential fatty acids and other polyunsaturated fatty acids (PUFAs) are able to heal wounds and increase immunity (Harbige, Layward, Morris-Downes, Dumonde, & Amor, 2000), and they play a vital role in the development of membrane phospholipid bilayers of cellular and organelle membranes (Oliwiecki, Burton, Elles, & Horrobin, 1991).

Oxidative reactions limit the shelf life of fresh and processed foodstuff and lower the nutritive value of the oil (Koski et al., 2002). The oxidation of edible fats and oils can be controlled by using processing techniques that reduce loss of tocopherols and other antioxidants (Miraliakbari & Shahidi, 2008). Owing to their antioxidant properties, carotenoids are widely used in the food, medical, pharmaceutical, and cosmetic industries. These compounds are characterized by high activity against both reactive oxygen species and free radicals (Siger, Dwiecki, et al., 2017; Siger, Józefiak, et al., 2017). The content of carotenoids in *F. sylvatica* seed oils and very effective photo-oxidation inhibitors is at a high level compared to other cold pressed oils (Siger, Dwiecki, et al., 2017; Siger, Józefiak, et al., 2017). Antioxidants are well-recognized for their potential in health promotion and prevention of aging-related diseases, including cancer and heart disease (Yu et al., 2002). Cold pressed *F. sylvatica* seed oils may serve as novel dietary sources of natural antioxidants desired by consumers and food manufacturers for health promotion through improving nutrition, and as a food additive for rising food quality and stability and disease prevention (Yu, Zhou, & Parry, 2005). Antimicrobial properties of cold pressed oils are also utilized (Ramadan, Asker, & Tadros, 2012). In food production, the cold pressed oils from unconventional sources are used for preparation of blended oils with those commonly produced in the food industry like canola or sunflower oils (Ramadan, 2013; Rudzińska, Hassanien, Abdel-Razek, Ratusz, & Siger, 2016).

3.1 Other uses

A smoke flavoring, the Primary Product AM 01, is produced from *F. sylvatica* and applied in and on foods for preservation purposes (EFSA, 2010). The edible nuts, or masts, were once used to feed pigs, and in France, the nuts are roasted and used as a coffee substitute (Durrant Houston et al., 2016). Despite the seeds being toxic because of tannins and alkaloids, they are a source of food for birds, rodents, and humans (Lyle, Prahl, & Sparrow, 2006). Additionally the seeds may be ground to make flour, which could be eaten after leaching out the tannins via soaking (Fergus, 2002, Fergus & Hansen, 2005; Lyle et al., 2006). Compared to many other broadleaved trees, the beech tree is also used for pulp, while its high cellulose content can also be spun into modal, which is used as a textile akin to cotton (Durrant Houston et al., 2016).

Beech is the most diversely used tree species in Europe owing to its properties; the tree is hard and strong, its wear-resistance and valuable and excellent bending capabilities making it ideal for a variety of purposes, including fuel, furniture, cooking utensils, tool handles, flooring, stairs, plywood, panels, veneering, musical instruments, boat building, and sports equipment, as well as for pulp and firewood (Durrant Houston et al., 2016). Fine-grained and knot-free, the wood is hard, making it ideal for producing wooden mallets and workbench tops. The wood is easy to soak, dye, varnish, and glue, and has a pale cream color and good workability (Goldstein, Simonetti, & Watschinger, 1995). It is fine and short-grained, making it an easy wood to work with hence it is well-suited for minor carpentry. Steaming makes the wood even easier to machine and it has a superb finish and is immune to compression and ripping, and it is stiff once flexed (Goldstein et al., 1995).

The common beech makes a standard hedging plant and if clipped it does not shed its leaves, which in turn offer a year-round dense screen, providing an excellent habitat for garden birds (Durrant Houston et al., 2016). Due to its relatively high energetic potential, the tree can be coppiced for firewood and charcoal, and it is also traditionally used to smoke herring (Goldstein et al., 1995; Horgan et al., 2003; Packham et al., 2012). The wood burns well and is considered one of the best firewood for fireplaces (Durrant Houston et al., 2016). The beech is considered to conserve the productive capacity of the soil better than several other species, owing to its capacity of its root system assisting in the circulation of air throughout the soil. Additionally, it has a significant amount of potash in its leaves, which may improve water retention (Durrant Houston et al., 2016).

4 Extraction and processing of cold pressed oil

4.1 Cold press method

Cold pressing, also known as the scarification method, is mainly used for extracting essential oils from plants, flower, seeds, lemon oils, and tangerine oils, wherein cold pressing does not require the use of organic solvent or heat (Arnould-Taylor, 1981; CAC, 2001). Cold pressed means that the oil is expeller-pressed at low temperatures and pressure; the method is ideal for acquiring nonconventional stable seed oil (Somesh, Rupali, Swati, Jose, & Manish, 2015). Consumers' desire and tendencies for natural and least-processed safe food products have made the method a substitute for conventional practices (Lutterodt et al., 2010; Parry et al., 2006). The production of cold pressed oils reduces chemical risk, extraction time, and high energy cost (Rassem, Nour, & Yunus, 2016).

The process of cold press production begins with the nuts being ground (mechanical extraction) into an even paste, then pressed to squeeze the material from the pulp, which is slowly stirred, encouraging the oil to separate from the solid parts and clump together (Rassem et al., 2016). Heat is reduced and minimized throughout the batching of the raw material and the pressure is applied, forcing the oil out by a hydraulic press, which produces more friction and heat. In addition, it takes a scrubbing and heating action to shear the oil cells, allowing the oil to release itself (Rassem et al., 2016). The oil rises to the surface of the material and is separated from the material by centrifugation. Cold pressing process guarantees that the resulting oil is 100% pure, and retains more of its original properties, including its flavor and aromas, uses minimal heat, and is free of chemical contamination (Rassem et al., 2016; Yu et al., 2005). Furthermore, cold pressing is able to preserve bioactive compounds such as essential fatty acids, phenolics, flavonoids, and tocopherol in the oils (Simopoulos, Leaf, & Salem, 2000). This inventive extraction process also brings with it health benefits, including a higher proportion of monounsaturated fatty acids (MUFAs), high amounts of polyphenols, β-sitosterol, vitamin E, lutein, and lycopene, which helps to reduce prostate cancer and macular degeneration (Rassem et al., 2016). These benefits make cold pressed oils an appealing choice since no solvents and no further processing other than filtering are involved (Koski et al., 2002). The method also has a higher smoking point, making it suitable for frying, baking, stir-frying, deep-frying, searing, barbecuing, and roasting and sautéing (Rassem et al., 2016).

The cold process extraction technology is suitable for producing vegetable oils from nonfood oilseed crops with high recovery under low temperature and the cold press machines can process vegetable oils continuously with minimal labor (Topare, Raut, Renge, et al., 2011). The technique requires low capital cost, can be suitable for small-scale extraction, and its residual meals could be easily used as feed for animals (Topare et al., 2011). Cold pressing has a similar concentration of fatty acids in oil extracted from seeds as solvent extraction using petroleum ether, aqueous extraction, and supercritical fluid extraction using carbon dioxide (SCFE CO_2) (Cenkowski, Yakimishen, Przybylski, et al., 2006). In order to obtain good-quality oil, it is vital to use ripe, clean, nondamaged, and correctly stored seeds (Sionek, 1997). In addition, the temperature and duration of pressing must be well-maintained as they determine the quality of oil; for instance, the quality of oil is reduced at higher temperatures such as above 40°C (Sionek, 1997). The process yields oil of good quality permitted for consumption when the temperatures are below 40°C. The quality and the sturdiness of the obtained oil rely upon several factors, including the composition of fatty acids, the conditions of production and storage, and the presence of prooxidants and antioxidants (Sionek, 1997). At least 6%–15% of oil remains in the pressed residues, hence a limitation of the use of this method is the low efficiency of the process (Sionek, 1997).

4.2 Cold pressed oil extraction

According to Siger, Kaczmarek, and Rudzińska (2015), *F. sylvatica* seeds were pressed using a Farmet Uno cold pressing machine (Farmet, Czech Republic). The temperature inside the press is 60°C for cold pressing and the temperature of the oil produced was 39°C. The oil was pressed under ambient conditions of about 18°C. Once produced, the oil is centrifuged at 5000 rpm for 15 min, and transferred to small, dark bottles and stored at 4°C. At low temperatures during oil extraction machine running, mechanical aspects may have no or less effects on the pH values of oil (Tesoro, 2012).

The color of the oils is determined by CIE *L/a/b/* color scales using a ColorTec-PCM plus 30 mm Benchtop Colorimeter (Color Tec, USA). The analyzed oil was characterized by a golden-yellow color and liquid form at 4°C with the refraction index at 40°C of 1.4725 (Siger, Dwiecki, et al., 2017; Siger, Józefiak, et al., 2017). Density is related to the chemical structure and composition of the oils, and is measured by weighing the mass of the sample in a certain volume, then dividing the oil mass by the volume (Qu, Wei, & Julson, 2013). The density of the oil was measured with respect to water density at a temperature of 20°C and amounted to 0.916 g/cm^3; these values are in good agreement with the requirements of the Codex Alimentarius (CODEX-STAN: 210-1999, 1999).

4.3 Water content in oil

Water plays a pivotal role in the hydrolysis of oil throughout the numerous handling and processing steps, which produces free fatty acids and glycerol products (Choo, Birch, & Dufour, 2007). However, it is desirable to have low moisture content in oils because the presence of water may influence lipid oxidation rate and must be controlled (Budilarto & Kamal-Eldin, 2015). The content of water in *F. sylvatica* seed oil amounted to 911 ppm, which is comparatively higher than that produced by other cold pressed oils, such as milk thistle oil (779 ppm) or black poppy (831 ppm) (Budilarto & Kamal-Eldin, 2015). The analyzed *F. sylvatica* seed moisture content was 8.65% (Siger, Dwiecki, et al., 2017; Siger, Józefiak, et al., 2017).

5 Lipid composition and extraction

The triacylglycerols, diacylglycerols, monoacylglycerols, free fatty acids, and sterols and phospholipids constitute the total lipid extract. Based on a dry basis, the total lipid extract obtained was 40.7% of *F. sylvatica* seeds (Prasad & Gülz, 1989). With such a high content of oil (40.7%), these seeds are considered an economically good source for edible or industrial uses (Prasad & Gülz, 1989). The physicochemical properties of *F. sylvatica* seed oil are presented in Table 1. The dominant component in the oil is triacylglycerols (TG) with 94.8%, followed by sterols (SL) (0.9%), diacylglycerols (DG) (0.8%),

TABLE 1 Physicochemical properties of *F. sylvatica* seed oil (Siger, Dwiecki, et al., 2017; Siger, Józefiak, et al., 2017).

CIE LAB L^* a^* b^* value[a]	$L^*=65.54$; $a^*=10.28$; $b^*=31.14$
Color	Golden yellow
Physical state at 4°C	Liquid
Index of refraction at 40°C	1.4725
Density at 20°C (g/cm^3)	0.916
β-Carotene (mg/kg)	10.68
Chlorophyll (mg/kg)	2.56
Water content (ppm)	911
Iodine value (g/100 g)	116.52
Peroxide value (mEq O_2/kg)	1.11
Acid value (mg KOH/g)	1.68
p-Anisidine value	1.28
Total oxidation value	3.50
Moisture content	8.65%

a^* = redness by positive or greenness by negative; b^* = yellowness by positive or blueness by negative.
[a]*L^* = lightness of the sample (0 = black, 100 = white).*

TABLE 2 Composition and yield of *F. sylvatica* seed oil (Prasad & Gülz, 1989).

Lipid class	mg	% of oil
Triacylglycerols	4950	94.8
Diacylglycerols	42	0.8
Monoacylglycerols	16	0.3
Free fatty acids	26	0.5
Phospholipids	37	0.7
Sterols	47	0.9
Unidentified and lost in column	104	2.0
	5222	100.0

phospholipids (PL) (0.7%), free fatty acids (FFA) (0.5%), and monoacylglycerols (MG) (0.3%), as given in Table 2. The TG was fractionated in a pure form with chloroform (Prasad & Gülz, 1989).

5.1 Acyl lipids and sterols

The acyl lipids and sterols are qualitatively identified by thin-layer chromatography (TLC) and the sterol fraction constituted β-sitosterol and stigmasterol with 89.3% and 10.7%, respectively (Prasad & Gülz, 1989). The presence of sterols was further confirmed by GC-MS data (Gülz, Scora, Müller, & Marner, 1987). The qualitative analysis based on the response to specific spray reagents and comparison of R_f values with authentic samples specified the presence of phosphatidylethanolamine with (PE) (R_f3, 0.45; 7R_f4, 0.45); phosphatidylinositol (PI) had (R_f3, 0.15; R_f4, 0.25) and phosphatidylcholine (PC) was found only in traces with undetectable amounts. The quantification of phospholipids was carried out on a colorimetric estimation of phosphorus, and phosphatidylinositol (PI) and phosphatidylethanolamine (PE) were found to be present in 55.1% and 44.9%, respectively (Prasad & Gülz, 1989).

5.2 Fatty acid composition

The sum of monounsaturated, polyunsaturated, and saturated fatty acids were 39.7%, 42.2%, and 18.1%, respectively (Siger, Dwiecki, et al., 2017; Siger, Józefiak, et al., 2017). The fatty acid methyl esters from *F. sylvatica* seeds are provided in Table 3. The content of oleic acid (C18:1n9) equal to 38.6% and linoleic acid (C18:2n6) equal to 38% was more or less the same, followed by the content of γ-linolenic acid (C18:3n6) with 4.2%, palmitic acid (C16:0) representing 8.3%, stearic acid (C18:0) 2.5%, and arachidic acid (C20:0) 7.3% (Siger, Dwiecki, et al., 2017; Siger, Józefiak, et al., 2017). *F. sylvatica* seed oil tocopherols and fatty acids content are relatively similar to soybean oil and apricot kernels oil. *F. sylvatica* seed oil might be classified as oleic-linoleic acids oil because fats from this group are the most abundant (Siger, Dwiecki, et al., 2017; Siger, Józefiak, et al., 2017). The major fatty acids present in triacylglycerols were linoleic (18:2) and oleic (18:1), followed by small quantities of palmitic (16:0), gadoleic (20:1), linolenic (18:3), and stearic (18:0) acids (Prasad & Gülz, 1989).

5.3 Chlorophyll and β-carotene

The cold pressed oils had significantly different levels of bioactive compounds such as tocopherol, β-carotene, chlorophyll, phenolics, and flavonoids. Total phenolic contents of cold pressed oils were evaluated since phenolics may act as antioxidants and protect lipids from peroxidation (Ramadan, 2013). The highly reactive free radicals which are present in biological systems may cause reversible or irreversible damage to nucleic acids, proteins, lipids, or DNA (Goldberg, 2003) and may initiate degenerative diseases (cancer, heart disease, and arthritis) (Cadenas & Davies, 2000). However, antioxidant compounds like phenolic acids, phenolics, and flavonoids have the ability to trap these free radicals such as peroxide, hydroperoxide, or lipid peroxyl, and thus inhibit the oxidative mechanisms that lead to chronic diseases. In comparison with other cold pressed oils, *F. sylvatica* seeds contain high levels of carotenoids (lutein and zeaxanthin), which are considered as the most effective compounds in inhibition of photooxidation (Siger, Dwiecki, et al., 2017; Siger, Józefiak, et al., 2017).

TABLE 3 Fatty acid composition of *F. sylvatica* seeds oil (Siger, Dwiecki, et al., 2017; Siger, Józefiak, et al., 2017).

Fatty acid		%
Saturated		
Palmitic acid	C16:0	8.3
Stearic acid	C18:0	2.5
Arachidic acid	C20:0	7.3
Monounsaturated		
Oleic acid (n-9)	C18:1	38.6
Eruic acid (n-9)	C22:1	1.1
Polyunsaturated		
Linoleic acid (n-6)	C18:2	38.0
γ-Linolenic acid (n-6)	C18:3	4.2
SAFAs (saturated fatty acids)	18.1 ± 0.65	
MUFAs (monounsaturated fatty acids)	39.7 ± 1.30	
PUFAs (polyunsaturated fatty acids)	42.2 ± 1.48	
U/Ss (unsaturated/saturated fatty acids)	4.52	
Cox value	5.21	

The analyzed oil is characterized by a golden-yellow color, which may be associated with the existence of carotenoids (Siger, Dwiecki, et al., 2017; Siger, Józefiak, et al., 2017). A relatively low chlorophyll content of 2.56 mg/kg, was detected in the oil. The carotenoids content, expressed as β-carotene, was 10.68 mg/kg, which is higher than that of other cold pressed oils. The same concentration of β-carotene was spotted in Japanese quince oil (10.7 mg/kg), while pumpkin seed oil contains 6.84 mg/kg, and hazelnut and walnut oils more than 2 mg/kg (Górnaś et al., 2014). For comparison, the content of carotenoids in rapeseed oil was 14.7 mg/kg, in sunflower oil it was 0.7 mg/kg, and in safflower oil it was 2.2 mg/kg (Franke, Fröhlich, Werner, Böhm, & Schöne, 2010).

5.4 Tocopherols

Tocols in oils are believed to protect PUFAs from peroxidation and improve the oxidative stability (OS) of the oil (Ramadan, Zayed, & El-Shamy, 2007). A small quantity of tocols (200 mg) was detected in *F. sylvatica* seeds oil wherein three homologues of tocopherol (α-T, δ-T, and γ-T) were detected, as shown in Table 4. The highest tocopherol homologue

TABLE 4 Tocopherols composition of *F. sylvatica* seed oil (Siger, Dwiecki, et al., 2017; Siger, Józefiak, et al., 2017).

Compound	mg/100 g oil	%
α-tocopherol	0.78	0.70
β-tocopherol	–	–
γ–tocopherol	75.40	68.41
δ-tocopherol	34.05	30.89
Total tocopherols	110.2	100

was γ-T (75.4 mg/100 g), which is more than 68% of the total content of tocopherols, while the minimum homologue was α-T (0.78 mg/100 g). The δ-T (34.05 mg/10 g) had a high content of more than 30% of the total tocopherols (Table 4). *F. sylvatica* seeds oil is comparable to soybean oil, which also has a high content of γ-T (8.9–230 mg/100 g) and δ-T (15.4–93.2 mg/100 g) when considering the presence of individual homologues and the concentration of tocopherols (CODEX-STAN: 210-1999, 1999).

5.5 Phytosterols

According to recent research by Siger, Dwiecki, et al. (2017) and Siger, Józefiak, et al. (2017), the contents of phytosterols and cholesterol are reported in *F. sylvatica* seed oil. In line with the procedure described by the AOCS method (Ch6-91, 2011), sterol contents and composition are determined by GC. Phytosterols and cholesterol were identified by comparing retention data with the appropriate standards. The total sterols content amounted to 1497.5 μg/g oil. Seven phytosterols were identified in *F. sylvatica* seed oil: campesterol, D5-stigmasterol, β-sitosterol, D5-avenasterol, cycloartenol, 24-methylenecycloartenol, and citrostadienol, as well as one phytostanol, namely sitostanol. The main species was β-sitosterol, with 1070.9 μg/g oil, which represents more than 71% of the total content of sterols while other compounds did not exceed 2.5% of the total concentration of sterols. For comparison, the soybean oil, which is related to *F. sylvatica* seed oil by means of the homologous profile of tocopherols, contains twice as many sterols (267.1–326.6 mg/100 g) (Verleyen et al., 2002). *F. sylvatica* seed oil has a similar qualitative composition of phytosterols in oils as in viper's bugloss (7.44 mg/g) and sea buckthorn (10.4 mg/g), though the quantities of sterols were several times higher (Nogala-Kalucka, Rudzinska, Zadernowski, Siger, & Krzyzostaniak, 2010). The content of phytosterols in *F. sylvatica* seed oil tallies approximately to the total amount of sterols in olive oil (176.3 mg/100 g) and refined walnut oil (144 mg/100 g) (Siger, Dwiecki, et al., 2017; Siger, Józefiak, et al., 2017). The *F. sylvatica* seed oil contained also higher sterols level than palm oil (59.9–79.4 mg/100 g) and coconut oil (68.9 mg/100 g). A high cholesterol content in vegetable oil was presented by Hassanien et al. (2014) as 591.7 μg/g oil in tomato seed oil (15% of total sterols), and this corresponds to Siger, Dwiecki, et al. (2017) and Siger, Józefiak, et al. (2017) where cholesterol (57.04 mg/g oil, 4% of total sterols) was reported (Table 5).

5.6 Calculated oxidizability (Cox) value

The Cox value shows a relatively high resistance to oxidation, and the Cox value of the oils is calculated by using the formula proposed by Fatemi and Hammond (1980). The calculated oxidizability (Cox) value in *F. sylvatica* seeds oil was 5.21 (Siger, Dwiecki, et al., 2017; Siger, Józefiak, et al., 2017), which confirms a relatively high resistance to oxidation of *F. sylvatica* seed oil. The Cox value in *F. sylvatica* seed oil is comparable with 3.3 in apricot kernels, 6.6 in black cumin seeds, 7.3 in grape seeds, 6.5 in tomato seeds, and 7.8 in wheat germ oils (Siger, Dwiecki, et al., 2017; Siger, Józefiak, et al., 2017).

TABLE 5 Phytosterols profile of *F. sylvatica* seed oil (Siger, Dwiecki, et al., 2017; Siger, Józefiak, et al., 2017).

Compound	μg/g lipids	%
Cholesterol	57.04	3.81
Campesterol	148.7	9.93
D5-stigmasterol	7.34	0.49
β-Sitosterol	1070.9	71.51
Sitostanol	36.21	2.42
D5-avenasterol	86.86	5.80
Cycloartenol	25.67	1.71
24-Methylenecycloartanol	29.04	1.94
Citrostadienol	35.78	2.39
Total	1497.5	100

5.7 Peroxide value (PV) and acid value (AV)

Good quality of the oil is shown by the content of peroxides and acid values. According to the Codex Alimentarius (CODEX-STAN: 210-1999, 1999), cold pressed oil should have an acid value <4 mg KOH/g, while its peroxide value should be <15 meq O_2/kg. The PV of oil is an empirical measure of oxidation that is useful for samples that are oxidized to relatively low levels (Frankel, 2005). The PV and AV of the *F. sylvatica* seed oil were calculated using the standard methods from ISO 3960 (2007) and ISO 660 (2009), and the PV amounted to 1.11 meq O_2/kg, whereas the AV was 1.68 mg KOH/g (Siger, Dwiecki, et al., 2017; Siger, Józefiak, et al., 2017). Anisidine and calculated TOTOX (total oxidation value) values were reported for *F. sylvatica* seed oil as 1.28 and 3.50, respectively. Previous researches by Górnaś et al. (2014) indicated that PVs of cold pressed oils from varied sources ranged from 0.59 meq O_2/kg for Japanese quince to 3.10 meq O_2/kg for peanut oil.

6 Conclusion

The physicochemical analysis presented by several studies has indicated that *F. sylvatica* seeds are a promising raw material for obtaining cold pressed oil with good quality. The cold pressing technique serves as a milestone toward the development of healthy blended oils with improved and high amounts of linoleic and oleic acid, as well as tocopherols, squalene, and sterols, ultimately providing added excellent utility value to the oil and nutritional value. Cold pressed oils are health-promoting products and their consumption may improve human health and prevent certain diseases. *F. sylvatica* seed oil could be successfully applied in the cosmetics industry, pharmaceuticals, and agricultural products. The need for widely usable and easily available *F. sylvatica* seed oil continues to grow, and its use in several applications would allow conservation of *F. sylvatica* forests and management of tons of seeds. In terms of tocopherols and fatty acids content, it could be concluded that *F. sylvatica* seed oil corresponds approximately to soybean oil and apricot kernels oil, and the content of phytosterols in beech seeds oil is most comparable to the total amount of those compounds in olive oil.

References

Arnould-Taylor, W. E. (1981). *Aromatherapy for the whole person* (pp. 22–26). UK: Stanley Thornes.

Augusto, L., Ranger, J., Binkley, D., & Rothe, A. (2002). Impact of several common tree species of European temperate forests on soil fertility. *Annals of Forest Science*, *59*, 233–253.

Bonner, F. T. (1990). Storage of seeds: Potential and limitations for germplasm conservation. *Forest Ecology and Management*, *35*, 35–43.

Budilarto, E. S., & Kamal-Eldin, A. (2015). The supramolecular chemistry of lipid oxidation and antioxidation in bulk oils. *European Journal of Lipid Science and Technology*, *117*, 1095–1137.

CAC (2001). *Codex Alimentarius fats, oils and related products. Codex standard for edible fats and oils not covered by individual standards codex Stan 19-1981 (REV. 2–1999)*. Codex Alimentarius Commission.

Cadenas, E., & Davies, K. J. A. (2000). Mitochondrial free radical generation, oxidative stress, and aging. *Free Radical Biology & Medicine*, *29*, 222–230.

Cenkowski, S., Yakimishen, R., Przybylski, R., et al. (2006). Quality of extracted sea buckthorn seed and pulp oil. *Canadian Biosystems Engineering*, *48*, 3.9–3.16.

Ch6-91 (2011). *AOCS official method. Determination of the composition of the sterol fraction of animal and vegetable oils and fats by TLC and capillary GLC*. AOCS.

Choo, W. S., Birch, J., & Dufour, J. P. (2007). Physicochemical and quality characteristics of cold-pressed flaxseed oils. *Journal of Food Composition and Analysis*, *20*, 202–211.

Christen, S., Woodall, A. A., Shigenaga, M. K., Southwell-Keely, P. T., Duncan, M. W., & Ames, B. N. (1997). C-tocopherol traps mutagenic electrophiles such as NOx and complements a-tocopherol: Physiological implications. *Proceedings of the National Academy of Sciences*, *94*, 3217–3222.

CODEX-STAN: 210-1999 (1999). *Joint FAO/WHO Codex Alimentarius Commission. Codex standard for named vegetable oils*.

Cornea, I., & Rudenco, A. (1969). *Bull, etudes et recherches tech. Vol. 1* (p. 169).

Durrant Houston, T., de Rigo, D., & Caudullo, G. (2016). *Fagus sylvatica* and other beeches in Europe: Distribution, habitat, usage and threats. In J. San-Miguel-Ayanz, D. de Rigo, G. Caudullo, T. Houston Durrant, & A. Mauri (Eds.), *European atlas of forest tree species* (p. e012b90). Luxembourg: Publications Office of the European Union.

European Food Safety Authority (EFSA) (2010). *Scientific opinion on safety of smoke flavour—primary product—AM* (p. 01). European Food Safety Authority.

Fang, J., & Lechowicz, M. J. (2006). Climatic limits for the present distribution of beech (*Fagus* L.) species in the world. *Journal of Biogeography*, *33*, 1804–1819.

Fatemi, S. H., & Hammond, E. G. (1980). Analysis of oleate, linoleate and linolenate hydroperoxides in oxidized ester mixtures. *Lipids*, *15*, 379–385.

Fergus, C. (2002). *Trees of Pennsylvania and the northeast*. Stackpole Books. ISBN 978-0-8117-2092-2.

Fergus, C., & Hansen, A. (2005). *Trees of New England: A natural history*. Globe Pequot. ISBN 978-0-7627-3795-6.

Franke, S., Fröhlich, K., Werner, S., Böhm, V., & Schöne, F. (2010). Analysis of carotenoids and vitamin E in selected oilseeds, press cakes and oils. *European Journal of Lipid Science and Technology*, *112*, 1122–1129.

Frankel, E. N. (2005). *Lipid oxidation*. Bridgwater, England: The Oily Press.

Geßler, A., Keitel, C., Kreuzwieser, J., Matyssek, R., Seiler, W., & Rennenberg, H. (2007a). Potential risks for European beech (*Fagus sylvatica* L.) in a changing climate. *Trees*, *21*, 1–11.

Geßler, A., Keitel, C., Kreuzwieser, J., Matyssek, R., Seiler, W., & Rennenberg, H. (2007b). Potential risks for European beech (*Fagus sylvatica* L.) in a changing climate. *Trees*, *21*, 1–11.

Goldberg, Z. (2003). Clinical implications of radiation-induced genomic instability. *Oncogene*, *22*, 7011–7017.

Goldstein, M., Simonetti, G., & Watschinger, M. (1995). *Alberi d'Europa Guide pratiche Mondadori illustrati*. Guide pratiche manuali. ISBN 8804395559, 256 p., 9788804395553.

Górnaś, P., Siger, A., Juhneviča, K., Lācis, G., Šnē, E., & Seglina, D. (2014). Cold-pressed Japanese quince (*Chaenomeles japonica* (Thunb.) Lindl. Ex Spach) seed oil as a rich source of a-tocopherol, carotenoids and phenolics: A comparison of the composition and antioxidant activity with nine other plant oils. *European Journal of Lipid Science and Technology*, *116*, 563–570.

Gosling, P. (1991). Beechnut storage: A review and practical interpretation of the scientific literature. *Forestry*, *64*, 51–59.

Granier, A., Reichstein, M., Bréda, N., Janssens, I. A., Falge, E., Ciais, P., et al. (2007). Evidence for soil water control on carbon and water dynamics in European forests during the extremely dry year: 2003. *Agricultural Meteorology*, *143*, 123–145. https://doi.org/10.1016/j.agrformet.2006.12.004.

Grime, J. P., Hodgson, J. G., & Hunt, R. (2007). *Comparative plant ecology: A functional approach to common British species* (2nd ed.). Dalbeattie, UK: Castlepoint Press.

Gülz, P.-G., Scora, R. W., Müller, E., & Marner, F.-J. (1987). Epicuticular leaf waxes of *Citrus halimii* stone. *Journal of Agricultural and Food Chemistry*, *35*, 716–720.

Gutfinger, T. (1981). Polyphenols in olive oils. *Journal of the American Oil Chemists Society*, *58*, 966–968.

Harbige, L. S., Layward, L., Morris-Downes, M. M., Dumonde, D. C., & Amor, S. (2000). The protective effects of omega-6 fatty acids in experimental autoimmune encephalomyelitis (EAE) in relation to transforming growth factor-beta 1 (TGF-beta1) up-regulation and increased prostaglandin E2 (PGE2) production. *Clinical and Experimental Immunology*, *122*(3), 445–452.

Harmer, R. (1994). Natural regeneration of broadleaved trees in Britain. 2. Seed production and predation. *Forestry*, *67*, 275–286.

Hassanien, M. M. M., Abdel-Razek, A. G., Rudzińska, M., Siger, A., Ratusz, K., & Przybylski, R. (2014). Phytochemical contents and oxidative stability of oils from non-traditional sources. *European Journal of Lipid Science and Technology*, *116*, 1563–1571.

Horgan, T., et al. (2003). *A guide to forest tree species selection and silviculture in Ireland*. National Council for Forest Research and Development (COFORD).

Hultén, E., & Fries, M. (1986). *Atlas of north European vascular plants: North of the tropic of Cancer I-III*. Königstein, Germany: Koeltz Scientific Books.

Johnson, O., & More, D. (2006). *Collins tree guide: The most complete field guide to the trees of Britain and Europe*. Harper Collins. ISBN: 9780007207718.

Kok, F. J., & Kromhout, D. (2004). Atherosclerosis-epidemiological studies on the health effects of a Mediterranean diet. *European Journal of Nutrition*, *43*, i2–i5.

Koski, A., Psomiadou, E., Tsimidou, M., Hopia, A., Kefalas, P., Wahala, K., et al. (2002). Oxidative stability and minor constituents of virgin olive oil and cold-pressed rapeseed oil. *European Food Research and Technology*, *214*, 294–298.

Kramer, K., Degen, B., Buschbom, J., Hickler, T., Thuiller, W., Sykes, M. T., et al. (2010). Modelling exploration of the future of European beech (*Fagus sylvatica* L.) under climate change-range, abundance, genetic diversity and adaptive response. *Forest Ecology and Management*, *259*, 2213–2222.

Kristensen, L. B., Larsen, P. O., & Sorensen, H. (1974). Free amino acids and I'-glutamyl peptides in seeds of *Facus silvatica*. *Phytochemistry*, *13*, 2803.

Krüssmann, G. (1977). *Fagus, in: Handbuch der Laubgehölze. Vol. 2* (p. 68). Verlag Paul Parey.

León-Lobos, P., & Ellis, R. H. (2002). Seed storage behaviour of *Fagus sylvatica* and *Fagus crenata*. *Seed Science Research*, *12*, 31–37.

Lutterodt, H., Luther, M., Slavin, M., Yin, J.-J., Parry, J., Gao, J.-S. M., et al. (2010). Fatty acid profile, thymoquinone content, oxidative stability, and antioxidant properties of cold-pressed black cumin seed oils. *LWT—Food Science and Technology*, *43*, 1409–1413.

Luzia, D. M., & Jorge, N. (2013). Bioactive substance contents and antioxidant capacity of the lipid fraction of *Annona crassiflora* Mart. seeds. *Industrial Crops and Products*, *42*, 231–235.

Lyle, M. W., Prahl, F. G., & Sparrow, M. A. (2006). *Organic carbon, freely extractable lipids and ketone recovery of ODP Leg 199 sediments*. PANGAEA. https://doi.org/10.1594/PANGAEA.777323.

Magri, D. (2008). Patterns of post-glacial spread and the extent of glacial refugia of European beech (*Fagus sylvatica*). *Journal of Biogeography*, *35*, 450–463.

Materac, E., Marczyński, Z., & Bodek, K. H. (2013). The role of long-chain fatty acids omega-3 and omega-6 in human body. *Bromatologia i Chemia Toksykologiczna*, *2*, 225–233.

Miraliakbari, H., & Shahidi, F. (2008). Lipid class compositions, tocopherols and sterols of tree nut oils extracted with different solvents. *Journal of Food Lipids*, *15*, 1–16.

Mitchell, A. F. (1974). *A field guide to the trees of Britain and Northern Europe*. London: Collins.

Neubeller, J., & Buchloh, G. M. (1971). *Klosterneuburg, Rebe und Wein, Obstbau und Früchteverwertung. 21* (p. 469).

Nogala-Kalucka, M., Rudzinska, M., Zadernowski, R., Siger, A., & Krzyzostaniak, I. (2010). Phytochemical content and antioxidant properties of seeds of unconventional oil plants. *Journal of the American Oil Chemists Society*, *87*, 1481–1487.

Oliwiecki, S., Burton, J. L., Elles, K., & Horrobin, D. F. (1991). Levels of essential and other fatty acids in plasma and red cell phospholipids from normal controls and patients with atopic eczema. *Acta Dermato-Venereologica*, *71*(3), 224–228.

Oomah, B. D., Ladet, S., Godfrey, D. V., Liang, J., & Girard, B. (2000). Characteristics of raspberry (*Rubus idaeus* L.) seed oil. *Food Chemistry*, *69*, 187–193.

Packham, J. R., Thomas, P. A., Atkinson, M. D., & Degen, T. (2012). Biological flora of the British Isles: *Fagus sylvatica*. *Journal of Ecology*, *100*, 1557–1608.

Parry, J., Su, L., Moore, J., Cheng, Z., Luther, M., Rao, J. N., et al. (2006). Chemical compositions, antioxidant capacities, and antiproliferative activities of selected fruit seed flours. *Journal of Agricultural and Food Chemistry*, *54*, 3773–3778.

Paule, L. (1995). Gene conservation in European beech (*Fagus sylvatica* L.). *Forest Genetics*, *2*, 161–170.

Pekkarinen, S., Hopia, A., & Heinonen, M. (1998). Effect of processing on the oxidative stability of low erucic acid turnip rapeseed oil. *Fett-Lipid*, *100*, 69–74.

Pidek, I. A., Svitavská-Svobodová, H., van der Knaap, W. O., Noryśkiewicz, A. M., Filbrandt-Czaja, A., Noryśkiewicz, B., et al. (2010). Variation in annual pollen accumulation rates of Fagus along a N-S transect in Europe based on pollen traps. *Vegetation History and Archaeobotany*, *19*, 259–270.

Praciak, A., et al. (2013). *The CABI encyclopedia of forest trees*. Wallingford, Oxfordshire, UK; Boston, Massachusetts: CABI.

Prasad, R., & Gülz, P.-G. (1989). Composition of lipids of beech (*Fagus sylvatica* L.) seed oil. *Zeitschrift für Naturforschung Section C*, *44*, 735–738.

Przybylski, R., & Mag, T. (2002). Canola/rapeseed oil. In F. D. Gunstone (Ed.), *Vegetable oils in food technology: Composition, properties and uses* (pp. 98–127). CRC Press.

Pukacka, S., Hoffmann, S. K., Goslar, J., Pukacki, P. M., & Wójkiewicz, E. (2003). Water and lipid relations in beech (*Fagus sylvatica* L.) seeds and its effect on storage behaviour. *Biochimica et Biophysica Acta*, *1621*, 48–56.

Pukacka, S., & Wójkiewicz, E. (2003). The effect of temperature of drying on viability and some factors affecting storability of *Fagus sylvatica* seeds. *Acta Physiologiae Plantarum*, *25*, 163–169.

Qu, W., Wei, L., & Julson, J. (2013). An exploration of improving the properties of heavy bio-oil. *Energy & Fuels*, *27*, 4717–4722.

Ramadan, M. F. (2013). Healthy blends of high linoleic sunflower oil with selected cold pressed oils: Functionality, stability and antioxidative characteristics. *Industrial Crops and Products*, *43*, 65–72.

Ramadan, M. F., Amer, M. M. A., & Awad, A. (2008). Coriander (*Coriandrum sativum* L.) seed oil improves plasma lipid profile in rats fed diet containing cholesterol. *European Food Research and Technology*, *227*, 1173–1182.

Ramadan, M. F., Asker, M. M. S., & Tadros, M. (2012). Antiradical and antimicrobial properties of cold-pressed black cumin and cumin oils. *European Food Research and Technology*, *234*, 833–844.

Ramadan, M. F., Zayed, R., & El-Shamy, H. (2007). Screening of bioactive lipids and radical scavenging potential of some solanaceae plants. *Food Chemistry*, *103*, 885–890.

Rankoff, D. G., & Popow, A. (1941). Untersuchungen übr das Öl aus den Samen von Wassermelonen. *Fette und Seifen*, *48*(8), 489–491.

Rassem, H. H. A., Nour, A. H., & Yunus, R. M. (2016). Techniques for extraction of essential oils from plants: A review. *Australian Journal of Basic and Applied Sciences*, *10*(16), 117–127.

Roberts, E. H. (1973). Predicting the storage life of seeds. *Seed Science and Technology*, *1*, 499–514.

Rudzińska, M., Hassanien, M. M. M., Abdel-Razek, A. G., Ratusz, K., & Siger, A. (2016). Blends of rapeseed oil with black cumin and rice bran oils for increasing the oxidative stability. *Journal of Food Science and Technology*, *53*, 1055–1062.

Salvador, M. D., Aranda, F., Gómez-Alonso, S., & Fregapane, G. (1999). Contribution of chemical components of Cornicabra virgin olive oils to oxidative stability. A study of three successive crop seasons. *Journal of the American Oil Chemists' Society*, *76*, 427–432. https://doi.org/10.1007/s11746-999-0020-8.

Schaffalitzky de Muckadell, M. (1955). A development stage in *Fagus sylvatica* L. characterized by abundant flowering. *Physiologia Plantarum*, *8*, 370–373.

Siger, A., Dwiecki, K., Borzyszkowski, W., Turski, M., Rudzińska, M., & Nogala-Kałucka, M. (2017). Physicochemical characteristics of the cold-pressed oil obtained from seeds of *Fagus sylvatica* L. *Food Chemistry*, *225*, 239–245. https://doi.org/10.1016/j.foodchem.2017.01.022.

Siger, A., Józefiak, M., & Górnaś, P. (2017). Cold-pressed and hot-pressed rapeseed oil: The effects of roasting and seed moisture on the antioxidant activity, canolol, and tocopherol level. *Acta Scientiarum Polonorum Technologia Alimentaria*, *16*(1), 69–81.

Siger, A., Kaczmarek, A., & Rudzińska, M. (2015). Antioxidant activity and phytochemical content of cold-pressed rapeseed oil obtained from roasted seeds. *European Journal of Lipid Science and Technology*, *117*, 1225–1237.

Simopoulos, A. P., Leaf, A., & Salem, N., Jr. (2000). Workshop statement on the essentiality of and recommended dietary intakes for omega-6 and omega-3 fatty acids. *Prostaglandins, Leukotrienes, and Essential Fatty Acids*, *63*(3), 119–121.

Sionek, B. (1997). Cold crushed oils. *Roczniki PZH*, *48*(3), 283–293.

Somesh, M., Rupali, S., Swati, S., Jose, M., & Manish, M. (2015). In-vitro comparative study on antimicrobial activity of five extract of few Citrus fruit: Peel & Pulp vs gentamicin. *Australian Journal of Basic and Applied Sciences*, *9*(1), 165–173.

Tesoro (2012). *Safety data sheet-jet fuel: 1–8*. San Antonio, TX: Tesoro Refining & Marketing Co. Available at: http://www.tsocorp.com/stellent/groups/corpcomm/documents/tsocorp_documents/msdsjetfuel.pdf. (Accessed 30 September 2019).

Topare, N. S., Raut, S. J., Renge, V. C., et al. (2011). Extraction of oil from algae by solvent extraction and oil expeller method. *International Journal of Chemical Sciences*, *9*, 1746–1750.

Tutin, T. G. (1984). Fagus. In T. G. Tutin, et al. (Ed.), *Vol. 1. Flora EuropaeaCambridge: Cambridge University Press.*

Vavpot, V. J., Williams, R. J., & Williams, M. A. (2013). Extrusion/expeller pressing as a means of processing green oils and meals. Green vegetable oil processing. In W. E. Farr, & A. Proctor (Eds.), *Green vegetable oil processing* (pp. 1–18). USA: AOCS Press. Champaign.

Vavpot, V. J., Williams, R. J., & Williams, M. A. (2014). In W. E. Farr, & A. Proctor (Eds.), *Green vegetable oil processing*. Champaign: AOCS.

Verleyen, T., Forcades, M., Verhé, R., Dewettinck, K., Huyghebaert, A., & De Greyt, W. (2002). Analysis of free and esterified sterols in vegetable oils. *Journal of the American Oil Chemists' Society*, *79*, 117–122.

Wühlisch, V. G. (2008). *Euforgen technical guidelines for genetic conservation and use for European beech (*Fagus sylvatica*)* (p. 6). Rome, Italy: Biodiversity International.

Yu, L., Haley, S., Perret, J., Harris, M., Wilson, J., & Qaian, M. (2002). Free radical scavenging properties of wheat extracts. *Journal of Agricultural and Food Chemistry*, *50*, 1619–1624.
Yu, L., Zhou, K., & Parry, J. (2005). Antioxidant properties of cold pressed black caraway, carrot, cranberry, and hemp seed oils. *Food Chemistry*, *91*, 723–729.

Further reading

AOCS Official Method Ce 2-66 (1997). *Preparations of methyl esters of fatty acids. Official methods and recommended practices of the AOCS* (6th ed.). Urbana, IL: AOCS.
Cc13i-96 (2013). *AOCS official method. Determination of chlorophyll pigments in crude vegetable oils.* AOCS.
Poulsen, K. M., & Knudsen, H. (1999). Viability constants based on eight years storage of beech nuts (*Fagus sylvatica* L.). *Seed Science Technology*, *27*, 1037–1039.
Siger, A., Kachlicki, P., Czubiński, J., Polcyn, D., Dwiecki, K., & Nogala-Kałucka, M. (2014). Isolation and purification of plastochromanol-8 for HPLC quantitative determinations. *European Journal of Lipid Science and Technology*, *116*, 413–422.
Siger, A., Michalak, M., & Rudzińska, M. (2016). Canolol, tocopherols, plastochromanol-8, and phytosterols content in residual oil extracted from rapeseed expeller cake obtained from roasted seed. *European Journal of Lipid Science and Technology*, *118*, 1358–1367.

Chapter 14

Cold pressed lemon (*Citrus limon*) seed oil

Ambrogina Albergamo[a,b], Rosaria Costa[a], and Giacomo Dugo[a,b]

[a]*Department of Biomedical, Dental, Morphological and Functional Imaging Sciences (BIOMORF), University of Messina, Messina, Italy*, [b]*Science4Life s.r.l., spin-off of the University of Messina, Messina, Italy*

Abbreviations

DPPH assay	2,2-diphenyl-1-picryl-hydrazyl-hydrate assay
FA	fatty acid
FAME	fatty acid methyl ester
FRAP assay	ferric reducing antioxidant power assay
GAE	gallic acid equivalents
GC-FID	gas chromatography coupled to flame ionization detector
REACH regulation	Registration, Evaluation, Authorisation and restriction of Chemicals regulation
RP HPLC-DAD	reversed phase high performance liquid chromatography coupled to a diode array detector
SPME-GC–MS	headspace solid-phase microextraction followed by gas chromatography coupled to mass spectrometry
TEAC assay	trolox equivalent antioxidant capacity assay
TLC	thin-layer chromatography

1 Introduction

1.1 Citrus: Industry, processing, and by-products

Citrus (*Rutaceae* family) represents one of the world's major fruit crops, produced in worldwide tropical and subtropical climates, and marked by global availability and popularity (Malacrida, Kimura, & Jorge, 2012). Due to complex biology explaining the presence of hybrid lines and apomictic clones, and to the wide geographical distribution allowing the crossing between species, the taxonomy of the *Citrus* genus has not yet been fully resolved, representing a continuing challenge for botanists (de Araújo, de Queiroz, & Machado, 2003; Moore, 2001). The most commonly industrialized fruits belong to species including *C. sinensis* (sweet orange), *C. reticulata* (tangerine), *C. limon* (lemon), *C. aurantifolia* (lime), and *C. paradisi* (grapefruit).

During the 2008–16 period, global citrus fruit production remained fairly stable, oscillating between a maximum of 132,002.3 thousand tons (2012) and a minimum of 115,541.8 thousand tons (2009) (Fig. 1A). In 2016, it was around 124,246.0 thousand tons, confirmed to be more significant in the Northern hemisphere (97,848.9 thousand tons) than in the Southern (26,397.1 thousand tons) (Fig. 1A). In particular, China (32,705.9 thousand tons) and the Mediterranean area (25,216.0 thousand tons) led production in the boreal hemisphere, whereas Brazil was the biggest producing hub in the austral hemisphere, with 16,555.1 thousand tons of citrus fruits (Fig. 1B). In the same year, sweet oranges ere confirmed to be the most produced fruit (66,974.1 thousand tons), followed by tangerines (32,968.5 thousand tons), lemons and limes (15,981.8 thousand tons), and grapefruit (8321.6 thousand tons) (Fig. 1C) (FAO, 2017).

In the past, citrus fruits were marketed and consumed exclusively as fresh fruits, due to exceptional postharvest stability encouraging the trade, even on an international scale. However, as time has gone on, new economic realities have been faced, and the processing of fruits has become a necessity, not only to meet the increasingly differentiated consumer demand, but also to keep alive farmers' businesses, especially at a local level (Berk, 2016). Consequently, the production of canned citrus fruits, marmalades, jams, liqueurs, and (not least) essential oils has become increasingly popular thanks to small and medium-sized businesses, distributed worldwide. However, larger-scale processing started at the beginning of the 20th century in certain US states, such as California and Florida, with the establishment of industries dedicated to citrus juice production (Berk, 2016). Nowadays, the amount of fruit intended for processing has oscillated between 29,400.0

Cold Pressed Oils. https://doi.org/10.1016/B978-0-12-818188-1.00014-1
Copyright © 2020 Elsevier Inc. All rights reserved.

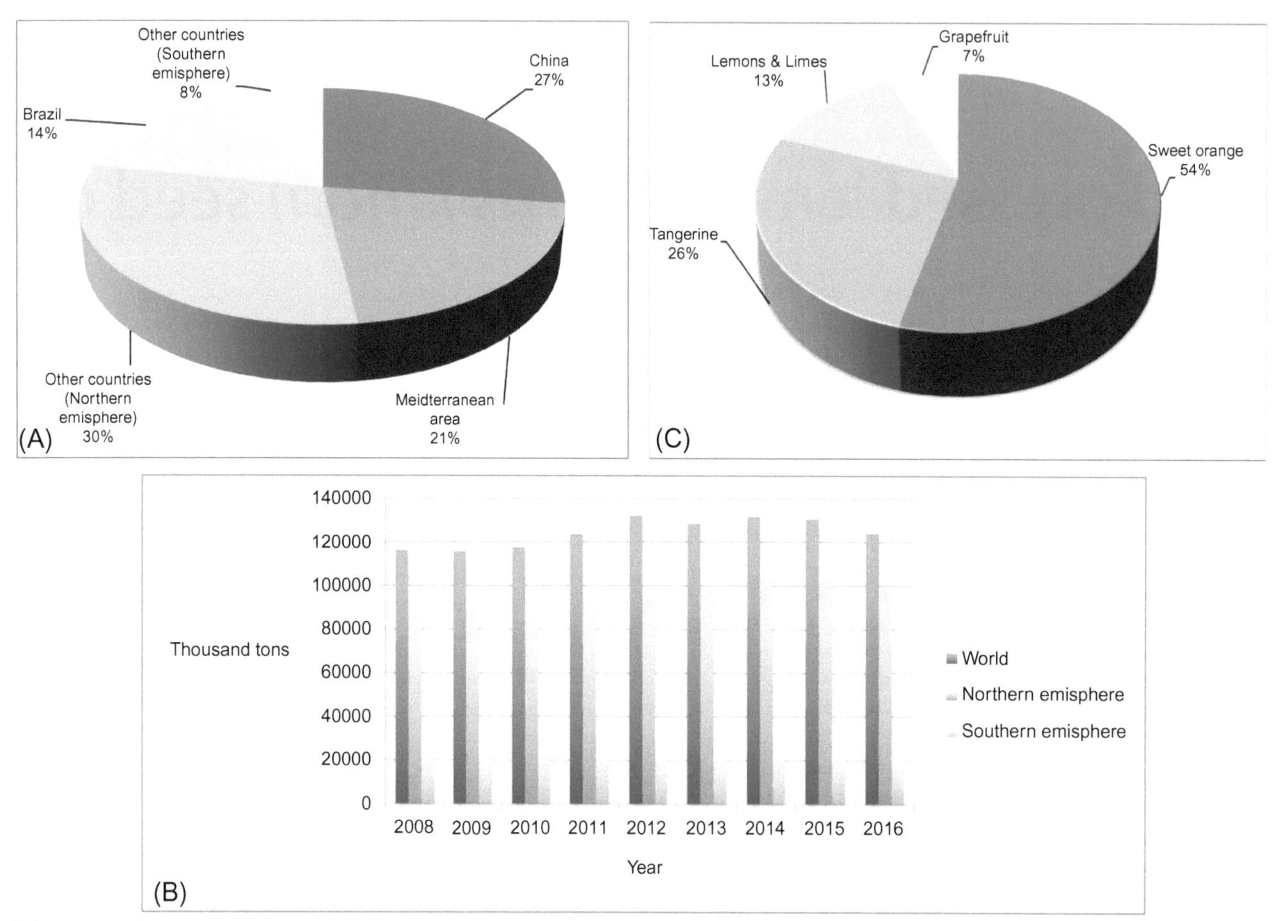

FIG. 1 The global overview of citrus fruit production. (A) Citrus production 2008–16, with particular focus on distribution between Northern and Southern hemispheres. (B) Distribution of citrus production according to the leading producer countries of Northern and Southern hemispheres in 2016. (C) Percentage amounts of the various citrus fruits produced in 2016. Data were retrieved by the FAO (2017).

thousand tons and 23,538.9 thousand tons (2008–16), representing, respectively, from 25% to 20% of the total citrus fruit production (Fig. 2A). In 2016, in particular, the sweet orange was the fruit variety most processed at global level (18,460.9 thousand tons), followed by lemons and limes (1821.5 thousand tons), tangerines (2469.0 thousand tons), and grapefruits (787.5 thousand tons) (Fig. 2B) (FAO, 2017).

Clearly, as a result of the impressive quantity of citrus fruits processed, large volumes of waste, representing roughly 50% of the raw fruits (Marín, Soler-Rivas, Benavente-García, Castillo, & Pérez-Alvarez, 2007), are generated (Berk, 2016). Overall, citrus waste is represented by solid and semisolid residues (fresh peel, pulp portions, membranes, seeds, dried peel and pulp, citrus meal and fines, and sludge), liquids (generally, citrus sludge from liquid wastes from citrus processing plants), and distillery effluents (effluents from citric acid and pectin production and essential oil plants, citrus peel liquor, and molasses) (Bampidis & Robinson, 2006; Sharma, Mahato, Cho, & Lee, 2017). The management of these materials usually leads to legal restriction problems (e.g., mandate reduction of the generation of waste), in addition to the significant costs required for the waste treatment. Conventional management of citrus waste has included: (i) direct release of treated effluents into lakes; (ii) dumping in waste ponds; (iii) dumping into groves/wells; and (iv) disposal in city systems (e.g., composting, anaerobic, digestion, incineration, thermolysis, and gasification) (Sharma et al., 2017). These methods may lead to potentially severe pollution issues, and, not least, to a loss of valuable biomass and nutrients.

Hence, in the perspective of preventing the environmental pollution and developing sustainable circular economies, similarly to a number of food wastes (Chandrasekaran, 2012; Lin et al., 2014), citrus by-products have widely demonstrated remarkable recovery potential in recent decades (Bampidis & Robinson, 2006; Mamma & Christakopoulos, 2014; Marín et al., 2007).

Citrus by-product valorization may represent an opportunity to expand the citrus industry, or even to develop another relevant and often independent business (Laufenberg, Kunz, & Nystroem, 2003). In this respect, one of the first plants for

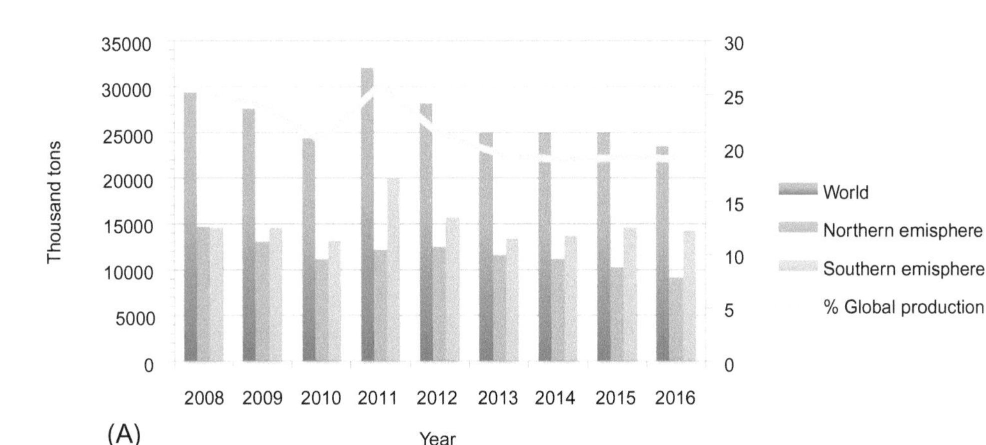

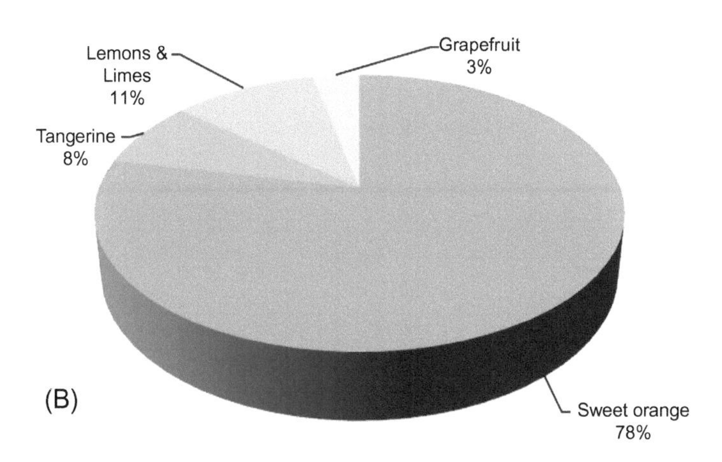

FIG. 2 The global overview of citrus fruit production intended for processing. (A) Citrus fruit processed 2008–16, with particular focus on the fruit amounts processed in Northern and Southern hemispheres. (B) Percent amounts of the various citrus fruits processed in 2016. Data were retrieved by the FAO (2017).

recovering by-products was formally opened in Ontario (California) in 1927, as a result of the large volumes of oranges and lemons that were not only discarded as cull fruit, but had to be disposed at a cost of at least $1 a ton. At that time, a plant in which ~10,000 tons of fruits were processed produced about 50,000 pounds of essential oil from the citrus peel that could be recovered, having a wholesale value of approximately $100,000 (AOCS, 1927).

Within this context, the present chapter aims to review comprehensively the extraction technologies, the nutritional and functional properties, and the practical applications of a high value-added product obtained from the citrus processing industry, especially cold pressed lemon (*C. limon*) seed oil. The summarized knowledge in the review may be helpful to encourage a greener and more eco-friendly utilization of lemon seeds for the production of cold pressed oil for its high nutritional and/or functional values.

1.2 Lemon seeds: Proximate composition, oil recovery, and yield

Similarly to other *Citrus* seeds, lemon seeds are obtained in abundance from the solid waste generated after juice production (Berk, 2016). One of the first scientific reports on lemon seed oils (Hendrickson & Kesterson, 1963) started with the study of the basic characteristics of seeds for getting into a reasonable physicochemical characterization of the final oil product. An average of 2.1–17.8 seeds per fruit, corresponding to 0.29%–2.06% of the fresh fruit weight, were present, irrespective of the lemon variety. In addition, fresh seeds showed moisture oscillating between 46% and 55%. In more recent works (Reda et al., 2005; Yilmaz & Güneşer, 2017), consistent moisture percentages were described for seeds from Sicilian and Turkish lemons (respectively, 48.30% and 41.94%). Additionally, information on mean size (length: 11.29 mm, width: 5.36 mm, height: 3.96 mm), skin/flesh ratio (0.65), and color (moderately light and yellow) was provided (Table 1).

In addition to basic information, the proximate composition of lemon seeds has been investigated in relation to fruit variety and production area (Juhaimi et al., 2016; Reda et al., 2005; Yilmaz & Güneşer, 2017). Overall, seeds showed a crude lipid and protein content ranging between 20.99% and 38.30% and 16.17% and 19.41%, respectively (Juhaimi

TABLE 1 General characteristics and proximate composition of lemon (*C. limon*) seeds.

	Lemon seeds				
Parameter	**(1)**	**(2)**	**(3)**	**(4a)**	**(4b)**
Number of seeds per fruit	2.1–17.8	–	–	–	–
Weight percent	0.29%–2.06%	–	–	–	–
Moisture	46%–55%	48.30%	41.94	–	–
Length (mm)	–	–	11.29	–	–
Width (mm)	–	–	5.36	–	–
Height (mm)	–	–	3.96	–	–
Skin/flesh ratio	–	–	0.65	–	–
Color	–	–	Moderately light and yellow	–	–
Crude lipid (%)[a]	–	38.30	34.55	20.99	30.21
Crude protein (%)[a]	–	–	19.41	16.17	16.21
Ash (%)[a]	–	–	1.41	3.24	3.63
Crude fiber (%)[a]	–	–	–	29.97	19.17

[a]*Values are reported on a dry weight basis*
– = not available.
(1) Hendrickson & Kesterson (1963) (lemon (*C. limon*) seeds from 39 varieties).
(2) Reda et al. (2005) (purchased Sicilian lemon (*C. limon*) seeds).
(3) Yilmaz & Güneşer (2017) (Turkish lemon (*C. limon* var. Kütdiken) seeds).
(4a) Juhaimi, Matthäus, Özcan, & Ghafoor (2016) (Saudi Arabian lemon (*C. limon* var. Eureka) seeds).
(4b) Juhaimi et al. (2016) (Turkish lemon (*C. limon* var. Kütdiken) seeds).

et al., 2016; Reda et al., 2005; Yilmaz & Güneşer, 2017); whereas crude fiber and ash varied from 19.17% to 29.17% and from 1.41% to 3.63%, respectively (Juhaimi et al., 2016; Yilmaz & Güneşer, 2017).

The oil yield of lemon seeds varied on average between 20% and 45%, although values out of range were occasionally reported (Table 2). Scientific evidence explained the diverse oil yields mainly irrespective of lemon variety (Matthaus & Özcan, 2012) and geographical area of production (Juhaimi et al., 2016). Only one recent study deepened the effects of different extraction methods, namely cold pressing and hexane extraction, on the oil yield (Yilmaz & Güneşer, 2017), obtaining a higher yield for the solvent-extracted oil (24.63%) than the cold pressed counterpart (12.73%) (Table 2). According to the available literature data, not considering extrinsic factors (e.g., plant variety and production area), the yield of cold pressed lemon seed oil was reduced by ~61% when compared with the mean yield of solvent-extracted oils (33.13%) (Table 2). This could be due to a lower efficiency of cold pressing (Moreau, Johnston, & Hicks, 2005; Yilmaz, Sevgi Arsunar, Aydeniz, & Güneşer, 2015), and may also be affected by the peculiar properties of seed material (Yilmaz et al., 2015). The lower yield of cold pressing explains why it is usually preferred for the production of oils intended for direct consumption, without any need for refining procedures.

However, the mean oil yield of *Citrus* seeds, including the lemon ones, is relatively higher than that of other seeds traditionally used as a source of edible oils (Saidani et al., 2004; Tarandjiiska & Nguyen, 1989).

1.3 Lemon seeds: Oil cake

In the perspective of sustaining circular economies, the dried cake or meal obtained after oil extraction, whether solvent-based or cold pressing, may be valorized by reutilization, thanks to valuable chemical, bioactive, and functional properties. The first insights on citrus meal suggested that it could contain 2%–3% of seed oil, which could be further recovered by solvent extraction (Ranganna, Govindarajan, Ramana, & Kefford, 1983). Additionally, once defatted, the press cake could serve as fertilizer or as an adjunct to cannery waste for cattle feed, due to the high content of protein (14%–21%) and fiber (26.5%) (Ranganna et al., 1983).

More recently, cakes from lemon seeds have been investigated in-depth, independent of the oil extraction method. Different studies (Karabiber, Zorba, & Yılmaz, 2018; Karaman, Yılmaz, & Tuncel, 2017; Yilmaz & Güneşer, 2017) have

TABLE 2 Oil yield of lemon (*C. limon*) seeds.

Seed source	Extraction method	Oil content (%)	References
Tunisian lemon (*C. limon*)	Soxhlet-extractor with petroleum ether	78.9	Saidani, Dhifi, & Marzouk (2004)
Purchased Sicilian lemon (*C. limon*)	Soxhlet-extractor with *n*-hexane	38.3	Reda et al. (2005)
Purchased lemon (*C. limon* var. Rosa)	Soxhlet-extractor with petroleum ether	34.9	Malacrida et al. (2012)
Turkish lemon (*C. limon* var. Interdonato)	Twisselmann-extractor with petroleum ether	45.7	Matthaus & Özcan (2012)
Turkish lemon (*C. limon* var. Kütdiken)		45.1	
–	Soxhlet-extractor with acetone	36.03	Qiubing, Jinghong, & Bi (2012)
Iranian lemon (*C. limon*)	Soxhlet-extractor with *n*-hexane	40.3–41.9	Reazai, Mohammadpourfard, Nazmara, Jahanbakhsh, & Shiri (2014)
Saudi Arabian lemon (*C. limon* var. Kütdiken)	Twisselmann-extractor with petroleum ether	30.21	Juhaimi et al. (2016)
Turkish lemon (*C. limon* var. Eureka)		20.99	
Turkish lemon (*C. limon* var. Kütdiken)	Cold pressing	12.73	Yilmaz & Güneşer (2017)
	n-hexane	24.63	

– = not available.

compared the proximate composition of cakes from cold pressed and solvent-extracted seeds. They found that the cold pressed cake was marked by greater levels of ash (2.79% vs. 2.57%), oil (17.18% vs. 5.25%), protein (24.59%–27.27% vs. 23.72%–20.98%), and soluble (7.55% vs. 5.95%) and insoluble (80% vs. 75.95%) fiber, with respect to the solvent-extracted counterpart (Table 3).

Lemon seed meals were also demonstrated to be rich in flavonoids (e.g., eriocitrin, rutin, naringenin, etc.), phenolic acids (Karaman, Karabiber, & Yılmaz, 2018), dietary fiber (Karaman et al., 2017), and proteins (Karabiber et al., 2018). In particular, Karaman et al. (2017) highlighted that fibers of lemon seed meals obtained via cold pressing and solvent extraction were respectively characterized by 2.44 and 2.54 mg/g phytate, and 1.38% and 1.36% ash, and that their main components were cellulose and lignin. Karabiber et al. (2018) pointed out interesting antimicrobial activity of the protein fraction, which proved active against some foodborne pathogens. Consequently, these meal components could be properly isolated for the production of functional foods or food supplements.

Finally, fibers and proteins from lemon seed cakes showed good functional properties, in terms of water and oil holding capacity, emulsification, and foaming properties. As a result, they could be employed as natural food additives (Karabiber et al., 2018; Karaman et al., 2017; Karaman et al., 2018)

2 Cold pressed lemon seed oil

According to the literature produced over the last 50 years, lemon seed oil has been predominantly extracted by solvents with the main aim of studying the effect of geographical provenance and/or cultivar on its chemical composition. The exclusive employment of this extraction protocol could be explained by the fact that organic solvents have been routinely adopted on an industrial scale for oilseed extraction, due to the apparent economic and practical advantages (e.g., easy oil recovery, good solvent solubilizing ability, etc.) (Kumar et al., 2017). However, it has been established that during the extraction and recovery processes, the organic solvent is released into the surrounding environment, with severe toxicological and polluting implications (Kumar et al., 2017). It is enough to mention that *n*-hexane, the most common solvent

TABLE 3 Proximate composition of the lemon (*C. limon*) seed cake.

Parameter (%)	Cold pressed seed cake (1)	Solvent-extracted seed cake	Cold pressed seed cake (2)	Solvent-extracted seed cake	Cold pressed seed cake (3)	Solvent-extracted seed cake
Moisture	10.53	11.91	–		–	
Ash[a]	2.79	2.57	–		–	
Oil[a]	17.18	5.25	–		–	
Protein[a]	24.59	23.72	27.27	20.98	–	
Soluble fiber[a]	–		–		7.55	5.95
Insoluble fiber[a]	–		–		80	75.95

[a]*Values are reported on a dry weight basis.*
– = not available.
(1) Yilmaz & Güneşer (2017) (oil cake from Turkish lemon (*C. limon* var. Kütdiken) seeds).
(2) Karabiber et al. (2018) (oil cake from Turkish lemon (*C. limon* var. Kütdiken) seeds).
(3) Karaman et al. (2017) (oil cake from Turkish lemon (*C. limon* var. Kütdiken) seeds).

used for oilseed extraction, is obtained from fossil resources, and registered under the Registration, Evaluation, Authorisation and restriction of Chemicals (REACH) regulations as a category 2 reprotoxic, and a category 2 aquatic chronic toxic substance (European Chemical Agency, 2019).

In the era of "green food processing" based on green chemistry and green engineering, the oil industry is increasingly moving toward the development of alternative extraction methodologies, which not only could ensure a significant extraction yield, but also could boost the health benefits and sensory properties of the final product, while preserving the environment. Additionally, in the perspective of concrete and virtuous sustainability, combining green technologies with the reclaim and valorization of food waste may lead to further environmental benefits and economic gains, and, not least, improve consumer health and lifestyle.

Coherently, in the last two decades, research efforts have increasingly focused on in-depth characterization of a variety of oils from seed waste and extracted by green technologies—particularly cold pressing (Cicero et al., 2018; Siger, Nogala-Kalucka, & Lampart-Szczapa, 2008)—and on improving their quality by optimizing the extraction procedure. However, the application of cold pressing procedures on citrus seeds and, consequently, the development of strategies for producing high-quality cold pressed oils are as yet poorly explored, and the lack of effective extraction methods may seriously compromise the utilization of citrus residues (Putnik et al., 2017). In temporal terms, the first scientific evidence on the cold pressing of citrus seeds occurred during the 1970s and 1980s (Braddock & Kesterson, 1973; Ranganna et al., 1983). A huge literature gap is then observed up to the last decade, with just a few works reported on the chemical characterization of cold pressed oils from different *Citrus* species (Güneşer & Yilmaz, 2017b; Terada et al., 2010), and in particular from *C. limon* (Güneşer & Yilmaz, 2017a; Yilmaz & Güneşer, 2017).

According to Braddock and Kesterson (1973), the processing of citrus seeds intended for cold pressing starts with the separation of seeds from the peel, rag, and pulp. Such separation may occur by a rotating reel with perforations allowing the finer material (e.g., seeds, juice sacs, and small rag pieces) to fall through them. Juice sacs and rag pieces are then separated by a paddle finisher. An alternative collection method is to lime the citrus solid residue containing seeds and to dry it, so that the dried seeds have higher densities than the other dried material, and can be separated by cyclone separators, screens, or winnowing systems. In any case, citrus seeds should achieve a moisture content $\sim$10% prior to oil extraction. In the next step, seeds are preferably decorticated for removing the fibrous envelopes, also called "hulls," and for obtaining the "kernels," concentrated in oil and proteins. Removing hulls becomes necessary as they could reduce the oil yield by absorbing the oil in the press cake, and compromise the performance of the extraction equipment as well. Once kernels are obtained, they are reduced to small particles or flakes and eventually roasted or cooked. Similarly to dehulling, the primary objective of cooking/roasting seeds is to increase the oil yield, by ensuring a moisture content between 6%

and 10%. Finally, the preprocessed seeds are oil-extracted by a hydraulic or mechanic press, as long as the temperature remains below 80.6°F.

In such a procedure, dehulling and cooking/roasting represent two of the critical steps that may be optimized depending on the physicochemical properties of seeds, to maximize the oil yield and quality (Hamm, Hamilton, & Calliauw, 2013). Considering the dehulling, the oil yield may be maximized by balancing the ratio of skin:flesh, in other words, the degree of separation of hulls from kernels. In fact, if a great portion of hulls is removed by the available seeds, the flesh could be carried over into the hulls and a loss of oil could occur. Similarly, if a great proportion of hulls is left among the kernels, there may be a loss of oil due to hull fiber absorption (Braddock & Kesterson, 1973). Additionally, dehulling procedures may improve the taste and smell, generating a milder oil than that of nondecorticated seeds (Koubaa, Mhemdi, & Vorobiev, 2016; Zhou, Yang, Huang, Zheng, & Deng, 2013).

As aforementioned, cooking or roasting procedures reduce the moisture content of seeds, so that the oil recovery from the meal is enhanced (Rostami, Farzaneh, Boujmehrani, Mohammadi, & Bakhshabadi, 2014). However, the heating process needs to be strictly controlled as, with increasing temperatures, the nontriglyceride fraction and acidity of the oil tend to increase (Prior, Vadke, & Sosulski, 1991). In addition, several studies reported that microwave cooking and roasting may induce the Maillard reaction of seed, so that heterocyclic compounds responsible of pleasant flavors (e.g., nutty and roasted odors) may be produced (Liu et al., 2011).

To the best knowledge of the authors, no attempts have been yet carried out to optimize the yield or quality of cold pressed lemon seed oil. Instead, a recent study (Juhaimi, Özcan, Uslu, & Ghafoor, 2018) evaluated the effect of different drying temperatures on total phenolics content, single polyphenols and tocopherols, and antioxidant activity of *C. limon* seeds, as well as the fatty acid profile of the derived oil. Overall, with the exception of an unaffected fatty acid profile, the increase of the drying temperature from 60°C to 80°C led to a corresponding reduction of investigated bioactive compounds and, consequently, of the antioxidant activity of such seeds. However, quite a few shortcomings are noticeable in such a study. In fact, the seeds dried at different temperatures were compared with an undefined control condition. Additionally, no mention was made of the extraction method of the oil.

On the other hand, another recent work (Güneşer & Yilmaz, 2017b) investigated the effects of roasting on the yield and chemical composition of cold pressed orange (*C. sinensis*) seed oil. Overall, the yield of roasted cold pressed oil increased by ~10% compared to that of the cold pressed one. The fatty acid composition, sterols, and α-tocopherol did not vary significantly between the two cold pressed oils (Güneşer & Yilmaz, 2017b). Hence, this study may represent a good starting point to deepen understanding of the effects of seed pretreatment (e.g., dehulling, drying, etc.) on cold pressed lemon seed oil.

3 Chemical characterization of cold pressed lemon seed oil

Oil is a natural component of various seeds, including lemon ones, and is roughly constituted by a major lipid (saponifiable) fraction (95%–99%) and a complex (unsaponifiable) mixture of minor compounds (5%–1%), with diverse chemical natures (Aluyor, Ozigagu, Oboh, & Aluyor, 2009; Giacomelli, Mattea, & Ceballos, 2006). Considering the saponifiable matter, only previous work on solvent-extracted lemon seed oil (Saidani et al., 2004) confirmed the largest abundance of triacylglycerols (98.20%), followed by phospholipids (3.30%) and diacylglycerols (2.70%). On the other side, available data on cold pressed oil suggested that the unsaponifiable fraction accounted for 1.08% (Yilmaz & Güneşer, 2017), and included sterols, tocopherols, pigments, polyphenols, and, not least, aroma compounds (Duda-Chodak & Tarko, 2007; Güneşer & Yilmaz, 2017a; Yilmaz & Güneşer, 2017). Similarly to other vegetable oils, minor compounds of lemon seed oil may vary consistently independent of several intrinsic and extrinsic factors, such as fruit variety and degree of maturation, growing area, cultivation techniques, soil management, climate, and, not least, oil production/processing method (Cert, Moreda, & Pérez-Camino, 2000; Fine et al., 2016).

3.1 Fatty acid composition

As mentioned above, vegetable oil is mainly constituted by triacylglycerols, derived from the esterification of three fatty acids (FAs) on the glycerol skeleton. The three FAs are not equivalent in terms of bioavailability. Indeed, the sn-2 position is conserved during the whole digestive process, which explains why in vegetable oils the most physiologically important FAs, namely unsaturated FAs, are esterified in the sn-2 position (Hunter, 2001).

Due to the growing evidence of the deleterious effect of dietary FAs on plasma lipid composition, and due to the growing awareness of the need to balance FAs for obtaining the most appropriate ω-6/ω-3 ratio, the fatty acid (FA) composition of an oil is primarily investigated for evaluating the goodness of its fat profile (Dubois, Breton, Linder,

Fanni, & Parmentier, 2007). Additionally, providing a fingerprint of the vegetable oil, the FA composition becomes a valuable tool for studying its genuineness (Aparicio & Aparicio-Ruı´z, 2000; Gunstone, 2011).

To the best knowledge of the authors, only a recent work (Yilmaz & Güneşer, 2017) characterized and compared the FA composition of lemon (*C. limon* var. Kütdiken) seed oil produced by cold pressing and extraction with solvent (*n*-hexane). In this study, perfectly overlapping FA profiles were described, as palmitic acid (C16:0) resulted between 20.88% and 20.90%, oleic acid (C18:0) between 30.86% and 30.27%, whereas linoleic acid (C18:2n-6) was between 33.77% and 33.9%, respectively, in cold pressed and solvent-extracted oils. In addition, less abundant FAs, such as palmitoleic acid (C16:1, 0.27% vs. 0.26%), stearic acid (C18:0, 4:28% vs. 4.32%), and linolenic acid (C18:3n-3, 8.35% vs. 8.65%), were comparable (Table 4).

Other previous works focused on the same lemon variety (*C. limon* var. Kütdiken), and highlighted that the seed oil extracted by solvent was characterized by palmitic acid ranging from 9% (Matthaus & Özcan, 2012) to 19.6% (Juhaimi et al., 2016), oleic acid oscillating between 31.4% (Juhaimi et al., 2016) and 38.5% (Matthaus & Özcan, 2012), and linoleic acid varying from 32.3% (Juhaimi et al., 2016) to 44.5% (Matthaus & Özcan, 2012) (Table 4). Assuming that the extraction method may affect to a small extent the FA profile of the oil (Yilmaz & Güneşer, 2017), the different FA contents could be attributable to other extrinsic factors such as the cultivation region of lemon orchards, characterized by peculiar geopedoclimatic conditions, and, not least, the harvest season considered for the production of the oil (Yilmaz & Güneşer, 2017).

Differently from cold pressing, the lemon seed oil extracted by solvent was elucidated in more detail, also independent of fruit variety and cultivation area (Hendrickson & Kesterson, 1963; Juhaimi et al., 2016, 2018; Malacrida et al., 2012; Matthaus & Özcan, 2012; Reazai et al., 2014; Reda et al., 2005; Saidani et al., 2004). Overall, the seed oil was confirmed to be mainly composed of palmitic (21.40%–29.40%), oleic (20.80%–36.60%), linoleic (31.40%–44.31%), and linolenic (6.20%–10.00%) acids, and proved to be a good source of essential fatty acids, such as linoleic and linolenic acids, which represent together 40%–50% of the total FA composition (Table 4). In particular, the content of linolenic acid proved to be higher than that of conventional and unconventional cold pressed and not seed oils, such as peanut (0.4%), soybean (7.8%) (Dubois et al., 2007), grapeseed (5.6%) (Cicero et al., 2018), caraway (0.24%), and carrot (0.28%) (Parker, Adams, Zhou, Harris, & Yu, 2003). Having high levels of linoleic acid associated with low amounts of palmitic acid, lemon seed oil could belong to the polyunsaturated fatty acid (42.2%–53.2%) class, according to the classification of vegetable oils proposed by Dubois et al. (2007).

However, the FA composition of cold pressed lemon seed oil needs to be elucidated further in terms of other minor FAs (e.g., arachidic acid, behenic acid, etc.), total saturated, mono- and poly-unsaturated FAs, and ω-6/ω-3 ratio as well, in order to evaluate thoroughly the effect of cold pressing on single FAs and FA classes as well, and to have a reliable indication of its nutritional value.

3.2 Minor bioactive compounds

3.2.1 *Phytosterols*

Phytosterols are cholesterol-like triterpenes with a wide range of chemical structures and related physical properties and biological activities. They are typically exploitable in the pharmaceutical, health-care, and nutrition sectors. In fact, they supply the majority of intermediates and precursors for the production of therapeutic steroids and serve as natural emulsifiers for a variety of cosmetic products. Additionally, a variety of plant sterols act as natural antioxidant of vegetable oils, and their hypocholesterolemic effects on human health have been largely documented (Abidi, 2001; Fernandes & Cabral, 2007).

In a recent study conducted by Yilmaz and Güneşer (2017), a variety of sterols were screened in Turkish lemon (*C. limon* var. Kütdiken) seed oils by thin-layer chromatography (TLC) followed by gas chromatography coupled to a flame ionization detector (GC-FID), for evaluating the effect of the extraction method, namely cold pressing and solvent-based extraction, on such a compound class. In cold pressed and solvent-extracted oils, Δ-5-avenesterol, stigmasterol, campesterol, and β-sitosterol were characterized by the highest contents, as they oscillated between 3.24% and 3.19%, 4.52% and 5.04%, 10.71% and 12.57%, and 76.55% and 76.10%, respectively. Next up, cholesterol showed intermediate concentrations (cold pressed oil: 1.23% vs. solvent-extracted oil: 1.12%), while the other sterols were $<1\%$ in both types of oil (Table 5). Among the most abundant sterols, significantly different contents of campesterol and stigmasterol were revealed, as they were significantly more concentrated in the oil extracted by solvent rather than the one produced by cold pressing. The content of almost every other compound was not significantly different in both seed oils (Yilmaz & Güneşer, 2017).

TABLE 4 Comparison of fatty acid (FA) compositions between cold pressed and solvent-extracted lemon (*C. limon* var. Kütdiken) seed oils.

FAME	Cold pressed oil	Solvent-extracted oil	Solvent-extracted oil									
(%)	(1)		(2a)	(2b)	(3a)	(3b)	(4)	(5)	(6)	(7)	(8)	(9)
C8:0	–		–	–	–	–	–	0.04	1	–	–	–
C10:0	–		–	–	–	–	–	0.05	–	–	–	–
C12:0	–		–	–	–	–	–	0.04	–	–	–	–
C14:0	–		–	–	–	–	–	0.10	0.10	–	–	0.143–0.155
C14:1	–		–	–	–	–	–	0.10	–	–	–	–
C16:0	20.88	20.90	9.0-	5.4	19.6	24.0	23.2	21.40	19.60	21.03	23.50–29.40	22.897–23.242
C16:1	0.27	0.26	–	–	–	–	–	–	–	0.65	0.60–0.90	–
C18:0	4.28	4.32	5	1.4	4.5	3.8	3.7	2.30	3.0	3.67	4.10–4.70	3.483–3.688
C18:1*n*–9	30.86	30.27	38.5	63.6	31.4	23.0	29.8	36.60	28.60	20.80	24.80–28.50	23.220–23.897
C18:1*n*–11	–		1.1	1.3	0.9	1.3	–	–	–	–	–	–
C18:2*n*–6	33.77	33.99	44.5	26.8	32.3	37.6	32.5	31.40	34.40	44.31	33.70–35.70	39.039–39.634
C18:3*n*–3	8.35	8.65	0.4	0.1	9.3	8.2	10.9	6.90	10.00	8.96	6.20–7.80	7.841–8.061
C20:0	–		0.5	0.1	0.4	0.4	–	nd	0.20	0.31	–	0.325–0.351
C20:1*n*–9	–		–	–	0.2	0.1	–	nd	–	–	–	–
C20:4*n*–6	–		–	–	–	–	–	–	–	–	–	0.170–0.256
C22:0	–		–	–	0.1	0.1	–	–	–	0.10	–	0.088–0.121
C24:0	–		–	–	–	–	–	–	–	0.17	–	–
SFA	–		–	–	–	–	–	–	–	25.28	–	–

Continued

TABLE 4 Comparison of fatty acid (FA) compositions between cold pressed and solvent-extracted lemon (*C. limon* var. Kütdiken) seed oils—cont'd

FAME	Cold pressed oil	Solvent-extracted oil	Solvent-extracted oil									
(%)	(1)		(2a)	(2b)	(3a)	(3b)	(4)	(5)	(6)	(7)	(8)	(9)
MUFA	–		–	–	–	–	–	75	–	21.45	–	–
PUFA	–		–	–	–	–	–		–	53.27	–	–

For a given FA, contents marked by the asterisk are significantly different at a level of confidence of 95%. Further comparative data on oil extracted by solvent-based procedures, are reported. FAME = Fatty Acid Methyl Ester; SFA = Saturated Fatty Acid content; MUFA = Monounsaturated Fatty Acid content; PUFAs = Polyunsaturated Fatty Acid content; nd = not detected; – = not available.
(1) Yilmaz & Güneşer (2017) (Turkish lemon (*C. limon* var. Kütdiken) seed oil).
(2a) Matthaus & Özcan (2012) (Turkish lemon (*C. limon* var. Kütdiken) seed oil).
(2b) Matthaus & Özcan (2012) (Turkish lemon (*C. limon* var. Interdonato) seed oil).
(3a) Juhaimi et al. (2016) (Turkish lemon (*C. limon* var. Kütdiken) seed oil).
(3b) Juhaimi et al. (2016) (Saudi Arabian lemon (*C. limon* var. Eureka) seed oil).
(4) Hendrickson & Kesterson (1963) (lemon (*C. limon*) seeds from 39 varieties). Data are expressed as mean value of the different varieties.
(5) Saidani et al. (2004) (Tunisian lemon (*C. limon*) seed oil).
(6) Reda et al. (2005) (Sicilian lemon (*C. limon*) seed oil).
(7) Malacrida et al. (2012) (lemon (*C. limon* var. Rosa) seed oil).
(8) Reazai et al. (2014) (Iranian lemon (*C .limon*) seed oil).
(9) Juhaimi et al. (2018) (Saudi Arabian lemon (*C. limon* var. Eureka) seed oil).
(Adapted from Yilmaz, E. & Güneşer, B.A. (2017). Cold pressed versus solvent extracted lemon (*Citrus limon* L.) seed oils: Yield and properties. *Journal of Food Science and Technology, 54*, *7*, 1891–1900.)

TABLE 5 Comparison of the sterol profile of lemon (*C. limon* var. Kütdiken) seed oils obtained by cold pressing and solvent-based extraction.

Sterol	Cold pressed oil (1)	Solvent-extracted oil	Solvent-extracted oil (2)
Cholesterol	1.23	1.12	2.7
Brassicasterol	0.05	0.05	–
24-Methylen cholesterol	0.10	0.03	0.1
Campesterol	10.71[a]	12.57[a]	10.4
Campestanol	0.18	0.19	0.0
Stigmasterol	4.52[a]	5.04[a]	4.4
Δ-7 Campesterol	0.32	0.34	0.2
Δ-5,23 Stigmastadienol	0.06	0.04	0.0
Chlerosterol	0.98	0.94	1.1
β-Sitosterol	76.55[a]	75.10[a]	75.6
Sitostanol	0.79	0.60	0.4
Δ-5 Ave-sterol	3.24	3.19	2.5
Δ-5,24 Stigmastadienol	0.21[a]	0.09[a]	0.8
Δ-7 Stigmastenol	0.69[a]	0.48[a]	1.1
Δ-7 Ave-sterol	0.34[a]	0.20[a]	0.7

[a]For a given sterol, amounts are significantly different at a level of confidence of 95%. Comparative literature on solvent-extracted oils from lemon seeds is also shown. –=not available.
(1) Yilmaz & Güneşer (2017) (Turkish lemon (*C. limon* var. Kütdiken) seed oil). Values are expressed on a % basis.
(2) Matthaus & Özcan (2012) (Turkish lemon (*C. limon* var. Kütdiken) seed oil). Values are expressed in terms of mg/Kg.
(Adapted from Yilmaz, E. & Güneşer, B.A. (2017). Cold pressed versus solvent extracted lemon (*Citrus limon* L.) seed oils: Yield and properties. *Journal of Food Science and Technology, 54*, 7, 1891–1900.)

Yilmaz and Güneşer (2017) commented that their data highlighted a noticeable similarity with the phytosterol contents revealed through TLC followed by gas-liquid chromatography (GLC), in the solvent-extracted lemon seed oil by Matthaus and Özcan (2012) (Table 5). However, the comparison was wrongly made, as the different units of measure of the two studies (percentage and mg/kg, respectively) were not considered. Indeed, assuming the correctness of the percentage measure, the cold pressed and solvent-extracted oils would be characterized by every sterol at exceptionally high content (i.e., cholesterol at 12,300 and 11,200 mg/Kg, respectively). This is rather unusual for a typical seed oil, independent of the seed species or extraction method. Conversely, Matthaus and Özcan (2012), although confirming campesterol, stigmasterol, and Δ-5-avenesterol as the sterols with the highest amounts in the oil, reported much lower (and reliable) contents (10.4, 4.4, and 2.5 mg/Kg, respectively) (Table 5).

Hence, it may be concluded that future research should be oriented toward a reliable study of the sterol profile of cold pressed lemon seed oil, while also deepening knowledge of the nutritional and pharmaceutical value of such oil.

3.2.2 *Tocols*

Tocopherols and tocotrienols, together abbreviated as tocols and commonly summarized under the term vitamin E, are a group of lipophilic antioxidants with a chromanol ring and a hydrophobic side chain (phytyl in the case of tocopherols, and isoprenyl in the case of tocotrienols). Single tocopherols (α-, β-, γ-, and δ-tocopherol) and single tocotrienols (α-, β-, γ-, and δ-tocotrienol) differ by the number and position of methyl substituents on the phenolic part of the chromanol ring. In particular, among tocols, α-tocopherol is characterized by the highest antioxidant activity, not only preventing the lipid oxidation of food, vegetable oils included, but also reducing the risk of cardiovascular diseases and of certain types of cancer (Schwartz, Ollilainen, Piironen, & Lampi, 2008; Seppanen, Song, & Saari Csallany, 2010).

Recently, the α-tocopherol content was determined in Turkish lemon (*C. limon* var. Kütdiken) seed oils by reversed-phase high-performance liquid chromatography coupled to a diode array detector (RP HPLC-DAD), for studying its variation independence of the oil production method (Yilmaz & Güneşer, 2017). It was pointed out that the cold pressed oil had significantly higher α-tocopherol content than the solvent-extracted counterpart (155.00 and 110.20 mg/Kg, respectively) (Table 6).

Although no further references on the presence of such antioxidant in cold pressed lemon seed oil was found, several correspondences could be highlighted for the oil extracted by solvent, as α-tocopherol ranged from 102 mg/Kg (Malacrida et al., 2012) to 130 mg/Kg (Matthaus & Özcan, 2012) (Table 6). However, available data must always be critically interpreted, as they underline a number of intrinsic and extrinsic factors that should not lead to an immediate comparison. Coherently, a previous work evaluating the effect of the drying temperature on the chemical composition of lemon (*C. limon* var. Eureka) seeds reported much higher—and not comparable—levels of α-tocopherol in the derived oil, as it ranged from 13,808 to 19,213 mg/Kg, considering seeds predried respectively at 80°C and 60°C (Juhaimi et al., 2018) (Table 6). The α-tocopherol content of cold pressed lemon oil was generally higher than that of other cold pressed seed oils, such as grapeseed (139.2 mg/Kg), peanut (73.2 mg/Kg), pumpkin (71.4 mg/Kg), soybean oil (92.1 mg/kg), rapeseed (90.9 mg/Kg), and flaxseed (9.5 mg/kg) (Tuberoso, Kowalczyk, Sarritzu, & Cabras, 2007). However, other common cold pressed oils, such as sunflower and olive, showed an opposite trend, reporting higher contents of such antioxidants (494.2 and 212.1 mg/Kg, respectively) (Tuberoso et al., 2007).

Other tocopherols and tocotrienols were characterized in the lemon seed oil obtained by solvent, but not in the cold pressed counterpart. With respect to the predominant α-tocopherol, much lower amounts of β-tocopherol (0.3–2.20 mg/Kg) and γ-tocopherol (0.3–1.33 mg/Kg) were revealed (Malacrida et al., 2012; Matthaus & Özcan, 2012). The same could apply for single tocotrienols, generally ranging from 0.1 to 0.8 mg/Kg (Juhaimi et al., 2016; Matthaus & Özcan, 2012) (Table 6).

Single tocopherols and single tocotrienols, as well as total tocopherols and tocotrienols, should be elucidated in more detail in cold pressed lemon seed oil for contributing to define its antioxidant capacity and nutritional value as well.

TABLE 6 **Comparison of the α-tocopherol content of lemon (*C. limon* var. Kütdiken) seed oils in dependence of the extraction method.**

	Cold pressed oil	Solvent-extracted oil	Solvent-extracted oil					
Tocols (mg/Kg)	(1)		(2)	(3a)	(3b)	(4a)	(4b)	(5)
α-Tocopherol	155.00[a]	110.20[a]	102.49	130	109	5.5	Nd	13,808–19,213
β-Tocopherol	–		2.20	0.4	0.3	–	–	–
γ-Tocopherol	–		1.33	0.3	0.6	nd	Nd	10,402–11,048
δ-Tocopherol	–		18.98	–	–	–	–	–
α-Tocotrienol	–		–	0.1	0.1	–	–	–
γ-Tocotrienol	–		–	0.3	0.2	0.7	Nd	–
Plastochromanol-8	–		–	0.1	0.2	0.8	Nd	–
Total	–		125.01	14.3	12.3	7.0	0.0	–

[a]For a given compound, contents are significantly different at a level of confidence of 95%. Comparative literature on the tocols of oils extracted by solvent from lemon seeds is also shown. nd = not detected; – = not available.
(1) Yilmaz & Güneşer (2017) (Turkish lemon (*C. limon* var. Kütdiken) seed oil).
(2) Malacrida et al. (2012) (lemon (*C .limon* var. Rosa) seed oil).
(3a) Matthaus & Özcan (2012) (Turkish lemon (*C. limon* var. Kütdiken) seed oil).
(3b) Matthaus & Özcan (2012) (Turkish lemon (*C. limon* var. Interdonato) seed oil).
(4a) Juhaimi et al. (2016) (Turkish lemon (*C. limon* var. Eureka) seed oil).
(4b) Juhaimi et al. (2016) (Turkish lemon (*C. limon* var. Kütdiken) seed oil).
(5) Juhaimi et al. (2018) (Saudi Arabian lemon (*C. limon* var. Eureka) seed oil).
(Adapted from Yilmaz, E. & Güneşer, B.A. (2017). Cold pressed versus solvent extracted lemon (*Citrus limon* L.) seed oils: Yield and properties. *Journal of Food Science and Technology, 54*, 7, 1891–1900.)

3.2.3 *Phenolics*

Phenolic compounds are known to enhance the fineness of vegetable oil, as they may improve its stability, and its sensory and nutritional characteristics as well (Fine et al., 2016). Additionally, due to strong antioxidant activity related to their oxygen-sparing, peroxide-destroying, metal-chelating, and free radical inhibitor properties (Dimitrios, 2006; van Acker, van Balen, van den Berg, Bast, & van der Vijgh, 1998), many phenolics confer on the oil a valuable functionality, which may be exploited in diverse sectors, such as pharmaceuticals, cosmetics, and nutraceuticals. In this sense, the characterization of the polyphenol fraction of the oil has often been associated with the study of its antioxidant activity (Bendini et al., 2007; Siger et al., 2008). However, the transfer of these antioxidants compounds into the oil during the extraction is quite low, as it mainly depends on their predominant hydrophilic nature and enzymatic activities of the matrix that they come from (Cecchi, Migliorini, Zanoni, Breschi, & Mulinacci, 2018). As a result, the seeds and the derived oil cake may be characterized by a more complex polyphenol fingerprint, and, consequently, by higher antioxidant activity than the relative oil (İnan, Özcan, & Aljuhaimi, 2018; Karaman et al., 2018).

In order to retain to major extent polyphenols in the oil, the cold pressing has commonly been preferred to solvent-based procedures for a variety of seeds (Siger et al., 2008). Additionally, the optimization of certain parameters of the pressing procedure demonstrated to improve the phenolic content of oil (Rombaut et al., 2015). Nevertheless, similarly to other compound classes, the effect of cold pressing on the polyphenol fraction of lemon seed oil has barely been addressed.

To the best knowledge of the authors, the total phenolic content and the antioxidant activity of cold pressed lemon seed oil have been not yet investigated. Conversely, previous works addressed such aspects on lemon seeds and solvent-extracted oils belonging to diverse fruit varieties and coming from different production areas (Duda-Chodak & Tarko, 2007; İnan et al., 2018; Juhaimi et al., 2018; Malacrida et al., 2012; Sultana, Anwar, Mushtaq, & Alim, 2015) (Table 7).

Although very different total phenolic contents were revealed (188.20–98,230 mg of Gallic Acid Equivalents (GAE)/Kg), the seeds proved to be a good source of such bioactives, which, in turn, contributed to a valuable antioxidant activity (evaluated in terms of TEAC, DPPH, and FRAP assays) (Duda-Chodak & Tarko, 2007; İnan et al., 2018; Juhaimi et al., 2018; Sultana et al., 2015) (Table 7). The lemon oil also proved to have a peculiar total phenolic content and antioxidant activity (Table 7). In particular, a slight decrease of total phenolic content and related antioxidant activity was observed between the lemon seeds and the derived oil (from 188.20 to 82.2 mg GAE/Kg, and from 402.96 to 232.76 mg of Trolox Equivalent Antioxidant Capacity (TEAC)/Kg, respectively), probably due to a suboptimal transfer of bioactives from the seeds into the oil (İnan et al., 2018). However, comparable antioxidant activity was revealed by the DPPH assay, as seeds and derived oil showed that 1.05 and 3.27 μg of the sample, respectively were able to inhibit 50% of the DPPH· radical (IC50) (İnan et al., 2018) (Table 7).

TABLE 7 Comparison of the total phenolic content and antioxidant activity (TEAC, DPPH and FRAP assays) of lemon (*C. limon*) seeds and solvent-extracted oil.

		Seeds		Seeds	Solvent-extracted oil	Solvent-extracted oil
Parameter	(1)	(2)	(3)	(4)		(5)
Total phenolic content (mg GAE/Kg)	1588.0	98,230	1102.43–1194.38	188.20	82.2	1196.71
TEAC assay (mg TEAC/Kg)	<2000	–	–	402.96	232.76	–
DPPH· assay (%)	–	39.98	43–58	1.05[a]	3.27[a]	29.25
FRAP assay (at 10 mg/mL extract)	–	0.63	–	–	–	–

Values of lemon seeds are reported on a dry weight basis. TEAC assay = Trolox Equivalent Antioxidant Capacity assay; DPPH assay = 2,2-diphenyl-1-picryl-hydrazyl-hydrate assay; FRAP assay = Ferric Reducing Antioxidant Power assay; – = not available.
(1) Duda-Chodak & Tarko (2007) (imported lemon (*C. limon* var. Primofiore) seeds), Value are expressed as mg catechin/Kg.
(2) Sultana et al. (2015) (Pakistani lemon (*C. limon*) seeds).
(3) Juhaimi et al. (2018) (Saudi Arabian lemon (*C. limon* var. Eureka) seeds).
(4) İnan et al. (2018) (Turkish lemon seeds and derived oils from six varieties). Data are expressed as mean value of the different varieties.
(5) Malacrida et al. (2012) (lemon (*C .limon* var. Rosa) seed oil).
[a]*Data are expressed as IC50 (μg).*

Overall, the total phenolics reported for the solvent-extracted lemon oils were higher than those verified by Siger et al. (2008) for a variety of cold pressed seed oils, such as soybean, sunflower, corn, canola, and rice (from 12.6 to 14.8 mg of caffeic acid equivalents/Kg). The same could apply for the antioxidant activity evaluated by DPPH· assay, as cold pressed soybean, sunflower, corn, canola, and rice oil reported a percentage of scavenged DPPH· oscillating from 11.1% to 51.2%, corresponding to IC_{50} values ranging from 9.7 to 34.0.

A lack of scientific information is evident also for the characterization of the phenolic profile, as only Güneşer and Yilmaz (2017a) screened single phenolics of cold pressed and hexane-extracted oils from *C. limon* var. Kütdiken, by RP HPLC-DAD. In this study, the authors proved that both oils had an identical qualitative profile, although every compound of the cold pressed oil had a higher, or at least comparable, content with respect to the solvent-extracted counterpart. Among phenolic acids, gallic acid (cold pressed: 93.42 mg/Kg vs. solvent-extracted: 43.96 mg/Kg) and *trans*-ferulic acid (cold pressed: 85.13 mg/Kg vs. solvent-extracted: 63.39 mg/Kg) were at the highest levels in both types of oil, while the most abundant flavonoids were eriocitrin (cold pressed: 1051.60 mg/Kg vs. solvent-extracted: 1007.00 mg/Kg), hesperidin (cold pressed: 907.39 mg/Kg vs. solvent-extracted: 868.64 mg/Kg), and naringin (cold pressed: 389.79 mg/Kg vs. solvent-extracted: 202.60 mg/Kg) (Güneşer & Yilmaz, 2017a) (Table 8).

TABLE 8 Comparison of the polyphenol profile between cold pressed and hexane-extracted lemon (*C. limon* var. Kütdiken) seed oils.

	Cold pressed oil	Solvent-extracted oil	Seeds	
Polyphenol (mg/Kg)	(1)		(2)	(3)
Gallic acid	93.42[a]	43.96[a]	49.38–112.21	–
Caffeic acid	–		11.34–53.23	84.7
p-Hydroxybenzoic acid	–		–	153.6
trans-Ferulic acid	85.13	63.39	6.82–26.73	–
p-Coumaric acid	–		3.81–12.02	–
Rosmarinic acid	16.64	27.25	–	–
trans-2-Hydroxycinnamic acid	10.99	13.26	0.30–2.34	–
Syringic acid	8.46[a]	6.99[a]	2.58–59.58	–
Apigenin 6,8-di-*C*-glucoside	–		–	32.3
Apigenin7-(malonylpiosyl)-glucoside	–		–	36.0
Eriocitrin	1051.60	1007.00	–	747.0
Hesperidin	907.39	868.64	–	40,399.7
Naringin	389.79[a]	202.60[a]	–	–
Narirutin	–		–	4704.0
Didimin	–		–	1584
Neohesperidin	111.06	198.19	13,808–19,213	–
Rutin	76.80	52.31	–	–
Naringenin	22.82	10.62	1.15–5.47	–
Catechin	15.79	16.63	44.59–125.77	–
Kaempherol	15.12[a]	27.83[a]	6.14–12.37	–

[a]For a given compound, contents are significantly different at a level of confidence of 95%. Comparative literature on the polyphenol profile of lemon seeds is also reported. – = not available.
(1) Yilmaz & Güneşer, 2017 (Turkish lemon (*C. limon* var. Kütdiken) seed oil).
(2) Juhaimi et al., 2018 (Saudi Arabian lemon (*C. limon* var. Eureka) seeds). Data of this work are expressed on a dry weight basis.
(3) Russo et al., 2014 (Sicilian lemon (*C. limon*) seeds). Data of this work are expressed on a dry weight basis.
(Adapted from Güneşer, B.A. & Yilmaz, E. (2017a). Bioactives, aromatics and sensory properties of cold-pressed and hexane-extracted lemon (*Citrus Limon* L.) seed oils. *Journal of the American Oil Chemists' Society, 94*(5), 723–731.)

Comparing the outlined phenolics profile with other literature data becomes challenging, as no studies on single phenolics of lemon seed oil—even extracted by solvent—are available in the literature. Conversely, the phenolics of lemon seeds of different variety and origin (Juhaimi et al., 2018; Russo et al., 2014) were screened and, although no comparable profiles were reasonably reported, valuable contents of bioactive compounds, in particular flavonoids, were reported (Table 8).

3.2.4 Carotenoids

Among secondary plant metabolites, carotenoids are a subgroup of isoprenoid compounds usually consisting of a 40-carbon atoms (tetraterpenes) skeleton, with a structure characterized by a complex conjugated double-bond system responsible for the color of vegetable in which they are contained, and for their functionality. Such pigments are present in deoxygenated (carotenes) and oxygenated (xanthophylls) forms. α-carotene and β-carotene are the most common carotene compounds, whereas, among xanthophylls, the main representatives are lutein, zeaxanthin, and β-cryptoxanthin (Franke, Fröhlich, Werner, Böhm, & Schöne, 2010). Provitamin A (in simple terms, the precursor of vitamin A) and carotenoids (namely α-carotene, β-carotene, and β-cryptoxanthin) are the major sources of vitamin A in the human diet. Of the three carotenoids, β-carotene is the most widely distributed molecule in edible plant sources, and usually it is characterized by the highest concentrations in human blood. Additionally, these pigments seem to be involved in the prevention of several disorders, such as cancer, heart disease, macular degeneration, and cataracts (Oliver & Palou, 2000). The positive health effects of carotenoid pigments through the consumption of vegetable oils are already well-established. In fact, if the oil color is acceptable from a sensory perspective, more pigments are retained in the oil, shelf life is longer, and improved health benefits are guaranteed (Franke et al., 2010). Similarly to other bioactives, cold pressed oils may contain higher pigment contents than the solvent-extracted counterpart, thus enhancing the beneficial effects of such molecules. From this perspective, Güneşer and Yilmaz (2017a) elucidated carotenoids in cold pressed and solvent-extracted lemon (*C. limon* var. Kütdiken) oils from Turkey, via UV–visible spectrophotometry. Overall, slightly higher total carotenoid contents and levels of β-carotene and lutein were found in the cold pressed oil (10.43, 10.06, and 10.03 mg/Kg, respectively) than the solvent-extracted one (10.01, 9.66, and 9.63 mg/Kg, respectively). However, these data were not significantly different from a statistical point of view (Güneşer & Yilmaz, 2017a) (Table 9).

In another work, Malacrida et al. (2012) explored total and single carotenoids in petroleum ether-extracted lemon seed oil, and found lower total carotenoid content and amounts of β-carotene and lutein (4.36, 0.63, and 3.53 mg/Kg, respectively). In addition, differently from Güneşer and Yilmaz (2017a), they detected β-cryptoxanthin, although at very low levels (0.20 mg/Kg) (Malacrida et al., 2012) (Table 9).

3.2.5 Flavor compounds

The flavor of cold pressed oil, meaning a combination of its aroma and taste, has proved to contribute significantly to the overall quality and acceptability of the oil itself (Aydeniz, Güneşer, & Yılmaz, 2014; Yilmaz et al., 2015). Indeed, cold pressed oil is characterized by a variety of odor- and taste-active compounds, which, not being removed or reduced by

TABLE 9 Comparison of total and single carotenoids of cold pressed and hexane-extracted lemon (*C. limon* var. Kütdiken) seed oils.

	Cold pressed oil	Solvent-extracted oil	Solvent-extracted oil
Carotenoid (mg/Kg)	**(1)**		**(2)**
Total carotenoid content	10.43	10.01	4.36
β-Cryptoxanthin	–	–	0.20
β-Carotene	10.06	9.66	0.63
Lutein	10.03	9.63	3.53

For a given compound, contents marked by the asterisk are significantly different at a level of confidence of 95%. Comparative literature is also reported. - = not available.
(1) Yilmaz & Güneşer, 2017 (Turkish lemon (*C. limon* var. Kütdiken) seed oil).
(2) Malacrida et al., 2012 (lemon (*C. limon* var. Rosa) seed oil).
(Adapted from Güneşer, B.A. & Yilmaz, E. (2017a). Bioactives, aromatics and sensory properties of cold-pressed and hexane-extracted lemon (*Citrus Limon* L.) seed oils. *Journal of the American Oil Chemists' Society, 94*(5), 723–731.)

conventional refining, are responsible for its sensorial properties, whether they are pleasant or not (Güneşer & Yilmaz, 2017a). Generally, volatiles are responsible for the aroma of oil, whereas nonvolatile molecules including (but not limited to) phenolics, develop its taste (Moreau & Kamal-Eldin, 2015). According to the characteristics of the pressing method and of the seed pretreatments, such compounds may incur qualitative and quantitative changes, thus affecting most of the sensorial aspects of an oil (Zhou et al., 2013). Up to now, limited information on aroma- and taste-active compounds of cold pressed lemon seed oil has been available.

Concerning the aroma, only Güneşer and Yilmaz (2017a) elucidated by headspace solid-phase microextraction followed by gas chromatography coupled to mass spectrometry (SPME-GC–MS) the variation of the volatile profile of the lemon (*C. limon* var. Kütdiken) oils independent of the oil isolation procedure. Around 30 aroma compounds were detected in both types of oil, although most of them were more concentrated in the cold pressed sample (Table 10). In quantitative terms, the most abundant compounds of cold pressed oil were D-limonene (8529.0 mg/Kg), γ-terpinene (655.8 mg/Kg), β-pinene (472.4 mg/Kg), β-cimene (224.5 mg/Kg), β-mircene (213.7 mg/Kg), and α-pinene (98.26 mg/Kg), while for the solvent-extracted oil, D-limonene (574.8 mg/Kg), γ-terpinene (53.94 mg/Kg), α-terpineol (26.86 mg/Kg), and β-cimene (22.94 mg/Kg) were highlighted (Güneşer & Yilmaz, 2017a) (Table 10).

TABLE 10 Aromatic profile of the lemon seed oil (*C. limon* var. Kütdiken) isolated by cold pressing and solvent-based procedures.

		Cold pressed oil	Hexane-extracted oil
Aroma compound	Aroma description[a]	Concentration (mg/Kg)	
Acetoin	Creamy, buttery	23.32	nd
Hexanal	Green, grass	10.65	5.89
Furfural	Sweet, caramel, baked bread	3.68	nd
Isoamyl acetate	Banana, fruity	7.49	nd
Xylene	Sweet	nd	16.34
Isobutyl isobutyrate	Pear, pineapple	nd	0.62
α-Thujene	Woody, green herb	28.15	0.57
α-Pinene	Herbal, woody	98.26	1.90
Isopropyl valerate	Fruity, pineapple	14.68	12.28
Benzaldehyde	Bitter almond	6.84	1.99
Butyl isobutanoate	Fruity, pineapple mixture in water	nd	1.87
Isobutyl butanoate	Fruity, pineapple	nd	2.57
β-Pinene	Woody, green pine	472.49	19.39
6-Methyl-5-hepten-2-One	Citrus, green	16.42	nd
β-Mircene	Spicy, terpenic, herbal	213.72	4.47
α-Phellandrene	Terpenic	12.11	0.34
3-Pentanol	Sweet, herbal	nd	1.41
4-Carene	Citrus fruit, citrus peel	23.58	nd
β-Cimene	Terpenic	224.51	22.94
D-Limonene	Fresh citrus	8529.0	574.83
β-Ocimene	Floral, green	10.27	0.69
γ-Terpinene	Terpenic	655.89	53.94
Terpinolene	Herbal, woody, citrus	28.07	2.26
Linalool	Citrus, floral	9.28	1.97

TABLE 10 Aromatic profile of the lemon seed oil (*C. limon* var. Kütdiken) isolated by cold pressing and solvent-based procedures—cont'd

		Cold pressed oil	Hexane-extracted oil
Aroma compound	**Aroma description**	**Concentration (mg/Kg)**	
Nonanal	Citrus peel, rose	1.44	0.90
Limonene oxide	Citrus	4.38	1.30
4-Carvomenthanol	Wood, minty	35.63	16.53
α-Terpineol	Oil, mint	46.13	26.86
Nerol	Floral, sweet rose	7.16	4.51
β-Citronellol	Floral	2.36	1.45
Geraniol	Floral, rose, fruity	10.54	9.42
Thymol	Herbal, thyme	0.85	1.58
Neryl acetate	Floral, fresh rose, citrus	1.98	2.56
Caryophyllene	Spicy, woody	1.64	1.08
(*E*)-α-Bergamotol	Woody, tea leaf	2.11	1.44

nd = not detected.
[a]*Aroma descriptions of volatiles were retrieved from http://www.thegoodscentscompany.com/index.html and from http://www.flavornet.org/flavornet.html.*
(Adapted from Güneşer, B.A. & Yilmaz, E. (2017a). Bioactives, aromatics and sensory properties of cold-pressed and hexane-extracted lemon (*Citrus Limon* L.) seed oils. *Journal of the American Oil Chemists' Society, 94*(5), 723–731.)

Clearly, the perception of any aromatic chemical is strictly linked to its peculiar odor threshold value, namely the minimum compound concentration that can be perceived by humans (Leão, Sampaio, Pagani, & Da Silva, 2014). As a result, an aromatic compound may show a very intense aroma at very low contents and vice versa (Czerny et al., 2008). Hence, the volatile concentrations reported in Table 10 do not necessarily correspond to the aromatic notes perceived in the oil samples; rather they were indicative of the fact that cold pressing, contrary to solvent-based methods, retained a major concentration of volatiles in the lemon seed oil. However, looking at the aroma description reported for each compound in Table 10, the lemon seed oil was presumably characterized by a majority of citrus, woody, terpenic, and herbal notes, which would encourage its employment especially in cosmetic and pharmaceutical areas.

Similarly to the volatile profile, only Güneşer and Yilmaz (2017a) reported a descriptive sensory analysis of cold pressed and solvent-extracted lemon (*C. limon* var. Kütdiken) oils carried out with the help of volunteer panelists, which evaluated the seed oils by means of appropriate sensory descriptors (Table 11).

Overall, the cold pressed oil showed slightly higher aroma and taste notes (e.g., roasted, nutty, and spicy aromas, and raw vegetable taste) when compared with the solvent-extracted counterpart (Table 10). In particular, the higher roasted (cold pressed: ~6 vs. solvent extracted: ~2), and nutty (cold pressed: ~6 vs. solvent extracted: ~4) descriptors could be related to the heat, not exceeding 40°C, generated during the cold pressing procedure (Güneşer & Yilmaz, 2017a) (Table 11). Hence, it was pointed out that cold pressing may affect to a certain extent the sensorial properties of an oil, conferring on it a slightly more intense aroma than the procedures involving the use of solvents. However, as shown in Table 11, a very marked bitter taste (>12), high astringent taste (~6), and throat-catching sensation (~7) were characteristic of both types of oil (Güneşer & Yilmaz, 2017a).

Bitterness and astringency, as well as a throat-catching sensation, could be related to flavonoids, such as the flavanone glycosides naringin and neohesperidin, naturally present in the lemon seeds (Russo et al., 2014), and passed into the oil, through either cold pressing or solvent-based isolation (Güneşer & Yilmaz, 2017a). In addition, the bitter note could be generated by limonoids, such as limonin and nomilin, mostly occurring in seeds in the form of aglycone (Gualdani, Cavalluzzi, Lentini, & Habtemariam, 2016; Russo et al., 2014), but not investigated in the derived oil, whether extracted by pressing or by solvent.

Overall, the high scores of bitterness, astringency, spicy, and raw vegetable of both oils were perceived as unpleasant notes by the trained panel and may explain why such seed oil, if not refined, is not suitable for human consumption (Güneşer

TABLE 11 Descriptive sensory analysis of cold pressed and hexane-extracted lemon (*C. limon* var. Kütdiken) seed oils.

Sensory attribute	Min. and max. scale values	Cold pressed oil	Solvent-extracted oil
		Indicative values	
Clarity	From 1 = detergent solution to 15 = pure water	~4	~3
Lemon peel aroma	From 1 = absent to 15 = fresh lemon peel	~4	~4
Roasted aroma	From 1 = absent to 15 = roasted wafer	~6	~2
Raw vegetable taste	From 1 = absent to 15 = fresh green bean	~5	<4
Grassy aroma	From 1 = absent to 15 = fresh cut grass	~2	<2
Bitter taste	From 1 = absent to 15 = 0.5% caffeine solution in water	>12	>12
Astringent taste	From 1 = absent to 15 = kaki fruit	~6	~6
Waxy aroma	From 1 = absent to 15 = paraffin	>4	>4
Nutty aroma	From 1 = absent to 15 = roasted peanut and hazelnut	~6	~4
Fatty taste and aroma	From 1 = absent to 15 = refined sunflower oil	<10	~8
Spicy aroma	From 1 = absent to 15 = thyme, peppermint and rosemary mixture in water	~8	~7
Menthol taste	From 1 = absent to 15 = menthol candy	~5	~5
Throat-catching burning sensation)	Intensity after 30 s of swallowing the oil from 1 to 15	~7	~7

(Adapted from Güneşer, B.A. & Yilmaz, E. (2017a). Bioactives, aromatics and sensory properties of cold-pressed and hexane-extracted lemon (*Citrus Limon* L.) seed oils. *Journal of the American Oil Chemists' Society, 94*(5), 723–731.)

& Yilmaz, 2017a). Conversely, the lemon peel aroma, equally perceived in both oils (~ 4) as a pleasant attribute, may support the employment of such oil in the cosmetic sector (Table 11) (Güneşer & Yilmaz, 2017a).

Clearly, further research on the lemon seed oil flavor may be useful, not only to provide comparative data but also to optimize cold pressing and storage conditions of the oil, by avoiding the potential development of off-flavors. From this perspective, greater consumer acceptance of such oil might be achieved.

4 Applications of lemon seed oil

Unfortunately, no application of cold pressed lemon seed oil has been yet reported in the literature. Conversely, few scientific and not evidences are present on the use of solvent-extracted lemon seed oil. In this respect, transesterified lemon seed oil has recently proved to serve as a valuable biodiesel source when blended with conventional diesel. In fact, thanks to appropriate density, viscosity, and calorific value, a specific blend of lemon oil and diesel proved not only to improve the performance of a single cylinder four-stroke diesel engine, but also to reduce fuel consumption, and HC and CO_2 emissions as well (Dhivagar, Sundararaj, & Vignesh, 2018). Additionally, according to the information reported from certain websites, lemon seed oil would seem to be used after a massive refining procedure, in the cosmetic sector, as a valuable carrier of diverse cosmetic formulations (The Soap Kitchen, 2010), due to its skin rejuvenating, antiinflammatory, and moisturizing effects (Naturally Thinking, 2016). Clearly, future studies are encouraged to validate scientifically the beneficial aspects of lemon seed oil, and to address in-depth other potential employments of such oil obtained, preferably by cold pressing procedures.

5 Issues with cold pressed lemon seed oil: Consumer acceptance

A multimillion dollar problem of the citrus industry has always been related to citrus bitterness, representing a limiting factor for the salable processing of fruits, and consumer acceptance as well. The literature has already thoroughly established that the bitterness of citrus products is imparted by peculiar compounds, namely limonoids and flavonoids

(Drewnowski & Gomez-Carneros, 2000; Hasegawa, Berhow, & Fong, 1996; Puri, Marwaha, Kothari, & Kennedy, 1996), and, as mentioned in Sections 3.2.3 and 3.2.5, lemon seeds have proved to be marked by discrete contents of flavonoids, and by the highest levels of limonoids when compared with other lemon by-products (Russo et al., 2014; Xi, Lu, Qun, & Jiao, 2017). In particular, the presence of bitter flavanone glycosides, such as naringin and neohesperidin (Güneşer & Yilmaz, 2017a; Xi et al., 2017), and bitter limonoids, such as limonin and nomilin, mostly occurring in seeds in the form of aglycone (Gualdani et al., 2016; Russo et al., 2014), was observed.

During cold pressing, such compounds pass into the oil, so that, as seen in Section 3.2.5, "bitter" is typically the dominant character perceived by the consumer (Güneşer & Yilmaz, 2017a). The bitterness of lemon seed oil has always limited its direct consumption and could explain why in the past it was exploited for the production of soap, or was heavily refined to obtain an edible oil (Berk, 2016). However, despite having bitter and astringent flavors, flavanones and limonoids are relevant bioactives (Drewnowski & Gomez-Carneros, 2000; González-Molina, Domínguez-Perles, Moreno, & García-Viguera, 2010; Mir & Tiku, 2015), and cold pressed lemon seed oil as such could serve as a functional ingredient for animal feed, manufactured foods, and health care.

In spite of the numerous approaches developed for minimizing the bitterness of citrus juices (Shaw, Baines, Milnes, & Agmon, 2000; Sun-Waterhouse & Wadhwa, 2013), to the best knowledge of the authors, no literature has been yet reported on the debitterizing of cold pressed lemon seed oil. Consequently, in the perspective of improving the downstream chain of the citrus industry, and increase the commercial value of the lemon seed oil, research efforts should be devoted to the setup of a method removing, or at least reducing, the strong bitterness, and, at the same time, having no impact on the positive flavor notes and nutritional value of seed oil.

References

Abidi, S. L. (2001). Chromatographic analysis of plant sterols in foods and vegetable oils. *Journal of Chromatography A*, *935*(1–2), 173–201.

van Acker, S. A., van Balen, G. P., van den Berg, D. J., Bast, A., & van der Vijgh, W. J. (1998). Influence of iron chelation on the antioxidant activity of flavonoids. *Biochemical Pharmacology*, *56*(8), 935–943.

Aluyor, E. O., Ozigagu, C. E., Oboh, O. I., & Aluyor, P. (2009). Chromatographic analysis of vegetable oils: A review. *Scientific Research and Essays*, *4*(4), 191–197.

AOCS-American Oil Chemist's Society (1927). Orange and lemon oil by-products bring prosperity to citrus fruit growers: One plant alone nets $100,000 from cull organges formerly of little value. *Journal of the American Oil Chemists' Society*, *4*, 53–54. https://doi.org/10.1007/BF02662903.

Aparicio, R., & Aparicio-Ruı´z, R. (2000). Authentication of vegetable oils by chromatographic techniques. *Journal of Chromatography A*, *881*(1–2), 93–104.

de Araújo, E. F., de Queiroz, L. P., & Machado, M. A. (2003). What is *Citrus*? Taxonomic implications from a study of cp-DNA evolution in the tribe Citreae (Rutaceae subfamily Aurantioideae). *Organisms, Diversity and Evolution*, *3*(1), 55–62.

Aydeniz, B., Güneşer, O., & Yılmaz, E. (2014). Physico-chemical, sensory and aromatic properties of cold press produced safflower oil. *Journal of the American Oil Chemists' Society*, *91*(1), 99–110.

Bampidis, V. A., & Robinson, P. H. (2006). Citrus by-products as ruminant feeds: A review. *Animal Feed Science and Technology*, *128*(3–4), 175–217.

Bendini, A., Cerretani, L., Carrasco-Pancorbo, A., Gómez-Caravaca, A., Segura-Carretero, A., Fernández-Gutiérrez, A., et al. (2007). Phenolic molecules in virgin olive oils: A survey of their sensory properties, health effects, antioxidant activity, and analytical methods. An overview of the last decade. *Molecules*, *12*(8), 1679–1719.

Berk, Z. (2016). *Citrus fruit processing*. Massachusetts: Academic Press.

Braddock, R. J., & Kesterson, J. W. (1973). *'Citrus seed oils', bulletin of institute of food and agricultural sciences (University of Florida). (Vol. 756)* (pp. 1–30). Available from: http://ufdc.ufl.edu/UF00027159/00001/1. [10 March 2019].

Cecchi, L., Migliorini, M., Zanoni, B., Breschi, C., & Mulinacci, N. (2018). An effective HPLC-based approach for the evaluation of the content of total phenolic compounds transferred from olives to virgin olive oil during the olive milling process. *Journal of the Science of Food and Agriculture*, *98*(10), 3636–3643.

Cert, A., Moreda, W., & Pérez-Camino, M. C. (2000). Chromatographic analysis of minor constituents in vegetable oils. *Journal of Chromatography A*, *881*(1–2), 131–148.

Chandrasekaran, M. (2012). *Valorization of food processing by-products*. Boca Raton: CRC Press.

Cicero, N., Albergamo, A., Salvo, A., Bua, G. D., Bartolomeo, G., Mangano, V., et al. (2018). Chemical characterization of a variety of cold-pressed gourmet oils available on the Brazilian market. *Food Research International*, *109*, 517–525.

Czerny, M., Christlbauer, M., Christlbauer, M., Fischer, A., Granvogl, M., Hammer, M., et al. (2008). Re-investigation on odour thresholds of key food aroma compounds and development of an aroma language based on odour qualities of defined aqueous odorant solutions. *European Food Research and Technology*, *228*(2), 265–273.

Dhivagar, R., Sundararaj, S., & Vignesh, V. R. (2018). Biodiesel from lemon and lemon grass oil and its effect on engine performance and exhaust emission. In *IOP Conference Series: Materials Science and Engineering: Proceedings of ICRAMMCE-2017*. Bristol: IOP Publishing. vol. 330, no. 1, p. 012103.

Dimitrios, B. (2006). Sources of natural phenolic antioxidants. *Trends in Food Science & Technology*, *17*(9), 505–512.

Drewnowski, A., & Gomez-Carneros, C. (2000). Bitter taste, phytonutrients, and the consumer: A review. *The American Journal of Clinical Nutrition*, *72*(6), 1424–1435.

Dubois, V., Breton, S., Linder, M., Fanni, J., & Parmentier, M. (2007). Fatty acid profiles of 80 vegetable oils with regard to their nutritional potential. *European Journal of Lipid Science and Technology*, *109*(7), 710–732.

Duda-Chodak, A., & Tarko, T. (2007). Antioxidant properties of different fruit seeds and peels. *Acta Scentiarium Polonorum Technologia Alimentaria*, *6*(3), 29–36.

European Chemical Agency. (2019). *C&L Inventory*. Available from: https://echa.europa.eu/information-on-chemicals/cl-inventory-database. [13 March 2019].

FAO (2017). *Citrus fruit fresh and processed: Statistical bulletin 2016*. Available from: http://www.fao.org/3/a-i8092e.pdf. [14 February 2019].

Fernandes, P., & Cabral, J. M. S. (2007). Phytosterols: Applications and recovery methods. *Bioresource Technology*, *98*(12), 2335–2350.

Fine, F., Brochet, C., Gaud, M., Carre, P., Simon, N., Ramli, F., et al. (2016). Micronutrients in vegetable oils: The impact of crushing and refining processes on vitamins and antioxidants in sunflower, rapeseed, and soybean oils. *European Journal of Lipid Science and Technology*, *118*(5), 680–697.

Franke, S., Fröhlich, K., Werner, S., Böhm, V., & Schöne, F. (2010). Analysis of carotenoids and vitamin E in selected oilseeds, press cakes, and oils. *European Journal of Lipid Science and Technology*, *112*(10), 1122–1129.

Giacomelli, L. M., Mattea, M., & Ceballos, C. D. (2006). Analysis and characterization of edible oils by chemometric methods. *Journal of the American Oil Chemists' Society*, *83*(4), 303–308.

González-Molina, E., Domínguez-Perles, R., Moreno, D. A., & García-Viguera, C. (2010). Natural bioactive compounds of *Citrus limon* for food and health. *Journal of Pharmaceutical and Biomedical Analysis*, *51*(2), 327–345.

Gualdani, R., Cavalluzzi, M., Lentini, G., & Habtemariam, S. (2016). The chemistry and pharmacology of *Citrus limonoids*. *Molecules*, *21*(11), 1530.

Güneşer, B. A., & Yilmaz, E. (2017a). Bioactives, aromatics and sensory properties of cold-pressed and hexane-extracted lemon (*Citrus Limon* L.) seed oils. *Journal of the American Oil Chemists' Society*, *94*(5), 723–731.

Güneşer, B. A., & Yilmaz, E. (2017b). Effects of microwave roasting on the yield and composition of cold pressed orange seed oils. *Grasas y Aceites*, *68*(1), 175.

Gunstone, F. (2011). *Vegetable oils in food technology: Composition, properties and uses*. Hoboken: John Wiley & Sons.

Hamm, W., Hamilton, R. J., & Calliauw, G. (2013). *Edible oil processing*. Hoboken: John Wiley & Sons.

Hasegawa, S., Berhow, M. A., & Fong, C. H. (1996). Analysis of bitter principles in citrus. In H. F. Linskens, & J. F. Jackson (Eds.), *Fruit analysis* (pp. 59–80): Springer, Berlin, Heidelberg.

Hendrickson, R., & Kesterson, J. W. (1963). Florida lemon seed oil. In *76. Proceedings of Florida state horticultural Society, The Society, University of Wisconsin* (pp. 249–253).

Hunter, J. E. (2001). Studies on effects of dietary fatty acids as related to their position on triglycerides. *Lipids*, *36*(7), 655–668.

İnan, Ö., Özcan, M. M., & Aljuhaimi, F. (2018). Effect of location and *Citrus* species on total phenolic, antioxidant, and radical scavenging activities of some *Citrus* seed and oils. *Journal of Food Processing and Preservation*, *42*(3), e13555.

Juhaimi, F. A., Matthäus, B., Özcan, M. M., & Ghafoor, K. (2016). The physico-chemical properties of some citrus seeds and seed oils. *Zeitschrift für Naturforschung. Section C*, *71*(3–4), 79–85.

Juhaimi, F. A., Özcan, M. M., Uslu, N., & Ghafoor, K. (2018). The effect of drying temperatures on antioxidant activity, phenolic compounds, fatty acid composition and tocopherol contents in citrus seed and oils. *Journal of Food Science and Technology*, *55*(1), 190–197.

Karabiber, E. B., Zorba, N. N., & Yılmaz, E. (2018). Antimicrobial and functional properties of the proteins extracted from lemon, orange and grapefruit seeds press meals. *Quality Assurance & Safety of Crops and Food*, *10*(2), 145–153.

Karaman, E., Karabiber, E. B., & Yılmaz, E. (2018). Physicochemical and functional properties of the cold press lemon, orange, and grapefruit seed meals. *Quality Assurance & Safety of Crops and Food*, *10*(3), 233–243.

Karaman, E., Yılmaz, E., & Tuncel, N. B. (2017). Physicochemical, microstructural and functional characterization of dietary fibers extracted from lemon, orange and grapefruit seeds press meals. *Bioactive Carbohydrates and Dietary Fibre*, *11*, 9–17.

Koubaa, M., Mhemdi, H., & Vorobiev, E. (2016). Influence of canola seed dehulling on the oil recovery by cold pressing and supercritical CO_2 extraction. *Journal of Food Engineering*, *182*, 18–25.

Kumar, S. J., Prasad, S. R., Banerjee, R., Agarwal, D. K., Kulkarni, K. S., & Ramesh, K. V. (2017). Green solvents and technologies for oil extraction from oilseeds. *Chemistry Central Journal*, *11*(1), 9.

Laufenberg, G., Kunz, B., & Nystroem, M. (2003). Transformation of vegetable waste into value added products: (A) the upgrading concept; (B) practical implementations. *Bioresource Technology*, *87*(2), 167–198.

Leão, K. M., Sampaio, K. L., Pagani, A. A., & Da Silva, M. A. A. (2014). Odor potency, aroma profile and volatiles composition of cold pressed oil from industrial passion fruit residues. *Industrial Crops and Products*, *58*, 280–286.

Lin, C. S. K., Koutinas, A. A., Stamatelatou, K., Mubofu, E. B., Matharu, A. S., Kopsahelis, N., et al. (2014). Current and future trends in food waste valorization for the production of chemicals, materials and fuels: A global perspective. *Biofuels, Bioproducts and Biorefining*, *8*(5), 686–715.

Liu, X., Jin, Q., Liu, Y., Huang, J., Wang, X., Mao, W., et al. (2011). Changes in volatile compounds of peanut oil during the roasting process for production of aromatic roasted peanut oil. *Journal of Food Science*, *76*(3), C404–C412.

Malacrida, C. R., Kimura, M., & Jorge, N. (2012). Phytochemicals and antioxidant activity of citrus seed oils. *Food Science and Technology Research*, *18*(3), 399–404.

Mamma, D., & Christakopoulos, P. (2014). Biotransformation of citrus by-products into value added products. *Waste and Biomass Valorization*, *5*(4), 529–549.

Marín, F. R., Soler-Rivas, C., Benavente-García, O., Castillo, J., & Pérez-Alvarez, J. A. (2007). By-products from different citrus processes as a source of customized functional fibers. *Food Chemistry*, *100*, 736–741.

Matthaus, B., & Özcan, M. M. (2012). Chemical evaluation of citrus seeds, an agro-industrial waste, as a new potential source of vegetable oils. *Grasas y Aceites*, *63*(3), 313–320.

Mir, I. A., & Tiku, A. B. (2015). Chemopreventive and therapeutic potential of "Naringenin", a flavanone present in citrus fruits. *Nutrition and Cancer*, *67*, 27–42.

Moore, G. A. (2001). Oranges and lemons: Clues to the taxonomy of citrus from molecular markers. *Trends in Genetics*, *17*(9), 536–540.

Moreau, R. A., Johnston, D. B., & Hicks, K. B. (2005). The influence of moisture content and cooking on the screw pressing and pre-pressing of corn oil from corn germ. *Journal of the American Oil Chemists' Society*, *82*(11), 851–854.

Moreau, R., & Kamal-Eldin, A. (2015). *Gourmet and health-promoting specialty oils*. Amsterdam: Elsevier.

Naturally Thinking. (2016). *The benefits of lemon seed carrier oil*. Available from: https://www.naturallythinking.com/blog/the-benefits-of-lemon-seed-carrier-oil/. [25 March 2019].

Oliver, J., & Palou, A. (2000). Chromatographic determination of carotenoids in foods. *Journal of Chromatography A*, *881*(1–2), 543–555.

Parker, T. D., Adams, D. A., Zhou, K., Harris, M., & Yu, L. (2003). Fatty acid composition and oxidative stability of cold-pressed edible seed oils. *Journal of Food Science*, *68*(4), 1240–1243.

Prior, E. M., Vadke, V. S., & Sosulski, F. W. (1991). Effect of heat treatments on canola press oils. Non-triglyceride components. *Journal of the American Oil Chemists' Society*, *68*(6), 401–406.

Puri, M., Marwaha, S. S., Kothari, R. M., & Kennedy, J. F. (1996). Biochemical basis of bitterness in citrus fruit juices and biotech approaches for debittering. *Critical Reviews in Biotechnology*, *16*(2), 145–155.

Putnik, P., Bursać Kovačević, D., Režek Jambrak, A., Barba, F., Cravotto, G., Binello, A., et al. (2017). Innovative "green" and novel strategies for the extraction of bioactive added value compounds from citrus wastes-a review. *Molecules*, *22*(5), 680.

Qiubing, C., Jinghong, L., & Bi, W. (2012). Study on the solvent-extraction method of lemon seed oil. *Journal of the Chinese Cereals and Oils Association*, *10*, 13.

Ranganna, S., Govindarajan, V. S., Ramana, K. V. R., & Kefford, J. F. (1983). Citrus fruits. Part II. Chemistry, technology, and quality evaluation. B. Technology. *Critical Reviews in Food Science & Nutrition*, *19*(1), 1–98.

Reazai, M., Mohammadpourfard, I., Nazmara, S., Jahanbakhsh, M., & Shiri, L. (2014). Physicochemical characteristics of citrus seed oils from Kerman, Iran. *Journal of Lipids*, *2014*, 1–3.

Reda, S. Y., Leal, E. S., Batista, E. A. C., Barana, A. C., Schnitze, E., & Carneiro, P. I. B. (2005). Caracterização dos óleos das sementes de limão rosa (*Citrus limonia* Osbeck) e limão siciliano (*Citrus limon*), um resíduo agroindustrial. *Ciência e Tecnologia de Alimentos*, *25*(4), 672–676.

Rombaut, N., Savoire, R., Thomasset, B., Castello, J., Van Hecke, E., & Lanoisellé, J. L. (2015). Optimization of oil yield and oil total phenolic content during grape seed cold screw pressing. *Industrial Crops and Products*, *63*, 26–33.

Rostami, M., Farzaneh, V., Boujmehrani, A., Mohammadi, M., & Bakhshabadi, H. (2014). Optimizing the extraction process of sesame seed's oil using response surface method on the industrial scale. *Industrial Crops and Products*, *58*, 160–165.

Russo, M., Bonaccorsi, I., Torre, G., Sarò, M., Dugo, P., & Mondello, L. (2014). Underestimated sources of flavonoids, limonoids, and dietary fibre: Availability in lemon's by-products. *Journal of Functional Foods*, *9*, 18–26.

Saidani, M., Dhifi, W., & Marzouk, B. (2004). Lipid evaluation of some Tunisian citrus seeds. *Journal of Food Lipids*, *11*, 242–250.

Schwartz, H., Ollilainen, V., Piironen, V., & Lampi, A. M. (2008). Tocopherol, tocotrienol and plant sterol contents of vegetable oils and industrial fats. *Journal of Food Composition and Analysis*, *21*(2), 152–161.

Seppanen, C. M., Song, Q., & Saari Csallany, A. (2010). The antioxidant functions of tocopherol and tocotrienol homologues in oils, fats, and food systems. *Journal of the American Oil Chemists' Society*, *87*(5), 469–481.

Sharma, K., Mahato, N., Cho, M. H., & Lee, Y. R. (2017). Converting citrus wastes into value-added products: Economic and environmently friendly approaches. *Nutrition*, *34*, 29–46.

Shaw, P. E., Baines, L., Milnes, B. A., & Agmon, G. (2000). Commercial debittering processes to upgrade quality of citrus juice products. In M. A. Berhow, S. Hasegawa, & G. D. Manners (Eds.), *Citrus limonoids: Functional chemicals in agriculture and foods*. Michigan: ACS Symposium Series.

Siger, A., Nogala-Kalucka, M., & Lampart-Szczapa, E. (2008). The content and antioxidant activity of phenolic compounds in cold-pressed plant oils. *Journal of Food Lipids*, *15*(2), 137–149.

Sultana, B., Anwar, F., Mushtaq, M., & Alim, M. (2015). Citrus residues: A potential source of phenolics with high antioxidant values. *International Food Research Journal*, *22*(3), 1163–1168.

Sun-Waterhouse, D., & Wadhwa, S. S. (2013). Industry-relevant approaches for minimising the bitterness of bioactive compounds in functional foods: A review. *Food and Bioprocess Technology*, *6*(3), 607–627.

Tarandjiiska, R., & Nguyen, H. (1989). Triglyceride composition of seed oils from Vietnamese citrus fruits. *Rivista Italiana delle Sostanze Grasse*, *66*, 99–102.

Terada, A., Kitajima, N., Machmudah, S., Tanaka, M., Sasaki, M., & Goto, M. (2010). Cold-pressed yuzu oil fractionation using countercurrent supercritical CO_2 extraction column. *Separation and Purification Technology*, *71*(1), 107–113.

The Soap Kitchen (2010). *Material safety data sheet of refined lemon seed oil*. Available from: https://www.thesoapkitchen.co.uk/media/custom_file_uploads/MSDS%20and%20Technical/base%20oils/PDF_MSDS_Lemon_Seed_Oil_Refined.pdf [25 March 2019].

Tuberoso, C. I., Kowalczyk, A., Sarritzu, E., & Cabras, P. (2007). Determination of antioxidant compounds and antioxidant activity in commercial oilseeds for food use. *Food Chemistry*, *103*(4), 1494–1501.

Xi, W., Lu, J., Qun, J., & Jiao, B. (2017). Characterization of phenolic profile and antioxidant capacity of different fruit part from lemon (*Citrus limon* Burm.) cultivars. *Journal of Food Science and Technology*, *54*(5), 1108–1118.

Yilmaz, E., & Güneşer, B. A. (2017). Cold pressed versus solvent extracted lemon (*Citrus limon* L.) seed oils: Yield and properties. *Journal of Food Science and Technology*, *54*(7), 1891–1900.

Yilmaz, E., Sevgi Arsunar, E., Aydeniz, B., & Güneşer, O. (2015). Cold pressed capia pepperseed (*Capsicum annuum* L.) oils: Composition, aroma, and sensory properties. *European Journal of Lipid Science and Technology*, *117*(7), 1016–1026.

Zhou, Q., Yang, M., Huang, F., Zheng, C., & Deng, Q. (2013). Effect of pretreatment with dehulling and microwaving on the flavor characteristics of cold-pressed rapeseed oil by GC-MS-PCA and electronic nose discrimination. *Journal of Food Science*, *78*(7), C961–C970.

Chapter 15

Cold pressed chia (*Salvia hispanica* L.) seed oil

Akeem Omolaja Akinfenwa[a], Ahmad Cheikhyoussef[b], Natascha Cheikhyoussef[c], and Ahmed A. Hussein[a]
[a]*Chemistry Department, Faculty of Applied Sciences, Cape Peninsula University of Technology, Bellville, South Africa,* [b]*Science and Technology Division, Multidisciplinary Research Centre, University of Namibia, Windhoek, Namibia,* [c]*Ministry of Higher Education, Training and Innovation, Windhoek, Namibia*

List of abbreviations

CSO	chia seed oil
ALA	α-linolenic acid
CP	cold pressed
CP-CSO	cold pressed chia seed oil
DHA	docosahexaenoic acid
EFA	essential fatty acids
EPA	eicosapentaenoic acid
OA	oleic acid
OC-PLS	one class partial least squares
PA	palmitic acid
PUFA	polyunsaturated fatty acid
ROS	reactive oxygen species
SA	stearic acid
SFE	supercritical fluid extraction
SIMCA	soft independent modeling by class analogy
ω-3 ALA	omega-3 α-linolenic acid
ω-6 LA	omega-6 linoleic acid

1 Introduction

Chia (*Salvia hispanica* L.), also known as Mexican chia and Spanish sage, is a flowering green plant endemic to Latin America; Mexico, Guatemala, and Argentina (Anacleto et al., 2016). The flowers have purple, blue, or white colors, and seeds are an oval shape with dark brown to gray-white colors. Chia is a member of the family *Lamiaceae*, genus *Salvia*, and species of *hispanica* adapted to semiarid to arid land with sandy loam soil, thus capable of surviving long-term drought (Ayerza & Coates, 2004; Ullah et al., 2015). Other species of the Chia plants include *S. columbriae B.*, *S. polystachya O.*, and S. tiliifolia *V.y* (Anacleto et al., 2016). The history of growing and popularity of chia dates back to the 15th century in Mexico by the Aztec Alliance Nations who used it as a high-energy food, a drug for the treatment of diseases, and as an offering to the Aztec gods during religious rites (Cahill, 2003, 2005). It is reported to have a long history of survival of extinction in Mexico following a ban on its cultivation and its various uses for more than two centuries by Spanish conquerors (Ortiz de Montellano, 1978). The Spanish colonizers, in a bid to establish control of the Aztecs, attempted to clean up the cultural and religious significance of the seed, and replaced it with foreign grains such as wheat and barley to serve the same purpose as chia seed (Anacleto et al., 2016). Chia's name can be traced to Aztecs' Nahuatl and Yucatan Maya ancestral languages, commonly spoken in Mexico, meaning "oily," having high oil content, and "strength," being a high-energy food, respectively (Anacleto et al., 2016). In recent times, the crop has been widely cultivated in Latin America and Australia as a staple food and for its oils that are known for health-promoting traits attributed to the high content of omega-3 α-linolenic acid (ω-3 ALA) and other polyunsaturated fatty acids (PUFAs). The seed is the most valuable part of the plant for its high oil content. Different studies in the literature indicate that cold pressed chia seed oil (CP-CSO) contains essential

Cold Pressed Oils. https://doi.org/10.1016/B978-0-12-818188-1.00015-3
Copyright © 2020 Elsevier Inc. All rights reserved.

fatty acids (EFA), mainly ω-3 ALA and ω-6 linoleic acid (LA). In addition, components such as phenolics, tocopherols, and phytosterols, which are important natural sources of antioxidant, antiinflammatory, and anticancer activities, were reported (Ixtaina et al., 2011). According to the United States Pharmacopeia (USP), chia seed oil (CSO) is defined as the oil extracted from the seeds by the cold press process and excludes the use of solvents or external heat used in other extraction processes. Research studies on extraction of CP-CSO have increased in recent years to raise awareness of this underutilized oil with excellent nutraceutical and nutritive value. This chapter aims to cover the cold pressed methods of extraction, phytochemicals and bioactive compounds, health-promoting traits, and application of the oil in functional foods.

2 Phytochemical of chia seed oil

Phytochemicals are secondary metabolites (bioactive compounds) that occur naturally in plants in small quantities with potential physiological functions in animals. The chemistry of CP-CSO showed an interesting profile in the literature as it contains phytochemicals (Figs. 1–4), triglycerides and EFA, ω-3 ALA (60%–65%), and ω-6 LA (20%) found as major constituents (Luz Magali, Valdivia-López, Aburto-Juárez, & Tecante, 2008). In addition, other secondary metabolites such as tocopherols, phytosterols, phenolics, sterols, and squalene are found as minor constituents (Martínez-Cruz & Paredes-lópez, 2014; Yingbin et al., 2018). These phytochemicals have been reported with activities that classified the oil among nontraditional oil with health-promoting traits and potential for use in functional food products (Dąbrowski, Konopka, & Czaplicki, 2018; Ixtaina et al., 2011; Marineli et al., 2014; Yingbin et al., 2018).

2.1 Triglycerides and fatty acids

The glycerides are esters of fatty acid with glycerol. CP-CSO glycerides consist of a combination of any three fatty acids from ALA, LA, OA, PA, and SA (Yingbin et al., 2018). Marineli et al. (2014), in a study of characterization of triglycerides of Chilean CP-CSO, found that ALA was predominant in most triglycerides of the oil. The fatty acids content of CP-CSO is rich in EFA, namely ω-3 ALA and ω-6 LA (Fig. 1) necessary for the neurodevelopment of the human brain and prevention of chronic heart diseases (Simopoulos, 2010). Because the human body cannot synthesize these fatty acids, they are obtained from seeds, seed oil, and fish. The ω-3 ALA and ω-6 LA represent approximately 64% and 20%, respectively, of CP-CSO, which is significantly higher than those percentages obtained from cold pressed flax, hemp, and canola lemon seed oil (Teh & Birch, 2013; Yilmaz & Güneşer, 2017). Other fatty acids reported of lesser quantities are oleic acid (OA), palmitic acid (PA), stearic acid (SA), and arachidonic acid (AA) in decreasing order (Mehmet, Fahad, Isam, Magdi, &

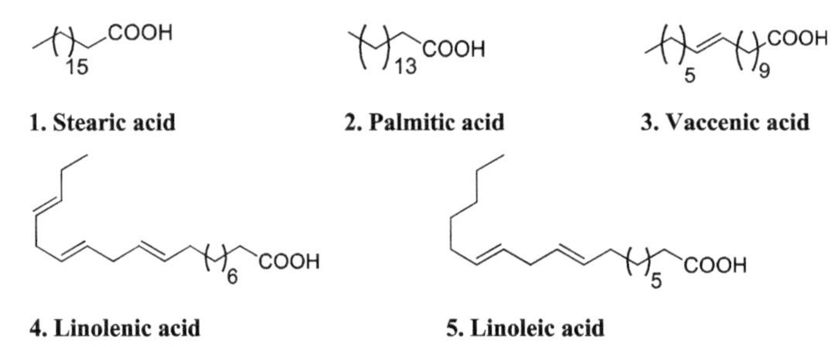

FIG. 1 Major fatty acid constituents of CP-CSO.

a-Flavonoids

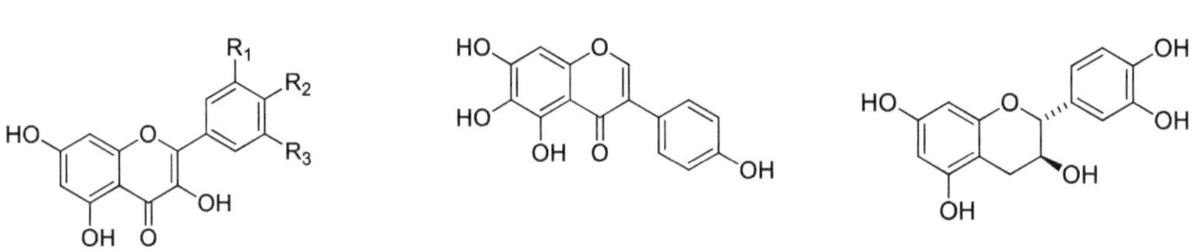

HO, O, OH, OH O, R_1, R_2, R_3

6. Kaempferol, R_1 = OH, R_2=R_3= H

7. Quercetin, R_1 =R_2= OH, R_3= H

8. Myricetin, R_1 =R_2=R_3= OH

9. Genistein

10. Catechin

FIG. 2 Flavonoids compounds structures from CP-CSO.

b-Phenolic acids

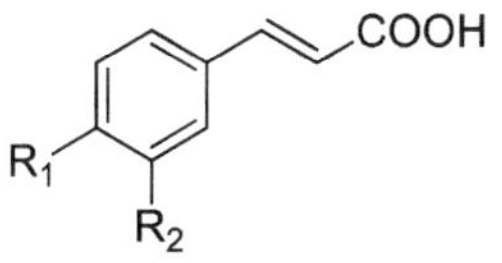

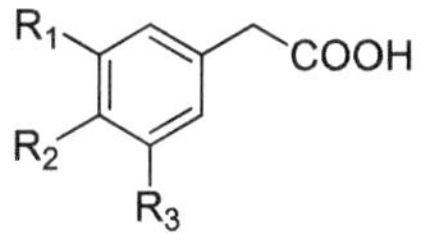

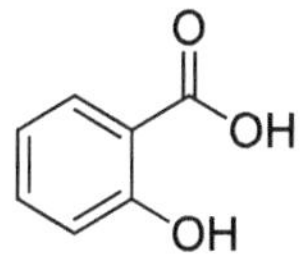

11. Caffeic acid, $R_1=R_2=OH$

12. Cumaric acid, $R_1=R_2=H$

13. Ferulic acid, $R_1=OH$, $R_2=OMe$

14. Gallic acid, $R_1=R_2=R_3=OH$

15. Syringic acid, $R_1=R_3=OH$, $R_2=H$

16. Protocatechuic acid, $R_2=R_3=OH$, $R_1=H$

17. Salicylic acid

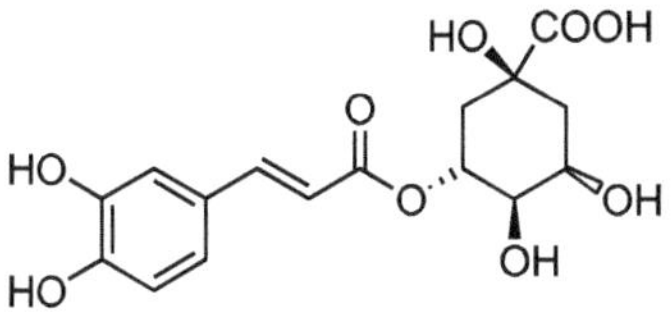

18. Chlorogenic acid

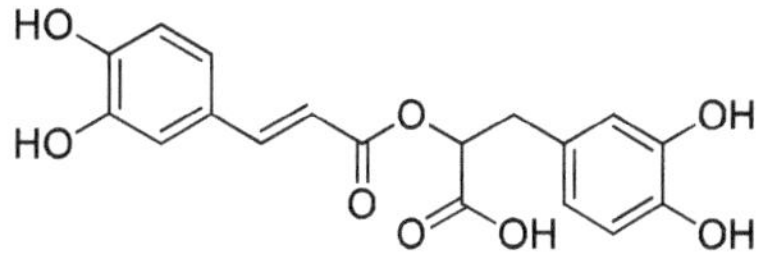

19. Rosemarinic acid

FIG. 3 Phenolic constituents of CP-CSO.

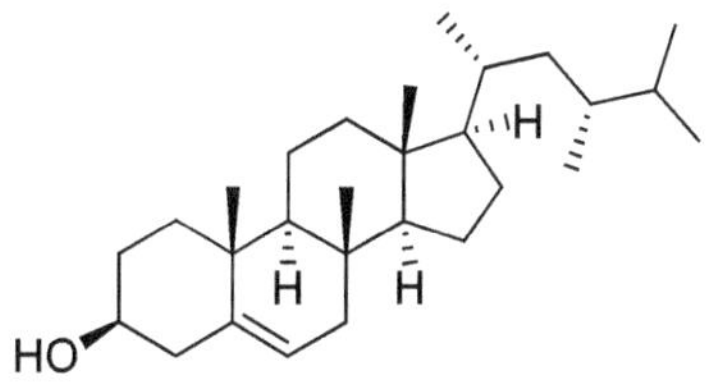

24. Campesterol

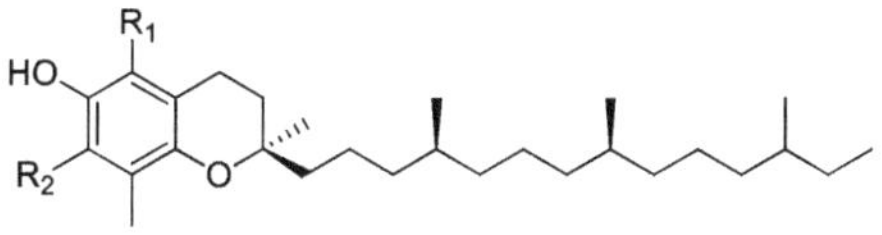

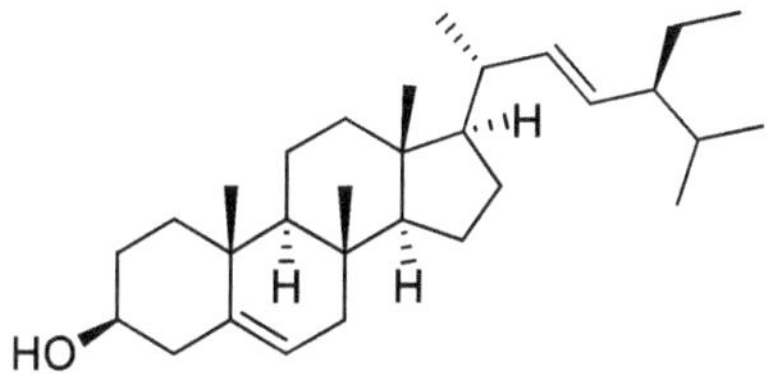

25. Stigmasterol

26. Stigmastanol

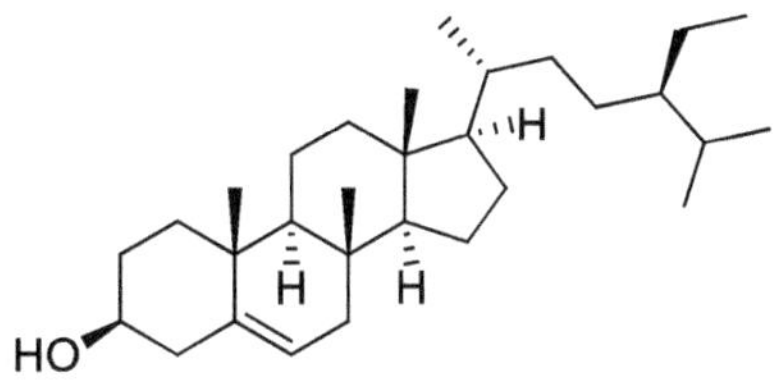

20. α-Tocopherol, $R_1=R_2=CH_3$

21. β-Tocopherol, $R_1=CH_3$, $R_2=H$

22. δ-Tocopherol, $R_1=R_2=H$

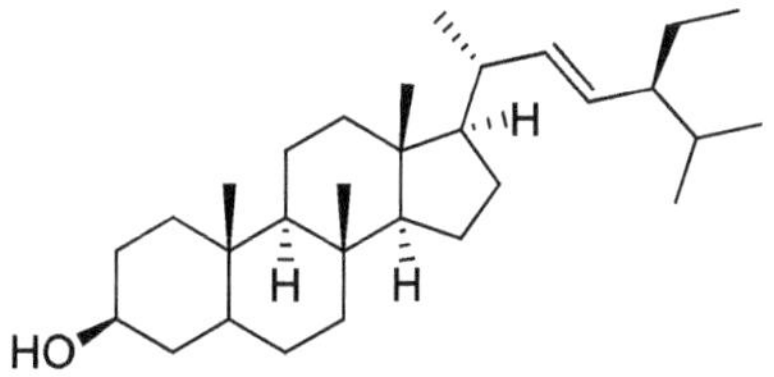

23. β- sitosterol

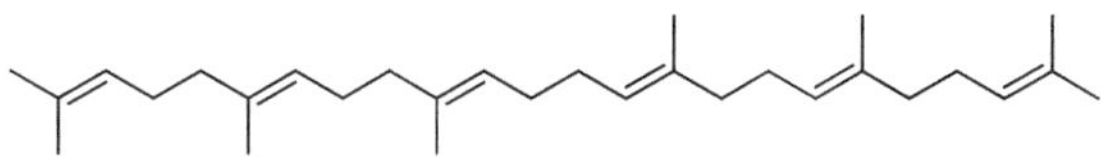

27. Squalene

FIG. 4 Tocopherol, phytosterol, and squalene of CP-CSO.

Mustafa, 2018). ω-3 ALA and ω-6 LA are 18 carbon fatty acids and differ only in the number and position of unsaturation in the carbon chain. While ω-3 ALA has three unsaturated double bonds from the third carbon of the methyl group end (18:3), ω-6 LA has two unsaturated double bonds from the sixth carbon of the methyl group (18:2).

2.2 Polyphenols

The polyphenols consist of large molecules with multiple hydroxyl groups attached to an aromatic system. They are good sources of natural antioxidant against auto-oxidation of fatty acid by reactive oxygen species (ROS) in oil. The presence of a double bond in CP-CSO fatty acids makes it prone to auto-oxidation, leading to the deterioration of the oil quality and short shelf life (Martínez-Cruz & Paredes-lópez, 2014). Antioxidants help to retain the wholesomeness of oil by prolonging shelf life and keeping the peroxide index at a low level for the oxidative stability of the oil. The literature reports of polyphenol compounds from CP-CSO seem inconclusive as there is a contrasting report regarding identification and type from chia oil where some of the identified phenolics vary slightly from reports, while some others were not detected. However, most reports indicated the presence of different types of polyphenols from CP-CSO as minor components. Flavonoids such as quercetin, kaempferol, myricetin, catechin, isorhamnetin, and genistein were observed, and phenolic acids such as chlorogenic, caffeic, rosemarinic, syringic, cinnamic, gallic, and cinnamic acids have also been reported (Ixtaina et al., 2011; Marineli et al., 2014; Mehmet et al., 2018). The chemical structures for the flavonoids and phenolic acids that have been isolated and identified from CP-CSO are shown in Figs. 2 and 3. The abovementioned phenolics were reported predominantly from different geographical origins. However, Yingbin et al.'s (2018) study of CP-CSO did not indicate the presence of phenolic compounds among minor bioactive components of the oil, but attributed antioxidant activity of Mexican CP-CSO to the tocopherols content.

2.3 Tocopherols and phytosterols

Tocopherols are natural antioxidants synthesized by plants that help protect against lipid peroxidation. They are amphipathic, having hydrophilic and lipophilic ends that act as antioxidants by scavenging peroxyl radicals of polyunsaturated fatty acids. Four homologous types (α-, β-, γ-, and δ-) of tocopherols were identified from CP-CSO by Mehmet et al. (2018), who reported that γ-tocopherol (85%) was the most abundant, followed by δ-tocopherol. Studies by Dąbrowski, Konopka, Czaplicki, and Tańska (2016) and Mehmet et al. (2018) found that the antioxidant activity of CSO is better retained through CP extraction compared to solvent extraction. The results of the cited studies reflect strong agreement for higher γ-tocopherol content in CSO obtained by CP extraction (Dąbrowski et al., 2016; Mehmet et al., 2018). However, Ixtaina et al. (2011) observed that the γ-tocopherol content of chia oil from Soxhlet extraction is significantly higher than that from CP. A comparison of the cold press method and the solvent method of CSO with some seed oils showed a higher value of α-tocopherol content for CP-CSO than that obtained for hemp oil and canola oil, but significantly less than sunflower and lemon seed oil (Teh & Birch, 2013; Mehmet et al., 2018; Yilmaz & Güneşer, 2017). This suggests that chia seed oil has better antioxidant properties and could be applied to functional foods (Table 1).

The literature is limited, with varying results on determination of the type and the exact amount of phytosterols from CP-CSO; however, there is a consensus that it is a minor constituent. Dąbrowski et al. (2018) reported the presence of the average amount of phytosterols such as β-sitosterol (55% of total sterols), campesterol (10.8%), 25-hydroxy-24-methylcholesterol (8.9%), and stigmasterol (4.2%) for CSO from Peru and Paraguay (Dąbrowski et al., 2018), while Luz Magali et al. (2008) found that Mexican CSO had predominant β-sitosterol (74%), stigmasterol, and stigmastanol (Luz Magali et al., 2008).

3 Extraction methods for chia seed oil

Chia seed oil is obtained from the seed through extraction methods that result in varying quality parameters of the oil: purity, percentage yield, fatty acids content, preservation of antioxidant content, and functionality of the oil. Different methods have been used to extract the oil from seeds over the years, such as cold pressing, conventional solvent extraction, supercritical fluid extraction, the ultrasound-assisted method, etc. (Ixtaina et al., 2010; Martínez et al., 2012). Each method produces oil of varying quality parameters: percentage purity and yield, fatty acids content, preservation of antioxidant content, organoleptic character, and wholesomeness of the oil. Hence the method selection depends on factors such as quality and quantity priorities, environmental considerations, and the cost of the process.

TABLE 1 α-Tocopherol content (mg/kg) of some cold-pressed seed oil.

Cold-pressed seed oil	α-tocopherol content
Chia	138 ± 1.89
Hemp	2.78 ± 01
Canola	11.99 ± 0.02
Lemon	155.00 ± 28.8
Sunflower	494.2 ± 15.1
Grape	139.2 ± 6.2
Rape	90.9 ± 3.4
Flax	9.5 ± 0.5

Adapted from Tuberoso et al., 2007; Teh, S. & Birch, J. (2013). Physicochemical and quality characteristics of cold-pressed hemp, flax and canola seed oils. *Journal of Food Composition and Analysis, 30*(1), 26–31. https://doi.org/10.1016/j.jfca.2013.01.004.; Yilmaz, E., & Güneşer, B. A. (2017). Cold pressed versus solvent extracted lemon (*Citrus limon* L.) seed oils: Yield and properties. *Journal of Food Science and Technology, 54*(7), 1891–1900; Mehmet, M., Fahad, Y., Isam A., Magdi, A. & Mustafa A. (2018). Effect of soxhlet and cold press extractions on the Physico-chemical characteristics of roasted and non-roasted chia seed oils. *Journal of Food Measurement and Characterization.* Springer US. https://doi.org/10.1007/s11694-018-9977-z.

3.1 Cold pressed extraction

Cold pressed extraction is a traditional method carried out by a screw expeller starting from seed separation to gray-white, where the oilseeds, loaded in a raffia sack, are pressed with a mechanical lever. The crude oil obtained is further purified by sedimentation and sieving. Pretreatments such as roasting and soaking in boiling water are usually employed to increase yield and reduce extraction time (Smain & Da-Wen, 2017). Like the age-old screw, the expeller does not involve organic treatment nor heat. It is a green method, cheap, nontoxic, and environmentally friendly. Martínez et al. (2012) developed a pilot design of a Box-Behnken screw press for extraction of CSO, and reported optimal conditions for oil yield comparable with a similar study by Ixtaina et al. (2011) using a Komet screw press. Ixtaina suggested that optimization conditions of oil yield increase screw press speed (20 rpm), moderate seed moistening to increase seed plasticity (0.101 g/g seed), and increase operating temperature (30 °C) and restriction die (6 mm) (Ixtaina et al., 2011; Martínez et al., 2012). The cold pressed extraction method was used to profile the physicochemical characteristic of hemp, flax, and canola seed oil by Teh and Birch (2013), who found that the cold pressed method had significant effects on the bioactive constituents. Hemp seed oil had the highest total phenolics, flavonoids, and γ-tocopherol (Teh & Birch, 2013). Several other literature reports proved that higher oil quality is obtained with the cold pressing method when compared with conventional solvent extraction techniques, even though oil recovery and time are low, however, However, the time involved is compensated for by the extraction of a purer oil and better retention of natural flavor, with a higher amount of total phenolics and antioxidant capacity for use as functional foods. This is also consistent with the report of the cold pressed method for lemon seed oil (Mehmet et al., 2018; Yilmaz, & Güneşer, 2017).

3.2 Soxhlet (solvent) extraction

CSO, due to its health-promoting traits and nutritional value, in recent times has experienced advancements in methods of extraction that influence the components and oxidative stability (Dąbrowski et al., 2016). Soxhlet extraction is a conventional method that involves the use of an organic solvent under controlled conditions. This conventional method relies on the use of solvent polarity and the hydrophilic or lipophilic character of the oil constituents. Solvents for extraction are varied between nonpolar to mid-polar and polar solvent (*n*-hexane < chloroform < dichloromethane < ethylacetate < acetone < methanol < water). Two advantages of this method over the traditional screw expeller are higher oil yield and better recovery (≈30% higher) in a short time (Dąbrowski et al., 2018). However, obtaining CSO totally free of the remnant of toxic and flammable solvents used requires rigorous procedures, posing a significant challenge for the process (Mehmet et al., 2018).

3.3 Supercritical fluid extraction (SFE)

This method was developed to address the drawbacks of the solvent method. It is based on the solubilizing power of fluid to exhibit properties of both liquid and gas simultaneously at critical temperature and pressure. At higher (supercritical) condition, it shows high solvation power with lower viscosity and higher diffusion rate, which is required for an ideal solvent for extraction of oils. Among gases, carbon dioxide is preferentially used as it is nontoxic, nonflammable, works at ambient temperature, is relatively cheap, and can be easily removed from the oil extract (Ixtaina et al., 2010). Antonio et al. (2011) and Ixtaina et al. (2010) reported the use of carbon dioxide gas for extraction of CSO with SFE techniques. Their results showed that the percentage of oil yield is higher than the oil yield from the solvent method. Oil recovery was reported at 92.8% after 300 min by Ixtaina et al. (2010). In addition, acidity and peroxide value were lower due to the absence of an organic solvent, suggestive of an advantage of SFE over solvent extraction (Antonio et al., 2011; Ixtaina et al., 2010).

3.4 Physicochemical properties of cold pressed chia oil

Properties such as peroxide value, moisture, residual oil and protein content, total dietary fiber, water holding capacity, tocopherols content, and antioxidant activity are important in determining the functional usage of CSO. The recorded values of these properties are found relative to the extraction method. Mehmet et al., in a study of the physicochemical properties of roasted and nonroasted chia seed oil using the cold pressed and Soxhlet methods, reported lower peroxide values (low rancidity), higher values of linolenic acid, and tocopherols being retained in the recovered oil with cold pressing over Soxhlet extraction (Mehmet et al., 2018), suggesting the advantage of the cold pressing method over the traditional Soxhlet method (Table 2). Similar results were reported by Capitani, Spotorno, Nolasco, and Tomás (2012) and Ixtaina et al. (2011), which attempt to justify the cold press method as an efficient method of extraction that maintains the physicochemical and functional properties of chia seed oil with respect to higher content of residual oil, moisture, and tocopherols (Capitani et al., 2012; Ixtaina et al., 2011).

3.5 Oxidative stability

Oxidative rancidity is a common spoilage process that occurs due to lipid auto-oxidation of oils containing monounsaturated and PUFAs exposed to light, oxygen, and high temperatures, resulting in undesirable changes such as off-odor, loss of nutritional qualities, and phase change. Yingbin et al. (2018) determined the stability of CSO by the response to forced oxidation and reported the induction period of CSO at 0.68 h, which was similar to the induction period (0.63 h) for camelina

TABLE 2 Comparison of physicochemical properties of chia seed oil extracted from different methods.

Physicochemical properties	Cold press method	Soxhlet method
Peroxide value	3.65 ± 0.29 meq O_2/kg	5.84 ± 1.56 meq O_2/kg
α-tocopherols	138.4 ± 1.89 mg/kg	135.6 ± 2.7 mg/kg
Linolenic acid	67.84%	66.75%
Antioxidant activity	0.54 ± 0.09%	0.47 ± 0.13%
Phenolic (rosemarinic acid)	2.17 ± 0.5 mg/g	2.67 ± 0.32 mg/g
Crude fiber	23.81 ± 0.34 g/100 g	27.57 ± 0.07 g/100 g
Total wax	109.21 ± 3.38 mg/kg	138.87 ± 5.36 mg/kg
Residual oil	11.39 g/100 g	10.85 g/100 g
Protein content	35 g/100 g	41.36 g/100 g
Water holding capacity	10.58 ± 0.55 (g/g)	10.64 ± 0.60 (g/g)
Moisture	10.84 ± 0.21 g/100 g	10.47 ± 0.16 g/100 g

Adapted from Capitani, M., Spotorno, V., Nolasco, S. & Tomás, M. (2012). Physicochemical and functional characterization of by-products from chia (*Salvia hispanica* L.) seeds of Argentina. *LWT- Food Science and Technology, 45*(1), 94–102. https://doi.org/10.1016/j.lwt.2011.07.012; Mehmet, M., Fahad, Y., Isam A., Magdi, A. & Mustafa A. (2018). Effect of soxhlet and cold press extractions on the Physico-chemical characteristics of roasted and non-roasted chia seed oils. *Journal of Food Measurement and Characterization*. Springer US. https://doi.org/10.1007/s11694-018-9977-z.

oil, but lower than for soybean oil; this was possibly due to the higher lipid content of the oil. Dąbrowski et al. (2016) found the stability of CSO to be dependent on the extraction method, wherein solvent extraction of CSO (with acetone) was reported as having higher stability than the SFE and CP methods (Dąbrowski et al., 2016). Although CSO contains tocopherols as natural antioxidants, it is relatively unstable with a short shelf life due to the higher contents of ALA and LA (Bodoira, Penci, Ribotta, & Martínez, 2017). This is similar to the report of oxidative stability of sunflower oil with a higher tocopherol content (Table 1) by Ramadan (2013). It was suggested that the stability of oils can be improved to retain their wholesomeness by storage conditions (in an airtight and lightproof container) that limit exposure of the oils to light and by a healthy blend with other cold pressed oils (Bodoira et al., 2017; Ramadan, 2013).

4 Edible and nonedible applications of cold pressed chia seed oil

4.1 Functional food materials

Vegetable oils play an important role in sensory assessment of diets and food products formulation. CP-CSO contains a high nutrient value that offers the highest amount of ω-3 ALA from seed plants. Extensive reports in the literature have detailed the high nutritional value and superiority of CSO over other seed oils as an excellent source of ALA, natural antioxidants, and minerals for use as a food supplement and in processed foods (Montrimaitė & Moščenkova, 2018). Although several seed oils, including rapeseed, sunflower seed, flaxseed, soybean oil, and CSO, have been consumed directly or infused as ingredients in functional foods, the increasing awareness of the health benefits of ALA from CP-CSO in relation to reduced risk of cardiovascular diseases and type 2 diabetes, and as a source of natural antioxidants, raises concern for the selection of oil to use in functional foods. CP-CSO might offer a healthy blend or a good replacement for vegetable oils for use as a leavening and emulsifying agent in the confectionery and bakery industries (Martínez-Cruz & Paredes-lópez, 2014; Monroy-Torres et al., 2008).

4.2 Natural source of fat replacers in foods

Obesity and overweight are major risk factors for chronic diseases such as diabetes and heart failure due to excess body fat from fatty food intake. The World Health Organization (WHO) in 2016 decried the alarming rate of obesity and overweight among adults (WHO, 2016); hence there is a need for food producers and consumers to reduce fat and identify alternative sources of low-fat foods. CP-CSO and cake possess low fat with lower triglyceride and cholesterol, which reduces the risk of obesity and heart-related disease. Akcicek and Karasu (2018) reported insights in the application of chia seed waste as a low-fat dressing for a salad that mimics the conventional fat oil-in-water emulsion of mayonnaise.

4.3 Cosmetics and skin care

Curatives of skin disorder are among the concerns of cosmetics industries in the manufacture of skin care products. A topical formulation of skin care product containing 4% CSO has been indicated to improve skin hydration, reduce intense itching sensation, and reduce *trans*-dermal water loss from healthy volunteers with xeroderma skin disorder (Jeong, Park, Park, & Hwan, 2010). Hyperpigmentation is another skin disorder related to tyrosinase enzymes by excess melanin production, UV-radiation damage, and inflammation of skin melanocytes. Research has been targeted toward the importance of seed oils rich in ω-3 ALA for reducing the prevalence of skin diseases by control of activities of tyrosinase enzymes in melanin production (Athar & Nasir, 2005). CSO is rich in ω-3 ALA, absorbs quickly into skin pores without irritation, and shows potential for application in inhibition of hyperpigmentation and tyrosinase trafficking through epidermal cells of the skin (Ligęza, Wyglądacz, & Tobiasz, 2016). An *in vitro* experiment carried out by Ando, Ryu, Hashimoto, Oka, and Ichihashi (1998) on lightening the effect of ALA, LA, and OA on UV-induced hyperpigmentation of dorsal skin of brownish guinea pigs indicated that the effect of ALA in the correction of the pigmentation was highest, followed by LA and last OA. Ando et al. (2006) further reported the mechanism of action as the ability of intracellular fatty acids (ω-3 ALA and ω-6 LA) in the endoplasmic reticulum to affect processing and function of tyrosinase through the expression of melanocyte genes in melanogenesis. In a similar report by Diwakar, Rana, Saito, Vredeveld, and Scholten (2014), the effect of the combination of α-linolenic and linoleic acids of chia seed with fruit extract from *Punica granatum* confirmed that the activity of the combination inhibits melanogenesis.

4.4 Health-promoting traits of cold pressed oil and oil constituents

CP-CSO is becoming popular among consumers for its nutraceutical values and prevention of chronic diseases. The oil provides tocopherols and polyphenols as natural sources of antioxidants for the prevention of stress-related disease caused by oxidative damage of ROS (Ixtaina et al., 2011). Studies of the oil from cold pressed extraction show health benefits of ω-3 ALA and ω-6 LA, with interconversion in relation to a balanced ratio of ω-6 LA to ω-3 ALA to mitigate the risk of cardiovascular diseases (Galli & Marangoni, 2006; Simopoulos, 2010). In a metabolic breakdown involving desaturation and chain elongation, ω-3 ALA produces eicosapentaenoic acid (EPA) and docosahexaenoic acid (DHA) essential for brain, retinal, and nervous functions, and ω-6 LA produces arachidonic acid, which is involved in the contraction and relaxation of muscles, regulation of receptor muscles, and blood vessel dilation.

Recent research discussed health-promoting traits of the oil in the prevention of risk factors of cardiovascular diseases: antihypertensive, antidiabetic, antiobesity, and anticholesterol. In animal studies, Wister rats fed with chia milled supplement as ALA were observed to have a significant reduction of triglycerides and cholesterol (Ayerza & Coates, 2007). Toscano et al. (2014) reported human studies of a group of 26 hypertensive subjects placed on daily consumption of 35 g/day of chia flour for 12 weeks; the subjects showed a significant reduction in blood pressure and systolic pressure when compared to a group administered with an antihypertensive drug (Toscano et al., 2014).

5 Market potential, adulteration, and authenticity

Historically, the market potential for chia seed and the oil started with domesticated chia by the Aztec Empire in the pre-Columbia age (3500 BCE–1000 CE), with a yearly consumption of 4000–15,000 tones in the forms of food, beverages, and medicine, and as a market commodity for the exchange of gold and corn (Anacleto et al., 2016). In recent times, there has been huge cultivation of the seed from more than 13 countries globally for the export of the oil. This is attributed to the growing popularity of the oil as the highest source of ω-3 ALA (PUFAs), health benefits, and application in functional foods. Countries like Mexico, Bolivia, Argentina, Paraguay, and Australia are the leading economies of the chia plant (Ayerza & Coates, 2004). In addition, the market potential for export to Europe has been reported to have increased between 2012 and 2016, with a significant increase of 27% from 3485 tons to 16,182 tons; this is probably because it is not grown in Europe and is also due to the fact that it is gluten and allergen free. Germany, the Netherlands, Scandinavia, the United Kingdom, and Denmark are the highest importers of chia seed for use as an ingredient in beverages, additives in food, and as a natural food supplement (Anacleto et al., 2016).

CSO is a pale yellow to golden liquid with a nutty flavor and high health value, thus having a retail market price higher than that of other oils (Rodríguez, Gagneten, Farroni, Percibaldi, & Buera, 2019). As with other cold pressed oils, it is prone to adulteration by fraudulent marketers and suppliers, and as such requires high quality standards and procedures for production (Smain & Da-Wen, 2017). The European Union's chia legislation in 2013 set out quality compliance standards for chia seed exported to Europe, to prevent adulteration and misbranding (Borg, 2013). The conventional method for determining the level of adulteration of oils relies on liquid or gas chromatography, involving solvents and the quantification of the fatty acids, which is time-consuming. Rodríguez et al. (2019) reported the used of Fourier transform infra-red (FTIR) spectroscopy using the untargeted methods of soft independent modeling by class analogy and one class partial least squares (SIMCA and OC-PLS) methods for the determination of adulterated CSO with corn, soybean peanut, and sunflower oils. Both methods were found to be effective for determining adulterants with acceptable prediction error.

6 Conclusion

Evaluation of CP-CSO showed it is a highly valorized oil with growing market potential due to the numerous health benefits of its constituents and applications in novel functional foods. Optimization of process parameters to increase oil yield and retain its natural aroma and phytochemicals comparable to other extraction methods, as well as maintaining the oxidative stabilities, will be a challenge for emerging researchers in cold pressing extraction of the oil.

References

Akcicek, A., & Karasu, S. (2018). Utilization of cold-pressed chia seed oil waste in a low-fat salad dressing as natural fat replacer. *Journal of Food Process Engineering*, 1–10. https://doi.org/10.1111/jfpe.12694.

Anacleto, S., Ruiz, G., Rana, J., Gordillo, G., West, H., Sharma, M., et al. (2016). Chia crop (*Salvia hispanica* L.): Its history and importance as a source of polyunsaturated fatty acids omega-3 around the world: A review. *Journal of Crop Research and Fertilizers*, *1*(104), 1–9. https://doi.org/10.17303/jcrf.2016.104.

Ando, H., Ryu, A., Hashimoto, A., Oka, M., & Ichihashi, M. (1998). Linoleic acid and α-linolenic acid lightens ultraviolet-induced hyperpigmentation of the skin. *Archives of Dermatological Research*, *290 (7)*, 375–381. https://doi.org/10.1007/s004030050320.

Ando, H., Wen, Z., Kim, H., Valencia, J., Costin, G., Watabe, H., et al. (2006). Intracellular composition of fatty acid affects the processing and function of tyrosinase through the ubiquitin-proteasome pathway. *The Biochemical Journal*, *394*, 43–50. https://doi.org/10.1042/BJ20051419.

Antonio, J., Uribe, R., Ivan, J., Perez, N., Castillo, H., Rosado, G., et al. (2011). Extraction of oil from chia seeds with supercritical CO_2. *The Journal of Supercritical Fluids*, *56*(2), 174–178. https://doi.org/10.1016/j.supflu.2010.12.007.

Athar, M., & Nasir, S. (2005). Taxonomic perspective of plant species yielding vegetable oils used in cosmetics and skincare products. *African Journal of Biotechnology*, *4*(1), 36–44.

Ayerza, R., & Coates, J. (2004). Composition of chia (*Salvia hispanica*) grown in six tropical and subtropical ecosystems of South America. *Tropical Science*, *44*(3), 131–135. https://doi.org/10.1002/ts.154.

Ayerza, R., & Coates, J. (2007). Effect of dietary α-linolenic fatty acid derived from chia when fed as ground seed, whole seed, and oil on lipid content and fatty acid composition of rat plasma. *Annals of Nutrition and Metabolism*, *85637*, 27–34. https://doi.org/10.1159/000100818.

Bodoira, R., Penci, M., Ribotta, P., & Martínez, M. (2017). Chia (*Salvia hispanica* L.) oil stability: Study of the effect of natural antioxidants. *LWT - Food Science and Technology*, *75*, 107–113. https://doi.org/10.1016/j.lwt.2016.08.031.

Borg, T. (2013). Commission implementing decision authorising an extension of the use of Chia (*Salvia hispanica*) seed as a novel food ingredient under Regulation (EC) No 258/97 of the European Parliament and Council (notified under document C). *Official Journal of the European Union*, *336*, 49–51.

Cahill, J. P. (2003). Ethnobotany of chia *Salvia hispanica* L. *Journal of Economic Botany*, *57*(4), 604–618.

Cahill, J. P. (2005). Human selection and domestication of chia (*Salvia hispanica* L.). *Journal of Ethnobiology*, *25*(2), 155–174. https://doi.org/10.2993/0278-0771.

Capitani, M., Spotorno, V., Nolasco, S., & Tomás, M. (2012). Physicochemical and functional characterization of by-products from chia (*Salvia hispanica* L.) seeds of Argentina. *LWT- Food Science and Technology*, *45*(1), 94–102. https://doi.org/10.1016/j.lwt.2011.07.012.

Dąbrowski, G., Konopka, I., & Czaplicki, S. (2018). Variation in oil quality and content of low olecular lipophilic compounds in chia seed oils. *International Journal of Food Properties*, *21*(1), 2016–2029. https://doi.org/10.1080/10942912.2018.1501699.

Dąbrowski, G., Konopka, I., Czaplicki, S., & Tańska, M. (2016). Composition and oxidative stability of oil from *Salvia hispanica* L. seeds in relation to extraction method. *European Journal of Lipid Science and Technology*, *119*(5), 1–26. https://doi.org/10.1002/ejlt.201600209.

Diwakar, G., Rana, J., Saito, L., Vredeveld, D., & Scholten, J. (2014). Inhibitory effect of a novel combination of *Salvia hispanica* (chia) seed and *Punica granatum* (pomegranate) fruit extracts on melanin production. *Fitoterapia*, *97*, 164–171. https://doi.org/10.1016/j.fitote.2014.05.021.

Galli, C. Ã., & Marangoni, F. (2006). N-3 fatty acids in the Mediterranean diet. *Prostaglandins, Leukotrienes and Essential Fatty Acids*, *75*, 129–133. https://doi.org/10.1016/j.plefa.2006.05.007.

Ixtaina, V., Martınez, M., Spotorno, V., Mateo, C., Diehl, B., Nolasco, S., et al. (2011). Characterization of chia seed oils obtained by pressing and solvent extraction. *Journal of Food Composition and Analysis*, *24*, 166–174. https://doi.org/10.1016/j.jfca.2010.08.006.

Ixtaina, V., Vega, A., Nolasco, S., Tomás, M., Gimeno, M., Bárzana, E., et al. (2010). Supercritical carbon dioxide extraction of oil from Mexican chia seed (*Salvia hispanica* L.): Characterization and process optimization. *Journal of Supercritical Fluids*, *5*(1), 192–199. https://doi.org/10.1016/j.supflu.2010.06.003.

Jeong, S., Park, H., Park, B., & Hwan, K. H. (2010). Effectiveness of topical chia seed oil on pruritus of end-stage renal disease (ESRD) patients and healthy volunteers. *Annals of Dermatology*, *22*(2), 2010. https://doi.org/10.5021/ad.2010.22.2.143.

Ligęza, M., Wyglądacz, D., & Tobiasz, A. (2016). Natural cold pressed oils as cosmetic products. *Family Medicine & Primary Care Review*, *18*(4), 443–447. https://doi.org/10.5114/fmpcr.2016.63699.

Luz Magali, A., Valdivia-López, M., Aburto-Juárez, M., & Tecante, A. (2008). Chemical characterization of the lipid fraction of Mexican chia seed (*Salvia hispanica* L.). *International Journal of Food Properties*, *11*(3), 687–697. https://doi.org/10.1080/10942910701622656.

Marineli, S., Aguiar, É., Alves, S., Teixeira, A., Nogueira, M., Roberto, M., et al. (2014). Chemical characterization and antioxidant potential of Chilean chia seeds and oil (*Salvia hispanica* L.). *LWT- Food Science and Technology*, *59*(2), 1304–1310. https://doi.org/10.1016/j.lwt.2014.04.014.

Martínez, M., Marín, M., Salgado, C., Revol, J., Penci, M., & Ribotta, P. (2012). Chia (*Salvia hispanica* L.) oil extraction : Study of processing parameters. *LWT- Food Science and Technology*, *47*(1), 78–82. https://doi.org/10.1016/j.lwt.2011.12.032.

Martínez-Cruz, O., & Paredes-lópez, O. (2014). Phytochemical profile and nutraceutical potential of chia seeds (*Salvia hispanica* L.) by ultra-high-performance liquid chromatography. *Journal of Chromatography A*, *1346*, 43–48. https://doi.org/10.1016/j.chroma.2014.04.007.

Mehmet, M., Fahad, Y., Isam, A., Magdi, A., & Mustafa, A. (2018). Effect of soxhlet and cold press extractions on the Physico-chemical characteristics of roasted and non-roasted chia seed oils. *Journal of Food Measurement and Characterization*. https://doi.org/10.1007/s11694-018-9977-z. Springer US.

Monroy-Torres, R., Mancilla-Escobar, M. L., Gallaga-Solórzano, J. C., & Santiago-García, E. J. (2008). Protein digestibility of chia seed Salvia hispanica L. *Revista Salud Pública y Nutrición*, *9*(1), 1–9.

Montrimaitė, K., & Moščenkova, E. (2018). Possibilities of usage of oilcake from non-traditional oil plants for the development of health-friendly functional food products. *Food Science and Applied Biotechnology*, *1*, 154–164. https://doi.org/10.30721/fsab2018.v1.i2.

Ortiz de Montellano, B. R. (1978). Aztec cannibalism: An ecological necessity? *Science*, *200*(4342), 611–617. (1978). https://www.jstor.org/stable/1746929.

Ramadan, M. F. (2013). Healthy blends of high linoleic sunflower oil with selected cold-pressed oils: Functionality, stability and antioxidative characteristics. *Industrial Crops and Products*, *43*, 65–72. https://doi.org/10.1016/j.indcrop.2012.07.013.

Rodríguez, S., Gagneten, M., Farroni, A., Percibaldi, N., & Buera, M. (2019). FT-IR and untargeted chemometric analysis for adulterant detection in chia and sesame oils. *Food Control*, *105*, 78–85. https://doi.org/10.1016/j.foodcont.2019.05.025.

Simopoulos, A. P. (2010). The omega-6/omega-3 fatty acid ratio : Health implications. *Oléagineux, Corps Gras, Lipides*, *17*, 267–275. https://doi.org/10.1684/ocl.2010.0325.

Smain, C., & Da-Wen, S. (Eds.). (2017). *Edible oils extraction, processing and applications*. Broken Sound Parkway NW/Boca Raton, FL: CRC Press Taylor & Francis Group.

Teh, S., & Birch, J. (2013). Physicochemical and quality characteristics of cold-pressed hemp, flax and canola seed oils. *Journal of Food Composition and Analysis*, *30*(1), 26–31. https://doi.org/10.1016/j.jfca.2013.01.004.

Toscano, L. T., da Silva, C. S. O., Toscano, L. T., de Almeida, A. E. M., da Cruz Santos, A., & Silva, A. S. (2014). Chia flour supplementation reduces blood pressure in hypertensive subjects. *Plant Foods for Human Nutrition*, *69*(4), 392–398.

Ullah, R., Nadeem, M., Khalique, A., Imran, M., Mehmood, S., Javid, A., et al. (2015). Nutritional and therapeutic perspectives of Chia (*Salvia hispanica* L.): A review. *Journal of Food Science and Technology*. https://doi.org/10.1007/s13197-015-1967-0.

WHO (2016). https://www.who.int/news-room/fact-sheets/detail/obesity-and-overweight. (Accessed22 July 2019).

Yilmaz, E., & Güneşer, B. A. (2017). Cold pressed versus solvent extracted lemon (Citrus limon L.) seed oils: Yield and properties. *Journal of Food Science and Technology*, *54*(7), 1891–1900.

Yingbin, S., Liyou, Z., Jun Jin, X., Junning Fu, M., Yifu, G., & Xun, S. (2018). Phytochemical and biological characteristics of Mexican chia seed oil. *Molecules*, *23*(3219), 16. https://doi.org/10.3390/molecules23123219.

Further reading

Carlo, T., Adam Kowalczyk, S., & Paolo, C. (2007). Determination of antioxidant compounds and antioxidant activity in commercial oilseeds for food use. *Food Chemistry*, *103*, 494–1501. https://doi.org/10.1016/j.foodchem.2006.08.014.

Emin, Y., & Buket, A. (2017). Cold-pressed versus solvent-extracted lemon (*Citrus limon* L.) seed oils : Yield and properties. *Journal of Food Science and Technology*, *54*, 891–1900. https://doi.org/10.1007/s13197-017-2622-8. June.

Grancieri, M., Stampini, H., Martino, D., & Mejia, E. (2019). Chia seed (*Salvia hispanica* L.) as a source of proteins and bioactive peptides with health benefits. *Comprehensive Reviews in Food Science and Food Safety*, 1–20. https://doi.org/10.1111/1541-4337.12423.

John, P., Amanda, N., Amy, L., Hellen, O., & Angela, I. (2018). Therapeutic perspectives on chia seed and its oil: A review. *Planta Medica*, *84*(9–10), 606–612. https://doi.org/10.1055/a- 0586-4711.

Luciana, T., Cássia, O., Lydiane, T., Antônio, M., Amilton, S., & Alexandre, S. (2014). Chia flour supplementation reduces blood pressure in hypertensive subjects. *Plant Foods for Human Nutrition*. https://doi.org/10.1007/s11130-014-0452-7.

Sheila, O., Débora, V., Cinthia, C., Mário, M., João, B., Andreia, S., et al. (2017). Characterization of phenolic compounds in chia (*Salvia hispanica* L.) seeds, fibre flour and oil. *Food Chemistry*, *232*, 295–305. https://doi.org/10.1016/j.foodchem.2017.04.002. Elsevier Ltd.

Sielicka, M., Małecka, M., & Purłan, M. (2014). Comparison of the antioxidant capacity of lipid-soluble compounds in selected cold-pressed oils using photochemiluminescence assay (PCL) and DPPH method. *European Journal of Lipid Science and Technology*, *116*, 388–394. https://doi.org/10.1002/ejlt.201300356.

Chapter 16

Cold pressed corn (*Zea mays*) oil

Zarina Mushtaq[a], Muhammad Imran[b], Nazir Ahmad[b], Muhammad Kamran Khan[b], and Nosheen Asghar[a]
[a]*Department of Food Science, Nutrition and Home Economics, Government College University, Faisalabad, Pakistan,* [b]*Institute of Home and Food Sciences, Faculty of Life Sciences, Government College University, Faisalabad, Pakistan*

1 Introduction

The cold press method for oilseeds and kernels is currently gaining attention for its special oil. There are supercritical extraction and solvent extraction methods in addition to regular expeller pressing to obtain oil, but each technique has its own advantages and drawbacks. The most important differences among the extraction techniques are related to oil yield and the minor component compositions of the oil. Thereby, oils extracted by different techniques usually have different end points for use; some must be fully refined, while some could be used directly for edible, cosmetic, or other purposes.

Corn is primarily grown for its starch and protein, while corn oil (extracted from corn germs) is a by-product of processing. Corn is composed of 61%–78% starch, 6%–12% protein, and 3.1%–5.7% oil. During the dry or wet milling process, corn germ as oil source is separated from the kernel for starch production. A wet milling process is usually preferred for corn germ separation to extract the oil. Corn oil is usually extracted with expeller pressing together with hexane or isohexane extraction to achieve around 97%–99% yield. Refining of crude corn oil is usually carried out by alkaline refining, bleaching, dewaxing, and deodorization processes (Stan, 2015). Corn oil is highly rich in essential fatty acids like omega-6 and omega-9. The omega-6 content was found to be 52.6%. Corn oil effluent is a source of polyunsaturated lipids (53.8%) and saturated lipids (14.8%) (Carrillo et al., 2017).

A comparison was made between cold pressed and fully refined winterized corn oils. The results showed that cold pressed corn oils have more pronounced free fatty acidity (FFA), peroxide (PV), and *p*-anisidin (*p*-AV) values, and higher saponification number, total carotenoids, and phenolic contents than those of the refined winterized corn oils. Some phytosterols (β-sitosterol, campesterol, and stigmasterol) were analyzed in cold pressed corn oil as well as various flavonoids (hesperidin and rutin) and phenolic acids (gallic, syringic, rosmarinic, and *trans*-ferulic) were detected in the cold pressed oil samples (Güneşer, Yılmaza, & Ok, 2017). Monounsaturated fatty acids (MUFAs):polyunsaturated fatty acids (PUFAs), saturated:unsaturated, as well as linoleic:linolenic acid ratios were improved on receiving a 75% recommended dose of fertilizer (Ray, Banerjee, Dutta, Hazra, & Majumdar, 2019).

Cold pressing is an easy, rapid, environmentally friendly, and cheap process, which yields good-quality oil but it is poor in overall oil yield. Hence, it is only preferable when special purpose oils are demanded. Furthermore, there are strict requirements for producing an acceptable grade of cold press oil. First, the seed or kernel must be very clean, homogeneous, and free from all unacceptable contaminations. The extraction process must be under mild conditions (no heating, solvent, or other chemicals). Finally, the cold pressed oils must be of edible grade regarding physicochemical and sensory properties, and must be safe. Under these circumstances, cold pressed oils could be very beneficial for human health, and could have exceptional aroma. Cold pressed oils do not contain traces of hexane or other solvents, and retain most of their bioactive components (tocopherols, sterols, phenols, etc.), making them highly valued products (Khoddami, Che Man, & Roberts, 2014; Yılmaz, Aydeniz, Güneşer, & Arsunar, 2015).

2 History of corn oil

Corn or maize (*Zea mays* L.) botanically belongs to the grass family gramineae (Poaceae). The cultivation of this plant started in Mesoamerica, and it is currently grown worldwide. The major corn producing countries are America, China, Brazil, Mexico, Indonesia, India, France, and Argentina. In 2016, 1.01 billion metric tons of corn were produced and more than 159 million hectares were used for corn cultivation. Production is increasing each year (Anonymous, 2015).

Cold Pressed Oils. https://doi.org/10.1016/B978-0-12-818188-1.00016-5
Copyright © 2020 Elsevier Inc. All rights reserved.

3 Fatty acid and acyl lipids profile of cold pressed corn oil

The major fatty acids of corn oil are specified as 59.8% linoleic acid, 25.8% oleic acid, 11.0% palmitic acid, 1.7% stearic acid, and 1.1% linolenic acid. A more detailed fatty acid composition range data is also available: myristic acid <0.1%, palmitic acid 8.0%–19.0%, palmitoleic acid <0.5%, stearic acid 0.5%–4.0%, oleic acid 19.0%–50.0%, linoleic acid 34.0%–62.0%, linolenic acid 0.1%–2.0%, arachidic acid <1.0%, gadoleic acid <0.5%, and behenic acids <0.5%. Deodorized corn oil contains 0.08%–0.12% of total tocopherols, which consist of 70%–80% g-, 20%–25% a-, and 3%–5% d-tocopherols (Table 1). Likewise, around 1.3%–2.3% unsaponifiable matter is present and around 60% of this is sterols. The predominant sterols were β-sitosterol, campesterol, and stigmasterol. The ranges of tocopherols for corn oil are given as: α-tocopherol 116–172 ppm, β-tocopherol 0–22 ppm, γ-tocopherol 1119–1401 ppm, and δ-tocopherol 59–65 ppm. The most important physical properties of corn oil were given as: iodine value of 118–133, saponification number of 187–193, refractive index of 1.470–1.474, unsaponifiable content of 1.3%–2.3%, titer of 14–20 °C, melting point of −12 °C to −10 °C, and solidification point of −1.0 °C to −20.0 °C (O'Brien, 2004).

TABLE 1 Bioactive compounds and antioxidant activities of cold pressed corn (*Zea mays*) oil.

Compound	Amount	Reference
Antioxidant		
β-Carotene	3.56 mg/100 g	Rafalowski et al. (2008)
Chlorophylls content	4.9 mg/kg	Tuberoso et al. (2007)
TEAC activity	2.30 mmol/L	Tuberoso et al. (2007)
EC_{50} value	14.8 mg	Siger et al. (2008)
DPPH$^{\bullet}$ radical scavenging activity	11.1%	Siger et al. (2008)
Total phenolic content	1.26 mg CAE/100 g	Siger et al. (2008)
Sterol and tocopherol		
β-Sitosterol	60%–62%	Stan (2015)
Campesterol	19%–22%	Stan (2015)
Stigmasterol	6%–7%	Stan (2015)
Sitosterol	2%–4%	Stan (2015)
α-Tocopherol	248.8 mg/kg	Rafalowski et al. (2008)
γ-Tocopherol	91.7 mg/100 g	Rafalowski et al. (2008)
δ-Tocopherol	5.93 mg/100 g	Rafalowski et al. (2008)
Phenolic compound		
Rutin	14.1 mg/kg	Pandey et al. (2013)
Hesperidin	23.8 mg/kg	Pandey et al. (2013)
Ferulic acid	0.5 ppm	Tuberoso et al. (2007)
trans-cinnamic acid	0.9 ppm	Tuberoso et al. (2007)
Vanillin	2.8 ppm	Tuberoso et al. (2007)
p-hydroxybenzoic acid	1.7 mg/100 g	Siger et al. (2008)
p-coumaric acid	1.9 mg/100 g	Siger et al. (2008)
Ferulic acid	5.8 mg/100 g	Siger et al. (2008)
Sinapic acid	0.6 mg/100 g	Siger et al. (2008)

4 Purification of cold pressed oil

In oil-bearing seeds, rendering cannot liberate oil from the cellular structure. The cell wall of seeds is broken by grinding, flaking, rolling, or pressing under high pressure to expel oil. The moisture content of the corn germs is adjusted to 12% by water conditioning before cold pressing, since that moisture level is determined as the best point for maximum oil yield and ease of the operation by preexperiments. Generally, the sequence of modern operation in cold pressing seeds is as follows: (1) to remove any stray bits of metal, seeds are passed over magnetic separators; (2) the hulls or shells are removed; (3) kernels are ground to convert them to coarse meal by hammer milling or by grinding them between grooved rollers; (4) finally, the material is pressed in screw or hydraulic presses without preliminary heating. Oil expelled without heating is often of edible quality without refining or further processing, and contains the least amount of impurities. Such oils are known as virgin, cold pressed, or cold drawn. Heating pressing of coarse meal removes more oil and also removes higher quantities of nonglyceride impurities like color bodies, phospholipids, and unsaponifiable matter. These oils are more highly colored than cold pressed oils (Aydeniz, Güneşer, & Yılmaz, 2014; Yılmaz, Arsunar, Aydeniz, & Güneşer, 2015).

5 Sterols and tocopherols composition of cold pressed corn oil

Table 1 presents the main bioactive compounds and antioxidant activities of cold pressed corn oil (CPCO). Some sterols (24-methylene cholesterol, campesterol, D-7-campesterol, D-5,23-stigmasterol, cholesterol, and D-5,24-stigmasterol) were present in trace amounts in the CPCO. A study of 15 samples found that the chief compounds present were β-sitosterol (60%–62%), campesterol (19%–22%), and stigmasterol (6%–7%). These values of given sterols agreed with the vegetable oils samples (Stan, 2015). The chief sterols of corn oils have been found to be β-sitosterol, campesterol, and stigmasterol (Strecker, Bieber, Maza, Grossberger, & Doskoczynski, 1996), whereas the quantification of tocopherol was as α-tocopherol in these samples. CPCO has a higher amount of α-tocopherol (248.8 mg/kg). A study was carried out to estimate the tocopherol from refined and CPCO with other oils (Rafalowski, Zegarska, Kuncewicz, & Borejszo, 2008). The results showed in cold pressed samples as 27.22, 91.76, and 5.93 mg/100 g oil of α-tocopherol, γ-tocopherol, and δ-tocopherol, respectively. In refined oils, these tocopherols were shown as 26.11, 44.31, and 0.0 mg/100 g oil, respectively. According to the findings of Strecker et al. (1996), the levels of tocopherols were 19.1, 94.2, and 4.2 mg/100 g for α-, γ-, and δ-tocopherols, respectively.

6 Phenolics composition

Studies did not report any phenolic compound in refined oil. This may be lost during reefing or processing from corn oil. Eight different samples (baby corn, raw, boiled, baked, fried, ripe corns, popcorn, and stylar) were analyzed for phenolic content (Pandey, Singh, Maurya, Singh, & Singh, 2013). Only rutin (14.17 mg/kg) and hesperidin (23.81 mg/kg) were quantified among eight samples of oils. The amounts of phenolics in raw maize were found to be 4.6, 3.2, 0.99, and 0.57 mg/g of gallic, caffeic, ferulic, and salicylic acids, respectively. Tuberoso, Kowalczyk, Sarritzu, and Cabras (2007) reported that the maize oil contained 0.5 ppm ferulic acid, 0.9 ppm *trans*-cinnamic acid, and 2.8 ppm vanillin. There was quantification of phenolic acids in CPCO, and 1.7, 1.9, 5.8, and 0.6 mg/100 g oil of *p*-hydroxybenzoic, *p*-coumaric, ferulic, and sinapic acids, respectively, was found (Siger, Nogala-Kalucka, & Lampart-Szczapa, 2008). It was concluded from the previous literature that CPCO contained a number of phenolic contents. These phenolics are antioxidants, which may help to improve health status through antioxidant, antibacterial, antiinflammatory, antiallergic, antithrombotic, and anticarcinogenic activities, and hepatoprotective and vasodilatory actions (Pandey et al., 2013).

7 Volatiles and aromatics composition

There are 22 volatiles found in cold pressed corn oil and seven volatiles in hexane-extracted corn oil. Obviously, maximum volatiles are missing in the pure sample, probably through the alkali refining and deodorization processes. The rich volatiles found in the CPCO sample were hexanal, acetic acid, 3-methylbutanal, hexanoic acid, and butanoic acid. It has been identified that in any food product, the characteristic smell is provided by the volatiles found in combination, but not in ranges of maximum amounts. The definition of aroma threshold value is most minimal concentration of a substance that can be detected by a human nose. The aroma threshold of certain volatiles is considerably lower, which can account for the perception of aroma at lower amounts relative to other volatiles found in similar foods at maximum concentrations (Meilgaard, Civille, & Carr, 1991).

The descriptive characteristics of aromatic components in samples were green, fresh, herbal, and sour. Limited literature is available regarding volatile compounds of corn oil. There were 29 volatile components present in corn oil (Tuan, 2011). In another study, octane, 2,6-dimethyl-3-heptanone, 2-heptenal, 2-pentylfuran, nonanal, hexedecanoic acid, 9-octadecenoic acid ethyl ester, acetic acid, α-pinene, benzaldehyde, limonene, and 3-carene were determined.

Cold pressing corn germ contained many volatile compounds, wherein Solid Phase Microextraction (SPME) with different resins was used to quantify the volatiles in corn (Steenson, Lee, & Min, 2002). The most abundant available volatile components identified at room temperature (storage 12–103 months) of corn oil were 2-heptenal, trans-2-octenal, pentane, and 2,4-heptedienal (Goicoechea & Guillén, 2014). Many secondary oxidized components such as alkanals, 2-alkenals, 2,4-alkadienals, acids, alcohols, ketones, alkylfurans, lactones, and a,b-unsaturated aldehydes were detected. With long duration, the type and concentration of volatile components were improved. At the initiation of the storage study, various resemblances between analyzed volatile components like hexanoic and butanoic acid, hexenal, nonanal, limonene, and some furans were determined.

8 Antioxidant activity of cold pressed corn oil

Table 1 presents the main bioactive compounds and antioxidant activities of cold pressed *Zea mays* oil. Various chemical compounds have been found in corn oils that showed strong importance in its nutritional potential. Various colored pigments and phenolics present in Zea oils have been lost during processing and during refining. The amount of the β-carotene content was found to be 91 ppm in commercial corn oil (Dauqan, Sani, Abdullah, Muhamad, & Top, 2011), while this content was present as 3.56 mg/100 g oil in CPCO, which is a negligible amount in processed or refined oil (Rafalowski et al., 2008). Furthermore, compounds like β-carotene (content 0.9 mg/kg), chlorophylls (content 4.9 mg/kg), and TEAC activity of 2.30 mmol/L were exploited for corn oil (Tuberoso et al., 2007). EC_{50} values, DPPH·radical scavenging activity, and total phenolic content (as caffeic acid equivalent) were 14.8 mg, 11.1%, and 1.26 mg CAE/100 g oil, respectively (Siger et al., 2008).

9 Health benefits of cold pressed corn oil

Potential health benefits of corn oil include its capacity to lower inflammation, reduce oxidative stress, lower allergic reactions, reduce "bad" cholesterol, improve eye health, and protect the skin. Consumers usually use corn oil for cooking due to its flavor. However, some health-conscious consumers use organic corn oil in medicinal use and topical application due to presence of various health components. Cold pressed organic oil delivers many health benefits. Polyunsaturated and monounsaturated fatty acids are both required to reduce excess cholesterol and reduce inflammation. However, neither of these acids are balanced in refined oils (Rose, Thomson, & Williams, 1965). Monounsaturated fats and polyunsaturated fats are found in high quantities in unrefined corn oil, which includes oleic and linoleic acids, although when it comes to the balance of omega-6 and omega-3 to fatty acids in corn oil, the ratio is far from ideal. Omega-6 fatty acids were 50 times greater than omega-3 fatty acids in oil, whereas the recommended ratio is 1:1. Omega-6 and omega-3 are considered to be effective against inflammation and relieve symptoms of headaches, arthritis, and gastrointestinal problems.

Several compounds in corn oil are considered to lower allergic reactivity in the body, also helping to reduce the symptoms of rhinitis and asthma. Flavonoids, like lutein, can lower free radical activity in the body (Mulrooney & Grimble, 1993). Lutein is also recommended to improve eye health and reduce the development of cataracts. The tocopherols and antioxidants in corn oil could help to reduce skin infections and to prevent blemishes, irritation, psoriasis, and eczema. High levels of vitamin as well as limited levels of xanthins, vitamin A, and trace amounts of other antioxidant compounds can be beneficial for health (Staughton, 2019).

References

Anonymous (2015). (2015). http://www.worldofcorn.com/#world-cornproduction. Accessed 26 August 2019.

Aydeniz, B., Güneşer, O., & Yılmaz, E. (2014). Physico-chemical, sensory and aromatic properties of cold press produced safflower oil. *Journal of the American Oil Chemists' Society*, *91*, 99–110. https://doi.org/10.1007/s11746-013-2355-4.

Carrillo, W., Carpio, C., Morales, D., Vilcacundo, E., Álvarez, M., & Silva, M. (2017). Content of fatty acids in corn (*Zea mays* L.) oil from Ecuador. *Asian Journal of Pharmaceutical and Clinical Research*, *10*(8), 150–153.

Codex Stan. (2015). Codex Standard for named vegetable oils. pp. 1–13. http://www.fao.org/input/.../standards/336/CXS_210e_2015.pdf. Accessed 13 September 2019.

Dauqan, E., Sani, H. A., Abdullah, A., Muhamad, H., & Top, A. G. (2011). Vitamin E and beta carotene composition in four different vegetable oils. *American Journal of Applied Sciences*, *8*, 407–412. https://doi.org/10.3844/ajassp.2011.407.412.

Goicoechea, E., & Guillén, M. D. (2014). Volatile compounds generated in corn oil stored at room temperature. Presence of toxic compounds. *European Journal of Lipid Science and Technology*, *116*, 395–406.

Güneşer, B. A., Yılmaza, E., & Ok, S. (2017). Cold pressed versus refined winterized corn oils: Quality, composition and aroma. *Grasas y Aceites*, *68*(2) e194.

Khoddami, A., Che Man, Y. B., & Roberts, T. H. (2014). Physicochemical properties and fatty acid profile of seed oils from pomegranate (*Punica granatum* L.) extracted by cold pressing. *European Journal of Lipid Science and Technology*, *116*, 553–562. https://doi.org/10.1002/ejlt.201300416.

Meilgaard, M., Civille, G. V., & Carr, B. T. (1991). *Sensory evaluation techniques*. Boca Raton, FL: CRC Press ISBN 10: 0849342805/ISBN 13: 9780849342806.

Mulrooney, H. M., & Grimble, R. F. (1993). Influence of butter and of corn, coconut and fish oils on the effects of recombinant human tumor necrosis factor-α in rats. *Clinical Science*, *84*(1), 105–112.

O'Brien, R. D. (2004). *Fats and oils: Formulating and processing for applications*. New York: CRC Press. ISBN 9780203483664.

Pandey, R., Singh, A., Maurya, S., Singh, U. P., & Singh, M. (2013). Phenolic acids in different preparations of maize (*Zea mays*) and their role in human health. *International Journal of Current Microbiology and Applied Sciences*, *2*, 84–92 (ISSN: 2319-7706).

Rafalowski, R., Zegarska, Z., Kuncewicz, A., & Borejszo, Z. (2008). Fatty acid composition, tocopherol and β-carotene content in Polish commercial vegetable oils. *Pakistan Journal of Nutrition*, *7*, 278–282.

Ray, K., Banerjee, H., Dutta, S., Hazra, A. K., & Majumdar, K. (2019). Macronutrients influence yield and oil quality of hybrid maize (*Zea mays* L.). *PLoS One*, *14*(5)e0216939.

Rose, G. A., Thomson, W. A., & Williams, R. T. (1965). Corn oil in treatment of Ischaemic heart disease. *British Medical Journal*, *1*(5449), 1531–1533.

Siger, A., Nogala-Kalucka, M., & Lampart-Szczapa, E. (2008). The content and antioxidant activity of phenolic compounds in cold pressed plant oils. *Journal of Food Lipids*, *15*, 137–149.

Staughton, J. (2019). Benefits & side effects of corn oil. https://www.organicfacts.net/health-benefits/oils/corn-oil.html.

Steenson, D. F., Lee, J. H., & Min, D. B. (2002). Solid phase microextraction of volatile soybean oil and corn oil compounds. *Journal of Food Science*, *67*, 71–76. https://doi.org/10.1111/j.1365-2621.2002. tb11361.x.

Strecker, L. R., Bieber, M. A., Maza, A., Grossberger, T., & Doskoczynski, W. J. (1996). Corn oil. In Y. H. Hui (Ed.), *Bailey's industrial oil and fat products* (pp. 125–158). New York: Wiley ISBN-13: 978-0471594307/ISBN-10: 047159430X.

Tuan, H. Q. (2011). *Characterization of natural edible oils regarding their quality and safety related constituents*. Master ThesisDivision of Food Chemistry, Department of Food Sciences and Technology, University of Natural Resources and Life Sciences, Vienna, Austria.

Tuberoso, C. I. G., Kowalczyk, A., Sarritzu, E., & Cabras, P. (2007). Determination of antioxidant compounds and antioxidant activity in commercial oilseeds for food use. *Food Chemistry*, *103*, 1494–1501.

Yılmaz, E., Arsunar, E. S., Aydeniz, B., & Güneşer, O. (2015). Cold pressed capia pepperseed (*Capsicum annuum* L.) oils: Composition, aroma and sensory properties. *European Journal of Lipid Science and Technology*, *117*, 1016–1026.

Yılmaz, E., Aydeniz, B., Güneşer, O., & Arsunar, E. S. (2015). Sensory and physico-chemical properties of cold press produced tomato (*Lycopersicon esculentum* L.) seed oils. *Journal of the American Oil Chemists' Society*, *92*, 833–842. https://doi.org/10.1007/s11746-015-2648-x.

Chapter 17

Cold pressed sunflower (*Helianthus annuus* L.) oil

Ranko Romanić
Faculty of Technology Novi Sad, University of Novi Sad, Novi Sad, Serbia

List of abbreviations

CPSO	cold pressed sunflower oil
FAO	Food and Agricultural Organization
HDL	high-density lipoprotein
HOSO	high oleic sunflower oil
HP	high palmitic
HPHOSO	high palmitic high oleic sunflower oil
HPSO	high palmitic sunflower oil
HS	high stearic
HSHOSO	high stearic high oleic sunflower oil
HSSO	high stearic sunflower oil
LDL	low-density lipoprotein
MOSO	mid oleic sunflower oil
NCEP	National Cholesterol Education Program
PET	polyethylene terephthalate
PUFA	polyunsaturated fatty acid
USDA	United States Department of Agriculture

1 Introduction

Sunflower oil is available as edible refined, and increasingly as edible unrefined, cold pressed, or virgin oil. Cold pressed oil contains minor components such as tocopherols, phytosterols, and pigments that are desirable in the oil, as well as those that are not desirable, including phospholipids and waxes. In order to remove unwanted components from the oil and to obtain products that are more acceptable to consumers, the oil is subjected to a refining process. The refining process involves several phases, and in addition to the undesirable ones, these processes remove the mentioned desirable components.

Sunflower breeding constantly creates different hybrids that favor certain characteristics of the plant, seed, and oil. In oilseeds and sunflowers, breeding is mainly aimed at increasing the oil content of the seeds as a basic technological requirement. However, there are a number of other requirements. On the basis of the available genetic variability of sunflowers in the world, in addition to hybrids with high yields of seeds and oils, resistant to dominant diseases, insects, and droughts, a large number of hybrids for specific (special) purposes have been developed, namely: productive hybrids with different oil quality, hybrids resistant to different herbicides, productive hybrids with increased protein content, dedicated hybrids for poultry and bird nutrition, and decorative sunflowers.

Edible unrefined oils include cold pressed and virgin oils. Due to the absence of refining, edible unrefined oils have specific sensory properties including the smell and taste of the original raw material (Dimić, 2005). Compared to the oil obtained from the traditional "standard" sunflower seeds of the "linoleic type," the "oleic type" oil contains about four to six times less linoleic acid, and three to four times more the oleic acid content (Cvejić, Jocić, Radeka, Balalić, & Miklič, 2014; Škorić, Demurin, Jocić, Lečić, & Verešbaranji, 1994).

Production of cold pressed oils takes place in two stages. First, the raw material (seed) is prepared to extract the oil. In the second phase, the edible unrefined cold pressed oil is mechanically extracted from the seed and remains of the cake as a

Cold Pressed Oils. https://doi.org/10.1016/B978-0-12-818188-1.00017-7
Copyright © 2020 Elsevier Inc. All rights reserved.

by-product (Dimić, 2005). In addition to the extracted cold pressed oil, a significant amount of oil remains in the cake, which is still mainly used in the production of animal feed. In addition to affecting the quality and oxidative stability of the oil (Romanić & Grbić, 2016), different proportions of hull and impurities in the seed to be pressed also influence the residual oil content of the cake (Romanić, 2015, 2016). Romanić (2015) found that the dependence of the oil content remaining in the cake (oil cake) on the contents of the hull (hull content), and impurities (impurities content) in the seed being pressed can be represented by the following equation: oil cake = − 1.0733·hull content −0.5813·impurities content +37.0500 (R^2 = 0.8659), and the dependence of the efficiency of pressing (yield) on the hull and the impurity content in the starting material is shown graphically in Fig. 1.

If, after (first) pressing, the seed of the cake contains significant amounts of "residual" oil, it can be pressed again: the second pressing (Fils, 2000). The utilization of pressing depends primarily on the type of raw material, that is, on the content of oil in the starting material for pressing, then on the preparation of the starting material, particle size composition, hull content, and residual impurity content. The design and strength of the press, as well as the pressing conditions (pressure, temperature, length of process, oil content in the cake, moisture content of the starting material, etc.), also affect the effectiveness of the pressing (Matthäus & Brühl, 2004; Zheng, Wiesenborn, Tostenson, & Kangas, 2003).

The Food and Agricultural Organization (FAO) has announced that, according to the United States Department of Agriculture (USDA), sunflower seeds have been ranked fourth or third in the last 4 years, with an average share of 7.92%, of the total quantities of oilseeds produced (FAO, 2018). Sunflower oil production in the world in 2016/2017 was 18,180 mmt, in 2017/2018 it was 18,207 mmt, while in 2018/2019 it was estimated to be 19,259 mmt. The largest producers of sunflower oil in the previous 2 years were Ukraine, with a share of an average of 33.64%, and Russia, with an average of 22.81% of the total world production. It was projected that in 2018/2019, Ukraine would produce 32.37%, and Russia 23.16% of the total amount of sunflower oil in the world, as can be seen in Fig. 2 (FAO, 2018).

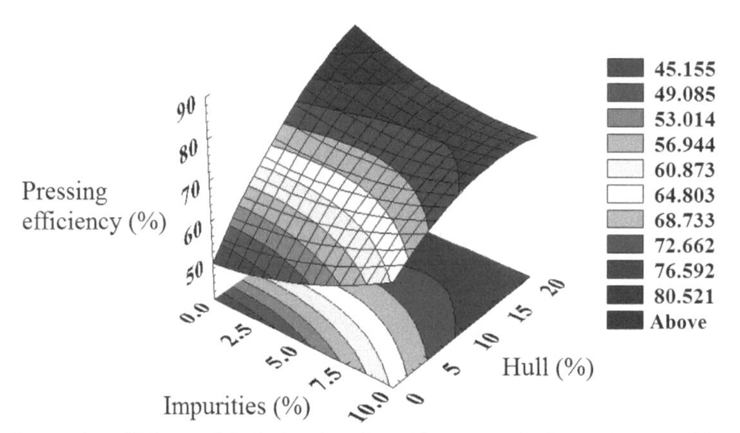

FIG. 1 Graphic depending on the pressing efficiency of the hull and the impurities content in the starting material: sunflower seeds.

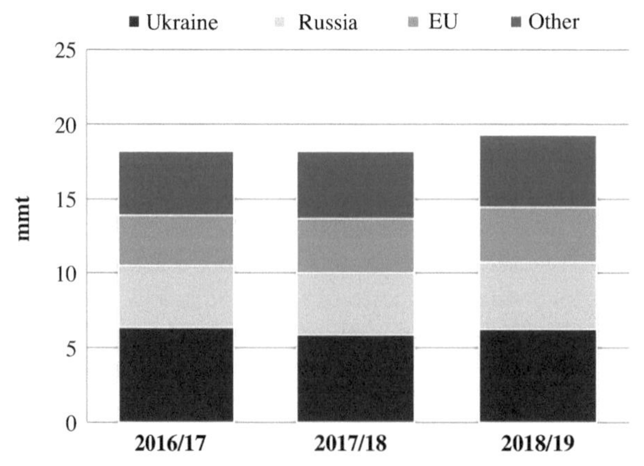

FIG. 2 The worldwide sunflower oil production.

Sunflower oil consumption is increasing at a fast pace in Europe. A growing segment in the European market is high oleic sunflower oil, originating from hybrid sunflower seeds. Demand for high oleic sunflower oil is increasing in Europe, but supplies remain limited. This points toward future opportunities for exporters from a developing country (CBI, 2016). The price of sunflower seeds over the past 11 years has averaged $492.45 per metric ton (US $/Mt). The maximum value of the price of sunflower oil was recorded in 2007/2008, when it was as high as 1824.5 US $/Mt. The lowest value of this oil in the last 11 years was recorded immediately following 2008/2009, when the price was 972.5 US $/Mt. When it comes to sunflower oil, prices differ significantly in the US and European markets; in the US market the price of this oil has risen on average by 40.49% over the last 11 years (FAO, 2018).

Although the production of cold pressed oil of sunflower seeds appears simple, this is in many ways not so. Manufacturers of these oils often do not notice all potential problems. Securing raw materials of high quality is one of the biggest obstacles. This chapter is intended to give at least a small positive contribution in this respect.

2 Sunflower (*Helianthus annuus* L.) seed and processing of cold pressed sunflower oil

The sunflower (*Helianthus annuus* L.) is an annual plant. It originates from the temperate climates of North America (Schilling, 2006), from where it was brought to the soil of Europe by the Spanish conquerors in the early 16th century. This plant had various uses, namely used as an ornamental plant, balm, for the production of makeup, body and hair care creams, etc. In Russia, it became increasingly important as a raw material for food production during Orthodox fasting that allowed the consumption of rare foods, including sunflowers (Dimić, 2005; Grompone, 2005). The "squeezing" of sunflower seed oil was suggested in the 18th century in Russia, after which mass cultivation of sunflowers began. Sunflower seed was the main oilseed in Russia in the early 20th century, which is why the scientist Pustavoit (Пуставойт) began researching the selection of sunflowers in 1912 in terms of increasing the quantity of seeds and oil, that is, their yield and quality (Grompone, 2005). Since then, scientists around the world have been concerned with increasing yields and improving sunflower quality (Miklič, Hladni, Jocić, Marinković, & Atlagić, 2008). The result of many years of work is a large number of different varieties and hybrids of sunflower.

The creation of hybrids in the direction of increasing the oil content led to the appearance of the so-called sunflower "oily type," mainly used for oil production, while the creation of hybrids aimed at increasing the protein content led to the emergence of a "protein (confectionary) type" sunflower, whose primary purpose is the production of an edible kernel. Owing to further intensive work on selection, two types of oil were distinguished in oil sunflower (Codex, 1999; Cvejić et al., 2014; Dimić, 2005; Romanić, 2015):

- standard linoleic type, in which the composition of the oil contains linoleic acid in amounts of about 55% to 75%; and
- oleic type:
  - high oleic, in which the oleic acid content is high in the oil composition, from about 80% to more than 90%; and.
  - mid oleic, dominated by oleic acid with a content of about 60% to 65%.

2.1 Seed characteristics of different sunflower hybrids

Hybrids with higher oil content in the seeds are characterized mainly by black seeds with a thin hull (which usually represents 20%–25% of the total mass of the seeds), and the oil content is above 40%. Hybrid seed with a lower oil content is somewhat different: it has larger seeds with a thicker, mostly black/white hull (representing 30%–45% of the total seed mass) (Jovanović, 2001), which is poorly attached to the kernel and easily removed (Gonzalez-Perez & Vereijken, 2007). This seed contains less oil (about 30%). Photographs of examples of seeds of hybrids with higher oil content (oil hybrids), and hybrids with lower oil content (confectionary-protein hybrids) are shown in Fig. 3A and B.

In "oily hybrids," the mass of 1000 seeds ranges from 30 to 80 g. The seeds of high oleic sunflower are slightly smaller than the linoleic type, having a smaller proportion of thinner hull, which is manifested by a smaller specific mass and a smaller mass of 1000 seeds (Brkić, 2004; Todorović & Komljenović, 2010). With oleic sunflowers, there are some differences depending on the place of cultivation. The average values of bulk density are 0.43 kg/dm^3 and true density are 0.69 kg/dm^3, while the weight of 1000 seeds is 47.82 g (Brkić, 2004).

Romanić et al. (2018a, 2018b) examined seed dimensions, hull, and kernel content of seeds, as well as the mass of 1000 seeds, of the latest sunflower hybrids of the "confectionary-protein" type. Seed length ranged from 14.79 to 20.68 mm and width from 7.01 to 7.99 mm, while seed thickness ranged from 3.91 to 4.24 mm. The hull content ranged from 41.91% to 46.74%, and the kernel content from 53.26% to 58.09%. The weight of 1000 seeds expressed on dry matter ranged from 114.22 to 125.56 g.

FIG. 3 Seeds of sunflower hybrids: (A) "oily type" and (B) "confectionary-protein" type.

TABLE 1 Fatty acid compositions of different types of sunflower oil (Codex, 1999; Salas et al., 2015; Salas, Bootello, Martínez-Force, & Garcés, 2011; Serrano-Vega, Martínez-Force, & Garcés, 2005).

	Fatty acid[a]								
Type of sunflower oil	C16:0	C16:1	C16:2	C18:0	C18:1 (n-9)	C18:1 (n-7)	C18:2	C20:0	C22:0
Linoleic	6.3	0.1	–	4.6	26.7	–	61.1	0.3	0.9
MOSO	4.9	–	–	3.8	57.9	–	32.3	0.3	0.8
HOSO	3.8	–	–	4.1	82.1	–	8.7	0.4	0.9
HSSO	7.4	–	–	27.1	16.1	–	46.3	1.5	1.6
HSHOSO	5.4	–	–	24.9	57.8	–	8.2	1.8	1.9
HPSO	34.7	5.1	0.6	2.6	6.9	3.4	45.1	0.5	1.1
HPHOSO	31.7	7.4	–	2.0	50.5	3.9	2.7	0.5	1.3

[a]*% m/m determined by gas chromatography.*

2.2 Sunflower oil with different fatty acid compositions

In order to increase oxidative stability and the possible applications of sunflower oil, breeders started working on changes in the composition of fatty acids in newly created hybrids as early as the 1970s (Salas, Bootello, & Garcés, 2015). Today, there are different types of hybrids in terms of fatty acid composition:

- standard sunflower oil (linoleic type);
- mid oleic sunflower oil (MOSO);
- high oleic sunflower oil (HOSO);
- high stearic sunflower oil (HSSO);
- high stearic high oleic sunflower oil (HSHOSO);
- high palmitic sunflower oil (HPSO); and
- high palmitic high oleic sunflower oil (HPHOSO).

The fatty acid compositions of the above types of sunflower oil are shown in Table 1.

Standard sunflower oil is rich in linoleic acid (C18:2), with a content of 48% to 74%. It contains small amounts of saturated fatty acids, mainly palmitic (C16:0) and stearic (C18:0). Unlike other oils such as soy and rapeseed, sunflower oil contains negligible amounts of linoleic acid (C18:3). The presence of long-chain fatty acids, such as arachidonic (C20:0) and behenic (C22:0), is also significant.

High stearic (HS) and high palmitic (HP) sunflower hybrids are created by breeding standard sunflower seeds. The stearic acid content of HSSO is from 20% to 30% (Table 1), which is up to five times higher than standard sunflower oil (linoleic type). Increasing the content of stearic acid decreases the oleic acid content, which also affects the physical characteristics of oil such as increasing the melting point. Another significant change in the fatty acid composition of high stearin sunflower oil is the increased content of long-chain fatty acids (arachidonic and behenic) of 3%, as opposed to standard sunflower oil, where their total content is below 1%.

The palmitic acid content of high palmitic sunflower oil is up to 30% due to the reduction of oleic and stearic fatty acid content. The content of long-chain fatty acids is slightly higher than their content in standard sunflower oil. However, the most significant aspect of HPSO is the presence of n-7 fatty acids, palmitoleic (C16:1), palmitolinoleic (C16:2), and *cis*-vaccenic (C18:1, n-7), which are not present in other sunflower oil types.

HSHOSO and HPHOSO are different from standard high-saturated sunflower oils in that they contain more oleic and less linoleic acid, which is even lower than in HOSO. With HSHOSO, oleic acid has replaced part of the stearic acid, so its content in these oils ranges from 16% to 25%. The fatty acid composition of sunflower seed oils is also greatly influenced by location and weather conditions during vegetation (Lajara, Diaz, & Quidiello, 1990; Marinković, Dozet, & Vasić, 2003; Mijić et al., 2011; Pospišil, Pospišil, & Antunović, 2006; Liović et al., 2010). If temperatures are high during the period of seed development and oil synthesis, the oleic content increases and the linoleic acid content decreases (Onemli, 2012). Increasing temperature differences between day and night result in an increase in linoleic acid accumulation, as confirmed by Krizmanić et al. (2013). In addition to the average daily temperatures and humidity, the oil content is influenced by the soil type as well as the application of agro-technical measures.

2.3 Mechanical extraction and processing of cold pressed oil

Edible unrefined oils include cold pressed (also called "cold pressed oils") and virgin oils. Unrefined oils are a category covered by technical regulations in the field of edible oils (Codex, 1999): “Cold pressed unrefined edible vegetable oil is produced without heating, pressing, precleaning (removing impurities), dehulling and milling mechanically (with certain raw materials). Cold pressed unrefined edible oil can only be purified by washing with water, precipitating, filtrating and centrifuging.”

The technological scheme for the production of virgin and cold pressed oil is shown in Fig. 4. The main difference between unrefined and refined oils is that the unrefined oils are used in the “raw” (original, unprocessed) state. The quality of the oil obtained depends directly on the quality of the raw material. The quality of the raw material is directly related to the conditions of cultivation, the method of storage—that is, the method of storage of the raw material—and the oil extraction processes themselves. There is no phase in the production of edible unrefined oils in the production of edible crude oils to allow the removal of undesirable components from the oil. This means that for the production of cold pressed and virgin oils, the raw materials must be of extremely high quality (Dimić, 2005; Grompone, 2005).

The preparation of raw material (sunflower seeds) for the extraction of oils is specific in the production of edible unrefined oils. The seeds are cleaned (Fig. 5) then partially or completely dehulled (Fig. 6), and ground and/or milled (Fig. 7).

The impurities from the seeds are removed either by sieving on different sieves or by aspiration by means of an air current. Sieves are used to remove coarse impurities, while fine impurities are removed by aspiration, Fig. 5 (Dimić, 2005). Seed dehulling uses different methods, but the hull is usually removed mechanically by using two basic operations: breaking the hull and releasing the kernel, and separating the hull from the kernel. Separation of the hull from the kernel is carried out by using a sieve (averaging), a wind current (aspiration) or by using an electric field as shown in Fig. 6 (Dimić, 2005).

Whether the raw material will be ground, and to what extent, in the production of cold pressed oil depends, first of all, on the type and characteristics of the press. If the material is ground, this is called coarse grinding. Coarse milling is practiced on rollers having different profiles, or on plate mills as shown in Fig. 7 (Dimić, 2005).

The oil is immediately extracted from the prepared material by means of a screw press (Fig. 8). In both cases, the precipitate is then removed, further processed, and the separated oil packaged in glass or, less often, in polyethylene terephthalate (PET) packaging. The resulting oil should have specific raw material properties and quality that depends directly on the preparation method and the pressing conditions (Grompone, 2005; Romanić, Grbić, & Savić, 2017). The use of organic solvents is not allowed in the production of these oils. The type of raw material, the method of extraction and the absence of solvent extraction and refining, have the consequence that the essential bioactive components are preserved in cold-pressed vegetable oils. The main characteristics of cold pressed oils, as well as their differences with respect

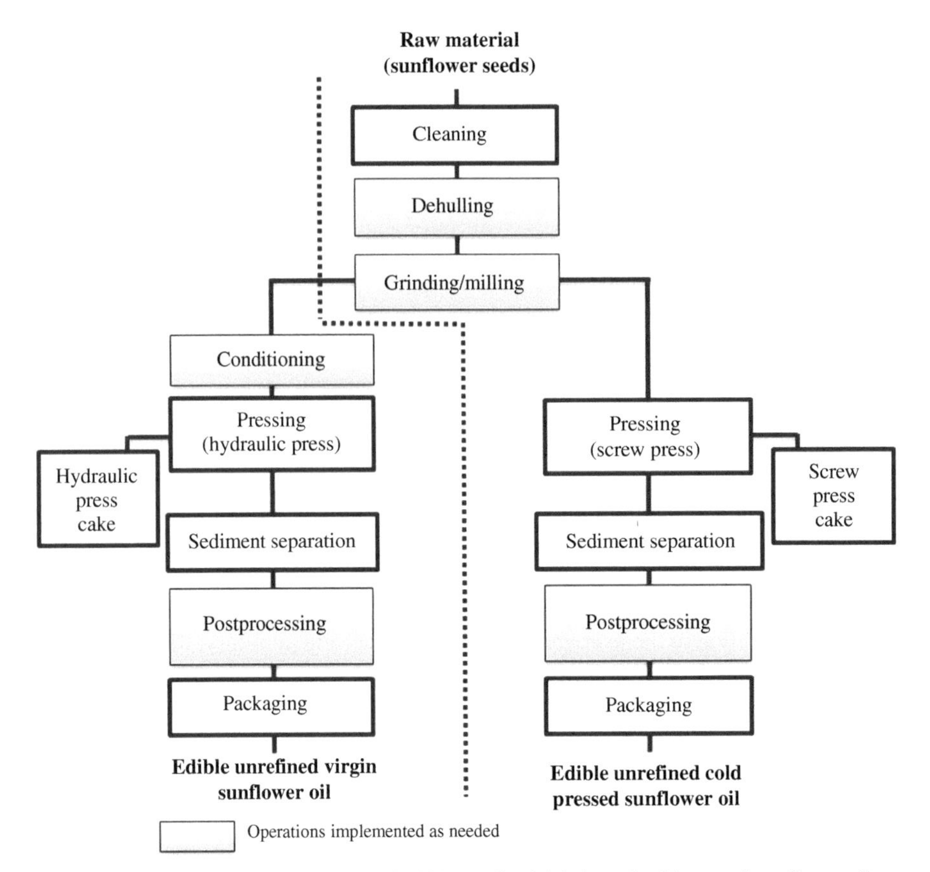

FIG. 4 Block diagram of technological process for production of edible unrefined (virgin and cold pressed) sunflower oil.

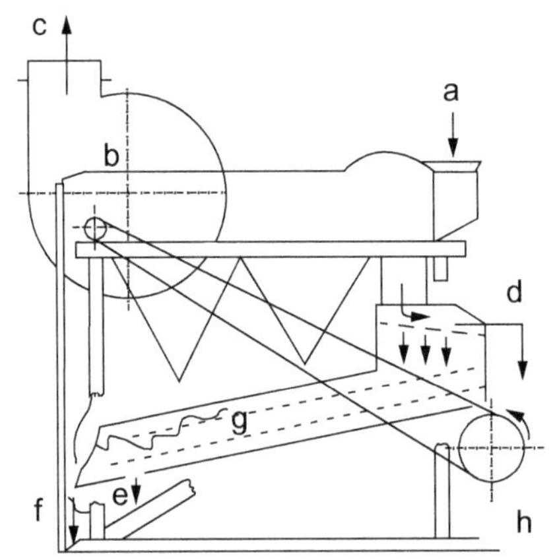

FIG. 5 Seed cleaner (aspirator): seed inlet (a), fan (b), dust outlet (c), coarse impurities (d), fine impurities (e), purified seed (f), sieve and grate (g), and eccentric drive shaft (h).

to refined oils, are as follows: specific sensory properties (appearance, color, odor, and taste), chemical composition, nutritive value, and shelf life (Ayerdi-Gotor & Rhazi, 2016; Dimić, 2005; Raß, Schein, & Matthäus, 2008).

As already pointed out, cold pressed sunflower oil is produced solely by "mechanical extraction," that is, pressing with screw presses. Mechanical pressing and solvent extraction are the most commonly used methods for commercial oil extraction (Beerens, 2007; Ionescu et al., 2014). Mechanical oil extraction (also known as pressing) is based on mechanical compression of oilseed materials. Through pressing, oil is separated from the oleaginous material under the action of compressive external forces that arise in special machines called presses. The first screw oil press was developed in 1900 by V. D. Anderson in the United States. This press allowed continuous operation through hydraulic pressure, which resulted in greater capacity with smaller equipment and less labor (Beerens, 2007).

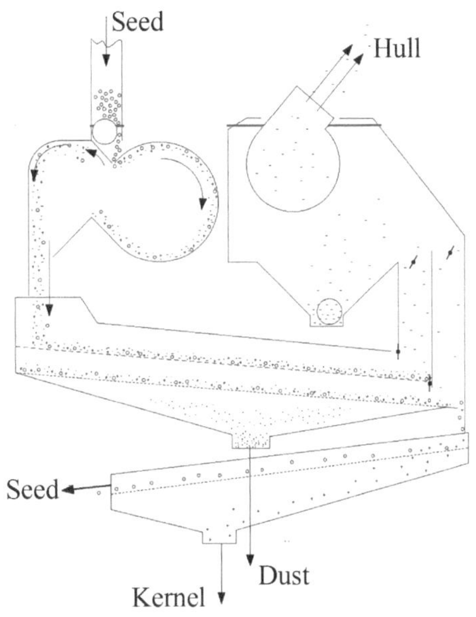

FIG. 6 Scheme of separation of individual seed fractions in the dehuller during sunflower dehulling.

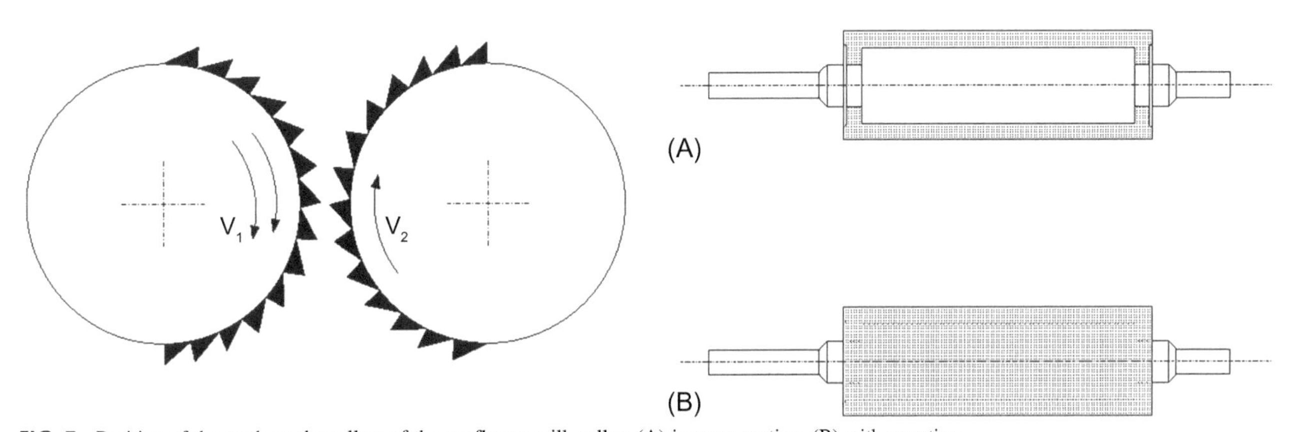

FIG. 7 Position of the teeth on the rollers of the sunflower mill, roller: (A) in cross section, (B) with serration.

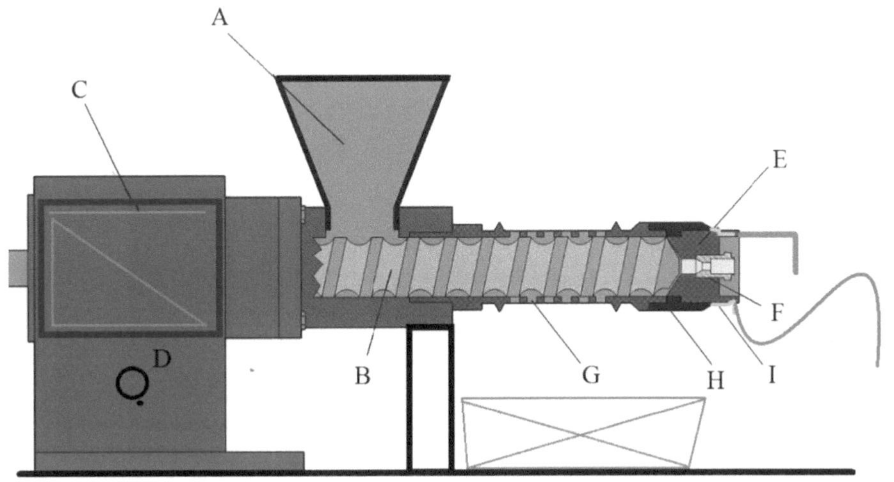

FIG. 8 Basic components of oilseed screw presses.

This method ensures the extraction of a noncontaminated, protein-rich, low-fat cake at relatively low cost. The disadvantage of this method is that the mechanical presses do not have high extraction efficiencies; about 8%–14% of the available oil remains in the press cake (Bamgboye & Adejumo, 2007; Ionescu et al., 2014). Pressing is an operation during which the oil is extracted (pressed, "squeezed out") from the prepared seed, only mechanically, under the influence of pressure. In spite of its slightly lower yield, screw pressing is the most popular oil extraction method as the process is simple, continuous, flexible, and safe (Beerens, 2007; Ionescu et al., 2014). Pressing is done using hydraulic or, most commonly, screw presses (Dimić, 2005; Faugno, del Piano, Crimaldi, Ricciardiello, & Sannino, 2016; Sivakumaran & Goodrum, 1988). Friction during pressing between the raw material and the material of the inner parts of the press is inevitable, which inevitably increases the temperature of the material from which the oil is pressed (Bockisch, 1998). In the production of cold pressed oils, the oil temperature immediately at the outlet of the press should not exceed 50 °C (120 °F) (De Panfilis, Gallina Toschi, & Lercker, 1998). The temperature of the output oil can be lowered by the specific design of the press, or the pressing can be carried out at lower pressures, which leaves more oil behind in the cake.

The mechanical screw press consists of a vertical feeder and a horizontal screw with increasing body diameter to exert pressure on the oilseeds as it advances along the length of the press. The barrel surrounding the screw has slots along its length, allowing the increasing internal pressure first to expel air then to drain the oil through the barrel. Oil is collected in a trough under the screw and the de-oiled cake is discharged at the end of the screw. The main advantage of the screw press is that large quantities of oilseeds can be processed with minimal labor, and it allows continuous oil extraction (Dunford, 2007; Ionescu et al., 2014). Other improvements on oil screw presses were made for the design of the presses and for the material of presses construction (Bamgboye & Adejumo, 2007; Dimić, 2005; Romanić, 2015).

Most commercially available oilseed presses have the same basic components (Fig. 8): A—feeding hopper, B—screw, C—electric motor, D—speed regulator, E—press head, F—clearance, orifice or nozzle for cake outlet, G—oil outlet holes, H—head temperature regulator, and I—temperature sensors.

Fig. 9A and B show the different screws and set nozzles of the screw presses, and in Fig. 10A–C, the sunflower cake at the exit of the screw presses. In cold pressing, the seed goes into a central hopper of the press, and is moved through

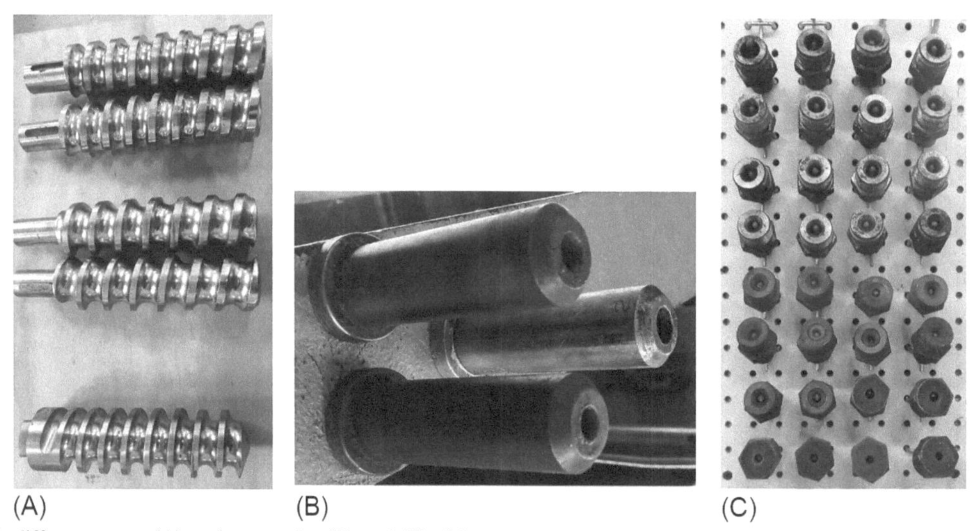

FIG. 9 Multiple different screws (A), and set nozzles (B), and (C) of the screw presses.

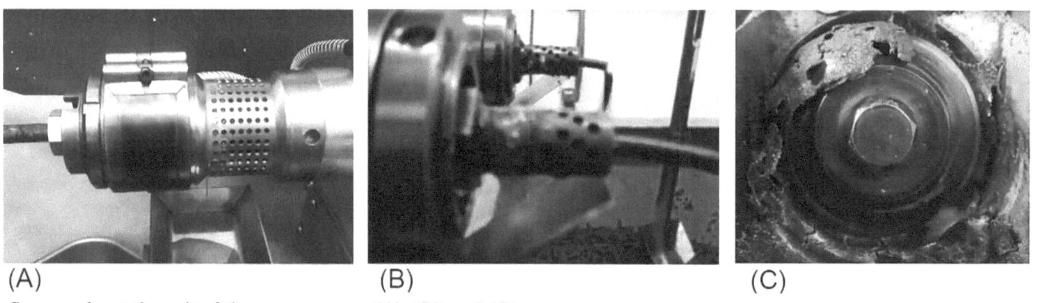

FIG. 10 Sunflower cake at the exit of the screw presses (A), (B), and (C).

TABLE 2 Specifications for six screw presses (Callahan, Harwood, Darby, Schaufler, & Elias, 2014).

Press, model	Estimated capacity (kg/h)	Approximate purchase cost	Power source	Rated load	Adjustable speed	Heated barrel/ head	Multiple screws available	Adjustable nozzle diameter
KernKraft 40	20	14,000 EUR	220 V	3.0 kW	X	X	X	X
Oil Prince (KernKraft 20F)	35	5700 EUR	220 V	2.2 kW	X	X	X	X
Komet CA59G3	5	7700 EUR	115 V AC	1.1 kW	X	X		X
Keller KEK P0020	20	7700 EUR	230 V	2.2 kW	X			
AgOil M70	15	8200 EUR	240 V	1.5 kW	X	X		X
Täby 70	30	6700 EUR	220–240 V	2.2 kW	X	X		X

one or more screws, crushed against a nozzle and screens to extract oil without heating the seed above a temperature of 50°C (120°F). Oil and cake are separated. Presses vary in the number and breadth of adjustment that each needs and is capable of. Some presses have multiple nozzles, with differing diameter holes for different raw materials. Some have heads that are heated to a given temperature before operation.

Unfortunately, manufacturers' instructions and customer service can vary greatly. Many presses are made overseas, and it can be difficult to get guidance in installing, wiring, and operating a press, due to either language barriers or geographical distance. In addition, the operational guidelines for different oilseeds can vary.

Table 2 lists the specifications of six different small screw presses.

When pressing seeds, the press, together with the oil, leaves a certain proportion of the solid particles that make up the so-called mechanical impurities. Mechanical impurities consist of dust, small or large parts of seeds (kernel, and hull), and many other undesirable impurities, which need to be separated from the oil. Mechanical impurities can be separated from the oil by the following processes: precipitation or filtration by means of vibrating sieves, filter presses, or separators. During sedimentation, mechanical impurities are deposited at the bottom of the tank. The separation of mechanical impurities by filtration is much more efficient and consists of passing crude oil through the filter on which they are retained. The following can be used as filters: filter cloths made of cotton, linen, and other textile fibers, filter paper, and filters in the form of sieves made of metal. Subsequent treatment, which can only be applied to edible unrefined oils, involves only washing the oil with water. This removes mucilages that can adversely affect the appearance and sensory properties of the oil. Separation can be done by decantation or centrifugal separators. Oil washing with water is done only as needed. Unless necessary, any treatment of oil other than separation of mechanical impurities should be avoided (Dimić, 2005).

2.4 Neutral lipids and phosphoacylglycerols (phospholipids) profile of cold pressed sunflower oil

Triacylglycerol, formerly called triglyceride, is a glycerol molecule esterified with three molecules of fatty acids. Since glycerol contains two primary hydroxyl groups (OH groups), the central carbon atom requires a chiral position if both primary hydroxyl groups are esterified with various fatty acids. Triacylglycerols in the body create energy depots from which, depending on the needs of the body, fatty acids are released, and from them occurs the process of oxidation of energy essential for the life of all cells and the organism in general (Grompone, 2011).

Phospholipids, as a very significant indicator of quality, appear only in crude-unrefined oils. Inedible refined oils are of no importance since they are completely removed at the prerefining stage—that is, during refining. The phospholipid complex of vegetable oils has a very complex composition. The basis of phosphatide is phosphatidic acid, which consists of glycerol in the position *sn*-1 and *sn*-2, which are fatty acids esterified with -OH groups, and of position 3, which is an esterified phosphoric acid molecule. To the phosphorus reactive group, acids also bind some basic amino or other ester bonds group. Thus, phospholipids or phosphatides are phosphatidic acid esters with amino alcohols (phosphatidylaminoalcohols) or with polyalcohols (phosphatidylpolyalcohols). In the processing of oilseeds by pressing or extracting, under the influence of heat, of moisture, or solvent, phospholipids turn into oil. Their content in crude oil depends on the amount of phospholipid in the seed, on the degree of maturity and storage conditions of the seed, as well as the mode and technological regime of oil extraction. The phospholipid content of crude regular sunflower oil varies between 0.5% and 1.2%. Solvent extracted oils generally have a higher phospholipid content than pressed oils. The phospholipid content of sunflower oil, expressed as phosphorus, is usually 200–400 ppm for crude oil (Gupta, 2002) and below 1 ppm for refined oil. Phosphatidylcholines (PC), phosphatidylethanolamines (PE), phosphatidylinositols (PI), and phosphatidic acids (PA) are major phospholipids in regular sunflower oil, most of which are hydratable and can be removed from the crude oil by degumming with water (Grompone, 2011).

3 Minor components of cold pressed sunflower (*Helianthus annuus* L.) oil

Cold pressed (or crude) sunflower oil consists mainly of triacylglycerol (>98%) together with a small fraction of minor constituents such as diacylglycerols, phospholipids, glycolipids, free fatty acids, aliphatic alcohols, waxes, hydrocarbons, phenolic compounds, carotenoids, chlorophyllides, chlorophyllides, chlorophylls, chlorophylls, and phytosterols (Grompone, 2005; Sánchez-Muniz & Cuesta, 2003). It may also contain pesticides (Gunstone, 2005). The major part of the minor components of sunflower oil is so-called unsaponifiable matter, which includes those minor components that cannot be saponified by bases but are soluble in organic solvents in which fats and oils are dissolved (AOCS, 2011). The unsaponifiable substances of cold pressed sunflower oil are tocopherols, phytosterols, alcohols, hydrocarbons, and phenols, which usually make up 0.5%–1.5% *w*/w of the oil. According to the Codex Alimentarius, the maximum amount of unsaponifiable matter in crude sunflower oil should be 15 g/kg (Codex, 1999). Table 3 shows the composition of the minor components of standard type crude sunflower oil.

Some minor components, such as tocopherols and phytosterols, have a significant positive effect on the nutritional and/or technological properties of the oil (Fernández-Martínez, Pérez-Vich, & Velasco, 2009). Others, however, have a negative impact on the quality and stability of the oil. Phospholipids, free fatty acids, and waxes are major components of this undesirable group of compounds (Kamal-Eldin, 2005). Accordingly, during extraction of oil, the extraction of desirable minor components should be favored, while the process of oil refining should preserve these desirable components while reducing the content of unwanted minor constituents (Anderson, 2005). Some of these minor constituents (e.g., phospholipids) are coproducts with important industrial applications (Salas, Martínez-Force, & Garcés, 2006).

TABLE 3 Content of minor components in crude sunflower oil of the standard type.

Minor component	Content (mg/kg of oil)	Source
Diacylglycerols	10,000–18,900	Sánchez-Muniz and Cuesta (2003)
Phospholipids	6000–12,000	Carelli, Ceci, and Crapiste (2002)
Free fatty acids	3700–4500	Sánchez-Muniz and Cuesta (2003)
Phytosterols	2400–4600	Grompone (2005)
Hydrocarbons	1000	Baštić, Baštić, Jovanović, and Spiteller (1978)
Tocopherols	403–935	Grompone (2005)
Waxes	200–3500	Carelli, Frizzera, et al. (2002)
Aliphatic alcohols	59–63	Sánchez-Muniz and Cuesta (2003)
Traces of metal	45–95	Gupta (2002)
Carotenoids	6.5–15.3	Premović, Dimić, Takači, and Romanić (2010)
Phenols	4.8–16.4	Perretti, Finotti, Adamuccio, Della Sera, and Montanari (2004)
Chlorophylls	0–1	Premović et al. (2010)

3.1 Tocopherols

Tocopherols are a group of liposoluble compounds consisting of a polar part derived from tyrosine, a chromanol ring, and a hydrophobic side, a derivative of the phytyl chain (Fig. 11). In nature, they occur in four isomeric forms: α-, β-, γ-, and δ-tocopherol, which differ in the number and position of methyl substituents on the chromanol ring (Mène-Saffrané & DellaPenna, 2010). Tocopherols are the main compounds with antioxidant activity in oilseeds, as well as in oils extracted from oilseeds, and have a great influence on their nutritional and technological characteristics. Tocopherols, together with other related compounds, such as tocotrienols (collectively known as tocochromanols), have E vitamin activity. They prevent oxidative, metabolic, and inflammatory damage at the cellular level and the subsequent development of various chronic diseases (Eitenmiller & Lee, 2004; Galli & Azzi, 2010). From a technological perspective, tocopherols protect the oil from oxidative changes (Shahidi & Zhong, 2010).

α: $R_1 = R_2 = CH_3$; β: $R_1 = CH_3$; $R_2 = H$; γ: $R_1 = H$; $R_2 = CH_3$; δ: $R_1 = R_2 = H$

Sunflower oil has a relatively high content of tocopherol and tocotrienol, consisting mainly of α-tocopherol, which makes sunflower oil the richest source of vitamin E among commercially available vegetable oils and is one of the foods rich in vitamin E (USDA, 2012). Other important vegetable oils contain mainly γ-tocopherol (soybean, rapeseed) or balanced amounts of α- and γ-tocopherol (peanuts and cottonseed). The only dandelion seed contains significant amounts of α-tocopherol. Olive oil has a similar tocochromanol profile, although the total content is much lower than that in sunflower oil.

Triacylglycerols and tocopherols accumulate in sunflower seed cells as a result of separate biosynthetic pathways. Because tocopherols are soluble in fats, they turn into oil during extraction (mechanical or chemical). Thus, the content of tocopherols in crude sunflower oil depends on the content of tocopherols in the seed, the oil content in the seed, and the oil extraction process (Velasco & Fernández-Martínez, 2012). The oil and tocopherols content of the seeds is a result of genetic and environmental factors as well as their interactions (Marquard, 1990; Velasco & Fernández-Martínez, 2012). Therefore, understanding the variations in the content of total tocopherols in sunflower oil requires a thorough analysis of all the factors that affect it. This explains the huge variations in tocopherols content in crude sunflower oil reported in the literature, with extreme values varying from 389 to 1873 mg/kg (Dolde, Vlahakis, & Hazebroek, 1999; Nolasco, Aguirrezábal, & Crapiste, 2004; Velasco, Fernández-Martínez, García-Ruíz, & Domínguez, 2002).

The total amount of tocopherols present in the oil and their composition is important for the nutritional and technological properties of the oil. α-tocopherol is a form of tocopherol with higher biological value, such as vitamin E. If a vitamin E value of 1.0 is assigned to α-tocopherol, it is generally accepted that the vitamin E values of other tocopherols are 0.5 for β-tocopherol, 0.1 for γ-tocopherol, and 0.01 for δ-tocopherol (Eitenmiller & Lee, 2004). Standard sunflower oil has a high tocopherol content, with a high content of α-tocopherol, making it the commercial vegetable oil with the highest E vitamin value. However, there is increasing evidence to suggest that the role of other forms of tocopherols, such as γ-tocopherol, in the promotion of health and prevention of disease is underestimated.

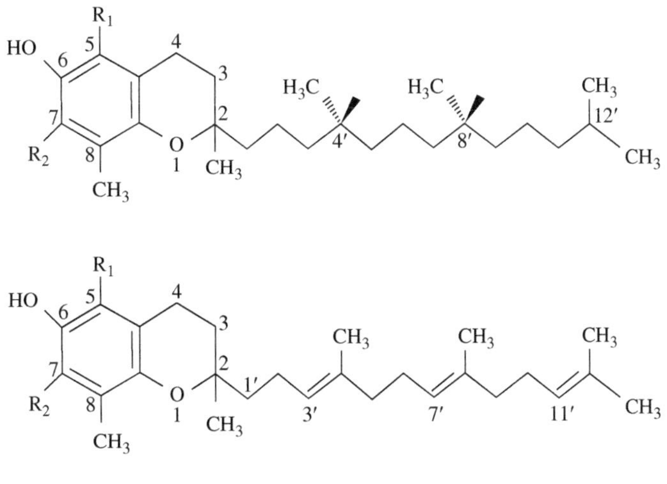

α: $R_1 = R_2 = CH_3$; β: $R_1 = CH_3$; $R_2 = H$; γ: $R_1 = H$; $R_2 = CH_3$; δ: $R_1 = R_2 = H$

FIG. 11 Chemical structure of tocopherols and tocotrienols (Velasco & Ruiz-Méndez, 2015).

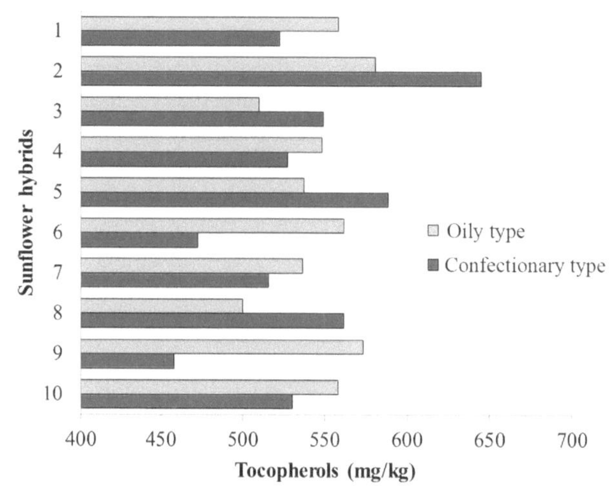

FIG. 12 Variations in the content of total tocopherols in cold pressed seed of different sunflower seed hybrids.

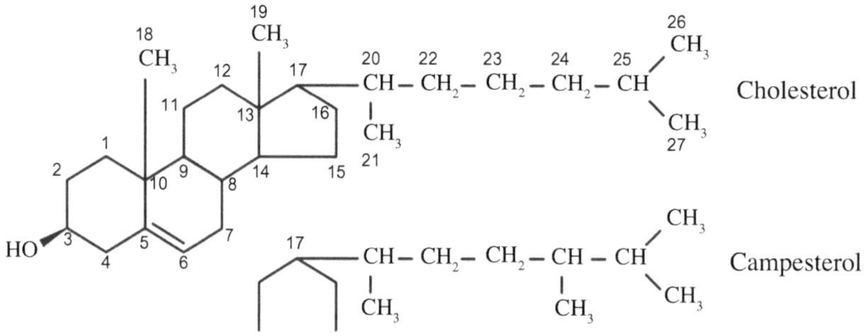

FIG. 13 Chemical structure of cholesterol and campesterol.

From a technological point of view, tocopherols protect the oil from oxidative reactions that occur mainly on double bonds of unsaturated fatty acids (Barrera-Arellano, Ruiz-Méndez, Velasco, Márquez-Ruiz, & Dobarganes, 2002). The effect of the type of tocopherols on the stability of the oils has been studied by adding tocopherols at various concentrations to refined oils (oils without natural tocopherols). The results show that the efficacy of tocopherols depends on their concentration as well as the temperature to which the oil is exposed. At moderately elevated temperatures, α-tocopherol has better antioxidant activity than γ-tocopherol when tocopherol concentrations are low, while the opposite occurs at high tocopherol concentrations (Fuster, Lampi, Hopia, & Kamal-Eldin, 1998; Lampi & Kamal-Eldin, 1998; Seppanen, Song, & Csallany, 2010). At high temperatures, γ- and δ-tocopherols more effectively inhibit polymerization reactions than α-tocopherol (Barrera-Arellano, Ruiz-Méndez, Márquez-Ruiz, & Dobarganes, 1999; Lampi & Kamal-Eldin, 1998; Seppanen et al., 2010; Warner & Moser, 2009).

According to our research (Romanić, 2015), the variations in the content of total tocopherols in cold pressed seed of different "oily" and "confectionary" type sunflower seed hybrids range from a minimum of 472 mg/kg to a maximum of 645 mg/kg, with a mean of 542 mg/kg (see Fig. 12).

3.2 Phytosterols

Phytosterols or plant sterols belong to a broad group of steroids of great physiological importance in both plants and animals. In plants, this group includes important compounds such as triterpenoids (of great importance in defense against pathogens and herbivores) (Flores-Sánchez, Ortega-López, Montes-Horcasitas, & Ramos-Valdivia, 2002), brassinosteroids (plant hormones essential for normal growth and development) (Bishop & Yokota, 2001), and numerous secondary plant metabolites such as glycoalkaloids and saponins (Hartmann, 1998). Plant sterols have a similar structure to mammalian cholesterol. They have a tetracyclic cyclopentanofenanthrene structure and an alkyl side chain of 8–10 carbon atoms (Hartmann, 1998). Fig. 13 shows a close structural similarity between cholesterol and campesterol, one of the most important phytosterols.

Studies of phytosterols content in sunflower oil and in sunflower seeds have shown significant variations. Most results on sunflower oil phytosterols include free and esterified desmethylsterols. Therefore, the total phytosterols content shown here applies to both phytosterol groups unless otherwise indicated. According to Piironen, Lindsay, Miettinen, Toivo, and Lampi (2000), the total phytosterols content of crude sunflower oil ranged from 3740 to 7250 mg/kg. A larger range of variations, from 1250 to 7650 mg/kg, was found by Ayerdi-Gotor et al. (2007) when examining a broad group of sunflower hybrids and inbred lines.

The most abundant phytosterols in standard sunflower oil are β-sitosterol (60% of total desmethylsterols), Δ7-stigmastenol (14%), campesterol (8%), and stigmasterol (8%). Vlahakis and Hazebroek (2000) found maximum concentrations of 20.4% campesterol, 17.9% stigmasterol, and 81.5% β-sitosterol. Maximum values of 23.9% for stigmasterol, 20.5% for Δ5-avenasterol, and 10.6% for Δ7-avenasterol were detected in wild *Helianthus* spp. (Fernández-Cuesta, Velasco, & Fernández-Martínez, 2011).

Phytosterols are important nutritional ingredients of vegetable oils. They are structurally cholesterol-like and phytosterols compete with cholesterol on absorption in the gut, which contributes to the reduction of serum cholesterol levels (Plat & Mensink, 2005). This action of phytosterols stimulated their use, and food is enriched with phytosterols (Zawistowski, 2010). Due to their poor solubility and bioavailability, phytosterols are used as phytostanol esters after hydrogenation and esterification with fatty acids (García-Llatas & Rodríguez-Estrada, 2011). Margarine was the first commercial phytosterols-enriched food, and today, phytosterols are added to mayonnaise, salads, dressings, dairy and non-dairy beverages, chocolates, meat, cheeses, and baked goods (MacKay & Jones, 2011). In addition to lowering blood cholesterol levels, there are studies indicating that phytosterols may have anticancer, antiatherosclerotic, antiinflammatory, and antioxidant properties (Berger, Jones, & Abumweis, 2004). The United States National Cholesterol Education Program (NCEP) recommends phytosterols supplementation of 2 g per day to reduce low-density lipoprotein (LDL cholesterol) in the blood and the risk of coronary heart disease (NCEP, 2002). However, the Federal Institute for Risk Assessment in Germany has found indications that stanols and sterols may be a problem for persons without hypercholesterolemia (BfR, 2011). Therefore, the European Union requires that products containing plant sterols and stanols have clear warnings that their use is "Not intended for persons who do not need to control their cholesterol level" (European Commission, 2013).

From a technological point of view, phytosterols increase the thermal stability of oils with a high content of polyunsaturated fatty acids (Winkler & Warner, 2008a). This effect has been found to be more pronounced for phytosterols such as Δ5-avenasterol containing the ethylene group in the side chain, as they protect the oil from polymerization and slow down the loss of tocopherols (Gordon & Magos, 1983). However, other studies have shown that the protective effect of phytosterols on oil oxidation was more related to the degree of phytosterol unsaturation than to the presence of an ethylidene group (Winkler & Warner, 2008b). Even though there is sufficient evidence to support the positive impact of phytosterols on the thermal stability of oils, more research needs to be done to understand fully the effect of phytosterol structure on oil stability.

Phytosterols are partially removed during the refining process. The degree of loss of phytosterol depends on the conditions applied at different stages of the process. Most authors agree that the greatest decrease in total sterols occurs during neutralization and deodorization, although a slight decrease in total phytosterol content can occur during the bleaching process (Ferrari, Schutte, Esteves, Brühl, & Mukherjee, 1996; Ortega-García et al., 2006; Verleyen et al., 2002). This loss can vary between 10% and 70% of the total sterols present in crude oil (Ferrari et al., 1996; Naz, Sherazi, & Talpur, 2011).

During the refining of vegetable oils, steradiens are produced during the bleaching process, on the surface of activated bleaching earth or during deodorization, under the influence of elevated temperature. Steradienes are steroidal hydrocarbons produced by the sterol dehydration reaction. The most abundant steradienes in vegetable oil are 3,5-stigmastadiene, obtained by the dehydration of β-sitosterol (Verleyen et al., 2002). In addition to steradiene, other products of sterol transformation are formed during refining. In addition, an increase in the esterified ratio, relative to free sterols, has been observed during refining (Ortega-García et al., 2006; Sciancalepore, 1981; Verleyen et al., 2002). Thanh et al. (2006) examined the effect of elevated temperature and storage time on the phytosterols content of the oil. Oil heating at 50 °C for several weeks and at 100 °C for 1 h did not show any significant variation in phytosterols content. In contrast, heating at 200 °C for 1 h resulted in a 50%–60% reduction in phytosterols relative to their starting content in the oil.

3.3 Other minor components

Squalene is one of the most biologically effective ingredients in vegetable oils. Chemically, squalene is a natural terpenoid hydrocarbon produced by all plants and animals, including humans. In plants, squalene is synthesized as a biochemical intermediate in the initial step of the phytosterol biosynthetic pathway. Similarly, in animals, it is an intermediate in the cholesterol biosynthetic pathway. In humans, squalene is one of the main constituents of lipids on the surface of the skin, making it important in the cosmetics industry (Huang, Lin, & Fang, 2009). It is also widely used in the preparation of vaccines and emulsions for drug delivery in the pharmaceutical industry (Fox, 2009). Shark liver oil has traditionally been a major source of squalene. Of the commercially important vegetable oils, virgin olive oil is the richest source of squalene, in quantities of 200 to 7500 mg/kg (Boskou, 2009). In crude sunflower oil, the squalene content is between 150 and 200 mg/kg (Grompone, 2005), although higher values have been reported, up to 271 mg/kg (Kalogeropoulos & Andrikopoulos, 2004). Squalene has a great influence on the nutritional properties of vegetable oils, and according to some authors, it has anticarcinogenic properties (Sotiroudis & Kyrtopoulos, 2008). It is very stable during oil storage as well as in technological processes. However, the results of its antioxidant activity are inconsistent (Amarowicz, 2009).

Carotenoids and chlorophylls are the two most common pigments present in sunflower oil. Carotenoids are tetraterpenoids, made up of eight isoprene molecules, and contain 40 carbon atoms. They are classified into two types: xanthophylls, containing oxygen, and carotene, which represents pure, hydrocarbons that do not contain oxygen. Xanthophylls are the major carotenoids in sunflower oil, with dihydroxy carotenoids (mainly lutein) ranging from 76% to 81% of total carotenoids (Rade, Mokrovčak, Štrucelj, Škevin, & Neđeral, 2004). The addition of β-carotene to sunflower oil increases oxidative stability at room temperature and in daylight, due to synergistic effects with tocopherols (Yanishlieva, Raneva, & Marinova, 2001). There is a great loss of carotenoids during the oil refining process. Maximum reduction occurs during the bleaching process, in which about 77% of the carotenoids are removed. During deodorization, the carotenoids are further removed to a final reduction of about 85% (Rade et al., 2004). Franke, Frölich, Werner, Böhm, and Schöne (2010) found carotenoids in cold pressed sunflower oil, but not in refined ones. The high proportion of removed carotenoids during refining is attributed to their thermolability (Ouyang, Daun, Chang, & Ho, 1980). Chlorophylls act as promoters of photooxidation in oils. Crude sunflower oil has a low chlorophyll content, which is removed during bleaching and during refining. Fig. 14 shows the content of total carotenoids and the content of total chlorophylls in cold pressed oils obtained by screw pressing sunflower seeds with different presence of hull and impurities (Romanić, 2015).

Crude sunflower oil is rich in phospholipids. Phospholipids form the major components of biological membranes. They consist of fatty acid esterified diacylglycerols at positions 1 and 2 of the glycerol skeleton, a phosphate group at position 3, and a variable hydrophilic group is linked to a phosphate group. Phospholipids have a strong antioxidant effect due to their synergistic effect with tocopherols; they have the ability to bind metals, as well as catalytic activity in the degradation of hydroperoxide (Carelli, Brevedan, & Crapiste, 1997; Smouse, 1995). However, phospholipids show greater sensitivity to oxidation than triacylglycerols because they possess a higher degree of fatty acid unsaturation, required to maintain membrane fluidity (Shahidi & Zhong, 2010). For this reason, they give the oil an unpleasant odor and taste during storage, reducing the shelf life of the oil (Desai, Mehta, Dave, & Mehta, 2002). The major phospholipid groups in sunflower oil are phosphatidylcholines, phosphatidylethanolamines, phosphatidylinositols, and phosphatidic acids (Gupta, 2002). Phospholipids are removed during prerefining, which is a refining phase, in the form of a mixture of polar lipids known as lecithin. Due to its emulsifying properties, lecithin has great commercial value in a wide range of applications for both nutritional and nonnutritional uses (Szuhaj, 2005).

Sunflower seeds contain about 30% of the shell and the shell is characterized by a high wax content, up to 3% (Carelli, Frizzera, Forbito, & Crapiste, 2002). Therefore, the seeds are partially peeled before oil extraction to a shell fraction below 15% (Grompone, 2005). Depending on the final hull content, the wax content in the oil may be between 200 and 3500 mg/kg (Carelli, Frizzera, et al., 2002). Chemically, waxes are esters of fatty acids and long-chain alcohols. Waxes in crude sunflower oil have 36–50 carbon atoms. The content and composition of waxes in sunflower oil depends on the mode of extraction, with more waxes being transferred to the oil during chemical extraction than mechanical pressing (Carelli, Ceci, & Crapiste, 2002). Waxes are removed during refining, in the process of winterization, which is a common

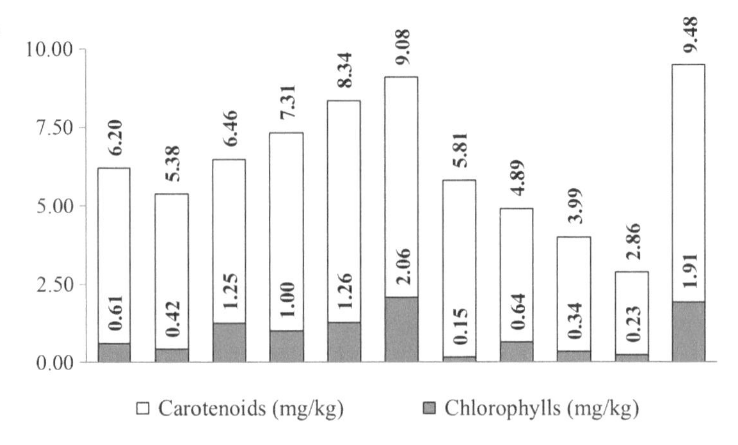

FIG. 14 Contents of total carotenoids and total chlorophylls in cold pressed sunflower oils.

practice in the sunflower oil industry. The wax concentration increases during storage depending on the conditions (time and temperature).

Martini and Añón (2005) found that the composition of sunflower seed waxes from C35–C39 was approximately constant during storage, while a significant change was observed for C40–C48 components. Waxes with more carbon atoms cause a higher degree of turbidity of crude sunflower oil due to its higher melting point, which affects the oil refining conditions.

4 Nutritive importance of cold pressed sunflower oil

Standard sunflower oil, like other vegetable oils, is an integral part of a healthy diet and a source of unsaturated fatty acids and liposoluble vitamins. Sunflower oil is rich in linoleic acid, which is an essential n-6 polyunsaturated fatty acid (PUFA). The biological effects of n-3 fatty acids are reflected in their conversion to n-6 eicosanoids, n-6 prostaglandins, and leukotrienes, which are hormones that act at different levels in human metabolism, especially in response to inflammatory processes (Simopoulos, 2002). In addition, the intake of PUFA has a positive effect on cholesterol buildup, which affects the incidence of cardiovascular disease (Erkkilä, De Mello, Risérus, & Laaksonen, 2008). Sunflower oil affects the reduction of total cholesterol by acting on low-density lipoprotein (LDL). High-density lipoprotein (HDL) remains unchanged. This indicates that sunflower oil intake is one of the main causes of atherosclerosis and is also recommended in a healthy diet rich in saturated fatty acids originating from foods of animal origin such as butter or tropical fats such as palm oil/fat (Katan, Zock, & Mensink, 1995). Oleic type sunflower oil (HOSO, MOSO) is very important from a nutritional and technical point of view. The FAO (Food and Agricultural Organization) recommends the intake of oils and fats with a high oleic acid content for its stability and ability to reduce blood cholesterol levels and thus prevent atherosclerosis. Thus, oleic type sunflower oil contains nutrients and vitamins, as does standard sunflower oil, but a higher level of oleic acid makes it fit better with healthy eating standards than many other oils.

High stearic sunflower oil is a healthier alternative to other fats rich in saturated fatty acids. Intake of medium-chain fatty acids, which are present in palm kernel and coconut oil, increases the level of total cholesterol leading, to atherosclerosis, and increases the risk of cardiovascular disease (Katan et al., 1995). A similar effect is caused by the intake of palmitic acid-rich fats such as palm oil/fat. Therefore, the intake of these fatty acids should be reduced in the diet. Stearic acid is the only saturated fatty acid that does not affect the level of cholesterol in the blood and, therefore, does not affect the onset of cardiovascular disease (Elson, 1992). There are not many stearic acid-rich fats, and they are often very expensive, such as cocoa butter or the stearin fraction of shea butter. High stearic sunflower oils can be a healthier and more convenient alternative to such fats. The advantage of this oil is the richness of tocopherols. High stearic high oleic sunflower oils are a source of stearic fatty acids, characterized by both a low palmitic content and a high oleic acid content. The saturated fatty acids in such fats are located at positions 1 and 3 within triacylglycerol, which enables the fractionation to produce fats enriched with saturated fatty acids that have a wide range of applications (Salas et al., 2015).

5 Applications of cold pressed sunflower oil

Due to its specific properties, the application of cold pressed oils including sunflower oil can be diverse, including:

1. direct consumption as salad oils or when preparing special dishes;
2. blending with other oils;
3. in the pharmaceutical and cosmetic industries;
4. in the chemical industry, such as oils and lubricants; and
5. directly as biofuel and/or for biodiesel production.

Cold pressed oils, which can be made from a wide variety of raw materials, are consumed directly as salad oils. These oils are especially favored by those who require oils with an intense aroma and flavor to the raw material. In addition to gourmets, there has also been an increasing number of consumers in the last years who are interested in "healthy foods." These consumers consume cold pressed oils because of their nutritional value. Since these oils are much more expensive, they are usually consumed in smaller quantities, which is also beneficial for people who take care of their fat intake.

Good food connoisseurs' kitchens and gourmet serving departments are constantly exploring new uses for these oils. The requirements are addressed primarily to special oils, which are blended with extracts of various herbs, spices, or essential oils, for example, oil blended with garlic, rosemary, or basil, oils containing dried hot peppers, ginger, pepper, or blends thereof. The enrichment of oils by spice plants or their extracts contributes to the accentuation or modification of sensory properties. Often, the addition of these components for flavor or aroma has antioxidant effects as well, which is certainly positive (Turkulov & Đurđev, 1995). Cold pressed sunflower oil, which is so far the most represented in some markets, except for salad oil, can also be used in daily culinary processing of food such as cooking, stewing, baking, salads, sauces, spreads, etc.

The cold pressed oils are very important for the blending process. The primary reason is to obtain a product of desirable flavor and standard quality, regardless of the quality of the raw material. Sensory features, which should evoke a gourmet experience in consumers, are usually crucial in evaluating the quality of specialty oils. This is the reason why these oils are often blended with different oils of different origins, or of the same origin but from different batches, in order to lessen or enhance a particular aroma. Consumers in some countries require a delicate aroma of the oil, but also some special oils, such as sesame seed oil or pumpkin seed, the aroma of which is very strong. The aroma of these oils can also be mitigated by adding a certain amount of refined oil with a neutral taste and aroma.

The use of oils in the pharmaceutical and cosmetics industries has been known since ancient times. However, in this field, the possibilities of applying new special oils are constantly being explored, for which special properties have been established, which can have a beneficial or functional effect from a certain aspect. Vegetable oils are widely used for pharmaceutical purposes (Chazan, 1996).

The cold pressed seed oils market can be segmented into types: applications, packaging, distribution channel, and region (Report, 2018). By applications, the cold pressed seed oils market can be segmented into the food industry, agriculture, cosmetics, and the personal care industry. In the food industry, it is used in cooking, mainly, in marinating salads, and via dietary supplements. In agriculture, cold pressed oils can be used as bio-pesticides. In cosmetics and the personal care industry, cold pressed seed oil is used for antiaging natural oils, oils for hair and skin, body lotions, face cream, body wash, gels, and scrubs. By packaging, the cold pressed seed oils market can be segmented into the glass bottle, PET bottles, pouches, cans, and others. Segmentation of sunflower oil in Europe is shown in Table 4.

TABLE 4 Segmentation of sunflower oil in Europe (CBI, 2016).

Food industry segmentation	
Bottling industry:	
Multipurpose cooking oil: frying, roasting, salad dressings, etc.	
Food processing industry:	
Margarine, sauces, mayonnaise, frying, spray oil on cereals and biscuits, etc.	
← Commodity	Specialty →
Conventional sunflower oil:	Specialty sunflower oil:
1. Futures market/international prices	1. Premium quality: virgin, cold pressed, high-oleic
2. Standard quality: refined, heat extracted	2. Organic certification

6 Adulteration and authenticity of cold pressed sunflower oil

Edible unrefined cold pressed oils are, due to their significantly higher prices, very attractive for blending with lower-cost oils, whether refined or unrefined. Unfortunately, despite the capabilities of modern analytical methods, checking and proving the authenticity of cold pressed oils is still quite problematic. Several methods have been proposed to authenticate edible unrefined, cold pressed, and virgin sunflower oils:

a. Gertz and Klostermann (2000) proposed an analytical procedure based on the determination of the triacylglycerol dimer content to distinguish unrefined from refined oils. Namely, the dimer content of most vegetable oils labeled as unrefined does not exceed 0.1%, while their content in virgin oils does not exceed 0.05%.
b. Marigheto, Kemsley, Defernez, and Wilson (1998) and McDowell, Osorio, Elliott, and Koidis (2018) proposed a procedure based on the comparative analysis of mid-infrared spectroscopy and Raman spectroscopy for the authentication of edible oils.
c. To test for cold pressed oils, Brühl (1996) proposed the determination of *trans* fatty acids. Namely, the falsification of cold pressed oils by refining or mixing with refined oils leads to an increase in the *trans* fatty acid content. The presence of *trans* fatty acids at a concentration above 0.1% is a sure proof of inadequate oil treatment during production.
d. In order to check the cold pressed oils, Brühl (1998) also proposed the determination of the ratio of lactic fatty acids to diacylglycerol, to determine whether neutralization has been carried out in the oil.

Edible unrefined oils can more easily prove their authenticity than organic oils. However, despite this, the presence of a small percentage of refined oil in such a product is very difficult to prove. In cases where small quantities of refined oil of the same type are added to unrefined oil, a complete database of the expected range of indicators for the refined and unrefined oils should be compiled to prove its authenticity. The physical and chemical characteristics that should be considered when authenticating cold pressed oils are as follows:

1. Color: The color of refined oils is less intense than the color of unrefined sunflower oils.
2. Acidity: rafination significantly affects the reduction of free fatty acids in oil.
3. Presence of stigmastadiene: these dienes are not present in unrefined oils, but are present in considerable quantities in refined oils.
4. Tocopherols: neutralization, bleaching, and deodorization affect the loss of tocopherols.
5. Fatty acid composition: high process temperatures can cause the formation of *trans* fatty acids, especially *trans*-octadecenic (linoleic and linolenic) acids.
6. Dimensions of triglycerides in oil: dimers ($<0.5\%$), together with polymers, are formed during the heating of oil. This method can only be applied in cases where a considerable amount of refined oil has been added to the unrefined.
7. Conjugated fatty acids: formed during bleaching from linoleic or linolenic acid, and their detection by UV absorption, or by HPLC. The method can only be used when large quantities ($>30\%$) of the oil have been refined.
8. Sterol dimer formation: these are formed during the bleaching process and can be used to prove this process.

In addition, there are tests that indicate that an extraction process with an organic solvent was used to extract the oil from the feedstock. Since many crude oils are extracted, and oils obtained by this process are not suitable for culinary use without first refining, tests to determine the following indicators are also significant: wax content, aliphatic alcohols, terpenes, and others (Jee, 2002).

Undoubtedly, the number and sensitivity of techniques available to analysts to determine the authenticity of the oil will increase. Variations in trace constituents (such as polyphenols, hydrocarbons, and terpenes) that are not normally present will be expected to provide new information, primarily concerning the origin of the oil.

One of the aspects of authentication is the question "What is authentic oil?" The 1970 Codex Alimentarius proposal implied that the erucic acid content of rapeseed oil used for human consumption was 30%–60%. Today's edible rapeseed oil should contain less than 1% erucic acid. So, if we were to present an oil that meets all the criteria of the 1970s, it would be rejected today. Similarly, high oleic sunflower and safflower oil would not be recognizable. Today, these two oils are well-known and included in the ordinance as two different oils. With the ability to produce oils of the desired composition by genetic modification, and given the aforementioned facts, the detection of oils obtained by genetic modification is very difficult and even impossible. This suggests that it may not even be possible to determine the authenticity of the oil in the traditional sense, and that it will be possible to "replace" the characteristics of a given oil with oil of other types (Jee, 2002).

7 Conclusion

Cold pressed sunflower oil is edible unrefined oil and is characterized by its sensory properties: the aroma and taste of the original raw material. The use of organic solvents in the production of this type of oil is not allowed, which means that the extraction of the oil is not allowed by extraction with the use of solvents, but is solely a mechanical process of extracting the oil without any processing charge. The most common problems that occur when producing this type of oil are related to the quality of the raw material or occur during the cleaning, dehulling, and screw pressing, that is, preparation of raw materials for the production of cold pressed oil.

The most important indicators of the chemical quality of raw materials intended for the production of cold pressed oils are those that indicate the degree of hydrolytic and oxidative changes: the acidity and oxidative state of the oil. These indicators are directly crucial in deciding whether or not the raw material is suitable for pressing. The uncharacteristic and unpleasant taste of both seeds and oils is most often the result of various spoilage processes: oxidative, hydrolytic, enzymatic, microbiological, resulting in the breakdown of lipids, proteins, phosphatides and other components of seeds and oils. Raw materials for the production of cold pressed oils must be of very high quality, as there is no way of subsequently adjusting the quality parameters in the finished product of edible unrefined cold pressed oil.

Based on the above conclusions, it can be generally concluded that the production of cold pressed sunflower oil is possible with the already existing technological production processes, with particular attention to the quality of the raw material, that is, efficiency and process utilization, and the yield of cold pressed oil.

Cold pressed sunflower oil is still a new "type" of superior quality oil, with high nutritive value, relatively good shelf life, and high yield. Although at first glance, similar products are currently on the market, edible sunflower seed oils, especially cold pressed oil are a rarity, and any new knowledge contributes to a better understanding of the technological process of production and quality of this oil. In this way, a new product of pronounced functional properties is developed, very up to date, in terms of both healthy nutrition and the shelf life of the oil itself, or the product in which it is composed. Its production would be very popular, especially in "mini-oil mills," but on the other hand, the use and consumption of this type of oil would also be popular.

Extensive knowledge of the technology of producing edible unrefined cold pressed oils is greatly needed. First of all, spreading the knowledge about the advantages of linoleic and high oleic sunflower type, present in the existing sunflower hybrid assortment from the aspect of cold pressed oil production. This would contribute to the expansion of the range of these oils with new very "attractive" oils.

In addition, the acquired knowledge could be, to a greater or lesser extent, expanded and "transferred" to the production of edible unrefined, primarily cold pressed oils from other raw materials. In order to overcome the potential problems of cold pressing and to ensure the desired quality of the resulting cold pressed sunflower oil, it is necessary to determine the optimal technological procedure for their production. By examining the changing technological parameters of raw material quality, the aim is to determine how they are affected and what their optimal values are for obtaining cold pressed oil of superior sensory and chemical quality, as well as a favorable oxidative state and high nutritive value. The aim is also to test, on the basis of relevant indicators, the oxidative stability and shelf life of cold pressed sunflower oil at higher as well as a lower and moderately high temperatures, and possibly obtain information on the manner, time, and conditions of rapid determination of the recommended shelf life of this type of oil.

Since sunflower seed dehulling is extremely important for obtaining high-quality crude and especially cold pressed oil, optimal conditions for the seed dehulling process are necessary. Extensive research of certain technological parameters in the production of cold pressed oil provides data on the optimal values of certain parameters for the dehulling of sunflower seeds, in terms of the content of seed moisture and air pressure in the dehuller.

In this way, consumers, as well as manufacturers, will be better acquainted with the characteristics of linoleic and high oleic sunflower oil and will receive answers to a number of concerns they have or are not aware of during its production. This would enable the continued production of high-quality cold pressed sunflower oil, and help "mini-oil mills" manufacturers overcome the difficulties they face, and provide consumers with a recognizable and stable product of high to excellent quality.

References

Amarowicz, R. (2009). Squalene: A natural antioxidant. *European Journal of Lipid Science and Technology, 111*, 411–412.

Anderson, D. (2005). A primer on oils processing technology. F. Shahidi (Ed.), *Edible oil and fat products: Edible oils* (pp. 1–56). *Bailey's industrial oil and fat products: Vol. 5*(pp. 1–56). New Jersey: John Wiley & Sons.

AOCS (2011). Official method Ca 6a-40. In *Official methods and recommended practices of the American Oil Chemists' Society*. (6th ed.). Champaign, IL: AOCS Press second printing.

Ayerdi-Gotor, A., Farkas, E., Berger, M., Labalette, F., Centis, S., Dayde, J., et al. (2007). Determination of tocopherols and phytosterols in sunflower seeds by NIR spectrometry. *European Journal of Lipid Science and Technology*, *109*, 525–530.

Ayerdi-Gotor, A., & Rhazi, L. (2016). Effects of refining process on sunflower oil minor components: A review. *Oilseeds and Fats, Crops and Lipids*, *23*(2), D207.

Bamgboye, A., & Adejumo, A. (2007). Development of a sunflower oil expeller. *Agricultural Engineering International: CIGR Journal*, *9*. Manuscript EE 06 015.

Barrera-Arellano, D., Ruiz-Méndez, V., Márquez-Ruiz, G., & Dobarganes, C. (1999). Loss of tocopherols and formation of degradation compounds in triacylglycerol model systems heated at high temperature. *Journal of the Science of Food and Agriculture*, *79*, 1923–1928.

Barrera-Arellano, D., Ruiz-Méndez, V., Velasco, J., Márquez-Ruiz, G., & Dobarganes, C. (2002). Loss of tocopherols and formation of degradation compounds at frying temperatures in oils differing in degree of unsaturation and natural antioxidant content. *Journal of the Science of Food and Agriculture*, *82*, 1696–1702.

Baštić, M., Baštić, L., Jovanović, J. A., & Spiteller, G. (1978). Hydrocarbons and other weakly unsaponifiables in some vegetable oils. *Journal of the American Oil Chemists' Society*, *55*, 886–891.

Beerens, P. (2007). *Screw-pressing of Jatropha seeds for fuelling purposes in less developed countries*. Ph.D. Thesis. The Netherlands: Eindhoven University of Technology.

Berger, A., Jones, P. J. H., & Abumweis, S. S. (2004). Plant sterols: Factors affecting their efficacy and safety as functional food ingredients. *Lipids in Health and Disease*, *3*, 5.

Bishop, G. J., & Yokota, T. (2001). Plant steroid hormones, brassinosteroids: Current highlights of molecular aspects on their synthesis/metabolism, transport, perception and response. *Plant & Cell Physiology*, *42*, 114–120.

Bockisch, M. (1998). *Fats and oils handbook*. Champaign, IL: AOCS Press.

Boskou, D. (2009). Other important minor constituents. In D. Boskou (Ed.), *Olive oil: Minor constituents and health* (pp. 45–54). Boca Raton, FL: CRC Press.

Brkić, S. (2004). *Quality and technical and technological characteristics of oleic sunflower seeds*. Graduate thesis. University of Novi Sad, Faculty of Technology Novi Sad.

Brühl, L. (1996). Determination of *trans* fatty acids in cold pressed oils and in dried seeds. *Fett-Lipid*, *98*(11), 380–383.

Brühl, L. (1998). Determination of free fatty acids and diacylglycerols in native vegetable oils. *Fett-Lipid*, *99*(12), 428–432.

Bundesinstitut für Risikobewertung (BfR). (2011). Addition of plant sterols and stanols to food: Assessment of a new study from the Netherlands. BfR opinion 006/2012, December 2011. Available at: http://www.bfr.bund.de/cm/349/addition-of-plant-sterols-and-stanols-to-foodassessment-of-a-new-study-from-the-netherlands.pdf. [Acessed at 18 August 2018].

Callahan, C., Harwood, H., Darby, H., Schaufler, D., & Elias, R. (2014). *Small-scale oilseed presses: An evaluation of six commercially-available designs*. Pennsylvania: Pennsylvania State University.

Carelli, A. A., Brevedan, M. I. V., & Crapiste, G. H. (1997). Quantitative determination of phospholipids in sunflower oil. *Journal of the American Oil Chemists' Society*, *74*, 511–514.

Carelli, A. A., Ceci, L. N., & Crapiste, G. H. (2002). Phosphorus-to-phospholipid conversion factors for crude and degummed sunflower oils. *Journal of the American Oil Chemists' Society*, *79*, 1177–1180.

Carelli, A. A., Frizzera, L. M., Forbito, P. R., & Crapiste, G. H. (2002). Wax composition of sunflower seed oils. *Journal of the American Oil Chemists' Society*, *79*, 763–768.

CBI (2016). *Product factsheet: Sunflower oil in Europe*. Practical market insights into your product CBI Market Information Database. Ministry of Foreign Affairs, CBI.

Chazan, J. B. (1996). Pharmacological and dietetic properties of oils and fats. In A. Karleskind (Ed.), *Vol. 1 Oils and fats manuel* (pp. 675–702). Vol. 1 (pp. 675–702). Andover: Intercept Ltd.

Codex. (1999). Codex Standards for named vegetable oils, Codex-Stan 210, Amended 2005, Joint FAO/WHO Food Standards Programme.

Cvejić, S., Jocić, S., Radeka, I., Balalić, I., & Miklič, V. (2014). Stvaranje i korišćenje visoko-oleinskih hibrida suncokreta. Jubilarno 55. Savetovanje industrije ulja: Proizvodnja i prerada uljarica, Zbornik radova, str. 113-119, Herceg Novi, Crna Gora.

De Panfilis, F., Gallina Toschi, T., & Lercker, G. (1998). Quality control for cold-pressed oils. *Inform*, *9*, 212–221.

Desai, N. C., Mehta, M. H., Dave, A. M., & Mehta, J. N. (2002). Degumming of vegetable oil by membrane technology. *Indian Journal of Chemical Technology*, *9*, 529–534.

Dimić, E. (2005). *Cold pressed oils*. Novi Sad: Faculty of Technology Novi Sad, University of Novi Sad.

Dolde, D., Vlahakis, C., & Hazebroek, J. (1999). Tocopherols in breeding lines and effects of planting location, fatty acid composition, and temperature during development. *Journal of the American Oil Chemists' Society*, *76*, 349–355.

Dunford, N. T. (2007). *Oil and oilseed processing II*. Robert M. Kerr Food & Agricultural Products Center. FAPC-159.

Eitenmiller, R., & Lee, J. (2004). *Vitamin E: Food chemistry, composition, and analysis*. New York: Marcel Dekker.

Elson, C. E. (1992). Tropical oils: Nutritional and scientific issues. *Critical Reviews in Food Science and Nutrition*, *31*, 79–102.

Erkkilä, A., De Mello, V. D., Risérus, U., & Laaksonen, D. E. (2008). Dietary fatty acids and cardiovascular disease: An epidemiological approach. *Progress in Lipid Research*, *47*, 172–187.

European Commission. (2013). *Commission Regulation (EU) (2013) No 718/2013 of 25 July 2013 Amending Regulation (EC) No 608/2004 Concerning the labeling of foods and food ingredients with added phytosterols, phytosterol esters, phytostanols and/or phytostanol esters. L 201* (pp. 49–50). Official Journal of the European Union.

Faugno, S., del Piano, L., Crimaldi, M., Ricciardiello, G., & Sannino, M. (2016). Mechanical oil extraction of *Nicotiana tabacum* L. seeds: Analysis of main extraction parameters on oil yield. *Journal of Agricultural Engineering*, *XLVII:539*, 142–147.

Fernández-Cuesta, A., Velasco, L., & Fernández-Martínez, J. M. (2011). Phytosterols in the seeds of wild sunflower species. *Helia, 55*, 31–38.
Fernández-Martínez, J. M., Pérez-Vich, B., & Velasco, L. (2009). Sunflower. In J. Vollmann, & I. Rajcan (Eds.), *Oil crops* (pp. 155–232). New York: Springer.
Ferrari, R. A., Schutte, E., Esteves, W., Brühl, L., & Mukherjee, K. D. (1996). Minor constituents of vegetable oils during industrial processing. *Journal of the American Oil Chemists' Society, 73*, 587–591.
Fils, J. M. (2000). The production of oils. In W. Hamm, & R. J. Hamilton (Eds.), *Edible oil processing* (pp. 47–48). Sheffield, England: Sheffield Academic Press.
Flores-Sánchez, I. J., Ortega-López, J., Montes-Horcasitas, M. D. C., & Ramos-Valdivia, A. C. (2002). Biosynthesis of sterols and triterpenes in cell suspension cultures of *Uncaria tomentosa. Plant & Cell Physiology, 43*, 1502–1509.
Food and Agriculture Organization of the United Nations (FAO). (2018). The State of Food Security and Nutrition in the World 2018. Building climate resilience for food security and nutrition. Rome: FAO. Retrieved from: http://www.fao.org/3/i9553en/i9553en.pdf.
Fox, C. B. (2009). Squalene emulsions for parenteral vaccine and drug delivery. *Molecules, 14*, 3286–3312.
Franke, S., Frölich, K., Werner, S., Böhm, V., & Schöne, F. (2010). Analysis of carotenoids and vitamin E in selected oilseeds, press cakes and oils. *European Journal of Lipid Science and Technology, 112*, 1122–1129.
Fuster, M. D., Lampi, A. M., Hopia, A., & Kamal-Eldin, A. (1998). Effects of α- and γ-tocopherols on the autooxidation of purified sunflower triacylglycerols. *Lipids, 33*, 715–722.
Galli, F., & Azzi, A. (2010). Present trends in vitamin E research. *BioFactors, 36*, 33–42.
García-Llatas, G., & Rodríguez-Estrada, M. T. (2011). Current and new insights on phytosterol oxidesin plant sterol-enriched food. *Chemistry and Physics of Lipids, 164*, 607–624.
Gertz, C., & Klostermann, S. (2000). A new analytical procedure to differentiate virgin or nonrefined from refined vegetable fats and oils. *European Journal of Lipid Science and Technology, 102*, 329–336.
Gonzalez-Perez, S., & Vereijken, J. M. (2007). Sunflower proteins: Overview of their physicochemical, structural and functional properties. *Journal of the Science of Food and Agriculture, 87*, 2173–2191.
Gordon, M. H., & Magos, P. (1983). The effect of sterols on the oxidation of edible oils. *Food Chemistry, 10*, 141–147.
Grompone, M. A. (2005). Sunflower oil. In F. Shahidi (Ed.), *Bailey's industrial oil and fat products: Vol. 1. Edible oil and fat products: Chemistry, properties, and health effects* (pp. 655–730). Hoboken, NJ: John Wiley & Sons.
Grompone, M. A. (2011). Sunflower oil. In F. D. Gunstone (Ed.), *Vegetable oils in food technology, composition, properties and uses* (2nd ed., pp. 137–167). Chichester, UK: Blackwell Publishing Ltd. and John Wiley & Sons, Ltd.
Gunstone, F. D. (2005). Vegetable oils. In F. Shahidi (Ed.), *Bailey's industrial oil and fat products: Vol. 1. Edible oil and fat products: Chemistry, properties, and health effects* (pp. 213–267). Hoboken, NJ: John Wiley & Sons.
Gupta, M. K. (2002). Sunflower oil. In F. D. Gunstone (Ed.), *Vegetable oils in food technology: Composition, properties and uses*. Oxford: Blackwell Publishing. Chapter 5.
Hartmann, M. A. (1998). Plant sterols and the membrane environment. *Trends in Plant Science, 3*, 170–175.
Huang, Z. R., Lin, Y. K., & Fang, J. Y. (2009). Biological and pharmacological activities of squalene and related compounds: Potential uses in cosmetic dermatology. *Molecules, 14*, 540–554.
Ionescu, M., Voicu, G., Biris, S.-S., Covaliu, C., Dincă, M., & Ungureanu, N. (2014). Parametars influencing the screw pressing process of oilseed materials. In *3rd International Conference of Thermal Equipment, Renewable Energy and Rural Development TE-RE-RD 2014, Proceedings* (pp. 243–248). Mamaia, Romania.
Jee, M. (2002). Adulteration and authentication of oils and fats: An overview. In M. Jee (Ed.), *Oils and fats authentical* (pp. 1–25). Oxford, Boca Raton, FL: Blackwell Publishing, CRC Press.
Jovanović, D. (2001). Possibility of using sunflower and breeding for special purposes. In *35. Proceedings of the Institute of Field and Vegetable Crops* (pp. 209–221). Novi Sad.
Kalogeropoulos, N., & Andrikopoulos, N. K. (2004). Squalene in oils and fats from domestic and commercial fryings of potatoes. *International Journal of Food Sciences and Nutrition, 55*, 125–129.
Kamal-Eldin, A. (2005). Minor components of fats and oils. In F. Shahidi (Ed.), *Bailey's industrial oil and fat products: Vol. 3. Edible oil and fat products: Specialty oils and oil products* (pp. 319–359). Hoboken, NJ: John Wiley & Sons.
Katan, M. B., Zock, P. L., & Mensink, R. P. (1995). Dietary oils, serum lipoproteins, and coronary heart disease. *The American Journal of Clinical Nutrition, 61*, 1368–1373.
Krizmanić, M., Mijić, A., Liović, I., Sudarić, A., Sudar, R., Duvnjak, T., et al. (2013). Environmental impact on oil content and fatty acid composition of new OS-hybrid combinations of sunflower. *Agriculture, 19*(1), 41–47.
Lajara, J. R., Diaz, U., & Quidiello, R. D. (1990). Definite influence of location and climatic conditions on the fatty acid composition of sunflower seed oil. *Journal of the American Oil Chemists' Society, 67*(10), 618–623.
Lampi, A. M., & Kamal-Eldin, A. (1998). Effect of α- and γ-tocopherol on thermal polymerisation of purified high-oleic sunflower triacylglycerols. *Journal of the American Oil Chemists' Society, 75*, 1699–1703.
Liović, I., Martinović, J., Bilandžić, M., Krizmanić, M., Mijić, A., & Šimić, B. (2010). Desiccation in regular and post-sowing sunflower. *Agriculture, 16*(1), 13–19.
MacKay, D. S., & Jones, P. J. H. (2011). Phytosterols in human nutrition: Type, formulation, delivery, and physiological function. *European Journal of Lipid Science and Technology, 113*, 1427–1432.

Marigheto, N. A., Kemsley, E. K., Defernez, M., & Wilson, R. K. (1998). A comparison of midinfrared and Raman spectroscopies for the authentication of edible oils. *Journal of the American Oil Chemists' Society*, *75*(8), 987–992.

Marinković, R., Dozet, B., & Vasić, D. (2003). *Sunflower breeding*. Monograph Novi Sad: Školska knjiga.

Marquard, R. (1990). Untersuchungen über dem einfluss von sorte und standort auf den tocopherolgehalt verschiedener pflanzenöle. *Fat Science and Technology*, *92*, 452–455.

Martini, S., & Añón, M. C. (2005). Storage of sunflower seeds: Variation on the wax content of the oil. *European Journal of Lipid Science and Technology*, *107*, 74–79.

Matthäus, B., & Brühl, L. (2004). Cold-pressed edible rapeseed oil production in Germany. *Inform*, *15*, 266–268.

McDowell, D., Osorio, M. T., Elliott, C. T., & Koidis, A. (2018). Detection of refined sunflower and rapeseed oil addition in cold pressed rapeseed oil using mid infrared and Raman spectroscopy. *European Journal of Lipid Science and Technology*, *120*, 1700472.

Mène-Saffrané, L., & DellaPenna, D. (2010). Biosynthesis, regulation and functions of tocochromanols in plants. *Plant Physiology and Biochemistry*, *48*, 301–309.

Mijić, A., Sudarević, A., Krizmić, M., Duvnjak, T., Bilandžić, M., Zdunić, Z., et al. (2011). Grain and oil yield of single-cross and three-way cross OS sunflower hybrids. *Agriculture*, *17*(1), 3–8.

Miklič, V., Hladni, N., Jocić, S., Marinković, R., & Atlagić, J. (2008). Oplemenjivanje suncokreta u Institutu za ratarstvo i povrtarstvo. *Zbornik radova Instituta za ratarstvo i povrtarstvo*, *45*(1), 31–36.

National Cholesterol Education Program (NCEP). (2002). Third report of the National Cholesterol Education Program (NCEP) expert panel on detection, evaluation, and treatment of high blood cholesterol in adults (Adult Treatment Panel III). National Cholesterol Education Program, National Heart, Lung, and Blood Institute, National Institutes of Health, NIH Publication No. 02-5215, September 2002. Available at: http://www.nhlbi.nih.gov/guidelines/ cholesterol/atp3full.pdf (Accessed 28 August 2018).

Naz, S., Sherazi, S. T. H., & Talpur, F. N. (2011). Changes of total tocopherol and tocopherol species during sunflower oil processing. *Journal of the American Oil Chemists' Society*, *88*, 127–132.

Nolasco, S. M., Aguirrezábal, L. A. N., & Crapiste, G. H. (2004). Tocopherol oil concentration in field-grown sunflower is accounted for by oil weight per seed. *Journal of the American Oil Chemists' Society*, *81*, 1045–1051.

Onemli, F. (2012). Changes in oil fatty acid composition during seed development of sunflower. *Asian Journal of Plant Sciences*, *11*, 241–245.

Ortega-García, J., Gámez-Meza, N., Noriega-Rodriguez, J. A., Dennis-Quiñonez, O., García-Galindo, H. S., Angulo-Guerrero, J. O., et al. (2006). Refining of High Oleic Safflower Oil: Effect on the Sterols and Tocopherols Content. *European Food Research and Technology*, *223*, 775–779.

Ouyang, J. M., Daun, H., Chang, S. S., & Ho, C. T. (1980). Formation of carbonyl compounds from β-carotene during palm oil deodorization. *Journal of Food Science*, *45*, 1214–1217.

Perretti, G., Finotti, E., Adamuccio, S., Della, Sera R., & Montanari, L. (2004). Composition of organic and conventionally produced sunflower seed oil. *Journal of the American Oil Chemists' Society*, *81*, 1119–1123.

Piironen, V., Lindsay, D. G., Miettinen, T. A., Toivo, J., & Lampi, A. M. (2000). Plant sterols: Biosynthesis, biological function and their importance to human nutrition. *Journal of the Science of Food and Agriculture*, *80*, 939–966.

Plat, J., & Mensink, R. P. (2005). Plant stanol and sterol esters in the control of blood cholesterol levels: Mechanism and safety aspects. *The American Journal of Cardiology*, *96*, 15–22.

Pospišil, M., Pospišil, A., & Antunović, M. (2006). Seed and oil yield of sunflower hybrids investigated as a function of weather. *Agriculture*, *12*(2), 11–16.

Premović, T., Dimić, E., Takači, A., & Romanić, R. (2010). Influence of impurities and hull content in material for pressing on sensory quality cold-pressed sunflower oil. *Acta Periodica Technologica*, *41*, 69–76.

Rade, D., Mokrovčak, Ž., Štrucelj, D., Škevin, D., & Neđeral, S. (2004). The effect of processing conditions on the nontriacylglycerol constituents of sunflower oil. *Acta Alimentaria*, *33*, 7–18.

Raß, M., Schein, C., & Matthäus, B. (2008). Virgin sunflower oil. *European Journal of Lipid Science and Technology*, *110*, 618–624.

Report. (2018). Cold pressed seed oils market: Global industry trend analysis 2012 to 2017 and forecast 2017–2025. Available at: https://www.persistencemarket (web archive link, 28 August 2019) research.com/market-research/cold-pressed-seed-oils-market.asp. (Accessed 28 August 2019).

Romanić, R. (2015). *Chemometric approach to the optimization of the production of cold pressed oil from high-oleic sunflower seeds*. Novi Sad: Faculty of Technology, The University of Novi Sad.

Romanić, R. (2016). Effect of hull and moisture and oil content in the press cake and on the yield of cold pressed high-oleic sunflower oil. In *FoodTech 2016—III International Congress "Food Technology, Quality and Safety"* (p. 187). Abstract Book, Novi Sad, Serbia.

Romanić, R., & Grbić, N. (2016). Pressing efficiency of cold pressed oil of high-oleic type sunflower seed with a different proportions of hull and impurities. In *57th Conference: Production and processing of oilseeds, Proceedings* (pp. 95–105). Herceg Novi.

Romanić, R., Grbić, N., & Savić, S. (2017). Utilization of cold pressed high-oleic sunflower seed oil with different proportion of hull and impurities. In *58th Oil Industry Conference: Oilseed Production and Processing, Proceedings* (pp. 131–140). Herceg Novi, Montenegro.

Romanić, R., Lužaić, T., Grahovac, N., Hladni, N., Kravić, S., & Stojanović, Z. (2018a). Composition investigation of the sunflower seeds of the latest NS confectionary hybrids. In *International GEA (Geo Eco-Eco Agro) Conference* (pp. 68–72). Podgorica, Montenegro. Book of Proceedings.

Romanić, R., Lužaić, T., Grahovac, N., Hladni, N., Kravić, S., & Stojanović, Z. (2018b). Study on dimensions of the sunflower seeds of the latest NS confectionary hybrids. In *International GEA (Geo Eco-Eco Agro) Conference* (pp. 73–77). Podgorica, Montenegro. Book of Proceedings.

Salas, J. J., Bootello, M., & Garcés, R. (2015). Food uses of sunflower oils. In E. Martínez-Force, N. T. Dunford, & J. J. Salas (Eds.), *Sunflower chemistry, production, processing, and utilization* (pp. 441–464). Urbana, IL: AOCS Press.

Salas, J. J., Bootello, M. A., Martínez-Force, E., & Garcés, R. (2011). Production of stearate-rich butters by solvent fractionation of high stearic-high oleic sunflower oil. *Food Chemistry*, *124*, 450–458.

Salas, J. J., Martínez-Force, E., & Garcés, R. (2006). Accumulation of phospholipids and glycolipids in seed Kernels of different sunflower mutants (*Helianthus annuus*). *Journal of the American Oil Chemists' Society*, *83*, 539–545.

Sánchez-Muniz, F. J., & Cuesta, C. (2003). Sunflower oil. In B. Caballero (Ed.), *Encyclopedia of food sciences and nutrition* (2nd ed., pp. 5672–5680). New York: Academic Press.

Schilling, E. (2006). Helianthus. Flora of North America Editorial Committee (Eds.)In *Flora of North America, North of Mexico* (pp. 141–169). New York: Oxford University Press.

Sciancalepore, V. (1981). The influence of processing on the content and composition of free and esterified sterols in sunflower seed oil. *Oli and Grassi Derivatives*, *17*, 11–12.

Seppanen, C. M., Song, Q., & Csallany, A. S. (2010). The antioxidant functions of tocopherol and tocotrienol homologues in oils, fats, and food systems. *Journal of the American Oil Chemists' Society*, *87*, 469–481.

Serrano-Vega, M. J., Martínez-Force, E., & Garcés, R. (2005). Lipid characterization of seeds oils from high-palmitic, and very high-stearic acid sunflowers lines. *Lipids*, *40*, 369–374.

Shahidi, F., & Zhong, Y. (2010). Lipid oxidation and improving the oxidative stability. *Chemical Society Reviews*, *39*, 4067–4079.

Simopoulos, A. P. (2002). Polyunsaturated fatty acids in biology and diseases. the importance of the ratio of omega-6/omega-3 essential fatty acids. *Biomedicine & Pharmacotherapy*, *56*, 365–379.

Sivakumaran, K., & Goodrum, J. W. (1988). Laboratory oilseed processing by a small screw press. *Journal of the American Oil Chemists' Society*, *65*, 932–935.

Škorić, D., Demurin, J., Jocić, S., Lečić, N., & Verešbaranji, I. (1994). Stvaranje hibrida suncokreta sa različitim kvalitetom ulja. 35. Savetovanje o unapređenju uljarstva Jugoslavije, Zbornik radova, Bar, str. 8–15.

Smouse, T. H. (1995). Factors affecting oil quality and stability. In K. Warner & N. A. M. Eskin (Eds.), *Methods to assess quality and stability of oils and fat-containing foods* (pp. 17–36). Urbana, IL: AOCS Press.

Sotiroudis, T. G., & Kyrtopoulos, S. A. (2008). Anticarcinogenic compounds of olive oil and related biomarkers. *European Journal of Nutrition*, *47*(2), 69–72.

Szuhaj, B. F. (2005). Lecithins. In F. Shahidi (Ed.), *Bailey's industrial oil and fat products: Vol. 5. Edible oil and fat products: Processing technologies* (pp. 361–456). Hoboken, NJ: John Wiley & Sons.

Thanh, T. T., Vergnes, M. F., Kaloustian, J., El-Moselhy, T. F., Amiot-Carlin, M. J., & Portugal, H. (2006). Effect of storage and heating on phytosterol concentrations in vegetable oils determined by GC/MS. *Journal of the Science of Food and Agriculture*, *86*, 220–225.

Todorović, J., & Komljenović, I. (2010). *Oilseeds (sunflower)—A vegetable and vegetable manual*. Part 7 Banja Luka: Faculty of Agriculture, University of Banja Luka.

Turkulov, J., & Đurđev, S. (1995). Antioxidant activity of herbs. In *36th Oil Industry Conference: Oilseed Production and Processing, Proceedings* (pp. 13–18). Budva, Yugoslavia.

USDA. (2012). *Oilseeds: World markets and trade*. [On line]. United States Department of Agriculture—Foreign Agricultural Service. Available at: http://www.fas.usda.gov. (Accessed 16 January 2019).

Velasco, L., & Fernández-Martínez, J. M. (2012). Environmental stability of contrasting seed tocopherol profiles in sunflower. *Crop Science*, *52*, 2446–2452.

Velasco, L., Fernández-Martínez, J. M., García-Ruíz, R., & Domínguez, J. (2002). Genetic and environmental variation for tocopherol content and composition in sunflower commercial hybrids. *The Journal of Agricultural Science*, *139*, 425–429.

Velasco, L., & Ruiz-Méndez, M. V. (2015). Sunflower oil minor constituents. In E. Martínez-Force, N. T. Dunford, & J. J. Salas (Eds.), *Sunflower chemistry, production, processing, and utilization* (pp. 297–329). Urbana, IL: AOCS Press.

Verleyen, T., Sosinska, U., Ioannidou, S., Verhe, R., Dewettinck, K., Huyghebaert, A., et al. (2002). Influence of the vegetable oil refining process on free and esterified sterols. *Journal of the American Oil Chemists' Society*, *79*, 947–953.

Vlahakis, C., & Hazebroek, J. (2000). Phytosterol accumulation in canola, sunflower, and soybean oils: Effects of genetics, planting location, and temperature. *Journal of the American Oil Chemists' Society*, *77*, 49–53.

Warner, K., & Moser, J. (2009). Frying stability of purified mid-oleic sunflower oil triacylglycerols with added pure tocopherols and tocopherol mixtures. *Journal of the American Oil Chemists' Society*, *86*, 1199–1207.

Winkler, J. K., & Warner, K. (2008a). The effect of phytosterol concentration on oxidative stability and thermal polymerization of heated oils. *European Journal of Lipid Science and Technology*, *110*, 455–464.

Winkler, J. K., & Warner, K. (2008b). Effect of phytosterol structure on thermal polymerization of heated soybean oil. *European Journal of Lipid Science and Technology*, *110*, 1068–1077.

Yanishlieva, N. V., Raneva, V. G., & Marinova, E. M. (2001). β-Carotene in sunflower oil oxidation. *Grasas y Aceites*, *52*, 10–16.

Zawistowski, J. (2010). Tangible health benefits of phytosterol functional foods. In J. Smith & E. Chartes (Eds.), *Food product development* (pp. 362–387). Chichester: Wiley-Blackwell.

Zheng, Y., Wiesenborn, P., Tostenson, K., & Kangas, N. (2003). Screw pressing of whole and dehulled flaxseed for organic oil. *Journal of the American Oil Chemists' Society*, *80*, 1039–1045.

Chapter 18

Cold pressed pumpkin seed oil

Zeynep Aksoylu Özbek and Pelin Günç Ergönül
Department of Food Engineering, Faculty of Engineering, Manisa Celal Bayar University, Manisa, Turkey

Abbreviations

CAC	Codex Alimentarius Commission
DPPH	1,1-diphenyl-2-picrylhydrazyl
FAO	Food and Agriculture Organization
GAE	gallic acid equivalent
LDL	low-density lipoprotein
LLL	trilinolein
LLLn	dilinoleinolinolenin
LLnLn	linoleinodilinolenin
LLO	linoleindiolein
LnPO	linoleninopalmitooleoin
MRL	maximum residue limit
OLL	oleodilinolein
OLLn	oleolinoleolinolenin
OOL	dioleolinolein
OOO	triolein
PAH	polycyclic aromatic hydrocarbon
PCB	polychlorinated biphenyl
PLL	palmitodilinolein
PLLn	palmitolinoleolinolenin
PLSt	palmitolinoleostearin
POL	palmitooleolinolein
PPL	dipalmitolinolein
PUFA	polyunsaturated fatty acid
StOL	stearooleolinolein
TBARS	thiobarbituric acid reactive substances
TPC	total phenolic content

1 Introduction

Pumpkin (*Cucurbita pepo* L.) is one of the most cultivated cucurbit crops in the world, along with cucumber, melon, and watermelon (Weng & Sun, 2012). It is believed that this annual crop was domesticated in South American and Mexican region, and then spread to Europe, Asia, and North America (Singh, Singh, Chand, & Nair, 2016; Yadav, Jain, Tomar, Prasad, & Yadav, 2010). The edible parts of the pumpkin are fruits, flowers, and seeds. Additionally, this plant is grown for decorative purposes (Ferriol & Picó, 2008). Due to its unique composition, seeds are considered the most important component of the pumpkin. Besides their high fat and protein contents with essential amino acids, pumpkin seeds contain substantial amounts of minerals like potassium, magnesium, selenium, zinc, copper, molybdenum, and chromium, and bioactive compounds such as tocopherols (in particular γ-tocopherol), and carotenoids (mainly lutein) (Glew et al., 2006; Jeznach, Danilcenko, Jariene, Kulaitiene, & Černiauskiene, 2012; Kim, Kim, Kim, Choi, & Lee, 2012). For a long time, pumpkin seeds have been consumed as a salted snack after roasting (Patel, 2013). However, the seeds are also referred to as oilseeds due to their very high oil content.

Cold Pressed Oils. https://doi.org/10.1016/B978-0-12-818188-1.00018-9
Copyright © 2020 Elsevier Inc. All rights reserved.

At the end of the 19th century, a natural mutation resulted in a pumpkin variety with naked seeds in central Europe (Pitrat, Chauvet, & Foury, 1999). This mutant is called Styrian oil pumpkin (*Cucurbita pepo* subsp. *pepo* var. Styriaca). Today, the Styrian pumpkin seed oil is predominantly produced in Styria, the southeastern province of Austria. The geographic origin and quality of this local oil are protected by the European Union under the regulation of protected geographical indications (Fruhwirth & Hermetter, 2008). In addition to Austria, Slovenia, Croatia, and Hungary are the other producer countries of similar types of pumpkin seed oil (Murkovic, 2009).

With increasing evidence of health-promoting effects of various bioactives, the interest of consumers in minimally processed foods like cold pressed oils has increased in recent years. The physicochemical properties of cold pressed pumpkin seed oil can be summarized as follows: refractive index at 25°C between 1.472 and 1.480; viscosity at 25°C between 78.1 and 78.6cP; saponification value between 184 and 290.7mg KOH/g oil; unsaponifiable matters between 0.6% and 0.9%; iodine value between 103 and 121.6g I_2/100g oil; acid value between 0.9 and 3.3mg KOH/g oil; free fatty acid 0.2%–0.6% oleic acid; peroxide value between 0.1 and 9meq O_2/kg; *p*-anisidine value between 0.2 and 1.9; totox value between 1.6 and 9.4 (Aktaş, Gerçekaslan, & Uzlaşır, 2018; Arslan, Akın, & Yılmaz, 2017; Gorjanović et al., 2011; Poiana, Alexa, Moigradean, & Popa, 2009; Vujasinovic, Djilas, Dimic, Basic, & Radocaj, 2012; Vujasinovic, Djilas, Dimic, Romanic, & Takaci, 2010; Vukša, Dimić, & Dimić, 2003). In summary, cold pressed pumpkin seed oils have lower acid and peroxide values than the maximum limits (4mg KOH/g oil and 15meq O_2/kg oil, respectively) set by the Codex Alimentarius (Codex Alimentarius, 1999). The formation of free fatty acids may be associated with the hydrolysis of triacylglycerols prior to pressing. Moreover, pressing of seeds with hulls promotes the increase of peroxide, acid, and *p*-anisidine values of cold pressed pumpkin seed oils (Vujasinovic et al., 2010). The higher iodine values of the oils indicate higher degrees of unsaturation.

The defatted oil cake obtained after pressing of oil is an excellent source of protein (more than 60%) with a balanced essential amino acid content, some vitamins (niacin, thiamine, pyridoxine, and pantothenic acid), minerals (phosphorus, potassium, magnesium, calcium, iron, and zinc), lignans (secoisolariciresinol and lariciresinol), and phenolic acids (esterified *p*-hydroxy benzaldehyde, free *p*-hydroxybenzoic, and vanillic acid) (Krimer-Malesevic, Posa, Stolic, Popovic, & Pericin-Starcevic, 2016; Mansour, Dworschák, Lugasi, Barna, & Gergely, 1993; Peričin, Krimer, Trivić, & Radulović, 2009; Sharma, Lal, Madaan, & Chatterjee, 1986; Sicilia, Niemeyer, Honig, & Metzler, 2003). This defatted seed cake has been successfully used as a high-nutritive value feed for chickens, pigs, dairy goats, lambs, and fish (Antunović et al., 2018; Greiling, Schwarz, Gierus, & Rodehutscord, 2018; Klir et al., 2017; Martínez, Valdivie, Martínez, Estarrón, & Córdova, 2010; Wafar, Hannison, Abdullahi, & Makinta, 2017; Zucker, Hays, Speer, & Catron, 1958). The authors stated that oil pumpkin seed cake could be utilized as a cheaper alternative to soybean cake in animal nutrition without any change in reproduction ability and milk production yield of animals or quality of the meat and milk. On the other hand, this valuable by-product of pumpkin seed oil production is the raw material of plant-based protein isolates and concentrates, which can be consumed as a dietary supplement or incorporated into other food products (Bučko et al., 2015). Additionally, recent studies have shown that bioactive peptides can be produced by hydrolyzation of the major storage protein of defatted pumpkin cake, namely cucurbitin (Popović et al., 2013, 2017). These biologically active peptides have antioxidant properties as well as in vitro angiotensin-I converting enzyme and α-amylase enzyme inhibitory activities; thus they act as a promising natural alternative for hypertension and diabetes treatment (Vaštag, Popović, & Popović, 2014; Vaštag, Popović, Popović, Krimer, & Peričin, 2011). Up to now, many attempts have been made to evaluate the feasibility of pumpkin seed cake in bread, biodegradable composite films, and fat-based spread, and as a substrate for enzyme production (*exo*-pectinase) (Filbrandt, 2012; Pericin, Madjarev, Radulovic, & Skrinjar, 2007; Popović, Lazić, Popović, Vaštag, & Peričin, 2010; Popović, Peričin, Vaštag, Lazić, & Popović, 2012; Popović, Peričin, Vaštag, Popović, & Lazić, 2011; Radocaj, Dimic, & Vujasinovic, 2011). The statistics about pumpkin seed production have not been reported by the FAO. The production quantity has been declared for pumpkins along with squash and gourds.

2 Extraction and processing of cold pressed oil

The well-known pumpkin seed oil, Styrian pumpkin seed oil, is classified as virgin oil due to its traditional production method. Unlike cold pressed pumpkin seed oil, virgin pumpkin seed oil is pressed from the roasted seeds (Vujasinovic et al., 2010). The roasting process contributes to the formation of the desired aroma characteristic of pumpkin seed oil, which is described as roasty, nutty, coffee-like, fatty, malty, and green (Matsui, Guth, & Grosch, 1998; Siegmund & Murkovic, 2004). It should be noted that this thermal treatment causes an undesirable increase in quality indicators of oil such as peroxide, *p*-anisidine, TBARS, and totox values (Aktaş et al., 2018; Vujasinovic et al., 2012). A new study demonstrated that keeping the pumpkin seeds in a water/salt mixture—called the wet-salting process—for one night before roasting may serve as an effective way to prevent oxidative changes from occurring throughout roasting (Aktaş et al., 2018).

However, in terms of oil stability, scientific evidence suggests the opposite. Vujasinovic et al. (2012) and Naziri, Mitić, and Tsimidou (2016) reported higher oxidative stability for the oils of roasted pumpkin seeds than that of unroasted ones due to antioxidative activity of Maillard reaction products formed during roasting, together with increased contents of phospholipids, tocopherols, and phenolics in the oil.

According to the Codex Alimentarius, cold pressed pumpkin seed oil is directly obtained from raw and dried seeds using a continuous screw press (Codex Alimentarius, 1999). The outlet temperature of the oil should be kept below 50°C for the oil to be labeled as cold pressed pumpkin seed oil (Vujasinovic et al., 2010).

3 Fatty acids and acyl lipids profile of cold pressed oil

The cold pressed pumpkin seed oil belongs to the oleic-linoleic acid group oils including corn, sesame, sunflower, soya, and cottonseed oils (Başoğlu, 2014). The proximate fatty acid composition of the cold pressed pumpkin seed oil is summarized in Table 1. The main fatty acids of the oil are linoleic, oleic, palmitic, and stearic acids. Owing to its high linoleic acid content, cold pressed pumpkin seed oil serves as a reliable source of essential fatty acids. Trace amounts of myristic (C14:0), palmitoleic (C16:1), margaric (C17:0), linolenic (C18:3), arachidic (C20:0), eicosenoic (C20:1), eicosatetraenoic (C20:4), behenic (C22:0), docosatetraenoic (C22:4), docosapentaenoic (C22:5), docosahexaenoic (C22:6), and lignoceric (C24:0) acids were identified in cold pressed pumpkin seed oil (Kulaitienė et al., 2018; Rabrenović et al., 2014).

By now, numerous studies have indicated the strong relationship between dietary linoleic acid and the blood cholesterol level. A linoleic acid-enriched diet contributes to the lowering of plasma cholesterol and reduces the risk of cardiovascular diseases (Heine, Mulder, Popp-Snijders, van der Meer, & van der Veen, 1989; Horrobin & Huang, 1987). Furthermore, linoleic acid is necessary for the production of vitamin D, cellular membranes, and some hormones (Fruhwirth & Hermetter, 2007). On the other hand, Louheranta, Porkkala-Sarataho, Nyyssönen, Salonen, and Salonen (1996) suggested that high linoleic acid intake increases the susceptibility of atherogenic lipoproteins against oxidation in men.

Acylglycerols constitute the neutral lipids together with fatty acids, alcohols, and sterols (O'Keefe & Sarnoski, 2017). The stereospecific location of fatty acids within acylglycerols determines the nutritive value, physical properties, and oxidative stability of any oil (McClements & Decker, 2017). The dominant triacylglycerols of cold pressed pumpkin seed oil are oleodilinolein (OLL), trilinolein (LLL), palmitodilinolein (PLL), and linoleindiolein (LLO) (Nederal et al., 2012). In addition to these, Arslan et al. (2017) identified linoleinodilinolenin (LLnLn), dilinoleinolinolenin (LLLn), oleolinoleolinolenin (OLLn), palmitolinoleolinolenin (PLLn), linoleninopalmitooleoin (LnPO), dioleolinolein (OOL), palmitooleolinolein (POL), dipalmitolinolein (PPL), triolein (OOO), stearooleolinolein (StOL), and palmitolinoleostearin (PLSt) in cold pressed pumpkin seed oils grown in the central Anatolia region of Turkey.

4 Minor bioactive lipids

The unsaponifiable part of lipids is comprised of sterols, terpenic alcohols, aliphatic alcohols, hydrocarbons, and squalene (Finley & DeMan, 2018). The plant sterols, namely phytosterols, attract consumers' attention due to the fact that they reduce the risk of cardiovascular diseases by lowering low-density lipoprotein (LDL) cholesterol levels (Marangoni & Poli, 2010). The phytosterol fraction of cold pressed pumpkin seed oil essentially consists of Δ7,22,25-stigmastatrienol (19.4–33.9 mg/100 g); campesterol (16.5 mg/100 g); Δ7,25-stigmastadienol (4.9–26.4 mg/100 g); Δ7-stigmasterol

TABLE 1 Fatty acid composition (%) of cold pressed pumpkin seed oil.

C16:0	C18:0	C18:1	C18:2	References
10.21–11.88	4.45–5.15	30.35–42.07	43.68–52.15	Vujasinovic et al. (2010)
11.60–11.80	5.10–5.30	28.90–35.20	46.70–53.20	Nederal et al. (2012)
11.20–15.50	5.20–6.20	37.10–43.60	37.30–44.50	Rabrenović, Dimić, Novaković, Tešević, and Basić (2014)
12.50	4.60	28.60	52.80	Konopka, Roszkowska, Czaplicki, and Tańska (2016)
11.83–11.87	3.18–3.58	16.01–18.33	64.65–67.24	Kulaitienė, Černiauskienė, Jarienė, Danilčenko, and Levickienė (2018)
11.90–11.99	5.26–5.29	27.52–27.59	53.20–53.28	Akin, Arslan, Karuk Elmas, and Yilmaz (2018)

(12.5–20.1 mg/100 g); spinasterol together with β-sitosterol (42.4–50.4 mg/100 g), and Δ7-avenasterol (2.1–9.4 mg/100 g). These studies revealed that the total sterol content of cold pressed pumpkin seed oil ranged between 71.8 and 180.6 mg/100 g (Akin et al., 2018; Konopka et al., 2016; Rabrenović et al., 2014).

Cold pressed pumpkin seed oil is characterized by relatively higher total phenolic content (TPC). It has higher TPC than cold pressed oils of soybean, sunflower, rapeseed, corn, grapeseed, hemp, flax, and rice bran (Siger, Nogala-Kalucka, & Lampart-Szczapa, 2008). Depending on the cultivar, the cold pressed oils of unroasted pumpkin seeds have a wide range of TPC, ranging from 4.63 mg gallic acid equivalent (GAE)/kg to 2240 mg GAE/kg (Aktaş et al., 2018; Vujasinovic et al., 2012). The phenolic compounds are responsible for the antioxidant activity of the pumpkin seed oil. A strong correlation ($r = 0.823$, $p < .05$) between TPC and 1,1-diphenyl-2-picrylhydrazyl (DPPH) radical· scavenging activity of cold pressed pumpkin seed oil was observed (Kulaitienė et al., 2018). In general, phenolic acids are found in cold pressed pumpkin seed oil. Siger et al. (2008) identified vanillic acid (11.4 μg/100 g), *p*-coumaric acid (3.8 μg/100 g), ferulic acid (3.8 μg/100 g), and protocatechuic acid (3.1 μg/100 g) in cold pressed pumpkin seed oil. Additionally, *trans*-cinnamic (1 mg/kg) acid was detected in the oil (Tuberoso, Kowalczyk, Sarritzu, & Cabras, 2007). Recently, the presence of caffeic acid (3.4–3.8 mg/100 g) and syringic acid (7.6–8 mg/100 g) in cold pressed pumpkin seed oils from Turkey was confirmed (Akin et al., 2018).

Besides phenolics, tocopherols are the other contributors to the antioxidant activity of this oil. The total tocols content of pumpkin seed oil varies from 265.7 mg/kg to 977.9 mg/kg (Akin et al., 2018; Vujasinovic et al., 2012). The major tocopherol isomer is γ-tocopherol, representing almost 90% of total tocopherols in cold pressed pumpkin seed oil. The content of this tocopherol isomer may range between 251 mg/kg and 775 mg/kg. The second most abundant tocol is α-tocopherol, with concentrations ranging from 10.1 mg/kg to 353 mg/kg (Broznić, Čanadi Jurešić, & Milin, 2016; Naziri et al., 2016). Cold pressed pumpkin seed oil contains minor amounts of β- and δ-tocopherols (Rabrenović et al., 2014; Vujasinovic et al., 2012). In addition to tocopherols, α-, β-, and γ-tocotrienols were identified in cold pressed oils of pumpkin seeds grown in Turkey (Akin et al., 2018). While Naziri et al. (2016) suggested that α- and γ-tocopherol increase the oxidative stability of pumpkin seed oil, Gorjanović et al. (2011) observed strong positive correlations between δ-tocopherol content and DPPH· and H_2O_2-scavenging activities of cold pressed pumpkin seed oil ($r = 0.89$ and $r = 0.87$, respectively).

Photooxidation of oils, which is induced by singlet oxygen, may be prevented by carotenoids via their quenching singlet oxygen mechanism (Elias & Decker, 2017). Cold pressed pumpkin seed oil contains mainly lutein (up to 8.8 mg/100 g) and γ-carotene (5.8 mg/100 g) along with lower amounts of β-carotene (0.5–0.6 mg/100 g), α-carotene (1.7 mg/100 g), zeaxanthin (2.6–2.9 mg/100 g), cryptoxanthin (0.4–0.5 mg/100 g), and 9-*cis*-γ-carotene (1.8 mg/100 g). Tuberoso et al. (2007) reported that cold pressed pumpkin seed oil contained higher levels of β-carotene than cold pressed oils of flaxseed, grapeseed, maize, peanut, rapeseed, soybean, and sunflower, but lower levels than virgin olive oil. Total carotenoid concentration was reported to be between 6.9 mg/100 g and 22.8 mg/100 g (Akin et al., 2018; Konopka et al., 2016). Chlorophylls, one of the most frequent pigments of edible oils, can serve as an antioxidant in lipid auto-oxidation, whereas they may act as a sensitizer and produce singlet oxygen molecules in the presence of light (Choe, 2017). Cold pressed pumpkin seed oil contains comparable amounts of chlorophylls to those of virgin olive oil (30.8 mg/kg and 33.9 mg/kg, respectively) (Tuberoso et al., 2007).

The dominant hydrocarbon of pumpkin seed oil is squalene. Its well-known health benefits are a precursor of steroids, regulation of vitamin D and cholesterol biosynthesis, and antitumor and antioxidant activity, as well as being a chemoprotective and radioprotective agent (Kim & Karadeniz, 2012). Naziri et al. (2016) suggested that squalene acts as a primary antioxidant against photosensitized oxidation together with α- and γ-tocopherols in cold pressed pumpkin seed oil. Squalene levels of 952.3–7880 mg/kg of oil have been reported (Gorjanović et al., 2011; Konopka et al., 2016). Cold pressed pumpkin seed oil (3529.9 mg/kg) is considered to be a rich source of plant-based squalene along with virgin olive oil (5990 mg/kg), cold pressed peanut (1276 mg/kg), rapeseed (437.4 mg/kg), and maize oils (338.7 mg/kg) (Tuberoso et al., 2007).

In summary, all of these minor phytochemicals directly contribute to the superior antioxidant effect of cold pressed pumpkin seed oil. This oil has the second highest DPPH· scavenging activity (65.3%), after hemp (76.2%), among the other cold pressed plant oils including rapeseed (51.2%), sunflower (23.8%), rice bran (23.7%), flax (19.3%), soybean (17.4%), grapeseed (13.4%), and corn (11.1%) (Siger et al., 2008). Contrary to this, Tuberoso et al. (2007) ranked cold pressed pumpkin seed oil as the seventh highest antioxidant capacity edible oil after cold pressed oils of corn, soybean, grapeseed, olive, sunflower, and flaxseed, respectively. Finally, cold pressed pumpkin seed oil (36.2%) was classified as "medium DPPH· radical scavenging activity oil" together with cold pressed oils of chia seed (30.4%), apricot seed (38.8%), sunflower seed (25.7%–39.3%), linseed (32.7%), argan seed (31.5%), mustard seed (34.6%), red grape seed (33.7%), almond seed (25.3%), rapeseed (35.1%), walnut (36.2%), negrilic (33%), and sesame seed (36.9%) (Casoni, Simion, & Sârbu, 2019). However, it should be kept in mind that these differences observed by various researchers might be caused by different chemical compositions of oilseeds.

The sensory properties of cold pressed pumpkin seed oil were defined as "unique" and "specific" with a mild and fruity taste by (Vujasinovic et al., 2010). The principal volatile compounds identified in cold pressed oil of unroasted pumpkin seeds are hexanol, 1-pentanol, hexanal, 3-methylbutanal, 2-methylbutanal, pentanal, and benzaldehyde (Potočnik & Košir, 2017). Hexanol gives a sweet and green fruity odor and aromatic flavor, whereas hexanal contributes to the fatty, green, grassy, and fruity odor of cold pressed pumpkin seed oil. On the other hand, benzaldehyde is responsible for the sweet, oily, almond-like, nutty, and woody taste of the oil (Burdock, 2010). Unlike cold pressed pumpkin seed oil, commercial pumpkin seed oil is commonly obtained from roasted seeds. The roasting process produces a number of aroma compounds such as furan derivatives and pyrazines, which lead to the characteristic roast and nutty aroma of the oil (Siegmund & Murkovic, 2004).

5 Health-promoting traits of cold pressed oil and oil constituents

Various parts of the pumpkin have been used as traditional medicine for centuries, mainly in America, Argentina, Brazil, Mexico, China, and India (Caili, Huan, & Quanhong, 2006). The beneficial effects of cold pressed pumpkin seed oil on human health are attributed to substantial amounts of bioactive components naturally found in the oil. In recent years, the oil has attracted great attention due to its restorative impacts on men's health. The best-known health benefit of cold pressed pumpkin seed oil is the prevention of testosterone-induced hyperplasia of the prostate (Gossell-Williams, Davis, & O'Connor, 2006). Phytosterols, mainly β-sitosterol, are considered to be the active components for improvement of prostate health (Klippel, Hiltl, & Schipp, 1997). Moreover, Cho et al. (2014) claimed that pumpkin seed oil is an effective agent for boosting hair growth in men with androgenetic alopecia. These positive effects on men's health have been associated with the inhibition of 5-α-reductase by phytosterols that catalyzes the production of a potent androgen, known as testosterone.

Similar to men's health, pumpkin seed oil maintains women's health by promoting estrogenic activity through its natural phytoestrogens. Gossell-Williams et al. (2011) reported that pumpkin seed oil intake prominently decreases menopausal symptoms in postmenopausal women. Furthermore, the abundant presence of lignans in the oil, especially secoisolariciresinol, is highly correlated with the reduced risk of breast cancer (Richter et al., 2013). However, the roasting process destroys secoisolariciresinol (Murkovic, 2009). Hence, cold pressed pumpkin seed oil is a valuable source of this kind of lignan.

Pumpkin seed oil has antihypertensive and cardioprotective properties (El-Mosallamy, Sleem, Abdel-Salam, Shaffie, & Kenawy, 2012). Landeka, Đikić, Radišić, Teparić, and Bačun-Družina (2011) suggested that diets rich in pumpkin seed oil decrease blood LDL cholesterol levels and the risk of cardiovascular diseases. This effect of the oil is associated with the hypocholesterolemic activity of linoleic acid (Landeka et al., 2011). On the other side, the antidiabetic effect of pumpkin seed oil is explained by the existence of tocopherol compounds (Bharti et al., 2013).

Another significant characteristic of the pumpkin seed oil is its strong antioxidant activity. By this mechanism, the oil has been shown to prevent diabetes, reduce lead-induced testicular toxicity, promote the action of antihypertensive agents, protect the lung tissue against acid-induced injury, protect the liver from the hepatotoxic effect caused by alcohol, decrease DNA damage by reducing toxic effect generated by sodium nitrite, protect against toxic insults, and prevent cytotoxicity of azathioprine in sperm (Abou Seif, 2014; Abou-Zeid, AbuBakr, Mohamed, & El-Bahrawy, 2018; Al Zuhair, Abd El-Fattah, & El-Sayed, 2000; Ali & Abdelzaher, 2017; Al-Masri, 2015; Elfiky, Elelaimy, Hassan, Ibrahim, & Elsayad, 2012; Eraslan, Kanbur, Aslan, & Karabacak, 2013; Makni et al., 2010; Omar & Sarhan, 2017).

Experimental studies have shown that cold pressed pumpkin seed oil acts as a promising agent for wound healing. Researchers correlated this effect with a balance between the proinflammatory activity of polyunsaturated fatty acids (PUFAs) and their antiinflammatory effect as a consequence of antibacterial and antioxidant activities (Bardaa et al., 2016; Bardaa, Moalla, Ben Khedir, Rebai, & Sahnoun, 2016).

6 Edible and nonedible applications of cold pressed oil

Pumpkin seed oil is commonly consumed as a salad oil in Austria, Slovenia, and Hungary (Kochhar, 2011; Murkovic, Hillebrand, Winkler, Leitner, & Pfannhauser, 1996). Due to its high unsaturation degree, it is not recommended to use pumpkin seed oil as a cooking or frying oil. Besides being a salad oil, it is also used for medical purposes.

7 Other issues

The major analytical parameters used for authenticity and adulteration assessment of cold pressed oils are fatty acid and/or sterol compositions, sterenes, and squalene (Kamm, Dionisi, Hischenhuber, & Engel, 2001). Butinar, Bučar-Miklavčič, Krumpak, and Raspor (2009) suggested that sterol and tocopherol compositions are the most confidential markers for adulteration, whereas fatty acid composition along with *trans*-isomers content may assist to evaluate the authenticity of pumpkin seed oils. Due to its simplicity, many authors commonly attempted to use fatty acid compositions of oils for detection of adulterations. However, the similarities in fatty acid profile of oils and the effects of environmental conditions on fatty acid profile may cause misinterpretations (Wenzl, Prettner, Schweiger, & Wagner, 2002). Hence, researchers have been investigating possible techniques that may be combined with fatty acid identification of oils. In accordance with this purpose, Potočnik, Ogrinc, Potočnik, and Košir (2016) successfully applied stable isotope ratio analysis accompanied by fatty acid characterization for verification of authenticity and geographical origin of pumpkin seed oils. The authors stated that the combination of fatty acid compositions and their $\delta^{13}C$ values is a reliable indicator of authenticity of pumpkin seed oils. On the other hand, stereospecific analysis of triacylglycerols can be also coupled with other chemical parameters such as fatty acid or tocopherol profile, to track adulteration. The genuineness of pumpkin seed oils produced in Slovenia was assessed based on presence and quantity of nine main triacylglycerols (LLLn, LLL, PLL, LOO, PLO, OOO, SPL, POO, and SLS). Additionally, the authors emphasized that roasting processes affect the triacylglycerol profile of oils (Butinar, Bučar-Miklavčič, Valenčič, & Raspor, 2010). Therefore, cold pressed pumpkin seed oils have unique chemical properties.

Similar to fatty acids, sterol composition can be used as an effective marker to evaluate the authenticity and adulteration of edible oils. The content of δ5-sterols, including stigmasterol, campesterol, and β-sitosterol, which is generally found in trace amounts in pumpkin seed oil, is considered a useful marker to detect the adulteration of the cold pressed pumpkin seed oils (Dulf, Bele, Ungureşan, Parlog, & Socaciu, 2009).

Determination of minerals can be an alternative approach for classification of edible oils. A study showed that pumpkin seed oils contain very low levels of potassium and nickel, while higher amounts of magnesium and calcium are present. These indicators may serve as an aid to track adulteration of pumpkin seed oil (Cindric, Zeiner, & Steffan, 2007). Joebstl, Bandoniene, Meisel, and Chatzistathis (2010) used rare earth elements patterns to identify geographical origin of pumpkin seed oils. In addition to adulteration detection, the elemental profile provides valuable insight on chemical stability of oils as well as demonstrating heavy metal contamination. Cold pressed pumpkin seed oil contains sodium (1.85–13.2 mg/kg), calcium (47.7–65.5 mg/kg), magnesium (20.0–38.5 mg/kg), potassium (2.31–28.6 mg/kg), phosphorus (0.32–67.0 mg/kg), aluminum (0.56 mg/kg), iron (3.74–14.1 mg/kg), zinc (0.31–9.88 mg/kg), tin (0.52 mg/kg), sulfur (1.08–6.07 mg/kg), copper (0.66–0.99 mg/kg), manganese (0.19–0.21 mg/kg), and selenium (0.047–0.058 mg/kg) (Aktaş et al., 2018; İmer & Taşan, 2018). The existence of copper and iron has been linked with oil oxidation (Benjellourr, Talou, Delmas, & Gaset, 1991; Knothe & Dunn, 2003). The maximum limits for these prooxidant metals are different depending on the type of oil. The maximum levels of iron and copper should be 5 mg/kg and 0.4 mg/kg, respectively, in cold pressed oils, whereas refined, bleached, and deodorized oils should have iron content lower than 0.1 mg/kg and copper content less than 0.02 mg/kg for maintaining oxidative stability (Codex Alimentarius, 1981; Smouse, 1995). It is obvious that the refining process will decrease the amount of these prooxidant metals, yet the iron and copper levels are comparatively higher in cold pressed pumpkin seed oil in terms of oxidation. On the other side, as contaminants, maximum limits for only lead and arsenic in edible fats and oils were established by the Codex Alimentarius Commission.

Apart from the intentional adulteration, cross contamination may occur throughout the production of several vegetable oils. In Latvia, the presence of sesamolin (7.17 mg/100 g in 2011 and 2.49 mg/100 g in 2012) and sesamin (20.54 mg/100 g in 2011 and 7.22 mg/100 g in 2012) lignans in cold pressed pumpkin seed oil were revealed (Górnaś, Siger, Pugajeva, & Segliņa, 2014). These results indicated the improper and inadequate cleaning procedures carried out in plants. Contamination of oils with any allergic substances is a serious health concern.

The most widely distributed environmental carcinogens are polycyclic aromatic hydrocarbons (PAH) (Harrison, 2001). Foods may be contaminated with PAHs through combustion of fossil fuels such as coal, oil, wood, or car exhaust emissions, as well as various food processes like cooking or smoking (Park & Penning, 2009). Due to their lipophilic character, vegetable oils are one of the most important source of PAHs (Shibamoto & Bjeldanes, 2009). Phenanthrene (13.6 μg/kg), fluoranthene (7.32 μg/kg), naphtalene (3.98 μg/kg), pyrene (3.91 μg/kg), fluorine (3.19 μg/kg), acenapthylene and acenaphtene (0.88 μg/kg), anthracene (0.70 μg/kg), benzo(b)fluoranthene (0.67 μg/kg), benzo(a)anthracene (0.44 μg/kg), benzo(a)pyrene (0.42 μg/kg), benzo[*g,h,i*]perylene (0.40 μg/kg), dibenz[*a,h*]anthracene (0.35 μg/kg), benzo(k)fluoranthene (0.28 μg/kg), and indeno[1,2,3-*cd*]pyrene (0.06 μg/kg) are the detected PAHs in cold pressed pumpkin seed oil (Roszko, Szterk, Szymczyk, & Waszkiewicz-Robak, 2012).

Polychlorinated biphenyls (PCB), including 209 congeners, are synthetic chemical pollutants distributed in the environment by human activity. Their high thermal and chemical stability turn the PCBs into a severe environmental concern (Arnold & Feeley, 2003; Schrenk & Chopra, 2017). PCBs are regarded as a probable carcinogen (La Rocca & Mantovani, 2006). Six nondioxin-like PCBs, namely congeners 28 (59.2 pg/g), 52 (25.9 pg/g), 101 (9.90 pg/g), 138 (21.1 pg/g), 153 (6.40 pg/g), and 180 (4.60 pg/g), and one dioxin-like PCB, namely congener 77 (1 pg/g), were identified in cold pressed pumpkin seed oil (Roszko et al., 2012).

Despite the popularity of pesticides in conventional farming, concerns have been raised about the health hazards of pesticide residues (Damalas & Eleftherohorinos, 2011). Pesticide exposure has been associated with reproductive, endocrine, immune, and nervous system disorders, as well as the formation of various cancers like leukemia/lymphoma, and kidney, brain, pancreas, liver, prostate, breast, skin, and lung cancer (Gilden, Huffling, & Sattler, 2010). Roszko et al. (2012) detected the presence of DDT (0.02 mg/kg) and trifluralin (0.137 mg/kg) in cold pressed pumpkin seed oil. In a recent study, the existence of chlorpyrifos in the range of 0.01–0.018 mg/kg in cold pressed oils of Turkish pumpkin seed varieties was confirmed (Arslan et al., 2017). The Codex Alimentarius Commission (CAC) has established maximum residue limits (MRLs) for pesticides in various plant and animal origin food commodities. The group of vegetable oils was divided into two subgroups by the CAC: crude vegetable oils and edible (refined) vegetable oils. However, data for crude vegetable oils are extremely limited. Only the MRLs of some pesticides has been determined in virgin olive oil, crude rapeseed oil, crude soybean oil, crude maize oil, crude cotton seed oil, crude linseed oil, crude peanut oil, and crude sunflower seed oil. Moreover, MRLs of different pesticides have been established in many edible (refined) oils. Nonetheless, no MRL for DDT in any crude vegetable oil has been proposed up to now. On the other hand, MRLs of chlorpyrifos has been adopted for edible cotton seed oil (0.05 mg/kg), edible maize oil (0.2 mg/kg), and refined soybean oil (0.03 mg/kg) (FAO/WHO, 2003). Taking into account all of these, cold pressed pumpkin seed oil complies with the MRLs for chlorpyrifos established by the CAC for edible cotton seed oil, edible maize oil, and refined soybean oil.

References

Abou Seif, H. S. (2014). Ameliorative effect of pumpkin oil (*Cucurbita pepo* L.) against alcohol-induced hepatotoxicity and oxidative stress in albino rats. *Beni-Suef University Journal of Basic and Applied Sciences*, *3*, 178–185.

Abou-Zeid, S. M., AbuBakr, H. O., Mohamed, M. A., & El-Bahrawy, A. (2018). Ameliorative effect of pumpkin seed oil against emamectin induced toxicity in mice. *Biomedicine & Pharmacotherapy*, *98*, 242–251.

Akin, G., Arslan, F. N., Karuk Elmas, S. N., & Yilmaz, I. (2018). Cold-pressed pumpkin seed (*Cucurbita pepo* L.) oils from the central Anatolia region of Turkey: Characterization of phytosterols, squalene, tocols, phenolic acids, carotenoids and fatty acid bioactive compounds. *Grasas y Aceites*, *69*, 232.

Aktaş, N., Gerçekaslan, K. E., & Uzlaşır, T. (2018). The effect of some pre-roasting treatments on quality characteristics of pumpkin seed oil. *OCL—Oilseeds fats, Crop and Lipids*, *25*, A301.

Al Zuhair, H., Abd El-Fattah, A. A., & El-Sayed, M. I. (2000). Pumpkin-seed oil modulates the effects of felodipine and captopril in spontaneously hypertensive rats. *Pharmacological Research*, *41*, 555–563.

Ali, D. M., & Abdelzaher, W. Y. (2017). Possible protective effect of pumpkin seed oil against sodium nitrite in rats; a biochemical and genetic study. *International Journal of Clinical Pharmacology & Toxicology*, *6*, 262–269.

Al-Masri, S. A. (2015). Effect of pumpkin oil and vitamin E on lead induced testicular toxicity in male rats. *Journal of Animal and Plant Sciences*, *25*, 72–77.

Antunović, Z., Klir, Ž., Šperanda, M., Sičaja, V., Čolović, D., Mioč, B., et al. (2018). Partial replacement of soybean meal with pumpkin seed cake in lamb diets: Effects on carcass traits, haemato-chemical parameters and fatty acids in meat. *South African Journal of Animal Science*, *48*, 695–704.

Arnold, D. L., & Feeley, M. (2003). Polychlorinated biphenyls. In J. P. F. D'Mello (Ed.), *Food safety: Contaminants and toxins* (pp. 125–152). Wallingford, UK: CABI Publishing.

Arslan, F. N., Akın, G., & Yılmaz, İ. (2017). Physicochemical characteristics, pesticide residue and aflatoxin contamination of cold pressed pumpkin seed (*Cucurbita pepo* L.) oils from Central Anatolia region of Turkey. *Anadolu University Journal of Science and Technology A—Applied Sciences and Engineering*, *18*, 468–483.

Bardaa, S., Ben Halima, N., Aloui, F., Ben Mansour, R., Jabeur, H., Bouaziz, M., et al. (2016). Oil from pumpkin (*Cucurbita pepo* L.) seeds: Evaluation of its functional properties on wound healing in rats. *Lipids in Health and Disease*, *15*, 73.

Bardaa, S., Moalla, D., Ben Khedir, S., Rebai, T., & Sahnoun, Z. (2016). The evaluation of the healing proprieties of pumpkin and linseed oils on deep second-degree burns in rats. *Pharmaceutical Biology*, *54*, 581–587.

Başoğlu, F. (2014). *Yemeklik Yağ Teknolojisi* (4th ed.). Bursa, Turkey: Dora Yayıncılık.

Benjellourr, B., Talou, T., Delmas, M., & Gaset, A. (1991). Oxidation of rapeseed oil: Effect of metal traces. *Journal of the American Oil Chemists' Society*, *68*, 210–211.

Bharti, S. K., Kumar, A., Sharma, N. K., Prakash, O., Jaiswal, S. K., Krishnan, S., et al. (2013). Tocopherol from seeds of *Cucurbita pepo* against diabetes: Validation by in vivo experiments supported by computational docking. *Journal of the Formosan Medical Association*, *112*, 676–690.

Broznić, D., Čanadi Jurešić, G., & Milin, Č. (2016). Involvement of α-, γ-, and δ-tocopherol isomers in the biphasic DPPH• disappearance kinetics of pumpkin (*Cucurbita pepo* L.) seed oil or oil mixtures. *Food Technology and Biotechnology*, *54*, 200–210.

Bučko, S., Katona, J., Popović, L., Vaštag, Ž., Petrović, L., & Vučinić–Vasić, M. (2015). Investigation on solubility, interfacial and emulsifying properties of pumpkin (*Cucurbita pepo*) seed protein isolate. *LWT—Food Science and Technology*, *64*, 609–615.

Burdock, G. A. (2010). *Fenaroli's handbook of flavor ingredients* (6th ed.). Boca Raton, FL: CRC Press.

Butinar, B., Bučar-Miklavčič, M., Krumpak, A., & Raspor, P. (2009). Experiences in olive oil purity and quality assessment as a tool for pumpkin seed oil evaluation. What can consumers benefit? *Acta Alimentaria*, *38*, 219–227.

Butinar, B., Bučar-Miklavčič, M., Valenčič, V., & Raspor, P. (2010). Stereospecific analysis of triacylglycerols as a useful means to evaluate genuineness of pumpkin seed oils: Lesson from virgin olive oil analyses. *Journal of Agricultural and Food Chemistry*, *58*, 5227–5234.

Caili, F., Huan, S., & Quanhong, L. (2006). A review on pharmacological activities and utilization technologies of pumpkin. *Plant Foods for Human Nutrition*, *61*, 70–77.

Casoni, D., Simion, I. M., & Sârbu, C. (2019). A comprehensive classification of edible oils according to their radical scavenging spectral profile evaluated by advanced chemometrics. *Spectrochimica Acta Part A: Molecular and Biomolecular Spectroscopy*, *213*, 204–209.

Cho, Y. H., Lee, S. Y., Jeong, D. W., Choi, E. J., Kim, Y. J., Lee, J. G., et al. (2014). Effect of pumpkin seed oil on hair growth in men with androgenetic alopecia: A randomized, double-blind, placebo-controlled trial. *Evidence-Based Complementary and Alternative Medicine*, *2014*, 549721.

Choe, E. (2017). Effects and mechanisms of minor compounds in oil on lipid oxidation. In C. C. Akoh (Ed.), *Food lipids: Chemistry, nutrition, and biotechnology* (pp. 567–588). Boca Raton, FL: CRC Press.

Cindric, I. J., Zeiner, M., & Steffan, I. (2007). Trace elemental characterization of edible oils by ICP–AES and GFAAS. *Microchemical Journal*, *85*, 136–139.

Codex Alimentarius (1981). Standard for edible fats and oils not covered by individual standards. *Codex Stan*, *19*, 1–4.

Codex Alimentarius (1999). Standard for named vegetable oils. *Codex Stan*, *210*, 1–13.

Damalas, C. A., & Eleftherohorinos, I. G. (2011). Pesticide exposure, safety issues, and risk assessment indicators. *International Journal of Environmental Research and Public Health*, *8*, 1402–1419.

Dulf, F. V., Bele, C., Ungureşan, M., Parlog, R., & Socaciu, C. (2009). Phytosterols as markers in identification of the adulterated pumpkin seed oil with sunflower oil. *Bulletin of University of Agricultural Sciences and Veterinary Medicine Cluj-Napoca Agriculture*, *66*, 301–307.

Elfiky, S. A., Elelaimy, I. A., Hassan, A. M., Ibrahim, H. M., & Elsayad, R. I. (2012). Protective effect of pumpkin seed oil against genotoxicity induced by azathioprine. *The Journal of Basic & Applied Zoology*, *65*, 289–298.

Elias, R. J., & Decker, E. A. (2017). Antioxidants and their mechanisms of action. In C. C. Akoh (Ed.), *Food lipids: Chemistry, nutrition, and biotechnology* (pp. 543–566). Boca Raton, FL: CRC Press.

El-Mosallamy, A. E. M. K., Sleem, A. A., Abdel-Salam, O. M. E., Shaffie, N., & Kenawy, S. A. (2012). Antihypertensive and cardioprotective effects of pumpkin seed oil. *Journal of Medicinal Food*, *15*, 180–189.

Eraslan, G., Kanbur, M., Aslan, Ö., & Karabacak, M. (2013). The antioxidant effects of pumpkin seed oil on subacute aflatoxin poisoning in mice. *Environmental Toxicology*, *28*, 681–688.

FAO/WHO (2003). *Codex pesticides residues in food online database-chlorpyrifos [WWW Document]*. http://www.fao.org/fao-who-codexalimentarius/codex-texts/dbs/pestres/pesticide-detail/en/?p_id=17 (accessed 6.3.19).

Ferriol, M., & Picó, B. (2008). Pumpkin and winter squash. In J. Prohens, & F. Nuez (Eds.), *Vegetables I: Asteraceae, Brassicaceae, Chenopodicaceae, and Cucurbitaceae* (pp. 317–349). New York, NY: Springer Science+Business Media.

Filbrandt, K. R. (2012). *Effect of pumpkin seed oil cake on the textural and sensory properties of white wheat bread*. University of Wisconsin-Stout.

Finley, J. W., & DeMan, J. M. (2018). Lipids. In J. M. DeMan, J. W. Finley, W. J. Hurst, & C. Y. Lee (Eds.), *Principles of food chemistry* (pp. 39–116). Cham, Switzerland: Springer International Publishing AG.

Fruhwirth, G. O., & Hermetter, A. (2007). Seeds and oil of the Styrian oil pumpkin: Components and biological activities. *European Journal of Lipid Science and Technology*, *109*, 1128–1140.

Fruhwirth, G. O., & Hermetter, A. (2008). Production technology and characteristics of *Styrian pumpkin* seed oil. *European Journal of Lipid Science and Technology*, *110*, 637–644.

Gilden, R. C., Huffling, K., & Sattler, B. (2010). Pesticides and health risks. *Journal of Obstetric, Gynecologic, and Neonatal Nursing*, *39*, 103–110.

Glew, R. H., Glew, R. S., Chuang, L.-T., Huang, Y.-S., Millson, M., Constans, D., et al. (2006). Amino acid, mineral and fatty acid content of pumpkin seeds (*Cucurbita* spp.) and *Cyperus esculentus* nuts in the Republic of Niger. *Plant Foods for Human Nutrition*, *61*, 49–54.

Gorjanović, S. Ž., Rabrenović, B. B., Novaković, M. M., Dimić, E. B., Basić, Z. N., & Sužnjević, D. Ž. (2011). Cold-pressed pumpkin seed oil antioxidant activity as determined by a DC polarographic assay based on hydrogen peroxide scavenge. *Journal of the American Oil Chemists' Society*, *88*, 1875–1882.

Górnaś, P., Siger, A., Pugajeva, I., & Segliņa, D. (2014). Sesamin and sesamolin as unexpected contaminants in various cold-pressed plant oils: NP-HPLC/FLD/DAD and RP-UPLC-ESI/MS n study. *Food Additives & Contaminants Part A, Chemistry, Analysis, Control, Exposure & Risk Assessment*, *31*, 567–573.

Gossell-Williams, M., Davis, A., & O'Connor, N. (2006). Inhibition of testosterone-induced hyperplasia of the prostate of sprague-dawley rats by pumpkin seed oil. *Journal of Medicinal Food*, *9*, 284–286.

Gossell-Williams, M., Hyde, C., Hunter, T., Simms-Stewart, D., Fletcher, H., McGrowder, D., et al. (2011). Improvement in HDL cholesterol in postmenopausal women supplemented with pumpkin seed oil: Pilot study. *Climacteric*, *14*, 558–564.

Greiling, A. M., Schwarz, C., Gierus, M., & Rodehutscord, M. (2018). Pumpkin seed cake as a fishmeal substitute in fish nutrition: Effects on growth performance, morphological traits and fillet colour of two freshwater salmonids and two catfish species. *Archives of Animal Nutrition*, *72*, 239–259.

Harrison, N. (2001). Environmental organic contaminants in food. D. H. Watson (Ed.), *Contaminants: Vol. 1. Food chemical safety* (pp. 169–192). Cambridge, UK: Woodhead Publishing Limited

Heine, R. J., Mulder, C., Popp-Snijders, C., van der Meer, J., & van der Veen, E. A. (1989). Linoleic-acid-enriched diet: Long-term effects on serum lipoprotein and apolipoprotein concentrations and insulin sensitivity in noninsulin-dependent diabetic patients. *The American Journal of Clinical Nutrition*, *49*, 448–456.

Horrobin, D. F., & Huang, Y.-S. (1987). The role of linoleic acid and its metabolites in the lowering of plasma cholesterol and the prevention of cardiovascular disease. *International Journal of Cardiology*, *17*, 241–255.

İmer, Y., & Taşan, M. (2018). Determination of some micro and macronutrient elements in various cold press vegetable oils. *Journal of Tekirdag Agricultural Faculty*, *15*, 14–25 (in Turkish).

Jeznach, M., Danilcenko, H., Jariene, E., Kulaitiene, J., & Černiauskiene, J. (2012). Accumulation of antioxidative vitamins and minerals in seeds of oil pumpkin (*Cucurbita pepo* L. var. styriaca) cultivars. *Journal of Food, Agriculture and Environment*, *10*, 245–247.

Joebstl, D., Bandoniene, D., Meisel, T., & Chatzistathis, S. (2010). Identification of the geographical origin of pumpkin seed oil by the use of rare earth elements and discriminant analysis. *Food Chemistry*, *123*, 1303–1309.

Kamm, W., Dionisi, F., Hischenhuber, C., & Engel, K.-H. (2001). Authenticity assessment of fats and oils. *Food Review International*, *17*, 249–290.

Kim, S.-K., & Karadeniz, F. (2012). Biological importance and applications of squalene and squalane. In S.-K. Kim (Ed.), *Advances in food and nutrition research volume 65—Implications and applications: Animals and microbes* (pp. 223–233). Waltham, MA, USA: Academic Press.

Kim, M. Y., Kim, E. J., Kim, Y.-N., Choi, C., & Lee, B.-H. (2012). Comparison of the chemical compositions and nutritive values of various pumpkin (*Cucurbitaceae*) species and parts. *Nutrition Research and Practice*, *6*, 21–27.

Klippel, K. F., Hiltl, D. M., & Schipp, B. (1997). A multicentric, placebo-controlled, double-blind clinical trial of beta-sitosterol (phytosterol) for the treatment of benign prostatic hyperplasia. *BJU International*, *80*, 427–432.

Klir, Z., Castro-Montoya, J. M., Novoselec, J., Molkentin, J., Domacinovic, M., Mioc, B., et al. (2017). Influence of pumpkin seed cake and extruded linseed on milk production and milk fatty acid profile in Alpine goats. *Animal*, *11*, 1772–1778.

Knothe, G., & Dunn, R. O. (2003). Dependence of oil stability index of fatty compounds on their structure and concentration and presence of metals. *Journal of the American Oil Chemists' Society*, *80*, 1021–1026.

Kochhar, S. P. (2011). Minor and speciality oils. In F. D. Gunstone (Ed.), *Vegetable oils in food technology-composition, properties and uses* (pp. 291–341). Oxford, UK: Wiley-Blackwell.

Konopka, I., Roszkowska, B., Czaplicki, S., & Tańska, M. (2016). Optimization of pumpkin oil recovery by using aqueous enzymatic extraction and comparison of the quality of the obtained oil with the quality of cold-pressed oil. *Food Technology and Biotechnology*, *54*, 413–420.

Krimer-Malesevic, V., Posa, M., Stolic, Z., Popovic, L., & Pericin-Starcevic, I. (2016). Determination of phenolic acids in seeds of black cumin, flax, pomegranate and pumpkin and their by-products. *Acta Periodica Technologica*, *265*, 19–28.

Kulaitienė, J., Černiauskienė, J., Jarienė, E., Danilčenko, H., & Levickienė, D. (2018). Antioxidant activity and other quality parameters of cold pressing pumpkin seed oil. *Notulae Botanicae Horti Agrobotanici Cluj-Napoca*, *46*, 161–166.

La Rocca, C., & Mantovani, A. (2006). From environment to food: The case of PCB. *Annali dell'Istituto Superiore di Sanità*, *42*, 410–416.

Landeka, I., Đikić, D., Radišić, I., Teparić, R., & Bačun-Družina, V. (2011). The effects of olive and pumpkin seed oil on serum lipid concentrations. *Croatian Journal of Food Technology, Biotechnology and Nutrition*, *6*, 63–68.

Louheranta, A. M., Porkkala-Sarataho, E. K., Nyyssönen, M. K., Salonen, R. M., & Salonen, J. T. (1996). Linoleic acid intake and susceptibility of very-low-density and low density lipoproteins to oxidation in men. *The American Journal of Clinical Nutrition*, *63*, 698–703.

Makni, M., Sefi, M., Fetoui, H., Garoui, E. M., Gargouri, N. K., Boudawara, T., et al. (2010). Flax and pumpkin seeds mixture ameliorates diabetic nephropathy in rats. *Food and Chemical Toxicology*, *48*, 2407–2412.

Mansour, E. H., Dworschák, E., Lugasi, A., Barna, É., & Gergely, A. (1993). Nutritive value of pumpkin (*Cucurbita pepo* Kakai 35) seed products. *Journal of the Science of Food and Agriculture*, *61*, 73–78.

Marangoni, F., & Poli, A. (2010). Phytosterols and cardiovascular health. *Pharmacological Research*, *61*, 193–199.

Martínez, Y., Valdivie, M., Martínez, O., Estarrón, M., & Córdova, J. (2010). Utilization of pumpkin (*Cucurbita moschata*) seed in broiler chicken diets. *Cuban Journal of Agricultural Sciences*, *44*, 387–392.

Matsui, T., Guth, H., & Grosch, W. (1998). A comparative study of potent odorants in peanut, hazelnut, and pumpkin seed oils on the basis of aroma extract dilution analysis (AEDA) and gas chromatography-olfactometry of headspace samples (GCOH). *Fett-Lipid*, *100*, 51–56.

McClements, D. J., & Decker, E. A. (2017). Lipids. In S. Damodaran, & K. L. Parkin (Eds.), *Fennema's food chemistry* (pp. 171–233). Boca Raton, FL: CRC Press.

Murkovic, M. (2009). Pumpkin seed oil. In R. A. Moreau, & A. Kamal-Eldin (Eds.), *Gourmet and health-promoting specialty oils* (pp. 345–358). Urbana, IL: AOCS Press.

Murkovic, M., Hillebrand, A., Winkler, J., Leitner, E., & Pfannhauser, W. (1996). Variability of fatty acid content in pumpkin seeds (*Cucurbita pepo* L.). *Zeitschrift für Lebensmittel-Untersuchung und -Forschung*, *203*, 216–219.

Naziri, E., Mitić, M. N., & Tsimidou, M. Z. (2016). Contribution of tocopherols and squalene to the oxidative stability of cold-pressed pumkin seed oil (*Cucurbita pepo* L.). *European Journal of Lipid Science and Technology*, *118*, 898–905.

Nederal, S., Škevin, D., Kraljić, K., Obranović, M., Papeša, S., & Bataljaku, A. (2012). Chemical composition and oxidative stability of roasted and cold pressed pumpkin seed oils. *Journal of the American Oil Chemists' Society*, *89*, 1763–1770.

O'Keefe, S. F., & Sarnoski, P. J. (2017). Nomenclature and classification of lipids. In C. C. Akoh (Ed.), *Food lipids—Chemistry, nutrition, and biotechnology* (pp. 1–36). Boca Raton, FL: CRC Press.

Omar, N. M., & Sarhan, N. R. (2017). The possible protective role of pumpkin seed oil in an animal model of acid aspiration pneumonia: Light and electron microscopic study. *Acta Histochemica*, *119*, 161–171.

Park, J.-H., & Penning, T. M. (2009). Polyaromatic hydrocarbons. In R. H. Stadler, & D. R. Lineback (Eds.), *Process-induced food toxicants: Occurrence, formation, mitigation, and health risks* (pp. 243–282). Hoboken, NJ, USA: John Wiley & Sons, Inc.

Patel, S. (2013). Pumpkin (*Cucurbita* sp.) seeds as nutraceutic: A review on status quo and scopes. *Mediterranean Journal of Nutrition and Metabolism*, *6*, 183–189.

Peričin, D., Krimer, V., Trivić, S., & Radulović, L. (2009). The distribution of phenolic acids in pumpkin's hull-less seed, skin, oil cake meal, dehulled kernel and hull. *Food Chemistry*, *113*, 450–456.

Pericin, D., Madjarev, S., Radulovic, L. M., & Skrinjar, M. (2007). Production of exo-pectinase by *Penicillium roqueorti* using pumpkin oil cake. *Zbornik Matice Srpske za Prirodne Nauke*, *113*, 313–320.

Pitrat, M., Chauvet, M., & Foury, C. (1999). Diversity, history and production of cultivated cucurbits. *Acta Horticulturae*, *21-28*.

Poiana, M.-A., Alexa, E., Moigradean, D., & Popa, M. (2009). The influence of the storage conditions on the oxidative stability and antioxidant properties of sunflower and pumpkin oil. In S. Maric, & Z. Loncaric (Eds.), *Proceedings of 44th Croatian and 4th international symposium on agriculture* (pp. 449–453). Croatia: Opatija.

Popović, S., Lazić, V., Popović, L., Vaštag, Ž., & Peričin, D. (2010). Effect of the addition of pumpkin oil cake to gelatin to produce biodegradable composite films. *International Journal of Food Science and Technology*, *45*, 1184–1190.

Popović, S., Peričin, D., Vaštag, Ž., Lazić, V., & Popović, L. (2012). Pumpkin oil cake protein isolate films as potential gas barrier coating. *Journal of Food Engineering*, *110*, 374–379.

Popović, L., Peričin, D., Vaštag, Ž., Popović, S., Krimer, V., & Torbica, A. (2013). Antioxidative and functional properties of pumpkin oil cake globulin hydrolysates. *Journal of the American Oil Chemists' Society*, *90*, 1157–1165.

Popović, S., Peričin, D., Vaštag, Ž., Popović, L., & Lazić, V. (2011). Evaluation of edible film-forming ability of pumpkin oil cake; effect of pH and temperature. *Food Hydrocolloids*, *25*, 470–476.

Popović, L., Stolić, Ž., Čakarević, J., Torbica, A., Tomić, J., & Šijački, M. (2017). Biologically active digests from pumpkin oil cake protein: Effect of cross-linking by transglutaminase. *Journal of the American Oil Chemists' Society*, *94*, 1245–1251.

Potočnik, T., & Košir, I. J. (2017). Influence of roasting temperature of pumpkin seed on PAH and aroma formation. *European Journal of Lipid Science and Technology*, *119*, 1500593.

Potočnik, T., Ogrinc, N., Potočnik, D., & Košir, I. J. (2016). Fatty acid composition and δ 13 C isotopic ratio characterisation of pumpkin seed oil. *Journal of Food Composition and Analysis*, *53*, 85–90.

Rabrenović, B. B., Dimić, E. B., Novaković, M. M., Tešević, V. V., & Basić, Z. N. (2014). The most important bioactive components of cold pressed oil from different pumpkin (*Cucurbita pepo* L.) seeds. *LWT—Food Science and Technology*, *55*, 521–527.

Radocaj, O., Dimic, E., & Vujasinovic, V. (2011). Optimization of the texture of fat-based spread containing hull-less pumpkin (*Cucurbita pepo* L.) seed press-cake. *Acta Periodica Technologica*, *42*, 131–143.

Richter, D., Abarzua, S., Chrobak, M., Vrekoussis, T., Weissenbacher, T., Kuhn, C., et al. (2013). Effects of phytoestrogen extracts isolated from pumpkin seeds on estradiol production and ER/PR expression in breast cancer and trophoblast tumor cells. *Nutrition and Cancer*, *65*, 739–745.

Roszko, M., Szterk, A., Szymczyk, K., & Waszkiewicz-Robak, B. (2012). PAHs, PCBs, PBDEs and pesticides in cold-pressed vegetable oils. *Journal of the American Oil Chemists' Society*, *89*, 389–400.

Schrenk, D., & Chopra, M. (2017). Dioxins and polychlorinated biphenyls in foods. In D. Schrenk, & A. Cartus (Eds.), *Chemical contaminants and residues in food* (pp. 69–89). Duxford, UK: Woodhead Publishing.

Sharma, P. B., Lal, B. M., Madaan, T. R., & Chatterjee, S. R. (1986). Studies on the nutritional quality of some cucurbit kernel proteins. *Journal of the Science of Food and Agriculture*, *37*, 418–420.

Shibamoto, T., & Bjeldanes, L. (2009). *Introduction to food toxicology*. Burlington, MA: Academic Press.

Sicilia, T., Niemeyer, H. B., Honig, D. M., & Metzler, M. (2003). Identification and stereochemical characterization of lignans in flaxseed and pumpkin seeds. *Journal of Agricultural and Food Chemistry*, *51*, 1181–1188.

Siegmund, B., & Murkovic, M. (2004). Changes in chemical composition of pumpkin seeds during the roasting process for production of pumpkin seed oil (Part 2: Volatile compounds). *Food Chemistry*, *84*, 367–374.

Siger, A., Nogala-Kalucka, M., & Lampart-Szczapa, E. (2008). The content and antioxidant activity of phenolic compounds in cold-pressed plant oils. *Journal of Food Lipids*, *15*, 137–149.

Singh, K., Singh, B., Chand, P., & Nair, R. (2016). Cucurbits: Assortment and therapeutic values. In M. Pessarakli (Ed.), *Handbook of cucurbits—Growth, cultural practices, and physiology* (pp. 405–416). Boca Raton, FL: CRC Press.

Smouse, T. H. (1995). Factors affecting oil quality and stability. In K. Warner, & N. A. M. Eskin (Eds.), *Methods to assess quality and stability of oils and fat-containing foods* (pp. 17–36). Champaign, Illinois: AOCS Press.

Tuberoso, C. I. G., Kowalczyk, A., Sarritzu, E., & Cabras, P. (2007). Determination of antioxidant compounds and antioxidant activity in commercial oilseeds for food use. *Food Chemistry*, *103*, 1494–1501.

Vaštag, Ž., Popović, L., & Popović, S. (2014). Bioactivity evaluation of cucurbitin derived enzymatic hydrolysates. *International Journal of Agricultural and Biosystems Engineering*, *8*, 445–448.

Vaštag, Ž., Popović, L., Popović, S., Krimer, V., & Peričin, D. (2011). Production of enzymatic hydrolysates with antioxidant and angiotensin-I converting enzyme inhibitory activity from pumpkin oil cake protein isolate. *Food Chemistry*, *124*, 1316–1321.

Vujasinovic, V., Djilas, S., Dimic, E., Basic, Z., & Radocaj, O. (2012). The effect of roasting on the chemical composition and oxidative stability of pumpkin oil. *European Journal of Lipid Science and Technology*, *114*, 568–574.

Vujasinovic, V., Djilas, S., Dimic, E., Romanic, R., & Takaci, A. (2010). Shelf life of cold-pressed pumpkin (*Cucurbita pepo* L.) seed oil obtained with a screw press. *Journal of the American Oil Chemists' Society*, *87*, 1497–1505.

Vukša, V., Dimić, E., & Dimić, V. (2003). Characteristics of cold pressed pumpkin seed oil. In R. Schubert, G. Flachowsky, G. Jahreis, & R. Bitsch (Eds.), *9th symposium micronutrients 2003: Vitamins and additives in the nutrition of man and animal* (pp. 493–496). Germany: Jena/Thüringen.

Wafar, R., Hannison, M., Abdullahi, U., & Makinta, A. (2017). Effect of pumpkin (*Cucurbita pepo* L.) seed meal on the performance and carcass characteristics of broiler chickens. *Asian Journal of Advances in Agricultural Research*, *2*, 1–7.

Weng, Y., & Sun, Z. (2012). Major cucurbit crops. In H.-Y. Wang, T. K. Behera, & C. Kole (Eds.), *Genetics, genomics and breeding of cucurbits* (pp. 1–16). Enfield, USA: Science Publishers.

Wenzl, T., Prettner, E., Schweiger, K., & Wagner, F. S. (2002). An improved method to discover adulteration of *Styrian pumpkin* seed oil. *Journal of Biochemical and Biophysical Methods*, *53*, 193–202.

Yadav, M., Jain, S., Tomar, R., Prasad, G. B. K. S., & Yadav, H. (2010). Medicinal and biological potential of pumpkin: An updated review. *Nutrition Research Reviews*, *23*, 184–190.

Zucker, H., Hays, V. W., Speer, V. C., & Catron, D. V. (1958). Evaluation of pumpkin seed meal as a source of protein for swine using a depletion-repletion technique. *The Journal of Nutrition*, *65*, 327–334.

Chapter 19

Cold pressed poppy seed oil

Zeynep Aksoylu Özbek and Pelin Günç Ergönül
Department of Food Engineering, Faculty of Engineering, Manisa Celal Bayar University, Manisa, Turkey

Abbreviations

DDVP	2,2-dichlorovinyl dimethyl phosphate
DPPH	1,1-diphenyl-2-picrylhydrazyl
LDL	low-density lipoprotein
LLL	trilinolein
LLLn	dilinoleolinolenin
LLnP	lineolinolenopalmitin
LLP	dilinoleopalmitin
LnLnLn	trilinolenoin
MUFA	monounsaturated fatty acid
OLL	oleodilinolein
OLP	oleolinoleopalmitin
OOL	dioleolinolein
OOP	dioleopalmitin
PAH	polycyclic aromatic hydrocarbon
PANH	polycyclic aromatic nitrogen hydrocarbon
PBDE	polybrominated biphenyl ether
PCB	polychlorinated biphenyl
POL	palmitooleolinolein
PUFA	polyunsaturated fatty acid
SOL	stearooleolinolein
SOO	stearodiolein
TAG	triacylglycerol

1 Introduction

Opium or oil poppy (*Papaver somniferum*) belongs to the Papaveraceae family and has been cultivated since Neolithic times (Küster, 2000). It was primarily grown in the Mediterranean region. Due to the high adaptative capability of the plant, its cultivation spread throughout temperate and tropical areas (Bernáth, 1999; Németh, 1999; Tétényi, 1997). The poppy serves as an important industrial crop because of its high content of alkaloids such as morphine, codeine, thebaine, narcotine, and papaverine, as well as being an ornamental plant. On the other hand, poppy seeds are a good source of edible oil and contain more than 30% oil (Levy & Milo, 1999; Özcan & Atalay, 2006). Today, the top producer countries of poppy seed are Czechia, Turkey, Spain, France, Hungary, Croatia, Germany, Slovakia, Austria, and Romania as shown in Table 1 (Food and Agriculture Organization of the United Nations, 2019).

The mean values for some physicochemical properties of cold pressed poppy seed oils are summarized in Table 2. Here, the most striking feature of the poppy seed oil is its relatively high iodine value. Poppy seed oil is classified as a drying oil and can be used in the production of oil paints and varnishes (Duke, 1973). Oil cakes, a by-product of edible oil processing, have a wide range of applications such as energy production, fertilizer, animal feed, and raw material for production of enzymes, vitamins, antibiotics, and biopesticides (Kachel-Jakubowska, Kraszkiewicz, & Krajewska, 2016). The studies have been concentrated on the use of poppy seed meal as animal feed. Dündar Emir et al. (2015) and Yilmaz and Dündar Emir (2016) suggested that the solvent-free poppy seed meal that emerges throughout cold pressing is an excellent protein source for human and animal nutrition. Up to now, incorporation of poppy seed meal into quail and laying hen

Cold Pressed Oils. https://doi.org/10.1016/B978-0-12-818188-1.00019-0
Copyright © 2020 Elsevier Inc. All rights reserved.

TABLE 1 The production quantities of leading poppy seed producer countries in 2017.

Country	Quantity (tonnes)
Czech	20,048
Turkey	15,244
Spain	11,900
France	5204
Germany	3104
Hungary	3000
Croatia	2681
Slovakia	1958
Romania	1849
Austria	1799

Adopted from Food and Agriculture Organization of the United Nations (2019). FAOSTAT [WWW Document]. http://www.fao.org/faostat/en/#data/QC/visualize (Accessed 4 September 2019).

TABLE 2 Physicochemical properties of cold pressed poppy seed oils.

Parameter	Value	Reference
Refractive index	1.474–1.476	Dündar Emir, Aydeniz, and Yılmaz (2015)
Viscosity (cP)	42.80–44.30	
Specific gravity (g/cm^3)	0.9194–0.9207	
Turbidity (NTU)	1.00–181.5	
Free fatty acids (% linoleic acid)	1.30–6.90	
Acid value (mg KOH/g)	0.35–4.82	Czaplicki, Ogrodowska, Derewiaka, Tańska, and Zadernowski (2011) and Symoniuk, Ratusz, Ostrowska-Ligęza, and Krygier (2018)
Peroxide value (mEq O_2/kg)	0.68–4.37	Czaplicki et al. (2011) and Dündar Emir et al. (2015)
P-Anisidine value	0.41–3.97	Czaplicki et al. (2011) and Symoniuk et al. (2018)
Totox value	2.45–4.77	Symoniuk et al. (2018)
Iodine number (g/100 g)	132.5–142	Dündar Emir et al. (2015)
Saponification number (mg KOH/g)	178.6–196	Dündar Emir et al. (2015)
Unsaponifiable matter (%)	0.48–1.02	Czaplicki et al. (2011) and Dündar Emir et al. (2015)
Total phenolics (mg GAE/100 g)	1.01–7.59	Dündar Emir et al. (2015)
Antioxidant capacity (mmol Trolox/g)	20.2–43.9	

rations and cattle and pig feed has been investigated (Akinci & Bayram, 2003; Annett & Sen, 1919; Bayram & Akıncı, 1998; Küçükersan, Yeşilbağ, & Küçükersan, 2009; Statham, 1984). However, Bayram, Akıncı Bozkurt, and Şehu (2005) stated that poppy seed meal levels higher than 15% in the laying hen rations decreased the yield and quality of egg as well as leading to liver problems in the animals. Therefore, poppy seed meal should be incorporated into animal feed formulations together with other oil cakes like soybean meal.

2 Extraction and processing of cold pressed oil

In the conventional cold pressing technique, oil is directly extracted from raw poppy seeds using a screw press at temperatures below 40°C. However, the main drawback of cold pressing is the lower yield compared to solvent extraction. In recent years, only a few attempts have been made to increase the oil recovery rate. Dündar Emir et al. (2015) reported that roasting at 150°C for 30 min or enzyme (hemicellulase and protease) treatment of poppy seeds before cold pressing increased oil yield.

3 Fatty acids and acyl lipids profile of cold pressed oil

Poppy seed oil is characterized by a high unsaturation degree. Cold pressed poppy seed oil is composed of palmitic (8.7%–11.2%), stearic (2.2%–13.9%), oleic (13.6%–19.6%), linoleic (62.8%–74.5%), and linolenic acids (0.5%–1.2%) (Bialek, Bialek, Jelinska, & Tokarz, 2016; Dündar Emir et al., 2015; Symoniuk et al., 2018). In terms of its fatty acid composition, poppy seed oil belongs to the oleic-linoleic fatty acid group along with sunflower oil, sesame oil, olive oil, corn oil, safflower oil, and palm oil (Başoğlu, 2014). Its low linolenic acid content improves the stability of the oil during storage (Özcan & Atalay, 2006). Although a maximum of 3% linolenic acid is desired by the food industry, lower than 1% linolenic acid is recommended for maintenance of acceptable flavor of any oil (Singh, Shukla, Khanna, Dixit, & Banerji, 1998). Hence, poppy seeds could be regarded as a suitable edible oil source.

Triacylglycerol (TAG) composition of the oil is commonly used for the detection of adulteration or authenticity as well as identification of geographical origin (Köseoğlu, Sevim, & Özdemir, 2017). Krist, Stuebiger, Unterweger, Bandion, and Buchbauer (2005) reported that cold pressed oils of different-colored poppy seeds (gray, white, and blue) include uniform TAG units principally composed of linoleic, oleic, and palmitic acids. LLL (19.1%–24.6%), LLP (16.3%–18.0%), OLL (13.2%–13.3%), OOL (7.3%–9.0%), OLP (7.2%–9.2%), SOL (1.8%–2.6%), OOP (1.0%–2.1%), LLnP (0.8%), LLLn (0.6%–0.8%), LnLnLn (0.3%–0.9%), SOO (0.4%–0.5%), and POL (0.3%–0.4%) were identified in cold pressed poppy seed oils (Krist et al., 2005). Chander (2010) stated that linoleic acid constituted approximately 72% of TAG composition of cold pressed poppy seed oil. The *sn*-1 position contained mostly palmitic acid and lower levels of linoleic acid (27%), while the *sn*-2 position involved 40% linoleic acid and 26% oleic acid and the *sn*-3 position was dominated by oleic acid (more than 50%).

4 Minor bioactive lipids

The main antioxidants of vegetable oils are tocopherols. Czaplicki et al. (2011) reported that cold pressed poppy seed oil contained significantly lower amounts of total tocopherols (292.7 mg/kg) than cold pressed oils of borage (1603 mg/kg), pumpkin seed (848.3 mg/kg), sesame (837 mg/kg), camelina (673.7 mg/kg), evening primrose (662.7 mg/kg), walnut (634.2 mg/kg), amaranth (617.9 mg/kg), and linseed (562.8 mg/kg). In contrast, Ayyildiz (2015) declared that cold pressed poppy seed oil contains total tocopherols (833.4 mg/kg) in amounts comparable to oils of safflower (917.3 mg/kg), walnut (878.6 mg/kg), canola (828.5 mg/kg), flaxseed (827.8 mg/kg), and more than extra virgin olive oil (249.8 mg/kg). On the other hand, it is reported (Ying, Wojciechowska, Siger, Kaczmarek, & Rudzińska, 2018) that storage of cold pressed poppy seed oil at high temperatures (60°C) reduced tocopherol amounts considerably (38% of them in the first 6 days and 64% in 12 days). γ-tocopherol (178–279 mg/kg) is the dominant isomer in the oil along with traces of α-tocopherol (3.0–91.7 mg/kg), γ-tocotrienol (2.10 mg/kg), and plastochromanol-8 (1.10–1.20 mg/kg) (Czaplicki et al., 2011; Dündar Emir et al., 2015; Górnaś et al., 2014; Gruszka & Kruk, 2007; Ying et al., 2018). In the conventional cold pressing method, the raw seeds are directly pressed. However, the feasibility of some pretreatments and their effects on quality characteristics of cold pressed oils have been evaluated in recent years. Dündar Emir et al. (2015) reported that roasting and enzyme pretreatments (hemicellulase and protease) affected the cold pressed oils of some poppy seed (blue, yellow, and white) varieties differently. While enzyme treatment caused dramatic reductions in the amounts of α-tocopherol (9.56%–42.1%), γ-tocopherol (0.69%–37.5%), and total tocopherol (2.95%–37.9%), roasting led to smaller decreases or increases in tocopherol levels of poppy seed oils (Dündar Emir et al., 2015).

The other biologically active components of edible oils, namely phytosterols, constitute the major part of the unsaponifiable fraction of poppy seed oils. Czaplicki et al. (2011) classified the cold pressed poppy seed oil as low-phytosterol oil together with camelina and evening primrose oils. Cycloartenol, campesterol, avenasterol, stigmasterol, and 24-methylene cycloartenol were identified with the β-sitosterol as the dominant sterol (Ying et al., 2018). In addition to these, high levelsof δ-5-avenasterol (114.2–266.2 ppm) and lower levels of 24-methylene-cholesterol (35.1–59.8 ppm), clerosterol

(21.7–27.6 ppm), δ-7-campesterol (3.62–14.6 ppm), δ-7-avenasterol (2.06–16.3 ppm), δ-5,24-stigmastadienol (0.83–13.8 ppm), δ-5,23-stigmastadienol (0–12.6 ppm), cholesterol (1.28–4.02 ppm), brassicasterol (0–1.81 ppm) and δ-7-stigmastadienol (0–1.19 ppm) were detected in cold pressed poppy seed oils (Dündar Emir et al., 2015).

The presence of various pigments that adsorb light, mainly chlorophyll and carotenoids, influences the color of oils (Wijaya, Wijaya, & Mehta, 2015). On the other hand, it should be remembered that chlorophyll concentrations higher than 50 μg/kg may trigger oil oxidation in the presence of light (Mag, 1990). Cold pressed poppy seed oil (0.17 mg/kg) is considered as a poor source of chlorophylls contrary to cold pressed oils of black cumin (62.7–67.1 mg/kg), pumpkin seed (6.04–30.5 mg/kg), evening primrose (14.4–15.9 mg/kg), milk thistle (3.93–6.46 mg/kg), hempseed (3.30–4.02 mg/kg), walnut (3.97 mg/kg), sunflower (0.09–1.73 mg/kg), hazelnut (1.38 mg/kg), peanut (0.69 mg/kg), almond (0.4 mg/kg), and flaxseed (0.35 mg/kg) (Górnaś, Siger, Juhņeviča, et al., 2014; Symoniuk et al., 2018). On the other side, another pigment group, carotenoids, enhance the antioxidative potential of food items. Hence, oils containing high levels of carotenoid isomers are preferable by both producers and consumers. In addition to being potent antioxidants, various carotenoids improve eye, heart, and skin health, the immune system, fertility, and brain-cognitive functions as well as protecting against certain cancer types (breast, ovarian, colorectal, cervical) and some chronic diseases such as obesity, diabetes, and cardiovascular diseases (Eggersdorfer & Wyss, 2018; Rodriguez-Concepcion et al., 2018). During refining, the majority of carotenoids are removed (up to 99%), but cold pressed oils do not require further purification treatments, and they thus serve as a rich natural source of carotenoids (Szydłowska-Czerniak, Trokowski, Karlovits, & Szłyk, 2011). However, as is the case with chlorophylls, cold pressed poppy seed oil involves trace amounts of carotenoids. Among all the carotenoid isomers, β-carotene is the most studied one at the time of writing. It is reported that cold pressed poppy seed oil (1.04 mg/kg) contains considerably lower amounts of β-carotene than cold pressed oils of Japanese quince (10.7 mg/kg), pumpkin seed (6.84 mg/kg), hazelnut (2.74 mg/kg), walnut (2.16 mg/kg), flaxseed (1.87 mg/kg), and sunflower (1.19 mg/kg), and higher β-carotene levels than cold pressed oils of sesame (0.37 mg/kg) and almond (0.49 mg/kg) (Górnaś, Siger, Juhņeviča, et al., 2014). Similarly, a recent study also showed that the carotenoid content of cold pressed poppy seed oil (2.17–2.32 mg/kg) is less than cold pressed oils of pumpkin seed (142.2–342.3 mg/kg), black cumin (28.3–40.5 mg/kg), linseed (18.4–38.7 mg/kg), hempseed (12.8–18.2 mg/kg), rapeseed (10.2–12.8 mg/kg), evening primrose (6.60–10.50 mg/kg), milk thistle (4.66–9.57 mg/kg), and sunflower (5.67–6.49 mg/kg) (Symoniuk et al., 2018).

The most well-known characteristic of biologically active substances is antioxidant activity. As listed above, cold pressed poppy seed oil is a poor source of tocopherols, carotenoids, and phytosterols. Hence, it was classified (Casoni, Simion, & Sârbu, 2019) as low 1,1-diphenyl-2-picrylhydrazyl (DPPH) radical scavenging activity oil (16.4%), together with cold pressed oils of peanut (4.94%), coconut (0.46% and 2.26%), and macadamia (2.07%).

The sensorial attributes of edible oils are influenced by the type and extent of pretreatments, since these practices contribute to the formation of several volatile compounds. In a research carried out by Krist et al. (2005), cold pressed oil of untreated poppy seeds were characterized by a mild, fatty, nutty smell and flavor, whereas cold pressed oil of preheated poppy seeds (at 60°C for 30 min) was described as having a stronger fatty, green, and roasted flavor and smell. The fatty and nutty perceptions were attributed to 1-pentanal, 1-hexanal, and caproic acid, while the sweet and fruity senses were correlated with 1-pentanol, 3-octen-2-one, and γ-hexalactone. The green notes were associated with the occurrence of trans-2-heptenal, 1-hexanal, and 2-pentylfuran compounds. Heat pretreatment caused the formation of strong roasted and nutty flavor in the oils, which was accompanied by development of 2-methylbutanal, 1-heptanal, 2-methylpyrazine, 2,5-dimethylpyrazine, and 2-ethyl-3,5- dimethylpyrazine compounds. Krist et al. (2005) concluded that pretreatments influence the type of volatile components, but long-term storage does not affect the volatile composition of cold pressed poppy seed oils. Emir, Güneşer, and Yılmaz (2014) mainly determined 1-hexanol, 2-heptanone, 2-pentanone, 2-pentyl furan, mercaptoacetic acid, 3-ethyl-2-methyl 1,3-hexadiene, 2-(dimethylamino)-3-phenylbenzo[*b*]thiophene, 3-octen-2-one, 4-hydroxyphenylacetic acid, α-pinene, limonene, dimethyl sulfone, hexanal, and nonanal in the cold pressed oils of raw, roasted and enzyme-assisted poppy seeds. These volatiles give sweet, fruity, green, creamy-cheese, green waxy, fermented, woody, soil, caramel, decayed onion, grassy, hazelnut, peanut popcorn, honey, spicy, citrus, terpenic, sulfur-burnt, and oily odor characteristics to cold pressed poppy seed oils. Emir et al. (2014) reported that poppy seed oil could be classified as a specialty oil due to its strong aromatic and rich flavor. On the other hand, in the same study, the cold pressed oils obtained from roasted poppy seeds were the most preferred samples by the consumers. Hence, the authors stated that roasting of seeds is desirable due to improvement of sensory characteristics of poppy seed oils.

5 Health-promoting traits of cold pressed oil and oil constituents

Monounsaturated fatty acids (MUFAs) and polyunsaturated fatty acids (PUFAs) constitute the major part of cold pressed poppy seed oil. Oleic and linoleic acids are the dominant fatty acids. Researchers have concentrated on primarily the effects of unsaturated fatty acids on lipoprotein metabolism. However, they are in contradiction. Some of them suggest that diets with higher PUFA content lead to greater reduction in low-density lipoprotein (LDL) cholesterol than do the diets with higher MUFA content (Howard et al., 1995). Likewise, Harris (2008) drew attention to the relationship between ω-6 PUFA consumption and cardiovascular disease and recommended intake of at least 5%–10% of energy from ω-6 PUFA, in particular linoleic acid, for reduction of risk. On the other side, Mensink and Katan (1989), Mata, Alvarez-Sala, Rubio, Nuño, and De Oya (1992), and Hodson, Skeaff, and Chisholm (2001) stated that MUFAs are as effective as PUFAs for lowering LDL and total cholesterol levels. Consequently, all researchers agree that both MUFAs and PUFAs are strong cholesterol-lowering agents. Considering the fact that poppy seed oil contains high amounts of MUFAs and PUFAs, its consumption may improve cholesterol metabolism in the human body and reduce the risk of cardiovascular diseases.

In general, oilseeds are rich sources of linoleic acid, which is one of the essential fatty acids for human. Linoleic acid together with linolenic acid is the precursor of some biologically active eicosanoids such as prostaglandin, thromboxane, and leukotrienes (Youdim, Martin, & Joseph, 2000). As is the case with the cholesterol-lowering effect of MUFAs and PUFAs, controversy exists about the effects of these eicosanoids. Patterson, Wall, Fitzgerald, Ross, and Stanton (2012) claimed that eicosanoids originated from ω-3 PUFA (i.e., linolenic acid) mostly exhibit antiinflammatory effects, whereas eicosanoids of ω-6 PUFA (i.e., linoleic acid) have pro-inflammatory effects. Minor levels of eicosanoids derived from arachidonic acid are desirable and biologically active; however, larger quantities of them assist the production of thrombi and atheroma that develop inflammatory and allergic disorders (Simopoulos, 1999). Fritsche (2008) declared that high ω-6 PUFA, particularly linoleic acid, intake does not correlate with in vivo ad ex vivo pro-inflammatory effects. Hence, there is still no consensus on whether eicosanoids of ω-6 PUFA promote the prevalence of chronic inflammatory diseases such as coronary heart disease, obesity, Alzheimer's, rheumatoid arthritis, nonalcoholic fatty liver disease, and inflammatory bowel disease or not. Nevertheless, it should be kept in mind that the dietary ω-6:ω-3 ratio of 1:1–2:1 is recommended to maintain the health status of the human body (Saini & Keum, 2018).

The fatty acids influence cancer and tumor formation in different ways depending on their chemical structure. It is thought that ω-3 fatty acids have preventive effect against various cancer types, while high intake of ω-6 fatty acids, in particular arachidonic acid, are associated with increased cancer incidences including prostate, breast, and colon cancers. However, linoleic acid, one of the ω-6 fatty acid and the dominant fatty acid of poppy seed oil, possess both pro- and anticancer properties (Huerta-Yépez, Tirado-Rodriguez, & Hankinson, 2016; Saini & Keum, 2018). Therefore, in vitro and in vivo studies should be carried out to reveal the exact effect of poppy seed oil consumption on development of different cancer types.

Numerous studies are available about the health benefits of α-tocopherol, but recent studies indicate that γ-tocopherol is a potent health-promoting agent as well. γ-tocopherol inhibits oxidative stress, reduces inflammation by decreasing formation of pro-inflammatory eicosanoids, acts as a chemopreventive agent, decreases coronary cardiovascular diseases, and prevents asthma symptoms such as mucous production and airway inflammation (Burbank et al., 2017, 2018; Campbell, Stone, Whaley, & Krishnan, 2003; Devaraj, Leonard, Traber, & Jialal, 2008; Dey et al., 2018; Hernandez et al., 2013; Jiang & Ames, 2003; Jiang, Christen, Shigenaga, & Ames, 2001; Wagner, Kamal-Eldin, & Elmadfa, 2004; Wiser et al., 2008). The most well-known effect of γ-tocopherol is to protect against prostate cancer in men. Helzlsouer et al. (2000) reported that high amounts of γ-tocopherol suppress prostate cancer development. Additionally, higher concentration of γ-tocopherol stimulates the protective properties of α-tocopherol and selenium against prostate cancer. Researchers concluded that the antioxidative strength of γ-tocopherol is higher than that of the α-tocopherol isomer (Li, Saldeen, Romeo, & Mehta, 1999; Saldeen, Li, & Mehta, 1999).

A very recent study showed that all the potentially beneficial compounds found in seed oils might contribute to their antimicrobial and antifungal activities. Kostadinović Veličkovska et al. (2018) reported that cold pressed poppy seed oil had low antibacterial against *Listeria monocytogenes*, while had moderate antifungal activity against *Candida albicans*. Compared to cold pressed oils of almond, walnut, and wheat germ, poppy seed oil had the highest antibacterial and antifungal activity. The researchers attributed this finding to its high linoleic acid content combined with other bioactives like phytosterols and phenolics (Kostadinović Veličkovska et al., 2018).

In summary, the bioactive micronutrients of cold pressed poppy seed oil perform different functions in the human body. Moreover, several studies have been demonstrated the synergistic interactions of these components. Therefore, more long-term studies about the effect of poppy seed oil consumption on various biological mechanisms should be carried out to clarify its beneficial and harmful health effects.

6 Edible and nonedible applications of cold pressed oil

Cold pressed poppy seed oil is generally used as salad or cooking oil without further refining and raw material for margarine production (Bozan & Temelli, 2003). In some countries where iodized salt is not available, the iodized poppy seed oil is used for the treatment of iodine deficiency-related disorders (Benmiloud et al., 1994; Huda, Grantham-McGregor, &

Tomkins, 2001; Zimmermann, Adou, Torresani, Zeder, & Hurrell, 2000). However, the high price of poppy seed oil forced researchers to find new and cheaper oil alternatives for iodine supplementation. Untoro, Schultink, West, Gross, and Hautvast (2006) evaluated the suitability of peanut oil for this purpose and reported that peanut oil may serve as a low-cost substitute. In addition to edible uses, its drying character makes it suitable for the manufacturing of paints, varnishes, and coatings as well as soaps (Bozan & Temelli, 2003; Duke, 1973; Özcan & Atalay, 2006). The popularity of poppy seed oil in the cosmetics industry results from its high moisturizing ability of skin and it is used as massage oil in aromatherapy (Baser & Arslan, 2014). Another use of poppy seed oil is in biodiesel. Aksoy (2011) declared that poppy seed oil is a suitable raw material for biodiesel production if the optimum conditions are provided. In addition, it causes lower oxidative damage (Aksoy, 2013). Nonetheless, cold pressed poppy seed oil is mostly used for edible or pharmaceutical purposes due to the lower oil yield of cold pressing procedure and limited and controlled cultivation of poppy plant.

7 Other issues

The authenticity of edible oils can be divided into three main groups:

a. determination of geographical origin;
b. adulteration-blending of cheaper oils with more expensive ones; and
c. adulteration-blending of lower quality oils (i.e., refined) with minimally processed ones (i.e., cold pressed) (Ulberth & Buchgraber, 2000).

The main limitation of the poppy seed oil is its higher price due to the lower production yield of seeds. Hence, the addition of another edible oil which exhibits identical properties to poppy seed oil has been practiced by the producers. For this purpose, sunflower oil is commonly used as an adulterant in poppy seed oil production. Due to the similar fatty acid compositions of sunflower and poppy seed oils, Krist, Stuebiger, Bail, and Unterweger (2006) assessed the feasibility of other markers for detection of adulteration. The authors stated that the presence of α-pinene, a volatile compound unique to sunflower oil, is a more reliable indicator than triacylglycerol composition to distinguish adulterated poppy seed oil. However, more practical approaches like sensory evaluation have been evaluated, owing to the complexity of these analytical methods. Bail, Majchrzak, Krist, Elmadfa, and Buchbauer (2008) reported that if the cold pressed and untreated sunflower oil is blended with cold pressed poppy seed oil at a minimum level of 30%, adulteration can be detected by sensory evaluation carried out by trained and experienced panelists. Nevertheless, the requirement of high levels of adulterant is the major drawback of sensory evaluation to detect adulteration in poppy seed oil. Differently from the adulteration carried out intentionally, cross contamination may occur due to improper cleaning of equipment in oil plants. For instance, sesamin and sesamolin were surprisingly detected in cold pressed poppy seed oils produced by a company which also produces sesame oil (Górnaś, Siger, Pugajeva, & Segliņa, 2014). These lignans, together with sesamol, are responsible for the allergenicity of sesame oil. Hence, contamination of other oils with these allergens is considered as a health concern and should be prevented.

Chemical risk assessment of foods is crucial in order to ensure food safety. Polycyclic aromatic hydrocarbons (PAHs) are one of the most prevalent toxic environmental and food processing contaminants (Purcaro, Moret, & Conte, 2013). Cold pressed vegetable oils may be contaminated through atmospheric deposition onto plant material, directly use of combustion smoke in the drying of oilseeds, and/or cultivation of oilseed plants in contaminated soils (Pandey, Mishra, Khanna, & Das, 2004). Due to the lipophilic property of PAHs, high levels of these toxic contaminants accumulate in edible oils. Cold pressed poppy seed oil was characterized by its comparatively lower total PAHs content (23.4 μg/kg) than cold pressed oils of amaranth (101.6 μg/kg), linseed (115 μg/kg), common flax (170.8 μg/kg), camelina (41.3 μg/kg), pumpkin seed (234.3 μg/kg), sesame (30.09 μg/kg), mustard (35.04 μg/kg), safflower (53.7 μg/kg), blackseed (221.2 μg/kg), walnut (45.9 μg/kg), borage (66.2 μg/kg), and evening primrose (68.9 μg/kg). The quantified PAHs in poppy seed oil were phenanthrene, anthracene, fluoranthene, pyrene, benzo[*a*]anthracene, benzo[*b*]fluoranthene, and benzo[*k*]fluoranthene. The sum of benzo[*a*]pyrene, chrysene, benzo[*a*]anthracene, and benzo[*b*]fluoranthene (1.71 μg/kg) in cold pressed poppy seed oil was below the maximum tolerable limit of 10 μg/kg set by Commission Regulation (EU) No 835/2011 (Ciecierska & Obiedziński, 2013). On the other hand, naphthalene, chrysenes, fluorene, acenaphtene, acenapthylene, benzo[*g*,*h*,*i*]perylene, and dibenz[*a*,*h*]anthracene were also detected in the oil (Roszko, Szterk, Szymczyk, & Waszkiewicz-Robak, 2012). In addition to these PAHs, benz[*a*]acridine, dibenz[*a*,*j*]acridine, and dibenz[*a*,*c*]acridine were the identified polycyclic aromatic nitrogen hydrocarbons (PANHs) in cold pressed poppy seed oil (Szterk, Roszko, & Cybulski, 2012). Moreover, residues of other contaminants such as dioxin-like and nondioxin-like polychlorinated biphenyls (PCBs), polybrominated biphenyl ethers (PBDEs), and pesticides such as 2,2-dichlorovinyl dimethyl phosphate (DDVP) have been isolated from the oil so far (Roszko et al., 2012).

References

Akinci, Z., & Bayram, I. (2003). Effects of poppy seed meal on egg production and hatching results in quail (*Coturnix coturnix japonica*). *Research in Veterinary Science*, *75*, 141–147.

Aksoy, L. (2011). Opium poppy (*Papaver somniferum* L.) oil for preparation of biodiesel: Optimization of conditions. *Applied Energy*, *88*, 4713–4718.

Aksoy, L. (2013). Oxidant/antioxidant equilibrium in rats supplemented with diesel fuel or with opium poppy (*Papaver somniferum* L.) seed oil biodiesel. *Revue de Médecine Vétérinaire (Toulouse)*, *164*, 34–38.

Annett, H. E., & Sen, J. N. (1919). The use of poppy seed cake as a cattle food and its effect on yield of milk and composition of the butter fat. *Journal of Agricultural Science*, *9*, 416–429.

Ayyildiz, H. F. (2015). Evaluation of new silica-based humic acid stationary phase for the separation of tocopherols in cold-pressed oils by normal-phase high-performance liquid chromatography. *Journal of Separation Science*, *38*, 813–820.

Bail, S., Majchrzak, D., Krist, S., Elmadfa, I., & Buchbauer, G. (2008). Sensory evaluation of poppy seed oil blended with sunflower seed oil. *Ernahrung/Nutrition*, *32*, 8–15.

Baser, K. H. C., & Arslan, N. (2014). Opium poppy (*Papaver somniferum*). In Z. Yaniv, & N. Dudai (Eds.), *Medicinal and aromatic plants of the middle-east* (pp. 305–332). Dordrecht: Springer.

Başoğlu, F. (2014). *Yemeklik Yağ Teknolojisi* (4th ed.). Bursa, Turkey: Dora Yayıncılık.

Bayram, İ., & Akıncı, Z. (1998). Bıldırcın rasyonlarına katılan haşhaş küspesinin besi performansı üzerine etkisi. *Ankara Üniversitesi Veteriner Fakültesi Dergisi*, *45*, 305–311.

Bayram, İ., Akıncı Bozkurt, Z., & Şehu, A. (2005). Haşhaş tohumu küspesinin yumurta tavuğu rasyonlarında kullanılmasının yumurta verimi ve bazı kan parametreleri üzerine etkisi. In *3. Ulusal Hayvan Besleme Kongresi* (pp. 176–181). Adana, Turkey.

Benmiloud, M., Chaouki, M. L., Gutekunst, R., Teichert, H. M., Wood, W. G., & Dunn, J. T. (1994). Oral iodized oil for correcting iodine deficiency: Optimal dosing and outcome indicator selection. *Journal of Clinical Endocrinology and Metabolism*, *79*, 20–24.

Bernáth, J. (1999). Cultivation of poppy under tropical conditions. In J. Bernáth (Ed.), *Poppy—The genus papaver* (pp. 271–283). London, UK: CRC Press.

Bialek, A., Bialek, M., Jelinska, M., & Tokarz, A. (2016). Fatty acid composition and oxidative characteristics of novel edible oils in Poland. *CyTA Journal of Food*, *15*, 1–8.

Bozan, B., & Temelli, F. (2003). Extraction of poppy seed oil using supercritical CO_2. *Journal of Food Science*, *68*, 422–426.

Burbank, A. J., Duran, C. G., Almond, M., Wells, H., Jenkins, S., Jiang, Q., et al. (2017). A short course of gamma-tocopherol mitigates LPS-induced inflammatory responses in humans ex vivo. *Journal of Allergy and Clinical Immunology*, *140*, 1179–1181.e4.

Burbank, A. J., Duran, C. G., Pan, Y., Burns, P., Jones, S., Jiang, Q., et al. (2018). Gamma tocopherol-enriched supplement reduces sputum eosinophilia and endotoxin-induced sputum neutrophilia in volunteers with asthma. *Journal of Allergy and Clinical Immunology*, *141*, 1231–1238.e1.

Campbell, S., Stone, W., Whaley, S., & Krishnan, K. (2003). Development of gamma (γ)-tocopherol as a colorectal cancer chemopreventive agent. *Critical Reviews in Oncology/Hematology*, *47*, 249–259.

Casoni, D., Simion, I. M., & Sârbu, C. (2019). A comprehensive classification of edible oils according to their radical scavenging spectral profile evaluated by advanced chemometrics. *Spectrochimica Acta Part A: Molecular and Biomolecular Spectroscopy*, *213*, 204–209.

Chander, A. K. (2010). *Characterisation and oxidative stability of specialty plant seed oils*. Aston University.

Ciecierska, M., & Obiedziński, M. W. (2013). Polycyclic aromatic hydrocarbons in vegetable oils from unconventional sources. *Food Control*, *30*, 556–562.

Czaplicki, S., Ogrodowska, D., Derewiaka, D., Tańska, M., & Zadernowski, R. (2011). Bioactive compounds in unsaponifiable fraction of oils from unconventional sources. *European Journal of Lipid Science and Technology*, *113*, 1456–1464.

Devaraj, S., Leonard, S., Traber, M. G., & Jialal, I. (2008). Gamma-tocopherol supplementation alone and in combination with alpha-tocopherol alters biomarkers of oxidative stress and inflammation in subjects with metabolic syndrome. *Free Radical Biology & Medicine*, *44*, 1203–1208.

Dey, P., Mah, E., Li, J., Jalili, T., Symons, J. D., & Bruno, R. S. (2018). Improved hepatic γ-tocopherol status limits oxidative and inflammatory stress-mediated liver injury in db/db mice with nonalcoholic steatohepatitis. *Journal of Functional Foods*, *40*, 670–678.

Duke, J. A. (1973). Utilization of papaver. *Economic Botany*, *27*, 390–400.

Dündar Emir, D., Aydeniz, B., & Yılmaz, E. (2015). Effects of roasting and enzyme pretreatments on yield and quality of cold-pressed poppy seed oils. *Turkish Journal of Agriculture and Forestry*, *39*, 260–271.

Eggersdorfer, M., & Wyss, A. (2018). Carotenoids in human nutrition and health. *Archives of Biochemistry and Biophysics*, *652*, 18–26.

Emir, D. D., Güneşer, O., & Yılmaz, E. (2014). Cold pressed poppy seed oils: Sensory properties, aromatic profiles, and consumer preferences. *Grasas y Aceites*, *65*, e029.

Food and Agriculture Organization of the United Nations (2019). *FAOSTAT [WWW Document]*. http://www.fao.org/faostat/en/#data/QC/visualize (Accessed 4 September 2019).

Fritsche, K. L. (2008). Too much linoleic acid promotes inflammation-doesn't it? *Prostaglandins, Leukotrienes and Essential Fatty Acids*, *79*, 173–175.

Górnaś, P., Siger, A., Juhņeviča, K., Lācis, G., Šnē, E., & Segliņa, D. (2014). Cold-pressed Japanese quince (*Chaenomeles japonica* (Thunb.) Lindl. ex Spach) seed oil as a rich source of α-tocopherol, carotenoids and phenolics: A comparison of the composition and antioxidant activity with nine other plant oils. *European Journal of Lipid Science and Technology*, *116*, 563–570.

Górnaś, P., Siger, A., Pugajeva, I., & Segliņa, D. (2014). Sesamin and sesamolin as unexpected contaminants in various cold-pressed plant oils: NP-HPLC/FLD/DAD and RP-UPLC-ESI/MS n study. *Food Additives & Contaminants: Part A*, *31*, 567–573.

Gruszka, J., & Kruk, J. (2007). RP-LC for determination of plastochromanol, tocotrienols and tocopherols in plant oils. *Chromatographia*, *66*, 909–913.

Harris, W. S. (2008). Linoleic acid and coronary heart disease. *Prostaglandins, Leukotrienes and Essential Fatty Acids*, *79*, 169–171.

Helzlsouer, K. J., Huang, H.-Y., Alberg, A. J., Hoffman, S., Burke, A., Norkus, E. P., et al. (2000). Association between alpha-tocopherol, gamma-tocopherol, selenium, and subsequent prostate cancer. *Journal of the National Cancer Institute*, *92*, 2018–2023.

Hernandez, M. L., Wagner, J. G., Kala, A., Mills, K., Wells, H. B., Alexis, N. E., et al. (2013). Vitamin E, γ-tocopherol, reduces airway neutrophil recruitment after inhaled endotoxin challenge in rats and in healthy volunteers. *Free Radical Biology & Medicine*, *60*, 56–62.

Hodson, L., Skeaff, C., & Chisholm, W.-A. (2001). The effect of replacing dietary saturated fat with polyunsaturated or monounsaturated fat on plasma lipids in free-living young adults. *European Journal of Clinical Nutrition*, *55*, 908–915.

Howard, B. V., Hannah, J. S., Heiser, C. C., Jablonski, K. A., Paidi, M. C., Alarif, L., et al. (1995). Polyunsaturated fatty acids result in greater cholesterol lowering and less triacylglycerol elevation than do monounsaturated fatty acids in a dose-response comparison in a multiracial study group. *The American Journal of Clinical Nutrition*, *62*, 392–402.

Huda, S. N., Grantham-McGregor, S. M., & Tomkins, A. (2001). Cognitive and motor functions of iodine-deficient but euthyroid children in Bangladesh do not benefit from iodized poppy seed oil (lipiodol). *Journal of Nutrition*, *131*, 72–77.

Huerta-Yépez, S., Tirado-Rodriguez, A. B., & Hankinson, O. (2016). Role of diets rich in omega-3 and omega-6 in the development of cancer. *Boletín Médico Del Hospital Infantil de México (English Edition)*, *73*, 446–456.

Jiang, Q., & Ames, B. N. (2003). γ-Tocopherol, but not α-tocopherol, decreases proinflammatory eicosanoids and inflammation damage in rats. *FASEB Journal*, *17*, 816–822.

Jiang, Q., Christen, S., Shigenaga, M. K., & Ames, B. N. (2001). γ-Tocopherol, the major form of vitamin E in the US diet, deserves more attention. *The American Journal of Clinical Nutrition*, *74*, 714–722.

Kachel-Jakubowska, M., Kraszkiewicz, A., & Krajewska, M. (2016). Possibilities of using waste after pressing oil from oilseeds for energy purposes. *Agricultural Engineering*, *20*, 45–54.

Köseoğlu, O., Sevim, D., & Özdemir, D. (2017). Determination of triacylglycerol composition of Ayvalık and Memecik olive oils during storage by chemometric methods. *Sakarya University Journal of Science*, *21*, 1497–1504.

Kostadinović Veličkovska, S., Naumova Letia, G., Čočevska, M., Brühl, L., Silaghi-Dumitrescu, R., Mirhosseini, H., et al. (2018). Effect of bioactive compounds on antiradical and antimicrobial activity of extracts and cold-pressed edible oils from nutty fruits from Macedonia. *Journal of Food Measurement and Characterization*, *12*, 2545–2552.

Krist, S., Stuebiger, G., Bail, S., & Unterweger, H. (2006). Detection of adulteration of poppy seed oil with sunflower oil based on volatiles and triacylglycerol composition. *Journal of Agricultural and Food Chemistry*, *54*, 6385–6389.

Krist, S., Stuebiger, G., Unterweger, H., Bandion, F., & Buchbauer, G. (2005). Analysis of volatile compounds and triglycerides of seed oils extracted from different poppy varieties (*Papaver somniferum* L.). *Journal of Agricultural and Food Chemistry*, *53*, 8310–8316.

Küçükersan, S., Yeşilbağ, D., & Küçükersan, K. (2009). Using of poppy seed meal and yeast culture (*Saccharomyces cerevisiae*) as an alternative protein source for layer hens Seher. *Kafkas Universitesi Veteriner Fakultesi Dergisi*, *15*, 971–974.

Küster, H. (2000). Spices and flavorings. In K. F. Kiple, & K. C. Ornelas (Eds.), *The Cambridge world history of food* (pp. 431–436). Cambridge, UK: Cambridge University Press.

Levy, A., & Milo, J. (1999). Genetics and breeding of *Papaver somniferum*. In J. Bernáth (Ed.), *Poppy—The genus Papaver* (pp. 93–103). Amsterdam, Netherlands: Harwood Academic Publishers.

Li, D., Saldeen, T., Romeo, F., & Mehta, J. L. (1999). Relative effects of α- and γ-tocopherol on low-density lipoprotein oxidation and superoxide dismutase and nitric oxide synthase activity and protein expression in rats. *Journal of Cardiovascular Pharmacology and Therapeutics*, *4*, 219–226.

Mag, T. K. (1990). Bleaching-theory and practice. In D. R. Erickson (Ed.), *Edible fats and oils processing: Basic principles and modern practices* (pp. 107–116). Champaign, IL: American Oil Chemists' Society.

Mata, P., Alvarez-Sala, L. A., Rubio, M. J., Nuño, J., & De Oya, M. (1992). Effects of long-term monounsaturated- vs polyunsaturated-enriched diets on lipoproteins in healthy men and women. *The American Journal of Clinical Nutrition*, *55*, 846–850.

Mensink, R. P., & Katan, M. B. (1989). Effect of a diet enriched with monounsaturated or polyunsaturated fatty acids on levels of low-density and high-density lipoprotein cholesterol in healthy women and men. *New England Journal of Medicine*, *321*, 436–441.

Németh, É. (1999). Cultivation of poppy in the temperate zone. In J. Bernáth (Ed.), *Poppy—The genus Papaver* (pp. 252–270). London, UK: CRC Press.

Özcan, M. M., & Atalay, Ç. (2006). Determination of seed and oil properties of some poppy (*Papaver somniferum* L.) varieties. *Grasas y Aceites*, *57*, 169–174.

Pandey, M. K., Mishra, K. K., Khanna, S. K., & Das, M. (2004). Detection of polycyclic aromatic hydrocarbons in commonly consumed edible oils and their likely intake in the Indian population. *Journal of the American Oil Chemists' Society*, *81*, 1131–1136.

Patterson, E., Wall, R., Fitzgerald, G. F., Ross, R. P., & Stanton, C. (2012). Health implications of high dietary omega-6 polyunsaturated fatty acids. *Journal of Nutrition and Metabolism*, *2012*, 1–16.

Purcaro, G., Moret, S., & Conte, L. S. (2013). Overview on polycyclic aromatic hydrocarbons: Occurrence, legislation and innovative determination in foods. *Talanta*, *105*, 292–305.

Rodriguez-Concepcion, M., Avalos, J., Bonet, M. L., Boronat, A., Gomez-Gomez, L., Hornero-Mendez, D., et al. (2018). A global perspective on carotenoids: Metabolism, biotechnology, and benefits for nutrition and health. *Progress in Lipid Research*, *70*, 62–93.

Roszko, M., Szterk, A., Szymczyk, K., & Waszkiewicz-Robak, B. (2012). PAHs, PCBs, PBDEs and pesticides in cold-pressed vegetable oils. *Journal of the American Oil Chemists' Society*, *89*, 389–400.

Saini, R. K., & Keum, Y.-S. (2018). Omega-3 and omega-6 polyunsaturated fatty acids: Dietary sources, metabolism, and significance—A review. *Life Sciences*, *203*, 255–267.

Saldeen, T., Li, D., & Mehta, J. L. (1999). Differential effects of α- and γ-tocopherol on low-density lipoprotein oxidation, superoxide activity, platelet aggregation and arterial thrombogenesis. *Journal of the American College of Cardiology*, *34*, 1208–1215.

Simopoulos, A. P. (1999). Essential fatty acids in health and chronic disease. *The American Journal of Clinical Nutrition*, *70*, 560s–569s.

Singh, S. P., Shukla, S., Khanna, K. R., Dixit, B. S., & Banerji, R. (1998). Variation of major fatty acids in F8 generation of opium poppy (*Papaver somniferum* × *Papaver setigerum*) genotypes. *Journal of the Science of Food and Agriculture*, *76*, 168–172.

Statham, M. (1984). Poppy seed meal (*Papaver somniferum*) as a protein source for growing pigs. *Australian Journal of Experimental Agriculture*, *24*, 170.

Symoniuk, E., Ratusz, K., Ostrowska-Ligęza, E., & Krygier, K. (2018). Impact of selected chemical characteristics of cold-pressed oils on their oxidative stability determined using the Rancimat and pressure differential scanning calorimetry method. *Food Analytical Methods*, *11*, 1095–1104.

Szterk, A., Roszko, M., & Cybulski, A. (2012). Determination of azaarenes in oils using the LC-APCI-MS/MS technique: New environmental toxicant in food oils. *Journal of Separation Science*, *35*, 2858–2865.

Szydłowska-Czerniak, A., Trokowski, K., Karlovits, G., & Szłyk, E. (2011). Effect of refining processes on antioxidant capacity, total contents of phenolics and carotenoids in palm oils. *Food Chemistry*, *129*, 1187–1192.

Tétényi, P. (1997). Opium poppy (*Papaver somniferum*): Botany and horticulture. In J. Janick (Ed.), *Horticultural reviews* (pp. 373–408). New York, NY: John Wiley & Sons, Inc..

Ulberth, F., & Buchgraber, M. (2000). Authenticity of fats and oils. *European Journal of Lipid Science and Technology*, *102*, 687–694.

Untoro, J., Schultink, W., West, C. E., Gross, R., & Hautvast, J. G. (2006). Efficacy of oral iodized peanut oil is greater than that of iodized poppy seed oil among Indonesian schoolchildren. *The American Journal of Clinical Nutrition*, *84*, 1208–1214.

Wagner, K.-H., Kamal-Eldin, A., & Elmadfa, I. (2004). Gamma-tocopherol—An underestimated vitamin? *Annals of Nutrition & Metabolism*, *48*, 169–188.

Wijaya, C. H., Wijaya, W., & Mehta, B. M. (2015). General properties of major food components. In P. C. K. Cheung, & B. M. Mehta (Eds.), *Handbook of food chemistry* (pp. 15–54). Berlin Heidelberg, Berlin, Heidelberg: Springer.

Wiser, J., Alexis, N. E., Jiang, Q., Wu, W., Robinette, C., Roubey, R., et al. (2008). In vivo γ-tocopherol supplementation decreases systemic oxidative stress and cytokine responses of human monocytes in normal and asthmatic subjects. *Free Radical Biology & Medicine*, *45*, 40–49.

Yilmaz, E., & Dündar Emir, D. (2016). Extraction and functional properties of proteins from pre-roasted and enzyme treated poppyseed (*Papaver somniferum* L.) press cakes. *Journal of Oleo Science*, *65*, 319–329.

Ying, Q., Wojciechowska, P., Siger, A., Kaczmarek, A., & Rudzińska, M. (2018). Phytochemical content, oxidative stability, and nutritional properties of unconventional cold-pressed edible oils. *Journal of Food and Nutrition Research*, *6*, 476–485.

Youdim, K. A., Martin, A., & Joseph, J. A. (2000). Essential fatty acids and the brain: Possible health implications. *International Journal of Developmental Neuroscience*, *18*, 383–399.

Zimmermann, M., Adou, P., Torresani, T., Zeder, C., & Hurrell, R. (2000). Low dose oral iodized oil for control of iodine deficiency in children. *The British Journal of Nutrition*, *84*, 139–141.

Chapter 20

Cold pressed hazelnut (*Corylus avellana*) oil

Veysel Umut Celenk[a], Zeliha Ustun Argon[b,c], and Zinar Pinar Gumus[d]
[a]*Drug Research and Pharmacokinetic Development and Applied Center, ARGEFAR, Ege University, İzmir, Turkey,* [b]*Department of Biosystems Engineering, Eregli Faculty of Engineering and Natural Sciences, Necmettin Erbakan University, Konya, Turkey,* [c]*Medical and Cosmetic Plants Application and Research Center, Necmettin Erbakan University, Konya, Turkey,* [d]*Central Research Testing and Analysis Laboratory Research and Application Center, EGE-MATAL, Ege University, İzmir, Turkey*

Abbreviations

CHD	coronary heart disease
CLnA	conjugated linolenic acid
FAME	fatty acid methyl esters
FAO	Food and Agriculture Organization
FDA	Food and Drug Administration
FT-IR	Fourier transform infrared spectroscopy
FT-MIR	Fourier transform mid-infrared spectroscopy
GC	gas chromatography
GC-FID	gas chromatography-flame ionization detector
HPLC-FLD	high-performance liquid chromatography-fluorescence detector
LDL	low-density lipoprotein
LOQ	limit of quantification
N.D.	not detected
NIR	near-infrared spectroscopy
NMR	nuclear magnetic resonance
PFG-NMR	high power pulse-field gradient-nuclear magnetic resonance
PS	pressing speed
PTV	programmable temperature vaporizing
Rd.	restriction dye
RPLC-GC	reversed-phase liquid chromatography and gas chromatography
rpm	rotation per minute
SDS-PAGE	sodium dodecyl sulfate polyacrylamide gel electrophoresis
SFE	supercritical fluid extraction
SMC	seed moisture content
TAG	triacylglycerol
WHO	World Health Organization

1 Introduction

The hazelnut (*Corylus avellana*, Betulaceae) is an important tree in Mediterranean countries especially in Turkey, which has the lion's share (72%) of hazelnut production (Eldin & Moreau, 2009; Gumus, Yorulmaz, & Aziz, 2016). Turkey is the main producer, with about 675,000 t hazelnuts in-shell followed by Italy with about 131,281 t, in 2017. The other European countries, like France and Spain, have much smaller production capacity, 10,833 t, and 10,487 t, respectively. Total hazelnut production capacity in European Union countries is 160,683 t. On the other hand, Germany, Italy, and France are very important importers in Europe. In 2017, the total production of hazelnut passed 1 million tons, which shows the popularity of hazelnuts around the world (FAOStat, 2019). The name of the tree comes from the Italian town of Avella,

Cold Pressed Oils. https://doi.org/10.1016/B978-0-12-818188-1.00020-7
Copyright © 2020 Elsevier Inc. All rights reserved.

TABLE 1 Oil contents of nuts.

Nut kernel	Oil (%)
Almonds	41–60
Brazil nuts	61–69
Cashew nuts	40–49
Hazelnuts	49–67
Macadamia nuts	59–78
Pecan nuts	58–74
Pine nuts	59–71
Pistachio nuts	45–59
Walnuts	51–65

when Carl Linnaeus (1707–78) described it as "*Avellana nux sylvestris*" meaning the wild nut of Avella. The tree, 3–8 m high, lives 75–100 years and produces 8–10 kg of nuts per year. Each nut is covered by a husk and a shell, and ripens around October, when the husks dry and release the kernels to fall on the ground. The hazelnut kernels include almost 60% of oil, 15% of carbohydrates, and 18% of protein. Corylus maxima, commonly called the fibert or filbert nut, is sometimes confused with hazelnuts. In recent years, hazelnut consumption has increased and many techniques have been developed to determine its chemical composition as a result of its high market value. Before the extraction of oil, the shell (pericarp or the fruit coat) and the husk (epicarp or the seed coat) must be removed from the nuts. The oil of hazelnuts can be consumed crude or refined. The oil contents in tree nut kernels are shown in Table 1 (Eldin & Moreau, 2009).

A remarkable amount of research has been performed on the nutritional values and physicochemical characteristics of hazelnut oil. Hazelnut oil consists of more than 90% unsaturated fatty acids, especially oleic (80%) and linoleic (6%–12%) acids. Hazelnut oil is a valuable edible oil and a good source of vitamin E (α-tocopherol) that improves shelf life by its antioxidant capacity. The cosmetics industry applies this nondrying oil to formulations because of its excellent shelf stability. Unsaturated fatty acids, sterols, and tocopherols in hazelnut oil have important roles in preventive medicine. Currently, hazelnut oil is used mainly in salad dressings and cosmetic and pharmaceutical products (Jokic et al., 2016).

The solvent extraction methods of oils have many disadvantages such as toxicity, high-cost solvents, and being harmful to the environment. Therefore, finding alternative eco-friendly methods like mechanical techniques become the current issue to obtain high-quality oil. Simple usage, rapid realization, short duration of the process, and high-quality oils are the main advantage of mechanical applications. On the other hand, low yield is the biggest disadvantage of mechanical techniques, but this technique can be optimized to reduce the yield up to 4%–6% by preheating the seeds. Oil extraction applications have some basic and unalterable rules such as not damaging the oil during operation, least impurity, minimum oil content in the cake, and maximum oil yield. The basic methods for oil extraction are schematized in Fig. 1 (Ionescu et al., 2016).

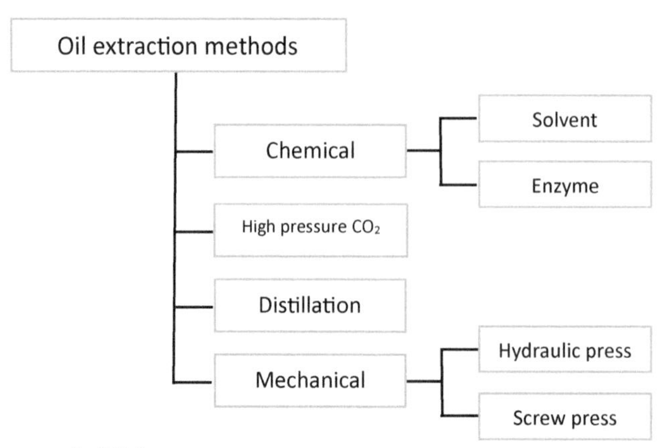

FIG. 1 Oil extraction methods (Ionescu et al., 2016).

2 Cold press extraction and processing of oil

Physical and chemical processes for the extraction of oil from crude material are performed successfully in the oil industry. The physical processes, such as hydraulic pressing and screw pressing, are based on mechanical power to obtain oil from the material. Chemical extraction involves organic solvents depending on the chemical characteristics of the crude material. The conventional chemical process based on solvent extraction produces low-quality oil that requires extensive purification operations. On the other hand, physical processes like screw pressing do not require the use of organic solvent and are able to retain bioactive compounds such as essential fatty acids, phenolics, flavonoids, and tocopherols in the oils, as well as the possibility of using cake free of toxic solvents in other processes. However, the primary problem of screw pressing is low oil extraction yield. The emerging technology of supercritical fluid extraction (SFE) has become a useful and remarkable technique for the extraction of fatty oils. Continuous mechanical pressing can be combined in commercial operations with supercritical CO_2 extraction, which is an environmentally friendly solvent. In the 1980s, SFE started to replace the use of organic solvents in applications of fatty oil extraction finitely. Nowadays, SFE on a commercial scale is limited to decaffeination, extraction of certain petroleum products, high-value compounds from spices, herbs, and other vegetable material, animal tissue, and microalgae (Jokic et al., 2016).

The cold pressing technique is an application based on gentle pressure on crude material, and due to the low yield of oil, this technique is usually expensive. The most important point in the pressing operation is that the temperature that does not exceed 30°C. The superior nutty flavor and fresh taste are results of low yield, nonchemical process, and limited temperature (Jokic et al., 2016). First-pressed oil can be classified as "extra-virgin nut oil," and the remaining part of the hazelnut is used for special baking, energy bars, and animal feeds. Because of the limited shelf life of cold pressed hazelnut oil, storage conditions must be at low temperatures. Packing under nitrogen to eliminate oxygen and supplemented with antioxidants are other stabilization conditions of cold pressed oil. Expeller-pressed oils are of lower quality than cold pressed oils that are carried out at higher temperatures may also extract the paste. Sometimes this lower quality oil is mixed with certain flavor compounds and sold as cold pressed oil. Cold pressed and expeller-pressed oils usually contain water, resins, color and flavor compounds, residual proteins and fiber that may make them dark and opaque. To remove these compounds, a clarification process is applied to the oils. The oils are allowed to stand undisturbed for a few days to yield a clear upper layer. Gentle heating follows to remove residual water and to destroy enzymes and any bacteria before filtration, and another brief heating occurs at c.100°C to remove the remaining traces of moisture. On the other hand, solvent extraction of kernels provides a higher yield than cold pressing, but the oils obtained are of lower quality compared to pressed oils and are generally refined by steps including neutralization, bleaching, degumming, and deodorization. The refined oils lose their nutty flavor but are better suited for high-temperature uses. The other technique, SFE, uses carbon dioxide and performs in wide temperature and pressure ranges. The oil yield of the SFE technique is lower than extraction with *n*-hexane but no significant differences were noted between the two oils regarding the composition of acyl lipids and sterols. Oils obtained by carbon dioxide extraction contain more tocopherols and are slightly more stable than oils extracted with *n*-hexane. However, the oxidative stability of carbon dioxide-extracted oils is lower than that of solvent-extracted oil despite the high content of tocopherols in carbon dioxide-extracted oils (Eldin & Moreau, 2009).

The cold press machine has a simple working scheme whereby seeds are fed to one inlet and two exits provide oil and a nonoiled cake. The oil yield depends on pretreatment and process parameters applied to the raw material. Pretreatment refers to peeling, drying, solvent, or enzymatic treatment of raw material and the process parameters mentioned as feeding rate, temperature, and rotation speed. Cold presses can be investigated under three main systems: expellers, expanders, and twin-cold systems. Twin-cold systems are for pilot- or laboratory-scale production. The other press expeller is the common type of cold press made by Anderson in 1902. In this technique, the cold or hot press can be applied. The oil separated from the cake is removed from the slot between the metal bars placed at regular intervals with the rotating cold (Cakaloglu, Ozyurt, & Otles, 2018).

All processes to reduce or eliminate the use of hazardous chemicals refer to green chemistry. This definition can be explained comprehensibly as follows: "Green Extraction is based on the discovery and design of extraction processes which will reduce energy consumption, allows use of alternative solvents and renewable natural products, and ensure a safe and high-quality extract/product." The listing of the "Six Principles of Green Extraction of Natural Products" should be viewed by industry and scientists as a direction to establish an innovative and green label, charter and standard, and as a reflection to innovate not only in process but in all aspects of solid-liquid extraction (Chemat, Vian, & Cravotto, 2012).

Principle 1: Innovation by selection of varieties and use of renewable plant resources.
Principle 2: Use of alternative solvents and principally water or agro-solvents.
Principle 3: Reduce energy consumption by energy recovery and using innovative technologies.

Principle 4: Production of co-products instead of waste to include the bio- and agro-refining industry.
Principle 5: Reduce unit operations and favor safe, robust and controlled processes.
Principle 6: Aim for a nondenatured and biodegradable extract without contaminants.

Extraction, according to the Six Principles of Green Extraction of Natural Products, is a new concept to meet the challenges of the 21st century, to protect both the environment and consumers, and in the meantime enhance the competitiveness of industries to be more ecologic, economic, and innovative. Within this green extraction approach, the concept of the green extract is introduced, which is an extract obtained in such a way to have the lowest possible impact on the environment (less energy and solvent consumption), and whose eventual recycling would have been planned for (coproducts, biodegradability, etc.). This green extract should be the result of a whole chain of values in both senses of the term: economic and responsible, starting from the production and harvesting of the plant, the transformation processes of extraction and separation, together with formulation and marketing. This "green extract" could be identifiable in the future by a European or International Label or Standard (Chemat et al., 2012).

3 Cold pressed oil recovery and content

The yield of the cold press technique depends on the specialty of crude material such as shell-shelled, moisture content, oil content, and type of raw material. The other benchmarks of yield are feed rate, temperature, rotation speed, the diameter of restriction dye, and pretreatment. Burg, Masan, and Rutkowski (2017) aimed to develop a cold press extraction process of grape seed oil with three white grape seeds. Experiment variances were grape varieties, rotational speed (20, 40, 60, 80 rpm), press performance, and oil yield. The results of this report showed that when the rotational speed increased, press performance increased but oil yield decreased. The yield of cold press technique was 6.75%–9.85% for grape seeds.

Singh, Wiesenborn, Tostenson, and Kangas (2002) investigated the influence of pretreatment as cooking and moisture content over the crambe seed oil by cold press extraction. In this study, first cooked then dried-up samples and only dried-up samples were compared. Heating ring degree and rotational speed were reported as 120°C and 20 rpm, respectively. The results showed that the cooking pretreatment has a positive effect on oil yield, but moisture content has a negative effect.

Martinez, Bordon, et al. (2017) used a 32-factor experimental design for walnut and almond oil extraction. Seed moisture content and restriction die were tested as yield parameters. Results showed that seed moisture content has a positive effect on oil yield, but restriction die has a negative effect. Martinez, Bordon, Lallana, Ribotta, and Maestri (2017) used Box-Behnken design to perform sesame oil extraction. Optimization of the cold press extraction method was based on seed moisture content (SMC), pressing speed (PS), and restriction dye (Rd) as the main processing parameters. The extraction procedure was carried out on a pilot and industrial scale. Oil yield was similar for both scales. The pilot-scale extraction resulted in 71.1% at 12.3% SMC, 4 mm Rd., and 20 rpm PS. On the other hand, industrial-scale extraction resulted in 74.4% at 8,03% SMC, 10 mm Rd., and 20 rpm PS. Quality parameters of both pilot-scale extraction and industrial-scale extraction oils were in the ranges of declared FAO/WHO standards for sesame oil.

Singh, Jhamb, and Kumar (2010) investigated the pretreatment effect on the yield of linseed oil. In this study, some of the linseeds were only established and some of them were subjected to steam pretreatment and some to enzyme pretreatment. The results showed that the seed with low moisture content had a better yield. The parameters studied within the study were preheating conditions, particle sizes, cold rotation speed, moisture content of seeds, and diameter of restriction dye. Whole seeds were given more yield than other particle sizes. At cold low rotational speeds, due to contact time with the machine, oil yield was found to be higher. The other results showed that moisture content affected oil yield negatively.

4 Chemical composition of cold pressed nuts oil

Triacylglycerols are the major component (96%–98%) in hazelnut oils like the other vegetable oils. Unsaponifiable components, diacylglycerols, monoacylglycerols, free fatty acids, phospholipids, sphingolipids, sterols, and sterol esters are the minor compounds in hazelnut oil. The fatty acid and triacylglycerol compositions depend on the geographic origin and variety of hazelnut (Miraliabari & Shahidi, 2007). Relative ratios of triacylglycerols, sterols, sterol esters, phosphatidylinositol, phosphatidylserine, phosphatidyl-choline, sphingolipids, and fatty acid composition for some selected nuts are presented in Table 2.

As shown in Table 2, there is no major difference between the nut oils in terms of triacylglycerol. All results are very similar to each other. Triacylglycerol content of hazelnut (98%) and almonds oils (98.2%) is higher than other nut oils. Focusing on hazelnut oil, free sterols (0.21%), sterol esters (0.04%), phosphatidylserine (0.27%), phosphatidylinositol

TABLE 2 Lipid-class and fatty acids distribution of selected nuts (N.D. = not detected).

Compound	Hazelnuts	Almonds	Brazil nuts	Pecan nuts	Pine nuts	Pistachio nuts	Walnuts
Triacylglycerols	98.0	98.2	96.7	96.4	97.2	96.2	97.2
Sterols (free)	0.21	0.22	0.18	0.26	0.13	0.19	0.26
Sterol esters	0.04	0.05	0.05	0.07	0.06	0.03	0.09
Phosphatidylserine	0.27	0.21	0.26	0.39	0.23	0.47	0.37
Phosphatidylinositol	0.06	0.11	0.09	0.15	0.14	0.21	0.25
Phosphatidylcholine	0.24	0.21	0.34	0.21	0.19	0.52	0.34
Sphingolipids	0.26	0.53	0.83	0.45	0.73	0.73	0.54
Myristic acid, 14:0	0.13	0.06	0.06	0.09	N.D.	0.09	0.13
Palmitic acid, 16:0	5.82	6.85	13.50	4.28	6.87	7.42	6.70
Palmitoleic acid, 16:1	0.29	0.63	0.33	0.09	0.14	0.70	0.23
Stearic acid, 18:0	2.74	1.29	11.77	1.80	4.48	0.86	2.27
Oleic acid, 18:1	79.30	69.24	29.09	40.63	39.55	58.19	21.00
Linoleic acid, 18:2	10.39	21.52	42.80	50.31	45.41	30.27	57.46
Linolenic acid,18:3	0.46	0.16	0.20	0.65	0.63	0.44	11.58
Arachidic acid, 20:0	0.16	0.16	0.54	<0.05	1.04	0.59	0.08
Eicosenoic acid, 20:1	N.D.	N.D.	0.21	1.21	1.06	0.60	N.D.
Ratio of unsaturated/ saturated fatty acids	9.90	10.80	2.79	13.54	6.81	9.75	9.50

(0.06%), phosphatidylcholine (0.24%) and sphingolipids (0.26%) were the other components. Hazelnut oil has the most oleic acid (79.3%) compounds in terms of fatty acid composition. The other fatty acids were found as follows: myristic acid (0.13%), palmitic acid (5.82%), palmitoleic acid (0.29%), stearic acid (2.74%), linoleic acid (10.3%), linolenic acid (0.46%), and arachidic acid (0.16%). Gondoic acid, behenic acid, and erucic acid were not detected. The ratio of unsaturated fatty acids to saturated fatty acids was 9.90 (Imbs, Nevshupova, & Pham, 1998; Maguire, O'Sullivan, Galvin, O'Connor, & O'Brien, 2009; Ryan, Galvin, O'Connor, Maguire, & O'Brien, 2009).

Other valuable and substantial compounds such as tocopherols, squalenes, and sterols were investigated. In terms of α-tocopherol, hazelnut oil has the richest content as 1218.8 μg/g. Another tocopherol isomer, γ-tocopherol, was reported as 320.4 μg/g in hazelnut oil (Table 3). In addition, Celenk, Gumus, Argon, Buyukhelvacigil, and Karasulu (2017) reported other tocopherol isomers, β-tocopherol (42.2 μg/g) and δ-tocopherol (13.9 μg/g), in cold pressed hazelnut oil.

TABLE 3 Squalene and tocopherol content of oil (μg/g oil) extracted from edible nuts (Amaral, Casal, Pereira, Seabra, & Oliveira, 2003; Imbs et al., 1998; Maguire et al., 2009; Ryan et al., 2009).

Compound	Hazelnuts	Almonds	Brazil nuts	Pecan nuts	Pine nuts	Pistachio nuts	Walnuts
α-Tocopherol	1218.8	439.5	82.9	12.2	124.3	15.6	20.6
γ-Tocopherol	320.4	12.5	116.2	168.5	105.2	275.4	300.5
Squalene	186.4	95.0	1377.8	151.7	39.5	91.4	9.4

As shown in Table 4, β-sitosterol, campesterol, and stigmasterol concentrations of hazelnut oil were reported as 991.2, 66.7, and 38.1 μg/g, respectively. The β-sitosterol and stigmasterol contents of hazelnut oil were the minimum amount of all nut oils. Relative percentages of triacylglycerol (TAG) compounds in tree nut oils are listed in Table 5.

TABLE 4 Campesterol, stigmasterol, and β-sitosterol content of oil (μg/g oil) extracted from edible nuts (Imbs et al., 1998; Maguire et al., 2009; Ryan et al., 2009).

Compound	Hazelnuts	Almonds	Brazil nuts	Pecan nuts	Pine nuts	Pistachio nuts	Walnuts
β-Sitosterol	991.2	2071.7	1325.4	1572.4	1841.7	4685.9	1129.5
Campesterol	66.7	55.0	26.9	52.2	214.9	236.8	51.0
Stigmasterol	38.1	51.7	577.5	340.4	680.5	663.3	55.5

TABLE 5 Triacylglycerol (TAG) molecular species (relative %) in tree nut oils (Amaral et al., 2003; Holcepek, Lisa, Jandera, & Kabatova, 2005; Imbs et al., 1998).

Compounds	Hazelnuts	Almonds	Brazil nuts	Pine nuts	Pistachio nuts	Walnuts
LLnLn	–	–	–	–	–	2.10
LLLn	0.10	–	0.10	–	0.60	20.00
LLL	3.70	8.70	14.80	5.00	11.70	34.60
PoLPo	–	–	–	–	–	–
PoPoPo	–	–	–	–	–	–
OLLn	0.50	0.10	0.20	–	1.00	4.60
LnLP	–	–	0.10	–	0.30	3.10
LnOP	–	–	–	–	–	–
OLL	12.30	27.60	16.70	14.70	24.80	17.20
PoLL	–	–	–	–	–	–
OLnO	0.70	–	–	5.40	0.90	–
LLP	1.60	4.80	13.00	3.80	7.10	10.60
OLO	28.20	28.00	13.10	7.40	25.20	3.10
LOP	5.20	11.30	16.70	8.30	11.80	4.20
PLP	0.20	0.50	2.60	0.60	1.10	–
OOO	36.50	13.30	4.60	7.60	8.90	–
SLO	1.40	1.80	10.00	3.50	1.30	–
OOP	6.10	2.70	–	2.60	3.40	–
SOO	2.80	0.60	2.30	1.30	0.40	–
SLL	–	–	–	6.70	–	–
SLP	–	–	–	0.50	–	–
POP	–	–	–	0.10	–	–
SOP	–	–	–	0.10	–	–

5 Comparison of chemical composition

Cold pressed oils are health-promoting products that improve health and prevent several diseases. A comparison of cold pressed oils in terms of chemical composition provides valuable information about the human diet. Celenk et al. (2017) reported a study into 15 cold pressed oils: black cumin seed (*N. sativa*), wheat germ (*Triticum* L.), sesame seed (*S. İndicum*), sunflower seed (*Helianthus* L.), poppy seed (*P. somniferum*), pomegranate seed (*P. granatum*), nettle seed (*U. dioica*), flax seed (*L. usitatissimum*), peanut (*A. hypogaea*), pumpkin seed (*Cucurbita pepo* L.), walnut (*Juglans* L.), hazelnut (*C. avellana*), grape seed (*Vitis vinifera* L.), safflower seed (*Carthamus tinctorium* L.), and canola seed (*Brassica napus* L.). The cold pressed oils were compared in terms of fatty acid composition and tocopherols content. The study deals with the sample preparation and validation of tocopherols using an HPLC-FLD method for simultaneous determination of α-, β-, γ-, and δ-tocopherols, and analysis of fatty acid methyl esters (FAME) using GC-FID. During the cold press process, the temperature was kept below 40°C. Neither solvent nor heating process was used to protect the physicochemical properties of the oil. The cold press machine was used by 3–4 kg seed/h capacity depending on the seeds' differences. Settling out the sediments has been done by storing the cold pressed oil in stainless steel intermediate tanks for 1 day. The next day, filtration of oil was completed by using 1 μm pores filtration paper. In the last step, the cold pressed oils were filled into 200 mL amber glass bottles and kept at 25 °C and at 60% relative humidity until the analysis was conducted. The tocopherol contents of 15 cold pressed oils are shown in Table 6.

According to Table 6, wheat germ oil is the richest oil (2556.1 μg/g) in terms of α-tocopherol. Hazelnut oil is the second richest oil with 1218.8 μg/g α-tocopherol concentration. Safflower and sunflower seed oils were reported as 1063.3 and 920.3 μg/g, respectively. In addition, β-tocopherol content of wheat germ oil was 1061.5 μg/g. Hazelnut oil is the second richest oil with a 42.2 μg/g β-tocopherol concentration. In terms of β-tocopherol concentration, sunflower seed oil and safflower seed oil contained 37.3 and 18.4 μg/g, respectively. Pomegranate seed oil is the richest oil (2625.0 μg/g) in terms of γ-tocopherol. The only nettle seed oil has a γ-tocopherol concentration (573.0 μg/g) between 500 and 1000 μg/g. The γ-tocopherol concentrations of other oils are smaller than 500 μg/g. However, only sunflower seed has a content close to the Limit of Quantification (LOQ) (0.08 μg/g). In terms of δ-tocopherol concentration, walnut oil and pomegranate seed oil was reported as 374.0 and 61.3 μg/g, respectively. Hazelnut oil has one of the lowest δ-tocopherol concentrations.

The fatty acid compositions of 15 cold pressed oils are set out in Table 7. The results show that hazelnut oil is the richest oil among the others in terms of oleic acid (77.0%). Canola oil was reported as the second richest oil (61.8%). Oleic acid was

TABLE 6 Amount of tocopherols in cold pressed seed oils (μg/g) (Celenk et al., 2017).

Cold pressed oil	α-Tocopherol	β-Tocopherol	γ-Tocopherol	δ-Tocopherol
Sunflower seed	920.3	37.3	1.2	N.D.
Poppy seed	32.9	N.D.	276.1	1.9
Sesame seed	53.7	N.D.	482.3	6.9
Pomegranate seed	72.4	N.D.	2625.0	61.3
Nettle seed	19.4	N.D.	573.0	9.2
Flax seed	7.5	N.D.	434.0	4.3
Peanut	114.7	1.4	45.5	9.9
Pumpkin seed	25.7	N.D.	678.0	10.0
Walnut	27.2	N.D.	417.7	374.0
Wheat germ	2556.1	1061.5	89.5	N.D.
Hazelnut	1218.8	42.2	320.4	13.9
Grape seed	82.4	0.3	83.8	20.2
Black cumin seed	58.1	N.D.	185.3	N.D.
Canola	235.1	N.D.	412.9	N.D.
Safflower seed	1063.3	18.4	45.7	N.D.

TABLE 7 Comparison of cold pressed oil in term of relative fatty acid composition (Celenk et al., 2017).

Cold pressed oil	Palmitic acid (%)	Stearic acid (%)	Oleic acid (%)	Linoleic acid (%)	Linolenic acid (%)	Punicic acid (%)	Arachidic acid (%)
Sunflower seed	6.414	4.476	21.316	64.952	0.114	N.D.	0.326
Poppy seed	8.954	2.440	15.793	71.920	0.637	N.D.	0.379
Sesame seed	8.882	5.671	38.374	45.350	0.311	N.D.	0.675
Pomegranate seed	2.508	2.127	5.307	5.187	0.857	80.923	0.520
Nettle seed	7.105	4.002	20.541	66.624	0.578	N.D.	0.427
Flax seed	5.905	3.977	22.703	17.988	54.771	N.D.	0.164
Peanut	9.404	3.329	45.191	30.668	0.251	N.D.	1.239
Pumpkin seed	11.083	6.901	36.481	47.457	0.287	N.D.	0.510
Walnut	6.694	2.877	19.489	59.288	11.339	N.D.	0.149
Wheat germ	16.603	0.869	15.781	55.429	7.532	N.D.	0.252
Hazelnut	5.087	2.346	77.043	14.965	0.265	N.D.	0.151
Grape seed	8.753	4.928	21.070	63.455	0.288	N.D.	0.261
Black cumin seed	11.918	3.635	24.665	55.250	0.329	N.D.	0.234
Canola	4.544	1.866	61.854	20.704	8.374	N.D.	0.590
Safflower seed	5.973	2.334	30.763	58.217	1.005	N.D.	0.566

found to be less than 50% in the remaining cold pressed oils. On the other hand, linoleic acid percentages were as follows: poppy seed oil (71.9%), nettle seed oil (66.6%), sunflower seed oil (64.9%), grape seed oil (63.4%), walnut oil (59.2%), safflower seed oil (58.2%), and black cumin seed oil (55.2%). In addition, punicic acid, known as a conjugated linolenic acid (CLnA), was reported in pomegranate seed oil (80.9%).

Juhaimi, Ozcan, Ghafoor, Babiker, and Hussain (2018) compared cold pressed and Soxhlet extraction systems for antioxidants, phenolics, fatty acids, and tocopherols. Antioxidant activities were 11.4% (peanut) to 67.5% (pistachio) after the cold pressed technique. However, Soxhlet extraction resulted in the range of 11.3% (hazelnut) to 51.2% (pistachio). The β-carotene contents of oils obtained by cold pressing and Soxhlet extraction changed between 7.53 (almond) and 13.84 μg/100 g (peanut). The total phenol content in cold pressed pistachio oil was 2.36 mg gallic acid equivalent/100 g. The catechin contents of cold pressed nut oils were between 0.56 (cashew) and 3.76 μg/100 g (pistachio), whereas Soxhlet-extracted oil varied between 0.64 (cashew) and 3.82 μg/100 g (pistachio). Flavonoid contents of oils obtained by cold pressing changed between 0.09 (peanut) and 3.89 mg/100 g (walnut), and Soxhlet extraction changed between 0.61 (peanut) and 6.37 mg/100 g (pecan). The carotenoid contents of cold pressed and Soxhlet-extracted oils were between 0.26 (cashew) and 1.18 mg/100 g (pecan) to 0.65 (cashew) and 1.97 mg/100 g (pecan), respectively (Table 8).

Anthocyanin content was found as 0.07 (hazelnut) to 0.36 mg/100 g (pistachio) after the cold pressed technique. Soxhlet extraction resulted in the range of 0.16 (apricot) to 0.69 mg/100 g (pistachio). Gallic concentrations in cold pressed oils varied between 0.13 (cashew) and 0.89 μg/100 g (almond). Gallic contents of nut oils obtained by Soxhlet extraction ranged between 0.19 (cashew) and 1.13 μg/100 g (almond). Protocatechuic contents of cold pressed and Soxhlet extracted oils were between 0.24 (almond) and 0.67 μg/100 g (pistachio) to 0.30 (cashew) and 0.74 μg/100 g (pistachio), respectively. The caffeic contents of cold pressed nut oils were between 0.06 (pecan) and 1.13 μg/100 g (pistachio), whereas the contents of Soxhlet-extracted oil varied between 0.14 (pecan) and 1.28 μg/100 g (pistachio). Ferulic concentrations in cold pressed oils ranged between 0.09 (peanut) and 0.43 μg/100 g (pecan). Ferulic contents of nut oils extracted by Soxhlet extraction varied between 0.18 (peanut) and 0.51 μg/100 g (pecan). Sinapic contents of cold pressed and Soxhlet-extracted oils were between 0.11 (peanut) and 0.83 μg/100 g (pistachio) to 0.19 (peanut) and 0.89 μg/100 g (pistachio), respectively. The naringenin contents of cold pressed nut oils were between 0.54 (cashew) and 2.75 μg/100 g (pistachio), whereas

TABLE 8 Bioproperties and phenolics of cold pressed oils (Juhaimi et al., 2018).

	Cold pressed oil							
Bioproperties and phenolics	Almond	Hazelnut	Peanut	Pecan	Walnut	Apricot	Pistachio	Cashew
β-Carotene (μg/100 g)	7.53	8.61	12.47	11.89	10.47	9.61	13.58	10.62
Total phenol (mg GAE/100 g)	1.56	1.29	0.98	0.78	0.84	1.32	2.36	1.13
Antioxidant activity (%)	27.81	13.56	11.43	17.58	14.61	32.45	67.58	17.43
Flavonoid (mg/100 g)	1.13	1.21	0.09	5.27	3.89	1.36	1.17	0.97
Carotenoid (mg/100 g)	0.68	0.56	0.42	0.53	0.76	0.57	1.18	0.26
Anthocyanin (mg/100 g)	0.09	0.07	0.13	0.21	0.19	0.11	0.36	0.05
Gallic (μg/100 g)	0.89	0.21	0.18	0.47	0.36	0.51	0.68	0.13
Protocatechuic (μg/100 g)	0.24	0.47	0.32	0.53	0.42	0.32	0.67	0.28
Catechin (μg/100 g)	1.17	0.89	1.34	2.38	1.67	1.26	3.76	0.56
Caffeic (μg/100 g)	0.43	0.67	0.21	0.06	0.18	0.15	1.13	0.09
Ferulic (μg/100 g)	0.27	0.38	0.09	0.43	0.23	0.31	0.13	0.16
Sinapic (μg/100 g)	0.73	0.42	0.11	0.38	0.41	0.68	0.83	0.23
Naringenin (μg/100 g)	1.38	1.13	0.98	2.17	3.86	0.77	2.75	0.54
Chlorogenic (μg/100 g)	1.41	0.65	0.74	0.53	0.61	1.07	1.45	0.32
p-coumaric (μg/100 g)	0.19	0.17	0.15	0.11	0.22	0.13	0.34	0.19
Rutin (μg/100 g)	0.15	0.09	0.05	0.07	0.09	0.11	0.17	0.14
Resveratrol (μg/100 g)	0.37	0.05	0.09	0.03	0.07	0.29	0.31	0.09
Vanillic (μg/100 g)	0.17	0.03	0.05	0.11	0.18	0.23	0.34	0.16
Kampferol (μg/100 g)	0.32	0.21	0.13	0.15	0.21	0.28	0.45	0.11
Quercetin (μg/100 g)	0.07	0.03	0.12	0.17	0.27	0.09	0.29	0.03
Luteolin (μg/100 g)	0.78	0.97	0.68	0.81	0.55	0.65	1.17	0.21
Pinocembrin (μg/100 g)	0.03	0.05	0.07	0.12	0.16	0.09	0.21	0.18

Soxhlet-extracted oil varied between 0.63 (cashew) and 3.94 μg/100 g (walnut). The chlorogenic contents of cold pressed nut oils were between 0.32 (cashew) and 1.45 μg/100 g (pistachio), whereas Soxhlet-extracted oil varied between 0.39 (cashew) and 1.58 μg/100 g (walnut). *p*-Coumaric contents of cold pressed and Soxhlet-extracted oils ranged from 0.11 (pecan) and 0.34 μg/100 g (pistachio) to 0.17 (apricot) and 0.40 μg/100 g (pistachio), respectively. Rutin concentrations in cold pressed oils changed between 0.05 (peanut) and 0.17 μg/100 g (pistachio) (Table 9).

Rutin contents of nut oils extracted by Soxhlet extraction varied between 0.11 (peanut), and 0.23 μg/100 g (almond and pistachio). Resveratrol contents of cold pressed and Soxhlet-extracted oils ranged from 0.03 (pecan) and 0.37 μg/100 g (almond) to 0.11 (hazelnut) and 0.39 μg/100 g (pistachio), respectively. The vanillic contents of cold pressed nut oils were between 0.03 (hazelnut) and 0.34 μg/100 g (pistachio), whereas Soxhlet-extracted oil varied between 0.13 (peanut) and 0.42 μg/100 g (pistachio). The kampferol contents of cold pressed nut oils were between 0.11 (cashew) and 0.45 μg/100 g (pistachio), whereas Soxhlet-extracted oil varied between 0.19 (pecan and cashew) and 0.51 μg/100 g (pistachio). Quercetin concentrations in cold pressed oils changed between 0.03 (hazelnut and cashew) and 0.29 μg/100 g (pistachio). Quarcetin contents of nut oils obtained by Soxhlet extraction changed between 0.11 (almond) and 0.43 μg/100 g (pistachio). Leteolin contents of cold pressed and Soxhlet-extracted oils ranged from 0.21 (cashew) and 1.17 μg/100 g (pistachio) to 0.36 (cashew) and 1.24 μg/100 g (pistachio), respectively. Pinocembrin contents of cold pressed and Soxhlet-extracted oils ranged from 0.03 (almond) and 0.21 μg/100 g (pistachio) to 0.09 (almond) and 0.36 μg/100 g (pistachio), respectively.

TABLE 9 Bioproperties and phenolics of Soxhlet-extracted oils (Juhaimi et al., 2018).

	Soxhlet extraction oil							
Bioproperties and phenolics	Almond	Hazelnut	Peanut	Pecan	Walnut	Apricot	Pistachio	Cashew
β-Carotene (μg/100g)	8.13	9.27	13.84	13.72	11.61	9.94	15.64	12.49
Total phenol (mg GAE/100g)	1.13	0.95	0.57	0.45	0.77	0.88	1.34	1.03
Antioxidant activity (%)	18.48	11.32	9.63	13.44	12.39	19.86	51.28	12.27
Flavonoid (mg/100g)	1.49	1.87	0.61	6.37	4.89	2.43	2.81	1.66
Carotenoid (mg/100g)	0.76	0.69	0.73	0.88	0.97	0.81	1.97	0.65
Anthocyanin (mg/100g)	0.21	0.33	0.19	0.38	0.27	0.16	0.69	0.18
Gallic (μg/100g)	1.13	0.38	0.27	0.55	0.43	0.59	0.73	0.19
Protocatechuic (μg/100g)	0.35	0.59	0.47	0.62	0.58	0.41	0.74	0.30
Catechin (μg/100g)	1.26	1.09	1.47	2.51	1.78	1.33	3.82	0.64
Caffeic (μg/100g)	0.52	0.78	0.34	0.14	0.27	0.23	1.28	0.17
Ferulic (μg/100g)	0.32	0.45	0.18	0.51	0.34	0.39	0.22	0.21
Sinapic (μg/100g)	0.84	0.51	0.19	0.43	0.41	0.77	0.89	0.34
Naringenin (μg/100g)	1.44	1.22	1.09	2.35	3.94	0.78	2.87	0.63
Chlorogenic (μg/100g)	1.49	0.71	0.83	0.59	0.67	1.23	1.58	0.39
p-Coumaric (μg/100g)	0.24	0.28	0.21	0.19	0.27	0.17	0.40	0.28
Rutin (μg/100g)	0.23	0.16	0.11	0.19	0.21	0.15	0.23	0.21
Resveratrol (μg/100g)	0.42	0.11	0.17	0.19	0.17	0.36	0.39	0.15
Vanillic (μg/100g)	0.24	0.28	0.13	0.23	0.29	0.34	0.42	0.28
Kampferol (μg/100g)	0.41	0.37	0.21	0.19	0.28	0.34	0.51	0.19
Quercetin (μg/100g)	0.11	0.13	0.19	0.23	0.38	0.17	0.43	0.15
Luteolin (μg/100g)	0.89	1.15	0.81	0.98	0.76	0.73	1.24	0.36
Pinocembrin (μg/100g)	0.09	0.13	0.17	0.24	0.28	0.19	0.36	0.27

6 Edible and nonedible applications

Edible cold pressed hazelnut oil is a functional product since the ingredients are expected to have additional health functions over basic nutritional properties. Hazelnut oil is used as salad oil, cooking oil, and for food preparation because of its unique flavor and specific characteristics. On the other hand, hazelnut oil has great potential in the pharmaceuticals and cosmetics industries due to its valuable bioactive substances such as tocopherols and tocotrienols, polyunsaturated fatty acids, free and esterified sterols, phenolics, lignans, squalene, triterpene alcohols, carotenoids, phospholipids, potassium, and magnesium (Boskou, 2017; Eldin & Moreau, 2009). Masson, Mrott, and Bardot (1990) reported that virgin hazelnut oil showed greater and longer moisturizing effects than refined hazelnut oil in a cosmetic emulsion. In particular, the phospholipid concentration of hazelnut oil is directly associated with a skin moisturizing effect.

7 Health-promoting traits

The beneficial effects of nuts on coronary heart disease (CHD) risk has been a very common research subject in recent years. Reports provide sufficient data that the consumption of nut has beneficial effects on CHD risk. These studies included the Adventist Health Study, the Nurses' Health Study, the Cholesterol and Recurrent Events Study, the Iowa Women's Health Study, and the Physicians' Health Study, all of which have provided valid evidence to the science world.

Common outputs of this research are that the consumption of nuts and nut lipids, despite their tremendous variability with respect to fatty acid composition, is beneficial for health, with a special impact on cardiovascular disease risk (Hu & Stampfer, 1999; Kris-Etherton et al., 1999). Many epidemiological studies have shown that consumption of nuts (hazelnuts, almonds, macadamia nuts, pecans, and pistachios) reduces low-density lipoprotein (LDL) and cholesterol concentrations in normo- and hypercholesterolemic people. In 2003, the FDA allowed the following health claim: "Scientific evidence suggests but does not prove that eating 1.5 ounces per day of most nuts as part of a diet low in saturated fat and cholesterol may reduce the risk of heart disease." Without excepting fatty acid composition, other bioactive compounds, such as fiber, phytosterols, antioxidants, folate, copper, magnesium, potassium, boron, and some amino acids contribute to the health-promoting effects of nuts (Abbey, Noakes, Belling, & Nestel, 1994; Almario, Vonghavaravat, Wong, & Kasim-Karakas, 2001; Chisholm et al., 1998; Curb, Wergowske, Dobbs, Abbott, & Huang, 2000; Durak et al., 1999; Edwards, Kwaw, Matud, & Kurtz, 1999; Garg, Blake, & Wills, 2003; Griel et al., 2008; Hyson, Schneeman, & Davis, 2002; Jenkins et al., 2002; Morgan et al., 2002; Morgan & Clayshulte, 2000; Sabaté et al., 1993; Sabaté, Haddad, Tanzman, & Jambazian, 2003; Sheridan, Cooper, Erario, & Cheifetz, 2007; Spiller et al., 1998; Zambon et al., 2000).

The influence of hazelnut oil on lipoprotein and peroxidation status of erythrocytes in rabbits was reported. Hazelnut oil reduced lipid-peroxide levels in the plasma and cholesterol-induced hemolytic anemia (Balkan, Hatipoglu, Aykac-Toker, & Uysal, 2003). The effect of hazelnut oil on rabbits fed a high-cholesterol diet was studied by Hatipoglu et al. (2004). Their study reported a reduction of lipid peroxidation, and no effect on the levels of cholesterol in any of the serum lipoproteins (Hatipoglu et al., 2004). Another study showed that feeding hazelnut oil to laying quail reduced their levels of serum triglycerides. However, no effects on the levels of serum LDL-cholesterol or egg-quality parameters were reported (Guclu, Uyanik, & Iscan, 2008).

Unrefined oils, especially cold pressed, contain variable peptides and proteins. Because of possible allergic reactions, products of nuts are remarked on the EU directive. Hazelnuts contain major allergens such as Cor a 1 (PR-10, accession), Cor a 2 (profilin), Cor a 8 (PR-14, lipid-transfer protein), and Cor a 9 (11S globulin-like protein, accession) (Jin et al., 2008; Robotham et al., 2005).

8 Authenticity and adulteration

Adulteration of oils has become a major problem for importers, producers, and consumers. In particular, adulteration of olive oil with hazelnut oil is a challenge for regulatory agencies and food scientists, because hazelnut oil and olive oil are very similar in terms of triacylglycerols and fatty acids. Many chemical approaches have been experimented with (Marco, Jean, Bordiga, Travaglia, & Garino, 2009). Flores, Castillo, Herraiz, and Blanch (2006) attempted to detect filbertone (*R*-filbertone and *S*-filbertone) in olive oils adulterated with both virgin and refined hazelnut oils by direct analysis (i.e., without sample preparation) using online coupled reversed-phase liquid chromatography and gas chromatography (RPLC-GC). The limit of detection for both enantiomers was 0.03 mg/L. This technique, reported as the proposed procedure based on the use of RPLC-GC coupled by means of a horizontally positioned PTV, may be of great usefulness as a rapid screening method for the detection of olive oils adulterated with 5% or 12% of some virgin and refined hazelnut oils, respectively.

Determination of flavoring components of hazelnut oil by GC, RPLC-GC, and isotopic assays, in addition to assessment of sterols and triacylglycerols in hazelnut oil with NMR and GC, are the proposed methods for the detection of olive oil adulteration with hazelnut oil. Ozen and Mauer (2002) reported FT-IR as a useful instrument for detecting the difference between pure hazelnut oil and other oils as well as to detect hazelnut oil adulteration by sunflower oil. Hazelnut oil adulterated by sunflower oil was detected as accurate to a 2% level. However, FT-IR was unable to detect olive oil adulteration with 5%–20% hazelnut oil. Another study showed that adulteration of extra virgin olive oil with solvent-extracted hazelnut oil can be traced by simple SDS-PAGE (sodium dodecyl sulfate polyacrylamide gel electrophoresis) analysis. This analysis confirmed the presence of hazelnut proteins in solvent-extracted hazelnut oil with molecular masses ranging from 10 to 60 kDa (Marco et al., 2009).

Several authors have described the application of vibrational spectroscopy (near-infrared spectroscopy (NIR), Fourier transform mid-infrared spectroscopy (FT-MIR), Fourier transform Raman spectroscopy (FT-Raman), NMR, and visible absorption spectroscopy) to investigate oil adulteration (Jiménez-Sanchidrián & Ruiz, 2016; Nunes, 2014; Ok, 2016; Sun, Lin, Li, Shen, & Luo, 2015). Dominguez, Sayago, Morales, and Recamales (2019) developed a rapid and simple luminescent method, in combination with advanced chemometric tools, to characterize edible oils and to detect adulterations.

PFG-NMR (high power pulse-field gradient-nuclear magnetic resonance) was employed for analyzing olive oil with adulterated oils of soybean, sunflower, peanut, and hazelnut. The results showed that the PFG-NMR method resulted

TABLE 10 Vitamin E and fatty acid compositions of hazelnut oils from different countries (ND: not detected; <0.1%).

Compound	Croatia	France	Italy	Spain	Turkey
α-Tocopherol (mg/kg)	340	343–545	118–440	314–590	319–492
β-Tocopherol (mg/kg)	ND	ND–10	ND–10	ND–15	ND–10
γ-Tocopherol (mg/kg)	65	ND–10	ND–10	ND–42	ND–130
δ-Tocopherol (mg/kg)	ND	ND–10	ND–10	ND–10	ND
α-Tocotrienol (mg/kg)	ND	ND	ND–209	ND	ND
β-Tocotrienol (mg/kg)	ND	ND	ND	ND–22	ND
γ-Tocotrienol (mg/kg)	ND	ND	ND	ND–34	ND
δ-Tocotrienol (mg/kg)	ND	ND	ND	ND	ND
Myristic acid, 14:0 (%)	ND	ND	ND	ND	ND
Palmitic acid, 16:0 (%)	0.7	0.2–0.8	ND–0.3	0.3–0.7	0.3–1.3
Palmitoleic acid, 16:1 (%)	0.2	0.1–0.2	0.1–0.2	0.1–0.2	0.1
Stearic acid, 18:0 (%)	0.3	0.1–0.4	0.1–0.2	0.1–0.3	0.2–0.7
Oleic acid, 18:1 (%)	86.1	75.6–85.3	84.3–89.6	76.5–84.5	78.7–88.2
Linoleic acid, 18:2 (%)	12.5	13.9–23.6	9.0–14.9	14.5–22.1	11.0–20.0
Linolenic acid, 18:3 (%)	0.1	0.1–0.3	0.1–1.0	0.1–0.2	0.1–0.2
Arachidic acid, 20:0 (%)	ND	ND	ND	ND	ND
Eicosenoic acid, 20:1 (%)	ND	ND–0.4	ND–0.3	ND–0.1	ND

in successful discrimination of authentic and highly adulterated extra virgin olive oil with hazelnut oil (as low as 30%), peanut oil (as low as 30%), with soybean oil (as low as 10%), and sunflower oils (as low as 10%) (Ok, 2016).

Crews et al. (2005) described the composition of authentic hazelnut oils obtained from five major supplier countries. The samples analyzed, in terms of fatty acids, fatty acids in the triacylglycerol 2-position, tocopherols and tocotrienols, triacylglycerols, sterols, steradienes, and iodine value by guiding standard methods. There were no major differences in the composition of oils from different countries. Vitamin E and fatty acid compositions of hazelnut oils from different countries are shown in Table 10.

References

Abbey, M., Noakes, M., Belling, G. B., & Nestel, P. J. (1994). Partial replacement of saturated fatty acids with almonds or walnuts lowers total plasma cholesterol and low-density-lipoprotein cholesterol. *The American Journal of Clinical Nutrition*, *59*, 995–999.

Almario, R. U., Vonghavaravat, V., Wong, R., & Kasim-Karakas, S. E. (2001). Effects of walnut consumption on plasma fatty acids and lipoproteins in combined hyperlipidemia. *The American Journal of Clinical Nutrition*, *74*, 72–79.

Amaral, J. S., Casal, S., Pereira, J. A., Seabra, R. M., & Oliveira, B. P. P. (2003). Determination of sterol and fatty acid compositions, oxidative stability, and nutritional value of six walnut (*Juglans regia* L.) cultivars grown in Portugal. *Journal of Agricultural and Food Chemistry*, *51*, 7698–7702.

Balkan, J., Hatipoglu, A., Aykac-Toker, G., & Uysal, M. (2003). Influence of hazelnut oil administration on peroxidation status of erythrocytes and apolipoprotein B 100-containing lipoproteins in rabbits fed a high cholesterol diet. *Journal of Agricultural and Food Chemistry*, *51*, 3905–3909.

Boskou, D. (2017). Edible cold pressed oils and their biologically active components. *Journal of Experimental Food Chemistry*, *17*(3).

Burg, P., Masan, V., & Rutkowski, K. (2017). Evaluation of the pressing process during oil extraction from grape seeds. *Potravinarstvo Slovak Journal of Food Sciences*, *11*, 1–6.

Cakaloglu, B., Ozyurt, V. H., & Otles, S. (2018). Cold press in oil extraction. A review. *Ukrainian Food Journal*, *7*(4), 640–654.

Celenk, V. U., Gumus, Z. P., Argon, Z. U., Buyukhelvacigil, M., & Karasulu, E. (2017). Analysis of chemical compositions of 15 different cold-pressed oils produced in Turkey: A case study of Tocopherol and fatty acid analysis. *Journal of the Turkish Chemical Society, Section A: Chemistry*, *5*(1), 1–20.

Chemat, F., Vian, M. A., & Cravotto, G. (2012). Green extraction of natural products: Concept and principles. *International Journal of Molecular Sciences*, *13*, 8615–8627.

Chisholm, A., Mann, J., Skeaff, M., Frampton, C., Sutherland, W., Duncan, A., et al. (1998). A diet rich in walnuts favourably influences plasma fatty acid profile in moderately hyperlipidaemic subjects. *European Journal of Clinical Nutrition*, *52*, 12–16.

Crews, C., Hough, P., Godward, J., Brereton, P., Lees, M., Guiet, S., et al. (2005). Study of the main constituents of some authentic hazelnut oils. *Journal of Agricultural and Food Chemistry*, *53*, 4843–4852.

Curb, J. D., Wergowske, G., Dobbs, J. C., Abbott, R. D., & Huang, B. (2000). Serum lipid effects of a high monounsaturated fat diet based on macadamia nuts. *Archives of Internal Medicine*, *160*, 1154–1158.

Dominguez, R. G., Sayago, A., Morales, M. T., & Recamales, A. F. (2019). Assessment of virgin olive oil adulteration by a rapid luminescent method. *Foods*, *8*, 287–296.

Durak, I., Koksal, J., Kacmaz, M., Buyukkocak, S., Cimen, B. M., & Ozturk, H. S. (1999). Hazelnut supplementation enhances plasma antioxidant potential and lowers plasma cholesterol levels. *Clinica Chimica Acta*, *284*, 113–115.

Edwards, K., Kwaw, I., Matud, J., & Kurtz, I. (1999). Effect of pistachio nuts on serum lipid levels in patients with moderate hypercholesterolemia. *Journal of the American College of Nutrition*, *18*, 229–232.

Eldin, A. K., & Moreau, R. A. (2009). *Gourmet and health-promoting specialty oils* (Chapter 3). Urbana, IL: AOCS Press.

FAOStat (2019). *Food and agriculture organization of the united nations, agricultural data.* http://www.fao.org/faostat/en/#data/QC. Accessed 20 September 2019.

Flores, G., Castillo, M. L. R., Herraiz, M., & Blanch, G. P. (2006). Study of the adulteration of olive oil with hazelnut oil by on-line coupled high performance liquid chromatographic and gas chromatographic analysis of filbertone. *Food Chemistry*, *97*, 742–749.

Garg, M. L., Blake, R. J., & Wills, R. B. (2003). Macadamia nut consumption lowers plasma total and LDL cholesterol levels in hypercholesterolemic men. *Journal of Nutrition*, *133*, 1060–1063.

Griel, A. E., Cao, Y., Bagshaw, D. D., Cifelli, A. M., Holub, B., & Kris-Etherton, P. M. (2008). A macadamia nut-rich diet reduces total and LDL-cholesterol in mildly hypercholesterolemic men and women. *The Journal of Nutrition*, *138*, 761–767.

Guclu, B. K., Uyanik, F., & Iscan, K. M. (2008). Effects of dietary oil sources on egg quality, fatty acid composition of eggs and blood lipids in laying quail. *South African Journal of Animal Science*, *38*, 91–100.

Gumus, C. E., Yorulmaz, A. Y., & Aziz, T. (2016). Differentiation of mechanically and chemically extracted hazelnut oils based on their sterol and wax profiles. *Journal of the American Oil Chemists' Society*, *93*, 1625–1635.

Hatipoglu, A., Kanbagli, O., Balkan, J., Kucuk, M., Cevikbas, U., Aykak-Toker, G., et al. (2004). Hazelnut oil administration reduces aortic cholesterol accumulation and lipid peroxides in the plasma, liver and aorta of rabbits fed a high-cholesterol diet. *Bioscience, Biotechnology, and Biochemistry*, *68*, 2050–2057.

Holcepek, M., Lisa, M., Jandera, P., & Kabatova, N. (2005). Quantitation of triacylglycerols in plant oils using HPLC with APCI-MS, evaporative light-scattering, and UV detection. *Journal of Separation Science*, *28*, 1315–1333.

Hu, F. B., & Stampfer, M. J. (1999). Nut consumption and risk of coronary heart disease: A review of epidemiologic evidence. *Current Atherosclerosis Reports*, *1*, 204–209.

Hyson, D. A., Schneeman, B. O., & Davis, P. A. (2002). Almonds and almond oil have similar effects on plasma lipids and LDL oxidation in healthy men and women. *The Journal of Nutrition*, *132*, 703–707.

Imbs, A. B., Nevshupova, N. V., & Pham, L. Q. (1998). Triacylglycerol composition of *Pinus koraiensis* seed oil. *Journal of the American Oil Chemists' Society*, *75*(7), 865–870.

Ionescu, M., Vladut, V., Ungureanu, N., Dibca, M., Zabava, B. S. T., & Stefan, M. (2016). Methods for oil obtaining from oleaginous materials. *Annals of the University of Craiova-Agriculture, Montanology, Cadastre Series*, *46*.

Jenkins, D. J., Kendall, C. W., Marchie, A., Parker, T. L., Connelly, P. W., Qian, W., et al. (2002). Dose response of almonds on coronary heart disease risk factors: Blood lipids, oxidized low-density lipoproteins, lipoprotein(a), homocysteine, and pulmonary nitric oxide: A randomized, controlled, crossover trial. *Circulation*, *106*, 1327–1332.

Jiménez-Sanchidrián, C., & Ruiz, J. R. (2016). Use of Raman spectroscopy for analyzing edible vegetable oils. *Applied Spectroscopy*, *51*, 417–430.

Jin, T., Albillos, S. M., Chen, Y.-W., Kothary, M. H., Fu, T.-J., & Zhang, Y.-Z. (2008). Purification and characterization of the 7S Vicilin from Korean pine (*Pinus koraiensis*). *Journal of Agricultural and Food Chemistry*, *56*, 8159–8165.

Jokic, S., Moslavac, T., Aladic, K., Bilic, M., Ackar, D., & Subaric, D. (2016). Hazelnut oil production using pressing and supercritical CO_2 extraction. *Hemijska Industrija*, *70*(4), 359–366.

Juhaimi, F. A., Ozcan, M. M., Ghafoor, K., Babiker, E. E., & Hussain, S. (2018). Comparison of cold-pressing and soxhlet extraction systems for bioactive compounds, antioxidant properties, polyphenols, fatty acids and tocopherols in eight nut oils. *Food Science and Technology*, *55*(8), 3163–3173.

Kris-Etherton, P. M., Yu-Poth, S., Sabate, J., Ratcliffe, H. E., Zhao, G., & Etherton, T. D. (1999). Nuts and their bioactive constituents: Effects on serum lipids and other factors that affect disease risk. *The American Journal of Clinical Nutrition*, *70*, 504S–511S.

Maguire, A. R., O'Sullivan, M., Galvin, K., O'Connor, T. P., & O'Brien, N. M. (2009). Fatty acid profile, tocopherol, squalene and phytosterol content of walnuts, almonds, peanuts, hazelnuts and the macadamia nut. *International Journal of Food Sciences and Nutrition*, *55*(3), 171–178.

Marco, A., Jean, D. C., Bordiga, M., Travaglia, F., & Garino, C. (2009). Olive oil adulterated with hazelnut oils: Simulation to identify possible risks to allergic consumers. *Food Additives and Contaminants*, *27*(01), 11–18.

Martinez, M. L., Bordon, M. G., Bodoria, R. M., Penci, M. C., Ribotta, P. D., & Maestri, D. M. (2017). Walnut and almond oil screw-press extraction at industrial scale: Effects of process parameters on oil yield and quality. *Grasas y Aceites*, *68*(4), 1–10.

Martinez, M. L., Bordon, M. G., Lallana, R. L., Ribotta, P. D., & Maestri, D. M. (2017). Optimization of sesame oil extraction by screw-pressing at low temperature. *Food and Bioprocess Technology*, *10*, 1113–1121.

Masson, P., Mrott, F., & Bardot, J. (1990). Influence of hazelnut on the skin moisturizing effect of a cosmetic emulsion. *International Journal of Cosmetic Science*, *12*(6), 243–251.

Miraliabari, H., & Shahidi, F. (2007). Lipid class compositions, tocopherols and sterols of tree nut oils extracted with different solvents. *Journal of Food Lipids*, *15*, 81–96.

Morgan, W. A., & Clayshulte, B. J. (2000). Pecans lower low-density lipoprotein cholesterol in people with normal lipid levels. *Journal of the American Dietetic Association*, *100*, 312–318.

Morgan, J. M., Horton, K., Reese, D., Carey, C., Walker, K., & Capuzzi, D. M. (2002). Effects of walnut consumption as part of a low-fat, low-cholesterol diet on serum cardiovascular risk factors. *International Journal for Vitamin and Nutrition Research*, *72*, 341–347.

Nunes, C. A. (2014). Vibrational spectroscopy and chemometrics to assess authenticity, adulteration and intrinsic quality parameters of edible oils and fats. *Food Research International*, *60*, 255–261.

Ok, S. (2016). Detection of olive oil adulteration by low-field NMR relaxometry and UV-Vis spectroscopy upon mixing olive oil with various edible oils. *Grasas y Aceites*, *68*(1), 173–186.

Ozen, B. F., & Mauer, L. J. (2002). Detection of hazelnut oil adulteration using FT-IR spectroscopy. *Journal of Agricultural and Food Chemistry*, *50*, 3898–3901.

Robotham, J. M., Wang, F., Seamon, V., Teuber, S. S., Sathe, S. K., Sampson, H. A., et al. (2005). Ana o 3, an important cashew nut (*Anacardium occidentale* L.) allergen of the 2S albumin family. *Journal of Allergy and Clinical Immunology*, *115*, 1284–1290.

Ryan, E., Galvin, K., O'Connor, T. P., Maguire, A. R., & O'Brien, N. M. (2009). Fatty acid profile, tocopherol, squalene and phytosterol content of brazil, pecan, pine, pistachio and cashew nuts. *International Journal of Food Sciences and Nutrition*, *57*(3–4), 219–228.

Sabaté, J., Fraser, G. E., Burke, K., Knutsen, S. F., Bennett, H., & Lindsted, K. D. (1993). Effects of walnuts on serum lipid concentrations and blood pressure in normal men. *New England Journal of Medicine*, *328*, 603–607.

Sabaté, J., Haddad, E., Tanzman, J. S., & Jambazian, P. S. (2003). Serum lipid response to the graduated enrichment of a step I diet with almonds: A randomized feeding trial. *The American Journal of Clinical Nutrition*, *77*, 1379–1384.

Sheridan, M. J., Cooper, J. N., Erario, M., & Cheifetz, C. E. (2007). Pistachio nut consumption and serum lipid levels. *Journal of the American College of Nutrition*, *26*, 141–148.

Singh, K. K., Jhamb, S. A., & Kumar, R. (2010). Effect of pretreatments on performance of screw pressing for flaxseed. *Journal of Food Process Engineering*, *35*, 543–556.

Singh, K. K., Wiesenborn, D. P., Tostenson, K., & Kangas, N. (2002). Influence of moisture content and cooking on screw pressing of Crambe seed. *Journal of the American Oil Chemists' Society*, *79*, 165–170.

Spiller, G. A., Jenkins, D. A., Bosello, O., Gates, J. E., Cragen, L. N., & Bruce, B. (1998). Nuts and plasma lipids: An almond-based diet lowers LDL-C while preserving HDL-C. *Journal of the American College of Nutrition*, *17*, 285–290.

Sun, X. D., Lin, W. Q., Li, X. H., Shen, Q., & Luo, H. Y. (2015). Detection and quantification of extra virgin olive oil adulteration with edible oils by FT-IR spectroscopy and chemometrics. *Analytical Methods*, *7*, 3939–3945.

Zambon, D., Sabate, J., Muno, S., Campero, B., Casals, E., Merlos, M., et al. (2000). Substituting walnuts for monounsaturated fat improves the serum lipid profile of hypercholesterolic men and women. *Annals of Internal Medicine*, *7*, 538–546.

Chapter 21

Cold pressed camelina (*Camelina sativa* L.) seed oil

Pelin Günç Ergönül and Zeynep Aksoylu Özbek
Department of Food Engineering, Faculty of Engineering, Manisa Celal Bayar University, Manisa, Turkey

Abbreviations

ALA	α-linolenic acid
BHA	butylated hydroxyanisol
BHT	butylated hydroxytoluene
DHA	docosahexaenoic
DM	dry matter
EDCL	enzyme-degummed camelina lecithin
EPA	eicosapentaenoic acid
FFA	free fatty acid
GSLs	glucosinolates
LC-MS	liquid chromatography mass spectroscopy
LDL	low-density lipoprotein
MUFA	monounsaturated fatty acid
PDSC	pressure differential scanning calorimetry
PG	propyl gallat
PUFA	polyunsaturated fatty acid
PV	peroxide value
SC-CO_2	supercritical CO_2
SFA	saturated fatty acid
SPE	solid phase extraction
TBHQ	tertbutylhydrooquinone
TPC	total polyphenol content
WDCL	water degummed camelina lecithin

1 Introduction

Camelina can be easily grown with low inputs of water, pesticides, and fertilizers. Camelina are early maturing crops requiring shorter growing days from planting to maturity and show higher tolerance to the cold. Camelina can either be grown as a spring or winter crop (Obour, Sintim, Obeng, & Jeliazkov, 2015). Many archeological excavations have revealed that camelina was grown at least 3000 years ago in Europe and it is also known as a natural plant of Northern Europe and Central Asia. After the 1960s, agricultural researches on camelina plants were carried out in both Europe and North America and, as a result of these studies, camelina has been reported to be a suitable plant for sustainable agricultural systems. In recent years, the importance of camelina has become more important with the idea of supplying omega-3 fatty acids from plant sources. Cold pressed camelina seed oil appears very attractive due to its unique fatty acid profile, particularly essential fatty acids such as linoleic (omega-6), and α-linolenic (omega-3) fatty acids, its high content of tocopherol isomers, phytosterols, and phenolic compounds, and it has taken its place among the alternative oil plants, which can contribute to the reduction of health problems caused by unbalanced nutrition, especially in developed industrial countries. The cold pressed camelina oil has clarity and pale straw color relative to canola oil and it is suitable for the production of cold pressed unbleached healthy oils (Francis & Campbell, 2003). Depending on the cold pressing procedure, cold pressed camelina seed oil may comprise some phytochemicals including natural antioxidants. Due to the high oil content,

Cold Pressed Oils. https://doi.org/10.1016/B978-0-12-818188-1.00021-9
Copyright © 2020 Elsevier Inc. All rights reserved.

omega-3 fatty acid, and crude protein, finding an alternative use of camelina meal (a coproduct obtained from camelina seed after oil extraction) in animal diets will increase the market value of the crop (Cherian, 2012). After oil extraction, cold pressed meal contains 100–150 g/kg of oil and 400 g/kg of protein with a low fiber content, making it an attractive animal feed (Berti, Russ, Eynck, Anderson, & Cermak, 2016).

Camelina meal is rich in protein, lipids, and essential n-3 and n-6 fatty acids, and could be incorporated into poultry rations as a source of energy, protein, and essential n-3 and n-6 fatty acids (Ibrahim & El Habbasha, 2015). Camelina meal consists of 13% residual oil, 6% ash, 12% crude fiber, 30% crude protein, 27% nonnitrogenous matter, and other substances such as vitamins. The gross energy is 4600–4800 kcal/kg. The oil content, fatty acid composition, and other nutrient profiles of the meal can vary due to cultivar, season, processing method, and other agronomic factors (e.g., soil type) (Cherian, 2012). Cold pressed camelina cake may contain 10%–30% oil, and its oil contains 35% α-linolenic acid (Karvonen et al., 2002).

The fatty acid composition of camelina meal has received considerable attention due to its high content of essential fatty acids. The meal is rich in omega-3 and omega-6 essential fatty acids (also known as n-3 and n-6 fatty acids). The omega-6:omega-3 fatty acid ratio is 0.90 to 0.70. α-Linolenic acid (18:3n3) is the major omega-3 fatty acid, constituting over 29%, with linoleic acid (18:2n-6) constituting up to 23%. Oleic acid is the major monounsaturated fatty acid, followed by eicosenoic acid (20:1). Other monounsaturated fatty acids (MUFAs) include palmitoleic (16:1) and erucic acids (22:1, $<2\%$). Altogether, total MUFAs constitute more than 32%. Saturated fatty acids (SFAs) in the meal include palmitic acid (9%) and stearic acid (2.5%) (Cherian, 2012).

In camelina meal, protein content is about 30%–35% dry matter (DM) basis. A large part of this percentage is seed storage proteins (Ibrahim & El Habbasha, 2015). The protein of camelina oil cakes, a coproduct of camelina seed pressing, is characterized by attractive amino acids, such as arginine, histidine, cystine, glycine, lysine, methionine, and threonine. The composition of the amino acids indicates that the protein is especially suitable for poultry feed (Cherian, 2012; Zubr, 1997). For these reasons, cold pressed camelina cake is becoming increasingly available for animal feeding (Woyengo, Patterson, Slominski, Beltranena, & Zijlstra, 2016). Lawrence, Anderson, and Clappert (2016) found that the percentage of total digestibility of cold pressed camelina meal protein is comparable with linseed and soyabean meals, which makes it valuable as an alternative protein source for dairy cattle. The presence of camelina meal or oil in animal diets also raised milk yield and changed milk composition in dairy cattle. Butter and other dairy products produced from milk of cows fed with camelina oil also had an increased content of conjugated linoleic acid, and better spreadability (Hurtaud & Peyraud, 2007; Ionescu, Ivopol, Neagu, Popescu, & Meghea, 2015). Many studies indicate that fish, chicken, swine, cattle, and sheep fed with camelina oil or meal have lower blood plasma cholesterol and higher contents of α-linolenic acid (ALA) or eicosapentaenoic acid (EPA), and docosahexaenoic (DHA) fatty acids (Berti et al., 2016).

Camelina meal is characterized by the presence of antinutritional compounds that affect the value of this by-product. In particular, plant secondary metabolites such as glucosinolates (GSLs), sinapine, inositol phosphates, erucic acid, and condensed tannins belong to widespread antinutritive compounds, which are generally present in oilseeds. Some studies have shown that high contents of camelina meal in fish diets decreases fish weight due to these antinutritional compounds in the meal (Ye, Anderson, & Lall, 2016). GSLs and sinapine have usually been associated with members of Brassicaceae, whereas inositol phosphates and condensed tannins are more generally distributed in flora (Ibrahim & El Habbasha, 2015). Camelina contains three main glucosinolates: glucoarabin (9-methyl-sulfinylnonylglucosinolate; GS9), glucocamelinin (10-methylsulfinyl-decyl glucosinolate; GS10), and 11-methyl-sulfinyl-undecyl glucosinolate (GS11). Compared to monogastric animals, ruminants are more tolerable to glucosinolates. Nevertheless, the federal regulation recommends not feeding meals containing glucosinolates in excess of 10% in the diet (Benz, 2010). Hurtaud and Peyraud (2007) was found that feeding lactating dairy cattle with camelina meal resulted in decreased milk fat yield and content, and changed fatty acid composition. Camelina meal is also evaluated as a fish meal replacement because of its high content of crude protein (45%) (Hixson, Parrish, & Anderson, 2014).

In camelina meal, carbohydrates of camelina include monosaccharides, disaccharides, oligosaccharides, polysaccharides, and fiber. Monosaccharides and disaccharides are easily digestible and in the human body provide easily metabolizable energy (Ibrahim & El Habbasha, 2015). The content of crude fiber in camelina meal is about 15% dry matter basis. The substantial part of crude fiber is cellulose. The proportionally high content of mucilage, crude fiber, and lignin indicates that camelina meal, when incorporated in food, can exert positive effects on gastrointestinal processes. A long-term human consumption of bread with added camelina meal confirmed the beneficial role of the ingredient in digestion (Ibrahim & El Habbasha, 2015).

Camelina meal is a good source of vitamins B1 (thiamin), B3 (niacin), and B5 (pantothenic acid). Thiamin in nature exists as thiamine pyrophosphate. It functions as a coenzyme in transketolation and is important in neural transmission. It is directly involved in maintenance of normal appetite and healthy attitude (Berdanier, 2002). Fat-soluble vitamin (tocopherols) content of the meal is over 200 μg/g (Cherian, 2012).

Analyses of camelina seed reveal a prevalently low content of macrominerals. The highest contents, between 1.0% and 1.6%, are calcium, potassium, and phosphorus. Among microminerals, camelina presents markedly high contents of iron (329 μg/g), manganese (40 μg/g), and zinc (69 μg/g) (Zubr, 2010). Among the minerals in camelina meal, potassium is the major mineral, followed by sulfur, phosphorus, magnesium, and calcium (Cherian, 2012). Mieriòa et al. (2017) determined total polyphenol content (TPC) in the meal extracts varied from 500 mg to more than 1500 mg GAE/100 g meal after cold pressing of oil. The greatest TPC occurred for ethanol extract of meal that was not defatted prior to the extraction.

The high protein, energy, omega-3 and omega-6 fatty acid, and essential amino acid content of camelina meal makes it a potentially suitable source of plant protein, and of essential fatty acids and amino acids for use in poultry rations (Cherian, 2012; Zubr, 1997). Recently, these oil cakes were used with success in a fodder mixture for egg-laying hens. A feeding experiment has proven that camelina oil cakes in fodder mixture (15%, w/w) had a similar biological value to extracted soybean meal. No negative effect of camelina oil cakes on the organoleptic quality of eggs and meat was observed. Nevertheless, camelina oil cakes in excess of 15% in the fodder mixture affected the flavor of eggs negatively (Zubr, 1997).

Boyle, Hansen, Hinnenkamp, and Ismail (2018) determined the effect of oil pressing conditions and protein extraction methods on protein yield and content from cold pressed defatted camelina meal. They obtained the oil by cold and hot pressing then they produced camelina protein concentrate from cold pressed and hot pressed defatted camelina meals. They found that the oil pressing conditions did not significantly affect protein yield and content. Nevertheless, with this research, the potential of camelina as a novel source of functional plant protein at a concentrate of ~70%–80% was revealed.

2 Extraction and processing of cold pressed oil

The cold pressing technique is applied in two ways: screw press and/or hydraulic press. It is an alternative technique to conventional solvent extraction method because it does not require the use of organic solvent or heat. This technique is affected by temperature, frequency, and press nozzles size (Popa, Jurcoane, & Dumitriu, 2017). Some researchers have attempted to optimize the camelina oil extraction by cold pressing (Moslavac et al., 2014). The best results were obtained at a temperature of 52°C, a frequency of 20 Hz, and a nozzle size of 9 mm. Cold pressing is capable of preserving bioactive compounds such as phenolics, essential fatty acids, flavonoids, and tocopherols in the oil (Moslavac et al., 2014).

Supercritical CO_2 (SC-CO_2) fluid extraction is an alternative technique to cold pressing. Bozan and Temelli (2002) observed higher ALA content in oil obtained from SC-CO_2 extraction compared to Soxhlet extraction. In contrast, tocopherol content was lower, suggesting the potential for lower oil stability. Temperatures (50°C and 70°C) and pressures (35 and 55 MPa) had little effect on fatty acid profiles. The stability of oil was not reported, thus further research is required regarding shelf life and comparisons to the cold pressing operation. When processed correctly, cold pressed oil has low peroxide values (PV) and low free fatty acid (FFA) contents. In a study carried out on the physicochemical properties and oxidative stability of cold pressed camelina seed oils sold commercially in Poland, the PV was found to be approximately 1.53 meq O_2/kg, while the FFA value was 0.64% (Raczyk, Popis, Kruszewski, Ratusz, & Rudzinska, 2016). The acid content and PV in tested cold pressed camelina oils were similar to results presented by Ratusz, Popis, Zytkiewicz, and Wroniak (2016). They found these values to be 0.14%–0.86% and 0.94–2.04 meq O_2/kg, respectively.

It is very important that cold pressed oils are low in moisture content and FFA to maintain the quality and shelf life of the oil (Teh & Birch, 2013). Limited data are available on the effects of storage conditions on oil stability. However, the oxidation of camelina seed oil was affected by the temperature of storage, with greater effects by light, especially at the initial stage of storage. Moslavac et al. (2014) determined oxidative stability of natural antioxidant added to cold pressed camelina oil by the Schall oven test at 63°C for 4 days. The addition of natural antioxidants of less than 0.1% in camelina oil had no effect on the stability of the oil (Gramza et al., 2006), but addition in concentration higher than 0.3% changed the organoleptic properties of the oil. The natural antioxidants used were green tea extract, pomegranate extract, rosemary extract, and olive leaf extract. Rosemary extract showed the highest protection efficiency, resulting in lower PV (2.46 mmol O_2/kg) after 4 days at concentration 0.3%. The pomegranate extract (PV 15.19 mmol O_2/kg) did not dissolve well in the oil. Adding olive leaf extract did not show an antioxidant effect on camelina oil. As for green tea extract, it showed the lowest effect on the oxidative stability of cold pressed camelina seed oil, with a PV of 19.80 mmol O_2/kg. Raczyk et al. (2016) compared Rancimat and pressure differential scanning calorimetry (PDSC) methods in determining oxidative stability of cold pressed camelina oil. Concerning the Rancimat method, the oxidative stability index of camelina oils ranged from 4.58 to 6.18 h, while in the thermoanalytical PDSC method, it was 30.6 min. In these two methods, the results showed similar rank values. The important correlation was also revealed between two methods by Dantas et al. (2011) and Kowalski, Gruszczynska, and Maciaszek (2000).

Raczyk et al. (2016) reported the anisidine value of cold pressed camelina oils, which indicated the secondary oxidation products content at a range of 0.46–1.60. It is typical for fresh cold pressed oils, due to no heat treatment being applied during extraction. Terpinc, Polak, Makuc, Ulrih, and Abramovic (2012) investigated the phenolic compounds and their occurrence in camelina seed cake and camelina seed oil after solvent extraction and solid phase extraction (SPE) purification steps by liquid chromatography mass spectroscopy (LC-MS). They found that after oil was pressed from the seeds, almost all phenolic compounds detected remain in the seed residues. Among the phenolic acids that are well-known for their antioxidant behavior, three hydroxybenzoic acids (protocatechuic, *p*-hydroxybenzoic, and salicylic) and one hydroxycinnamic acid (sinapic) were identified. Rutin, catechin, and quercetin are flavonoids widespread in nature that act as powerful antioxidants and possess antimicrobial, anticarcinogenic, and antimutagenic properties. Onyilagha et al. (2003) reported the presence of quercetin in camelina leaves. Salminen, Estevez, Kivikari, and Heinonen (2006) found that in camelina cake, the concentration of flavonols predominated over that of hydroxycinnamic acids, and they also contribute to its antioxidant activity, especially the quercetin glycosides.

Belayneh, Wehling, Cahoon, and Ciftci (2015) compared the oil yield and oil composition (fatty acids, tocopherols, and phytosterols) of camelina seed oil obtained by different extraction techniques such as SC-CO_2, cold press, and hexane extraction. They found that the SC-CO_2 extraction technique was more efficient to extract oil from camelina seed than conventional methods (*n*-hexane extraction and cold pressing). The total oil yield obtained with Soxhlet (*n*-hexane) extraction for 6h was 35.9%, whereas it was 31.6% and 29.9% with SC-CO_2 extraction and cold pressing, respectively (Berti et al., 2016). In respect to oil composition, the extraction method did not have a significant effect on the fatty acid composition of the oils, but it affected tocopherol content. The tocopherol content of SC-CO_2-extracted, Soxhlet-extracted, and cold pressed camelina oils were 801, 682, and 695 mg/kg, respectively. The higher phytosterol content was found from SC-CO_2-extracted oil to be 1419 mg/kg, whereas it was 1408 mg/kg and 568 mg/kg in the Soxhlet and cold pressed oils, respectively. SC-CO_2 extraction has the facility to extract more phytosterols than other conventional techniques at high pressures and temperatures, as reported by some researchers (Ciftci, Calderon, & Temelli, 2012; Shukla, Dutta, & Artz, 2002; Szterk, Roszko, Sosinska, Derewiaka, & Lewicki, 2010).

3 Fatty acids and acyl lipids profile of cold pressed oil

Camelina seed oil contains 35%–40% of α-linolenic acid (C18:3 ω3, ALA), which is an essential omega-3 fatty acid (Belayneh et al., 2015). ALA can be converted to eicosapentaenoic acid (EPA) and docosahexaenoic acid (DHA); however, the conversion of ALA to EPA and DHA is limited and affected by the level of the dietary intake of ALA and linoleic acid. The optimal intake of ALA was reported as 2 g/day or 0.6%–1% of the total energy intake. Camelina seed oil can be used as a source of ALA as an alternative to EPA and DHA from marine sources (Belayneh, Wehling, Cahoon, & Ciftci, 2017). Some researchers reported that camelina seed oil is composed of palmitic acid (5.3%–6.4%), stearic acid (1.4%–3.0%), oleic acid (14.0%–18.7%), linoleic acid (13.5%–19.0%), α-linolenic acid (27.1%–38.9%), arachidic acid (0.4%–1.5%), gondoic acid (11.6%–16.2%), eicosadienoic acid (1.7%–2.1%), eicosatrienoic acid (1.3%–1.7%), and erucic acid (1.6%–4.0%) (Abramovic & Abram, 2005; Eidhin, Burke, & O'Beirne, 2003).

Camelina seed oil composition change depends on the cultivar, location, environment, and extraction method. Generally, the linolenic acid content varies with temperature during seed development in all oilseeds (Berti et al., 2016). The fatty acid composition of cold pressed camelina oil is given and compared with the values determined by some studies (Berti et al., 2016; Raczyk et al., 2016; Wu & Leung, 2011; Yang, Claude, Corscadden, Sophia He, & Li, 2016) in Table 1. Linolenic acid was the major fatty acid, followed by linoleic acid, oleic acid, gadoleic acid, palmitic acid, erucic acid, stearic acid, and other fatty acids with trace amounts.

Due to the high content of unsaturated fatty acids in camelina oil, its oxidative stability should be an important factor. Camelina oil was found to be more stable toward oxidation than highly unsaturated linseed oil, but less stable than rapeseed, olive, corn, sesame, and sunflower oils. Camelina oil has an unusual fatty acid profile, consisting of higher levels of α-linolenic acid and comparatively low concentrations of erucic acid. Camelina oil can serve as **an interesting** source of n-3 (omega-3) fatty acid due to its cholesterol-lowering properties in the human diet (Karvonen et al., 2002).

According to recent research, camelina seed oil should become a new potential source of lecithin (Belayneh et al., 2017; Belayneh, Wehling, Cahoon, & Ciftci, 2018). Belayneh et al. (2018) investigated the chemical composition and emulsifying properties of lecithin obtained from cold pressed camelina seed oil by water and enzymatic degumming for the first time. They found that ALA (C18:3) was the predominant fatty acid in both water-degummed camelina lecithin (WDCL), and enzyme-degummed camelina lecithin (EDCL), followed by linoleic acid (C18:2). ALA content in the EDCL and WDCL was 28.3% and 34.3%, respectively. Fatty acid results indicate that camelina lecithin is an omega-3 rich emulsifier for food and pharmaceutical industries. EDCL contained 17.4% linolenic acid, 16.5% eicosenoic acid (C20:1), and 15.5% oleic acid (C18:1), whereas WDCL contained 19.6% linolenic acid, 12.5% eicosenoic acid (C20:1), and 15.5% oleic acid (C18:1). The content of saturated fatty acids in EDCL was higher than in WDCL and cold pressed camelina seed oil

TABLE 1 The fatty acid composition (%) of cold pressed camelina seed oil from different published studies.

	Yang et al. (2016)	Wu et al. (2011)	Berti et al. (2016)	Raczyk et al. (2016)
Miristic acid	0.1	0.2	–	–
Palmitic acid	5.5	5.1	5.7	6.1
Palmitoleic acid	0.1	0.3	–	0.1
Stearic acid	2.4	2.4	2.4	2.6
Oleic acid	14.4	17.6	15.7	16.1
Linoleic acid	19.1	18.7	18.5	18.5
Linolenic acid	33.5	28.6	32.8	35.8
Arachidic acid	1.5	1.8	1.7	1.3
Gadoleic acid	15.0	11.9	15.1	14.0
Eicosadienoic acid	2.2	1.9	1.8	1.8
Arachidonic acid	1.4	–	–	1.4
Behenic acid	0.3	0.8	0.4	–
Erucic acid	3.1	4.2	3.5	2.3
Clupanodinic acid	0.2	0.4	0.5	–
Docosatrienoic acid	0.4	–	–	–
Lignoseric acid	0.2	–	–	–
Nervonic acid	0.6	–	0.7	–

(Belayneh et al., 2018). The emulsion stability of camelina lecithins (obtained from both techniques) was poor when compared with soy lecithin at high pH (7.5). However, camelina lecithins can be used as an omega-3 rich alternative emulsifier and functional ingredient for different food systems due to its precise chemical composition.

4 Health-promoting traits of cold pressed oil and oil constituents

Some studies in human nutrition and health have determined the relationship between the diet and the occurrence of various diseases among the population in industrialized countries. The nutritional deficiency due to the disproportion of polyunsaturated fatty acids (PUFAs) can be alleviated by the addition of n-3 fatty acid-rich oils in the diet. In such a situation, camelina oil can be an excellent source of PUFAs and n-3 fatty acids in particular. Camelina oil can enhance the biological value of diet by changing the proportion of n-6/n-3 fatty acids (Skjervold, 1993). Camelina is being marketed in Europe in salad dressing and as cooking oil, but is not suitable as deep-fat frying oil (Ibrahim & El Habbasha, 2015).

4.1 Health-promoting effects of camelina oil

The human body cannot synthesize ALA and its deficiency may result in clinical symptoms including neurological abnormalities and poor growth. Therefore, ALA should be included in the diet. ALA can be elongated to EPA and DHA, the metabolic products of which have beneficial effects that help in preventing coronary heart disease, arrhythmias, and thrombosis. The consumption of camelina oil can help to improve the general health of the population to the desired level (Rokka, Alen, Valaja, & Ryhanen, 2002; Zubr, 1997). Camelina oil is helpful in the regeneration of cells, skin elasticity, and slenderness recovery (Vollmann, Damboeck, Eckl, Schrems, & Ruckenbauer, 1996). The preventive and curative effects were ascribed primarily to a reinforced immunity of the human organism. The immunity was apparently deriving from biochemical processes in which linoleic acid, α-linolenic acid, tocopherols, and phytosterols were involved. Both the linoleic acid and α-linolenic acid are precursors for pure unsaturated fatty acids and are the substrates for important hormones with various functions in the human organism (Calder, 2001). Motivated by the unique nutritional quality and beneficial properties of camelina oil, flax oil in the diet was replaced with camelina oil. The major ingredients (c.80%, *v*/v fermented milk and 20% v/v camelina oil) were mixed to obtain emulsion. The fermented milk was a source of essential amino acids and microorganisms (*Lactobacillus acidophilus*), with well-known positive dietary effects. The oil provided linoleic acid, α-linolenic acid, tocopherols, phenols, and phytosterols. During testing with adults, eight tablespoons of fermented milk and two tablespoons of camelina oil were used per serving. To improve the nutritional value and taste, oats flakes and seasonally available small fruits, minced fruits or vegetables, dry fruits, grape raisins, jam, and sugar were mixed with the emulsion. The complex mixture was consumed with two or three slices of toast. The most convenient was the consumption of the diet at breakfast (Ibrahim & El Habbasha, 2015).

4.2 Cholesterol-reducing effect of camelina oil

The high contents of ALA, tocopherols and other antioxidants make camelina oil nutritionally very attractive. Besides being a substrate in human metabolism, ALA is capable of improving the n-6/n-3 fatty acid ratio in food (Simopoulos, 1999). A mixed fat product consisting of butterfat and camelina oil (1:1) was tested by mildly hypercholesterolemic subjects. The volunteers, aged 25–75 years (14 males and 31 females), during 4 weeks consumed 50–60 g/d of the mixed fat. Their habitual diet was maintained, only fats (butter, margarine, and oil) were substituted with the tested product. Blood analyses were performed during morning hours at 2-week intervals. The initial mean content of total cholesterol in blood serum of the subjects was 6.3 mmol/L. After 2 weeks on the diet, the volunteers experienced decreasing cholesterol levels. At the end of the trial, their cholesterol in blood serum was reduced to 5.8 mmol/L. The cholesterol-reducing effect was ascribed to camelina oil. A similar cholesterol-reducing effect of camelina oil was achieved in a test with mildly and moderately hypercholesterolemic subjects. The volunteers consumed 33 mL camelina oil per day for 6 weeks. Their total cholesterol in blood serum was reduced from 5.9 to 5.6 mmol/L and LDL (low-density lipoprotein) decreased by 12.2%. In another trial, volunteers during 4 weeks consumed 12 g/d α-linolenic acid in the form of ground flax seed (50 g/d) or flax oil (20 g/d).

The content of long chain n-3 fatty acids and erythrocyte lipids in blood serum of the subjects increased significantly. Simultaneously serum total cholesterol was lowered by 9% and LDL (low-density lipoprotein) cholesterol by 18%. A provision of functional oil with flax oil reduced the total cholesterol by 12.5% and LDL by 13.9% (Onge, Lamarche, Mauger, & Jones, 2003). Meanwhile, unusually high content of cholesterol in camelina oil was reported by Shukla et al. (2002) as 188 mg/g. Karvonen et al. (2002) determined the cholesterol-reducing effect of camelina oil in a test with mildly and moderately hypercholesterolemic subjects. The volunteers consumed 33 mL camelina oil per day for 6 weeks. Their total cholesterol in blood serum was reduced from 5.9 to 5.6 mmol/L and LDL decreased by 12.2%. Experiments showed that camelina oil possesses a cholesterol-reducing property. In addition to the effects of α-linolenic acid and tocopherols, phytosterols were also found to be effective in lowering cholesterol (Ortega, Palencia, & Lopez-Sobaler, 2006). Preliminary unpublished experimental results indicate that the amount of cholesterol in blood serum is not correlated to the dietary intake. The major determinants of cholesterol level in blood serum are saturated fatty acids (C12:0–C16:0). The development of atherosclerosis actually stems from the oxidation of LDL (Ibrahim & El Habbasha, 2015).

5 Minor bioactive lipids in cold pressed oil

The oil is very rich in natural antioxidants, such as tocopherols, making this highly stable oil very resistant to oxidation and rancidity. The vitamin E content of camelina oil is approximately 110 mg/100 g oil. It is well-suited for use as a cooking oil (Ibrahim & El Habbasha, 2015). Mieriòa et al. (2017) advocated that the oxidative stability of camelina oil was even higher than that of rapeseed or flaxseed oil despite its fatty acid composition. This argument is consistent with findings that camelina oil in comparison with other vegetable oils is rich in natural antioxidant tocopherols (Belayneh et al., 2015): the amount of total tocopherols varies from 700 mg/kg oil (Belayneh et al., 2015) to 830 mg/kg (Zubr & Matthäus, 2002). γ-Tocopherol has been identified as the major tocopherol in camelina oil by Zubr and Matthäus (2002) and Abramovic, Butinar, and Nikolicˇ (2007). Similarly, Grajzer et al. (2015) and Ergönül and Özbek (2018) confirmed that γ-tocopherol was the main tocopherol in camelina oil, with its amount as 654.5 and 1254.9 mg/kg, respectively. These results were in line with data obtained by other researchers (Abramovic et al., 2007; Zubr & Matthäus, 2002). Carotenoids, which can prevent PUFA oxidation and inactivate photooxidation as well as radical scavenging (Fromm, Bayha, Kammerer, & Carle, 2012), were detected at levels below 20 mg/kg by Grajzer et al. (2015). Raczyk et al. (2016) determined total carotenoids and chlorophyll pigments of cold pressed camelina oils at a range of 78–112 mg of β-carotene/kg oil and 50–117 mg of pheophytin a/kg oil by UV spectrophotometer.

Additionally, the polyphenol contents of camelina oil could reach up to 130 mg/kg (Grajzer et al., 2015). According to Grajzer et al. (2015), sinapic acid was a predominant phenolic acid in camelina oil, with a content of 155.72 μg/kg. However, Ergönül and Özbek (2018) reported that luteolin represented the major part of phenolic compounds in cold pressed camelina oils obtained from 10 camelina seed varieties, followed by apigenin, sinapic acid, and 3-hydroxytyrosol.

According to one report, camelina seed oil contained 0.54% of unsaponifiable materials, mainly composed of the following desmethyl sterols (0.36%), cholesterol (188 ppm), brassicasterol (133 ppm), campesterol (893 ppm), stigmasterol (103 ppm), sitosterol (1884 ppm), and Δ^5-avenasterol (393 ppm) (Shukla et al., 2002). According to the study carried out by Grajzer et al. (2015), the total sterol content of cold pressed camelina seed oil was 2312 mg/kg.

The content of carbohydrates in camelina is very small; for example, sucrose is about 5.5%, which was twice as high as flaxseed (2.8%) but lower than rapeseed (6.8%) (Knudsen & Betty, 1991). Oligosaccharides (raffinose and stachyose) are very low in camelina (below 1%). Starch is a polysaccharide consisting of different chain lengths and straight-chained amylase and branch chained amylopectin. Starch levels in camelina are 1% (Zubr, 2010). Starch is incompletely digestible in the small intestine, but it is fermented by microbes in the large intestine. Pectin is a heteropolysaccharide consisting mainly of d-galacturonic acid linked with fucose, xylose, and galactose. This fermentable fiber is very low in camelina—less than 1% (Kitts, 2007). Mucilage is a water-soluble fiber that forms gel. Soluble fibers delay gastric emptying and transit through the colon. These fibers also interfere with the absorption of sugars and fats, and absorb potentially noxious carcinogenic compounds of the ingesta. The content of mucilage in camelina is 6.7%, which is lower than flaxseed (8%) (Zubr, 2010). Crude fiber includes cellulose and hemicelluloses. Cellulose is a nondigestible glucose polymer. It is found in the cell walls of all vegetation. Hemicellulose fibers are cellulose molecules substituted with other sugars such as xylan, galactan, and mannan. Cellulose and hemicelluloses are microbially fermented in the large intestine. A mixture of short-chain fatty acids, such as acetate, butyrate, and propionate, is produced. Lignin is a polyphenolic compound associated with dietary fiber. It is water-insoluble and in the gastrointestinal system, it increases the amount of stool and absorption of water. Camelina contains 7.4% of lignin (Zubr, 2010).

Other natural plant products found in camelina are terpenes, which are used in the flavor, fragrance, and medical industries, and, moreover, have the potential to be developed as drop-in biofuels (Augustin, Higashi, Feng, & Kutchan, 2015). Augustin et al. (2015) observed that accumulation of terpenes in camelina seed increases the calorific value of the cold pressed oil.

6 Contribution of bioactive compounds in cold pressed oil to organoleptic traits and functions in food or nonfood products

Cold pressed camelina oil has an attractive yellow color, mustard-like taste, and a characteristic but not unpleasant odor. However, flavor and odor are subjective, and the oil should be refined to a neutral product for use as a commercial edible oil (Crowley & Fröhlich, 1998).

Due to high levels of essential fatty acids, particularly omega-3, the fatty acid of camelina oil is also being investigated as a food ingredient (Abramovic & Abram, 2005; Eidhin & O'Beirne, 2010). In 2010, Health Canada approved the use of cold pressed, unrefined camelina oil as a food ingredient in Canada. In some eastern European countries, camelina oil is used in folk medicine for the treatment of burns, wounds, and eye inflammations, as well as to cure stomach ulcers, and as a tonic Ibrahim & El Habbasha, 2015).

Omega-3 fatty acids are one of the main components of functional food ingredients, because they play a positive role against many degenerative and inflammatory complications such as heart disease, stroke, rheumatoid arthritis, asthma, and some types of cancer, and contribute toward development and functioning in infants' brains and retinas (Belayneh et al., 2015). This makes them an attractive product. In Europe, omega-3 fatty acids are used in salad oils and margarine formulations (Francis & Campbell, 2003). Camelina oil and meal are potential replacements of fish meal and oil in aquaculture feeds because of their higher ALA (30%) and crude protein level (45%) (Hixson et al., 2014).

Dietary supplementation of ALA in European countries, the USA, and Canada was estimated to be between 0.8 and 2.2 g/d per person (Burdge & Calder, 2005). The dietary provision of ALA was a subject of numerous investigations. Recommendations for dietary intake of ALA, however, are somewhat inconsistent. A supplementation of c.2 g ALA (20 g rape oil) and 7–10 g linoleic acid for the daily intake by healthy persons was suggested. The conversion of ALA to EPA was found to be about 10% (Metzner & Lüder, 2007). Other studies show that intake of 3.5 g/d ALA increased the proportion of EPA but not DHA in plasma phospholipids. Supplementation of 4.5 and 9.5 g/d ALA was used experimentally (Finnegan, Hawarth, et al., 2003). On the basis of studies with volunteers, the intake of 12 g α-linolenic acid was proposed, corresponding to 20 g flax oil per day. The exploitation of ALA as a substrate for EPA depends on the n-6/n-3 fatty acids ratio.

The conversion of ALA to EPA can be inhibited by the excess of linoleic acid. Appropriate supplementation of ALA is needed to ensure the conversion of ALA to EPA (Emken, Adlof, et al., 1994). Enrichment of food with ALA appears to be extraordinarily important for infants and children. Dietary ALA promotes healthy growth as well as optimal neurological development (Simopoulos, 1991). The incorporation of camelina oil in diet for children appears to be a promising health-promoting measure.

The health-promoting potential of camelina oil is due to high contents of ALA, tocopherols, and other antioxidants, which make camelina oil nutritionally very attractive. Besides being a substrate in human metabolism, ALA is capable of improving the n-6/n-3 fatty acid ratio in food. Experimental documentation showed that in the human metabolism, linoleic acid and ALA are convertible to pure unsaturated fatty acids through desaturation and chain elongation metabolic pathways. ALA is a precursor for prostaglandins and other eicosanoids and hormones involved in a wide range of body functions, including the immune system (Calder, 2001). Additional health effects may be ascribed to antioxidants and phytosterols in camelina oil. Specific studies disclosed that phytosterols, incorporated in functional fat with flax oil, had beneficial effects on lipids and cholesterol in blood serum. Dietary intake of 13.7 g/d ALA from flax oil significantly increased the content of ALA in blood serum. The concentration of ALA increased approximately eightfold in the serum lipid fractions (phospholipids, cholesteryl esters, and triglycerides) and 50% in the neutrophil phospholipids. The concentration of EPA in plasma phospholipids increased 2.5-fold. A supplementation of ALA from vegetable oils can elevate EPA in tissues to concentrations comparable to those achieved with fish oil (Onge et al., 2003). Another investigation showed the conversion of linoleic acid to n-6 metabolites ranging from 1.0% to 2.2%. The conversion of ALA to n-3 metabolites was 11.0%–18.5% and to DHA was 3.8%. A significant increase of ALA, EPA, and DHA in blood serum in a trial with volunteers was reported. At the same time, the saturated fatty acids (C14:0, C15:0, and C16:0) decreased. The trial demonstrated that the supplementation of camelina oil had about the same effects as flax oil (Ibrahim & El Habbasha, 2015).

The high content of ALA in linseed and camelina oils makes them highly sensitive to heat, oxygen, and light (Choo, Birch, & Dufour, 2007; Nag, 2000; Rubilar, Gutierrez, Verdugo, Shene, & Sineiro, 2010). Due to their low stability, linseed and camelina oils are intended for low temperature processing (Nag, 2000). Deep frying with the use of linseed and camelina oils can quickly produce undesirable degradation products, including polymers and cyclic fatty acids (Brinkmann, 2000). Some investigations showed camelina oil to be more stable toward oxidation than linseed oil, but less stable than rapeseed, olive, corn, sesame, and sunflower oils (Eidhin et al., 2003). This is caused not only by lower levels of PUFAs, but also by a high content of natural antioxidants, predominantly from the group of tocopherols (Raczyk et al., 2016).

7 Edible and nonedible applications of cold pressed oil

The possible industrial applications of camelina include its use in environmentally safe paints, coatings, cosmetics, and low emission biodiesel fuels (Bonjean & Le Goffic, 1999). Although the presence of PUFAs makes camelina oil susceptible to lipid oxidation, it remains sufficiently stable during storage due to the presence of antioxidants in the seed (Abramovic & Abram, 2005). Camelina seed contains oil contents between 32% and 46%, and concentration of ALA is in the broad range from 28% to 43% of total fatty acids (Budin, Breene, & Putnam, 1995). Seed quality characteristics of camelina are important features for processing and marketing of the crop in competition with other oilseeds. Several reports have suggested camelina as one of the most cost-effective oilseed crops to produce, due to the search for new sources of essential fatty acids, particularly n-3 (omega-3) fatty acids and multiple-use values (Ibrahim & El Habbasha, 2015).

The specific dermatological effects of PUFA make camelina oil suitable for cosmetic applications, such as cosmetic oils, skin creams, and lotions (Ibrahim & El Habbasha, 2015). *Camelina sativa* oil is also applied for the production of potential adhesives containing epoxide (Kim, Li, & Sun, 2015) or acrylate and hydroxyl functionalities (Li & Sun, 2015). Besides the abovementioned synthetic modifications, transgenic camelina oils have been used for production of acetyl triacylglycerides, which possess lower viscosity and have improved cold temperature properties (Liu et al., 2015), and sciadonic acid (Gonzalez-Thuillier, Napier, & Sayanova, 2016).

Camelina is also considered as a promising raw material for biodiesel production. Yang et al. (2016) investigated the fatty acid composition of cold pressed camelina oil, transesterification process of oil by alkali, and fuel properties of camelina biodiesel to evaluate the suitableness of biodiesel production. Under controlled reaction conditions, the maximum conversion was found to be approximately 96%. However, similar to derived biodiesel from other vegetable oils, oxidative stability is not acceptable, due to the high percentage of unsaturated fatty acids (about 90%) (Yang et al., 2016; Yang, He, Corscadden, & Caldwell, 2017). From this point of view, Yang et al. (2017) examined the effect of some antioxidants such as butylated hydroxytoluene (BHT), butylated hydroxyanisol (BHA), tertbutylhydrooquinone, and propyl gallat (PG) on the oxidative and storage stabilities of camelina biodiesel obtained from cold pressed camelina seed oil. The rate of biodiesel oxidation depends on the number of allylic and bis-allylic positions in the unsaturated fatty acid ester chains. Oxygen can bind to these reactive allylic and bis-allylic sites and thus initiates the primary free radical chain reaction and forms primary oxidation products (hydroperoxides). In the secondary oxidation step, the hydroperoxides broke up to volatile organic acids, alcohols, ketones, and aldehydes (Zuleta, Libia, Luis, & Jorge, 2012). According to Yang et al. (2017), TBHQ showed the highest antioxidant activity, followed by PG, BHT, and BHA. The smaller amount of TBHQ resulted in an oxidative stability index of 8.8 h.

In another study carried out by Patil, Gude, Camacho, and Deng (2010), microwave radiation was estimated as a nonconventional heat source for transesterification of cold pressed *C. sativa* oil. Three types of catalyst (two homogeneous catalysts, two heterogeneous metal oxide catalysts, and two sol-gel derived catalysts) were used to determine their effect on biodiesel production from camelina oil. It was compared with conventional heat treatment. Catalyst concentration is a critical factor to be determined in the transesterification process. Low and high catalyst concentrations may result in undesired biodiesel yield as well as high production costs. The homogeneous catalysts resulted in high biodiesel yield with short reaction times. The reason for high reaction times can be explained as the inability of the microwaves to influence the solid materials in the same way as the liquid solvents. The microwaves have a poor ability to penetrate through solid materials. In the case of homogeneous catalysts, this is entirely different because the catalyst is completely dissolved in the solvent and the microwave effect is higher in liquid solvents, resulting in higher biodiesel yield. From the information above, it could be concluded that the homogeneous catalysts will, in general, require shorter reaction times compared to heterogeneous catalysts for microwave-assisted transesterification. Another result from this study is that 30–60 s were sufficient for microwave heating while 30–60 min were required for conventional heating to achieve comparable biodiesel yields. This large difference in reaction time is attributed to the limitations of conventional heating, in which the energy is first utilized to increase the temperature of the reaction vessel and the higher temperature of the reaction vessel results in higher heat losses to the ambient (Patil et al., 2010).

Although there is limited research on nonfood type uses for camelina meal, some of the applications included bioherbicides, soil fungicides, adhesives, and biooils (Berti et al., 2016). Cao, Gu, Muthukumarappan, and Gibbons (2015) extracted glucosinolates from the meal by the membrane separation technique. Glucosinolates can produce bioherbicide, named ionic thiocyanates (SCN), against redroot pigweed (*Amaranthus retroflexus* L.) and wild oat (*Avena fatua* L.). Li, Qi, Sun, Xu, and Wang (2015) remarked that the protein composition of cold pressed camelina meal is similar to that of canola meal, which has shown potential as an alternative to conventional petroleum-based adhesives. They found that albumin, globulin, and glutelin are more extractable protein fractions from camelina meal. Among them, globulins have better adhesion performance. In some parts of Europe and North America, *C. sativa* is grown as a nonedible grade crop. Cold pressed camelina oil is used to produce polyols due to its high percentage of unsaturates, as an alternative to castor oil (Omonov, Kharraz, & Curtis, 2017).

References

Abramovic, H., & Abram, V. (2005). Physiochemical properties, composition and oxidative stability of *Camelina sativa* oil. *Food Technology and Biotechnology*, *43*, 63–70.

Abramovic, H., Butinar, B., & Nikolicˇ, V. (2007). Changes occurring in phenolic content, tocopherol composition and oxidative stability of *Camelina sativa* oil during storage. *Food Chemistry*, *104*, 903–909.

Augustin, J. M., Higashi, Y., Feng, X., & Kutchan, T. M. (2015). Production of mono- and sesquiterpenes in *Camelina sativa* oilseed. *Planta*, *242*, 693–708.

Belayneh, H., Wehling, R., Cahoon, E., & Ciftci, O. N. (2015). Extraction of omega-3-rich oil from *Camelina sativa* seed using supercritical carbon dioxide. *Journal of Supercritical Fluids*, *104*, 153.

Belayneh, H. D., Wehling, R. L., Cahoon, E. B., & Ciftci, O. N. (2017). Effect of extraction method on the oxidative stability of camelina seed oil studied by differential scanning calorimetry. *Journal of Food Science*, *82*(3), 632–637.

Belayneh, H. D., Wehling, R. L., Cahoon, E., & Ciftci, O. N. (2018). Lipid composition and emulsifying properties of *Camelina sativa* seed lecithin. *Food Chemistry*, *242*, 139–146.

Benz, S. A. (2010). *Food and drug administration-department of health and human services*. http://agr.mt.gov/agr/Producer/Checkoff/Camelina/pdf/FDA_exception_9-2-10.Pdf (Accessed 15 July 2014).

Berdanier, C. D. (2002). Food constituents. In C. D. Berdanier (Ed.), *Handbook of nutrition and food* (pp. 3–95). Boca Raton, FL: CRC Press.

Berti, M., Russ, G., Eynck, C., Anderson, J., & Cermak, S. (2016). Camelina uses, genetics, genomics, production, and management. *Industrial Crops and Products*, *94*, 690–710.

Bonjean, A., & Le Goffic, F. (1999). *Camelina sativa* (L.) Crantz: An opportunity for European agriculture and industry. *Oleagineux, Corps Gras, Lipides*, *6*, 28–34.

Boyle, C., Hansen, L., Hinnenkamp, C., & Ismail, B. P. (2018). Emerging camelina protein: Extraction, modification, and structural/functional characterization. *Journal of the American Oil Chemists' Society*, *95*, 1049–1062.

Bozan, B., & Temelli, F. (2002). Supercritical CO_2 extraction of flaxseed. *Journal of the American Oil Chemists' Society*, *79*(3), 231–235.

Brinkmann, B. (2000). Quality criteria of industrial frying oils and fats. *European Journal of Lipid Science and Technology*, *102*, 539–541.

Budin, J. T., Breene, W. M., & Putnam, D. H. (1995). Some compositional properties of camelina (*Camelina sativa* (L.) Crantz) seeds and oils. *Journal of the American Oil Chemists' Society*, *72*, 309–315.

Burdge, G. C., & Calder, P. C. (2005). α-Linolenic acid metabolism in adult humans: The effects of gender and age on conversion to longer-chain polyunsaturated fatty acids. *European Journal of Lipid Science and Technology*, *107*(6), 426–439.

Calder, P. C. (2001). The ratio of n-6 to n-3 fatty acids in the diet: Impact on T lymphocyte function. *European Journal of Lipid Science and Technology*, *103*(6), 390–398.

Cao, Y., Gu, Z., Muthukumarappan, K., & Gibbons, W. (2015). Separation of glucosinolates from camelina seed meal via membrane and acidic aluminumoxide column. *Journal of Liquid Chromatography and Related Technologies*, *38*, 1273–1278.

Cherian, G. (2012). Biofuel co-products as livestock feed-opportunities and challenges. In H. P. S. Makkar (Ed.), Camelina sativa *in poultry diets: Opportunities and challenges*. Rome: FAO (Chapter 17).

Choo, W. S., Birch, E. J., & Dufour, J. P. (2007). Physiochemical and stability characteristics of flaxseed oils during pan heating. *Journal of the American Oil Chemists' Society*, *84*, 735–740.

Ciftci, O. N., Calderon, J., & Temelli, F. (2012). Supercritical carbon dioxide extraction of corn distiller's dried grains with solubles: Experiments and mathematical modeling. *Journal of Agricultural and Food Chemistry*, *60*, 482–490.

Crowley, J. G., & Fröhlich, A. (1998). Factors affecting the composition and use of camelina. In *End of project reports*. Teagasc. ISBN: 1 901138 66 6.

Dantas, M. B., Albuquerque, A. R., Barros, A. K., Rodrigues Filho, M. G., Antoniosi Filho, N. R., Sinfronio, F. S. M., et al. (2011). Evaluation of the oxidative stability of corn biodiesel. *Fuel*, *90*, 773–778.

Eidhin, N. D., Burke, J., & O'Beirne, D. (2003). Oxidative stability of ω3-rich camelina oil and camelina oil-based spread compared with plant and fish oils and sunflower spread. *Journal of Food Science: Sensory and Nutritive Qualities of Food*, *68*(1), 345–353.

Eidhin, D. N., & O'Beirne, D. (2010). Oxidative stability of camelina oil in salad dressings, mayonnaises and during frying. *International Journal of Food Science and Technology*, *45*, 444–452.

Emken, E. A., Adlof, R. O., et al. (1994). Dietary linoleic acid influences desaturation and acylation of deuterium-labeled linoleic and linolenic acids in young adult males. *Biochimica et Biophysica Acta*, *1213*, 277–288.

Ergönül, P. G., & Özbek, Z. A. (2018). Identification of bioactive compounds and total phenol contents of cold pressed oils from safflower and camelina seeds. *Journal of Food Measurement and Characterization*, *12*, 2313–2323.

Finnegan, Y. E., Hawarth, D., et al. (2003). Plant and marine derived (n-3) polyunsaturated fatty acids do not affect blood coagulation and fibrinolytic factors in moderately hyperlipidemic humans. *Journal of Nutrition*, *133*, 221–2213.

Francis, C. M., & Campbell, M. C. (2003). *New high quality oil seed crops for temperate and tropical Australia: A report for the Rural Industries Research and Development Corporation*. RIRDC Publication No 03/045.

Fromm, M., Bayha, S., Kammerer, D. R., & Carle, R. (2012). Identification and quantitation of carotenoids and tocopherols in seed oils recovered from different Rosaceae species. *Journal of Agricultural and Food Chemistry*, *60*(43), 10733–10742.

Gonzalez-Thuillier, I., Napier, J., & Sayanova, O. (2016). Engineering synthetic pathways for the production of pharmaceutical sciadonic acid in transgenic oilseed *Camelina sativa*. *New Biotechnology*, *33*, S46.

Grajzer, M., Prescha, A., Korzonek, K., Wojakowska, A., Dziadas, M., Kulma, A., et al. (2015). Characteristics of rose hip (*Rosa canina* L.) cold-pressed oil and its oxidative stability studied by the differential scanning calorimetry method. *Food Chemistry*, *188*, 459–466.

Gramza, A., Khokhar, S., Yoko, S., Gliszczynska-Swiglo, A., Hes, M., & Korczak, J. (2006). Antioxidant activity of tea extracts in lipids and correlation with polyphenol content. *European Journal of Lipid Science and Technology*, *108*, 351–362.

Hixson, S. M., Parrish, C. C., & Anderson, D. M. (2014). Full substitution of fish oil with camelina (*Camelina sativa*) oil, with partial substitution of fish meal with camelina meal, in diets for farmed Atlantic salmon (*Salmo salar*) and its effect on tissue lipids and sensory quality. *Food Chemistry*, *157*, 51–61.

Hurtaud, C., & Peyraud, J. (2007). Effects of feeding camelina (seeds or meal) on milk fatty acid composition and butter spreadability. *Journal of Dairy Science*, *90*, 5134–5145.

Ibrahim, F. M., & El Habbasha, S. F. (2015). Chemical composition, medicinal impacts and cultivation of camelina (*Camelina sativa*): Review. *International Journal of PharmTech Research*, *8*(10), 114–122.

Ionescu, N., Ivopol, G.-C., Neagu, M., Popescu, M., & Meghea, A. (2015). Fatty acids and antioxidant activity in vegetable oils used in cosmetic formulations. *UPB Scientific Bulletin, Series B: Chemistry and Materials Science*, *77*(3), 39–48.

Karvonen, H. M., Aro, A., Tapola, N. S., Salminen, I., Uusitupa, M. I. J., & Sarkkinen, E. S. (2002). Effect of α-linolenic acid-rich *Camelina sativa* oil on serum fatty acid composition and serum lipids in hypercholesterolemic subjects. *Metabolism, Clinical and Experimental*, *51*, 253–1260.

Kim, N., Li, Y., & Sun, X. S. (2015). Epoxidation of *Camelina sativa* oil and peel adhesion properties. *Industrial Crops and Products*, *64*, 1–8.

Kitts, D. D. (2007). Carbohydrates and mineral metabolism. In *Functional food carbohydrates* (pp. 413–433). Boca Raton, FL: CRC Press.

Knudsen, K. E. B., & Betty, W. L. (1991). Determination of oligosaccharides in protein-rich feedstuffs by gas-liquid chromatography and high-performance liquid chromatography. *Journal of Agricultural and Food Chemistry*, *39*, 689–694.

Kowalski, B., Gruszczynska, E., & Maciaszek, K. (2000). Kinetics of rapeseed oil oxidation by pressure differential scanning calorimetry measurements. *European Journal of Lipid Science and Technology*, *102*, 337–341.

Lawrence, R. D., Anderson, J. L., & Clappert, J. A. (2016). Evaluation of camelina meal as a feedstuff for growing dairy heifers. *Journal of Dairy Science*, *99*, 6215–6228.

Li, N., Qi, G., Sun, X. S., Xu, F., & Wang, D. (2015). Adhesion properties of camelina protein fractions isolated with different methods. *Industrial Crops and Products*, *69*, 263–272.

Li, Y., & Sun, X. S. (2015). Camelina oil derivatives and adhesion properties. *Industrial Crops and Products*, *73*, 73–80.

Liu, J., Tjellström, H., McGlew, K., Shaw, V., Rice, A., Simpson, J., et al. (2015). Field production, purification, and analysis of high-oleic acetyl-triacylglycerols from transgenic *Camelina sativa*. *Industrial Crops and Products*, *65*, 259–268.

Metzner, C., & Lüder, W. (2007). Pflanzlische w-3 und w-6 Fedttsäuren: Wissenswertes zu Pflanzenölen (German). *Pharmazie in Unserer Zeit*, *36*(2), 134–141.

Mieriòa, I., Adere, L., Krasauska, K., Zoltnere, E., Skrastiòa, D. Z., & Jure, M. (2017). Antioxidant properties of *Camelina sativa* oil and press cakes. *Proceedings of The Latvian Academy of Sciences Section B*, *71*(6 (711)), 515–521.

Moslavac, T., Jokic, S., Subari, D., Aladi, D., Vukoja, J., & Prce, N. (2014). Pressing and supercritical CO_2 extraction of *Camelina sativa* oil. *Industrial Crops And Products*, *54*, 122–129.

Nag, A. (2000). Stabilization of flaxseed oil with capsicum antioxidant. *Journal of the American Oil Chemists' Society*, *77*, 799–800.

Obour, A. K., Sintim, H. Y., Obeng, E., & Jeliazkov, V. D. (2015). Oilseed camelina (*Camelina sativa* L Crantz): Production systems, prospects, and challenges in the USA Great Plains. *Advances in Plants & Agriculture Research*, *2*(2), 1–10.

Omonov, T. S., Kharraz, E., & Curtis, J. M. (2017). Camelina (*Camelina sativa*) oil polyols as an alternative to Castor oil. *Industrial Crops and Products*, *107*, 378–385.

Onge, M. P., Lamarche, B., Mauger, J. F., & Jones, P. J. H. (2003). Consumption of a functional oil rich in phytosterols and medium-chain triglyceride oil improves plasma lipid profiles in men. *Journal of Nutrition*, *133*, 1815–1820.

Onyilagha, J., Bala, A., Hallett, R., Gruber, M., Soroka, J., & Westcott, N. (2003). Leaf flavonoids of the cruciferous species, *Camelina sativa*, *Crambe* spp., *Thlaspi arvense* and several other genera of the family Brassicaceae. *Biochemical Systematics and Ecology*, *31*, 1309–1322.

Ortega, R. M., Palencia, A., & Lopez-Sobaler, A. M. (2006). Improvement of cholesterol levels and reduction of cardiovascular risk via the consumption of phytosterols. *British Journal of Nutrition*, *96*(S1), 89–93.

Patil, P. D., Gude, V. G., Camacho, L. M., & Deng, S. (2010). Microwave-assisted catalytic transesterification of *Camelina sativa* oil. *Energy & Fuels*, *24*, 1298–1304.

Popa, A. L., Jurcoane, S., & Dumitriu, B. (2017). *Camelina sativa* oil—A review. *Scientific Bulletin Series F. Biotechnologies*, *XXI*, ISSN 2285-1364, ISSN 2285-5521, ISSN online 2285-1372, ISSN-L 2285-1364.

Raczyk, M., Popis, E., Kruszewski, B., Ratusz, K., & Rudzinska, M. (2016). Physicochemical quality and oxidative stability of linseed (*Linum usitatissimum*) and camelina (*Camelina sativa*) cold-pressed oils from retail outlets. *European Journal of Lipid Science and Technology*, *118*, 834–839.

Ratusz, K., Popis, E., Zytkiewicz, H. C., & Wroniak, M. (2016). Oxidative stability of camelina (*Camelina sativa* L.) oil using pressure differential scanning calorimetry and Rancimat method. *Journal of Thermal Analysis and Calorimetry*, *126*, 343–351.

Rokka, T., Alen, K., Valaja, J., & Ryhanen, E. L. (2002). The effect of a Camelina sativa enriched diet on the composition and sensory quality of hen eggs. *Food Research International*, *35*, 253–256.

Rubilar, M., Gutierrez, C., Verdugo, M., Shene, C., & Sineiro, J. (2010). Flaxseed as a source of functional ingredients. *Journal of Soil Science and Plant Nutrition*, *10*, 373–377.

Salminen, H., Estevez, M., Kivikari, R., & Heinonen, M. (2006). Inhibition of protein and lipid oxidation by rapeseed, camelina and soy meal in cooked pork meat patties. *European Food Research and Technology*, *223*, 461–468.

Shukla, V. K. S., Dutta, P. C., & Artz, W. (2002). Camelina oil and its unusual cholesterol content. *Journal of the American Oil Chemists' Society*, *79*, 965–969.

Simopoulos, A. P. (1991). Omega-3 fatty acids in health and disease and in growth and development. *The American Journal of Clinical Nutrition*, *54*, 438–463.

Simopoulos, A. P. (1999). New products from the agri-food industry: The return of n-3 fatty acids into the food supply. *Lipids*, *34*, 297–301.

Skjervold, H. (1993). Lifestyle diseases and human diet. In: *Abstracts to Minisymposium-lifestyle diseases and the human diet—A challenge to future food production, National Institute of Animal Science Denmark, 16–19 August, Aarhus, Denmark.*

Szterk, A., Roszko, M., Sosinska, E., Derewiaka, D., & Lewicki, P. P. (2010). Chemical composition and oxidative stability of selected plant oils. *Journal of the American Oil Chemists' Society*, *87*, 637–645.

Teh, S. S., & Birch, J. (2013). Physicochemical and quality characteristics of cold-pressed hemp, flax, and canola seed oils. *Journal of Food Composition and Analysis*, *30*, 26–31.

Terpinc, P., Polak, T., Makuc, D., Ulrih, N. P., & Abramovic, H. (2012). The occurrence and characterisation of phenolic compounds in *Camelina sativa* seed, cake and oil. *Food Chemistry*, *131*, 580–589.

Vollmann, J., Damboeck, A., Eckl, A., Schrems, H., & Ruckenbauer, P. (1996). Improvement of Camelina sativa, an underexploited oilseed. In J. Janick (Ed.), *Progress in new crops* (pp. 357–362). Alexandria: American Society of Horticultural Science Press.

Woyengo, T. A., Patterson, R., Slominski, B. A., Beltranena, E., & Zijlstra, R. T. (2016). Metabolism and Nutrition. Nutritive value of cold-pressed camelina cake with or without supplementation of multi-enzyme in broiler chickens. *Poultry Science*, *95*, 2314–2321.

Wu, X., & Leung, D. Y. C. (2011). Optimization of biodiesel production from camelina oil using orthogonal experiment. *Applied Energy*, *88*, 3615–3624.

Yang, J., Claude, C., Corscadden, K., Sophia He, Q., & Li, J. (2016). An evaluation of biodiesel production from *Camelina sativa* grown in Nova Scotia. *Industrial Crops and Products*, *81*, 162–168.

Yang, J., He, Q. S., Corscadden, K., & Caldwell, C. (2017). Improvement on oxidation and storage stability of biodiesel derived from an emerging feedstock camelina. *Fuel Processing Technology*, *157*, 90–98.

Ye, C. L., Anderson, D. M., & Lall, S. P. (2016). The effects of camelina oil and solvent extracted camelina meal on the growth: Carcass composition and hindguthistology of Atlantic salmon (*Salmo salar*) parr in freshwater. *Aquaculture*, *450*, 397–404.

Zubr, J. (1997). Oil-seed crop: *Camelina sativa*. *Industrial Crops and Products*, *6*(1997), 113–119.

Zubr, J. (2010). Carbohydrates, vitamins and minerals of *Camelina sativa* seed. *Nutrition and Food Science*, *40*, 523–531.

Zubr, J., & Matthäus, B. (2002). Effects of growth conditions on fatty acids and tocopherols in *Camelina sativa* oil. *Industrial Crops and Products*, *15*(2), 155–162.

Zuleta, E. C., Libia, M. B., Luis, A. R., & Jorge, A. C. (2012). The oxidative stability of biodiesel and its impact on the deterioration of metallic and polymeric materials: A review. *Journal of the Brazilian Chemical Society*, *23*, 2159–2175.

Chapter 22

Cold pressed pistachio (*Pistacia vera*) oil

Engy Shams-Eldin[a] and Mariam Abdur-Rahman[b]

[a]*Special Food and Nutrition Department, Food Technology Research Institute, Agriculture Research Center, Giza, Egypt,* [b]*Chemistry Department, Faculty of Science, Cairo University, Giza, Egypt*

Abbreviations

DPKF defatted pistachio kernel flour
FAO Food and Agriculture Organization
FTIR Fourier transform infrared spectroscopy
IR infrared
LDL low-density lipoprotein
MUFA monounsaturated fatty acid
PKO pistachio kernel oil
PUFA polyunsaturated fatty acid
SFA saturated fatty acid

1 Introduction

Pistacia is a genus of aromatic trees in the cashew family. It includes pistachio (*Pistacia vera*) and several medicinal species. Pistachio acquired its economic importance from being one of the most favored nuts around the world. Moreover, pistachio has a high oil content consisting of about 50%–62% of its weight. The industrial extraction of pistachio oil involves using either supercritical fluids, organic solvents, or pressure methods (Catalán et al., 2017). However, only mechanical methods, including the screw press (oil expeller) and the hydraulic press methods, are considered as cold-extraction methods.

The by-product remaining after pistachio oil extraction is called "oil cake," or defatted pistachio kernel flour (DPKF). It is a valuable source of protein that is readily used as a food ingredient in human nutrition and animal feeding (Pardo-Giménez et al., 2016). It can also be used as a supporting matrix for various biotechnological processes, including the production of enzymes, antibiotics, and mushrooms. In addition, oil cake is used as a source for the production of vitamins and antioxidants (Ramachandran, Singh, Larroche, Soccol, & Pandey, 2007).

Studies were carried out to evaluate the nutritional value, functional properties, bioactivity, and microstructure of DPKF prepared by cold pressing. DPKF was found to be rich in protein and carbohydrate. Furthermore, DPKF showed a low Na content coupled with high concentrations of K, P, Ca, Mg, Zn, and Se. All essential amino acids, except sulfur-containing ones, were found to be significantly higher than the reference values recommended by the Food and Agriculture Organization (FAO). DPKF also exhibited appreciably high total phenolics and flavonoid contents as well as antioxidant capacities. Therefore, DPKF may have potential applications such as functional food ingredients and supplement (Ling, Yang, Li, & Wang, 2016).

2 Extraction and processing of cold pressed oil

Pistachio oil is one of the most valuable products from a nutritional point of view. It can be produced with high quality by controlling the extraction conditions to improve its organoleptic properties and maintain its nutritional value (Catalán et al., 2017). Several factors may affect the nutritional value of pistachio cold pressed oil as well as its quality, in terms of flavor and color. These factors include drying parameters of nuts and oil extraction conditions.

Cold Pressed Oils. https://doi.org/10.1016/B978-0-12-818188-1.00022-0
Copyright © 2020 Elsevier Inc. All rights reserved.

Sena-Moreno et al. (2015) used different extraction techniques to study the effects of different drying temperatures (30°C, 50°C, and 70°C) on oxidative stability, chemical composition, and sensory evaluation. A significant increment of peroxide levels was observed either at 70°C drying temperature or when the screw press extraction technique was used. However, oils extracted via hydraulic press remain at lower peroxide levels. The polyphenolic content of pistachio oil increased at higher temperatures (50°C and 70°C), which is accompanied by a decrease in stigmasterol. No significant change was observed with respect to fatty acids composition using any of the different drying temperatures and extraction techniques.

Meeting market needs is crucial in the pistachio oil industry. It was noted, for example, in a color sensory test that consumers preferred the greenish pistachio oil dried at 70°C. Therefore, the best conditions of pistachio oil extraction methods, that is, hydraulic and screw presses, were further studied with respect to yield, sensory, and physicochemical parameters. There were no differences with respect to yield and physicochemical parameters in hydraulic press extraction of pistachio oil, while sensory analysis recommended using roasted pistachios at 100°C for 30 min. However, in the case of screw press extraction, there were differences in both yield and sensory characteristics, in which lower speed was associated with higher yields and the oil was more valued by consumers (Rabadán, Álvarez-Ortí, Gómez, Alvarruiz, & Pardo, 2017).

In the pistachio oil industry, the moisture content must not exceed 5%, in order to obtain higher yields and avoid the microbial growth (Catalán et al., 2017). Recent work by Rabadán, Álvarez-Ortí, Pardo, and Alvarruiz (2018) and Rabadán, Gallardo-Guerrero, Gandul-Rojas, Álvarez-Ortí, and Pardo (2018) reported that roasting temperature, applied before oil extraction, could affect the physical and chemical parameters of the produced pistachio oil. Density value was reduced after higher drying conditions at 125°C, while viscosity was slightly increased. These effects were accompanied by a significant increase in chlorophyll content and carotenoid pigments, in addition to higher oxidative stability. Collectively, increasing drying temperature had a clear impact on color, that is, from yellow to brilliant green, due to the transfer of the favored pigments into pistachio oil. This confirms that roasting treatment, done prior to oil extraction, has a great effect on sensory properties of pistachio oil. On the other hand, the ratio of oil recovery depends on the extraction method; it reached up to 30% in case of hydraulic press, and 40% in the screw press (Álvarez-Ortí et al., 2012).

3 Fatty acids and acyl lipids

Although pistachio oils have different nutritional values according to the nut origin source, pistachio cold pressed oil has generally a great amount of unsaturated fatty acids and bioactive phytochemicals such as phenolics, stilbenes, carotenoids, phytosterols, and proanthocyanidins (Bolling, Chen, McKay, & Blumberg, 2011; Catalán et al., 2017). Pistachio oil contains ~30% polyunsaturated, ~53% monounsaturated, and ~17% saturated fatty acids (Alizadeh, Alizadeh, Amari, & Zare, 2013). Palmitic and stearic acids are among saturated fatty acids found in pistachio oil, while the most abundant unsaturated fatty acids are oleic acid (~51%) followed by linoleic acid (~30%). A recent study (Rabadán, Álvarez-Ortí, et al., 2018; Rabadán, Gallardo-Guerrero, et al., 2018) summarized the composition of freshly pressed pistachio oil (Table 1). Oleic acid has fungistatic activity against a wide spectrum of saprophytic molds and yeasts. It may be used as an edible cooking oil or for margarine manufacture (Givianrad, Saffarpour, & Beheshti, 2011).

4 Minor bioactive lipids

Studies on pistachio oil samples from Turkey and Italy have identified 21 polyphenols via mass spectra, in which benzoic acid derivatives such as 4-hydroxybenzoic, protocatechuic, and gallic acids were the main components (Saitta, La Torre, Potortì, Di Bella, & Dugo, 2014; Tsantili, Konstantinidis, & Christopoulos, 2011). Polyphenols increase the antioxidant activity, protecting cells from free radical damage. They have a protective effect against different diseases such as breast cancer, colorectal cancers, and heart diseases (Saitta, Giuffrida, Di Bella, La Torre, & Dugo, 2011). The anthocyanins content in pistachios represents the larger proportion of total polyphenols concentration (Ballistreri, Arena, & Fallico, 2009). In addition, pistachio oil contains higher values of phytosterols compared to other nuts oils (Bolling et al., 2011; Kornsteiner-Krenn, Wagner, & Elmadfa, 2013). Major pistachio oil sterols, including β-sitosterol, stigmasterol, campesterol, and δ5-avenasterol could be used to identify oil origin with better results than fatty acids constituents (Arena, Campisi, Fallico, & Maccarone, 2007). Tocopherols, especially γ-tocopherol, are found in pistachio oil in considerable amounts, which are however lower than in other nuts, such as walnuts and almonds (Delgado-Zamarreño, Bustamante-Rangel, Sánchez-Pérez, & Carabias-Martínez, 2004). Miraliakbari and Shahidi (2008) reported that notable amounts of other tocopherols (α-, β-, and δ-tocopherols) were also detected.

TABLE 1 Fatty acid and sterol constituents of freshly pressed pistachio oil (Rabadán, Álvarez-Ortí, et al., 2018; Rabadán, Gallardo-Guerrero, et al., 2018).

Fatty acid and sterol	(%)
Myristic, C14:0	0.08 ± 0.00
Palmitic, C16:0	10.76 ± 0.02
Palmitoleic, C16:1	1.01 ± 0.01
Margaroleic, C17:1	0.09 ± 0.00
Stearic, C18:0	1.05 ± 0.00
Oleic, C18:1	54.59 ± 0.02
Linoleic, C18:2	31.33 ± 0.03
Linolenic, C18:3	0.50 ± 0.00
Arachidic, C20:0	0.12 ± 0.00
Gadoleic, C20:1	0.33 ± 0.01
Behenic, C22:0	0.08 ± 0.00
SFA[a]	12.09 ± 0.02
MUFA[a]	56.02 ± 0.02
PUFA[a]	31.83 ± 0.03
Campesterol	4.2 ± 0.1
Stigmasterol	0.6 ± 0.0
Apparent β-sitosterol	94.1 ± 0.2
Δ-7-Stigmastenol	0.4 ± 0.0

[a]SFA, *saturated fatty acid;* MUFA, *monounsaturated fatty acid;* PUFA, *polyunsaturated fatty acid.*

Flavor and aroma compounds in pistachio oil were investigated by Ling, Zhang, Li, and Wang (2016). They analyzed the volatile compounds of pistachio kernel oil (PKO) either raw or in roasted kernels, prepared by conventional or microwave roasting. Among 43 volatile identified compounds, the main compounds were limonene, β-myrcene, α-pinene, nonanal, and hexanoic acid. Concentrations of these compounds were decreased as a result of the heat-labile effect caused by roasting.

5 Contribution of bioactive compounds in cold pressed oil to organoleptic traits and functions in food or nonfood products

Production of high-quality pistachio oil as a novel healthy oil depends on several factors, including immediate drying of nuts prior to the extraction step until reaching the appropriate moisture content (5%–8% on dry basis) (Catalán et al., 2017), as well as the extraction method itself. Moreover, with respect to consumer preference, attractive and sensory characteristics can add higher value of less-processed virgin oils compared to refined oils (Kamal-Eldin & Moreau, 2010). Cold pressed oil, as defined by the WHO-FAO Codex Stan 2010, is the one with the highest quality among oils, and it is usually produced on a small scale to be sold in health markets. Commercial products of pistachio oil have appeared in some European as well as Middle Eastern countries. They are used essentially as salad dressing, in addition to therapeutic and cosmetic products (Ojeda-Amador, Fregapane, & Salvador, 2018).

Polar bioactive compounds of pistachio nut remain in the defatted by-product portion, called "residual cake," after the oil extraction step (Ling, Yang, et al., 2016). However, bioactive constituents that are present in pistachio oil include anacardic acids, tocopherols, carotenoids, flavonoids, isoflavones, anthocyanins, chlorophylls, proanthocyanidins, cardanols, and resveratrol (Saitta et al., 2011). They could contribute effectively to the nutritional value of this pistachio oil.

Organoleptic properties of pistachio oil have been usually affected by various factors, such as plant origin and roasting of nuts before extraction, as well as the extraction method.

6 Health-promoting traits of cold pressed oil and oil constituents

Pistachio nuts have several different classes of antioxidants. These antioxidants protect cells against oxidative damage through scavenging free radicals and decomposing peroxides. Several bioactive constituents have been identified in pistachio oil such as tocopherols, anacardic acids, carotenoids, anthocyanins, flavonoids, isoflavones, chlorophylls, cardanols, and proanthocyanidins, as well as vitamin C and resveratrol (Saitta et al., 2011). Concerning eight different nut oils including walnut, pecan, hazelnut, apricot, pistachio, peanut, cashew, and almond, the literature showed a variety of antioxidant activities that were changed from 11.4% to 65.5% in peanut and pistachio, respectively. Moreover, β-carotene and total phenolics levels of different oils were observed, where catechin contents varied from 0.56% to 3.76% in cashew and pistachio, respectively (Al Juhaimi, Özcan, Ghafoor, Babiker, & Hussain, 2018).

β-Sitosterol is considered as the major sterol in Iranian pistachio oil, accounting for around 87% of the lipid fraction sterols. Phytosterols have a major role in reducing serum low-density lipoprotein cholesterol (LDL), and hence can help protect the cardiovascular system against harmful diseases. Saber-Tehrani, Givianrad, Aberoomand-Azar, Waqif-Husain, and Jafari Mohammadi (2013) determined a large amount of tocopherols in cold pressed pistachio oil, about 409 mg tocopherol/kg oil, in which α-tocopherol represented the major form compared to β-, γ-, and δ-tocopherols.

7 Edible and nonedible applications of cold pressed oil

Despite being an exceptional oil in both health and sensory terms, pistachio oil is not considered as a commercial oil in Europe, due to its high market prices. Therefore, efforts should be taken to improve its sensory parameters, in order to attract more consumers to use this healthy product. In this context, factors including roasting and extraction methods can profoundly affect pistachio oil color, varying from yellow to rich green, and taste, varying from a natural to a roasted flavor. From a commercial point of view, pistachio oil production can ideally solve the problem of having low-quality pistachios with damaged, stained kernels that are not adequate for fresh use. In other words, oil production can transform discards into valuable stuff.

Having special sensory properties, comprising rich green color and strong taste and odor, pistachio oil is highly suitable for culinary use and high cuisine. It can substitute butter and margarine in confectionery. Moreover, it could be used as a delicious steamed vegetable dressing. However, it is not preferable to use in stews and fried dishes (Catalán et al., 2017).

Because of its special chemical composition, pistachio oil also plays an important role in the industrial production of cosmetics, with positive gentle effects on skin and hair. Moreover, it can contribute significantly to the pharmaceuticals industry, as it is rich in bioactive ingredients, having both antioxidant and antiinflammatory properties (Zhang, Kris-Etherton, Thompson, & Vanden Heuvel, 2010). On the other hand, after oil extraction, a solid paste remains as a by-product, which can be ground, homogenized, and used as an additive or condiment in human nutrition, for example, creams, sauces, stews, mixing pasta, pastries, biscuits, and crunchy confectionery. This paste can also be used as a nutritional supplement in animal feeding and agriculture, for example, edible cultivated fungi such as mushrooms (Pardo-Giménez et al., 2016).

8 Oil adulteration

Adulteration of high-cost pistachio oil, including partial replacement with lower-priced oils, can unfortunately be performed. This could be controlled using several analytical methods. NMR, HPLC, UV, GC, chemiluminescence, and fiber optic sensors techniques can readily identify processes of mixing original oils with cheaper ones (Cordella, Moussa, Martel, Sbirrazzuoli, & Lizzani-Cuvelier, 2002; Indelicato et al., 2017; Rohman et al., 2016).

Edible oils consist of a complex matrix of di- and triacylglycerols, phospholipids, free fatty acids, as well as other constituents (Obeidat & Khanfar, 2009). Infrared (IR) and Raman spectroscopy techniques have been used for years to distinguish between oils (Sinclair, McKay, Myers, & Jones, 1952). These techniques have the advantage of being rapid, and analyses require only a short time to take place. Moreover, Fourier transform infrared spectroscopy (FTIR) has been used for quantitative analysis of pistachio oil. Quantification is done by comparing the band intensity in the spectrum, which is, in turn, proportional to the amount of the considered compounds (Sheibani, Bahraman, & Sadeghi, 2014; Zhang et al., 2012).

References

Al Juhaimi, F., Özcan, M. M., Ghafoor, K., Babiker, E. E., & Hussain, S. (2018). Comparison of cold-pressing and soxhlet extraction systems for bioactive compounds, antioxidant properties, polyphenols, fatty acids and tocopherols in eight nut oils. *Journal of Food Science and Technology*, *55*(8), 3163–3173.

Alizadeh, A., Alizadeh, O., Amari, G., & Zare, M. (2013). Essential oil composition, total phenolic content, antioxidant activity and antifungal properties of Iranian *Thymus daenensis* subsp. *daenensis* Celak. as in influenced by ontogenetical variation. *Journal of Essential Oil Bearing Plants*, *16*(1), 59–70. https://doi.org/10.1080/0972060X.2013.764190.

Álvarez-Ortí, M., Quintanilla, C., Sena, E., Alvarruiz, A., Pardo, J. E., & Pardo, J. E. (2012). The effects of a pressure extraction system on quality the parameters of different virgin pistachio (*Pistacia vera* L. var. Larnaka) oils. *Grasas y Aceites*, *63*(3), 260–266. (2012). http://grasasyaceites.revistas.csic.es/index.php/grasasyaceites/article/view/1377/1374.

Arena, E., Campisi, S., Fallico, B., & Maccarone, E. (2007). Distribution of fatty acids and phytosterols as a criterion to discriminate geographic origin of pistachio seeds. *Food Chemistry*, *104*(1), 403–408. (2007). https://www.sciencedirect.com/science/article/pii/S0308814606007321.

Ballistreri, G., Arena, E., & Fallico, B. (2009). Evaluation of different degree of ripeness and sun-drying process on triacylglycerols composition of Pistacia vera L oil. In: *Vol. 912. V International symposium on pistachios and almonds* (pp. 791–794). https://www.actahort.org/books/912/912_120.htm.

Bolling, B., Chen, C., McKay, D., & Blumberg, J. (2011). Tree nut phytochemicals: Composition, antioxidant capacity, bioactivity, impact factors. A systematic review of almonds, brazils, cashews, hazelnuts, macadamias, pecans, pine nuts, pistachios and walnuts. *Nutrition Research Reviews*, *24*(2), 244–275.

Catalán, L., Alvarez-Ortí, M., Pardo-Giménez, A., Gómez, R., Rabadán, A., & Pardo, J. E. (2017). Pistachio oil: A review on its chemical composition, extraction systems, and uses. *European Journal of Lipid Science and Technology*, *119*(5), 1–8.

Cordella, C., Moussa, I., Martel, A.-C., Sbirrazzuoli, N., & Lizzani-Cuvelier, L. (2002). Recent developments in food characterization and adulteration detection: Technique-oriented perspectives. *Journal of Agricultural and Food Chemistry*, *50*(7), 1751–1764. https://doi.org/10.1021/jf011096z.

Delgado-Zamarreño, M. M., Bustamante-Rangel, M., Sánchez-Pérez, A., & Carabias-Martínez, R. (2004). Pressurized liquid extraction prior to liquid chromatography with electrochemical detection for the analysis of vitamin e isomers in seeds and nuts. *Journal of Chromatography A*, *1056*(2), 249–252.

Givianrad, M., Saffarpour, S., & Beheshti, P. (2011). Fatty acid and triacylglycerol compositions of *Capparis spinosa* seed oil. *Chemistry of Natural Compounds*, *47*(5), 798–799. (2011). http://www.springerlink.com/index/J1517X668230U519.pdf.

Indelicato, S., Bongiorno, D., Pitonzo, R., Di Stefano, V., Calabrese, V., Indelicato, S., et al. (2017). Triacylglycerols in edible oils: Determination, characterization, quantitation, chemometric approach and evaluation of adulterations. *Journal of Chromatography A*, *1515*, 1–16. (2017). https://08101zqdr-1106-y-https-www-sciencedirect-com.mplbci.ekb.eg/science/article/pii/S0021967317311445.

Kamal-Eldin, A., & Moreau, R. A. (Eds.), (2010). Tree nut oils. In *Gourmet and health-promoting specialty oils*: Elsevier.

Kornsteiner-Krenn, M., Wagner, K.-H., & Elmadfa, I. (2013). Phytosterol content and fatty acid pattern of ten different nut types. *International Journal for Vitamin and Nutrition Research*, *83*(5), 263–270. https://doi.org/10.1024/0300-9831/a000168.

Ling, B., Yang, X., Li, R., & Wang, S. (2016). Physicochemical properties, volatile compounds, and oxidative stability of cold-pressed kernel oils from raw and roasted pistachio (*Pistacia vera* L. Var Kerman). *European Journal of Lipid Science and Technology*, *118*(9), 1368–1379.

Ling, B., Zhang, B., Li, R., & Wang, S. (2016). Nutritional quality, functional properties, bioactivity, and microstructure of defatted pistachio kernel flour. *Journal of the American Oil Chemists' Society*, *93*(5), 689–699.

Miraliakbari, H., & Shahidi, F. (2008). Oxidative stability of tree nut oils. *Journal of Agricultural and Food Chemistry*, *56*(12), 4751–4759. https://doi.org/10.1021/jf8000982.

Obeidat, S., & Khanfar, M. (2009). Classification of edible oils and uncovering adulteration of virgin olive oil using FTIR with the aid of chemometrics. *Australian Journal of Basic and Applied Sciences*, *3*(3), 2048–2053. https://www.researchgate.net/profile/Mai_Khanfar/publication/279554122_Classification_of_edible_oils_and_uncovering_adulteration_of_virgin_olive_oil_using_FTIR_with_the_aid_of_chemometrics/links/02bfe5106e6662d4a2000000/Classification-of-edible-oils-and-u.

Ojeda-Amador, R. M., Fregapane, G., & Salvador, M. D. (2018). Composition and properties of virgin pistachio oils and their by-products from different cultivars. *Food Chemistry*, *240*, 123–130. https://08101y288-1106-y-https-www-sciencedirect-com.mplbci.ekb.eg/science/article/pii/S0308814617312414.

Pardo-Giménez, A., Catalán, L., Carrasco, J., Álvarez-Ortí, M., Zied, D., & Pardo, J. (2016). Effect of supplementing crop substrate with defatted pistachio meal on *Agaricus bisporus* and *Pleurotus ostreatus* production. *Journal of the Science of Food and Agriculture*, *96*(11), 3838–3845.

Rabadán, A., Álvarez-Ortí, M., Gómez, R., Alvarruiz, A., & Pardo, J. E. (2017). Optimization of pistachio oil extraction regarding processing parameters of screw and hydraulic presses. *LWT—Food Science and Technology*, *83*, 79–85.

Rabadán, A., Álvarez-Ortí, M., Pardo, J. E., & Alvarruiz, A. (2018). Storage stability and composition changes of three cold-pressed nut oils under refrigeration and room temperature conditions. *Food Chemistry*, *259*, 31–35.

Rabadán, A., Gallardo-Guerrero, L., Gandul-Rojas, B., Álvarez-Ortí, M., & Pardo, J. E. (2018). Effect of roasting conditions on pigment composition and some quality parameters of pistachio oil. *Food Chemistry*, *264*, 49–57. https://doi.org/10.1016/j.foodchem.2018.05.030.

Ramachandran, S., Singh, S. K., Larroche, C., Soccol, C. R., & Pandey, A. (2007). Oil cakes and their biotechnological applications—A review. *Bioresource Technology*, *98*(10), 2000–2009.

Rohman, A., Windarsih, A., Riyanto, S., Sudjadi, Shuhel Ahmad, S. A., Rosman, A. S., et al. (2016). Fourier transform infrared spectroscopy combined with multivariate calibrations for the authentication of avocado oil. *International Journal of Food Properties*, *19*(3), 680–687. https://doi.org/10.1080/10942912.2015.1039029.

Saber-Tehrani, M., Givianrad, M. H., Aberoomand-Azar, P., Waqif-Husain, S., & Jafari Mohammadi, S. A. (2013). Chemical composition of irans *Pistacia atlantica* cold-pressed oil. *Journal of Chemistry*, *2013*.

Saitta, M., Giuffrida, D., Di Bella, G., La Torre, G. L., & Dugo, G. (2011). In V. R. Preedy, R. R. Watson, & V. B. Patel (Eds.), *Nuts and seeds in health and disease prevention* (pp. 909–918). London: Academic Press.

Saitta, M., La Torre, G. L., Potortì, A. G., Di Bella, G., & Dugo, G. (2014). Polyphenols of pistachio (*Pistacia vera* L.) oil samples and geographical differentiation by principal component analysis. *Journal of the American Oil Chemists' Society*, *91*(9), 1595–1603. https://doi.org/10.1007/s11746-014-2493-3.

Sena-Moreno, E., Pardo, J. E., Catalán, L., Gómez, R., Pardo-Giménez, A., & Alvarez-Ortí, M. (2015). Drying temperature and extraction method influence physicochemical and sensory characteristics of pistachio oils. *European Journal of Lipid Science and Technology*, *117*(5), 684–691.

Sheibani, A., Bahraman, N., & Sadeghi, F. (2014). FT-IR application for the detection of pistachio oil adulteration. *Oriental Journal of Chemistry*, *30*(3), 1205–1209. http://www.orientjchem.org/vol30no3/ft-ir-application-for-the-detection-of-pistachio-oil-adulteration.

Sinclair, R. G., McKay, A. F., Myers, G. S., & Jones, R. N. (1952). The infrared absorption spectra of unsaturated fatty acids and esters. *Journal of the American Chemical Society*, *74*(10), 2578–2585. https://doi.org/10.1021/ja01130a035.

Tsantili, E., Konstantinidis, K., & Christopoulos, M. (2011). Total phenolics and flavonoids and total antioxidant capacity in pistachio (*Pistachia vera* L.) nuts in relation to cultivars and storage conditions. *Scientia Horticulturae*, *129*(4), 694–701. (2011). https://www.sciencedirect.com/science/article/pii/S0304423811002676.

Zhang, J., Kris-Etherton, P. M., Thompson, J. T., & Vanden Heuvel, J. P. (2010). Effect of pistachio oil on gene expression of IFN-induced protein with tetratricopeptide repeats 2: A biomarker of inflammatory response. *Molecular Nutrition & Food Research*, *54*(S1), S83–S92. https://doi.org/10.1002/mnfr.200900244.

Zhang, Y., Smuts, J. P., Dodbiba, E., Rangarajan, R., Lang, J. C., & Armstrong, D. W. (2012). Degradation study of carnosic acid, carnosol, rosmarinic acid, and rosemary extract (*Rosmarinus officinalis* L.) assessed using HPLC. *Journal of Agricultural and Food Chemistry*, *60*(36), 9305–9314.

Chapter 23

Cold pressed clove (*Syzygium aromaticum*) oil

Hari Prasad Devkota and Anjana Adhikari-Devkota
Graduate School of Pharmaceutical Sciences, Kumamoto University, Kumamoto, Japan

1 Introduction

The flowering buds of *Syzygium aromaticum* (L.) Merr. & L.M. Perry. (synonym: *Eugenia caryophyllata* Thunb.; family: Myrtaceae), commonly known as clove (Fig. 1), are widely used as a spice all over the world due to their strong aromatic smell (Cortés-Rojas, de Souza, & Oliveira, 2014). The clove is a native of some islands of Malay Archipelago, especially the Moluccas. It is cultivated in Zanzibar and Pemba (Tanzania), Indonesia, Penang, and Malagasy, and to a lesser extent in the Seychelles, Mauritius, and Sri Lanka. In India, it is reported to be grown in Tamil Nadu and Kerala (Sharma, Yelne, & Dennis, 2000). Known by many names such as Devapuspa in Sanskrit, Lwang in Nepali, and Lavanga in Hindi, it is an important component of Ayurveda and used in formulations such as "Lavangadi Vati" and "Lavangadi Churna" (Ministry for Health and Family Welfare, 2010). It has aromatic, stimulant, and carminative properties, and is used to combat gastric irritation, dyspepsia, nausea, vomiting, and flatulence (Rajbhandari, Joshi, Shrestha, Joshi, & Acharya, 1995). The main chemical constituents of cloves are phenylpropanoids, sesquiterpenoids, tannins, and triterpenoids. The extracts obtained from clove have shown various biological activities including antifungal, antibacterial, anticarcinogenic, antioxidant, antibacterial, antiviral, anticonvulsant, histamine release inhibitory, and tyrosinase inhibitory activities (Adhikari et al., 2008; Chaieb et al., 2007; Cortés-Rojas et al., 2014; Devkota et al., 2006; El-Maati, Mahgoub, Labib, Al-Gaby, & Ramadan, 2016; Sharma et al., 2000).

Clove essential oil is one of the most widely used processed product of clove. It is traditionally used in dental care to relieve toothache as an antiseptic and analgesic agent (Chaieb et al., 2007). Clove oil has rubefacient, carminative, and antispasmodic activities (Rajbhandari et al., 1995). The oil is a light yellow liquid with a characteristic aroma taste (The Ministry of Health Labor and Welfare of Japan, 2016). Eugenol (Fig. 2) is the main constituent of clove oil. The content of eugenol in clove oil has been reported to vary from about 45% to almost 90% in different studies (Chaieb et al., 2007). The Japanese Pharmacopeia describes clove oil as the volatile oil distilled with steam from the flower buds or leaves of *Syzygium aromaticm* and the oil contains not less than 80% of eugenol (The Ministry of Health Labor and Welfare of Japan, 2016). The United States Food and Drug Administration (USFDA) has classified eugenol as safe. Similarly, the clove oil below the concentration of 1500 ppm is classified as "Generally Regarded as Safe" (GRAS) (Ibrahim, Attia, Maklad, Ahmed, & Ramadan, 2017). Other chemical constituents of clove oil include eugenyl acetate, carvacol, thymol, β-caryophyllene, cinnamaldehyde, 2-heptanone, and calacorene among others (Chaieb et al., 2007). Clove oil has been reported to possess strong antibacterial, antifungal, antiviral, antitumor, anesthetic, and insecticidal properties (Chaieb et al., 2007; Cortés-Rojas et al., 2014; de Meneses et al., 2019; Dorman, Figueiredo, Barroso, & Deans, 2000; Issac, Gopakumar, Kuttan, Maliakel, & Krishnakumar, 2015; Ogata, Hoshi, Urano, & Endo, 2000; Tennyson, Samraj, Jeyasundar, & Chalieu, 2013; Zhu, Henderson, Chen, Fei, & Laine, 2001).

In recent years, there has been increasing interest in cold pressed oils. They are obtained by a technique known as cold pressing, which involves no heat and chemical treatments, and is thought to be safer and rich in antioxidant lipophilic compounds than other techniques (Ibrahim et al., 2017; Parker, Adams, Zhou, Harris, & Yu, 2003; Prescha, Grajzer, Dedyk, & Grajeta, 2014; Ramadan, 2013; Siger, Nogala-Kalucka, & Lampart-Szczapa, 2008). This chapter aims to provide an overview of the chemistry and reported biological activities of cold pressed clove oil.

Cold Pressed Oils. https://doi.org/10.1016/B978-0-12-818188-1.00023-2
Copyright © 2020 Elsevier Inc. All rights reserved.

FIG. 1 Photograph of clove (flowering buds of *Syzygium aromaticum* (L.) Merr. & L.M. Perry).

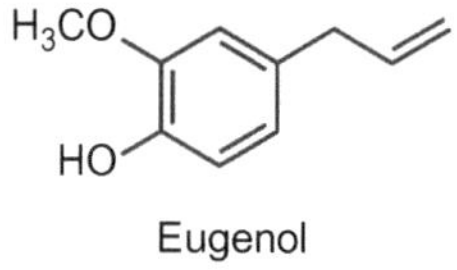

H_3CO

HO

Eugenol

FIG. 2 Structure of eugenol, a principal component of clove oil.

2 Chemical composition of cold pressed clove oil

There have been various studies on the chemical composition of clove oils obtained using different extraction methods. For example, Guan, Li, Yan, Tang, and Quan (2007) studied the effects of supercritical fluid extraction method and three traditional methods of extraction (hydrodistillation, steam distillation, and Soxhlet extraction), and reported that the temperature has a significant effect on the content of eugenol in the essential oil. Cold pressing might be a method of choice for the extraction of clove oil, as no heat or chemical treatment is applied.

Ramadan, Asker, and Tadros (2013) studied the lipid profiles of cold pressed clove oils purchased from a local market in Egypt. The characterization of lipid subclasses, analysis of methyl esters of fatty acids, high-performance liquid chromatography (HPLC) analysis of tocols, and extraction and quantification of phenolic compounds were performed. The authors reported that the level of neutral lipids (NL) was the highest (about 94.7%), followed by glycolipids (GL, 0.76%) and phospholipids (PL, 0.39%). The constituents of neutral fatty acids were triacylglycerols, free fatty acids, diacylglycerols, esterified fatty acids, and monoacyl glycerols. Linoleic and oleic acids were the main unsaturated fatty acids, whereas stearic acid and palmitic acid were the main saturated fatty acids. Among tocols, α-, β-, and γ-tocopherols were reported to be higher than α-, β-, and γ-tocotrienols. The cold pressed clove oil had higher content to total phenolics (4.6 mg/g) as compared to virgin clove oil (3.4 mg/g). Similar results were reported in another paper (Assiri & Hassanien, 2013). In a separate study, Ramadan (2013) reported the total phenolic content of cold pressed oil to be 5.9 mg gallic acid equivalents (GAE) per gram of oil.

3 Biological activities of cold pressed clove oil

Various biological activities have been reported for the cold pressed clove oil. Ramadan et al. (2013) evaluated the free radical scavenging activity of cold pressed clove oil and virgin clove oil. It was reported that the cold pressed oil had deactivated 70% of 1,1-diphenyl-2-picrylhydrazyl (DPPH·) radical, while the virgin oil had deactivated 45% of DPPH· radicals. The stronger antiradical activity of cold pressed clove oil was suggested to be due to higher content of phenolics in cold pressed clove oil. Cold pressed oil also showed potent antibacterial activity against various Gram-positive and Gram-negative bacteria (Assiri & Hassanien, 2013; Ramadan et al., 2013).

El-Hadary and Ramadan Hassanien (2016) evaluated the hepatoprotective activity of cold pressed clove oil against carbon tetrachloride-induced hepatotoxicity in healthy male albino rats. Carbon tetrachloride in olive oil (1:1, w/w) was used to induce hepatotoxicity. Clove oil at a dose of 100 or 200 mg/kg was administered orally (three times a week) for 8 weeks and carbon tetrachloride in olive oil (1:1 w/w, 1 mL/kg, three times a week) was administered during the last 4 weeks. The administration of clove oil attenuated the increased levels of liver enzymes alanine transaminase (ALT), aspartate transaminase (AST), and alkaline phosphatase (ALP) induced by carbon tetrachloride. The levels of serum total bilirubin (TB) and direct bilirubin (B) were also decreased. Administration of clove oil also reduced the levels of kidney function indicators such as the levels of urea, creatinine, and uric acids as compared to the control groups. Histopathological examinations revealed the reduced fatty degenerations, cytoplasmic vacuolization, and necrosis in clove oil-treated groups as compared to the controls. Based on a 24-h oral toxicity test, the lethal dose (LD_{50}) was found to be 5950 mg/kg. During the long-term toxicity study, 400 mg/kg caused a decrease in body weight.

Ibrahim et al. (2017) studied the effects of different cold pressed oils including clove oil for acute toxicity and antiinflammatory and ulcerogenic activity in female albino rats. From the acute toxicity test, the dose of 400 mg/kg was found to be safe for administration. Clove oil administration was found to reduce the rat paw edema in the carrageenan-induced edema model in rats. During ulcerogenic testing, clove oil did not show any ulcerogenic activity.

Hussein, Abd El-Hack, Mahgoub, Saadeldin, and Swelum (2019) studied the effects of cold pressed clove oil on the growth, carcass traits, blood components, meat quality, and intestinal microbiota in Japanese quails. Administration of clove oil (1.5 mL/kg) resulted in an increase in live body weight and daily body weight. The activities of antioxidant enzymes and lipid profiles were also improved. Similarly, the population of intestinal bacteria such as *Escherichia coli* and *Salmonella* spp. was also lower compared to the control group.

4 Conclusions

Clove oil is a popular remedy for toothache due to its potent antiseptic and analgesic activity. It also has strong antioxidant and antiviral activities. Eugenol is the main component of clove oil. Various studies reported the fatty acid and tocols composition of cold pressed clove oil; however, the quantification of major constituents such as eugenol, eugenyl acetate, and other compounds was not performed in detail. Cold pressed clove oil was reported to possess stronger radical scavenging and antimicrobial activity compared to virgin oil. Similarly, the cold pressed clove oil was also reported to possess hepatoprotective activity in experimental animals along with other biological activities. Future studies should focus on the comparison of the activities of cold pressed oil with other oils obtained from different conventional using advanced methods. The relationship between their bioactivity and the chemical composition should be studied in detail.

References

Adhikari, A., Devkota, H. P., Takano, A., Masuda, K., Nakane, T., Basnet, P., et al. (2008). Screening of Nepalese crude drugs traditionally used to treat hyperpigmentation: In vitro tyrosinase inhibition. *International Journal of Cosmetic Science*, *30*, 353–360. https://doi.org/10.1111/j.1468-2494.2008.00463.x.

Assiri, A. M. A., & Hassanien, M. F. R. (2013). Bioactive lipids, radical scavenging potential, and antimicrobial properties of cold pressed clove (*Syzygium aromaticum*) oil. *Journal of Medicinal Food*. https://doi.org/10.1089/jmf.2012.0288.

Chaieb, K., Hajlaoui, H., Zmantar, T., Kahla-Nakbi, A. B., Rouabhia, M., Mahdouani, K., et al. (2007). The chemical composition and biological activity of clove essential oil, Eugenia caryophyllata (*Syzygium aromaticum* L. Myrtaceae): A short review. *Phytotherapy Research*. https://doi.org/10.1002/ptr.2124.

Cortés-Rojas, D. F., de Souza, C. R. F., & Oliveira, W. P. (2014). Clove (*Syzygium aromaticum*): A precious spice. *Asian Pacific Journal of Tropical Biomedicine*, *4*, 90–96. https://doi.org/10.1016/S2221-1691(14)60215-X.

de Meneses, A. C., Sayer, C., Puton, B. M. S., Cansian, R. L., Araújo, P. H. H., & de Oliveira, D. (2019). Production of clove oil nanoemulsion with rapid and enhanced antimicrobial activity against gram-positive and gram-negative bacteria. *Journal of Food Process Engineering*, 42. https://doi.org/10.1111/jfpe.13209.

Devkota, H. P., Adhikari, A., Paudel, S., G.C., S., Takano, A., & Basnet, P. (2006). Antioxidative activity of common natural medicines in Nepal. *Journal of Nepal Pharmaceutical Association*, *26*, 39–46.

Dorman, H. J. D., Figueiredo, A. C., Barroso, J. G., & Deans, S. G. (2000). In vitro evaluation of antioxidant activity of essential oils and their components. *Flavour and Fragrance Journal*, *15*, 12–16. https://doi.org/10.1002/(SICI)1099-1026(200001/02)15:1<12::AID-FFJ858>3.0.CO;2-V.

El-Hadary, A. E., & Ramadan Hassanien, M. F. (2016). Hepatoprotective effect of cold-pressed *Syzygium aromaticum* oil against carbon tetrachloride (CCl_4)-induced hepatotoxicity in rats. *Pharmaceutical Biology*, *54*, 1364–1372. https://doi.org/10.3109/13880209.2015.1078381.

El-Maati, M. F. A., Mahgoub, S. A., Labib, S. M., Al-Gaby, A. M. A., & Ramadan, M. F. (2016). Phenolic extracts of clove (*Syzygium aromaticum*) with novel antioxidant and antibacterial activities. *European Journal of Integrative Medicine*, *8*, 494–504. https://doi.org/10.1016/j.eujim.2016.02.006.

Guan, W., Li, S., Yan, R., Tang, S., & Quan, C. (2007). Comparison of essential oils of clove buds extracted with supercritical carbon dioxide and other three traditional extraction methods. *Food Chemistry*, *101*, 1558–1564. https://doi.org/10.1016/j.foodchem.2006.04.009.

Hussein, M. M. A., Abd El-Hack, M. E., Mahgoub, S. A., Saadeldin, I. M., & Swelum, A. A. (2019). Effects of clove (*Syzygium aromaticum*) oil on quail growth, carcass traits, blood components, meat quality, and intestinal microbiota. *Poultry Science*. https://doi.org/10.3382/ps/pey348.

Ibrahim, F. M., Attia, H. N., Maklad, Y. A. A., Ahmed, K. A., & Ramadan, M. F. (2017). Biochemical characterization, anti-inflammatory properties and ulcerogenic traits of some cold-pressed oils in experimental animals. *Pharmaceutical Biology*. https://doi.org/10.1080/13880209.2016.1275705.

Issac, A., Gopakumar, G., Kuttan, R., Maliakel, B., & Krishnakumar, I. M. (2015). Safety and anti-ulcerogenic activity of a novel polyphenol-rich extract of clove buds (*Syzygium aromaticum* L). *Food & Function*, *6*, 842–852. https://doi.org/10.1039/c4fo00711e.

Ministry for Health and Family Welfare (2010). *The ayurvedic pharmacopoeia of India*.

Ogata, M., Hoshi, M., Urano, S., & Endo, T. (2000). Antioxidant activity of eugenol and related monomeric and dimeric compounds. *Chemical & Pharmaceutical Bulletin*, *48*, 1467–1469. https://doi.org/10.1248/cpb.48.1467.

Parker, T. D., Adams, D. A., Zhou, K., Harris, M., & Yu, L. (2003). Fatty acid composition and oxidative stability of cold-pressed edible seed oils. *Journal of Food Science*. https://doi.org/10.1111/j.1365-2621.2003.tb09632.x.

Prescha, A., Grajzer, M., Dedyk, M., & Grajeta, H. (2014). The antioxidant activity and oxidative stability of cold-pressed oils. *Journal of the American Oil Chemists' Society*. https://doi.org/10.1007/s11746-014-2479-1.

Rajbhandari, T., Joshi, M., Shrestha, T., Joshi, S., & Acharya, B. (1995). *Medicinal plants of Nepal for ayurvedic drugs*. Kathmandu: Department of Plant Resources.

Ramadan, M. F. (2013). Healthy blends of high linoleic sunflower oil with selected cold pressed oils: Functionality, stability and antioxidative characteristics. *Industrial Crops and Products*. https://doi.org/10.1016/j.indcrop.2012.07.013.

Ramadan, M. F., Asker, M. M. S., & Tadros, M. (2013). Lipid profile, antiradical power and antimicrobial properties of *Syzygium aromaticum* oil. *Grasas y Aceites*. https://doi.org/10.3989/gya.011713.

Sharma, P., Yelne, M., & Dennis, T. (2000). *Database on medicinal plants used in ayurveda*. New Delhi: Central Council for Research in Ayurveda and Siddha.

Siger, A., Nogala-Kalucka, M., & Lampart-Szczapa, E. (2008). The content and antioxidant activity of phenolic compounds in cold-pressed plant oils. *Journal of Food Lipids*. https://doi.org/10.1111/j.1745-4522.2007.00107.x.

Tennyson, S., Samraj, D. A., Jeyasundar, D., & Chalieu, K. (2013). Larvicidal efficacy of plant oils against the dengue vector *Aedes aegypti* (L.) (Diptera: Culicidae). *Middle-East Journal of Scientific Research*, *13*, 64–68. https://doi.org/10.5829/idosi.mejsr.2013.13.1.64107.

The Ministry of Health Labor and Welfare of Japan (2016). *The Japanese pharmacopoeia* (17th ed.). Tokyo.

Zhu, B. C. R., Henderson, G., Chen, F., Fei, H., & Laine, R. A. (2001). Evaluation of vetiver oil and seven insect-active essential oils against the formosan subterranean termite. *Journal of Chemical Ecology*, *27*, 1617–1625. https://doi.org/10.1023/A:1010410325174.

Chapter 24

Cold pressed berry seed oils

Ahmad Cheikhyoussef[a], Natascha Cheikhyoussef[b], Ateeq Rahman[c], and Alfred Maroyi[d]

[a]*Science and Technology Division, Multidisciplinary Research Centre, University of Namibia, Windhoek, Namibia,* [b]*Ministry of Higher Education, Training and Innovation, Windhoek, Namibia,* [c]*Department of Chemistry and Biochemistry, Faculty of Science, University of Namibia, Windhoek, Namibia,* [d]*Medicinal Plants and Economic Development (MPED) Research Centre, Department of Botany, University of Fort Hare, Alice, South Africa*

Abbreviations

BCPSO blackberry cold pressed seed oil
BUCPSO blueberry cold pressed seed oil
CCPSO cranberry cold pressed seed oil
CPBSO cold pressed berry seed oil
PUFA polyunsaturated fatty acid
RCPSO raspberry cold pressed seed oil
SCPSO strawberry cold pressed seed oil

1 Introduction

Cold pressing is a seed oil extraction process that does not involve chemicals or extreme heat prior to or during the procedure (Van Hoed, Clercq, Echim, Andjelkovic, et al., 2009). According to the Codex Alimentarius Standard 210 (1999), cold pressed oils are obtained, without altering the oil, by mechanical procedures such as expelling or pressing, without any heat application (Van Hoed et al., 2011). During cold pressing, temperature should not exceed 50°C (Moreau & Kamal-Eldin, 2009) depending on the characteristics of the fruits and seeds. Oils can then be purified by washing with water, settling, filtering, and centrifuging. Cold pressed seed oils are known to retain more phytochemicals, including natural antioxidants.

Cold pressed seed oils from berries have become commercially available due to their health-promoting effects. These oils are extracted from berry seeds, which are considered as agricultural by-products. Raspberry and blackberry are common names of edible fruits, which belong to the *Rubus* L. genus of the Rosaceae family. The raspberry (*Rubus idaeus* L.) and blackberry (*Rubus fruticosus* L.) belong to caneberries (bramble fruits) of Rosaceae that grow on a leafy cane and produce multiple small fruits (Bushman et al., 2004). During the processing of berries, the seed materials are removed as waste by-products. The seeds contain high levels of a diverse range of phytochemicals and bioactive compounds such as vitamins C and E, fatty acids, dietary fibers (Nile & Park, 2014), and phenolics such as anthocyanins, phenolic acids, flavonols, and tannins (Bowen-Forbes, Zhang, & Nair, 2010; Szajdek & Borowska, 2008), which are known for their health benefits. Blackberry and raspberry pomace is a by-product of pressing fruit juice that mostly consists of the seeds. These seeds contain lipids and two main important fatty acids, namely linoleic (C18:2ω-6) and α-linolenic (C18:3ω-3) acid. These oils are rich in tocopherols, phytosterols, phenolics, and other micro-constituents that have antioxidant properties and are considered as value-added products (Radočaj et al., 2014). This chapter aims to present a summary of the cold pressed method of extraction and processing of berry seed oils, their composition of bioactive compounds, their health effects, and their potential applications.

2 Extraction, processing, and physicochemical characteristics of cold pressed berry seed oils

Recently, scientists have become interested in unconventional sources of plant oils and their nutritional and health impact. Berry oils are recovered from almost all types of berry seeds and found to contain a high content of PUFA (Ahmad, Anwar, &

Cold Pressed Oils. https://doi.org/10.1016/B978-0-12-818188-1.00024-4
Copyright © 2020 Elsevier Inc. All rights reserved.

Abbas, 2019). Seeds from berries contain between 11% and 23% oil, which can be extracted by cold pressing. Cold pressed berry seed oils (CPBSO) are processed between 40°C and 60°C depending on the composition of the different seeds. Bushman et al. (2004) used a screw-press (Botanic Oil Innovation, Spooner, WI) to extract the oil from cranberries (*Rubus* spp.) seeds. The extrusion of the oil took place at the press head under pressure up to 9652 kPa and temperatures ranged from 27°C to 38°C.

Oomah, Ladet, Godfrey, Liang, and Girard (2000) extracted the raspberry cold pressed oil from raspberry seeds hydraulically using a Carver Press at 280 kg/cm^2. High-grade food-quality cranberry seed oil is produced by cold pressing (Thyagarajan, 2012; Van Hoed et al., 2009). Raspberry (*Rubus idaeus*), blackberry (*Rubus fruticosus*), cranberry (*Vaccinium macrocarpon* Aiton), blueberry (*Vaccinium corymbosum* L.), and strawberry (*Fragaria* × *ananassa* (Weston) Duchesne) oils were prepared by pressing the berries to make juice (within 24 h) after the pomace was collected and then it was dried within 10 h of collection (from about 80% to 8% moisture) (Van Hoed et al., 2009). The seeds were mechanically separated (approx. 99.9% pure seed) and double wrapped in heavy-duty polyethylene bags; after that, they were stored in dry to ambient-temperature cardboard boxes before being pressed. The oils were cold pressed in a Täby Type 40a expeller press at temperatures ranging from 40.6°C to about 60.0°C, depending on the seed type and quality. In the final step of the CPBSO preparation, the oils were cold filtered at a temperature below 10°C and the raw oils were drawn through tubes immersed in ice-water before passing through a plate filter Model MFP320-10 with 1%–1.5% of diatomaceous earth (grades FW 60 and FP 4) for polishing (at or below 10°C) (Van Hoed et al., 2009). Cold pressing extraction that is maintained at room temperatures retain the natural antioxidants and provide nutritional quality, desirable flavor, and oil stability of berry seed oils (Van Hoed et al., 2009). Expelled pressing uses a press to physically squeeze the oil out of the seed without use of chemical solvents. A typical expeller press is a screw-type machine that presses oil through a caged barrel-like-cavity, using friction and continuous pressure. During processing, the screw drives forward to squeeze the oil from the compressed seeds (Broaddus, 2017). Strawberry (*Fragaria* × *ananassa*) seeds constitute 1% of the fruit; however, their content in the fruit depends on its species (Cheel, Theoduloz, Rodríguez, et al., 2007). It is one of the most interesting raw materials whereby after the pressing process, the oilcake is almost completely defatted (Kowalczewski et al., 2019). The content of seeds in raspberries is higher than strawberry (9%–12%) of the fruit weight and is a by-product formed mainly in the production of wine and juices (Kowalczewski et al., 2019). Due to the high recovery rate of raspberry seed oil (10%–23%), it is also increasingly used for the production of cold-pressed oil and more widely used in cosmetics (Teng, Chen, Huang, et al., 2016).

Parry and Yu (2004) reported that color perception is critical for potential food applications of edible oils. They further confirmed that cold pressed black raspberry seed oil had very high intensity in yellowness (6.08–16.26) and redness (3.93–7.73) (Table 1). The cold pressed black raspberry seed oils also exhibited significantly more redness than commercial corn and soybean oils and cold pressed black caraway, carrot, hemp, and cranberry seed oils (Parker et al., 2003). In addition, the two preparations of the cold pressed black raspberry seed oils significantly differed in their *L*, *a*, and *b* values, suggesting that the processing, growing, and harvesting conditions of the seeds may alter the color of the cold pressed black raspberry oils (Parry & Yu, 2004). These conditions may also partially explain why the two oil samples significantly differed in their DPPH· (2,2-diphenyl-1-picryhydrazyl) radical scavenging capacities. Van Hoed et al. (2009, 2011) reported that blackberry cold pressed seed oil (BCPSO) had free fatty acid values of 0.96% and 0.80% (Table 1), while RCPSO had free fatty acid values of 0.49% and 0.69% (expressed as oleic acid). However, the oils in their study were obtained from the seeds originating from fresh berries.

3 Fatty acids and acyl lipids profile of cold pressed berry seed oils

Edible oils are primarily composed of triacylglycerols (95%), and minor amounts of other components such as diacylglycerols, monoacylglycerols, free fatty acids, phospholipids, tocols (tocopherols and tocotrienols), sterols, and pigments (Shahidi & Shukla, 1996). Berry lipids are found as berry waxes (epicuticular and cuticular waxes), cytoplasmatic lipids, and seed lipids (Klavins & Klavina, 2019). Raspberry cold pressed seed oil (RCPSO) consists primarily of neutral lipid (93.8%) with minor amounts of free fatty acid (3.5%) and phospholipids (2.7%) of the total crude oil (Oomah et al., 2000). Similar high levels of neutral lipids (95.7%) have been reported for raspberry seed oil (Winton & Winton, 1935) and other berry fruit oils, such as sea buckthorn seed oil (92%) (Zadernowski, Nowak-Polakowska, Lossow, & Nesterowicz, 1997). The most abundant fatty acids in RCPSO are linoleic (C18:2ω-6) (PubChem CID: 5280450), α-linolenic acid (C18:3ω-3) (PubChem CID: 5280934), and oleic acid (OA, C18:1,n-9) (PubChem CID: 445639), which together comprise 96% of the total fatty acid. The phospholipid fraction of RCPSO contains saturated fatty acids (11%) and monounsaturated fatty acids (19%), and contains much lower polyunsaturated compared to RCPSO neutral lipid fraction. The polyunsaturated fraction of the free fatty acid fraction in RCPSO can reach up to 61% of the total fatty acids,

TABLE 1 Physicochemical characteristics of CPBSO.

Characteristic	Blackberry	Blueberry	Cranberry	Raspberry	Strawberry
Oil yield (%) dry matter	13.30–15.00	NR	NR	10.70–18.70	NR
Acid value (mg KOH/g)	6.85 ± 0.11	NR	NR	17.18 ± 0.43	5.0
Free fatty acid (as oleic acid, %)	0.79–3.43	0.59–0.67	0.70–1.10	0.49–8.59	1.54–1.59
Peroxide value (meq O_2/kg of oil)	0.59–34.41	24.62–47.31	8.41–42.75	8.25–50.45	26.25–30.31
Iodine value (g/I_2/100 g oil)	NR	153.1	NR	194.00–195.30	NR
Saponification number (mg KOH/g oil)	NR	NR	NR	191 ± 0.1	NR
p-Anisidine value	8.66–10.80	2.28–10.30	6.18–61.49	14.30–16.00	5.86–22.74
Viscosity (mPa s)	NR	NR	NR	26 ± 1.1	NR
Diene value	NR	NR	NR	0.837 ± 0.0003	NR
Carotenoid content (mg/100 g)	NR	NR	NR	23 ± 0.04	NR
UV absorption at 232	1.88–3.49	2.34–2.67	0.593–0.600	4.76	3.47–5.52
UV absorption at 270	2.50–3.11	2.13–2.49	0.511–1.660	7.37	1.19–2.45
Color (red, yellow)	11.09–12.09	9.05–11.28	1.22–16.30	2.55–16.26	12.85–14.43
Oxidative stability index (at 97.8°C)	8.23–8.77	2.55–8.24	6.33–7.90	3.43–6.54	4.28–5.00
Origin	USA	USA	USA	Canada USA	USA
Reference	Van Hoed et al. (2009, 2011), Dimić, Vujasinović, Radočaj, and Pastor (2012), and Bushman et al. (2004)	Van Hoed et al. (2009, 2011)	Van Hoed et al. (2009, 2011)	Oomah et al. (2000), Van Hoed et al. (2009), Dimić et al. (2012), Bushman et al. (2004), Parry and Yu (2004), and Parker, Adams, Zhou, Harris, and Yu (2003)	Van Hoed et al. (2009)

NR, not reported.

meanwhile, the monounsaturated and saturated fatty acids reach 27% and 12%, respectively. Hence, the ratio of polyunsaturated to monounsaturated to saturated fatty acid varies from 84:12:4 to 61:27:12, depending on lipid fractions, which have a strong impact on the quality of the oil and its characteristics such as oxidation (Oomah et al., 2000).

Cranberry cold pressed seed oil (CCPSO) is characterized by higher contents of polyunsaturated such as α-linolenic acid (C18:3ω-3) (30%–35%) and linoleic acid (C18:2ω-6) (35%–40%), followed by oleic acid (OA, C18:1,n-9) (20%–25%) (Liangli, Zhou, & Parry, 2005). CPBSO is an excellent natural source of dietary α-linolenic (about 35%) and about 55%–58% linoleic fatty acid with high phenolic content. These compounds are well-known for their significant contribution to the antioxidant properties and antioxidative stability of these oils (Parry & Yu, 2004; Van Hoed et al., 2009). Oils derived from diverse berry seeds exhibited impressive/exciting ratios of n-6/n-3 essential fatty acids (Table 2) in comparison to conventional vegetable fixed oils (Parker et al., 2003; Parry, Su, Moore, Cheng, & Luther, 2006). Van Hoed et al. (2009) reported that seed oils within the same genus such as *Rubus* spp. (BCPSO and RCPSO) have obvious differences for all fatty acids; meanwhile, regarding the oils of the *Vaccinium* L. spp. (BUCPSO and CCPSO), only the amounts of C18:2 and C18:3 were different (Table 2). Li et al. (2016) reported that palmitic (C16:0), stearic (C18:0), oleic (18:1), linoleic (C18:2ω-6), and α-linolenic (C18:3ω-3) acids were detected in cold pressed seed oils of BCPSO, RCPSO, and blueberry

TABLE 2 Fatty acid composition of CPBSO.

Fatty acid (fractions %)	Blackberry	Blueberry	Cranberry	Raspberry	Strawberry
Free fatty acid (as oleic acid %)	0.96	0.67	0.70	0.49–3.50	1.54
C12:0	0.04	0.02	0.14	ND	ND
C14:0	0.05	0.09	0.08	0.07	0.05
C16:0	3.3–3.71	5.20–5.66	5.38–7.83	1.22–10.46	4.32–4.70
C16:1	ND	ND	ND	0.1	ND
C16:2	ND	ND	ND	0.1	ND
C18:0	1.60–3.3 0	1.30–1.78	1.25–1.91	0.60–1.26	0.586–1.68
C18:1	14.40–14.72	21.42–22.20	22.69–25.30	6.15–26.62	14.55–16.90
C18:2	61.22–63.40	41.90–42.51	37.68–44.31	47.28–57.59	42.22–47.20
C18:3	16.50–17.60	28.10–28.28	22.28–30.09	14.35–35.33	29.30–36.48
C20:0	0.47–0.80	0.25	0.07	0.37–0.40	0.71–1.29
C20:1	0.30	ND	ND	0.20	ND
C20:2	ND	ND	0.98±0.00	0.10	ND
C22:0	ND	ND	ND	0.10	ND
Sat	4.9–6.45	6.50–7.80	6.88–9.74	1.62–3.82	6.85
Mono	14.40–14.81	21.43–22.20	22.69–25.14	6.15–11.02	14.71
n-3	16.50	28.10	22.28±0.01	33.70–35.33	ND
n-6	ND	ND	45.29±0.03	55.85–57.59	ND
n-6/n-3	3.48–3.80	1.50	1.25	1.60–1.69	1.16
PUFA	78.74–79.90	70.00–70.77	67.57–67.98	85.16–92.92	78.44
Origin	USA	USA	USA	Canada/USA	USA
Reference	Van Hoed et al. (2009), Li, Wang, and Shahidi (2016), and Bushman et al. (2004)	Van Hoed et al. (2009) and Li et al. (2016)	Parker et al. (2003) and Van Hoed et al. (2009)	Bushman et al. (2004), Li et al. (2016), Oomah et al. (2000), Parry and Yu (2004), and Van Hoed et al. (2009)	Van Hoed et al. (2009) and Jurgoński, Fotschki, and Juśkiewicz (2015)

ND, not detected.

cold pressed seed oil (BUCPSO). BCPSO has PUFAs (87.2%), while BUCPSO contained monounsaturated fatty acids (22.2%) and saturated fatty acids (6.5%) (Table 2). Linoleic acid (C18:2ω-6) as reported by Li et al. (2016) was the most prevalent fatty acid in all tested cold pressed seed oils, contributing 41.3%–63.4% to the total fatty acids whereby BCPSO had 63.6% of this important fatty acid. Li et al. (2016) reported that α-linolenic acid led to favorable ω-6/ω-3 ratios, of which BUCPSO demonstrated the lowest ω-6/ω-3 ratio of 1.5, while BCPSO exhibited the highest ratio (3.8). Nevertheless, this ratio is still considered as being favorable because it is lower than that for most vegetable oils.

4 Minor bioactive lipids in cold pressed berry seed oils

Berries are a rich source of a wide variety of and bioactive compounds such as flavonoids, phenolics, anthocyanins, phenolic acids, stilbenes, and tannins, essential oils, carotenoids, vitamins, and minerals. Desmethylsterols and squalene were quantified in the unsaponifiable fraction of the berry seed oils as reported by Van Hoed et al. (2009). They further reported that total desmethylsterol content varied widely between the different berry seed oils, ranging from 404 mg/100 g for blueberry cold pressed seed oil (BUCPSO) to 692 mg/100 g for cranberry cold pressed seed oil (CCPSO). Squalene is an important antioxidant, and is a precursor of phytosterols (Khallouki et al., 2003). Squalene contents in cold pressed berry oils have been reported to be varied widely. For example, a very high value (671.5 mg/100 g) was found for CCPSO, which had the highest sterol content, while blackberry cold pressed seed oil (BCPSO) has 17.0 mg/100 g), raspberry cold pressed seed oil (RCPSO) has 8.4 mg/100 g, and strawberry cold pressed seed oil (SCPSO) has 5.8 mg/100 g (Van Hoed et al., 2009).

The γ-isomer (272 mg/100 g) was reported to be the major tocopherol in RCPSO (75%) of the total tocopherols meanwhile, δ-tocopherol, α-tocopherol, and total vitamin E equivalent can reach 17.4, 71, and 97 mg/100 g, respectively (Oomah et al., 2000). A similar finding was reported by Van Hoed et al. (2009), where γ-tocopherol was the main tocol in BCPSO and RCPSO: 1311.7 and 1640.7 mg/kg, respectively. Exceptionally, CCPSO showed a unique composition with 1235.0 mg/kg of γ-tocotrienol, and 152.7 mg/kg of α-tocotrienol. β-tocopherol and β-tocotrienol were not detected (Table 3) in any of the tested oil.

The biologically active vitamin E content relative to that of α-tocopherol, calculated by using the formula proposed by McLaughlin and Weihrauch (1979), was 97.8 mg/100 g for hexane-extracted oil in comparison to 58.4 mg/100 g for the cold pressed oils. Van Hoed et al. (2009) reported that RCPSO had the highest tocopherol content (2113 mg/kg), followed by CCPSO (1532 mg/kg) and BCPSO (1388 mg/kg), while all other oils contained much lower amounts, as given in Table 3. It was reported that total tocopherol and tocotrienol contents may vary and depend on the different growing conditions, processing and storage conditions, and consequently may vary considerably within a certain oil class (Boskou, 2006).

TABLE 3 Sterols and tocopherols composition (mg/kg oil) of CPBSO.

Sterol/Tocol	Blackberry	Blueberry	Cranberry	Raspberries	Strawberry
Total desmethylsterols (mg/100 g)	403.7	580.2	692.3	493.7	489.4
Sum Δ5 sterols (%)	98.7	84.0	65.9	92.5	87.5
Sum Δ7 sterols (%)	1.3	16.0	34.1	7.5	12.5
Squalene (mg/100 g)	17.0	178.1	671.5	8.4	Tr
α-Tocopherol	25.4–52.4	4.4–30.4	42.4–148.7	136.4–461	ND
β-Tocopherol	ND	ND	ND	ND	ND
γ-Tocopherol	187.4–1311.7	34.4–55.8	59.7–91	144.0–1640.7	260.3–537.0
δ-Tocopherol	31.7–46.5	2.2–18.6	20.9–32.5	55.3–143.5	20.0
Total tocopherols	286.3–1388.7	375.2–1302.9	532.4	198–2112	280.3
α-Tocotrienol	19.9	ND	115.0–157	ND	18.2–20.1
β-Tocotrienol	ND	ND	ND	ND	ND
γ-Tocotrienol	20.0–305.6	330.4–622.2	1105.3–1235.0	7.2 ± 0.3	ND
δ-Tocotrienol	ND	6.0–1244.8	14.7	ND	ND
Total tocols	1584.8–1638.5	705.0–726.8	1353.5–1467.3	443.3	560.0–587.8
Origin	USA	USA	USA	Canada/ USA	USA
Reference	Van Hoed et al. (2009, 2011) and Li et al. (2016)	Van Hoed et al. (2009, 2011) and Li et al. (2016)	Van Hoed et al. (2009, 2011)	Oomah et al. (2000), Van Hoed et al. (2009, 2011), and Li et al. (2016)	Van Hoed et al. (2011)

ND, not detected; *Tr*, trace, detected but not quantified.

Parry and Yu (2004) suggested the possible effects of processing conditions for RCPSO and seed quality on antioxidant properties. They reported on the DPPH·, $ABTS^+$, and total phenolics contents of the cold pressed black raspberry seed oil to range from 10.6% to 49.3%, 0.28 to 0.71 (μmol/g) and 0.035 to 0.093 mg GE/g, respectively.

Parry, Su, Luther, Zhou, et al. (2005) have used the HPLC to determine total carotenoids content for the BCPSO (23.4 μmol/kg) and 12.5 μmol/kg for RCPSO. Van Hoed et al. (2011) using the same instrumentation, identified and quantified phenolic compounds in filtered BCPSO (91 mg/kg) and filtered SCPSO (15,810 mg/kg). They further emphasized that no correlation was found between the quantified phenolic compounds and the oxidative stability of these oils, confirming that the tocopherols were the main antioxidants protecting the lipids during storage in these oils. Another phenolic compound reported by Van Hoed et al. (2011) is *p*-coumaric acid found in CCPSO, BCPSO, and SCPSO. Tyrosol was reported from RCPSO and SCPSO, while homo-vanillic acid, vanillic acid, and vanillin were detected in BCPSO, BUCPSO, CCPSO, RCPSO, and SCPSO. Protocatechuic acid was found only in BCPSO and CCPSO, while ferulic acid was only found in SCPSO (Van Hoed et al., 2011).

Meanwhile, Oomah et al. (2000) reported that the total carotenoids content (expressed as β-carotene) of RCPSO (*Rubus idaeus*) was 23 mg/100 g, which is higher than the total reported carotenoids content by Dimić et al. (2012) of 40 mg/kg for RCPSO and 33 mg/kg for BCPSO. Several studies reported that CCPSO possesses significantly high amounts of tocols, tannins, anthocyanins, and flavonoid compounds (Neto, 2007; Ruel & Couillard, 2007). Van Hoed et al. (2009) reported the presence of campesterol (0.29 g/kg), stigmasterol (0.09 g/kg), β-sitosterol (4.18 g/kg), Δ7-sitosterol (2.19 g/kg), and Δ7-avenasterol (0.17 g/kg) in CCPSO (Table 3). Tocols (Table 3) are also present in large quantities in RCPSO (2.11 g/kg oil), and SCPSO has a significant amount of phenolic compounds (9 mg/g) oil of which *p*-coumaric acid has the highest content (6.6 mg/g) (Van Hoed et al., 2011).

5 Contribution of bioactive compounds in cold pressed oil to organoleptic traits and functions in food or nonfood products

Phytochemicals from phenolics and flavonoids classes have been reported to be strongly associated with the organoleptic properties of fruits (Ahmad et al., 2019). The oxidative processes in oils and fats are one of the main causes of the deterioration of the principal organoleptic and nutritional characteristics of foodstuffs (Van Hoed et al., 2009). The oxidation process is composed of two phases: first, fatty acids react with oxygen to produce peroxides as odorless compounds, which are then degraded into volatile aldehydes and other compounds, responsible for the rancid odor and flavor, and also a nonvolatile fraction. Oxidation primary products are measured as peroxide value and secondary products as *p*-anisidine value. The peroxide value indicates the actual oxidative status and p-anisidine value indicates its history from the oxidative point of view (Van Hoed et al., 2009). The oxidative stability of oil is influenced by its fatty acid composition, the oil quality parameters, and the tocopherol content (Van Hoed et al., 2009). The fatty acids present in triacylglycerols (TAG) are saturated, monounsaturated, or polyunsaturated, and the degree of unsaturation is the main factor that dictates the oxidative stability of the oils (Li et al., 2016). The oxidative stability is also influenced by the positional distribution of fatty acids in the TAG as well as the minor components in oils and their storage conditions. The minor components, such as tocols, including tocopherols and tocotrienols, may play a role in protecting oils from oxidation via several mechanisms of action such as neutralizing free radicals, forming chelation complexes with transition metals, reducing peroxides, and stimulating antioxidative defense enzymes in the body (Li et al., 2016; Parker et al., 2003). Pigments such as chlorophylls can act as photosensitizers when exposed to light, which leads to rapid oxidation of edible oils (Fakourelis, Lee, & Min, 1987). Li et al. (2016) reported that the oxidative stability of BCPSO, RCPSO, and BUCPSO was compromised after the removal of tocols under high-temperature-induced oxidation and the loss of chlorophylls led to weak oxidative stability when exposed to fluorescent lights. Oomah et al. (2000) reported that raspberry cold pressed oil has unique thermal characteristics that influence its physicochemical properties, with a solid-liquid ratio of approximately 1:1 at −23°C. They further added that the solid-liquid ratio, as well as the melting and recrystallization characteristics of raspberry seed oil, can impinge on its consistency, taste, and texture. CCPSO also has unique features such as high oxidative stability and long shelf life whereby its characteristics did not deteriorate easily as compared with other oils such as virgin olive, walnut, pistachio, and hazelnut (Nawar, 2004). The unique fatty acids composition of cold pressed cranberry seed oil enables it to exhibit a pleasant flavor and high oxidative stability (Ahmad et al., 2019). Cold pressed raspberry, blueberry, blackberry, and cranberry seed oils have been reported to contain antioxidants and possess a remarkable radical scavenging activity and oxygen radical absorption capacity due to the fact that during the preparation of these oils, they retain phenols present in the seed that may have potential for applications in the promotion of health and prevention against oxidation damages mediated by radicals (Dimitrios, 2006; Yu, Zhou, & Parry, 2005). Van Hoed et al. (2009) reported that berry seed oil samples were highly susceptible to oxidation, which is due to the high degree of unsaturation of these oils. The low oxidative stability values can be partially attributed to the exceptionally high PUFA content; therefore cold pressed oils from berries require refrigerated storage in well-sealed small bottles, protected from light. In addition, tocopherols, phenolics, and some other antioxidants (pigments, and squalene) or metal catalysts may contribute to the oxidative stability of these oils (Van Hoed et al., 2009). Bushman et al. (2004) evaluated the antioxidant activity using oxygen radical absorbance capacity (ORAC), of which the RCPSO ranged from 52.3 to 53.6 μmol Trolox equivalents (TE)/g oil basis, while, BCPSO ranged from 88.3 to 119.6 μmol TE/g oil. The authors attributed these values to the presence of other lipophilic antioxidants, like tocotrienols or carotenoids.

6 Health-promoting traits of cold pressed berry seed oil

Raspberry seed oil with high levels of δ-tocopherol may be as important as α-tocopherol in the prevention of degenerative diseases (Oomah et al., 2000). Bioactive compounds from BSCPO have potent antioxidant, anticancer, antimutagenic, antimicrobial, antiinflammatory, and antineurodegenerative properties. The special content of cranberry cold pressed oil gives it a vital role in lowering serum cholesterol, and thus helps to prevent heart-related problems, hypertension, certain types of cancer, and autoimmune diseases (Ahmad et al., 2019). Cranberry seed oil is marketed as a healthy oil for its balanced omega-3, omega-6, and omega-9 fatty acids ratio along with high contents of tocols (tocopherols and tocotrienols) (Ahmad et al., 2019). A high ratio of PUFA to saturated fatty acids is reported in cranberry seed oil, which is valuable in lowering serum cholesterol and atherosclerosis and thus helps to prevent heart-related disorders (Heeg, Lager, & Bernard, 2002). PUFAs are believed to act as potential mediators for the nervous, endocrine, and immune systems by different mechanisms including alteration of membrane structure, influence on gene expression, and alteration of prostaglandin production (Yehuda, Rabinovita, & Mostofsky, 2001).

Cranberry and cranberry seed oil has been utilized for the treatment of urinary disorders, wounds, diarrhea, blood poisoning, and diabetes (Ahmad et al., 2019). Cranberry seed oil and cranberry fruit have long been used by long ocean voyagers as protection from scurvy (Henig & Leahy, 2000). Cranberry fruits and cranberry seed oil have been found to contain biologically active components with potential anticancer, antioxidant, and antiinflammatory activities (Ahmad et al., 2019). Shahidi and Weerasinghe (2004) attributed the biological of cranberry products to the presence of phenolic acids, flavonoids, and proanthocyanidins. The presence of ellagic acid, *p*-hydroxybenzoic acid, and other phenolic acids in cranberry fruit and in the seed oil have been believed to prevent and cure urinary tract infections (Marwan & Nagel, 1986), and possess several biological properties such as anticancer, antimutagenic, anticarcinogenic, antioxidant, and antiviral (Daniel et al., 1989). Tocols and phenolics are antioxidants with potent preventive effects on cardiovascular disease, whereas phytosterols, such as β-sitosterol, inhibit cholesterol absorption in the intestine, thus exhibiting cholesterolemia-lowering properties (Jurgoński et al., 2018). Phytosterols are effectively able to reduce the LDL-cholesterol when given as supplements, and the smaller amounts present in natural foods also appear to be important (Ostlund Jr, 2004). Extracts of fruits from blackberry (*Rubus fruticosus*), black raspberry (*Rubus occidentalis* L.), red raspberry (*Rubus idaeus*), blueberry (*Vaccinium corymbosum* L.), and cranberry (*Vaccinium macrocarpon*) act effectively and are considered as inhibitors of free radicals. It has been reported that tocotrienols exhibit a higher bio-potency than tocopherols against different diseases (Nakagawa et al., 2007). The reduction in the ω-6/ω-3 ratio could help to prevent or slow cancer development (Aronson et al., 2001; Narayanan, Narayanan, & Reddy, 2001), reduce the risk of cardiovascular disease (Hu et al., 2002), and improve bone health (Watkins, Li, Allen, Hoffmann, & Seifert, 2000).

Tocopherols are common lipophilic antioxidants abundant in berries oils and nuts and their presence in red raspberry seed could provide vitamin E activity and antioxidant potential (Bramley et al., 2000). They are reported to inhibit lipid oxidation in ORAC assays (Huang, Ou, Hampsch-Woodill, Flanagan, & Deemer, 2002) and in radical trapping antioxidant parameter assays (Cabrini et al., 2001). α-linolenic acid and its derivatives, especially eicosapentaenoic acid (20:5n-3) and docosahexaenoic acid (22:6n-3), have received attention due to their effects on functions of the brain and retina (Fernstrom, 1999), suppressive effects on coronary heart disease (Okuyama, Fujii, & Ikemoto, 2000), antiinflammatory properties (KanKaanpää, Sütas, Salminen, Lichtenstein, & Isolauri, 1999), cancer (Aronson et al., 2001), hypertension and autoimmune disorders (Tapiero, Ba, Couvreur, & Tew, 2002), and their involvement in infant development (Xiang & Zetterstrom, 1999). A monounsaturated fatty acid-enriched diet can reduce the aortic accumulation of oxidized low-density lipoprotein, which may be positively associated with the formation of fatty streaks which is considered as the earliest identifiable lesions of atherosclerosis (Nicolosi et al., 2002).

Bushman et al. (2004) indicated that the range of 15%–31% α-linolenic acid in cranberry seed oils is high relative to the vegetable oils and more similar to other berry oils such as blackcurrant, blueberry, and cranberry seeds (Parker et al., 2003). Linoleic and α-linolenic acids constituted 74% and 84% of the majority of the polyunsaturated lipids in the cranberry seed oils, respectively. In humans, linoleic and α-linolenic are considered as substrates for oxidation, carbon recycling, or conversion in the liver to eicosapentaenoic acid (20:5n-3) and docosahexaenoic acid (22:6n-3) (Sinclair, Attar-Bashi, & Li, 2002). Linoleic acid can compete with α-linolenic acid for the same metabolism enzymes, as reported by Seppanen-Laakso, Laakso, and Hiltunen (2002). The abundance of linoleic and α-linolenic acids in tissue lipids can thus be affected by the choice of ingested fats (Keys, Anderson, & Grande, 1957) and the proportion of ω-6/ω-3 fatty acids has been proposed as an accurate method of predicting coronary heart disease (Okuyama et al., 2000). Most vegetable oils have ω-6/ω-3 ratios in excess of 6/1, whereas ratios of 2/1 are considered beneficial (Simopoulos, 1999). Cranberry seed oil ratios range from 1.7/1 in the black and red raspberries to 4/1 in Marion blackberries (Bushman et al., 2004). The biological activities of eicosapentaenoic acid (20:5n-3) and docosahexaenoic acid (22:6n-3) and their potential in disease prevention have led to recent recommendations for an increase in daily intake of n-3 unsaturated fatty acids from 0.1 g to 0.6 g (Simopoulos, 1999). A berry-based diet has been reported to increase the expression of a neuroprotective trophic factor, IGF-1, in rat brains, suggesting that berries are potent regulators of brain signaling that is correlated to enhancement in cognitive function (Shukitt-Hale et al., 2008). It was reported that raspberry seed oil decreases liver fat content and atherogenic index in healthy rats and rats with low-grade systemic inflammation (Jurgoński et al., 2015). Another study on strawberry seed oils confirmed that there was no significant effect on cholesterol, triglycerides, high-density lipoprotein (HDL), and low-density lipoprotein (LDL) levels in rats (Pieszka, Tombarkiewicz, Roman, Migdał, & Niedziółka, 2013).

7 Applications of cold pressed berry seed oils

Several applications of cold pressed berry seed oils have been reported (Fig. 1).

Berry seed oils, such as BCPSO, RCPSO, and BUCPSO, are considered specialty oils, which may be used in different products by the food, nutraceutical, and cosmetic industries (Li et al., 2016). Berry seed oils are used as important ingredients in cosmeceutical and aromatherapy applications due to their high contents of essential fatty acids and antioxidants (Bushman et al., 2004; Parry et al., 2005; Yu et al., 2005). Ellagic acid was reported to be more abundant in red raspberry and blackberry (*Rubus* spp.) than in other fruits and nuts (Daniel et al., 1989), and has shown chemopreventative activity in animal models (Xue et al., 2001). These characteristics of red raspberry seeds suggest possible roles in human nutritional products. O'lenick Jr. and Lavay (2006) reported a skin-protecting effect for raspberry seed oil which is mainly linked to its content of ellagic acid, a very important phenolic compound in the cosmetic industry. Cold pressed oils obtained from raspberry seed oil and strawberry seed oil were used by 80 adult volunteers to assess the cosmetic properties of these oils by applying them on the skin. Blackcurrant seed oil showed the best absorbency (85% positive opinions) with no irritation being observed within the analyzed oils (Ligęza, Wyglądacz, Tobiasz, Jaworecka, & Reich, 2016). Processing of raspberries and strawberries into juices and purees usually removes the seeds as a by-product (Brownmiller, Howard, & Prior, 2008), which can be used as a food additive for bread production to develop the growing amount of wasted seeds (Kowalczewski et al., 2019), and to re-utilize unused foodstuff with interesting nutritional values (Bushman et al., 2004; Parry & Yu, 2004). According to Korus, Juszczak, Ziobro, et al. (2012), after the oil pressing process, almost completely

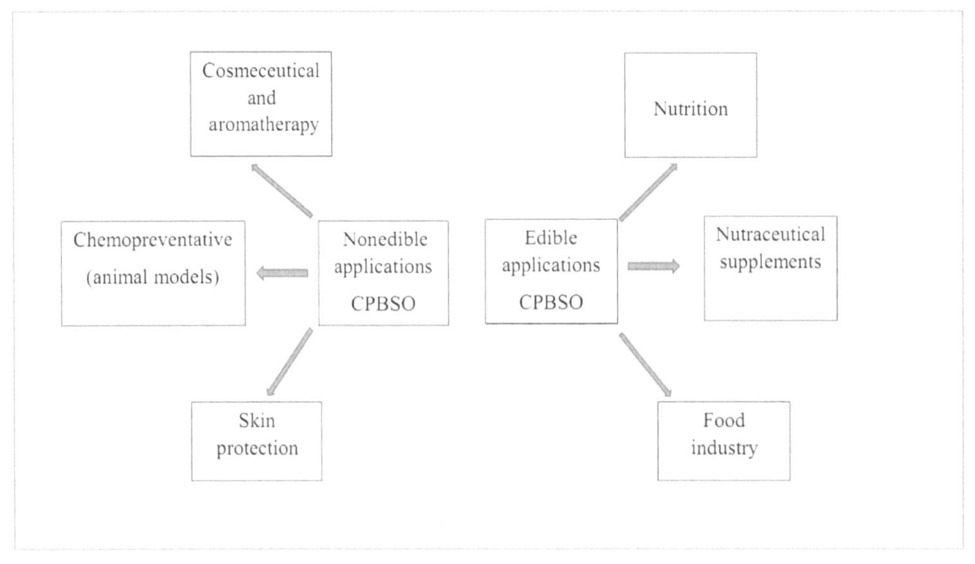

FIG. 1 Edible and nonedible applications of cold pressed berry seed oils.

fat-free oilcake was obtained, which can be used as a functional additive for the production of gluten-free bread. In addition, the use of by-products both after oil extraction as well as from fruit and vegetable processing can reduce the amount of waste generated (Nawirska & Kwaśniewska, 2005).

8 Conclusions

CPBSO considered specialty oils that can be used in different products by the food, nutraceutical, and cosmetic industries. BCPO has a unique character due to its high content of PUFAs, and low ω-6/ω-3 ratio (3/1–2/1) fatty acids. It has very high levels of tocopherols and tocotrienols, which explain the berries' cold pressed oil stability. Cranberry seed oils possess considerable amounts of tocopherols and tocotrienols, which are comparable to those of other berries such as blackberry, blueberry, and strawberry; however, they are rather lower than those of red raspberry. All berry seed oils have a high PUFA content and some are an excellent source of essential fatty acids. These berry seed oils are abundant in oleic, linoleic, and α-linolenic acids, they are rich sources of essential PUFAs, and their potential health benefits have led to rapid development and commercialization of products based on them in various applications. CPBSO is a good dietary source of essential fatty acids with favorable ω-6/ω-3 ratios, and could be incorporated into a daily diet to reduce the ratio of ω-6/ω-3 fatty acids for humans. CPBSO retains higher levels of natural antioxidants, which are responsible for a stable shelf life and safety without added synthetic antioxidants to these healthy oils.

References

Ahmad, N., Anwar, F., & Abbas, A. (2019). Cranberry seed oil. In M. Ramadan (Ed.), *Fruit oils: Chemistry and functionality*. Cham: Springer.

Aronson, W. J., Glaspy, J. A., Reddy, S. T., Reese, D., Heber, D., & Bagga, D. (2001). Modulation of omega-3/omega-6 polyunsaturated ratios with dietary fish oils in men with prostate cancer. *Urology*, *58*, 283–288.

Boskou, D. (2006). *Olive oil: Chemistry and technology* (2nd ed.). Champaign, IL: AOCS Press.

Bowen-Forbes, C. S., Zhang, Y., & Nair, M. G. (2010). Anthocyanin content, antioxidant, anti-inflammatory and anticancer properties of blackberry and raspberry fruits. *Journal of Food Composition and Analysis*, *23*, 554–560.

Bramley, P. M., Elmadfa, I., Kafatos, A., Kelly, F. J., Manios, Y., Roxborough, H. E., et al. (2000). Vitamin E. *Journal of Agricultural and Food Chemistry*, *80*(913), 938.

Broaddus, H. (2017). *The difference between solvent expelled, expeller pressed and cold pressed oil*. Available at http://www.centrafoods.com/blog/the-difference-between-solvent-expelled-expeller-pressed-and-cold-pressed-oil (Accessed 7 November 2019).

Brownmiller, C., Howard, L. R., & Prior, R. L. (2008). Processing and storage effects on monomeric anthocyanins, percent polymeric color, and antioxidant capacity of processed blueberry products. *Journal of Food Science*, *73*, 72–79.

Bushman, B. S., Phillips, B., Isbell, T., Ou, B., Crane, J. M., & Knapp, S. J. (2004). Chemical composition of cranberry (*Rubus* spp.) seeds and oils and their antioxidant potential. *Journal of Agricultural and Food Chemistry*, *52*, 7982–7987.

Cabrini, L., Barzanti, V., Cipollone, M., Fiorentini, D., Grossi, G., Tolomelli, B., et al. (2001). Antioxidants and total peroxyl radical-trapping ability of olive and seed oils. *Journal of Agricultural and Food Chemistry*, *49*, 6026–6032.

Cheel, J., Theoduloz, C., Rodríguez, J. A., et al. (2007). Free radical scavenging activity and phenolic content in achenes and thalamus from *Fragaria chiloensis* ssp. *chiloensis*, *F. vesca* and *F. x ananassa* cv. Chandler. *Food Chemistry*, *102*(1), 36–44.

Codex Alimentarius Standard 210 (1999). Codex standard for named vegetable oils. *Codex Alimentarius*, *8*, 11–25.

Daniel, E. M., Krupnick, A., Heur, Y. H., Blinzler, J. A., Nims, R. W., & Stoner, G. D. (1989). Extraction stability and quantification of ellagic acid in various fruits and nuts. *Journal of Food Composition and Analysis*, *2*, 338–349.

Dimić, E. B., Vujasinović, V. B., Radočaj, O. F., & Pastor, O. P. (2012). Characteristics of blackberry and raspberry seeds and oils. *Acta Periodica Technologica*, *43*, 1–9. https://doi.org/10.2298/APT1243001D.

Dimitrios, B. (2006). Sources of natural phenolic antioxidants. *Trends in Food Science & Technology*, *17*, 505–512.

Fakourelis, N., Lee, E. C., & Min, D. B. (1987). Effects of chlorophyll and β-carotene on the oxidation stability of olive oil. *Journal of Food Science*, *52*, 234–235.

Fernstrom, J. D. (1999). Effects of dietary polyunsaturated fatty acids on neurnal function. *Lipids*, *34*, 161–167.

Heeg, T., Lager, H., & Bernard, G. (2002). Cranberry seed oil, cranberry seed flour and a method for making. US patent, 6 391 345.

Henig, Y. S., & Leahy, M. M. (2000). Cranberry juice and urinary-tract health: Science supports folklore. *Nutrition*, *16*, 684–687.

Hu, F. B., Bronner, L., Willett, W. C., Stampfer, M. J., Rexrode, K. M., Albert, C. M., et al. (2002). Fish and omega-3 fatty acid intake and risk of coronary heart disease in women. *Journal of the American Medical Association*, *287*, 1815–1821.

Huang, D., Ou, B., Hampsch-Woodill, M., Flanagan, J. A., & Deemer, E. K. (2002). Development and validation of oxygen radical absorbance capacity assay for lipophilic antioxidants using randomly methylated cyclodextrin as the solubility enhancer. *Journal of Agricultural and Food Chemistry*, *50*, 1815–1821.

Jurgoński, A., Fotschki, B., & Juśkiewicz, J. (2015). Dietary strawberry seed oil affects metabolite formation in the distal intestine and ameliorates lipid metabolism in rats fed an obesogenic diet. *Food & Nutrition Research*, *59*, 26104. https://doi.org/10.3402/fnr.v59.26104.

Jurgoński, A., et al. (2018). Berry seed oils as potential cardioprotective food supplements. *Nutrire*, *43*(1), 26. https://doi.org/10.1186/s41110-018-0086-x.

KanKaanpää, P., Sütas, Y., Salminen, S., Lichtenstein, A., & Isolauri, E. (1999). Dietary fatty acids and allergy. *Annals of Medicine*, *31*, 282–287.

Keys, A., Anderson, J. T., & Grande, F. (1957). Prediction of serum cholesterol responses of man to changes in fats in the diet. *Lancet*, *273*, 959–966.

Khallouki, F., Younos, C., Soulimani, R., Oster, T., Charrouf, Z., Spiegelhalder, B., et al. (2003). Consumption of argan oil (Morocco) with its unique profile of fatty acids, tocopherols, squalene, sterols and phenolic antioxidants should confer valuable cancer chemopreventive effects. *European Journal of Cancer Prevention*, *12*, 67–75.

Klavins, M., & Klavina, L. (2019). *Vaccinium* genus berry waxes and oils. In M. Ramadan (Ed.), *Fruit oils: Chemistry and functionality*. Cham: Springer.

Korus, J., Juszczak, L., Ziobro, R., et al. (2012). Defatted strawberry and blackcurrant seeds as functional ingredients of gluten-free bread. *Journal of Texture Studies*, *43*, 29–39.

Kowalczewski, P. Ł., et al. (2019). Wheat bread enriched with raspberry and strawberry oilcakes: Effects on proximate composition, texture and water properties. *European Food Research and Technology*, *245*, 2591–2600.

Li, Q. Q., Wang, J. K., & Shahidi, F. (2016). Chemical characteristics of cold-pressed blackberry, black raspberry, and blueberry seed oils and the role of the minor components in their oxidative stability. *Journal of Agricultural and Food Chemistry*, *64*, 5410–5416.

Liangli, L. Y., Zhou, K. K., & Parry, J. (2005). Antioxidant properties of cold-pressed black caraway, carrot, cranberry and hemp seed oils. *Food Chemistry*, *91*, 723–729.

Ligęza, M., Wyglądacz, D., Tobiasz, A., Jaworecka, K., & Reich, A. (2016). Natural cold pressed oils as cosmetic products. *Family Medicine & Primary Care Review*, *18*(4), 443–447.

Marwan, A. G., & Nagel, C. W. (1986). Characterization of cranberry benzoates and their antimicrobial properties. *Journal of Food Science*, *51*, 1069–1070.

McLaughlin, P. J., & Weihrauch, J. L. (1979). Vitamin E content of foods. *Journal of the American Dietetic Association*, *75*, 647–665.

Moreau, R. A., & Kamal-Eldin, A. (2009). *Gourmet and health-promoting specialty oils. gourmet and health-promoting specialty oils*. Elsevier Inc https://doi.org/10.1016/C2015-0-02415-7.

Nakagawa, K., Shibata, A., Yamashita, S., Tsuzuki, T., Kariya, K., Oikawa, S., et al. (2007). In vivo angiogenesis is suppressed by unsaturated vitamin E Tocotrienol. *Journal of Nutrition*, *137*, 1938–1943.

Narayanan, B. A., Narayanan, N. K., & Reddy, B. S. (2001). Docosahexaenoic acid regulated genes and transcription factors inducing apoptosis in human colon cancer cells. *International Journal of Oncology*, *19*, 1255–1262.

Nawar, W., 2004. 'Cranberry seed oil extract and compositions containing components thereof', US Patent, 0 258 734 A1.

Nawirska, A., & Kwaśniewska, M. (2005). Dietary fibre fractions from fruit and vegetable processing waste. *Food Chemistry*, *91*, 221–225.

Neto, C. C. (2007). Cranberry and its phytochemicals: A review of in vitro anticancer studies. *Journal of Nutrition*, *137*, 186–193.

Nicolosi, R. J., Wilson, T. A., Handelman, G., Foxall, T., Keaney, J. F., & Vita, J. A. (2002). Decreased aortic early atherosclerosis in hypercholesterolemic hamsters fed oleic acid-rich Trisun oil compared to linoleic acid-rich sunflower oil. *Journal of Nutritional Biochemistry*, *13*(7), 392–402.

Nile, S. H., & Park, S. W. (2014). Edible berries: Review on bioactive components and their effect on human health. *Nutrition*, *30*, 134–144.

O'lenick, A. J., Jr., & Lavay, C. (2006). Raspberry amido amines and betaines as a delivery system for natural antioxidants. US patent no. 7,078,545 B1.

Okuyama, H., Fujii, Y., & Ikemoto, A. (2000). n-6/n-3 ratio of dietary fatty acids rather than hypercholesterolemia as the major risk factor for atherosclerosis and coronary heart disease. *Journal of Health Science*, *46*, 157–177.

Oomah, B. D., Ladet, S., Godfrey, D. V., Liang, J., & Girard, B. (2000). Characteristics of raspberry (*Rubus idaeus* L.) seed oil. *Food Chemistry*, *69*, 187–193.

Ostlund, R. E., Jr. (2004). Phytosterols and cholesterol metabolism. *Current Opinion in Lipidology*, *15*, 37–41.

Parker, T. D., Adams, D. A., Zhou, K., Harris, M., & Yu, L. (2003). Fatty acid composition and oxidative stability of cold-pressed edible seed oils. *Journal of Food Science*, *68*, 1240–1243.

Parry, J., Su, L., Luther, M., Zhou, K., et al. (2005). Fatty acid composition and antioxidant properties of cold-pressed marionberry, boysenberry, red raspberry, and blueberry seed oils. *Journal of Agricultural and Food Chemistry*, *53*, 566–573.

Parry, J., Su, L., Moore, J., Cheng, Z., & Luther, M. (2006). Chemical compositions, antioxidant capacities and antiproliferative activities of selected fruit seed lours. *Journal of Agricultural and Food Chemistry*, *54*, 3773–3778.

Parry, J., & Yu, L. (2004). Fatty acid content and antioxidant properties of cold-pressed black raspberry seed oil and meal. *Journal of Food Science*, *69*, 189–193.

Pieszka, M., Tombarkiewicz, B., Roman, A., Migdał, W., & Niedziółka, J. (2013). Effect of bioactive substances found in rapeseedraspberry and strawberry seed oils on blood lipid profile and selected parameters of oxidative status in rats. *Environmental Toxicology and Pharmacology*, *36*, 1055–1062.

Radočaj, O., et al. (2014). Blackberry (*Rubus fruticosus* L.) and raspberry (*Rubus idaeus* L.) seed oils extracted from dried press pomace after long-term frozen storage of berries can be used as functional food ingredients. *European Journal of Lipid Science and Technology*, *116*(8), 1015–1024.

Ruel, G., & Couillard, C. (2007). Evidences of the cardioprotective potential of fruits: The case of cranberries. *Molecular Nutrition & Food Research*, *51*, 692–701.

Seppanen-Laakso, T., Laakso, I., & Hiltunen, R. (2002). Analysis of fatty acids by gas chromatography, and its relevance to research on health and nutrition. *Analytica Chimica Acta*, *465*, 39–62.

Shahidi, F., & Shukla, V. (1996). Nontriacylglycerol constituents of fats and oils. *Inform*, *7*, 1227–1232.

Shahidi, F., & Weerasinghe, D. K. (2004). *Nutraceutical beverages: Chemistry, nutrition and health effects ACS Symposium Series 871 of the American Chemical Society, Washington, DC*.

Shukitt-Hale, B., Lau, F. C., Carey, A. N., Galli, R. L., Spangler, E. L., Ingram, D. K., et al. (2008). Blueberry polyphenols attenuate kainic acid-induced decrements in cognition and alter inflammatory gene expression in rat hippocampus. *Nutritional Neuroscience*, *11*, 172–182.

Simopoulos, A. P. (1999). Essential fatty acids in health and chronic disease. *American Journal of Clinical Nutrition*, *70*(Suppl), 560S–569S.
Sinclair, A. J., Attar-Bashi, N. M., & Li, D. (2002). What is the role of R-linolenic acid for mammals? *Lipids*, *37*, 1113–1123.
Szajdek, A., & Borowska, E. J. (2008). Bioactive compounds and health-promoting properties of berry fruits: A review. *Plant Foods for Human Nutrition*, *63*, 147–156.
Tapiero, H., Ba, G. N., Couvreur, P., & Tew, K. D. (2002). Polyunsaturated fatty acids (PUFA) and eicosanoids in human health and pathologies. *Biomedicine & Pharmacotherapy*, *56*(5), 215–222.
Teng, H., Chen, L., Huang, Q., et al. (2016). Ultrasonic-assisted extraction of raspberry seed oil and evaluation of its physico-chemical properties, fatty acid compositions and antioxidant activities. *PLoS ONE*. *11*, e0153457 https://doi.org/10.1371/journal.pone.0153457.
Thyagarajan, P. (2012). *Evaluation and optimization of cranberry seed oil extraction methods.* MSc thesis: McGill University.
Van Hoed, V., Clercq, N. D., Echim, C., Andjelkovic, M., et al. (2009). Berry seeds: A source of specialty oils with high content of bioactives and nutritional values. *Journal of Food Lipids*, *16*, 33–49.
Van Hoed, V., et al. (2011). Influence of filtering of cold pressed berry seed oils on their antioxidant profile and quality characteristics. *Food Chemistry*, *127*, 1848–1855.
Watkins, B. A., Li, Y., Allen, K. G., Hoffmann, W. E., & Seifert, M. F. (2000). Dietary ratio of (n-6)/(n-3) polyunsaturated fatty acids alters the fatty acid composition of bone compartments and biomarkers of bone formation in rats. *Journal of Nutrition*, *130*, 2274–2284.
Winton, A. L., & Winton, K. B. (1935). *The structure and composition of foods, Vol. II, vegetables, legumes, fruits.* New York, NY: John Wiley & Sons.
Xiang, M., & Zetterstrom, R. (1999). Relation between polyunsaturated fatty acids and growth. *Acta Paediatrica*, *88*, 78–82.
Xue, H., Aziz, R. M., Sun, N., Cassady, J. M., Kamendulis, L. M., Xu, Y., et al. (2001). Inhibition of cellular transformation by berry extracts. *Carcinogenesis*, *22*(351), 356.
Yehuda, S., Rabinovita, S., & Mostofsky, D. I. (2001). PUFA: Mediators for the nervous, endocrine, and immune systems. In D. I. Mostofsky, S. Yehuda, & N. Salem (Eds.), *Fatty acids: Physiological and behavioral functions* (pp. 403–420). Totowa, NJ: Humana Press.
Zadernowski, R., Nowak-Polakowska, H., Lossow, B., & Nesterowicz, J. (1997). Sea-buckthorn lipids. *Journal of Food Lipids*, *4*, 165–172.

Chapter 25

Cold pressed oregano (*Origanum vulgare*) oil

Ahmad Cheikhyoussef[a], Natascha Cheikhyoussef[b], and Mohamed Fawzy Ramadan[c]
[a]*Science and Technology Division, Multidisciplinary Research Centre, University of Namibia, Windhoek, Namibia,* [b]*Ministry of Higher Education, Training and Innovation, Windhoek, Namibia,* [c]*Agricultural Biochemistry Department, Faculty of Agriculture, Zagazig University, Zagazig, Egypt*

Abbreviations

CER cerebroside
CPO cold pressed oil
DG diacylglycerol
DGD digalactosyldiglyceride
ESG esterified sterylglycoside
EVOO extra virgin olive oil
FFA free fatty acid
GL glycolipid
MAG monoacylglycerol
MGD monogalactosyldiglyceride
MUFA monounsaturated fatty acid
NL neutral lipid
OCPO oregano cold pressed oil
PC phosphatidylcholine
PE phosphatidylethanolamine
PI phosphatidylinositol
PL phospholipid
PS phosphatidylserine
PUFA polyunsaturated fatty acid
SFA saturated fatty acid
SG sterylglycoside
SQD sulphoquinovosyldiacylglycerol
STE esterified sterol
TAG triacylglycerol
USFA unsaturated fatty acid

1 Introduction

Herbs and spices are among the most important targets to search for natural antioxidants from the point of view of safety (Yanishlieva, Marinova, & Pokorný, 2006). Among these herbs is the *Origanum* species (family Lamiaceae), which is a group of plants rich in bioactive compounds (Ortega-Nieblas et al., 2011). The genus *Origanum* is divided based on the morphological criteria into groups including 10 sections, 38 species, 6 subspecies, and 17 hybrids. Due to their high biodiversity, four groups of oregano have been clearly established; these are Greek oregano (*Origanum vulgare* subsp. *hirtus* (Link) letswaart), Spanish oregano (*Coridohymus capitatu* (L) Hoffmanns and Links), Turkish oregano (*Origanum onites* L), and Mexican oregano (*Lippia graveolens* Kunth or *Lippia berlandieri*) (Cid-Pérez et al., 2016; Ortega-Ramirez et al., 2016). *Origanum* species are commercially called oregano and have great economic importance as a natural source of active compounds. Oregano is one of the most traded and consumed spice plants in the world. There is an internationally growing market in plant-based natural antioxidants due to the worldwide trend toward the use of natural-based additives in

Cold Pressed Oils. https://doi.org/10.1016/B978-0-12-818188-1.00025-6
Copyright © 2020 Elsevier Inc. All rights reserved.

food, cosmetics, and health products. Oregano (*Origanum vulgare* L.) is a perennial herb commonly utilized as a spice in diets with a very pleasant flavor. The plant is used in traditional medicine to treat coughs, colds, and digestive ailments and disorders (Yin, Fretté, Christensen, & Grevsen, 2012). The herb and its essential oil are highly valued for its antioxidant and antimicrobial activities (Dutra et al., 2019; Yanishlieva et al., 2006). The aroma of *O. vulgare* is due to the presence of essential oil that accumulates in the leaf trichomes (Assiri, Elbanna, Al-Thubiani, & Ramadan, 2016). Very few studies are found in the literature on the composition and biological properties of cold pressed oregano oil, with *O. vulgare* being the only researched species for its cold pressed oil to date (Assiri et al., 2016; Elbanna & Ramadan, 2017). This chapter highlights the most important aspects of this important variety of herb-based oil and increases awareness of its potential for nutraceutical applications.

2 Extraction and processing of cold pressed oil

Cold pressed oils (CPOs) have been recognized for their high levels of lipid-soluble bio-actives with nutritive properties. CPOs have received increased interest and high demand in the past decades since they contain a high level of lipophilic phytochemicals including natural antioxidants. The technique is environmentally friendly since no organic solvents are involved in the extraction process. Therefore, this technology is becoming a potential substitute for conventional practices of oil extraction (Ramadan, 2013). Recently, CPOs have been introduced to the market internationally. However, few reports on their bio-information, chemical composition, and antioxidant and antimicrobial properties are available. Such bio-information is vital to evaluate the nutritional and health properties of these oils (Assiri et al., 2016).

3 Lipids and fatty acids composition

Assiri et al. (2016) have reported on the lipid classes and subclasses proportions found in OCPO, which include neutral lipid (NL) 91%, glycolipids (GL) 0.76% and phospholipids (PL) 0.5% (Fig. 1A). Subclasses of neutral lipid contained triacylglycerol (TAG), free fatty acids (FFAs), diacylglycerol (DG), esterified sterols (STEs) and monoacylglycerol (MAG) in decreasing order (Fig. 1B). A significant amount of TAG was found (96% of total NL) in OCPO followed by a lower level of FFA (1.7% of total NL). Meanwhile, DG and STE were recovered in lower levels. Subclasses of glycolipids (GL, Fig. 1C) found in OCPO were cerebrosides (CERs), sterylglycosides (SGs), and esterified sterylglycosides (ESGs), which contribute up to 89% of the total glycolipids, and the remaining 11% is formed of sulphoquinovosyldiacylglycerol (SQD), digalactosyldiglycerides (DGDs), and monogalactosyldiglycerides (MGDs) (Assiri et al., 2016). Thin-layer chromatography analysis revealed that the major subclasses of phospholipids (PLs) were phosphatidylcholine (PC) (50%) followed by phosphatidylethanolamine (PE) (25%), phosphatidylinositol (PI), and phosphatidylserine (PS), respectively (Fig. 1D).

The fatty acid profile of OCPO was reported by Assiri et al. (2016). Nine fatty acids were identified in OCPO, wherein linoleic (39.9%) and oleic acids (43.5%) were the main fatty acids. OCPO contained significant levels of monounsaturated fatty acids (MUFAs, 44%) and polyunsaturated fatty acids (PUFAs, 41.1%). OCPO contained about 14.8 g/100 g of saturated fatty acids (SFAs), whereby palmitic and stearic were the major SFAs, comprising together about 15% (Fig. 1E). The main fatty acids in OCPO were linoleic, oleic, stearic, and palmitic acids. Fatty acids of NL and polar lipids (GL and PL) were not significantly different from each other, wherein linoleic and oleic acids were the main fatty acids. The ratio of unsaturated fatty acids (USFAs) to SFAs was not significantly higher in NL than in the corresponding polar fractions (Assiri et al., 2016).

4 Minor bioactive lipids in oregano cold pressed oil

Assiri et al. (2016) reported that OCPO contains α-, β-, γ-, and δ-tocopherols at concentrations of 180.4, 60.4, 650, and 117.6 mg/100 g oil, respectively and α-, γ-, and δ-tocotrienols at 521, 58.9, and 430 mg/100 g oil, respectively. These levels of tocols detected in OCPO may contribute to the stability of the oil toward oxidation. It was further confirmed that OCPO contains high amounts of phenolic compounds (5.6 mg/g as GAE) and the radical scavenging activities of OCPO were compared with extra virgin olive oil (EVOO) (3.6 mg/g as GAE) using stable DPPH· and galvinoxyl radicals. The superiority of radicals scavenging activity of OCPO over EVOO was attributed to the differences in content and profile of unsaponifiable materials (unsaponifiable in OCPO 23.7 g/kg), the diversity in structural characteristics of potential antioxidants, the presence of synergism between antioxidants with other bioactive components, and the different kinetic behaviors of antioxidants in OCPO (Assiri et al., 2016). The antioxidant activity of OCPO was also reported by Assiri et al. (2016) via Rancimat assay in sunflower oil blend and found to be superior due to the presence of high levels of tocols and phenolic compounds.

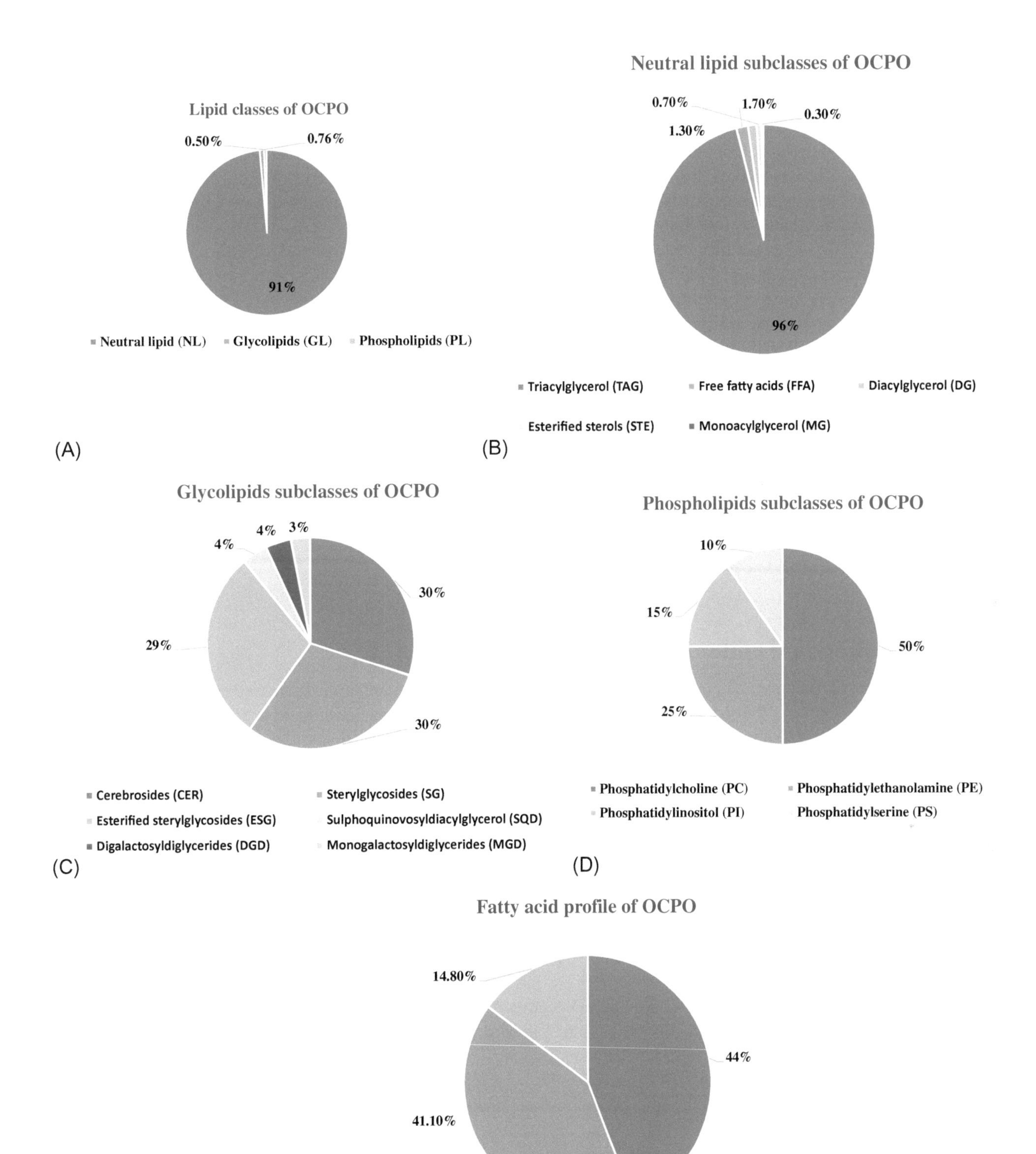

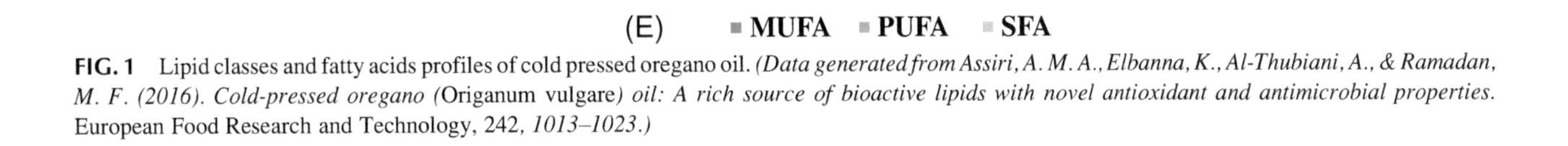

FIG. 1 Lipid classes and fatty acids profiles of cold pressed oregano oil. *(Data generated from Assiri, A. M. A., Elbanna, K., Al-Thubiani, A., & Ramadan, M. F. (2016). Cold-pressed oregano (*Origanum vulgare*) oil: A rich source of bioactive lipids with novel antioxidant and antimicrobial properties.* European Food Research and Technology, 242, *1013–1023.)*

5 Health-promoting traits of cold pressed oil and oil constituents

MUFAs and PUFAs have been reported to reduce LDL (low-density lipoprotein) cholesterol and retain HDL (high-density lipoprotein) cholesterol (Ramadan, Kinni, Seshagiri, & Mörsel, 2010). Several health-promoting benefits of PUFAs were reported in alleviating diseases including inflammatory diseases, heart diseases, cardiovascular, atherosclerosis, and diabetes (El-Hadary & Ramadan, 2016; Ibrahim, Attia, Maklad, Ahmed, & Ramadan, 2017). Edible oils rich in natural antioxidants play a vital role in promoting their health activities, whereby these oils are reported to be a rich source of phenolic compounds and play an important role in the oxidative stability of PUFAs in these oils (Assiri et al., 2016). The antioxidant effect of phenolic compounds is attributed to their redox properties and is a result of several activities including free radical scavenging activity, transition-metal-chelating activity, and/or singlet oxygen-quenching capacity (Bettaieb et al., 2010). Tocol levels in oils may have a great effect on their radical scavenging activities, whereby an increasing ring methyl substitution may lead to an increase in scavenging activity against the DPPH· radicals and leads to a decrease in oxygen radical absorbance capacity (Müller, Theile, & Böhm, 2010).

The antimicrobial activity of OCPO has been reported by Assiri et al. (2016). OCPO exhibited a wide spectrum of antimicrobial activity against foodborne pathogens of *Salmonella enteritidis* (30mm), *Listeria monocytogenes* (15 mm), *Escherichia coli* (33 mm), and *Staphylococcus aureus* (35 mm) and food spoilage fungi of *Candida albicans* (32 mm) and *Aspergillus lavus* (36 mm) as well as the dermatophyte fungi of *Trichophyton mentagrophytes* (42 mm) and *Trichophyton rubrum* (38mm) (Assiri et al., 2016). OCPO had a lethal effect since there was no growth observed when new agar plates or broth media were inoculated with a loop from the clear zone area. Very low MLC values (40 μg/mL) tested against dermatophytic fungi (*T. mentagrophytes* and *T. rubrum*) were observed. The antimicrobial activity was attributed to the presence of the outer membrane surrounding the cell wall in bacteria, which limits the diffusion of hydrophobic substances through its lipopolysaccharide covering (Assiri et al., 2016).

6 Adulteration and authenticity of cold pressed oil

Increasing concern on functional foods especially in developed countries has led to the increased consumption of CPO (El Makawy, Ibrahim, Mabrouk, Ahmed, & Ramadan, 2019; Ge, Chen, Liu, & Zhao, 2014; Kiralan et al., 2017; Parker, Adams, Zhou, Harris, & Yu, 2003; Ramadan, 2013; Yu, Zhou, & Parry, 2005). However, due to the high prices of CPO, it is considered as a target for adulteration with cheaper and lower-quality refined vegetable oils. Hence, the development of rapid analytical methods for detecting adulteration of CPO is of high interest to be targeted for both reasons of humans' health and products' quality (Ge et al., 2014; Li, Wang, Zhao, Ouyang, & Wu, 2015; Poulli, Mousdis, & Georgiou, 2006). Adulteration of spices and herbs is a highly dynamic fraudulent practice, exploiting a wide range of inorganic, synthetic organic, and plant-derived material adulterants (Osman et al., 2019). To combat the inefficient practices in adulteration of spices and herbs, a range of analytical techniques have been developed to aid in the detection of adulterants and establish quantifiable concentrations for the compounds in question in various matrixes including herbs, spices, and their CPO (Osman et al., 2019). Examples of the techniques currently in use include chromatographic methods such as HPLC, HPLC-MS, DART-ToF-MS, and GC-MS. Spectroscopic methods include NIR, Raman, and NMR spectroscopy (Osman et al., 2019). Fourier transform-infrared (FTIR) spectroscopy has emerged as an important tool for the quality control of agro-food products. It has become an attractive option because of its advantages of being simple, nondestructive, rapid, and less costly than other methods (Li et al., 2015; Ozulku, Yildirim, Toker, Karasu, & Durak, 2017). Recently, FTIR spectroscopic methods combined with multivariate data analyses have received great attention in adulteration analysis of different varieties of food products such as CPO, in terms of both qualitative (characterization, classification, and adulteration) and quantitative (determination of adulterant concentration) analysis (Arslan & Çağlar, 2019; Teresa, Koidis, Jim, & Gonz, 2017).

7 Conclusion

Oregano cold pressed oil contains high levels of MUFAs (44.4%) and PUFAs (41.1%). Alpha-tocopherol constituted 32.1% of total measured tocols followed by α-tocotrienol (25.8%) and γ-tocotrienol (21.3%). OCPO contains linoleic, oleic, stearic, and palmitic acids. The oil contains high amounts of phenolic compounds (5.6mg/g as GAE) and possesses strong antioxidant and antimicrobial activity. It has high economic value as a rich natural product with novel functional properties in the food, cosmetics, and pharmaceutical industries.

References

Arslan, F. N., & Çağlar, F. (2019). Attenuated total Reflectance-Fourier Transform Infrared (ATR FTIR) spectroscopy combined with chemometrics for rapid determination of cold-pressed wheat germ oil adulteration. *Food Analytical Methods*, *12*, 355–370.

Assiri, A. M. A., Elbanna, K., Al-Thubiani, A., & Ramadan, M. F. (2016). Cold-pressed oregano (*Origanum vulgare*) oil: A rich source of bioactive lipids with novel antioxidant and antimicrobial properties. *European Food Research and Technology*, *242*, 1013–1023.

Bettaieb, I., Bourgou, S., Wannes, W. A., Hamrouni, I., Limam, F., & Marzouk, B. (2010). Essential oils, phenolics, and antioxidant activities of different parts of cumin (*Cuminum cyminum* L.). *Journal of Agricultural Food Chemistry*, *58*, 10410–10418.

Cid-Pérez, T. S., et al. (2016). Mexican Oregano (*Lippia berlandieri* and *Poliomintha longiflora*) oils. In V. R. Preedy (Ed.), *Essential oils in food preservation, flavor and safety* (pp. 551–560). Academic Press.

Dutra, T. V., et al. (2019). Bioactivity of oregano (*Origanum vulgare*) essential oil against *Alicyclobacillus* spp. *Industrial Crops and Products*, *129*, 345–349.

El Makawy, A. I., Ibrahim, F. M., Mabrouk, D. M., Ahmed, K. A., & Ramadan, M. F. (2019). Effect of antiepileptic drug (Topiramate) and cold pressed ginger oil on testicular genes expression, sexual hormones and histopathological alterations in mice. *Biomedicine and Pharmacotherapy*, *110*, 409–419.

Elbanna, K., & Ramadan, M. F. (2017). The oil of oregano (*Origanum vulgare*). *Inform*, *28*, 18–20.

El-Hadary, A. E., & Ramadan, M. F. (2016). Hepatoprotective effect of cold-pressed *Syzygium aromaticum* oil against carbon tetrachloride (CCl_4)-induced hepatotoxicity in rats. *Pharmaceutical Biology*, *54*, 1364–1372.

Ge, F., Chen, C., Liu, D., & Zhao, S. (2014). Rapid quantitative determination of walnut oil adulteration with sunflower oil using fluorescence spectroscopy. *Food Analytical Methods*, *7*, 146–150.

Ibrahim, F. M., Attia, H. N., Maklad, Y. A. A., Ahmed, K. A., & Ramadan, M. F. (2017). Biochemical characterization, anti-inflammatory properties and ulcerogenic traits of some cold pressed oils in experimental animals. *Pharmaceutical Biology*, *55*, 740–748.

Kiralan, M., Ulaş, M., Özaydin, A. G., Özdemir, N., Özkan, G., Bayrak, A., et al. (2017). Blends of cold pressed black cumin oil and sunflower oil with improved stability: A study based on changes in the levels of volatiles, tocopherols and thymoquinone during accelerated oxidation conditions. *Journal of Food Biochemistry*, *41*, e12272.

Li, B., Wang, H., Zhao, Q., Ouyang, J., & Wu, Y. (2015). Rapid detection of authenticity and adulteration of walnut oil by FTIR and fluorescence spectroscopy: A comparative study. *Food Chemistry*, *181*, 25–30.

Müller, L., Theile, K., & Böhm, V. (2010). In vitro antioxidant activity of tocopherols and tocotrienols and comparison of vitamin E concentration and lipophilic antioxidant capacity in human plasma. *Molecular Nutrition Food Research*, *54*, 731–742.

Ortega-Nieblas, M. M., Robles-Burgueño, M. R., Acedo-Félix, E., González-León, A., Morales-Trejo, A., & Vázquez-Moreno, M. L. (2011). Chemical composition and antimicrobial activity of oregano (*Lippia palmeri* S. Wats) essential oil. *Revista Fitotecnia Mexicana*, *34*, 11–17.

Ortega-Ramirez, L. A., et al. (2016). Oregano (*Origanum* spp.) oils. In V. R. Preedy (Ed.), *Essential oils in food preservation, flavor and safety* (pp. 625–631). Academic Press.

Osman, A. G., et al. (2019). Overview of analytical tools for the identification of adulterants in commonly traded herbs and spices. *Journal of AOAC International*, *102*(2), 376–385.

Ozulku, G., Yildirim, R. M., Toker, O. S., Karasu, S., & Durak, M. Z. (2017). Rapid detection of adulteration of cold pressed sesame oil adultered with hazelnut, canola, and sunflower oils using ATR-FTIR spectroscopy combined with chemometric. *Food Control*, *82*, 212–216.

Parker, T. D., Adams, D. A., Zhou, K., Harris, M., & Yu, L. (2003). Fatty acid composition and oxidative stability of cold-pressed edible seed oils. *Journal of Food Science*, *68*, 1240–1243.

Poulli, K. I., Mousdis, G. A., & Georgiou, C. (2006). Synchronous fluorescence spectroscopy for quantitative determination of virgin olive oil adulteration with sunflower oil. *Analytical Biochemistry*, *386*, 1571–1575.

Ramadan, M. F. (2013). Healthy blends of high linoleic sunflower oil with selected cold pressed oils: Functionality, stability and antioxidative characteristics. *Industrial Crops and Products*, *43*, 65–72.

Ramadan, M. F., Kinni, S. G., Seshagiri, M., & Mörsel, J.-T. (2010). Fat soluble bioactives, fatty acid profile and radical scavenging activity of *Semecarpus anacardium* seed oil. *Journal of American Oil Chemists' Society*, *87*, 885–894.

Teresa, M., Koidis, A., Jim, A. M., & Gonz, A. (2017). Chemometric classification and quantification of olive oil in blends with any edible vegetable oils using FTIR-ATR and Raman spectroscopy. *LWT—Food Science and Technology*, *86*, 174–184.

Yanishlieva, N. V., Marinova, E., & Pokorný, J. (2006). Natural antioxidants from herbs and spices. *European Journal of Lipid Science and Technology*, *108*, 776–793.

Yin, H., Fretté, X. C., Christensen, L. P., & Grevsen, K. (2012). Chitosan oligosaccharides promote the content of polyphenols in Greek Oregano (*Origanum vulgare* ssp. *hirtum*). *Journal of Agricultural and Food Chemistry*, *60*, 136–143.

Yu, L. L., Zhou, K. K., & Parry, J. (2005). Antioxidant properties of cold-pressed black caraway, carrot, cranberry, and hemp seed oils. *Food Chemistry*, *91*, 723–729.

Chapter 26

Cold pressed onion (*Allium cepa* L.) seed oil

Yan Yi Sim and Kar Lin Nyam
Faculty of Applied Sciences, UCSI University, Kuala Lumpur, Malaysia

Abbreviations

CVD	cardiovascular disease
DNA	deoxyribonucleic acid
DPPH·	2,2-diphenyl-1-picrylhydrazyl
FRS	free radical scavengers
HDL-C	high-density lipoprotein cholesterol
LDL-C	low-density lipoprotein cholesterol
MUFA	monounsaturated fatty acid
OCS	organosulfur compound
ORAC	oxygen radical absorbance capacity
OSI	oxidative stability index
PUFA	polyunsaturated fatty acid
ROS	reactive oxygen species
SFAs	saturated fatty acid
SOD	superoxide dismutase
T	tocopherol
TPC	total phenolic content

1 Introduction

Onion (*Allium cepa*) is considered as one of the most important spice commodities grown all over the world and is consumed in various forms. It has been in cultivation for more than 4000 years (Peter, 2001). The five greatest producers of onion are China, Turkey, Russia, Egypt, and the United States (USDA, 2011). The extensive cultivation of onion is attributed to its distinctive flavor, which is widely used in food processing and as an essential part of cuisines of the world (Peter, 2001). The world total onion output has increased by at least 25% over the past 10 years to the current production of approximately 44 million tons, making it the second most important horticultural crop after tomatoes (Yang et al., 2018). The genus Allium includes cultivated and wild species. Onion does not exist as a wild species. The cultivated type of onion falls into two broad horticultural groups: the common onion group, and the *Aggregatum* group, which is distinguished by bulb dormancy, phenotype, and photoperiod requirement that are different from those of common onion (McCallum, 2007).

Onions have been used in traditional medical practice to prevent and treat several diseases and disorders. Modern scientific research has shown that onions have antioxidant, anticancer, antibacterial, hypolipidemic, hypoglycemic, and antiplatelet aggregating activities (Benkeblia, Shiomi, & Osaki, 2007; Rodríguez Galdón, Rodríguez Rodríguez, & Díaz Romero, 2008; Zhu & Li, 2016). According to Preedy (2016), one of the best-studied effects of onion is its antiasthmatic activity, which is attributed to its antiinflammatory activity. In addition, a patented pharmaceutical or veterinary mixture of onion and coconut (*Cocos nucifera*) has been created for controlling brain parasites (nematodes) and/or flatworms (Platyhelminthes) in humans or animals (Mehlhorn et al., 2011).

Onion seeds are also eaten, but their commercial availability is currently limited. Perhaps if consumers were more familiar with onion seed health, functional, and nutritional properties, there would be an increase in consumption of this

Cold Pressed Oils. https://doi.org/10.1016/B978-0-12-818188-1.00026-8
Copyright © 2020 Elsevier Inc. All rights reserved.

product (Dini, Tenore, & Dini, 2008a; Dini, Trimarco, Tenore, & Dini, 2005). According to a study carried out by Dini, Tenore, and Dini (2008b), the onion seed consists of 24.8% crude protein, 20.4% oil, and 10.5% moisture. The high cysteine content in the onion bulb makes it a functional food. However, onion seed contains only low concentrations of these cysteine components (Yalcin & Kavuncuoglu, 2014).

Industrially, there are different types of extraction methods for onion seed oil such as organic solvent expression, supercritical fluid expression, and the screw press method. The cold pressed onion seed oil method is less efficient than the other methods. However, cold pressed onion seed oil can retain more nutritional properties and chemical components compared to the other methods. As the cold press method does not expose the onion seed to organic solvent and high heat treatment, the natural phytochemicals and essential fatty acids, as well as an antioxidant compound, are protected at higher levels in cold pressed onion seed oils. According to the study carried out by Yalcin and Kavuncuoglu (2014), onion seed possesses a high amount of oil (21.8%–25.8%). Therefore, onion seed could be considered as a profitable oilseed crop.

1.1 Recovery and content of cold pressed onion seed oil

Cold pressed onion seed oil has a high polyunsaturated fatty acid (PUFA) content (65.3%), with linoleic acid (18:2n-6) as the primary fatty acid. Cold pressed onion seed oil consists of 64%–65% linoleic acid and 25%–26% oleic acid, with the ratio of oleic to linoleic acid being 0.4. The saturated fatty acid (SFA) and monounsaturated fatty acid (MUFA) content of cold pressed onion seed oil were 0.3% and 25.4%, respectively (Parry et al., 2006). In addition, Parry et al. (2006) also reported the cold pressed onion seed oil contained significantly higher ($P < .0001$) amounts of α-tocopherol and more than double the total tocopherols (1.8–2.0 mmol/kg oil) when compared with cold pressed red raspberry, boysenberry seed, parsley, cardamom, mullein, roasted pumpkin, and milk thistle seed oils.

The 460–634 mg/L concentrations of α-tocopherol in cold pressed onion seed oil are comparable with or higher than those reported for commercial extra virgin sunflower, corn, olive, and peanut seed oils (174–578 mg/L), and higher than soybean oil (89 mg/L), but much lower than that of 1300 mg/kg in wheat germ oil (Cabrini et al., 2001; Wagner, Kamal-Eldin, & Elmadfa, 2004). Cold pressed onion seed oil also has α-tocopherol concentration higher than cold pressed marionberry, blueberry, boysenberry, and red raspberry seed oils (21–151 mg/kg) (Parry et al., 2005). The ratio of α- to γ-tocopherol for cold pressed onion seed oil is 3.1, which makes it a preferred dietary source for total, α-, and γ-tocopherols, with significant levels of δ-tocopherols (Parry et al., 2005).

A significant level of carotenoids was detected in cold pressed onion seed oil, with total carotenoid 4.01 μmol/kg. Zeaxanthin was the primary carotenoids compound present (1.22–1.74 mg/kg), followed by cryptoxanthin (0.51–0.75 mg/kg), and lutein (17.3–17.9 μg/kg) (Parry et al., 2006). The cold pressed onion seed oil had higher carotenoids content than cardamom (0.05 μmol/kg) and milk thistle seed oil (2.30 μmol/kg) (Parry et al., 2006). These data suggest that cold pressed onion seed oil has the potential to serve as a dietary source of carotenoids, especially zeaxanthin.

For antioxidant properties, cold pressed onion seed oil showed total phenolic content TPC 3.35 mg GAE/g (Parry et al., 2006). This TPC value was higher than that of cold pressed black raspberry seed oil (0.04–0.09 mg GAE/g), parsley seed oil (2.27 mg GAE/g), and milk thistle seed oil (3.07 mg GAE/g) (Parry et al., 2006; Parry & Yu, 2004). Cold pressed onion seed oil also exhibited stronger diphenylpicrylhydrazyl (DPPH·) radical scavenging properties (75.8%–77.3%), than cardamom (41.8%), roasted pumpkin (35.9%), and milk thistle seed oil (32.7%). This showed that cold pressed onion seed oil could retain a higher level of beneficial compounds.

1.2 Physical properties of cold pressed onion seed oil

Color is one of the important characteristics for determining the visual acceptance of oil by suggesting the sensation of attractive appearance. The *L*, *a*, and *b* values of cold pressed onion seed oil were 3.50, −1.38, and 3.90, respectively (Parry et al., 2006). Cold pressed onion seed oil had the lightest and most yellow color compared with cold pressed parsley, pumpkin, and mullein seed oil (Parry et al., 2006). The oxidative stability of the oil could be determined by the oxidative stability index (OSI), and a higher OSI indicated a longer shelf life. The oxidative stability index of cold pressed onion seed oil was 31.4 h (Parry et al., 2006). The OSI of cold pressed onion seed oil was comparable with that of cold pressed milk thistle seed oil (13.3 h) and *Trichosanthes kirilowii* seed oil (0.58 h) (Jiang, Wu, Zhou, & Akoh, 2015; Parry et al., 2006). The reflective index and density of cold pressed onion seed oil were 1.4725^{25} D and 0.930 g/mL, respectively (Parry, Cheng, & Moore, 2008).

1.3 The economy of cold pressed onion seed oil

Onion has been used for flavoring throughout history and formed part of the economy of countries with growing populations. Onion is easily grown and adapted to a wide variety of soils and climatic conditions. Therefore, onions have long been cultivated on a varying scale for local and export use. The onion seed yield was 500–1200 kg/ha (Brewster, 1994), which was higher than that for carrot seed (600–1000 kg/ha) and tomato seed (250–400 kg hectare) (Welbaum, 2005). In addition, high amounts of oil (21.1%) can be extracted from onion seeds using cold press extraction (Topkafa, 2016).

Cold pressed onion seed oil is extensively and traditionally used to impart flavor to processed food (salad dressing, sauces, soup, and meat) without the difficulties of handling large amounts of fresh onion (Lawless, 2012). The edible cold pressed onion seed oil rich in MUFA has greater market potential due to the current consumer trend in reducing the intake of total, saturated, and *trans* fat, and relatively better oxidative stability during cooking (Moreau & Kamal-Eldin, 2015). In addition to its flavoring properties, it has been shown that cold pressed onion oil has antioxidant properties, antibrowning properties, hepatoprotective activity, and hypolipidemic activity (Sohail, Karim, Sarwar, & Alhasin, 2011; Yalcin & Kavuncuoglu, 2014). Cold pressed onion seed oil has high potential to experience a huge growth rate in the market due to its nutrition, flavoring, and health benefits.

2 Extraction and processing of cold pressed onion seed oil

2.1 Comparison of different extraction methods

There are several extraction and processing methods to extract the oil from the onion seed reported in the literature, which are cold pressed, Soxhlet extraction, and supercritical carbon dioxide (Golubkina et al., 2015; Hou, Zhang, & Wang, 2008; Parry et al., 2006; Topkafa, Kara, & Sherazi, 2015). Table 1 summarizes the onion seed oil content extracted by these three methods.

2.1.1 Soxhlet extraction

An oil yield of 13.5% was obtained from the onion seed by Soxhlet extraction with *n*-hexane. The onion seed oil was extracted with *n*-hexane at 60°C for 6 h. The *n*-hexane was removed from the onion seed oil using a rotary evaporator under reduced pressure of 20 mbar at 30°C (Golubkina et al., 2015). The onion seed oil extracted by Soxhlet extraction contained

TABLE 1 Onion seed oil content extracted by Soxhlet, supercritical fluid, and cold pressed extraction from previous studies.

Extraction methods	Soxhlet extraction (Golubkina et al., 2015)	Supercritical fluid extraction (Hou et al., 2008)	Cold pressed extraction (Parry et al., 2006; Topkafa, 2016)
Oil yield (%)	13.5	12.08	21.1
Fatty acid			
Monounsaturated fatty acid (MUFA) (%)	23.21	–	26.6
Polyunsaturated fatty acid (PUFA) (%)	66.89	–	58.7
Linoleic acid (%)	61.5–73.7	41.72	25–26
Oleic acid (%)	19.0–27.2	37.30	64%–65%
Total tocopherol content (mg/kg)	1666.85	–	2506.3
Antioxidant properties			
Diphenylpicrylhydrazyl (DPPH) radical scavenging activity (%)	13.58–38.26	–	75.8–77.3
Total phenolic content (mg GAE/g)	1.84–6.87	–	3.35

61.5%–73.7% of linoleic acid, and 19.0%–27.2% of oleic acid, with the ratio of oleic to linoleic acid being 0.36 (Golubkina et al., 2015). MUFA, PUFA, and total tocopherol content in the Soxhlet-extracted onion seed oil were 23.21%, 66.89%, and 3.87 mmol/kg, respectively (Golubkina et al., 2015). In terms of antioxidant properties, the Soxhlet-extracted onion seed oil exhibited 13.58%–38.26% DPPH· radical scavenging activity (Parry et al., 2006) and 1.84–6.87 mg GAE/g TPC (Golubkina et al., 2015).

For Soxhlet extraction, which involves the use of an organic solvent, the solvent needs to be removed using a solvent evaporator. However, there might be a risk of the existence of residual solvent if the solvent removal process is not complete. The mass transfer resistances that involve more than one phase within the system repeatedly limit the use of this extraction method (Jadhav, Rekha, Gogate, & Rathod, 2009). Furthermore, Soxhlet extraction requires a very long time and the thermally sensitive compounds are deteriorated, resulting in low extraction yield (Danlami, Arsad, Zaini, & Sulaiman, 2014).

2.1.2 Supercritical fluid extraction

For supercritical fluid extraction, the onion seed oil yield was 12.08%. The supercritical fluid extraction was carried out at the extraction temperature (35–45°C, and extraction pressure 15–25 MPa), and the extraction time adjusted to be between 30 and 90 min. The onion seed oil extracted by supercritical fluid extraction had 41.7% of linoleic acid and 37.3% of oleic acid, with the ratio of oleic to linoleic acid being 0.89 (Hou et al., 2008).

This extraction method is generally performed at low temperatures, with a short extraction time, and using little solvent compared to the traditional Soxhlet extraction method (Liza et al., 2010). In addition, carbon dioxide (CO_2) is a low-cost, inert, nontoxic, and environmentally friendly solvent that permits extraction at low pressures and temperatures (Jadhav et al., 2009). However, this extraction method has complicated system thermodynamics and high operation costs (Jose, Miguel, Alejandro, & Elena, 2007).

2.1.3 Cold pressed extraction

The oil yield of cold pressed onion seed oil was 21.1% (Topkafa, 2016), which is higher than the oil yield from Soxhlet extraction and supercritical fluid extraction. The cold pressed onion seed oil also had higher MUFA (26.6%), PUFA (58.7%), and total tocopherol content (2506.3 mg/kg) than the Soxhlet extraction method (Topkafa, 2016). For the two main fatty acids found in the onion seed oil, the amount of oleic acid (25%–26%) in cold press extraction was higher than for Soxhlet extraction (19%–27.2%), but lower than supercritical fluid extraction (37.0%). For the linoleic acid content in the onion seed oil, cold press extraction exhibited higher content than supercritical fluid extraction (41.7%), but lower than Soxhlet extraction (61.5%–73.7%) (Golubkina et al., 2015; Hou et al., 2008; Parry et al., 2006). In terms of antioxidant properties, the TPC in the cold pressed onion seed oil was 3.35 mg GAE/g. For the radical scavenging activity, the Soxhlet-extracted onion seed oil exhibited lower DPPH· radical scavenging activity (13.5%–38.2%) than the cold press extraction (75.8%–77.3%) (Parry et al., 2006).

Mechanical pressing (cold press) oil extraction is technically less expensive and less labor-intensive than the other extraction methods. The safety and simplicity of cold press extraction are advantageous over the more efficient solvent extraction equipment. Furthermore, products pressed out generally have better preserved native properties, end products are chemically contaminant free, and it is a safer process (Thanonkaew, Wongyai, McClements, & Decker, 2012). Moreover, this method is conducted at low temperatures to minimize the thermal degradation of the bioactive compounds present in the oil (Mariod, Matthäus, & Ismail, 2011) and to achieve complete removal of the fluid at the end of the extraction (Chan & Ismail, 2009).

3 Fatty acid and acyl lipids profile of cold pressed onion seed oil

3.1 Triglyceride composition in cold pressed onion seed oil

Triglycerides are major components of vegetable oil. Triglycerides act as one of the most concentrated sources of energy in the diet and function as carrier for fat-soluble vitamins such as A, D, E, and K (Topkafa et al., 2015). The most abundant triglycerides in cold pressed onion seed oil are OLL (28.3%), LLL (24.6%), and LLnLn (19.9%), as shown in Table 2 (Topkafa, 2016).

TABLE 2 Triglyceride composition in cold pressed onion seed oil (%).

Triglyceride (%)	Cold pressed onion seed oil
LnLnLn	ND[a]
LLnLn	19.9±1.1
LLLn	6.1±0.5
LLL	24.6±2.2
OLLn	ND[a]
PLLn	ND[a]
OLL	28.3±3.1
PLL	6.4±1.0
OOL	10.4±1.3
POL+SLL	4.3±0.3
PPL	ND[a]
OOO	ND[a]
SOL	ND[a]

[a]*ND = not detected.*

3.2 Fatty acid profile in cold pressed onion seed oil

Novel specialty cold pressed onion seed oils, which are rich in factors beneficial to health, are in high demand as there is increasing consumer interest in health-promoting diet and disease prevention. The beneficial factors in cold pressed onion seed oil include special fatty acid components such as high amounts of unsaturated fatty acids. The fatty acid composition of cold pressed onion seed oil has been reported in several studies. These reports vary considerably in fatty acid composition of cold pressed onion seed oil, as shown in Table 3. Parry et al. (2006) and Topkafa (2016) investigated the fatty acid composition of cold pressed onion seed oil. Fatty acid present in cold pressed onion seed oil was composed of approximately 30.3% MUFA and 58.7% PUFA. Linoleic acid (C18:2, 58.5%) was the predominant fatty acid, followed by oleic acid (C18:1, 29.3%) and palmitic acid (C16:0, 7.7%). Myristic, palmitoleic, stearic, α-linolenic, arachidonic, and eicosenoic acids were found in the cold pressed onion seed oil in minor percentages.

In some studies, saturated fatty acids (SFAs) were considered as one of the dietary risk factors of cardiovascular disease (CVD), and unsaturated fatty acids can act as CVD preventive factors (Li et al., 2015). Consumption of oil high in saturated fatty acids will increase the low-density-lipoprotein cholesterol (LDL-C) level in the blood and increase the risk of developing CVD. Consumption of oil rich in MUFA and PUFA can reduce the risk of high blood cholesterol levels and have other health benefits when they replace saturated fats in the diet. Oleic acid is more stable than linoleic acid to withstand high heat application such as oil processing or cooking. Thus, a high level of oleic acid (omega-9) makes the oil desirable as a cooking oil in terms of nutrition and stability (Nyam, Tan, Lai, Long, & Yaakob, 2009).

Linoleic acid is the most common type of omega-6 fatty acid—an essential PUFA that can increase the nutritional value of vegetable oils. Omega-6 fatty acids are an important part of a healthy diet, as they confer benefits on the immune system and metabolism. Palmitic acid is a saturated fatty acid that is suitable to make margarine, shortening, and other fat products (Nyam et al., 2009). A minor amount of stearic acid (2.4 mg/100 g oil) was identified in cold pressed onion seed oil, which is a neutral saturated fatty acid that does not increase the LDL-C level in human blood (Coetzee, Labuschagne, & Hugo, 2008). α-linolenic acid was present in a minor amount (0.1 mg/100 g) in the cold pressed onion seed oil, which is an essential omega-3 fatty acid that has neuroprotective, antiinflammatory, and antidepressant properties (Blondeau et al., 2015). The high amounts of MUFA and PUFA make cold pressed onion seed oil suitable to use as an edible oil for diet enrichment.

TABLE 3 Fatty acid composition of cold pressed onion seed oil reported from the previous studies.

Fatty acid	Parry et al. (2006) (mg/100g oil)	Topkafa (2016) (%)
C14:0	TR[a]	0.2±0.1
C16:0	6.4±0.01	7.7±0.2
C16:1	0.2±0.01	0.4±0.1
C18:0	2.4±0.00	2.5±0.1
C18:1	24.8±0.02	29.3±0.1
C18:2	65.2±0.03	58.5±0.1
C18:3	0.1±0.03	0.1±0.1
C20:0	0.3±0.00	0.3±0.1
C20:1	0.4±0.01	0.5±0.1
C20:2	ND[a]	0.3±0.1
C20:4	ND[a]	0.2±0.1
C23:0	ND[a]	0.1±0.1
C24:0	ND[a]	0.1±0.1
SAT	9.3	10.9±0.4
MUFA	25.4	30.3±0.2
PUFA	65.3	58.7±0.2

[a]TR, *trace;* ND, *not detected.*

4 Minor bioactive lipids in cold pressed onion seed oil

4.1 Tocopherols

Tocopherols, the major forms of vitamin E, are a group of fat-soluble phenolic compounds. Each tocopherol consists of a chromanol ring and a 16-carbon phytyl chain. Tocopherols are designated as α, β, δ, and γ, depending on the number and position of methyl groups on the chromanol ring. α-tocopherol (T) is trimethylated at the 5-, 7-, and 8-positions of the chromanol ring, while β-T is dimethylated at the 5- and 8-positions (Constantinou, Papas, & Constantinou, 2008). In addition, γ-T is dimethylated at the 7- and 8-positions and δ-T is methylated at the 8-position (Constantinou et al., 2008). All tocopherols are strong antioxidants that can interrupt the propagation stage of oxidation reaction by scavenging free radicals. However, δ-T and γ-T are more effective in trapping reactive nitrogen species than α-T (Patel, Liebner, Netscher, Mereiter, & Rosenau, 2007). Tocopherols have been suggested to reduce the risk of cancer as they possess strong antioxidant properties (Ju, Picinich, Yang, & Zhao, 2010).

According to a study carried out by Parry et al. (2006), cold pressed onion seed oil contains 1937.8 μmol/kg total tocopherols, 9±18.4 mg/g α-tocopherol, 28.6±0.1 mg/g β-tocopherol, and 219.2±4.6 mg/g γ-tocopherol. The highest total tocopherol content was also observed in the onion seed oil at a level of 1.8–2.0 mmol/kg, which was comparable to cold pressed boysenberry and red raspberry seed oils (2.3 and 2.1 mmol/kg, respectively).

4.2 Phenolic compounds

Phenolic compounds, commonly found in plants, are an important part of the human diet and are of considerable interest due to their antioxidant properties and potential beneficial health effects. There is much evidence indicating that consumption of a variety of phenolic compounds present in foods can lower the risk of developing health disorders due to their antioxidant activity (Shahidi & Ambigaipalan, 2015). Phenolic compounds are classified as primary antioxidants, which are mainly free radical scavengers (FRS) that inhibit or delay the initiation step or interrupt the propagation step of lipid oxidation (Alamed, Chaiyasit, McClements, & Decker, 2009). This will decrease the formation of volatile decomposition products, such as ketones and aldehydes, which cause rancidity (Naczk & Shahidi, 2004; Kil et al., 2009).

Different types of oils may present different types and content of phenolic compounds. Phenolics will influence the antioxidant activity and flavor of the oils (Nyam et al., 2009). The TPC in the cold pressed onion seed oil was 2.16–3.35 mg GAE/g oil. Phenolic compounds can enhance the oxidative stability of vegetable oil. In addition, plant phenolic compounds are also being considered as potential drugs in combating cancer or as diet enrichment to reduce cancer risk (Wong, Tan, Tan, Long, & Nyam, 2014).

4.3 Carotenoids

Carotenoids, which act as antioxidant micronutrients, are abundant in fruit and vegetables and have been known to contribute to the body's defense against reactive oxygen species (Stanner, Hughes, Kelly, & Buttriss, 2004). The majority of carotenoids accumulate in the liver and combine with lipoprotein for release into the blood circulation. Ingested carotenoids could participate in an antioxidant defense system when high concentrations of free radical species are present in the liver, and these physiological functions of carotenoids could inhibit the development of liver dysfunction (Suguira, 2013). A significant level of carotenoids was detected in cold pressed onion seed oil, with total carotenoids of 4.01 μmol/kg. Zeaxanthin was the primary carotenoid compound present (1.22–1.74 mg/kg), followed by cryptoxanthin (0.51–0.75 mg/kg) and lutein (17.3–17.9 μg/kg) (Parry et al., 2006).

5 Contribution of bioactive compounds in cold pressed onion seed oil to organoleptic traits and functions in food or nonfood products

5.1 Organosulfur compounds (OCS)

The characteristic flavor compounds occurring in the individual species of the genus *Allium* are dominated by various biologically active OCS (Gyawali et al., 2006). The OCS can be classified based on the functional group in which sulfur atoms are bound to a cyanate group or a carbon atom in a cyclic or noncyclic configuration (Abuaja, Ogbonna, & Osuji, 2014). The functional ingredients of foods containing OCS obtained only after cutting, chewing, or crushing have disrupted the cells to expose them. The volatile compounds are formed by the hydrolysis of nonvolatile alkyl and alkenyl-substituted L-cysteine sulfoxides due to the action of the enzyme alliinase following onion seed tissue disruption (Crozier, Clifford, & Ashihara, 2008).

OCS present in natural food are generally considered beneficial for health because of their anticarcinogenic and antioxidant properties. This has led to the excessive and long-term consumption of cold pressed onion seed oil. According to the study carried out by Yalcin and Kavuncuoglu (2014), onion seeds oil contained especially carbon disulfide, 4-phenyl-2-thiazolethiol, methyl sulfonyl-methane, and dipropyl disulfide. According to a study carried out by Vazquez-Armenta, Ayala-Zavala, Olivas, Molina-Corral, and Silva-Espinoza (2014), the dipropyl disulfide present in the onion seed oil had an antibrowning effect. The OCS have also proven to be the principal antimicrobial agent present in onions (Rose, Whiteman, Moore, & Zhu, 2005).

5.2 Hydrocarbons

Hydrocarbons are compounds comprised exclusively of carbon and hydrogen and they are by far the dominant components of crude oil. Terpenes are a large class of hydrocarbon compounds constructed from five-carbon isoprene units that are combined to generate a great variety of skeletons, which are then acted on by various enzymes to add functionality and altered oxidation (Liu & Mander, 2010). Terpenes are secondary metabolites, which provide the plant with its organoleptic characteristics (flavor and aroma) and that constitute most of the essential oil produced by plant seeds (Fundacion Canna, 2019). Monocyclic monoterpene (limonene), cyclic monoterpene (α-phellandrene), and bicyclic monoterpene (*p*-cymene) were also detected in onion seed oil (Yalcin & Kavuncuoglu, 2014).

5.3 Esters

Esters are very important flavor compounds and found widespread in nature in a great variety of foodstuffs. Esters are compounds with high perception thresholds and typically exhibit sweet, fruity, and green notes (Salas, 2004). According to a study carried out by Yalcin and Kavuncuoglu (2014), the esters detected in the onion seed oil were hexyl acetate, nonyl acetate, ethyl octanoate, and methyl hexanoate.

6 Health-promoting traits of cold pressed onion seed oil and oil constitute

6.1 Antioxidant activity

Antioxidants are substances that, at relatively low concentrations, inhibit the rate of oxidizing a substrate with oxygen by reacting with a free radical early in the oxidation process. Thus, the intermediate compound formed is unable to continue the free radical chain reaction. Free radicals will cause oxidation in the body, and subsequently disrupt the functions of DNA, lipids, proteins, lipids, and other biomolecules, which affect the metabolism and cause mutagenesis. Recently, there has been increasing demand in extracting antioxidant compound from natural materials like plants, and increasing interest in bioactives found among secondary metabolic products of organisms (Jin et al., 2009). Therefore, cold pressed onion seed oil is suggested for use as a functional oil as it contains high amounts of phenolics and tocopherols.

Different assays can use to assess the antioxidant activity of vegetable oil, such as DPPH· and oxygen radical absorbance capacity (ORAC) (Parry et al., 2006). DPPH· has been widely used to determine the free radical scavenging effect of various antioxidant substances (Ozcelik, Lee, & Min, 2003). DPPH· assay is widely used to assess the hydrophobic system, such as oil affected by lipophilic antioxidants, including carotenoids, tocopherols, chlorophyll, and phytosterols. The ORAC assay can be used to measure the radical chain-breaking ability of antioxidants by monitoring the inhibition of peroxyl radical-induced oxidation. Peroxyl radicals are the major free radicals found in lipid oxidation in foods and biological systems under physiological conditions. Therefore, ORAC values are of biological relevance as a reference for antioxidant effectiveness.

The antioxidant activity of cold pressed onion seed oil could be attributed to the high content of phenolic compounds present in the oil. Phenolic compounds consist of functional groups (hydroxyl, $OH^{\bullet}$), which are able to scavenge free radicals to disrupt the initiation and propagation phases of the lipid oxidation mechanism. Different types of phenolic compounds can exert their antioxidant activity with different pathways in different oil systems to delay the rate of the oxidation of lipids. Thus, the phenolic compounds can minimize rancidity, retard the formation of toxic oxidation products, maintain nutritional quality, and increase the shelf life of cold pressed onion seed oil (Fukumoto & Mazza, 2000).

Tocopherols are natural compounds, present in high amounts in cold pressed onion seed oil. Edible seed oils are an important dietary source of essential tocopherols and the observed highly significant correlation between tocopherol content and galvinoxyl reduction also emphasizes their important contribution to maintaining oxidative stability of the bulk food product (Duthie, Gardner, Morrice, & McPhail, 2016; Murphy, Subar, & Block, 1990). The biological activity of α-tocopherol is double that of β- and γ-, and is 100 times higher than that of δ-tocopherol, so the biological activity of α-tocopherol is the highest among the tocopherol isomers (Chew, Tan, & Nyam, 2017).

Carotenoids are lipophilic molecules, which tend to accumulate in lipophilic compartments like lipoprotein or membranes. Among the different type of defense strategies, carotenoids are usually involved in the scavenging of two of the reactive oxygen species: peroxyl radicals and singlet molecular oxygen ($_1O^2$). In addition, they can effectively deactivate the electronically excited sensitizer molecules, which are involved in the generation of radicals and singlet oxygen (Stahl & Sies, 2003). According to Parry et al. (2006), cold pressed onion seed oil exhibited comparable antioxidant activity with other cold pressed seed oils and has great potential for use as a high antioxidant edible seed oil.

6.2 Antipathogenic activity

Bacteria and fungi are two of the main cause of different infections in humans (Sohail et al., 2011). Antipathogenic compounds can prevent the production of toxins or abolish the ability of bacteria to adapt to the mammalian environment, which would give the host immune system competitive advantages to allow clearance of the infectious organism (González-Lamothe et al., 2009). Wilson and Adams reported that the organic sulfur compounds in the onion could inhibit the proinflammatory messengers and bacterial growth (Wilson & Adams, 2007). In addition, the sulfur compounds may be involved in the inhibition of the bacterial activity of key enzymes such as triose phosphate dehydrogenase, succinate dehydrogenase, and xanthine oxidase (Kim, Huh, Kyung, & Kyung, 2004). Hence, cold pressed onion seed oil possesses this advantage and may have potential as an antipathogenic agent.

6.3 Hepatoprotective activity

The liver is a key organ that regulates metabolism, secretion, detoxifying, and storage functions in the body, and hepatic damage is often associated with distortion of these functions (Wolf, 1994). Liver cells show a number of compensatory mechanisms to deal with reactive oxygen species (ROS) and its effects; among these are the induction of antioxidant proteins such as superoxide dismutase (SOD), catalase, and glutathione peroxidase (Chandrasekaran et al., 2012). Most hepatotoxic chemicals damage liver cells primarily by inducing lipid peroxidation and other oxidative damage (Chandrasekaran et al., 2012). Plant-derived phenolica, flavonoida, and polyphenolic compounds are considered to contribute to the prevention of diseases associated with oxidative stress. Vidyavati, Manjunatha, Hemavathy, and Srinivasan (2010) found increased vitamin E levels in rats' serum because of onion administration; this vitamin is reported to have a beneficial effect on renal oxidative damage in rats. The use of cold pressed onion seed oil in the daily human diet can reduce the occurrence of liver disease, so it should be adopted in routine life.

6.4 Hypolipidemic activity

Hyperlipidemia is a well-known risk factor for cardiovascular disease. In addition, hyperlipidemia was found to have a positive relationship with obesity (Dennis, Aziz, She, Faruqui, & Davis, 2006). By increasing the intake of total cholesterol, triglyceride, and lipoproteins such as LDL-C, very-low-density lipoprotein cholesterol, and decreasing high-density lipoprotein (HDL) are considered as the symptoms of hyperlipidemia (Tilak, Veeraiah, & Koteswara Rao, 2001). There are a few lines of evidence showing that HDL is inversely related to total body cholesterol and a decrease in plasma HDL concentration may accelerate the development of atherosclerosis leading to ischemic heart diseases, by impairing the clearing of cholesterol from the arterial wall (Kanungo, Panda, Swain, Barik, & Tripathi, 2007).

The commercial hypocholesterolemic drugs, statins, are highly effective drugs that have the ability to reduce the risk of major cardiovascular events by up to 10% in primary prevention and 5% in secondary prevention over 5 years (Pinal-Fernandez, Casal-Dominguez, & Mammen, 2018). However, there are side effects accompanied with the statin drug, such as abdominal pain, fatigue, headache, skin rashes, constipation, dizziness, muscle weakness, blurred version, and/or liver inflammation (Cheong, Tan, Tan, & Nyam, 2016). Thus, natural plants are recommended to be involved in the drug or food supplementation to avoid these side effects.

Linoleic acid, tocopherols, and phenolics in cold pressed onion seed oil are believed to play an important role in cholesterol-lowering activity (Sohail et al., 2011). Phenolic compounds possess antioxidant activity to scavenge the body of free radicals and reduce oxidative stress, thus decreasing the risk of getting hypercholesterolemia. α-tocopherol was identified as the primary antioxidant found in human lipoproteins; it received much attention as a suppressor of LDL lipid oxidation and as an epidemiological marker for ischemic heart disease (Catalgol & Ozer, 2011).

7 Edible and nonedible applications of cold pressed onion seed oil

7.1 Edible applications of cold pressed onion seed oil

The odor and flavor properties of cold pressed onion seed oil and its components (Table 4) have been used in the food industry as food additives to add traditional flavor in novel food. The OCS, the volatile compounds found in the oil, are perceived as odorants like fried onion, garlic, and/or cooked cabbage. Cold pressed onion seed oil can also be added into carriers such as salt, carbohydrates, or other edible material to get fat-based soluble, dry-soluble, liquid-soluble, and encapsulated spices (Farrell, 1998).

7.2 Nonedible applications of cold press onion seed oil

Cold pressed onion seed oil can be used as a new treatment for alopecia areata. This is an unpredictable, usually patchy, nonscarring hair loss condition (Madini & Shapiro, 2000). The sulfur and phenolic compounds in cold pressed onion seed oil may cause irritant contact dermatitis, which can stimulate hair regrowth through antigenic competition (Sharquie & Al-Obaidi, 2002).

TABLE 4 Main components in cold pressed onion seed oil.

Main components	Amount (g/100g oil)
Palmitic acid	6.4±0.01
Stearic acid	2.4±0.00
Oleic acid	24.8±0.02
Linoleic acid	65.2±0.03
SAT	9.3
MUFA	25.4
PUFA	65.3
α-Tocopherol	0.1±18.4
γ-Tocopherol	0.02±4.6
δ-Tocopherol	0.003±0.1
Total tocopherols	0.3

8 Other issues

8.1 Adulteration and authenticity of cold pressed onion seed oil

Previous literature reported that cold pressed onion seed oil is edible, with no toxicity. The edible cold pressed onion seed oil is rich in MUFA, high in linoleic acid content, and has large market potential compared to other oils rich in PUFA, as there is growing consumer demand for oil that has good oxidative stability during cooking, and is low in total, saturated, and *trans* fat. The commercially available cold pressed onion seed oil may contain more health-promoting phytochemicals, and may hold currently untapped commercial and nutritional value (Moreau & Kamal-Eldin, 2015). Cold pressed onion seed oil can serve as an alternative source of tocopherol, phenolic compounds, and MUFA in the human diet.

References

Abuaja, C. I., Ogbonna, A. C., & Osuji, C. M. (2014). Functional components and medicinal properties of food: A review. *Journal of Food Science and Technology*, *52*(5), 2522–2529.

Alamed, J., Chaiyasit, W., McClements, D. J., & Decker, E. A. (2009). Relationship between free radical scavenging and antioxidant activity in foods. *Journal of Agricultural and Food Chemistry*, *57*, 2969–2976.

Benkeblia, N., Shiomi, N., & Osaki, M. (2007). Kinetics and hydrolysis parameters of total fructooligosaccharides of onion bulbs: Effects of temperature regimes and cultivars. *Journal of Food Biochemistry*, *31*(1), 14–27.

Blondeau, N., Lipsky, R. H., Bourourou, M., Duncan, M. W., Gorelick, P. B., & Marini, A. M. (2015). Alpha-linolenic acid: An omega-3 fatty acid with neuroprotective properties ready for use in the stroke clinic? *Biomed Research International*. https://doi.org/10.1155/2015/519830.

Brewster, J. L. (Ed.). (1994). *Onions and other vegetables alliums*. Wallingford, United Kingdom: CAB International.

Cabrini, L., Barzanti, V., Cipollone, M., Fiorentini, D., Grossi, G., Tolomelli, B., et al. (2001). Antioxidants and total peroxyl radical-trapping ability of olive and seed oils. *Journal of Agricultural and Food Chemistry*, *49*, 6026–6032.

Catalgol, B., & Ozer, N. (2011). Protective effects of vitamin E against hypercholesterolemia-induced age-related diseases. *Genes & Nutrition*, *7*(1), 91–98.

Chan, K. W., & Ismail, M. (2009). Supercritical carbon dioxide fluid extraction of *Hibiscus cannabinus* L. seed oil: A+ potential solvent-free and high antioxidative edible oil. *Food Chemistry*, *114*, 970–975.

Chandrasekaran, C., Dethe, S., Mundkinajeddu, D., Pandre, M., Balachandran, J., Agarwal, A., et al. (2012). Hepatoprotective and antioxidant activity of standardized herbal extracts. *Pharmacognosy Magazine*, *8*(30), 116.

Cheong, A. M., Tan, K. W., Tan, C. P., & Nyam, K. L. (2016). Improvement of physical stability properties of kenaf (*Hibiscus cannabinus* L.) seed oil-in-water nanoemulsions. *Industrial Crops and Products*, *80*, 77–85.

Chew, S. C., Tan, C. P., & Nyam, K. L. (2017). Comparative storage of crude and refined kenaf (*Hibiscus cannabinus* L.) seed oil during accelerated storage. *Food Science and Biotechnology*, *26*, 63–69.

Coetzee, R., Labuschagne, M. T., & Hugo, A. (2008). Fatty acid and oil variation in seed from kenaf (*Hibiscus cannabinus* L.). *Industrial Crops and Products*, *27*, 104–109.

Constantinou, C., Papas, A., & Constantinou, A. I. (2008). Vitamin E and cancer: An insight into the anticancer activities of vitamin E isomers and analogs. *International Journal of Cancer*, *123*, 739–752.

Crozier, A., Clifford, M., & Ashihara, H. (2008). *Plant secondary metabolites*. New York: John Wiley & Sons.

Danlami, J. M., Arsad, A., Zaini, M. A. A., & Sulaiman, H. (2014). A comparative study of various extraction technique from plants. *Reviews in Chemical Engineering*, *30*(6), 605–626.

Dennis, B., Aziz, K., She, L., Faruqui, M., & Davis, C. E. (2006). High rates of obesity and cardiovascular disease risk factors in lower middle class community in Pakistan: The Metroville health study. *Journal of the Pakistan Medical Association*, *56*, 267–272.

Dini, I., Tenore, G. C., & Dini, A. (2008a). S-alkenyl cysteine sulfoxide and its antioxidant properties from *Allium cepa* var. tropeana (red onion) seeds. *Journal of Natural Products*, *71*, 2036–2037.

Dini, I., Tenore, G. C., & Dini, A. (2008b). Chemical composition. Nutritional value and antioxidant properties of *Allium cepa* L. var. *tropeana* (red onion) seeds. *Food Chemistry*, *107*, 613–621.

Dini, I., Trimarco, E., Tenore, G. C., & Dini, A. (2005). Furostanol saponins in *Allium caepa* L. var. tropeana seeds. *Food Chemistry*, *93*, 205–214.

Duthie, G. G., Gardner, P. T., Morrice, P. C., & McPhail, D. B. (2016). The contribution of dα-Tocopherol and dγ-tocopherol to the antioxidant capacity of several edible plant oils. *Natural Science*, *8*, 41–48.

Farrell, K. T. (1998). *Spices, condiments, and seasonings*. Germany: Springer.

Fukumoto, L. R., & Mazza, G. (2000). Assessing antioxidant and prooxidant activities of phenolic compounds. *Journal of Agricultural and Food Chemistry*, *48*, 3597–3604.

Fundacion Canna (2019). *Terpenes. (2019)*. https://www.fundacion-canna.es/en/terpenes (Accessed 10 March 2019).

Golubkina, N. A., Nadezhkin, S. M., Agafonov, A. F., Kosheleva, O. V., Molchanova, A. V., Russu, G., et al. (2015). Seed oil content, fatty acid composition, and antioxidant properties are affected by genotype in *Allium cepa* L. and perennial onion species. *Advances in Horticultural Science*, *29*(4), 199–206.

González-Lamothe, R., Mitchell, G., Gattuso, M., Diarra, M., Malouin, F., & Bouarab, K. (2009). Plant antimicrobial agents and their effects on plant and human pathogens. *International Journal of Molecular Sciences*, *10*(8), 3400–3419.

Gyawali, R., Seo, H. Y., Lee, H. J., Song, H. P., Kim, H. D., Byun, M. W., et al. (2006). Effect of γ-irradiation on volatile compounds of dried Welsh onion (*Allium fistulosum* L.). *Radiation Physics and Chemistry*, *75*, 322–328.

Hou, X. L., Zhang, Y. S., & Wang, J. Q. (2008). *Method for extracting onion seed oil by supercritical carbon dioxide*. 200810123769.

Jadhav, D., Rekha, B. N., Gogate, P. R., & Rathod, V. K. (2009). Extraction of vanillin from vanilla pods: A comparison study of conventional Soxhlet and ultrasonic assisted extraction. *Journal of Food Engineering*, *93*, 421–426.

Jiang, X. R., Wu, S. M., Zhou, Z. J., & Akoh, C. C. (2015). Physicochemical properties and volatile profiles of cold-pressed *Trichosanthes kirilowii* maxim seed oils. *International Journal of Food Properties*, *19*(8), 1765–1775.

Jin, C. W., Eom, S. H., Park, H. J., Ghimeray, A. K., Yu, C. Y., & Cho, D. H. (2009). Antioxidant activity of *Hibiscus cannabinus* L. leaves in different growth time. *Korean Journal of Medicinal Crop Science*, *17*, 21–25.

Jose, A. M., Miguel, H. M., Alejandro, C., & Elena, I. (2007). Use of compressed fluids for sample preparation: Food applications. *Journal of Chromatography A*, *1152*, 234–246.

Ju, J., Picinich, S. C., Yang, Z., & Zhao, Y. (2010). Cancer-preventive activities of tocopherols and tocotrienols. *Carcinogenesis*, *31*, 533–542.

Kanungo, S. K., Panda, D. S., Swain, S. R., Barik, B. B., & Tripathi, D. K. (2007). Comparative evaluation of hypolipidemic activity of some marketed herbal formulations in Triton induced hyperlipidemic rats. *Pharmacology Online*, *3*, 211–221.

Kil, H. Y., Seong, E. S., Ghimire, B. K., Chung, I. M., Kwon, S. S., Goh, E. J., et al. (2009). Antioxidant and antimicrobial activities of crude sorghum extract. *Food Chemistry*, *115*, 1234–1239.

Kim, J. W., Huh, J. E., Kyung, S. H., & Kyung, K. K. (2004). Antimicrobial activity of alk(*en*)yl sulfides found in essential oil of garlic and onion. *Food Science and Biotechnology*, *13*, 235–239.

Lawless, J. (2012). *Encyclopedia of essential oils: The complete guide to the use of aromatic oils in aromatherapy, herbalism, health and well-being* (1st ed.). USA: HarperCollins Publishers.

Li, Y., Hruby, A., Bernstein, A. M., Ley, S. H., Wang, D. D., Chiuve, S. E., et al. (2015). Saturated fats compared with unsaturated fats and sources of carbohydrates in relation to risk of coronary heart disease. *Journal of the American College of Cardiology*, *66*, 1538–1548.

Liu, H. W., & Mander, L. (2010). Comprehensive natural products II. In K. Tidgewell, B. Clark, & W. Gerwick (Eds.), *The natural products chemistry of cyanobacteria* (pp. 141–188). Netherlands: Elsevier Science.

Liza, M. S., Abdul Rahman, R. A., Mandana, B., Jinap, S., Rahmat, A., Zaidul, I., et al. (2010). Supercritical carbon dioxide extraction of bioactive flavonoid from *Strobilanthes crispus* (Pecah Kaca). *Food and Bioproducts Processing*, *88*, 319–326.

Madini, S., & Shapiro, J. (2000). Alopecia areata update. *Journal of the American Academy of Dermatology*, *42*, 549–566.

Mariod, A. A., Matthäus, B., & Ismail, M. (2011). Comparison of supercritical fluid and hexane extraction methods in extracting kenaf (*Hibiscus cannabinus*) seed oil lipids. *Journal of the American Oil Chemists' Society*, *88*, 931–935.

McCallum, J. (2007). Onion. In C. Kole (Ed.), *Vegetables* (pp. 331–347). Berlin, Heidelberg: Springer.

Mehlhorn, H., Schmidt, J., Abdel-Ghaffar, F., Al-Rasheid, K., Quraishi, S., Al-Farhan, A., Schmahl, G., 2011. Pharmaceutical or veterinary medical preparation obtained from coconut (*Cocos nucifera*) and Onion (*Allium cepa*) for controlling flatworms (Plathelminthes) and/or threadworms (Nematodes) in animals or humans. WO Patent 2011042054.

Moreau, R., & Kamal-Eldin, A. (2015). Gourmet and health-promoting specialty oils. In L. L. Yu, & J. J. Hao (Eds.), *Parsley, carrot, and onion seed oils* (pp. 479–490). Netherland: Elsevier Science.

Murphy, S. P., Subar, A. F., & Block, G. (1990). Vitamin E intakes and sources in the United States. *The American Journal of Clinical Nutrition*, *52*, 361–367.

Naczk, M., & Shahidi, F. (2004). Extraction and analysis of phenolics in food. *Journal of Chromatography A*, *1054*(1–2), 95–111.

Nyam, K. L., Tan, C. P., Lai, O. M., Long, K., & Yaakob, C. M. (2009). Physicochemical properties and bioactive compounds of selected seed oils. *LWT—Food Science and Technology*, *42*, 1396–1403.

Ozcelik, B., Lee, J. H., & Min, D. B. (2003). Effects of light, oxygen, and pH on the 2,2-diphenyl-1-picrylhydrazyl (DPPH) method to evaluate antioxidants. *Journal of Food Science*, *68*, 487–490.

Parry, J. W., Cheng, Z. H., & Moore, J. (2008). Fatty acid composition, antioxidant properties, and antiproliferative capacity of selected cold-pressed seed flours. *Journal of the American Oil Chemists' Society*, *85*, 457–464.

Parry, J., Hao, Z., Luther, M., Su, L., Zhou, K., & Yu, L. (2006). Characterization of cold-pressed onion, parsley, cardamom, mullein, roasted pumpkin, and milk thistle seed oils. *Journal of the American Oil Chemists' Society*, *83*, 847–854.

Parry, J. W., Su, L., Luther, M., Zhou, K., Yurawecz, M. P., Whittaker, P., et al. (2005). Fatty acid content and antioxidant properties of cold-pressed marionberry, boysenberry, red raspberry, and blueberry seed oils. *Journal of Agricultural and Food Chemistry*, *53*, 566–573.

Parry, J. W., & Yu, L. (2004). Fatty acid content and antioxidant properties of cold pressed black raspberry seed oil and meal. *Journal of Food Science*, *69*, 189–193.

Patel, A., Liebner, F., Netscher, T., Mereiter, K., & Rosenau, T. (2007). Vitamin E chemistry. Nitration of non-alpha-tocopherols: Products and mechanistic considerations. *Journal of Organic Chemistry*, *72*, 6504–6512.

Peter, K. V. (2001). *Handbook of herbs and spices*. (Vol. 1). United Kingdom: Woodhead Publishing.

Pinal-Fernandez, I., Casal-Dominguez, M., & Mammen, A. (2018). Statins: Pros and cons. *Medicina Clínica*, *50*(10), 398–402.

Preedy, V. (2016). Essential oil in food preservation, flavour, and safety. In F. J. Vazquez-Armenta, M. R. Cruz-Valenzuela, & J. F. Ayala-Zavala (Eds.), *Onion (*Allium cepa*) essential oil* (pp. 617–623). Cambridge, MA: Academic Press.

Rodríguez Galdón, B., Rodríguez Rodríguez, E., & Díaz Romero, C. (2008). Flavonoids in onion cultivars (*Allium cepa* L.). *Journal of Food Science*, *73*(8), 599–605.

Rose, P., Whiteman, M., Moore, P. K., & Zhu, Y. Z. (2005). Bioactive S-alk(en)yl cysteine sulfoxide metabolites in the genus allium: The chemistry of potential therapeutic agents. *Natural Product Reports*, *22*, 351–368.

Salas, J. J. (2004). Characterization of alcohol acyltransferase from olive fruit. *Journal of Agricultural and Food Chemistry*, *52*, 3155–3158.

Shahidi, F., & Ambigaipalan, P. (2015). Phenolics and polyphenolics in foods, beverages, and spices: Antioxidant activity and health effects—A review. *Journal of Functional Foods*, *18*, 820–897.

Sharquie, K. E., & Al-Obaidi, H. K. (2002). Onion juice (*Allium cepa* L.), a new tropical treatment for *Alopecia areata*. *Journal of Dermatology*, *29*, 343–346.

Sohail, M. N., Karim, A., Sarwar, M., & Alhasin, A. M. (2011). Onion (*Alium cepa* L.): An alternate medicine for Pakistan population. *International Journal of Pharmacology*, *7*(6), 736–744.

Stahl, W., & Sies, H. (2003). Antioxidant activity of carotenoids. *Molecular Aspects of Medicine*, *24*, 345–351.

Stanner, S. A., Hughes, J., Kelly, C. N., & Buttriss, J. (2004). A review of the epidemiological evidence for the 'antioxidant hypothesis'. *Public Health Nutrition*, *7*, 407–422.

Suguira, M. (2013). *Carotenoids: Liver diseases prevention* (1st ed.). USA: Academic Press (chapter 27).

Thanonkaew, A., Wongyai, S., McClements, D. J., & Decker, E. A. (2012). Effect of stabilization of rice bran by domestic heating on mechanical extraction yield, quality, and antioxidant properties of cold-pressed rice bran oil (*Oryza saltiva* L.). *LWT—Food Science and Technology*, *48*(2), 231–236.

Tilak, K. S., Veeraiah, K., & Koteswara Rao, D. K. (2001). Restoration on tissue antioxidants by fenugreek seeds (*Trigonella foenum* Graecum) in alloxan-diabetic rats. *Indian Journal of Physiology and Pharmacology*, *45*, 408–420.

Topkafa, M. (2016). Evaluation of chemical properties of cold pressed onion, okra, rosehip, safflower and carrot seed oils: Triglyceride, fatty acid and tocol compositions. *Analytical Methods*, *21*, 1–16.

Topkafa, M., Kara, H., & Sherazi, S. T. H. (2015). *Journal of the American Oil Chemists' Society*, *92*, 791–800.

USDA (2011). *U.S. onion statistics. (2011)*. http://usda.mannlib.cornell.edu/MannUsda/viewDocumentInfo.do?documentID=1396 (web archive link, 16 February 2019) (Accessed 16 February 2019).

Vazquez-Armenta, F. J., Ayala-Zavala, J. F., Olivas, G. I., Molina-Corral, F. J., & Silva-Espinoza, B. A. (2014). Antibrowning and antimicrobial effects of onion essential oil to preserve the quality of cut potatoes. *Acta Alimentaria*, *43*(4), 640–649.

Vidyavati, H. G., Manjunatha, H., Hemavathy, J., & Srinivasan, K. (2010). Hypolipidemia and antioxidant efficacy of dehydrated onion in experimental rats. *Journal of Food Science and Technology*, *47*, 55–60.

Wagner, K. H., Kamal-Eldin, A., & Elmadfa, I. (2004). Gamma-tocopherol an underestimated vitamin? *Annals of Nutrition & Metabolism*, *48*, 169–188.

Welbaum, G. (2005). *Vegetable seed production: Carrot. (2005). https://booksite.elsevier.com/9780081019375/content/Elsevier%20Standard%20Reference%20Styles.pdf (Accessed 08 March 2019).*

Wilson, E. A., & Adams, B. D. (2007). Antioxidant, anti-inflammatory, and antimicrobial properties of garlic and onion. *Nutrition & Food Science*, *37*, 178–183.

Wolf, P. L. (1994). Biochemical diagnosis of liver diseases. *Indian Journal of Clinical Biochemistry*, *14*, 59–90.

Wong, Y. H., Tan, W. Y., Tan, C. P., Long, K., & Nyam, K. L. (2014). Cytotoxic activity of kenaf (*Hibiscus cannabinus* L.) seed extract and oil against human cancer cell lines. *Asian Pacific Journal of Tropical Biomedicine*, (Suppl. 1), S510–S515.

Yalcin, H., & Kavuncuoglu, H. (2014). Physical, chemical and bioactive properties of onion (*Allium cepa* L.) seed and seed oil. *Journal of Applied Botany and Food Quality*, *87*, 87–92.

Yang, C., Li, L., Yang, L. G., Lu, H., Wang, S. K., & Sun, G. J. (2018). Anti-obesity and hypolipidemic effects of garlic oil and onion oil in rats fed a high-fat diet. *Nutrition and Metabolism*, *15*, 43.

Zhu, Y. B., & Li, J. S. (2016). Analysis on the harm and preventive of computer network viruses. *Journal of Mechanical Engineering Research and Developments*, *39*(2), 469–472.

Chapter 27

Cold pressed okra (*Abelmoschus esculentus*) seed oil

Farooq Anwar[a], Rahman Qadir[a], and Naveed Ahmad[b]
[a]*Department of Chemistry, University of Sargodha, Sargodha, Pakistan,* [b]*Department of Chemistry, University of Education, Lahore, Faisalabad Campus, Faisalabad, Pakistan*

1 Introduction

Okra (*Abelmoschus esculentus* L. Moench), is valued as a nutritious vegetable crop (Naveed, Khan, & Khan, 2009). A native to Africa, okra is now cultivated in several tropical regions across the world (Akassey & Daubrey, 1992; Sathish & Eswar, 2013). It is a flowering plant that has long been a part of the human diet in several countries around the world (Jarret, Wang, & Levy, 2011).

Okra seeds have been documented as a promising source of oil/oil cake and vegetable proteins (Akingbala, Akinwande, & Uzo-Peters, 2003; Ndangui et al., 2010). Over the years, okra has been cultivated in different parts of the world as a vegetable as well as to get edible quality oil (Sanjeet et al., 2010). Okra seeds have been reported to possess about 20%–40% oil (Benchasr, 2012; MEF, 2013). Andras et al. (2005) reported 15.9%–20.7% oil contents in okra seeds collected from different areas of Greece. In another study, using *n*-hexane, Acikgoz, Borazan, Andoglu, and Gokdai (2016) reported that okra seeds can yield oil ranging from 21% to 23%. The variation in the oil yield from okra seeds can be linked to extraction methods applied (Dong, Zhang, Tian, Pan, & Wei, 2014). The oil contents in okra seeds were found to be comparable to various oilseed crops except for palm oil and soybean oil (Sanjeet et al., 2010) (Fig. 1).

The analysis of okra seed revealed the occurrence of potent antioxidant compounds such as catechins (2.4 mg/g of seeds), flavonols (3.3 mg/g of seeds), hydroxycinnamic (0.2 mg/g of skins), and quercetin derivatives (0.3 mg/g of skins) (Arapitsas, 2008). The specialty oils, which contain some medicinally important minor bioactive, are sold as high-value oils due to their peculiar fatty acids and nutrients profile. As far as okra seed oil composition is concerned, it contains considerable levels of unsaturated fatty acids and tocols. Savello, Martins, and Hill (1980) investigated okra seed oil as one of the potential sources of unsaturated fatty acids. It has been suggested that several pathologies such as carcinogenic, neurodegenerative and cardiovascular disorders can be prevented using omega-3 (n-3) unsaturated fatty acids (Harel et al., 2002). The roasted okra seeds are a favorable candidate to be used in various industrial applications including confectionery and as a caffeine-free substitute for coffee (Adetuyi, Osagie, & Adekunle, 2011; Calisir & Yildiz, 2005). Furthermore, okra seed oil has a potential hypocholesterolemic effect.

2 Extraction and processing of okra seed oil

Due to the increasing human population and the expanding oleo-chemicals industrial sector, the exploration of some newer and under-utilized resources of vegetable oils is of vital importance (Schalau, 2002). Among different methods employed for oil extraction at commercial level, the extraction yield obtained by solvent extraction is higher; however, traces of volatile organic solvent may be left in the extracted oil (Liu, Xu, Hao, & Gao, 2009). The oil yields obtained using different extraction methods and techniques may vary depending upon efficacy of the method. When compared with other techniques, cold pressing is less efficient in terms of oil yield recovery; however, better nutritional, organoleptic, and nutraceutical attributes have been reported for cold pressed seed oils. In fact, cold pressing does not involve the heat treatment of raw material that minimizes the loss of nutrients. Moreover, the recovered cold pressed oil as against solvent extracted oil is healthier and beneficial as it is free of any traces of toxic solvent (Ramadan, 2013).

Cold Pressed Oils. https://doi.org/10.1016/B978-0-12-818188-1.00027-X
Copyright © 2020 Elsevier Inc. All rights reserved.

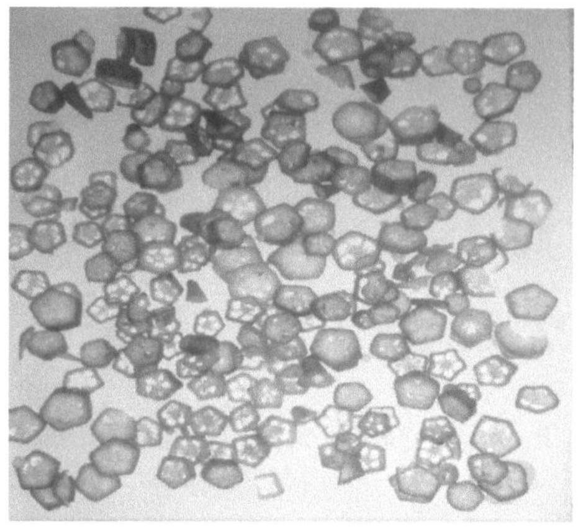

FIG. 1 Okra (*Abelmoschus esculentus*) fruits and green pods.

3 Fatty acids profile of cold pressed okra oil

Okra, mainly grown as a vegetable crop, contains considerable amounts of oil (20%–40%), and thus can be cultivated as an edible/vegetable oilseed crop (Benchasr, 2012; MEF, 2013). Among essential fatty acids, α-linolenic acid is one that humans need in the body and it has to be obtained through dietary sources. Fatty acids such as palmitic acid ($C_{16:0}$), oleic acid ($C_{18:1,\ n\text{-}9}$), and linoleic acid ($C_{18:2,\ n\text{-}9}$), being present in most of the common vegetable oils, are frequently available in human diet and nutrition. Previous studies showed that okra seed oil has appreciable amounts of oleic acid ($C_{18:1,\ n\text{-}9}$) and linoleic acid ($C_{18:2,\ n\text{-}9}$), along with a low level of linolenic acid ($C_{18:3,\ n\text{-}3}$) (Adelakun, Ade-Omowaye, Adeyemi, & Van De Venter, 2010; Adetuyi et al., 2011; Arapitsas, 2008; Benjawan, Chutichudet, & Kaewsit, 2007; Bozan & Temelli, 2008). Acikgoz et al. (2016) found that okra oil is a valuable source of palmitic (28.60%), stearic (3.57%), oleic (16.81%), linoleic (49.54%), and linolenic acids (1.48%). Andras et al. (2005) and Jarret et al. (2011) reported that okra seed oil is rich in important fatty acids (palmitic, oleic, and linoleic acids) and unsaturated, with linoleic acid (up to 47.4%) as a major constituent. The other fatty acids of the seed oil were noted to be myristic (0.19%–0.38%), palmitic (21.9%–30.4%), palmitoleic (0.31%–0.98%), stearic (2.64%–4.75%), oleic (16.8%–27.4%), linoleic (36.5%–49.5%), linolenic (0.17%–2.64%), and behenic (0.16%–0.52%) acids. Table 1 shows the comparative fatty acids composition of okra seed oils produced by different extraction techniques.

TABLE 1 Oil yield and fatty acids composition (g/100g) of okra seed oil using different extraction techniques.

Extraction method	SCF (CO_2)	SE	SPE	*n*-Hexane	Ethanol	Cold pressed
Myristic acid ($C_{14:0}$)	0.23	0.36	0.24	–	–	0.30
Palmitic acid ($C_{16:0}$)	28.9	28.5	24.1	32.5	29.9	30.4
Palmitoleic ($C_{16:1}$)	0.44	0.45	0.48	–	–	0.60
Heptadecanoic acid ($C_{17:0}$)	0.28	0.20	0.16	–	–	0.10
Heptadecenoic acid ($C_{17:1}$)	0.44	0.35	0.32	–	–	–
Stearic acid ($C_{18:0}$)	4.14	4.18	4.75	2.80	3.20	2.70
Oleic acid ($C_{18:1}$)	24.3	25.1	27.5	16.1	17.4	16.9
Linoleic acid ($C_{18:2}$)	38.3	36.5	37.8	47.4	47.5	47.7
Linolenic acid ($C_{18:3}$)	ND	0.89	ND	1.20	1.30	0.20
Cyclopaneoctanoic acid ($C_{19:0}$)	1.53	1.08	1.42	–	–	–
Arachidic acid ($C_{20:0}$)	0.53	0.66	0.77	0.50	0.20	0.30
Paullinic acid ($C_{20:1}$)	–	–	–	–	–	0.10

TABLE 1 Oil yield and fatty acids composition (g/100g) of okra seed oil using different extraction techniques—cont'd

Extraction method	SCF (CO_2)	SE	SPE	*n*-Hexane	Ethanol	Cold pressed
Arachidonic acid ($C_{20:4}$)	–	–	–	–	–	0.10
Behenic acid ($C_{22:0}$)	0.33	0.17	0.45		0.30	0.20
Erucic acid ($C_{22:1}$)	–	–	–	–	0.30	–
MUFA	25.2	25.7	28.3	16.3	20.7	17.6
PUFA	38.3	37.4	37.8	–	–	47.9
SFA	36.0	35.6	31.6	–	–	34.2
Oil yield (%)	7.17	18.2	15.9	–	–	13.8
Reference	Dong et al. (2014)			Andras et al. (2005)		Topkafa (2016)

MUFA, mono unsaturated fatty acid; *PUFA*, polyunsaturated fatty acid; *SCF*, supercritical fluid extraction; *SE*, solvent extraction; *SFA*, saturated fatty acid; *SPE*, screw press expression; *UFA*, unsaturated fatty acid.

4 Minor bioactive lipids in cold pressed okra oil

Cold pressed seed oils contain relatively higher amounts of natural beneficial minor components such as tocols, phytosterols, phospholipids, carotenoids, and antioxidant phenolic moieties (Parker, Adams, Zhou, Harris, & Yu, 2003). Tocols (α-, β-, γ-, and δ-tocopherols, and α-, β-, γ-, and δ-tocotrienols) are well-known antioxidants with vitamin E activity/potency. In addition to playing their antioxidant role in vegetable oils, tocols exhibit cholesterol-lowering, inhibition of lipid peroxidation (in biological membranes), and free radical chain-breaking properties. Like other seed oils, okra seed oil is reported to contain significant levels of tocols and phytosterols (Bozan & Temelli, 2008; Parry et al., 2006). Table 2 depicts tocopherols and β-sitosterol, while Table 3 shows triglycerides composition of okra seed oil, recovered by different extraction means. The presence of carotenoids and tocopherols in an unsaponifiable fraction of okra oil was confirmed by thin-layer chromatography (TLC). HPLC chromatograms revealed that the most abundant tocopherols are γ-tocopherol and α-tocopherol, while a β isomer is present in trace level. Furthermore, xanthophylls (lutein) and carotenes (α and β) were also identified using HPLC (Pham, Peralta, & Pham, 2003). According to another report, okra seeds are reported to contain 10 times higher quantity of flavonols and almost 15 times higher amounts of catechins compared to skins (Arapitsas, 2008). The phenolic contents in okra seeds were found to be higher (2.85 mg/kg) than in the skin part (0.20) (Arapitsas, 2008). The presence of considerable amount of tocopherols and phenolics with antioxidant activity advocates potential nutraceutical applications of cold pressed okra seed oil.

TABLE 2 Tocopherol (mg/kg) and sterol composition of okra seed oils produced by different extraction means.

Extraction method	Cold pressed	SCF	*n*-Hexane	Ethanol
α-Tocopherol	320.1	930	780	620
β-Tocopherol	24.4	ND	ND	ND
γ-Tocopherol	371.3	407	380	494
δ-Tocopherol	1	ND	ND	ND
β-Sitosterol (g/kg)	ND	15	12.3	12.9
Reference	Topkafa (2016)	Andras et al. (2005)		

ND, not detected; *SCF*, supercritical fluid extraction.

TABLE 3 Triglyceride composition of cold pressed okra seed oils (%) (Topkafa, 2016).

Sr. no.	Triglyceride	Okra seed oil (cold pressed)
1	LnLnLn	ND
2	LLnLn	0.4
3	LLLn	0.5
4	LLL	17.3
5	OLLn	ND
6	PLLn	0.4
7	OLL	22.6
8	PLL	27.4
9	OOL	4.3
10	POL+SLL	13.6
11	PPL	9.3
12	OOO	2.2
13	SOL	2.2

ND, not detected.

5 Organoleptic traits and applications of cold pressed okra oil in food/nonfood products

A good-quality vegetable oil is considerably resistant to oxidative stress, light, and heat. and thus maintains its characteristic organoleptic value and nutritional characteristics (Gunstone & Harwood, 2007). Cold pressed high-value oils produced using mechanical extraction have attracted special attention from consumers due to their functional food attributes. The consumers acknowledge superior characteristic taste and aroma of cold pressed oils compared with solvent-extracted oil (Ramadan, 2013). Okra seed flour has been used to fortify different cereal flours (Adelakun et al., 2010) due to its unique flavoring characteristics (Arapitsas, 2008; Kumar, Patil, Patil, & Paschapur, 2009). According to Acikgoz et al. (2016), okra seed oil has been commonly employed in cosmetics as it contains palmitic acid. Interestingly, okra seed oil, due to the presence of palmitic acid, can be explored as an essential ingredient for manufacture of esters and plasticizers (Ndangui et al., 2010). Furthermore, the linoleic acid, a constituent of okra seed oil, may be utilized in the preparation of dyes and resins on a commercial scale. Basco (1995) reported that linoleic acid, which is present in high amounts in okra seed oil, is important in the human diet and thus the oil could serve as a cooking oil and find applications in the shortening and margarine industries. Linoleic acid is an important fatty acid often used in cosmetics, plastic, paint, production of soap, and pharmaceutical industries, and has multiple health benefits as well (Dangarembizi, Chivandi, & Erlwanger, 2013; Ndangui et al., 2010).

6 Health-promoting traits of cold pressed okra oil constituents

Plants including herbs, vegetables, and fruits have long been recognized as an impressive source of phytomedicines and natural drugs to treat different ailments (Dai & Mumper, 2010; Qadir, Anwar, Batool, Mushtaq, & Jabbar, 2019). The health-promoting attributes of plant foods, especially fruits, seeds/grains, and vegetables, have been attributed to the presence of a wide range of functional bioactive such as polyphenols with multiple biological activities (Dai & Mumper, 2010; Qadir et al., 2019). It is believed that seed-derived oils enriched with tocopherols offer greater protection against various pathologies such as cardiovascular, inflammation, and certain type of cancer (Van Hoed et al., 2009; Wang & Jiao, 2000). In this context, the entire okra plant is edible and has been used in several nutritional formulations/products (Madison, 2008; Maramag, 2013). Pods and seeds of okra are rich in phenolic compounds such as catechin oligomers and hydroxycinnamic derivatives with medicinal value (Manach, Williamson, Morand, Scalbert, & Remesy, 2005).

A wide array of nutritional and pharmaceutical properties of the okra seeds and okra seed oil has been reported (Jarret et al., 2011; Kumar et al., 2009). Okra has also been employed in folk medicine as antiulcer, antispasmodic, gastroprotective, and diuretic agents (Gurbuz, 2003; Martin, 1982). Okra seed oil consumption can be associated with medicinal benefits due to the high degree of unsaturation with linoleic acid as the major constituent (Jarret et al., 2011).

7 Other issues

Natural antioxidants, due to their safer nature, are attracting greater attention from the food industry and consumers. Cold pressed oils possess higher amount of natural antioxidants, and thus exhibit improved safety and shelf life (Yu, Haley, Perret, & Harris, 2002). In fact, higher oxidative stability, as well as the longer shelf life, of cold pressed okra seed oil is one of its peculiar features. Pressurized differential scanning colorimetry also revealed that okra seed oil is quite stable even up to a temperature of 175.3°C (Jarret et al., 2011). The reason behind the higher oxidative stability of cold pressed okra seed oil may be linked to the presence of higher contents of tocopherols relative to other compounds, as reported in the literature by Pham et al. (2003).

References

Acikgoz, C., Borazan, A. A., Andoglu, E. M., & Gokdai, D. (2016). Chemical composition of Turkish okra seeds (*Hibiscus esculenta* L.) and the total phenolic content of okra seeds flour. *Anadolu University Journal of Science and Technology*, *17*(5), 766–774.

Adelakun, O. E., Ade-Omowaye, B. I. O., Adeyemi, I. A., & Van De Venter, M. (2010). Functional properties and mineral contents of a Nigerian okra seed (*Abelmoschus esculentus* Moench) flour as influenced by pretreatment. *Journal of Food Technology*, *8*(2), 39–45.

Adetuyi, F. O., Osagie, A. U., & Adekunle, A. T. (2011). Nutrient, antinutrient, mineral and zinc bioavailability of okra *Abelmoschus esculentus* (L) Moench variety. *The American Journal of Clinical Nutrition*, *1*(2), 4954.

Akassey, R. P., & Daubrey, A. (1992). *Le guide de l'agriculteur en Côte d'Ivoire*: (p. 282). Edition Prat/Europa, 3.

Akingbala, J. O., Akinwande, B. A., & Uzo-Peters, P. I. (2003). Effects of color and flavor changes on acceptability of ogi supplemented with okra seed meals. *Plant Foods for Human Nutrition*, *58*, 1–9.

Andras, C. D., Simandi, B., Orsi, F., Lambrou, C., Tatla, D. M., & Panayiotou, C. (2005). Supercritical carbon dioxide extraction of okra (*Hibiscus esculentus* L.) seeds. *Journal of the Science of Food and Agriculture*, *85*, 1415–1419.

Arapitsas, P. (2008). Identification and quantification of polyphenolic compounds from okra seeds and skins. *Food Chemistry*, *110*, 1041–1045.

Basco, M. H. A. (1995). *Physical and chemical properties of okra seeds: their possible uses for local industries.* (Unpublished MSc thesis) Manila, Philippines: Department of Chemistry, Centro Escolar University.

Benchasr, S. (2012). Okra (*Abelmoschus esculentus* (L.) Moench) as a valuable vegetable of the world. *Ratarstvo Povrtarstvo*, *49*, 105–112.

Benjawan, C., Chutichudet, P., & Kaewsit, S. (2007). Effect of green manures on growth yield and quality of green okra (*Abelmoschus esculentus* L.) harlium cultivar. *Pakistan Journal of Biological Sciences*, *10*, 1028–1035.

Bozan, B., & Temelli, F. (2008). Chemical composition and oxidative stability of flax, safflower and poppy seed and seed oils. *Bioresource Technology*, *99*, 6354–6359.

Calisir, S., & Yildiz, M. U. (2005). A study on some physico-chemical properties of Turkey okra (*Hibiscus esculentus*) seeds. *Journal of Food Engineering*, *68*, 73–78.

Dai, J., & Mumper, R. (2010). Plant phenolics: Extraction, analysis and their antioxidant and anticancer properties. *Molecules*, *15*(10), 7313–7352.

Dangarembizi, R., Chivandi, E., & Erlwanger, K. (2013). Aloe ferox seed: A potential source of oil for cosmetic and pharmaceutical use. *Natural Product Communications*, *8*(3), 411–414.

Dong, Z., Zhang, J. G., Tian, K. W., Pan, W. J., & Wei, Z. J. (2014). The fatty oil from okra seed: Supercritical carbon dioxide extraction, composition and antioxidant activity. *Current Topics in Nutraceutical Research*, *12*, 75–84.

Gunstone, F. D., & Harwood, J. L. (2007). *Occurrence and characterization of oils and fats. In the lipid handbook with CD-ROM* (3rd ed.). New York: LLC.

Gurbuz, I. (2003). Antiulcerogenic activity of some plants used as folk remedy in Turkey. *Journal of Ethnopharmacology*, *88*(1), 93–97.

Harel, Z., Gascon, G., Riggs, S., Vaz, R., Brown, W., & Exil, G. (2002). Supplementation with omega-3 polyunsaturated fatty acids in the management of recurrent migraines in adolescents. *Journal of Adolescent Health*, *31*(2), 154–161.

Jarret, R. L., Wang, M. L., & Levy, I. J. (2011). Seed oil and fatty acid content in okra (*Abelmoschus esculentus*) and related species. *Journal of Agricultural and Food Chemistry*, *59*, 4019–4024.

Kumar, R., Patil, M. B., Patil, S. R., & Paschapur, M. S. (2009). Evaluation of *Abelmoschus esculentus* mucilage as paracetamol suspension. *International Journal of PharmTech Research*, *1*, 658–665.

Liu, G., Xu, X., Hao, Q. F., & Gao, Y. X. (2009). Supercritical CO_2 extraction optimization of pomegranate (*Punica granatum* L.) seed oil using response surface methodology. *LWT—Food Science and Technology*, *42*, 1491–1495.

Madison, D. (2008). *Renewing America's food traditions*: (p. 167). Chelsea Green Publishing.

Manach, C., Williamson, G., Morand, C., Scalbert, A., & Remesy, C. (2005). Bioavailability and bioefficacy of polyphenols in humans. *American Journal of Clinical Nutrition*, *81*, 230–242.

Maramag, R. P. (2013). Diuretic potential of *Capsicum frutescens* L., *Corchorus oliturius* L., and *Abelmoschus esculentus* L. *Asian Journal of Natural and Applied Science*, *2*(1), 60–69.

Martin, F. W. (1982). Okra, potential multiple- purpose crop for the temperate zones and tropics. *Economic Botany*, *36*(3), 340–345.

MEF (2013). *Biology of okra. Series of crop specific biology document*: (pp. 1–8). Ministry of Environmental and Forest Government of India.

Naveed, A., Khan, A. A., & Khan, I. A. (2009). Generation mean analysis of water stress tolerance in okra (*Abelmoschus esculentus* L.). *Pakistan Journal of Botany*, *41*, 195–205.

Ndangui, C. B., Kimbonguila, A., Nzikou, J. M., Matos, L., Pambou, N. P. G., Abena, A. A., et al. (2010). Nutritive composition and physico-chemical properties of gumbo (*Abelmoschus esculentus* L.) seed and oil. *Research Journal of Environmental and Earth Sciences*, *2*(1), 49–54.

Parker, T. D., Adams, D., Zhou, K., Harris, M., & Yu, L. (2003). Fatty acid composition and oxidative stability of cold-pressed edible seed oils. *Journal of Food Sciences*, *68*, 1240–1243.

Parry, J., Hao, Z., Luther, M., Su, L., Zhou, K., & Yu, L. L. (2006). Characterization of cold-pressed onion, parsley, cardamom, mullein, roasted pumpkin, and milk thistle seed oils. *Journal of American Oil and Chemical Society*, *83*, 847–854.

Pham, J. P., Peralta, M. M., & Pham, J. L. (2003). Okra (*Hibiscus esculentus* L.) seed oil: Characterization and potential use for high value products. In *Advanced research on plant lipids* (pp. 35–38).

Qadir, R., Anwar, F., Batool, F., Mushtaq, M., & Jabbar, A. (2019). Enzyme-assisted extraction of *Momordica balsamina* L. fruit phenolics: Process optimized by response surface methodology. *Journal of Food Measurement and Characterization*, *13*, 697–706.

Qadir, R., Anwar, F., Gilani, M. A., Zahoor, S., Rehman, M., & Mustaqeem, M. (2019). RSM/ANN based optimized recovery of phenolics from mulberry leaves by enzyme- assisted extraction. *Czech Journal of Food Sciences*, *37*(2), 99–105.

Ramadan, M. F. (2013). Healthy blends of high linoleic sunflower oil with selected cold pressed oils: Functionality, stability and antioxidative characteristics. *Industrial Crops and Products*, *43*, 65–72.

Sanjeet, K., Sokona, D., Adamou, H., Alain, R., Dov, P., & Christophe, K. (2010). Okra (*Abelmoschus* spp.) in west and Central Africa: Potential and progress on its improvement. *African Journal of Agricultural Research*, *5*(25), 3590–3598.

Sathish, D., & Eswar, A. (2013). A review on: *Abelmoschus esculentus* (okra). *International Research Journal of Pharmaceutical and Applied Science*, *3*(4), 129–132.

Savello, P. A., Martins, F. W., & Hill, J. M. (1980). Nutritional composition of okra seed meals. *Journal of Agriculture and Food Chemistry*, *28*(6), 1163–1166.

Schalau, J. (2002). *Backyard gardener. Available at (2002).* http://ag.arizona.edu./yavapai/anr/hort/byg/.

Topkafa, M. (2016). Evaluation of chemical properties for cold pressed onion, okra, rosehip, safflower and carrot seed oils: Triglyceride, fatty acid and tocol compositions. *Analytical Methods*, *8*(21), 4220–4225.

Van Hoed, V., De Clercq, N., Echim, C., Andjelkovic, M., Leber, E., Dewettinck, K., et al. (2009). Berry seeds: A source of specialty oils with high content of bioactives and nutritional value. *Journal of Food Lipids*, *16*, 33–49.

Wang, S. Y., & Jiao, H. J. (2000). Scavenging capacity of berry crops on superoxide radicals, hydrogen peroxide, hydroxyl radicals and singlet oxygen. *Journal of Agriculture and Food Chemistry*, *48*, 5677–5684.

Yu, L., Haley, S., Perret, J., & Harris, M. (2002). Antioxidant properties of extracts from hard winter wheat. *Food Chemistry*, *78*(4), 457–461.

Chapter 28

Cold pressed rosehip seed oil

Rahman Qadir and Farooq Anwar
Department of Chemistry, University of Sargodha, Sargodha, Pakistan

Abbreviations

ALA α-linolenic acid
EFA essential fatty acid
FFA free (unesterified) fatty acid
LA linoleic acid
MIR mid-infrared
MUFA monounsaturated fatty acid
PUFA polyunsaturated fatty acid
PV peroxide value

1 Introduction

Since ancient times, it has been a common practice of human beings to use vegetables and fruits as an important source of therapeutic agents/products, which can cure various pathologies due to the presence of selected phytochemicals and bioactives' so-called secondary metabolites (Cai & Ding, 1995; Vossen, Utrera, de Smet, Morcuende, & Estevez, 2012). However, some parts of plants/fruits such as seeds are discarded every year as agro-waste regardless of their importance as a potential source of oil and other nutraceutical components. Such under-utilized seeds can be explored for value-addition by extracting oil mostly rich in polyunsaturated fatty acids (PUFAs). Typically, the quantity of oil in seeds may vary from 5% to 18% depending upon the species (Szentmihalyi, Vinkler, Lakatos, Illes, & Then, 2002).

Out of 150 species from the Rosacea family, rosehip (*Rosa canina*) is one of the most peculiar species widely distributed across Europe, the Middle East, Asia, and North America (Ercisli et al., 2007; Hosni et al., 2010). The color of rosehip flowers varies from red to orange. Rosehip fruits (Fig. 1) comprise of seed and pericarp ranging from 30% to 35% and 65% to 70%, respectively (Uggla, Gao, & Werlemark, 2003). Soxhlet extraction, involving heat treatment and an organic solvent, is a common practice to extract oils from seeds; however, the food industry and consumers have serious concerns about the quality of such extracted oils (Szentmihalyi et al., 2002). In this regard, cold pressing technique has been of keen interest for nutritionists and food scientists to be applicable for better-quality oil extraction due to its safer and environmentally friendly nature. An increase in the consumption of cold pressed edible oils has been observed, as these are relatively much more beneficial than the refined ones in terms of their functional food and nutraceutical perspectives/benefits (Salgın, Salgın, Ekici, & Uluda, 2016).

Mostly, the cold pressed oil contains considerably higher amounts of valuable components/bioactives such as n-3 and n-6 PUFAs, tocopherols, phenolics, and phytosterols having biological activities and therapeutic effects (Osmianski, Bourzeix, & Heredia, 1986). Regardless of extraction techniques, the concentration of such high-value compounds in oils is also dependent on the quality and variety of grain. Owing to the rich profile of bioactive compounds, rosehip seeds can be explored as an important source of functional food and nutraceuticals, and thus can be utilized for commercial applications and value-addition (Szentmihalyi et al., 2002). The main objective of this chapter is to highlight the utility of cold pressing as a viable approach to extract rosehip seed oil and thereby explore the potential functional food and nutra-pharmaceutical food science applications of cold pressed rosehip seed oil.

Cold Pressed Oils. https://doi.org/10.1016/B978-0-12-818188-1.00028-1
Copyright © 2020 Elsevier Inc. All rights reserved.

FIG. 1 Typical rosehip fruits.

2 Rosehip cold pressed oil: Recovery and contents

Cold pressed vegetable oils can be produced from grains (including seeds), fruit, nuts, or germs. Generally, cold pressed oils are derived from seeds and fruits of oil plants having more than 15% fat content. These cold pressed oils mainly contain triglycerides (approx. 95%) and a small amount of diglycerides, monoglycerides, and free (unesterified) fatty acids (FFAs). The value of cold pressed oils is mainly due to the fact that they are a potential source of monounsaturated fatty acids (MUFAs) and PUFAs (Olsson et al., 2004). PUFAs include essential fatty acids (EFAs) that are not synthesized by humans and must be provided through food. These are: α-linolenic acid (18:3, n-3; ALA), linoleic acid (LA, 18:2, n-6), and γ-linolenic (GLA) (C18:3; n-6) (Concha, Soto, Chamy, & Zuniga, 2006). Cold pressed oils containing high amounts of PUFAs (LA and ALA) are those extracted from flax (15.8% and 56.9%, respectively), *Camelina sativa* (13.8% and 38.8%), hippophae or sea-buckthorns (37.3% and 23.1%), evening primrose (14.0% and 75.0%), borage (16.0% and 23.0%), rosehip (54.4% and 7.13%), and raspberry (54.5% and 29.1%) (Denev et al., 2014; Gunstone, 2005). The variation in rosehip seed oil yield based upon origin is depicted in Table 1.

TABLE 1 Geographical variation in rosehip seed oil yield.

Species	Oil content (%)	Country of origin	Reference
Rosa canina	3.23–6.59	Hungary	Szentmihalyi et al. (2002)
Rosa canina	1.22–1.25	Turkey	Demir and Ozcan (2001)
Rosa canina	2.53–3.00	Poland	Grajzer et al. (2015)
Rosa canina	13.3–14.5	Turkey (Hadium)	Ozcan (2002)
Rosa canina	17.8–19.2	Turkey (Ermenek)	Ozcan (2002)
Rosa canina	6.29–7.45	Turkey (Gumushane)	Ilyasoglu (2014)
Rosa sp.	14.3–15.8	France	Machmudah, Kawahito, Sasaki, and Goto (2007)
Rosa canina	3.10–3.40	Poland	Topkafa (2016)

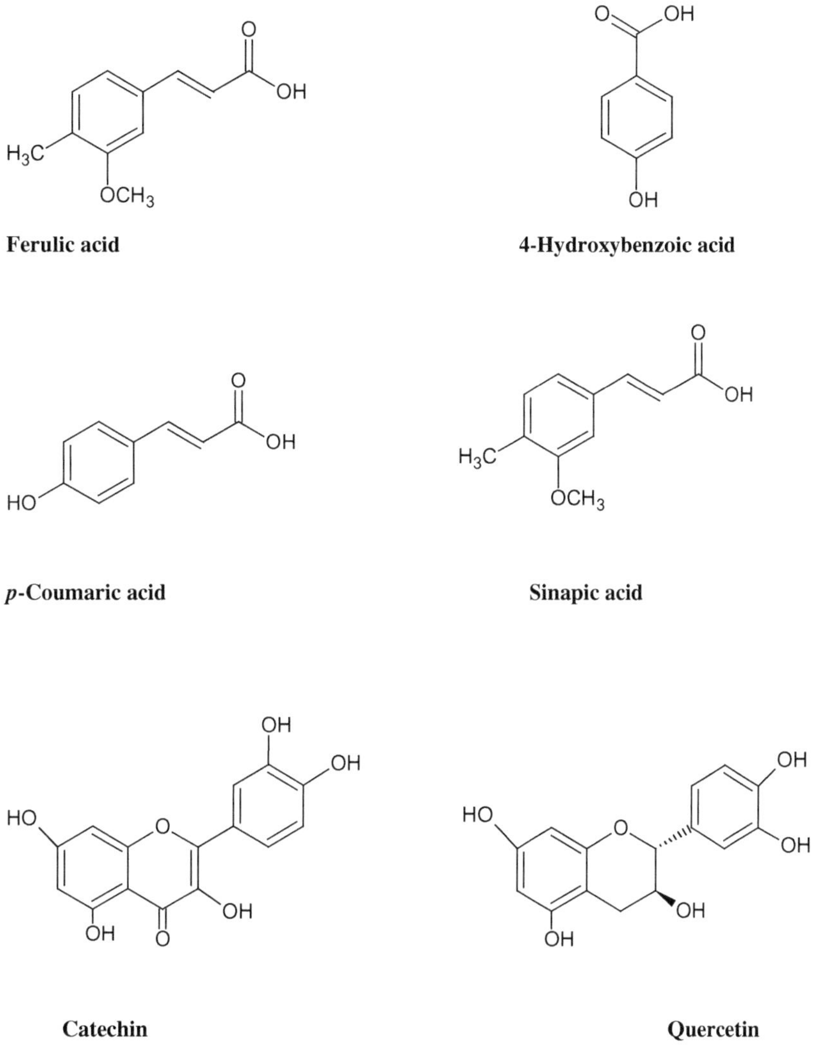

Ferulic acid

4-Hydroxybenzoic acid

p-Coumaric acid

Sinapic acid

Catechin

Quercetin

FIG. 2 Chemical structures of important phenolic compounds in rosehip seed oil.

Rosehip fruit and its pericarp are important due to their potential industrial applications. The by-products obtained from rosehip fruits such as seeds, which are generated in appreciable quantities, are often treated as under-utilized material and/or discarded as agro-waste. Generally, the rosehip seeds are crushed into a cake to be used for animal nutrition/animal feed (Szentmihalyi et al., 2002). However, rosehip seeds have been explored as a noticeable source of high-value oil with functional constituents such as essential fatty acids, tocopherols, phytosterols, phenolics, and phospholipids (Szentmihalyi et al., 2002; Zlatanov, 1999). The chemical structures of some important phenolics found in rosehip seed oil are given in Fig. 2.

Around 8%–10% of oil contents have been noted in rosehip seeds; however, the recovered oil is very valuable due to the rich profile of PUFAs and other medicinally important bioactives. Out of the most abundant fatty acids in rosehip seed oil, linoleic acid (41.1%–51.0%) comes at the top, followed by α-linolenic (19.6%–23.8%) and oleic (20.3%–23.0%) acids (Çelik, Balta, Ercişli, Kazankaya, & Javidipour, 2010). High nutritional value and health-contributing properties of edible plant oils such as rosehip seed oil, rich in omega-3 PUFAs, have prompted the need for production of cold pressed oils and oilcake (left over after oil extraction/cold pressing) for nutritional, nutraceutical, and industrial purposes.

3 Extraction and processing of rosehip seed cold pressed oil

Various conventional and nonconventional methods such as solvent, ultrasound, microwave, and supercritical fluid extraction have been devised by researchers to extract oil from rosehip seeds (Chrubasik, Roufogalis, Müller-Ladner, & Chrubasik, 2008; Machmudah et al., 2007; Maddocks-Jennings et al., 2005; Szentmihalyi et al., 2002).

As reported earlier, the supercritical fluid extraction technique yielded oil (6.68 wt/wt%) with high linoleic acid content (57.5%) compared to other methods employed. In another study, microwave was employed to extract rosehip seed oil with appreciable contents of unsaturated fatty acids (oleic, linoleic and linolenic acid) ranging from 35.9% to 54.7% (Szentmihalyi et al., 2002). Grajzer et al. (2015) employed cold pressing techniques to obtain two new commercially available high linolenic oils from rosehip seeds at lower temperatures. It has also been revealed that cold pressed oil has lower FFA content, peroxide value (PV), and acid value compared to the Soxhlet extraction technique (Paladines et al., 2014). The conventional process of oil extraction using solvents and accelerated temperature conditions has some drawbacks/limitations that can be overcome by using cold pressing techniques as green alternatives, thus producing oils with a longer shelf life, better quality, and improved stability due to the presence of higher amounts of antioxidants. Cold pressing is becoming a popular extraction process to yield nutritionally rich oils and allows the bioactive and PUFAs to remain intact (Salgın et al., 2016; Topkafa, 2016).

4 Fatty acids composition of rosehip seed oil

A number of researchers reported variations in the fatty acid composition of rosehip seed oils (Table 2). Almost 97% composition of rosehip seed oil is comprised of linoleic (41%–55%), oleic (14%–20%), and palmitic and stearic acid (20%–30%) (Franco, Pinelo, et al., 2007), while the remaining oil percentage is comprised of minor fatty acids and bioactive components. Likewise, researchers also isolated some galactolipids from rosehip seeds. Most galactolipids possess

TABLE 2 Variation in fatty acid composition of rosehip seed oils.

	Fatty acid composition (%)						
Fatty acid	Ercisli, Orhan, and Esitken (2007)	Kazaz, Baydar, and Erbas (2009)	Szentmihalyi et al. (2002)	Prescha, Grajzer, Dedyk, and Grajeta (2014)	Ilyasoglu (2014)	Turan, Solak, Kiralan, and Ramadan (2018)	Topkafa (2016)
C 12:0	4.80	–	–	–	–	–	
C 14:0	–	–	–	0.10	–	0.03	0.20
C 15:0	–	–	–	–	–	–	–
C 16:0	16.40	5.26	7.87	3.80	3.34	3.66	3.40
C 16:1	–	–	–	–	–	–	0.10
C 18:0	–	3.13	3.18	1.80	1.69	2.19	2.10
C 18:1	–	22.14	21.50	14.60	19.50	14.79	18.50
C 18:2	16.00	48.84	35.94	44.10	54.05	54.80	49.60
C 18:3	40.50	20.65	24.65	34.00	19.37	–	24.30
C 19:0	4.740	–	–	–	–	–	–
C 19:1	5.79	–	–	–	–	–	–
C 20:0	–	–	–	0.60	1.00	–	1.20
C 20:1	–	–	–	0.30	–	–	0.10
C 21:1	0.00	–	–	–	–	–	–
C 20:2	–	–	–	–	–	–	–
C 22:0	–	–	–	–	–	0.14	–
C 22:1	–	–	–	–	–	–	–
C 22:2	6.60	–	–	–	–	–	–

high levels of unsaturated fatty acids such as linolenic acid and impart antiinflammatory/antitumorigenic roles. Machmudah et al. (2007) extracted rosehip seed oil using supercritical CO_2 extraction, and found the oil to contain linoleic acid as the most abundant acid, followed by linolenic, palmitic, and stearic acids. Grajzer et al. (2015) demonstrated that among the selected cold pressed oils, rosehip seed oils proved to be a valuable source of nutrition with regard to the contents of tocopherols, which were as high as 1124 mg/kg. The results were in close agreement with the data obtained by Fromm, Bayha, Kammerer, and Carle (2012), who extracted oil at room temperature. While the amount of tocopherols was noted to be 10 times higher than reported by Zlatanov (1999) and Peredi, Facsar, Domokos, and Makk (1995) for Soxhlet-extracted oils. Schwager et al. (2014) compared two rosehip shells and found that fatty acids in the seed shell were almost higher than in the shells without seeds. In rosehip seeds, linoleic as well as α-linolenic acids are most important components of triglycerides.

5 Other bioactives in rosehip seed oil

Rosehip oils have been characterized by several other bioactives such as phenolics, carotenoids, lycopene, vitamin C, tocopherols, squalene, and chlorophyll (Roman, Stanila, & Stanila, 2013). However, the concentration of these bioactives is reported to be variable in different varieties depending upon the ripening stage, the geographical origin of the plant, the species, and ecological factors. The oil also contains high levels of carotenoids (i.e., lycopene, β-carotene, and rubixanthin) (Fig. 3) (Franco, Pinelo, et al., 2007; Franco, Sineiro, et al., 2007; Silva, Vandenabeele, Edwards, & Oliveira, 2008).

Among phenolics, quercetin and (+)-catechin were the main phenolics in *Rosa canina* L., as described by Turkben, Uylaser, Incedayi, and Çelikkol (2010). In another study, Demir, Yildiz, Alpaslan, and Hayaloglu (2014) observed vanillic acid, caffeic acid, and protocatechuic acid as major phenolics in *Rosa canina*. In a work conducted by Cunja, Mikulic-Petkovsek, Zupan, Stampar, and Schmitzer (2015), high concentrations of carotene and lycopene in *Rosa canina* were

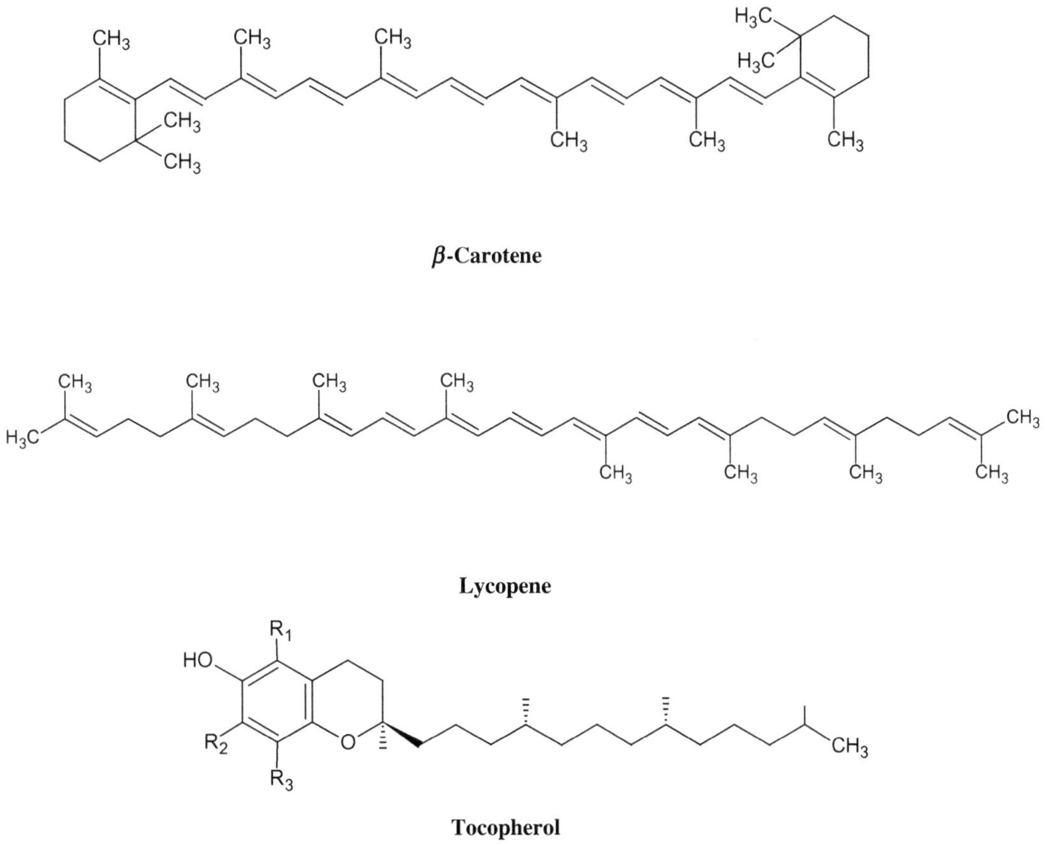

β-Carotene

Lycopene

Tocopherol

*R_1, R_2, R_3 = CH_3 or H

FIG. 3 Chemical structures of lipid-soluble antioxidants in rosehip seed oil.

detected; however, ascorbic acid and some quercetin glycosides were present in lesser amounts. Szentmihalyi et al. (2002) characterized rosehip oils by increased amounts (107.7 mg/kg) of carotene pigments, while in camelina and walnut oils these values were observed to be lower than 20 mg/kg. Furthermore, the carotenoids in rosehip oils were noted to be even higher (15 mg/kg) than in olive/pumpkin oils. Grajzer et al. (2015) investigated the individual phenolics such as *p*-coumaric, 4-hydroxybenzoic, ferulic, and sinapic acids in rosehip seed oils. A strong relationship may exist between organoleptic traits and health attributes of rosehip cold pressed oil to the combined effect of phenolics, carotenoids, and vitamins. Cold pressed rosehip oil, among various oils, might be a promising candidate to be utilized for functional food and nutra-pharmaceutical applications.

6 Organoleptic traits of bioactive compounds from rosehip seed oil

As a valuable source of bioactive compounds and PUFAs, rosehip seed oil is of particular importance due to its potential for functional food and cosmo-nutraceutical applications. Rosehip seed oils have been employed in cosmetic formulations due to its therapeutic properties against skin disorders (Franco, Pinelo, et al., 2007; Szentmihalyi et al., 2002). It could be used in the treatment of pigmentation, ulceration, and scarring problems (Concha et al., 2006). In another study, Martínez-Romero et al. (2017) added rosehip oil to *Aloe vera* gel to delay the ripening of stone-fruits and plums. Rosehip seed oil was also utilized in curing various skin pathologies such as scars, dermatitis, acne, eczema, and burns, pertaining to its potential bioactive components such as essential fatty acids, carotenoids, lycopene, tocopherols, and phenolics. In this regard, various herbal products and skin-caring lotions have been developed using rosehip seed oil that may act as moisturizers, scar healers, and antiaging agents (Concha et al., 2006). Based upon the multiple nutritional and therapeutic effects of rosehip oil, it might be employed as a valuable ingredient in several food/nonfood products. The organoleptic attributes and health traits of oils can be directly linked to the minor components; hence there is wider interest in the use of bioactive rich high-value oils for cosmo-nutraceutical applications. In another study conducted by Prescha et al. (2014) and Grajzer et al. (2015), rosehip oil offered higher antiradical and antioxidant traits than those of other cold pressed oils studied. This supports the strong relationship between the synergetic effects of rosehip oil bioactives and that of strong antioxidant activity (Prescha et al., 2014).

7 Health-promoting traits of cold pressed rosehip seed oil

Even though the oil content in rosehip seeds is quite low, the oil is rich in bioactives including tocols and phytosterols that contribute to the healthy traits of this special oil (Fromm et al., 2012; Grajzer et al., 2015; Ramadan, 2015). Due to increased levels of bioactives, rosehip oil has multiple bioactivities such as antibacterial, antifungal, and antiinflammatory, and thus could constrain cancer cell propagation (Olsson, Gustavsson, Andersson, Nilsson, & Duan, 2004; Silva et al., 2008). In an experiment on hamsters, researchers concluded that the hypolipidemic effect of rosehip oil might be linked to the presence of linoleic acid (Gonzalez, Francois, & Renaud, 1997). This acid has a protective effect against heart disease and is important in the development of brain and retinal cells (Connor, 1999). Linoleic and linolenic acids, essential fatty acids present in rosehip seed oil, are precursors of omega-3 and omega-6 fatty acids that are linked to reduced chances of chronic diseases (Aronson, Glaspy, Reddy, Reese, & Bagga, 2001). Furthermore, rosehip seed oil can be used together with an oral poly-vitamin that exhibits a synergistic effect (Iso et al., 2002; Jimenez et al., 2017; Patel, 2013; Wissemann, Gallenmüller, Ritz, Steinbrecher, & Speck, 2006). Lopes, Daletos, Proksch, Andrade, and Valentao (2014) found that some galactolipids in rosehip seed oil possessed antiinflammatory action and have antitumorigenic roles (Maeda, Hada, Yoshida, & Mizushina, 2007; Maeda, Kokai, Hada, Yoshida, & Mizushina, 2013).

8 Adulteration and authenticity

Rosehip oil, due to its remarkable nutritive quality, is a potential candidate for applications in the pharmaceutical and cosmetic industries. However, there are chances of adulteration of such high-value oils with a mixture of lower-grade oils. In this regard, the mid-infrared (MIR) method can effectively distinguish between the authentic and adulterated rosehip oil (De Santana et al., 2016). Rosehip oil consists of more than 77% PUFA, and hence there is much susceptibility of oxidation (Concha et al., 2006), but on the other hand, carotenoids and other antioxidant compounds present have the ability to effectively quench the oxygen radical species (Van den Bergh et al., 2000). Therefore, it can be concluded that a considerable amount of carotenoids, synergistically combined with other bioactives such as phenolic compounds, tocopherols, squalene, and chlorophyll, could contribute to the higher oxidative stability of rosehip oil, and thus demonstrated significant antiradical/antioxidant activities. Conclusively, it can be advocated that to preserve nutritional and nutraceutical benefits and attributes, extraction through cold pressing can be a promising means of production of high-quality rosehip seed oil.

References

Aronson, W. J., Glaspy, S. T., Reddy, D., Reese, D., & Bagga, D. (2001). Modulation of omega-3/omega-6 polyunsaturated ratios with dietary fish oils in men with prostate cancer. *Urology*, *58*, 283–288.

Cai, J. T., & Ding, Z. H. (1995). Nutrients composition of Rosa laevigata fruits. *Science and Technology of Food Industry*, *3*, 26–29.

Çelik, F., Balta, F., Ercişli, S., Kazankaya, A., & Javidipour, I. (2010). Seed oil profiles of five rosehip species (*Rosa* spp.) from Hakkari, Turkey. *Journal of Food, Agriculture and Environment*, *8*, 482–484.

Chrubasik, C., Roufogalis, B. D., Müller-Ladner, U., & Chrubasik, S. (2008). A systematic review on the *Rosa canina* effect and efficacy profiles. *Phytotherapy Research*, *22*(6), 725–733.

Concha, J., Soto, C., Chamy, R., & Zuniga, M. E. (2006). Effect of rose-hip extraction process on oil and defatted meal physico-chemical properties. *Journal of the American Oil Chemists' Society*, *83*, 771–775.

Connor, W. E. (1999). α-Linolenic acid in health and disease. *American Journal of Clinical Nutrition*, *69*, 827–828.

Cunja, V., Mikulic-Petkovsek, M., Zupan, A., Stampar, F., & Schmitzer, V. (2015). Frost decreases content of sugars, ascorbic acid and some quercetin glycosides but stimulates selected carotenes in *Rosa canina* hips. *Journal of Plant Physiology*, *178*, 55–63.

De Santana, F. B., Gontijo, L. C., Mitsutake, H., Mazivila, S. J., de Souza, L. M., & Neto, W. B. (2016). Non-destructive fraud detection in rosehip oil by MIR spectroscopy and chemometrics. *Food Chemistry*, *209*, 228–233.

Demir, F., & Ozcan, M. (2001). Chemical and technological properties of rose (*Rosa canina* L.) fruits grown wild in Turkey. *Journal of Food Engineering*, *47*, 333–336.

Demir, N., Yildiz, O., Alpaslan, M., & Hayaloglu, A. A. (2014). Evaluation of volatiles, phenolic compounds and antioxidant activities of rosehip (*Rosa* L.) fruits in Turkey. *Food Science and Technology*, *57*, 126–133.

Denev, P., Kratchanova, M., Ciz, M., Lojek, A., Vasicek, O., Nedelcheva, P., et al. (2014). Biological activities of selected polyphenol-rich fruits related to immunity and gastrointestinal health. *Food Chemistry*, *157*, 37–44.

Ercisli, S., Orhan, E., & Esitken, A. (2007). Fatty acid composition of *Rosa* species seeds in Turkey. *Chemistry of Natural Compounds*, *43*(5), 605–606.

Franco, D., Pinelo, M., Sineiro, J., & Nunez, M. J. (2007). Processing of *Rosa rubiginosa*: Extraction of oil and antioxidant substances. *Bioresource Technology*, *98*, 3506–3512.

Franco, D., Sineiro, J., Pinelo, M., & Nunez, M. J. (2007). Ethanolic extraction of *Rosa rubiginosa* soluble substances: Oil solubility equilibria and kinetic studies. *Journal of Food Engineering*, *79*, 150–157.

Fromm, M., Bayha, S., Kammerer, D. R., & Carle, R. (2012). Identification and quantitation of carotenoids and tocopherols in seed oils recovered from different *Rosaceae* species. *Journal of Agriculture and Food Chemistry*, *60*, 10733–10742.

Gonzalez, B., Francois, J., & Renaud, M. A. (1997). Rapid and reliable method for metabolite extraction in yeast using boiling buffered ethanol. *Yeast*, *13*(14), 1347–1355.

Grajzer, M., Prescha, A., Korzonek, K., Wojakowska, A., Dziadas, M., Kulma, A., et al. (2015). Characteristics of rosehip (*Rosa canina* L.) cold-pressed oil and its oxidative stability studied by the differential scanning calorimetry method. *Food Chemistry*, *188*, 459–466.

Gunstone, F. D. (2005). Vegetable oils. F. Shahidi (Ed.), *Bailey's industrial oil and fat products* (pp. 213–267). *Edible oil and fat products: Chemistry, properties, and health effects: Vol. 1*(pp. 213–267). Haboken, NJ: John Wiley & Sons, Inc.

Hosni, K., Chrif, R., Zahed, N., Abid, I., Medfei, W., Sebei, H., et al. (2010). Fatty acid and phenolic constituents of leaves, flowers and fruits of the dog rose (*Rosa canina* L.). *Rivista Italiana Delle Sostanze Grasse*, *87*, 117–123.

Ilyasoglu, H. (2014). Characterization of rosehip (*Rosa canina* L.) seed and seed oil. *International Journal of Food Properties*, *17*(7), 1591–1598.

Iso, H., Sato, S., Umemura, U., Kudo, M., Koike, K., Kitamura, A., et al. (2002). Linoleic acid, other fatty acids, and the risk of stroke. *Stroke*, *33*, 2086–2093.

Jimenez, S., Jimenez-Moreno, N., Luquin, A., Laguna, M., Rodriguez-Yoldi, M. J., & Ancín-Azpilicueta, C. (2017). Chemical composition of rosehips from different *Rosa* species: An alternative source of antioxidants for food industry. *Food Additives & Contaminants Part A, Chemistry, Analysis, Control, Exposure & Risk Assessment*, *34*(7), 1121–1130.

Kazaz, S., Baydar, H., & Erbas, S. (2009). Variations in chemical compositions of *Rosa damascena* Mill. and *Rosa canina* L. fruits. *Czech Journal of Food Sciences*, *27*, 178–184.

Lopes, G., Daletos, G., Proksch, P., Andrade, P. B., & Valentao, P. (2014). Anti-inflammatory potential of monogalactosyl diacylglycerols and a monoacylglycerol from the edible brown seaweed fucus *Spiralis linnaeus*. *Marine Drugs*, *12*, 1406–1418.

Machmudah, S., Kawahito, Y., Sasaki, M., & Goto, M. (2007). Supercritical CO_2 extraction of rosehip seed oil: Fatty acids composition and process optimization. *Journal of Supercritical Fluids*, *41*, 421–428.

Maddocks-Jennings, W., Wilkinson, J. M., & Shillington, D. (2005). Novel approaches to radiotherapy-induced skin reactions: A literature review. *Complementary Therapies in Clinical Practice*, *11*(4), 224–231.

Maeda, N., Hada, T., Yoshida, H., & Mizushina, Y. (2007). Inhibitory effect on replicative DNA polymerases, human cancer cell proliferation, and in vivo anti-tumor activity by glycolipids from spinach. *Current Medicinal Chemistry*, *14*, 955–967.

Maeda, N., Kokai, Y., Hada, T., Yoshida, H., & Mizushina, Y. (2013). Oral administration of monogalactosyl diacylglycerol from spinach inhibits colon tumor growth in mice. *Experimental and Therapeutic Medicine*, *5*, 17–22.

Martínez-Romero, D., Zapata, P. J., Guillen, F., Paladines, D., Castillo, S., Valero, D., et al. (2017). The addition of rosehip oil to Aloe gels improves their properties as postharvest coatings for maintaining quality in plum. *Food Chemistry*, *217*, 585–592.

Olsson, M. E., Gustavsson, K. E., Andersson, S., Nilsson, A., & Duan, R. D. (2004). Inhibition of cancer cell proliferation in vitro by fruit and berry extracts and correlations with antioxidant levels. *Journal of Agriculture and Food Chemistry*, *52*, 7264–7271.

Osmianski, J., Bourzeix, M., & Heredia, N. (1986). Les composes phenoliques du fruit de Ieglantier. *Bull Liaison Groupe Polyphenols*, *13*, 488–490.

Ozcan, M. (2002). Nutrient composition of rose (*Rosa canina* L.) seed and oils. *Journal of Medicinal Food*, *5*(3), 137–140.

Paladines, D., Valero, D., Valverde, J. M., Díaz-Mula, H. M., Serrano, M., & Martínez-Romero, D. (2014). The addition of rosehip oil improves the beneficial effect of *Aloe vera* gel on delaying ripening and maintaining postharvest quality of several stonefruits. *Postharvest Biology and Technology*, *92*, 23–28.

Patel, S. (2013). Rosehips as complementary and alternative medicine: Overview of the present status and prospects. *Mediterranean Journal of Nutrition and Metabolism*, *6*(2), 89–97.

Peredi, J., Facsar, G., Domokos, J., & Makk, A. (1995). Characteristics of seed oils of important central European wild rose species. *Olaj Szappan Kozm*, *44*, 142–145.

Prescha, A., Grajzer, M., Dedyk, M., & Grajeta, H. (2014). The antioxidant activity and oxidative stability of cold-pressed oils. *Journal of the American Oil Chemists Society*, *91*(8), 1291–1301.

Ramadan, M. F. (2015). Oxidation of β-sitosterol and campesterol in sunflower oil upon deep- and pan-frying of French fries. *Journal of Food Science and Technology*, *52*, 6301–6311.

Roman, I., Stanila, A., & Stanila, S. (2013). Bioactive compounds and antioxidant activity of *Rosa canina* L. biotypes from spontaneous flora of Transylvania. *Chemistry Central Journal*, *7*, 73.

Salgın, U., Salgın, S., Ekici, D. D., & Uluda, L. G. (2016). Oil recovery in rosehip seeds from food plant waste products using supercritical CO_2 extraction. *Journal of Supercritical Fluids*, *118*, 194–202.

Schwager, J., Richard, N., Schoop, R., & Wolfram, S. (2014). A novel rosehip preparation with enhanced anti-inflammatory and chondroprotective effects. *Mediators of Inflammation*, *2014*, 10–57.

Silva, C. E., Vandenabeele, P., Edwards, H. G., & Oliveira, L. F. (2008). NIR-FT-Raman spectroscopic analytical characterization of the fruits, seeds, and phytotherapeutic oils from rose-hips. *Analytical and Bioanalytical Chemistry*, *392*, 1489–1496.

Szentmihalyi, K., Vinkler, P., Lakatos, B., Illes, V., & Then, M. (2002). Rosehip (*Rosa canina* L.) oil obtained from waste hip seeds by different extraction methods. *Bioresource Technology*, *82*(2), 195–201.

Topkafa, M. (2016). Evaluation of chemical properties of cold pressed onion, okra, rosehip, safflower and carrot seed oils: Triglyceride, fatty acid and tocol compositions. *Analytical Methods*, *8*(21), 4220–4225.

Turan, S., Solak, R., Kiralan, M., & Ramadan, M. F. (2018). Bioactive lipids, antiradical activity and stability of rosehip seed oil under thermal and photo-induced oxidation. *Grasas y Aceites*, *69*(2), 1–9.

Turkben, C., Uylaser, V., Incedayi, B., & Çelikkol, I. (2010). Effects of different maturity periods and processes on nutritional components of rosehip (*Rosa canina* L.). *Journal of Food, Agriculture and Environment*, *8*, 26–30.

Uggla, M., Gao, X., & Werlemark, G. (2003). Variation among and within dogrose taxa (*Rosa sect. caninae*) in fruit weight, percentages of fruit flesh and dry matter, and vitamin C content. *Acta Agriculturae Scandinavica, Section B-Plant Soil Science*, *53*(3), 147–155.

Van den Bergh, J. P., ten Bruggenkate, C. M., Disch, F. J., & Tuinzing, D. B. (2000). Anatomical aspects of sinus floor elevations. *Clinical Oral Implants Research*, *11*(3), 256–265.

Vossen, E., Utrera, M., de Smet, S., Morcuende, D., & Estevez, M. (2012). Dog rose (*Rosa canina* L.) as a functional ingredient in porcine frankfurters without added sodium ascorbate and sodium nitrite. *Meat Science*, *92*, 451–457.

Wissemann, V., Gallenmüller, F., Ritz, C., Steinbrecher, T., & Speck, T. (2006). Inheritance of growth form and mechanical characters in reciprocal polyploid hybrids of rosa section caninae-implications for the ecological niche differentiation and radiation process of hybrid offspring. *Trees*, *20*, 340–347.

Zlatanov, M. D. (1999). Lipid composition of Bulgarian chokeberry, black currant and rosehip seed oils. *Journal of the Science of Food and Agriculture*, *79*(12), 1620–1624.

Chapter 29

Cold pressed safflower (*Carthamus tinctorius* L.) seed oil

Pelin Günç Ergönül and Zeynep Aksoylu Özbek
Department of Food Engineering, Faculty of Engineering, Manisa Celal Bayar University, Manisa, Turkey

Abbreviations

3-MCPD	3-monochloro-propane-diol
CD	conjugated diene
CT	conjugated triene
FA	fatty acid
MUFA	monounsaturated fatty acid
PAH	polycyclic aromatic hydrocarbon
PUFA	polyunsaturated fatty acid
SPC	safflower seed press cake
TG	triglyceride

1 Introduction

Safflower (*Carthamus tinctorius* L.) is one of the most ancient oil crops in the world. Today, Kazakhstan, Mexico, India, the USA, Russia, Turkey, China, Argentina, and Uzbekistan are the major producing countries of safflower. With a share of 93% total production, America and Asia are the major production regions, as shown in Fig. 1 (Khalid et al., 2017). Due to its high tolerance to drought and salt, safflower may be well adapted to poor and dry areas. The plant varies in length from 0.3 to 1.5 m, with globular flower heads of different color fractions ranging from yellow to orange and red. There is a growing trend to improve new safflower seed varieties with higher oil contents (Khalid et al., 2017). The world production of safflower increased promptly in the early 1960s as a result of the demand for highly unsaturated oil as an edible oil. It has been observed that the trend of safflower production has been augmented over the previous few years due to the crop area increasing at the rate of 4.9% per annum. An average global yield of safflower seeds has varied from 805 to 872 kg/ha with an annual growth rate of 0.97% (Khalid et al., 2017). The hull fraction of safflower seeds is 35%–45%. According to hull types, the oil content of the seed ranges between 20% and 47% (Cosge, Gurbuz, & Kıralan, 2007). The whole seeds in normal hull types contain 27%–32% oil of very high quality, 5%–8% moisture, 14%–15% protein, 2%–7% ash, and 32%–40% crude fiber (Gecgel, Demırcı, & Esendal, 2007). The main characteristics of the oil including specific gravity of 0.919–0.924, refractive index of 1.473–1.476, titer of 15–17°C, flash point of 121–149°C, free acidity of 0.15%–0.60%, saponification value of 186–194 mg KOH/g oil, iodine value of 141–147 g/100 g oil, unsaponifiable matter of 0.3%–0.6%, peroxide value of 0–1.0 mequiv. O_2/kg oil (fresh oil), and moisture and volatile matter content of 0.03%–0.1% have been reported (Gunstone, Harwood, & Dijkstra, 2007; Smith, 1996; Velasco & Fernandez-Martinez, 2001).

Depending on the genetics of the safflower line, two different types of oil can be produced. One type is high in linoleic acid (>70%), and the other is high in oleic acid (>70%), which is stable toward oxidation. A third experimental type of safflower oil contains almost equal amounts of oleic and linoleic acids (Ahmadzadeh, Kadıvar, & Saeıdı, 2014). Safflower oil contents can also vary depending upon the process and technique of oil extraction (Han, Cheng, Zhang, & Bi, 2009). Conjugated diene (CD) and conjugated triene (CT) values are good indicators to follow lipid oxidation in oils and fats. Kıralan and Ramadan (2016) identified K232 and K270 values of cold pressed safflower oil during storage at 60°C. The initial value for safflower oil increased from 1.96 to 11.97 at the end of the storage time. Regarding CT values, an increase from 0.01 to 1.28 was observed after 12 days of storage. Smaller increases in K232 and K270 values were observed

Cold Pressed Oils. https://doi.org/10.1016/B978-0-12-818188-1.00029-3
Copyright © 2020 Elsevier Inc. All rights reserved.

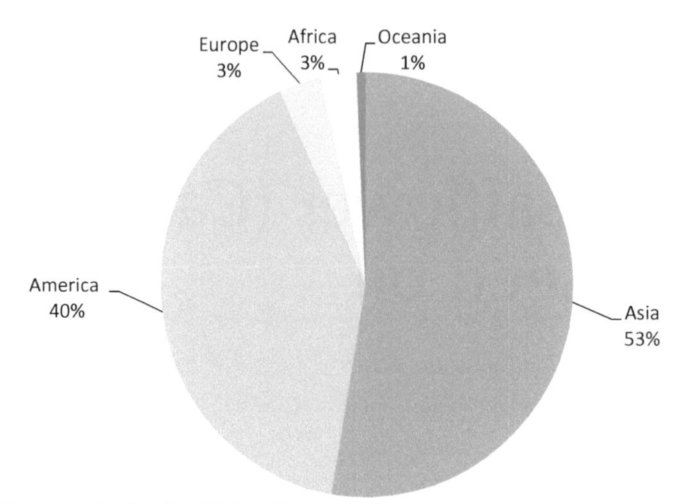

FIG. 1 Region distribution of safflower production (FAOSTAT, 2015).

in microwave-treated cold pressed safflower oil samples, reaching up to 4.22 and 0.64, respectively. It has been confirmed that the effect of microwave treatment on K232 and K270 values of cold pressed oil is more efficient than that of conventional heating.

In recent years, interest in varieties with high-linoleic acid content has increased, due to its beneficial effects on human health (Kıralan & Ramadan, 2016). On the other hand, an oleic-rich variety (>70%), which is suitable as a heat-tolerant cooking oil, was developed. Currently, the oil is preferred to produce margarines, salad dressings, infant formulas, and food coatings in the food industry. Safflower oil can be used for industrial purposes, and safflower seed cake is only used as animal feed (Stanford, Wallins, Lees, & Mundel, 2001). Angın (2013) studied the effects of the pyrolysis temperatures and heating rates on the pore structure, yield, and physicochemical properties of biochar obtained from safflower seed cold pressed cake. It was stated that a more valuable and functional product (biochar) can be obtained from safflower seed press cake (SPC). The SPC biochars can be effectively used as a raw material for the preparation of activated carbon. Biochars obtained at high pyrolysis temperatures (600°C) are suitable for direct use in fuel applications due to their high fixed carbon content, higher heating value, and low volatile matter content. Biochars can also be valorized as a chemical feedstock for industrial purpose (Angın, 2013).

2 Extraction and processing of cold pressed oil (developments in extraction and isolation of high-value lipid compounds and phytochemicals)

Vegetable oils can be obtained from oilseeds using different types of press systems, solvent extractors, or a combination of both. Generally, oilseeds containing high levels of oil are first prepressed then solvent extracted, or direct solvent extraction can be applied to materials with a lower oil level. The technology of extraction can be selected depending on the manufacturing cost, availability, material properties, the usage goals of the cake (meal), and environmental factors (Ghazani, Garcia Llatas, & Marangoni, 2014). Based on the seed composition and structure, some parts of the oil may always remain in the cake or meal. This should be taken into account when comparing the industrial press and cold press for the oil yield and meal composition values. The crude lipids, crude protein, crude fiber, moisture, and ash contents of safflower meal obtained from industrial type expeller presses are 6.6%, 21.0%, 32.2%, 9.0%, and 3.7%, respectively. While prepress solvent extraction reduced the crude oil level to 0.5%–1.5%, it did not change other components significantly (Smith, 1996).

With mechanical pressing, which is a simple and safe technique, more oil remains in the meal than by solvent extraction. Heating is not applied to oilseeds during cold pressing; moreover, the oilseeds processed by this technique must be very clean, uniform, and have an appropriate moisture level (Aydeniz, Güneser, & Yılmaz, 2014). By this technique a very pure yield of safe, nutritionally rich, and sensorially acceptable virgin oils can be obtained, which do not require refining and are consumable directly. Nevertheless, the oil yield is lower than with hot pressing and solvent extraction. In order to increase the oil yield of cold pressing systems, some pretreatments are applied to oil seeds before pressing, such as microwave treatment, steaming, enzyme application, and preroasting. It is also indicated that special cold pressed oils are high-demand products in world markets, not only for food usage but also cosmetic, medicinal, and other uses. On the other hand, cold press-produced virgin seed oils do not need expensive refining procedures. Only centrifugation or filtration is necessary to obtain edible quality oils. In addition, minor nutrients, which are mostly lost during chemical refining, are retained in cold pressed oils. A producer will prefer a production type depending on the aim of production, end uses of both oil and meal, and the amount of the seed that must be processed. Unquestionably, solvent extraction and/or full expeller pressing are the major oil production techniques worldwide. Cold pressed oils are produced only for unique uses and demands (Aydeniz et al., 2014).

Traditionally, safflower oil has been extracted using local mills called "ghani" (mainly in Indian regions), which involve the pestle and mortar technique. Approximately 10%–15% oily cake or meal obtained from ghani is partially dehulled before milling. By expeller pressing, the residual oil can decrease to 6%–8% (Khalid et al., 2017). According to the pre-experiments, the optimum seed moisture content for cold pressing is 12%, and for all succeeding operations, the seed moisture level was constantly measured and adjusted by the addition of water (Aydeniz et al., 2014). The majority of the moisture has been retained by the cake.

Cold pressing of safflower seeds was performed on a cold press machine at constant parameters of 10 m exit die, 40-rpm screw rotation speed, and a maximum 40°C exit temperature. When oil (liquid phase) and oily cake-meal (solid phase) were collected and weighed after each cold pressing, the oil was immediately filtered through a 40 μm screen to separate suspended materials. The oil was centrifuged in a refrigerated centrifuge at 10°C (Aydeniz et al., 2014; Ergönül & Özbek, 2018). Cold pressed safflower oil has a relatively high phenolic content, particularly in terms of apigenin and luteolin. Additionally, this specialty oil contains significant amounts of α-tocopherol. The antioxidative characteristic of safflower oil is derived from phenolic compounds and tocopherol isomers. Other biologically active components of the oil are phytosterols such as β-sitosterol, δ-7-stigmasterol, campesterol, stigmasterol, δ-7-campesterol, δ-7-avenasterol, and δ-5-avenasterol. In general, cold pressed oils have higher phytochemical contents compared to refined oils, and retain their natural flavor and aroma.

3 Fatty acids and acyl lipids profile of cold pressed oil

Safflower oil is constituted of 6%–8% palmitic, 2%–3% stearic, 16%–20% oleic, and 71%–75% linoleic acids, and contains the highest linoleic acid content among all the commercial oils (Aydeniz et al., 2014). Safflower seed oils had significant amounts of oleic acid (18:1n-9), linoleic acid (18:2n-9), and linolenic acid (18:3n-3), which reduced the risk of cardiovascular disease (Topkafa, 2016). PUFAs were found to be dominant in cold pressed safflower seed oil (73.8%). They comprised the highest amount of linoleic acid (73.7%) from the total FA (Topkafa, 2016). Oleic (15.2%) and palmitic acids (7.1%) were established to be distinctive in cold pressed safflower seed oil (Ergönül & Özbek, 2018; Topkafa, 2016). The Tre-Aso 12/08 variety was high in oleic acid (62.6%) among all varieties, whereas other varieties were high in linoleic acid (51.2%–76.8%). Samancı and Özkaynak (2003) identified linoleic acid, oleic acid, and palmitic acid in the cold pressed oils of Yenice and Dinçer safflower varieties as 55.3%–66.6%, 21.2%–27.4%, and 45.0%–13.87%, respectively. Celenk, Gumuş, Argon, Buyukhelvacigil, and Karasulu (2018) demonstrated that the linoleic acid content of cold pressed safflower seed oil is 58.2%.

Cold pressed safflower seed oil contains high amounts of TGs that include monounsaturated fatty acids (MUFAs) and polyunsaturated fatty acids (PUFAs), which have important nutritional value and act as carriers for fat-soluble vitamins (A, D, E, and K) (Topkafa, Kara, & Sherazi, 2015). LLL (57.3%) was found to be the dominant triglyceride in cold pressed safflower seed oil (Topkafa, 2016). OLL, PLL, OLL, and POL + SOL containing linoleic acid accounted for 20.3%, 9.5%, 5.2%, and 4.3%, respectively (Fig. 2).

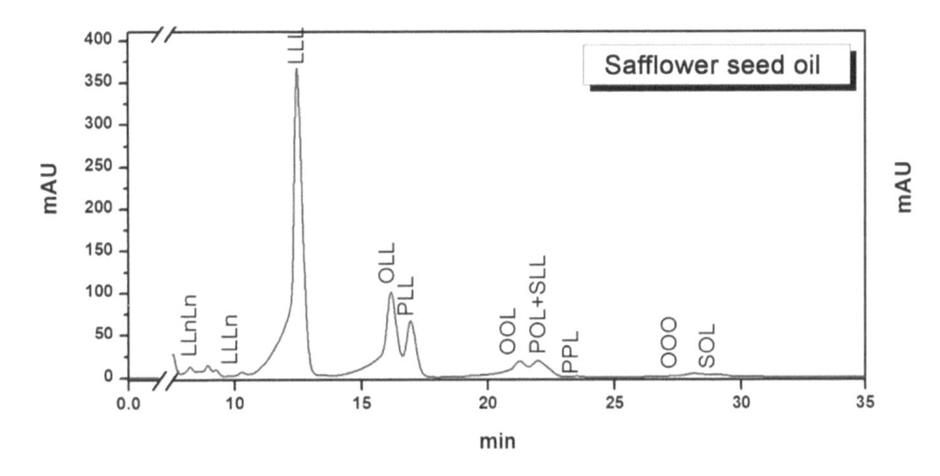

FIG. 2 A sample of triglyceride chromatogram of cold pressed safflower oil (Topkafa, 2016).

4 Minor bioactive lipids in cold pressed oil

Tocols are antioxidants that inhibit the oxidation of edible oils, influence shelf life of edible oil during storage, and delay lipid peroxidation associated with the pathogenesis of numerous human diseases/disorders (Gimeno, Castellote, Lamuela-Raventos, De La Torre, & Lopez-Sabater, 2000). Cold pressed safflower seed oil had α-tocopherol at 376.6 mg/kg, as the main component. This high content of α-tocopherol makes safflower oil an excellent dietary source of vitamin E, but confers on it low thermostability for high-temperature applications such as deep frying or lubrication (Cuesta, Velasco, & Méndez, 2014). It also contains β-tocopherol (4.4 mg/kg), γ-tocopherol (25.9 mg/kg), γ-tocotrienol (9.7 mg/kg), and δ-tocopherol (1.0 mg/kg) (Topkafa, 2016). The total tocopherol content in cold pressed safflower seed oil (407.9 mg/kg) was lower than that reported by Celenk et al., 2018 (1126.9 mg/kg). In another study, three types of tocopherols were found in safflower oil in various amounts: α-tocopherol, β-tocopherol, and γ-tocopherol ranging from 46.05 to 70.93 mg/100 g, 0.85 to 2.16 mg/100 g, and trace amounts 0.45 mg/100 g of oils, respectively (Matthaus, Özcan, & Al Juhaimi, 2015). α-Tocopherol was found to be the dominant tocopherol in cold pressed safflower seed oils, amounting to between 112.6 and 131.8 mg/100 g (Ergönül & Özbek, 2018).

Table 1 shows the differences between average tocopherol contents in different varieties of safflower seed oil according to the extraction techniques. α-Tocopherol was the dominant tocopherol, accounting for 94%–96% of total tocopherols in safflower seed oil, which is higher in the oil obtained by cold pressing when compared with the oils obtained by solvent extraction. According to Aydeniz et al. (2014), the highest contents of α-tocopherol (502.1 mg/kg) were found in roasted safflower oil using different cold press techniques, whereas microwave extraction technique resulted in decreased α-tocopherol contents (366.2 mg/kg).

Cuesta et al. (2014) revealed the effect of extraction methods such as solvent extraction and cold pressed on the tocopherol contents and oxidative stability of safflower seed oil. They observed that pressed oil had more tocopherols than the solvent-extracted oil. The main difference between both oils was found to be oil stability. In pressed oil, the induction time was 38.0 h, while it was 32.0 h in the solvent-extracted oil. The results demonstrated that the antioxidative compounds are more protected in cold pressed oil. According to Franke, Frohlich, Werner, Bohm, and Schone (2010), the safflower oil contained trace amounts of carotenoids such as 0.08 mg/100 g of lutein, and 0.15 mg/100 g of zeaxanthine. There are no more data on the carotenoids content of cold pressed safflower seed oil.

Phenolic compounds are one of the bioactive compounds that protect the seeds from oxidative deterioration even at low concentrations. Ergönül and Özbek (2018) revealed the effect of variety on bioactive compounds of cold pressed safflower seeds. They identified the highest level of total phenolic compounds in the Balcı variety (5393.6 μg/100 g), which is locally developed. Apigenin was the main phenolic between safflower varieties followed by luteolin, and their amounts ranged between 40.08 and 2222.1 μg/100 g (Ergönül & Özbek, 2018). Apart from these, the authors determined some phenolics as *p*-coumaric, ferulic, sinapic, vanillic, veratric, and caffeic acids in trace amounts. Although there is no further study on determination of phenolic compounds in cold pressed safflower seed oil, Moumen et al. (2015) determined them in safflower seed oil extracted by the Soxhlet method. In the four oil varieties, they identified two hydroxybenzoic acids (vanillic acid and syringic acid), four hydroxycinnamic acids (*p*-coumaric acid, cinnamic acid, sinapic acid, and ferulic acid), and

TABLE 1 Amount of minerals in cold pressed safflower oil (Alam et al., 2006).

Ingredient	Amount (ppm) (Mean ± SD)	Ingredient	Amount (ppm) (Mean ± SD)
Br	5.15 ± 0.05	K	15,500 ± 600
Ca	3260 ± 168	Mg	2980 ± 170
Cl	2096 ± 17	Mn	27.6 ± 0.2
Co	45.7 ± 2	Na	126 ± 2
Cr	0.17 ± 0.04	Rb	7.1 ± 0.3
Cu	51.4 ± 14.2	Ru	28.1 ± 10.1
Fe	350 ± 6	Sb	22.4 ± 4.6
Sc	1.5 ± 0.3	Zn	61.7 ± 0.3

seven compounds from other phenolic classes (1–3 DHN, pinoresinol, vanillin, rutin, *trans*-chalcone, naringin, and tyrosol). They showed that the major phenolic compounds in Soxhlet-extracted safflower seed oils are *trans*-chalcone (13.4%), and naringin (26.8%). In another study, the major phenolic compounds in a hot water extract of safflower seeds were (−)-epigallocatechin (109.6 mg/g), a 4-hydroxy benzohydrazide derivative (18.28 mg/g), and gallocatechin (17.02 mg/g) (Yu, Lee, Kim, et al., 2013). As can be seen from the results in all studies, the major phenolic compounds differ depending on the seed varieties and oil extraction techniques.

The studies on sterol contents are also very limited. Ergönül and Özbek (2018) determined mainly β-sitosterol, δ-7-stigmasterol, campesterol, stigmasterol, δ-7-campesterol, δ-7-avenasterol, and δ-5-avenasterol, as well as negligible amounts of δ-5,24-stigmastadienol, sitostanol, and cholesterol in cold pressed safflower seed oils from different varieties. According to their results, the Tre-Aso 12/08 variety contained the highest amounts of campesterol and δ-5-avenasterol compared with other varieties, while the Balcı variety had the highest β-sitosterol content. According to the results of the study done by Aydeniz et al. (2014), the concentrations of some phytosterols (campesterol, stigmasterol, D-5-23-stigmastadienol, β-sitosterol, and D-7-stigmastenol) decreased significantly after roasting or microwaving. It should be kept in mind that the concentrations of sterols obtained from solvent extraction may yield more sterols than cold pressing. Plant sterols are very important bioactive compounds for their cholesterol-lowering effect. Additionally, antioxidant effects of phytosterols such as stigmasterol, campesterol, and β-sitosterol against lipid peroxidation were reported by Yoshida and Niki (2003).

Kıralan and Ramadan (2016) identified the major aroma compounds as hexanal, 2,4-heptadienal, hexanal, and 2-heptenal in cold pressed safflower oil. They subjected cold pressed oils to conventional heating (oven test) using an air-forced oven at 60°C and microwave heating for 2 and 4 min. 2,4-Decadienal was present at the maximum level at the end of storage. They pointed out that 2-heptenal was the major volatile aldehyde after 10 days of storage at 60°C. Hexanal rises from linoleic acid and is a good indicator of lipid oxidation (Kıralan & Ramadan, 2016). Jelen, Obuchowska, Zawirska-Wojtasiak, and Wasowicz (2000) stated that hexanal was the most abundant volatile oxidation compound in fresh cold pressed oil. 2-Heptenal is a well-known and important volatile aldehyde induced during linoleic acid oxidation, wherein the odor threshold of this compound is very low, at 0.001 mg/kg (Morales, Luna, & Aparicio, 2005). 2-Hexenal, α-thujene, 2-heptenal, nonanal, and 2,4-decadienal were found in the samples treated with microwave heating. Hexanal content increased with the exposure time and 2-heptenal was determined in only one sample after 4 min of application.

Aydeniz et al. (2014) determined the effect of pretreatment on the physicochemical, sensory, and aromatic properties of cold pressed safflower oil. The volatile compounds measured in all samples (control, microwaved, and roasted) were 2,3-butanediol, hexanal, ethylbenzene, *p*-xylene, heptanal, α-pinene, *p*-cymene, D-limonen, 1,5-octadien-3-ol, 2-octenal, nonanal, phenylethyl alcohol, 2-nonenal, 2,4-nonadienal, octanoic acid, naphthalene, *E,E*-2,4-nonadienal, 3-dodecen-1-al, 2,4-decadienal, undecanal, *E,E*-2,4-decadienal, 2-dodecenal, 2-cyclohexen-1-one, 5-tetradecene, 1-tetradecene, 2 (3*H*)-furanone-dihydro-5,5-dimethyl4-3-oxo-butyl, methyl eugenol, *trans*-caryophyllene, tetradecenal, 2,4-dodecadienal, benzene-nonyl, 17-octadecenal, *c*-dodecalactone, and 9-octadecanoic acid methyl ester. 2-Methyl butanol, pentanal, pyrazine, methyl pyrazine, furfural, heptanone, pyrazine-2,5-dimethyl, hexanoic acid methyl ester, 2-ethyl-pyrazine, 1-ethyl-2-formyl-pyrrol, 3,5-dimethyl-2-ethylpyrazine, 1-propyl pentyl ester butyric acid, 2-acetyl-6-methylpyrazine, *E*-3-nonene-2-one, 2-methyl-5*H*-6,7 dihydrocyclopentapyrazine, and tridecanoic acid were the volatiles present only in microwaved samples, but not in the control and roasted samples. Likewise, volatiles found only in the roasted samples were isoamyl alcohol, 1-phellandrene, isophytol, and 2-*N*-heptyl furan. The results showed that safflower oil is rich in aromatic compounds. Aromatic compounds associated with oily, buttery, creamy, fruity, green, plant, waxy, wood, citrus, sweet, herbal, earthy, hay, pungent, bitter, spicy, and pepper sensory definitions were determined in almost all samples. Microwave treatment was more effective than roasting in the production of the pyrazine, furfural, and similar volatiles, which are mostly responsible for the roasted, nutty, caramel, and similar aroma/flavor descriptions. Aromatics with fruity, herbal, and floral definitions were present in roasted samples in addition to common descriptors (Aydeniz et al., 2014).

5 Health-promoting traits of cold pressed safflower oil and oil constituents

Safflower oil has many beneficial health effects according to the studies performed. A balanced fatty acid profile found in safflower oil has shown to decrease fat accumulation in rats when compared to a beef tallow diet (Shimomaoera, Tamaoera, & Suzuki, 1990). The presence of conjugated linoleic acid in safflower oil has effectively indicated to decrease body weight and fat, as demonstrated in clinical trials (Norris et al., 2009). Additionally, safflower oil has been found to be effective in fat-induced insulin resistance, which is a common problem (Neschen et al., 2002).

Bone and joint health is mainly affected by degenerative diseases. Safflower oils have been reported to diminish these degenerative diseases. Not many studies have been designed to develop functional foods using safflower oils (Choi, Kim, & Im, 2011; Zhou, Tang, Xu, Zhou, & Wang, 2014). There has been an effort to design dietary supplements containing higher amounts of PUFAs from various vegetable sources. India, the USA, and Japan have used safflower oil in controlling blood parameters such as high-density lipoprotein (HDL) levels and cholesterol (Choi et al., 2011).

As shown as in Table 2, cold pressed safflower oil is rich in minerals that prevent bone loss in osteoporosis induced-ovariectomized rats caused by estrogen inadequacy. Estrogen plays an important role in mineral homeostasis and its deficiency is known as a major factor in the loss of bone minerals in postmenopausal osteoporosis (Alam et al., 2006). In particular, Ca and Mg in safflower seed oils may affect bone metabolism.

The fatty acid composition of safflower oil is quite similar to that of olive oil, especially in oleic and linoleic acid contents, with a much lower price for consumers. The PUFA in safflower oil inhibits the level of low-density lipoproteins (LDLs, bad cholesterol). Therefore, in North America, Germany, and Japan, high amounts of safflower oil are consumed. Another advantage of safflower oil is minimal allergic responses, which make it suitable for many cosmetic products when compared to other functional oils.

In high-cholesterol-fed rats, safflower seeds and oils caused damage/toxicity to the liver by reducing plasma and hepatic total cholesterol, plasma triglycerides, and atherogenic index. The supplementation also extended the activity of hepatic HMG-CoA reductase activity (Moon et al., 2001). Another study by Rahimi, Asgary, and Kabiri (2014) reported that the treatment of rats with 200 mg/kg safflower seed oil in the diet caused decreases in the levels of blood glucose, triglycerides LDLs, total cholesterol, alanine aminotransferase (ALT), alkaline phosphatase (ALP), and aspartate aminotransferase (AST).

The lignans and flavonoids found in safflower seeds have physiological effects similar to that of estrogen's so-called phytoestrogens. These phytoestrogens have anticarcinogenic properties, bone health improvement, antioxidant activities, and regulation of serum cholesterol (Draper et al., 1997). Similarly, their extraction from defatted safflower oil has been shown to improve blood lipid status and cholesterol elimination without significant uterotropic action in estrogen-deficient animal models (Cho et al., 2004).

The arachidonic acid concentration in a Sprague-Dawley rats model increased after safflower oil supplementation. The increased arachidonic acid resulted in a reduced rate of diabetic embryopathy (Reece et al., 1996). Recently, Higa et al. (2010) showed that safflower supplementation decreased the malformation rates in maternal diabetes. The reduction of resorption was due to regulation of nitric oxide homeostasis and arachidonic acid in embryos, which prevented developmental damage during organogenesis. Safflower oil showed significant hypoglycemic and hypolipidemic effects in hyperglycemic rats after supplementation at a rate of 200 mg per kg body weight for a period of 30 days (Rahimi et al., 2014).

TABLE 2 Tocopherol contents in different varieties of safflower oil as affected by the extraction technique (Aydeniz et al., 2014; Lee, Kim, et al., 2004; Lee, Oh, Chang, & Kim, 2004; Matthaus et al., 2015; Vosoughkia, Ghavamib, Gharachorloo, Sharrifmoghaddasi, & Omidi, 2011).

	Compound	Content	Method
Tocopherols (%)	α-Tocopherol	0.0554	Solvent extraction
	β-Tocopherol	0.001	Solvent extraction
	γ-Tocopherol	0.0002	Solvent extraction
Tocopherols (Iran; mg/kg)	α-Tocopherol	192.05–439.64	Solvent extraction
	β-Tocopherol	Not detected	Solvent extraction
	γ-Tocopherol	5.59–14.68	Solvent extraction
	δ-Tocopherol	3.06–11.50	Solvent extraction
Tocopherols (mg/kg)	α-Tocopherol	366.24–502.12	Cold pressed
Tocopherols (mg/kg)	α-Tocopherol	386–520	Roasted seeds and expeller press
	β-Tocopherol	8.90–12.40	Roasted seeds and expeller press
	γ-Tocopherol	2.40–7.70	Roasted seeds and expeller press

Asp et al. (2011) carried out a study on a randomized, double-masked crossover with 55 postmenopausal, obese women with type 2 diabetes; the authors concluded that the addition of 8 g safflower oil in diet improved glycemia, inflammation, and blood lipids over a period of 16 weeks. Zhang, Li, Liu, Sun, and Zhang (2010) showed that a safflower oil supplemented diet can powerfully alter adipocytic adiposity-related gene expression and resulted in effective amelioration of diet-induced obesity. Shimomaoera et al. (1990) maintained that consumption of a safflower oil diet increased lipoprotein lipase activity in the heart and skeletal muscle, resulting in the elevation of lipids oxidation rate and the depression of serum triacylglycerol levels.

The high content of linoleic acid in safflower oil improves the quality and appearance of skin. Therefore, it is frequently used in formulations of skin conditioners and other cosmetics for handled acne vulgaris (Toombs, 2005). Safflower oil also promotes hair and scalp health, wherein numerous patents describe the usage of safflower oil in combination with other oils or as a separate ingredient in the formulation of many cosmetics and herbal products. The growth-promoting effect of safflower oil on hair and healthy skin is due to vitamin E content and its light texture, which allows its easy absorption into the scalp (Maru, 2014). The *N*-feruloylserotonin, and *N*-*p*-coumaroyl serotonin derivatives and acacetin from ethyl acetate extracted safflower seed inhibited the formation of melanin in the skin and stimulated a skin-whitening effect in *Streptomyces bikiniensis* and B16 melanoma cell lines (Roh, Han, Kim, & Hwang, 2004). Similarly, safflower oil has potential to improve the symptoms of meningitis (encephalomyelitis) (Harbige, Yeatman, Amor, & Crawford, 1995). The oil extracted from the seeds contains alkane-6,8-diols, which inhibits the activity of 12-*O*-tetradecanoylphorbol-13-acetate-induced tumor promotion in mouse skin (Loo, Cheung, & Chow, 2004). *N*-Feruloylserotonin and *N*-(*p*-coumaroyl) serotonin from seeds strongly inhibited melanin production in *Streptomyces bikiniensis* and B16 melanoma cell lines (Roh et al., 2004). *N*-(*p*-Coumaroyl) tryptamine and *N*-(*p*-coumaroyl) serotonin from seed extracts inhibit the production of proinflammatory cytokines (IL-1a, IL-1b, IL-6, IL-8, and TNF-a) from lipopolysaccharide-stimulated human monocytes (Takii et al., 2003). Safflower seed extracts together with a dendritic cell (DC)-based vaccine increased the levels of TNF-a and IL-1b in Mouse CD117 (c-kit)-derived DCs and showed strong antitumor activity (Chang, Hung, Chyan, Cheng, & Wu, 2011).

6 Edible and nonedible applications of cold pressed oil

Safflower oil has common applications in the food industry due to the high MUFA and PUFA content. It may also be used entirely or in combination with other oils as biodiesel. The utilization of safflower oil in combination with castor oil to produce lower viscosity biodiesel was reported (Thomas, Birney, & Auld, 2012).

Safflower oil has a high nutritional value and many biological activities. The incorporation of safflower oil will increase the biological activities of the diet, thereby helping to reduce animal and human alimentations (Alizadeh et al., 2012). As a potential bioactive food ingredient, it can be integrated into various cosmetics, health supplements, pharmaceuticals, food products, and beverages, and used as a fortificant for feed. Because of the incidents of increase in ruminant body mass and milk yield, safflower seeds are used in ruminants' diets in more than 60 countries (Alizadeh et al., 2012). Moreover, feeding safflower oil to lambs increased the content of unsaturated fatty acids in muscle tissues (Kott et al., 2003). Supplementation of 10–30 g safflower oil in goat diets has been shown to ameliorate the milk fat composition (Shi, Luo, Zhang, & Sheng, 2015). Furthermore, the combination of safflower oil and monensin increased the concentration of conjugated linoleic acid in bovine milk (Bell, Griinari, & Kennelly, 2006).

Safflower oil has the capacity to serve as a natural vehicle for stabilizing and storing biological lipophilic compounds for developing functional foods or for use as a vehicle for drug distribution. To increase the oxidative stability of PUFAs, cod liver oil can be encapsulated in safflower oil (Fischer et al., 2014). Due to the good oxidative stability of safflower oil along with its diverse nature, it can be textured into different products like organogels (semisolids systems with self-assembled gelation capacities). The safflower oil containing organogels are noncytotoxic and capable of holding lipophilic compounds for longer periods with good delivery at the target site (Morales-Rueda, Dibildox-Alvarado, Charo, Weiss, & Toro-Vazquez, 2009). Safflower oil emulsions can be a good source of dietary supplements. These supplements have the capacity to increase metabolism rate and support a healthy weight program containing linoleic fatty acid diets (Maru, 2014). Vitamin D can also be encapsulated in safflower oil using microchannel emulsification. These oil-in-water (*O/W*) emulsions could remain stable for more than 30 days without any increase in droplet size (Khalid et al., 2017). Safflower oil can also be used in *O/W* emulsions for nutritional and medical purposes (Floyd, 1999; Yeh, Chao, Lin, & Chen, 2000).

7 Other issues

Cold pressed oils have more nutritional advantages when compared with refined oils, through increased compounds such as tocopherols, phenolic, and sterols, which provide health benefits and prevent disease (Teh & Birch, 2013). Cold pressed oils may be subjected to adulteration. For example, cold pressed vegetable oils have a lower efficiency process than solvent-extracted oil. Therefore, cold pressed oils have higher prices and thus may be subject to adulteration. Oils may be substituted or added to other oils of lower value (Aued-Pimentel, Castro, de Sousa, Mello, & Abe-Matsumoto, 2015; Azadmard-Damirchi & Torbati, 2015; Hirashima, Silva, Caruso, & Aued-Pimentel, 2013; Rohman & Man, 2012; Stankova, Kremmyda, Tvrzicka, & Žák, 2013). There are very limited studies about the adulteration and authenticity of cold pressed oils. One of these has been carried out by Silva, da Silva Torres, de Almeida, and Sampaio (2018), who identified the fatty acid composition of 19 commercial cold pressed safflower oils. According to the results of FAME contents, 47% (9/19) of the samples were authentic with high linoleic values, a characteristic of this oil (Codex Alimentarius, 2015). In some samples, α-linolenic levels were higher (maximum of 0.2%) than the legislation (<0.1%); however, levels were very close to it and similar to those found in the literature (Oz, 2016). Ten samples were adulterated, suggesting a possible mixture of vegetable oils with high levels to 18:1c (27.3%–69.0%) and 18:3, n-3 (1.0%–7.0%) and low to 18:2c (20.2%–55.1%), in dissimilarity with the range allowed by the Codex Alimentarius (2015).

In order to ensure food safety, it is essential to know contamination sources and take the necessary precautions. Polycyclic aromatic hydrocarbons (PAHs) are a group of organic chemical contaminants. Vegetable oils can be contaminated with PAHs by environmental pollution of the vegetable raw material, during industrial processing as heating and seed drying, the extraction with solvent, burning of soil, the material of packing, residues of mineral oils, and migration from contaminated water or soils (Bansal & Kim, 2015; Camargo, Antoniolli, & Vicente, 2011; Ciecierska & Obiedziński, 2013). The presence of PAHs in vegetable oils have been investigated, but there are very limited works on the identification of PAHs contents in cold pressed oils (Camargo et al., 2011; Roszko, Szterk, Szymczyk, & Waszkiewicz-Robak, 2012; Ciecierska & Obiedziński, 2013). Silva et al. (2018) examined the contamination with four PAHs—benzo[*a*] anthracene (BaA), chrysene (Chr), benzo[*b*]fluoranthene (BbF), and benzo[*a*]pyrene (BaP)—in cold pressed safflower oils marketed in São Paulo, Brazil. Authentic cold pressed safflower oil contained 0.68 μg/kg, 1.23 μg/kg, <LOQ, 0.40 μg/kg, and 2.94 μg/kg of BaA, Chr, BbF, and BaP, and total four PAHs, respectively. The adulterated cold pressed safflower oils contained 2.07, 3.76, 1.23, 0.78, and 7.84 μg/kg (Silva et al., 2018). According to Regulation No. 835/2011 by the European Union, the limit for PAHs and the maximum value for BaP is 2.0 μg/kg, and a limit for the total of four PAHs is 10.0 μg/kg for vegetable oils (Commission of the European Communities, 2011). The results indicate that authentic and adulterated safflower oils did not exceed the recommended limits.

In another study, the researcher determined 3-chloropropane-1,2-diol (3-MCPD) and some other chloropropanols in cold pressed safflower oils. 3-MCPD are known as food-processing contaminants with carcinogenic and genotoxic effects (Weißhaar, 2008). Out of 11 samples of cold pressed safflower oils, there were two samples of safflower oil labeled as cold pressed that had high levels of 3-MCPD (2450 and 2461 μg/kg). A significant amount of 3-MCPD was formed because of the presence of chloride ions. According to the revised regulation of EFSA, the daily intake of 3-MCPD was recommended as 2 μg/kg body weight (EFSA, 2018).

References

Ahmadzadeh, S., Kadıvar, M., & Saeıdı, G. (2014). Investıgatıon of oil properties and seed composıtıon in some safflower lınes and cultıvars. *Journal of Food Biochemistry*, *38*, 527–532.

Alam, M. R., Kim, S. M., Lee, J., Chon, S. K., Choi, S. J., Choi, H., et al. (2006). Effects of safflower seed oil in osteoporosis induced ovariectomized rats. *The American Journal of Chinese Medicine*, *34*(4), 601–612.

Alizadeh, A. R., Alikhani, M., Ghorbani, G. R., Rahmani, H. R., Rashidi, L., & Loor, J. J. (2012). Effects of feeding roasted safflower seeds (variety IL-111) and fish oil on dry matter intake, performance and milk fatty acid profiles in dairy cattle. *Journal of Animal Physiology and Animal Nutrition*, *96*, 466–473.

Angın, D. (2013). Effect of pyrolysis temperature and heating rate on biochar obtained from pyrolysis of safflower seed press cake. *Bioresource Technology*, *128*, 593–597.

Asp, M. L., Collene, A. L., Norris, L. E., Cole, R. M., Stout, M. B., Tang, S. Y., et al. (2011). Time-dependent effects of safflower oil to improve glycemia, inflammation and blood lipids in obese, post-menopausal women with type 2 diabetes: A randomized, double-masked, crossover study. *Clinical Nutrition*, *30*(4), 443–449.

Aued-Pimentel, S., Castro, F. D., de Sousa, R. J., Mello, M. R. P. A., & Abe-Matsumoto, L. T. (2015). Commercialised coconut oil in São Paulo City, Brazil: Evaluation of authenticity and nutritional labelling. *Journal of Agriculture and Life Sciences*, *2*(1), 76–83.

Aydeniz, B., Güneser, O., & Yılmaz, E. (2014). Physico-chemical, sensory and aromatic properties of cold press produced safflower oil. *Journal of the American Oil Chemists' Society*, *91*(1), 99–110.

Azadmard-Damirchi, S., & Torbati, M. (2015). Adulterations in some edible oils and fats and their detection methods. *Journal of Food Quality and Hazards Control*, *2*(2), 38–44.

Bansal, V., & Kim, K. H. (2015). Review of PAH contamination in food products and their health hazards. *Environment International*, *84*, 26–38.

Bell, J. A., Griinari, J. M., & Kennelly, J. J. (2006). Effect of safflower oil, flaxseed oil, Monensin, and vitamin E on concentration of conjugated linoleic acid in bovine milk fat. *Journal of Dairy Science*, *89*, 733–748.

Camargo, M. C., Antoniolli, P. R., & Vicente, E. (2011). HPLC-FLD simultaneous determination of 13 polycyclic aromatic hydrocarbons: Validation of an analytical procedure for soybean oils. *Journal of the Brazilian Chemical Society*, *22*(7), 1354–1361.

Celenk, V. U., Gumuş, Z. P., Argon, Z. U., Buyukhelvacigil, M., & Karasulu, E. (2018). Analysis of chemical compositions of 15 different cold-pressed oils produced in Turkey: A case study of tocopherol and fatty acid analysis. *Journal of the Turkish Chemical Society Section A: Chemistry*, *5*(1), 1–18.

Chang, J. M., Hung, L. M., Chyan, Y. J., Cheng, C. M., & Wu, R. Y. (2011). *Carthamus tinctorius* enhances the antitumor activity of dendritic cell vaccines via polarization toward Th1 cytokines and increase of cytotoxic T lymphocytes. *Journal of Evidence Based Complementary Medicine*, *2011*, 274858.

Cho, S. H., Lee, H. R., Kim, T. H., Choi, S. W., Lee, W. J., & Choi, Y. (2004). Effects of defatted safflower seed extract and phenolic compounds in diet on plasma and liver lipid in ovariectomized rats fed high-cholesterol diets. *Journal of Nutrition Science Vitaminology*, *50*(1), 32–37.

Choi, C.-H., Kim, H.-D., & Im, E.-B. (2011). Reviews of research trends on safflower seed (*Carthamus tinctorius* L.). *Journal of Korean Medical Classics*, *24*(6), 63–90.

Ciecierska, M., & Obiedziński, M. W. (2013). Polycyclic aromatic hydrocarbons in vegetable oils from unconventional sources. *Food Control*, *30*(2), 556–562.

Codex Alimentarius (2015). *Codex standard for named vegetable oils (Codex Stan 210–1999)*: (pp. 1–13).

Commission of the European Communities (2011). Commission regulation (EC) no 835/2011, of 19 August 2011 amending regulation (EC) no 1881/2006 as regards maximum levels for polycyclic aromatic hydrocarbons in foodstuffs. *Official Journal of the European Union*, *L215*, 4–8.

Cosge, B., Gurbuz, B., & Kıralan, M. (2007). Oil content and fatty acid composition of some safflower (*Carthamus tinctorius* L.) varieties sown in spring and winter. *International Journal of Natural and Engineering Sciences*, *1*, 11–15.

Cuesta, A. F., Velasco, L., & Méndez, M. V. R. (2014). Novel safflower oil with high γ-tocopherol content has a high oxidative stability. *European Journal of Lipid Science and Technology*, *116*, 832–836.

Draper, C. R., Edel, M. J., Dick, I. M., Randall, A. G., Martin, G. B., & Prince, R. L. (1997). Phytoestrogens reduce bone loss and bone resorption in oophorectomized rats. *Journal of Nutrition*, *127*(9), 1795–1799.

EFSA (2018). *Revised safe intake for 3-MCPD in vegetable oils and food*. EFSA Panel on Contaminants in the Food Chain (CONTAM). https://www.efsa.europa.eu/en/press/news/180110 (Accessed 03 March 2018).

Ergönül, P. G., & Özbek, Z. A. (2018). Identification of bioactive compounds and total phenol contents of cold pressed oils from safflower and camelina seeds. *Journal of Food Measurement and Characterization*, *12*, 2313–2323.

FAOSTAT (2015). FAOSTAT online database at http://faostat3.fao.org/saffower

Fischer, J. J., Nykiforuk, C. L., Chen, X., Shen, Y., Zaplachinksi, S., Murray, E. W., et al. (2014). Delayed oxidation of polyunsaturated fatty acids encapsulated in safflower (*Carthamus tinctorius*) oil bodies. *International Journal of Engineering Science and Innovation Technologies*, *3*, 512–522.

Floyd, A. G. (1999). Top ten considerations in the development of parenteral emulsions. *Pharmaceutical Science & Technology Today*, *2*(4), 134–143.

Franke, S., Frohlich, K., Werner, S., Bohm, V., & Schone, F. (2010). Analysis of carotenoids and vitamin E in selected oilseeds, press cakes, and oils. *European Journal of Lipid Science and Technology*, *112*, 1122–1129.

Gecgel, U., Demırcı, M., & Esendal, E. (2007). Fatty acid composition of the oil from developing seeds of different varieties of safflower (*Carthamus tinctorius* L.). *Journal of the American Oil Chemists' Society*, *84*, 47–54.

Ghazani, S. M., Garcia Llatas, G., & Marangoni, A. G. (2014). Micronutrient content of cold-pressed, hot-pressed, solvent extracted and RBD canola oil: Implications for nutrition and quality. *European Journal of Lipid Science and Technology*, *116*, 380–387.

Gimeno, E., Castellote, A., Lamuela-Raventos, R., De La Torre, M., & Lopez-Sabater, M. (2000). *Journal of Chromatography A*, *881*, 251–254.

Gunstone, F. D., Harwood, J. L., & Dijkstra, A. J. (2007). *The lipid handbook* (3rd ed.). USA: CRC Press.

Han, X., Cheng, L., Zhang, R., & Bi, J. (2009). Extraction of safflower seed oil by supercritical CO_2. *Journal of Food Engineering*, *92*, 370–376.

Harbige, L., Yeatman, N., Amor, S., & Crawford, M. (1995). Prevention of experimental autoimmune encephalomyelitis in Lewis rats by a novel fungal source of g-linolenic acid. *British Journal of Nutrition*, *74*(05), 701–715.

Higa, R., White, V., Martinez, N., Kurtz, M., Capobianco, E., & Jawerbaum, A. (2010). Safflower and olive oil dietary treatments rescue aberrant embryonic arachidonic acid and nitric oxide metabolism and prevent diabetic embryopathy in rats. *Molecular Human Reproduction*, *16*(4), 286–295.

Hirashima, K., Silva, S. A. D., Caruso, M. S. F., & Aued-Pimentel, S. (2013). Encapsulated specialty oils commercialized in São Paulo state, Brazil: Evaluation of identity (fatty acid profile) and compliance of fatty acids and vitamin E contents with nutrition labeling. *Food Science and Technology*, *33*(1), 107–115.

Jelen, H. H., Obuchowska, M., Zawirska-Wojtasiak, R., & Wasowicz, E. (2000). Headspace solid-phase microextraction use for the characterization of volatile compounds in vegetable oils of different sensory quality. *Journal of Agricultural and Food Chemistry*, *48*, 2360–2367.

Khalid, N., Khan, R. S., Hussain, M. I., Farooq, M., Ahmad, A., & Ahmed, I. (2017). A comprehensive characterisation of safflower oil for its potential applications as a bioactive food ingredient—A review. *Trends in Food Science & Technology*, *66*(2017), 176–186.

Kıralan, M., & Ramadan, M. F. (2016). Volatile oxidation compounds and stability of safflower, sesame and canola cold-pressed oils as affected by thermal and microwave treatments. *Journal of Oleo Science*, *65*(10), 825–833.

Kott, R. W., Hatfield, P. G., Bergman, J. W., Flynn, C. R., Wagoner, H. V., & Boles, J. A. (2003). Feedlot performance, carcass composition, and muscle and fat CLA concentrations of lambs fed diets supplemented with safflower seeds. *Small Ruminant Research*, *49*, 11–17.

Lee, Y., Kim, I.-H., Chang, J., Rhee, Y., Oh, H., & Park, H. (2004). Chemical composition and oxidative stability of safflower oil prepared with expeller from safflower seeds roasted at different temperatures. *Journal of Food Science*, *69*(1).

Lee, Y.-C., Oh, S.-W., Chang, J., & Kim, I.-H. (2004). Chemical composition and oxidative stability of safflower oil prepared from safflower seed roasted with different temperatures. *Food Chemistry*, *84*(1), 1–6.

Loo, W. T., Cheung, M. N., & Chow, L. W. (2004). The inhibitory effect of a herbal formula comprising ginseng and *Carthamus tinctorius* on breast cancer. *Life Sciences*, *76*(2), 191–200.

Maru, R. (2014). Safflower oil emulsion as dietary supplement and preparation thereof. Patent No. US8623433 B1.

Matthaus, B., Özcan, M. M., & Al Juhaimi, F. Y. (2015). Fatty acid composition and tocopherol profiles of safflower (*Carthamus tinctorius* L.) seed oils. *Natural Product Research*, *29*(2), 193–196.

Moon, K.-D., Back, S.-S., Kim, J.-H., Jeon, S.-M., Lee, M.-K., & Choi, M.-S. (2001). Safflower seed extract lowers plasma and hepatic lipids in rats fed high cholesterol diet. *Nutrition Research*, *21*(6), 895–904.

Morales, M., Luna, G., & Aparicio, R. (2005). Comparative study of virgin olive oil sensory defects. *Food Chemistry*, *91*, 293–301.

Morales-Rueda, J. A., Dibildox-Alvarado, E., Charo, M., Weiss, R. G., & Toro-Vazquez, J. F. (2009). Thermo-mechanical properties of candelilla wax and dotriacontane organogels in safflower oil. *European Journal of Lipid Science and Technology*, *111*(2), 207–215.

Moumen, A. B., Mansouri, F., Richard, G., Abid, M., Fauconnier, M. L., Sindic, M., et al. (2015). Biochemical characterisation of the seed oils of four safflower (*Carthamus tinctorius*) varieties grown in North-Eastern of Morocco. *International Journal of Food Science and Technology*, *50*, 804–810.

Neschen, S., Moore, I., Regittnig, W., Yu, C. L., Wang, Y., Pypaert, M., et al. (2002). Contrasting effects of fish oil and safflower oil on hepatic peroxisomal and tissue lipid content. *American Journal of Physiology-Endocrinology and Metabolism*, *282*(2), 395–401.

Norris, L. E., Collene, A. L., Asp, M. L., Hsu, J. C., Liu, L.-F., Richardson, J. R., et al. (2009). Comparison of dietary conjugated linoleic acid with safflower oil on body composition in obese postmenopausal women with type 2 diabetes mellitus. *The American Journal of Clinical Nutrition*, *90*(3), 468–476.

Oz, M. (2016). Relationship between sowing time, variety, and quality in safflower. *Journal of Chemistry*, *2016*, 1–8 Article ID 9835641.

Rahimi, P., Asgary, S., & Kabiri, N. (2014). Hepatoprotective and hypolipidemic effects of *Carthamus tinctorius* oil in alloxan-induced type 1 diabetic rats. *Journal of HerbMed Pharmacology*, *3*(2), 107–111.

Reece, E. A., Wu, Y.-K., Wiznitzer, A., Homko, C., Yao, J., Borenstein, M., et al. (1996). Dietary polyunsaturated fatty acid prevents malformations in offspring of diabetic rats. *American Journal of Obstetrics and Gynecology*, *175*(4, Part 1), 818–823.

Roh, J. S., Han, J. Y., Kim, J. H., & Hwang, J. K. (2004). Inhibitory effects of active compounds isolated from safflower (*Carthamus tinctorius* L.) seeds for melanogenesis. *Biological and Pharmaceutical Bulletin*, *27*(12), 1976–1978.

Rohman, A., & Man, Y. C. (2012). Application of Fourier transform infrared spectroscopy for authentication of functional food oils. *Applied Spectroscopy Reviews*, *47*(1), 1–13.

Roszko, M., Szterk, A., Szymczyk, K., & Waszkiewicz-Robak, B. (2012). PAHs, PCBs, PBDEs and pesticides in cold-pressed vegetable oils. *Journal of the American Oil Chemists' Society*, *89*(3), 389–400.

Samancı, B., & Özkaynak, E. (2003). Effect of planting date on seed yield, oil content and fatty acid composition of safflower (*Carthamus tinctorius*) cultivars grown in the mediterranean region of Turkey. *Journal of Agronomy and Crop Science*, *189*, 359–360.

Shi, H., Luo, J., Zhang, W., & Sheng, H. (2015). Using safflower supplementation to improve the fatty acid profile in milk of dairy goat. *Small Ruminant Research*, *127*, 68–73.

Shimomaoera, Y., Tamaoera, T., & Suzuki, M. (1990). Less body fat accumulation in rats fed a safflower oil diet than in rats fed a beef tallow diet. *Journal of Nutrition*, *120*, 1291–1296.

Silva, S. A., da Silva Torres, E. A. F., de Almeida, A. P., & Sampaio, G. R. (2018). Polycyclic aromatic hydrocarbons content and fatty acids profile in coconut, safflower, evening primrose and linseed oils. *Food Chemistry*, *245*(2018), 798–805.

Smith, J. (1996). Safflower oil. Y. H. Hui (Ed.), *Bailey's industrial oil and fat products* (pp. 411–456). Vol. 2(pp. 411–456). New York: John Wiley.

Stanford, K., Wallins, G. L., Lees, B. M., & Mundel, H. H. (2001). Immature safflower forage as a feed for ewes. In: J. W. Bergman, & H. H. Mundel (Eds.), *Vth International Safflower Conference, Williston, North Dakota, Sidney, Montana, USA*, pp. 29–32.

Stankova, B., Kremmyda, L. S., Tvrzicka, E., & Žák, A. (2013). Fatty acid composition of commercially available nutrition supplements. *Czech Journal of Food Science*, *31*(3), 241–248.

Takii, T., Kawashima, S., Chiba, T., Hayashi, H., Hayashi, M., Hiroma, H., et al. (2003). Multiple mechanisms involved in the inhibition of proinflammatory cytokine production from human monocytes by N-(p-coumaroyl)serotonin and its derivatives. *International Journal of Immunopharmacology*, *3*(2), 273–277.

Teh, S. S., & Birch, J. (2013). Physicochemical and quality characteristics of cold-pressed hemp, flax and canola seed oils. *Journal of Food Composition and Analysis*, *30*(1), 26–31.

Thomas, T. P., Birney, D. M., & Auld, D. L. (2012). Viscosity reduction of castor oil esters by the addition of diesel, safflower oil esters and additives. *Industrial Crops and Products*, *36*(1), 267–270.

Toombs, E. L. (2005). Cosmetics in the treatment of acne vulgaris. *Dermatologic Clinics*, *23*(3), 575–581.

Topkafa, M. (2016). Evaluation of chemical properties of cold pressed onion, okra, rosehip, safflower and carrot seed oils: Triglyceride, fatty acid and tocol compositions. *Analytical Methods*, *8*, 4220.

Topkafa, M., Kara, H., & Sherazi, S. T. H. (2015). Evaluation of the triglyceride composition of pomegranate seed oil by RP-HPLC followed by GC-MS. *Journal of the American Oil Chemists' Society*, *92*, 791–800.

Velasco, L., & Fernandez-Martinez, J. M. (2001). Breeding for oil quality in safflower. In: J. W. Bergman, & H. H. Mündel (Eds.), *7th International safflower conference, Williston, North Dakota and Sidney, Montana*, pp. 133–137.

Vosoughkia, M., Ghavamib, M., Gharachorloo, M., Sharrifmoghaddasi, M., & Omidi, A. H. (2011). Lipid composition and oxidative stability of oils in safflower (*Carthamus tinctorius* L.) seed varieties grown in Iran. *Advances in Environmental Biology*, 897e903.

Weißhaar, R. (2008). Determination of total 3-chloropropane-1,2-diol (3-MCPD) in edible oils by cleavage of MCPD esters with sodium Methoxide. *European Journal of Lipid Science and Technology*, *110*, 183–186.

Yeh, S., Chao, C., Lin, M., & Chen, W. (2000). Effects of parenteral infusion with medium-chain triglycerides and safflower oil emulsions on hepatic lipids, plasma amino acids and inflammatory mediators in septic rats. *Journal of Clinical Nutrition*, *19*(2) 115e120.

Yoshida, Y., & Niki, E. (2003). Antioxidant effects of phytosterol and its components. *Journal of Nutritional Science and Vitaminology*, *49*, 277.

Yu, S. Y., Lee, Y. J., Kim, J. D., et al. (2013). Phenolic composition, antioxidant activity and anti-adipogenic effect of hot water extract from safflower (*Carthamus tinctorius* L.) seed. *Nutrients*, *5*, 4894–4907.

Zhang, Z., Li, Q., Liu, F., Sun, Y., & Zhang, J. (2010). Prevention of diet-induced obesity by safflower oil: Insights at the levels of PPARa, orexin, and ghrelin gene expression of adipocytes in mice. *Acta Biochimica et Biophysica Sinica*, *42*(3), 202–208. https://doi.org/10.1093/abbs/gmq010.

Zhou, X., Tang, L., Xu, Y., Zhou, G., & Wang, Z. (2014). Towards a better understanding of medicinal uses of *Carthamus tinctorius* L. in traditional Chinese medicine: A phytochemical and pharmacological review. *Journal of Ethnopharmacology*, *151*(1), 27–43.

Chapter 30

Cold pressed carrot (*Daucus carota* subsp. *sativus*) seed oil

Mehmet Aksu[a], Yasemin Incegul[b], Sündüz Sezer Kiralan[c], Mustafa Kiralan[c], and Gulcan Ozkan[b]

[a]*Agriculture and Rural Development Support Institution, Isparta Provincial Coordination Unit, Isparta, Turkey,* [b]*Department of Food Engineering, Faculty of Engineering, Suleyman Demirel University, Isparta, Turkey,* [c]*Department of Food Engineering, Faculty of Engineering, Balıkesir University, Balikesir, Turkey*

Abbreviations

ABTS 2,2′-azino-bis-3-ethylbenzthiazoline-6-sulphonic acid
DIC instant controlled pressure drop
DNA deoxyribonucleic acid
DPPH 2,2-diphenyl-1-picrylhydrazyl
LLL trilinolein
MUFA monounsaturated fatty acid
OLL oleodilinolein
OOL dioleolinolein
OOO triolein
ORAC oxygen radical absorbance capacity
OSI oxidative stability index
PLL palmitodilinolein
POL palmitooleolinolein
PUFA polyunsaturated fatty acid
SLL stearodilinolein
SOL stearooleolinolein
SPF sun protection factor
TEAC trolox equivalent antioxidant capacity

1 Introduction

Daucus carota L., commonly known as carrot, is a member of the *Apiaceae* family (Kataria, Chahal, Kaur, & Kaur, 2016). The carrot is a root vegetable, which has not only orange but also black, red, purple, white, and yellow colors (Iorizzo et al., 2013). It can be consumed as fresh and dried forms or evaluated as a raw material of the fruit juice and concentrate industry. The world's most significant carrot-growing areas are in South and Southeast Asia, Middle Asia, and European countries, in addition to the USA. The production quantities and harvested areas of carrots are listed in Table 1. China is the world's primary carrot-growing country in terms of both quantity and area.

The carrot (*Daucus carota* L.) species contains both cultivated (*Daucus carota* L. ssp. *sativus*) and wild (*Daucus carota* L. ssp. *carota*) subspecies (Iorizzo et al., 2013). The cultivated carrot has more economic importance. Cultivated ones become popular in recent decades because of increasing interest for their nutritive value. The chemical composition of carrot tubers has been reported in different references and their main chemical constituents are moisture (86%–89%), carbohydrates (6.0%–10.6%), protein (0.7%–1.1%), lipids (0.2%–0.5%), and crude fiber (1.2%–2.8%) (Nguyen & Nguyen, 2015). Minor chemical components found in carrots such as carotenoids and phenolics are important for nutritional and functional properties (Sharma, Karki, Thakur, & Attri, 2012). Orange-colored carrots are often described as good for the eyes thanks to their considerable carotenoid content. Predominating carotenoid fractions in carrots are α- and β-carotene. Lycopene and lutein are other important carotenoids (Arscott & Tanumihardjo, 2010). Yellow-colored carrots are rich in

Cold Pressed Oils. https://doi.org/10.1016/B978-0-12-818188-1.00030-X
Copyright © 2020 Elsevier Inc. All rights reserved.

TABLE 1 Production quantity and harvested area of carrot in 2017.[1]

Country	Production quantity (tons)	Area harvested (ha)
China	20,374,421	405,620
Uzbekistan	2,249,733	30,978
Russian Federation	1,805,787	66,309
United States of America	1,540,280	31,850
United Kingdom	957,036	14,629
Ukraine	839,010	42,700
Poland	827,138	22,118

[1] *Food and Agriculture Organization (2017).*

lutein, which has a protective effect against macular degeneration. The red carrot color comes from its high quantity of lycopene (da Silva Dias, 2014).

Carrots can be consumed as fresh or processed (i.e., juice), dried, or fried products (Arscott & Tanumihardjo, 2010). As a new and alternative consumption type, the carrot seed oil is produced by the cold pressing method (Yetim, Sagdic, & Ozturk, 2008). Carrot seeds are formed on umbels after flowering and anthesis. Maturation of seeds takes approximately 30–60 days after anthesis. Moisture content and seed color are generally accepted maturation indicators. Mature carrot seeds have about 10% moisture content and a brown color (de Miranda, De Augusto, Picoli, & Nascimento, 2017). Reserve oil (triglycerides) is located in the seeds while essential oil is generally located in the mericarp of the fruit (Glišić et al., 2007). The approximate chemical composition of mature seeds is 17% lipids, 14% protein, and 11% sugar (Panayotov, 2010). For lipid extraction of carrot seeds, cold pressing techniques are suitable and valuable due to transferring of nutritional and bioactive compounds from seeds and ensuring high-quality products (Parker, Adams, Zhou, Harris, & Yu, 2003).

Carrot seed oil consists of unsaturated fatty acids, tocopherols, phenolics, sterols, and aroma compounds (Topkafa, 2016; Yu, Zhou, & Parry, 2005). Fatty acids of oils from cultivated and wild carrots seeds differ from each other. Seed oils of sativus cultivars contain oleic acid (*cis*-9-octadecenoic acid) (Bialek, Bialek, Jelinska, & Tokarz, 2016), and the main fatty acid of wild cultivar's seed oil is petroselinic acid (*cis*-6-octadecenoic acid) (Dutta & Appelqvist, 1989). For cultivated carrot (*Daucus carota* subsp. *sativus*) seed oil, oleic acid is followed by linoleic acid and their total levels comprise more than 90% of total fatty acids (Gao, Yang, & Birch, 2016; Parker et al., 2003; Topkafa, 2016). Apart from fatty acids, triacylglycerides and phenolic acids are mentioned as important phytochemicals (Gao et al., 2016; Gao & Birch, 2015; Yu et al., 2005).

High unsaturated fatty acids content of cold pressed carrot seed oil shows similarity to other common edible oils, and this similarity may explain its popularity for culinary consumption. Antimicrobial (Jasicka-Misiak et al., 2004), antioxidative (Yu et al., 2005), and health-promoting (Shakheel, Saliyan, Satish, & Hedge, 2017) effects of carrot seed oil and seed extracts have been reported. Carrot seed oil is used in cosmetic applications to enhance the protective effect of cosmetic emulsions against sunlight (Singh, Lohani, Mishra, & Verma, 2018; Suryawanshi, 2016). Additionally, essential oil of carrot seed is discussed from the point of antimicrobial (Gupta, Rath, Dash, & Mishra, 2004; Imamu et al., 2009; Kaur, Chahal, Kataria, & Kumar, 2018), antioxidant (Alves-Silva et al., 2016) cytotoxic (Sieniawska et al., 2016), and health-protective (da Silva Dias, 2014) effects.

2 Cold press extraction and processing of oil

Oils in nuts and seeds could be extracted in many ways. Conventional extraction methods are high capacity systems and generally use heat and chemicals in order to simplify and accelerate the process. The solvent extraction method is the most common conventional extraction system (Cravotto, Binello, & Orio, 2011). Conventional extraction systems are commonly applicable and feasible for industrial-scale production. However, they have some disadvantages such as negative environmental effects, excessive refining process, and high energy consumption (Savoire, Lanoisellé, & Vorobiev, 2013).

Green extraction techniques are alternative extraction methods, which provide acquisition of low-cost, more safe and pure products by means of consuming less energy and chemical substances. In addition, they need less time than commercial extraction systems and are environmentally friendly. Some of these methods are instant controlled pressure drop (DIC), subcritical water extraction, supercritical fluid extraction, ultrasound and/or microwave-assisted extraction, pulsed electric field assisted extraction, pressurized liquid extraction, enzymatic hydrolysis, and mechanic expression (Chemat, Vian, & Cravotto, 2012; Cravotto et al., 2011; Soquetta, Terra, & Bastos, 2018).

Mechanical pressing (cold pressing) is an old method with several advantages such as low cost, effective transferring of nutritional and bioactive compounds from seeds, and ensuring the high quality of products (Cravotto et al., 2011). "Cold pressed oil" is a term used for plant seed oils that are extracted by screw or hydraulic press. The leading aim of this method is to maintain bioactive constituents such as essential fatty acids, phenolics, flavonoids, and tocols (Teh & Birch, 2013). It should be taken into consideration that the capacity and productivity of the mechanical pressing technique are lower than those of commercial systems (Uluata, 2016).

As a result of consumer demand, it is possible to obtain natural and safe food products by means of the cold extraction method without heat effect (Yetim et al., 2008). Due to lack of chemical use, the cold pressing method allows seed oils to be pure and have more natural antioxidants than those obtained via other extraction methods (Yu et al., 2005). In addition, these techniques do not affect the environment and natural resources (Chemat et al., 2012).

It is significant that cold pressed oils may have a longer shelf life due to lack of refining processes, wherein natural antioxidants are separated from oil (Prescha, Grajzer, Dedyk, & Grajeta, 2014). Natural antioxidants also show health benefits in preventing diseases and contributing to human health (Parker et al., 2003). Carotenoids and tocols retained in the cold pressed oils may be responsible for their ultimate oxidative stability (Yetim et al., 2008). However, it should be noted that cold pressed oils are rich in polyunsaturated fatty acids (PUFAs) and therefore tend to experience lipid oxidation (Teh & Birch, 2013).

Carrot seed oil could be extracted by cold pressing or solvent extraction (Yildirim & Kostem, 2014). Supercritical carbon dioxide extraction has been also reported (Gao et al., 2016). Carrot seed oil is generally obtained from discarded seeds and could be evaluated in the food and cosmetics industries (Kula, Izydorczyk, Czajkowska, & Bonikowski, 2006). Studies on cold pressed carrot seed oil are very limited. Available articles on carrot seed oil have generally been conducted with final products, which are sold in local markets.

3 Fatty acids and acyl lipids (neutral lipids, glycolipids, and phospholipids) profile of cold pressed carrot seed oil

In a similar manner to many plant-sourced oils, the main fatty acid composition of carrot seed oil is composed of unsaturated fatty acids. Detailed fatty acids distribution of carrot seed oil is given in Table 2. Carrot seed oil comprises mainly monounsaturated fatty acids (MUFAs), followed by PUFAs. Total unsaturated fatty acids constitute about 90%–95% of total fatty acids. Oleic acid was the major fatty acid, accounting for nearly 80% of total fatty acids. In terms of PUFAs, linoleic acid had the highest level (13%). The saturated fatty acids of carrot seed oil are palmitic and stearic acids (Bialek et al., 2016; Gao & Birch, 2015; Parker et al., 2003; Topkafa, 2016; Yetim et al., 2008). Carrot seed oil has a remarkable amount of oleic acid (18:1n-9), while the total saturated fatty acid content is very low. Carrot seed oil may be a possible nutritional source of MUFAs (Parker et al., 2003). Although petroselinic acid has been mentioned as another important

TABLE 2 Main fatty acids (%) of carrot seed oil.

Palmitic (C16:0)	Stearic (C18:0)	Oleic (C18:1n9)	Linoleic (C18:2n6)	Linolenic (C18:3n3)	Reference
3.71	0.42	82.08	13.19	0.28	Parker et al. (2003)
10.21	0.40	75.61	13.29	0.20	Yetim et al. (2008)
4.20	3.1	79.20	7.70	0.10	Bialek et al. (2016)
3.75	0.87	81.45	13.32	0.48	Gao and Birch (2015)
3.70	0.90	80.70	12.20	0.20	Topkafa (2016)

TABLE 3 Triglyceride profile of cold pressed carrot seed oil.[1]

Triglyceride type[2]	%
OOL	29.5
POL + SLL	20.4
OOO	19.4
SOL	11.8
PLL	9.7
OLL	5.3
LLL	1.9

[1]*Topkafa (2016).*
[2]*Triglyceride types:* O, *oleic;* P, *palmitic;* S, *stearic;* L, *linoleic acid.*

unsaturated fatty acid in a few works (Dutta & Appelqvist, 1989; Ozcan & Chalchat, 2007), it was only found in wild cultivars (Dutta & Appelqvist, 1989).

Topkafa (2016) investigated triglyceride profile of some cold pressed oils including carrot seed oil. He detected seven different triglyceride forms (Table 3). The main triglycerides were OOL, OOO, and POL + SLL mixture. They comprised nearly 70% of the total triglycerides. The remaining triglycerides were SOL, PLL, OLL, and LLL, according to the quantities, respectively.

4 Minor bioactive lipids (sterols, tocols, carotenoids, phenolic compounds, hydrocarbons, flavor and aroma compounds) in cold pressed carrot seed oil

Table 4 summarizes the tocopherol and tocotrienol compositions of cold pressed carrot seed oil. The total tocopherol and tocotrienol contents were 19.7 and 238.4 mg/kg, respectively. Alpha (α) and gamma (γ) isomers have been determined as the main compounds for both tocopherols and tocotrienols (Topkafa, 2016).

5 Composition of cold pressed carrot seed oil in comparison to other cold pressed oils

Carrot seed oil is mainly composed of unsaturated fatty acids (90% or more). The most dominant unsaturated fatty acid is oleic acid (80%) and the next is linoleic acid. A detailed comparison of different cold pressed seed oils in terms of fatty acid distribution is given in Table 5. It can be concluded that carrot seed oil is an excellent source for oleic acid when compared to other cold pressed seed oils. Apricot seed oil showed the closest similarity to carrot seed oil in terms of fatty acid ratios.

TABLE 4 Tocopherols and tocotrienols of cold pressed carrot seed oil.[1]

Isomer	Tocopherol (mg/kg)	Tocotrienol (mg/kg)
Alpha (α)	5.6	86.4
Beta (β)	4.8	ND
Gamma (γ)	8.2	143.2
Delta (δ)	1.0	8.8
Total	19.7	238.4

ND, not detected.
[1]*Topkafa (2016).*

The triglyceride contents of some cold pressed seed oils are given in Table 6. Cold pressed oils comprise about 95% triglycerides (Makala, 2015). Since the main fatty acid of carrot seed oil is oleic acid, the triglyceride profile is mainly composed of oleic acid-containing triglycerides. About half of the total triglycerides contain two or more oleic acids (OOL + OOO). When triglycerides of some cold pressed seed oils were compared, the triglyceride compositions of the oils showed parallelism with their fatty acid contents.

TABLE 5 Comparison of the main fatty acids in some cold pressed seed oils.

Seed oil	Fatty acid (%)				
	Palmitic (C16:0)	Stearic (C18:0)	Oleic (C18:1n9)	Linoleic (C18:2n6)	Linolenic (C18:3n3)
Carrot[1,2]	3.70–3.71	0.42–0.90	80.70–82.08	12.20–13.19	0.20–0.28
Safflower[2,4]	6.2–7.1	2.4–2.88	15.2–20.6	67.3–73.7	0.1–0.6
Sunflower[3]	6.29	3.85	31.86	56.30	0.27
Flax[3,4]	5.20–5.59	4.20–4.35	19.30–20.72	13.37–14.00	51.20–55.14
Sesame[3,4]	9.28–10.40	5.30–5.64	39.50–40.57	42.72–42.90	0.31–0.50
Rapeseed[3,6]	4.84	1.71	59.16–63.40	20.19–21.70	9.29–9.60
Hemp[1,3,4]	6.10–6.55	2.60–3.09	11.72–13.20	55.20–59.96	17.06–19.33
Apricot[3,5]	4.25–6.10	1.20–1.36	70.40–70.90	20.93–21.70	0.74

[1]*Parker et al. (2003).*
[2]*Topkafa (2016).*
[3]*Veličkovska, Brühl, Mitrev, Mirhosseini, and Matthäus (2015).*
[4]*Prescha et al. (2014).*
[5]*Uluata (2016).*
[6]*Makala (2015).*

TABLE 6 Comparison of triglyceride profile (%) of some cold pressed seed oils.[1]

Triglyceride type[2]	Carrot	Onion	Okra	Rosehip	Safflower
LnLnLn	ND[3]	ND	ND	5.9	ND
LLnLn	ND	19.9	0.4	18.9	0.5
LLLn	ND	6.1	0.5	27.3	0.3
LLL	1.9	24.6	17.3	15.0	57.3
OLLn	ND	ND	ND	8.2	ND
PLLn	ND	ND	0.4	2.7	ND
OLL	5.3	28.3	22.6	9.3	20.3
PLL	9.7	6.4	27.4	5.5	9.5
OOL	29.5	10.4	4.3	3.0	5.2
POL + SLL	20.4	4.3	13.6	2.2	4.3
PPL	ND	ND	9.3	ND	0.2
OOO	19.4	ND	2.2	0.9	1.1
SOL	11.8	ND	2.2	0.7	0.8

[1]*Topkafa (2016).*
[2]*Triglyceride types:* Ln, *linolenic;* L, *linoleic;* O, *oleic;* P, *palmitic;* S, *stearic acid.*
[3]*ND, not detected.*

Phospholipids, tocopherols, tocotrienols, sterols, hydrocarbons, triterpene alcohols, and carotenoids are among the nonglycerol fraction of cold pressed oils (Makala, 2015). These bioactive compounds are more abundant in cold pressed oils due to a lack of refining process. Table 7 shows a comparative summary of tocopherol and tocotrienol contents of some cold pressed oils. Carrot seed oil has a remarkable amount of tocols. Its total tocol amount is close to that of sunflower oil.

Oxidative stability and antioxidant properties of different kinds of cold pressed oils have been reported, including oxidative stability index, oxygen radical absorbing capacity (ORAC), radical cation $ABTS^+$ scavenging activity, radical scavenging activity by DPPH· and TEAC assay, chelating capacity, and total phenolic content (Gao & Birch, 2015; Parker et al., 2003; Prescha et al., 2014; Uluata, 2016; Veličkovska et al., 2015; Yu et al., 2005). Table 8 presents the results of some antioxidant for selected cold pressed oils. The antioxidant properties of cold pressed carrot seed oil were investigated by comparisons with other cold pressed seed oils (Yu et al., 2005). In order to evaluate the antioxidant properties of different cold pressed seed oils (black caraway, carrot, cranberry, hemp seed), radical scavenging activity values ($ABTS^+$ and DPPH

TABLE 7 Tocopherol and tocotrienol content (mg/kg) of some cold pressed seed oils.

Type	Isomer	Carrot[1]	Safflower[1]	Sunflower[2]	Flaxseed[2]	Sesame[2]	Rapeseed[2]	Apricot[3]
Tocopherol	Alpha (α)	5.6	376.6	207	2	–	74	39.6
	Beta (β)	4.8	4.4	21	–	–	–	9.2
	Gamma (γ)	8.2	25.9	51	404	516	338	498.5
	Delta (δ)	1.0	1.0	2	4	5	7	15.1
	Total	19.7	407.9	281	410	521	419	562.4
Tocotrienol	Alpha (α)	86.4	ND	–	–	–	–	–
	Beta (β)	ND	ND	–	–	–	–	–
	Gamma (γ)	143.2	9.7	–	–	–	–	–
	Delta (δ)	8.8	ND	–	–	–	–	–
	Total	238.4	9.7	–	–	–	–	–
Total tocols (tocopherols + tocotrienols)		258.0	417.5	281	410	521	419	562.4

ND, not detected.
[1]*Topkafa (2016).*
[2]*Veličkovska et al. (2015).*
[3]*Uluata (2016)*

TABLE 8 Antioxidant properties of cold pressed carrot oil against other oils (Yu et al., 2005).

Oil	ORAC[1]	ABTS[2]	DPPH[3]	Chelating[4]	TPC[5]
Carrot seed	160	8.90	10.9	25.5	1.98
Caraway	220	30.8	2.7	12.6	3.53
Cranberry	NA	22.5	11.3	NA	1.61
Hemp	28.2	11.4	10.5	10.5	0.44

NA, data not available.
[1]*μmol TE/g.*
[2]*μmol TE/g.*
[3]*mg oil equivalent/mL.*
[4]*EDTAE mg/g.*
[5]*mg GE/g.*

TABLE 9 Total phenolic content of some cold pressed oils.

Cold pressed oil	Total phenolic content[1]	Reference
Carrot	391.4	Gao and Birch (2015)
Apricot	24.9	Uluata (2016)
Sunflower	60.8	Veličkovska et al. (2015)
Flax	72.5	Veličkovska et al. (2015)
Sesame	133.4	Veličkovska et al., 2015
Hemp	64.2	Veličkovska et al. (2015)
Black cumin	295.8	Veličkovska et al. (2015)
Rapeseed	74.1	Veličkovska et al. (2015)

[1]*mg gallic acid equivalent/kg oil.*

•), oxygen radical absorbing capacity (ORAC), chelating activity, and total phenolic contents were investigated. According to Yu et al., cold pressed carrot seed oil has become prominent for preventing lipids, proteins, and DNA from oxidative damage by means of natural antioxidant contents, and the oil is suggested for potential novel food applications.

Total phenolic content of vegetable oils is another indicator in terms of important bioactive constituents, and they have been mentioned as hydrophilic antioxidants (Prescha et al., 2014). The total phenolic contents of some cold pressed oils are given in Table 9. Compared to other oils, carrot seed oil has a considerable amount of phenolic substances.

6 Edible and nonedible applications of cold pressed carrot seed oil

Vegetable oils are widely used in cosmetic formulations due to their antioxidant content. The degree of fatty acid unsaturation directly affects properties of oils, wherein most seed oils contain high amounts of unsaturated fatty acids. Vegetable oils rich in unsaturated fatty acids are both more homogeneous and better absorbed in the skin compared to those rich in saturated fatty acids (Yildirim & Kostem, 2014). Adding vegetable and seed oils into cosmetic formulae may contribute to better antioxidative and sun-protecting effects of these products. Sun protection factor (SPF) is a specific criterion for determining the efficiency of dermatological products. Carrot seed oil has been found to be effective in SPF for skin (Suryawanshi, 2016). Cosmetic emulsions prepared from carrot seed oil were investigated and it was concluded that formulations including carrot seed oil have a renewing effect in cases of skin damage (Singh et al., 2018).

A medical soap that includes carrot seed oil was found to have a therapeutic effect on fungal infections (Abdulrasheed, Aroke, & Sani, 2015). Ürüşan, Erhan, and Bölükbaşı (2018) investigated the effect of cold pressed carrot seed oil on the performance, carcass properties, and shelf life of broiler chickens. The authors found that broilers fed with diets containing carrot seed oil gained more weight than control broilers. While intake of carrot seed oil increased hot carcass weight and hot carcass yield of broilers, feed intake and feed conversation ratio did not affected to gain weight.

7 Health-promoting traits of cold pressed carrot seed oil and oil constituents

To our knowledge, there is no study on the health effects of cold pressed carrot seed oil hot carcass yield on broilers. However, aqueous, ethanol, and methanol extracts of carrot seeds have been mentioned as having antiinflammatory (Vasudevan, Gunnam, & Parle, 2006), antisteroidogenic (Majumder, Dasgupta, Mukhopadhaya, Mazumdar, & Gupta, 1997), and hepatoprotective (Singh, Singh, Chandy, & Manigauha, 2012) influence. In addition, essential oil from carrot seeds has been investigated in terms of antifungal (Jasicka-Misiak et al., 2004) and antibacterial (Gupta et al., 2004; Imamu et al., 2009) properties and found to be effective.

8 Oxidative stability of cold pressed carrot seed oil

It has been mentioned by different researchers that some important factors affecting oxidative stability of oils include the extraction method (Uluata, 2016), fatty acid composition (Kiralan, Özkan, Bayrak, & Ramadan, 2014; Parker et al., 2003;

Ramadan & Asker, 2009), and the ratio of hydrophilic and lipophilic fractions of oils (Prescha et al., 2014; Ramadan, 2008; Ramadan, Amer, & Sulieman, 2006). In a comparative study, the oxidative stability index (OSI) value of carrot seed oil was found to be about 48 h. This value is close to that of soybean oil and higher than that of hemp seed oil. The shelf life of carrot seed oil may be evaluated as similar to tested commercial soybean oil and this result could indicate potential culinary applications of carrot seed oil (Parker et al., 2003).

The oxidative stability and thermal response of carrot seed oil extracted by supercritical fluid extraction have been compared with those of cold pressed flax, hemp, and canola seed oils. It has been concluded that the positional distribution of fatty acids influenced the oxidative stability of oils, wherein carrot seed oil showed higher thermal stability than other tested oils (Gao & Birch, 2015).

References

Abdulrasheed, A., Aroke, U. O., & Sani, I. M. (2015). Parametric studies of carrot seed oil extract for the production of medicated soap. *International Journal of Recent Development in Engineering and Technology*, *4*(1), 1–5.

Alves-Silva, J. M., Zuzarte, M., Gonçalves, M. J., Cavaleiro, C., Cruz, M. T., Cardoso, S. M., et al. (2016). New claims for wild carrot (*Daucus carota* subsp. *carota*) essential oil. *Evidence-Based Complementary and Alternative Medicine*, *2016*, 1–10.

Arscott, S. A., & Tanumihardjo, S. A. (2010). Carrots of many colors provide basic nutrition and bioavailable phytochemicals acting as a functional food. *Comprehensive Reviews in Food Science and Food Safety*, *9*, 223–239.

Bialek, A., Bialek, M., Jelinska, M., & Tokarz, A. (2016). Fatty acid profile of new promising unconventional plant oils for cosmetic use. *International Journal of Cosmetic Science*, *38*(4), 382–388.

Chemat, F., Vian, M. A., & Cravotto, G. (2012). Green extraction of natural products: Concept and principles. *International Journal of Molecular Sciences*, *13*(7), 8615–8627.

Cravotto, G., Binello, A., & Orio, L. (2011). Green extraction techniques: For high-quality natural products. *Agro Food Industry Hi Tech*, *22*(6), 57–59.

da Silva Dias, J. C. (2014). Nutritional and health benefits of carrots and their seed extracts. *Food and Nutrition Sciences*, *5*, 2147–2156.

de Miranda, R. M., De Augusto, E., Picoli, D. T., & Nascimento, W. M. (2017). Physiological quality, anatomy and histochemistry during the development of carrot seeds (*Daucus carota* L.). *Ciência e Agrotecnologia*, *41*(2), 169–180.

Dutta, P. C., & Appelqvist, L. (1989). The effects of different cultural conditions on the accumulation of depot lipids notably petroselinic acid during somatic embryogenesis in *Daucus carota* L. *Plant Science*, *64*, 167–177.

Food and Agriculture Organization (2017). *Food and agriculture data*. Available from: http://www.fao.org/faostat/en/?#data/QC (Accessed 1 March 2019).

Gao, F., & Birch, J. (2015). Oxidative stability, thermal decomposition, and oxidation onset prediction of carrot, flax, hemp, and canola seed oils in relation to oil composition and positional distribution of fatty acids. *European Journal of Lipid Science and Technology*, *117*(7), 1042–1052.

Gao, F., Yang, S., & Birch, J. (2016). Physicochemical characteristics, fatty acid positional distribution and triglyceride composition in oil extracted from carrot seeds using supercritical CO_2. *Journal of Food Composition and Analysis*, *45*, 26–33.

Glišić, S. B., Mišić, D. R., Stamenić, M. D., Zizovic, I. T., Ašanin, R. M., & Skala, D. U. (2007). Supercritical carbon dioxide extraction of carrot fruit essential oil: Chemical composition and antimicrobial activity. *Food Chemistry*, *105*(1), 346–352.

Gupta, R., Rath, C. C., Dash, S. K., & Mishra, R. K. (2004). In-vitro antibacterial potential assessment of carrot (*Daucus carota*) and celery (*Apium graveolens*) seed essential oils against twenty one bacteria. *Journal of Essential Oil-Bearing Plants*, *7*(1), 79–86.

Imamu, X., Yili, A., Aisa, H. A., Maksimov, V. V., Veshkurova, O. N., & Salikhov, S. I. (2009). Chemical composition and antimicrobial activity of essential oil from seeds of *Anethum graveolens* growing in Uzbekistan. *Chemistry of Natural Compounds*, *45*(2), 280–281.

Iorizzo, M., Senalik, D. A., Ellison, S. L., Grzebelus, D., Cavagnaro, P. F., Allender, C., et al. (2013). Genetic structure and domestication of carrot (*Daucus carota* subsp. *sativus*) (Apiaceae). *American Journal of Botany*, *100*(5), 930–938.

Jasicka-Misiak, I., Lipok, J., Nowakowska, E. M., Wieczorek, P. P., Młynarz, P., & Kafarski, P. (2004). Antifungal activity of the carrot seed oil and its major sesquiterpene compounds. *Zeitschrift Fur Naturforschung—Section C Journal of Biosciences*, *59*(11–12), 791–796.

Kataria, D., Chahal, K. K., Kaur, P., & Kaur, R. (2016). Carrot plant—A potential source of high value compounds and biological activities: A review. *Proceedings of the Indian National Science Academy*, *82*(4), 1237–1248.

Kaur, A., Chahal, K. K., Kataria, D., & Kumar, A. (2018). Assessment of carrot seed essential oil and its chemical constituents against *Meloidogyne incognita*. *Journal of Pharmacognosy and Phytochemistry*, *7*(1), 896–903.

Kiralan, M., Özkan, G., Bayrak, A., & Ramadan, M. F. (2014). Physicochemical properties and stability of black cumin (*Nigella sativa*) seed oil as affected by different extraction methods. *Industrial Crops and Products*, *57*, 52–58.

Kula, J., Izydorczyk, K., Czajkowska, A., & Bonikowski, R. (2006). Chemical composition of carrot umbel oils from *Daucus carota* L. ssp. *sativus* cultivated in Poland. *Flavour and Fragrance Journal*, *21*(4), 667–669.

Majumder, P. K., Dasgupta, S., Mukhopadhaya, R. K., Mazumdar, U. K., & Gupta, M. (1997). Anti-steroidogenic activity of the petroleum ether extract and fraction 5 (fatty acids) of carrot (*Daucus carota* L.) seeds in mouse ovary. *Journal of Ethnopharmacology*, *57*(3), 209–212.

Makala, H. (2015). Cold pressed oils as functional food. In G. Budryn, & D. Zyzelewicz (Eds.), *Plant Lipids Science, Technology, Nutritional Value and Benefits to Human Health* (pp. 185–200). Kerala: Research Signpost.

Nguyen, H. H. V., & Nguyen, L. T. (2015). Carrot processing. In Y. H. Hui, & E. Ö. Evranuz (Eds.), *Handbook of vegetable preservation and processing* (pp. 449–466). Florida: CRC Press.

Ozcan, M. M., & Chalchat, J. C. (2007). Chemical composition of carrot seeds (*Daucus carota* L.) cultivated in Turkey: Characterization of the seed oil and essential oil. *Grasas y Aceites*, *58*(4), 359–365.

Panayotov, N. (2010). Heterogeneity of carrot seeds depending on their position on the mother plant. *Folia Horticulturae*, *22*(1), 25–30.

Parker, T. D., Adams, D. A., Zhou, K., Harris, M., & Yu, L. (2003). Fatty acid composition and oxidative stability of cold-pressed edible seed oils. *Food Chemistry and Toxicology*, *68*(4), 1240–1243.

Prescha, A., Grajzer, M., Dedyk, M., & Grajeta, H. (2014). The antioxidant activity and oxidative stability of cold-pressed oils. *Journal of American Oil Chemists' Society*, *91*, 1291–1301.

Ramadan, M. F. (2008). Quercetin increases antioxidant activity of soy lecithin in a triolein model system. *LWT—Food Science and Technology*, *41*, 581–587.

Ramadan, M. F., Amer, M. M. A., & Sulieman, A. M. (2006). Correlation between physicochemical analysis and radical scavenging activity of vegetable oil blends as affected by frying of French fries. *European Journal of Lipid Science and Technology*, *108*, 670–678.

Ramadan, M. F., & Asker, M. M. S. (2009). Antimicrobical and antivirial impact of novel quercetin-enriched lecithin. *Journal of Food Biochemistry*, *33*, 557–571.

Savoire, R., Lanoisellé, J. L., & Vorobiev, E. (2013). Mechanical continuous oil expression from oilseeds: A review. *Food and Bioprocess Technology*, *6*(1), 1–16.

Shakheel, M., Saliyan, T., Satish, S., & Hedge, K. (2017). Therapeutic uses of *Daucus carota*: A review. *International Journal of Pharma and Chemical Research*, *3*(2), 138–143.

Sharma, K. D., Karki, S., Thakur, N. S., & Attri, S. (2012). Chemical composition, functional properties and processing of carrot—A review. *Journal of Food Science & Technology*, *49*(1), 22–32.

Sieniawska, E., Świątek, Ł., Rajtar, B., Kozioł, E., Polz-Dacewicz, M., & Skalicka-Woźniak, K. (2016). Carrot seed essential oil-source of carotol and cytotoxicity study. *Industrial Crops and Products*, *92*, 109–115.

Singh, S., Lohani, A., Mishra, A. K., & Verma, A. (2018). Formulation and evaluation of carrot seed oil-based cosmetic emulsions. *Journal of Cosmetic and Laser Therapy*, *21*(2), 99–107.

Singh, K., Singh, N., Chandy, A., & Manigauha, A. (2012). In vivo antioxidant and hepatoprotective activity of methanolic extracts of *Daucus carota* seeds in experimental animals. *Asian Pacific Journal of Tropical Biomedicine*, *2*(5), 385–388.

Soquetta, M. B., Terra, L. M., & Bastos, C. P. (2018). Green technologies for the extraction of bioactive compounds in fruits and vegetables. *CYTA—Journal of Food*, *16*(1), 400–412.

Suryawanshi, J. A. S. (2016). In-vitro determination of sun protection factor and evaluation of herbal oils. *International Journal of Pharmacology Research*, *6*(1), 37–43.

Teh, S. S., & Birch, J. (2013). Physicochemical and quality characteristics of cold-pressed hemp, flax and canola seed oils. *Journal of Food Composition and Analysis*, *30*(1), 26–31.

Topkafa, M. (2016). Evaluation of chemical properties of cold pressed onion, okra, rosehip, safflower and carrot seed oils: Triglyceride, fatty acid and tocol compositions. *Analytical Methods*, *8*(21), 4220–4225.

Uluata, S. (2016). Effect of extraction method on biochemical properties and oxidative stability of apricot seed oil. *Akademik Gida*, *14*(4), 333–340.

Ürüşan, H., Erhan, M. K., & Bölükbaşı, S. C. (2018). Effect of cold-press carrot seed oil on the performance, carcass characteristics, and shelf life of broiler chickens. *Journal of Animal & Plant Sciences*, *28*(6), 1662–1668.

Vasudevan, M., Gunnam, K. K., & Parle, M. (2006). Antinociceptive and anti-ınflammatory properties of *Daucus carota* seeds extract. *Journal of Health Science*, *52*(5), 598–606.

Veličkovska, S. K., Brühl, L., Mitrev, S., Mirhosseini, H., & Matthäus, B. (2015). Quality evaluation of cold-pressed edible oils from Macedonia. *European Journal of Lipid Science and Technology*, *117*, 1–13.

Yetim, H., Sagdic, O., & Ozturk, I. (2008). Fatty acid compositions of cold press oils of seven edible plant seeds grown in Turkey. *Chemistry of Natural Compounds*, *44*(5), 634–636.

Yildirim, K., & Kostem, A. M. (2014). A technical glance on some cosmetic oils. *European Scientific Journal*, *2*, 425–435.

Yu, L. L., Zhou, K. K., & Parry, J. (2005). Antioxidant properties of cold-pressed black caraway, carrot, cranberry, and hemp seed oils. *Food Chemistry*, *91*(4), 723–729.

Chapter 31

Cold pressed coriander (*Coriandrum sativum* L.) seed oil

Rizwan Ashraf[a], Saba Ghufran[a], Sumia Akram[b], Muhammad Mushtaq[c], and Bushra Sultana[a]

[a]*Department of Chemistry, University of Agriculture, Faisalabad, Pakistan,* [b]*Division of Science and Technology, University of Education Lahore, Lahore, Pakistan,* [c]*Department of Chemistry, GC University, Lahore, Pakistan*

Abbreviations

MUFA monounsaturated fatty acid
PUFA polyunsaturated fatty acid
QS quorum sensing
SFA saturated fatty acid
SFE supercritical fluid extraction
SWE subcritical water extraction
UAE ultrasound-assisted extraction

1 Introduction

Coriandrum sativum L.; an annual herbaceous plant of the *Apiaceae* family, is traditionally known by various names like parsley in China, cilantro in America, and dhania in Pakistan (Mhemdi, Rodier, Kechaou, & Fages, 2011). The available history lacks authentic data regarding coriander origin; however, this herb is considered to be well-adapted in fertile areas of tropical regions, so the Mediterranean, Middle East (Iran), Southern Europe, and Western Australia are all suitable for coriander growth and cultivation (Seidemann, 2005).

The use of coriander seeds can be traced back to 1000 BC, but it has been supposed that they were cultivated for the first time over 2000 BC in the Hanging Gardens of Babylon. The Egyptians carried the seeds to England as a meat preserver and later on to other parts of the world. Now, coriander is cultivated almost everywhere in the world where the climate favors its growth, particularly in Iran, Indonesia, Russian, Afghanistan, China, India, Tanzania, Turkey, and Bulgaria. Pakistan falls among small coriander producers with an estimated 29,000 tons of coriander seeds covering an area of 5600 ha.

The coriander plant belongs to upright branched herbaceous species that can grow to a height of 50 cm having a green stem and leaves that turn an intense violet or reddish color at flowering time. The young herb of coriander has a characteristic odor, especially as an immature herb, that could be associated with the aroma of essential oil (Rajeshwari & Andallu, 2011; Small, 1997). The coriander plant carries white to pink flowers arranged in umbels (Fig. 1), which provides the basis for its former family *Umbelliferae*. The appearance of this plant varies with ecotype and origin; the most common varieties of coriander, that is, Vulgare, Alef, and Microcarpum DC, bear smaller fruit about 1.50–3.0 mm in diameter (Small, 1997).

The coriander foliage and seeds have their own demand for a wide range of applications in the food, pharmaceutical, and medicinal industries (Bhat, Kaushal, Kaur, & Sharma, 2014). It seems that almost all parts of coriander plants have secured a special place in folk kitchens. For example, the leaves and foliage work as flavor and aroma enhancers in a variety of food products (Carrubba, 2009). The green coriander leaves, when incorporated as an essential part of salads, pickles, soups, and cooked foods, especially meats, pulses, and vegetables, enhance the flavor, beauty, and fragrance of these dishes. In China and many other parts of the world, coriander seeds or their powder find application in flavoring and thickening of soups, cousins, liquors, and curries. The coriander roots have little importance in foodstuffs, but Thais greatly relish these roots in their cousins. In addition to food applications, coriander plant parts have been used in folk health care and herbal medicine, to treat digestion-related complications and to facilitate appetite (Maroufi, Farahani, & Darvishi, 2010). Coriander leaves

Cold Pressed Oils. https://doi.org/10.1016/B978-0-12-818188-1.00031-1
Copyright © 2020 Elsevier Inc. All rights reserved.

FIG. 1 Different parts (roots, leaves, seeds, and flower) of coriander plant.

are very effective to make up for vitamin deficiencies (Emamghoreishi, Khasaki, & Aazam, 2005). Mani, Parle, Ramasamy, and Abdul Majeed (2011) reported that coriander leaves have the potential to cope with Alzheimer's disease as well as control cholesterol levels to get rid of anxiety and insomnia (Cioanca, Hritcu, Mihasan, & Hancianu, 2013).

Coriander seeds, the fruit of coriander, are soft, tiny, and weightless, but are valuable parts of this plant due to the presence of characteristic aroma compounds (about 1.8% essential oil). Coriander seed oil has been frequently applied for skin care, hair conditioning, and as a deodorant. Consumption of coriander seed oil may provide benefits against diabetes, digestion complications, and heart issues depending upon the active ingredients present in it (Matasyoh, Maiyo, Ngure, & Chepkorir, 2009; Nambiar, Daniel, & Guin, 2010; Sriti et al., 2010).

The abovementioned benefits of coriander seed oils or their products have been ascribed to the presence of active aroma and lipophilic compounds, whose concentration and activities depend upon the extent and nature of postharvest processing. In general, the majority of these compounds are either volatile or thermally labile; therefore, a thorough understanding of the behavior of active ingredients and their fate during postharvest processing has been a challenging and unavoidable task. This chapter will provide a comprehensive understanding of frequently adopted storage and extraction technologies with special emphasis on comparison of quality attributes of final products.

2 Extraction and processing of coriander seeds

2.1 Seed harvest and storage

In Pakistan and many other parts of the world, coriander is cultivated for the production of green leaves and seeds/fruit during mild temperature days (October–November in Pakistan). The plant normally matures within 90–120days and assessment of proper maturity state is crucial for both essential oil production and confectionery products. For oil extraction, it is necessary to harvest plants when almost 60% of fruits turn brown or central umbels attain a yellow color, because at this maturity stage the contents of the volatile oil are at their maximum level. The immature seeds develop a fetid aroma due to the presence of aldehydes and may not be suitable for confectionery uses. The overripe fruits, on the other hand, may shred or split to cause a loss in seed and oil yield. Once harvested, the coriander seeds meant for essential oil production should be dried in shade, whereas those meant for the production of spices are usually dried in sunlight. In general, sun drying may cause loss of peculiar seed texture or quality; therefore, in both cases, it is safer to dry plants in shade or bundled to avoid direct sunlight. Finally, the dried plants are shredded or beaten with light sticks to collect the seeds, which should be stored in a zone free of moisture and air to prevent deterioration and pest attacks.

2.2 Extraction of seed oil

The extraction of phytochemicals involves a set of activities adopted to isolate/recover a particular class of bioactive compounds from the complex biological matrix of the plant. Following agro-climate conditions, in terms of seed maturity/health, extraction plays the most vital role in the quality and composition of end products. Therefore, careful selection of extraction technique/methods activities, thorough knowledge of how to use them, and understanding of which extraction conditions/parameters are more sensitive are required. A large number of extraction techniques are available for the recovery of oils, aroma, and other bioactive compounds (Mushtaq, 2018). These extraction techniques are often divided into two categories, viz., conventional and nonconventional techniques. Conventional techniques include commonly hydro-distillation, cold pressed extraction, and sometimes solvent-based extraction (Giray et al., 2008). These techniques are considered more eco-friendly and cost-effective to extract some fragile components, but at the cost of extraction yield as well as involving a long extraction time (De Castro & Garcıa-Ayuso, 1998). However, more advanced and preferable techniques have been introduced to reduce these limitations, and are known as nonconventional or advanced techniques. These techniques include pulsed electric field extraction, ohmic heating, enzyme digestion, microwave-assisted extraction, pressurized liquid extraction, ultrasound-assisted extraction, supercritical fluid extraction, and enzyme-assisted extraction (Mushtaq, Akram, & Adnan, 2018). Overall, modern extraction techniques use either energy (microwave, ultrasound, or ohmic radiations), matter (solvent), or both of these simultaneously to extract the target compounds. The use of energy, however, may cause detrimental effects on the quality of the extracted compounds (Azmir et al., 2013). Conventional methods, such as cold press, in this context, do not deteriorate the thermally labile bioactives.

2.2.1 Nonconventional extraction techniques

Ultrasound-assisted extraction (UAE)

In this extraction technique, ultrasound radiations are used in a controlled manner in order to extract the components of interest from desired samples (Priego-Capote & de Castro, 2007). When sound-waves strike the cell walls, this ruptures the cell wall make it lenient for mass ejection through continuously contraction and expansion of cell wall (Toma, Vinatoru, Paniwnyk, & Mason, 2001). Ultrasound-assisted extraction (UAE) is a simple and low-cost extraction technique with promising efficiency (Pingret, Fabiano-Tixier, & Chemat, 2013). This technique has been frequently used for the extraction of phenolic compounds and anthocyanins from various vegetables like raspberry (Teng, Lee, & Choi, 2013), grapes (Carrera, Ruiz-Rodríguez, Palma, & Barroso, 2012), sugar beet (Chen, Zhao, & Yu, 2015), and onion (Katsampa, Valsamedou, Grigorakis, & Makris, 2015). Rahmani, Khorrami, and Mizani (2015) compared ultrasonic-based extraction and microwave-based extraction of coriander oil. The results showed that although ultrasound extraction provided a comparable yield, it lost some of the essential components in coriander oil. Zeković et al. (2015) also reported the extraction of coriander seed oil using UAE with a major proportion of polar compounds. In general, UAE is suitable for extraction of polar compounds and offers low extraction yield along with the risk of valuable components degradation, particularly when applied at higher frequencies. Similar concerns have been raised by Giray et al. (2008) during UAE of essential oil from *Lavandula stoechas* flowers.

2.2.2 Supercritical fluid extraction (SFE)

SFE uses a special phase of gases known as the "supercritical phase" which distributes like gas but has solubilities comparable to those of liquids. This phase has been found to be tunable with thermodynamic conditions. Hannay and Hogarth (1879) were first who introduced this phase followed by Zosel who employed SFE for the decaffeination of coffee. This technique has enormous applications nowadays in the pharmaceutical, food, and environmental industries (Zougagh, Redigolo, Ríos, & Valcárcel, 2004). The most widely used supercritical fluid, CO_2, is considered to be cheap, nontoxic, highly stable, and noninflammable. In contrast to ultrasonic extraction, SFE is highly preferable and viable for the extraction of nonpolar compounds, and for polar analytes, various polar solvents can be used as a cosolvent (Mhemdi et al., 2011).

The coriander essential oil obtained by SFE presents some quantitative and qualitative differences that give a superior aroma compared with commercial oil obtained by hydro-distillation. First, the application of supercritical fluid for extraction of coriander seeds oil yields relatively higher amounts (4.55%–6.9%) of essential oil (Anitescu, Doneanu, & Radulescu, 1997; Reverchon, 1997; Shrirame et al., 2018). In another study, the extraction composition was monitored by changing the instrumental conditions and supercritical fluid nature from CO_2 to propane; this adversely affected yield as well as the composition of extracted oil (Illés, Daood, Perneczki, Szokonya, & Then, 2000).

2.3 Conventional techniques

2.3.1 *Hydro-distillation*

Hydro-distillation uses water (steam) as an extraction solvent to recover volatile or polar components of plant materials. Hydro-distillation is often carried out using a setup known as Clevenger apparatus or simple steam distillation. In the Clevenger apparatus, sample mixed water is boiled to evaporate volatile components while in the steam distillation approach, steam is passed through a bed of the sample. In both approaches, two layers (aqueous and oil-rich) are obtained and oil can be further separated via separating funnels. In this technique, separation of aroma compounds from water, particularly polar ones, is a challenging task and perhaps the greatest limitation of hydro-distillation type extraction. Many researchers have attempted to produce coriander seed oil through hydro-distillation (Sovová & Aleksovski, 2006) and almost all of these studies faced limited recovery of essential oils comparative to other extraction techniques (Eikani, Golmohammad, & Rowshanzamir, 2007; Zeković et al., 2015).

2.3.2 *Subcritical water extraction (SWE)*

Subcritical water extraction (SWE) is commonly known as superheated water extraction or pressurized hot water extraction. Among the other techniques for the extraction of coriander seed oil, SWE has been applied for coriander seed extraction. In this technique, hot water (100–374°C) is applied as an extraction medium that also supports the extraction of less polar components. Although most polar components are soluble in water at ambient conditions, higher temperatures alter the properties of water and solubilize even mild polar constituents. When SWE was applied for coriander oil extraction, it did not produce a significant yield compared to other studied techniques. However, the extraction of polar as well as less polar solvent through a green approach distinguishes it in terms of coriander oil extraction (Saim, Osman, Yasin, & Hamid, 2008). The extraction yield could be enhanced through optimum conditions (extraction time, temperature, and pressure) up to 14.1% (Eikani et al., 2007). Moreover, supercritical carbon dioxide flow rate also affect the quantity of essential oil (Eikani et al., 2007; Zeković et al., 2016).

2.3.3 *Soxhlet extraction*

This technique uses a particular type condenser known as the Soxhlet apparatus, which is filled with a solid sample of homogenous porosity and mounted over a flask containing a boiling extraction solvent. The solvent vapors enter the plant matrix, solubilize compatible components, hit the colder tubes of a condenser, and condense back to the solvent reservoir along with the extracted components. This technique has been applied for the recovery of fixed oil from coriander seeds and was able to produce a 19.4% yield on a dry mass basis (Eikani et al., 2007; Ramadan, Kroh, & Moersel, 2003; Ramadan & Moersel, 2003, 2004). Although this technique offers a good quantity of lipids, its use is limited to fixed type oils, and contamination of the final extract with cuticle wax and other fatty content deteriorates the quality of coriander seed oil. The other major disadvantage is the degradation of some useful components upon the continuous heating of the solution (Zeković et al., 2015).

2.3.4 *Cold pressing technique*

The cold pressers simply apply pressure on ground plant material to squeeze out what they contain. If the plant material is properly dried and ground, this technique offers a simple, viable, and ecofriendly solution for the extraction of oils, but again cold pressing type extractions are not suitable for the production of essential oils, and the final product may contain other chemicals or contaminants. The oil glands in plants are ruptured mechanically, which helps in squeezing out the oil they contain independent of the polarity of the components rich in a number of lipophilic compounds. Plant oils obtained by this technique are more stable to oxidative stress than refined or processed oils are (Ramadan et al., 2003). When coriander seed oil is extracted through cold presser type extractors, the oil has been found to be good in quantity and quality (Sriti et al., 2010). For simplicity and easy scalability, this technique was commercialized for the production of vegetable and seed oil, but as mentioned above, only a fixed type of oil.

Cold pressing has less detrimental effects on the extracted components of coriander oils. Proteins extracted from coriander seed by cold pressing showed greater thermal stability than that reported for acid-precipitated proteins. Protein thermal stability resulting from cold pressed fractions has been attributed to the presence of greater hydrophobic amino acid residues, specifically leucine, isoleucine, and phenylalanine, which remain inviable during the harsh extraction conditions of other techniques. Cold pressed protein extracts also exhibited higher protein purity (>84% crude protein) than that of whole fruit (67%). Cold pressing of coriander prove it to be the most suitable, high yield, and eco-friendly technique for the extraction of coriander oil and proteins.

Cold pressed coriander seed oil mainly contains nonvolatile bioactive with traces of aroma compounds. Fixed oil fraction of coriander seed oil normally contains fatty acids, phospholipids, pigments, tocopherols, and traces of metals. The percentage of fixed lipophilic in coriander seed oil obtained via this technique may range from 9.9% to 27.7%, while the essential oil content may vary from 0.03 to 2.6 (Eidi & Eidi, 2011). The concentration of various bioactive in the extracted oil can be varied for different varieties of coriander, but petroselinic acid and linalool are present in dominant amounts. It has been observed that coriander essential oil may contain 41%–80% of linalool: a monoterpene alcohol with characteristic lemon flavor (Asgarpanah & Kazemivash, 2012).

3 Fatty acid and acyl lipids profile of cold pressed coriander seed oil

Coriander seed is reported to contain about 1% of fatty acids (Uitterhaegen et al., 2018). The identification and distribution of fatty acids in the coriander seed oil are frequently assessed through chromatographic techniques (GC-MS) and their profiles are presented in Figs. 2 and 3. The most probable component in the fatty acid profile is petroselinic acid

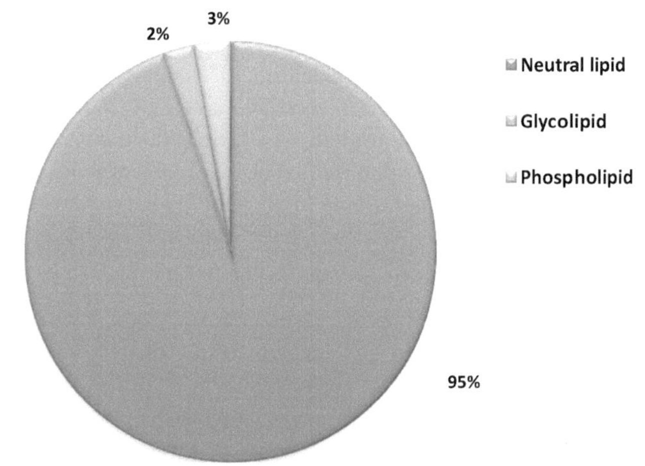

FIG. 2 Distribution of lipids in coriander seed oil (Ramadan & Mörsel, 2002).

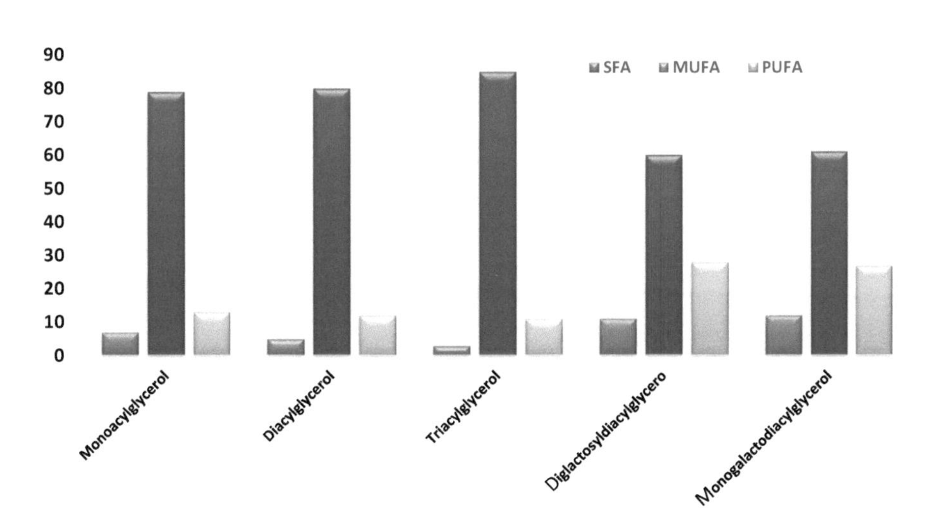

FIG. 3 Percentages of SFA (saturated fatty acid), MUFAs (monounsaturated fatty acids), and PUFAs (polyunsaturated fatty acids) in the neutral lipid classes of coriander seed oil (Sriti et al., 2019).

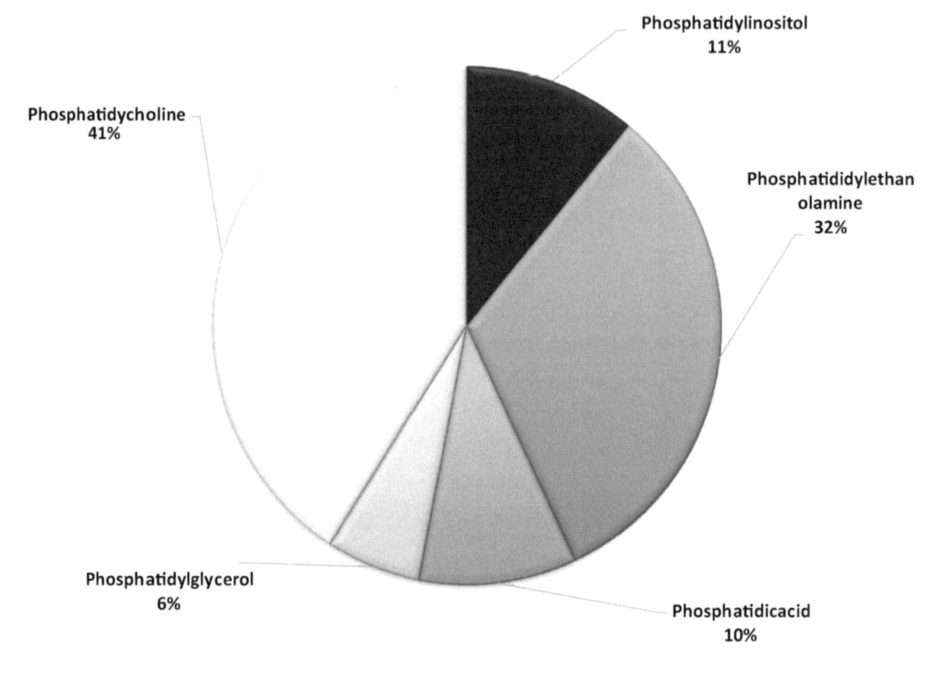

FIG. 4 The relative abundance of phospholipid classes in coriander seed oil (Sriti et al., 2010; Uitterhaegen et al., 2016).

(65%–75%), followed by linoleic acid (13%–16%) and oleic acid (5%–7%) (Ramadan, Amer, & Awad, 2008; Ramadan & Mörsel, 2002; Ramadan & Wahdan, 2012; Sriti, Bettaieb, Bachrouch, Talou, & Marzouk, 2019).

There is a significant amount of total lipids measured in coriander seeds (Neffati & Marzouk, 2008). These lipids are mainly in three classes: neutral lipid, glycolipids, and phospholipids. Among these constituents, neutral lipid contributes the highest percentage, followed by glycolipids and phospholipids. Phospholipids are also present in significant amounts in coriander oil. The most abundant components that contribute to phospholipids are phosphatidylcholine, phosphatidylglycerol, phosphatidic acid, phosphatidylethanolamine, and phosphatidylinositol, as shown in Fig. 4 (Ramadan & Moersel, 2003; Sriti et al., 2010; Uitterhaegen et al., 2016).

4 Minor bioactive lipids

4.1 Sterols

Coriander seed oil is a rich source of phytosterols (Fig. 5), which are highly beneficial for health due to their antioxidant potential (Ramadan & Mörsel, 2002; Sriti et al., 2010, 2019). According to a seminal report, coriander seed oil may contain

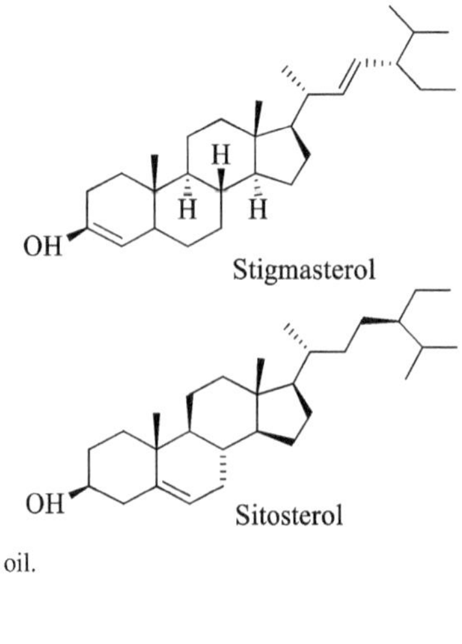

FIG. 5 Structure of sterols found in coriander seed oil.

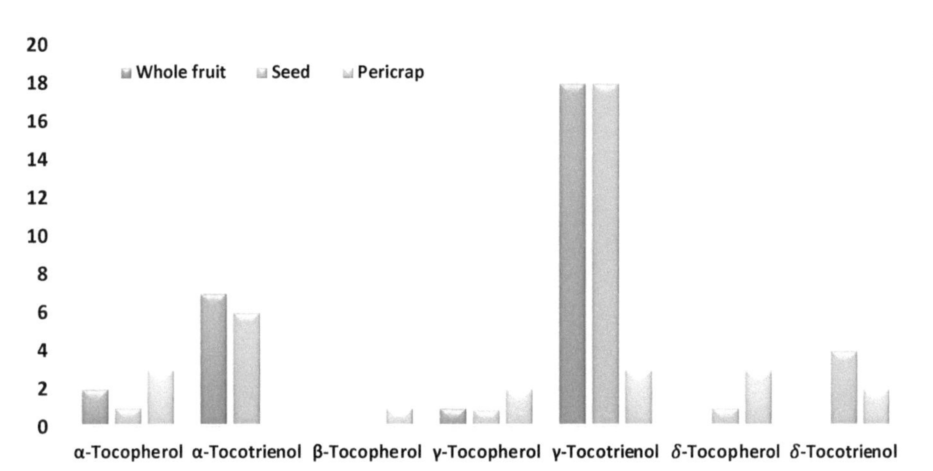

FIG. 6 Levels of tocochromanols (mg/100mL oil) in different parts of coriander.

phytosterols in a range of 36.9–91.8 mg/g oil. Phytosterols usually compete with the absorption of dietary cholesterol to make a balance of low-density lipid profile. The most distinct sterol found in coriander has been labeled as β-sitosterol (25%–37%), followed by stigmasterol (22%–30%).

4.2 Tocochromanols

Coriander oil is considered a medium source of tocochromanols, which are present in amounts of 280–330 ppm. Total tocochromanols may contain various individual compounds in quite different ratios, as shown in Fig. 6. The most abundant tocochromanol in coriander seed oil is γ-tocotrienol, which may present in a quantity of about 200 ppm (Sriti et al., 2010).

4.3 Other bioactive compounds

Coriander seed oil also contains minor amounts of well-known phenolic antioxidants. These compounds may include but are not limited to caffeic, ferulic, gallic, and chlorogenic acid (Fig. 7). In addition to phenolic acids, some flavonoids (Fig. 8) like luteolin, kaempferol, and quercetin have also been found in coriander seed oil or its products (Nambiar et al., 2010). Oganesyan, Nersesyan, and Parkhomenko (2007) found tannins and anthocyanins in coriander oil.

4.4 Essential oil composition and functionality

The essential oil extracted from coriander is characterized as a colorless or pale yellow liquid with a specific odor and mild sweet aromatic in pharmaceuticals flavor (Burdock, 2002). Approximately 20% of extracted essential oil is employed and

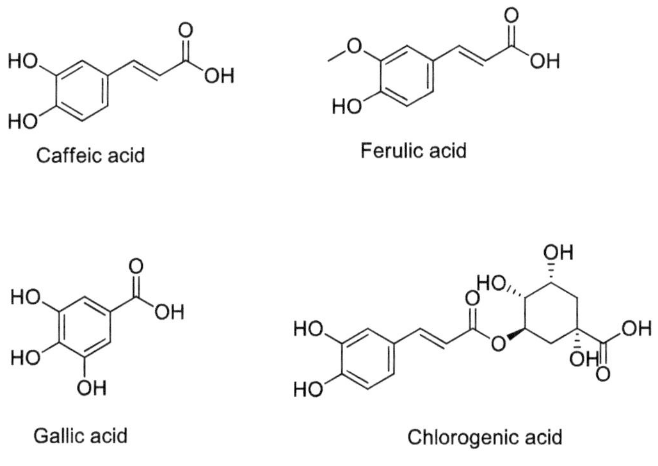

Caffeic acid
Ferulic acid
Gallic acid
Chlorogenic acid

FIG. 7 Phenolic acids found in coriander seed oil.

FIG. 8 Chemical structures of selected flavonoids.

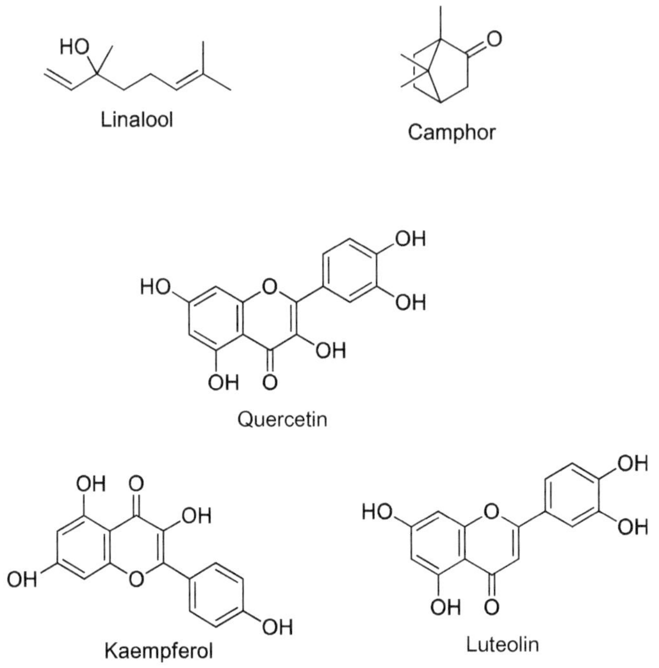

Linalool

Camphor

Quercetin

Kaempferol

Luteolin

similar fractions are being used as flavoring agents in the food industry. The major volume of coriander seed oil is used for perfumery, cosmetics, and aromatherapy (Douglas, Heyes, & Smallfield, 2005). As far as its composition is concerned, coriander essential oil may contain geranyl acetate (4.99%), linalool (3.2%), γ-terpinene (4.17%), α-pinen (1.63%), anethol (1.15%), and *p*-cymen (1.12%) (Anwar et al., 2011). Linalool, the major component of coriander seed oil, has many uses in the food and pharmaceuticals industries. It works as a good alternative to synthetic antioxidants due to its strong radical scavenging activities and lipid peroxidation inhibition activity. Linalool also has the ability to enhance the shelf life of food as well as food safety for storage purposes (Duarte, Luís, Oleastro, & Domingues, 2016).

5 Health-promoting traits of cold pressed coriander seed oil and its constituents

Coriander seed has a folk history for the treatment of various diseases like dysentery, indigestion, and nausea (Al-Mofleh et al., 2006). Chronic disorders such as cancer and heart-related diseases are strongly linked with oxidative damages in DNA caused by reactive oxygen species. Coriander, being rich in antioxidants, can retard oxidative stress and damage or related diseases. These antioxidant traits could be linked with the presence of fatty acids, tocochromanols, sterols, terpenoids, and volatile or aroma compounds of high medicinal and nutritional potential (Reddy, Subba Rao, & Chowdari, 2013).

5.1 Antioxidant-related activities

Antioxidants are basically the compounds capable of delaying or retarding the process of oxidation of lipids and other biomolecules, thereby preventing the production of reactive species. Although many synthetic antioxidants like butylated hydroxytoluene (BHT) and butylated hydroxyanisole (BHA) are available, concerns exist regarding their carcinogenic character. Hence, natural and nontoxic antioxidant resources like coriander seed oil are gaining far more focus for the application in the food, cosmetics, and pharmaceuticals industries. Coriander oil contains many of those valuable antioxidants like phenolics, flavonoids, PUFAs, linalool, and phytosterols. The cold pressed extracts of coriander seeds have been found to have ample levels of antioxidant compounds (Wangensteen, Samuelsen, & Malterud, 2004).

Coriander seed oil contains monoterpenes, which bear hypoglycemic potential and help in modulating the enzymes to metabolize the carbohydrates. Furthermore, these bioactives enhance insulin sensitivity and activate the insulin receptor cells and signaling. Monoterpenes also inhibit the absorption of glucose by gastro-intestine by inhibiting the enzymes in the gastrointestinal tract. Likewise, the presence of linalool and PUFA in coriander seed oil renders it a potent hypolipidemic agent (Ramadan et al., 2008). These biomolecules decline the cholesterol level by inhibiting the 3-hydroxy-3-methyl-glutaryl-CoA reductase (HMG-CoA-reductase) or block the synthesis of fatty acids by inhibiting the lipogenic enzymes. In addition, these bioactives may act as a peroxisome proliferator-activated receptor alpha (PPARα) agonist or retard low-density lipid (LDL) oxidation (Duarte et al., 2016).

Coriander oil has been known to have hypotensive potential; it thus acts as a vasorelaxant and modulates the NOS pathways that enhance the release of NO species. It also inhibits Ca^{2+} influx or acts as a Ca^{2+} channel antagonist, enhancing the K+ channel opening and hyperpolarization. It also increases NO availability as an antioxidant. In addition, it increases urine output as a diuretic (Jabeen, Bashir, Lyoussi, & Gilani, 2009). When the composition of coriander seed oil is keenly observed, it may be seen that linalool, monoterpene alcohol, is a very important bioactive compound that is useful in performing the analgesic activity as well as insecticidal activity by modulating the multiple pain pathways.

5.2 AntiQS potential

Coriander oil also shows anti quorum sensing (QS) activity. It is a sort of intercellular communication mechanism named quorum sensing that makes the bacteria able to direct some physiological activities such as gaining antibiotic resistance to perform a bactericidal activity in food products (Gölz, Sharbati, Backert, & Alter, 2012). Coriander oil has the capability to inhibit this activity. Duarte et al. (2016) studied the antiQS activity of coriander oil against a Gram-negative bacterium (*C. violaceum*) that is a biosensor strain and produces violacein (a purple pigment), which is responsible for food damage. Hence, the violacein production is inhibited by coriander oil, which leads to better preservation of food.

5.3 Antiinflammatory potential

The antiinflammatory potential of coriander oil is highly significant as it contains bioactive compounds such as rutin, chlorophyllin, β-sitosterol, and β-sitosterolin, which showed antiinflammatory activity and inhibit the NF-κB pathway in macrophages and suppress the pro-inflammatory mediators. In addition, they inhibit macrophage activation, infiltration, and aggregation (Sabogal-Guáqueta, Osorio, & Cardona-Gómez, 2016).

5.4 Antimicrobial potential

The antibacterial activity of coriander oil against Gram-negative and Gram-positive bacteria has been linked with the presence of bioactive compounds (Laribi, Kouki, M'Hamdi, & Bettaieb, 2015). In this context, an antimicrobial peptide named "Plantaricin CS" with wide antibacterial activity has been reported in coriander (Zare-Shehneh et al., 2014). Some silver nanoparticles are formed through a simple reaction of coriander extract with silver nitrate. These silver nanoparticles act as a proficient antimicrobial because of their smaller surface area. These small nanoparticles not only interact with the bacterial external membrane, but also have great potency to collapse bacterial DNA and inhibit their reproductive system as well (Ashraf et al., 2019). Coriander oil also comprises several polar and moderately polar secondary metabolites such as phenolics and flavonoids (Bhat et al., 2014), which play a major role in the human diet as well as combating or inhibiting various microbial and enzymatic activities like human immunodeficiency virus (HIV) replication, glucosyl transferases of *Streptococcus* mutants, human simplex virus (HSV), cytotoxic effects, tumor promotion, ascorbate auto-oxidation and xanthine, and monoamine oxidases (Havsteen, 2002).

6 Edible and nonedible applications of cold pressed oil

Apparently, the food that we usually consume is of low quality with a very minimal life span. Coriander seed oil has the potential to increase the shelf life of the food when incorporated in food as a preservative. According to an international survey of coriander oil supply, it is sold under various brands by different companies throughout the world that indicates its commercial applicability (Sasaki et al., 2002). Coriander oil has fatty acid content mainly of petroselinic acid (up to 75%). Petroselinic acid assists in the manufacturing of useful *trans*-isomers without any saturated fatty acid by-products (Kalyna, Lutsenko, & Kharytonov, 2018). This acid also plays a significant role in the food, chemical, and cosmetics industries (Uitterhaegen et al., 2016). When petroselinic acid undergoes oxidative cleavage, it yields two industrially important compounds: adipic acid, which is used in synthesizing Nylon 66 polymers, and lauric acid, which is a commercial surfactant (Cermak, Isbell, Evangelista, & Johnson, 2011). Coriander oil has also contributed to the generation of alternative sustainable energy resources. The use of coriander oil in biodiesel fuel production is extensively reported due to its unique fatty acid profile (Moser & Vaughn, 2010). The important components that contribute to biodiesel production are 18 carbon fatty acids like oleic acid, stearic acid, and linoleic acid, which are abundantly present in coriander oil. The coriander oil properties like an acid number, iodine number, total glycerol component, free glycerol content, sulfur number, and phosphorus number are also suitable for biodiesel fuel production (Moser & Vaughn, 2010). Interestingly, coriander oil has been used as a finishing material for wood products such as furniture and building material for indoor applications, where it not only protects the structure from pests but also improves the air quality and environment (Uitterhaegen et al., 2018).

Although there is a huge demand for coriander oil, it can be fulfilled with simple procedural extraction of drying and pressing without the use of any preservatives. In this way, these products can be used throughout the year without any fear of wilting or degradation. In addition, these preserved herbs can be marketed all around the world to various process industries and consumers (Pirbalouti, Salehi, & Craker, 2017). However, most consumers have shown concern for quality in the case of dried herbs (Díaz-Maroto, Pérez-Coello, & Cabezudo, 2002). By using the drying procedure, these herbs enable an enhanced shelf life that ultimately slows down microbial degradation and other biochemical reactions that can change the organoleptic properties in these plants (Sellami et al., 2011).

References

Al-Mofleh, I., Alhaider, A., Mossa, J., Al-Sohaibani, M., Rafatullah, S., & Qureshi, S. (2006). Protection of gastric mucosal damage by *Coriandrum sativum* L. pretreatment in wistar albino rats. *Environmental Toxicology and Pharmacology*, *22*(1), 64–69.

Anitescu, G., Doneanu, C., & Radulescu, V. (1997). Isolation of coriander oil: Comparison between steam distillation and supercritical CO_2 extraction. *Flavour and Fragrance Journal*, *12*(3), 173–176.

Anwar, F., Sulman, M., Hussain, A. I., Saari, N., Iqbal, S., & Rashid, U. (2011). Physicochemical composition of hydro-distilled essential oil from coriander (*Coriandrum sativum* L.) seeds cultivated in Pakistan. *Journal of Medicinal Plant Research*, *5*(15), 3537–3544.

Asgarpanah, J., & Kazemivash, N. (2012). Phytochemistry, pharmacology and medicinal properties of *Coriandrum sativum* L. *African Journal of Pharmacy and Pharmacology*, *6*(31), 2340–2345.

Ashraf, A., Zafar, S., Zahid, K., Shah, M. S., Al-Ghanim, K. A., Al-Misned, F., et al. (2019). Synthesis, characterization, and antibacterial potential of silver nanoparticles synthesized from *Coriandrum sativum* L. *Journal of Infection and Public Health*, *12*, 275–281.

Azmir, J., Zaidul, I., Rahman, M., Sharif, K., Mohamed, A., Sahena, F., et al. (2013). Techniques for extraction of bioactive compounds from plant materials. *Journal of Food Engineering*, *117*, 426–436.

Bhat, S., Kaushal, P., Kaur, M., & Sharma, H. (2014). Coriander (*Coriandrum sativum* L.): Processing, nutritional and functional aspects. *African Journal of Plant Science*, *8*, 25–33.

Burdock, G. A. (2002). Regulation of flavor ingredients. In *Nutritional toxicology* (pp. 316–339). New York, NY: Taylor and Francis.

Carrera, C., Ruiz-Rodríguez, A., Palma, M., & Barroso, C. G. (2012). Ultrasound assisted extraction of phenolic compounds from grapes. *Analytica Chimica Acta*, *732*, 100–104.

Carrubba, A. (2009). Nitrogen fertilisation in coriander (*Coriandrum sativum* L.): A review and meta-analysis. *Journal of the Science of Food and Agriculture*, *89*(6), 921–926.

Cermak, S. C., Isbell, T. A., Evangelista, R. L., & Johnson, B. L. (2011). Synthesis and physical properties of petroselinic based estolide esters. *Industrial Crops and Products*, *33*(1), 132–139.

Chen, M., Zhao, Y., & Yu, S. (2015). Optimisation of ultrasonic-assisted extraction of phenolic compounds, antioxidants, and anthocyanins from sugar beet molasses. *Food Chemistry*, *172*, 543–550.

Cioanca, O., Hritcu, L., Mihasan, M., & Hancianu, M. (2013). Cognitive-enhancing and antioxidant activities of inhaled coriander volatile oil in amyloid β(1-42) rat model of Alzheimer's disease. *Physiology and Behavior*, *120*, 193–202.

De Castro, M. L., & Garcıa-Ayuso, L. (1998). Soxhlet extraction of solid materials: An outdated technique with a promising innovative future. *Analytica Chimica Acta*, *369*, 1–10.

Díaz-Maroto, M., Pérez-Coello, M., & Cabezudo, M. (2002). Effect of different drying methods on the volatile components of parsley (*Petroselinum crispum* L.). *European Food Research and Technology*, *215*(3), 227–230.

Douglas, M., Heyes, J., & Smallfield, B. (2005). *Herbs, spices and essential oils: Post-harvest operations in developing countries*: (p. 61). UNIDO and FAO.

Duarte, A., Luís, Â., Oleastro, M., & Domingues, F. C. (2016). Antioxidant properties of coriander essential oil and linalool and their potential to control *Campylobacter* spp. *Food Control*, *61*, 115–122.

Eidi, M., & Eidi, A. (2011). Effect of coriander (*Coriandrum sativum* L.) seed ethanol extract in experimental diabetes. In *Nuts and Seeds in Health and Disease Prevention* (pp. 395–400). Elsevier.

Eikani, M. H., Golmohammad, F., & Rowshanzamir, S. (2007). Subcritical water extraction of essential oils from coriander seeds (*Coriandrum sativum* L.). *Journal of Food Engineering*, *80*(2), 735–740.

Emamghoreishi, M., Khasaki, M., & Aazam, M. F. (2005). Coriandrumsativum: Evaluation of its anxiolytic effect in the elevated plus-maze. *Journal of Ethnopharmacology*, *96*(3), 365–370.

Giray, E. S., Kırıcı, S., Kaya, D. A., Türk, M., Sönmez, Ö., & Inan, M. (2008). Comparing the effect of sub-critical water extraction with conventional extraction methods on the chemical composition of *Lavandula stoechas*. *Talanta*, *74*(4), 930–935.

Gölz, G., Sharbati, S., Backert, S., & Alter, T. (2012). Quorum sensing dependent phenotypes and their molecular mechanisms in *campylobacterales*. *European Journal of Microbiology and Immunology*, *2*(1), 50–60.

Hannay, J., & Hogarth, J. (1879). On the solubility of solids in gases. *Proceedings of the Royal Society of London*, *29*, 324.

Havsteen, B. H. (2002). The biochemistry and medical significance of the flavonoids. *Pharmacology and Therapeutics*, *96*(2–3), 67–202.

Illés, V., Daood, H., Perneczki, S., Szokonya, L., & Then, M. (2000). Extraction of coriander seed oil by CO_2 and propane at super-and subcritical conditions. *Journal of Supercritical Fluids*, *17*(2), 177–186.

Jabeen, Q., Bashir, S., Lyoussi, B., & Gilani, A. H. (2009). Coriander fruit exhibits gut modulatory, blood pressure lowering and diuretic activities. *Journal of Ethnopharmacology*, *122*(1), 123–130.

Kalyna, V., Lutsenko, M., & Kharytonov, M. (2018). Feasibility study of the technology of fatty coriander oil complex processing. *Annals of Agrarian Science*, *16*(2), 95–100.

Katsampa, P., Valsamedou, E., Grigorakis, S., & Makris, D. P. (2015). A green ultrasound-assisted extraction process for the recovery of antioxidant polyphenols and pigments from onion solid wastes using Box-Behnken experimental design and kinetics. *Industrial Crops and Products*, *77*, 535–543.

Laribi, B., Kouki, K., M'Hamdi, M., & Bettaieb, T. (2015). Coriander (*Coriandrum sativum* L.) and its bioactive constituents. *Fitoterapia*, *103*, 9–26.

Mani, V., Parle, M., Ramasamy, K., & Abdul Majeed, A. B. (2011). Reversal of memory deficits by *Coriandrum sativum* leaves in mice. *Journal of the Science of Food and Agriculture*, *91*(1), 186–192.

Maroufi, K., Farahani, H. A., & Darvishi, H. H. (2010). Importance of coriander (*Coriandrum sativum* L.) between the medicinal and aromatic plants. *Advances in Environmental Biology*, *4*, 433–436.

Matasyoh, J., Maiyo, Z., Ngure, R., & Chepkorir, R. (2009). Chemical composition and antimicrobial activity of the essential oil of *Coriandrum sativum*. *Food Chemistry*, *113*, 526–529.

Mhemdi, H., Rodier, E., Kechaou, N., & Fages, J. (2011). A supercritical tuneable process for the selective extraction of fats and essential oil from coriander seeds. *Journal of Food Engineering*, *105*, 609–616.

Moser, B. R., & Vaughn, S. F. (2010). Coriander seed oil methyl esters as biodiesel fuel: Unique fatty acid composition and excellent oxidative stability. *Biomass and Bioenergy*, *34*(4), 550–558.

Mushtaq, M. (2018). Extraction of fruit juice: An overview. In *Fruit juices* (pp. 131–159). Elsevier.

Mushtaq, M., Akram, S., & Adnan, A. (2018). Novel extraction technologies. In *Fruit Juices* (pp. 161–181). Elsevier.

Nambiar, V. S., Daniel, M., & Guin, P. (2010). Characterization of polyphenols from coriander leaves (*Coriandrum sativum*), red amaranthus (*A. paniculatus*) and green amaranthus (*A. frumentaceus*) using paper chromatography and their health implications. *Journal of Herbal Medicine and Toxicology*, *4*, 173–177.

Neffati, M., & Marzouk, B. (2008). Changes in essential oil and fatty acid composition in coriander (*Coriandrum sativum* L.) leaves under saline conditions. *Industrial Crops and Products*, *28*(2), 137–142.

Oganesyan, E., Nersesyan, Z., & Parkhomenko, A. Y. (2007). Chemical composition of the above-ground part of *Coriandrum sativum*. *Pharmaceutical Chemistry Journal*, *41*(3), 149–153.

Pingret, D., Fabiano-Tixier, A.-S., & Chemat, F. (2013). Degradation during application of ultrasound in food processing: A review. *Food Control*, *31*(2), 593–606.

Pirbalouti, A. G., Salehi, S., & Craker, L. (2017). Effect of drying methods on qualitative and quantitative properties of essential oil from the aerial parts of coriander. *Journal of Applied Research on Medicinal and Aromatic Plants*, *4*, 35–40.

Priego-Capote, F., & de Castro, L. (2007). Ultrasound-assisted digestion: A useful alternative in sample preparation. *Journal of Biochemical and Biophysical Methods*, *70*(2), 299–310.

Rahmani, S., Khorrami, A. R., & Mizani, F. (2015). A survey of the effects of coriander seed essential oil by solvent-free microwave extraction and ultrasonic waves on control of the growth of *Salmonella typhimurium* bacteria. *Journal of Applied Environmental and Biological Sciences*, *5*(4S), 65–71.

Rajeshwari, U., & Andallu, B. (2011). Medicinal benefits of coriander (*Coriandrum sativum* L.). *Spatula DD*, *1*(1), 51–58.

Ramadan, M. F., Amer, M. M. A., & Awad, A. (2008). Coriander (*Coriandrum sativum* L.) seed oil improves plasma lipid profile in rats fed diet containing cholesterol. *European Food Research and Technology*, *227*, 1173–1182.

Ramadan, M. F., Kroh, L. W., & Moersel, J.-T. (2003). Radical scavenging activity of black cumin (*Nigella sativa* L.), coriander (*Coriandrum sativum* L.) and niger (*Guizotia abyssinica* Cass.) crude seed oils and oil fractions. *Journal of Agricultural and Food Chemistry*, *51*, 6961–6969.

Ramadan, M. F., & Moersel, J.-T. (2003). Analysis of glycolipids from black cumin (*Nigella sative* L.), coriander (*Coriandrum sativum* L.) and niger (*Guizotia abyssinica* Cass.) oilseeds. *Food Chemistry*, *80*, 197–204.

Ramadan, M. F., & Moersel, J.-T. (2004). Oxidative stability of black cumin (*Nigella sativa* L.), coriander (*Coriandrum sativum* L.) and niger (*Guizotia abyssinica* Cass.) upon stripping. *European Journal of Lipid Science and Technology*, *106*(1), 35–43.

Ramadan, M., & Mörsel, J.-T. (2002). Oil composition of coriander (*Coriandrum sativum* L.) fruit-seeds. *European Food Research and Technology*, *215* (3), 204–209.

Ramadan, M. F., & Wahdan, K. M. M. (2012). Blending of corn oil with black cumin (*Nigella sativa*) and coriander (*Coriandrum sativum*) seed oils: Impact on functionality, stability and radical scavenging activity. *Food Chemistry*, *132*, 873–879.

Reddy, M., Subba Rao, G., & Chowdari, B. (2013). Metal oxides and oxysalts as anode materials for Li ion batteries. *Chemical Reviews*, *113*(7), 5364–5457.

Reverchon, E. (1997). Supercritical fluid extraction and fractionation of essential oils and related products. *The Journal of Supercritical Fluids*, *10*(1), 1–37.

Sabogal-Guáqueta, A. M., Osorio, E., & Cardona-Gómez, G. P. (2016). Linalool reverses neuropathological and behavioral impairments in old triple transgenic Alzheimer's Alzheimer's mice. *Neuropharmacology*, *102*, 111–120.

Saim, N., Osman, R., Yasin, W., & Hamid, R. D. (2008). Subcritical water extraction of essential oil from coriander (*Coriandrum sativum* L.) seeds. *The Malaysian Journal of Analytical Sciences*, *12*(1), 22–24.

Sasaki, Y. F., Kawaguchi, S., Kamaya, A., Ohshita, M., Kabasawa, K., Iwama, K., et al. (2002). The comet assay with 8 mouse organs: Results with 39 currently used food additives. *Mutation Research, Genetic Toxicology and Environmental Mutagenesis*, *519*, 103–119.

Seidemann, J. (2005). World spice plants. Springer. Carrubba, A. 2009. Nitrogen fertilisation in coriander (*Coriandrum sativum* L.): A review and meta-analysis. *Journal of the Science of Food and Agriculture*, *89*, 921–926.

Sellami, I. H., Wannes, W. A., Bettaieb, I., Berrima, S., Chahed, T., Marzouk, B., et al. (2011). Qualitative and quantitative changes in the essential oil of *Laurus nobilis* L. leaves as affected by different drying methods. *Food Chemistry*, *126*(2), 691–697.

Shrirame, B., Geed, S., Raj, A., Prasad, S., Rai, M., Singh, A., et al. (2018). Optimization of supercritical extraction of coriander (*Coriandrum sativum* L.) seed and characterization of essential ingredients. *Journal of Essential Oil-Bearing Plants*, *21*(2), 330–344.

Small, E. (1997). *Culinary herbs.* Ottawa: NRC Research Press.

Sovová, H., & Aleksovski, S. A. (2006). Mathematical model for hydrodistillation of essential oils. *Flavour and Fragrance Journal*, *21*(6), 881–889.

Sriti, J., Bettaieb, I., Bachrouch, O., Talou, T., & Marzouk, B. (2019). Chemical composition and antioxidant activity of the coriander cake obtained by extrusion. *Arabian Journal of Chemistry*, *17*, 1765–1773.

Sriti, J., Wannes, W. A., Talou, T., Mhamdi, B., Cerny, M., & Marzouk, B. (2010). Lipid profiles of Tunisian coriander (*Coriandrum sativum*) seed. *Journal of the American Oil Chemists' Society*, *87*(4), 395–400.

Teng, H., Lee, W. Y., & Choi, Y. H. (2013). Optimization of microwave-assisted extraction for anthocyanins, polyphenols, and antioxidants from raspberry (*Rubus coreanus* Miq.) using response surface methodology. *Journal of Separation Science*, *36*(18), 3107–3114.

Toma, M., Vinatoru, M., Paniwnyk, L., & Mason, T. (2001). Investigation of the effects of ultrasound on vegetal tissues during solvent extraction. *Ultrasonics Sonochemistry*, *8*(2), 137–142.

Uitterhaegen, E., Burianová, K., Ballas, S., Véronèse, T., Merah, O., Talou, T., et al. (2018). Characterization of volatile organic compound emissions from self-bonded boards resulting from a coriander biorefinery. *Industrial Crops and Products*, *122*, 57–65.

Uitterhaegen, E., Sampaio, K., Delbeke, E., De Greyt, W., Cerny, M., Evon, P., et al. (2016). Characterization of French coriander oil as source of petroselinic acid. *Molecules*, *21*(9), 1202.

Wangensteen, H., Samuelsen, A. B., & Malterud, K. E. (2004). Antioxidant activity in extracts from coriander. *Food Chemistry*, *88*(2), 293–297.

Zare-Shehneh, M., Askarfarashah, M., Ebrahimi, L., Kor, N. M., Zare-Zardini, H., Soltaninejad, H., et al. (2014). Biological activities of a new antimicrobial peptide from *Coriandrum sativum. International Journal of Bioscience*, *4*(6), 89–99.

Zeković, Z., Bušić, A., Komes, D., Vladić, J., Adamović, D., & Pavlić, B. (2015). Coriander seeds processing: Sequential extraction of non-polar and polar fractions using supercritical carbon dioxide extraction and ultrasound-assisted extraction. *Food and Bioproducts Processing*, *95*, 218–227.

Zeković, Z., Kaplan, M., Pavlić, B., Olgun, E. O., Vladić, J., Canlı, O., et al. (2016). Chemical characterization of polyphenols and volatile fraction of coriander (*Coriandrum sativum* L.) extracts obtained by subcritical water extraction. *Industrial Crops and Products*, *87*, 54–63.

Zougagh, M., Redigolo, H., Ríos, A., & Valcárcel, M. (2004). Screening and confirmation of pahs in vegetable oil samples by use of supercritical fluid extraction in conjunction with liquid chromatography and fluorimetric detection. *Analytica Chimica Acta*, *525*(2), 265–271.

Chapter 32

Cold pressed peanut (*Arachis hypogaea* L.) oil

Seok Shin Tan[a], Chin Xuan Tan[b], and Seok Tyug Tan[c]

[a]*Department of Nutrition and Dietetics, School of Health Sciences, International Medical University, Kuala Lumpur, Malaysia,* [b]*Department of Allied Health Sciences, Faculty of Science, Universiti Tunku Abdul Rahman, Kampar Perak, Malaysia,* [c]*Department of Healthcare Professional, Faculty of Health and Life Sciences, Management and Science University, Shah Alam, Selangor, Malaysia*

Abbreviations

CCPO commercial crude peanut oil
CPMO cold pressed *Moringa oleifera* seed oil
HDL high-density lipoprotein
LDL low-density lipoprotein
MUFA monounsaturated fatty acid
OSI oxidative stability index
PPC peanut protein concentrate
PPI peanut protein isolate
PUFA polyunsaturated fatty acid
SFA saturated fatty acid

1 Introduction

The peanut (*Arachis hypogaea* L.), also known as the groundnut or earthnut, is a legume crop that belongs to the family of *Fabaceae*. The crop is native to South America and was cultivated as early as 2000–3000 BC (Dean, Davis, & Sanders, 2011). In the 16th century, the peanut was introduced by the Portuguese from Brazil to West Africa and then later on to Southwestern India (Sahdev, 2015). The peanut crop grows well on light, sandy loam soil in tropical and subtropical regions (Sahdev, 2015; Suri, Singh, Kaur, & Singh, 2019). The peanut kernel is commonly used in the production of cooking oil, peanut butter, flour, snacks, soups, and desserts.

Global peanut production generally showed a growing trend between 2007 and 2017, reaching more than 45.3 million metric tons in 2017 (Fig. 1). Between 2013 and 2017, the average productions of the four main peanut-producing countries were 17,245,200 metric tons in China, 5,624,000 metric tons in India, 3,000,000 metric tons in Nigeria, and 2,557,600 metric tons in the USA (International Nut and Dried Fruit, 2018). These countries accounted for more than 60% of the total world peanut production. Approximately two-thirds of the world peanut production is crushed for oil, while the remaining is used in food preparations (Akcali, Ince, & Guzel, 2006; Wang et al., 2016). Peanut accounts for about 5.77 million metric tons of plant oil production worldwide (Suri et al., 2019).

2 Composition of the peanut

About 68%–72% of the peanut is constituted by the edible kernel; the shell (28%–32%) constitutes the rest (Wang, Liu, Wang, Guo, & Wang, 2016). At 4.6% moisture content, the average thickness, width, and length of the peanut kernel were 6.9, 3.6, and 8.5 mm, respectively (Olajide & Igbeka, 2003). The main nutrients of peanut kernels were crude lipids (32.7%–53.9%), crude protein (25.9%–32.4%), and total carbohydrate (9.89%–23.62%) (Musa, 2011; Wang, Liu, Wang, et al., 2016). In addition, peanut kernels are also rich in vitamins (e.g., niacin, folate, and vitamin E), minerals (e.g., copper, manganese, iron, and phosphorus), phytosterols (e.g., β-sitosterol, campesterol, and stigmasterol), and phenolics

Cold Pressed Oils. https://doi.org/10.1016/B978-0-12-818188-1.00032-3
Copyright © 2020 Elsevier Inc. All rights reserved.

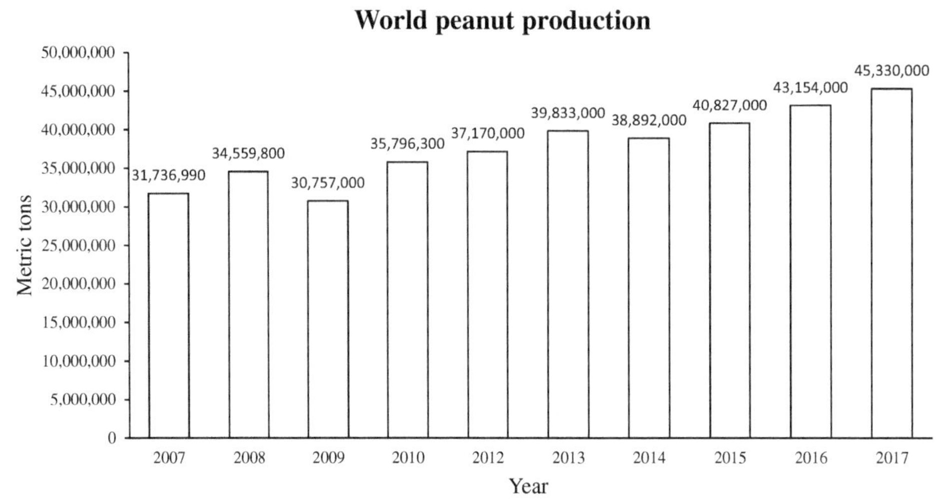

FIG. 1 World peanut production from 2007 to 2017. *(Source: International Nut and Dried Fruit (2018). Nuts and dried fruits statistical year book 2017/2018.)*

(e.g., resveratrol and *p*-coumaric acid) (Arya, Salve, & Chauhan, 2016). The chemical composition of the peanut kernel is affected by its maturity stage (Kim & Hung, 1991), environmental conditions (Bansal, Satija, & Ahuja, 1993), geographical origin (Grosso, Lamarque, Maestri, Zygadlo, & Guzman, 1994), and variety (Musa, 2011). Nutritionists categorize peanut kernels as an A+ grade crop and they are the fourth largest source of oil-bearing crop in the world (Wang, Liu, Wang, et al., 2016).

3 Extraction and processing of cold pressed peanut oil

Initially, the impurities on the surface of peanut are cleaned using water. Unremoved organic (e.g., leaves and stems) and inorganic (e.g., dust and sand) impurities can affect the yield and quality of oil produced (Wang, Liu, Hu, et al., 2016). The peanut is dehulled using manual or mechanical methods to obtain the kernel. Due to time and labor savings, mechanical shelling is typically adopted during the production of cold pressed peanut oil. Subsequently, the red skin-coated peanut kernel is dehydrated to moisture contents of 4%–8% in order to allow the skin coat to be easily removed from the kernel (Wang, Liu, Hu, et al., 2016). Color-sorting machine is then used to ensure all the peanut kernels are free from shriveled, rotten, and moldy. After color sorting, the peanut is placed into the conditioning tank, where both the pressing temperature and moisture are conditioned to enhance the oil yield (Wang, Liu, Hu, et al., 2016). Commercial cold pressed extraction of peanut oil is carried out using a twin-screw press at temperatures below 60°C (Wang, Liu, Hu, et al., 2016; Zheng, Ren, Su, Yang, & Zhao, 2013). Lastly, a frame filter is used to press and filter three times under 30°C to obtain the cold pressed peanut oil (Wang, Liu, Hu, et al., 2016). Cold pressing of peanut kernel generates peanut oil and pressed peanut meal. Commercial crude peanut oil (CCPO) is usually produced employing the cold pressing technique (Dun et al., 2019; Ogunsina et al., 2014; Wang, Liu, Wang, et al., 2016).

4 Utilization of cold pressed peanut meal

Cold pressed peanut meal, also known as defatted peanut meal or pressed peanut meal, is a by-product of peanut oil extraction. It contains a high protein level (>25%), and is white to creamy in color due to the mild thermal (<60°C) treatment (Zheng et al., 2013). Cold pressing of 100kg peanut kernels can generate 70kg of defatted peanut meal (Arya et al., 2016). In comparison with defatted soybean meal, the defatted peanut meal has several benefits, such as low level of sulfur-containing amino acids, no bean smell, inexpensiveness, and easy extrusion (Liu et al., 2016). Seven nonessential amino acids (aspartate, serine, glutamate, proline, alanine, cysteine, and tyrosine) and 10 essential amino acids (threonine, valine, methionine, isoleucine, leucine, phenylalanine, histidine, lysine, arginine, and tryptophan) were detected in defatted peanut meal (Batal, Dale, & Café, 2005). The total concentration of essential amino acids (18.4%) was lower than the total concentration of nonessential amino acids (21.0%).

The effect of roasting on the functional properties of peanut flour produced from the defatted meal was examined by Yu, Ahmedna, and Goktepe (2007). Roasting reduced the oil binding capacity and water binding capacity of defatted peanut flour (Table 1); this was probably due to irreversible denaturation caused by high-temperature roasting, which might have

TABLE 1 Functional properties of raw and roasted defatted peanut flours.

Functional properties (mL/g)	Raw	Roasted
Oil binding capacity	2.67	1.67
Water binding capacity	1.67	1.00
Emulsifying capacity	87.08	87.50
Foaming capacity	0.06	0.03

Source: Yu, J., Ahmedna, M., & Goktepe, I. (2007). Peanut protein concentrate: Production and functional properties as affected by processing. *Food Chemistry, 103*, 121–129.

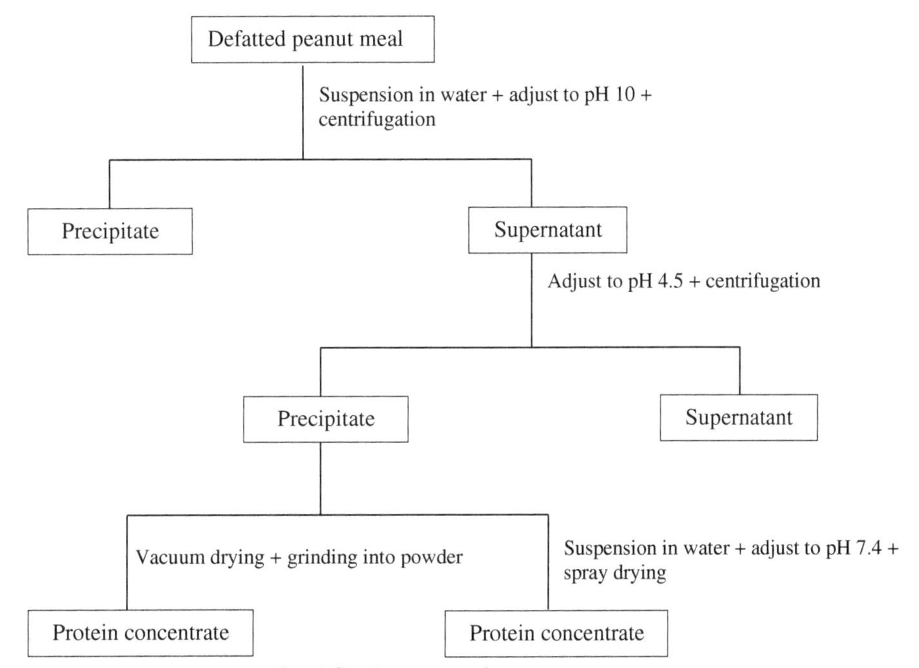

FIG. 2 Production of peanut protein concentrate using defatted peanut meal.

decomposed the hydrophobic and hydrophilic groups of peanut proteins. Both raw and roasted defatted peanut flours contained high emulsifying capacity (>87 mL/g) and low foaming capacity (<0.06 mL/g), indicating each flour's suitability to be used as a food emulsifier. In addition, Yu et al. (2007) suggested the feasibility of peanut protein concentrate (PPC) production (Fig. 2) using the defatted meal. The authors concluded that the PPC could be an ideal source ingredient for protein fortification and product formulation.

On the other hand, peanut protein isolate (PPI) is obtained by removing the nonprotein components of the defatted peanut meal mainly using the alkali solution and acid isolation method (Liu et al., 2016). Compared with the PPC, the insoluble glycan and soluble saccharides are removed in the PPI method. Study of Shafiqur, Islam, Rahman, Uddin, and Mazumder (2018) showed that PPI contained high protein solubility, which could be used as a raw ingredient to formulate protein-rich beverages and bakery food products.

5 Fatty acids and acyl lipids profile

The major fatty acids in cold pressed peanut oil were oleic (35.7%–55.3%), linoleic (20.4%–39.1%), and palmitic (9.4%–15.0%) acids (Table 2). The fatty acid composition of peanut oil is highly variable according to the environmental conditions, variety, and peanut maturity level (Akhtar, Khalid, Ahmed, Shahzad, & Suleria, 2014). Cold pressed peanut oil contained less than 27% of saturated fatty acids (SFAs) (Table 2), which are beneficial to human health. There is strong evidence to suggest that prolonged consumption of diets rich in SFAs (e.g., palmitic, lauric, and myristic acids) increase the risk of insulin resistance, coagulation, and inflammation, leading to the development of hypercholesterolemia,

TABLE 2 Fatty acid composition of cold pressed peanut oil.

Fatty acid	Veličkovska, Mitrev, and Mihajlov (2016)	Suri et al. (2019)	Ananth, Deviram, Mahalakshmi, Sivasudha, and Tietel (2019)	Konuskan, Arslan, and Oksuz (2019)
C12:0	NR	NR	0.03	NR
C14:0	NR	NR	0.15	NR
C16:0	10.06	11.65	14.87	9.37
C16:1	NR	NR	0.13	NR
C17:0	NR	NR	0.14	NR
C18:0	4.40	3.35	6.26	3.73
C18:1	35.68	38.65	49.60	55.33
C18:2	36.13	39.08	20.38	23.69
C18:3	0.33	NR	0.31	NR
C20:0	0.10	1.28	3.44	1.83
C20:1	1.37	1.17	1.80	1.57
C20:2	0.16	NR	NR	NR
C21:0	NR	NR	1.89	NR
C22:0	NR	2.82	NR	3.25
C24:0	3.93	0.89	NR	1.62
SFA	18.49	19.99	26.78	19.80
MUFA	37.05	39.82	51.53	56.90
PUFA	36.62	39.08	20.69	23.69

MUFAs, monounsaturated fatty acids; *NR*, not reported; *PUFAs*, polyunsaturated fatty acids; *SFAs*, saturated fatty acids.

cardiovascular diseases, and type 2 diabetes mellitus (Calder, 2015; Tan, Chong, Hamzah, & Ghazali, 2018a). In a study conducted by Juhaimi, Özcan, Ghafoor, Babiker, and Hussain (2018), the polyunsaturated fatty acid (PUFA) content of cold pressed plant oils followed the order of peanut (30.3%) > pistachio (28.9%) > almond (21.8%) > pecan (20.1%) > cashew (19.9%) > apricot (18.6%) > hazelnut (13.0%). High PUFA content in cold pressed peanut oil may not be advantageous in terms of oxidative stability.

More than 96% of the peanut oil is contributed by the triacylglycerols (Akhtar et al., 2014). The triacylglycerol composition of cold pressed peanut oil purchased at local grocery stores and factories in Hungary was reported by Jakab, Heberger, and Forgacs (2002). Their study showed the main triacylglycerols of peanut oil were triolein (OOO: 34.3%–46.1%), linoleoyl-dioleoyl-glycerol (LOO: 13.1%–17.5%), palmitoyl-dioleoyl-glycerol (POO: 10.1%–11.8%), and dilinoleoyl-oleoyl-glycerol (LLO: 6.7%–8.5%).

6 Minor bioactive lipid components

Cold pressed plant oil contains high amounts of essential fatty acids and bioactive lipid components (Bail, Stuebiger, Krist, Unterweger, & Buchbauer, 2008). Cold pressing and Soxhlet extraction are common methods used for oil production from oleaginous plants. The performance of an alternative (e.g., cold pressing) can be evaluated by comparison with a standard reference technique, which is Soxhlet extraction (Tan, Chong, Hamzah, & Ghazali, 2018b). Table 3 shows the phenolics and tocol contents of cold pressed and Soxhlet-extracted peanut oil. Phenolic compounds can act as antioxidants by scavenging of free radicals. In addition, the oxidative stability of unsaturated fatty acids, particularly the PUFAs of plant oils, are strongly affected by its phenolic concentrations (Siger, Nogala-Kalucka, & Lampart-Szczapa, 2008). Regardless of the extraction methods, the main phenolics in peanut oil were catechin (1.3–1.5 μg/100 g), naringenin (1.0–1.1 μg/100 g), and chlorogenic acid (0.7–0.8 μg/100 g). The total phenolics of peanut oil obtained by Soxhlet extraction

TABLE 3 Phenolics and tocols profile of peanut oil obtained by cold pressed and Soxhlet extraction.

Component	Cold pressed	Soxhlet extraction
Phenolic (μg/100g)		
Catechin	1.34	1.47
Caffeic	0.21	0.34
Chlorogenic	0.74	0.83
Ferulic	0.09	0.18
Gallic	0.18	0.27
Kampferol	0.13	0.21
Luteolin	0.68	0.81
Naringenin	0.98	1.09
p-Coumaric	0.15	0.21
Pinocembrin	0.07	0.17
Protocatechuic	0.32	0.47
Quercetin	0.12	0.19
Rutin	0.05	0.11
Resveratol	0.09	0.17
Sinapic	0.11	0.19
Vanillic	0.05	0.13
Total	5.31	6.84
Tocol (mg/kg)		
α-Tocopherol	98.65	82.49
β-Tocopherol	65.93	60.57
γ-Tocopherol	47.52	42.61
δ-Tocopherol	8.33	7.22
Total	220.43	192.89

Source: Juhaimi, F. A., Özcan, M. M., Ghafoor, K., Babiker, E. E., Hussain, S. (2018). Comparison of cold-pressing and soxhlet extraction systems for bioactive compounds, antioxidant properties, polyphenols, fatty acids and tocopherols in eight nut oils. *Journal of Food Science and Technology, 55,* 3163–3173.

(6.84 μg/100 g) were slightly greater than by cold pressing (5.31 μg/100 g); this was probably due to the good solubility of phenolic compounds in the extraction solvent used.

Plant oil is notably rich in tocopherol contents (Ananth et al., 2019). Veličkovska et al. (2016) reported that the total tocopherol content of cold pressed peanut oil was 295.6 mg/kg. Their study showed that cold pressed peanut oil contained almost the same amounts of α- and γ-tocopherol content (143.9 and 145.1 mg/kg, respectively). Another study, conducted by Juhaimi et al. (2018), indicated that the total tocopherols of peanut oil extracted using cold pressed extraction (220.4 mg/kg) was greater than by using Soxhlet extraction (192.9 mg/kg) (Table 3). A recent study published by Ananth et al. (2019) showed that the α-tocopherol level of cold pressed peanut oil was greater than that of other cold pressed plant (sesame, castor, coconut, neem, and iluppai) oils. Compared to other tocopherol isomers, α-tocopherol is the most effective free radical reducing scavenger in oil oxidation (Ananth et al., 2019). Generally, the tocopherol content of peanut oil is affected by analytical methods and conditions used to analyze the tocopherol components as well as the varietal difference of peanuts and the oil manufacturing conditions.

Phytosterols belong to the triterpene family of natural products and contribute the greatest proportion of the unsaponifiable fraction of oil. The chemical structures of phytosterols and cholesterol are similar, except for the side chain.

TABLE 4 Free and esterified phytosterols composition of cold pressed peanut oil.

Component (mg/100 g)	Free	Esterified
Sitosterol	83.9	15.1
Campesterol	11.9	3.4
Stigmasterol	13.4	ND
Δ5-Avenasterol	13.9	3.1
Sitostanol	2.0	ND
Campestanol	1.2	ND
Total	126.3	21.6

ND, nondetectable.
Source: Phillips, K. M., Ruggio, D. M., Toivo, J. I., Swank, M. A., & Simpkins, A. H., 2002. Free and esterified sterol composition of edible oils and fats. *Journal of Food Composition and Analysis, 15*, 123–142..

The phytosterols of plant oil exist mainly in free and esterified forms. Both free and esterified phytosterols reduce cholesterol absorption, resulting in a reduction of low-density lipoprotein (LDL)-cholesterol level (Richelle et al., 2004). The predominant phytosterols in peanut oil were sitosterol, campesterol, stigmasterol, and Δ5-avenasterol (Table 4).

7 Volatile components

The quantitative and qualitative of volatile components in plant oil is closely related to its aroma and flavor. Oil extraction techniques can greatly influence the levels of volatile components, thereby affecting the aroma and flavor of plant oil (Tan, 2019; Tan et al., 2018b). By using the headspace solid-phase microextraction technique, Dun et al. (2019) identified 64 volatile components in cold pressed peanut oil. These components included aldehydes, hydrocarbons, alcohols, acids, ketones, esters, furan, and pyrrole. Aldehydes (e.g., hexanal, heptanal, octanal, pentanal, nonanal, benzaldehyde, (*E*)-2-heptenal, (*E*)-2-octenal, (*E*)-2-nonenal, (*E*)-2-decenal, (*E*,*E*)-2,4-decadienal, and (*E*,*E*)-2,4-nonadienal), which are responsible for the fatty, nutty, and grassy aroma, were the main volatiles (50.9%–70.1%) detected in cold pressed peanut oil.

8 Health-promoting traits of peanut oil

A limited number of studies have been carried out on the health-promoting characteristics of peanut oil to date. Thus, much potential is waiting to be explored in this field. Peanut oil contains a high percentage of unsaturated fatty acids, which is associated with improvement of cardiovascular biomarkers such as increment of high-density lipoprotein (HDL)-cholesterol level and reduction in LDL-cholesterol and total triacylglycerol levels (Dean et al., 2011; Kris-Etherton, 1999; Kris-Etherton, Hu, Ros, & Sabaté, 2008). Pelkman et al. (2004) found that overweight subjects with moderate consumption of oil rich in unsaturated fatty acids (e.g., peanut oil) showed improvement in the cardiovascular disease risk profile. On the other hand, several studies have demonstrated that regular consumption of peanut and other oilseeds is associated with bodyweight management (Coelho et al., 2006; Dean et al., 2011; Fraser, Sabate, Beeson, & Strahan, 1992; Traoret et al., 2008), possibly due to the satiety enhancement.

9 Edible and nonedible applications of peanut oil

Cold pressed peanut oil is often used in the preparation of mayonnaise, shortening, and margarine because of its pleasant nut-like aroma. Owing to a high smoke point (>150°C), crude peanut oil is suitable for deep-fat frying (BaselineFoundation, 2019). Vanaspati, a vegetable ghee substitute used in India, is produced using peanut oil (Dean et al., 2011). Crude or refined peanut oil is widely used in creating soft conditioning bar soap with long-lasting lather properties. In addition, crude peanut oil contains a high amount of tocopherols, which is suitable as a cosmetic formulation for nourishing and moisturizing the skin.

10 Other issues

10.1 Oxidative stability

Peanut oil is commonly used in India to prepare various domestic cooking applications like frying and baking. The oxidative stability index (OSI) of solvent-extracted and cold pressed peanut oils at temperatures of 120°C and an airflow rate of 20 L/h was reported by Suri et al. (2019). Their study showed that the OSI of cold pressed peanut oil (4.3 h) was lower than solvent-extracted peanut oil (4.8 h), probably due to the presence of greater antioxidant bioactive compounds in cold pressed peanut oil. In another study, Ogunsina et al. (2014) compared the oxidative stability of CCPO and cold pressed *Moringa oleifera* seed oil (CPMO). After incubation in lab incubator at 37 °C with 55% relative humidity for 42 days, the incremental percentages of peroxide value for CCPO and CPMO were 88.86% and 78.57%, respectively. The authors' study indicates that CPMO has more oxidative stability than CCPO, which can be attributed to the presence of high amounts of unsaturated fatty acids, particularly PUFAs, in CCPO. According to Sunil, Reddy, Krishna, and Urooj (2013), the oxidative stability of high-PUFA CCPO can be enhanced by blending it with SFA- or MUFA-rich plant oils.

10.2 Adulteration and authenticity

Cold pressed peanut oil is rich in unsaturated fatty acids, and bioactive compounds, hence adulteration could occur. A novel technique to identify the adulterants in peanut oil by monitoring the change of absorbance value with refrigeration time using a spectrophotometer was reported by Su et al. (2012). Their study showed the absorbance value of peanut oil adulterated with another plant (e.g., palm olein, corn, cottonseed, soybean, or rapeseed) oils reduced appreciably when refrigerated at the freezing point. In another study, Sunil et al. (2013) indicated that incorporation of 20% of red palm olein oil or rice bran oil shifted the proportional distributions of the fatty acids in CCPO, resulting in SFA content increment in the range of 2.1%–6.9%.

References

Akcali, I. D., Ince, A., & Guzel, E. (2006). Selected physical properties of peanuts. *International Journal of Food Properties*, *9*, 25–37.

Akhtar, S., Khalid, N., Ahmed, I., Shahzad, A., & Suleria, H. A. R. (2014). Physicochemical characteristics, functional properties, and nutritional benefits of peanut oil: A review. *Critical Reviews in Food Science and Nutrition*, *54*, 1562–1575.

Ananth, D. A., Deviram, G., Mahalakshmi, V., Sivasudha, T., & Tietel, Z. (2019). Phytochemical composition and antioxidant characteristics of traditional cold pressed seed oils in South India. *Biocatalysis and Agricultural Biotechnology*, *17*, 416–421.

Arya, S. S., Salve, A. R., & Chauhan, S. (2016). Peanuts as functional food : A review. *Journal of Food Science and Technology*, *53*, 31–41.

Bail, S., Stuebiger, G., Krist, S., Unterweger, H., & Buchbauer, G. (2008). Characterisation of various grape seed oils by volatile compounds, triacylglycerol composition, total phenols and antioxidant capacity. *Food Chemistry*, *108*, 1122–1132.

Bansal, U. K., Satija, D. R., & Ahuja, K. L. (1993). Oil composition of diverse groundnut (*Arachis hypogaea* L.) genotypes in relation to different environments. *Journal of the Science of Food and Agriculture*, *63*, 17–19.

BaselineFoundation (2019). *Healthiest cooking oil comparison chart with smoke points and omega 3 fatty acid ratios* [WWW document]. *(2019)*. https://www.jonbarron.org/diet-and-nutrition/healthiest-cooking-oil-chart-smoke-points (Accessed 7 February 2019).

Batal, A., Dale, N., & Café, M. (2005). Nutrient composition of peanut meal. *Journal of Applied Poultry Research*, *14*, 254–257.

Calder, P. C. (2015). Functional roles of fatty acids and their effects on human health. *Journal of Parenteral and Enteral Nutrition*, *39*, 18S–32S.

Coelho, S. B., de Sales, R. L., Iyer, S. S., Bressan, J., Costa, N. M. B., Lokko, P., et al. (2006). Effects of peanut oil load on energy expenditure, body composition, lipid profile, and appetite in lean and overweight adults. *Nutrition*, *22*, 585–592.

Dean, L. L., Davis, J. P., & Sanders, T. H. (2011). Groundnut (peanut) oil. In F. D. Gunstone (Ed.), *Vegetable oils in food technology: Composition, properties, and uses* (pp. 225–242). West Sussex, UK: Blackwell Publishing Ltd.

Dun, Q., Yao, L., Deng, Z., Li, H., Li, J., Fan, Y., et al. (2019). Effects of hot and cold-pressed processes on volatile compounds of peanut oil and corresponding analysis of characteristic flavor components. *LWT—Food Science and Technology*, *112*, 107648.

Fraser, G. E., Sabate, J., Beeson, W. L., & Strahan, T. M. (1992). A possible protective effect of nut consumption on risk of coronary heart disease: The Adventist health study. *Archives of Internal Medicine*, *152*, 1416–1424.

Grosso, N. R., Lamarque, A., Maestri, D. M., Zygadlo, J. A., & Guzman, C. A. (1994). Fatty acid variation of runner peanut (*Arachis hypogaea* L.) among geographic localities from Córdoba (Argentina). *Journal of the American Oil Chemists' Society*, *71*, 541–542.

International Nut and Dried Fruit (2018). *Nuts and dried fruits statistical year book 2017/2018*. .

Jakab, A., Heberger, K., & Forgacs, E. (2002). Comparative analysis of different plant oils by high-performance liquid chromatography-atmospheric pressure chemical ionization mass spectrometry. *Journal of Chromatography A*, *976*, 255–263.

Juhaimi, F. A., Özcan, M. M., Ghafoor, K., Babiker, E. E., & Hussain, S. (2018). Comparison of cold-pressing and soxhlet extraction systems for bioactive compounds, antioxidant properties, polyphenols, fatty acids and tocopherols in eight nut oils. *Journal of Food Science and Technology*, *55*, 3163–3173.

Kim, N. K., & Hung, Y. C. (1991). Mechanical properties and chemical composition of peanuts as affected by harvest date and maturity. *Journal of Food Science*, *56*, 1378–1381.

Konuskan, D. B., Arslan, M., & Oksuz, A. (2019). Physicochemical properties of cold pressed sunflower, peanut, rapeseed, mustard and olive oils grown in the Eastern Mediterranean region. *Saudi Journal of Biological Sciences*, *26*, 340–344.

Kris-Etherton, P. M. (1999). Monounsaturated fatty acids and risk of cardiovascular disease. *Circulation*, *100*, 1253–1258.

Kris-Etherton, P. M., Hu, F. B., Ros, E., & Sabaté, J. (2008). The role of tree nuts and peanuts in the prevention of coronary heart disease: Multiple potential mechanisms. *Journal of Nutrition*, *138*, 1746–1751.

Liu, H., Shi, A., Liu, L., Wu, H., Ma, T., He, X., et al. (2016). Peanut protein processing technology. In Q. Wang (Ed.), *Peanuts: Processing technology and product development* (pp. 83–209). London, UK: Academic Press.

Musa, Ö. M. (2011). Some nutritional characteristics of kernel and oil of peanut (*Arachis hypogaea* L.). *Journal of Oleo Science*, *59*, 1–5.

Ogunsina, B. S., Indira, T. N., Bhatnagar, A. S., Radha, C., Debnath, S., & Gopala Krishna, A. G. (2014). Quality characteristics and stability of *Moringa oleifera* seed oil of Indian origin. *Journal of Food Science and Technology*, *51*, 503–510.

Olajide, J. O., & Igbeka, J. C. (2003). Some physical properties of groundnut kernels. *Journal of Food Engineering*, *58*, 201–204.

Pelkman, C. L., Fishell, V. K., Maddox, D. H., Pearson, T. A., Mauger, D. T., & Kris-Etherton, P. M. (2004). Effects of moderate-fat (from monounsaturated fat) and low-fat weight-loss diets on the serum lipid profile in overweight and obese men and women. *The American Journal of Clinical Nutrition*, *79*, 204–212.

Richelle, M., Enslen, M., Hager, C., Groux, M., Tavazzi, I., Godin, J. P., et al. (2004). Both free and esterified plant sterols reduce cholesterol absorption and the bioavailability of β-carotene and α-tocopherol in normocholesterolemic humans. *The American Journal of Clinical Nutrition*, *80*, 171–177.

Sahdev, R. K. (2015). Present status of peanuts and progression in its processing and preservation techniques. *Agricultural Engineering International: CIGR Journal*, *17*, .

Shafiqur, R., Islam, A., Rahman, M. M., Uddin, M. B., & Mazumder, A. R. (2018). Isolation of protein from defatted peanut meal and characterize their nutritional profile. *Chemistry Research Journal*, *3*, 187–196.

Siger, A., Nogala-Kalucka, M., & Lampart-Szczapa, E. (2008). The content and antioxidant activity of phenolic compounds in cold-pressed plant oils. *Journal of Food Lipids*, *15*, 137–149.

Su, R., Wang, X. H., Zhao, T. Q., Yu, W. Z., Feng, X. D., Zhang, H. Q., et al. (2012). Analysis of peanut oil adulterated with other edible oils by spectrophotometry. *Chemical Research in Chinese Universities*, *28*, 14–18.

Sunil, L., Reddy, P. V., Krishna, A. G. G., & Urooj, A. (2013). Retention of natural antioxidants of blends of groundnut and sunflower oils with minor oils during storage and frying. *Journal of Food Science and Technology*, *52*, 849–857.

Suri, K., Singh, B., Kaur, A., & Singh, N. (2019). Impact of roasting and extraction methods on chemical properties, oxidative stability, and Maillard reaction products of peanut oils. *Journal of Food Science and Technology*, *56*, 2436–2445.

Tan, C. X. (2019). Virgin avocado oil: An emerging source of functional fruit oil. *Journal of Functional Foods*, *54*, 381–392.

Tan, C. X., Chong, G. H., Hamzah, H., & Ghazali, H. M. (2018a). Hypocholesterolaemic and hepatoprotective effects of virgin avocado oil in diet-induced hypercholesterolaemia rats. *International Journal of Food Science and Technology*, *53*, 2706–2713.

Tan, C. X., Chong, G. H., Hamzah, H., & Ghazali, H. M. (2018b). Characterization of virgin avocado oil obtained via advanced green techniques. *European Journal of Lipid Science and Technology*, *120*, e1800170.

Traoret, C. J., Lokko, P., Cruz, A. C. R. F., Oliveira, C. G., Costa, N. M. B., Bressan, J., et al. (2008). Peanut digestion and energy balance. *International Journal of Obesity*, *32*, 322–328.

Veličkovska, S. K., Mitrev, S., & Mihajlov, L. (2016). Physicochemical characterization and quality of cold-pressed peanut oil obtained from organically produced peanuts from Macedonian "Virginia" variety. *Grasas y Aceites*, *67*, e118.

Wang, Q., Liu, H., Hu, H., Mzimbiri, R., Yang, Y., & Chen, Y. (2016). Peanut oil processing technology. In Q. Wang (Ed.), *Peanuts: Processing technology and product development* (pp. 63–81). London, UK: Academic Press.

Wang, Q., Liu, L., Wang, L., Guo, Y., & Wang, J. (2016). Introduction. In Q. Wang (Ed.), *Peanuts: Processing technology and product development* (pp. 1–22). London, UK: Academic Press.

Yu, J., Ahmedna, M., & Goktepe, I. (2007). Peanut protein concentrate: Production and functional properties as affected by processing. *Food Chemistry*, *103*, 121–129.

Zheng, L., Ren, J., Su, G., Yang, B., & Zhao, M. (2013). Comparison of in vitro digestion characteristics and antioxidant activity of hot- and cold-pressed peanut meals. *Food Chemistry*, *141*, 4246–4252.

Further reading

Phillips, K. M., Ruggio, D. M., Toivo, J. I., Swank, M. A., & Simpkins, A. H. (2002). Free and esterified sterol composition of edible oils and fats. *Journal of Food Composition and Analysis*, *15*, 123–142.

Chapter 33

Cold pressed pequi (*Caryocar brasiliense* Camb.) almond oil

Onur Ketenoglu[a], Mustafa Kiralan[b], and Mohamed Fawzy Ramadan[c,d]

[a]*Department of Food Engineering, Faculty of Engineering, Cankiri Karatekin University, Cankiri, Turkey,* [b]*Department of Food Engineering, Faculty of Engineering, Balıkesir University, Balikesir, Turkey,* [c]*Agricultural Biochemistry Department, Faculty of Agriculture, Zagazig University, Zagazig, Egypt,* [d]*Deanship of Scientific Research, Umm Al-Qura University, Makkah, Saudi Arabia*

Abbreviations

CPPO cold pressed pequi oil
DPPH 2,2-diphenyl-1-picrylhydrazyl
MUFA monounsaturated fatty acid
ORAC oxygen radical absorbance capacity
PUFA polyunsaturated fatty acid
SFA saturated fatty acid
SFEAP supercritical fluid extraction assisted by pressing
TE trolox equivalents
TEAC trolox equivalent antioxidant capacity

1 Introduction

Pequi (*Caryocar brasiliense*) is a typical fruit found in Brazil (Cerrado region), which belongs to the *Caryocaraceae* family (Machado, Mello, & Hubinger, 2015). The fruits have been used in various food applications (juice, ice cream, liquor, jelly, and traditional dishes), cosmetics, and traditional medicine (Faria, Damasceno, & Ferrari, 2014; Faria-Machado et al., 2015; Guedes, Antoniassi, & de Faria-Machado, 2017). In addition to their usage in oil production, the leaves and the fruit pulp of pequi have been used as tannin sources (De Araujo, 1995).

The peel and residual parts are the majority of fresh fruit mass (approximately 90% of its total mass). Pulp and almond parts of the fruit represent 10% of fruit mass. However, these parts are valuable for their nutritional value (Roesler et al., 2007). The proximate composition of pequi almonds is given in Table 1. As seen in this table, the pulp of pequi is rich in lipids (Cardoso, Reis, Hamacek, & Sant'ana, 2013; Macedo et al., 2011). In addition, almonds and pulp contain high amounts of phenolics and carotenoids (De Lima, Silva, Trindade, Torres, & Mancini-Filho, 2007).

The highest level of MUFA was 58.5% in cold pressed pequi oil (CPPO), followed by SFA (38.8%). The PUFA content in CPPO was reported as 2.63%. Among MUFAs, oleic acid ($C_{18:1}$) was the most abundant one, and it was also the major fatty acid in the fatty acid composition of CPPO (56.4%). The main SFAs of CPPO were palmitic ($C_{16:0}$) and stearic ($C_{18:0}$) acids. In particular, palmitic acid was the main acid abundant in pequi oil (35.2%). The major PUFA, linoleic acid (C18:2), was found at a level of 2.25% (Cicero et al., 2018).

As for minor compounds, CPPO contains tocopherols, phytosterols, and carotenoids which have biological activity on human health. Among tocopherols, the main isomer was α-tocopherol in CPPO, with a range between 79.7 and 100 mg/kg. The total phytosterol content of CPPO is similar to conventional edible oils, ranging between 743 mg/kg and 964 mg/kg. Stigmasterol was the predominant sterol fraction found in CPPO (482–653 mg/kg) (Torres et al., 2016). Azevedo-Meleiro and Rodriguez-Amaya (2004) reported that such carotenoids as violaxanthin, lutein, and zeaxanthin were the major carotenoids of pequi, while the fruit contained lesser amounts of β-carotene, β-cryptoxanthin, and neoxanthin.

The pequi fruit has also been reported as a good source of dietary fiber and vitamins. Cardoso et al. (2013) reported that the cooked fruit contained 9.9 g/100 g of dietary fiber and 14.3 mg/100 g of vitamin C. De Oliveira Sousa et al. (2011)

Cold Pressed Oils. https://doi.org/10.1016/B978-0-12-818188-1.00033-5
Copyright © 2020 Elsevier Inc. All rights reserved.

TABLE 1 Nutritional value of different parts of pequi.

Macronutrient	Pulp[a] (%, w/w)	Kernel[a] (%, w/w)	Almond[b] (g/100 g)	Pulp-cooked[c] (g/100 g)
Moisture	51.3	31.7	4.97	51.7
Ash	0.55	3.02	4.54	0.5
Fiber	3.68 (crude)	1.05 (crude)	10.4 (total)	9.9 (total)
Protein	3.39	20.79	29.65	2.2
Lipid	29.6	32.5	50.0	33.1
Carbohydrate	11.4	10.9	0.40	2.7

[a]*Adapted from Macedo, Santos, Pantoja, and Santos (2011).*
[b]*Adapted from De Oliveira Sousa, Fernandes, Alves, De Freitas, and Naves (2011).*
[c]*Adapted from Cicero et al. (2018).*

reported that the total dietary fiber content of pequi almond was 10.4 g/100 g, most of which were constituted by insoluble fibers (6.82 g/100 g). Other reported vitamin contents (mg/100 g) for the respective internal and the external mesocarps of the fruit are vitamin C (12.0, 6.12), vitamin B1 (29.7, 0.81), vitamin B2 (463, 360), and niacin (387, 346) (Carvalho & Burger, 1960).

Different parts of pequi are also good sources for mineral compounds. Some of the reported mineral compositions (g/100 g) of pequi are calcium (0.05), phosphorus (0.21), iron (0.83–1.39), and copper (0.24) (Carvalho & Burger, 1960). In another report by De Oliveira Sousa et al. (2011), pequi almond was defined as rich in minerals, particularly in terms of phosphorus, potassium, and magnesium. In addition, the presence of selenium and zinc was stated by the researchers.

The amino acid composition of pequi almond was compared to that of such exotic fruits as baru almond, cerrado cashew nut, and peanut (De Oliveira Sousa et al., 2011). The researchers reported that the total essential amino acid content of pequi almond was 352 mg amino acid/g protein, and this value was found to be lower than those of other tested fruits. According to the study, this was due to the low lysine content of pequi almond fruit.

2 Cold press extraction and processing of oils

In recent years, consumers have tended to prefer natural, nutritional, and safe food. Therefore, cold pressed oils are popular among food products. The cold pressing extraction method is used to obtain oil from plant seed instead of traditional solvent extraction because cold pressing does not involve the use of toxic solvents or heat. These oils are therefore safer due to their lack of solvent content. In addition, most bioactive compounds such as essential fatty acids, flavonoids, phenolics, and tocols remain in cold pressed oils (Parker, Adams, Zhou, Harris, & Yu, 2003; Teh & Birch, 2013). In a traditional method called a handmade process, an intensive cooking in hot water and subsequent removal of the supernatant is carried out (Torres, Shinagawa, et al., 2016). Low yield and high temperature are the major disadvantages of this traditional extraction method (Facioli & Gonçalves, 1998).

Johner, Hatami, and Meireles (2018) studied a new technique called supercritical fluid extraction assisted by pressing (SFEAP), and evaluated this technique on the extraction of oil from pequi. At the optimum working condition, the yield in the SFEAP method was eight times greater than that of supercritical fluid extraction during the first minute of extraction. Pessoa et al. (2015) investigated the subcritical propane extraction of pequi pulp oil. Their applied pressure and temperature for the subcritical propane extraction ranged from 5 to 15 MPa and 303.15 to 333.15 K, respectively. The researchers also investigated the effects of different extraction techniques such as ultrasound-assisted extraction and Soxhlet extraction on the properties of pequi pulp oil. The highest yield was obtained when ethanol was used as the solvent. Their research also indicated that the fatty acid composition of pequi pulp oil was not significantly affected by the extraction technique. The overall oleic and palmitic acid contents ranged between 58.7% and 60.5% and 33.6% and 35.4%, respectively. Araújo, Oliveira, Carvalho, Menezes, and Queiroz (2019) studied the ethanol extraction of oil from pequi seeds using cosolvents. They built a model using the central composite rotational design of the response surface methodology (RSM). They achieved to recover the highest amount of soluble solids (84.4%) from pequi at 53.9°C when they used *n*-hexane and isopropanol as cosolvents, with mass fractions of 15.9% and 4.05%, respectively.

3 Cold pressed oil recovery, content, uses of oil cake, and economy

Pequi cake contains a high level of starch, and this component could therefore be considered for the production of bioethanol. Macedo et al. (2011) used the remaining oil cake after extracting the oil. Carbohydrates present in the pequi cake were acid-hydrolyzed to obtain fermentable sugars and the conversion of these substances to bioethanol was evaluated. Pequi cake could be used in bioethanol production, yielding a 5.3% (v/p) dry basis under fermentative conditions.

Pequi by-products were also studied in terms of their recoverable dietary fiber and antioxidant compounds. Leão, Franca, Oliveira, Bastos, and Coimbra (2017) produced flour from the wastes of pequi and determined the antioxidant capacity and dietary fiber content of the flours. According to researchers, an average of approximately 40g/100g of total dietary fibers was determined, where most of the dietary fibers were the insoluble fraction. The total extractable phenolics of the by-products were averaged as 16g GAE/100g.

Monteiro, da Silva, da S. Martins, Barin, and da Rosa (2015) investigated the antioxidant activity and chemical composition of pequi peels processed by different methods. Their highest inhibitory concentration (IC_{50}) was 69.1 μg/mL when the DPPH· method was used. The researchers also stated that the total phenolic content was the highest (78.58 g GAE/kg dry weight) when hydro-ethanolic extraction was applied by shaking, followed by hydroethanolic extraction using a microwave.

4 Fatty acid composition of cold pressed oil

The fatty acid composition of the CPPO is presented in Table 2. CPPO contains a high proportion of MUFA followed by SFA and a low proportion of PUFA. Oleic ($C_{18:1}$) and linoleic ($C_{18:2}$) acids as major unsaturated fatty acids are identified in

TABLE 2 Fatty acid composition of CPPO (%).

Fatty acid	Almond[a]	Almond[b]	Fruit[c]
$C_{14:0}$	0.36–0.38	0.36	0.01
$C_{16:0}$	33.3–35.8	29.4	35.2
$C_{16:1}$	0.52–0.85	0.66	1.65
$C_{17:0}$	–	–	0.54
$C_{17:1}$	–	–	0.28
$C_{18:0}$	2.20–2.63	2.44	2.11
$C_{18:1}$	53.4–55.3	59.9	56.4
$C_{18:2}$	6.39–7.23	6.44	2.25
$C_{18:3}$	–	–	0.38
$C_{20:0}$	–	–	0.36
$C_{20:1}$	–	–	0.12
$C_{22:0}$	–	–	0.56
$C_{22:1}$	–	0.63	–
$C_{23:0}$	0.51–0.81	–	–
$C_{24:0}$	–	–	0.01
SFAs	36.8–39.2	32.2	38.8
MUFAs	54.3–55.9	61.2	58.5
PUFAs	6.39–7.23	6.44	2.63

[a]*Adapted from Torres, Shinagawa, et al. (2016).*
[b]*Adapted from Torres, de Santana, et al. (2016).*
[c]*Adapted from Cicero et al. (2018).*

CPPO. The oleic acid content of CPPO was within 53.4%–59.9% of total fatty acids, while linoleic acid was found at lower levels (2.25%–7.23%). Palmitic acid is the predominant SFA (29.4%–35.8% of total fatty acids) in CPPO.

The nature of the fruit may differ regarding different parts of the plant. According to a report by Passos et al. (2003), a difference in the essential oil composition of the seeds and the leaves was observed. Ethyl hexanoate was the most abundant compound in the oil obtained from the pequi seed, while octacosane, heptadecane, and hexadecanol were the major components in the oil extracted from the leaves. Maia, Andrade, and da Silva (2008) reported the composition of aroma compounds present in the fruit. According to their report, the high amount of ethyl hexanoate (52.9%) was found to be liable for the aromatic character of the fruit. Some other compounds such as ethyl octanoate, tetrahydro furfuryl alcohol, ethyl butanoate, butyl palmitate, isobutyl stearate, and 3-methyl valeric acid also played roles as minor aroma components of the pequi.

5 Minor bioactive lipids in cold pressed oil

The tocopherol contents in CPPO are shown in Table 3. The most abundant tocopherol in CPPO was α-tocopherol, followed by γ-tocopherol. Total tocopherol content varied from 133.2 to 191.9 mg/kg. Torres, Shinagawa, et al. (2016) revealed that α-tocopherol and γ-tocopherol accounted for 58.9% and 41.0% of the average total tocopherols in CPPO, respectively. The α-tocopherol contents were between 79.7 and 100 mg/kg, while the γ-tocopherol level changed between 46.6 and 91.3 mg/kg.

Phytosterols are important in human health due to their LDL cholesterol-reducing effect and antioxidant activity; levels of phytosterols in edible oils are also used for oil identification and determination of oil quality (Köhler, Teupser, Elsässer, & Weingärtner, 2017; Wester, 2000). The levels of total plant sterols ranged from 743.8 to 964.5 mg/kg. β-Sitosterol is predominant in many vegetable oils. In contrast, stigmasterol was the major sterol found in CPPO (482.0–653.4 mg/kg), followed by β-sitosterol with a range of 220.9–273.1 mg/kg.

According to Torres, Shinagawa, et al. (2016), the total phenolic content in CPPO is between 87.56 and 155.0 mg GAE/100 g. Cicero et al. (2018) identified five phenolic compounds. Vanillin (0.69 μg/g), *p*-hydroxybenzoic acid

TABLE 3 Minor bioactive components of CPPO.

	Compound	
Tocopherol (mg/kg)[1]	α-Tocopherol	79.7–100.0
	γ-Tocopherol	46.6–91.3
	Total tocopherols	133.2–191.9
Sterol (mg/kg)[a]	Campesterol	38.0–48.1
	Stigmasterol	482.0–653.4
	β-sitosterol	220.9–273.1
	Total sterols	743.8–964.5
Total phenolics (mg GAE/100 g)[a]		87.6–155.0
Total carotenoids (μg/100 g)[a]		41.6–159.8
Squalene (μg/g)[b]		181.0
Phenolic compound (μg/g)[b]	*p*-Coumaric acid	0.59
	Luteolin	0.08
	Vanillin	0.69
	Caffeic acid	0.18
	p-Hydroxybenzoic acid	0.61

[a]*Adapted from Torres, Shinagawa, et al. (2016).*
[b]*Adapted from Cicero et al. (2018).*

(0.61 μg/g), and *p*-coumaric acid (0.59 μg/g) were the most abundant phenolic compounds in CPPO, followed by smaller quantities of luteolin and caffeic acid. The range of the total carotenoids was reported as between 41.58 and 159.8 μg/100 g in CPPO.

Squalene is a triterpene and found at higher levels in olive oil, peanut oil, and pumpkin oil (Amarowicz, 2009). Squalene also plays a significant role in human health such as preventing some cancer types (Newmark, 1997). Cicero et al. (2018) demonstrated that the squalene content in CPPO was 181.0 μg/g.

6 Composition of cold pressed oil in comparison to other cold pressed oils

The two major fatty acids of CPPO are oleic ($C_{18:1}$) and palmitic ($C_{16:0}$) acids, constituting approximately 90% of total fatty acids. The remaining fatty acid composition includes myristic ($C_{14:0}$), stearic ($C_{18:0}$), palmitoleic ($C_{16:1}$), linoleic ($C_{18:2}$), and erucic acid ($C_{22:1}$). The reported oleic and palmitic acid concentrations of CPPO are (in respective order): 54.3% (cold pressed)–54.9% (handmade), 34.7% (cold pressed)–34.92% (handmade) (Torres, de Santana, et al., 2016), 59.9% (cold pressed)–56.3% (handmade), and 29.4% (cold pressed)–33.7% (handmade) (Torres, Shinagawa, et al., 2016). A comparison of the fatty acid compositions of some cold pressed oils versus cold pressed and handmade pequi almond oils is given in Table 4.

TABLE 4 Comparison of fatty acids in some selected cold pressed oils versus pequi fruit (%).

Fatty acid	Avocado oil[a]	Moringa oleifera[b]	Moringa peregrina[b]	Pumpkin seed oil[c]	Pequi oil (handmade)[d]	Pequi oil (cold pressed)[d]
$C_{12:0}$	–	2.17	1.93	–	–	–
$C_{14:0}$	0.01	0.78	0.61	0.08–0.1	0.35	0.36
$C_{16:0}$	14.2	9.58	11.7	10.2–11.8	33.7	29.4
$C_{16:1}$	7.06	3.17	1.83	–	0.59	0.66
$C_{17:0}$	0.01	–	–	–	–	–
$C_{17:1}$	0.08	–	–	–	–	–
$C_{18:0}$	2.15	2.81	4.76	4.45–5.15	2.62	2.44
$C_{18:1}$	59.4	75.4	74.6	30.3–42.0	56.3	59.9
$C_{18:2}$	14.6	1.69	1.84	43.6–52.1	5.74	6.44
$C_{18:3}$	1.3	1.87	1.97	0.16–0.26	–	–
$C_{20:0}$	0.41	5.71	3.84	0.28–0.33	–	–
$C_{20:1}$	0.51	1.43	1.13	–	–	–
$C_{22:0}$	0.08	5.83	2.74	–	–	–
$C_{22:1}$	–	–	–	–	0.60	0.63
$C_{24:0}$	0.06	–	–	–	–	–
SFAs	16.9	26.8	25.6	15.1–16.7	36.7	32.2
MUFAs	67.1	80.0	77.5	30.3–43.6	57.5	61.2
PUFAs	15.9	3.56	3.81	42.2–52.4	5.74	6.44

[a]*Adapted from Cicero et al. (2018).*
[b]*Adapted from Özcan, Ghafoor, Al Juhaimi, Ahmed, and Babiker (2019).*
[c]*Adapted from Vujasinovic, Djilas, Dimic, Romanic, and Takaci (2010).*
[d]*Adapted from Torres, de Santana, et al. (2016).*

7 Edible and nonedible applications of cold pressed oil

Caryocar brasiliense oil contains a high level of MUFA, approximately 60% of oleic acid, making this fruit interesting for food, cosmetic, and oleochemical uses (Guedes, Antoniassi, & de Faria-Machado, 2017). Pequi oil could be reacted with a chemical interesterification process to produce a *trans*-free fat that could be used in different food industries (Guedes et al., 2017). *Caryocar brasiliense* oil is also used in cosmetic products. Faria et al. (2014) investigated the effects of *Caryocar brasiliense* oil on stratum corneum hydration and skin barrier function. The stratum corneum is a barrier toward physical and chemical influences, and also against dehydration. This study revealed that formulations containing pequi oil improved skin hydration.

Borges et al. (2012) studied the physical and chemical characteristics of biodiesel produced by the transesterification of pequi oil. The authors emphasized that pequi oil could be used as an alternative resource for obtaining biodiesel, and many of the physical and chemical characteristics such as acidity, density, viscosity, and carbon residue of this biodiesel meet the NP (Brazilian National Petroleum Agency) standards. In a study by Silva, de Assunção, Vieira, de Oliveira, and Batista (2014), results demonstrated that the biodiesels obtained from pequi oil exhibited satisfactory thermal stability and qualified as potential substituents of traditional mineral diesel.

8 Health-promoting traits of cold pressed oil and oil constituents

Torres, de Santana, et al. (2016) demonstrated that CPPO attenuated the carbon tetrachloride-induced variations in serum and hepatic tissue in rats due to its antioxidant and antiinflammatory properties. Traesel et al. (2016) investigated the toxicological potential of pequi oil through acute and subchronic toxicity methods. Their results revealed the low acute and subchronic toxicity of pequi oil in rats.

Palmeira et al. (2016) investigated the potential of pequi oil in the prevention of liver cancer in mice. Results demonstrated that this oil showed a hepatoprotective effect against preneoplastic liver lesions induced by the carcinogen diethylnitrosamine in mice. Roll, Miranda-Vilela, Longo, Agostini-Costa, and Grisolia (2018) emphasized that pequi oil could be used as a natural dietary supplementation, and therefore it could be a good strategy for the elderly, mainly for females, to protect against some diseases such as anemia, inflammation, and oxidative stress-related to aging. Nascimento-Silva and Naves (2019) stated that the pequi pulp oil exhibited such health effects as antioxidant, cardioprotective, antiinflammatory, antigenotoxic, and anticarcinogenic on animals. They also emphasized that oil could also benefit human health.

9 Oxidative stability and antiradical and antimicrobial activity of CPPO

The oxidative stability of CPPO was determined using Rancimat, thermogravimetric, and differential scanning calorimetry. The oil is reported to exhibit good oxidative and thermal stability. Induction periods of handmade and CPPO varied between 7.33 and 15.9h (Torres, Shinagawa, et al., 2016). Pequi oils exhibit antiradical activity according to DPPH·, ABTS, and ORAC methods. IC_{50} values varied between 58.48mg/mL and 76.46mg/mL in a DPPH· test. ABTS results showed that antioxidant capacity ranged from 10.6μmol TE/g to 40.4μmol TE/g. The lipophilic ORAC values changed between 113.9μmol TE/100g and 280.8μmol TE/100g (Torres, Shinagawa, et al., 2016). Roesler, Catharino, Malta, Eberlin, and Pastore (2008) stated that the maximum concentration for half inhibition (IC_{50}) against lipid peroxidation was 0.78μg/mL for the ethanol extracts of the fruit's peel. In addition, they noted that the high antioxidant activity of the fruit might be due to its contents of such bioactive components as gallic acid, quinic acid, quercetin, and quercetin 3-O-arabinose.

In a study by Machado et al. (2015), the FRAP values (mM/mL) of ultrafiltration and two nanofiltration retentates ranged from 3.16 to 3.71, while the FRAP value of nonfiltrated feed was 2.62. Their study also revealed that the highest total phenolics in retentates was 338.4μg/mL, while it was 216.3μg/mL in the feed.

The antifungal activity of pequi essential oils was reported by Passos et al. (2003). The authors stated that the minimal inhibitory concentration (MIC) of pequi essential oil was 500μg/mL for the inhibition of *Paracoccidioides brasiliensis*. The study also revealed the antifungal activity of the oil on *Cryptococcus neoformans*.

References

Amarowicz, R. (2009). Squalene: A natural antioxidant? *European Journal of Lipid Science and Technology*, *111*(5), 411–412.

Araújo, A. C. M. A., Oliveira, É. R., Carvalho, G. R., Menezes, E. G. T., & Queiroz, F. (2019). Modeling of cosolvent and temperature effects on ethanolic extraction of Pequi and Murici seeds. *Emirates Journal of Food and Agriculture*, *31*(3), 153–163.

Azevedo-Meleiro, C. H., & Rodriguez-Amaya, D. B. (2004). Confirmation of the identity of the carotenoids of tropical fruits by HPLC-DAD and HPLC-MS. *Journal of Food Composition and Analysis*, *17*(3–4), 385–396.

Borges, K. A., Batista, A. C. F., Rodrigues, H. D. S., Terrones, M. H., Vieira, A. T., & de Oliveira, M. F. (2012). Production of methyl and ethyl biodiesel fuel from pequi oil (*Caryocar brasiliensis* Camb.). *Chemistry and Technology of Fuels and Oils*, *48*(2), 83–89.

Cardoso, L. D. M., Reis, B. D. L., Hamacek, F. R., & Sant'ana, H. M. P. (2013). Chemical characteristics and bioactive compounds of cooked pequi fruits (*Caryocar brasiliense* Camb.) from the Brazilian Savannah. *Fruits*, *68*(1), 3–14.

Carvalho, M. C., & Burger, O. N. (1960). *Contribuição ao estudo do pequi de Brasília*: (pp. 7–15). (Vol. 50). Brasília: Coleção Estudo e Pesquisa Alimentar-SAPS.

Cicero, N., Albergamo, A., Salvo, A., Bua, G. D., Bartolomeo, G., Mangano, V., et al. (2018). Chemical characterization of a variety of cold-pressed gourmet oils available on the Brazilian market. *Food Research International*, *109*, 517–525.

De Araujo, F. D. (1995). A review of *Caryocar brasiliense* (caryocaraceae)—An economically valuable species of the central Brazilian cerrados. *Economic Botany*, *49*, 40–48.

De Lima, A. D., Silva, A. M. O., Trindade, R. A., Torres, R. P., & Mancini-Filho, J. (2007). Composição química e compostos bioativos presentes na polpa e na amêndoa do pequi (*Caryocar brasiliense*, Camb.). *Revista Brasileira de Fruticultura*, *29*(3), 695–698.

De Oliveira Sousa, A. G., Fernandes, D. C., Alves, A. M., De Freitas, J. B., & Naves, M. M. V. (2011). Nutritional quality and protein value of exotic almonds and nut from the Brazilian savanna compared to peanut. *Food Research International*, *44*(7), 2319–2325.

Facioli, N. L., & Gonçalves, L. A. G. (1998). Modificação por via enzimática da composição triglicerídica do óleo de piqui (*Caryocar brasiliense* Camb). *Química Nova*, *21*(1), 16–19.

Faria, W. C. S., Damasceno, G. A. D. B., & Ferrari, M. (2014). Moisturizing effect of a cosmetic formulation containing pequi oil (*Caryocar brasiliense*) from the Brazilian cerrado biome. *Brazilian Journal of Pharmaceutical Sciences*, *50*(1), 131–136.

Faria-Machado, A. F., Tres, A., van Ruth, S. M., Antoniassi, R., Junqueira, N. T. V., Lopes, P. S. N., et al. (2015). Discrimination of pulp oil and kernel oil from pequi (*Caryocar brasiliense*) by fatty acid methyl esters fingerprinting, using GC-FID and multivariate analysis. *Journal of Agricultural and Food Chemistry*, *63*(45), 10064–10069.

Guedes, A. M. M., Antoniassi, R., & de Faria-Machado, A. F. (2017). Pequi: A Brazilian fruit with potential uses for the fat industry. *Oilseeds and fats, Crops and Lipids*, *24*(5), D507.

Guedes, A. M. M., Antoniassi, R., Galdeano, M. C., Grimaldi, R., de Carvalho, M. G., Wilhelm, A. E., et al. (2017). Length-scale specific crystalline structural changes induced by molecular randomization of pequi oil. *Journal of Oleo Science*, *66*(5), 469–478.

Johner, J. C. F., Hatami, T., & Meireles, M. A. A. (2018). Developing a supercritical fluid extraction method assisted by cold pressing for extraction of pequi (*Caryocar brasiliense*). *The Journal of Supercritical Fluids*, *137*, 34–39.

Köhler, J., Teupser, D., Elsässer, A., & Weingärtner, O. (2017). Plant sterol enriched functional food and atherosclerosis. *British Journal of Pharmacology*, *174*(11), 1281–1289.

Leão, D. P., Franca, A. S., Oliveira, L. S., Bastos, R., & Coimbra, M. A. (2017). Physicochemical characterization, antioxidant capacity, total phenolic and proanthocyanidin content of flours prepared from pequi (*Caryocar brasilense* Camb.) fruit by-products. *Food Chemistry*, *225*, 146–153.

Macedo, A. L., Santos, R. S., Pantoja, L., & Santos, A. S. (2011). Pequi cake composition, hydrolysis and fermentation to bioethanol. *Brazilian Journal of Chemical Engineering*, *28*, 9–15.

Machado, M. T. C., Mello, B. C. B. S., & Hubinger, M. D. (2015). Evaluation of pequi (*Caryocar Brasiliense* Camb.) aqueous extract quality processed by membranes. *Food and Bioproducts Processing*, *95*, 304–312.

Maia, J. G. S., Andrade, E. H. A., & da Silva, M. H. L. (2008). Aroma volatiles of pequi fruit (*Caryocar brasiliense* Camb.). *Journal of Food Composition and Analysis*, *21*(7), 574–576.

Monteiro, S. S., da Silva, R. R., da S. Martins, S. C., Barin, J. S., & da Rosa, C. S. (2015). Phenolic compounds and antioxidant activity of extracts of pequi peel (*Caryocar brasiliense* Camb.). *International Food Research Journal*, *22*(5), 1985–1992.

Nascimento-Silva, N. R. R. D., & Naves, M. M. V. (2019). Potential of whole Pequi (*Caryocar* spp.) fruit-pulp, almond, oil, and shell-as a medicinal food. *Journal of Medicinal Food*, *22*(9), 952–962.

Newmark, H. L. (1997). Squalene, olive oil, and cancer risk: A review and hypothesis. *Cancer Epidemiology, Biomarkers and Prevention*, *6*(12), 1101–1103.

Özcan, M. M., Ghafoor, K., Al Juhaimi, F., Ahmed, I. A. M., & Babiker, E. E. (2019). Effect of cold-press and soxhlet extraction on fatty acids, tocopherols and sterol contents of the Moringa seed oils. *South African Journal of Botany*, *124*, 333–337.

Palmeira, S. M., Silva, P. R. P., Ferrão, J. S. P., Ladd, A. A. B. L., Dagli, M. L. Z., Grisolia, C. K., et al. (2016). Chemopreventive effects of pequi oil (*Caryocar brasiliense* Camb.) on preneoplastic lesions in a mouse model of hepatocarcinogenesis. *European Journal of Cancer Prevention*, *25*(4), 299–305.

Parker, T. D., Adams, D. A., Zhou, K., Harris, M., & Yu, L. (2003). Fatty acid composition and oxidative stability of cold-pressed edible seed oils. *Journal of Food Science*, *68*(4), 1240–1243.

Passos, X. S., Castro, A. C. M., Pires, J. S., Garcia, A. C. F., Campos, F. C., Fernandes, O. F. L., et al. (2003). Composition and antifungal activity of the essential oils of *Caryocar brasiliensis*. *Pharmaceutical Biology*, *41*(5), 319–324.

Pessoa, A. S., Podesta, R., Block, J. M., Franceschi, E., Dariva, C., & Lanza, M. (2015). Extraction of pequi (*Caryocar coriaceum*) pulp oil using subcritical propane: Determination of process yield and fatty acid profile. *The Journal of Supercritical Fluids*, *101*, 95–103.

Roesler, R., Catharino, R. R., Malta, L. G., Eberlin, M. N., & Pastore, G. (2008). Antioxidant activity of *Caryocar brasiliense* (pequi) and characterization of components by electrospray ionization mass spectrometry. *Food Chemistry*, *110*(3), 711–717.

Roesler, R., Malta, L. G., Carrasco, L. C., Holanda, R. B., Sousa, C. A. S., & Pastore, G. M. (2007). Antioxidant activity of cerrado fruits. *Food Science and Technology*, *27*(1), 53–60.

Roll, M. M., Miranda-Vilela, A. L. M., Longo, J. P. F., Agostini-Costa, T. D. S., & Grisolia, C. K. (2018). The pequi pulp oil (*Caryocar brasiliense* Camb.) provides protection against aging-related anemia, inflammation and oxidative stress in Swiss mice, especially in females. *Genetics and Molecular Biology*, *41*(4), 858–869.

Silva, T. A., de Assunção, R. M. N., Vieira, A. T., de Oliveira, M. F., & Batista, A. C. F. (2014). Methylic and ethylic biodiesels from pequi oil (*Caryocar brasiliense* Camb.): Production and thermogravimetric studies. *Fuel*, *136*, 10–18.

Teh, S. S., & Birch, J. (2013). Physicochemical and quality characteristics of cold-pressed hemp, flax and canola seed oils. *Journal of Food Composition and Analysis*, *30*(1), 26–31.

Torres, L. R. O., de Santana, F. C., Torres-Leal, F. L., de Melo, I. L. P., Yoshime, L. T., Matos-Neto, E. M., et al. (2016). Pequi (*Caryocar brasiliense* Camb.) almond oil attenuates carbon tetrachloride-induced acute hepatic injury in rats: Antioxidant and anti-inflammatory effects. *Food and Chemical Toxicology*, *97*, 205–216.

Torres, L. R. O., Shinagawa, F. B., Santana, F. C., Araújo, E. S., Oropeza, M. V. C., Macedo, L. F. L., et al. (2016). Physicochemical and antioxidant properties of the pequi (*Caryocar brasiliense* Camb.) almond oil obtained by handmade and cold-pressed processes. *International Food Research Journal*, *23*(4), 1541–1551.

Traesel, G. K., Menegati, S. E. L. T., dos Santos, A. C., Souza, R. I. C., Boas, G. R. V., Justi, P. N., et al. (2016). Oral acute and subchronic toxicity studies of the oil extracted from pequi (*Caryocar brasiliense*, Camb.) pulp in rats. *Food and Chemical Toxicology*, *97*, 224–231.

Vujasinovic, V., Djilas, S., Dimic, E., Romanic, R., & Takaci, A. (2010). Shelf life of cold-pressed pumpkin (*Cucurbita pepo* L.) seed oil obtained with a screw press. *Journal of the American Oil Chemists' Society*, *87*, 1497–1505.

Wester, I. (2000). Cholesterol-lowering effect of plant sterols. *European Journal of Lipid Science and Technology*, *102*(1), 37–44.

Chapter 34

Cold pressed *Pistacia lentiscus* seed oils

Moncef Chouaibi[a], Leila Rezig[a], Nesrine Gaout[a], Khaled Ben Daoued[a], Kamel Msaada[b], and Salem Hamdi[a]
[a]*Food Preservation Laboratory, High Institute of Food Industry, Tunis, Tunisia,* [b]*Laboratory of Medicinal and Aromatic Plants, Biotechnology Center of Borj-Cedria Technopole, Hammam-Lif, Tunisia*

Abbreviations

ABTS 2,2′-azino-bis-3-ethylbenzthiazoline-6-sulphonic acid
DPPH· 2,2-diphenyl-1-picrylhydrazyl
FRAP ferric reducing antioxidant power
MUFA monounsaturated fatty acid
PUFA polyunsaturated fatty acid
RDA recommended dietary allowance
SFA saturated fatty acid

1 Introduction

Due to the growing tendency of replacing synthetic ingredients with their natural counterparts, changing consumer expectations, and environmental protection concerns, a number of industries nowadays are returning to the use of natural additives extracted from natural raw materials (Chouaibi, Rezig, Hamdi, & Ferrari, 2019; Chouaibi, Rezig, Lakoud, et al., 2019). The supreme challenge for most commercial-scale producers is connected with the management of by-products and reducing rising production costs. A good sample of the management of by-products produced by the fruit industry is the valorization of seeds in oil production (Chouaibi et al., 2012a, 2013, Chouaibi, Rezig, Hamdi, and Ferrari, 2019; Chouaibi, Rezig, Lakoud, et al., 2019; Ben Daoued, Chouaibi, Gaout, Bel Haj, & Hamdi, 2016; Rezig, Chouaibi, Meddeb, et al., 2019).

In recent years, an increasing quantity of publications has been released regarding numerous unconventional sources of seed oil and its physicochemical characteristics and compositions (Ben Daoued et al., 2016; Rezig et al., 2019; Górnas, Siger, & Seglin, 2013). Because of the fact that unconventional seed oils have valuable natural compounds such as unsaturated fatty acids, triacylglycerols, vitamins, phytosterols, micronutrients, pigments, and phenolics, they could be incorporated in the food, medicine, and cosmetics industries (Chouaibi et al., 2013; Chouaibi, Rezig, Hamdi, & Ferrari, 2019; Górnas et al., 2013). Both saturated and unsaturated fatty acids are vital ingredients of biological membranes, where their influence on the hydrophobic-hydrophilic balance (HLB) in the bilayer is indispensable. Indeed, saturated acyl chains provide a rigid membrane structure, whereas unsaturated acyl chains provide its liquid character. Nevertheless, the existence of embedded specific biological compounds in the membrane, such as α-tocopherol, can change that balance (Dwiecki, Górnas, Jackowiak, Nogala-Kałucka, & Polewski, 2007).

Pistacia lentiscus L. is an evergreen member of the Anacardiaceae family consisting of more than 11 species. It is largely distributed in the "extreme" ecosystems of the Mediterranean basin (Ben Daoued et al., 2016; Gaout, 2012). In Tunisia, it has a large geographical and bioclimatic distribution, extending from humid to arid areas (Pottier & Alapetite, 1981). It is cultivated in the wild also in Algeria, Turkey, Morocco, France, Spain, Italy, and Greece (Le Floc'h & Nabli, 1995). The fruit, galls, resin, and leaves of *Pistacia lentiscus* have a long tradition in folk medicine dating from the time of the ancient Greeks. Despite its limited distribution in the world, this plant is known internationally for several therapeutic properties such as its antifungal, antimicrobial, and antioxidant activities (Ben Daoued et al., 2016; Gaout, 2012).

Conventionally, oils and lipids are extracted using *n*-hexane as a solvent. Nonetheless, this flammable solvent, derived from petroleum, has been considered as a reprotoxic category 2 substance under the REACH regulation (European

Cold Pressed Oils. https://doi.org/10.1016/B978-0-12-818188-1.00034-7
Copyright © 2020 Elsevier Inc. All rights reserved.

Directives and Registration, Evaluation, Authorization and Restriction of Chemicals), as category 2 aquatic chronic toxic, and listed as a forbidden substance in cosmetic products (Rapinel et al., 2016). Nowadays, the development of green chemistry has prompted chemists to find alternative solvents to hazardous ones (Kerton & Marriott, 2013).

After seed separation from fruits (by drying and screening), oil recovery from plant seeds is realized. At an industrial scale, the plant seed oil is extracted by continuous mechanical expression (screw pressing) and/or by a solvent process (Chouaibi, Rezig, Hamdi, & Ferrari, 2019; Zuin & Ramin, 2018). To ensure a higher oil quality, mechanical pressing is preferred (lower process temperature, no solvent) although lower yields are achieved (from 55% to 95% according to the process draw material) (Singh & Bargale, 2000). Screw pressing has been studied for a variety of oilseeds including red pepper, linseed, canola, *Zizyphus lotus*, crambe, and chia seeds (Savoire, Lanoisellé, & Vorobiev, 2013). Considering oil yield, studies have proved that factors contributing to increasing pressure and temperature in the screw barrel have a major positive influence on oil yield. Those factors can be modulated by impacting on the diameter of the restriction die located at the meal discharge and by screw rotation speed (the lower the speed and the smaller the restriction opening, the higher the yield) (Savoire et al., 2013). Suitable pretreatments of seeds such as flaking, cooking, and moisture conditioning can also enhance oil yield. Moisture conditioning has a key role in expression as an optimal moisture content regarding oil yield depends on seed variety (e.g., linseeds 9%–11% (db), rapeseed 5% (db), crambe seed 3%–4% (db)) (Savoire et al., 2013). An oilseed thermal treatment such as cooking, although it allows an increase in oil yield, could be responsible for oil degradation. Thus cold pressing is usually preferred to retain more health beneficial components such as antioxidants increasing oil nutritive value (Lutterodt, Slavin, Whent, Turner, & Yu, 2011; Maier, Schieber, Kammerer, & Carle, 2009; Siger, Nogala-Kalucka, & Lampart-Szczapa, 2008). Insufficient works describe the effect of screw pressing parameters on oil quality (i.e., free fatty acids content, peroxide value, and oxidative stability) (Chouaibi, Rezig, Hamdi, & Ferrari, 2019). Nevertheless, relatively high temperatures can be reached during the process: 60–68°C according to Maier et al. (2009), which may impact the oil composition and quality.

There is no generalized technique that can be used to all feedstocks for converting to biodiesel. This is a major technical challenge (Greenwell, Laurens, Shields, Lovitt, & Flynn, 2010; Santacesaria, Vicente, Di Serio, & Tesser, 2012). For instance, different equipment and testing facilities are required in the biodiesel production processes and some of that equipment may not be needed for all feedstocks; thus, production cost per unit of biodiesel can be higher than for ordinary techniques (Goldemberg, 2008; Mettler, Vlachos, & Dauenhauer, 2012; Vasudevan & Briggs, 2008). Some feedstocks, such as bio-oils, may have toxic acids that require advanced techniques and longer processing times, and this would lead to higher costs for fuel processing (Liu, Wang, & Guo, 2012). To overcome technical challenges, advanced and organized study and development efforts are required among various universities, biodiesel producing companies, and state governments. Strong and favorable government policy and end user (public) support can help produce and maintain production efficiencies (Azad, Rasul, Khan, Sharma, & Hazrat, 2015).

2 Oil content of *Pistacia lentiscus* fruits

Fruits of *Pistacia lentiscus* give an edible oil, which is rich in unsaturated fatty acids such as oleic and linoleic (Ben Daoued et al., 2016; Gaout, 2012; Trabelsi, Sakouhi, et al., 2012; Trabelsi, Cherif, et al., 2012). In Algeria, the oil of the fruit is used by the population in traditional medicine in many ways, as an antidiarrheal and as a constituent of cattle feed. The oil amounts of *Pistacia lentiscus* fruits extracted by cold pressed machines are recapitulated in Table 1. The highest lipid amount (42.5%) was found in the Ain Draham region, followed by that in Nefza (39.1%), values two times higher than that observed in Sajnene (21.0%). Therefore, the obtained contents were smaller than those reported by Trabelsi, Cherif, et al. (2012), of 1.85%–42.54% for *Pistacia* fruits grown in northeast Tunisia (Rimel). Furthermore, our results were in concordance with those found by Ben Daoued et al. (2016). These researchers reported a higher oil amount, extracted with *n*-hexane, of 38.8% for *Pistacia lentiscus* fruits collected from the Tabarka region. In the current study, we reinforced that all fruits were collected in the same method, where all seed samples were extracted and stored under the same procedure and environmental conditions. Thus, the observed differences related only to geographic factors.

The statistical analysis showed that significant correlations were found between lipid yield in *Pistacia lentiscus* fruits collected in different regions: a positive correlation for oleic acid content ($r = 0.788$, $P < .01$) and a negative one for linolenic acid ($r = 0.943$, $P < .05$) (Gaout, 2012). In practice, this means that with an increasing oil yield in *Pistacia lentiscus* fruits, a lower percentage of linolenic acid and a higher content of oleic acid in the total fatty acid amounts may be expected.

The oil yield of seed oils extracted by the Soxhlet method from *Pistacia lentiscus* varied from 19.4% for immature fruits (red) to 42.5% for mature fruits (black), as shown by Trabelsi, Cherif, et al. (2012). Total lipid accumulation showed an increasing trend as maturity progressed. In good agreement with our results, Charef, Yousfi, Saidi, and Stoccker (2008) showed that the crude lipid amount of the *Pistacia lentiscus* fruit differs from 32.8% for black fruits to 11.7% for red fruits.

TABLE 1 Oil yield of cold pressed *Pistacia lentiscus* fruits collected at different regions of Tunisia (Gaout, 2012).

Region	Oil yield (%)
Tabarka	34.7
Ain Draham	42.5
Ouechteta	36.3
Nefza	39.1
Zaghouan	28.4
Sejnene	21.0
Jendouba	30.9
Tabouba	33.5
Tibar	28.7
Amdoun	28.5

Trabelsi found that the oil contents of Tunisian *Pistacia lentiscus* fruits were function of harvest dates. Indeed, the transition of the fruit from immature (green) to mature (red) was described by significant growth of total lipid amounts of *Pistacia lentiscus* fruit, as indicated by Trabelsi, Cherif, et al. (2012).

3 Mineral composition

Mineral levels are naturally present in the crude oil extracted from the seeds despite the variations in concentration depending on many factors, such as the soil used for the cultivation, irrigation water, and maturity stage. The cold pressed lentisc seed oils included calcium and to a lesser extent magnesium as the main minerals (9.20–19.4 and 5.10–13.7 mg/kg oil, respectively). In contrast, the potassium levels were only 0.47–4.31 mg/kg of oil, whereas sodium was not detected in extracted seed oils collected from Zaghouan, Tabouba, and Jendouba. It has been reported that calcium is essential in organisms particularly in cell physiology, where the movement of calcium ions into and out of the cytoplasm function is a signal for many cellular processes (Remington, 2005).

Heavy metals are an important factor for the assessment of seed oil quality since they affect several properties such as the rate of oil oxidation, its nutritional value, and storability (Murillo et al., 1999). *Pistacia lentiscus* seed oils are rich in iron, manganese, and copper (1.90–9.39, 0.06–3.12, and 0.31–2.65 mg/kg, respectively). Copper is an essential element, but it may be toxic to both human beings and animals when its level exceeds the safe limits (Yaman, Akdeniz, Bakidere, & Atici, 2005). Nickel, cobalt, chromium, and vanadium were also present but at low levels. Interestingly, other metals were also present in cold pressed lentisc seed oil depending on regions. The results indicated that cold pressed lentisc seed oil provides most of the minerals for human nutrition and health. Iron-containing enzymes, usually heme prosthetic groups, participate in the catalysis of oxidation reactions in biology as well as the transport of a number of soluble gazes (Pineda & Ashmead, 2001). It has been reported that cadmium and lead are nonessential elements in plant nutrition. In addition, they are highly toxic elements that accumulate in biological systems and have a long half-life (Mena, Cobrera, Lorenzo, & Lopez, 1996). Hence, based on the findings of this investigation, it appears that cold pressed lentisc seed oil, with its enhanced levels of minerals, could be considered a potentially successful functional food.

4 Physicochemical properties of cold pressed *Pistacia lentiscus* seed oil

The physicochemical characteristics of cold pressed *Pistacia lentiscus* seed oils are displayed in Table 2. The refractive index of oil revealed that the chain length and unsaturation degree of the fatty acids (Shahidi, 2005). The refractive index values of the cold pressed lentisc seed oils ranged from 1.459 to 1.467, and were similar to those of soybean oil (1.477), mustard oil (1.466), sunflower oil (1.467), *Zizyphus lotus*, pumpkin, and red pepper seed oils (Bachheti, Rai, Joshi, & Rana, 2012; Nehdi, 2011, Rezig, Chouaibi, Msaada, et al., 2012, Rezig, Chouaibi, & Fregapane, 2018; Chouaibi et al., 2012a, 2012b, Chouaibi, Rezig, Hamdi, & Ferrari, 2019; Chouaibi, Rezig, Lakoud, et al., 2019), indicating that the fatty

TABLE 2 Physicochemical characteristics of cold pressed *Pistacia lentiscus* seed oils collected at different regions (Gaout, 2012).

Region	Refractive index	Density	Acid value (mg/g)	Peroxide value (mmol/kg)	Iodine value (I_2/100g)	Saponification value (mg/g)	*L*	*a*	*b*
Tabarka	1.463	0.9021	13.8	9.30	140.3	183.5	50.23	−5.50	22.39
Ain Draham	1.467	0.9120	16.2	8.65	143.1	185.2	52.19	−4.93	21.50
Ouechteta	1.464	0.9126	10.4	10.4	137.5	185.4	45.30	−4.47	28.63
Nefza	1.460	0.9130	13.5	7.56	138.4	186.5	48.33	−5.56	29.04
Zaghouan	1.459	0.9120	11.4	8.20	136.2	189.1	55.68	−2.46	20.22
Sejnene	1.462	0.9100	9.10	8.97	142.4	185.5	49.44	−3.20	26.88
Jendouba	1.460	0.9122	8.80	12.0	140.2	187.6	47.69	−2.50	29.18
Tabouba	1.465	0.9118	8.30	10.2	141.4	185.7	51.40	−5.46	22.45
Tibar	1.466	0.9119	12.6	8.50	139.5	191.3	47.70	−3.10	26.66
Amdoun	1.462	0.9112	15.3	7.93	142.7	185.5	53.80	−6.56	22.40

acid compositions of the cold pressed lentisc seed oils were similar to those of common vegetable oils. It is worth noting that the seed oils from Ain Draham exhibited a higher refractive index than those collected in Zaghouan, Jendouba, and Nefza, demonstrating that the unsaturation degree of the seed oils was much higher than their counterparts. The variations of the relative density among the oil samples were not obvious, which ranged from 0.9021 to 0.9130. On the other hand, acid values, peroxide values, iodine values, and saponification values of the cold pressed lentisc seed oils were quite different, which varied depending on the source of the oils. The acid values and peroxide values of the oils obtained from the cold pressed lentisc seed oils were higher than those of the other vegetable oils, suggesting that the latter had relatively good storage stability and high degree of edibility (Chouaibi et al., 2012b). Fruit oil displayed the lowest acid values (8.30 mg/g and 16.20 mg/g) and peroxide values (7.56 mmol/kg and 12.06 mmol/kg).

Generally, the iodine value usually reflected the unsaturation degree of the fatty acids and the saponification value was a ruler to evaluate the chain length of all fatty acids present in the triacylglycerols (Shahidi, 2005). The iodine values of the cold pressed lentisc seed oils (136.2–140.1 g I_2/100 g) were much higher than those of the vegetable oils (Chouaibi et al., 2012b) such as pumpkin seed oil (116.2 g I_2/100 g), soybean oil (122.5 g I_2/100 g), and sunflower oil (128.0 g I_2/100 g) (Bachheti et al., 2012; Jiao et al., 2014; Nehdi, 2011), indicating the high content of unsaturated fatty acids in the cold pressed lentisc seed oils. The saponification values of the cold pressed lentisc seed oils varied from 183.5 to 191.3 mg/g, wherein the Tibar seed oil had the highest value, followed by the Zaghouan and Jendouba samples, indicating that a relative percentage of long-chain fatty acids existed in the triacylglycerols of the seed oils.

The color parameters (L, a, and b) of the cold pressed *Pistacia lentiscus* seed oils were: Luminosity (L) ranged between 45.3 and 55.6, a parameter varied from –6.56 and –2.46, and b parameter was between 20.22 and 29.18 (Gaout, 2012). The results are different from the CIELAB obtained for the most vegetable oils (Chouaibi et al., 2012a, 2012b). The higher b value in cold pressed lentisc seed oils designates a more intense yellow color, and hence the presence of increased amounts of carotenoids (Chouaibi, Rezig, Hamdi, & Ferrari, 2019). Interestingly, the cold pressed lentisc seed oils had a negative value for a (greenness), which can be attributed to the presence of chlorophylls or any other green pigments extracted from the fruits. In addition, oxidative stability was assessed and the results revealed that their values ranged from 10.09 to 15.90 h.

5 Composition of cold pressed lentisc seed oil

5.1 Fatty acid composition of cold pressed *Pistacia lentiscus* seed oils

The fatty acids profile is a main determinant of the oil quality. In Tunisia, the oil of lentisc is used by population in traditional medicine in many ways, as an antidiarrheal, antiinflammatory, and for asthma treatment. In addition to fatty acids, vegetable oils contain phytosterols, which are the major components of the unsaponifiable matter. The fatty acid composition of cold pressed lentisc seed oils consisted principally of oleic acid (C18:1) (42.4%–57.1%), followed by linoleic acid (C18:2), palmitic acid (C16:0), and stearic acid (C18:0) (Gaout, 2012). In addition to these main fatty acids, others were identified and quantified, namely: myristic (C14:0), heptadecanoic (C17:0), arachidic (C20:0), behenic (C22:0), palmitoleic (C16:1), heptadecenoic (C17:1), eicosenoic (C20:1), α-linolenic (C18:3), and eicosadienoic (C20:2n6) acids. Minor *trans* fatty acids were detected (Gaout, 2012). The data were comparable to those obtained by other researchers that indicated that oleic acid was the most abundant fatty acid in *Pistacia lentiscus* seed oils (Trabelsi, Cherif, et al., 2012; Trabelsi, Sakouhi, et al., 2012). Similarly, Ben Daoued et al. (2016) found content for oleic acid of 50.4% in *Pistacia* oils from the Tabarka region; Trabelsi, Cherif, et al. (2012) and Trabelsi, Sakouhi, et al. (2012) also reported levels between 41.7% (Tebaba) and 51.7% (Rimel). The high content of linoleic acid in *Pistacia lentiscus* seed oils is interesting, once linoleic acid is the major dietary fatty acid that regulates the low-density lipoprotein (LDL)-C metabolism by downregulating LDL-C production and enhancing its clearance (Wijendran & Hayes, 2004).

When considering the general classification of the fatty acids, it was found that the cold pressed *Pistacia lentiscus* seed oils had the following sequence: PUFA > MUFA > SFA, which is in concordance with other research (Ben Daoued et al., 2016; Trabelsi, Cherif, et al., 2012; Trabelsi, Sakouhi, et al., 2012). By analyzing the lipid profiles of the 10 cold pressed *Pistacia lentiscus* collected at different regions, minor changes were found. The region with the highest levels of SFA and MUFA was Ain Draham, followed by Nefza. The highest values of PUFA were detected in oils from Tibar, Zaghouan, and Amdoun, followed by Tabouba, Ouichteta, and Sejnene. In addition, the amounts of SFA (ranging between 23.1% and 28.4%), MUFA (between 53.3% and 60.3%), and PUFA (between 25.2% and 28.0%) were analogous to those found in other works (Gaout, 2012; Trabelsi, Sakouhi, et al., 2012). The PUFA/SFA ratio varied between 0.95 (Tabarka), and 1.25 (Zaghouan). Our ratios were similar to the maximum of 0.92 reported by Trabelsi, Sakouhi, et al. (2012) and Trabelsi, Cherif, et al. (2012) for *Pistacia lentiscus* collected in northern Tunisia (Rimel). However, the unsaturated fatty acids (MUFAs and PUFAs) determined in the regions were 70.4% (Tibar) and 75.2% (Ain Draham) of the total fatty acids. Taking into account the high proportion of unsaturated fatty acids, the seed oils of 10 *Pistacia lentiscus* fruits were highly recommended for human consumption, presenting a more favorable fatty acid profile than other vegetable oils. Moreover, concerning *trans* fatty acids that might raise LDL cholesterol, and lower HDL cholesterol concentrations, the oils analyzed contained very low percentages of these fatty acids. Unsaturated fatty acids perform various vital functions in biological membranes and considered as precursors of diverse lipid regulators of cellular metabolism (Ray, Kassan, Busija, Rangamani, & Patel, 2016). Although PUFAs are scientifically proven to be healthy fatty acids, the human body is often unable to synthesize them efficiently de novo to a sufficient level. Therefore, they must be supplied through diet or dietary supplements. Although fish oils are currently the main industrial sources of PUFAs, they are not suitable for vegetarian and vegan consumers and they cannot meet the increasing global demand of omega-3 PUFA from natural fish sources.

A cluster analysis was achieved to classify the numerous cold pressed *Pistacia lentiscus* according to their seed oils' fatty acid compositions. Four clusters were detected with an R-squared of 0.83. The first cluster included five regions, Jendouba, Tibar, Tabouba, Ouechteta, and Sejenene, because the fatty acid composition was similar among them. The second and third clusters included Tabarka and Ain Draham and Nefza, which were the regions with the highest levels of SFA and MUFA. However, Ain Draham had a higher SFA content than Nefza seed oil, whereas their MUFA contents were quite similar. Finally, the fourth cluster included Zaghouan and Amdoun because it showed the highest PUFA and the lowest MUFA levels.

The variations in fatty acid composition observed in cold pressed *Pistacia lentiscus* oil samples obtained from both regions were probably related to both genetic factors and environmental conditions during fruit development and maturity (Breton, Souyris, Villemur, & Bervillé, 2009). Some authors have used the total fatty acid composition to differentiate *Pistacia lentiscus* fruit oils according to the geographical area of production for oils from Algeria (Charef et al., 2008) and Tunisia (Gaout, 2012). The observed differences in fatty acid composition suggest strong influence of geographical location, climatic condition, growing practice, and variety of plants on the fatty acid profile of *Pistacia lentiscus* (Gaout, 2012).

5.2 Sterol composition

Phytosterols are steroid compounds, and their structure and functions are similar to those of cholesterol. In the pharmaceuticals and cosmetics industries, phytosterols are recovered from cellulose processing of vegetable oil waste and they are utilized in the production of therapeutic steroids, creams, or lipsticks (Fernandes & Cabral, 2007). The total sterol content of cold pressed lentisc seed oils ranged from 205.7 to 482.3 mg/100 g oil. These results showed that there were no significant variations between the Tabarka and Ain Draham regions in terms of total sterol content (Gaout, 2012).

These data are in concordance with those reported by Trabelsi, Sakouhi, et al. (2012). Six compounds were found, where β-sitosterol was the major sterol, ranging from 147.8 to 183 mg/100 g oil. The highest and lowest contents of β-sitosterol were detected in samples from the Ain Draham and Sejnene regions, respectively. β-sitosterol was the predominant phytosterol present in *Pistacia lentiscus* seed oils (Ben Daoued et al., 2016). The next major component was Δ5-avenasterol and then campesterol, representing 60.2%–92.0% and 25.3%–42.6% of total sterols, respectively. Other components, citrostadienol and stigmasterol, were also reported (Gaout, 2012). Additionally, cholesterol was at lower level ranging from 0.02 to 0.15 mg/100 g oil, which is relatively higher than that in olive and soybean oils (0.40%), but lower than that in palm and tomato seed oils (2.3% and 20%), respectively (Chouaibi et al., 2012b). From the results of the present investigation, it appears that cold pressed lentisc seed oil, with its high amounts of sterols, can be considered a potentially successful functional food.

5.3 Tocol composition

Tocopherols in vegetable oils are believed to protect polyunsaturated fatty acids (PUFAs) from peroxidation. Tocopherols are widely used for food, cosmetics, and resins. In food, they are used as an antioxidant for frying oil and margarine (Ben Daoued et al., 2016). Seven compounds were determined in almost all cold pressed lentisc seed oils, including three tocopherols (α-, γ-, δ-) and four tocotrienols (α-, β-, γ-, δ-). All seed oils had higher tocotrienol levels than tocopherols. γ-tocotrienol was the most abundant, followed by α-tocotrienol and α tocopherol. β-tocopherol was not detected in any of the seed oils, whereas δ-tocopherol was only detected in small amounts in all regions, namely Nefza, Tabouba, Ain Draham, Tabarka, and Ouechteta (Gaout, 2012). The seed oil of the Tabarka region was the one that contained the highest vitamin E content (Gaout, 2012). The high content observed for α-tocopherol was comparable to that reported by Ben Daoued et al. (2016) for *Pistacia lentiscus* oils (383.7 mg/kg oil). However, lower amounts were found in Algerian (8.54 mg/g) regions. Ben Daoued et al. (2016) and Gaout (2012) found that γ-tocotrienol was the highest tocol-like compound, which is in line with the reports in the present work. Concerning the total content of tocopherols and tocotrienols, the values ranged between 749 and 2192 mg/kg oil, which were higher than those reported by Ben Daoued et al. (2016). α-tocopherol and γ-tocotrienol were those compounds that presented the highest variability between varieties, the Tabarka region being the one that presented the highest values of both compounds. Regarding γ-tocopherol, lower values and lower variability were observed when compared to α-tocopherol. It should be mentioned that for γ-tocopherol, Ouechteta was the region that presented the highest concentration. In terms of α-tocotrienol, a lower variability and lower values were obtained when compared to γ-tocotrienol. Sejnene was the region that presented the highest α-tocotrienol concentration. The data also indicates that cold pressed lentisc seed oils from 10 regions in the present study are an excellent source of vitamin E (187–280 α-tocopherol equivalents). The inclusion of 13 g of Tabarka *Pistacia* or 5.10 g of Nefza *Pistacia* seed oil in the human diet per day will supply 10% of the recommended dietary allowance (RDA) of vitamin E for adults. The RDA for both men and women is 15 mg (35 μmol)/day of α-tocopherol (National Research Council, 1989). Vitamin E brings several health benefits, as it is a powerful lipid-soluble antioxidant. It is able to protect the human body's cells from the damage caused by free radicals. The cold pressed lentisc seed oils from 10 regions of Tunisia might be used to delay the body's aging process and to prevent the occurrence of some chronic diseases.

After performing a dendrogram to vitamin E contents of the 10 cold pressed *Pistacia lentiscus* seed oils examined in this study (Fig. 1), four clusters were found with R-squared equal to 0.85, through which a significant proportion of the total variation was retained. The first cluster was constituted by four *Pistacia lentiscus* seed oils, namely the Jendouba, Zaghouan, Tibar, and Tabouba regions, because all of them had comparable tocol levels. Ouechteta, Sejnene, and Amdoun were involved in cluster 2, because both regions had comparatively lower α- and γ-tocotrienol contents. The *Pistacia lentiscus* seed oil of Nefza region was included in cluster 3 due to its high γ-tocopherol amount (20.0 mg/kg). The last cluster was represented by Tabarka. This region presented the highest levels of γ-tocotrienol, α-tocotrienol, α-tocopherol, β-tocotrienol, and δ-tocotrienol (Gaout, 2012). Such a high tocopherol content was expected to contribute good oxidative stability and protection to cold pressed lentisc seed oil during its storage and processing (Chouaibi et al., 2012b).

5.4 Carotenoids composition

The composition of the total natural carotenoid pigments in cold pressed *Pistacia lentiscus* seed oils were described and determined by Gaout (2012). Seven compounds were found, where β-carotene was the major carotenoid, constituting 1.11–5.13 mg $100\,g^{-1}$ oil, and the next major component was α-carotene (0.87–1.64 mg $100\,g^{-1}$ oil). Other components including γ-carotene and lutein formed 0.05–0.22 and 0.18–0.86 mg $100\,g^{-1}$ oil, respectively. Cryptoxanthine and echinenone were found at lower levels (0.20–0.52 and 0.12–0.38 mg $100\,g^{-1}$ oil, respectively). Lycopene, which is specific

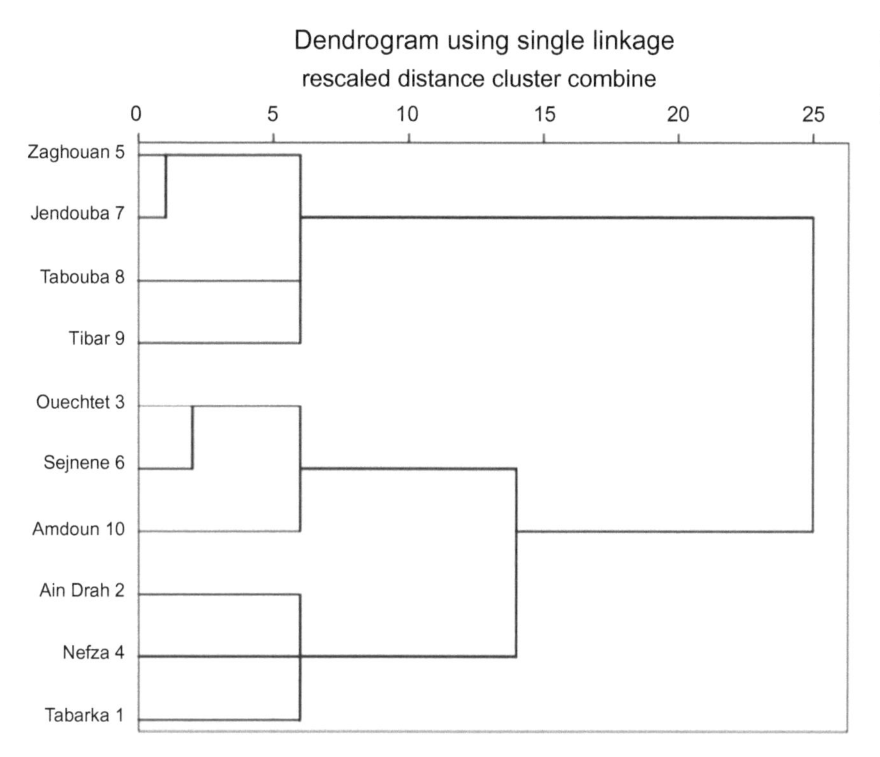

FIG. 1 Dendrograms of cold pressed lentisc seed oils from different regions of Tunisia taking into account their vitamin E compositions (Gaout, 2012).

to vegetal lipids, is present at low levels in most vegetable oils. Although, lycopene content was detected in lower amounts in cold pressed *Pistacia lentiscus* seed oil. It appears that cold pressed *Pistacia lentiscus* seed oil, with its high amounts of carotenoids, can be considered a potentially successful functional food. Obviously, the difference in concentrations among the seed oils might be attributed to differences between cultivars, degree of ripeness, latitude, environmental conditions, processing techniques, and storage conditions.

In general, the composition of the carotenoids pigment of oils is an important quality parameter, because they correlate with color, which is a basic attribute for the evaluation of the quality of oil. It is generally known that pigment concentration in olives decreases during ripening (Gandul-Rojas and Minguez-Mosquera, 1996). In fact, as ripening progresses, photosynthetic activity decreases, and the level of both chlorophylls and carotenoids decreases progressively (Criado, Motilva, Goñi, & Romero, 2007). The first three principal components, accounting for 81.7% of the variance in the *Pistacia lentiscus* seed oils, were examined. The first, second, and three principal components (F1, F2, and F3) explained 47.2%, 22.2%, and 12.2% of the total variance, respectively. According to the loading matrix, it is clear that factor 1 was positively correlated with total carotenes and carotenoids, α- and β-carotene (Gaout, 2012). The second factor was positively correlated with lutein and gamma carotene and negatively correlated with cryptoxanthine. The third factor was positively and negatively correlated with echinenone and lycopene, respectively. Fig. 2 presents the principal component of carotenoid composition of cold pressed lentisc seed oils collected from different regions of Tunisia.

6 Antioxidant activities of cold pressed *Pistacia lentiscus* seed oil

Due to the presence of inherent powerful antioxidants, cold pressed lentisc seed oils displayed excellent in vitro antioxidant capacities. Ben Daoued et al. (2016) pointed out that the seed oils from fruits exhibited strong DPPH· and ABTS radical scavenging activity and pancreatic lipase inhibitory activity in vitro. Meanwhile, Gaout (2012) reported that the strong antioxidant activities of the cold pressed lentisc seed oils were positively correlated with phenolics content.

To depict the antioxidant potential of the seed oils of the 10 cold pressed *Pistacia lentiscus* from different regions, DPPH· and ABTS radical scavenging activities were determined. It is clear that DPPH· radical scavenging activities varied between 38.6% and 69.8%. Regarding the values of the antioxidant capacity of the *Pistacia lentiscus* seed oils obtained by ABTS assay, similar results (0.322–0.457 μmol Trolox/ml oil) were obtained for the 10 seed oils from different regions of Tunisia. The lowest antioxidant capacity was determined for Sejenene, whereas the highest values were found for Ain Draham, Nefza, and Ouechteta.

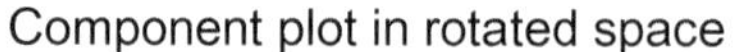

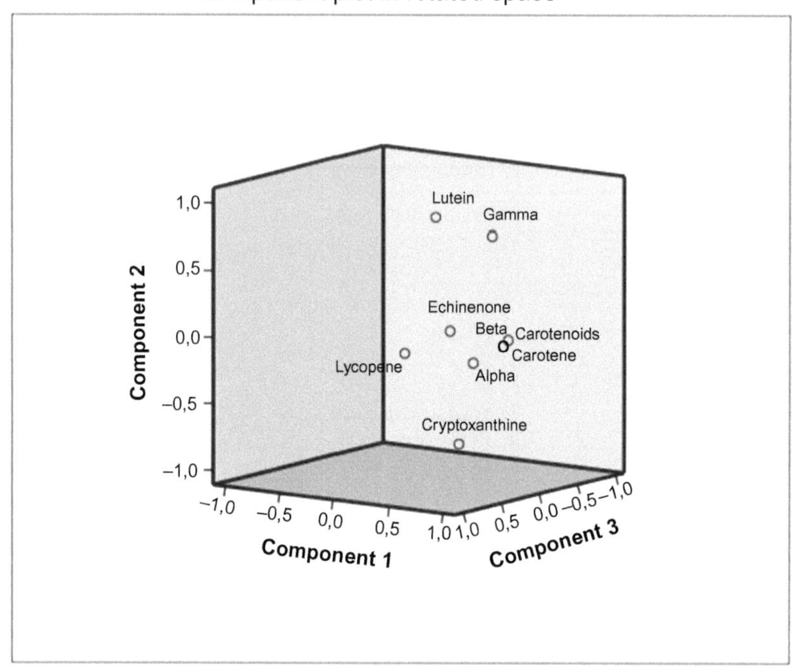

FIG. 2 Principal component of carotenoid composition of cold pressed lentisc seed oils collected from different regions of Tunisia (Gaout, 2012).

The relatively high antioxidant activity of the oil may be due to its high content of total phenolics and flavonoids. Phenolic compounds are known to react with hydroxyl radicals, peroxyl radicals, and superoxide anions, and inhibit the lipid oxidation at an early stage (Chang, Yen, Huang, & Duh, 2002). Many researchers have demonstrated that phospholipids such as phosphatidylcholine, phosphatidylethanolamine, and phosphatidylserine are responsible for the high antioxidant activity of vegetable oils (Segwa, Kamata, & Totani, 1995). Other studies have suggested that some fatty acids such as linoleic acid were important in relation to the antioxidant activity of oils (Ramaprasad, Srinivasan, Baskaran, Sambaiah, & Lokesh, 2006). All of these activities are represented in *Pistacia lentiscus* seed oil and may explain the high antioxidant activities. Yet the presence of many antioxidants, such as pigments, occasionally leads to a decrease in the activity of components like tocopherols and phenolic compounds (Chouaibi et al., 2013). It has been shown that antioxidant systems based on α-tocopherol, ascorbic acid, citric acid, and phospholipids can reduce peroxidation activity (Olsen, Vogt, Saarem, Greibrokk, & Nillsson, 2005). In addition, Henry, Catignani, and Schwartz (1998) found that in safflower seed oil, a pro-oxidant effect of carotenoids was observed at concentrations above 500 ppm, and that carotenoids decreased the antioxidant capacity of tocopherols.

7 Physical properties

7.1 Rheological properties

It is important to be able to predict the oil viscosity at each purification step because this step of the vegetable oil involves different temperature conditions. All cold pressed lentisc seed oils presented Newtonian behavior at shear rates ranging from 10 to 500 s^{-1} at 25°C. There is linear relationship between shear stress and shear rate. This relation is in agreement with Newton's law of viscosity, as expressed by the following equation:

$$\sigma = \eta \gamma^{\cdot},$$

where σ is the shear stress (Pa), $\gamma^{\cdot}$ is the shear rate (s^{-1}), and η is the viscosity (Pa.s). The viscosity at each temperature can be obtained from this curve using the slope from fits of the experimental shear stress-shear rate data. Similar findings had been found previously by other researchers (Chouaibi et al., 2012a, 2012b; Kim, Kim, Lee, Yoo, & Lee, 2010; Santos, Dos Santos, & De Souza, 2005). They showed that vegetable oils are Newtonian liquids having high viscosity due to their long-chain structure. However, apparent viscosity values varied distinctly depending on the regions of oil samples. The viscosity of Tabarka seed oil was the highest, followed by these collected in Nefza, Amdoun, Ouechteta, Sejnene, Tabouba, and

Zaghouan oils. No significant difference was observed in Tabarka and Ain Draham oils in terms of viscosity. Oil viscosities were positively correlated with the amount of oleic acid (i.e., viscosity increased with an increase in this type of fatty acid), and negatively correlated with the amount of linoleic acid (i.e., viscosity decreased with an increase in this fatty acid) (Chouaibi et al., 2012b). This trend is in line with the reported correlation of viscosity with the iodine value, which describes that the oil viscosity almost linearly decreases as the iodine value increases (Maskan, 2003; Kim et al., 2010). This logic justifies the high viscosity of cold pressed *Pistacia lentiscus* seed oil. It has been shown that high viscosity can allow oils to be accumulated more easily on the surface of fried foods and to enter inside during the cooling period (Kalogianni, Karapantsios, & Miller, 2011).

Detailed knowledge of the rheological properties of natural vegetable oils is of particular importance whenever lubrication is their intended usage. The apparent viscosity vs temperature data of the vegetable oils, along with cold pressed lentisc seed oil, was employed (Gaout, 2012). There was a reduction in apparent viscosity η_a as temperature increased (20–100°C). The apparent viscosity (η_a) of all oils exhibited an exponential decrease with temperature, as illustrated by the data. *Pistacia lentiscus* seed oil from Tabarka had higher viscosities value at temperatures ranging from 20°C to 100°C. Heating did not change the Newtonian behavior. Decreasing of vegetable oil viscosity is due to high thermal movement among molecules, reducing intermolecular forces, making flow among easier and reducing viscosity. This phenomenon is in agreement with various studies (Abramovic & Kloufutar, 1998; Eromosele & Paschal, 2003; Maskan & Gogus, 2000; Santos et al., 2005; Wan Nik, Ani, Masjuki, & Eng Giap, 2005).

The overall results suggested some potential applications for cold pressed lentisc seed oil. Interestingly, from the viscosity-temperature profiles, the oil might be utilized as lubricating oil base stock at or above ambient temperature if only the viscosity were taken into account (and with the aid of antioxidant/antihydrolysis/antirust packages). Temperature effect on oil viscosity was evaluated by applying the Arrhenius model: ($\eta = Aexp\left(\frac{Ea}{RT}\right)$) where A is the frequency factor or viscosity coefficient at a reference temperature (Pa.s), E_a is the activation energy (kJ/mol), R is the gas constant (kJ/mol K), and T is the absolute temperature (K). Activation energy can be determined from the slope of ln η vs $1/T$ plot. Generally, the high activation energy reflects the viscosity-temperature stability of oils. Activation energy values E_a for cold pressed lentisc seed oils varied from 16.15 to 28.44 kJ/mol (Gaout, 2012). The activation energy value of Tabarka lentisc seed oil was significantly high. Frequency factors A increased significantly from 1.30×10^{-5} to 2.05×10^{-5} (Pa.s) as a function of oil region.

7.2 Thermal properties

DSC and TGA have been applied to study the thermo-oxidative processes that occur in the oil as a function of temperature and time. The characteristics DSC thermogram of cold pressed lentisc seed oils showing the variation of enthalpy values with the increase of heating temperature (Gaout, 2012). A crystallization peak was not prominent when cold pressed lentisc seed oil was cooled to −80°C, but when the oil was heated, the melting curve showed a single prominent phase transition starting from −50°C, and ending at −10°C, with a peak ranging from −40°C to −36.2°C with corresponding melting enthalpy of 62.40–72.30 J/g. These thermal properties can be observed during the melting of more than one group of triglycerides with melting points too close to be differentiated by DSC experiments (Ferrari et al., 2007). It has been reported that melting of the least stable polymorphic forms of triacylglycerols such as a-forms occurs at low temperatures. The melting of a triglyceride depends on its chain length, extent of branching of the chain, and degree of unsaturation of its constituent fatty acids as well as their stereo-specific distribution along the glycerol molecules (Dickinson & McClements, 1995; Ferrari et al., 2007). In general, vegetable oils contain a mixture of triglycerides, and thus their melting occurs over a wide range of temperatures rather than at a single distinct temperature (Dickinson & McClements, 1995). Thus, the shoulder-like feature in a DSC curve is observed in most vegetable oils because they contain a mixture of triacylglycerols. Generally, the melting point of an oil increases with the increase in chain length and degree of saturation of constituent fatty acids. The oils rich in unsaturated fatty acids normally melt at a lower temperature, which is corroborated by the melting curve of CSO. This curve is relatively flat above 0°C, indicating that cold pressed lentisc seed oils are purely liquid without any detectable solid crystals at ambient temperature.

Thermogravimetric and its derivative profile of cold pressed lentisc seed oils exhibited negligible loss of oil mass (<1% of the initial mass), when seed oils were heated up to 180°C from the ambient temperature. This means that cold pressed lentisc seed oils can be heated up to 230°C with negligible degradation. In other words, it can be used as a frying oil. However, the loss of mass increased rapidly afterward reaching up to 60% when cold pressed lentisc seed oils were heated to 500°C. The degradation of oil was more pronounced when it was heated above 500°C. Approximately 75% of the oil was degraded when it was heated to 500°C. The degradation continued upon further heating and nearly all of the oil was volatilized when the temperature reached 550°C.

8 Conclusion

The present work demonstrated that seed oils recovered from *Pistacia lentiscus* fruits from different regions from Tunisia are a promising source of bio-compounds, particularly unsaturated fatty acids such as oleic and linoleic acids, which represent approximately 70% of the total fatty acids, as well as phytosterols, mainly β-sitosterol. The results demonstrated that *Pistacia lentiscus* fruit and seed oils have high nutritive value and great health benefits and are good candidates for consumption as new resource edible oils or functional ingredients in the food industry, which would be further contributable to the industrial exploitation and application of *Pistacia lentiscus* fruit resources. Nevertheless, a high variation in oil yield as well as contents of phytosterols and squalene, and lower for fatty acids composition, might be found in seed oils recovered from different regions. Moreover, the content of phytosterols, as well as the composition of fatty acids, was more associated with oil yield. Nevertheless, appropriate utilization of *Pistacia lentiscus* fruits would facilitate management of tons of the by-product generated during lentiscus harvesting. Hence, it may contribute to environmental sustainability, more effective use of harvested plant material, and increased economic profits. Cold pressed lentisc seed oils could be successfully used in the pharmaceuticals and cosmetics industries, e.g., in the production of natural cosmetics, such as creams, hair care products, or lipstick.

Acknowledgments

The authors are grateful for the financial support of the Ministry of Higher Education, Scientific Research and Technology of Tunisia.

References

Abramovic, H., & Kloufutar, C. (1998). The temperature dependence of dynamic viscosity for some vegetable oils. *Acta Chimica Slovaca*, *45*(1), 69–77.

Azad, A. K., Rasul, M. G., Khan, M. M. K., Sharma, S. C., & Hazrat, M. A. (2015). Prospect of biofuels as an alternative transport fuel in Australia. *Renewable and Sustainable Energy Reviews*, *43*(Supplement C), 331–351.

Bachheti, R., Rai, I., Joshi, A., & Rana, V. (2012). Physico-chemical study of seed oil of *Prunus armeniaca* L. grown in Garhwal region (India) and its comparison with some conventional food oils. *International Food Research Journal*, *19*, 577–581.

Ben Daoued, K., Chouaibi, M., Gaout, N., Bel Haj, O., & Hamdi, S. (2016). Chemical composition and antioxidant activities of cold pressed lentisc (*Pistacia lentiscus* L.) seed oil. *Rivista Italiana Delle Sostanze Grasse*, *93*, 31–38.

Breton, C. M., Souyris, I., Villemur, P., & Bervillé, A. (2009). Oil accumulation kinetic along ripening in four olive cultivars varying for fruit size. *Oilseeds and fats, Crops and Lipids*, *16*, 1–7.

Chang, L. W., Yen, W. J., Huang, S. C., & Duh, P. D. (2002). Antioxidant activities of sesame coat. *Food Chemistry*, *70*, 347–354.

Charef, M., Yousfi, M., Saidi, M., & Stoccker, P. (2008). Determination of the fatty acid composition of acorn (*Quercus*), *Pistacia lentiscus* seeds growing in Algeria. *Journal of the American Oil Chemists' Society*, *85*, 921–924.

Chouaibi, M., Mahfoudhi, N., Rezig, L., Donsì, F., Ferrari, G., & Hamdi, S. (2012a). Nutritional composition of *Zizyphus lotus* L. seeds. *Journal of the Science of Food and Agriculture*, *92*, 1171–1177.

Chouaibi, M., Mahfoudhi, N., Rezig, L., Donsì, F., Ferrari, G., & Hamdi, S. (2012b). A comparative study on physicochemical, rheological, and surface tension properties of Tunisian jujube (*Zizyphus lotus* L.) seed and vegetable oils. *International Journal of Food Engineering*, *8*, 11–18.

Chouaibi, M., Rezig, L., Hamdi, S., & Ferrari, G. (2019). Chemical characteristics and compositions of red pepper seed oils extracted by different methods. *Industrial Crops and Products*, *128*, 363–370.

Chouaibi, M., Rezig, L., Lakoud, A., Boussaid, A., Hassouna, M., Ferrari, G., et al. (2019). Exploring potential new galactomannan source of *Retama reatam* seeds for food, cosmetic and pharmaceuticals: Characterization and physical, emulsifying and antidiabetic properties. *International Journal of Biological Macromolecules*, *124*, 1167–1176.

Chouaibi, M., Rezig, L., Mahfoudhi, N., Arafa, S., Donsì, F., Ferrari, G., et al. (2013). Physicochemical characteristics and antioxidant activities of *Zizyphus lotus* L. seed oil. *Journal of Food Biochemistry*, *37*, 554–563.

Criado, M. N., Motilva, M. J., Goñi, M., & Romero, M. P. (2007). Comparative study of the effect of the maturation process of the olive fruit on the chlorophyll and carotenoid fractions of drupes and virgin oils from Arbequina and Farga cultivars. *Food Chemistry*, *100*(2), 748–755. https://doi.org/10.1016/j.foodchem.2005.10.035.

Dickinson, E., & McClements, D. J. (1995). *Advances in food colloids*. Springer Science & Business Media.

Dwiecki, K., Górnas, P., Jackowiak, H., Nogala-Kałucka, M., & Polewski, K. (2007). The effect of d-alpha-tocopherol on the solubilization of dipalmitoylphosphatidyl-choline membrane by anionic detergent sodium dodecyl sulfate. *Journal of Food Lipids*, *14*, 50–61.

Eromosele, C. O., & Paschal, N. H. (2003). Characterization and viscosity parameters of seed oils from wild plants. *Bioresource Technology*, *86*, 203–205.

Fernandes, N., & Cabral, J. M. (2007). Phytosterols: Applications and recovery methods. *Bioresource Technology*, *98*(12), 2335–2350.

Ferrari, C., Angiuli, M., Tombari, E., Righetti, M. C., Matteoli, E., & Salvetti, G. (2007). Promoting calorimetry for olive oil authentication. *Thermochimica Acta*, *459*(1–2), 58–63.

Gandul-Rojas, B., & Mínguez-Mosquera, M. I. (1996). Chlorophyll and carotenoid composition in the virgin olive oil from different Spanish olive varieties. *Journal of the Science Food and Agriculture*, *72*, 31–39.

Gaout, N. (2012). *Etude des propriétés physico chimiques, biologiques et fonctionnelles de l'huile de* Pistacia lentiscus *extraite par la pression au froid.* Tunisie: Thèse doctorat à l'université de Carthage.

Goldemberg, J. (2008). The challenge of biofuels. *Energy & Environmental Science*, *1*(5), 523–525.

Górnas, P., Siger, A., & Seglin, D. (2013). Physicochemical characteristics of the cold-pressed Japanese quince seed oil: New promising unconventional bio-oil from by-products for the pharmaceutical and cosmetic industry. *Industrial Crops and Products*, *48*, 178–182.

Greenwell, H. C., Laurens, L. M. L., Shields, R. J., Lovitt, R. W., & Flynn, K. J. (2010). Placing microalgae on the biofuels priority list: A review of the technological challenges. *Journal of the Royal Society Interface*, *7*(46), 703–726.

Henry, L. K., Catignani, G. L., & Schwartz, S. J. (1998). The influence of carotenoids and tocopherols on the stability of safflower seed oil during heat-catalyzed oxidation. *Journal of the American Oil Chemists' Society*, *75*(10), 1399–1402.

Jiao, J., Li, Z.-G., Gai, Q.-Y., Li, X.-J., Wei, F.-Y., Fu, Y.-J., et al. (2014). Microwave assisted aqueous enzymatic extraction of oil from pumpkin seeds and evaluation of its physicochemical properties, fatty acid compositions and antioxidant activities. *Food Chemistry*, *147*, 17–24.

Kalogianni, E. P., Karapantsios, T. D., & Miller, R. (2011). Effect of repeated frying on the viscosity, density and dynamic interfacial tension of palm and olive oil. *Journal of Food Engineering*, *105*, 169–179.

Kerton, F. M., & Marriott, R. (2013). *Alternative solvents for green chemistry*. Croydon: Royal Society of Chemistry.

Kim, J., Kim, D. N., Lee, S. H., Yoo, S. H., & Lee, S. (2010). Correlation of fatty acid composition of vegetable oils with rheological behavior and oil uptake. *Food Chemistry*, *118*, 398–402.

Le Floc'h, E., & Nabli, M. A. (1995). In M. A. Nabli (Ed.), *Essai de Synthèse sur la Végétation et la Phyto-Ecologie Tunisiennes II & III. Le Milieu Physique et la Végétation, Ecologie Végétale Appliquée* (pp. 523–536). Tunisie.

Liu, C.-Z., Wang, A. R., & Guo, C. (2012). Ionic liquids for biofuel production: Opportunities and challenges. *Applied Energy*, *92*, 406–414.

Lutterodt, H., Slavin, M., Whent, M., Turner, E., & Yu, L. (2011). Fatty acid composition, oxidative stability, antioxidant and antiproliferative properties of selected cold-pressed grape seed oils and flours. *Food Chemistry*, *128*, 391–399. https://doi.org/10.1016/j.foodchem.2011.03.040.

Maier, T., Schieber, A., Kammerer, D. R., & Carle, R. (2009). Residues of grape (*Vitis vinifera* L.) seed oil production as a valuable source of phenolic antioxidants. *Food Chemistry*, *112*, 551–559. https://doi.org/10.1016/j.foodchem.2008.06.005.

Maskan, M. (2003). Change in colour and rheological behavior of sunflower seed oil during frying and after adsorbent treatment of used oil. *European Food Research and Technology*, *218*, 20–25.

Maskan, M., & Gogus, F. (2000). Effect of sugar on the rheological properties of sunflower oil-water emulsions. *Journal of Food Engineering*, *43*, 173–177.

Mena, C., Cobrera, C., Lorenzo, M. L., & Lopez, M. C. (1996). Cadmium levels in wine, beer and other alcoholic beverages: Possible sources of contamination. *Science of the Total Environment*, *181*, 201–208.

Mettler, M. S., Vlachos, D. G., & Dauenhauer, P. J. (2012). Top ten fundamental challenges of biomass pyrolysis for biofuels. *Energy and Environmental Science*, *5*(7), 7797–7809.

Murillo, M., Benzo, Z., Marcano, E., Gomez, C., Garaboto, A., & Marin, C. (1999). Determination of copper, iron and Nikel in edible oils using emulsified solutions by ICP-AES. *Journal of Analytical Atomic Spectrometry*, *14*, 815–820.

National Research Council. (1989). *Recommended dietary allowances* (10th ed.). Washington: National Academy Press.

Nehdi, I. (2011). Characteristics, chemical composition and utilisation of *Albizia julibrissin* seed oil. *Industrial Crops and Products*, *33*, 30–34.

Olsen, E., Vogt, G., Saarem, K., Greibrokk, T., & Nillsson, A. (2005). Autoxidation of cod liver oil with tocopherol and ascorbyl palmitate. *Journal of the American Oil Chemists' Society*, *82*(2), 97–103.

Pineda, O., & Ashmead, H. D. (2001). Effectiveness of treatment of iron-deficiency anemia in infants and young children with ferrous bis-glycinate chelate. *Nutrition*, *17*, 381–384.

Pottier, P., & Alapetite, G. (1981). *Programme Flore et Végétation Tunisiennes*. Tunis: Flore de la Tunisie Publications Scientifiques Tunisiennes.

Ramaprasad, T. R., Srinivasan, K., Baskaran, V., Sambaiah, K., & Lokesh, B. R. (2006). Spray-dried milk supplemented with a-linoleic acid and eicosapentaenoic acid and docasahhexaenoic acid decrease HMG Co-A reductase activity and increases biliary secretion of lipids of rats. *Steroids*, *71*, 409–415.

Rapinel, V., Rombaut, N., Rakotomanomana, N., Vallageas, A., Cravotto, G., & Chemat, F. (2016). An original approach for lipophilic natural products extraction: Use of liquefied *n*-butane as alternative solvent to *n*-hexane. *LWT-Food Science and Technology*, *85*, 524–533.

Ray, S., Kassan, A., Busija, A. R., Rangamani, P., & Patel, H. H. (2016). The plasma membrane as a capacitor for energy and metabolism. *American Journal of Physiology Cell Physiology*, *310*(3), C181–C192.

Remington, J. (2005). *Remington: The science and practice of pharmacy*. Williams & Wilkins: Lippincott.

Rezig, L., Chouaibi, M., & Fregapane, G. (2018). *Profile characterisation of pumpkin (Cucurbita maxima) seed oils*. Lambert Academic Publishing, ISBN: 978-613-9-87846-8. 53p.

Rezig, L., Chouaibi, M., Meddeb, W., et al. (2019). Chemical composition and bioactive compounds of Cucurbitaceae seeds: Potential sources for new trends of plant oils. *Process Safety and Environmental Protection*. https://doi.org/10.1016/j.psep.2019.05.005.

Rezig, L., Chouaibi, M., Msaada, K., et al. (2012). Chemical composition and profile characterization of pumpkin (*Cucurbita maxima*) seed oil. *Industrial Crops and Products*, *37*(82–87), 2012.

Santacesaria, E., Vicente, G. M., Di Serio, M., & Tesser, R. (2012). Main technologies in biodiesel production: State of the art and future challenges. *Catalysis Today*, *195*(1), 2–13.

Santos, J. C. O., Dos Santos, I. M. G., & De Souza, A. G. (2005). Effect of heating and cooling on rheological parameters of vegetable edible oils. *Journal of Food Engineering*, *67*, 401–405.

Savoire, R., Lanoisellé, J.-L., & Vorobiev, E. (2013). Mechanical continuous oil expression from oilseeds, a review. *Food and Bioprocess Technology*, *6*, 1–16. https://doi.org/10.1007/s11947-012-0947-x.

Segwa, T., Kamata, M., & Totani, H. Y. (1995). Antioxidant activity of phospholipids for polyunsaturated fatty acids of fish oil. III. Synergism of nitrogen-containing phospholipids with tocopherols. *Journal of Japan Oil Chemists' Society*, *44*, 36–42.

Shahidi, F. (2005). *Baley's industrial oil and fat products* (6th ed.). New York: Wiley.

Siger, A., Nogala-Kalucka, M., & Lampart-Szczapa, E. (2008). The content and antioxidant activity of phenolic compounds in cold-pressed plant oils. *Journal of Food Lipids*, *15*, 137–149. https://doi.org/10.1111/j.1745-4522.2007.00107.x.

Singh, J., & Bargale, P. C. (2000). Development of a small capacity double stage compression screw press for oil expression. *Journal of Food Engineering*, *43*, 75–82. https://doi.org/10.1016/S0260 8774(99)00134-X.

Trabelsi, H., Cherif, O. A., Sakouhi, F., Villeneuve, P., Renaud, J., Barouh, N., et al. (2012). Total lipid content, fatty acids and 4-desmethylsterols accumulation in developing fruit of *Pistacia lentiscus* L. growing wild in Tunisia. *Food Chemistry*, *131*, 2–6.

Trabelsi, H., Sakouhi, F., Renaud, J., Villeneuve, P., Khouja, M. L., Mayer, P., et al. (2012). Fatty acids, 4-desmethylsterols, and triterpene alcohols from Tunisian lentisc (*Pistacia lentiscus*) fruits. *European Journal of Lipid Science and Technology*, *114*(8), 968–973.

Vasudevan, P. T., & Briggs, M. (2008). Biodiesel production-current state of the art and challenges. *Journal of Industrial Microbiology & Biotechnology*, *35*(5), 421–427.

Wan Nik, W. B., Ani, F. N., Masjuki, H. H., & Eng Giap, S. G. (2005). Rheology of bio-edible oils according to several rheological models and its potential as hydraulitic fluid? *Industrial Crops and Products*, *22*, 249–255.

Wijendran, V., & Hayes, K. C. (2004). Dietary n-6 and n-3 fatty acid balance and cardiovascular health. *Annual Review of Nutrition*, *24*, 597–615.

Yaman, M., Akdeniz, I., Bakidere, S., & Atici, D. (2005). Comparison of trace metal concentrations in malign and benign human prostate. *Journal of Medicinal Chemistry*, *48*, 630–634.

Zuin, V. G., & Ramin, L. Z. (2018). Green and sustainable separation of natural products from agro-industrial waste: Challenges, potentialities, and perspectives on emerging approach. *Topics in Current Chemistry*, *376*, 3. https://doi.org/10.1007/s41061-017-0182-z.

Chapter 35

Cold pressed niger (*Guizotia abyssinica* Cass.) seed oil

Ali Osman
Agricultural Biochemistry Department, Faculty of Agriculture, Zagazig University, Zagazig, Egypt

Abbreviations

ASG acylated steryl glucoside
CER glucocerebroside
CHD coronary heart disease
CM chloroform/methanol
GL glycolipid
H *n*-hexane
HDL high density lipoprotein
HPLC high-performance liquid chromatography
IC_{50} the half maximal inhibitory concentration
LDL low-density lipoprotein
NL neutral lipid
PL phospholipids
PUFA polyunsaturated fatty acid
RSA radical scavenging activity
SG steryl glucoside
ST sterol
TAG triacylglycerol
TGLs total glycolipids
TL total lipid

1 Introduction

Niger (*Guizotia abyssinica* (L.f.) Cass.) is an oil crop planted in India and Ethiopia for its edible oil. *G. abyssinica* belongs to the same Compositae family as safflower and sunflower seeds (Ramadan, 2009; Ramadan, 2012). Niger seeds yield niger oil and niger oil cake. Niger oil cake is a feed rich in protein, oil, fiber, free from any toxic substance, and suited to all classes of livestock that can digest fibrous feeds. The niger plant is a stout, erect annual herb, that grows up to a height of 2 m (Mariod, Mirghani, & Hussein, 2017). The root system is well developed, with a taproot that has many lateral roots, particularly in the upper 5 cm. The stems are soft, hairy, hollow with a diameter up to 2 cm, and branched. Their color is pale green, often stained or dotted with purple, and they become yellow with age. The leaves are opposite, sometimes alternate at the apices of the stems. The limbs are simple and sessile. The leaf blade is lanceolate to obovate, 3–23 cm × 1–6 cm, variable in shape, with a margin that is either entire or toothed, ciliate, softly hairy on both surfaces. Leaves are usually dark green, but the lower ones have a distinct yellow color. The inflorescences are arranged in apical or axillary cymes surrounded by leafy bracts up to 3 cm long. The flowers are capitula, ranging from 15 to 50 mm in diameter, bright yellow becoming golden yellow as they mature. Each flower produces about 50 seeds. The seeds are small achenes (actually a fruit), 3–6 mm long × 1.5–4 mm broad, glossy black in color (Mariod et al., 2017).

In Ethiopia, niger is a major source of edible oil, and provides about 50% of the country's oilseed production. Ethiopian niger seeds contain about 40% oil (Dutta, Helmersson, Kebedu, Alemaw, & Appelqvist, 1994). In India, niger oil is only 3% of oilseed production (Mariod et al., 2017). The average oil yield of niger seed has been reported to be between 40% and

Cold Pressed Oils. https://doi.org/10.1016/B978-0-12-818188-1.00035-9
Copyright © 2020 Elsevier Inc. All rights reserved.

44% (Ramadan, 2012). Niger oil is pale yellow, with a nutty taste and a pleasant odor. It is mostly used for cooking, as well as in paints and the extraction of perfume from flowers (Dutta et al., 1994). Niger seeds are used as food in many Ethiopian dishes, condiments, and snacks such as "litlit" or "chibito" when niger seeds are roasted and ground with salt, then mixed with roasted cereals (Mariod et al., 2017). In Ethiopia, niger oil cake is the main protein supplement for livestock (Dutta et al., 1994). In Western countries, niger seeds are important components of birdseed mixtures (Lin, 2005).

2 Oil content

The main components in cold pressed niger oil are presented in Table 1. Niger seed yields about 30%–35% of its weight oil, which is clear, slow-drying, and edible. Niger seed oil is a polyunsaturated semi-drying oil. It has a pale yellow or orange color with a nutty taste and sweet odor. The crude oil has low acidity, and can be used directly for cooking. Normally it has a poor shelf life, and become rancid when stored for a long period. Its fatty acid composition is similar to that of sunflower oil, and has high content of linoleic acid. It is used as a substitute for olive oil and can be mixed with

TABLE 1 The main components in the cold pressed niger oil.

Lipid class	References
1. Neutral lipids	Bhatnagar and Gopala Krishna (2014)
1.1. Monoacylglycerols	
1.2. Diacylglycerols	
1.3. Triacylglycerols	
1.3.1. Trilinolein	
2. Glycolipids	Bhatnagar and Gopala Krishna (2014)
2.1. Acylated steryl glucoside	
2.2. Steryl glucoside	
2.3. Cerebroside	
3. Phospholipids	Bhatnagar and Gopala Krishna (2014)
3.1. Phosphatidylcholine	
3.2. Phosphatidylethanolamine	
3.3. Phosphatidylinositol	
3.4. Phosphatidylserine	
3.5. Lysophosphatidylcholine	
Fatty acid	
Myristic acid (C14:0)	Ramadan and Mörsel (2003)
Palmitic acid (C16:0)	Dutta et al. (1994)
Stearic acid (C18:0)	Dutta et al. (1994)
Oleic acid (C18:1)	Ramadan and Mörsel (2003)
Linoleic acid (C18:2)	Ramadan (2009)
Arachidic acid (C20:0)	Ramadan (2009)

linseed oil. The presence of linoleic acid varies from 45% to 65% depending on harvested soil conditions and seed variety (Chimdessa, Petros, & Wakuma, 2017).

Niger seeds contain 2.8%–7.8% moisture, 483 cal, 34%–39% carbohydrates, 9%–13% fiber, 17%–30% protein, 1.8–9.9 g ash, 180–800 mg/100 g phosphorus, 50–587 mg/100 g calcium, 0.22–0.55 mg/100 g riboflavin, 0.43 mg/100 g thiamine, and 3.66 mg/100 g niacin (Rao, 1994). Reports in the literature showed the oil content in niger seed to be in the range 30%–50% (Dagne & Jonsson, 1997), wherein there was a variation in oil content (28.5%–38.8%) of different samples collected from Ethiopia (Dutta et al., 1994). With high levels of linoleic acid, it is very similar to sunflower and safflower oils (Francis & Campbell, 2003).

3 Lipid classes and fatty acid profile of niger seed oil

Concerning fatty acid composition, niger seed oil resembles that of sunflower safflower and with a high level of linoleic acid (*c.*85%) depending on the origin (Riley & Belayneh, 1989). In addition to the main fatty acids, the seed oil can contain up to 1% of arachidic acid, and in certain varieties up to 3% of linolenic acid (Ramadan, 2009). Seed oils of niger grown in different regions of Ethiopia are typically high in linoleic acid (54%–73%), but with varying levels of oleic (5%–27%), palmitic (8%–10%), and stearic acids (5.5%–8.1%) (Dutta et al., 1994). Palmitoleic, linolenic, arachidic, eicosenoic, behenic, erucic and lignoceric acids account for about 2%–3% of the total fatty acids present in the oil (Fig. 1).

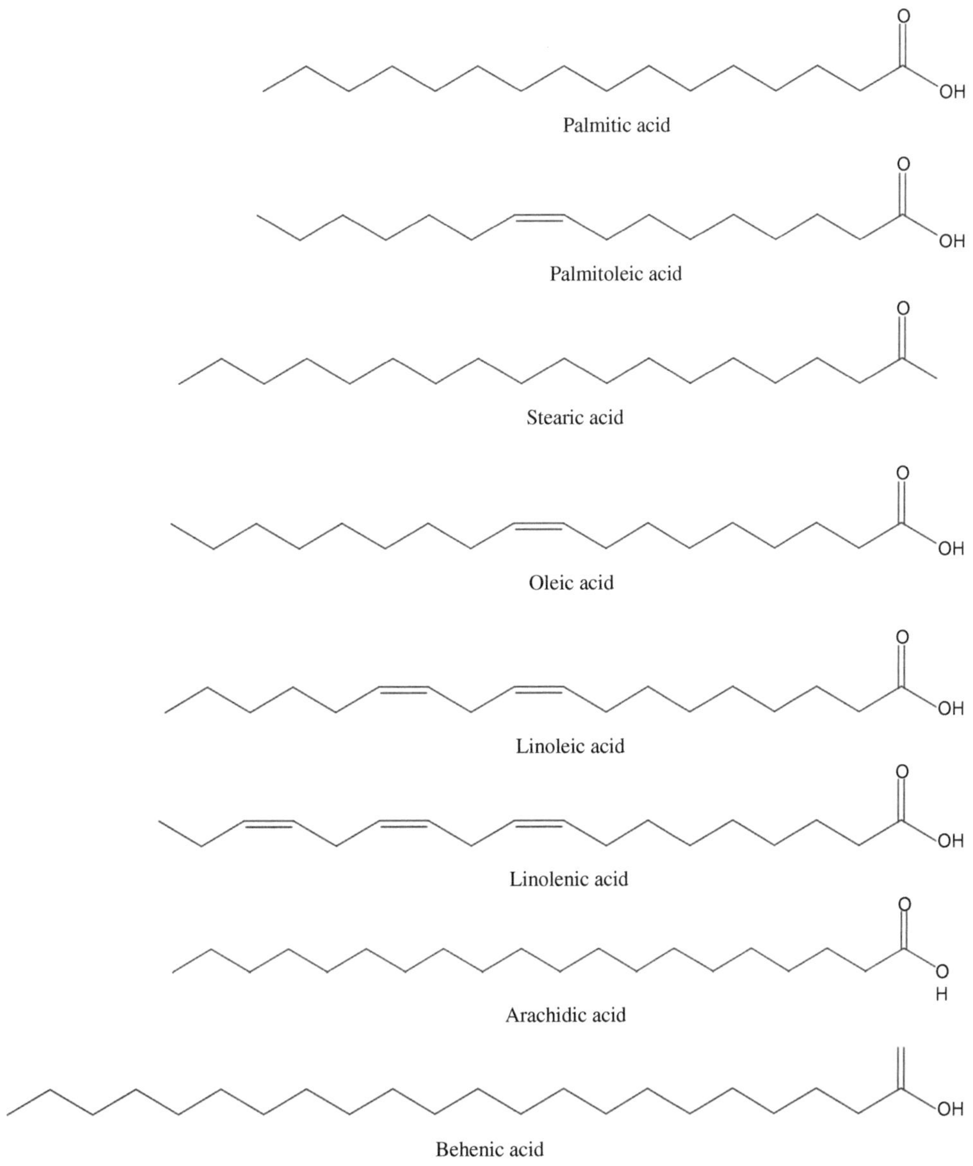

O
OH
Palmitic acid
O
OH
Palmitoleic acid
O
Stearic acid
O
OH
Oleic acid
O
OH
Linoleic acid
O
OH
Linolenic acid
O
O
H
Arachidic acid
OH
Behenic acid

FIG. 1 The major fatty acids in niger seed oil.

Besides triacylglycerols, niger oil contains other bioactive classes such as polar lipids, and in particular phospholipids. Phospholipids are widely distributed in food, wherein pro- as well as antioxidant effects have been attributed to them (Mcdonald & Mossoba, 1997). They have significant potential as multifunctional additives to food (because of their technological functions and nutritional benefit) as well as pharmaceutical and industrial applications (Ramadan & Mörsel, 2002).

4 Effect of extraction methods on lipid classes and bioactive composition

Bhatnagar and Gopala Krishna (2013) investigated commercial *G. abyssinica* seed to study the effect of extraction solvent on oil and bioactive lipids profile. Niger seeds were extracted using solvents of different polarities including petroleum ether, *n*-hexane, chloroform, acetone, methanol, and ethanol. The oil recovery after extraction with solvents of different polarities was between 31.8 and 41.3 g/100 g. The solvent-extracted oil was characterized by the following parameters: free fatty acids (3.6–12.3 g/100 g), color (40.0–95.0 Lovibond units), iodine value (137.6–140.3 g I_2/g), peroxide value (3.2–7.8 mequiv O_2/kg), saponification value (177.3–185.9 mg KOH/g), and unsaponifiable matter (1.3–4.3 g/100 g). Linoleic acid (69%–73%) was the main fatty acid, and trilinolein (31%–33%) was the main triacylglycerol. The composition of active molecules was 247–647 ppm of total phenolics, 171–345 ppm of total tocols, 18.9–181 ppm of total carotenoids, and 1.249–6.309 ppm of total sterols. α-tocopherol was the main component (154–276 ppm), and vanillic acid (1.70–176 ppm) was the main phenolic compound in the oil extracted using different solvents. Ethanol-extracted oil exhibited a 13.9-fold higher oxidative stability and a stronger antiradical activity (IC_{50} of 9.2 mg/mL) compared to *n*-hexane-extracted oil (IC_{50} of 40.3 mg/mL).

Bhatnagar and Gopala Krishna (2014) studied that lipid classes of solvent-extracted (*n*-hexane and ethanol) and cold pressed oils from commercial *G. abyssinica* seeds. The oil recovery of niger seeds obtained by cold press was 28.3 g/100 g, while the oil contents obtained by *n*-hexane and ethanol extractions were 38.3 and 29.7 g/100 g, respectively. The lipid fractions of the niger seed oils contained neutral lipids (c.65%–95.5%), glycolipids (c.2.7%–24.6%), and phospholipids (c.1.8%–9.5%). The acylglycerol profile of neutral lipids of niger seed oils contained triacylglycerols (c.76.9%–91.6%), diacylglycerols (c.3.9%–7.3%), and monoacylglycerols (c.0.6%–2.5%). The fatty acid profile of tri-, di-, and monoacylglycerols of seed oils contained linoleic acid (c.66.7%–71.6%) as the main fatty acid. The triacylglycerol profile of *G. abyssinica* seed oils showed trilinolein (c.39%–40%) as the main triacylglycerol. The extracted *G. abyssinica* seed oils contained 1289–6215 ppm of total sterols with β-sitosterol (c.42%–43%) as the main phytosterol. Acylated sterylglucoside was the major glycolipid (c.39%–52%) in extracted *G. abyssinica* seed oils. Phosphatidylcholine (c.47%) was the main phospholipid in cold pressed and *n*-hexane-extracted *G. abyssinica* seed oils, while phosphatidyl ethanolamine (57%) was the main subclass in ethanol-extracted *G. abyssinica* seed oil.

Ramadan and Mörsel (2003) extracted niger seeds (G. abyssinica Cass.), which are of interest as a new source of vegetable oils, using Soxhlet extraction with *n*-hexane and the oil was analyzed using a combination of chromatographic methods. The total lipid content was c.300 mg/g seed material, and the fatty acid profile showed a high content of linoleic acid (up to 63%) together with palmitic acid (17%), oleic acid (11%), and stearic acid (7%). Column chromatography separation over silica gel eluted with solvents of increasing polarity yielded 291 mg/g of neutral lipids, 5.76 mg/g of glycolipids, and 0.84 mg/g of phospholipids. The major fatty acid present in all lipid classes was linoleic acid together with minor amounts of palmitic, oleic, and stearic acids. Polar lipids were characterized by higher levels of palmitic acid and a lower content of linoleic acid. Phospholipid classes separated by normal-phase HPLC consisted of phosphatidylcholine (49%), phosphatidylethanolamine (22%), phosphatidylinositol (14%), phosphatidylserine (8%), and minor amounts (2%–3%) of phosphatidylglycerol and lysophosphatidylcholine.

Ramadan and Mörsel (2003) extracted niger (*G. abyssinica* Cass.) seeds with chloroform:methanol (2:1, *v*/v), and the total lipid content was 50% of seed weight. The major unsaturated fatty acids were linoleic and oleic acids. Palmitic acid was the major saturated fatty acid, followed by stearic acid. Column chromatography with solvent of increasing polarity yielded 93.0% neutral lipids, 4.90% glycolipids, and 0.60% phospholipids (PLs). Polar lipids were characterized by lower linoleic acid and higher palmitic acid levels. PL subclasses were separated via high-performance liquid chromatography (HPLC). The major individual PL subclasses were found to be phosphatidylcholine followed by phosphatidyl ethanolamine, phosphatidyl inositol, and phosphatidyl serine, respectively. The predominant PL-bound fatty acids presented were linoleic, palmitic, and oleic acids.

Niger (*G. abyssinica* Cass.) seeds were extracted with *n*-hexane (H) and chloroform/methanol (CM) (2:1, v/v) to determine the effect of the solvent on the amount and composition of the recovered lipid (Ramadan & Morsel, 2002). The amount of the total lipid (TL) extracted with H was 29.6% of the seed weight, while that extracted with CM was 49.9%. Column and thin-layer chromatographic procedures on silica gel were performed to fractionate the main neutral lipid (NL) subclasses. Fatty acid compositions of the NL subclasses, triacylglycerol (TAG) molecular species, and sterol (ST) content were estimated. The NL amount was found to be higher (97% of TL) in the H extract than in the CM extract (93% of TL). TAG appeared to be the predominant NL subclass (89.7%–91.9% of total NL). Linoleic acid (C18:2n-6) was the main fatty acid, followed by oleic acid (C18:1n-9) as the second main unsaturated fatty acid, while palmitic acid (C16:0) was the main saturated fatty acid. With high temperature gas liquid chromatography using a flame ionization detector (HTGLC/FID) and H_2 as the carrier gas, the actual TAG molecular species according to their carbon numbers were separated. The TAGs of even carbon numbers 16 and 18 were contained in six TAG molecular species. The major peaks occurred at C54:3 and C54:6, corresponding to triolein and trilinolein. TAG molecular species, expected to contain two molecules of linoleic acid (C54:5), were detected at a high level. Phytosterol pattern was determined without derivatization in the unsaponifiable fractions by HTGLC/FID. Sterols (ST) profile was characterized by a high total amount accounting for 4.22 g/kg of total H extract, and 4.00 g/kg of total CM extract. β-sitosterol (2.035–1.929 g/kg TL) was the main component in both extracts. The major ST found were, in order of decreasing levels, β-sitosterol > campesterol > stigmasterol > △5-avenasterol > △7-avenasterol > lanosterol.

Ramadan and Mörsel (2003) reported that edible plant glycolipids (GLs) are anticipated to play a role in human nutrition. Total glycolipids (TGLs) were separated from niger (*G. abyssinica* Cass.) seed oils by silica gel chromatography. Different GL subclasses were identified and separated using high-performance liquid chromatography with ultraviolet adsorption (HPLC/UV). The separation was accomplished using Zorbax-Sil (5 μm) column with an isocratic elution using isooctane:2-propanol (1:1, *v*/v) and detection at 206 nm. Methods are described for the analysis of GL constituents, sugar and sterols (ST), using gas–liquid chromatography equipped with flame ionization detector (GLC/FID). A high level of TGL was detected. Among the TGL from niger seed oil, acylated steryl glucoside (ASG), steryl glucoside (SG), and glucocerebroside (CER) were detected. The fatty acid profiles of GL fractions from niger seed oils were linoleic acid C18:2n-6, followed by oleic acid C18:1n-9.

5 Health benefits of niger oil

G. abyssinica seed oil is considered nutritionally valuable, as the high amount of linoleic acid could prevent cardiovascular diseases, and a precursor of some metabolic regulatory compounds and structural components of plasma membranes (Vles, 1989). Even less information is available on the associations of linoleic acid with coronary heart disease (CHD) than is available for PUFA intake overall. Linoleic acid could lower low-density lipoprotein (LDL)-cholesterol with minimal impact on high density lipoprotein (HDL)-cholesterol (Mensink & Katan, 1992). In vitro studies showed that linoleic acid and other n-6 fatty acids have been purported to be pro-thrombotic or pro-inflammatory (Calder, 2001). Linoleic acid might also decrease arrhythmias (Charnock, Sundram, Abeywardena, Mclennan, & Tan, 1991), and improve insulin sensitivity (Erkkilä, De Mello, Risérus, & Laaksonen, 2008). Rao (1994) reported that niger seed oil can be used in rheumatism. In addition, niger sprouts mixed with garlic are used to treat coughs (Getinet & Teklewold, 1995). Antiradical tests were used to evaluate the health impact of bioactive compounds found in foods. The antiradical properties of *G. abyssinica* seed oil was compared with common and nonconventional edible oils toward different stable free radicals (Ramadan & Moersel, 2006).

References

Bhatnagar, A. S., & Gopala Krishna, A. G. (2013). Effect of extraction solvent on oil and bioactives composition of commercial Indian Niger (*Guizotia abyssinica* (L.F.) Cass.) seed. *Journal of the American Oil Chemists' Society*, *90*, 1203–1212.

Bhatnagar, A. S., & Gopala Krishna, A. G. (2014). Lipid classes and subclasses of cold-pressed and solvent-extracted oils from commercial Indian Niger (*Guizotia abyssinica* (L.F.) Cass.) seed. *Journal of the American Oil Chemists' Society*, *91*, 1205–1216.

Calder, P. C. (2001). Polyunsaturated fatty acids, inflammation, and immunity. *Lipids*, *36*, 1007–1024.

Charnock, J., Sundram, K., Abeywardena, M., Mclennan, P., & Tan, D. (1991). Dietary fats and oils in cardiac arrythmia in rats. *The American Journal of Clinical Nutrition*, *53*, 1047s–1049s.

Chimdessa, M., Petros, Y., & Wakuma, D. A. (2017). *Influence of nitrogen and phosphorus fertilization on growth performance and seed yields of Niger* Guizotia abyssinica *(Lf) Cass*. Haramaya University.

Dagne, K., & Jonsson, A. (1997). Oil content and fatty acid composition of seeds of *Guizotia* Cass (Compositae). *Journal of the Science of Food and Agriculture*, *73*, 274–278.

Dutta, P. C., Helmersson, S., Kebedu, E., Alemaw, G., & Appelqvist, L. Å. (1994). Variation in lipid composition of Niger seed (*Guizotia abyssinica* Cass.) samples collected from different regions in Ethiopia. *Journal of the American Oil Chemists' Society*, *71*, 839–843.

Erkkilä, A., De Mello, V. D., Risérus, U., & Laaksonen, D. E. (2008). Dietary fatty acids and cardiovascular disease: An epidemiological approach. *Progress in Lipid Research*, *47*, 172–187.

Francis, C., & Campbell, M. (2003). *New high quality oil seed crops for temperate and tropical Australia.* Canberra: Rural Industries Research And Development Corporation.

Getinet, A., & Teklewold, A. (1995). An agronomic and seed-quality evaluation of Niger (*Guizotia abyssinica Cass.*) germplasm grown in Ethiopia. *Plant Breed*, *114*(4), 375–376.

Lin, E. (2005). *Production and processing of small seeds for birds.* Food and Agriculture Organization of The United Nations.

Mariod, A. A., Mirghani, M. E. S., & Hussein, I. H. (2017). *Unconventional oilseeds and oil sources.* Academic Press.

Mcdonald, R. E., & Mossoba, M. M. (1997). *New techniques and applications in lipid analysis.* The American Oil Chemists Society.

Mensink, R. P., & Katan, M. B. (1992). Effect of dietary fatty acids on serum lipids and lipoproteins. A meta-analysis of 27 trials. *Arteriosclerosis and Thrombosis: A Journal of Vascular Biology*, *12*, 911–919.

Ramadan, M. F. (2009). Niger seed oil. In *Gourmet and health-promoting specialty oils.* Elsevier.

Ramadan, M. F. (2012). Functional properties, nutritional value, and industrial applications of Niger oilseeds (*Guizotia abyssinica* Cass.). *Critical Reviews in Food Science and Nutrition*, *52*, 1–8.

Ramadan, M. F., & Moersel, J.-T. (2006). Screening of the antiradical action of vegetable oils. *Journal of Food Composition and Analysis*, *19*, 838–842.

Ramadan, M. F., & Morsel, J. (2002). Proximate neutral lipid composition of Niger. *Czech Journal of Food Science*, *20*, 98–104.

Ramadan, M. F., & Mörsel, J. T. (2002). Characterization of phospholipid composition of black cumin (*Nigella sativa* L.) seed oil. *Food/Nahrung*, *46*, 240–244.

Ramadan, M. F., & Mörsel, J.-T. (2003). Determination of the lipid classes and fatty acid profile of Niger (*Guizotia abyssinica* Cass.) seed oil. *Phytochemical Analysis*, *14*, 366–370.

Rao, P. U. (1994). Nutrient composition of some less-familiar oil seeds. *Food Chemistry*, *50*, 379–382.

Riley, K., & Belayneh, H. (1989). Niger seed: *Guizotia abyssinica* Cass. In *Oil crops in the world*: *Their breeding and utilization.* McGraw-Hill.

Vles, R. (1989). Nutritional characteristics and food uses of vegetable oils. In *Oil crops of the world, their breeding and utilization* (pp. 63–86). McGraw-Hill.

Chapter 36

Cold pressed rice (*Oryza sativa*) bran oil

Tossaton Charoonratana
Department of Pharmacognosy, College of Pharmacy, Rangsit University, Muang Pathum Thani, Thailand

Abbreviations

CPRBO cold pressed rice bran oil
CVD cardiovascular disease
FFA free fatty acids
GABA γ-aminobutyric acid
HDL-C high-density lipoprotein cholesterol
LDL-C low-density lipoprotein cholesterol
MUFA monounsaturated fatty acid
PUFA polyunsaturated fatty acid
RBO rice bran oil
RCTs randomized controlled trials
SFA saturated fatty acid
TC total cholesterol

1 Introduction

Rice is generally known as *Oryza sativa* L. (Asian rice); however, *Oryza glaberrima* is also a species of African rice, which is not widely grown nowadays, even in Africa. Asian rice is one of the major crops of the world; the annual production is more than 700 million metric tons, and China accounts for 30% of this production (Shahbandeh, 2019). The whole rice grain comprises of endosperm, hull, bran, and embryo. Rice bran is a by-product from rice milling, representing about 10% of the grain by weight, and constitutes only about 12%–32% of rice bran oil (RBO) (da Silva, Sanches, & Amante, 2006; Marshall & Wadsworth, 1994; Samad, 2015). In 2015–2016, RBO production was estimated at more than 700,000 metric tons annually, which were 12–32 folds far from its potential since the rice bran has been mainly used as a feedstock. In China, a leader in rice production, 90,000 metric tons of RBO were produced, which was only 1.7%–5.0% of its potential. Likewise, in a list of the top 10 rice producers, only Japan and India have greatly utilized their rice bran as RBO, totaling 65,000 metric tons (22%–65% of its potential), and 430,000 metric tons (8%–25% of its potential), respectively (Pal & Pratap, 2017). RBO content depends on rice genotype, the degree of milling, and the downstream process of bran. For example, the RBO content of 204 rice accessions grown in Texas (USA) was about 17%–27% (Goffman, Pinson, & Bergman, 2003). Increasing a degree of milling from 6%–9% to 9%–10% caused a reduction in RBO content (Saunders, 1985).

RBO is commonly used as a cooking medium and cosmetics ingredients. In commercial RBO production, the solvent extraction technique, using *n*-hexane, is mainly used to extract rice bran since it is relatively cheap and provides a high RBO yield (Sparks, Hernandez, Zappi, Blackwell, & Fleming, 2006). However, it has drawbacks in the form of residual chemicals and a high temperature required for the process leading to some detrimental components in RBO (Amarasinghe, Kumarasiri, & Gangodavilage, 2009). Since RBO has high nutritional value, it is also used for human health promotion (Nayik, Majid, Gull, & Muzaffer, 2015). To preserve its nutritive properties for pharmaceutical use, cold pressed extraction is a good option. Cold pressed rice bran oil (CPRBO) is produced by mechanical compressing of the rice bran using a screw or hydraulic press under mild heating (lower than 60°C), then the oil is refined by filtration. Therefore, the process involves neither heat nor chemical treatment, which is a soft process. However, the extraction yield was about 3%–10% of the rice bran weight by pressing (Nakagawa, 1983; Srikaeo & Pradit, 2011).

Cold Pressed Oils. https://doi.org/10.1016/B978-0-12-818188-1.00036-0
Copyright © 2020 Elsevier Inc. All rights reserved.

FIG. 1 CPRBO from Hom-Mali rice bran (left) and Rice Berry rice bran (right).

CPRBO has a brownish color, unique odor, and a characteristic earthy flavor. The color can be varied to rice varieties (Fig. 1). The major chemical composition of CPRBO is triacylglycerols, which account for up to 80% of the oil weight. Its bioactive constituent is γ-oryzanol, the strongest antioxidant in rice bran, and the content was about 0.3%–2.5% of CPRBO by weight (Charoonratana, Songsak, Sakunpak, Pathompak, & Charoenchai, 2015; Pengkumsri et al., 2015; Sakunpak, Suksaeree, Monton, Pathompak, & Kraisintu, 2014). It was reported to possess a hypolipidemic effect, improving symptoms of dementia and menopause (Lai, Jacoby, Leong, & Lai, 2019). Moreover, a significant amount of by-product is generated during the CPRBO production process. These by-products include γ-oryzanol, γ-aminobutyric acid (GABA), vitamin E, phytosterols, and policosanol, which can be used as constituents for pharmaceutical manufacture (Liu, Yu, & Wang, 2008).

RBO contains high free fatty acids (FFA) content, so tremendous oil losses occur during the refined process. The esterification process was optimized to reduce oil losses, but this approach has not been generally adopted by the industry. For this reason, RBO has low market availability and higher costs compared to palm and soybean oils. It is doubtful whether RBO could compete with those vegetable oils as a feedstock for industrial manufacturing. Targeting a niche market such as high-value products in cosmetics and pharmaceuticals appears to be a more reasonable way for RBO market expansion (Pal & Pratap, 2017). Cold press, a simple green technology, can be used to produce RBO with its full nutritional value at lower cost. The cold pressed oil machine and other necessary facilities are cheap, suitable for small and medium-sized enterprises, and increase opportunities in business competition. For these reasons, cold press technology is fitting to the situation on RBO utilization in the past, current, near future, and possibly beyond.

2 Extraction and processing of cold pressed rice bran oil

The extraction steps of CPRBO are very easy and simple, starting with compression of the rice bran; the recovered oil is then filtered to obtain functional oil.

First, the material for CPRBO extraction (rice bran) has a very short shelf life, wherein milling of rice bran for more than 4 h must be avoided since lipase, an enzyme inside the rice bran, converts triacylglycerol into FFA (rising 5%–7% per day), which in high levels are not appropriate for human consumption. Oxidation of FFA into free radical compounds is also a cause of human health problems. It was reported that FFA content in RBO should not exceed 10% (Saunders, 1985).

In addition to using fresh rice bran, adding a stabilization step is useful to preserve its nutrition. One option is to stabilize rice bran by physical methods. For example, domestic heating can improve oil extraction content and quality. When comparing unstabilized rice bran and stabilized rice bran through hot air heating, roasting, steaming, and microwave heating methods, it was found that all methods could be used for rice bran stabilization. Moreover, it was found that rice bran treated with hot air heating at 150°C contained the highest oil content and γ-oryzanol with low FFA. Interestingly, microwave heating was the best method to reduce FFA content, while retaining the highest amount of total phenolics and flavonoids. It is possible to use both methods; at the end of the process, the CPRBO color will become slightly redder but the lightness is almost the same (Thanonkaew, Wongyai, McClements, & Decker, 2012). This is related to a previous finding, which reported how the effects of microwave and air drying of parboiled rice can increase oil extraction and reduce FFA. It was shown that FFA of microwave-stabilized rice bran increases less than that in air drying-stabilized rice bran during 10 weeks storage at 25°C (Rizk, Basyony, & Doss, 1995). Another study compared ohmic (electric) and microwave heating. Microwave heating was better in FFA reduction, and both methods were effective for rice bran stabilization with the addition of moisture (Lakkakula, Lima, & Walker, 2004). Theoretically, microwave heating can cause dipole rotation of free water molecules; friction occurs and heats the rice bran consistently through the microwave energy penetrating depth. Although, nowadays, the price of microwave equipment has reduced, for the productivity scale, the price is still much higher than the oven treatment. However, air-dried rice bran needs to be stored in sealed containers to preserve stability, which causes poststabilization and transportation problems (Tao, 1989). From another study, it was noted that heating the rice bran up to 70°C was not possible since it will become very soft and cannot be fed to the screw (Srikaeo & Pradit, 2011). This is contrary to the first example when the temperature was 150°C. This may be due to the differences in heating duration, screw press machine ability, and work experience from both studies. Stabilization using a conveyor was also studied and this was reported to reduce FFA generation after 2 weeks of storage (Kurniawati, Yuliana, & Budijanto, 2014). Another option is to stabilize rice bran by chemical methods, which is not preferable for green extraction like CPRBO production.

Second, the rice bran is sieved through 150–300 mesh sizes to remove foreign matter, such as grains, husks, and stones, which have a negative effect on machine pressing efficacy and oil quality (Fig. 2). It was shown that rice bran obtained from each mesh size comprised of different components and oil content (Singanusong, 2015).

Third, CPRBO is produced by compressing rice bran using a screw press or hydraulic press, which generate only mild heat. These mechanical pressing methods have been used in small and medium-scale industries for CPRBO production. They involve feasible green technology since the machine is less expensive, requires small quantities of physical effort to

FIG. 2 Rice bran.

FIG. 3 Screw press machine with a 2 hp motor.

complete a task, and is free of organic solvent (Cousins, Fore, & Janssen, 1953; Gopala Krishna, 1993). The screw press was continuous and produced a slightly higher yield than the hydraulic press (Gul, Yousuf, Singha, Singh, & Wani, 2015). The screw press (expeller) consists of a hopper (feed), gearbox, discharge of by-product, a screw rotating inside a cylindrical cage (barrel), and an electric motor (Fig. 3). The process is begun by feeding rice bran to the gap between the barrel and the screw via the hopper, where it is propelled by a rotating screw in a direction parallel to the axis and forces oil through the slits along the barrel length. The product from this step is called crude RBO. The defatted rice bran, a by-product from this step, is discharged through a tray on the periphery of the barrel. Crude RBO recovery from the rice bran by this method could be increased by adjusting the speed of the screw using an inverter or gear changing (Sayasoonthorn, Kaewrueng, & Patharasathapornkul, 2012). Using this method, 9%–10% of the rice bran weight could be extracted (Nakagawa, 1983).

The hydraulic press technique is classified as batch mechanical pressing (Sivala, Mukherjee, & Bhole, 1993). The machine consists of a boom, handle, distribution, pressure gauge, barrel, cylinder, and tank, and a heater and a temperature controller that could be installed for pressing at a higher temperature (Fig. 4). The process was started by placing rice bran packed in cloth into the metallic porous cylinder, then it was pressed using a hydraulic press machine. The pressure was increased gradually and kept constant until the crude RBO was forced through the holes or slits of the cylinder into the tank. After a batch was finished, the defatted rice bran could be easily collected since it was in cloth. There is a report indicating that using hydraulic press could recover oil with 11.8% weight of rice bran (Shukla & Pratap, 2017).

In the last step, the crude RBO is subjected to a refined process. To make the whole process green extraction, physical refining without hazardous chemicals is preferable. The crude RBO is left for sedimentation for a single night. After that, filtration through filter papers occurs three times to remove suspended particles. The crude RBO is regularly filtered through filter papers of 10, 5, and 0.25 μm to obtain CPRBO, which is sterilized using UV irradiation (Fig. 5). The by-product of the refined process is called filter cake (Wongwaiwech, Weerawatanakorn, Tharatha, & Ho, 2019). Moreover, adding the sedimentation step showed a reduction of gum and wax in CPRBO from 3.6% to 1.3%. The reduction of gum, especially hygroscopic gum (phosphatides), causes fewer sediments during RBO storage, which improves CPRBO product appearance (Srikaeo & Pradit, 2011). CPRBO should be stored in sealed containers, protected from light at room

FIG. 4 Hydraulic press machine.

temperature with a 1-year shelf life (Fig. 6). Unsealed CPRBO should be kept for less than 1 year since 12-month opened CPRBO product significantly increased in peroxide value when compared with freshly opened or 12-month sealed at room temperature products (Ponchurdchai & Singkhonrat, 2016).

During the whole process, significant amounts of by-products is produced; all of the are rich sources of valuable substances. To decide whether a by-product can be of interest to recover the bioactive substances, the oil content, recovery, substance concentration, and total amount of by-product per batch need to be considered. For CPRBO, there is a study on the amount of nutraceuticals in defatted rice bran and filter cake. The total amount of bioactive compounds of the by-products (100g) was surprisingly high; the highest one was γ-oryzanol (1288mg), followed by phytosterol (249mg), vitamin E (141mg), policosanol (122mg), and GABA (29mg), respectively. In Thailand in 2017 (Wongwaiwech et al., 2019), defatted rice bran was discarded from manufacture of about 640,000tons and the total amount of bioactive constituent found in defatted rice bran from CPRBO production was 321mg/100g, which contained γ-oryzanol (230mg/100g). Consequently, the total amount of nutraceuticals remaining in defatted rice bran is estimated at around 3200tons per year (containing 2300tons of γ-oryzanol). Combining γ-oryzanol and filter cake, this unique compound can be extracted and purified to add value to the product.

γ-oryzanol, the molecule of interest, is found only in rice species. There are several reports about its isolation from crude RBO. First, preparative column chromatography can be used under the normal phase silica-based mode. From 2.4g crude RBO, around 400mg γ-oryzanol can be isolated with the productivity of 10mg/20min, while the purity was 90%–98% (Lai et al., 2019). Second, countercurrent chromatography could provide 261mg γ-oryzanol in 200min from 8g crude RBO with 95% purity (Angelis, Urbain, Halabalaki, Aligiannis, & Skaltsounis, 2011). Third, with two-step crystallization, γ-oryzanol was recovered at around 59% with 93%–95% purity (Zullaikah, Melwita, & Ju, 2009). In addition to pure compound utilization, CPRBO itself is also valuable from its components. Good-quality CPRBO should be considered in terms of the varieties and contents of bioactive constituents that have positive effects on humans.

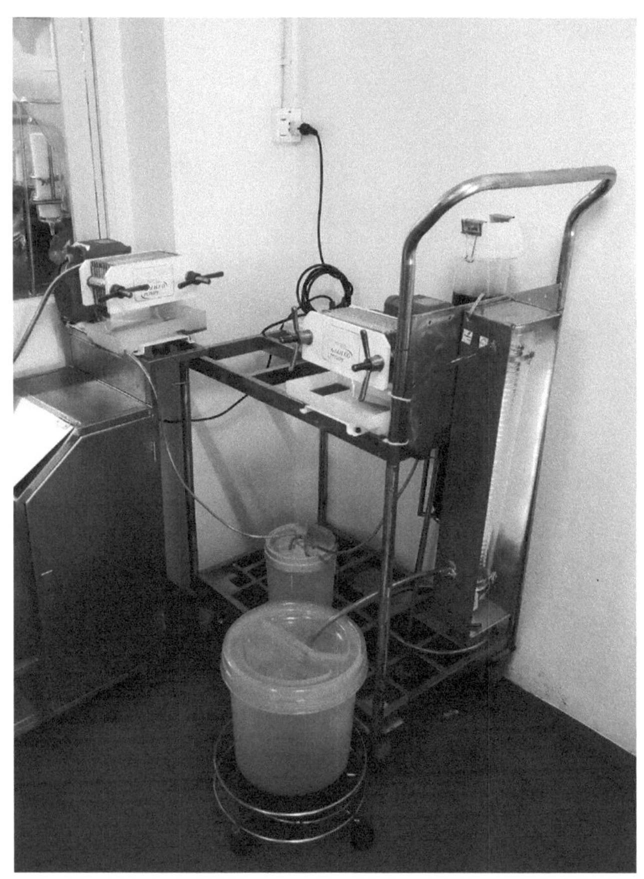

FIG. 5 Membrane filters and UV lamp.

FIG. 6 Sealed Hom Pathum CPRBO ready to store in dark container.

3 Fatty acids and acyl lipids profile of cold pressed rice bran oil

The chemical and fatty acid constituents of rice vary significantly with genetics, ecological conditions, and agronomic practices, which affect chemical composition in rice bran. Studies on 24 nonglutinous rice varieties grown in the Hiroshima Agricultural Experiment Station found that variety had a considerable effect on linoleic, oleic, and stearic acids content in rice bran (Taira, Taira, & Maeshige, 1979). Generally, up to 80% of RBO weight is triacylglycerols, while diacylglycerols and monoacylglycerols are 3%–5% and 1%–2%, respectively. About 49% and 40% of the triacylglycerols in RBO were triunsaturated and diunsaturated, meaning that around 76% of fatty acids on the glycerol backbone was unsaturated. This value is also related to free fatty acids composition in RBO (Jin et al., 2016).

FFA is reported to be in a range of 0.9%–10.2% of CPRBO weight (Charoonratana et al., 2015; Pal & Pratap, 2017; Srikaeo & Pradit, 2011). Generally, the saturated fatty acid composition of CPRBO is quite high, in a range of 19%–24%. Palmitic acid is the major saturated fatty acid (16%–21%), while stearic acid is the minor constituent (2.5%–3.8%). Unsaturated fatty acids contain most of the fatty acids in CPRBO (76%–81%). Monounsaturated fatty acid (MUFA), oleic acid, is the richest unsaturated fatty acid (37%–44%), followed by polyunsaturated fatty acid (PUFA), linoleic acid (30%–37%), and linolenic acid (1.4%–2.2%), respectively (Charoonratana et al., 2015; Katsri, Noitup, Junsangsree, & Singanusong, 2014; Singanusong, 2015).

Compared to other studies, the fatty acid composition of CPRBO is relatively similar to RBO (Latha & Nasirullah, 2014; Rukmini & Raghuram, 1991). RBO has high oxidative stability since it contains natural antioxidants such as γ-oryzanol and vitamin E. It was found that only 9% of PUFA was lost through degradation to oxidation products when RBO was heated at 180°C for 8h. No *trans*-unsaturated fatty acid was found in the heated oil (Latha & Nasirullah, 2014).

4 Minor bioactive lipids in cold pressed rice bran oil

CPRBO is thought to be the best product that preserves the full nutrition of rice bran. Considering reported metabolomics data in rice bran from three different US rice varieties, 453 compounds were found including amino acids, cofactors, vitamins, and secondary metabolites. Five major secondary metabolites from this finding were gluconate, 2-piperidinone, ferulate, feruloylputrescine, and stachydrine. Other interesting minor bioactive compounds were 4-guanidinobutanoate, taurine, tocopherols, tocotrienols, ergothioneine, and quinate. They were identified as rice bran antioxidants, which could be used to inactivate free radicals, reducing oxidative stress that may cause some chronic diseases. Luteolin, α-amyrin, and leucic acid were also found in rice bran. Luteolin, a plant flavonoid, possesses antimicrobial activity, while triterpene α-amyrin has antiinflammatory activity. Leucic acid has also been identified in fermented food and possesses fungicidal activity (Zarei, Brown, Nealon, & Ryan, 2017).

To date, in CPRBO, not more than 20 minor compounds were analyzed compared to their reference standards. γ-oryzanol is the main target in every study. Tocopherols and tocotrienols are the next compounds of interest. Other compounds are squalene and groups of phytosterols, while policosanol and GABA are detected in CPRBO by-products.

γ-oryzanol is one of the most important compounds in CPRBO. It is a mixture of ferulic acid esters of phytosterols and triterpenoids, particularly cycloartenyl ferulate, 24-methylenecycloartanyl ferulate, campsteryl ferulate, and β-sitosteryl ferulate. Pure γ-oryzanol is white, odorless, and crystalline (Kangcare, 2012). The hydroxyl group in the ferulic acid portion of γ-oryzanol caused its relatively high polarity and possible solubility in polar and nonpolar solvents. It is soluble in isopropanol but insoluble in water. The content was about 0.28%–2.64% of CPRBO by weight. A wide range of γ-oryzanol content in CPRBO could be due to the difference in rice varieties and refined methods. One study was conducted on CPRBO from three different rice varieties, Khao Dawk Mali, Red Jasmine, and Hom-nin; after a refining process using phosphoric acid, bleaching earth, activated carbon, and very high temperature (220°C, 1h), it was found that γ-oryzanol content was 0.33% *w*/w, 0.28% *w*/w, and 0.54% *w*/w, respectively (Mingyai et al., 2018). This harsh refining process may reduce the amount of γ-oryzanol. A study from India (Shukla & Pratap, 2017) used a soft process, comprising a hydraulic press and filter to produce CPRBO, wherein the γ-oryzanol content was surprisingly low (0.17% *w*/w). However, these data are excluded from this chapter since the authors used the spectrophotometer to estimate γ-oryzanol content, which was different from other kinds of literature that used more accurate detectors. Another study was conducted to determine four ferulic esters from CPRBO. While using the soft refining process, γ-oryzanol content reached 2.5% *w*/w in Khao-Hom-Pathum CPRBO from Chainat province, while the most abundant ester was 24-methylenecycloartanyl ferulate (Charoonratana et al., 2015). γ-oryzanol content in CPRBO from Sung Yod rice was also reported but the value was low (0.31% *w*/w) (Uttama et al., 2014). γ-oryzanol content was reported in RBO in a range of 1.2–1.7 (Pal & Pratap, 2017). Finally, a study showed that unidentified rice variety CPRBO possessed 2.64% *w*/w γ-oryzanol (Pokkanta et al., 2019).

Tocopherols, along with tocotrienols, are a member of the vitamin E family. There are four naturally occurring tocopherols: α-, β-, γ-, and δ-tocopherol. A pure form of tocopherols is a yellow viscous liquid, easily degraded under exposure to light, oxygen, heat, and alkali. It is soluble in an organic solvent but not in water (Sarmento, Ferreira, & Hense, 2006). Tocopherols consist of six chromanol nuclei with the addition of methyl groups at positions five, seven, and eight. It has a C16 saturated side chain at position two (Ahsan, Ahad, & Siddiqui, 2015). Tocopherol esters are considered more stable than tocotrienols. α-tocopherol is the form of vitamin E that is preferably absorbed and accumulated in humans. Its stability decreases by half with every increase of 10°C when exposed to light (Lampi, Kamal-Eldin, & Piironen, 2002). The tocopherols content was about 0.003%–0.024% of CPRBO by weight. The reason behind a wide range of content is the same as that of γ-oryzanol. Tocopherols content was lowest in CPRBO from Khao Dawk Mali rice bran, while the value was highest in Hom-nin CPRBO from the same study with a harsh refining process (Mingyai et al., 2018). It was found that tocopherols were present in a wide variety of rice types at different levels. Tocopherols contents in CPRBO produced from Chiang Mai black rice, Suphanburi-1 brown rice, and Mali red rice were 0.016% *w*/w, 0.021% *w*/w, and 0.013% *w*/w, respectively (Pengkumsri et al., 2015). Compared to RBO extracted by other methods, refined RBO was reported to contain tocopherols up to 0.08% *w*/w, which was higher than that of CPRBO (Pal & Pratap, 2017).

Tocotrienols, along with tocopherols, are known as tocols, consisting of four isomers of different analogs (α, β, γ, and δ). A pure form of tocotrienols is a yellow viscous liquid, soluble in an organic solvent but insoluble in water. They are not stable to oxygen, light, heat, and alkali. Tocotrienols differ from tocopherols in the number and position of methyl groups on the benzene ring and the farnesyl side chain (Aggarwal, Sundaram, Prasad, & Kannappan, 2010). Tocotrienols are only in R-configuration and *trans*-geometry, with three double bonds at positions three, seven, and eleven. The tocotrienols content was about 0.014%–0.139% of CPRBO by weight and they varied according to rice species (Mingyai et al., 2018; Pengkumsri et al., 2015). Compared to other literature, tocotrienols content of CPRBO is relatively similar to RBO (Pal & Pratap, 2017).

A conclusion on the amounts of γ-oryzanol, tocopherols, and tocotrienols in CPRBO was drawn from limited available resources. It was found that the refining process involving dewaxing, degumming, bleaching, and deodorizing strongly decreased those metabolites content (Mingyai et al., 2018). Excluding harsh refining, the ranges of γ-oryzanol, tocopherols, and tocotrienols were 0.61%–2.64%, 0.013%–0.024%, and 0.034%–0.139% of CPRBO by weight, respectively (Charoonratana et al., 2015; Pengkumsri et al., 2015; Pokkanta et al., 2019).

Phytosterols or plant sterols are a family of compounds related to cholesterol. They are found mainly in plant cell walls and membranes. The most common phytosterols are campesterol, stigmasterol, and β-sitosterol (Dunford & King, 2000). To utilize phytosterols in applications, esterification is required since all of them in the free form are poorly soluble in water or oil. Two studies in phytosterols content in CPRBO were carried out, both from Thailand. It was found that without a harsh refining process, the content of sterols was 2.53% of CPRBO by weight. Values fell into a range of 0.99%–1.14% *w*/w after refining (Mingyai et al., 2018; Pokkanta et al., 2019). Values were close to those of phytosterols found in commercially refined RBO, which are 0.86%–1.03% *w*/w (Niawanti & Zullaikah, 2017). One-half of phytosterol in CPRBO is β-sitosterol.

Squalene is in the triterpene family, which is intermediate for phytosterols or triterpenoid saponins biosynthesis. It is a colorless and odorless liquid at room temperature. It is highly soluble in oil. Due to its unsaturated structure, squalene is not very stable and is easily oxidized (Rosales-García, Jimenez Martinez, & Davila-Ortiz, 2017). In contrast, squalene in CPRBO was reported as 0.32% *w*/w, which is the same value as in commercially refined RBO (Pokkanta et al., 2019; Popa, Bəbeanu, Popa, Niţă, & Dinu-Parvu, 2015). This may be because the CPRBO and RBO were from different rice varieties. The chemical compositions of CPRBO and RBO were compared, as shown in Table 1.

5 Health-promoting traits of cold pressed rice bran oil and oil constituents

The concept of a single nutrient affecting human health has been accepted and maintained for a long time. However, much evidence indicates that the sample matrix also has influence on health promotion. Therefore, it is necessary to evaluate the health promotion effect of CPRBO as it is, and as its individual component. To consider the health effect of CPRBO, firm clinical evidence should be provided; however, only two studies in rodents have been reported about the impact of CPRBO on the memory-enhancing effect and antihypertensive effect, respectively. All results showed positive effects (Choowong-in & Sattayasai, 2017; Jan-on et al., 2019). Due to lacking evidence and as CPRBO composition is relatively similar to RBO in some studies, meta-analysis and randomized controlled trials (RCTs) of RBO are selected and discussed in this section.

From 11 RCTs studying the effects of RBO on cholesterol in humans, it was found that consumption of RBO could reduce low-density lipoprotein cholesterol (LDL-C) and total cholesterol (TC) (Jolfaie, Rouhani, Surkan, & Azabdakht, 2016). This may lead to the prevention of cardiovascular disease (CVD) since 10 mg/L reduction in LDL-C

TABLE 1 Main chemical composition of CPRBO compared to RBO.

Composition	Content in oil (% *w/w*)	
	CPRBO	**RBO**
Saturated fatty acids		
Palmitic acid	16.0–21.0	17.0–21.5
Stearic acid	2.5–3.8	1.0–3.0
MUFA		
Oleic acid	37.0–44.0	38.4–42.3
PUFA		
Linoleic acid	30.0–37.0	33.1–37.0
Linolenic acid	1.4–2.2	0.5–2.2
Other bioactive compounds		
γ-oryzanol	0.61–2.64	1.20–1.70
Tocopherols	0.013–0.024	0.020–0.080
Tocotrienols	0.034–0.139	0.025–0.170
Phytosterols	0.99–1.14	0.86–1.03
Squalene	0.32	0.30–0.40

levels can lead to a 1%–2% decrease in the risk of CVD (Zavoshy, Noroozi, & Jahanihashemi, 2012). In addition, a 10% decrease in TC concentration results in a 30% reduction of CVD incidence (Lichtenstein et al., 2006). Besides reductions in LDL-C and TC values, it was found that RBO consumption could increase high-density lipoprotein cholesterol (HDL-C) (good cholesterol) in men. Moreover, in two studies, the LDL-C/HDL-C ratio showed a significant reduction following RBO intake, and change in this ratio is a predictor of CVD (Kinosian, Glick, Preiss, & Puder, 1995). Although there were different control groups in the studies—for example, corn, palm, peanut, soybean, and sunflower—compared with these oils, RBO could have favorable effects on LDL-C and TC levels. The main difference between RBO and control oils is related to its high level of unsaponifiable compounds. It is difficult to draw a conclusion about the ideal amount of RBO to consume per day to promote a benefit to lipid profiles; as detailed in the literature, five studies were not blinded, so both performance and detection biases can occur. The duration of trials ranged between 21 and 90 days. The number and age of volunteers ranged from 9 to 60 volunteers and 34 to 61 years, respectively. Eight studies were conducted in hyperlipidemic patients, while the others focused on healthy volunteers.

For the antihypertensive studies, the results were mixed. Even there is a study which demonstrated that a blend of RBO and sesame oil can lower blood pressure in mild-to-moderate hypertensive patients, there is another study found no a significant impact of RBO on blood pressure in postmenopausal type 2 diabetic women (Devarajan, Singh, Chatterjee, Zhang, & Ali, 2016; Salar, Faghih, Pishdad, & Rezaie, 2015). The positive effect of the first study may be from sesame oil or the mixtures to a specific group of people.

To date, among oil produced from higher plants, the ratio of fatty acids of CPRBO (1.3:1.5:1; PUFA:MUFA:SFA) is the closest ratio to those recommended by the World Health Organization for edible oil. Moreover, CPRBO had no *trans*-unsaturated fatty acid, which is safe with regard to heart disease and cancer caused by this compound (Singanusong, 2015). However, considering individual fatty acids, the result may be completely different. For example, it was found that increasing dietary linoleic acid intake has no significant effect on inflammatory markers in the blood concentrations (Su, Liu, Chang, Huang, & Wang, 2017).

Considering a single-nutrient approach, taking too little or too much of fatty acids daily may cause a nutritional deficiency or undesirable effect in humans. For example, the adequate intake for omega-3 s for male adults (19–50 years old) is 1.60 g daily (Institute of Medicine, Food and Nutrition Board, 2005). By averaging the fatty acids value content mentioned earlier, CPRBO contains free omega-3 s around 0.10% of CPRBO weight. Adding this to the data provided before that 87% *w*/w of RBO is glycerol, and 76% of fatty acids on the glycerol backbone was unsaturated, it means that 100 g CPRBO contains 66 g unsaturated fatty acids from glycerol, which is omega-3 s around 2.3% (1.52 g). Combining 1.52 g with 0.10 g, it is found that there is 1.62 g linolenic acid in 100 g CPRBO. This means that at least 100 g of CPRBO can be taken by men per day. However, it should be noted that not all linolenic acid can be converted to docosahexenoic acid or eicosapentaenoic acid, because of enzymatic limitations in the human body.

Similarly, estimated free and unfree saturated fatty acids in CPRBO are 1.19% w/w and 34.00% w/w, respectively. It is recommended to reduce the intake of total saturated fatty acids to less than 10% of total energy consumption (Astrup et al., 2019). Men need not more than 1050 kJ per day from SFA, while 100 g of CPRBO gives around 577 kJ from SFA (calculated based on the ratio of palmitic acid and stearic acid; 1 g palmitic acid gives 19.46 kJ based on its molar heat combustion). It is recommended for men not to consume more than 182 g or 200 mL CPRBO per day, while it is 160 mL in women. Along with other food and product intake, an appropriate amount of CPRBO can be roughly calculated and it will be more accurate if those products contain ingredient labels.

There is very limited evidence of pure γ-oryzanol for human health promotion. One review in the literature showed that it possessed great potential for lowering blood cholesterol (Cicero & Gaddi, 2001). Later, it was found that the human intestinal load of 453–740 mg of de-ferulated plant sterol improved lipoprotein pattern in mildly hypercholesterolemic patients (Berger et al., 2005). Moreover, the reason for the beneficial effect of γ-oryzanol was because it can directly be absorbed by the intestine and is present in the intact form in plasma (Kobayashi et al., 2016). γ-oryzanol is safe in rats after 90 days of oral administration at doses of 1000 and 2000 mg/kg body weight/day (Moon et al., 2017).

Many trials were conducted concerning the effects of vitamin E on the human body, and some relating to CVD are exemplified here. The direct effect on CVD was studied from 16 RCTs. It was found that 33–800 IU vitamin E alone can reduced myocardial infarction in patients who have a follow-up ranging from 0.5 to 9.4 years (Loffredo et al., 2015). Circuitous studies were conducted to determine the effects of vitamin E on cholesterol behavior. Tocotrienols can suppress 3-hydroxy-3-methyl-glutaryl-coenzyme A (HMG-COA) reductase activity, resulting in lower endogenous cholesterol synthesis (Minhajuddin, Beg, & Iqbal, 2005). Moreover, it is thought that tocotrienols components can increase the expression of cholesterol 7-alpha-hydroxylase (CYP7A1), which is the important enzyme in maintaining stable TC levels in the body through its conversion to bile acid (Pullinger et al., 2002; Zavoshy et al., 2012). However, it was not recommended to take a high dose (>399 IU per day) of vitamin E, since it increased all-cause mortality in humans (Miller III et al., 2005).

Several studies have demonstrated that the consumption of phytosterols can lead to a significant reduction in serum cholesterol. In 2017, it was shown from a meta-analysis from 12 RCTs that phytosterols can reduce LDL-C levels in healthy volunteers and patients with hypercholesterolemia (Ying & Kang, 2017). Later clinical trials also supported the benefit of phytosterols supplement effectively decreasing LDL-C in healthy adults within 1 month (Reaver et al., 2019). Phytosterols also play a role in the inhibition of cholesterol absorption into enterocytes by reducing its micelles solubility, and may compete in binding activity of cholesterol due to having a similar structure to it (Abumweis, Barake, & Jones, 2008).

The study design in a randomized, double-blind, placebo-controlled trial was conducted to evaluate the effect of ferulic acid, minor constituent (4×10^{-7}% *w*/w) in CPRBO (Siger, Nogala-Kalucka, & Lampart-Szczapa, 2007), on many parameters in hyperlipidemia. It was found that 1000 mg/day ferulic acid can improve lipid profiles and oxidative stress, oxidized LDL-C, and inflammation in hyperlipidemic patients. Therefore, a sufficient amount of ferulic acid has the potential to reduce cardiovascular disease risk (Bumrungpert, Lilitchan, Tuntipopipat, Tirawanchai, & Komindr, 2018). However, ferulic content in CPRBO was extremely low, suggesting that it is not a good source for a pure compound extraction.

6 Other applications of cold pressed rice bran oil

The unique characteristics of CPRBO have rendered its benefit in the human cholesterol-lowering effect, the complimentary supplement is suitable for its application. Interesting other applications should be focused on the by-products. First, there is a study using the particles extracted from CPRBO by-products as the ingredients for body mask and source of dietary fiber (Srikaeo, Poungsampao, & Phuong, 2017). Another attempt was also made to prepare a mask using CPRBO as the ingredient (Suksaeree et al., 2013). There is also growing interest in using CPRBO niosomes to produce materials for drug delivery of oil-soluble bioactive compounds for water-soluble functional ingredient applications (Hunthayung, Klinkesorn, Hongsprabhas, & Chanput, 2019). Concerning food packaging applications, RBO can be used as a material to produce composite biodegradable polymer films in packaging (Kale et al., 2018). There is 141 mg vitamin E in 100 g by-product. This vitamin E can be extracted and purified to use in advanced applications. For example, it is utilized as the coating material in quality dialysis membrane, which can reduce the erythropoietin resistance in hemodialysis patients (Huang, Yi, Li, & Zhang, 2015). For application in construction, the major secondary metabolite in rice bran was gluconate as suggested from a rice bran metabolomics study (Zarei et al., 2017). Instead of discarding this by-product, the gluconate can be used as a retarder to slow down cement hydration reactions and to delay cement setting time (Ramachandran & Lowery, 1992).

7 Conclusion

Rice bran oil's reputation as a "heart oil" is well-earned since it can reduce LDL-C and total cholesterol, which are risk factors of cardiovascular disease. Cold pressing is a simple green technology for the production of a fully nutritional rice bran oil. Its feasibility in small-scale enterprise provides opportunities for local people with limited budgets to achieve market penetration, while the CPRBO price is up to 600 US dollars per kilogram. In terms of CPRBO issues, much more information is still waiting for us to discover it; uncharted rice varieties richer in bioactive composition should be explored, along with higher efficient extraction, and a preserved simple method could be developed to recover a greater oil yield. Clinical studies data in various health promotions can also be added to support an increase of the product value.

Acknowledgments

The author thanks Mr. Pornthep Thanakulrungsarit and Mr. Natawat Chankana for kindly providing suggestions, materials, and field visits.

References

Abumweis, S. S., Barake, R., & Jones, P. J. (2008). Plant sterols/stanols as cholesterol lowering agents: A meta-analysis of randomized controlled trials. *Food & Nutrition Research*, *52*, 1–17.

Aggarwal, B. B., Sundaram, C., Prasad, S., & Kannappan, R. (2010). Tocotrienols, the vitamin E of the 21st century: Its potential against cancer and other chronic diseases. *Biochemical Pharmacology*, *80*(11), 1613–1631.

Ahsan, H., Ahad, A., & Siddiqui, W. A. (2015). A review of characterization of tocotrienols from plant oils and foods. *Journal of Chemical Biology*, *8*(2), 45–59.

Amarasinghe, B. M. W. P. K., Kumarasiri, M. P. M., & Gangodavilage, N. C. (2009). Effect of method of stabilization on aqueous extraction of rice bran oil. *Food and Bioproducts Processing*, *87*(2), 108–114.

Angelis, A., Urbain, A., Halabalaki, M., Aligiannis, N., & Skaltsounis, A. L. (2011). One-step isolation of γ-oryzanol from rice bran oil by non-aqueous hydrostatic countercurrent chromatography. *Journal of Separation Science*, *34*, 2528–2537.

Astrup, A., Bertram, H. C. S., Bonjour, J. P., de Groot, L. C. P., de Oliveira Otto, M. C., Feeney, E. L., et al. (2019). WHO draft guidelines on dietary saturated and trans fatty acids: Time for a new approach? *British Medical Journal*, *366*, 1–6.

Berger, A., Rein, D., Schafer, A., Monnard, L., Gremaud, G., Lambelet, P., et al. (2005). Similar cholesterol-lowering properties of rice bran oil, with varied γ-oryzanol in mildly hypercholesterolemic men. *European Journal of Nutrition*, *44*(3), 163–173.

Bumrungpert, A., Lilitchan, S., Tuntipopipat, S., Tirawanchai, N., & Komindr, S. (2018). Ferulic acid supplementation improves lipid profiles, oxidative stress, and inflammatory status in Hyperlipidemic subjects: A randomized, double-blind, placebo-controlled clinical trial. *Nutrients*, *10*(6), 713–720.

Charoonratana, T., Songsak, T., Sakunpak, A., Pathompak, P., & Charoenchai, L. (2015). Using liquid chromatography-mass spectrometry based metabolomics to discriminate between cold pressed rice bran oils produced from two different cultivars of *Oryza sativa* L. ssp. *indica* in Thailand. *Chinese Journal of Chromatography*, *33*(9), 966–973.

Choowong-in, P., & Sattayasai, J. (2017). Memory enhancing effects of virgin rice bran oil derived from Thai brown rice in mice. *Journal of Thai Traditional & Alternative Medicine*, *14*(1), 37–45.

Cicero, A. F. G., & Gaddi, A. (2001). Rice bran oil and γ-oryzanol in the treatment of hyperlipoproteinaemias and other conditions. *Phytotherapy Research*, *15*(4), 277–289.

Cousins, E., Fore, S., & Janssen, H. (1953). Rice bran oil. VIII. Tank settlings from crude rice bran oil as a source of wax. *Journal of the American Oil Chemists' Society*, *30*, 9–14.

Devarajan, S., Singh, R., Chatterjee, B., Zhang, B., & Ali, A. (2016). A blend of sesame oil and rice bran oil lowers blood pressure and improves the lipid profile in mild-to-moderate hypertensive patients. *Journal of Clinical Lipidology*, *10*(2), 339–349.

Dunford, N. T., & King, J. W. (2000). Phytosterol enrichment of rice bran oil by a supercritical carbon dioxide fractionation technique. *Journal of Food Science*, *65*(8), 1395–1399.

Goffman, F., Pinson, S. R. M., & Bergman, C. (2003). Genetic diversity for lipid content and fatty acid profile in rice bran. *Journal of the American Oil Chemists' Society*, *80*(5), 485–490.

Gopala Krishna, A. (1993). Influence of viscosity on wax settling and refining loss in rice bran oil. *Journal of the American Oil Chemists' Society*, *70*, 895–898.

Gul, K., Yousuf, B., Singha, A., Singh, P., & Wani, A. (2015). Rice bran: Nutritional values and its emerging potential for development of functional food—A review. *Bioactive Carbohydrates and Dietary Fibre*, *6*, 24–30.

Huang, J., Yi, B., Li, A. M., & Zhang, H. (2015). Effects of vitamin E-coated dialysis membranes on anemia, nutrition and dyslipidemia status in hemodialysis patients: A meta-analysis. *Renal Failure*, *37*(3), 398–407.

Hunthayung, K., Klinkesorn, U., Hongsprabhas, P., & Chanput, W. (2019). Controlled release and macrophage polarizing activity of cold-pressed rice bran oil in noisome system. *Food & Function*, *10*, 3272–3281.

Institute of Medicine, Food and Nutrition Board (2005). *Dietary reference intakes for energy, carbohydrate, fiber, fat, fatty acids, cholesterol, protein, and amino acids (macronutrients)*. Washington, DC: National Academy Press.

Jan-on, G., Sangartit, W., Pakdeechote, P., Kukongviriyapan, V., Sattayasai, J., Senaphan, K., et al. (2019). Virgin rice bran oil alleviates hypertension through the upregulation of eNOS, and reduction of oxidative stress and inflammation in L-NAME-induced hypertensive rats. *Nutrition*, *69*, 110575.

Jin, J., Xie, D., Chen, H., Wang, X., Jin, Q., & Wang, X. (2016). Production of rice bran oil with light color and high oryzanol content by multi-stage molecular distillation. *Journal of the American Oil Chemists' Society*, *93*(1), 145–153.

Jolfaie, N. R., Rouhani, M. H., Surkan, P. J., & Azabdakht, L. (2016). Rice bran oil decreases total and LDL cholesterol in humans: A systematic review and meta-analysis of randomized controlled clinical trials. *Hormone and Metabolic Research*, *48*(7), 417–426.

Kale, R. D., Gorade, V. G., Madye, N., Chaudhary, B., Bangde, P. S., & Dandekar, P. P. (2018). Preparation and characterization of biocomposite packaging film from poly(lactic acid) and acylated microcrystalline cellulose using rice bran oil. *International Journal of Biological Macromolecules*, *118*, 1090–1102.

Kangcare (2012). *Oroyal TM Gamma Oryzanol 99% (Vol. August 16)*. Retrieved from http://www.kangcare.com/editor/attached/file/20131030/20131030191112_51946.pdf.

Katsri, K., Noitup, P., Junsangsree, P., & Singanusong, R. (2014). Physical, chemical and microbiological properties of mixed hydrogenated palm kernel oil and cold-pressed rice bran oil as ingredients in non-dairy creamer. *Songklanakarin Journal of Science and Technology*, *36*(1), 73–81.

Kinosian, B., Glick, H., Preiss, L., & Puder, K. L. (1995). Cholesterol and coronary heart disease: Predicting risks in men by changes in levels and ratios. *Journal of Investigative Medicine*, *43*, 443–450.

Kobayashi, E., Ito, J., Kato, S., Sawada, K., Matsuki, M., Hashimoto, H., et al. (2016). Presence of orally administered rice bran oil γ-oryzanol in its intact form in mouse plasma. *Food & Function*, *7*(7), 4816–4822.

Kurniawati, M., Yuliana, N. D., & Budijanto, S. (2014). The effect of single screw conveyor stabilization on free fatty acids, α-tocopherol, and γ-oryzanol content of rice bran. *International Food Research Journal*, *21*(3), 1201–1205.

Lai, O. M., Jacoby, J. J., Leong, W. F., & Lai, W. T. (2019). Nutritional studies of rice bran oil. In L. Z. Cheong, & X. Xu (Eds.), *Chemistry, processing and utilization* (pp. 19–53). Illinois: AOCS Press. Available from: Elsevier E-book. [26 September 2019].

Lakkakula, N. R., Lima, M., & Walker, T. (2004). Rice bran stabilization and rice bran oil extraction using ohmic heating. *Bioresource Technology*, *92*, 157–161.

Lampi, A., Kamal-Eldin, M., & Piironen, V. (2002). Functional foods and nutraceutical series. In *Functional foods biochemical and processing aspects* (pp. 1–29). (2nd ed.). London: CRC Press.

Latha, R. B., & Nasirullah, D. R. (2014). Physico-chemical changes in rice bran oil during heating at frying temperature. *Journal of Food Science and Technology*, *51*(2), 335–340.

Lichtenstein, A. H., Appel, L. J., Brands, M., Carnethon, M., Daniels, S., Franch, H. A., et al. (2006). Diet and lifestyle recommendations revision 2006: A scientific statement from the American Heart Association nutrition committee. *Circulation*, *114*, 82–96.

Liu, Y., Yu, J., & Wang, X. (2008). Extraction of policosanols from hydrolysed rice bran wax by high-intensity ultrasound. *International Journal of Food Science & Technology*, *43*, 763–769.

Loffredo, L., Perri, L., Castelnuovo, A. D., Lacoviello, L., Gaetano, G. D., & Violi, F. (2015). Supplementation with vitamin E alone is associated with reduced myocardial infarction: A meta-analysis. *Nutrition Metabolism and Cardiovascular Diseases*, *25*(4), 354–363.

Marshall, W., & Wadsworth, J. (1994). *Rice science and technology*: (pp. 421–438). New York: Marcel Dekker.

Miller, E. R., III, Pastor-Barriuso, R., Dalal, D., Riemersma, R. A., Appel, L. J., & Guallar, E. (2005). Meta-analysis: High-dosage vitamin E supplementation may increase all-cause mortality. *Annals of Internal Medicine*, *142*(1), 37–46.

Mingyai, S., Srikaeo, K., Kettawan, A., Singanusong, R., Nakagawa, K., Kimura, F., et al. (2018). Effects of extraction methods on phytochemicals of rice bran oils produced from colored rice. *Journal of Oleo Science*, *67*(2), 135–142.

Minhajuddin, M., Beg, Z. H., & Iqbal, J. (2005). Hypolipidemic and antioxidant properties of tocotrienol rich fraction isolated from rice bran oil in experimentally induced hyperlipidemic rats. *Food and Chemical Toxicology*, *43*, 747–753.

Moon, S. H., Kim, D., Shimizu, N., Okada, T., Hitoe, S., & Shimoda, H. (2017). Ninety-day oral toxicity study of rice-derived γ-oryzanol in Sprague-Dawley rats. *Toxicology Reports*, *4*, 9–18.

Nakagawa, K. (1983). *Apparatus for pretreatment for extracing crude oil from rice bran*. US Patent. 4,384,837.

Nayik, G. A., Majid, I., Gull, A., & Muzaffer, K. (2015). Rice bran oil, the future edible oil of India: A mini review. *Rice Research*, *3*(4), 1–3.

Niawanti, H., & Zullaikah, S. (2017). Removal of bioactive compound (γ-oryzanol) from rice bran oil-based biodiesel using deep eutectic solvent. *Chemical Engineering Transactions*, *56*, 1513–1518.

Pal, Y. P., & Pratap, A. P. (2017). Rice bran oil: A versatile source for edible and industrial applications. *Journal of Oleo Science*, *66*(6), 551–556.

Pengkumsri, N., Chaiyasut, C., Sivamaruthi, B. S., Saenjum, C., Sirilun, S., Peerajan, S., et al. (2015). The influence of extraction methods on composition and antioxidant properties of rice bran oil. *Food Science and Technology*, *35*(3), 493–501.

Pokkanta, P., Sookwong, P., Tanang, M., Setchaiyan, S., Boontakham, P., & Mahatheeranont, S. (2019). Simultaneous determination of tocols, γ-oryzanol, phytosterols, squalene, cholecalciferol and phylloquinone in rice bran and vegetable oil samples. *Food Chemistry*, *271*, 630–638.

Ponchurdchai, C., & Singkhonrat, J. (2016). Monitoring the effect of different storage conditions of cold-pressed rice bran oil. *International Journal of Chemical Engineering and Applications*, *7*(4), 254–258.

Popa, O., Bəbeanu, N. E., Popa, I., Niţă, S., & Dinu-Parvu, C. E. (2015). Methods for obtaining and determination of squalene from natural sources. *BioMed Research International*, *2015*, 1–16.

Pullinger, C. R., Eng, C., Salen, G., Shefer, S., Batta, A. K., Erickson, S. K., et al. (2002). Human cholesterol 7alpha-hydroxylase (CYP7A1) deficiency has a hypercholesterolemic phenotype. *Journal of Clinical Investigation*, *110*, 109–117.

Ramachandran, V. S., & Lowery, M. S. (1992). Conduction calorimetric investigation of the effect of retarders on the hydration of Portland cement. *Thermochimica Acta*, *195*, 373–387.

Reaver, A., Hewlings, S., Westerman, K., Blander, G., Schmeller, T., Heer, M., et al. (2019). Controlled, double-blind crossover study to assess a unique phytosterol ester formulation in lowering LDL virtual tracking tool. *Nutrients*, *11*(9), 2108.

Rizk, L. F., Basyony, A. E., & Doss, H. A. (1995). Effect of microwave and air drying of parboiled rice on stabilization of rice bran oil. *Grasas y Aceites*, *46* (3), 158–163.

Rosales-García, T., Jimenez Martinez, C., & Davila-Ortiz, G. (2017). Squalene extraction: Biological sources and extraction methods. *International Journal of Environment, Agriculture and Biotechnology*, *4*, 1662–1670.

Rukmini, C., & Raghuram, T. C. (1991). Nutritional and biochemical aspects of the hypolipidemic action of rice bran oil: A review. *Journal of the American College of Nutrition*, *10*(6), 593–601.

Sakunpak, A., Suksaeree, J., Monton, C., Pathompak, P., & Kraisintu, K. (2014). Quantitative analysis of γ-oryzanol content in cold pressed rice bran oil by TLC-image analysis method. *Asian Pacific Journal of Tropical Biomedicine*, *4*(2), 119–123.

Salar, A., Faghih, S., Pishdad, G., & Rezaie, P. (2015). Comparison of the effects of canola oil and rice bran oil consumption on oxidative stress and blood pressure in postmenopausal type 2 diabetic women: A randomized controlled clinical trial. *Journal of Nutritional Science and Diabetics*, *1*(4), 199–205.

Samad, N. (2015). Rice bran oil prevents neuroleptic-induced extrapyramidal symptoms in rats: Possible antioxidant mechanisms. *Journal of Food and Drug Analysis*, *23*, 370–375.

Sarmento, C. M. P., Ferreira, S. R. S., & Hense, H. (2006). Supercritical fluid extraction (SFE) of rice bran oil to obtain fractions enriched with tocopherols and tocotrienols. *Brazilian Journal of Chemical Engineering*, *23*(2), 243–249.

Saunders, R. M. (1985). Rice bran: Composition and potential food uses. *Food Reviews International*, *1*, 465–495.

Sayasoonthorn, S., Kaewrueng, S., & Patharasathapornkul, P. (2012). Rice bran oil extraction by screw press method: Optimum operating settings, oil extraction level and press cake appearance. *Rice Science*, *19*(1), 75–78.

Shahbandeh, M. (2019). *Paddy rice production worldwide in 2017 and 2018, by country (in million metric tons)*. Statista Available from: https://www.statista.com/statistics/255937/leading-rice-producers-worldwide/. [20 September 2019]. 9 August 2019.

Shukla, H. S., & Pratap, A. (2017). Comparative studies between conventional and microwave assisted extraction for rice bran oil. *Journal of Oleo Science*, *66*(9), 973–979.

Siger, A., Nogala-Kalucka, M., & Lampart-Szczapa, L. (2007). The content and antioxidant activity of phenolic compounds in cold-pressed plant oils. *Journal of Food Lipids*, *15*, 137–149.

da Silva, M. A., Sanches, C., & Amante, E. R. (2006). Prevention of hydrolytic rancidity in rice bran. *Journal of Food Engineering*, *75*, 487–491.

Singanusong, R. (2015). Processing, nutrition and health of cold-pressed rice bran oil. In C. L. Hii, S. V. Jangam, S. Z. Ong, P. L. Show, & A. S. Mujumdar (Eds.), *Processing of foods, vegetables and fruits: Recent advances* (pp. 27–44). The University of Nottingham Press, Singapore, Available from: Arunmujumdar E-book. Accessed: 26 September 2019.

Sivala, K., Mukherjee, R. K., & Bhole, N. G. (1993). A preliminary study of rice bran oil expression in a manually operated hydraulic press. *Journal of Food Engineering*, *20*(3), 215–222.

Sparks, D., Hernandez, R., Zappi, M., Blackwell, D., & Fleming, T. (2006). Extraction of rice bran oil using supercritical carbon dioxide and propane. *Journal of the American Oil Chemists' Society*, *83*(10), 10–16.

Srikaeo, K., Poungsampao, P., & Phuong, N. T. (2017). Utilization of the fine particles obtained from cold pressed vegetable oils: A case study in organic rice bran, sunflower and sesame oils. *Journal of Oleo Science*, *66*(1), 21–29.

Srikaeo, K., & Pradit, M. (2011). Simple techniques to increase the production yield and enhance the quality of organic rice bran oils. *Journal of Oleo Science*, *60*(1), 1–5.

Su, H., Liu, R., Chang, M., Huang, J., & Wang, X. (2017). Dietary linoleic acid intake and blood inflammatory markers: A systematic review and meta-analysis of randomized controlled trial. *Food & Function*, *8*, 3091–3103.

Suksaeree, J., Charoenchai, L., Sakunpak, A., Monton, C., Chusut, T., Saingam, W., et al. (2013). Preliminary formulations study of cold pressed rice bran oil mask. *Thai Journal of Pharmaceutical Sciences*, *38*, 188–191.

Taira, H., Taira, H., & Maeshige, M. (1979). Influence of variety and crop year on lipid content and fatty acid composition of lowland non-Glutinus brown rice. *Japanese Journal of Crop Science*, *48*(2), 220–228.

Tao, J. (1989). *Rice bran stabilization by improved internal and external heating methods*. Louisiana, US: UMI.

Thanonkaew, A., Wongyai, S., McClements, D. J., & Decker, E. A. (2012). Effect of stabilization of rice bran by domestic heating on mechanical extraction yield, quality, and antioxidant properties of cold-pressed rice bran oil (*Oryza saltiva* L.). *LWT-Food Science and Technology*, *48*, 231–236.

Uttama, S., Itharat, A., Rattarom, R., Makchuchit, S., Panthong, S., & Sakpakdeejaroen, I. (2014). Biological activities and chemical content of Sung Yod rice bran oil extracted by expression and Soxhlet extraction methods. *Journal of the Medical Association of Thailand*, *97*(8), s125–s132.

Wongwaiwech, D., Weerawatanakorn, M., Tharatha, S., & Ho, C. T. (2019). Comparative study on amount of nutraceuticals in by-products from solvent and cold pressing methods of rice bran oil processing. *Journal of Food and Drug Analysis*, *27*, 71–82.

Ying, J., & Kang, Y. (2017). Low density lipoprotein-cholesterol lowering mechanism and blood lipid profile managing effect of phytosterol in healthy normal and hypercholesterolemia: A meta-analysis of randomly controlled trial. *Chinese Journal of Clinical Nutrition*, *25*(6), 335–349.

Zarei, I., Brown, D. G., Nealon, N. J., & Ryan, E. P. (2017). Rice bran metabolome contains amino acids, vitamins & cofactors, and phytochemicals with medicinal and nutritional properties. *Rice*, *10*, 24–44.

Zavoshy, R., Noroozi, M., & Jahanihashemi, H. (2012). Effect of low calorie diet with rice bran oil on cardiovascular risk factors in hyperlipidemic patients. *Journal of Research in Medical Sciences*, *17*, 626–631.

Zullaikah, S., Melwita, E., & Ju, Y. H. (2009). Isolation of oryzanol from crude rice bran oil. *Bioresource Technology*, *100*, 299–302.

Chapter 37

Cold pressed avocado (*Persea americana* Mill.) oil

Valeria da Silva Santos and Gabriel Deschamps Fernandes
Independent Consultant on Fats and Oils Technology, Pelotas, Rio Grande do Sul, Brazil

Abbreviations

BHA	butylated hydroxyanisole
FA	fatty acid
FAEES	fatty acid ethyl esters
FAMES	fatty acid methyl esters
FFA	free fatty acid
LDL-C	low-density lipoprotein cholesterol
MUFA	monounsaturated fatty acids
PUFA	polyunsaturated fatty acid
SC-CO_2	supercritical carbon dioxide
SFE	supercritical fluid extraction
UAAE	ultrasound-assisted aqueous extraction

1 Introduction

The avocado tree is taxonomically referred to as the *Persea americana* species, which is separated into three distinct races, commonly known as Mexican, Guatemalan, and West Indian (Mohameed et al., 1997). Over many years of cultivation, the crossing of the three races resulted in the development of a large number of avocado cultivars; however, due to differences in the oil content, only a few of them are used as industrial oil sources (Tango, Carvalho, & Soares, 2004). In this way, the Hass (from the Guatemalan variety), and Fuerte (from the crossing between Guatemalan and Mexican) varieties are commonly applied for industrial oil extraction, with oil contents around 30%. Among other cultivars with potential applications in the oil industry, we can cite Pinkerton and Bacon varieties, both coming from Guatemalan and Mexican variety crossing (Ashworth & Clegg, 2003; Mohameed et al., 1997).

According to the FAO, Mexico is the world's largest producer of avocado. In 2017, this country produced 2 million tons of avocado fruits (Fig. 1), around 33% of the world's stock (FAOSTAT, 2019). In Fig. 2, you can see the world's biggest avocado producers. However, the avocado oil industry grows slowly because this is just a side-industry of the fresh fruit business (Woolf et al., 2009).

Regarding cultivars, Hass is the most produced cultivar, accounting for more than 90% of total production in many large and small avocado-producing countries. Hass is also the most likely cultivar to be used for avocado oil production.

Avocado oil is extracted from the pulp of *Persea americana* fruits. In the fruit, the oil is stored as a single oil droplet in the idioblast cells, which are dispersed in the mesocarp (pulp). The idioblast wall structure is very complex, and is basically composed of two cellulose layers separated by a suberin layer (Platt & Thomson, 1992). This morphology hampers oil extraction.

Cold pressed avocado oil is aligned to the minimally processed products tendency. It has been compared to olive oil over the years, mainly due to the high amounts of monounsaturated fatty acids (MUFAs) as well as high levels of pigments (chlorophylls and carotenoids), which act as antioxidants. In this way, the culinary uses of avocado oil have been widespread around the globe. Avocado oil consumers are looking for a delicate buttery flavor without the pungent notes of extra virgin olive oil (Woolf et al., 2009).

Cold Pressed Oils. https://doi.org/10.1016/B978-0-12-818188-1.00037-2
Copyright © 2020 Elsevier Inc. All rights reserved.

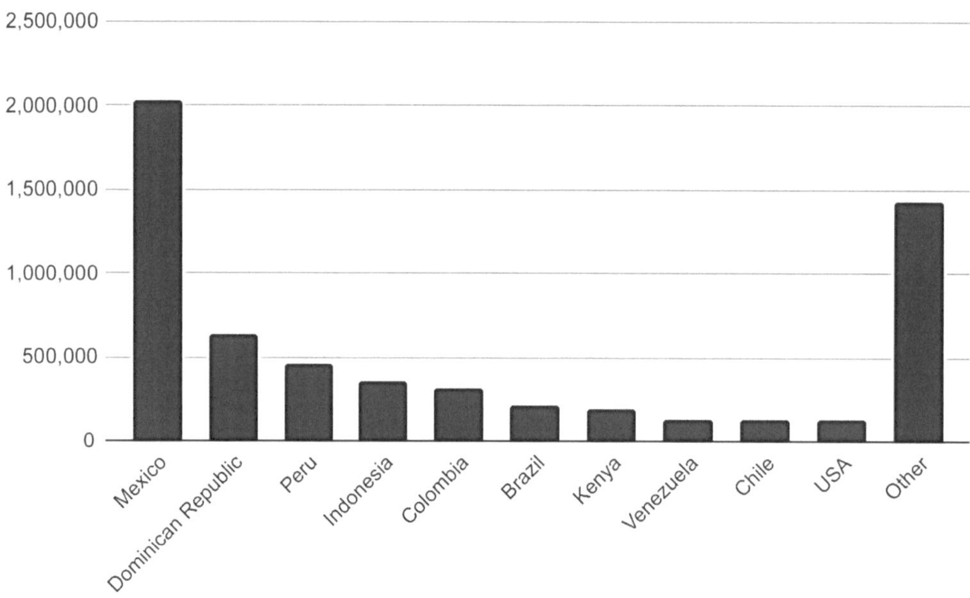

FIG. 1 Production (tons) of the main producer countries in the world in 2017 (FAOSTAT, 2019).

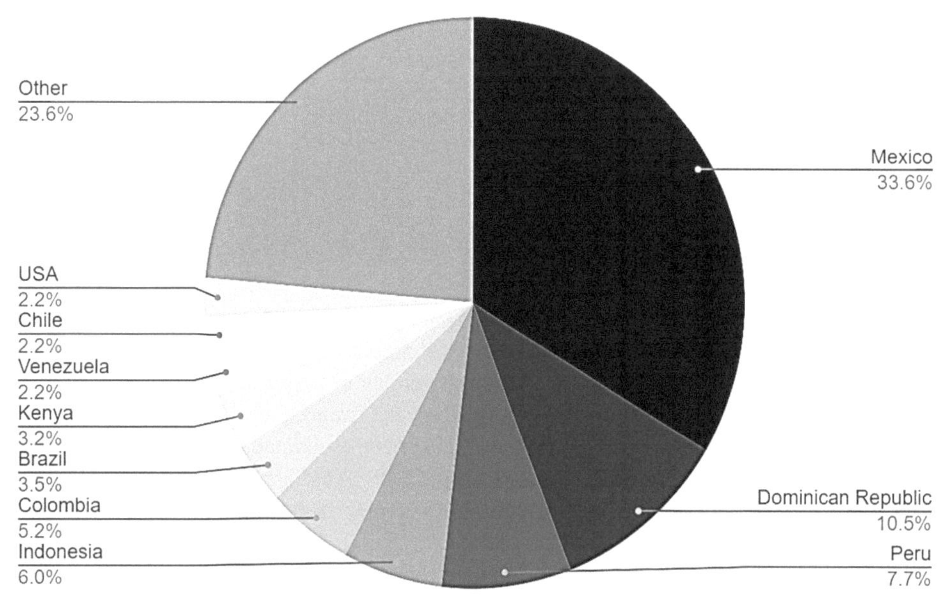

FIG. 2 Biggest avocado producers in the world (FAOSTAT, 2019).

It is clear that the main fraction of avocado production is designated for fresh fruit consumption. However, the avocado oil industry takes the opportunity to use "rejected grade" fruits. These are commonly low-price products since they do not meet consumer requirements. In addition, the recent movements of the consumer market are driving an increase in consumption and a willingness to pay a premium for alternatives to standard oils. Avocado oil production is limited and specialized; therefore, the oil is considered a specialty. This means that avocado oil enjoys a high market value (Requejo-Jackman, 2009; Woolf et al., 2009).

2 Extraction and processing of avocado oil

The avocado oil is extracted from the pulp of *Persea americana* fruits. As previously described, the oil is stored as a single oil droplet in the idioblast cells, which are dispersed in the fruit mesocarp (pulp). The idioblast wall structure is very complex, and it is basically composed of two cellulose layers separated by a suberin layer; this morphology hampers oil extraction. The structure of the idioblast cells also makes the oil extraction of avocado oil more difficult than in the case of the olive mesocarp (Wong, Requejo-Jackman, & Woolf, 2010). As a result, the establishment of extraction methods for avocado oil has been the focus of many studies (Abreu & Pinto, 2009; Botha, 2004; Fernandes, Gómez-Coca, Pérez-Camino, Moreda, & Barrera-Arellano, 2018; Freitas, Lago, Jablonka, & Hartman, 1993; Moreno, Dorantes, Galindez, & Guzman, 2003; Mostert, Botha, Plessis, & Duodu, 2007; Ortiz, Dorantes, Gallndez, & Cardenas, 2004; Salgado, Danieli, Regitano-D'arce, Frias, & Mansi, 2008; USDA, 2015).

The importance of studies of different extraction methods of avocado oil is related to the preparation of fruit pulp, extraction yield, and physical and chemical characteristics of the oil obtained, as well as the fatty acids and volatile compounds profiles. Several methods have been proposed for avocado oil recovery, including extraction with solvents, pressing, centrifugation, ultrasound, supercritical fluid extraction, and the use of technological aids to increase the extraction yield, like salt, microwave, microtalc, and enzymes.

Solvent extraction methods are very important to quantify the oil content in the avocado fruit. Because of the polarity of the solvent, it is more effective in extracting the oil. Thus, it is widely used for characterization of the total content of oil in the fruit, as well as to compare with the extraction yields of other methods. However, solvents are not used in industrial extraction, due to the necessity of removal and recovery of the solvent.

Nowadays, producers and researchers are looking for methods that are green, economically viable, and that allow production of oils of good quality, preserving the functional and sensorial properties. In this way, methods that use lower temperatures have been gaining prominence, such as the centrifugation method.

Another point to highlight is that the industrial extraction of avocado oil has become quite similar to that of olive oil. Thus, olive extraction plants have been used for the production of avocado oil, avocado only being different in relation to skin removal, stoning, and malaxation (Fig. 3). In this next step, the authors suggest the use of technological aids to improve the effectiveness of avocado oil extraction.

On this topic, we reviewed the methods that have been used over time for the extraction of avocado oil, highlighting its main characteristics in terms of physical–chemical properties and process yield (Table 1).

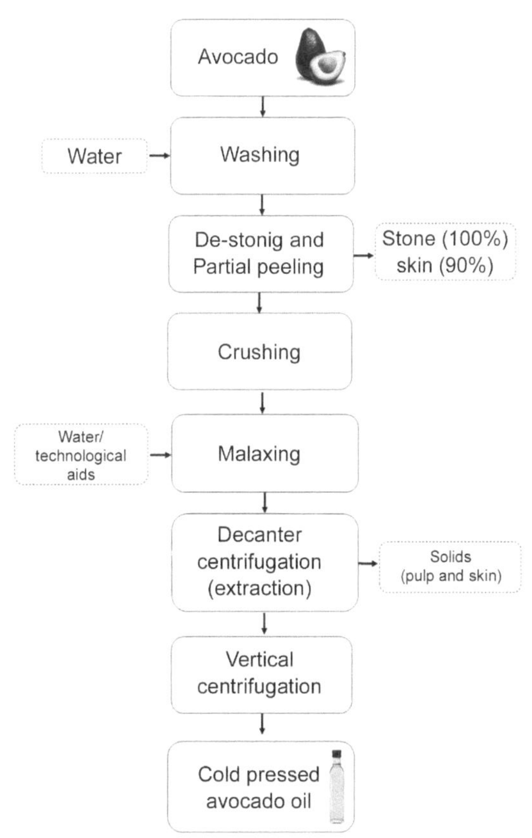

FIG. 3 Avocado oil cold process flow chart.

TABLE 1 Overview of various extraction methods for crude avocado oil.

Extraction method	Variety/ sampling place	Pretreatment of avocado pulp	Observations	References
Solvent extraction				
Hexane	*Persea americana* Mill./Mexico City	*Dehydration* Avocado manually peeled➔seed removed➔ground pulp in mortar➔paste➔dehydrated in vacuum oven (at 70°C, ≤100 mmHg)➔pulp with 27% moisture➔Soxhlet extraction method➔crude oil	Extraction yield of ~54% Official method used to quantify the oil content and to characterize the sample The content of free fatty acids, peroxide value and trans fatty acid of crude oil were 0.65%, 10.68 meq/kg and 0.3%, respectively	Moreno et al. (2003)
Hexane	Persea americana Mill./México City	*Microwave* Avocado manually peeled➔seed removed➔ground pulp in mortar➔paste➔heated by microwave (11 min)➔Soxhlet extraction method➔crude oil	Extraction yield of ~97% Microwave-assisted hexane extraction method provided the high yield and slight modification to the characteristic of the oil when compared to extraction with only hexane The content of free fatty acids, peroxide value, and *trans* fatty acid of crude oil were 0.277%, 9.55 meq/kg, and 0.09%, respectively	Moreno et al. (2003)
Acetone	Persea americana Mill./México City	*Heating* Avocado manually peeled➔seed removed➔ground pulp in mortar➔paste-acetone extraction at 25°C (first extraction)➔ paste-acetone extraction at 55°C (second extraction) ➔crude oil	Extraction yield of ~12%. Most of the oil remains inside the cells because the stronger changes in the idioblastic cellular structure were observed, and this effect may be responsible for the low oil extraction yield The content of free fatty acids, peroxide value and *trans* fatty acid of crude oil were 2.84% 12.74 meq/kg, and 0.45%, respectively	Moreno et al. (2003)
Hexane	Varieties Fuerte and Hass/South Africa	*Ultra-turrax and hexane* Ripe pulp➔skinned➔deseeded➔dried in oven (45°C)➔milled➔pulp and hexane (50:50 m/m)➔ ultra-turrax treatment (10 min)➔Soxhlet extraction (24 h)➔crude oil	Extraction yield of ~63% (Hass) and ~64% (Fuerte) The extraction yield by ultra-turrax treatment was similar to tradition Soxhlet (Hass = ~65% and Fuerte = ~64%)	Reddy, Moodley, and Jonnalagadda (2012)

TABLE 1 Overview of various extraction methods for crude avocado oil—cont'd

Extraction method	Variety/ sampling place	Pretreatment of avocado pulp	Observations	References
Hexane	Varieties Fuerte and Hass/South Africa	*Microwave* Ripe pulp➔skinned➔deseeded➔paste➔heated by microwave (11 min) ➔Soxhlet extraction method➔crude oil	Extraction yield of ~70% (Hass) and ~ 61% (Fuerte) Microwave treatment promoted the highest yield of oil from the Hass variety when compared to traditional extraction only by Soxhlet (~65%), while for the Fuerte variety it was the opposite (~64%).	Reddy et al. (2012)
Cold pressed extraction				
Pressing	Uninformed variety/ Cameroon	*Water and heating* Fresh avocado flesh➔pounded and wrapped paste➔boiled in water (15–20 min) ➔put in a wooden trough and pressed➔oil/water mixture➔12 h of standing to form oil layer➔scooped into a heating aluminum pot➔crude oil	Oil yield of ~2–12% (yield in wet basis). Acid and peroxide values were 4.5 mg KOH/g and 7.3 meqO_2.kg^{-1}, respectively. The avocado oil obtained using this method would be suitable for cosmetic use because has a bitter taste	Kameni and Tchamo (2003)
Pressing	Persea americana Mill./México City	*Microwave* Avocado manually peeled➔seed removed➔ground pulp in mortar➔paste➔ microwave treatment➔extraction squeezing/ pressing the pulp in cloth mesh➔crude oil	Extraction yield of ~65%. Optimized conditions of microwave treatment: microwave energy of 1.8 KJ/g and exposure time of 11 min. The content of free fatty acids, peroxide value and trans fatty acid of crude oil were 0.144%, 3.77 meq.kg^{-1} and 0.29%, respectively. In general, a lesser deterioration of the oils was produced with microwaves rather than solvents	Moreno et al. (2003)
Centrifugation	Variety Hass/ California	*Inorganic salt, pH and heating* Fresh avocado➔flesh➔diluted and pounded paste➔held at suitable conditions➔centrifugation➔crude oil (upper layer)	Extraction yield of 75.30%. Optimized conditions: 75°C, pH of 5.5 and 5% of NaCl. NaC1 may cause corrosion effect, especially on metallic equipment parts	Werman and Neeman (1987)

Continued

TABLE 1 Overview of various extraction methods for crude avocado oil—cont'd

Extraction method	Variety/ sampling place	Pretreatment of avocado pulp	Observations	References
Centrifugation	Variety Hass/ Minnesota	*Inorganic salt, pH and Heating* Fresh avocado➔flesh➔diluted and pounded paste➔held at suitable conditions➔centrifugation➔crude oil (upper layer)	Extraction yield of optimized conditions: 5:1 water:paste, pH 5.5 and centrifugal force of 12.300 rpm, 5% $CaCO_3$ or $CaSO_4$. Salt concentrations above 5% increase costs, increase corrosion of equipment, and limit the utilization of the extracted residue	Bizimana, Breene, and Csallany (1993)
Centrifugation	Variety Hass/ México	*Enzymes* Fresh avocado pulp➔paste➔diluted paste with water➔enzymatic reaction➔centrifugation (12.300 rpm/ 10 min)➔crude oil (upper layer)	Extraction yield of ~78%. Optimized conditions of enzyme treatment: 1% of α-amylase, enzymatic reaction time of 1 h, temperature of 65 °C, dilution rate of 1:5 (paste: water). Oil extraction yield significantly increases from paste treated with the enzyme compared to the control paste (Soxhlet extraction)	Buenrostro and López-Munguia (1986)
Centrifugation	Uninformed variety/ Fortaleza, Brazil	*Enzymes* Fresh avocado pulp➔peeled➔seed removed mechanically➔paste➔diluted paste with water ➔enzymatic reaction➔centrifugation (950 rpm/30 min)➔cooling (12 h)➔crude oil (upper layer)	Extraction yield of ~93%. Optimized conditions of enzyme treatment: 0.1% of carbohydrase mix, enzymatic reaction time of 3 h, temperature of 30°C, dilution rate of 1:3 (paste: water) and 150 rpm stirring in orbital shaker. Oil extraction yield significantly increases (~10%) from paste treated with the enzyme compared to the control paste (Soxhlet extraction)	Abreu and Pinto (2009)
Centrifugation	Varieties Bacon, Fuerte, Hass, and Pinkerton/ Málaga-Spain	*Talc* Ripe avocado pulp➔milled➔paste➔ talc addition➔Abencor® system malaxer➔Abencor® system centrifuge➔bench centrifuge➔filtration➔ filtered oil	Extraction yield uninformed. Abencor® system is a laboratory scaling system very similar to the processes used on an industrial scale	Fernandes et al. (2018)

TABLE 1 Overview of various extraction methods for crude avocado oil—cont'd

Extraction method	Variety/ sampling place	Pretreatment of avocado pulp	Observations	References
Centrifugation	Margarida and Hass/ Brazil	*Heating* Ripe avocado pulp ➔ peeled➔seed removed➔ blender homogenization➔thermal mixer (40 °C/ 40 min)➔ horizontal centrifuge➔filtration➔centrifugation➔crude oil	Extraction yield of ~50% (Margarida) and ~65% (Hass). Commercial extra virgin cold pressed avocado oil (New Zealand). Extraction process by centrifugation obtained a high yield in oil, which attests that it is a viable process, applicable in food industries	Jorge, Polachini, Dias, Jorge, and Telis-Romero (2015)
Unconventional extraction technologies				
Supercritical fluid SC-CO_2	Variety Fuerte/South Africa	*Heating drying* Ripe avocado pulp➔unpeeled➔ oven drying (80 °C/24 h)➔sliced (2 mm)➔ supercritical fluid extraction ➔vacuum evaporation of CO_2 and water (30 min) ➔crude oil	Extraction yield of ~60%. Optimized conditions of supercritical fluid extraction: 2 h, 37°C, 3.5×10^7 Pa and fluid flow rate of 4.5 mL min^{-1} (pump head). Higher oil yield was obtained with hexane extraction (~93%). However, lower chlorophyll content was obtained by SC-CO_2 extraction. It also resulted in shorter extraction times and with the absence of organic solvents	Botha (2004)
Supercritical fluid SC-CO_2	Variety Fuerte/South Africa	*Heating drying* Avocado pulp*➔unpeeled➔deep-frozen at −20°C➔ oven drying (80°C/24 h)➔sliced (10 mm)➔ supercritical fluid extraction➔crude oil *Unripe and Ripe pulps were evaluated.	Extraction yield of ~52% (unripe) and ~60% (ripe). Optimized conditions of SC-CO_2 extraction: 37 °C, 3.5×10^7Pa, 1.7 mL min^{-1} fluid flow rate, terminated after oil yield less than 1% was obtained in 1 h. Higher oil yield was obtained with hexane extraction (unripe = ~63% and ripe = ~71%), respectively	Mostert et al. (2007)

Continued

TABLE 1 Overview of various extraction methods for crude avocado oil—cont'd

Extraction method	Variety/ sampling place	Pretreatment of avocado pulp	Observations	References
Supercritical fluid SC-CO_2	Variety Fuerte/South Africa	*Freeze-drying* Avocado pulp*➔unpeeled➔deep-frozen at −20°C➔freeze-drying➔sliced (10 mm)➔ supercritical fluid extraction➔crude oil *Unripe and ripe pulps were evaluated.	Extraction yield of ~59% (unripe) and ~65% (ripe). Optimized conditions of SC-CO_2 extraction: 37°C, 3.5×10^7 Pa, 1.7 mL min^{-1} fluid flow rate, terminated after oil yield less than 1% was obtained in 1 h. Higher oil yield was obtained with hexane extraction (unripe = ~68% and ripe = ~72%)	Mostert et al. (2007)
Supercritical fluid SC-CO_2	Varieties Fuerte and Hass/South Africa	*Heating drying* Ripe avocado➔skinned➔deseeded➔dried in oven (45°C)➔milled➔ supercritical fluid extraction➔ oven (150°C to constant mass)➔ vacuum evaporation (30 min)➔crude oil	Extraction yield of ~63% (Hass) and ~60% (Fuerte) Optimized conditions of SC-CO_2 extraction: 2 h, 37°C, 3.5×10^7Pa, 2.8–3.5 mL min^{-1} fluid flow rate. Soxhlet extraction yields were ~65% (Hass) and ~64% (Fuerte) SC-CO_2 extraction co-extracted the lowest amount of oxidizing metals. SC-CO_2 is more favorable to produce oils for food/ pharmaceutical industries because the method is free from solvent and is reported to be feasible on a large-scale operation	Reddy et al. (2012)
Ultrasound water bath	Varieties Fuerte and Hass/ South Africa	*Heating drying* Ripe avocado➔skinned➔deseeded➔dried in oven (45°C)➔milled➔sonicated in a water bath (60°C) with hexane for 1 h➔suction filtration➔evaporation➔ crude oil	Extraction yield of ~55% (Hass) and ~59% (Fuerte). Soxhlet ~65% (Hass) and Fuerte). Ultrasound extraction showed high dissolution percentages for most metals, should be avoided to prevent possible oxidation and destabilization of avocado oils.	Reddy et al. (2012)

TABLE 1 Overview of various extraction methods for crude avocado oil—cont'd

Extraction method	Variety/ sampling place	Pretreatment of avocado pulp	Observations	References
Ultrasound water bath	Uninformed variety/ Malaysia	*Tray-type dryer* Ripe avocado➔skinned➔deseeded➔tray-type dryer (35°C)/3 days➔ milled (360 mm)➔ sonicated in a water bath➔centrifugation (8000 rpm/20 min)➔ crude oil (upper layer)	Extraction yield of ~73%. Optimized conditions UAAE: 6 mL/g water-to-powder ratio, 30 min of sonication time and at 35°C. The Soxhlet extraction yield was ~21%. UAAE-extracted oil is more suitable for edible and industrial usages than Soxhlet-extracted oil because the "solvent-free" extraction and the high extraction yield.	Tan, Gun Hean, Hamzah, and Ghazali (2018)
Homogenization pressure	Uninformed variety/ Indonesia	*Water and heating* Avocado flesh➔water (1:1)➔ phosphate acid (0.05%)➔blender homogenization (5000 rpm)➔ homogenization pressure (7, 71, and 176 kg/cm^2)➔heating (100°C)➔ natural separation➔crude oil (upper layer)	Extraction yield of wet and pressure method combined ~21% (7 kg/cm^2), ~23% (71 kg/cm^2), and ~23% (176 kg/cm^2) (yield in wet basis). The wet extraction yield was ~20% (wet basis). The Soxhlet extraction yield was ~30% (wet basis). The use of homogenization for better breakdown O/W emulsion leads to an increase in oil yield	Hamzah (2013)

(Adapted from Qin, X. and Zhong, J. (2016). A review of extraction techniques for avocado oil. *Journal of Oleo Science, 65*, 881–888.)

2.1 Solvent extraction of avocado oil

2.1.1 Hexane extraction

The *n*-hexane method of extraction is considered an official method of analysis by the Association of Official Analytical Chemists (AOAC), and is used to characterize the oil content of different samples, according to the Soxhlet method AOAC 963.15 (Cunniff & Association of Official Analytical Chemists, 1995). It is used for avocado oil extraction, in order to quantify the total oil content in the sample, as well as to compare the result with the other extraction methods. According to this standardized method, the sample—in this case the avocado pulp—was previously dehydrated, usually in a vacuum oven at 70°C, and at a working pressure of 100 mmHg, until the sample reached 27% moisture. After dehydration, pulp is submitted to the Soxhlet extraction system, which consists of exposure of the sample in a refluxed system containing *n*-hexane under heating. The sample is refluxed for 4 h, with the heat adjusted so the extractor siphons ≥30 times. In this method, the solvent is generally evaporated on a steam bath and is reused in other extractions.

Moreno et al. (2003) used the Soxhlet systems to obtain avocado oil to investigate the detailed changes that the oil undergoes when different pretreatments of the pulp and extraction methods are applied. The extraction yield was 54% of avocado oil, with a predominance of oleic acid (C18:1; 60.2 g/100 g oil) followed by linoleic acid (C18:2; 13.6 g/100 g oil), palmitic acid (C16:0; 15.7 g/100 g oil), and palmitoleic acid (C16:1; 7.26 g/100 g oil). Corresponding to the fraction of volatile compounds, the presence of propionic acid, decanol, 2,4-decadienal, and aromatic hydrocarbons such as tridecane and undecane, as well as terpenoids, was reported.

2.1.2 Acetone extraction

Avocado oil was extracted using acetone as a solvent, according to a patented method (Curiel, Neve, & Petach, 1985). The method consists of grinding the avocado to small particles of about 3–5 mm, after peeling and seed removal. Firstly, the fruit particles were brought into contact with acetone at 25°C, at which most of the water present in the avocado fruit was extracted into the acetone phase. Then, another extraction operation of substantially water-free avocado with acetone solution was carried out at 55°C. After that, all liquid solutions obtained from both extractions were admixed and cooled to ambient temperature, as well as obtaining two main fractions, which were separated: (1) an upper layer of oil-acetone-water, which contained mainly the unsaponifiable matter; and (2) a lower phase of oil-acetone-water, which contained the main fraction of avocado oil. The acetone was removed by distillation and recycled to the extraction operation. The oil was separated out as the upper phase above the aqueous phase. Moreno et al. (2003) used the patented method in their studies. These authors observed that the extraction yield was 12% of avocado oil, which was much lower than when extracted with hexane (54%). The author reported that, when the acetone method is used, most of the oil remains inside the cells. Besides that, stronger changes in the idioblastic cellular structure were observed, and this effect may be responsible for the low oil extraction yield. The fatty acids predominating in the oil extracted by acetone method were oleic acid (C18:1; 60.2 g/100 g oil), linoleic acid (C18:2; 15.3 g/100 g oil), palmitic acid (C16:0, 14.9 g/100 g oil), and palmitoleic acid (C16:1, 5.88 g/100 g oil). Many volatile compounds were found, mainly terpenoids, aldehydes, and short-chain fatty acids. Some of the terpenoids were α-bergamotene, α-humulene, α-copaene, α-cubebene, α-farnesane, β-caryophyllene, and β-bisabolene.

2.2 Cold pressed avocado oil

2.2.1 Pressing

The pressing method was used for a long time to obtain olive oil. The method starts by grinding the olives (with stone); this usually occurs in stone mills. This type of mill consists of large round or conical granite stones that roll over the olives, crushing them and releasing the olive oil from the interior of the vacuoles. In the mill itself, the malaxation stage occurs, which consists of joining the small drops of oil dispersed in the olive paste, forming larger droplets, and improving the extraction yields. The separation of the solid and liquid fractions from the pulp takes place by pressing the olive paste in a discontinuous system of hydraulic presses, where the mass is disposed in the middle of cast discs, through which the olive oil flows when the pressing force is exerted. The liquid fraction is then collected in a settling tank to separate the oil and water (naturally present in the olives). However, this method has been discarded because of the deposition of olive mass in the guts of the pressing discs, resulting in the occurrence of fermentative processes and reducing the quality of the oil produced (Petrakis, 2006).

The extraction of avocado oil using the pressing method is initiated with manual separation of the kernel and skin from the pulp. The kernel and skin do not participate in the extraction method; both are discarded (Kameni & Tchamo, 2003; Moreno et al., 2003). Kameni and Tchamo (2003) reported in their studies that after kernel and skin separation, the oil was extracted by cooking in water, manual pressing, and refining by heating. Thus, the pulp was pounded in a mortar with eucalyptus sawdust (10%, *w*/w), and the mixture was wrapped in muslin cloth and put in boiling water for 15–20 min. The mixture was then put in a wooden trough with a drain hole, and pressed increasingly with a wooden board. When the flow of the oil stopped, the cake was loosened, and put back in boiling water for 10 min and then pressed again. The oil/water mixture was left to stand for 12 h, and the oil layer was scooped into an aluminum pot, which was heated until foaming occurred, and dried particles were deposited on the side of the pot. When heating was stopped, the particles settled at the bottom and the oil could be drained off. The yields of oil were 1.8%–11.6% of the weight of the fruits, with an extraction rate of 82%–85%. The values of acidity level and peroxide were 4.5 mg KOH/g and 7.3 meq O_2.kg^{-1}, respectively. The authors concluded that the avocado oil obtained using this method would only be suitable for cosmetic use because it has a bitter taste.

Moreno et al. (2003) developed a combined extraction method, denominated microwave-squeezing. The avocado pulp was spread on the rotary plate of a microwave oven to form a uniform layer that was $\pm$5 mm thick. The sample was placed in the oven, and heated at a high power level. The sample was removed from the plate and the oil extracted by squeezing, pressing the pulp manually through a cloth mesh. For process optimization, the authors used results from a central compound experimental design using Design-Expert software. The highest extraction yield (65.2%) from the microwave-squeezing combined method was obtained when the sample weighed 300 g, and the exposure time was 11 min. Oleic acid (18:1) was the major component in the microwave-squeezing processed oil (52.04 g/100 g oil), followed by palmitic acid (C16:0), linoleic acid (C18:2), and palmitoleic acid (C16:1), having concentrations of 21.1, 14.9, and 8.90 g/100 g oil, respectively. Four volatile compounds in the avocado oil extracted by microwave-squeezing were detected: hexanal, octanal, nonanal, and β-caryophyllene. The authors concluded that the microwave-squeezing method caused the slightest modification to the oil quality compared to other methods, of microwave combined with hexane extraction, hexane extraction, and acetone extraction.

2.2.2 Centrifugation

In order to meet market demands, the appearance of continuous operating plants contributed to the reduction of costs and to the increase of processing capacity; in addition, to improve the yield and quality of the olive oil produced, the centrifugation process was developed. In this process, as with the pressing process, the olive is used whole, without seed removal. Basically, the flow sheet of the used extraction plant comprises four main operations: fruit cleaning (defoliation, washing), preparation of the paste (crushing, malaxation), horizontal decanter centrifugation (separation of the solid, pomace, and liquid phases, oily must, and wastewater), and vertical centrifugation (separation of the liquid phases, oil/waste/water) (Martinez Nieto, Barranco Barranco, & Moreno, 1992; Petrakis, 2006). The obtained oil can be decanted into storage tanks and filtered before the potting process.

The same process has been used to obtain avocado oil (Martinez Nieto et al., 1992). Commonly, the avocado fruit is cleaned, and the skin and seed removed manually. The malaxation step aims to break up the oil/water emulsion, so that the droplets of oil join together to form larger drops. It is a fundamental step for increasing extraction yields and is designed to enhance the effect of crushing and to make the paste uniform (Martinez Nieto et al., 1992; Petrakis, 2006). In this step, process assistants might be used to improve extraction efficiency, such as enzymes and talc (Abreu & Pinto, 2009; Buenrostro & López-Munguia, 1986; Fernandes et al., 2018), which will be discussed later.

The horizontal decanter centrifugation step comprises a centrifugal force that operates by sucking out the liquid from the paste. This is done with rotary machines whose speed and separation efficiency are directly proportional to the angular speed and rotation radius, as well as to the difference in the density of the liquids that have to be separated. The machines applied in this kind of process are horizontal centrifuges that operate at an angular speed producing up to 3000 times greater acceleration than natural gravitational acceleration. In this way, by promoting the separation of the solid part as pomace and liquid phases, oily must, and wastewater (Petrakis, 2006). Petrakis (2006) further noted that decanter centrifuges are composed of four important parts:

- *Inlet zone*: this is responsible for accelerating the feed slurry up to the speed of the bowl, designed to keep any degradation of the feed solids to a minimum so that disturbance of the sediment in the bowl is avoided;
- *Screw conveyor*: scrolling of the sedimented solids occurs here, wherein the design of the screw conveyor is crucial to efficient and effective decanter performance;
- *Solid discharge section*: this allows the output of solids as pomace; and
- *Liquid discharge section*: in a two-phase decanter, the liquid level is regulated by dam plates. When operating in a three-phase mode, each phase discharges over a set of dam plates into separate baffled compartments in the casing.

For many years, three-phase centrifugation was used to obtain olive oil. It consists of the addition of lukewarm water, equivalent to approximately 40%–100% of the weight of the olive fruits (depending on the ratio of initial moisture/work humidity) in the malaxation step. The water-thinned paste is then centrifuged in the decanter. The pastes undergoing centrifugal extraction have to be relatively fluid to facilitate separation of fractions with different specific weights (Paraskeva & Diamadopoulos, 2006; Petrakis, 2006). It is important to note that phenolic compounds migrate to the water fraction, resulting in its lower concentration of oil obtained. In addition, the water resulting from this extraction process carries large amounts of organic matter. In this way, the volume of waste that needs treatment is very large, becoming a disadvantage of the system (Paraskeva & Diamadopoulos, 2006). Due to the environmental issues involved in the three-stage process, the two-phase system was developed, which does not include the addition of water to the batter. In this case, technological aids, such as enzymes and microtalc, can be added in the malaxation step, facilitating the emulsion breaking, and consequently increasing the extraction yields of the process (Abreu & Pinto, 2009; Buenrostro & López-Munguia, 1986; Freitas et al., 1993). After the malaxation step, the paste is centrifuged in a decanter centrifuge, separating the pomace (containing the talc or enzymes) from the liquid fraction. Despite the generation of water as waste at this stage, the quantities are much smaller than in the three-phase system, facilitating waste treatment (Aparicio, 2000; Di Giovacchino, Marsilio, Costantini, & Di Serio, 2005; Petrakis, 2006).

Silva, Freitas, Cabrita, and Garcia (2012) compared monovarietal virgin oils obtained by both processes, wherein the oils from two-phase decanters have a higher content of *E*-hex-2-enal and total aroma substances, but lower values of aliphatic and triterpenic alcohols.

In both the two- and three-phase systems, the resulting liquid fraction needs an additional centrifugation process. Thus, the vertical centrifuge is applied to effectively separate the oil and vegetation water (intrinsic water from the fruit). The vertical centrifugation is a system that consists of a disk stack centrifuge for final cold pressed avocado oil purification to remove residual water and solids to avoid degradation of oil by contact with water and pomace residual (Costagli & Betti, 2015). After the avocado oil is obtained by vertical centrifugation, it is decanted and can be filtered to separate possible suspended solids and water droplets in the lipid fraction. These processes aim to reduce the possibility of hydrolysis and fermentation processes during storage, which contribute to the reduction of the quality of the final product.

Jorge et al. (2015) physicochemically evaluated avocado oils (Margarida and Hass varieties obtained in Brazil) obtained by the two-phase cold press process. The pulp was removed manually and homogenized in a blender. The resulting paste was sifted and kept in a thermal mixer at 40°C for 40min. Finally, the paste was placed in a horizontal centrifuge in order to obtain the oil. The extracted oils were filtered, centrifuged, stored in amber glass bottles, and inerted with gaseous nitrogen. The obtained oils were compared to commercial extra virgin cold pressed avocado oil from New Zealand. Margarida pulp showed 7.70% of oil, based on initial paste weight, remaining as residual oil (2.25%) in avocado paste after extraction process. Hass variety fruits obtained higher oil yield, 8.80%, but suffered higher losses during the extraction, since the resulting paste contained 14.2%, a higher content than that extracted by centrifugation. The Margarida variety showed the highest acidity level, 3.6mg KOH/g, and Hass oil presented the highest peroxide index, 5.54meq O_2/kg. These value are in accordance with international regulations for crude, cold pressed oil, whose threshold is 4mg KOH/g for acidity and a peroxide index of 15.0meq O_2/kg oil (CODEX, 2009). The commercial avocado oil showed the highest chlorophyll content (18.96 $mg\,kg^{-1}$) compared to the other oils analyzed, which grants it a darker green color. The authors concluded that the extraction process by centrifugation recovered high oil yield, which attests that it is a viable process, applicable in the food industry.

In view of the above, the centrifugation continuous operating plants can be used in between the harvest of the olive to extract avocado oil, thus, making the investment in the equipment even more profitable. In addition, as for olive oil, the yield and quality of avocado oil can be improved. Another point to highlight is that the use of technological aids in the stage of malaxation also helps to increase the extraction yield, as will be discussed later.

2.3 Technological aids in cold pressed avocado oil extraction

2.3.1 *Inorganic salt*

The addition of inorganic salt (such as NaCl, $CaCO_3$, and $CaSO_4$) assists oil separation from the emulsion by physical means. Werman and Neeman (1987) studied the influences of temperature, pH, and NaC1 concentration on avocado oil extraction efficiency by centrifugal processing. The authors observed that NaC1 increases the difference in specific gravity between the oil and aqueous phase, and also suppresses stability of the double layer at O/W interface. Optimal separation conditions were obtained at 5% NaC1 and pH 5.5 at 75°C, with extraction efficiency of 75.3%. The efficiency without added salt was somewhat lower, at 72.7%. The authors concluded that since this improvement was not significant, it might be better to dispense with NaC1 in the future and not risk a possible corrosion effect, especially on metallic equipment parts. Bizimana et al. (1993) verified the interplay among factors such as centrifugation extraction, pH, pulp/water ratios, inorganic salt type ($CaCO_3$ and $CaSO_4$), concentration, temperature, and time factors for optimizing a procedure for extracting avocado oil. The results of this study suggested for industrial use the addition of 5% of $CaCO_3$ and $CaSO_4$ increased oil recovery, but with little, if any, benefit above 5% was observed. The authors emphasize that salt concentrations above 5% increase costs, increase corrosion of equipment, and limit the utilization of the extracted residue.

2.3.2 *Enzymes*

Enzyme-assisted extraction of avocado oil has emerged as an ecological process (Abreu & Pinto, 2009; Buenrostro & López-Munguia, 1986; Freitas et al., 1993). Buenrostro and López-Munguia (1986) studied the efficiency of different enzymes, as pectinases, α-amylase, proteases, and cellulases, in the extraction of avocado oil. The avocado paste was conditioned in a stirred tank with temperature control with sodium sulfite and butylated hydroxyanisole (BHA) to avoid oxidation, and different dilution rates for paste:water (1:3, 1:4, 1:5, and 1:6, *w*/w) were employed before the addition of enzymes. After reaction, the solution was centrifuged at 12.300rpm/10min. The best result was obtained for extraction using α-amylase during 1h at 65°C, with a dilution rate of 1:5, reaching an extraction yield of about 78% of avocado oil. Abreu and Pinto (2009) used a carbohydrases enzymatic complex composed of arabanase, cellulase, β-glucanase, hemicellulase, and xylanase. The avocado pulp samples were treated with 1000 ppm of the enzymatic complex, with incubations of 1, 2, and 3 h and dilutions (substrate/water ratio) in 1:1, 1:2, and 1:3 (*w*/w) proportions, under shaking (150 rpm) at 30°C, followed by centrifugation and cooling (12*h*) for the separation of the oil fraction. The highest extraction yield was obtained with 3 h incubation, and 1:3 dilution, with an extraction yield of 92.7%. The authors concluded that the avocado oil extraction yield significantly increased from paste treated with the enzyme compared to the control paste. The higher yield can be explained by the enzyme's greater access to the pulp, by destructuring the cell wall, by breaking the carbohydrates, facilitating the release of the oil. The lipid profile and volatile compounds of the avocado oils obtained were not studied by these authors.

2.4 Microtalc

Research has been conducted on the extraction of olive oil and avocado oil using microtalc as a technological aid. This kind of material has lipophilic properties, and has presented positives results for the reduction of the oil/water emulsion, resulting in a consequent increase of free oil during extraction (Petrakis, 2006).

Fernandes et al. (2018) characterized chemically single varieties of avocado oils (Bacon, Fuerte, Hass, and Pinkerton) obtained by extraction using natural microtalc (Mg_3 (Si_4O_{10}) (OH)). Ripe avocados, without seeds, were milled in a knife mill and the paste was taken for extraction by means of an Abencor® system malaxer, and centrifuge, followed by an additional centrifugation step. Malaxation was carried out below 40°C for 40 min, with talc addition (~10 g,100 g^{-1} paste). Distilled water (20 mL,100 g^{-1} paste) was added after 10 min of starting the malaxation process. The first centrifugation was carried out in the Abencor® system centrifuge at 3000 rpm for 60 s. The paste was then poured out and the liquid phase was further centrifuged in a bench centrifuge (5000 rpm, 10 min), followed by filtration. The squalene concentrations were higher in the Bacon, Fuerte, and Pinkerton oils than in the Hass and commercial oils used to compare the results. Those oils obtained might be suggested as potential squalene sources. Tocopherols (but not tocotrienols) were found and were mainly α- and γ-tocopherol. The authors highlighted that the presence of ω7, ω9, and ω11 isomers of oleic and palmitoleic acids in avocado oil were described for the first time in this work, also the minor compounds like methyl- and dimethylsterols, terpenic alkenols, and aliphatic alcohols. The authors did not report the yields of extractions using microtalc as a technological aid.

2.5 Unconventional extraction technologies

2.5.1 *Supercritical fluid*

Supercritical fluid extraction (SFE) has been used in separating desired compounds from solid matrices in the pharmaceutical and food industries. Supercritical carbon dioxide (SC-CO_2) is mainly used, as it is a green solvent, ensuring biological safety, and leaves no solvent residue in the final product, compared with organic solvents used in traditional oil extractions (Qin & Zhong, 2016). The oil extractions by SFE presents some peculiarities, such as differences in SC-CO_2 solubility to bioactive compounds, which can be controlled by operating conditions, such as pressure and temperature. For example, Botha (2004) showed that lower temperature and pressure favor reducing coextraction of chlorophyll when extracting oil from dried and ground avocado, compared with *n*-hexane extraction. Another very important point to address about SFE is that the method is economical, since it uses half the energy of the distillation extraction processes and has been reported to be feasible for large-scale operation in the food sector (Qin & Zhong, 2016; Reddy et al., 2012). However, the SFE technique is considered to be under-exploited to obtain avocado oil. Some researchers have studied the influence of various process parameters (temperature, pressure, fluid flow rate, and extraction time) on the extraction yield and on characteristics of the oils obtained (fatty acids composition, free fatty acid content, pigments, and others). In addition, the influence of the treatment of the avocado pulp, mainly in the yield of the processing, will be discussed in more detail in the following topic.

Botha (2004) studied avocado oil extraction by SC-CO_2 using different extraction conditions. The fruit was allowed to ripen before processing. Dried avocado slices were prepared by oven drying, destoned, unpeeled, fresh fruit at 80 °C for 24 h. Dried avocado was ground to less than 2 mm by means of a food processor. Supercritical fluid extractions were performed on 4 g avocado samples in a 10 mL extractor, during 2 h with a fluid flow rate of 4.5 mL min^{-1}, measured at the pump head, at 3.5×10^7 and 5.4×10^7 Pa, with temperatures of 37°C and 81°C. Prior to any analysis, the extracted oil was subjected to vacuum evaporation to remove any water and dissolved carbon dioxide. The author compared the oil obtained by supercritical extraction with *n*-hexane extraction. The two extraction methods were shown to be statistically equivalent with regard to the acid value, but the color and unsaponifiable matter content were significantly different. The colors of oils obtained at 3.5×10^7 and 5.4×10^7 Pa at 37°C were evaluated to be straw-yellow and straw-yellow with a greenish tint, while at 81°C they were straw-yellow with a strong greenish tint and green with a yellow tint, respectively. The oil extracted by *n*-hexane was a darker green than the oils extracted by SC-CO_2. The unsaponifiable content decreases during SC-CO_2 extraction (20 min = 13.9%, 40 min = 9.6%, and 60 min = 5.6%). The results indicated that these compounds might have higher solubility in the carbon dioxide than the triglycerides. This implies that it might be possible to further enrich or fractionally extract the unsaponifiable matter.

Some years later, this author and his colleagues verified the effect of fruit ripeness (unripe and ripe) and method of fruit drying (freeze-dried and oven-dried at 80°C) on the extractability of avocado oil with *n*-hexane and supercritical carbon dioxide (Mostert et al., 2007). The average oil yield from ripe fruit (freeze-dried and oven-dried combined) was 72 g kg^{-1} higher than from unripe fruit for SC-CO_2 extracts and 61 g kg^{-1} higher for *n*-hexane extracts. This may be due to enzymatic degradation of parenchyma cell walls during ripening, thus making the oil more available for extraction. Freeze-dried samples had a mean oil yield 55 g kg^{-1} greater than oven-dried samples for SC-CO_2 extracts and 31 g kg^{-1} higher for *n*-hexane extracts. However, oil yields from ripe fruit (freeze-dried and oven-dried) subjected to *n*-hexane extraction were not significantly different. All *n*-hexane extracts combined had a mean oil yield 93 g kg^{-1} higher than SC-CO_2 extracts. In both studies, it was found that SC-CO_2 was better than *n*-hexane as a solvent for the extraction of avocado oil, mainly due to shorter extraction times and absence of organic solvents.

2.5.2 Ultrasound water bath extraction

The application of ultrasound approach in extracting plant oils has received increasing attention in recent years. It utilizes the cavitation forces produced by acoustic waves to decompose the cell walls of the oil cells and the structure of the oil emulsion, thereby releasing these intracellular components into the solvent.

Ultrasound-assisted aqueous extraction (UAAE) is a simple and inexpensive substitute for traditional oil extraction practices. However, there are few studies available on the production of edible oil using this technique (Tan et al., 2018). Reddy et al. (2012) studied the influence of ultrasound on fatty acid profile and elemental content of avocado oil from two avocado varieties (Hass and Fuerte). The authors studied the extraction during 1 h, wherein the avocado pulp was subjected to sonication in a water bath at 60°C with *n*-hexane as a solvent, in a 1:1 ratio (pulp: hexane). The resultant mixture was filtered by suction and filtrate evaporated. The authors compared this method with Soxhlet extraction. The extraction yield using ultrasound method was 54.6% for Hass, and 58.7% for Fuerte, while the values obtained by the Soxhlet method were 64.7% and 63.6%, respectively. Element dissolution was relatively low for traditional Soxhlet extraction and relatively high for the ultrasound water bath. For example, the concentration of Fe in oils were 3.34% for Hass (1.80 μg g^{-1}) and 4.12% for Fuerte (1.82 μg g^{-1}). Although these concentrations are not high enough to affect oil stability, ultrasound extraction showed high dissolution percentages for most metals, which should be avoided to prevent possible oxidation and destabilization of avocado oils.

Tan et al. (2018) optimized the UAAE parameters such as sonication time, sonication temperature, and water-to-powder ratio on the recovery of virgin avocado oil with low free fatty acid (FFA) content. The physicochemical properties of UAAE-extracted virgin avocado oil were compared with the Soxhlet-extracted avocado oil. The avocado powder was sonicated in an ultrasonic bath (Thermo-10D; 40 kHz frequency; 240 W ultrasonic output power; $500 \times 300 \times 150$ internal dimensions). When the required sonication temperature and time were reached, a laboratory-scale screw press was used to press the mixture to obtain an aqueous-oil mixture. The aqueous-oil mixture was centrifuged at 8000 rpm for 20 min at room temperature to separate the oil from the water layer. A Pasteur pipette was used to remove the top oil layer, which was weighed. Response surface methodology with a three-factor, three-level, Box-Behnken design (BBD) was utilized to study the effect of UAAE parameters on the recovery of virgin avocado oil with low FFA content. The dependent variables used were sonication time (min), sonication temperature (°C), and water-to-powder ratio (mL/g).

Within the experimental conditions, the recovery of low FFA-virgin avocado oil was highest at the optimum conditions of 6 mL/g water-to-powder ratio as well as sonication time and sonication temperature of 30 min and 35°C, respectively. The authors reported that the oil extraction yield and FFA level at these optimum conditions for UAAE were 72.9% and 0.29%, respectively, while for Soxhlet extraction the oil yield was 20.7%. UAAE extracted the oil appeared as liquid, while Soxhlet-extracted oil appeared as semisolid. This implies that the physical state of lipids is influenced by extraction methods. Regardless of the extraction methods, oleic (40.6%–41.6%), palmitic (28.0%–34.5%), linoleic (15.5%–19.0%), and palmitoleic (6.64%–8.37%) acids were the major fatty acids of avocado oil. The unsaturated fatty acids of UAAE-extracted oil (71.2%) were higher than Soxhlet-extracted oil (65.4%). The authors explained that this variation could be attributed to the selectivity of specific extracting solvent for certain fatty acids because the interaction between unsaturated fatty acids with a polar solvent (e.g., water) is stronger than a nonpolar solvent (e.g., *n*-hexane). In addition, the high temperatures used in the Soxhlet extraction could possibly lead to the decomposition of unsaturated fatty acids. The authors concluded that, in general, the UAAE-extracted oil is more suitable for edible and industrial usages than Soxhlet-extracted oil. As the environmental safety and health concerns arise, the use of "solvent-free" extraction is highly warranted.

2.5.3 Homogenization pressure extraction

Hamzah (2013) studied the effect of homogenization pressures on the extraction of avocado oil. The author stated that only heating is not enough to break down emulsion and extract the avocado oil. In the avocado pulp, there was not only water but also proteins and carbohydrates. Linkage of oil with proteins and carbohydrates made the emulsion and the linkage slightly harder to break down during heating. In this way, the author decided to apply a similar method used in the extraction of coconut oil. The avocado pulp was added water, in the proportion 1:1 (*w*/w), and 0.05% of phosphate acid. The mixture was then heated at a temperature of 105°C, and homogenized by a blender (5000 rpm for 3 min), followed by homogenization pressure of 7, 71, and 176 kg/cm^2 in order to be much easier to break down water in oil emulsion, then heated at 100 °C until the emulsion of water-oil broke down and the avocado oil could be separated. The results showed that the wet method presents a yield of extraction of 20.0% and when homogenization pressures were used, the yield increased from 7 to 71 kg/cm^2, being 21.4% and 22.8%, respectively. However, when homogenization pressures increased from 71 to 176 kg/cm^2, no significant increase was observed. When compared to soxhlet method (solvent) the wet method combined to homogenization pressure delivered 28%, 23%, and 21% less oil yield for 7, 71, and 176 kg/cm^2, respectively.

3 Major components of cold pressed avocado oil—Fatty acids and acyl lipids (neutral lipids, glycolipids, and phospholipids) profile

Fatty acids are the main fundamental molecule of oils and fats. In regular fats and oils, the fatty acids are bound to a glycerol molecule forming a triacylglycerol, which is the main molecule composed of fatty acids in regular oils. The fatty acid composition is considered one of the most widely studied parameters of oils; it is present in a large number of regulations and laws. When comparing to other oils, cold pressed avocado oil is a new trend in oil market; for this reason, there is no specific regulation for it.

Several scientific studies have reported the fatty acid composition of avocado oil. In most cases, they have compared different extraction process or different varieties. In general, avocado oil is composed of unsaturated fatty acids, mainly oleic (18:1) and palmitoleic (16:1) acids. Considering a general approach, without discrimination of process and/or variety, we found in the scientific bibliography the following ranges of the main fatty acids in avocado oil: palmitic acid (16:0) 10.0%–35.2%, palmitoleic acid (16:1) 2.8%–16.1%, stearic acid (18:0) 0.2%–1.5%, oleic acid (18:1) 36.9%–74%, linoleic acid (18:2) 6.1%–21.2%, and linolenic acid (18,3) 0.3%–2.1% (Werman & Neeman, 1987; Martinez Nieto et al., 1992; Moreno et al., 2003; Salgado et al., 2008; Fernandes et al., 2018; Tan et al., 2018; Tango et al., 2004; Rueda et al., 2014; Rueda et al., 2016).

The cold extracted avocado oil market is considered to be new; however, regarding the fatty acid composition of cold extracted avocado oil, we have about 30 years of scientific studies. From the first study until now, oleic acid has been described as the main fatty acid, followed by palmitic, linoleic, and palmitoleic acids. Table 2 provides the fatty acid composition described by many authors for cold pressed avocado oil. The detailed composition has increased over time, and certainly with the complexity and goals of each work.

Werman and Neeman (1987) described the composition of commercial avocado oil from the Hass variety from California. They described more than 74% of oleic acid and similar amounts of palmitic and linoleic acids (~12%). Moreno et al. (2003) characterized the composition of pressed avocado oil after a microwave pretreatment. The authors did not describe the analyzed variety, but they have found over around 52% of oleic acid and higher amounts of palmitic acid compared to linoleic acid: ~21% and ~15%, respectively. Regarding the work of Jorge et al. (2015), we find in Table 2 ranges of the characterization of oils from the Margarida and Hass varieties. The range of palmitic acid was again higher than the range of linoleic acid; however, the authors particularly found higher amounts of palmitoleic acids than those described before. In this work, the authors also analyzed a commercial cold extracted avocado oil sample (Table 2). In this case, the commercial sample presented higher amounts of oleic acid and lower amounts of palmitic and linoleic acid compared to the extracted samples from the same study.

Krumreich, Borges, Mendonça, Jansen-Alves, and Zambiazi (2018) analyzed avocado oils obtained by different press temperatures (40°C and 60°C). The authors described the main fatty acids of each sample and, for the first time, they described the presence of docosadienoic (22:2), and tricosanoic (23:0) acids. In 2018, Fernandes and coworkers analyzed the chemical composition of avocado oils from Hass, Fuerte, Bacon, and Pinkerton planted in Malaga (Spain), in addition,

TABLE 2 Fatty acid composition of cold extracted avocado oil.

Fatty acid profile (%)	Werman and Neeman (1987) Commercial	Moreno et al. (2003)	Jorge et al. (2015)	Jorge et al. (2015) Commercial	Krumreich et al. (2018)	Fernandes et al. (2018)	Fernandes et al. (2018) Commercial	Total range
Miristic—14:0	–	–	–	–	–	0.02–0.03	0.03–0.06	0.02–0.06
Pentadecanoic—15:0	–	–	–	–	–	0.02–0.02	0–0.04	0–0.04
Palmitic—16:0	12.20	21.10	19.43–23.28	11.74	19.90–21.20	12.16–18.17	11.64–21.05	11.64–23.8
Palmitoleic—16:1	3.50	8.90	2.65–11.35	4.80	2.70–6.50	4.22–7.8	3.43–11.73	2.65–11.73
Margaric—17:0	–	–	–	–	–	0.01–0.08	0.1–0.15	0.01–0.15
Margaroleic—17:1	–	–	–	–	–	0.09–0.11	0.1–0.15	0.09–0.15
Stearic—18:0	–	0.50	–	–	–	0.37–0.51	0.37–0.79	0.37–0.79
Oleic—18:1	74.10	52.04	54.72–57.33	73.88	58.60–64.50	61.56–73.57	53.24–71.62	52.04–74.1
Elaidic—18:1t		0.29	–	–	–	–	–	0.29
Linoleic—18:2	12.70	14.89	13.22–14.84	9.51	10.60–10.70	8.25–11.12	8.8–16.5	8.25–16.5
Linoleic trans—18:2t	–	0.04	–	–	–	–	0–0.04	0–0.04
Linolenic—18:3	–	1.83	0.83–1.25	0.32	0.40–0.60	0.56–0.59	0.49–1.15	0.32–1.83
Arachidic—20:0	–	–	–	–	–	0.04–0.05	0.04–0.11	0.04–0.11
Eicosaenoic—20:1	–	0.22	–	–	–	0.16–0.19	0.12–0.22	0.12–0.22
Arachidonic—20:4	–	0.09	–	–	–	–	–	0.09
Behenic—22:0	–	–	–	–	–	0–0.01	0–0.06	0–0.06
Docosadienoic—22:2	–	–	–	–	0.20–0.80	–	–	0.20–0.80
Erucic—22:1	–	0.09	–	–	–	–	–	0.09
Tricosanoic—23:0	–	–	–	–	0.9–2.40	–	–	0.9–2.4
Lignoceric—24:0	–	–	–	–	–	–	0–0.05	0–0.05

Absolute value and ranges of scientific works from 1987 to 2018.

TABLE 3 Fatty acid composition of monovarietal cold extracted avocado oil.

	Jorge et al. (2015)		Fernandes et al. (2018)			
Fatty acid profile % area	**Margarida**	**Hass**	**Fuerte**	**Hass**	**Pinkerton**	**Bacon**
Miristic—14:0	–	–	0.03	0.03	0.03	0.02
Pentadecanoic—15:0	–	–	0.02	0.02	0.02	0.02
Palmitic—16:0	23.28	19.43	12.37	18.17	16.93	12.16
Palmitoleic—16:1	2.65	11.35	4.22	7.80	7.46	6.76
Margaric—17:0	–	–	0.02	0.01	0.08	0.02
Margaroleic—17:1	–	–	0.09	0.10	0.11	0.11
Stearic—18:0	–	–	0.51	0.37	0.43	0.38
Oleic—18:1	57.33	54.72	73.57	61.56	65.93	71.55
Linoleic—18:2	14.84	13.22	8.46	11.12	11.12	8.30
Linolenic—18:3	1.25	0.83	0.47	0.59	0.59	0.44
Arachidic—20:0	–	–	0.05	0.04	0.04	0.04
Gondoic—20:1	–	–	0.18	0.18	0.16	0.19
Behenic—22:0	–	–	0.00	0.00	0.00	0.01
Others	0.66	0.45	–	–	–	–

they analyzed commercial samples from Brazil, New Zealand, Chile (2), and Ecuador. For both groups of samples, the fatty acid composition was very detailed, evaluating fatty acids from 14 (myristic acid) to 24 (lignoceric) carbons, including the minor fatty acids. As expected, due to the pedoclimatic differences among countries, the authors found a broad range of fatty acid composition in commercial samples.

In Table 2 we can find a general range of fatty acid composition for cold extracted avocado oil. Considering a general consensus, it is possible to describe the oleic acid range from 52% to 74%, palmitic acid varies from 11% to 23%, linoleic acid from 8% to 16%, and the most variable one, palmitoleic acid, from 2% to 11%.

One of the main factors that have influence on the fatty acid composition is the avocado variety. Two recent scientific works have analyzed monovarietal cold extracted avocado oils (Table 3) from Brazil (Jorge et al., 2015) and Spain (Fernandes et al., 2018).

Regarding fruits from Brazil, the main difference between the Margarida and Hass varieties is the amount of palmitoleic acid, which is four times higher in the Hass than in the Margarida variety. In the Spanish samples, the amount of palmitoleic acid was around 7% in the Hass, Pinkerton, and Bacon varieties and around 4% in the Fuerte variety. Comparing the Hass variety harvested in Brazil and Spain, palmitoleic acid was almost two times higher in Brazil than in Spain. Comparing both studies, the Fuerte variety (from Spain) showed a higher amount of oleic acid (73.5%), and Hass (from Brazil) the lowest amount (54.7%). Palmitic acid was detected in higher amounts in Margarida (from Brazil), at 23.2%, and in the lowest amount in Bacon (from Spain, 12.1%). The highest amount of linoleic acid was found in Margarida (from Brazil, 14.8%): in this way, this could be the least stable sample. The Hass sample from Brazil showed a higher amount of palmitoleic acid (11.3%). The same observation was made by Fernandes et al. (2018) when analyzing a commercial sample of cold extracted avocado oil from Brazil as well.

During their study, Fernandes and coworkers first determined the monounsaturated fatty acids isomers, basically from palmitoleic (16:1; ω7, ω9, and ω11), margaroleic (17:1; two unidentified isomers), and oleic (18:1; ω7 and ω9), from monovarietal cold extracted avocado oils, Fig. 4. The distribution (% of total 16:1) of palmitoleic isomers 16:1 ω7, 16:1 ω9, and 16:1 ω11 were quite similar for all varieties. For margaroleic acid, the Hass variety showed a higher ratio of the second isomer. The Hass variety was also the most different sample, showing lower amounts of the 18:1 ω9 isomer.

The fatty acids can be analyzed from a different perspective: triacylglycerol (TAG) composition. In this case, it is possible to analyze the distribution of fatty acids in the TAG molecule. The TAG composition of cold extracted avocado oil can

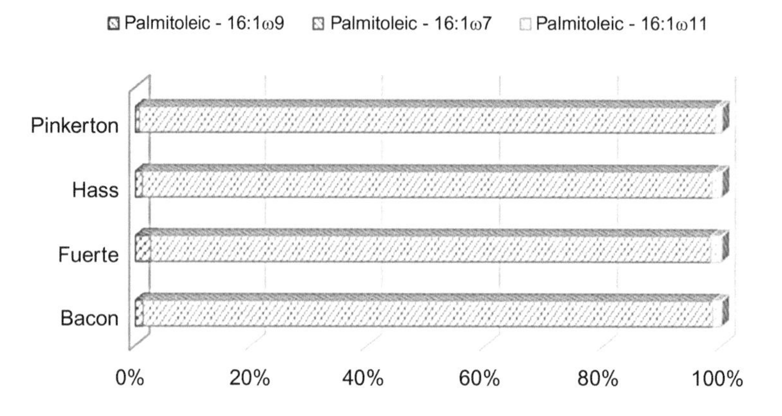

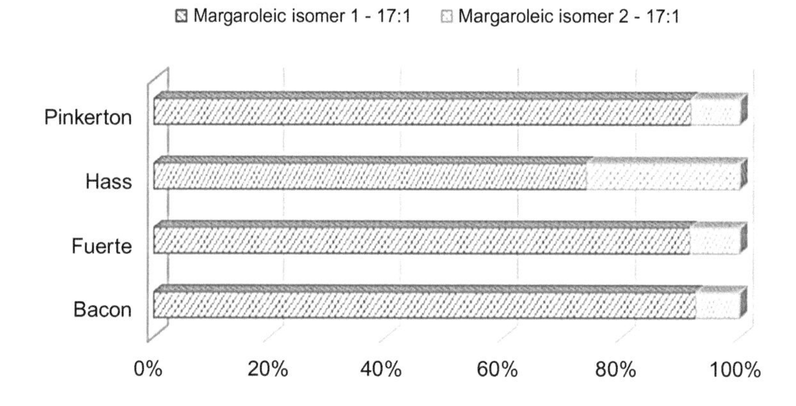

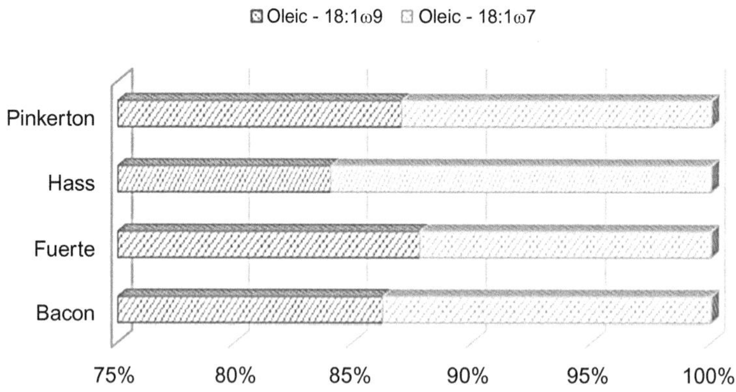

FIG. 4 Distribution (% of the total fatty acid amount, for each 16:1, 17:1, and 18:1) of isomers of monounsaturated fatty acids from monovarietal cold extracted avocado oil, according to Fernandes et al. (2018).

be found in two scientific works, from Hierro, Tomás, Fernández-Martín, and Santa-María (1992) and Fernandes et al. (2018), as presented in Table 4. Unfortunately, Hierro et al. (1992) presented the TAG composition in graphical view, which makes it difficult to obtain the values for this discussion; however, there is overall agreement with the results obtained by Fernandes et al. (2018).

TABLE 4 Triacylglycerol composition of monovarietal and commercial cold extracted avocado oil.

TAG profile %		Fernandes et al. (2018) Monovarietal	Fernandes et al. (2018) Commercial
ECN 42	Unknown	0.03–0.05	0.00–0.16
	LLL	0.24–0.40	0.37–1.09
	OLLn	0.20–0.50	0.33–1.17
	PoLL	0.06–0.12	0.07–0.28
	PLLn	0.00–0.09	0.00–0.24
ECN 44	OLL	0.83–1.58	1.67–5.60
	OOLn + PoOL	2.04–3.40	2.45–4.41
	PLL + PoPoO	0.70–1.71	1.40–3.45
	POLn + PPoPo + PPoL	0.91–2.46	1.08–3.51
ECN 46	Unknown II	0.00–0.73	0.00–0.99
	OOL + LnPP	8.72–12.76	8.34–16.98
	PoOO	5.80–8.88	4.40–9.31
	SLL + PLO	6.02–9.81	7.16–10.44
	PoOP + SPoL + POLn + SPoPo	3.16–6.70	2.14–10.48
ECN 48	PLP + OOO + PoPP	27.61–40.67	18.47–32.84
	SOL + POO	21.09–22.57	17.72–24.98
ECN 50	POP	2.97–5.01	2.89–5.73
	SOO	0.00–0.61	0.00–1.13
ECN 52	POS + SSL	–	0.00–0.32

Due to the degradation process, fatty acids can be found in different forms in regular oils such as FFA, fatty acid methyl esters (FAMES), and fatty acid ethyl esters (FAEES). Free fatty acids are formed by the hydrolysis of triacylglycerols by an enzymatic process. In addition, the fatty acids are totally related to the oxidation process of oils since the oxidation occurs in the neighborhood carbon of double bonds of unsaturated fatty acids. Both hydrolytic and oxidative degradation are related to the quality parameters of cold pressed avocado oil. Hydrolysis can be monitored by the percentage of free fatty acids or the acidity index, while oxidation is commonly monitored by the peroxide value. As expected, all quality parameters suffer constantly the influence of storage conditions, temperature, time, and raw material quality.

4 Minor bioactive lipids

The unsaponifiable matter is basically composed of nonfatty acids related molecules; in regular oils, these are the minor compounds. In general, these minor compounds are closely related to the biological activity as well as the identity of oils (Gómez-Coca, Pérez-Camino, & Moreda, 2015). In regular oils, the minor compounds are related to some lipid classes such as sterols, tocopherols, hydrocarbons, and terpenic molecules. Sterols are one of the most representative classes of unsaponifiable components and among them we can find the desmethyl, methyl, and dimethylsterols, being the desmethylsterols, generally known as sterols (Gómez-Coca et al., 2015).

Fernandes et al. (2018) determined the three molecule groups of the sterol fraction of avocado oils. As expected, desmethylsterols were the most representative group, wherein the total amount of desmethyl-sterols in avocado oil was very high, ranging from 3828.7 mg kg^{-1} in Chilean samples to 7611.8 mg kg^{-1} in Hass monovarietal oil. β-sitosterol was the most abundant desmethyl sterols, with concentrations between 80.5% in a Fuerte monovarietal oil and 86.0% in a New Zealand commercial oil. The amount of the major sterols is very close in many studies, where the avocado oil contained β-sitosterol as the main sterol (71.8%–93.0%) with low amounts of clerosterol, avenasterol, campesterol, and sitostanol (Hierro et al., 1992; Salgado et al., 2008). A more detailed composition was reported by Fernandes et al. (2018), who reported that olive oil has around 1500 $mg\,kg^{-1}$, while the amount of desmethylsterols (phytosterols) in avocado oils is very high, sometimes five times higher.

Desmethyl sterols are popularly known as phytosterols; their consumption is related to the reduction of the absorption of cholesterol. Phytosterols have been used to enhance many functional foods, which are indicated for the prevention of cardiovascular diseases. In this way, the consumption of avocado oil has also been related to these healthy effects (Carvajal-Zarrabal et al., 2014; Ortiz-Avila et al., 2013).

For methyl and dimethylsterols groups, Fernandes and coworkers have found a large variation for the profile and total amounts. As presented in Table 5, amounts of methyl-sterols from 287.9 to 1091.6 $mg\,kg^{-1}$ were found. Within this group,

TABLE 5 Ranges of minority bioactive compounds presents in avocado oils, according to Fernandes et al. (2018), Hierro et al. (1992), Salgado et al. (2008), and Gómez-Coca et al. (2015).

Desmethylsterols % (profile)		**Dymethylsterols % (profile)**	
Cholesterol	0.15–2.30	Butyrospermol	0–6.63
Brassicasterol	0–0.03	Cicloartenol	24.57–70.43
24-methylene-cholesterol	0.3–1.94	24-methylencycloartenol	28.82–74.47
Campesterol	3.71–6.60	*Total mg kg^{-1}*	39.68–545.33
Campestanol	0.04–0.7	*Terpenic alkenols % (profile)*	
Stigmasterol	0.11–1.15	Phytol	20.56–50.72
△7-campesterol	0.1–13.40	Geranylgeraniol	49.28–79.44
△5,23-stigmastadienol + clerosterol	1.72–2.08	*Total mg kg^{-1}*	41.68–92.36
β-sitosterol	71.60–86.03	*Aliphatic alcohol % (profile)*	
Sitostanol	0.41–0.81	C22-OH	18.61–43.75
△5-avenasterol	1.8–9.16	C23-OH	0–3.6
△5,24-stigmastadienol	0.07–0.75	C24-OH	0–28.31
△7-stigamastenol	0.06–0.36	C25-OH	2.88–9.62
△7-avenasterol	0.13–0.34	C26-OH	14.4–31.59
Total mg kg^{-1}	*1300.0–7611.88*	C28-OH	6.77–27.28
Methylsterols % (profile)		*Total mg kg^{-1}*	2.77–14.41
Obtusifoliol	5.48–10.98	*Tocopherols mg kg^{-1}*	
Gramisterol	11.48–20.24	α-tocoferol	0.15–103.11
Citrostadienol	71.02–83.04	β-tocoferol	0–8.07
Total mg kg^{-1}	287.98–1091.66	γ-tocoferol	2.42–71.61
Squalene mg kg^{-1}		δ-tocoferol	0–44.1
Squalene mg kg^{-1}	190.52–1366.64	*Total mg kg^{-1}*	4.99–141.5

citrostadienol was the predominant compound, reaching 83.0% in Bacon monovarietal oil. The amounts of dimethylsterols were the lowest amount in the sterol class; the highest value found was 545.33 mg kg^{-1} in Pinkerton monovarietal oil. There is no solid information on the health effects of methyl and dimethylsterols; the full profile of these compounds in many avocado oils can be found in Table 5.

An important bioactive compound of avocado oil is squalene. The main studies of the health effects of this molecule come mainly from olive oil consumption (Newmark, 1997). However, the high amounts of squalene in avocado oil can also be related to the health benefits of avocado oil (Gómez-Coca et al., 2015). In the samples under study by Fernandes et al. (2018), the squalene amounts reach 1366.6 mg kg^{-1} in Fuerte monovarietal oil.

Phytol and geranylgeraniol are primary alkenols with terpenic skeletons with total amounts below 100 mg kg^{-1} in avocado oil, as reported by Fernandes et al. (2018). The authors also reported the profile of aliphatic alcohols, which was dominated by molecules with even carbon numbers, like C_{22}-OH, C_{24}-OH, C_{26}-OH, and C_{28}-OH.

Tocopherols have antioxidant activity in oils as well as important vitamin E activity for consumers. In avocado oil, the amount of tocopherols is not as high as in other vegetable oils, such as soybean, but it is still important for the oil oxidative stability and vitamin E intake. In the reported studies, monovarietal avocado oil samples showed up to 141.5 mg kg^{-1}. Generally, α-tocopherol was the main species found, which is very good for the health benefits of avocado oil since α-tocopherol has higher vitamin E activity.

5 Health-promoting traits of cold extraction oil and oil constituents

Avocado oil has attracted growing interest in terms of human nutrition, mainly because the lipid content is rich in monounsaturated fatty acids, which could exert many cardiovascular benefits, and also other important functions, including the prevention of cancer, increasing the absorption of lipophilic functional compounds in the body, antioxidant activity, and wound healing (Forero-Doria, García, Vergara, & Guzman, 2017; Oliveira et al., 2013; Unlu, Bohn, Clinton, & Schwartz, 2004).

Avocado consumption was reported to be associated with cardiovascular health, because of the fatty acid composition, rich in MUFA and polyunsaturated fatty acid (PUFA), and the high level of phytosterols (Dreher & Davenport, 2013). Forero-Doria et al. (2017) reported the high concentration of oleic acid and highlighted the increased of omega-3 PUFA incorporation into the cell membranes, which decreases the possibility that LDL-C oxidizes, producing beneficial effects on health, such as reducing the risk of cardiovascular disease. In addition, avocado oil has a high concentration of components in the unsaponifiable fraction, such as phytosterols that can decrease the low-density lipoprotein cholesterol (LDL-C) in blood (Forero-Doria et al., 2017). Because of their similarity to cholesterol molecules, phytosterols are known to compete with cholesterol for absorption, leading to lower blood cholesterol levels. Studies indicate that the consumption of 2 g of phytosterols per day has a significant effect on the reduction of cholesterol levels and, consequently, the prevention of coronary heart disease (Moruisi, Oosthuizen, & Opperman, 2013; Wu, Fu, Yang, Zhang, & Han, 2009).

The ingestion of phytosterols also has been associated with cancer prevention. Recent studies indicate that the prevention of cancer by the consumption of phytosterols is related to the modulation of sterol biosynthesis, improvement of the immune response, and induction of tumor metastases (Shahzad et al., 2017). In this way, the high amounts of phytosterols in avocado oil bring the possibility to these oils becoming a natural food source of consumers' diets.

Unlu et al. (2004) highlighted that carotenoid (α-carotene, β-carotene, and lutein) absorption from salad by humans was enhanced by the addition of avocado or avocado oil. The hypothesis of these authors was that a frequently consumed lipid-rich fruit, avocado, can increase carotenoid absorption when consumed together with carotenoid-rich foods and that this increase is comparable to the effect of adding equivalent amounts of fat/oil alone. The experiment was conducted by addition of 75 g and 150 g of avocado fruit or 24 g of avocado oil in salads. The salads were offered to men and women over 2, 4, 6, and 8 weeks. The addition of 24 g avocado oil to the salad significantly enhanced the absorption of lutein, α-carotene, and β-carotene to an extent that was not significantly different from the addition of 150 g of avocado fruit. The authors considered that avocados and avocado oils contain a large variety of nutrients, and that adding avocado fruit to carotenoid-containing meals as a lipid source can facilitate carotenoid absorption while offering additional nutritional benefits.

Oliveira et al. (2013) evaluated the wound-healing activity of a semisolid formulation of avocado oil (composed of 50% avocado oil and 50% petroleum jelly as a vehicle) and avocado oil in nature, on incisional and excisional cutaneous wound models in Wistar rats. The oil from the Margarida variety avocado, in Brazil, was extracted using *n*-hexane as a solvent using the Soxhlet extraction method. The authors highlighted that avocado oil is a rich source of oleic acid and contains essential fatty acids. In avocado oil, the main fatty acids were oleic fatty acid (47.2%), followed by palmitic (23.6%), linoleic (13.4%), docosadienoic (8.88%), palmitoleic (3.58%), linolenic (1.60%), eicosenoic (1.29%), and myristic acids

(0.33%). The authors observed a significant increase, on the 14th day, in percentage wound contraction and reepithelialization in the groups treated with a semisolid formulation of avocado oil or avocado oil *in natura* compared to the control. Antiinflammatory activity increase in density of collagen and tensile strength were observed in a semisolid formulation of avocado oil or avocado oil *in natura* groups, when compared to control groups. The authors concluded that, when used *in natura* or in pharmaceutical formulations for topical use, avocado oil can promote the increase in collagen synthesis and the decrease in the numbers of inflammatory cells during the wound-healing process, and may thus be considered a new option for treating skin wounds.

6 Edible and nonedible applications of cold pressed oil

6.1 Culinary applications

As we have been discussing during this chapter, cold pressed avocado oil has similar properties to olive oil in both chemical and healthy characteristics. Regarding these characteristics, we can highlight that the high amount of oleic acid and antioxidant compounds could provide high shelf-life stability for avocado oil. In the same way as olive oil, avocado oil fits consumer expectation regarding less processed food products. On the other hand, the main difference between these two oils resides in the sensory characteristics, wherein olive oil is known by its bitter and pungent characteristics, avocado oil is known by its butter-like and softer flavors. In this way, cold pressed avocado oil appeals to the consumer as a delicate oil to add in culinary preparations from salads and fresh foods to cooked foods. In terms of gastronomy, avocado oil has been used from the simplest to the most sophisticated preparation (Woolf et al., 2009). Regarding the application of avocado oil in other foods, we have not enough data on the industrial application of avocado oil in foods, only a few home (sometimes even artisanal) applications.

6.2 Cosmetic applications

In the cosmetic market, the use of avocado oil has a significant history. According to cosmetic producers and users, avocado oil is one of the most penetrating oils for both skin and hair hydration. In addition, from the perspective of cosmetic products, it forms a fine emulsion due to the reduction of superficial tension, as well as providing high oxidative stability for the system. For some cosmetic uses, crude avocado oil can be refined, resulting in a pale yellow oil with low or mule avocado taste. In general, avocado oil is claimed to be good for tissue and massage creams, muscle oils, soaps, shampoo, hair treatments, and other products (Woolf et al., 2009).

7 Other issues

There is no global law or recommendation regarding avocado oil identity and quality. Some countries have established their own regulations based on internal product characteristics. The lack of regulations makes it difficult to detect fraud and mislabeling of products. Avocado oil could be classified according to processing and quality parameters: "extra virgin," with high quality and low sensory defects; "virgin," with low quality features and some sensory defects; "pure," a kind of refined avocado oil; and "blend," when the oil is mixed with other vegetable oils (Woolf et al., 2009). A typical fatty acid (FA) profile is composed of palmitic (C16:0) 10%–25%, palmitoleic (C16:1) 2%–8%, stearic (C18:0) 0.1%–1.5%, oleic (C18:1) 60%–80%, linoleic (C18:2) 7%–20%, and linolenic (C18:3) 0.2%–1% acids. In addition, the tocopherol content was established to be between 70 and 190 $mg\,kg^{-1}$. Returning to the major and minor compounds session, we can easily see that some of the proposed values are out of even monovarietal oils; thus, based on recent studies by Woolf et al. (2009), proposals could be updated and proposed as a legal standard for avocado oil.

8 Conclusion

The avocado fruit is outstanding for its high content of monounsaturated oil and is an excellent source of important lipid-soluble compounds such as tocopherols, carotenoids, and phytosterols, which play a significant role in reducing the risk of cardiovascular diseases and in providing health benefits. With increasing production of this fruit, one of the industrial processes is the production of oil from the fruit. The extraction efficiency and quality of avocado oil depend on various extraction methods including conventional and nonconventional extraction methods.

References

Abreu, R. F. A., & Pinto, G. A. S. (2009). Extração do Óleo da Polpa de Abacate Assistida por Enzimas em Meio Aquoso. In: *XVII Simpósio Nacional de bioprocessos. Natal, RN, Brasil: Embrapa Agroindústria Tropical (CNPAT).*

Aparicio, R. (2000). Authentication. In J. Harwood, & R. Aparicio (Eds.), *Handbook of olive oil: Analysis and properties*. Gaithersburg, MD: Aspen Publishers.

Ashworth, V. E. T. M., & Clegg, M. T. (2003). Microsatellite markers in avocado (Persea americana Mill.): Genealogical relationships cultivated avocado genotypes. *The Journal of Heredity*, *94*, 407–415.

Bizimana, V., Breene, W. M., & Csallany, A. S. (1993). Avocado oil extraction with appropriate technology for developing countries. *Journal of the American Oil Chemists' Society*, *70*, 821–822.

Botha, B. M. (2004). Supercritical fluid extraction of avocado oil. *South African Avocado Growers' Association*, *27*, 24–27.

Buenrostro, M., & López-Munguia, A. C. (1986). Enzymatic extraction of avocado oil. *Biotechnology Letters*, *8*, 505–506.

Carvajal-Zarrabal, O., Nolasco-Hipolito, C., Aguilar-Uscanga, M. G., Melo-Santiesteban, G., Hayward-Jones, P. M., & Barradas-Dermitz, D. M. (2014). Avocado oil supplementation modifies cardiovascular risk profile markers in a rat model of sucrose-induced metabolic changes. *Disease Markers*, *2014*, 1–8. https://doi.org/10.1155/2014/386425.

CODEX (2009). *Standard for named vegetable oils. Current official standards—CODEX STAN*: (p. 210).

Costagli, G., & Betti, M. (2015). Avocado oil extraction processes: Method for cold-pressed high-quality edible oil production versus traditional production. *Journal of Agricultural Engineering*, *46*, 115.

Cunniff, P., & Association of Official Analytical Chemists (1995). *Official methods of analysis of AOAC international*. Washington, DC: Association of Official Analytical Chemists.

Curiel, M., Neve, O., & Petach, T. (1985). *Process for the recovery of oil from avocados fruit*. USA patent application.

Di Giovacchino, L., Marsilio, V., Costantini, N., & Di Serio, G. (2005). Use of olive mill wastewater (OMW) as fertiliser of the agricultural soil: Effects on crop production and soil characteristics. *European Bioremediation Conference*, *3*, 165.

Dreher, M. L., & Davenport, A. J. (2013). Hass avocado composition and potential health effects. *Critical Reviews in Food Science and Nutrition*, *53*, 738–750.

FAOSTAT (2019). *(Food and Agriculture Organization of United Nations, Statistic Division). Production/Crops/Avocado 2014*. URL http://faostat.fao.org/, Accessed 27.08.19.

Fernandes, G. D., Gómez-Coca, R. B., Pérez-Camino, M. C., Moreda, W., & Barrera-Arellano, D. (2018). Chemical characterization of commercial and single-variety avocado oils. *Grasas y Aceites*, *69*, 256.

Forero-Doria, O., García, M. F., Vergara, C. E., & Guzman, L. (2017). Thermal analysis and antioxidant activity of oil extracted from pulp of ripe avocados. *Journal of Thermal Analysis and Calorimetry*, *130*, 959–966.

Freitas, S. P., Lago, R., Jablonka, F. H., & Hartman, L. (1993). Enzimatic aqueous extraction of avocado oil from fresh pulp. *Revue Francaise des Corps Gras*, *40*, 365–371.

Gómez-Coca, R. B., Pérez-Camino, M. C., & Moreda, W. (2015). Analysis of neutral lipids: Unsaponifiable. In L. Nollet, & F. Toldrá (Eds.), *Handbook of food analysis*. Boca Raton: CRC Press.

Hamzah, B. (2013). The effect of homogenization pressures on extraction of avocado oil by wet method. *Advance Journal of Food Science and Technology*, *5*, 1666–1668.

Hierro, M. T., Tomás, M. C., Fernández-Martín, F., & Santa-María, G. (1992). Determination of the triclyceride compositon of avocado oil by high-peerformance liquid chromatography using a light-scattering detector. *Journal of Chromatography*, *607*, 329–338.

Jorge, T. D. S., Polachini, T. C., Dias, L. S., Jorge, N., & Telis-Romero, J. (2015). Physicochemical and rheological characterization of avocado oils. *Ciência e Agrotecnologia*, *39*, 390–400.

Kameni, A., & Tchamo, P. (2003). Water extraction of avocado oil in the high lands of Cameroon. *Tropical Science*, *43*, 10–12.

Krumreich, F. D., Borges, C. D., Mendonça, C. R. B., Jansen-Alves, C., & Zambiazi, R. C. (2018). Bioactive compounds and quality parameters of avocado oil obtained by different processes. *Food Chemistry*, *257*, 376–381.

Martinez Nieto, L., Barranco Barranco, R., & Moreno, M. V. (1992). Extracción de aceite de aguacate: Un experimento industrial. *Grasas y Aceites*, *43*, 11–15.

Mohameed, S., Sharon, D., Kaufman, D., Lahav, E., Hillel, J., Degani, C., et al. (1997). Genetic relationships within avocado (*Persea americana* Mill) cultivars and between *Persea* species. *Theoretical and Applied Genetics*, *94*, 279–286.

Moreno, A. O., Dorantes, L., Galindez, J., & Guzman, R. I. (2003). Effect of different extraction methods on fatty acids, volatile compounds, and physical and chemical properties of avocado (*Persea americana* Mill.) oil. *Journal of Agricultural and Food Chemistry*, *51*, 2216–2221.

Moruisi, K. G., Oosthuizen, W., & Opperman, A. M. (2013). Phytosterols/stanols lower cholesterol concentrations in familial Hypercholesterolemic subjects: A systematic review with meta-analysis. *Journal of the American College of Nutrition*, *25*, 41–48.

Mostert, M. E., Botha, B. M., Plessis, L. M. D., & Duodu, K. G. (2007). Effect of fruit ripeness and method of fruit drying on the extractability of avocado oil with hexane and supercritical carbon dioxide. *Journal of the Science of Food and Agriculture*, *87*, 2880–2885.

Newmark, H. L. (1997). Squalene, olive oil, and cancer risk: A review and hypothesis. *Cancer Epidemiology, Biomarkers & Prevention*, *6*, 1101–1103.

Oliveira, A. P., Franco, E. D. S., Rodrigues Barreto, R., Cordeiro, D. P., Melo, R. G., Aquino, C. M. F., et al. (2013). Effect of semisolid formulation of *Persea americana* mill (avocado) oil on wound healing in rats. *Evidence-based Complementary and Alternative Medicine: Ecam*, *2013*, 472382.

Ortiz, M. A., Dorantes, A. L., Gallndez, M. J., & Cardenas, S. E. (2004). Effect of a novel oil extraction method on avocado (*Persea americana* Mill) pulp microstructure. *Plant Foods for Human Nutrition*, *59*, 11–14.

Ortiz-Avila, O., Sámano-García, C. A., Calderón-Cortés, E., Pérez-Hernández, I. H., Mejía-Zepeda, R., Rodríguez-Orozco, A. R., et al. (2013). Dietary avocado oil supplementation attenuates the alterations induced by type I diabetes and oxidative stress in electron transfer at the complex II-complex III segment of the electron transport chain in rat kidney mitochondria. *Journal of Bioenergetics and Biomembranes*, *45*, 271–287.

Paraskeva, P., & Diamadopoulos, E. (2006). Technologies for olive mill wastewater (OMW) treatment: A review. *Journal of Chemical Technology & Biotechnology*, *81*, 1475–1485.

Petrakis, C. (2006). Olive oil extraction. In D. Boskou (Ed.), *Olive oil chemistry and technology*. Champaign: AOCS Press.

Platt, K. A., & Thomson, W. W. (1992). Idioblast oil cells of avocado: Distribution, isolation, ultrastructure, histochemistry, and biochemistry. *International Journal of Plant Sciences*, *153*, 301–310.

Qin, X., & Zhong, J. (2016). A review of extraction techniques for avocado oil. *Journal of Oleo Science*, *65*, 881–888.

Reddy, M., Moodley, R., & Jonnalagadda, S. B. (2012). Fatty acid profile and elemental content of avocado *(Persea americana* Mill.) oil-effect of extraction methods. *Journal of Environmental Science and Health. Part. B*, *47*, 529–537.

Requejo-Jackman, E. C. (2009). Avocado oil. From cosmetic to culinary oil. In R. Moreau, & A. Kamal-Eldin (Eds.), *Gourmet and health-promoting specialty oils*. Urbana: AOCS.

Rueda, A., Samaniego-Sánchez, C., Olalla, M., Giménez, R., Cabrera-Vique, C., Seiquer, I., & Lara, L. (2016). Combination of analytical and chemometric methods as a useful tool for the characterization of extra virgin argan oil and other edible virgin oils. Role of polyphenols and tocopherols. *Journal of AOAC International*, *99*, 489–494.

Rueda, A., Seiquer, I., Olalla, M., Giménez, R., Lara, L., & Cabrera-Vique, C. (2014). Characterization of fatty acid profile of argan oil and other edible vegetable oils by gas chromatography and discriminant analysis. *Journal of Chemistry*, *2014*, 1–8. https://doi.org/10.1155/2014/843908.

Salgado, J. M., Danieli, F., Regitano-D'arce, M. A. B., Frias, A., & Mansi, D. N. (2008). O óleo de abacate (*Persea americana* Mill) como matéria-prima para a indústria alimentícia. *Food Science and Technology*, *28*, 20–26.

Shahzad, N., Khan, W., Md, S., Ali, A., Saluja, S. S., Sharma, S., et al. (2017). Phytosterols as a natural anticancer agent: Current status and future perspective. *Biomedicine & Pharmacotherapy*, *88*, 786–794.

Silva, M. D. R. G., Freitas, A. M. C., Cabrita, M. J., & Garcia, R. (2012). Olive oil composition: Volatile compounds. In D. Boskou (Ed.), *Olive oil—Constituents, quality, health properties and Bioconversions*: InTech.

Tan, C. X., Gun Hean, C., Hamzah, H., & Ghazali, H. M. (2018). Optimization of ultrasound-assisted aqueous extraction to produce virgin avocado oil with low free fatty acids. *Journal of Food Process Engineering*, *41*, e12656.

Tango, J. S., Carvalho, C. R. L., & Soares, N. B. (2004). Physical and chemical characterization of avocado fruits aiming its potential for oil extraction. *Revista Brasileira de Fruticultura*, *26*, 17–23.

Unlu, N. Z., Bohn, T., Clinton, S. K., & Schwartz, S. J. (2004). Carotenoid absorption from salad and salsa by humans is enhanced by the addition of avocado or avocado oil. *The Journal of Nutrition*, *135*(3), 431–436. https://doi.org/10.1093/jn/135.3.431.

USDA (2015). *Dietary guidelines for Americans 2015–2020*. United States: United States Department of Agriculture.

Werman, M. J., & Neeman, I. (1987). Avocado oil production and chemical characteristics. *Journal of the American Oil Chemists' Society*, *64*, 229–232.

Wong, M., Requejo-Jackman, C., & Woolf, A. (2010). What is unrefined, extra virgin cold-pressed avocado oil? In *Vol. 21 International news on fats, oils, and related materials* (pp. 189–260). INFORM AOCS.

Woolf, A., Wong, M., Eyres, L., Mcghie, T., Lund, C., Olsson, S., et al. (2009). 2—Avocado oil. In R. A. Moreau, & A. Kamal-Eldin (Eds.), *Gourmet and health-promoting specialty oils*. AOCS Press.

Wu, T., Fu, J., Yang, Y., Zhang, L., & Han, J. (2009). The effects of phytosterols/stanols on blood lipid profiles: A systematic review with meta-analysis. *Asia Pacific Journal of Clinical Nutrition*, *18*, 179–186.

Chapter 38

Cold pressed colza oil

Zahra Piravi-Vanak
Food Technology and Agricultural Products Research Center, Standard Research Institute of Iran, Karaj, Iran

Abbreviations

AOF Australian Oilseeds Federation
CXC Codex Alimentarius Commission
GM Genetically modified

1 Introduction

Colza (*Brassica napus*) seed oil or *Brassica campestris* (formerly known as rapeseed oil), a member of the Cruciferae family, is an amphidiploid species resulting from a cross between certain forms of cabbage and turnip in nature (Małgorzata, Agnieszka, & Katarzyna, 2016). It is an annual plant, with spring and autumn types. Aside from *B. napus*, other *Brassica* species (*B. campestris* or field mustard, *Brassica nigra* or black mustard, and *Brassica carinata* or Ethiopian mustard) are also called colza in the international markets. Based on an Australian Oilseeds Federation (AOF) standard, canola is defined as a seed of the species *B. napus* or *Brassica rapa* but containing less than 30 μmol of specified glucosinolates per g of oil-free air-dry solids and not more than 2% erucic acid in the oil component, as a proportion of the total fatty acids content. The specified glucosinolates are any one or a mixture of 3-butenyl, 4-pentenyl, 2-hydroxy-3-butenyl, and 2-hydroxy-4-pentenyl glucosinolates (Australian Oilseeds Federation, 2018).

The seeds of rape consist of three basic components: the embryo, the endosperm, and the seed coat (hull). The endosperm degenerates during seed maturation and the seed coat enwraps the embryo tightly. The embryo contains two pieces of cotyledons (which serve as food reserve structures), a radicle and hypocotyl. The oil in this oilseed is distributed in spherosomes throughout the germ cell. Rapeseeds are composed of 38%–50% lipid, 20%–32% protein, and 10%–15% crude fiber. These major rapeseed constituents are not evenly distributed throughout the rapeseed. The respective oil contents for hulls and kernels range from 10.6 to 16.4% (dry basis) and from 47.1% to 59.6%. The protein content of the hulls ranges from 17% to 18% (defatted dry basis), while the protein content in defatted kernels ranges from 46% up to 79%. Crude fiber contents range from 27.0% to 44.1% (dry basis) in defatted hulls and 3.0% to 12% in de-oiled kernels (Rękasa et al., 2017).

Canola, known as low-glucosinolate, low-erucic acid rapeseed, is a major oilseed crop worldwide (Grageolaa et al., 2013). Low erucic acid rape is divided into two different seeds in terms of genetic manipulation: genetically modified (GM) canola and nonGM canola. In this regard, nonGM canola should contain low-level presence of up to 0.9% of GM events approved by the national government office of the Gene Technology Regulator (Australian Oilseeds Federation, 2018). Rapeseed is growing as a major source of vegetable oil and is now the second largest oilseed after soybeans around the world. It is mainly grown in Western Europe, China, Canada (which has a variety of canola production), and India. Due to the combination of rapeseed fatty acids, approximately 31 million hectares of land are allocated to rapeseed each year and about 60 million tons of oil are produced (Leckband et al., 2002).

According to (Codex Alimentarius commission) CXC 210, the production definition for colza oil is as follows:

Rapeseed oil (turnip rape oil; colza oil; ravison oil; sarson oil: toria oil) is produced from seeds of *Brassica napus* L., *Brassica rapa* L., *Brassica juncea* L., and *Brassica tournefortii* Gouan species.
Rapeseed oil-low erucic acid (low erucic acid turnip rape oil; low erucic acid colza oil; canola oil) is produced from low erucic acid oil-bearing seeds of varieties derived from the *Brassica napus* L., *Brassica rapa* L., and *Brassica juncea* L. species (CXS 210-1999, 2019).

Cold Pressed Oils. https://doi.org/10.1016/B978-0-12-818188-1.00038-4
Copyright © 2020 Elsevier Inc. All rights reserved.

Colza is also classified in various ways concerning erucic acid, glucosinolate, fiber, and genetic manipulation. According to these properties, there is a common classification for colza as follows:

Traditional colza cultivars: this type contains oil with 26%–60% erucic acid and 100–205 μmoles of glucosinolates/100 g canola meal. This fatty acid specification of extracted oil is similar to the codex limit for rapeseed oil.
Single-zero cultivars: these Canadian cultivars contain oil with less than 5% erucic acid and 100–205 μmoles of glucosinolates/100 g of canola meal.
Double zero cultivars: These contain oil with less than 2% erucic acid and 0–18 μmoles of glucosinolates/100 g of canola meal. The limit of erucic acid of its oil is within CXS (Codex Alimentarius Commission) 210 for rapeseed oil (low erucic acid).
Triple-zero cultivars: these improved colza cultivars are also called candle cultivars, and contain oil with minimal amounts of erucic acid, glucosinolate as well as fiber content. Canola is, in fact, the shortened form of Canada oil low erucic acid and refers to *B. napus* and *B. campestris* (Abrehdari, Ghavami, Gharachorloo, & Delkhosh, 2015).

Global production of canola oil is the third highest after palm and soybean oil. Consumer interest in canola oil has recently increased because of the unique characteristics of the oil (Ghazani, García-Llatas, & Marangoni, 2014). Canola oil is undergoing genetic modification and several rapeseed types have been modified with the structure of fatty acid. It is still unclear how many of these types are economically viable. Canola cultivars have been developed with oil that is rich in oleic acid (60%–85%), greatly resembles olive oil, and is very stable and has limited amounts of *trans* fatty acids (Choo, Birch, & Dufour, 2007).

There is a limited demand for rapeseed oil with high levels of erucic acid, which generally used to produce erucamide, one of the main components of the polyethylene packaging film. Rapeseed oil with less linolenic acid has been developed with increased levels of lauric acid, stearic acid, oleic acid, or unusual acids such as γ-linolenic acid, ricinoleic acid, or vernolic acid for commercial exploitation. The combination of modern biotechnology, such as genetic engineering, with classical plant breeding types of varieties efficient rapeseed oil, was reported (Leckband et al., 2002; Yang et al., 2013).

Based on the Codex Alimentarius commission in description section cold pressed oil are obtained, without altering the oil, by mechanical procedures only, e.g., expelling or pressing, without the application of heat. They may have been purified by washing with water, settling, filtering, and centrifuging only (CXS 19-1981, 2019; CXS 210-1999, 2019).

2 Cold pressed colza oil

Today, the oil extraction industry uses a variety of methods. However, the extraction by solvent and mechanical extraction (hot and cold press) is one of the most common. Solvent extraction is typically used for low oil seeds (<20%). Considering the disadvantages of solvent extraction, which includes high costs, safety concerns, leakage of organic matter, explosiveness, and low quality oil (Matthäus & Brühl, 2003) as well as the use of high temperatures and toxic substances that all cause great damage to the environment. Other methods of oil extraction with less damage and harm can be used (Çakaloğlu, Hazal Özyurt, & Ötleş, 2018). In cold pressed colza oil, the content of soluble protein (SP) in the meal is twice as high as in extraction with *n*-hexane (Ghazani et al., 2014).

Cold pressing is one of the oldest natural extraction methods. It is used instead of conventional solvent extraction to extract oil from vegetable seeds, fruits, nuts, or germs because cold pressing does not require the use of organic solvents or heat. The procedure is Generally Recognized As Safe (GRAS). Hence, cold pressing is capable of retaining bioactive compounds such as essential fatty acids, phenolics, flavonoids, and tocopherols in oils. For example, cold pressed rapeseed oil, which has a characteristic taste with nuts, a distinct flavor, and an intense color, is popular not only in Poland, but also in Germany, Switzerland, Austria, and the United Kingdom (Makała, 2015).

Cold pressed oils free from chemical activities characteristic of refining can be valuable edible oils providing that they do not contain harmful to humans chemicals and microbiological contamination including mycotoxins and metals (i.e., Fe, and Cu) accelerating the oxidation of oils. Modern technology of obtaining oils through cold pressing using a nitrogen atmosphere or supercritical carbon dioxide extraction allows them to be retained almost in an intact state. Their chemical composition is richer in by-products with a higher level of biological and antioxidative activities. The presence of antioxidants inhibiting deleterious changes and an insight into their activity and stability is of paramount importance not only for food technologists but also for nutritionists (Makała, 2015). Fig. 1 shows the specification of cold pressed colza oil.

Parameters that affect product yield in cold press extraction are; characteristics of the raw material (shell-shellless, moisture content, oil content, and type of raw material), feed rate, temperature (hot or cold), cold rotation speed, the diameter of restriction dye, and pretreatment. The studies that investigated the parameters affecting oil yield in the cold press method were reported (Çakaloğlu et al., 2018).

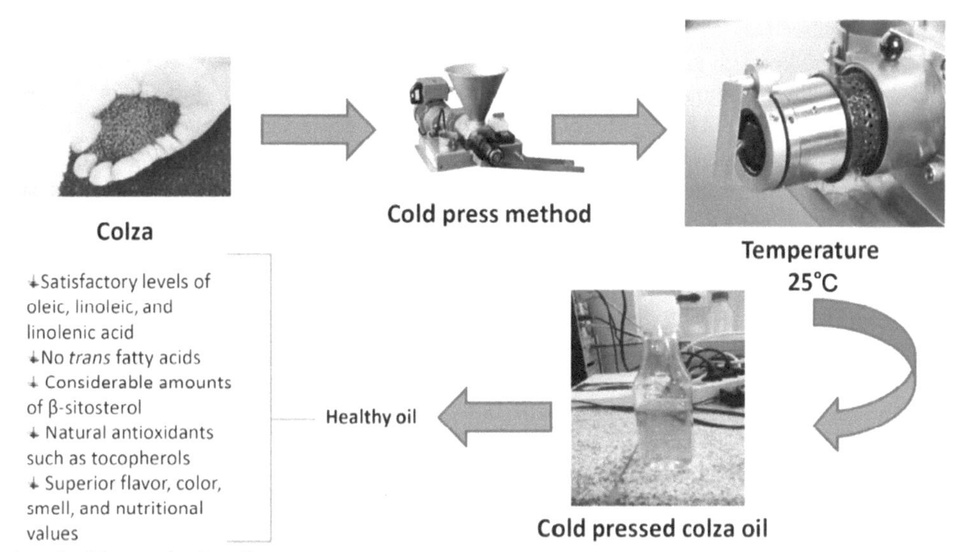

FIG. 1 Specification of cold pressed colza oil.

With the exception of cold pressing parameters, the technical characteristics of raw materials play an important role in obtaining the final quality of oils. The most valuable in terms of sensory properties are oils of high-quality raw materials, which are subjected to the least oxidation changes. Therefore, physical characteristics of the seed play an important role. The clean canola seeds were poured in a stainless-steel cylinder, which has a hole in the bottom, as shown in Fig. 2. When the bolt shown in Fig. 2 is screwed down, it presses canola seeds slowly to extract oil, which is collected in the beaker. The hole is covered with a stainless steel mesh to stop crushed residues coming out from the hole with oil. The extracted canola oil through the cold press method may be called virgin oil (Saleem & Naveed, 2018).

Cold pressed oils derived from seeds of oil plants contain more than 15% lipids content. Rapeseed seeds with a particle size greater than 2 mm were associated with higher technological values and increased extraction efficiency due to lower levels of contamination with higher lipids. The oils derived from these seeds are of higher quality due to the lower content of chlorophyll, phospholipids, and lipid hydrolysis and oxidation products; generally, expeller pressing is less complex, more cost-efficient, and safer than solvent extraction. Moreover, cold pressed oils retain their natural properties better. Cold pressed colza oil has been found to have numerous health benefits (Nasab & Vanak, 2015).

3 The effect of heat on the oil components

Saleem and Naveed (2018) reported that heating of cold pressed canola oil during extraction and refining processes at ultra-high temperatures leaves it with no natural color (β-carotene and chlorophyll), fatty acids, or vitamin E contents. Therefore,

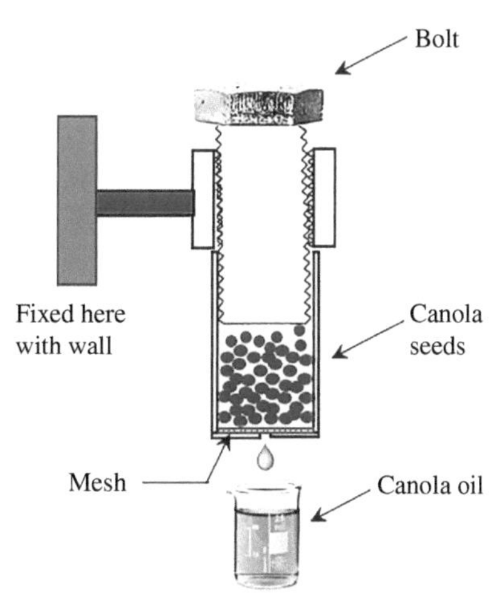

FIG. 2 Schematic diagram of the cold press oil extraction setup.

it was suggested that cold pressed canola oil is comparatively better than the one extracted chemically in the laboratory in terms of their nutritional ingredients. It has been investigated that heating of cold pressed canola oil up to 180°C does not cause it to lose most of its valuable ingredients. It could be used safely for the cooking of foods where usually the temperature remains between 100°C and 115°C. Therefore, canola oil extracted through cold pressing is far better to use directly for cooking and frying purposes as it is rich with fat-soluble vitamins, essential fatty acids, β-carotene, and chlorophylls (Saleem & Naveed, 2018). In addition, cold pressed canola oil can be used safely without destroying much of its valuable ingredients for frying eggs, which needs a temperature range of 140–160°C, and deep frying of meat and fish that needs the oil temperature to be around 180°C.

Opportunities exist to develop oil types for specific applications. High oleic acid oils with very low levels of linolenic acid are useful for frying; however, totally different fatty acid profiles are useful for baking. Increasing the tocopherol content in canola oil would increase the stability of the oil. During processing, the tocopherols stripped during steam distillation can be collected and sold to the cosmetics, animal feed, and human supplement industries, increasing the value of the product (NSW Department of Primary Industries, 2014).

4 Colza oil composition

Rapeseed oil is characterized by its unique chemical properties and the possibility of the oil to be used in the human diet. Canola oil is a relatively good source of fatty acid components such as oleic, linoleic, and linolenic acids. These fatty acids are considered essential fatty acids and their functions tend to relate to nutrition and oxidative stability issues. Rapeseed oil is characterized by the ideal polyunsaturated fatty acid (PUFA) content with a 2:1 ratio of linoleic (*n*-6) versus linolenic (*n*-3) fatty acid and high contents of active biological compounds, such as tocopherols, plastochromanol-8, phytosterols, and phenol compounds (Siger, Gawrysiak, & Bartkowiak-Broda, 2017).

The main nonsaponifiable components in vegetable oils are tocopherols and sterols, which are present in varying amounts depending on the oil. Tocopherols are natural antioxidants and their amount in the plant is governed by the content of unsaturated fatty acids. Rapeseed has the highest phenolic content compared to other oilseeds (e.g., soybean, and sunflower); about 10-fold, however, of phenolics remain in the meal after pressing (Siger, Gawrysiak, et al., 2017). According to CXC 19, cold pressed rapeseed oil should have an acid value <4 mg KOH/g oil, while its peroxide value should be <15 meq O_2/kg. The conditions and type of oil extraction process could greatly influence the peroxide value of the product. Cold pressed oil has a lower peroxide value compared to solvent extracted oil, which may be due to the latter's long exposure to air during the extraction process, and the subsequent evaporation of the solvent.

The high quality of cold pressed oils is affected by many factors. The most important raw materials quality is purity, uniformity, integrity, and maturity. The harvest time is another decisive factor. In addition, the quality of raw materials is influenced by prepressing activities including harvesting, drying, storage, and postharvest management.

5 Fatty acids profile of colza oil

From a nutritional point of view, canola's fatty acid composition can be considered extremely health-promoting as it serves as an excellent source of α-linolenic and oleic (50%–66%) acids. In addition, the ratio of ω-6 to ω-3 fatty acids is relatively low (about 2:1). It also contains a low amount of saturated fatty acids (<7%) compared to other vegetable oils. Rapeseed and canola oils contain low amounts of saturated fatty acids 4%–9%, high levels of unsaturated fatty acid (90%–95%), some linoleic acid (11%–30%), and α-linolenic acid (5%–14%). Therefore, the optimal ratio of omega-6 (linoleic acid) to omega-3 fatty acids (linolenic acid) (2:1) is naturally occurring in colza oil for human health.

Recently, essential fatty acids (EFAs) have been considered as functional food and nutraceuticals. Many research studies have documented their significant roles in many biochemical pathways resulting in cardioprotective effect because of their considerable antiatherogenic, antithrombotic, antiinflammatory, antiarrhythmic, and hypolipidemic effects, because of the potential of reducing the risk of serious diseases, especially cardiovascular diseases, cancer, osteoporosis, diabetes, and other health conditions. Promotion activities occur due to their complex influence on concentrations of lipoproteins, fluidity of biological membranes, function of membraned enzymes and receptors, modulation of eicosanoids production, blood pressure regulation, and, finally, on the metabolism of minerals (Orsavova, Misurcova, Vavra, Vicha, & Mlcek, 2015).

Traditional rapeseed oil is rich in erucic acid (22:1), and its seed meal has an unwanted level of glucosinolate. These components reduce the amount of oil and protein in the oilseed. However, both are out of modern rapeseed (Teh & Birch, 2013). Table 1 shows the fatty acid composition (expressed as percentage of total fatty acids) of rapeseed oil as determined by gas–liquid chromatography (CXS 210-1999, 2019).

TABLE 1 Fatty acid composition of different rapeseed oils (expressed as percentage of total fatty acids).

Fatty acid	Rapeseed oil	Rape seed oil (low erucic acid)
C6:0	ND	ND
C8:0	ND	ND
C10:0	ND	ND
C12:0	ND	ND
C14:0	ND–0.2	ND–0.2
C16:0	1.5–6.0	2.5–7.0
C16:1	ND–3.0	ND–0.5
C17:0	ND–0.1	ND–0.3
C17:1	ND–0.1	ND–0.3
C18:0	0.5–3.1	0.8–3.0
C18:1	8.0–60.0	51.0–70.0
C18:2	11.0–23.0	15.0–30.0
C18:3	5.0–13.0	5.0–14.0
C20:0	ND–3.0	0.2–1.2
C20:1	3.0–15.0	0.1–4.3
C20:2	ND–1.0	ND–0.1
C22:0	ND–2.0	ND–0.6
C22:1	>2.0–60.0	ND–2.0
C22:2	ND–2.0	ND–0.1
C24:0	ND–2.0	ND–0.3
C24:1	ND–3.0	ND–0.4

ND, nondetectable.

6 Sterol composition of colza oil

Plant sterols form the majority of the unsaponifiable components of most plant oils. Phytosterols are present in all plants and plant foods as free or esterified sterols. The overall quantity of sterols in "00" rapeseed oil varies between 0.45% and 1.13%. As shown in Table 2, the total sterols of rapeseed oil is 4500–11,300 mg/kg. The main phytosterols of rapeseed oil are β-sitosterol (45%–57%), campesterol (24%–38%), brassicasterol (5%–13%), stigmasterol (0.2%–1%), and D5-avenasterol. A characteristic feature of the Brassica genus is the level of brassicasterol, the presence of which allows an oil to be identified as rapeseed oil. Other sterols found in rape include β-sitosterol and campesterol, which together make up 80%–88% of all sterols (Siger, Gawrysiak, et al., 2017). The amounts of phytosterols in the oil depend on many factors including variety/cultivar, growth conditions, postharvest storage, oil production methods (press, refining), and storage conditions (Yang et al., 2013). Table 2 shows the sterol composition of rapeseed oil.

7 Tocols (tocopherols and tocotrienols) profile of colza oil

The main unsaponifiable components in rapeseed oils are tocopherols, plastochromanol-8, and sterols. The total tocopherol content of canola is close to corn, cottonseed, saffron, and sunflower oils (60–70 mg/100 g). Tocopherols are the most important natural antioxidants that inhibit the peroxidation of fatty acids in vegetable oils and act as free radicals. Rapeseed oil contains mostly α- and γ-tocopherols, and γ-tocopherols are usually in higher concentrations. The total amount of tocopherols in canola oil is higher than other common vegetable oils such as sunflower, safflower, palm oil, and coconut,

TABLE 2 Levels of desmethyl sterols in crude rapeseed oil from as a percentage of total sterols.

Sterol composition	Rapeseed (low erucic acid)
Cholesterol	ND–1.3
Brassicasterol	5.0–13.0
Campesterol	24.7–38.6
Stigmasterol	0.2–1.0
Beta-sitosterol	45.1–57.9
Delta-5-avenasterol	2.5–6.6
Delta-7-stigmastenol	ND–0.01
Delta-7-avenasterol	ND–0.8
Others	ND–4.2
Total sterols (mg/kg)	4500–11,300

ND, nondetectable.

but less than soybean and corn oil (Ratnayake & Daun, 2004; Yang et al., 2013). As Table 3 shows, the total concentration of tocols in rapeseed oil is 430–2680 mg/kg. The highest amount of tocopherol is related to γ-tocopherol (189–753 mg/kg) and α-tocopherol (100–386 mg/kg). Tocopherols are known to be very efficient natural antioxidants. The antioxidant activity of the homologous tocopherols in vivo is as follows: α-T > β-T > γ-T > δ-T. In vitro, the activity is the reverse: α-T < β-T ≈ γ-T < δ-T (Siger, Józefiak, & Górnaś, 2017).

8 Phenolic compounds of colza oil

Many phenolic compounds have antioxidant properties. Among commercial oilseeds, canola has the highest content of phenolic acids; its level is 10 times as high as other oilseeds. Defatted rapeseed meal may contain up to 2% phenolic acids. Canola oil contains high amounts of phenolic compounds, chiefly in the form of sinapic acid, and its derivatives. The main phenolic compound of rapeseed is sinapinic acid, which makes up 70% of the total content of free phenolic acids and their derivatives such as sinapine. Other derivatives of sinapinic acid that have been identified include 1-*O*-β-D-glucopyranosyl sinapate, 1,6-di-*O*-sinapoylglucose, 1,2,6′-tri-*O*-sinapoylgentiobiose, sinapic acid methyl ester, 3-dihexoside-7-sinapoylhexoside kaempferol, and 3-hexoside-7-sinapoyl hexoside kaempferol (Siger, Gawrysiak, et al., 2017; Siger, Józefiak, et al., 2017).

TABLE 3 Levels of tocopherols and tocotrienols in crude rapeseed oil (mg/kg).

Alpha-tocopherol	100–386
Beta-tocopherol	ND–140
Gamma-tocopherol	189–753
Delta-tocopherol	ND–22
Alpha-tocotrienol	ND
Gamma-tocotrienol	ND
Delta-tocotrienol	ND
Total (mg/kg)	430–2680

ND, nondetectable.

9 Pigments in colza oil

9.1 Carotenoids

Crud rapeseed and canola oil may contain up to 95 mg/kg of total carotenoids. They are mainly xanthophylls (about 85%–90%), including lutein (50%), 13-*cis*-lutein (15%), and 9-*cis*-lutein (20%). About 7%–10% of carotenoids are β-carotene. β-carotene has been shown to protect lipids against free radical auto-oxidation by reacting with radicals, therefore inhibiting propagation of the oxidation chain reaction (Nasab & Vanak, 2015;Ratnayake & Daun, 2004; Yang et al., 2013). Lutein has been reported to have pro-oxidant activity in triacylglycerol (TAG) both in darkness and in light (Yang et al., 2013).

9.2 Chlorophyll

The presence of chlorophyll in rapeseed seeds causes an undesirable green color in the oil, but the pigment is important for photosynthesis. In addition, chlorophyll stimulates photooxidation and inhibits the catalyst required for hydrogenation. Chlorophyll has a pro-oxidant effect in the presence of light, but it has been shown to work as an antioxidant in the dark (Ratnayake & Daun, 2004; Yang et al., 2013). A study reported that the chlorophyll content of the cold pressed canola oils was about 10 times lower than that of the solvent-extracted oil (Ghazani et al., 2014).

10 Phospholipids of colza oil

Research has mainly concerned the impact of extraction technologies on the phosphorus and/or total phospholipid content in oils. The method of oil extraction had a significant impact on the content of phosphorus and the profile of phospholipids, of which the most significant to the share of phosphatidic acid. It was found that the heat treatment of whole and milled seeds increased the content of phosphorus and the share of phospholipids (including phosphatidic acid) in cold pressed oils and petroleum ether extracted oils. The samples of cold pressed oil obtained from the unheated and heated seeds were characterized by a lower content of phospholipids (Ambrosewicz-Walacik, Tańska, & Rotkiewicz, 2015). Despite extracting by solvent, cold pressed canola oils were free of phosphorous. Based on the Codex Standard for Named Vegetable Oils, cold pressed oils may be purified by washing with water followed by centrifugation. Upon water washing, phosphatides would hydrate and be removed by centrifugation. This would explain the lack of phosphorous in cold pressed oils (Ghazani et al., 2014).

11 Bioactive compounds in colza oil

The chemical composition of oils determines their health beneficial capacities and their practical application. The consumers turn to natural food with health-promoting qualities, termed functional food. Food may be defined as functional if, beyond its nutritional values, it has been proven to have therapeutic potentials on one or more of the body functions leading to health promotion and disease prevention. Functional foods must have attributes of conventional foods and exhibit beneficial effects with values expected to be consumed with every day's diet being neither pills nor capsules, but an integral part of an appropriate diet. The health-contributing effect of functional food should be documented with scientific research (Makała, 2015).

Rapeseed oil is a rich source of bioactive compounds, such as phenolics, phytosterols, β-carotene, and lutein, which may prevent cardiovascular disease, cancer, diabetes, high blood pressure, neurodegenerative diseases (Alzheimer's and Parkinson's disease), and autoimmune diseases. Most of these beneficial effects are due to antioxidant activity (Yang et al., 2013).

Functional foods may incorporate cold pressed oils. Cold pressed oils provide a wide range of bioactive substances, such as tocopherols and tocotrienols, free and esterified sterols, hydrocarbons (squalene), triterpene alcohols, carotenoids, and chlorophylls; colorants are also valuable nutrients. They also contain ω-3 and ω-6 PUFA and sterols having biologically active effects. It should be mentioned the content of tocopherols in oils have powerful antioxidant, vitamin E, anticancer, and anticardiovascular effects. Tocochromanols show the ability to suppress reactive oxygen species (ROS) and free radicals. The concentration of these ingredients in oils is dependent on the quality, grain type, and its varieties that may serve as a functional food.

The intake of cold pressed edible oils can result in inhibition or retardation of diet-dependent lifestyle diseases, such as obesity, ischemic heart disease (IHD), and hypertension thanks to the presence of antioxidants like tocopherols and phenolic compounds (Makała, 2015). It is worth noting that sterols play a major role in edible oils. Phytosterols, a natural component of vegetable oil, can lower the level of LDL-cholesterol in the blood by reducing cholesterol absorption, and thus prevent the development of heart disease. By the consumption of 1.5–2 g of plant sterols and stanols, the concentration of LDL-cholesterol decreases by 9%–14%, without affecting the level of HDL and triglyceride.

12 Sensory evaluation of colza oil

It has been reported that both the presence of impurities and rapeseed's origin have an impact on the sensory characteristics and physicochemical quality of cold pressed oils. Rapeseeds with impurity content of <1% are most suitable for cold pressing. Their use ensures the production of oil with low degrees of hydrolysis and oxidation corresponding to those of refined oil. The presence of impurities exceeding 5% in the starting material for pressing have a negative effect on the sensory quality of cold pressed oil. The presence of impurities at such a level was associated with the appearance of woody, strawy, and fusty/musty off-flavors. However, no correlation between the impurity content and oxidative stability of the oils tested was found. The highest-quality cold pressed rapeseed oil was obtained by pressing the seeds originating from individual cultivars, characterized by the lowest impurity content. It seems advisable, therefore, to remove impurities from the raw material prior to pressing, in order to produce high-quality cold pressed rapeseed oil (Małgorzata et al., 2016).

13 Conclusions

In recent years, consumers have been paying more and more attention to the aspects of their life that improve its quality. Therefore, diet, in addition to the way and conditions of life, is one of the key factors influencing human health and well-being. Consumers fearing chemical remnants in the food and those who are environmentally conscious choose oils unaffected by drastic thermal treatment. Recently, an increase in the consumption of cold pressed edible oils has been observed. These oils are much more beneficial than the refined ones in terms of nutritional values. Such oils contain natural, nutritionally valuable components like tocopherols, sterols, carotenoids, and phospholipids with oxidative properties partly stripped from refined oils or destroyed during the refining process. The main reason for the increased demand for cold pressed canola oil could be the perception of improved nutritional properties compared to other common vegetable oils and highly refined canola oil. Extraction methods affected the quality and nutritional properties of canola oils by changing the fatty acid composition, tocopherols, free and total phytosterols, and total phenolic content. Cold pressed canola oils had more favorable quality indicators (lower FFA, PV, *p*-AV, chlorophyll, and almost zero phosphatides) compared to solvent-extracted and hot-pressed canola oils.

References

Abrehdari, S., Ghavami, M., Gharachorloo, M., & Delkhosh, B. (2015). Evaluation and chemical comparison of triple-zero canola cultivars with emphasis on the extracted oil. *Biological Forum—An International Journal*, *7*(1), 1037–1044.

Ambrosewicz-Walacik, M., Tańska, M., & Rotkiewicz, D. (2015). Effect of heat treatment of rapeseed and methods of oil extraction on the content of phosphorus and profile of phospholipids. *Polish Journal of Natural Science*, *30*(2), 123–136.

Australian Oilseeds Federation (AOF) (2018). *Quality standards, technical information & typical analysis*. Issue 17.

Çakaloğlu, B., Hazal Özyurt, V., & Ötleş, S. (2018). Cold press in oil extraction: A review. Food Technology *Ukrainian Food Journal*, *7*(4), 640–654.

Choo, W. S., Birch, J., & Dufour, J. P. (2007). Physicochemical and quality characteristics of cold-pressed flaxseed oils. *Journal of Food Composition and Analysis*, *20*(3–4), 202–211.

CXS 19-1981 (2019). Amended *Standard for edible fats and oils not covered by individual standards*. Codex Alimentarius Commission.

CXS 210-1999 (2019). Revised *Standard for named vegetable oil*. Codex Alimentarius Commission.

Ghazani, S. M., García-Llatas, G., & Marangoni, A. G. (2014). Micronutrient content of cold-pressed, hot-pressed, solvent extracted and RBD canola oil: Implications for nutrition and quality. *European Journal of Lipid Science and Technology*, *116*, 1–8.

Grageolaa, F., Landero, J. L., Beltranena, E. M., Cervantes, A., Araiza, R. T., & Zijlstra (2013). Energy and amino acid digestibility of expeller-pressed canola meal and cold-pressed canola cake in ileal-cannulated finishing pigs. *Animal Feed Science and Technology*, *186*, 169–176.

Leckband, G., Frauen, M., Friedt, W., & NAPUS (2002). Rapeseed (*Brassica napus*) breeding for improved human nutrition. *Food Research International*, *35*(2–3), 273–278.

Makała, H. (2015). Cold-pressed oils as functional food. In *Plant lipids science, technology, nutritional value and benefits to human health* (pp. 185–200).

Małgorzata, W., Agnieszka, R., & Katarzyna, R. (2016). Influence of impurities in raw material on sensory and physicochemical properties of cold-pressed rapeseed oil produced from conventionally and ecologically grown seeds. *Acta Scientiarum Polonorum Technologia Alimentaria*, *15*(3), 289–297.

Matthäus, B., & Brühl, L. (2003). Quality of cold-pressed edible rapeseed oil in Germany. *Food/Nahrung*, *47*(6), 413–419.

Nasab, Z., & Vanak, Z. (2015). A study on the specifications of cold pressed colza oil. *Recent Patents on Food, Nutrition & Agriculture*, *7*(1), 47–52.

NSW Department of Primary Industries (2014). *Variability of quality traits in canola seed, oil and meal—A review*: (pp. 1–26). NSW Department of Primary Industries.

Orsavova, J., Misurcova, L., Vavra, J., Vicha, R., & Mlcek, J. (2015). Fatty acids composition of vegetable oils and its contribution to dietary energy intake and dependence of cardiovascular mortality on dietary intake of fatty acids. *International Journal of Molecular Sciences*, *16*, 12,871–90.

Ratnayake, W. M., & Daun, J. K. (2004). *Chemical composition of canola and rapeseed oils. Rapeseed and canola oil: production, processing, properties and use*: (pp. 37–78). Wiley-Blackwell.

Rękasa, A., Sigerb, A., Wroniaka, M., Ścibisza, I., Derewiakac, D., & Andersd, A. (2017). Influence of de-hulled rapeseed roasting on the physicochemical composition and oxidative state of oil. *Grasas y Aceites*, *68*(1)e176.

Saleem, M., & Naveed, A. (2018). Characterization of canola oil extracted by different methods using fluorescence spectroscopy. *PLoS One*, *13*(12) e020864.

Siger, A., Gawrysiak, W., & Bartkowiak-Broda, I. (2017). Antioxidant (tocopherol and canolol) content in rapeseed oil: obtained from roasted yellow-seeded *Brassica napus*. *Journal of the American Oil Chemists' Society*, *94*, 37–46.

Siger, A., Józefiak, M., & Górnaś, P. (2017). Cold-pressed and hot-pressed rapeseed oil: The effects of roasting and seed moisture on the antioxidant activity, canolol, and tocopherol level. *Acta Scientiarum Polonorum Technologia Alimentaria*, *16*(1), 69–81.

Teh, S. S., & Birch, J. (2013). Physicochemical and quality characteristics of cold-pressed hemp, flax and canola seed oils. *Journal of Food Composition and Analysis*, *30*(1), 26–31.

Yang, M., Zheng, C., Zhou, Q., Huang, F., Liu, C., & Wang, H. (2013). Minor components and oxidative stability of cold-pressed oil from rapeseed cultivars in China. *Journal of Food Composition and Analysis*, *29*(1), 1–9.

Chapter 39

Cold pressed capia pepper (*Capsicum annuum* L.) seed oil

Nazish Jahan[a] and Khalil-ur-Rahman[b]

[a]Department of Chemistry, University of Agriculture, Faisalabad, Pakistan, [b]Department of Biochemistry, University of Agriculture, Faisalabad, Pakistan

1 Introduction

The term "cold pressed extraction" means the expeller-pressed oil at low pressures and temperatures. The cold pressed method is considered one of various green procedures to extract oil. This process is equally useful for vegetable and essential oils. The benefit of cold pressing is that it ensures high purity of the resulting oil and retains the total qualities of the plant. It is essentially mechanical extraction at low temperature for the whole batch of any food, feed, or seed materials (Rassem, Nour, & Yunus, 2016). In the cold pressing method, there is no need for organic solvents or heating for extraction of oil. Therefore, this technique is considered better than many as it retains the flavor, aroma, and beneficial value-added components with unique properties in seed oils that might be lost due to chemical modifications by organic solvents, during evaporation, and through conventional solvent extraction methods. Although both solvent-extracted and cold pressed oils are used in pharmaceuticals, beauty products, and skincare products, only cold pressed oils are more favored for cosmeceuticals due to being safer in nature and the presence of bioactive compounds (Sanati, Razavi, & Hosseinzadeh, 2018).

The procedure for extracting oil through cold pressing involves grinding of seeds or nuts materials and expelling out the oil. The seed materials are placed inside a container with a moving screw. This screw is used for grinding and crushing the seeds and ultimately for extracting the oil. The bottom of the cylinder contains small orifices, which allow the oil to exit into the collection container. During this process, a small amount of heat through friction is produced, but this is not sufficient to affect the quality of oil in terms of aroma, flavor, or nutritional contents. In the case of hot pressing, seeds are heated inside the cylinders to make them soft so that they can be broken down easily. During this process, a greater quantity of oil is extracted, but the heat may reduce the flavor, aroma, and nutritional components of oils. Olive, sesame, sunflower, canola, and coconut oil for cooking purposes can easily be extracted using the cold pressing procedure. The almond, jojoba, and rosehip seed oils extracted through cold pressing are used in aromatherapy and cosmetic products. The oils extracted from flaxseed and *Oenothera biennis* are available as nutritional supplements, while neem oil is used in biopesticides. Cold pressed neem oil protects crops from insects without any residual and harmful effect on plants and the ecosystem. Commercially available cold pressed oil also contains higher quantities of antioxidants. The contribution of cold pressed oils to a healthier life is due to its cholesterol-free nature and presence of tocopherols, vitamin C, and phosphates (such as lecithin). In these oils, there are no added chemicals and preservatives. These oils are never deodorized, refined, or processed in any way during extraction. Therefore, cold pressed oils are considered superior due to their potential health benefits.

2 Capia pepper (*Capsicum annuum* L.)

Capia pepper (Chili) is one of the oldest crops, and belongs to the plant genus *Capsicum* and family Solanaceae. Worldwide, more than 20 varieties of pepper belonging to the genus *Capsicum* are available, such as bell peppers, jalapeños, New Mexico Chile, and cayenne peppers with different taste, colors, sizes, shapes, and pungencies. *Capsicum annuum* L. is the most commonly cultivated species (Bosland & Votava, 2000). Fresh pepper fruits are an excellent source of vitamins C, E, and B6, carotenoids, carbohydrates, phenols, flavonoids, and pungent components as well as minerals like potassium, sodium, calcium, iron, copper, and manganese (Pundir, Rani, Tyagi, & Pundir, 2016). Pepper can be eaten raw (salads) or cooked (baked dishes, salsa, pizza), and used as a spice in dried and powdered form (the fruits with seeds are usually used),

Cold Pressed Oils. https://doi.org/10.1016/B978-0-12-818188-1.00039-6
Copyright © 2020 Elsevier Inc. All rights reserved.

or processed into oleoresins (Aybak, 2002). Peppers are mostly consumed as green vegetables and in powdered form as a flavoring spice. Moreover, chilies are utilized in many processed food products like sauces, salad dressings, pickles, meat, and bakery items (Yılmaz, Arsunar, Aydeniz, & Güneşer, 2015). The shape of pepper seeds is planar, disc-like with a deep chalazal depression. The color of seeds is yellow but depending upon various types, brown to black seeds can also occur. The seeds of capia pepper contain many essential fatty acids and rich in vitamins, nutrients and bioactive phytochemicals (Bosland & Votava, 2000; Yılmaz et al., 2015).

3 Composition of *C. annuum* seeds

Proximate analysis of any dietary material is actually a set of methods that used to collect information about macronutrients of the products. It has become mandatory almost for all food/feed products to have some standard nutritional labeling. Firstly, this enables consumers to be aware of the nutritional composition of their diet and make considered decisions about it, and secondly, it allows fair market competition. Proximate analysis of several varieties of pepper is given in Table 1. Proximate nutritional composition may also include amino acids, minerals, fatty acids, vitamins, and heavy metal contents of *C. annuum*. These are actually functional properties of any nutritional commodity. The amino acids and minerals found in the seeds of capia pepper are given in Table 2. The frequency distribution of amino acids in *C. annuum* indicated that glutamic acid was found in highest quantity while isoleucine was the lowest amino acid. The oil content of seeds varied from 13% to 27% in different Greek and Indian varieties of Capsicum and are averaged to be 24%. The frequency distribution of oil varies according to different varieties cultivated under different conditions (Jarret, Levy, Potter, & Cermak, 2013).

4 Fatty acid composition and bioactive components of cold pressed paprika seed oil

Research works about hot pepper seed oil are relatively limited, and even many common quality characteristics of the oil are unidentified. Although efforts have been made to extract the pepper seed oils by organic solvent and supercritical solvent extraction techniques, information about extraction of oil through cold press is very limited.

The saturated fatty acids in the seed of *C. annuum* range from 12% to 18.5% and total unsaturated fatty acids from 82% to 87.7% (Zaki, Hasib, Hakmaoui, Dehbi, & Ouatmane, 2013). The fatty acid composition depends upon the varieties of pepper and extraction procedure adopted. The average fatty acid composition of pepper seed oil is 7.5%–13.0% palmitic acid; 8.5%–14.0% oleic acid, 72%–80% linoleic, and 3%–4% stearic acid (Yılmaz et al., 2015, Koncsek, Helyes, & Daood, 2017). The significance of linolenic acid in human nutrition is well documented, and insufficiency in food leads to improper growth, reproductive disability, skin injuries, and many other symptoms. The estimated oil contents of various varieties of *Capsicum* including *C. tovarii*, *C. flexuosum*, *C. galapagoense*, *C. eximium*, *T. anomalum*, and *C. annuum* are 18%, 17%, 11%, 25%, 21%, and 26%, respectively. The range in percentage (g/100 g) of fatty acid naming palmitic acid, linolenic acid, stearic acid, and oleic acid in these varieties are 4–9, 71–80, 2–4, and 6–12, respectively (Jarret et al., 2013).

TABLE 1 Proximate analysis (g/100 g) of *Capsicum annuum* seeds.

Parameter	Hot pepper	Sweat pepper
Moisture	4–6	70.59
Dried extract meal	77	29
Oil	26	–
Ash	4.5–5.5	4.88
Crude fat	23–24	19.57
Crude protein	20.5–29	19.28
Total dietary fiber	29–39	–
Nitrogen free extract	36.37	–
Carbohydrate	–	56.28

TABLE 2 Amino acid and mineral composition (g/100g) of hot pepper seeds.

Alanine	0.63–0.71
Arginine	0.96–1.06
Aspartic acid	1.96–2.1
Cysteine	0.91–1.03
Glutamic acid	3.34–3.56
Glycine	0.78–0.86
Histidine	1.55–1.69
Isoleucine	0.46–0.50
Leucine	0.86–0.96
Lysine	0.97–1.06
Methionine	0.79–0.85
Phenylalanine	1.18–1.34
Proline	0.65–0.71
Serine	0.81–0.91
Threonine	0.76–0.86
Tryptophan	0.83–0.91
Tyrosine	0.76–0.84
Valine	0.47–0.51
Calcium	171.7–177.6
Iron	17.24–17.74
Magnesium	233.9–241.2
Manganese	2.11–2.21
Potassium	640.6–659.5
Sodium	12.26–12.50
Zinc	7.84–8.1

Capsicum annuum also contains myristic acid (0.21%), linolenic acid (0.35%), lauric acid (0.145%), and arachidonic acid (0.09%) (Azabau, Taheur, & Jridi, 2017, Zou, Ma, & Tian, 2015).

Paprika seed oils extracted through cold pressing are transparent with a bright red color. This sharp color is due to a considerable amount of carotenoids. In addition to β-carotene (free and mono- or di-esters), lutein, capsantin, and capsorubin are also the primary carotenoids of pepper seed oil. The fatty acid (FA) profile indicates the high amount 69%–75% of *cis*-linoleic acid. The most abundant saturated FA present in oil is palmitic acid (11%–12%). The prevalence of stearic acid is only 3%–4%. A very high amount of γ-tocopherol (58–84 mg/100 g) is present in paprika seed oils, whereas the level of α-tocopherol is very low (5%–15%) (Koncsek et al., 2017).

The physicochemical properties of the oils are summarized as free fatty acidity 0.71%–4.74%, pour point −6 to −9°C, viscosity 25.1–25.6 at 40 °C, acid value 1.43–9.44 mg/g, cloud point 16–17°C, and oxidation onset stability 153.54–156.08°C (Bosland & Votava, 2000; Yılmaz et al., 2015). Cold pressed oil is attracting more attention due to its technological and nutritional importance. Owing to high antioxidant levels in pepper seed oils, it has also become a sought-after product in the cosmetics and pharmaceuticals industries. The daily minimum consumption of pepper seed oil (10 g) can cover the daily requirement of 70%–74% of linoleic acid, 3.7%–13.6% of vitamin E (α-tocopherol), and 10.3%–14.2% of vitamin A (retinol equivalent) (Koncsek et al., 2017). The high nutritional values of oils extracted through cold pressing have made them part of a modern nutrition program. The presence of valuable bioactive phytochemicals like γ-tocopherol, lutein, and

TABLE 3 Fatty acids (g/100 g) in different *capsicum* varieties.

Variety of capsicum	Palmitic acid	Linoleic acid	Stearic acid	Oleic acid
Capsicum tovarii	13.0	76.0	2.1	6.3
Capsicum flexuosum	13.0	73.0	2.6	8.3
Capsicum galapagoense	14.0	71.0	3.3	8.5
Capsicum eximium	14.0	75.0	2.7	6.5
Tubocapsicum anomalum	9.0	80.0	2.2	6.3
Capsicum annum	11.2	70.9	3.54	12.18

TABLE 4 Fatty acid profile of cold pressed *Capsicum annuum* seed oil (g/100 g).

Fatty acid	Symbol	g/100 g
Lauric acid	C12:0	0.05–0.08
Myristic acid	C14:0	0.07–0.27
Palmitic acid	C16:0	11.0–12.1
Palmitoleic acid	C16:1	0.18–0.26
Stearic acid	C18:0	3.2–2–4
cis-Oleic acid (n9)	C18:1	7.90–11.1
cis-Linoleic acid (n6)	C18:2	70.7–74.3
Arachidic acid	C20:0	0.40–0.48
Eicosenoic acid	C20:1	0.11–0.54
α-Linolenic acid (n3)	C18:3	0.20–0.73
11,14-Eicosadienoic acid	C20:2	0.03–0.07
Behenic acid	C22:0	0.34–0.48
Arachidonic acid (n6)	C20:4	0.03–0.5
Lignoceric acid	C24:0	0.34–0.44
Total saturated		15.5–17.2
Total monounsaturated		8.5–10.1
Total polyunsaturated		71.2–74.8

zeaxanthin has made cold pressed oil more suitable for its protective effect against age-related macular degeneration. Fatty acid compositions of cold pressed paprika seed oils are given in Tables 3 and 4.

5 Bioactive compounds of pepper

Various parts of peppers, including seeds, are considered waste materials and removed during processing for marketing. Different waste parts like the stem, placenta, and seeds of the pepper are consumed as animal fodder and are not used as human food. The seeds of *C. annuum* are discarded from pepper during processing as a by-product and can potentially be used as high a value-added foodstuff or nutraceutical product. The seeds contain essential fatty acids, nutrients, and other bioactive compounds that are required for human health (Jarret et al., 2013). *C. annuum* is an excellent source of such compounds, which are nutrients, antioxidants, and biofunctional components including different flavonoids, polyphenols, carotenoids, and various vitamins like ascorbic acid, tocopherols, and retinol/al (Mendoza, Sanchez, Marquez, Arreola, & Cordova, 2015; Parrilla, Rosa, Amarowicz, & Shahidi, 2011; Yang et al., 2010). The paramount phytoactive components available in peppers like alkamide, flavonoids, and saponin glycosides are attributed to free radical scavenging and anti-inflammatory potential with other aspects of pharmacological importance. The pungency in hot peppers is due to capsaicinoids, while sweet peppers contain nonpungent compound capsinoids. Capsinoids are reported to possess antiinflammatory potential and accelerated basal metabolic rate (BMR), and also enhance the body's temperature by suppressing fat anabolic pathways. The structures of some important bioactive capsinoids present in various parts of *C. annuum* are presented in Fig. 1 The nonpungent nature of capsinoids makes them appropriate additives in food and pharmaceutical preparation. The culinary properties and presence of abundant antioxidants not only impart nutrition, but also make pepper seeds suitable for pharmacological products (Ezekiell & Oluwole, 2014).

The GC–MS spectroscopic analysis of organic solvent extracts reported the presence of variety of volatile biologically active components, like 3-carene ($C_{10}H_{16}$), hexadecane ($C_{16}H_{34}$), octadecane ($C_{18}H_{38}$), elicosane ($C_{20}H_{42}$), 10-heneicosene ($C_{21}H_{42}$), decosane ($C_{19}H_{34}O_2$), 9,12-octadecadienoic acid methyl ester ($C_{19}H_{34}O_2$), hexadecanoic acid ($C_{16}H_{32}O_2$), 1-decosene ($C_{22}H_{44}$), 9,12 octadecadienal ($C_{18}H_{32}O$), heptadec-8-ene 2,4-dione ($C_{17}H_{30}O_2$), tetracosane ($C_{24}H_{50}$), 2(3H) furanone, dihydro 5 (2-octenyl) ($C_{12}H_{20}O_2$), hexadec-8-ene 2,4 dione ($C_{16}H_{28}O_2$), pentadec-8-ene 2,4 dione ($C_{15}H_{26}O_2$), 1-phenyloctane ($C_{14}H_{22}$), phenol,2,4 bis (1,1-dimethylethyl) C14H22O, octadecane (C18H38), 1 hexadecen ($C_{16}H_{32}$), elicosane ($C_{20}H_{42}$), 5 eicosene ($C_{20}H_{40}$), hexadecanoic acid ethyl ester ($C_{18}H_{36}O_2$), decosane ($C_{22}H_{46}$), 1 dodecene ($C_{12}H_{24}$), 9 hexadecenoic acid methyl ester ($C_{17}H_{32}O_2$), and 9,12-octadecadienoic acid methyl ester ($C_{19}H_{34}O_2$) (Gurnania, Guptab, Mehtaa, & Mehta, 2016; Kevresan, Hrabovski, Kuhajda, Mimica-Dukić, & Sakac, 2009).

Paprika seed oil is transparent with a shiny red color, due to the presence of carotenoids (630–850 µg/g). The main carotenoids present are capsantin, lutein, capsorubin ($C_{40}H_{56}$), and β-carotene as free xanthophylls, or mono- or di-esters. γ-tocopherol (57.8–83.5 mg/100 g) is the main tocopherol, while the amount of α-tocopherol is relatively low (4.50–16.4 mg/100 g) in paprika seed oil (Koncsek et al., 2017). Two important capsicosides, A and G, are isolated from the ethanol extract of pepper seeds (Fig. 2).

The seeds of *C. annuum* also possess three novel furostanol saponins named capsicoside E, capsicoside F, and capsicoside G with other oligoglycosides. Through chemical and spectroscopic analysis, the structures of capsicoside E

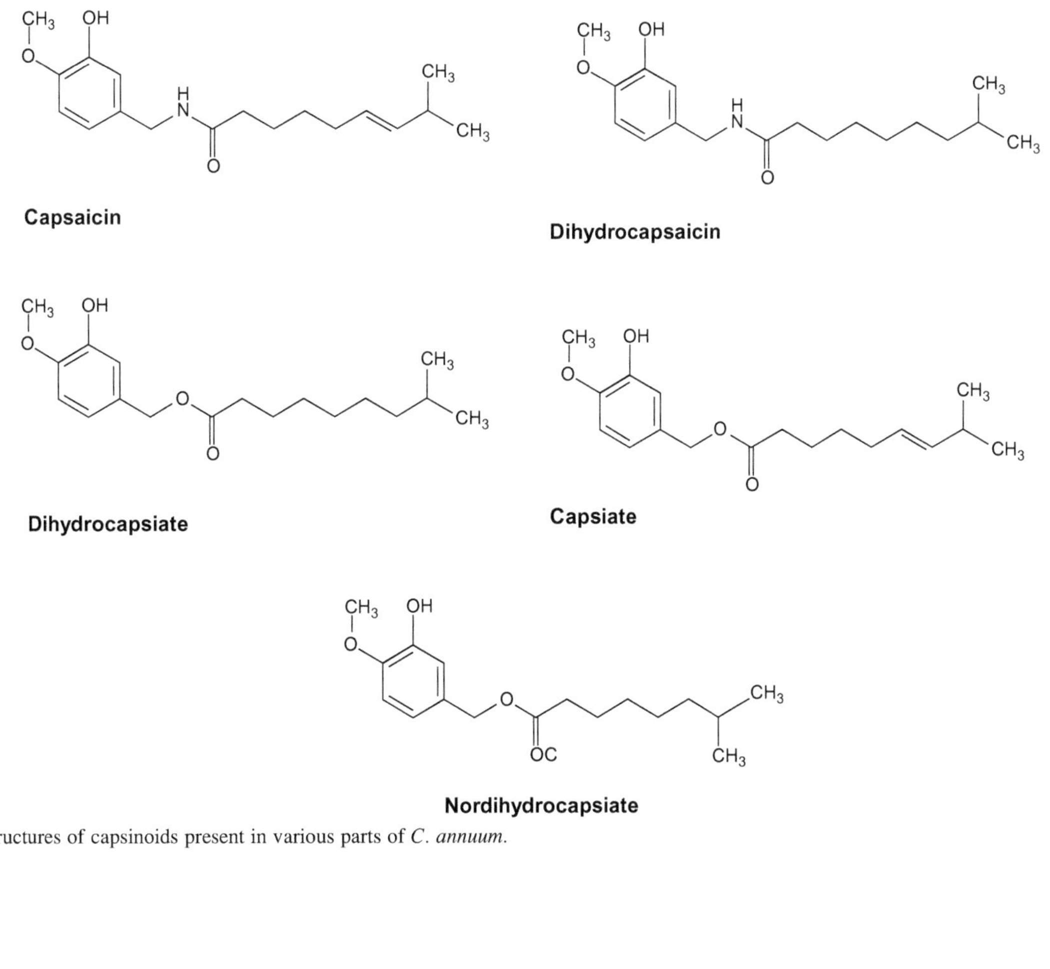

Capsaicin

Dihydrocapsaicin

Dihydrocapsiate

Capsiate

Nordihydrocapsiate

FIG. 1 Structures of capsinoids present in various parts of *C. annuum*.

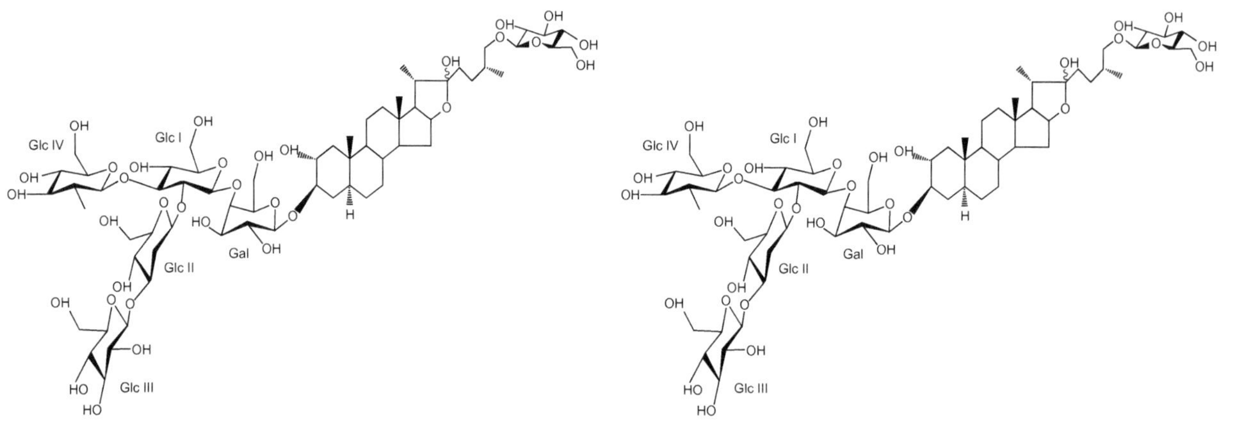

Capsicoside A Capsicoside G

FIG. 2 Structure of capsicoside A and G isolated from ethanol extract of pepper seeds.

oligoglycosides is elucidated as 26-*O*-β-D-glucopyranosyl-22-*O*-methyl-5α-furost-25 (27)-en-2α,3β,22ξ,26-tetraol-3-*O*-β-D-glucopyranosyl (1→3)-β-D-glucopyranosyl (1→2)-[β-D-glucopyranosyl (1→3)]-β-D-glucopyranosyl (1→4)-β-galactopyranoside. The structure of capsicoside F is 26-*O*-β-D-glucopyranosyl-(25*R*)-5α-furost-20 (22)-en-2α,3β, 26-triol-3-*O*-β-D-glucopyranosyl (1→3)-β-D-glucopyranosyl (1→2)-[β-D-glucopyranosyl (1→3)]-β-D-glucopyranosyl (1→4)-β-D-galactopyranoside. The structure capsicoside G is 26-*O*-β-D-gluco-pyranosyl-(25*R*)-5α-furosta-3β,22ξ,26-triol-3-*O*-β-D-glucopyranosyl (1→3)-β-D-glucopyranosyl (1→2)-[β-D-glucopyranosyl (1→3)]-β-D-glucopyranosyl (1→4)-β-D-galactopyranoside (Iorizzi, Lanzotti, Ranalli, De Marino, & Zollo, 2002). It may therefore be concluded that the seeds of *C. annuum* are an excellent source of biologically active compounds like alkamide, carotenoids, flavonoids, phenolic acids, and important vitamins. These phytoprotectants possess well-established free radical scavengers (Harvell & Bosland, 1997) with antiinflammatory (Sanati et al., 2018), antiobesity (Sung et al., 2016), antineoplastic, and antinociceptive effects (Amruthraj, Raj, Saravanan, & Lebel, 2014).

6 Health-promoting potential of red pepper (*Capsicum annuum*) seeds

C. annuum assumes an extraordinary role in human health because our body is unable to produce a variety of healthy essential compounds, which this plant possesses. The high contents of phytonutrients actually made pepper seed oil the best choice in recent diet programs as salad dressings and frying or cooking oil in food manufacturing units (Yang et al., 2010).

7 Antiobesity effect

Obesity has become a common metabolic syndrome, and it is considered globally a serious threat to human health. Owing to the severity of this problem, interest is growing in finding a green solution to obesity. It was reported that bioactive compounds from seeds of pepper like capsicoside G reduced diet-induced obesity in mice. Epididymal adipose tissue weight and adipocyte hypertrophy remained significantly lower in experimental mice treated with a capsinoid-rich fraction than in mice administered with a high-fat diet. The hepatic fat deposition is lower in mice fed on a capsicoside G and high-fat diet than mice fed with a high-fat diet only. The obesity-related abnormalities treated with capsicoside can reverse the fat-induced raised expression of adipocyte differentiation regulators, counting peroxisome proliferator-stimulated receptor γ, sterol regulatory binding protein, and their respective target genes. It has been proved that the capsicoside-rich fraction from the seed of pepper can be successfully used as dietary therapy for the management of obesity and its related metabolic ailments (Sung et al., 2016).

8 Antimicrobial activity

The bioactive compounds of capsicum have an inhibitory effect on bacterial growth. The important bioactive alkamides in *capsicum* are affinin and capsaicin, which impart antimicrobial activity against *P. solanacearum*, *E. coli*, *S. cerevisiae*, and *B. subtilis*. The affinin can inhibit the growth of *E. coli*, and *S. cerevisiae* at low concentration (25 μg/mL). The growth of *B.*

TABLE 5 Antimicrobial potential of *C. annuum* various parts.

	Zone of inhibition (mm)					
Microorganism	Callus	Leaf	Shoot	Fruit	Seed	Standard
E. coli	13	10	9	14	6	16
Klebsiella pneumoniae	11	6	9	15	8	16
Salmonella typhi	18	9	16	11	6	16
Staphylococcus aureus	19	15	15	21	9	18
Streptococcus pyogenes	15	20	19	8	12	18
Bacillus cereus	9	11	15	9	8	15
Aspergillus flavus	6	13	15	–	13	17
Candida albicans	5	6	10	8	12	21

subtilis and *P. solanacearum* was inhibited with high concentration of amides. The capsaicin can inhibit the growth of microbes at high concentration. In addition to natural capsaicin, various synthetic capsaicinoids derivatives are also effective antimicrobial agents (Molina-Torres, Garcia, & Ramirez, 1999). Antimicrobial activity of chili pepper is not bound to any part of the plant, but almost all parts exhibited antimicrobial activities as shown in Table 5 (Gurnania et al., 2016; Omolo et al., 2014; Santos et al., 2012). Many procedures are available to check the antimicrobial potential of capsicum species. The findings of different procedures may vary for many reasons, including inconsistency of analysis and plant materials. Most of the antimicrobial activities have been studied in vitro, and there is a significant need to conduct more and more in vivo experiments to ensure the antimicrobial activities of various bioactive molecules of chili pepper in the human body.

9 Antioxidant potential

The pungency of hot pepper is because of the presence of capsaicinoids, containing branched-chain fatty acids of C9 to 11. The most abundant capsaicinoids of chili peppers are capsaicin (E-N-4-hydroxy-3-methoxybenzyl-8-methyl-6-nonenamide) and dihydrocapsaicin (6,7-dihydro derivative of capsaicin), which account for about 90% of the total capsaicinoids. Chili pepper, being a rich source of vitamins and flavonoids like quercetin, luteolin, and carotenoids, has high antioxidant capacity (Caporaso, Paduano, Nicoletti, & Sacchi, 2013; Mansor, Kalita, Bartolo, & Jayanty, 2019; Mendoza et al., 2015). The most common assays for determination of antiradical potential are 1-1-diphenyl 2-picryl hydrazine (DPPH), 2,2′-azino-bis(3-ethylbenzothiazoline-6-sulphonic acid) ABTS, ferric reducing antioxidant power (FRAP), lipid peroxidation (LPO), and reducing power (RP) (Gurnania et al., 2016). The free radical scavenging potential of mixture of polyphenols in the seed oil of *C. annuum* is increased in a dose-dependent manner (Bae, Jayaprakashaa, Jifon, & Patil, 2012; Silva, Azevedo, Pereira, Valentão, & Andrade, 2013; Sim & Sil, 2008). The antioxidant potential determined through all the abovementioned assays evidenced a similar pattern of results. Pepper seed oil has significant antioxidant potential due to the presence of multiple antioxidants. There is a positive synergism toward the antioxidant potential, which ultimately has beneficial potential in human health. The 28 bioactive polyphenol components in three different-varieties of *C. annuum* are derivatives of cinnamic acid like feruloyl hexoside, sinapoyl hexoside, and feruloyl hexoside, as well as five flavonoid components including vicenin-2, orientin, isoscoparin, quercetin 3-*O*-hexoside, and luteolin malonylpentosyldihexoside (Sotto et al., 2018). The contents of important antioxidants, capsaicinoids, are affected by variables factors such as agricultural practices, amount of water, temperature, plant food, and length of exposure to light (Harvell & Bosland, 1997).

10 Anticancer potential

The human diet may contain a diversity of natural mutagens and cancer-causing agents, but at the same time it also contains various anticarcinogens and antimutagens that perform important roles to decrease or eradicate the effects of carcinogenic agents. Owing to the variety and high amounts of different anticancer compounds like dietary fiber, retinal/ol, ascorbic acid, tocopherol, and Se, Ca, carotenoids, and polyphenols, almost no human diet is free of antimutagenic or anticarcinogenic agents. These green antimutagenic factors exhibited excellent potential by inhibiting carcinogenicity and enhancing detoxification against carcinogens, neutralizing DNA reactive mediators, destroying any irregular proliferation of early preneoplastic lesions, and changing the nature of the tumor cells (Amruthraj et al., 2014). Bioactive capsaicin in *Capsicum annuum* exhibited significant potential in terms of anticarcinogenic and antimutagenic activities (Clark & Lee, 2016).

11 Antidiabetic potential

The extracts and seed oil of peppers significantly reduce the α-amylase activity. This amylase inhibition potential of the pepper seed oil was in a concentration-dependent manner. Both extracts of pericarp and seeds are the most active α-amylase inhibitory potential and may inhibit about 85%–86% enzyme activity at 500 μg/mL (Sotto et al., 2018). The aqueous extracts of yellow, green, and red long peppers can significantly inhibit α-glucosidase with very weak or no inhibitory potential against α-amylase. The lipophilic fractions obtained from different varieties of capsicum are very potent α-amylase inhibitors (Sotto et al., 2018). High temperatures during extraction of oils or drying and frying may reduce the antidiabetic potential of hot pepper extracts; therefore, extract obtained through cold pressing contain high amount of nutraceutical components of chilli peppers. The phenolics and carotenoids are actually potential bioactive candidates which inhibit the activity of α-amylase and α-glucosidase to ultimately control the hyperglycemia. In addition, the antidiabetic potential of pepper is linked with the presence of various flavonoids (Vinayagam & Xu, 2015).

12 Other health benefits

Due to the presence of a galaxy of important antioxidants, chili pepper has the potential to decrease the prevalence of thrombosis. It may also reduce hyperglycemia along with antithrombotic activity. The capsaicin analogs can inhibit the aggregation of platelets (Ezekiell & Oluwole, 2014). Metal toxicity due to environmental pollution has become very common in human beings. The bioactive molecules present in chili pepper seed oil impart many other physiological and pharmacological effects including analgesia, antiinflammatory, antiaging, and cardioprotective (Sanati et al., 2018).

References

Amruthraj, N. J., Raj, J. P. P., Saravanan, S., & Lebel, L. A. (2014). *In vitro* studies on anticancer activity of capsaicinoids from *Capsicum chinense* against human hepatocellular carcinoma cells. *International Journal of Pharmacy and Pharmaceutical Sciences*, *6*(4), 254–558.

Aybak, H. (2002). *Biber Yetiştiriciliği (Pepper cultivation)*. Istanbul (Turkey): Hasad Publishing Co.

Azabau, S., Taheur, F. B., & Jridi, M. (2017). Discarded seeds from red pepper (*Capsicum annum*) processing industry as a sustainable source of high added-value compounds and edible oil. *Environmental Science and Pollution Research International*, *24*(28), 22196–22203.

Bae, H., Jayaprakashaa, G. K., Jifon, J., & Patil, B. S. (2012). Variation of antioxidant activity and the levels of bioactive compounds in lipophilic and hydrophilic extracts from hot pepper (*Capsicum* spp.) cultivars. *Food Chemistry*, *134*, 1912–1918.

Bosland, P. W., & Votava, E. J. (2000). *Peppers: Vegetable and spice capsicums*. New York (USA): CABI Publishing.

Caporaso, N., Paduano, A., Nicoletti, G., & Sacchi, R. (2013). Capsaicinoids, antioxidant activity, and volatile compounds in olive oil flavored with dried chili pepper (*Capsicum annuum*). *European Journal of Lipid Science and Technology*, *115*, 1434–1442.

Clark, R., & Lee, S. H. (2016). Anticancer properties of capsaicin against human cancer. *Anticancer Research*, *36*, 837–843.

Ezekiell, J. A. T., & Oluwole, O. J. A. (2014). Effects of capsaicin on coagulation: Will this be the new blood thinner. *Clinical Medicine & Research*, *3*, 145–149.

Gurnania, N., Guptab, M., Mehtaa, D., & Mehta, B. K. (2016). Chemical composition, total phenolic and flavonoid contents, and *in vitro* antimicrobial and antioxidant activities of crude extracts from red chilli seeds (*Capsicum frutescens* L.). *Journal of Taibah University for Science*, *10*, 462–470.

Harvell, K. P., & Bosland, P. W. (1997). The environment produces a significant effect on pungency of chiles. *HortScience*, *32*, 1292.

Iorizzi, M., Lanzotti, V., Ranalli, G., De Marino, S., & Zollo, F. (2002). Antimicrobial furostanol saponins from the seeds of *Capsicum annuum* L. var. *acuminatum*. *Journal of Agricultural and Food Chemistry*, *50*, 4310–4316.

Jarret, R. L., Levy, I. J., Potter, T. L., & Cermak, S. C. (2013). Seed oil and fatty acid composition in *Capsicum* spp. *Journal of Food Composition and Analysis*, *30*, 102–108.

Kevresan, Z. S., Hrabovski, N. C., Kuhajda, K. N., Mimica-Dukić, N., & Sakac, M. B. (2009). Essential oil composition of fresh and dried pepper fruits (*Capsicum annuum* L.). *Food Processing, Quality and Safety*, *36*(12), 36–39.

Koncsek, A., Helyes, L. D., & Daood, H. (2017). Bioactive compounds of cold pressed spice paprika seeds oils. *Journal of Food Processing and Preservation*, *42*(3), e13403.

Mansor, H., Kalita, D., Bartolo, M. E., & Jayanty, S. S. (2019). Capsaicinoids, polyphenols and antioxidant activities of *Capsicum annuum*: Comparative study of the effect of ripening stage and cooking methods. *Antioxidants*, *8*, 364.

Mendoza, C. C., Sanchez, E., Marquez, E. M., Arreola, J. P. S., & Cordova, M. A. F. (2015). Bioactive compounds and antioxidant activity in different grafted varieties of bell pepper. *Antioxidants*, *4*, 427–446.

Molina-Torres, J., Garcia, A. C., & Ramirez, E. C. (1999). Antimicrobial properties of alkamides present in flavouring plants traditionally used in Mesoamerica: Affinin and capsaicin. *Ethnopharmacology*, *64*, 241–248.

Omolo, M. A., Wong, Z. Z., Mergen, A. K., Hastings, J. C., Le, N. C., Reiland, H. A., et al. (2014). Antimicrobial properties of chili peppers. *Journal of Infectious Diseases and Therapy*, *2*, 145.

Parrilla, E. A., Rosa, L. A. D., Amarowicz, R., & Shahidi, F. (2011). Antioxidant activity of fresh and processed Jalapeno and serrano peppers. *Journal of Agricultural and Food Chemistry*, *59*, 163–173.

Pundir, R., Rani, R., Tyagi, S., & Pundir, P. (2016). Advance review on nutritional phytochemical, pharmacological and antimicrobial properties of chili. *International Journal of Ayurveda and Pharma Research*, *4*, 53–59.

Rassem, H. H. A., Nour, A. H., & Yunus, R. M. (2016). Techniques for extraction of essential oils from plants. *Australian Journal of Basic and Applied Sciences*, *10*(16), 117–127.

Sanati, S., Razavi, B. M., & Hosseinzadeh, H. (2018). A review of the effects of *Capsicum annuum* L. and its constituent, capsaicin, in metabolic syndrome. *Iranian Journal of Basic Medical Sciences*, *21*, 439–448.

Santos, M. M. P., Vieira-da-Motta, O., Vieira, I. J. C., Braz-Filho, R., Alves, P. S. G., Maria, E. J., et al. (2012). Antibacterial activity of *Capsicum annuum* extract and synthetic capsaicinoid derivatives against *Streptococcus mutans*. *Journal of Natural Medicines*, *66*, 354–356.

Silva, L. R., Azevedo, J., Pereira, M. J., Valentão, P., & Andrade, P. B. (2013). Chemical assessment and antioxidant capacity of pepper (*Capsicum annuum* L.) seeds. *Food and Chemical Toxicology*, *53*, 240–248.

Sim, K. H., & Sil, H. Y. (2008). Antioxidant activities of red pepper (*Capsicum annuum*) pericarp and seed extracts. *International Journal of Food Science & Technology*, *43*, 1813–1823.

Sotto, A. D., Vecchiato, M., Abetea, L., Toniolob, C., Giustic, A. M., Manninad, L., et al. (2018). *Capsicum annuum* L. var. Cornetto di Pontecorvo PDO: Polyphenolic profile and in vitro biological activities. *Journal of Functional Foods*, *40*, 679–691.

Sung, J., Yang, J., Kim, Y., Kim, M., Jeong, H. S., & Lee, J. (2016). Effect of defatted pepper (*Capsicum annuum* L.) seed extracts on high-fat diet-induced obesity in C57BL/6J mice. *Food Science and Biotechnology*, *25*(5), 1457–1461.

Vinayagam, R., & Xu, B. (2015). Antidiabetic properties of dietary flavonoids: A cellular mechanism review. *Nutrition and Metabolism*, *12*, 60.

Yang, C. Y., Mandal, P. K., Han, K. H., Fukushima, M., Choi, K., Kim, C. J., et al. (2010). Capsaicin and tocopherol in red pepper seed oil enhances the thermal oxidative stability during frying. *Journal of Food Science and Technology*, *47*, 162–165.

Yılmaz, E., Arsunar, E. S., Aydeniz, B., & Güneşer, O. (2015). Cold pressed capia pepper seed (*Capsicum annuum* L.) oils: Composition, aroma, and sensory properties. *European Journal of Lipid Science and Technology*, *117*, 1016–1026.

Zaki, N., Hasib, A., Hakmaoui, A., Dehbi, F., & Ouatmane, A. (2013). Assessment of color, capsaicinoids, carotenoids and fatty acids composition of paprika produced from moroccan pepper cultivars (*Capsicum Annuum* L.). *Journal of Natural Science Research*, *3*, 111.

Zou, Y., Ma, K., & Tian, M. (2015). Chemical composition and nutritive value of hot pepper seed (*Capsicum annuum*) grown in northeast region of China. *Journal of Food Science and Technology*, *35*, 659–663.

Further reading

Castro, C. J. S., Morales, M. V., Oomah, B. D., Dorado, R. G., Godoy, S. M., & Alonso, L. G. E. (2017). Bioactive compounds and antioxidant activity in scalded Jalapeño pepper industrial byproduct (*Capsicum annuum*). *Journal of Food Science and Technology*, *54*, 1999–2010.

Embaby, H. E., & Mokhtar, S. M. (2011). Chemical composition and nutritive value of lantana and sweet pepper seeds and nabak seed kernels. *Journal of Food Science*, *76*, 736–741.

Manninad, L., Locatellie, M., Nicolettib, M., & Giacomoa, S. D. (2018). *Capsicum annuum* L. var. Cornetto di Pontecorvo PDO: Polyphenolic profile and in vitro biological activities. *Journal of Functional Foods*, *40*, 679–691.

Sottoa, A. D., Vecchiatoa, M., Abetea, L., Toniolob, C., Giustic, A. M., Wesołowska, A., et al. (2015). GC-MS analysis of essential oils isolated from fruits of chosen hot pepper (*Capsicum annuum* L.) cultivars. *Folia pomeranae universitatis technologiae stetinensis. agricultura, alimentaria, piscaria et zootechnica*, *320*, 95–108.

Chapter 40

Cold pressed tomato (*Lycopersicon esculentum* L.) seed oil

Zinar Pinar Gumus[a], Zeliha Ustun Argon[b,c], Veysel Umut Celenk[d], and Suna Timur[e]

[a]*Central Research Testing and Analysis Laboratory Research and Application Center, EGE-MATAL, Ege University, İzmir, Turkey,* [b]*Department of Biosystems Engineering, Eregli Faculty of Engineering and Natural Sciences, Necmettin Erbakan University, Konya, Turkey,* [c]*Medical and Cosmetic Plants Application and Research Center, Necmettin Erbakan University, Konya, Turkey,* [d]*Central Research Testing and Analysis Laboratory Research and Application Center, EGE-MATAL, Ege University, İzmir, Turkey,* [e]*Department of Biochemistry, Faculty of Science, Ege University, İzmir, Turkey*

Abbreviations

FA fatty acid
LDL low-density lipoprotein
PUFA polyunsaturated fatty acid
QDA quantitative descriptive analysis
QDA quantitative descriptive analysis
SFE supercritical fluid extraction
TSO tomato seed oil

1 Introduction

Tomatoes (*Lycopersicum esculentum*) are one of the most common agricultural crops used for human nutrition around the world (Bhowmik, Kumar, Paswan, & Srivastava, 2012; Khachik et al., 2002; Shao et al., 2015). Tomatoes are in great demand due to their nutritional value, while tomatoes are the basis of many processed products, such as tomato extract, juice, or ketchup. Therefore, tomatoes are consumed both fresh and processed. The processing of tomatoes produces a large amount of tomato pomace. Tomato pomace consists mainly of skins, seeds, fibrous substances, and cull tomatoes, wherein seeds make up 60% of the waste (Westphal, Bauerfeind, Rohrer, & Böhm, 2014. Therefore, the recycling of wastes after industrial tomato processing is of great importance for agriculture and industry. In addition, the increase of by-products and wastes from the plant processing industry is an important ecological and financial problem (da Silva & Jorge, 2014). The oils obtained using seeds and pulps, which are the main components of these wastes, are used for both food and industrial purposes (Botinestean, Gruia, & Jianu, 2014; Botinestean & Jianu, 2013; Kalogeropoulos, Chiou, Pyriochou, Peristeraki, & Karathanos, 2012).

The tomato processing industry generates high amounts of waste (Botinestean & Jianu, 2013). Tomato pomace has no commercial value and is currently disposed of as a solid waste or used for animal feed. However, a careful examination of this material reveals that it is a rich source of important nutrients and phytochemicals. In particular, tomato seeds contain oil of high nutritional quality, and significant amounts of carotenoids are present in the skins (Eller, Moser, Kenar, & Taylor, 2010; Khachik et al., 2002). The increase of waste quantities from the tomato processing industry is an important ecological and financial problem (Westphal et al., 2014). Tomato seed oil (TSO) is considered a good source of edible oil, and one of the major food ingredients across the world. It can be extracted from tomato seed, which is the major by-product of the tomato paste manufacturing industry. In order to make full use of TSO and to improve its economic value, it is very important to perform a full analysis of the oil (Ma et al., 2014).

Tomato wastes are a rich source of bioactive compounds like antioxidants (Machmudah et al., 2012; Westphal et al., 2014; Yilmaz, Aydeniz, Güneser, & Arsunar, 2015). Pure components derived from plant extracts, oils, or plants exhibit activity on a very broad scale, such as anticancer, antioxidant, antibacterial, antifungal, and wound healing (Evans & Johnson, 2010; Franko, Van De Bovenkamp, & Bicanic, 1998; Shao et al., 2013). The recovery of vegetable and fruit by-products ensures that certain bioactive compounds are applied in the health food, medical, cosmetics, and

Cold Pressed Oils. https://doi.org/10.1016/B978-0-12-818188-1.00040-2
Copyright © 2020 Elsevier Inc. All rights reserved.

pharmaceuticals industries. This is necessary to ensure the recovery of waste, increase benefit, and decrease pollution risk for the environment.

Extraction of TSO is of great importance. One of the extraction systems that maintains the properties of the components of oil is the cold pressing process, which also provides protection of the bioactive substances in the oil (Kiralan, Özkan, Bayrak, & Ramadan, 2014; Teh & Birch, 2013). The cold pressing technique is more advantageous than other oil extraction methods in that the bioactive components are obtained at higher concentrations, since this method is both environmentally friendly and practical, requiring no heating, organic solvent, or chemical treatment (Ramadan, 2013). Other advantages of this method are that it minimizes oxidation reaction and the oil components remain more stable (Van Hoed et al., 2011).

The importance of biologically active components obtained from natural sources is to increase the health of people and to prevent chronic diseases (Ma et al., 2014). Many plants are excellent resources for therapeutically active substances. Carotenoids found in tomatoes have an important place in the human diet (Singh & Goyal, 2008). Every year, the pulps of tomatoes are squeezed with the cold press method to obtain oils rich in lycopene and β-carotene in the tomato processing industry (Eller et al., 2010). Lycopene is the most abundant carotenoid and the compound responsible for the red color of tomato products (Baysal, Ersus, & Starmans, 2000). Epidemiological studies indicate that high consumption of tomatoes is related to a lower risk of cancer and is protective against a number of cardiovascular diseases (Britton, 1995). Lycopene is a carotenoid of industrial and commercial interest due to its potential health benefits (Ha et al., 2015; Sotomayor-Gerding et al., 2016).

Carotenoids are important compounds in the human diet as anticarcinogens, cardiovascular disease preventers (Shao et al., 2015), and immune system regulators (Singh & Goyal, 2008). They act as antioxidants and precursors of vitamin A. One of the most important carotenoids is lycopene ($C_{40}H_{56}$). Lycopene is widely used as a food additive as a colorant and antioxidant agent. A large amount of lycopene and other carotenoids contained in tomato processing by-products make their recycling desirable from both an environmental and a health viewpoint (Machmudah et al., 2012). Lycopene is also used for prevention of cutaneous signs of skin damage, and TSO has been suggested to protect from aging, eczema, psoriasis, and UV damage to the skin (Evans & Johnson, 2010). On the other hand, β-carotene plays a fundamental role in human nutrition as provitamin A, which is also an antioxidant agent (Sabio et al., 2003).

2 Cold press extraction and processing of oil

The cold press extraction technique, also called cold drawn extraction, produces a better flavor than heat-assisted oil extraction and is purer in terms of natural compounds. After pressing, the remaining cake includes 5%–15% oil. The remaining oil could be extracted by petroleum benzine (also known as petroleum ether, commercial *n*-hexane, or heptane). The solvent is recovered from the solvent-oil solution by evaporating to use over again. The mechanical process involves two techniques—hydraulic or screw presses—to force oil out of oil-containing material (Arisanu, 2013). Mechanical pressure, pressing time, and bulk product temperature directly affect the oil yield (Mwithiga & Moriasi, 2007). The screw press method has advantage over the hydraulic press method with regard to oil yield. In addition to a higher oil yield, the screw press method has a continuous mode of operation (Arisanu, 2013). Manual or powered mechanical presses are simple, safer, and have fewer steps compared to solvent extraction of oils. Because of the low operation cost and high quality flavored oil, mechanical press methods are often applied to oilseeds having an oil proportion higher than 20% (Sinha, Haddar, & Majumdar, 2015). Mechanical processes can be separated into two types: cold press and hot press methods. Cold press techniques are carried out below 50°C, while the hot press method is carried out at elevated temperatures and pressures. This difference in temperature induces decreased oxidative stability, degradation of valuable oil components, and reduced oil-keeping quality. Natural compounds, flavors, aromatic components, and physicochemical properties are preserved in cold pressed oils (Bhatol, 2013).

Time and solvent consuming conventional oil extraction methods give way to other alternative innovative methods such as microwave-assisted extraction (MAE), ultrasonic-assisted extraction (UAE), and supercritical fluid extraction (SFE) (Bampouli et al., 2014). Microwave-assisted extraction needs pretreatment of oilseed in a microwave oven (frequency range of about 300MHz to 300GHz). This process results in the rupture of cell membranes, making it possible to obtain a higher extraction yield (Azadmard-Damirchi, Alirezalu, & Achachlousi, 2011). Veldsink et al. (1999) reported that microwave-assisted extraction of oil from rapeseed showed a markedly improved oxidative stability, most likely due to an increase in phenolic antioxidants. This technique allows improving oil extraction yield and quality, lower energy consumption, faster processing time, reduced solvent consumption, and availability of desirable nutraceuticals such as phytosterols and tocopherols, canolol, and phenolic compounds in the extracted oil. Because of high antioxidant content, cold pressed oils have an improved shelf life. Microwave energy disrupts plant structure and degrades the polyunsaturated fatty acids (PUFAs) in vegetable oil (Uquiche, Jerez, & Ortiz, 2008).

Ultrasonic-assisted extraction is a new innovative technique that represents an innovative way of increasing extracted oil yield. Ultrasonic sound waves make plant cell walls thinner, thus enhancing the interaction of the solvent. There are many advantages of ultrasonic-assisted extraction, such as reduction in extraction time, lower energy consumption and being ecofriendly, increased extraction yield, and higher processing throughput (Takadas & Doker, 2017).

Supercritical fluid extraction (SFE) has become one of the most effective methods for the extraction of valuable compounds from oilseeds. In addition, SC-CO_2 has been used to extract minor constituents from tomatoes, including phytosterols, tocopherols, β-carotene, and lycopene, which have excellent antioxidant activity due to their ability to quench singlet oxygen and trap peroxyl radicals. The usage of CO_2 as a unique extraction solvent makes this technique nontoxic, nonflammable, noncorrosive, highly selective, safer, and more environmentally friendly compared to the use of most conventional organic solvents (Eller et al., 2010).

3 Cold pressed oil recovery and content, uses of oil cake, and economy

The by-products of the tomato industry's processing have been evaluated in a very wide range for different applications and the amount of disposals needs to be considered for environmentally sound and economical ways of recycling/reusing, sustainability and value-added products (Kalogeropoulos et al., 2012). In this respect, tomato pomace is important with its seeds, pulp, and skin content due to phytochemical components in it. The tomato seeds, which are mostly by-products of tomato sauce production, have been processed with different applications, such as hot and cold breaking processing to extract the seed oil and lycopene. The production of industrial tomato paste is carried out by two methods: cold and hot crushing. While cold crushing temperature is below 70 degrees, hot crushing temperatures between 85 and 90 degrees are used. These processes affect the physicochemical properties of the by-products. For this, it is also important to know the chemical properties of tomatoes by-products obtained by cold break (Goodman, Fawcett, & Barringer, 2002; Kelebek et al., 2017). Cold press and solvent extractions have been applied to by-products obtained by different processes and these extractions showed different advantages in many ways such as oil yield, antioxidant capacity, and better physicochemical properties. Among these methods, cold pressed extraction is mostly preferred due to its resultant high content of important phytochemicals and antioxidants, as mentioned earlier. Different studies show that tomato seeds contain approximately 17%–24% of oil, depending on the extraction methods (Giuffrè & Capocasale, 2015). Tomato seed oil (TSO) is considered a good edible oil source, with its fatty acid composition, especially with linoleic, oleic, and linolenic acids (Gómez-Romero, Segura-Carretero, & Fernández-Gutiérrez, 2010). Protein isolates of tomato seeds are also an important source of amino acids, as they contain all the essential amino acids and can supply the minimum level of protein requirements of children from 1 to 2 years old (Kowalska et al., 2017). Crude protein content from TSO cold pressed cake was found to be 23%–39%. Therefore, it can be considered for use for animals, especially in cattle feed formulations, or as a fertilizer (Cantarelli, Regitano-d'Arce, & Palma, 1993; Giannelos, Sxizas, Lois, Zannikos, & Anastopoulos, 2005; Giuffrè & Capocasale, 2016; Lazos, Tsaknis, & Lalas, 1998; Shen & Xu, 2005).

4 Fatty acids composition and acyl lipids

Vegetable oils are known to be natural products with a vegetable origin that contain mixtures of esters derived from glycerol that have chains of fatty acid with 14–20 carbon atoms that have different degrees of unsaturation. Vegetable oils are also sources of energy and essential fatty acids, like linoleic and linolenic, which are responsible for growth and the health of organisms (Botineştean, Hadaruga, & Jianu, 2012; Nesma, Hany, Mohammed, & Mohammed, 2010).

Yilmaz et al. (2015) cleaned and dried tomato seeds before the cold pressing of TSO with a group roasting process. Roasting was not applied to the control group. They analyzed sterol, fatty acids, tocopherols, aroma compounds, and minerals in the cold pressed TSOs obtained without roasting (control) and with roasting. The fatty acid, sterol, and policasonal contents of TSO were analyzed also by Giuffrè, Capocasale, and Zappia (2017).

According to a study by Yilmaz et al. (2015), there was no significant difference between the control and roasted cold pressed TSO in terms of fatty acids composition. The major fatty acids were linoleic acid (53%), oleic acid (22%), palmitic acid (14%), linolenic acid (1.92%), and stearic acid (6%). Other fatty acids (myristic, pentadecanoic, palmitoleic, heptadecanoic, arachidic, gadoleic, behenic, and lignoceric acid) were less than 1% in the cold pressed seed oil. Giuffrè, Capocasale, Zappia, and Poiana (2017) evaluated fatty acid compositions of TSOs as saturated fatty acids, unsaturated fatty acids, monounsaturated fatty acids, and PUFAs. The means of saturated fatty acids, unsaturated fatty acids, monounsaturated fatty acids, and PUFAs of TSO amounts are 20%, 80%, 21%, and 58.5%, respectively.

The major component of cold pressed TSO was linoleic acid (54.9%), followed by palmitic acid (13.5%) and oleic acid (18.8%). All the fatty acids were expressed in methyl esters using GC–MS, as reported by Botineştean et al. (2012). In addition, they found myristic acid (0.13%) and linolenic acid (1.08%) in cold pressed TSO.

5 Minor bioactive lipids

Depending on the type of tomato, the degree of maturity, and the characteristics of the processing plant, tomato waste cake is about 3.0% by weight of the total fruit content. Fresh tomato seeds contain 10%–12% lipids (Giuffrè et al., 2016). The effects of the cultivar, geographical conditions, and climate on the composition of vegetable oils are known (Rondanini, Castro, Searles, & Rousseaux, 2011).

Phytosterols have positive effects on human health. A high intake of plant sterols is inversely associated with lower concentrations of total and low-density lipoprotein (LDL) serum cholesterol (Andersson et al., 2004). The most commonly encountered phytosterols in plants are β-sitosterol, campesterol, and stigmasterol (García-Llatas & Rodriguez-Estrada, 2011). The total sterol composition of TSOs was found to range from 1546 to 2077 mg/kg oil (Giuffrè, Capocasale, Zappia, & Poiana, 2017). The difference in total sterol composition may vary depending on the type of tomato and agroclimatic properties. When the amount of major sterols contained in cold pressed TSO was compared to those of roasted seeds, no significant differences were found. The sterol content of cold pressed oil extracted from untreated seeds was found to be cholesterol 9.62%, campesterol 2.67%, stigmasterol 6.54%, and β-sitosterol 31.8%. The roasted TSO has cholesterol 8.34%, campesterol 3.19%, stigmasterol 5.63%, and β-sitosterol 25.44%. D-7-avenestrol was not detected in the untreated oil, while roasted seed oil has a proportion of 1.85% D-7-avenestrol (Yilmaz et al., 2015).

Gamma and alpha-tocopherols were quantified in TSO. Gamma-tocopherol is the major tocopherol in TSOs (0.761 and 1.295 mg/kg oil). Alpha-tocopherols accounted for 0.761–1.295 mg/kg in cold pressed TSO (Yilmaz et al., 2015).

Policosanol is the common name for a mixture of high molecular weight (20–36 carbon) aliphatic primary alcohols, which are constituents of plant epicuticular waxes (Irmak & Dunford, 2005). The amounts of policosanol contained in eight different TSOs obtained from cold-break extraction were 14.67–101.00 mg/kg docosanol, 3.67–31.33 mg/kg tetracosanol, 8.67–76.33 mg/kg hexacosanol, and 3.33–35.0 mg/kg octacosanol (Giuffrè, Capocasale, Zappia, & Poiana, 2017). Even odd-chain fatty alcohols in the policosanol mixture are important for human and plant health (Giuffrè & Capocasale, 2015).

Vegetable oils have a unique aroma and taste. A balanced distribution of these sensory properties determines the flavor characteristics of the oil. The aroma and volatile components generally comprise aldehydes, esters, alcohols, and ketones. The volatile compounds hexanal, hexanol, D-limonene, 2-isobutylthiazole, 2-acetyl pyrrole, tetramethyl pyrazine, nonanal, phenylethyl alcohol, methyl-2-hdroxybenzoate, decanal, beta-cyclocitral, phenyl ethyl acetate, (+)-cyclosa- tivene, alpha-copaene, tetradecane, and beta-gurjunene were determined in TSO. In addition, heptanal, furfural, 2,5-dimethyl pyrazine, benzaldehyde, benzene acetaldehyde, corylone pyrazine, 2-phenyl-2-butenal, and 3-butyl-2,5-dimethyl pyrazine have been detected in cold pressed TSO obtained from roasted seeds (Yilmaz et al., 2015).

Minerals play an important role in oil characterization and adulteration detection. The presence of metals in vegetable oils may be due to different factors. The minerals can be incorporated into the oil from the soil or be introduced during the manufacturing of the foodstuff. Zn, Pb, Cd, Ni, Ba, Fe, B, Mn, Cr, Mg, Ca, Cu, Al, Na, and K were detected in TSO (Yilmaz et al., 2015).

In a study performed by Kalogeropoulos et al. (2012) with a cold-break tomato paste process, cinnamic acid, phloretic acid, *p*-coumaric acid, ferulic acid, caffeic acid, sinapic acid, chlorogenic acid, benzoic acid, vannilic acid, syringic acid, chrysin, epicatechin, catechin, kaempferol, quercetin, and luteolin were found (Kalogeropoulos et al., 2012). Rutin, quercetin, kaempferol-3-*O*-rutinoside, naringenin, naringenin-7-*O*-glucoside, protocatechuic acid, ferulic acid, ferulic acid-*O*-hexose, caffeic acid-*O*-hexose, caffeic acid, chlorogenic acid, and *p*-coumaric acid were identified in cold-break tomato paste (Kelebek et al., 2017).

Kalogeropoulos et al. (2012) performed lycopene and β-carotene analysis on by-products obtained from the cold-break tomato paste process and whole tomatoes. Lycopene and β-carotene have been identified and quantified in TSO that was obtained using accelerated solvent extraction by Eller et al. (2010). They performed accelerated solvent extraction using *n*-hexane and ethanol; the results are shown in Table 1.

When compared to the amount of lycopene in the whole tomatoes, by-products of tomato industries, and TSO, the lycopene quantity was lower in the oil than the other products, although lycopene was found in all products.

6 Composition of cold pressed oil in comparison to other cold pressed oils

Cold pressed seed oils may retain natural beneficial components of the seeds. Seeds include natural antioxidants, and cold pressed seed oils are free of chemical contamination (Yu, Zhou, & Parry, 2005). Cold pressed extraction preserves bioactive components such as vitamins, provitamins, phytosterols, phospholipids, and squalene, which are, together with some fatty acids, key nutritional value factors (Rabrenović, Dimić, Novaković, Tešević, & Basić, 2014). The consumption of new and improved products such as cold pressed oils may improve human health and may prevent certain

TABLE 1 Lycopene and β-carotene content of tomatoes, by-products, and tomato seed oil.

Constituents	Whole tomatoes	By-products	Tomato seed oil (hexane) (*cis*+*trans*)	Tomato seed oil (ethanol) (*cis*+*trans*)
Lycopene (mg/kg)	1013.2	413.7	25.86	21.05
β-carotene (mg/kg)	86.1	149.8	4.53	4.06
References	Kalogeropoulos et al. (2012)	Kalogeropoulos et al. (2012)	Eller et al. (2010)	Eller et al. (2010)

TABLE 2 Fatty acid composition of different cold pressed oils.

Fatty acids (%)	Tomato	Pumpkin	Flax	Black cumin	Sesame	Wheat germ
C16:0 palmitic acid	14.4	11.8	6.1	7.0	8.9	16.6
C18:0 stearic acid	5.9	6.2	2.7	3.7	5.7	0.9
C18:1 oleic acid	22.2	40.7	15.2	24.7	38.4	15.8
C18:2 linoleic acid	53.6	40.8	16.7	55.8	45.4	55.4
C18:3 α-linolenic acid	1.94	0.2	59.3	0.94	0.3	7.5
C20:0 arachidic acid	0.4	–	–	–	0.7	0.3
References	Yilmaz et al. (2015)	Rabrenović et al. (2014)	Teh and Birch (2013)	Ramadan (2013)	Celenk, Gumus, Ustun, Buyukhelvacigil, and Karasulu (2018)	Celenk et al. (2018)

diseases (Van Hoed et al., 2011). The major fatty acid compositions of different cold pressed oils are compared in Table 2. The highest C18:3 ratio in terms of PUFAs is found in TSO, after flaxseed oil. A saturated palmitic acid amount of TSOs was less than wheat germ oil, while palmitic acid is higher than other oils.

The sterol contents of cold pressed TSO, black cumin seed oil, and pistachio oil are shown in Table 3. β-sitosterol was found to be the lowest amount in TSO, while cholesterol was the highest quantity of sterol in TSO.

Tocopherols are antioxidants that inhibit the oxidation of edible oils, influence the shelf life of edible oil during storage, and delay lipid peroxidation associated with the pathogenesis of numerous human diseases/disorders (Topkafa, 2016). The tocopherol compositions of different cold pressed seed oils are given in Table 4. Gamma-tocopherol is the most commonly found tocopherol in TSO, followed by alpha-tocopherol. Therefore, only alpha and gamma tocopherols in the oils are reported in the table. When evaluated in terms of tocopherols, alpha- and gamma-tocopherol ratios of TSO have average values compared to other oils. The amount of gamma-tocopherol is higher than safflower, hempseed, and canola oils.

7 Organoleptic properties of cold pressed tomato seed oil

Cold pressed oils are valuable due to their sensory, functional, and nutritional characteristics. It has been emphasized that this method is effective to obtain bioactive compounds such as flavonoids and phenolic acids. Pretreatments like pre-roasting, enzyme applications, and microwave heating affect the yield of cold pressed seed oils (Aydeniz, Güneser, & Yılmaz, 2014; Dundar, Güneser, & Yılmaz, 2014; Zuorro, Lavecchia, Medici, & Piga, 2013).

TABLE 3 Sterol compositions of cold pressed oil.

Sterols	Tomato	Black cumin	Pistacia
Campesterol (%)	2.67	0.89	4.35
β-sitosterol (%)	31.80	62.24	87.73
Stigmasterol (%)	5.63	1.10	0.98
Cholesterol (%)	9.62	–	0.44
References	Yilmaz et al. (2015)	Gumus and Celenk (2017)	Saber-Tehrani, Givianrad, Aberoomand-Azar, Waqif-Husain, and Jafari Mohammadi (2013)

TABLE 4 Tocopherol composition of different cold pressed oils.

Tocopherols (mg/kg)	Safflower	Rosehip	Hemp seed	Canola	Nettle seed	Wheat germ
Alpha tocopherols	376.6	57.5	2.8	11.9	19.4	2556.1
Gamma tocopherols	25.9	1058.1	56.4	21.3	5.7	573.0
References	Topkafa (2016)	Topkafa (2016)	Teh and Birch (2013)	Teh and Birch (2013)	Celenk et al. (2018)	Celenk et al. (2018)

Yilmaz et al. (2015) used a laboratory scale (12 kg seed/h capacity, single head, 2 hp., 1.5 kW power) cold press machine to obtain oil from tomato seeds. The cold pressing procedure was set as a 10-mm exit die, 40 rpm of screw rotation speed, and 40 °C of exit temperature to get the two products: the cold pressed oil (liquid part), and seed press cake or meal (solid part). Refractive indices, free fatty acidity, peroxide, iodine number values, mineral contents, volatile compounds, fatty acid profiles, tocopherol contents, and sterol compositions of TSOs were reported. In addition, sensory analysis (appearance, color, odor, and taste/flavor) of the oils was carried out. Quantitative descriptive analysis (QDA) was used to describe the sensory properties of the oil samples. According to Yilmaz et al.'s sensory panel study, TSOs have roasted, olive, green, tomato, hay, fruity, spicy, and bitter sensory properties (Yilmaz et al., 2015).

The roasting process prior to cold pressing enhanced the antioxidant activity and tocopherol contents whereas this pretreatment had no effect on the fatty acid and mineral content of the TSOs. In addition, the roasting process affected the sterol composition of the TSOs. Furthermore, in terms of sensory properties and volatile compositions, roasted seed oils were well-accepted oils.

8 Edible and nonedible applications of cold pressed oil

The tomato by-products for industrial applications were found to be significantly important, with bioactive phytochemicals such as tocopherols, carotenoids, polyphenols, some sterols, and terpenes, which show antioxidant activities. These bioorganic components can be used in the formulations of functional foods or supplements with their antioxidant properties. Additionally, these phytochemicals can act as protective agents within the food additives sector and can help to increase shelf life of the foods (Kalogeropoulos et al., 2012; Vidyarthi, Li, & Pan, 2019). Carotenoids and antioxidant components can be used to inhibit oxidation of refined oils. These properties can be increased by roasting the seeds prior to cold pressing, since, with pretreatment, tocopherols content and antioxidant activity increase. The roasting process can also improve the aromatic composition of TSO and help to develop and formulate new food products (Yilmaz et al., 2015).

The fatty acid composition of TSO is considered to be an important source for fatty acids, and can be included in essential fatty acids deficient diets as a dietary supplement (Cantarelli et al., 1993; Giuffrè & Capocasale, 2016), and with cholesterol-lowering effect it is also preferred in medical applications (Giannelos et al., 2005). Due to its organoleptic properties and high amounts of antioxidants and phytosterols, TSO is defined among the high-quality, stable edible oils, which are good for food applications and cooking (Eller et al., 2010; Kamazani, Tavakolipour, Hasani, & Amiri, 2014; Lazos et al., 1998). The oil also can be used in processing of food that are packed in oil, such as tuna and vegetables, or can be an ingredient of the formulations of dried tomatoes and sauces (Porretta, 2019).

TSO is an important ingredient for cosmetics, as a cream cleanser, and is classified irritation and sensitizer safe (da Silva & Jorge, 2014; Giannelos et al., 2005; Hassanien et al., 2014). TSO is also used in different applications such as lubricant, paint, varnish, and soap (Giannelos et al., 2005). Research has shown that TSO is a good source for biofuel production compared with other diesel fuels, since its sulfur and ash level is low, and its density and iodine values are high. Therefore, it can be considered as a renewable energy source and alternative raw material for diesel fuel (Fahimdanesh & Bahrami, 2013; Giuffrè, Capocasale, Zappia, & Poiana, 2017).

9 Health-promoting traits of cold pressed oil and oil constituents

Phytochemical components such as vitamins, carotenoids, and phenolic compounds in TSO show many health-promoting effects with their antioxidant properties. Studies showed that TSO has cholesterol-lowering effects in hamsters and guinea pigs fed with high-fat diets (Giannelos et al., 2005; Yilmaz et al., 2015). As TSO is rich in linoleic acid, many health effects are attributed to this oil, such as anticarcinogenic activity, reducing the catabolic effects of immune stimulation, and preventing atherosclerosis risk (Gómez-Romero et al., 2010; Shen & Xu, 2005). All these effects are primarily related to the strong antioxidant properties of TSO, but the stimulation of immune and hormonal systems and modulation of intercellular communication mechanisms can be added to these effects (Lavecchia & Zuorro, 2006). Lycopene is also an important component of TSO and shows a protective effect against cardiovascular disease, certain types of cancer, and skin damage. Its antioxidant impact, bioavailability, and the ability of singlet oxygen quenching are higher when it is combined with dietary lipids (Shao et al., 2013). TSO is preferred for nutraceuticals and antiobesity formulations with its essential fatty acid components and also recommended to use for eczema, aging, UV damage of skin, and psoriasis. Carotenoid components are also important for prevention of digestive tract, pancreas, and prostate cancer, inflammation, cardiovascular problems, and some chronic diseases (Banerjee et al., 2017; da Silva & Jorge, 2014; Muller et al., 2013; Vagi et al., 2007).

10 Adulteration and authenticity

The authenticity of foods has become a focal point that attracts the attention of producers and consumers. Controlling adulteration in oils is one of the main issues in terms of authentication and contamination of oil. Adding cheaper oil to expensive oil, adding refined oil, solvent-extracted oil to cold pressed oil, mixing old oil with new oil, and adding an external synthetic bioactive component to the oil are the most commonly used adulterants (Ulberth & Buchgraber, 2000). The oil yield of cold pressed oils is lower than processed oils. Therefore, a producer obtains high prices for cold pressed oils. Economic reasons constitute the basis of adulteration. There are many methods used to detect fraud, which are based on knowledge of the natural variations of the characteristic components of the commodity of interest and use complex statistical procedures to obtain indicators of adulteration (Aparicio & Aparicio-Ruiz, 2000; Aparicio, Morales, Aparicio-Ruiz, Tena, & García-González, 2013). A number of analytical methods to detect fraud should be continuously developed, modified, and revised, one step ahead of those who monitor these illegal activities (Kamm, Dionisi, Hischenhuber, & Engel, 2001). Knowing the compounds found in the natural structure of oils is important for understanding adulteration. Therefore, advanced analysis techniques and multivariate data analysis tools are used (Gumus, Ertas, Yasar, & Gumus, 2018).

The chemical content of cold pressed TSO should be known in detail. For this, studies with cold pressed TSOs are becoming more important. Adulteration of oils where oil yield is low is more common. Determination of both major and minor components of oils is necessary to reveal the characteristics of the oil. The properties of oils such as sterol composition, fatty acid composition, triacylglycerol composition, aroma components, mineral structure, tocopherols, phenols, and policosanol content give a lot of information about fat. It is also critical to know the amounts of bioactive components specific to the oil.

Plant sterols or phytosterols are components in different types of seed oils. A sterol profile is used for determining the adulteration. The quantitative analysis of fatty acids, triglycerides, or sterols can be useful for the detection of oil adulteration. Even so, these characteristics are known to be affected by many factors: agronomic climate, soil, water, geographic altitude, longitude, harvesting cultivar, ripeness, technological conservation of the fruit or of the oil, extraction systems, processing, refining, and solvent extraction.

Mineral composition analysis plays an important role in oil characterization and adulteration detection. The presence of metals in edible oils may be due to different factors. The metals can be incorporated into the oil from the soil or be introduced during the manufacturing of the foodstuff. Therefore, it can be assumed that the trace elemental distribution in oils varies according to their origin, and it can then be supposed that a suitable statistical treatment on trace element data could allow a geographical characterization of different oils.

Although tomatoes contain a lot of lycopene and carotene, they are present in smaller quantities in TSO. However, there are not many studies on lycopene and carotene in cold pressed TSO. More studies are needed to determine specific parameters that investigate the characteristic of cold pressed TSO in terms of tocopherols, composition of fatty acids, phenols, minerals, and aroma components. Furthermore, considering that tomato is one of the most consumed foods in the world, the remaining pulp from tomato processes can be evaluated and more research could be carried out with cold pressed TSO. In this way, the characteristics of the chemical content of cold pressed TSO can be determined. Also in this way, geographical origin studies could occur and regional differences could be revealed. With the increase of studies, the adulteration of cold pressed tomato oil might be prevented.

References

Andersson, S. W., Skinner, J., Ellegård, L., Welch, A. A., Bingham, S., Mulligan, A., et al. (2004). Intake of dietary plant sterols is inversely related to serum cholesterol concentration in men and women in the EPIC Norfolk population: A cross-sectional study. *European Journal of Clinical Nutrition*, *58*, 1378–1385.

Aparicio, R., & Aparicio-Ruiz, R. (2000). Authentication of vegetable oils by chromatographic techniques. *Journal of Chromatography A*, *881*, 93–104.

Aparicio, R., Morales, M. T., Aparicio-Ruiz, R., Tena, N., & García-González, D. L. (2013). Authenticity of olive oil: Mapping and comparing official methods and promising alternatives. *Food Research International*, *54*(2), 2025–2038.

Arisanu, A. O. (2013). Mechanical continuous oil expression from oil seeds: Oil yield and press capacity. In: *International conference "computational mechanics and virtual engineering" COMEC 2013, 24–25 October 2013, Brasov, Romania.*

Aydeniz, B., Güneser, O., & Yılmaz, E. (2014). Physico-chemical, sensory and aromatic properties of cold press produced safflower oil. *Journal of the American Oil Chemists' Society*, *91*, 99–110.

Azadmard-Damirchi, S., Alirezalu, K., & Achachlousi, B. F. (2011). Microwave pretreatment of seeds to extract high quality vegetable oil. *International Journal of Nutrition and Food Engineering*, *5*(9), 508–511.

Bampouli, A., Kyriakopoulou, K., Papaefstathiou, G., Lauli, V., Krokida, M., & Magoulas, K. (2014). Comparison of different extraction methods of *Pistacia lentiscus* var. Chia leaves: Yield, antioxidant activity and essential oil chemical composition. *Journal of Applied Research on Medicinal and Aromatic Plants*, *1*(3), 81–91.

Banerjee, J., Singh, R., Vijayaraghavan, R., Macfarlane, D., Patti, A. F., & Arora, A. (2017). Bioactives from fruit processing wastes : Green approaches to valuable chemicals. *Food Chemistry*, *225*, 10–22. Elsevier.

Baysal, T., Ersus, S., & Starmans, D. A. J. (2000). Supercritical CO_2 extraction of b-carotene and lycopene from tomato paste waste. *Journal of Agricultural and Food Chemistry*, *48*, 5507–5511.

Bhatol, K. (2013). *Castor oil obtained by cold press method.* Banaskantha, Gujarat, India: Shri Bhgwati Oil Mill (SBOM) manufacturer's. Info.

Bhowmik, D., Kumar, K. P. S., Paswan, S., & Srivastava, S. (2012). Tomato—A natural medicine and its health benefits. *Phytojournal*, *1*(1), 33–43. Retrieved from http://www.phytojournal.com/vol1Issue1/Issue_may_2012/3.pdf.

Botinestean, C., Gruia, A., & Jianu, I. (2014). Utilization of seeds from tomato processing wastes as raw material for oil production. *Journal of Material Cycles and Waste Management*, *17*, 1–7.

Botineștean, C., Hadaruga, D. I., & Jianu, I. (2012). Fatty acids composition by gas chromatography-mass spectrometry (GC-MS) and most important physical-chemicals parameters of tomato seed oil. *Journal of Agroalimentary Processes and Technologies*, *18*(1), 89–94.

Botinestean, C., & Jianu, I. (2013). Tomato seed oil for industrial use: Separation of seeds from tomato pomace using wet fermentation technology. *Current Opinion in Biotechnology*, *24*, 78.

Britton, G. (1995). Structure and properties of carotenoids in relation to function. *The FASEB Journal*, *9*(15), 1551.

Cantarelli, P. R., Regitano-d'Arce, M. A. B., & Palma, E. R. (1993). Physicochemical characteristics and fatty acid composition of tomato seed oils from processing wastes. *Science in Agriculture*, *50*(1), 117–120.

Celenk, V. U., Gumus, Z. P., Ustun, A. Z., Buyukhelvacigil, M., & Karasulu, E. (2018). Analysis of chemical compositions of 15 different cold-pressed oils produced in Turkey: A case study of tocopherol and fatty acid analysis. *Journal of the Turkish Chemical Society Section A: Chemistry*, *5*(1), 1–18.

Dundar, E. D., Güneser, O., & Yılmaz, E. (2014). Cold pressed poppyseed oils: Sensory properties, aromatic profiles and consumer preferences. *Grasas y Aceites*, *65*(3), e029.

Eller, F. J., Moser, J. K., Kenar, J. A., & Taylor, S. L. (2010). Extraction and analysis of tomato seed oil. *Journal of the American Oil Chemists' Society*, *87*, 755–762.

Evans, J., & Johnson, E. J. (2010). The role of phytonutrients in skin health. *Nutrients*, *2*(8), 903–928.

Fahimdanesh, M., & Bahrami, M. E. (2013). Evaluation of physicochemical properties of Iranian tomato seed oil. *Journal of Nutrition and Food Sciences*, *3*(3).

Franko, M., Van De Bovenkamp, P., & Bicanic, D. (1998). Determination of trans—β-carotene and other carotenoids in blood plasma using high-performance liquid chromatography and thermal lens detection. *Journal of Chromatography B: Biomedical Sciences and Applications*, *718*, 47–54.

García-Llatas, G., & Rodriguez-Estrada, M. T. (2011). Current and new insights on phytosterol oxides in plant sterol-enriched food. *Chemistry and Physics of Lipids*, *164*, 607–624.

Giannelos, P. N., Sxizas, S., Lois, E., Zannikos, F., & Anastopoulos, G. (2005). Physical, chemical and fuel related properties of tomato seed oil for evaluating its direct use in diesel engines. *Industrial Crops and Products*, *22*, 193–199.

Giuffrè, A. M., & Capocasale, M. (2015). Policosanol in tomato (*Solanum lycopersicum* L.) seed oil: The effect of cultivar. *Journal of Oleo Science*, *64*, 625–631.

Giuffrè, A., & Capocasale, M. (2016). Physicochemical composition of tomato seed oil for an edible use : The effect of cultivar. *International Food Research Journal*, *23*(2), 583–591.

Giuffrè, A. M., Capocasale, M., & Zappia, C. (2017). Tomato seed oil for edible use: Cold break, hot break, and harvest year effects. *Journal of Food Process Preservation*, *41*, e13309.

Giuffrè, A. M., Capocasale, M., Zappia, C., & Poiana, M. (2017). Biodiesel from tomato seed oil : Transesterification and characterisation of chemical-physical properties. *Agronomy Research*, *15*(1), 133–143.

Giuffrè, A. M., Capocasale, M., Zappia, C., Sicari, V., Pellican, O. T. M., Poiana, M., et al. (2016). Tomato seed oil for biodiesel production. *European Journal of Lipid Science and Technology*, *118*, 640–650.

Gómez-Romero, M., Segura-Carretero, A., & Fernández-Gutiérrez, A. (2010). Phytochemistry metabolite profiling and quantification of phenolic compounds in methanol extracts of tomato fruit. *Phytochemistry*, *71*, 1848–1864.

Goodman, C., Fawcett, S., & Barringer, S. A. (2002). Flavor, viscosity, and color analyses of hot and cold break tomato juices. *Journal of Food Science*, *67*, 404–408.

Gumus, Z. P., & Celenk, V. U. (2017). A case study on profile investigation of cold-pressed black cumin seed oil produced in Turkey. *Hacettepe Journal of Biology and Chemistry*, *45*(4), 475–484015.

Gumus, Z. P., Ertas, H., Yasar, E., & Gumus, O. (2018). Classification of olive oils using chromatography, principal component analysis and artificial neural network modelling. *Journal of Food Measurement and Characterization*, *12*, 1325–1333.

Ha, T. V. A., Kim, S., Choi, Y., Kwak, H.-S., Lee, S. J., Wen, J., et al. (2015). Antioxidant activity and bioaccessibility of size-different nanoemulsions for lycopene-enriched tomato extract. *Food Chemistry*, *178*, 115–121.

Hassanien, M. M. M., et al. (2014). Phytochemical contents and oxidative stability of oils from non-traditional sources. *European Journal of Lipid Science and Technology*, *116*, 1563–1571.

Irmak, S., & Dunford, N. (2005). Policosanol contents and compositions of wheat varieties. *Journal of Agricultural and Food Chemistry*, *53*(14), 5583–5586.

Kalogeropoulos, N., Chiou, A., Pyriochou, V., Peristeraki, A., & Karathanos, V. T. (2012). Bioactive phytochemicals in industrial tomatoes and their processing byproducts. *LWT-Food Science and Technology*, *49*(2), 213–216.

Kamazani, N. A., Tavakolipour, H., Hasani, M., & Amiri, M. (2014). Evaluation and analysis of the ultrasound-assisted extracted tomato seed oil. *Journal of Food Biosciences and Technology*, *4*(2), 57–66.

Kamm, W., Dionisi, F., Hischenhuber, C., & Engel, K. H. (2001). Authenticity assessment of fats and oils. *Food Review International*, *17*(3), 249–290.

Kelebek, H., Selli, S., Kadiroglu, P., Kola, O., Kesen, S., Uçar, B., et al. (2017). Bioactive compounds and antioxidant potential in tomato pastes as affected by hot and cold break process. *Food Chemistry*, *220*, 31–41.

Khachik, F., Carvalho, L., Bernstein, P. S., Muir, G. J., Zhao, D.-Y., & Katz, N. B. (2002). Chemistry, distribution, and metabolism of tomato carotenoids and their impact on human health. *Experimental Biology and Medicine (Maywood, NJ)*, *227*(10), 845–851.

Kiralan, M., Özkan, G., Bayrak, A., & Ramadan, M. F. (2014). Physicochemical properties and stability of black cumin (*Nigella sativa*) seed oil as affected by different extraction methods. *Industrial Crops and Products*, *57*, 52–58.

Kowalska, H., et al. (2017). Trends in food science and technology what's new in biopotential of fruit and vegetable by-products applied in the food processing industry. *Trends in Food Science and Technology*, *67*, 150–159. https://doi.org/10.1016/j.tifs.2017.06.016.

Lavecchia, R., & Zuorro, A. (2006). Thermal stability of tomato lycopene in vegetable oils. *Chemical Technology: An Indian Journal*, *1*(2–4), 80–87.

Lazos, E. S., Tsaknis, J., & Lalas, S. (1998). Characteristics and composition of tomato seed oil. *Grasas y Aceites*, *49*, 440–445.

Ma, Y., Ma, J., Yang, T., Cheng, W., Lu, Y., Cao, Y., et al. (2014). Components, antioxidant and antibacterial activity of tomato seed oil. *Food Science and Technology Research*, *20*(1), 1–6.

Machmudah, S., Zakaria, Winardi, S., Sasaki, M., Goto, M., Kusumoto, N., et al. (2012). Lycopene extraction from tomato peel by-product containing tomato seed using supercritical carbon dioxide. *Journal of Food Engineering*, *108*(2), 290–296.

Muller, L., Catalona, A., Simone, R., Cittadini, A., Fröhlich, K., Böhm, V., et al. (2013). Antioxidant capacity of tomato seed oil in solution and its redox properties in cultured macrophages. *Journal of Agricultural and Food Chemistry*, *61*, 346–354.

Mwithiga, G., & Moriasi, L. (2007). A study of yield characteristics during mechanical oil extraction of pretreated and ground soybeans. *Journal of Applied Sciences Research*, *3*(10), 1146–1151.

Nesma, E., Hany, A. A., Mohammed, H. A. E., & Mohammed, M. Y. (2010). Red palm olein: Characterization and utilization in formulating novel functional biscuits. *Journal of the American Oil Chemists' Society*, *87*(3), 295–304.

Porretta, S. (2019). Tomato seeds and skins as a source of functional compounds. In S. Porretta (Ed.), *Tomato chemistry, industrial processing and product development* (pp. 231–244). Croydon; UK: The Royal Society of Chemistry.

Rabrenović, B. B., Dimić, E. B., Novaković, M. M., Tešević, V. V., & Basić, Z. N. (2014). The most important bioactive components of cold pressed oil from different pumpkin (*Cucurbita pepo* L.) seeds. *LWT-Food Science and Technology*, *55*(2), 521–527.

Ramadan, M. F. (2013). Healthy blends of high linoleic sunflower oil with selected cold pressed oils: Functionality, stability and antioxidative characteristics. *Industrial Crops and Products*, *43*(1), 65–72.

Rondanini, D. P., Castro, D. N., Searles, P. S., & Rousseaux, M. C. (2011). Fatty acid profiles of varietal virgin olive oils (*Olea europaea* L.) from mature orchards in warm arid valleys of Northwestern Argentina (La Rioja). *Grasas y Aceites*, *62*(4), 399–409.

Saber-Tehrani, M., Givianrad, M. H., Aberoomand-Azar, P., Waqif-Husain, S., & Jafari Mohammadi, S. A. (2013). Chemical composition of Iran's *Pistacia atlantica* cold-pressed oil. *Journal of Chemistry*, *2013*, 126106. 6 pages.

Sabio, E., Lozano, M., de Espinosa, V. M., Mendes, R. L., Pereira, A. P., Palavra, A. F., et al. (2003). Lycopene and β-carotene extraction from tomato processing waste using supercritical CO_2. *Industrial and Engineering Chemistry Research*, *42*, 6641–6646.

Shao, D., Bartley, G. E., Yokoyama, W., Pan, Z., Zhang, H., & Zhang, A. (2013). Plasma and hepatic cholesterol-lowering effects of tomato pomace, tomato seed oil and defatted tomato seed in hamsters fed with high-fat diets. *Food Chemistry*, *139*(1–4), 589–596.

Shao, D., Venkitasamy, C., Li, X., Pan, Z., Shi, J., Wang, B., et al. (2015). Thermal and storage characteristics of tomato seed oil. *LWT- Food Science and Technology*, *63*(1), 191–197.

Shen, X., & Xu, S. (2005). Supercritical CO_2 extraction of tomato seed oil. *Journal of Food Technology*, *3*(2), 226–231.

da Silva, A. C., & Jorge, N. (2014). Bioactive compounds of the lipid fractions of agro-industrial waste. *Food Research International*, *66*, 493–500.

Singh, P., & Goyal, G. K. (2008). Dietary lycopene: Its properties and anticarcinogenic effects. *Comprehensive Reviews in Food Science and Food Safety*, *7*(3), 255–270.

Sinha, L. K., Haddar, S., & Majumdar, G. C. (2015). Effect of operating parameter on mechanical expression of solvent-soaked soybean grits. *Journal of Food Science and Technology*, *52*(5), 2942–2949.

Sotomayor-gerding, D., Oomah, B. D., Acevedo, F., Morales, E., Bustamante, M., Shene, C., et al. (2016). High carotenoid bioaccessibility through linseed oil nanoemulsions with enhanced physical and oxidative stability. *Food Chemistry*, *199*, 463–470.

Takadas, F., & Doker, O. (2017). Extraction method and solvent effect on safflower seed oil production. *Chemical and Process Engineering Research*, *51*, 9–17.

Teh, S. S., & Birch, J. (2013). Physicochemical and quality characteristics of cold-pressed hemp, flax and canola seed oils. *Journal of Food Composition and Analysis*, *30*(1), 26–31.

Topkafa, M. (2016). Evaluation of chemical properties of cold pressed onion, okra, rosehip, safflower and carrot seed oils: Triglyceride, fatty acid and tocol compositions. *Analytical Methods*, *8*, 4220–4225.

Ulberth, F., & Buchgraber, M. (2000). Authenticity of fats and oils. *European Journal of Lipid Science and Technology*, *102*(11), 687–694.

Uquiche, E., Jerez, M., & Ortiz, J. (2008). Effect of treatment with microwaves on mechanical extraction yield and quality of vegetable oil from Chilean hazelnuts (*Gevuina avellana* Mol). *Innovative Food Science & Emerging Technologies*, *9*(4), 495–500.

Vagi, E., et al. (2007). Supercritical carbon dioxide extraction of carotenoids, tocopherols and sitosterols from industrial tomato by-products. *Journal of Supercritical Fluids*, *40*, 218–226. https://doi.org/10.1016/j.supflu.2006.05.009.

Van Hoed, V., Barbouche, I., De Clercq, N., Dewettinck, K., Slah, M., Leber, E., et al. (2011). Influence of filtering of cold pressed berry seed oils on their antioxidant profile and quality characteristics. *Food Chemistry*, *127*(4), 1848–1855.

Veldsink, J. W., Muuse, B. G., Meijer, M. M. T., Cuperus, F. P., Van De Sande, R. L. K. M., & Van Putte, K. P. A. M. (1999). Heat pretreatment of oilseeds: Effect on oil quality. *Fett-Lipid*, *7*, 244–248.

Vidyarthi, S., Li, X., & Pan, Z. (2019). Peeling of tomatoes using infrared heating technology. In S. Porretta (Ed.), *Tomato chemistry, industrial processing and product development* (pp. 180–200). UK: The Royal Society of Chemistry.

Westphal, A., Bauerfeind, J., Rohrer, C., & Böhm, V. (2014). Analytical characterisation of the seeds of two tomato varieties as a basis for recycling of waste materials in the food industry. *European Food Research and Technology*, *239*(4), 613–620.

Yilmaz, E., Aydeniz, B., Güneser, O., & Arsunar, E. S. (2015). Sensory and physico-chemical properties of cold press-produced tomato (*Lycopersicon esculentum* L.) seed oils. *Journal of the American Oil Chemists' Society*, *92*, 833–842.

Yu, L. L., Zhou, K. K., & Parry, J. (2005). Antioxidant properties of cold-pressed black caraway, carrot, cranberry, and hemp seed oils. *Food Chemistry*, *91*(4), 723–729.

Zuorro, A., Lavecchia, R., Medici, F., & Piga, L. (2013). Enzyme assisted production of tomato seed oil enriched with lycopene from tomato pomace. *Food and Bioprocess Technology*, *6*, 3499–3509.

Chapter 41

Cold pressed argan (*Argania spinose*) oil

Massimo Lucarini[a], Alessandra Durazzo[a], Stefano Ferrari Nicoli[a], Antonio Raffo[a], Antonello Santini[b], Ettore Novellino[b], Eliana B. Souto[c,d], Annalisa Romani[e], Maria Francesca Belcaro[e], and Chiara Vita[e]
[a]*CREA—Research Centre for Food and Nutrition, Rome, Italy,* [b]*Department of Pharmacy, University of Napoli Federico II, Napoli, Italy,* [c]*Department of Pharmaceutical Technology, Faculty of Pharmacy, University of Coimbra, Coimbra, Portugal,* [d]*CEB—Centre of Biological Engineering, University of Minho, Campus de Gualtar, Braga, Portugal,* [e]*DiSIA-PHYTOLAB (Pharmaceutical, Cosmetic, Food supplement Technology and Analysis), University of Florence, Florence, Italy*

Abbreviations

COX cyclooxygenase
NO nitric oxide

1 Introduction

Among edible fats and oils, cold pressed oil is a derivative that has undergone a minimum of processing or none at all. In fact, the oil is extracted from its vegetal matrix without applying heat or stressing the pressure extraction by increasing the pressure. For this reason, it is generally considered able to achieve a higher quality. Many oils present a unique flavor, odor, and special characteristics related to their use addressed in a specific way for cosmetic, therapeutic, and dietary scope. They are promoted as specialty oils and usually available on the market at a higher price. Cold pressed oil products are reported to contain pro- and antioxidative compounds, which could affect the quality of these oils. In general, the consumer's perception is that foods subjected to minor or no industrial treatment or processing at all can be better and healthier. In this respect, cold pressed oil is becoming an appealing choice for consumers since no solvents and no further processing other than filtering is involved.

This chapter is addressed to cold pressed argan (*Argania spinose*) oil, the main product of the argan tree (Guillaume, Pioch, & Charrouf, 2019). The argan tree belongs to the family *Sapotaceae* del. As the latter has been known since the late Cretaceous period, it is agreed that the argan tree appeared in the tertiary. The argan tree is of great economic interest because it is a multipurpose tree. Each part of the tree is usable and it is a source of income or nourishment for the user: the wood is used as fuel, the leaves and the fruits constitute a fodder, and the oil extracted from the kernel is used in human nutrition and traditional medicine. In addition to this socio-economic role, the argan tree plays an irreplaceable role in the ecological balance and in preserving biodiversity. Thanks to its powerful root system, it contributes to the maintenance of the soil and it makes it possible to fight against the erosion of water and wind, which threat of desertification.

2 Extraction and processing of cold pressed oil

A recent review reported the advantages and disadvantages of the extraction by means of cold pressing in comparison with other commonly used extraction methods. Cold press extraction implies mechanical extraction and requires less energy than any other oil extraction techniques; it is also environmentally friendly since no chemicals are needed (e.g., organic solvents). Çakaloğlu, Özyurt, and Ötleş (2018) remarked that the critical parameters in extraction with cold press depend on the raw starting material (i.e., shell-shells, moisture content, oil content, and type of raw material), feed rate, temperature (hot or cold), cold rotation speed, diameter of restriction dye, and pretreatment. Hilali, Charrouf, El Aziz, Hachimi, and Guillaume (2005) highlighted how, in addition to the seed origin, the technology associated with argan oil extraction represents a parameter which can possibly modify the argan oil acidity. Other authors reported that in southwestern Marocco, women's cooperatives take care to increase and improve the production of high-quality argan oil

Cold Pressed Oils. https://doi.org/10.1016/B978-0-12-818188-1.00041-4
Copyright © 2020 Elsevier Inc. All rights reserved.

(Charrouf & Guillaume, 2008; Charrouf & Guillaume, 2010; Henry, Danoux, Moser, Charrouf, & Pauly, 2002), specifically to obtain the argan oil that is prepared by mechanically cold pressing argan kernels.

Generally, virgin edible argan oil is prepared by cold pressing of roasted argan kernels and includes at least six steps: fruit collection, sun drying, dehulling, nut breaking (or kernel collection), kernel roasting, and cold pressing (Kartah et al., 2015). The work of El Monfalouti, Guillaume, Denhez, and Charrouf (2010) specified that edible argan oil is prepared from roasted kernels, whereas unroasted kernels are employed in the production of argan oil for cosmetic uses. The same authors described the differences between cold pressed edible oil and cold pressed oil for cosmetic use: materials, hand-picked fruit, roasted kernels vs. hand-picked fruit, unroasted kernels; process, the use of press for both items; preservation for several months vs. up to 1 month; taste, hazelnut-like vs. bitter; quality, high for both items; antioxidants, high for both items; and moisture, low vs. discrete/high amount.

3 Composition of cold pressed oil

Males, Mišković, Bojić, and Ćurak (2018) showed that the major constituents of argan oil, obtained by cold press from argan (*Argania spinosa* L. Skeels) plant seeds, are triglycerides containing up to 80% of monounsaturated fatty acids (MUFAs), including oleic and linolenic acids along with phenolics, squalene, and tocopherols. The same authors explained how these constituents could be responsible for the biological properties of argan oil, which include antiinflammatory, cardioprotective, and antioxidant health beneficial effects. Madawalaa, Kochharb, and Duttaa (2012), by studying the lipid components and oxidative status of high-value specially selected oils, determined that argan oil contains, among sterols, also schottenol (35%), and spinasterol (32%).

Gharby, Harhar, Guillaume, Haddad, and Charrouf (2012) compared the physicochemical parameters of the two main usages of argan oil, namely edible and beauty oil (using roasted or unroasted argan kernels and cold pressed), immediately after preparation and after 2 years of storage. Their results indicated that phospholipids are a new and essential type of oil component contributing to the excellent oxidative stability of edible argan oil, in addition to the already suggested Maillard-reaction products, which can be formed during storage, namely phenolics and tocopherols.

Kartah et al. (2015) reported the chemical composition of crude cold pressed unfiltered and filtered argan oil. The filtering leads to significant changes in the free fatty acids content, in the peroxide value, in the total content of phospholipids, and in the oxidative stability as measured by the Rancimat test at 110°C. Moreover, the same authors concluded that the high level of tocopherols, fatty acids, and sterols composition were not significantly influenced by filtrations, but it contributes to the oxidative stability of the oils (Kartah et al., 2015).

Matthäus (2013), by investigating quality traits of cold pressed argan oil, concluded that after 35 days under accelerated storage, oil from roasted seeds obtained using a screw press showed the highest oxidative stability. A further study of the same author and a colleague (Matthäus & Brühl, 2015) evaluated 17 argan oils purchased from Swiss and German local markets to assess the sensory quality, identity, and oxidative state as well as the presence of contaminants. Only one oil exceeded the limit of the total oxidation value of 20, and the shelf life calculated from the induction period was 196–435 days. Moreover, the authors reported that adulterations of expensive argan oil with cheaper oils could be detected by the dominant occurrence of γ-tocopherol, α-spinasterol, and 7-stigmastenol (Schottenol) as characteristic tocopherol and phytosterols, respectively (Matthäus & Brühl, 2015). Indeed the level of γ-tocopherol can be seen as a potential biomarker of cold pressed argan oil. Madawalaa et al. (2012) reported values for α-tocopherol, γ-tocopherol, and δ-tocopherol as 90, 463, and 71 $\mu g\ g^{-1}$, respectively, in specialty argan oil.

In a current work, Matthäus, Willenberg, Engert, and Steinberg (2019a, 2019b) summarize the main results obtained on cold pressed argan oil with particular regard to adulteration: oleic acid and linoleic acid represent the predominant fatty acids, the low level of α-linolenic acid (0.1%) could be helpful to confirm the authenticity of argan oil. γ-tocopherol comprises about 90% of the total tocopherols. Moreover, Matthaus observed how the composition of the phytosterols can be another useful parameter, in addition to γ-tocopherol, to distinguish argan oil from other edible oils. The oil contains schottenol, which otherwise is only detected in pumpkin seed oil, tea seed oil, and shea butter, and does not contain β-sitosterol, the main phytosterol of most other edible oils. Matthäus et al. (2019a, 2019b) noted how the triacylglycerol composition, considering their stable characteristic feature, can also be used as marker to confirm the identity of pure argan oil. While the fatty acid composition of adulterated oils often falls into specified limits, the triacylglycerol composition shows conspicuous features.

4 Health-promoting traits of cold pressed oil and oil constituents

In addition to several studies on the benefits of argan oil consumption (El Monfalouti et al., 2010), only few studies have been devoted to researching cold pressed argan oil. Consumption of cold pressed oils rich in polyunsaturated fatty acids

(PUFAs) is considered healthy. However, these oils are not stable, and this leads to the accumulation of lipid oxidation products, with a cascade of related effects on health. Indeed, it is of great interest to identify new strategies to increase the oxidative stability of PUFA-rich oils. Some authors have proposed strategies to increase the oxidative stability of cold pressed oils, i.e., the enrichment of cold pressed oils with antioxidants or limiting oxidation factors, which can promote lipid oxidation. Grosshagauer, Steinschaden, and Pignitter (2019), in a recent work, remarked how many attempts have been performed to prolong the shelf life of edible oils by fortification with antioxidants and limiting factors, which promote lipid oxidation, whereas few studies have been carried out on enrichment of cold pressed oils with antioxidants together with modifying the production process. Moreover, the same authors, besides phenolics and vitamin E used as conventional antioxidants in oils, observed that advanced lipid oxidation end products (ALEs) can be used. ALEs have antioxidant activities and are formed by the nonenzymatic reaction between lipid aldehydes and amino phospholipids.

Generally, the interactions of bioactive components in a foodstuff define its potential benefit (Santini & Novellino, 2014; Andrew & Izzo, 2017; Santini, Tenore, & Novellino, 2017; Santini et al., 2018; Santini & Novellino, 2018; Daliu, Santini, & Novellino, 2018; Durazzo, 2018; Durazzo & Lucarini, 2019; Durazzo, D'Addezio, Camilli, et al., 2018). Berrougui et al. (2004) investigated the effects of 7 weeks of treatment with hand-pressed argan oil (10 mL kg^{-1}) on the blood pressure and endothelial function of spontaneously hypertensive rats (SHRs) and normotensive Wistar-Kyoto rats. The treatment of hypertensive animals with argan oil not only prevented an increase in blood pressure, but also improved endothelial function. Moreover, a high concentration of linoleic acid and α-tocopherol could contribute to explaining this effect, which was dependent on both cyclooxygenase (COX) products and nitric oxide (NO) synthase.

5 Volatile flavor compounds

The aroma of cold pressed argan oil is strongly dependent on the roasting process that is generally applied to kernels to obtain food-grade argan oil, as opposite to the cosmetic-grade oil obtained from raw kernels. During roasting an aroma develops, and this is retained in the oil during the extraction process (El Monfalouti, Charrouf, Guillaume, Denhez, & Zeppa, 2013). The main routes of flavor formation during argan kernel roasting are the Maillard reaction and lipid oxidation. As a result, the volatile profile changes dramatically when compared to the profile of raw kernels. Eighty volatiles have been identified in oils extracted from raw or roasted kernels (Gracka et al., 2018), belonging to the chemical groups of aldehydes, ketones, alcohols, esters, acids, terpenes, and N-heterocycles, such as pyrazines and pyrroles. The volatile profile of the oil extracted from raw kernels is dominated by the presence of lipid oxidation products, such as pentanal, hexanal, and 2-pentyl-furan, and also by N-heterocycles, such as 1-methyl-1-pyrrole, 2,6-dimethyl pyrazine and other pyrazines, alcohols (1-hexanol, 2,3-butanediol, 2-methyl-1-propanol), ketones (acetoin), and acetic acid (El Monfalouti et al., 2013; Gracka et al., 2018). The profile of oil obtained from roasted kernels is characterized by markedly increased levels of products of the Maillard reaction, such as the Strecker aldehydes 2-methyl propanal, 2- and 3-methyl butanal, and also the N-heterocycles 1-methyl-1-pyrrole, methyl pyrazine, 2,6-dimethyl pyrazine, 2-ethyl-5-methyl pyrazine, trimethyl pyrazine, 3-ethyl-2,5-dimethyl pyrazine, and furfural. On the contrary, products of lipid oxidation, such as hexanal and 2-pentyl furan, tend to decrease when the roasting process is relatively short ($<$20 min), whereas for longer roasting times an increase in their content was observed (El Monfalouti et al., 2013). El Monfalouti et al. (2013) investigated the effect of roasting time on volatile content in the extracted oil, highlighting that most of roasting related volatiles started to be formed after 15–25 min of roasting. In addition, Gracka et al. (2018) determined key odorants, by gas chromatography-olfactometry analysis, and sensory profile of both raw and roasted kernels oils. Raw kernel oil was characterized by oily, slightly acidic and green odor notes, with key odorants represented by 2- and 3-methyl butanal, hexanal, 2-acetyl pyrroline, phenyl ethanol, and other unknown compounds. Roasted kernel oil aroma, which was markedly more intense, was described as roasted and nutty, wherein key odorants were 3-isopropyl-2-methoxy pyrazine, 2-ethyl-3,5-dimethyl pyrazine, 2-*S*-butyl-3-methoxy pyrazine, 3-mercapto-3-methyl-1-butanol, 2-acetyl pyrroline, 2,3,5-trimethyl pyrazine, (Z)-2-nonenal, phenylacetaldehyde, 2- and 3-methyl butanal, 2,3-butanedione, 1-octen-3-one, and (E,Z)-2,4-nonadienal, some of them being present at very low levels. Both oils showed enhanced oily and rancid odor notes at the end of an accelerated storage experiment, whereas some aldehydes (propanal and pentanal in raw kernel oil, as well as pentanal, hexanal, heptanal, and nonanal in roasted kernel oil) were identified as potential oxidation markers during storage (Gracka et al., 2018). Results from another study confirmed that hexanal may be considered as a marker of oxidation of raw and roasted kernels argan oils. This is also related to the rise of off-odors in oils extracted by hand-pressing from roasted kernels and kernels coming from goat-digested fruits, and in oils extracted by mechanical pressing from unroasted kernels (Matthäus et al., 2010).

6 Edible and nonedible applications of cold pressed oil

Traditional use of argan oil is originally culinary, but it is used in traditional medicine for many purposes such as skin and joint issues. Charrouf and Guillaume (2008), in a review on the active components and potential benefits of argan oil, concluded how argan oil has become popular on the international edible oil market, thanks to its organoleptic properties and health-promoting properties. Moreover, the authors remarked that the increase of using argan oil has a key role in the maintenance of argan tree and forests, from the perspective of biodiversity (Guillaume, & Charrouf, 2016; Charrouf, & Guillaume, 2018).

7 Infrared spectroscopy combined with chemometrics applied to cold pressed oil

In recent years, the consumption of cold pressed oils has been increasing (Ramadan, 2013). However, due to the high prices of cold pressed oils, they are exposed to adulteration with cheaper and lower-quality refined oils of vegetal origin. Adulterate means to replace an original product by the addition or deletion of a compound, by a different one (usually cheaper). Several strategies have been applied to monitor and control the authenticity of argan oils for reasons of both human health and demand as well as with regard to controlling product quality in industrial laboratories. Among these, new frontiers are open toward the use of rapid analytical methods and chemometrics (Addou, Fethi, Chikri, & Rrhioua, 2016; Ge, Chen, Liu, & Zhao, 2014; Li, Wang, Zhao, Ouyang, & Wu, 2015). In this regard, it is worth mentioning the work of Casoni, Simion, and Sârbu (2019), where 30 cold pressed edible oils according to their UV–vis spectra and radical scavenging profiles using the 2,2-diphenyl-1-picrylhydrazyl (DPPH·) by chemometrics were characterized and classified. Addou et al. (2016), by studying the detection of argan oil adulteration with olive oil using fluorescence spectroscopy and chemometrics tools, developed a model with a regression coefficient of $r^2=0.992$, a Standard Error of Prediction (SEP) = 1.311, and bias = 0.31.

Fourier Transformed Infrared Spectroscopy (FTIR) spectroscopy can be defined as a "fingerprint analytical technique" for the structural identification of compounds considering that no two chemical structures will have the same FTIR spectrum (Yap, Chan, & Lim, 2007). FTIR provides a characteristic signature of chemical or biochemical substances in the sample by featuring their molecular vibrations (stretching, bending, and torsions of the chemical bonds) in specific infrared regions. FTIR is an example of a green and fast analytical technique that requires minimal or no sample preparation. Recently, FTIR spectroscopy has become an important analytical technique for the quality control analysis of agro-food products (Durazzo, D'Addezio, Camilli, et al., 2018; Durazzo, Kiefer, Lucarini, Camilli, et al., 2018; Durazzo, Kiefer, Lucarini, Marconi, et al., 2018). Moreover, this spectroscopic technique with chemometrics can be used for qualitative and quantitative analysis of food products (Rohman & Che Man, 2010; Lohumi, Lee, Lee, & Cho, 2015; Marikkar, Mirghani, & Jaswir, 2016; Lucarini, Durazzo, Sánchez Del Pulgar, Gabrielli, & Lombardi-Boccia, 2018). Recently, FTIR spectroscopic methods combined with multivariate data analyses have received a great deal of attention in adulteration analysis of different varieties of food products. The joined approach of advanced technologies with the multivariate statistical approach enlarges the possibilities to exploit a wide range of food traits and aspects (Durazzo, Kiefer, Lucarini, Camilli, et al., 2018; Durazzo, Kiefer, Lucarini, Marconi, et al., 2018; Durazzo, D'Addezio, Camilli, et al., 2018; Lucarini, Durazzo, Raffo, Giovannini, & Kiefer, 2019). In the literature, the adulteration of high-quality cold pressed oils with lower-quality refined vegetable oils, such as virgin coconut oil (Ge et al., 2014; Li et al., 2015; Ozulku, Yildirim, Toker, Karasu, & Durak, 2017), is well-documented.

For instance, Oussama, Elabadi, and Devos (2012), focused on the detection and quantification of argan oil adulteration with different edible oils, using mid-infrared spectroscopy with chemometrics. The PLS model has been established to predict the concentration of soybean and sunflower oil as adulterants in the calibration range between 0% and 30% (*w/w*) in argan oil with good prediction performances in the external validation. Generally, Reflectance–Fourier transform infrared (FTIR-ATR) spectroscopy combined with multivariate data analyses is an innovative green and rapid approach for determination of oil adulteration. The number of studies on the application of advanced analytical methods to cold pressed argan oil is rather limited. Fig. 1 shows the FTIR spectra of commercial cold-pressed argan oil, acquired using a Nicolet iS10 FT-IR spectrometer equipped with a diamond crystal ATR cell in the range of 4000–650 cm^{-1}, at a resolution of 4 cm^{-1}, in our laboratory. The spectra in Fig. 1 contain a multitude of bands that are more or less characteristic of food samples.

The peak at 3008 cm^{-1} is due to the stretching vibration of double bound =C—H (*cis-*). Asymmetric and symmetric stretching vibrations of CH_2 groups are found at 2923 and 2853 cm^{-1}, probably associated to hydrocarbures chains of lipids (Jović, Smolić, Jurišić, Meić, & Hrenara, 2013; Vlachos et al., 2006). The band at 1744 cm^{-1} correlated to the stretching vibration of the ester carbonyl functional groups of the triglycerides (Jović et al., 2013; Moharam & Abbas, 2010).

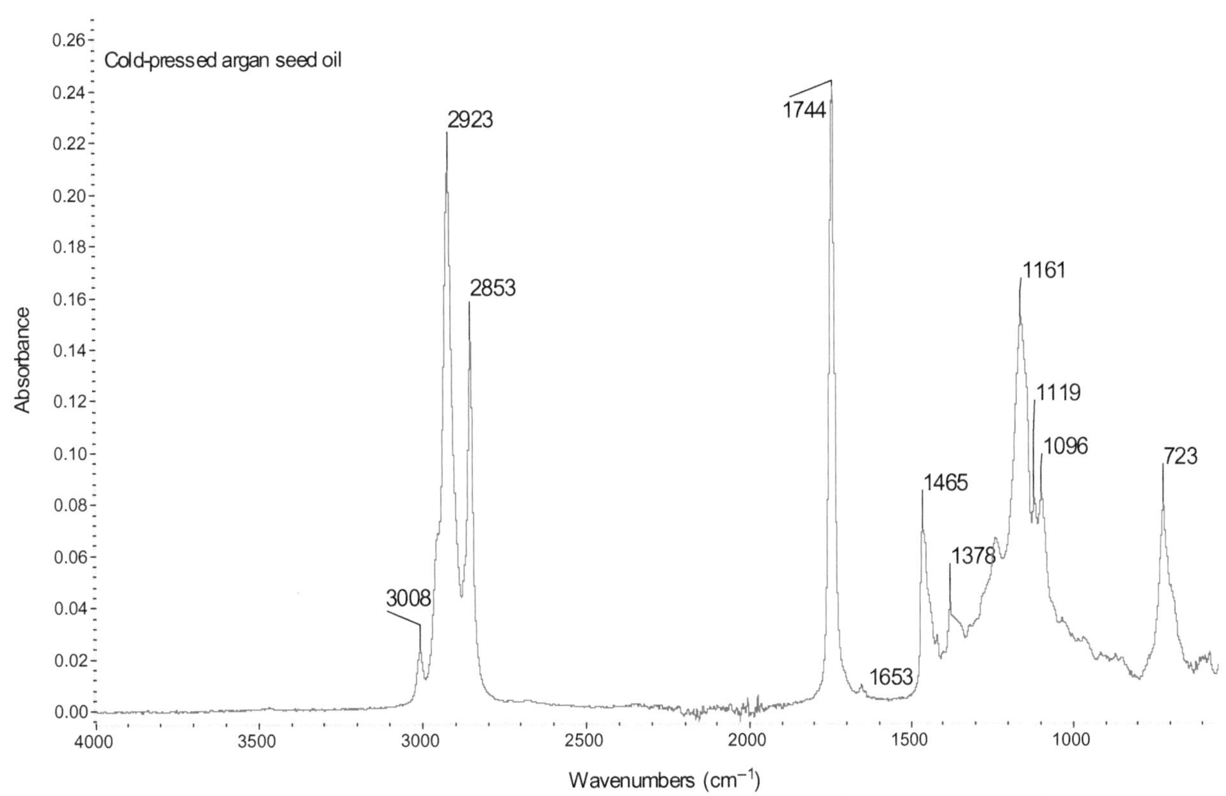

FIG. 1 FTIR spectra of a commercial cold-pressed argan oil acquired in the range of 4000–650 cm^{-1} at resolution of 4 cm^{-1}.

The peculiar band at 1653 cm^{-1} represents stretching vibrations of —C=C— of *cis*-olefins (of disubstituted olefins, RHC=CHR) (Guillen & Cabo, 1997; Vlachos et al., 2006). The IR region from 1500 to 700 cm^{-1} is referred as the "fingerprint" region, which includes bands corresponding to the vibrations of the C—O, C—C, C—H, and C—N bonds. This region is, on the one hand, very rich in information but, on the other, difficult to analyze due to its complexity. The bands at 1464 cm^{-1} and 1378 cm^{-1} are due to bending (scissoring) vibrations of bonds of aliphatic groups of CH_2 and CH_3. The bands at 1161 cm^{-1} and 109 cm^{-1} are correlated to stretching vibrations of the C—O bond of the ester group (Gouvinhas, de Almeida, Carvalho, Machado, & Barros, 2015; Guillen & Cabo, 1997; Jović et al., 2013; Ozulku et al., 2017). The band at 723 cm^{-1} is related to overlapping peaks of the CH_2 rocking vibration and the out-of-plane vibration of *cis*-disubstituted olefins, related to the presence of long-chain hydrocarbons (Vlachos et al., 2006).

References

Addou, S., Fethi, F., Chikri, M., & Rrhioua, A. (2016). Detection of argan oil adulteration with olive oil using fluorescence spectroscopy and chemometrics tools. *Journal of Materials and Environmental Science*, *7*(8), 2689–2698.

Andrew, R., & Izzo, A. A. (2017). Principles of pharmacological research of nutraceuticals. *British Journal of Pharmacology*, *174*, 1177–1194.

Berrougui, H., Alvarez de Sotomayor, M., Perez-Guerrero, C., der Ettaib, A., Hmamouchi, M., Marhuenda, E., et al. (2004). Argan (*Argania spinosa*) oil lowers blood pressure and improves endothelial dysfunction in spontaneously hypertensive rats. *British Journal of Nutrition*, *92*(6), 921–929.

Çakaloğlu, B., Özyurt, V. H., & Ötleş, S. (2018). Cold press in oil extraction: A review. *Ukrainian Food Journal*, *7*(4).

Casoni, D., Simion, I. M., & Sârbu, C. (2019). A comprehensive classification of edible oils according to their radical scavenging spectral profile evaluated by advanced chemometrics. *Spectrochimica Acta Part A, Molecular and Biomolecular Spectroscopy*, *213*, 204–209.

Charrouf, Z., & Guillaume, D. (2008). Argan oil: Occurrence, composition, and impact on human health. *European Journal of Lipid Science and Technology*, *110*, 632–636.

Charrouf, Z., & Guillaume, D. (2010). The rebirth of the argan tree or how to give a future to Amazigh women. In R. L. Harpelle, & B. Muirhead (Eds.), *Long-term solutions for a short-term world: Canada and research development*. Waterloo: WLU Press ISBN 13: 978-1-55458-223-225.

Charrouf, Z., & Guillaume, D. (2018). The argan oil project, going from utopia to reality in 20 years. *Oilseed and Fats, Crops and Lipids*, *25*, D209.

Daliu, P., Santini, A., & Novellino, E. (2018). From pharmaceuticals to nutraceuticals: Bridging disease prevention and management. *Expert Review of Clinical Pharmacology*, *28*, 1–7.

Durazzo, A. (2018). Extractable and non-extractable polyphenols: An overview. F. Saura-Calixto, & J. Pérez-Jiménez (Eds.), *Non-extractable polyphenols and carotenoids: Importance in human nutrition and health* (p. 37). *Food Chemistry, Functional and Analysis No 5*, London, UK: Royal Society of Chemistry.

Durazzo, A., D'Addezio, L., Camilli, E., Piccinelli, R., Turrini, A., Marletta, L., et al. (2018). From plant compounds to botanicals and back: A current snapshot. *Molecules*, *23*(8) Pii: E1844.

Durazzo, A., Kiefer, J., Lucarini, L., Camilli, E., Marconi, S., Gabrielli, P., et al. (2018). Qualitative analysis of traditional Italian dishes: FTIR approach. *Sustainability*, *10*(11), 4112.

Durazzo, A., Kiefer, J., Lucarini, M., Marconi, S., Lisciani, S., Camilli, E., et al. (2018). An innovative and integrated food research approach: Spectroscopy applications to milk and a case study of a milk-based dishes. *Brazilian Journal of Analytical Chemistry*, *5*, 12–27.

Durazzo, A., & Lucarini, M. (2019). A current shot and re-thinking of antioxidant research strategy. *Brazilian Journal of Analytical Chemistry*, *5*, 9–11.

El Monfalouti, H., Charrouf, Z., Guillaume, D., Denhez, C., & Zeppa, G. (2013). Volatile compound formation during argan kernel roasting. *Natural Product Communications*, *8*, 33–36.

El Monfalouti, H., Guillaume, D., Denhez, C., & Charrouf, Z. (2010). Therapeutic potential of argan oil: A review. *Journal of Pharmacy and Pharmacology*, *62*, 1669–1675.

Ge, F., Chen, C., Liu, D., & Zhao, S. (2014). Rapid quantitative determination of walnut oil adulteration with sunflower oil using fluorescence spectroscopy. *Food Analytical Methods*, *7*, 146–150. https://doi.org/10.1007/s12161-013-9610-z.

Gharby, S., Harhar, H., Guillaume, D., Haddad, A., & Charrouf, Z. (2012). The origin of virgin argan oil's high oxidative stability unraveled. *Natural Product Communications*, *7*(5), 621–624.

Gouvinhas, I., de Almeida, J. M. M. M., Carvalho, T., Machado, N., & Barros, A. I. R. N. A. (2015). Discrimination and characterisation of extra virgin olive oils from three cultivars in different maturation stages using Fourier transform infrared spectroscopy in tandem with chemometrics. *Food Chemistry*, *174*, 226–232.

Gracka, A., Majcher, M., Kludská, E., Hradecký, J., Hajšlová, J., & Jeleń, H. H. (2018). Storage-induced changes in volatile compounds in argan oils obtained from raw and roasted kernels. *Journal of the American Oil Chemists' Society*, *95*(12), 1475–1485.

Grosshagauer, S., Steinschaden, R., & Pignitter, M. (2019). Strategies to increase the oxidative stability of cold pressed oils. *LWT- Food Science and Technology*, *106*, 72–77.

Guillaume, D., & Charrouf, Z. (2016). Functional food and sustainable development once met in the argan forest: The tale of argan oil. In K. Kristbergsson, & S. Otles (Eds.), *Functional properties of traditional foods* (pp. 309–320). New York Inc.: Springer-Verlag.

Guillaume, D., Pioch, D., & Charrouf, Z. (2019). Argan [*Argania spinosa* (L.) skeels] oil. In M. Ramadan (Ed.), *Fruit oils: Chemistry and functionality*. Cham: Springer.

Guillen, M. D., & Cabo, N. (1997). Characterization of edible oils and lard by Fourier transform infrared spectroscopy. Relationships between composition and frequency of concrete bands in the fingerprint region. *Journal American Oil Chemists' Society*, *74*, 1281–1286.

Henry, F., Danoux, L., Moser, P., Charrouf, Z., & Pauly, G. (2002). New potential cosmetic active ingredient containing polyphenols from *Argania Spinosa* L. skeels leaves. In: *Polyphenols communications 2002 XXI international conference on polyphenols Marrakech-Morocco, September 9–12, 2002 Vol. 2 Ismail El Hadrami*.

Hilali, M., Charrouf, Z., El Aziz, S. A., Hachimi, L., & Guillaume, D. (2005). Influence of origin and extraction method on argan oil physico-chemical characteristics and composition. *Journal of Agricultural and Food Chemistry*, *53*, 2081–2087.

Jović, O., Smolić, T., Jurišić, Z., Meić, Z., & Hrenara, T. (2013). Chemometric analysis of Croatian extra virgin olive oils from Central Dalmatia region. *Croatica Chemica Acta*, *86*, 335–344.

Kartah, B. E., El Monfalouti, H., Harhar, H., Gharby, S., Charrouf, Z., & Matthaus, B. (2015). Effect of filtration on virgin argan oil: Quality and stability. *Journal of Materials and Environmental Science*, *6*, 2871–2877.

Li, B., Wang, H., Zhao, Q., Ouyang, J., & Wu, Y. (2015). Rapid detection of authenticity and adulteration of walnut oil by FTIR and fluorescence spectroscopy: A comparative study. *Food Chemistry*, *181*, 25–30.

Lohumi, S., Lee, S., Lee, H., & Cho, B. K. (2015). A review of vibrational spectroscopic techniques for the detection of food authenticity and adulteration. *Trends in Food Science & Technology*, *46*, 85–98.

Lucarini, M., Durazzo, A., Raffo, A., Giovannini, A., & Kiefer, J. (2019). Passion fruit (*Passiflora* spp.) seed oil. In M. Ramadan (Ed.), *Fruit oils: Chemistry and functionality* (pp. 577–603). Cham, Switzerland AG: Springer-Nature.

Lucarini, M., Durazzo, A., Sánchez Del Pulgar, J., Gabrielli, P., & Lombardi-Boccia, G. (2018). Determination of fatty acid content in meat and meat products: The FTIR-ATR approach. *Food Chemistry*, *267*, 223–230.

Madawalaa, S. R. P., Kochharb, S. P., & Duttaa, P. C. (2012). Lipid components and oxidative status of selected specialty oils. *Grasas y Aceites*, *63*(2), 143–151.

Males, Z., Mišković, G., Bojić, M., & Ćurak, I. (2018). Argan oil. *Farmaceutski glasnik*, *74*(11), 817–824.

Marikkar, J. N. M., Mirghani, M. E. S., & Jaswir, I. (2016). Application of chromatographic and infra-red spectroscopic techniques for detection of adulteration in food lipids: A review. *Journal of Food Chemistry and Nanotechnology*, *2*, 32–41.

Matthäus, B. (2013). Quality parameters for cold pressed edible argan oils. *Natural Product Communications*, *8*(1), 37–41.

Matthäus, B., & Brühl, L. (2015). Quality parameters for the evaluation of cold-pressed edible argan oil. *Journal für Verbraucherschutz und Lebensmittelsicherheit*, *10*, 143–154.

Matthäus, B., Guillaume, D., Gharby, S., Haddad, A., Harhar, H., & Charrouf, Z. (2010). Effect of processing on the quality of edible argan oil. *Food Chemistry*, *120*(2), 426–432.

Matthäus, B., Willenberg, I., Engert, S., & Steinberg, P. (2019a). The German National Reference Centre for authentic food (NRZ-Authent). *OCL*, *26*, 11.

Matthäus, B., Willenberg, I., Engert, S., & Steinberg, P. (2019b). The German National Reference Centre for authentic food (NRZ-Authent). *OCL*, *26*, 11.

Moharam, M. A., & Abbas, L. M. (2010). A study on the effect of microwave heating on the properties of edible oils using FTIR spectroscopy. *African Journal of Microbiological Research*, *4*, 1921–1927.

Oussama, A., Elabadi, F., & Devos, O. (2012). Analysis of Argan oil adulteration using infrared spectroscopy. *Spectroscopy Letters*, *45*(6), 458–463.

Ozulku, G., Yildirim, R. M., Toker, O. S., Karasu, S., & Durak, M. Z. (2017). Rapid detection of adulteration of cold pressed sesame oil adultered with hazelnut, canola, and sunflower oils using ATR-FTIR spectroscopy combined with chemometric. *Food Control*, *82*, 212–216.

Ramadan, M. F. (2013). Healthy blends of high linoleic sunflower oil with selected cold pressed oils: Functionality, stability and antioxidative characteristics. *Industrial Crops and Products*, *43*, 65–72.

Rohman, A., & Che Man, Y. B. (2010). FTIR spectroscopy combined with chemometrics for analysis of lard in the mixtures with body fats of lamb, cow, and chicken. *International Journal of Food Research*, *17*, 519–526.

Santini, A., Cammarata, S. M., Capone, G., Ianaro, A., Tenore, G. C., Pani, L., et al. (2018). Nutraceuticals: Opening the debate for a regulatory framework. *British Journal of Clinical Pharmacology*, *84*(4), 659–672.

Santini, A., & Novellino, E. (2014). Nutraceuticals: Beyond the diet before the drugs. *Current Bioactive Compounds*, *10*, 1–12.

Santini, A., & Novellino, E. (2018). Nutraceuticals: Shedding light on the grey area between pharmaceuticals and food. *Expert Review of Clinical Pharmacology*, *11*(6), 545–547.

Santini, A., Tenore, G. C., & Novellino, E. (2017). Nutraceuticals: A paradigm of proactive medicine. *European Journal of Pharmaceutical Sciences*, *96*, 53–61.

Vlachos, N., Skopelitis, Y., Psaroudaki, M., Konstantinidou, V., Chatzilazarou, A., & Tegou, E. (2006). Applications of Fourier transform-infrared spectroscopy to edible oils. *Analytica Chimica Acta*, *573–574*, 459–465.

Yap, K. Y. L., Chan, S. Y., & Lim, C. S. (2007). Infrared-based protocol for the identification and categorisation of ginseng and its products. *Food Research International*, *40*, 643–652.

Chapter 42

Cold pressed *Moringa oleifera* seed oil

Natascha Cheikhyoussef[a], Martha Kandawa-Schulz[b], Ronnie Böck[c], and Ahmad Cheikhyoussef[d]

[a]*Ministry of Higher Education, Technology and Innovation, Windhoek, Namibia,* [b]*Department of Chemistry and Biochemistry, University of Namibia, Windhoek, Namibia,* [c]*Department of Biological Sciences, University of Namibia, Windhoek, Namibia,* [d]*Science and Technology Division, Multidisciplinary Research Centre, University of Namibia, Windhoek, Namibia*

Abbreviations

DUFAs di-unsaturated fatty acids
HDL high-density lipoprotein
MUFAs monounsaturated fatty acids
PUFAs polyunsaturated fatty acids
SFAs saturated fatty acids
TUFAs tri-unsaturated fatty acids
USFAs unsaturated fatty acids

1 Introduction

Moringa oleifera Lam. belongs to the family *Moringaceae*, and is predominantly found in India, and tropical and arid countries. *M. oleifera* is commonly known and is the most researched plant among the other species of *Moringa* due to its wide distribution (Tsaknis, Lalas, Gergis, & Spiliotis, 1998). The drumstick tree, horseradish tree, and ben tree are some common names used to describe the tree. It is a deciduous, fast-growing tree that can reach a height of 10–12 m (Parrotta, 2009). The *Moringa* fruit contains 10–12 three-angled, globular seeds that are about 1 cm in diameter (Sánchez-Machado & Núñez-Gastélum, 2010; Leone et al., 2016). The various plant parts of *M. oleifera* such as the green immature pods, flowers, leaves, and roots are edible (Sánchez-Machado & Núñez-Gastélum, 2010). The different plant parts of the tree have been subject to extensive research for many years due to the beneficial properties and applications in the medicinal, food and fuel industries. Several reviews reported on the beneficial uses in food, cosmetics, and medicine of the *M. oleifera* and it being a multipurpose tree (Anwar, Latif, Ashraf, & Gilani, 2007; Brilhante et al., 2017; Fahey, 2005; Farooq, Rai, Tiwari, Khan, & Farooq, 2012; Fuglie, 1999).

Knowledge of the *M. oleifera* seed oil for use in food and as watch oil is existing since over a century (Brown, Kleiman, & Hill, 2003). Use of the oil in cosmetics by the ancient Egyptians is well documented (Ghazali & Mohammed, 2011; Kleiman, Brown, & Hill, 2006). As far back as 1848, the first compositional analysis of the oil was carried out that determined the presence of the saturated fatty acid, behenic acid/docosanoic acid ($C_{21}H_{43}COOH$), resulting in the oil generally being referred to as "Ben oil" (Brown et al., 2003). Ben oil is edible, has a nutty flavor (peanut-like), a pleasant aroma, and is nondrying and yellow in color (Abdulkarim, Long, Lai, Muhammad, & Ghazali, 2005; Brown et al., 2003; Tsaknis, Lalas, Gergis, Dourtoglou, & Spiliotis, 1999). Reviews on the properties, composition and application potentials of the *M. oleifera* seed oil obtained through various extraction methods (Soxhlet extraction, cold pressing, supercritical carbon dioxide extraction, microwave-assisted extraction) have been published by several researchers (Boukandoul, Casal, & Zaidi, 2018; Ghazali & Mohammed, 2011; Mariod, Mirghani, & Hussein, 2017; Nadeem & Imran, 2016). This chapter specifically summarizes the composition, physicochemical properties, health-promoting traits, and other applications of cold pressed *M. oleifera* seed oil.

Cold Pressed Oils. https://doi.org/10.1016/B978-0-12-818188-1.00042-6
Copyright © 2020 Elsevier Inc. All rights reserved.

2 Extraction and processing of cold pressed oil

Mature pods of the *M. oleifera* tree are collected and the seeds are de-hulled. Seeds are dried at 35–40°C before oil extraction (Barakat & Ghazal, 2016). The kernel is used for the oil extraction process. Seeds are generally milled into a fine paste with the addition of water (1 seed:2 water) (Tsaknis et al., 1998; Tsaknis, Lalas, et al., 1999; Tsaknis, Spiliotis, Lalas, Gergis, & Dourtoglou, 1999) or simply milled to fine particles (Barakat & Ghazal, 2016) before extraction with an oil extraction press. The oil yield after mechanical pressing is about 11%–62% (Table 1). Crude extracted oils with a cloudy appearance are generally degummed. Degumming, a refining process, of the cold pressed oil has been carried out by several researchers before analysis of the oil characteristics and composition (Lalas & Tsaknis, 2002; Tsaknis et al., 1998). Tsaknis et al. (1998) recovered 98.7% of oil after the degumming process of cold pressed *Moringa* seed oil. Anwar and Rashid (2007) reported that the concentration of the tocopherols in *M. oleifera* oil (from Pakistan) considerably decreased after degumming. Sánchez-Machado et al. (2015) determined the effect of refining (neutralization, degumming, bleaching) on *n*-hexane-extracted *M. oleifera* oil and reported that the process did not have a significant effect on the composition of the fatty acids, except on stearic acid and α-linolenic acid. The refining process improved the oxidative stability, increased the viscosity of the oil, and increased the saponification value, but had no effect on the iodine value (Sánchez-Machado et al., 2015). Fotouo-M, du Toit, and Robbertse (2016) have pressed de-hulled seeds with a screw press to obtain the oil and have additionally ground the pressed cake to a fine paste followed by Soxhlet extraction before analysis. Aptowitz (2014) through MIT's D-Lab and the start-up MoringaConnect have developed a device with a rotating chamber for farmers in Ghana that is operated via a foot pedal to cost-effectively deshell and winnow the seeds of *Moringa* (20 kg/h). The rotating mechanisms cause the shells to crack and to escape through the chamber's mesh system (Aptowitz, 2014).

Mechanical pressing or cold pressing is commonly done at the local/rural level (Boukandoul et al., 2018) and by small to medium enterprises. Although extraction of the *M. oleifera* seed oil by organic solvents produces higher yields of oil recovery as compared to the cold pressing method (Tsaknis, Lalas, et al., 1999; Ogunsina et al., 2014), this method is becoming more popular due to the need to curb the environmental impact of organic solvents and the need to preserve important bioactive compounds (Boukandoul et al., 2018). Research is underway toward optimizing the design and process parameters for oil extraction from *M. oleifera* seeds (Ajav & Fakayode, 2013; Aviara, Musa, Owolarafe, Ogunsina, & Oluwole, 2015; Fakayodea & Ajav, 2016). The optimization targets are generally to increase oil yield, reduce extraction time and associated extraction costs. The design of a suitable oil expresser requires knowledge of the mechanical properties of the respective oilseed of interest.

Krotowski, Yarnall, Lombardi, and Katenov (2014) have developed a seed press to extract *Moringa* oil that includes a mechanism for stripping the seeds from the pods (*Moringa* Pod Stripper) and a shell cracker to remove the layer encasing the seed pulp. Ajav and Fakayode (2013) have determined the mechanical properties such as deformation (5.099 mm), maximum strain (85%), crushing force (58.535 N), and rupture energy (0.1344 N.m) of the seeds of *M. oleifera* (from Nigeria). The oil point pressures (MPa) for *M. oleifera* (from Nigeria) at different conditions such as heating temperature (50–100°C) and time (15–30 min, 5 min interval) and moisture content (4.78%, 6%, 8%, and 10%) were reported by Aviara et al. (2015). Increased moisture content resulted in increased oil point pressure, whereby increasing heating temperature and time caused a decrease in oil point pressure (Aviara et al., 2015). The moisture content of 10% at 50°C for 15 min resulted in the highest oil point pressure (1.1239 MPa), while the lowest pressure (0.3164 MPa) was obtained with a moisture content of 4.78% at 100°C for 30 min. The model generated by Aviara et al. (2015) using Response Surface Methodology can be used for optimizing and controlling the oil expression from *M. oleifera* seeds. Fakayodea and Ajav (2016) have applied a 4 × 5 Central Composite Rotatable Design with factors such as pressure, heating temperature, time, and moisture content for the optimization of the oil expression (mechanically) from *M. oleifera* (from Nigeria) seeds. The moisture content of 11% at 80°C and 20 MPa for 30 min resulted in the highest oil yield (25.5%). The predicted values were oil yield (28.2%) at a moisture content of 11.3% at 85.5°C and 19.63 MPa for 27.17 min (Fakayodea & Ajav, 2016).

3 Physical and chemical characteristics of cold pressed oil

The reported physical and chemical properties of cold pressed *M. oleifera* are presented in Table 1. The oils are characterized by low acid values (1.20%–2.25% as oleic acid) and peroxide values (0.11–3.33 meq O_2/kg oil), and are within the acceptable standards for cold pressed oils according to the CODEX-STAN 210-1999. The saponification number has been reported to range from 172 to 199.32 mg KOH/g oil with an iodine value of 65–68.56 g/I_2/100 g oil. The saponification number is in the range for that of olive oil (184–196), soybean oil (189–195), and sunflower oil (188–194) (CODEX-STAN 210-1999, 2019). The iodine value for cold pressed *M. oleifera* oil is generally lower than the values reported for crude vegetable oils (CODEX-STAN 210-1999, 2019). The presence of polar compounds has been reported by

TABLE 1 **Physicochemical characteristics of cold pressed *Moringa oleifera* seed oil.**

Oil yield (%)	26.3	25.8	25.1	61.5	–	–	–	–	11.36
Acid value	1.20[b] % as oleic acid	1.01[b]–2.06% as oleic acid	–	–	2.25	–	0.29–0.37 mg/g	–	8.8 mg/g
Free fatty acid (% oleic acid)	–	1.01b	1.94b	3.5	–	1.33	–	1.75e	–
Peroxide value (meq O_2/kg of oil)	0.19	0.36	0.11	1.0	3.33	0.65	1.67–2.47	–	3.3
Iodine value (g/I_2/100 g oil)	65.83	66.81	65.73	67.8	68.56	–	65–67	66.43	68.9
Saponification number (mg KOH/g oil)	188.70	179.80	199.32	190.4	121.51	–	172–178	–	180.6
Specific gravity at 25°C	–	–	–	0.93	–	–	0.901–0.918	–	–
Density (mg/mL)	0.9016 (24°C)	0.9037 (24°C)	0.899 (24°C)	0.90 (25°C)	0.93	–	–	0.899d	0.9095 (20°C)
Viscosity (mPa.s)	91	103	80	43.8	–	84.4	–	40.5c	43.6
Color (red, yellow)	2.0, 28	1.90, 30	1.9, 30	1, 25	–	–	–	–	–
Refractive index (n_D 40°C)	1.4598	1.4591	1.460	1.47	–	–	1.4607–1.4613	–	–
Polar compounds (g/100 g oil)	–	3.58	4.37	3.1%	–	–	–	–	–
Smoke point (°C)	204	201	203	–	–	204.5	–	–	–
Induction time (h)	–	34.1	28.2	–	–	–	–	26.86	–
Origin	Malawi	Kenya	India	India	Thailand	India	Egyptc	South Africa	Brazil
Reference	Tsaknis et al. (1998)a	Tsaknis, Lalas, et al. (1999); Tsaknis, Spiliotis, et al. (1999)a	Lalas and Tsaknis (2002)a	Ogunsina et al. (2014)	Ruttarattanamongkol and Petrasch (2015)	Janaki (2015)	Barakat and Ghazal (2016)	Fotouo-M et al. (2016)	Pereira et al. (2016)

[a] *Degummed oil.*
[b] *Nondegummed oil.*
[c] *Data represented from three different regions of Egypt, namely Asuit, Ismalia, and Monofya governorates (Barakat & Ghazal, 2016).*
[d] *Screw press.*
[e] *Hexane (pressed cake + screw press).*

Tsaknis, Lalas, et al. (1999) and Ogunsina et al. (2014). Frying of cold pressed *M. oleifera* oil resulted in the lowest amount of polar compounds compared to *n*-hexane extracted *M. oleifera* oil (Lalas & Tsaknis, 2002). Smoke point and induction time for the oil are 201–204.5°C and 26.86–34.1°C, respectively. The oil viscosity ranges between 43.8 and 103 mPa.s. Lalas and Tsaknis (2002) found that cold pressed *M. oleifera* oil has the least change in viscosity after 10 periods of frying compared to *n*-hexane-extracted *M. oleifera* oil and olive oil.

4 Fatty acid composition of cold pressed oil

The reported fatty acids contained in cold pressed *M. oleifera* are presented in Table 2. The oil is composed of 15%–23% saturated fatty acids, 72%–83% monounsaturated fatty acids, 0.7%–2.8% polyunsaturated fatty acids, and 72%–83% total unsaturated fatty acids. The content of monounsaturated fatty acids is similar to that reported for cold pressed avocado oil

TABLE 2 Fatty acid compositions of cold pressed *Moringa oleifera* seed oil.

Caprylic acid (C8:0)	0.03	0.03	0.04	–	–	–	–	–
Myristic (C14:0)	0.10	0.11	0.13	0.24	–	–	–	0.2
Palmitic (C16:0)	5.40	5.73	6.34	5.8	12.97	6.25	5.66–6.44	5.8
Palmitoleic (C16:1)	0.11	0.10	0.10	1.2	–	1.18	1.43–1.92	1.4
C16:1 *cis*ω7	0.98	1.32	1.28					
Margaric (C17:0)	0.04	0.09	0.08	–	1.40	–	0.08–0.09	–
Heptadecenoic (C17:1)	–	0.04	–	–	–	–	0.06–0.10	–
Stearic (C18:0)	5.80	3.83	5.70	3.9	2.95	6.29	4.79–7.94	6.2
Oleic (C18:1)	67.65	75.39	71.60	79.5	77.40	71.87	73.30–79.58	70.2
Vaccenic acid (C18:1)	–	–	–	–	–	–	–	4.6
Linoleic (C18:2)	0.69	0.72	0.77	nd	1.40	0.50	0.58–0.59	0.4
α-Linolenic (C18:3)	0.20	0.20	0.20	2.2	1.39	0.20	0.15–0.17	–
Arachidic (C20:0)	3.72	2.52	3.52	2.2	–	3.85	1.57–5.10	–
Arachidonic acid (C20:0)	–	–	–	–	–	–	–	3.7
Gadoleic (C20:1)	2.64	2.54	2.24	nd	–	2.27	1.18–3.16	1.9
Behenic (C22:0)	6.74	5.83	6.21	5.1	–	6.00	2.62–3.62	5.6
Erucic acid (C22:1)	0.12	0.15	0.12	–	–	–	–	–
Lignoceric (C24:0)	nd	nd	nd	–	–	1.05	–	–
Cerotic acid (C26:0)	0.90	0.96	1.21	–	–	–	–	–
SFA	22.73	19.1	23.23	17.2	17.24	23.44	15.00–22.83	21.5
MUFAs	71.50	79.54	75.34	80.7	80.7	75.32	75.91–82.84	78.1
DUFAs	0.69	0.72	0.77	–	1.40	0.50	0.58–0.59	0.4

TABLE 2 Fatty acid compositions of cold pressed *Moringa oleifera* seed oil—cont'd

TUFAs	0.20	0.20	0.20	2.2	1.39	0.20	0.15–0.17	0.0
PUFAs	0.99	0.92	0.97	2.2	2.79	0.70	0.73–0.76	0.0
USFA	72.49	80.46	77.28	82.9	83.49	76.02	77.14–88.98	78.50
Origin	Malawi	Kenya	India	India	India	Thailand	Egypt	Brazil
Reference	Tsaknis et al. (1998)[a]	Tsaknis, Lalas, et al. (1999); Tsaknis, Spiliotis, et al. (1999)[a]	Lalas and Tsaknis (2002)[a]	Ogunsina et al. (2014)	Janaki (2015)	Ruttarattanamongkol and Petrasch (2015)	Barakat and Ghazal (2016)[b]	Pereira et al. (2016)

[a]*Degummed oil.*
[b]*Data represented from three different regions of Egypt, namely Asuit, Ismalia, and Monofya governorates (Barakat & Ghazal, 2016);* SFA, *saturated fatty acids;* MUFAs, *monounsaturated fatty acids;* DUFAs, *di-unsaturated fatty acids;* TUFAs, *tri-unsaturated fatty acids;* PUFAs, *polyunsaturated fatty acids;* USFAs, *unsaturated fatty acids.*

(74.7%), macadamia nut oil (81.3%), and olive oil (74.9%) (Diamante & Lan, 2014). Seventeen fatty acids have been detected in cold pressed *M. oleifera* oil originating from different countries. The dominant fatty acid is oleic acid (67.6%–79.5%), which is similar to olive oil (55.0%–83.0%) (Tsimidou, Blekas, & Boskau, 2003), which categorizes the oil into the high-oleic acid oil group. Oil high in monounsaturated fatty acids, such as oleic acid, impose oxidative stability on the oils (Sánchez-Machado et al., 2015). Linoleic acid (0.5%–0.77%) and α-linolenic acid (0.15%–1.39%) are present in low concentrations. Palmitic acid (5%–13%), stearic acid (3.8%–8.0%), and behenic acid (2.6%–6.74%) are found at higher concentrations, while arachidic acid and gadoleic acid are found at lower concentrations of 1.57%–5.10% and 1.18%–3.16%, respectively. Arachidic acid, gadoleic acid, and behenic acid are found in considerable higher amounts compared to some of the common vegetable oils (CODEX-STAN 210-1999, 2019). The presence of heptadecenoic acid, palmitoleic acid, arachidonic acid, lignoceric acid, erucic acid, and cerotic acid have been reported by few researchers (Table 2). Trace amounts of C16:1*cis*ω7 have been reported to be contained in cold pressed *M. oleifera* oil from Malawi, Kenya, and India (Lalas & Tsaknis, 2002; Tsaknis et al., 1998; Tsaknis, Spiliotis, et al., 1999).

5 Tocopherols and sterols profile of cold pressed oil

The reported tocopherol and sterol compositions in cold pressed *M. oleifera* are presented in Tables 3 and 4, respectively. The highest tocopherol content has been reported for the cold pressed *M. oleifera* oil from Malawi (Tsaknis et al., 1998). The concentration of α-, γ-, and δ-tocopherols in cold pressed *M. oleifera* oil from Malawi, Kenya, and India have been reported to be 5.05–226.9 mg/kg, 15.82–71.47 mg/kg, and 3.55–216.57 mg/kg, respectively (Table 3). Total tocopherol concentration reported for the oil was in the range of 34–514.94 mg/kg. The presence of □-tocopherol has not been

TABLE 3 Tocopherol compositions (mg/kg oil) of cold pressed *Moringa oleifera* seed oil.

α-tocopherol	226.9	101.46	42.61	5.06	–	5.05
γ-tocopherol	71.47	39.54	15.82	25.40	–	25.40
δ-tocopherol	216.57	75.67	24.21	3.55	–	3.55
Total tocopherols	514.94	216.67	82.64	34.01	95.5	34.00
Origin	Malawi	Kenya	Kenya	India	India	India
Reference	Tsaknis et al. (1998)b	Tsaknis, Lalas, et al. (1999)b	Tsaknis, Spiliotis, et al. (1999)a	Lalas and Tsaknis (2002)b	Ogunsina et al. (2014)	Janaki (2015)

[a]*Degummed oil.*
[b]*Nondegummed oil.*

TABLE 4 Sterol compositions (%) of cold pressed *Moringa oleifera* seed oil.

Cholesterol	0.19	0.13	0.18
Brassicasterol	0.12	–	0.06
$^{\Delta5,25-}$ergostadienol	0.08	–	–
24-methylene cholesterol	0.91	0.85	0.07
Campesterol	23.68	14.03	15.81
Campestanol	0.47	–	0.36
Stigmasterol	17.40	17.27	23.10
$^{\Delta5,24-}$ergostadienol	0.53	–	0.30
Clerosterol	0.65	0.95	2.08
□-sitosterol	46.73	49.19	45.58
Stigmastanol	0.87	1.05	0.76
$^{\Delta5-}$avenasterol	2.87	12.79	8.46
$^{\Delta5,23-}$stigmastadienol	1.23	–	–
$^{\Delta7,14-}$stigmastadienol	0.52	–	0.52
28-isoavenasterol	0.27	1.01	0.27
$^{\Delta7-}$avenasterol	0.45	0.94	0.53
$^{\Delta7-}$stigmastenol	0.81	–	–
$^{\Delta7,14-}$stigmastanol	–	0.83	0.35
Origin	Malawi	Kenya	India
Reference	Tsaknis et al. (1998)a	Tsaknis, Lalas, et al. (1999)a	Lalas and Tsaknis (2002)a

[a]*Degummed oil.*

reported. Higher concentrations of α-, γ-, and δ-tocopherols have been reported to be contained in cold pressed *M. oleifera* oil compared to solvent extracted oil (Tsaknis et al., 1998; Tsaknis, Lalas, et al., 1999). Cold pressed *M. oleifera* (variety Periyakulam) oil from India in comparison contained lower amounts of α- and δ-tocopherols, but had the highest amount of γ-tocopherol (Lalas & Tsaknis, 2002). The reported stability (resistance to oxidation) of cold pressed *M. oleifera* oil has been attributed to the presence of α-, γ-, and δ-tocopherols and the presence of the sterol, $^{\Delta5-}$avenasterol (Tsaknis et al., 1998; Tsaknis & Lalas, 2002). Tsaknis, Spiliotis, et al. (1999) and Tsaknis and Lalas (2002) have studied the changes in *M. oleifera* (Mbololo variety of Kenya) and *M. oleifera* (variety Periyakulum of India) cold pressed oil quality that occur during frying. Cold pressed *M. oleifera* oil experienced the least deterioration compared to the organic solvent extracted oil. The dominant sterols in cold pressed *M. oleifera* oil (Table 4) are reported to be campesterol, stigmasterol, and $^{\Delta5-}$avenasterol, with □-sitosterol being present at the highest concentrations. β-sitosterol, which is a main dietary phytosterol, has been reported to exhibit antiviral, antifungal, antiinflammatory (Malini & Vanithakumari 1990), anticarcinogenic, and anti-atherogenic properties (National Center for Biotechnology Information, 2020).

6 Health-promoting traits of *M. oleifera* cold pressed oil

Kusolrat and Kupittayanant (2013) have studied the effect of cold pressed *M. oleifera* seed oil on female ovariectomized rats and have found that the seed oil increases the serum estradiol, relative uterine weight, and serum high-density lipoprotein (HDL), a property that could be applied in the treatment of the symptoms of menopause. The microbial count of *Escherichia coli*, *Salmonella*, *Pseudomonas aeruginosa*, and *Staphylococcus aureus* on fried and deep-fried potatoes in cold pressed *M. oleifera* oil was investigated by Janaki and Yamuna Devi (2015) over a 15-day period. The results indicated that cold pressed *M. oleifera* oil has good resistance toward harmful bacteria, making the use of this oil especially suitable for food storage applications (Janaki & Yamuna Devi, 2015). The presence of a high concentration of monounsaturated fatty acids (MUFAs) in cold pressed *M. oleifera* oil could be functional in reducing cardiovascular disease due to the positive effects on low and high-density lipoprotein cholesterol and in the development of functional products as MUFAs have antithrombotic and antiinflammatory effects (Cicero & Gaddi, 2001). Abd-Rabou, Zoheir, Kishta, Shalby, and Ezzo (2016) have reported on the anticancer properties of *M. oleifera* seed oil in its pure form and as nano-formulations via mitochondrial-mediated apoptosis. The nano-form of the oil had a greater impact on the colorectal cancer Caco-2 and HCT 116 cytotoxicity via mitochondrial dysfunction triggering. The polyunsaturated fatty acids α-linolenic acid and linoleic acid were reported to reduce heavy cholesterol (low-density lipoprotein) (Chan et al. 1991, cited in Sánchez-Machado et al., 2015). Decreased saturated fatty acids and α-linolenic acid have been found to decrease blood pressure, triglyceride levels, and platelet aggregation (Wijendran & Hayes, 2004; Sánchez-Machado et al., 2015).

7 Applications of cold pressed *M. oleifera* oil

The cold pressed oil of *M. oleifera* can be used in both edible and nonedible (lubrication oil, cosmetic oil, biodiesel) applications due to the fatty acid composition, tocopherols composition, and its excellent resistance to autoxidation and thermal stability (Nadeem & Imran, 2016; Ogunsina et al., 2014; Rashid, Anwar, Moser, & Knothe, 2008). Since the stability (Tsaknis, Spiliotis, et al., 1999) is greater than that of some of the commercial oils such as groundnut oil (Ogunsina et al., 2014), soybean oil, and sunflower oil, the oil could be blended with these less stable oils, functioning as an antioxidant (Nadeem & Imran, 2016), and could be used in frying and cooking (Ogunsina et al., 2014). Cold pressed *M. oleifera* oil has been shown to have better organoleptic properties during repeated frying compared to organic solvent extracted oil (Tsaknis, Spiliotis, et al., 1999). Due to the oil's reported resistance to some harmful bacteria, applications in food preservation are possible (Janaki & Yamuna Devi, 2015), including applications as an antimicrobial agent (Atolani et al., 2016). The flow-ability of the oil, which is due to the presence of a high concentration of oleic acid, which also confers stability on the oil, allows it to be used for the production of biodiesel (Tsaknis, Spiliotis, et al., 1999). The search for the utilization of environmentally friendly and renewable raw materials in the industry is encouraging the use of vegetable oils for their chemical modification and product development. Oleic acid oils are being investigated in terms of their chemical modification and application in the bio-lubricant industry (Salimon & Salih, 2010). The oil is a good emollient (Rahman et al., 2009) and is free-spreading due to its low iodine value of 65–69 (Kleiman et al., 2006). The nondrying property of the oil allows it to be used in aromatherapy treatments and as massage oil (Ghazali & Mohammed, 2011; Kleiman et al., 2006). Cosmetic formulations containing *M. oleifera* oil have soothing properties on the skin and provide the unique skin feel property due to the presence of the behenic acid (3%–7%) found in the oil (Kleiman et al., 2006). This fatty acid has emollient and lubricant properties that contribute to skin hydration (L'ORÉAL Paris, 2009), making cold pressed *M. oleifera* oil unique for the development of skin formulations.

8 Conclusion

This chapter specifically described the physical and chemical properties, composition, health-promoting traits, and applications of cold pressed *M. oleifera* seed oil. This oil has excellent physical and chemical properties that make it suitable for various edible and nonedible applications. The high oxidative and thermal stability of the oil adds to its usefulness in the food, cosmetics, and fuel industries. In particular, the high content of MUFAs, the significant presence of behenic acid, tocopherols, campesterol, □-sitosterol, stigmasterol, and $^{\Delta5-}$avenasterol, allows this oil to be considered for further developments into nutraceutical and pharmaceutical products.

References

Abd-Rabou, A. A., Zoheir, K. M. A., Kishta, M. S., Shalby, A. B., & Ezzo, M. I. (2016). Nano-micelle of *Moringa oleifera* seed oil triggers mitochondrial cancer cell apoptosis. *Asian Pacific Journal of Cancer Prevention*, *17*(11), 4929–4933.

Abdulkarim, S. M., Long, K., Lai, O. M., Muhammad, S. K. S., & Ghazali, H. M. (2005). Some physico-chemical properties of *Moringa oleifera* seed oil extracted using solvent and aqueous enzymatic methods. *Food Chemistry*, *93*(2), 253–263.

Ajav, E. A., & Fakayode, O. A. (2013). Mechanical properties of Moringa (*Moringa oleifera*) seeds in relation to an oil expeller design. *Agrosearch*, *13*(3), 206–216.

Anwar, F., Latif, S., Ashraf, M., & Gilani, A. H. (2007). *Moringa oleifera*: A food plant with multiple medicinal uses. *Phytotherapy Research*, *21*(1), 17–25.

Anwar, F., & Rashid, U. (2007). Physico-chemical characteristics of *Moringa oleifera* seeds and seed oil from a wild provenance of Pakistan. *Pakistan Journal of Botany*, *39*(5), 1443–1453.

Aptowitz, R. (2014). *Treadle-powered seed sheller. Retrieved from (2014).* https://seelio.com/w/cbf/treadle_powered-seed-sheller>. Accessed 25 April 2019.

Atolani, O., Olabiyi, E. T., Issa, A. A., Azeez, H. T., Onoja, E. G., Ibrahim, S. O., et al. (2016). Green synthesis and characterisation of natural antiseptic soaps from the oils of underutilised tropical seed. *Sustainable Chemistry and Pharmacy*, *4*, 32–39.

Aviara, N. A., Musa, W. B., Owolarafe, O. K., Ogunsina, B. S., & Oluwole, F. A. (2015). Effect of processing conditions on oil point pressure of *Moringa oleifera* seed. *Journal of Food Science and Technology*, *52*(7), 4499–4506.

Barakat, H., & Ghazal, G. A. (2016). Physicochemical properties of *Moringa oleifera* seeds and their edible oil cultivated at different regions in Egypt. *Food and Nutrition Sciences*, *7*(6), 472–484.

Boukandoul, S., Casal, S., & Zaidi, F. (2018). The potential of some *Moringa* species for seed oil production. *Agriculture*, *8*(10), 150. https://doi.org/10.3390/agriculture8100150.

Brilhante, R. S. N., Sales, J. A., Pereira, V. S., Castelo-Branco, D. S. C. M., Cordeiro, R. A., de Souza Sampaio, C. M., et al. (2017). Research advances on the multiple uses of *Moringa oleifera*: A sustainable alternative for socially neglected population. *Asian Pacific Journal of Tropical Medicine*, *10*(7), 621–630.

Brown, J. H., Kleiman, R., & Hill, J. C. (2003). *Ultra-stable composition comprising Moringa oil and its derivatives and uses thereof.* US Patent 6667047 B2.

Cicero, A. F., & Gaddi, A. (2001). Rice bran oil and gamma-oryzanol in the treatment of hyperlipoproteinemias and other conditions. *Physiotherapy Research*, *15*(4), 277–286.

CODEX-STAN 210-1999 (2019) Codex standard for named vegetable oils. Retrieved from http://www.fao.org/3/y2774e/y2774e04.htm#bm4.1. [Accessed 6 May 2019].

Diamante, L. M., & Lan, T. (2014). Absolute viscosities of vegetable oils at different temperatures and shear rate range of 64.5 to 4835 s^{-1}. *Journal of Food Processing*. *2014*, 234583. https://doi.org/10.1155/2014/234583.

Fahey, J. W. (2005). *Moringa oleifera*: A review of the medical evidence for its nutritional, therapeutic, and prophylactic properties Part 1. *Trees Life Journal*, *1*(5) Retrieved from (2005). http://www.tfljournal.org/article.php/20051201124931586. Accessed 29 April 2019.

Fakayodea, O. A., & Ajav, E. A. (2016). Process optimization of mechanical oil expression from Moringa (*Moringa oleifera*) seeds. *Industrial Crops and Products*, *90*, 142–151.

Farooq, F., Rai, M., Tiwari, A., Khan, A. A., & Farooq, S. (2012). Medicinal properties of *Moringa oleifera*: An overview of promising healer. *Journal of Medicinal Plant Research*, *6*(27), 4368–4374. Retrieved from (2012). http://www.academicjournals.org/JMPR. Accessed 29 April 2019.

Fotouo-M, H., du Toit, E. S., & Robbertse, P. J. (2016). Effect of storage conditions on *Moringa oleifera* Lam. seed oil: Biodiesel feedstock quality. *Industrial Crops and Products*, *84*, 80–86.

Fuglie, L. J. (1999). *The miracle tree:* Moringa oleifera*: Natural nutrition for the tropics.* New York: Church World Service.

Ghazali, H. M., & Mohammed, A. S. (2011). Moringa (*Moringa oleifera*) seed oil: Composition, nutritional aspects, and health attributes, nuts and seeds. In V. R. Preedy, R. R. Watson, & V. B. Patel (Eds.), *Health and disease prevention* (pp. 787–793). Elsevier Life Sciences.

Janaki, S. (2015). Characterization of cold press *Moringa* oil. *International Journal of Science and Research*, *4*(4), 386–389.

Janaki, S., & Yamuna Devi, P. (2015). Microbial properties of cold press *Moringa* oil. *International Journal of Current Research*, *7*(4), 14953–14955.

Kleiman, R., Brown, J. H., & Hill, J. (2006). *Moringa* oil: A "new yet old" unique cosmetic emollient. *Information*, *17*, 739–740. Retrieved from:(2006). http://aocs.files.cms-plus.com/inform/2006/11/739.pdf. Accessed 29 April 2019.

Krotowski, M., Yarnall, M., Lombardi, J., & Katenov, M. (2014). *Moringa seed press: Final report*. Retrieved from (2014). https://marcyarnall.weebly.com/uploads/4/3/8/8/43880957/moringa_seed_press_final_report.pdf. Accessed 29 April 2019.

Kusolrat, P., & Kupittayanant, S. (2013). *The effects of* Moringa oleifera *Lam. seed oil in ovariectomized rats. In: 37th World congress of the International Union of Physiological Sciences, Birmingham, United Kingdom, Poster Communications* Retrieved from (2013). http://www.physoc.org/proceedings/abstract/Proc%2037th%20IUPSPCB290. Accessed 29 April 2019.

L'ORÉAL Paris (2009). *Behenic acid. Retrieved from (2009).* https://www.lorealparisusa.com/ingredient-library/behenic-acid.aspx> Accessed 29 April 2019.

Lalas, S., & Tsaknis, J. (2002). Characterization of *Moringa oleifera* seed oil variety "Periyakulam 1" *Journal of Food Composition and Analysis*, *15*(1), 65–77.

Leone, A., Spada, A., Battezzati, A., Schiraldi, A., Aristil, J., & Bertoli, S. (2016). *Moringa oleifera* seeds and oil: Characteristics and uses for human health. *International Journal of Molecular Sciences*, *17*(12), 2141 Retrieved from: https://doi.org/10.3390/ijms17122141. (Accessed 29 April 2019).

Malini, T., & Vanithakumari, G. (1990). Rat toxicity studies with β-sitosterol. *Journal of Ethnopharmacology*, *28*, 221–234. https://doi.org/10.1016/0378-8741(90)90032-O.

Mariod, A. A., Mirghani, M. E. S., & Hussein, I. (2017). *Unconventional oilseeds and oil sources.* Elsevier Science Publishing Co Inc.

Nadeem, M., & Imran, M. (2016). Promising features of *Moringa oleifera* oil: Recent updates and perspectives. *Lipids in Health and Disease*, *15*(212) Retrieved from (2016). https://doi.org/10.1186/s12944-016-0379-0 29 April 2019.

National Center for Biotechnology Information (2020). *PubChem Database. beta-Sitosterol.* CID=222284, Retrieved from:(2020). https://pubchem.ncbi.nlm.nih.gov/compound/222284. Accessed 10 June 2020.

Ogunsina, B. S., Indira, T. N., Bhatnagar, A. S., Radha, C., Debnath, S., & Gopala Krishna, A. G. (2014). Quality characteristics and stability of *Moringa oleifera* seed oil of Indian origin. *Journal of Food Science and Technology*, *51*, 503–510.

Parrotta, J. A. (2009). *Moringa oleifera*' Lam., 1785. In A. Roloff, H. Weisgerber, U. Lang, B. Stimm, & P. Schütt (Eds.), *Enzyklopädie der Holzgewächse Handbuch und Atlas der Dendrologie* (pp. 1–8). Weinheim: Wiley-VCH Verlag GmbH & Co. KGaA.

Pereira, F. S. G., Galvão, C. C., de Lima, V. F., da Rocha, M. F. A., Schuler, A. R. P., da Silva, V. L., et al. (2016). The versatility of the *Moringa oleifera* oil in sustainable applications. *Oilseeds and fats Crops and Lipids*, *23*(6), 1–7. Retrieved from (2016). https://www.ocl-journal.org/articles/ocl/full_html/2016/06/ocl160001/ocl160001.html Accessed 30 April 2019.

Rahman, I. M., Barua, S., Nazimuddin, M., Begum, Z. A., Rahman, M. A., & Hasegawa, H. (2009). Physicochemical properties of *Moringa oleifera* lam. Seed oil of the indigenous-cultivar of Bangladesh. *Journal of Food Lipids*, *16*(4), 540–553.

Rashid, U., Anwar, F., Moser, B. R., & Knothe, G. (2008). *Moringa oleifera* oil: A possible source of biodiesel. *Bioresource Technology*, *99*, 8175–8179.

Ruttarattanamongkol, K., & Petrasch, A. (2015). Antimicrobial activities of *Moringa oleifera* seed and seed oil residue and oxidative stability of its cold pressed oil compared with extra virgin olive oil. *Songklanakarin Journal of Science and Technology*, *37*(5), 587–594.

Salimon, J., & Salih, N. (2010). Chemical modification of oleic acid oil for biolubricant industrial applications. *Australian Journal of Basic and Applied Sciences*, *4*(7), 1999–2003.

Sánchez-Machado, D. I., López-Cervantes, J., Núñez-Gastélum, J. A., Servín de la Mora-López, G., López-Hernández, J., & Paseiro-Losada, P. (2015). Effect of the refining process on *Moringa oleifera* seed oil quality. *Food Chemistry*, *187*, 53–57.

Sánchez-Machado, D. I., & Núñez-Gastélum, J. A. (2010). Nutritional quality of edible parts of *Moringa oleifera*. *Food Analytical Methods*, *3*, 175–180.

Tsaknis, J., & Lalas, S. (2002). Stability during frying of *Moringa oleifera* seed oil variety "Periyakulam 1" *Journal of Food Composition and Analysis*, *15*(1), 79–101.

Tsaknis, J., Lalas, S., Gergis, V., Dourtoglou, V., & Spiliotis, V. (1999). Characterization of *Moringa oleifera* seed oil (Mbololo variety of Kenya). *Journal of Agricultural and Food Chemistry*, *47*, 4495–4499.

Tsaknis, J., Lalas, S., Gergis, V., & Spiliotis, V. (1998). A total characterisation of *Moringa oleifera* Malawi seed oil. *La Rivista Italiana Delle Sostanze Grasse*, *75*, 21–27.

Tsaknis, J., Spiliotis, V., Lalas, S., Gergis, V., & Dourtoglou, V. (1999). Quality changes of *Moringa oleifera*, variety Mbololo of Kenya, seed oil during frying. *Grasas y Aceites*, *50*, 37–48.

Tsimidou, M., Blekas, G., & Boskau, D. (2003). Olive oil. In B. Caballero, P. Finglas, & F. Toldra (Eds.), *Encyclopedia of food sciences and nutrition* (pp. 4252–4260). (2nd ed.). Academic Press.

Wijendran, V., & Hayes, K. C. (2004). Dietary n-6 and n-3 fatty acid balance and cardiovascular health. *Annual Review of Nutrition*, *24*, 597–615. https://doi.org/10.1146/annurev.nutr.24.012003.132106.

Chapter 43

Cold pressed garden cress (*Lepidium sativum* L.) seed oil

Collen Musara[a], Alfred Maroyi[a], Natascha Cheikhyoussef[b], and Ahmad Cheikhyoussef[c]
[a]*Medicinal Plants and Economic Development (MPED) Research Centre, Department of Botany, University of Fort Hare, Alice, South Africa,*
[b]*Ministry of Higher Education, Technology and Innovation, Windhoek, Namibia,* [c]*Science and Technology Division, Multidisciplinary Research Centre, University of Namibia, Windhoek, Namibia*

1 Introduction

Lepidium sativum L. is an annual herb, belonging to the Brassicaceae or mustard family, which is commonly known as garden cress. The species is native to Africa and Asia but widely cultivated throughout the world as a culinary, traditional medicine, and oilseed crop. The plant requires minimal agricultural resources, and grows well in semiarid regions and low fertility soils. The seeds of *L. sativum* can be harvested in 70–90 days and the yield is estimated to be 800–1000 kg/ha (Gupta, 2006). *Lepidium sativum* seeds have been used in traditional medicine since ancient times in India with various diseases such as asthma, coughs with expectoration, diabetes, diarrhea, dysentery, poultices for sprains, leprosy, skin disease, splenomegaly, dyspepsia, lumbago, leucorrhoea, scurvy, and seminal weakness (Singh & Paswan, 2017). The seeds are traditionally used in the diet of lactating women to induce milk secretion and the seed paste is applied to rheumatic joints to relieve pain and swelling (Zia-Ul-Haq et al., 2012). The seeds are also used as ethnoveterinary medicines as laxative, dressing for sores in camels and horses, and also considered galactagogue and emmenagogue (Datta, Diwakar, Viswanatha, Murthy, & Naidu, 2011). The seed oil of *L. sativum* is used as an illuminant and for making soap (Jansen, 2004; Manohar, Viswanatha, Nagesh, Jain, & Shivaprasad, 2012). Boiled seeds and the ground seeds are used in health drink formulations, either ground in honey or as an infusion in hot milk (Jansen, 2004; Mohit, Gharal, Ranveer, Sahoo, & Ghosh, 2012; Wadhwa, Panwar, Agrawal, Saini, & Patidar, 2012). Whole fruits or seeds of *L. sativum* are used, fresh or dried, as a seasoning, giving a peppery flavor, as a condiment, and in baking, although the odor of the oil is not pleasant (Jansen, 2004).

2 Physical properties of *L. sativum* seed oil

The physical properties of *L. sativum* seed oil extracted by different extraction methods are presented in Table 1. Extracted *L. sativum* seed oil by screw press, cold pressed, solvent extraction, Soxhlet extraction, and supercritical CO_2 extraction methods ranged from 9.33% to 24.7% (Diwakar et al., 2010; Yenge et al., 2017). These different extraction methods produced significant differences in yield and physical properties of *L. sativum* seed oil with the Soxhlet extraction method producing the maximum oil content, ranging from 21.5% to 24.7% (Table 1). Physical properties like color, odor, viscosity, specific gravity, and refractive index are important during the development of food products, because these properties may affect the different quality parameters of the developed products (Singh & Paswan, 2017). The oil yield of *L. sativum* seed is low when compared to the other oil seeds of Brassicaceae family members like *Brassica napus* L. (40%–45%), *Camelina sativa* (L.) Crantz (40%–45%), and *Sinapis alba* L. (25%–40%) (Budin, Breene, & Putnam, 1995; Singh & Paswan, 2017). Soxhlet extracted oil is dark yellowish, while cold pressed oil is light yellowish in color (Diwakar et al., 2010). Yenge et al. (2017) argued that solvent-extracted and cold pressed extracted oil are light yellowish in color, while Soxhlet extracted oil is dark yellow in color. The color of *L. sativum* oil is dirty yellow due to the presence of chlorophyll and carotenoids pigments, which are unintentionally coextracted during oil extraction. The seeds of the Brassicaceae family are known to contain chlorophyll and its derivatives that could be extracted with the oil, and these processes affect oil color, which is an important characteristic for determining the visual acceptance of the oil.

Cold Pressed Oils. https://doi.org/10.1016/B978-0-12-818188-1.00043-8
Copyright © 2020 Elsevier Inc. All rights reserved.

TABLE 1 Physical properties of *L. sativum* seed oil extracted by screw press, cold pressed, solvent extraction, soxhlet extraction, and supercritical CO_2 extraction (Diwakar, Dutta, Lokesh, & Naidu, 2010; Yenge et al., 2017).

Property	Cold pressed	Soxhlet extraction	Supercritical CO_2 extraction	Screw press	Solvent extraction
Oil yield (% dry weight)	12.6	21.5–24.7	18.15	9.33	14.73
Refractive index (nDt[a]) at 24°C	1.47	1.47	1.47	–	–
Specific gravity (g/mL)[b]	0.91	0.90	0.91	0.91	0.90
Viscosity (η)[c]	64.3	42.6–55.5	53.8	53.6	39.9
Peroxide value (m equiv. peroxide/kg oil)	0.70	2.53–4.09	2.63	0.83	1.27
Free fatty acid (% oleic)	0.28	0.39	1.52	–	–
Saponification value (mg KOH/g)	178.8	182.2	174.0	–	–
Unsaponifiable matter (g %)	1.65	1.39	1.16	–	–
Iodine value (g I_2 absorbed/100g)	122.0	131.0	123.0	97.27	134.2
Acid value (mg KOH/g)	–	0.69–0.74	–	0.61	0.69

[a]*nDt is the unit of refractive index (nD) for light with a wavelength equal to 589.3nm at temperature,* t = *24°C.*
[b]*The direct pycnometer determination at 33°C.*
[c]*Viscosity determined at 25°CMPa/s.*

The cold pressed, Soxhlet extraction, and supercritical CO_2 extraction methods yielded the same refractive index, wherein nonsignificant differences were observed for specific gravity of oil extracted by all extraction methods (Table 1). A high refractive index value of 1.47 exhibited by all extraction methods is an indication of substantial unsaturation and the presence of unusual components such as hydroxyl groups in *L. sativum* seed oil. The refractive index of *L. sativum* seed oil is within the range of edible oils; therefore, it can be a good fortifying agent for product development (Diwakar et al., 2010). The viscosity of *L. sativum* seed oil extracted with cold pressed was the highest with the value of 64.3, while the solvent extraction method yielded the least viscosity value of 39.94 (Table 1). Soxhlet extraction yielded the highest peroxide and acid values of 4.09m Equiv peroxide/kg oil and 0.74mg KOH/g, respectively (Table 1).

The saponification value indicates the amount of saponifiable units, that is, acyl groups per unit weight of extracted oil. A high saponification value indicates a higher proportion of low molecular weight fatty acids in the oil or vice versa. Diwakar et al. (2010) revealed that *L. sativum* seed oil showed saponification values ranging from 174.00 to 182.23mg KOH/g, indicating that the oil contains high molecular weight mass fatty acids (Table 1). The saponification value of *L. sativum* seed oil (178.3mg KOH/g) obtained via cold pressed method is lower than that of palm oil (196–205mg KOH/g), olive oil (188–196mg KOH/g), sunflower oil (186–196mg KOH/g), soybean oil (188–195mg KOH/g), and safflower oil (186–198mg KOH/g) (Diwakar et al., 2010). Unsaponifiable matter of *L. sativum* seed oil extracted by different methods varied between 1.16g/100g and 1.65g/100g (Table 1). The unsaponifiable matter shows the pigments, chlorophyll, and other heterocyclic compounds present in *L. sativum* seed oil. The unsaponifiable matter content is higher in *L. sativum* seed oil compared to other oils such as sesame (1.20%), white melon (1.10%), corn (0.92%), cotton (0.52), palm (0.34%), peanut (0.33), palm kernel (0.22), and coco kernel (0.09) (Diwakar et al., 2010). The iodine value is a measure of average unsaturation of an oil or fat and this value is affected by all unsaturated components in the oil or fat. The iodine value of solvent extracted oil is the highest, with a value of 134.0g I_2 absorbed/100g, and screw press method yielded the least iodine, with a value of 97.27g I_2 absorbed/100g (Table 1). Research by Diwakar et al. (2010) and Singh and Paswan (2017) showed that the iodine value of *L. sativum* seed oil is affected by the presence of many long-chain unsaturated components like olefins, including carotenoids and squalenes.

The characteristics of *L. sativum* seed oil were also investigated by Nehdi, Sbihi, Tan, and Al-Resayes (2012) aimed at determining whether it has the potential to be used as a raw material for biodiesel production. The kinematic viscosity (1.92 mm^2/s), cetane number (49.23), gross heat value (40.45), and other fuel properties were within the limits for biodiesel specified by the American Standard for Testing and Materials. Nehdi et al. (2012) argued that *L. sativum* seed oil methyl esters have the potential to supplement petroleum-based diesel and further investigation of biodiesel originating from *L. sativum* seed oil can be widened by the examination of different catalyst types and the impact of the oil on the exhaust emissions in comparison with other biodiesels.

3 Nutritional and chemical properties of *L. sativum* seed oil

Lepidium sativum seed oil is a good source of both macro minerals and trace elements such as calcium, copper, iron, magnesium, manganese, phosphorus, potassium, sodium, and zinc (Table 2). The nutritional content of *L. sativum* seed oil is comparable to seed oil of *Sinapis alba*, a well-known seed oil source (Abul-Fadl, El-Badry, & Ammar, 2011; Al-Jasass & Al-Jasser, 2012). Several amino acids and fatty acids (Table 3) have been identified from the seed oil of *L. sativum*, and these include essential amino acids such as histidine, isoleucine, leucine, lysine, methionine, phenylalanine, threonine, tryptophan, and valine (Zia-Ul-Haq et al., 2012). The primary fatty acids found in *L. sativum* seed oil are oleic acid (18.5%–30.6%), α-linoleic acid (29.3%–30.1%), and α-linolenic acid (32.1%–34.4%) (Moser et al., 2009; Nehdi et al., 2012; Yenge et al., 2017; Zia-Ul-Haq et al., 2012). The amino acid and fatty acid constituents and other physicochemical properties of *L. sativum* seed oil make the seed oil a valuable source of these nutrients, when compared with the nutritional value of *S. alba* and the FAO/WHO/UNU dietary reference intakes or recommended dietary allowance (RDA), required to meet essential nutrients for a healthy person (Tables 1 and 2). The varied bioactive profiles of *L. sativum* seed oils makes the species an excellent choice for functional food and nutraceutical applications. A diet rich in *L. sativum* seed oils will provide the much-required balance of omega-6:omega-3, thereby countering the effect of otherwise consumed proinflammatory diet (Naik & Lele, 2012). Similarly, Diwakar, Dutta, Lokesh, and Naidu (2008) and Umesha and Naidu (2015) argued that *L. sativum* seed oil is more stable due to the presence of a more balanced monounsaturated fatty acid (MUFA) and polyunsaturated fatty acid (PUFA) ratio and high concentration of antioxidants in the oil. Yenge et al. (2017) also argued that *L. sativum* seed oil has a linoleic acid: linolenic acid ratio in the range of 1:4 to 2:3, which could give the species nutritional advantages over the currently available linolenic acid-rich plant oils in altering the n-6/n-3 ratio.

Other phytochemical compounds that have been isolated from *L. sativum* seed oil include essential oils, flavonoids, isoflavonoids, and phytosterols (Table 4). According to Nayak, Upadhyaya, and Upadhyaya (2009), *L. sativum* seed oil also contains seven imidazole alkaloids: lepidine B, C, D, E, and F (dimeric), and monomeric alkaloids lipidene A and B. Some of the documented phytochemical compounds may be used to explain the functional food, nutraceutical applications, and pharmacological properties of the species.

4 Pharmacological properties of *L. sativum* seed oil

4.1 Antidiarrheal and antispasmodic activities

The methanol (70%) seed extract of *L. sativum* at 100 and 300 mg/kg inhibited castor oil-induced diarrhea in rats (Rehman, Mehmood, Alkharfy, & Gilani, 2012). The authors considered that antidiarrheal and antispasmodic activities of the seed extracts were mediated through dual blockade of muscarinic receptors and Ca^{++} channels. Similarly, Gilani, Rehman, Mehmood, and AlKharfy (2012) also observed that castor oil-induced diarrhea in mice was inhibited at doses of 300 and 1000 mg/kg. In the in vitro experiments utilizing two different segments of intestine (ileum and jejunum) from three different species (rat, guinea pig, and rabbit), it was demonstrated that the antidiarrheal and antispasmodic activities of the plant were mediated through multiple pathways including Ca^{++} antagonist, K^{+} channels opener, and inhibition of PDE enzyme (Gilani et al., 2012).

4.2 Antihypertensive activities

Maghrani, Zeggwah, Michel, and Eddouks (2005) investigated the antihypertensive and diuretic effect of aqueous extract of *L. sativum* seeds in normotensive Wistar Kyoto (WKY) rats, and spontaneously hypertensive rats (SHRs). After 3 weeks of oral administration at 20 mg/kg per day, the extract caused a significant decrease in blood pressure and increase electrolytes excretion in SHR (Maghrani et al., 2005). A similar result was obtained by Patel, Kulkarni, Undale, and Bhosale (2009), when investigating the diuretic effects of aqueous and methanol extracts of the seeds of *L. sativum* in normal rats. The extract caused a considerable increase in urine volume following two doses of extracts of 50 and 100 mg/kg in comparison to the control group of hydrochlorothiazide at 10 mg/kg.

TABLE 2 Nutritional composition of *L. sativum* seed oil compared with nutritional values of *Sinapis alba* and the recommended dietary allowance (RDA) (Al-Jasass & Al-Jasser, 2012; Diwakar et al., 2010; Karazhiyan et al., 2011; Razmkhah, Mohammadifar, Razavi, & Ale, 2016; Zia-Ul-Haq et al., 2012).

Nutritional composition	*L. sativum*	*S. alba*	RDA
Ash (%)	4.25–15.07	7.10	–
Calcium (mg/100 g)	266.3	105.0	1000.0–1300.0
Carbon (%)	27.56	–	–
β-Carotene (μg/100 g)	371.8	–	6.00–15.00
Copper (mg/100 g)	5.73	4.80	1.00–3.00
Crude fiber (%)	6.75	11.90	25.0–38.0
Crude protein (%)	2.45–24.18	24.19	34.00
Hydrogen (%)	4.62	–	–
Iron (mg/100 g)	8.31	21.0	8.00–15.0
Lutein (μg/100 g)	39.9	–	10.00
Magnesium (mg/100 g)	339.2	102.0	310.0–320.0
Manganese (mg/100 g)	2.00	1.10	1.00–5.00
Moisture (%)	3.92–13.45	4.36	–
Nitrogen (%)	1.81	–	–
Phosphorus (mg/100 g)	608.6	–	1250.0
Potassium (mg/100 g)	1236.5	663.00	4700.0
Sodium (mg/100 g)	19.65	–	2300.0
α-Tocopherol (mg/100 g)	14.64–17.19	–	20.00
β-Tocopherol (mg/100 g)	0.10	–	–
γ-Tocopherol (mg/100 g)	10.88–87.74	–	20.00
δ-Tocopherol (mg/100 g)	6.74–111.5	–	–
Total lipids (%)	28.03	–	300.00
Total carbohydrates (%)	32.8	30.74	45.0–65.0
Zinc (mg/100 g)	6.99	3.70	8.00–11.0
Constituent sugar			
Arabinose (%)	11.02–19.40	–	–
Fructose (%)	6.80	–	–
Fucose (%)	0.31	–	–
Galactose (%)	4.70–8.80	–	–
Galactose A (%)	6.95–8.00	–	–
Galactose A + glucose A (%)	8.23	–	–
Glucose (%)	1.00–2.41	–	–
Glucose A (%)	1.27–6.7	–	–
Mannose (%)	0.73–38.9	–	–
Rhamnose (%)	1.90–11.8	–	–
Xylose (%)	9.06	–	–

TABLE 3 Amino acids and fatty acids composition of *L. sativum* seed oil compared with the values of *Sinapis alba* seed oil and the recommended dietary allowance (RDA) (Abul-Fadl et al., 2011; Al-Jasass & Al-Jasser, 2012; Alqahtani et al., 2018; Moser, Shah, Winkler-Moser, Vaughn, & Evangelista, 2009; Naik & Lele, 2012; Nehdi et al., 2012; Yenge et al., 2017; Zia-Ul-Haq et al., 2012).

Acid	*L. sativum* (%)	*S. alba* (mg/100 g)	RDA
Essential amino acid			
Histidine	3.87	2.51–2.61	21.00–32.00
Isoleucine	4.19	4.28–4.54	20.00
Leucine	7.03	7.89–8.00	39.00
Lysine	5.98	5.54–6.53	30.00
Methionine	0.51	–	28.0–43.0
Methionine + cysteine	–	4.26–5.89	15.00
Phenylalanine	5.39	–	54.0–84.0
Phenylalanine + tyrosine	–	7.57–8.41	–
Threonine	3.76	4.04–4.36	16.0
Tryptophan	0.92	–	4.00
Valine	6.21	5.66–6.72	26.0
Conditionally essential amino acid			
Arginine	3.44	4.70–7.76	–
Cysteine	0.21	–	28.0–43.0
Glycine	5.08	4.47–5.04	–
Proline	4.63	4.19–8.07	–
Tyrosine	2.88	–	37.0–58.0
Nonessential amino acid			
Alanine	4.59	4.03–4.54	–
Aspartic acid	12.07	8.69–8.99	–
Glutamic acid	24.29	19.8–20.1	–
Serine	4.18	4.94–5.55	–
Fatty acid			
Arachidic acid	2.10–3.92	0.91–1.04	0.1
Behenic acid	0.60–0.99	2.23–2.34	–
Capric acid	0.10	–	–
Cyclopropanepentanoic acid, 2-undecyl	0.20	–	–
Decanoic acid	0.04	–	–
13-Docosenoic acid	0.10	–	–
Dodecanoic acid	0.04	–	–
Eicosenoic acid	11.07	–	–
Eicosaenoic acid	13.40	–	–
Eicosadienoic acid	0.30–0.71	–	–
Eicosatrienoic acid	0.45–0.62	–	–
Eicosenoic acid methylester	5.65	–	–
Erucic acid	3.00–4.51	23.9–37.8	–
Gondoic acid	11.1–12.01	–	–

Continued

TABLE 3 Amino acids and fatty acids composition of *L. sativum* seed oil compared with the values of *Sinapis alba* seed oil and the recommended dietary allowance (RDA) (Abul-Fadl et al., 2011; Al-Jasass & Al-Jasser, 2012; Alqahtani et al., 2018; Moser, Shah, Winkler-Moser, Vaughn, & Evangelista, 2009; Naik & Lele, 2012; Nehdi et al., 2012; Yenge et al., 2017; Zia-Ul-Haq et al., 2012)—cont'd

Acid	*L. sativum* (%)	*S. alba* (mg/100 g)	RDA
Heneicosanoic acid	1.30	–	–
Heptadecanoic acid	0.05	–	–
6-Heptenoic acid	0.05	–	–
Hexadecanoic acid methylester	9.83	–	–
9,12-Hexadecadienoic acid	1.10	–	–
7,10-Hexadecadienoic acid	44.4	–	–
7,10,13-Hexadecadienoic acid	9.90	–	–
Hexadecadienoic acid	0.04	–	–
Hexadecanoic acid, 15-methyl	4.30	–	–
Lignoceric acid	0.41	–	–
Linoleic acid	7.60–12.9	12.3–23.5	6.70
α-Linoleic acid	29.3–30.1	23.92	–
α-Linolenic acid	32.1–34.4	11.56–12.00	1.40
Myristic acid	0.11	–	–
Myristoleic acid	0.04	–	–
8-Nonynoic acid	0.08	–	–
9,12-Octadecadienoic acid	0.40	–	–
13,16-Octadecadienoic acid	0.20	–	–
10-Octadecenoic acid	1.40	–	–
11-Octadecenoic acid	15.5	–	–
7-Octadecynoic acid	0.20	–	–
Octadecadienoic acid methylester	12.4	–	–
Octadecanoic acid methylester	5.00–20.8	–	–
Octadecatrienoic acid methylester	26.71	–	–
Oleic acid	18.5–30.6	19.0–20.2	–
Palmitic acid	9.03–10.3	2.58–3.64	–
Palmitoleic acid	0.15–0.70	0.32–0.34	–
Pentadecanoic acid	0.05	–	–
Pentadecenoic acid	0.04	–	–
Stearic acid	1.90–3.28	1.13–1.46	–
15-Tetracosenoic acid	5.70	–	–
Tetracosanoic acid	0.90	–	–
Tetradecanoic acid, 12-methyl	0.07	–	–
Triacontanoic acid	0.09	–	–
4-Tridecen-6-yne	0.10	–	–
Undecanoic acid	0.07	–	–
Vaccenic acid	1.30–1.40	–	–

TABLE 4 Phytochemical compounds isolated from *L. sativum* seed oil (Ait-Yahia et al., 2018; Kimbaris, Koliopoulos, Michaelakis, & Konstantopoulou, 2012; Moser et al., 2009; Sakran, Selim, & Zidan, 2014; Zia-Ul-Haq et al., 2012).

Compound	%
Essential oil	
Eucalyptol	35.80
Linanool	4.30
Linalool propanoate	2.60
E-Nerolidol	2.60
α-Phellandrene	1.00
α-Pinene	0.70
Sabinene	1.00
γ-Terpinene	0.60
Terpinen-4-ol	1.80
α-Terpineol	4.30
α-Terpinyl acetate	44.10
Flavonoid	
Caffeic acid	–
Caffeic acid-hexoside	–
Caffeoylquinic acid	–
Coumaric acid	–
Coumaric acid-hexoside	–
Coumaroylquinic acid	–
Ferulic acid	0.10
Ferulic acid-hexoside	–
Gallic acid	–
Glucopaoline	67.5
Kaempferol di-hexose rhamnose	3.80
Kaempferol-glucuronide	–
Kaempferol hexose rhamnose 1	1.70
Kaempferol hexose rhamnose 2	0.80
Kaempferol hexose rhamnose 3	0.60
Kaempferol-hexoside	–
Kaempferol rhamnose	0.10
Kaempferol rhamnose (benzo) di-hexose 1	1.90
Kaempferol rhamnose (benzo) di-hexose 2	0.10
Kaempferol rhamnose (benzo) di-hexose 3	0.40
Protocatechuic acid	–
Quercetin	–
Quercetin di-hexose rhamnose	2.10
Quercetin di-hexose rhamnose	2.10
Quercetin-hexoside	–

Continued

TABLE 4 Phytochemical compounds isolated from *L. sativum* seed oil (Ait-Yahia et al., 2018; Kimbaris, Koliopoulos, Michaelakis, & Konstantopoulou, 2012; Moser et al., 2009; Sakran, Selim, & Zidan, 2014; Zia-Ul-Haq et al., 2012)—cont'd

Compound	%
Quercetin hexose rhamnose	0.10
Quercetin rhamnose	0.30
Sinapic acid	0.60
Sinapine	13.40
Sinapoyl di-glucose	2.70
Sinapoyl malate	1.70
Vanillic acid-hexoside	–
Isoflavonoid	
5,6-dimethoxy-2′,3′-methylenedioxy-7-C-β-D-gluco-pyranosyl isoflavone	–
7-hydroxy-4′,5,6-trimethoxyisoflavone	–
7-hydroxy-5,6-dimethoxy-2′,3′-methylenedioxyisoflavone	–
Phytosterol	
Cholesterol	0.5
Campesterol	3.95
Stigmasterol	0.30
Sitosterol	5.82
Avenasterol	3.44
Dihydrolanosterol	0.25
β-Amyrin	0.16

4.3 Antimicrobial activities

Adam, Salih, and Abdelgadir (2011) evaluated the antimicrobial activities of *L. sativum* seed extracts against *Staphylococcus aureus*, *Escherichia coli*, *Klebsiella pneumoniae*, *Proteus vulgaris*, *Pseudomonas aeruginosa*, and *Candida albicans*. The petroleum ether extract was found to be active against all the pathogens (Adam et al., 2011). Sharma, Vyas, and Manda (2012) evaluated the antifungal activities of ethanol extract of *L. sativum* seeds against *Fusarium equiseti*, *Aspergillus flavus*, and *Alternaria alternata*. The seed extract exhibited activities with the zone of inhibition ranging from 4 mm to 22 mm (Sharma et al., 2012). Ait-Yahia et al. (2018) evaluated antimicrobial activities of *L. sativum* seed oil extracts against *Staphylococcus aureus*, *Enterococcus faecalis*, *Escherichia coli*, and *Pseudomonas aeruginosa*. The extract exhibited activities with half-maximal inhibitory concentration (IC_{50}) values ranging from 0.40 to 0.47 μg/mL and the minimum inhibitory concentration (MIC) values ranging from 3.50 to 5.00 μg/mL.

4.4 Antioxidant activities

Bhasin, Bansal, Yadav, and Punia (2011) investigated the antioxidant activities of *L. sativum* seed extracts using reducing power assay, with ethanol extract exhibiting the strongest antioxidant activities. Yadav et al. (2009) evaluated antioxidant activities of *L. sativum* seed extracts using 2,2-dipheny-l-picrylhydrazyl (DPPH·) free radical scavenging, ferric chloride, and phosphormolybdenum assays. The IC_{50} values for DPPH·, ferric chloride, and phosphormolybdenum were 18.50, 9.10, and 18.40 μg/mL, respectively (Yadav et al., 2009). Zia-Ul-Haq et al. (2012) evaluated antioxidant capacity of *L. sativum* seed oil using 2,2′-azino-bis(3-ethylbenzthiazoline-6-sulphonic acid) radical-scavenging (ABTS), ferric reducing antioxidant power (FRAP), and total radical-trapping antioxidant parameter (TRAP) assays. The FRAP value was

1317.00 μmol Fe^{2+}/g, while TEAC and TRAP values were 168.10 μmol TE/g, and 506.40 μmol TE/g, respectively (Zia-Ul-Haq et al., 2012). Indumathy and Aruna (2013) evaluated antioxidant activities of *L. sativum* seed extracts using DPPH·, hydroxyl radical, superoxide anion scavenging, nitric oxide, and hydrogen peroxide scavenging assays, and the methanol extract exhibited activities that were stronger than the reference standards used. Umesha and Naidu (2015) evaluated the effect of *L. sativum* seed oil and its blends with n-6 PUFA-rich edible vegetable oils such as sunflower oil, rice bran oil, and sesame oil on antioxidant status of oils and antioxidative enzymes in Wistar rats. Blending of *L. sativum* seed oil with other vegetable oils decreased the n-6/n-3 PUFA ratio (>2.0) and dietary feeding of *L. sativum* seed oil blended oils increased the antioxidant status and activity of antioxidant enzymes (catalase and GPx) in experimental rats. Ait-Yahia et al. (2018) evaluated the antioxidant activities of *L. sativum* seed extracts through DPPH·, reducing power (RP), and β-carotene/linoleic acid assays. The extracts exhibited activities with IC_{50} values of 67.10 μg/mL (DPPH·), and 100.00 μg/mL (RP), which were comparable to IC_{50} values of 4.00 μg/mL to 20.00 μg/mL exhibited by ascorbic acid and quercetin, the two standards used in the study.

4.5 Cytotoxicity activities

Mahassni and Al-Reemi (2013) evaluated the ability of *L. sativum* seed extracts to induce apoptosis and necrosis in the human breast cancer cell line MCF-7 through comparison to normal human skin fibroblasts (HFS) by determining the morphological changes in the cells using light microscopy, DNA fragmentation assay, and fluorescent stains (Annexin V and propidium iodide) using flow cytometry and fluorescent microscopy. Apoptosis was induced in both cells, and more in MCF-7, when they were treated with 25% and 50% extract, while necrosis was observed mainly after exposure to elevated extract concentrations (75%). DNA fragmentation resulted in both cells, in a time- and dose-dependent manner. Both cells, at all extract concentrations, showed no significant differences in the number of living, dead, apoptotic, and necrotic cells (Mahassni & Al-Reemi, 2013). Sakran et al. (2014) evaluated the ability of isoflavonoids isolated from seeds of *L. sativum*, namely 5,6-dimethoxy-2′,3′-methylenedioxy-7-C-β-D-gluco-pyranosyl isoflavone, 7-hydroxy-4′,5,6-trimethoxyisoflavone, and 7-hydroxy-5,6-dimethoxy-2′,3′-methylenedioxyisoflavone to reduce the hepatotoxicity induced by paracetamol in adult Sprague Dawley male rats by reducing the damage and toxicity effects on liver cells. The compounds showed significant improvement of total antioxidant capacity, normalizing the levels of liver enzymes GSH, SOD, GPX, CAT, and GST compared to the control group (Sakran et al., 2014).

4.6 Diuretic activities

Patel et al. (2009) evaluated the diuretic activities of aqueous and methanolic seed extracts of *L. sativum* in adult male Wistar rats, which were administered orally at doses of 50 and 100 mg/kg body weight. The extracts showed a dose-dependent increase in urine excretion. The excretion of sodium was increased, while potassium excretion was increased only by the aqueous extract at a dose of 100 mg/kg, and the methanolic extract exhibited a potassium-conserving effect. Both aqueous and methanolic extracts of *L. sativum* exhibited diuretic activities, which were comparable to those produced by the reference diuretic hydrochlorothiazide (Patel et al., 2009).

4.7 Galactagogue activities

Galactagogue activities of *L. sativum* seeds were studied by Al-Yawer and Al-Khafaji (2006) in adult female virgin Norway rats. Each experimental rat was administered with 1.6 mg of *L. sativum* seed powder/g body weight per day for 14 days. Different parameters such as gross assessment, histological examination, enzymatic histochemical, and hormonal assay of follicle-stimulating hormone, luteinizing hormone, prolactin, estrogen, and progesterone were evaluated to assess the effect of *L. sativum* seeds on the mammary gland of young adult virgin rats. All the parameters significantly exhibited strong mammotrophic and lactogenic effects of *L. sativum* seeds on the nonprimed mammary gland of adult virgin rats.

4.8 Hepatoprotective activities

Abuelgasim, Nuha, and Mohammed (2008) evaluated the hepatoprotective activities of *L. sativum* seed extracts on carbon tetrachloride-induced liver damage in albino Wister rats. The severe changes in the livers of rats caused by carbon tetrachloride were decreased in rats treated with *L. sativum* seed extracts, implying that the seeds of *L. sativum* have hepatoprotective activities.

4.9 Hypoglycemic activities

Research by Eddouks, Maghrani, Zeggwagh, and Michel (2005) revealed that aqueous seed extracts of *L. sativum* at 20mg/kg in an acute single dose or chronic dose repeated daily for 15 days as oral administration resulted in a significant decrease in blood glucose levels in streptozotocin-induced diabetic rats. Two weeks after daily repeated oral administration, there was a marked normalization of glycemia. The basal plasma insulin concentrations did not differ considerably in the *L. sativum* extract-treated groups compared to the untreated group either in normal or diabetic rats when treated once daily at a dose of 20mg/kg. Shukla, Bigoniya, and Srivastava (2012) investigated the hypoglycemic activities of total alkaloids isolated from *L. sativum* seed on alloxan-induced diabetic rats. The diabetic rats were fed with total alkaloid of 50mg/kg of body weight, 150mg/kg of body weight, and 250mg/kg of body weight for 21 days. At the dose level of 250mg/kg, there was significant reduction in the blood glucose, cholesterol, triglyceride, and urea level in diabetic rats.

4.10 Hypolipidemic activities

Research by Al Hamedan (2010) revealed that at 8 weeks following aqueous *L. sativum* seed extract administration into hypercholesterolemic rats, the weight gain, feeding efficiency ratio, serum cholesterol, triglycerides, very low density lipoprotein cholesterol, low density lipoprotein cholesterol, high density lipoprotein cholesterol, serum, serum creatinine, urea, liver cholesterol, and total lipids levels all decreased in comparison with the control group (hypercholesterolemic rat), with a significant increase in both serum globulin and liver triglycerides.

4.11 Larvicidal activities

Kimbaris et al. (2012) evaluated the larvicidal activities of *L. sativum* essential oils against the late third to early fourth instar mosquito larvae of *Culex pipiens*. The essential oils of *L. sativum* were moderately active with the median lethal concentration (LC_{50}) value of >65 mg/L on mosquito larvae while eucalyptol and α-terpinyl acetate, its major components, displayed LC_{50} values above 100mg/L and less than 25 mg/L, respectively.

4.12 Nephroprotective and curative activities

Research by Yadav et al. (2009) revealed that significant nephroprotective and curative activities could be attained by using the ethanolic extract of *L. sativum* seeds at 200mg/kg for 16 days against nephrotoxicity damage induced by cisplatin. The extract inhibited the activities of antioxidant enzyme in renal tissue, glutathione depletion, and increase in thiobarbituric acid reactive substance. Saini, Yadav, Srivastava, and Seth (2010) found that 400mg/kg ethanolic extract of *L. sativum* seeds used against cisplatin (5 mg/kg, i.p.) induced nephrotoxicity and exhibited nephroprotective and curative activities. Al Hamedan (2010) evaluated the protective effect of *L. sativum* seed extracts on the nutritional status and some biochemical analyses of serum and liver in hypercholesterolemic adult albino male rats of the Sprague Dawley strain. The *L. sativum* extracts showed significantly lower values of weight gain, feed efficiency ratio, serum cholesterol, serum creatinine, urea, liver cholesterol, and total lipids, with significant increase in both serum globulin and liver triglycerides. These results imply that *L. sativum* seeds may have nephroprotective and curative activities.

4.13 Prokinetic and laxative activities

Research by Rehman, Mehmood, Alkharfy, and Gilani (2011) revealed that the aqueous-methanolic extract of *L. sativum* seeds at doses of 30 and 100mg/kg exhibited atropine-sensitive prokinetic and laxative action in mice. Further investigations on isolated jejunum and ileum preparations from different animals, such as a mouse, guinea pig, and rabbit, demonstrated that the extract and tissue-selectivity with extracts were more efficacious in gut preparations of the rabbit than in the guinea pig or mouse.

4.14 Protective activities

Al-Sheddi et al. (2016) evaluated the protective effects of chloroform extract of *L. sativum* seed against oxidative stress and cytotoxicity induced by hydrogen peroxide (H_2O_2) in human liver cells (HepG2). Preexposure of *L. sativum* seed extract significantly attenuated the loss of cell viability up to 48% at 25 mg/mL concentration against H_2O_2 with median lethal dose (LD_{50}) value of 2.5 mM. Preexposure of *L. sativum* seed extract at 25 mg/mL concentration significantly inhibited the induction of reactive oxygen species generation (45%) and lipid peroxidation (56%), and increased the mitochondrial membrane potential (55%) and reduced glutathione levels (46%).

4.15 Toxicity

Datta et al. (2011) evaluated the safety of *L. sativum* seeds by conducting acute and subchronic toxicity studies in adult Wistar rats. For the acute toxicity study, 0.50 g/kg body weight to 5.00 g/kg body weight of *L. sativum* seed powder was administered through diet to rats and obvious symptoms of toxicity and mortality were monitored for 72 h. Acute doses of *L. sativum* seed powder did not induce any symptoms of toxicity or mortality of rats. In the subchronic toxicity study, 1%–10% of *L. sativum* was administered to rats through diet for 14 weeks. Dietary feeding of *L. sativum* seed powder did not produce any mortality, and no significant changes in food intake, gain in body weight, the relative weight of organs, hematological parameters, or macroscopic and microscopic changes in vital organs, were observed between experimental and control groups. These results showed that acute and subchronic feeding of *L. sativum* seed for 14 weeks did not produce any toxic effects in male and female rats and, therefore, seeds of *L. sativum* can be considered nontoxic and safe.

5 Development of new food products using *L. sativum* seed oil

Since *L. sativum* seed oil is characterized by high nutritional and functional properties, it can be used for the fortification of drinks and foods. According to Singh and Paswan (2017), *L. sativum* seeds are used as an ingredient of dahiwala bread, iron-rich biscuits, omega-3 fatty acid-rich biscuits, and fortified health drinks. Addition of *L. sativum* seed flour results in increased amounts of protein, lipids, calcium, iron, and phosphorous, and decreases of antinutritional components like oxalate, total cyanogens, and phytic acid. Varsha and Rohini (2007) argued that the intake of *L. sativum* seeds characterized by high n-6 PUFAs results in changes of physiological conditions like prothrombotic and proaggregatory, characterized by increase in blood viscosity, vasospasm, and vasoconstriction and decrease in bleeding time. Similarly, Singh and Paswan (2017) argued that atopic dermatitis, rheumatoid arthritis, asthma, ulcerative colitis, and cancer are usually caused by the deficiency of n-3 PUFAs. However, sufficient intake of n-3 PUFAs alters membrane fluidity, downregulates inflammatory genes and lipid synthesis, and stimulates fatty acid degradation (Wien, Raiaram, Oda, & Sabate, 2010). Humans cannot synthesize PUFAs like n-3 and n-6, which are essential fatty acids; therefore, they need to be supplemented through food (Singh & Paswan, 2017).

References

Abuelgasim, A. I., Nuha, H. S., & Mohammed, A. H. (2008). Hepatoprotective effect of *Lepidium sativum* against carbon tetrachloride induced damage in rats. *Research Journal of Animal and Veterinary Sciences*, *3*, 20–23.

Abul-Fadl, M. M., El-Badry, N., & Ammar, M. S. (2011). Nutritional and chemical evaluation for two different varieties of mustard seeds. *World Applied Sciences Journal*, *15*(9), 1225–1233.

Adam, S. I. Y., Salih, S. A. M., & Abdelgadir, W. S. (2011). *In vitro* antimicrobial assessment of *Lepidium sativum* L. seeds extracts. *Asian Journal of Medical Sciences*, *3*(6), 261–266.

Ait-Yahia, O., Perreau, F., Bouzroura, S.-A., Benmalek, Y., Dob, T., & Belkebir, A. (2018). Chemical composition and biological activities of n-butanol extract of *Lepidium sativum* L. (Brassicaceae) seed. *Tropical Journal of Pharmaceutical Research*, *17*(5), 891–896.

Al Hamedan, W. A. (2010). Protective effect of *Lepidium sativum* L. seeds powder and extract on hypercholesterolemic rats. *The Journal of American Science*, *6*(11), 873–879.

Al-Jasass, F. M., & Al-Jasser, M. S. (2012). Chemical composition and fatty acid content of some spices and herbs under Saudi Arabia conditions. *The Scientific World Journal*, *2012*, Article ID 859892.

Alqahtani, F. Y., Aleanizy, F. S., Mahmoud, A. Z., Farshori, N. N., Alfaraj, R., Al-sheddi, E. S., et al. (2018). Chemical composition and antimicrobial, antioxidant, and anti-inflammatory activities of *Lepidium sativum* seed oil. *Saudi Journal of Biological Science*. https://doi.org/10.1016/j.sjbs.2018.05.007.

Al-Sheddi, E. S., Farshori, N. N., Al-Oqail, M. M., Musarrat, J., Al-Khedhairy, A. A., & Siddiqui, M. A. (2016). Protective effect of *Lepidium sativum* seed extract against hydrogen peroxide-induced cytotoxicity and oxidative stress in human liver cells (HepG2). *Pharmaceutical Biology*, *54*(2), 314–321.

Al-Yawer, M. A., & Al-Khafaji, H. M. A. (2006). Garden cress seed could be a factual galactagogue. The Iraqi postgrad. *Medizinhistorisches Journal*, *5*(1), 62–67.

Bhasin, P., Bansal, D., Yadav, O. P., & Punia, A. (2011). *In vitro* antioxidant activity and phytochemical analysis of seed extracts of *Lepidium sativum*: A medicinal herb. *Journal of Bioscience and Technology*, *2*(6), 410–415.

Budin, J. T., Breene, W. M., & Putnam, D. H. (1995). Some compositional properties of Camelina (*Camelina sativa* L. Crantz) seeds and oils. *Journal of the American Oil Chemists' Society*, *72*, 309–315.

Datta, P. K., Diwakar, B. K., Viswanatha, S., Murthy, K. N., & Naidu, K. A. (2011). Safety evaluation studies on garden cress (*Lepidium sativum* L.) seeds in Wistar rats. *International Journal of Applied Research in Natural Products*, *4*, 37–43.

Diwakar, B. T., Dutta, P. K., Lokesh, B. R., & Naidu, K. A. (2008). Bioavailability and metabolism of n-3 fatty acid rich garden cress (*Lepidium sativum*) seed oil in albino rats. Prostaglandins, leukotrienes Ess. *Fatty Acids*, *78*, 123–130.

Diwakar, B. T., Dutta, P. K., Lokesh, B. R., & Naidu, K. A. (2010). Physicochemical properties of garden cress (*Lepidium sativum* L.) seed oil. *Journal of the American Oil Chemists' Society*, *87*, 539–548.

Eddouks, M., Maghrani, M., Zeggwagh, N. A., & Michel, J. B. (2005). Study of the hypoglycaemic activity of *Lepidium sativum* L. aqueous extract in normal and diabetic rats. *Journal of Ethnopharmacology*, *97*, 391–395.

Gilani, A. H., Rehman, N. U., Mehmood, M. H., & AlKharfy, K. M. (2012). Species differences in the antidiarrheal and antispasmodic activities of *Lepidium sativum* and insight into underlying mechanisms. *Phytotherapy Research*, *27*(7), 1086–1094.

Gupta, S. (2006). *Effect of nutrients and plant density on growth and yield of garden cress (*Lepidium sativum *L.).* Ph.D. Thesis Bangalore: University of Agricultural Sciences.

Indumathy, R., & Aruna, A. (2013). Free radical scavenging activities, total phenolic and flavonoid content of *Lepidium sativum* (Linn.). *International Journal of Pharmacy and Pharmaceutical Sciences*, *5*(4), 634–637.

Jansen, P. C. M. (2004). *Lepidium sativum* L. In G. J. H. Grubben, & O. A. Denton (Eds.), *Plant resources ofTropical Africa 2: Vegetables* (pp. 365–367). Leiden, Netherlands: PROTA Foundation, Backhuys Publishers.

Karazhiyan, H., Razavi, S. M. A., Phillips, G. O., Fang, Y., Al-Assaf, S., & Nishinari, K. (2011). Physicochemical aspects of hydrocolloid extract from the seeds of *Lepidium sativum*. *International Journal of Food Science and Technology*, *46*, 1066–1072.

Kimbaris, A. C., Koliopoulos, G., Michaelakis, A., & Konstantopoulou, M. A. (2012). Bioactivity of *Dianthus caryophyllus, Lepidium sativum, Pimpinella anisum* and *Illicium verum* essential oils and their major components against the West Nile vector *Culex pipiens*. *Parasitology Research*, *111*, 2403–2410.

Maghrani, M., Zeggwah, N. A., Michel, J. B., & Eddouks, M. (2005). Antihypertensive effect of *Lepidium sativum* in spontaeneously hypertensive rats. *Journal of Ethnopharmacology*, *100*, 193–197.

Mahassni, S. H., & Al-Reemi, R. M. (2013). Apoptosis and necrosis of human breast cancer cells by an aqueous extract of garden cress (*Lepidium sativum*) seeds. *Saudi Journal of Biological Science*, *20*, 131–139.

Manohar, D., Viswanatha, G. L., Nagesh, S., Jain, V., & Shivaprasad, H. N. (2012). Ethnopharmacology of *Lepidium sativum* Linn (Brassicaceae): A review. *International Journal of Phytothearpy Research*, *2*, 1–7.

Mohit, S. Y., Gharal, D. B., Ranveer, R. C., Sahoo, A. K., & Ghosh, J. S. (2012). Development of health drink enriched with processed garden-cress (*Lepidium sativum* L.) seeds. *American Journal of Food Technology*, *7*(9), 571–576.

Moser, B. R., Shah, S. N., Winkler-Moser, J. K., Vaughn, S. F., & Evangelista, R. L. (2009). Composition and physical properties of cress (*Lepidium sativum* L.) and field pennycress (*Thlaspi arvense* L.) oils. *Industrial Crops and Products*, *30*, 199–205.

Naik, A. S., & Lele, S. S. (2012). Functional lipids and bioactive compounds from oil rich indigenous seeds. *International Journal of Nutrition and Food Sciences*, *6*(9), 722–730.

Nayak, P. S., Upadhyaya, S. D., & Upadhyaya, A. A. (2009). HTPLC densitometer determination of sinapic acid in chandrashur (*Lepidium sativum*). *Journal of Scientific Research*, *1*(1), 121–127.

Nehdi, I. A., Sbihi, H., Tan, C. P., & Al-Resayes, S. I. (2012). Garden cress (*Lepidium sativum* Linn.) seed oil as a potential feedstock for biodiesel production. *Bioresource Technology*, *126*, 193–197.

Patel, U., Kulkarni, M., Undale, V., & Bhosale, A. (2009). Evaluation of diuretic activity of aqueous and methanol extracts of *Lepidium sativum* "garden cress" (Cruciferae) in rats. *Tropical Journal of Pharmaceutical Research*, *8*, 215–219.

Razmkhah, S., Mohammadifar, M. A., Razavi, S. M. A., & Ale, M. T. (2016). Purification of cress seed (*Lepidium sativum*) gum: Physicochemical characterization and functional properties. *Carbohydrate Polymers*, *141*, 166–174.

Rehman, N. U., Mehmood, M. H., Alkharfy, K. M., & Gilani, A. H. (2011). Prokinetic and laxative activities of *Lepidium sativum* seed extract with species and tissue selective gut stimulatory actions. *Journal of Ethnopharmacology*, *134*, 878–883.

Rehman, N. U., Mehmood, M. H., Alkharfy, K. M., & Gilani, A. H. (2012). Studies on antidiarrhel and antispasmodic activities of *Lepidium sativum* crude extract in rats. *Phytotherapy Research*, *26*, 136–141.

Saini, V., Yadav, Y. C., Srivastava, D. N., & Seth, A. K. (2010). Nephroprotective and curative activity of *Lepidium sativum* L. seeds in Albino rats using cisplatin induced acute renal failure. *Pharmacology*, *2*(4), 57–64.

Sakran, M., Selim, Y., & Zidan, N. (2014). A new isoflavonoid from seeds of *Lepidium sativum* L. and its protective effect on hepatotoxicity induced by paracetamol in male rats. *Molecules*, *19*, 15440–15451.

Sharma, R. K., Vyas, K., & Manda, H. (2012). Evaluation of antifungal effect on ethanolic extract of *Lepidium sativum* L. seed. *International Journal of Phytopharmacology*, *3*(2), 117–120.

Shukla, A., Bigoniya, P., & Srivastava, B. (2012). Hypoglycemic activity of *Lepidium sativum* Linn. seed total alkaloid on alloxan induced diabetic rats. *Research Journal of Medicinal Plants*, *6*(8), 587–596.

Singh, C. S., & Paswan, V. K. (2017). *The potential of garden cress (*Lepidium sativum *L.) seeds for development of functional foods. In J. C. Jimenez-Lopez (Ed.), Seed biology* (pp. 279–294). IntechOpen. https://doi.org/10.5772/intechopen.70355. Chapter 14. Available from:(2017). https://www.intechopen.com/books/advances-in-seed-biology/the-potential-of-garden-cress-lepidium-sativum-l-seeds-for-development-of-functional-foods.

Umesha, S. S., & Naidu, K. A. (2015). Antioxidants and antioxidant enzymes status of rats fed on n-3 PUFA rich garden cress (*Lepidium sativum* L.) seed oil and its blended oils. *Journal of Food Science and Technology*, *52*(4), 1993–2002.

Varsha, S. Z., & Rohini, D. (2007). Biofortification of biscuits with garden cress seeds for prevention of anaemia. *Asian Journal of Home Science*, *2*, 1–5.

Wadhwa, S., Panwar, M. S., Agrawal, A., Saini, N., & Patidar, L. N. (2012). A review on pharmacognostical study of *Lepidium sativum*. *Advance Research in Pharmaceuticals and Biologicals*, *2*(4), 316–323.

Wien, M., Raiaram, S., Oda, K., & Sabate, J. (2010). Decreasing the linoleic acid to a linoleic acid diet ratio increase eicosapentaeonic acid in erythrocyte in adults. *Lipids*, *45*, 683–692.

Yadav, Y. C., Srivastav, D. N., Seth, A. K., Gupta, V. D., Yadav, K. S., & Kumar, S. (2009). Nephroprotective and curative activity of *Lepidium sativum* L. seeds in albino rats using cisplatin induced nephrotoxicity. *Pharmacology*, *3*, 640–646.

Yenge, G. B., More, H. G., Kenghe, R. N., Kanawade, V. L., Nimbalkar, C. A., & Patil, A. P. (2017). Effect of different extraction methods on yield and physico-chemical properties of garden cress (*Lepidium sativam* L.) oil. *Journal of Oilseed Brassica*, *8*(2), 138–142.

Zia-Ul-Haq, M., Ahmad, S., Calani, L., Mazzeo, T., Rio, D. D., Pellegrini, N., et al. (2012). Compositional study and antioxidant potential of *Ipomoea hederacea* Jacq. and *Lepidium sativum* L. seeds. *Molecules*, *17*, 10306–10321.

Chapter 44

Cold pressed walnut (*Juglans regia* L.) oil

Farooq Anwar[a], Rahman Qadir[a], and Ali Abbas[b]
[a]*Department of Chemistry, University of Sargodha, Sargodha, Pakistan,* [b]*Department of Chemistry, Government Postgraduate Taleem-ul-Islam College, Chiniot, Punjab, Pakistan*

1 Introduction

The walnut (*Juglans regia*) is the nut of a tree with genus *Juglans*, belonging to the family *Juglandaceae*. It is recognized as a nutritious edible seed of a drupe and commonly consumed as a nut. Around the world, China is the biggest walnut producer (with 46% production share), followed by Iran, the USA, Turkey, and Ukraine. Pakistan also contributes about 0.30% of the total world's production of walnuts (Huffman, Harika, Eilander, & Osendarp, 2011). In recent years, walnuts have gained special attention from nutritionists and food scientists due to the contents of their edible seed oil, which is composed of high-value fatty acids, such as ω-3 and ω-6 polyunsaturated fatty acids (PUFAs) (Rabrenovic, Dimic, Maksimovic, Sobajic, & Gajic-Krstajic, 2011). Interestingly, walnut seeds (Fig. 1) comprise 63%–71% of oil contents of their total weight. Almost 90% of walnut oil contains unsaturated fatty acids, such as oleic acid (12%–20%), linoleic acid (50%–58%), and linolenic acid (11%–17%). Minor components with medicinal and cosmo-nutraceutical features such as phytosterols and tocopherols are also present in appreciable amounts in walnut oil and thus favor its potential uses in the industrial sector (Savage, McNeil, & Dutta, 2001).

Recently, walnut oil/oil cake has gained much attraction of consumers and food scientists due to its high nutritional value, sensory qualities, and nutraceutical potential; it is thus becoming an important food commodity in several developing and under-developed countries. Being a dietary source of essential fatty acids, walnut oil (an expensive culinary oil) is employed in several food preparations such as cold dishes (Huffman et al., 2011). The presence of high unsaturated fatty acid contents especially linoleic acid, oleic acid, and linolenic acid renders walnut oil as an important component of the nutraceutical industry as an alternative health care medicine (Wan, Dong, & Li, 2001). Epidemiological studies revealed that walnut oil not only reduces serum cholesterol, but can also be employed as a nutrition health care oil (Zhao, Chen, & Ge, 2010).

Extraction of any oil from a vegetable oil seed material is a critical step as it drastically affects the quality of extracted oil and oilseed residues. Conventionally, oilseed extraction is based on the use of nonpolar organic solvents such as petroleum ether and *n*-hexane; however, such techniques have limitations with regard to process safety and quality of end-use products (Gecgel et al., 2015). Among different oil extraction techniques, cold pressing is considered to be a viable green process that can be used for the extraction of pure and high-quality vegetable oils, especially high-value oils (Radocaj & Dimic, 2013). With the growing trends, cold pressed oils are becoming more acceptable in the food industry and market. This chapter is mainly designed to reveal the usefulness of cold pressed walnut oil and present its nutritional and nutraceutical attributes along with the bioactives profile and food science uses of this high-value nut oil.

2 Extraction and processing of cold pressed walnut oil

Oil from the plant materials can be extracted by multiple extracting methods using different nonpolar solvents, e.g., hexane or petroleum ether (Sabudak, 2007). As the demand for organic food is increasing, the consumers and food industry are looking for the products that have not been produced by obtaining the oil from seeds using conventional solvent extraction methods or having undergone a chemical refining process (Gecgel et al., 2015). In line with the concept of optimal nutrition, consumers are now looking for oils that can boost human health, that are not contaminated with different chemicals, and that prevent certain diseases. In this regard, cold pressed oils are found to be an alternative high-value product (Radocaj & Dimic, 2013). Interest in obtaining the oil using a cold pressing method is increasing day by day due to its high nutrient and nutraceuticals density. Furthermore, the cold pressed oils obtained from nuts show antimicrobial activities and provide a natural way to control infectious diseases and decrease the pathogens load in food products (Gecgel et al., 2015).

Cold Pressed Oils. https://doi.org/10.1016/B978-0-12-818188-1.00044-X
Copyright © 2020 Elsevier Inc. All rights reserved.

FIG. 1 Walnut (*Juglans regia* L.) fruits with nuts inside.

Compared to other refined oils, simple cold pressed oil (mechanically pressed) preserves many valuable and health beneficial ingredients, such as phenolics, tocopherols, sterols, and carotenoids with medicinal potential (Kania, Michalak, Gogolewski, & Hoffmann, 2004).

3 Fatty acids and acyl lipids profile of cold pressed walnut oil

Walnuts contain triacylglycerol containing monounsaturated fatty acids (MUFAs) and polyunsaturated fatty acids (PUFAs) as a major component in seed oil. The compositions of fatty acids in walnut oil reported by different researchers are given in Table 1. Most of the nut oils have a high content of MUFAs (Robbins, Shin, Shewfelt, Eitenmiller, & Pegg, 2011). Walnuts oil has been shown to have high percentages of PUFAs including linoleic acid and linolenic acid (Arranz, Cert, Jimenez, Cert, & Saura-Calixto, 2008; Bujdoso et al., 2016; Popa, Boran, & Davidescu, 2016). More than 60% of walnut oil is composed of PUFAs, as noted for different walnut cultivars (Arranz et al., 2008). Martínez and Maestri (2008) investigated considerable variation in contents of oleic acid (16.1%–25.4%), linoleic acid (52.5%–58.9%), and linolenic acid (11.4%–16.5%) in different varieties of walnut from Argentina. Somewhat similar results have also been documented by Amaral, Casal, Pereira, Seabra, and Oliveira (2003) for six walnut cultivars grown in Portugal. Total PUFAs in walnut oil varied from 71% to 75%, while total MUFAs ranged from 16% to 20%. In a study on cold pressed walnut oil, eight different saturated and unsaturated fatty acids were separated, wherein linoleic acid (ω-6, C18:2n-9, 12) and α-linolenic acid (ω-3, C18:3n-9, 12, 15) were found in higher amounts in walnut oil (Copolovici et al., 2017). Another study revealed that total MUFAs in walnut oil were lesser in concentration (9.5%) compared to PUFAs (63.3%). Linoleic acid was the major fatty acid of walnut oil, followed by oleic, linolenic, palmitic, and stearic acids (Dogan, Elik, Balta, Javidipour, & Yavic, 2010). In fact, among vegetable oils, walnut oil has one of the highest amounts of PUFAs (up to 78% of the total fatty acids) (Amaral et al., 2003). Romanian walnut fruits are enriched in nutritional compounds such as high lipid content (50%–70%), proteins (10%–20%), and carbohydrates (15%). Almost 70% of the ratio of PUFAs is found in walnut oil, out of which linoleic acid was the major component (Patras & Dorobantu, 2010).

4 Minor bioactive lipids in cold pressed walnut oil

Among the minor bioactive components of vegetable oils, tocopherols are of greater importance; however, their concentration may vary depending upon the variety/species of oilseed crops as well as postharvest processing techniques. Likewise, walnut oil is a good source of tocopherols; the concentration of those generally varies within the range of 100–436 mg/kg. In walnut oil tocopherols fraction, γ-tocopherol is predominant, contributing 88% of total tocopherols

TABLE 1 Fatty acids composition of walnut oil extracted by different extraction techniques.

Reference		Copolovici, Bungau, Boscencu, Tit, and Copolovici (2017)	Bialek, Bialek, Jelinska, and Tokarz (2016)	Bujdoso et al. (2016)	Nikovska (2010)	Sabudak (2007)
Technique for oil extraction		Cold pressing	Solvent extraction	Solvent extraction	Cold pressing	*n*-Hexane extraction
Fatty acid (%)	Palmitic acid $C_{16:0}$	6.58	7.40	6.82	9.00	6.82
	Stearic acid $C_{18:0}$	2.16	2.60	4.60	2.80	3.74
	Arachidic acid $C_{20:0}$	0.06	0.10	0.09	–	0.13
	Oleic acid $C_{18:1}$	18.0	15.2	21.1	24.9	17.3
	Gondoic acid $C_{20:1}$	0.09	–	–	–	0.08
	Linoleic acid $C_{18:2}$	60.8	59.5	57.0	54.6	58.2
	Linolenic acid $C_{18:3}$	12.0	9.60	11.0	6.50	11.3

(Rabrenovic et al., 2011; Savage, McNeil, & Dutta, 1998). According to Lavendrine, Ravel, Poupard, and Alary (1997), β-tocopherol is rarely present in walnut oils. The presence of large amounts of γ-tocopherol provides some protection against oxidation (Dolde, Vlahakis, & Hazebroek, 1999). Phytosterols are physiologically active substances, which constitute the main part of the unsaponifiable matters. Among the total sterols tested in walnut oil, β-sitosterol constitutes up to 85%, followed by Δ5-avenasterol (7%) and campesterol (4%) (Amaral et al., 2003). Cholesterol, stigmasterol, Δ7-stigmasterol, and Δ7-avenasterol were also detected. Verleyen et al. (2002) investigated only campesterol and β-sitosterol in walnut oil. The amount of total sterols varies widely between 120 and 200 mg/100 g oil. High contents of tocopherols along with phytosterols in walnut oil can be linked to the nutraceutical properties of this important nut oil.

5 Organoleptic and health-promoting traits of cold pressed walnut oil

Studies on walnut oils have revealed that they could be used in cosmetics due to their exceptional lipid quality (Bialek et al., 2016). Depending upon their moisturizing properties and free radical scavenging activities, walnut oils are employed in dry skin creams as antiwrinkle and antiaging products (Espín, Soler-Rivas, & Wichers, 2000). In another study, Sabate et al. (1993) demonstrated that the inclusion of walnuts (around 80 g) in the daily diet of men for a period of 7–8 weeks can significantly reduce their blood cholesterol levels, provided that total fat intake is kept at a reasonable level. A higher concentration of linolenic acid gives walnut oils additional cosmeceutical properties and prompts their potential applications of the cosmo-nutraceutical industry.

Walnuts contain a high percentage of PUFAs such as omega-6 and omega-3, which are essential dietary fatty acids. Clinical and epidemiological trials suggest that omega-3 fatty acids might have a significant role in the prevention of coronary heart disease (Bucher, Hengstler, Schindler, & Meier, 2002; Harper & Jacobson, 2001). The consumption of walnut oil enhances endothelial function in diabetes type II, and may therefore help to reduce cardiovascular disease risk in this high-risk population (May et al., 2010). Recently, it has been demonstrated that walnut consumption could protect against colon cancer by changes in the gut microbiome (Nakanishi et al., 2016). Cold pressed oil obtained from walnuts could decrease the sugar concentrations in patients with diabetes mellitus type 2 (Copolovici et al., 2017). It has a high amount of omega-6 along with a reasonable amount of omega-3, which are suitable for salads or as a gourmet condiment (Boskou, 2017).

6 Other issues (stability and shelf life)

The fatty acids composition of walnut oil suggests that it is likely to be oxidized and become rancid fairly rapidly (Fox & Cameron, 1995; Savage et al., 1998; Zwarts, Savage, & McNeil, 1999). Oil stability is notably more influenced by antioxidants and the position of fatty acids within the triglycerides than by fatty acids composition (Kaijser, 1998; Neff et al., 1992). In this regard, Savage et al. (1998) proved that walnut oil is reasonably stable and does not undergo deterioration over 4 months, when stored under optimum conditions in the dark at room temperature. It is likely that the high natural antioxidant (vitamin E) contents of walnut oil play an important role in achieving high storage stability. Walnut oil has a reasonable shelf life and offers good organoleptic flavor and nutritional benefits.

References

Amaral, J. S., Casal, S., Pereira, J. A., Seabra, R. M., & Oliveira, B. P. (2003). Determination of sterol and fatty acid compositions, oxidative stability, and nutritional value of six walnut (*Juglans regia* L.) cultivars grown in Portugal. *Journal of Agricultural and Food Chemistry*, *51*(26), 7698–7702.

Arranz, S., Cert, R., Jimenez, P. J., Cert, A., & Saura-Calixto, F. (2008). Comparison between free radical scavenging capacity and oxidative stability of nut oils. *Food Chemistry*, *110*(4), 985–990.

Bialek, A., Bialek, M., Jelinska, M., & Tokarz, A. (2016). Fatty acid profile of new promising unconventional plant oils for cosmetic use. *International Journal of Cosmetic Science*, *38*(4), 382–388.

Boskou, D. (2017). Edible cold pressed oils and their biologically active components. *Journal of Experimental Food Chemistry*, *3*, 108.

Bucher, H. C., Hengstler, P., Schindler, C., & Meier, G. (2002). N-3 polyunsaturated fatty acids in coronary heart disease: A meta analysis of randomized controlled trials. *American Journal of Medicine*, *112*, 298–304.

Bujdoso, G., Konya, E., Berki, M., Nagy-Gasztonyi, M., Bartha-Szuegyi, K., Marton, B., et al. (2016). Fatty acid composition, oxidative stability, and antioxidant properties of some Hungarian and other Persian walnut cultivars. *Turkish Journal of Agriculture and Forestry*, *40*(2), 160–168.

Copolovici, D., Bungau, S., Boscencu, R., Tit, D. M., & Copolovici, L. U. C. I. A. N. (2017). The fatty acids composition and antioxidant activity of walnut cold press oil. *Revista de Chimie*, *68*(3), 507–509.

Dogan, A. C., Elik, F., Balta, F., Javidipour, I., & Yavic, A. (2010). Analysis of fatty acid profiles of pistachios (*Pistachio vera* L.) and native walnuts (*Juglans regia* L.). *Asian Journal of Chemistry*, *22*(1), 517–521.

Dolde, D., Vlahakis, C., & Hazebroek, J. (1999). Tocopherols in breeding lines and effects of planting location, fatty acid composition, and temperature during development. *Journal of the American Oil Chemists' Society*, *76*(3), 349–355.

Espín, J. C., Soler-Rivas, C., & Wichers, H. J. (2000). Characterization of the total free radical scavenger capacity of vegetable oils and oil fractions using 2,2-diphenyl-1-picrylhydrazyl radical. *Journal of Agricultural and Food Chemistry*, *48*(3), 648–656.

Fox, B. A., & Cameron, A. G. (1995). *Food science, nutrition & health* (6th ed.). London, UK: Edward Arnold.

Gecgel, U., Demirci, A. S., Dulger, G. C., Gecgel, U., Tasan, M., Arici, M., et al. (2015). Some physicochemical properties, fatty acid composition and antimicrobial characteristics of different cold-pressed oils. *La Rivista Italiana Delle Sostanze Grasse*, *92*, 187–200.

Harper, C. R., & Jacobson, T. A. (2001). The fats of life: The role of Omega-3 fatty acids in the prevention of coronary heart disease. *Archives of Internal Medicine*, *161*, 2185–2192.

Huffman, S. L., Harika, R. K., Eilander, A., & Osendarp, S. J. (2011). Essential fats: How do they affect growth and development of infants and young children in developing countries? A literature review. *Maternal & Child Nutrition*, *7*, 44–65.

Kaijser, A. (1998). *Stability and lipid composition of oils in macadamia nuts (*Macadamia tetraphylla *and* M. integrfolia*) grown in New Zealand*. Uppsala: Institutionen för Livsmedelsvetenskap, nr. 86, Swedish University of Agricultural Sciences.

Kania, M., Michalak, M., Gogolewski, M., & Hoffmann, A. (2004). Antioxidative potential of substances contained in cold pressed soybean oil and after each phase of refining process. *Acta Scientiarum Polonorum, Technologia Alimentaria*, *3*, 113–121.

Lavedrine, F., Ravel, A., Poupard, A., & Alary, J. (1997). Effect of geographic origin, variety and storage on tocopherol concentrations in walnuts by HPLC. *Food Chemistry*, *58*(1–2), 135–140.

Martínez, M. L., & Maestri, D. M. (2008). Oil chemical variation in walnut (*Juglans regia* L.) genotypes grown in Argentina. *European Journal of Lipid Science and Technology*, *110*(12), 1183–1189.

May, Y., Njike, V. Y., Mille, T. J., Dutta, S., Doughty, K., Treu, J., et al. (2010). Effects of walnut consumption on endothelial function in Type 2 diabetic subjects. *Diabetes Care*, *33*, 227–232.

Nakanishi, M., Chen, Y., Qendro, V., Miyamoto, S., Weinstock, E., Weinstock, G. M., et al. (2016). Effects of walnut consumption on colon carcinogenesis and microbial community structure. *Cancer Prevention Research*, *9*(8), 692–703.

Neff, W. E., Selke, E., Mounts, T. L., Rinsch, W., Frankel, E. N., & Zeitoun, M. A. M. (1992). Effect of triacylglycerol composition and structures on oxidative stability of oils from selected soybean germplasm. *Journal of the American Oil Chemists Society*, *69*(2), 111–118.

Nikovska, K. (2010). Oxidative stability and rheological properties of oil-in-water emulsions with walnut oil. *Advance Journal of Food Science and Technology*, *2*(3), 172–177.

Patras, A., & Dorobantu, P. (2010). Physical and chemical composition of some walnut (*Juglans regia* L) biotypes from Moldavia. *USAMV I I de la Brad" Iasi Scientific Paper Agronomy Series*, *53*(2), 57–60.

Popa, S., Boran, S., & Davidescu, C. (2016). Walnut food oil under UV radiation-influence upon CIELAB and thermal properties of anthocyanin addition. *Revista de Chimie*, *67*(5), 858–861.

Rabrenovic, B., Dimic, E., Maksimovic, M., Sobajic, S., & Gajic-Krstajic, L. (2011). Determination of fatty acid and tocopherol compositions and the oxidative stability of walnut (*Juglans regia* L.) cultivars grown in Serbia. *Czech Journal of Food Sciences*, *29*(1), 74–78.

Radocaj, O., & Dimic, E. (2013). Physico-chemical and nutritive characteristics of selected cold-pressed oils found in the European market. *Rivista Italiana Delle Sostanze Grasse*, *90*, 219–228.

Robbins, K. S., Shin, E. C., Shewfelt, R. L., Eitenmiller, R. R., & Pegg, R. B. (2011). Update on the healthful lipid constituents of commercially important tree nuts. *Journal of Agricultural and Food Chemistry*, *59*(22), 12083–12092.

Sabate, J., Fraser, G. E., Burke, K., Knutsen, S. F., Benett, H., & Linstead, K. D. (1993). Effects of walnuts on serum lipid levels and blood pressure in normal men. *New England Journal of Medicine*, *329*, 603–660.

Sabudak, T. (2007). Fatty acid composition of seed and leaf oils of pumpkin, walnut, almond, maize, sunflower, and melon. *Chemistry of Natural Compounds*, *43*(4), 465–467.

Savage, G. P., McNeil, D. L., & Dutta, P. C. (1998). Vitamin E content and oxidative stability of fatty acids in walnut oil. *Proceedings of the Nutrition Society of New Zealand*, *23*, 81–90.

Savage, G. P., McNeil, D. L., & Dutta, P. C. (2001). Some nutritional advantages of walnuts. In: *ISHS Acta Horticulturae, 544, IV International Walnut Symposium*.

Verleyen, T., Forcades, M., Verhé, R., Dewettinck, K., Huyghebaert, A., & De Greyt, W. (2002). Analysis of free and esterified sterols in vegetable oils. *Journal of the American Oil Chemists' Society*, *79*(2), 117–122.

Wan, B. Y., Dong, H. Z., & Li, H. (2001). Research on the properties and nutrition of walnut. *China Western Cereals & Oils Technology*, *26*, 18–19.

Zhao, S. L., Chen, C. Y., & Ge, F. (2010). Progressive study on functions and components. *Journal of Yunnan University of Traditional Chinese Medicine*, *33*, 71–74.

Zwarts, L., Savage, G. P., & McNeil, D. L. (1999). Fatty acid content of New Zealand grown walnuts (*Juglans regia* L.). *International Journal of Food Sciences and Nutrition*, *50*(3), 184–194.

Chapter 45

Cold pressed grapefruit (*Citrus paradisi* L.) oil

Buket Aydeniz-Guneser and Onur Guneser
Department of Food Engineering, Engineering Faculty, Usak University, Usak, Turkey

Abbreviations

FAO Food and Agriculture Organization
GC-MS gas chromatography-mass spectrometry
GC-O gas chromatography-olfactometry
GRAS generally recognized as safe
IR120 amberlite IR120 sodium form
IRA400 amberlite IRA-400 chloride form
QDA quantitative descriptive analysis
USDA United States Department of Agriculture
XAD7 amberlite XAD polymeric resins

1 Introduction

Citrus fruits belonging to the *Rutaceae* taxa cover a large botanical family including the sweet orange (*Citrus sinensis*), sour orange (*Citrus aurantium*), mandarin (*Citrus reticulata*), lemon (*Citrus limon*), lime (*Citrus aurantifolia*), pummelo (*Citrus maxima*), bergamot (*Citrus bergawia*), and grapefruit (*Citrus paradisi*) (Anwar et al., 2008; Chebrolu et al., 2012). Among the major citrus varieties, the grapefruit, which is believed to have originated in the West Indies (*Citrus* X *paradisi* Macfady), is a hybrid tropical fruit between the pummelo (or shaddock) and the sweet orange (Iwabuchi et al., 2010; Jian, Lee, & Binns, 2007). Grapefruits are classified according to their pulp colors from white to red. There are many grapefruit varieties including Star Ruby, Marsh Seedless, Flame, Oro Blanco, Shambar, Red Blush, Garner Seedless, Davis Seedless, Pink, Rio Star, White Marsh, Duncan, Foster, Flame, Thompson, and Reed (Iwabuchi et al., 2010; Ortuno et al., 1995).

Grapefruits are a rich source of vitamins A and C, dietary fiber, potassium, magnesium, phenolics, and limonoids. In particular, pink grapefruit (1.4 mg lycopene/100 g fruit) draws attention due to its characteristic color and it ranks third in the list of foods with the highest lycopene content (Whitbread, 2019). Moreover, grapefruit is accepted as a weight-loss friendly food due to its low calorific value (Kelebek & Selli, 2011).

A regular increase has occurred in grapefruit production (million metric tons) worldwide from 2012/13 to 2017/18 (USDA, 2019). The FAO's forecast for 2018/19 global grapefruit production was 7 million metric tonnes, up 2.85% (6.7 million metric tonnes) compared to the previous year (2017/18) (USDA, 2019). Global grapefruit production is intended for both fresh consumption and industrial processing, and their amounts were reported as 6.2 million metric tons and 0.5 million metrics tons, respectively. Due to suitable weather and growing conditions, China (4.9 million metric tonnes) is the leading grapefruit producing country, followed by the USA (606 metric tonnes) and South Africa (450 metric tonnes) (USDA, 2019). Grapefruit production in the main producing countries from the 2013/14 to 2017/18 season is shown in Table 1 (USDA, 2019).

Grapefruit oil is a valuable ingredient and its consumption rate and approximate value was reported as 700 metric tonnes and 9 million €, respectively (Brud, 2015). Soft drinks, fragrances, and cosmetics as well as alternative medicine are the major application areas of grapefruit essential oils obtained by peel expression or hydrodistillation (Brud, 2015; Lin & Rouseff, 2001). The conditions and techniques of oil extraction may have created the differences in extracted oils physicochemical, compositional, functional, and sensorial properties. Cold pressed grapefruit oils have a characteristic

Cold Pressed Oils. https://doi.org/10.1016/B978-0-12-818188-1.00045-1
Copyright © 2020 Elsevier Inc. All rights reserved.

TABLE 1 Grapefruit production for some countries.

	Year (1000 metric tonnes)					
Production	**2013/14**	**2014/15**	**2015/16**	**2016/17**	**2017/18**	**2018/19 (Feb)**
China	3717	4050	4350	4600	4800	4900
United States	950	826	728	633	469	606
South Africa	413	387	315	354	419	450
Mexico	424	424	438	442	445	445
Turkey	235	238	250	253	265	270
Israel	236	186	163	149	144	155
European Union	92	109	107	106	108	112
Other	0	26	25	25	22	24
Total	6067	6246	6376	6562	6672	6962
	Fresh domestic consumption					
	2013/14	**2014/15**	**2015/16**	**2016/17**	**2017/18**	**2018/19**
China	3578	3957	4224	4460	4670	4740
European Union	417	415	438	398	445	445
Mexico	328	323	331	334	333	333
United States	346	325	287	282	245	307
Russia	133	101	117	106	130	140
Japan	109	125	105	108	92	103
Turkey	63	96	62	131	76	71
Canada	42	40	39	37	36	38
Ukraine	27	15	18	15	25	30
Israel	24	8	10	8	8	8
Other	20	17	17	18	19	20
Total	5087	5422	5648	5897	6079	6235
	For processing					
	2013/14	**2014/15**	**2015/16**	**2016/17**	**2017/18**	**2018/19**
United States	470	370	333	268	184	54
South Africa	203	168	111	118	127	135
Mexico	84	84	87	88	93	93
Israel	134	117	92	80	68	72
European Union	16	18	20	19	20	20
Other	0	1	1	1	2	2
Total	907	758	644	574	494	576

TABLE 1 Grapefruit production for some countries—cont'd

	Export					
	2013/14	2014/15	2015/16	2016/17	2017/18	2018/19
South Africa	217	221	203	231	288	310
China	165	124	159	180	208	230
Turkey	177	145	190	125	190	200
Israel	78	61	61	61	68	75
United States	147	141	124	108	58	60
Hong Kong	8	10	11	15	30	40
Mexico	14	19	22	21	20	20
Other	19	15	15	15	17	17
Total	825	736	785	756	879	952

	Import					
	2013/14	2014/15	2015/16	2016/17	2017/18	2018/19
European Union	360	339	365	326	374	370
Russia	133	101	117	106	130	140
Japan	109	100	82	84	71	80
China	26	31	33	40	78	70
Hong Kong	16	15	16	20	35	45
Canada	42	40	39	37	36	38
Ukraine	27	15	18	15	25	30
United States	13	10	16	25	18	15
Switzerland	7	7	7	7	7	7
South africa	12	7	4	1	3	3
Other	7	5	4	4	3	3
Total	752	670	701	665	780	801

Adapted from USDA (2019). https://apps.fas.usda.gov/psdonline/circulars/citrus.pdf (Available at 07 March 2019).

bitter taste and distinct aroma that can be obtained from grapefruit peel, leaves, or seeds. Several researchers indicated the biological activities of cold pressed grapefruit oils due to their great distinctive and nutraceutical composition. This chapter presents the current scientific knowledge on the cold pressing technique of extraction of grapefruit oils, focusing on compositional properties, bioactive components, volatile compounds, sensory attributes, and health-promoting effects of cold pressed grapefruit oils.

2 Physicochemical properties of cold pressed grapefruit oils

There is much research in the literature on grapefruit oils and their physicochemical properties (Anwar et al., 2008; Habib et al., 1986; Kesterson & Braddock, 1975). In a study by Kesterson and Braddock (1975), peel oil levels of Duncan, Marsh, and Ruby Red grapefruit varieties grown in Florida were determined as 2.0–3.5 and 2.5–4.0 kg/metric tonnes, respectively.

TABLE 2 Some physicochemical properties of grapefruit seeds and seed oils (Anwar et al., 2008; Habib et al., 1986).

Property	Grapefruit seed
Oil yield (%)	36.54/41.36
Protein content (%, N*6.25)	16.60
Dietary fiber content (%)	8.50
Ash content (%)	2.60/5.03
	Grapefruit seed oil
Refractive index (25°C)	1.4639/1.4662
Specific gravity (g/mL) (25°C)	0.913/0.932
Acid value (mg KOH/g oil)	0.66/0.90
Iodine number (g I/100g oil)	91.4/101.50
Peroxide value (meq O_2/kg oil)	1.55
Saponification number (mg KOH/g oil)	189.6/198.85
Unsaponifiable matter (%)	0.39

Numerous researchers compared the seed oil characteristics of various citrus varieties including orange, mandarin, lime, and grapefruit (Anwar et al., 2008; Habib et al., 1986; Haze, Sakai, & Gozu, 2002; Saidani, Dhifi, & Marzouk, 2004). Orange (*Citrus sinensis*) seeds have a higher oil content (up to 52%) than lemon (*Citrus lemon*) (up to 45%) seeds and grapefruit (*Citrus paradisi*) (up to 49%) seeds. Kesterson and Braddock (1975) evaluated chemical properties of solvent-extracted oils from grapefruit seeds, which contained 41% of oil. Researchers informed that grapefruit seed oils, which is nondrying (91 g I/100 g iodine value), has low acid as 0.9 mg KOH/g oil and a saponification number of 189.6 mg KOH/g oil. Some physicochemical properties of fruit seed oils obtained from grapefruit processing are shown in Table 2.

Recently, the cold pressing technique has been located among environmentally friendly techniques for oil extraction from by-products obtained by food-processing operations (Aydeniz, Güneşer, & Yilmaz, 2014; Aydeniz-Güneşer & Yilmaz, 2018). Although there are several papers on grapefruits, oil produced by hydrodistillation, and/or solvent extraction, there is limited research in the literature on cold pressed grapefruit oils. Citrus peel essential oils produced by cold pressing have a yellow-pale green color and sweet, caramel, fresh citrus, and creamy aroma characteristics (Curtis, Thomas, & Johnson, 2016).

In a recent study by Aydeniz-Güneşer and Yilmaz (2018), the effects of the debittering enzymes on grapefruit seed oil's compositional and physicochemical properties and especially characteristic flavor was reported. Researchers observed that although enzyme treatment did not cause remarkable differences in physical properties like specific gravity, viscosity, turbidity, and L, a^*, b^* values between the control group (no enzyme treatment) and enzyme-treated group, the oil yield in enzyme-treated samples (65%) was significantly higher than that (55%) of the control group (Table 3). Iodine values, which indicate the unsaturation degree of the oil in enzyme-treated seed oils, were significantly lower than those of control seed oils. Free fatty acids, peroxide values, and iodine values of enzyme-treated seed oils were higher than in control samples. Another important result was that naringinase and hesperinidase enzyme treatments caused a statistically significant improvement in the total phenolics content and antioxidant capacity values of cold pressed grapefruit seed oils.

Valorization of oily cake or press meals gained after cold pressing is an increasingly important issue, because meals could be further utilized to extract protein, dietary fiber, and flavonoids. Yilmaz et al. (2018) investigated the physicochemical values of grapefruit seed press meal formed during the cold pressing technique. Debittering enzymes applications on the grapefruit seeds before cold pressing caused significant differences in meal samples. One of the striking and undesirable findings (Table 4) was that meal without enzyme treatment (control) has a higher remaining oil content than that of meal with enzyme treatment. In this respect, it is possible to say that enzyme application can enhance oil and protein yields extracted from seeds.

TABLE 3 Physicochemical properties of the cold pressed grapefruit seed oils (Yilmaz, Aydeniz-Guneser, & Ok, 2018).

	Grapefruit seed oil	
Property	Control	Enzyme treated
Oil yield (%)	55.15 ± 2.40^{B}	$65.35 \pm 1\ 51^{A}$
Specific gravity (25°C)	0.92 ± 0.01^{A}	0.91 ± 0.01^{A}
Refractive Index (25°C)	1.47 ± 0.01^{A}	1.47 ± 0.01^{A}
Viscosity (25°C, cP)	58.42 ± 0.10^{A}	58.17 ± 0.05^{A}
Turbidity (25°C, NTU)	8.75 ± 1.89^{A}	8.00 ± 0.58^{A}
Color		
L	35.83 ± 1.89^{A}	38.95 ± 0.93^{A}
*a**	-0.96 ± 0.30^{A}	-0.14 ± 0.06^{A}
*b**	17.19 ± 4.16^{B}	24.81 ± 3.03^{A}
Sediment content (%)	8.25 ± 1.69^{A}	5.83 ± 0.75^{A}
Free fatty acid (% linoleic acid)	0.37 ± 0.06^{B}	0.57 ± 0.04^{A}
Acid value (mg KOH/g oil)	0.74 ± 0.13^{B}	1.13 ± 0.08^{A}
Peroxide value (meq O_2/kg oil)	13.83 ± 1.73^{A}	14.84 ± 2.18^{A}
p-Anisidine value	0.73 ± 0.19^{A}	0.86 ± 0.39^{A}
Iodine number (g I/100 g oil)	126.60 ± 25.10^{A}	114.72 ± 0.54^{B}
Saponification number (mg KOH/g oil)	202.53 ± 2.43^{A}	204.61 ± 1.59^{A}
Unsaponifiable matter (%)	0.79 ± 0.06^{A}	0.89 ± 0.10^{A}
Total phenolics (µg GA/100 g)	6432 ± 968^{B}	9811 ± 340^{A}
TEAC (µmol Trolox/100 g oil)	15.709 ± 1800^{B}	24.345 ± 739^{A}

$^{A-B}$ Same rows followed by different superscript letters were significantly different ($P < .05$).

3 Fatty acids profile of cold pressed grapefruit oil

Due to a high interest in citrus seed oils for edible, pharmaceutical, and therapeutic purposes, the lipid profiles of citrus seeds have been well studied. Fatty acid and acyl lipid profiles of citrus oils vary widely depending on several factors such as citrus variety, climatic conditions, and the production processes. In general, citrus seed oils contain four main fatty acids at high levels: palmitic (16:0), oleic (C18:1n-9), linoleic (C18:2n-6), and linolenic (C18:3 n-3) (Ajewole & Adeyeye, 1993; Anwar et al., 2008; Matthaus & Özcan, 2012; Saidani et al., 2004). In a study by Teles et al. (1972), the major saturated fatty acids of Arizona grapefruit seed oil were determined as 28%–32% palmitic and 2.5%–4.5% stearic, while the major unsaturated fatty acids were 21%–22% of oleic and 36%–39% of linoleic acids. Anwar et al. (2008) reported that the fatty acid composition of grapefruits from Pakistan was 32% of palmitic acid, 3.7% of stearic acid, 22% of oleic acid, and 36% of linoleic acid. Similarly, Waheed, Mahmud, Saleem, and Ahmad (2009) reported that the contents of palmitic, oleic, linoleic, and linolenic acids in the Pakistan variety of grapefruit were 35.9%, 24.0%, 24.4%, and 3.97%, respectively. Matthaus and Özcan (2012) reported that oleic, linoleic, and palmitic acids were major fatty acids for the Turkish variety of grapefruit and their amounts in grapefruit were 70.1%, 19.5%, and 5.80%, respectively. In a study on grapefruit seed oils obtained from cold pressing (Yilmaz et al., 2018), similar results (Tables 5 and 6) were reported, wherein palmitic (28%) and linoleic (40%) acids, β-sitosterol (80%), and α-tocopherol (220 ppm) as dominant fatty acids, sterol, and tocopherol forms were determined, respectively.

TABLE 4 The proximate compositions of the grapefruit seeds and its cold press meals (Yilmaz et al., 2018).

			Grapefruit seed meal	
	Grapefruit seed		Control	Enzyme-treated
Seed size (mm)		Moisture (%)	7.74 ± 0.06^{A}	7.72 ± 0.09^{A}
Length	12.45 ± 0.34	Water activity (25°C)	0.51 ± 0.01^{B}	0.55 ± 0.01^{A}
Width	7.68 ± 0.16	Ash (%)	4.21 ± 0.45^{A}	4.23 ± 0.63^{A}
Height	4.08 ± 0.09	Oil[a](%)	16.55 ± 0.15^{A}	13.24 ± 0.13^{B}
1000-seed weight (g)	313.40 ± 13.60	Protein[a] (%)	22.10 ± 1.44^{A}	23.73 ± 0.74^{A}
Skin:flesh ratio	0.39 ± 0.04	Color		
Moisture (%)	50.73 ± 0.18	*L*	52.21 ± 0.57^{A}	49.14 ± 0.47^{B}
Water activity (25°C)	0.96 ± 0.01	*a**	8.47 ± 0.55^{A}	9.11 ± 0.11^{A}
Oil[a] (%)	45.72 ± 0.05	*b**	20.82 ± 0.48^{A}	19.45 ± 0.09^{B}
Protein[a] (%)	16.25 ± 0.44			
Ash (%)	1.28 ± 0.01			
Color				
L	58.94 ± 0.73			
*a**	0.75 ± 0.15			
*b**	14.85 ± 0.43			

A-B Same rows followed by different superscript letters were significantly different ($P<.05$).
[a]*Values on dry weight basis.*

TABLE 5 Fatty acid compositions of the cold pressed grapefruit seed oils (Yilmaz et al., 2018).

	Grapefruit seed oil	
Fatty acids (%)	Control	Enzyme-treated
Palmitic (C16:0)	28.19 ± 0.06^{A}	28.66 ± 0.01^{A}
Palmitoleic (C16:1)	0.75 ± 0.09^{A}	0.40 ± 0.15^{A}
Stearic (C18:0)	3.77 ± 0.12^{A}	3.81 ± 0.04^{A}
Oleic (C18:1 n-9)	20.78 ± 0.01^{A}	20.74 ± 0.13^{A}
Linoleic (C18:2 n-6)	40.78 ± 0.01^{A}	40.95 ± 0.03^{A}
Linolenic (C18:3 n-3)	5.71 ± 0.10^{A}	5.45 ± 0.02^{A}

A-B Same rows followed by different superscript letters were significantly different ($P < 0.05$).

4 Minor bioactive compounds in cold pressed grapefruit oils

The cold pressed technique is known to produce oils having superior sensorial and nutritional properties and aroma quality. While cold pressed grapefruit seed oils have come into prominence due to rich bioactive composition, their highly bitter taste can limit their consumption and these oils have therefore not been proposed for direct human consumption. When minor bioactive compounds profiles of cold pressed grapefruit oils were examined, it could be seen that there are several phytosterols, phenolic and flavonoid compounds, and color pigments that provide the characteristic and distinct taste of grapefruit seed oil.

TABLE 6 Sterol and tocopherol compositions of the cold pressed grapefruit seed oils (Yilmaz et al., 2018).

	Grapefruit seed oil	
	Control	Enzyme-treated
Sterols (%)		
Cholesterol	0.48 ± 0.01^{A}	0.45 ± 0.01^{A}
Brassicasterol	0.27 ± 0.01^{A}	0.13 ± 0.04^{B}
24-Methylen cholesterol	0.16 ± 0.06^{A}	0.19 ± 0.13^{A}
Campesterol	9.00 ± 0.06^{B}	9.75 ± 0.02^{A}
Campestanol	0.14 ± 0.01^{A}	0.16 ± 0.03^{A}
Stigmasterol	2.18 ± 0.08^{B}	2.71 ± 0.01^{A}
Delta-7-campesterol	0.52 ± 0.06^{A}	0.46 ± 0.11^{A}
Delta-5,23-stigmastadienol	nd	0.03 ± 0.02
Chlerosterol	0.87 ± 0.09^{A}	0.91 ± 0.08^{A}
Beta-sitosterol	81.74 ± 0.08^{A}	80.85 ± 0.22^{A}
Sitostanol	1.66 ± 0.19^{A}	1.55 ± 0.38^{A}
Delta-5-avenasterol	1.46 ± 0.44^{A}	1.71 ± 0.58^{A}
Delta-5,24-stigmastadienol	0.85 ± 0.09^{A}	0.51 ± 0.09^{B}
Delta-7-stigmastenol	0.52 ± 0.03^{A}	0.44 ± 0.09^{B}
Delta-7-avenasterol	0.12 ± 0.01^{A}	0.11 ± 0.04^{A}
Tocopherol (mg/kg oil)		
α-Tocopherol	221.81 ± 6.31^{B}	322.0 ± 9.23^{A}

$^{A-B}$ Same rows followed by different superscript letters were significantly different ($P < .05$). *nd*, not detected.

β-Sitosterol was determined as the predominant sterol, while naringin and hesperidin were the major flavonoids. Ferulic and gallic acids are the main phenolic acids in grapefruit seed oils (Adeyeye & Adesina, 2015; Matthaus & Özcan, 2012; Vanamala et al., 2005). In an early study by Barroso et al. (1972), it was revealed that grapefruit seed oil has three main plant sterols: campesterol, β-sitosterol, and stigmasterol. The content of these sterols were in the following order: β-sitosterol (90.1%) > campesterol (7.4%) > stigmasterol (2.5%). Similarly, Adeyeye and Adesina (2015) mentioned that the major sterols of grapefruit seed oil from Nigeria were campesterol (54.4 mg/kg), sitosterol (283 mg/kg), stigmasterol (12.7 mg/kg), and 5-avenasterol (9.93 mg/kg).

Evaluation of the effects of enzyme treatment on grapefruit seeds before cold pressing and quantification of the sterols and α-tocopherol contents of grapefruit seed oils produced by the cold pressing were studied by Yilmaz et al. (2018). It was apparent that naringinase/hesperinidase applications on grapefruit seed did not result in significant changes in sterol compounds, and remarkable differences were observed only between brassicasterol, campesterol, stigmasterol, Δ-5,24-stigmastadienol, and Δ-7-stigmasterol levels. In contrast to sterol composition, α-tocopherol content was significantly affected by enzyme applications and α-tocopherol content (322 ppm) in the enzyme-treated seed oils increased to 50% compared to the control (without enzyme treatment) seed oils (220 ppm) (Table 6).

In the same study (Yilmaz et al., 2018), the effects of the debittering enzymes on grapefruit seed oils' flavonoid and phenolic acids composition were determined. For that purpose, naringinase (0.06 U/g seed activity) and hesperinidase (0.033 U/g seed activity) enzymes were added to grapefruit seeds before cold pressing. Significant reductions (Table 7) occurred at phenolic compositions, especially flavanone glycosides naringin and hesperidin levels, in grapefruit seed oils, but this was not sufficient to remove all flavanone glycosides. Reduction levels for hesperidin, naringin, and rutin were 43.5%, 42.0%, and 33.0%, respectively.

TABLE 7 Flavonoids, phenolic acid composition of cold pressed grapefruit seed oils treated with debittering enzymes (Yilmaz et al., 2018).

	Grapefruit seed oil		
	Control	Enzyme-treated	Reduction (%)
Flavonoids (mg/kg oil)			
Eriocitrin	94.70 ± 18.40^{A}	97.85 ± 1.76^{A}	–
Rutin	443.0 ± 95.30^{A}	295.91 ± 2.44^{B}	33.2
Naringin	1324.60 ± 76.60^{A}	764.12 ± 9.60^{B}	42.3
Naringenin	57.45 ± 9.90^{A}	33.14 ± 0.09^{A}	42.3
Hesperidin	884.50 ± 17.50^{A}	498.89 ± 1.23^{B}	43.5
Neohesperidin	452.20 ± 63.60^{A}	310.60 ± 14.00^{A}	31.3
Kaempherol	18.55 ± 2.14^{A}	13.19 ± 0.43^{A}	28.8
Phenolic acids (mg/kg oil)			
Gallic	78.40 ± 12.20^{A}	77.60 ± 10.20^{A}	1.02
Syringic	7.23 ± 0.12^{A}	7.19 ± 0.01^{A}	0.55
tr-ferulic	206.10 ± 34.00^{A}	205.27 ± 2.97^{A}	0.40
Rosmaniric	16.00 ± 0.17^{A}	15.64 ± 0.09^{A}	2.25
tr-2-Hydrocinnamic	11.32 ± 0.55^{A}	10.99 ± 0.07^{A}	2.91

$^{A\text{-}B}$ Same rows followed by different superscript letters were significantly different ($P < .05$).

TABLE 8 Carotenoid and chlorophyll composition of cold pressed grapefruit seed oils (Yilmaz et al., 2018).

	Cold pressed grapefruit oil		
Pigments (mg/kg oil)	Control	Enzyme-treated	Reduction (%)
Total carotenoid	2.35 ± 0.16^{B}	2.99 ± 0.15^{A}	–
β-Carotene	2.27 ± 0.16^{B}	2.89 ± 0.15^{A}	–
Lutein	2.26 ± 0.16^{B}	2.88 ± 0.15^{A}	–
Total chlorophyll (pheophytin a)	0.08 ± 0.03^{A}	0.06 ± 0.02^{A}	25

$^{A\text{-}B}$ Same rows followed by different superscript letters were significantly different ($P < .05$).

Total carotenoid and chlorophyll contents (Table 8) of cold pressed grapefruit seed oils were measured wherein β-carotene was the major color pigment. β-Carotene and lutein contents in enzyme-treated oil samples were higher than those in the control oil despite pheophytin levels not differing between the oil samples (Yilmaz et al., 2018).

In another study by Aydeniz-Güneşer and Yilmaz (2018), amberlite resins such as XAD7, IR120, IRA400, and natural and low-cost adsorbent materials such as zeolite, sepiolite, and montmorillonite were evaluated to remove bitterness and reduce the throat-catching feeling caused by cold pressed grapefruit seed oil. All oil samples added with adsorption materials at 3% (w/w) level were analyzed in regard to physicochemical properties, and phenolic and phytosterol compositions. Furthermore, Quantitative Descriptive Analysis (QDA) and consumer acceptance tests were performed in order to observe sensorial changes in treated cold pressed grapefruit oils. Natural sepiolite, zeolite, and XAD7 amberlite have the largest surface area, pore radius, pore-volume, and the highest adsorption capacity, and these morphological properties were seen to play a remarkable role in reducing flavonoids such as rutin, naringin, and hesperidin. Lower free acidity, turbidity, viscosity, and soap contents in grapefruit oils treated with adsorbents were recorded. Sepiolite, followed by zeolite and amberlite XAD7 resin, was determined as the most successful material (Table 9) to reduced undesirable bitter, astringent taste/flavor, and throat-catching feeling.

Thermal behavior has critical importance on the oxidative stability of oils, as well as food and chemistry applications of vegetable oils. In addition, crystallization and melting temperatures of oils or fats have provided reliable and accurate information on lipid chemistry, authentication, and adulteration (Indriyani, Rohman, & Riyanto, 2016). Aydeniz-Güneşer (2016) evaluated the crystallization and melting characteristics (onset, crystallization and melting temperatures (T_c and T_m) and enthalpies of crystallization and melting (ΔH)) of grapefruit oil obtained by cold pressing (Table 10). Results indicated that the onset crystallization temperatures were at −3.2°C and −2.6°C for cold pressed oils and grapefruit oils produced from the seeds treated with debittering enzymes before pressing, respectively. Cold pressed grapefruit oils were completely crystallized between −6.2°C and −5.4°C, then grapefruit oils included the four different fractions (which have different melting temperatures from −22.7°C to 5.7°C) started to melt at around −24°C. Attention was drawn to the fact that the oxidative stability of enzyme-treated oils was higher than that of control (nonenzyme-treated) oils.

5 Flavor and sensory characterization of cold pressed grapefruit oils

Citrus essential oils are the most popular group of aromatics in the natural flavors and fragrances market. Many food products including beverages and confectioneries contain flavoring agents based on citrus essential oil due to their unique and attractive individual aromas (Sawamura, 2010). The key aroma compounds of various kinds of citrus species have been revealed by advanced chromatographic separations such as gas chromatography-olfactometry (GC-O) and gas chromatography-mass spectrometry (GC-MS) (Shahidi & Zhong, 2012). Aldehydes, terpene aldehydes, terpene alcohols, and ketones are major aromatic compounds for most citrus essential oils. The most typical aromatic compounds in citrus essential oils are limonene (citrus), terpinene, α-terpineol (floral), pinene (pine), linalool (woody), nerol, and geraniol. It was emphasized that grapefruit peel/seed oils have more than 200 flavor compounds including aldehydes, esters, alcohols, and free acids. Several studies have revealed that nootkatone is the most abundant and characteristic volatile compound in grapefruit peel/seed oils (Lin & Rouseff, 2001; Sun & Petracek, 1999). D-Limonene found in citrus has also important roles like a system regulation and reduction of cholesterol levels. Oranges (84%–96%) and grapefruit (85%–95%) were accepted as the richest sources of D-limonene (Zielisnki, 2017).

An early study by Wilson and Shaw (1980) identified that cold pressed oil of Florida grapefruit had 32 volatile compounds. Among these compounds, limonene, myrcene, and sabinene were found to compose up to 88% of volatiles in cold pressed grapefruit oil. Similarly, Viuda-Martos, Ruiz-Navajas, Fernández-López, and Pérez-Álvarez (2009) reported that essential oils of grapefruit extracted from peel contain 25 volatile compounds. Limonene (96.2%) and myrcene (1.5%) were the most abundant volatile compounds, while α-thujene, β-pinene, β-ocymene, nonanal, isogeraniol, *trans*-limonene oxide, α-terpineol, *trans*-carveol, carvone, citral, geranil acetate, germacrene D, and linalool in grapefruit essential oil were found to be at trace levels. In a study by Lin and Rouseff (2001), aroma compounds in cold pressed grapefruit oil were characterized using GC-O and GC-MS techniques. The researchers revealed that 22 volatile compounds had significant major aroma activity for grapefruit oil. Researchers identified 1,8-cineole (minty), octanal (fresh minty), dodecanal (citrus), trans-4,5-epoxy-(*E*)-2-decenal (green), 4-hydroxy-2,5-dimethyl-3(2*H*)-furanone (caramel), eugenol (sweet/honey), β-sinensal, and nootkatone as key volatiles in commercial grapefruit peel oil. In another study, volatile components of six different essential oils were analyzed by Haze et al. (2002) to examine the effect of fragrance inhalation on sympathetic activity in normal adults. Limonene (93%), myrcene (1.5%), sabinene (0.6%), and α-pinene (0.6%) were determined as major aromatic compounds. Tisserand and Young (2014) reported similar results on grapefruit essential oils composition, confirming that it consisted mainly of 90% limonene and bergamot (27%–52%). Ahmad et al. (2006) reported that essential oils of grapefruit from Pakistan contain 93.9% of monoterpenes, and 2.17% sesquiterpenes. Among the monoterpenes, 86.2% of limonene and 6.28% of myrcene were found to be major components, while α-terpinene was determined at the level of 2.1%. Similarly, Njoroge et al. (2005) found that four monoterpenes (limonene, α-terpinene, α-pinene, and sabinene) amounted to 93.3% of the total volatile compounds in essential oil of Red Blush grapefruit from Kenya.

Considering the sensory characteristics of grapefruit oil, it is assumed not to proper for direct human consumption. It is very bitter, astringent, and has a throat-catching property due to its flavonoids, phenolic acids, and volatile composition. In this context, recent research was conducted by Yilmaz et al. (2018) on the aromatic and sensory characteristics of enzyme-treated and untreated cold pressed grapefruit seed oils (Table 11). The researchers found that the major volatile compound in both oils was D-limonene, wherein enzymes (naringinase and hesperidinase) treatment reduced by almost half the content of D-limonene in grapefruit seed oil.

The researchers characterized the cold pressed grapefruit seed oils with 11 different sensory terms (Fig. 1). The most important sensory terms were expressed as bitter, fatty, spicy, and clarity for cold pressed grapefruit seed oils.

TABLE 9 Phenolic composition (mg/kg) of cold pressed grapefruit seed oils treated with different adsorbents (Aydeniz-Güneşer & Yilmaz, 2018).

	Control	Amberlite XAD7	Amberlite IR120	Amberlite IRA 400	Natural zeolite	Natural sepiolite	Natural montmorillonite
Flavonoid							
Eriocitrin (P = .475)	101.50 ± 0.05	101.57 ± 0.17	101.70 ± 0.02	101.56 ± 0.01	102.02 ± 0.15	101.49 ± 0.01	101.68 ± 0.0605
Rutin (P = .002)	313.10 ± 74.70^{a}	215.90 ± 12.0^{a,b} (−31.04%)‡	243.90 ± 14.30^{a} (−22.10%)	211.23 ± 4.70^{a,b} (−32.53%)	256.10 ± 25.70^{a} (−18.20%)	175.24 ± 1.19^{b} (−44.03%)	190.08 ± 8.32^{b} (−39.29%)
Naringin (P = .006)	1160.70 ± 39.20^{a}	1118.50 ± 0.16^{c} (−3.63%)	1120 ± 0.16^{a,b,c} (−3.50%)	1120.20 ± 0.25^{a,b,c} (−3.48%)	1121.20 ± 0.71^{a,b} (−3.40%)	1121.60 ± 0.53^{a,b} (−3.48%)	1118.90 ± 0.19^{b,c} (−3.60%)
Naringenin (P = .382)	45.35 ± 2.21	40.24 ± 4.27	30.20 ± 11.40	44.67 ± 0.32	44.54 ± 4.21	26.66 ± 1.02	43.53 ± 2.15
Hesperidin (P = .009)	883.60 ± 18.50^{a}	808 ± 134^{a,b} (−8.55%)	868.70 ± 16.80^{a} (−1.68%)	796 ± 13.80^{a} (−9.91%)	814.3 ± 63.4^{a} (−8.51%)	320.70 ± 22.00^{b} (−60.70%)	761 ± 108^{a,b} (−13.87%)
Neohesperidin (P = .398)	390.89 ± 2.33	354 ± 36.10	363.30 ± 22.40	403.10 ± 49.90	387 ± 46.10	185.84 ± 8.46	393.90 ± 18.30
Kaempherol (P = .037)	25.25 ± 0.20^{a,b}	19.88 ± 0.21^{b} (−21.26%)	23.05 ± 0.51^{a,b} (−8.71%)	23.73 ± 0.23^{a,b} (−6.01%)	24.08 ± 0.93^{a,b} (−4.63%)	24.44 ± 0.57^{a} (−3.20%)	21.88 ± 0.14^{a,b} (−13.34%)
Phenolic acid							
Gallic acid (P = .000)	71.40 ± 5.28^{a}	47.09 ± 6.19^{b} (−34.04%)	36.26 ± 1.48^{b,c} (−49.21%)	30.99 ± 0.52^{b,c} (−56.59%)	38.46 ± 4.28^{b,c} (−46.13%)	23.09 ± 0.45^{c} (−67.66%)	20.24 ± 0.39^{c} (−71.65%)
Syringic acid (P = .009)	6.97 ± 0.14^{a}	6.56 ± 0.05^{b} (−5.88%)	6.79 ± 0.01^{a,b} (−2.58%)	6.55 ± 0.01^{b} (−6.02%)	6.81 ± 0.04^{a,b} (−2.29%)	6.56 ± 0.01^{b} (−5.88%)	6.78 ± 0.02^{a,b} (−2.72%)
tr-Ferulik asit (P = .390)	170.81 ± 1.33	157.40 ± 17.70	172.60 ± 7.87	176.10 ± 4.71	170.20 ± 18.40	78.13 ± 4.41	181.08 ± 8.41
tr-2-Hydrocinnamic acid (P = .004)	11.45 ± 0.14^{a}	9.85 ± 0.14^{b} (−13.97%)	10.61 ± 0.14^{a} (−7.33%)	10.31 ± 0.01^{a,b} (−9.95%)	10.85 ± 0.17^{a} (−5.24%)	10.90 ± 0.05^{a} (−4.80%)	10.51 ± 0.09^{a} (−8.20%)

$^{a–c}$ Same rows followed by different superscript letters were significantly different (P < .05).

TABLE 10 Thermal properties of the grapefruit seed oils (Aydeniz-Güneşer, 2016).

	Cold pressed	Enzyme-treated
Crystallization		
$Onset_c$ (°C)	-3.23 ± 006^a	-2.68 ± 0.35^a
T_c (°C)	-6.28 ± 0.17^a	-5.43 ± 0.27^a
ΔH_c (J/g)	-30.86 ± 1.00^b	-2591 ± 0.42^a
Melting		
$Onset_m$ (°C)	-24.78 ± 0.57^a	-21.61 ± 0.38^b
T_m1 (°C)	-22.74 ± 0.42^b	-5.03 ± 0.05^a
T_m2 (°C)	-6.68 ± 0.05^b	-1.12 ± 0.43^a
T_m3 (°C)	-2.40 ± 0.01^b	4.98 ± 0.70^a
T_m4 (°C)	5.77 ± 0.06^a	nd
ΔH_m (J/g)	50.21 ± 2.80^a	42.08 ± 4.84^a
OIT (min)	36.30 ± 10.66^a	4392 ± 3.36^a

[a–b] Same rows followed by different superscript letters were significantly different ($P < .05$). *nd*, not detected.

TABLE 11 Volatile aromatics composition of cold pressed grapefruit seed oils (Yilmaz et al., 2018).

				Concentration (µg/kg oil)	
No	RI[a]	Volatile compound	Aroma/flavor description[b]	Control	Enzyme-treated
1	<700	3-Methylbutanal	Fruity, sweet	27.77 ± 8.48	28.46 ± 2.87
2	716	Acetoin	Creamy, butter	11.54 ± 2.63	nd
3	799	Hexanal	Green, grass	7.76 ± 1.81	6.86 ± 0.98
4	813	Furfural	Sweet, caramel, baked bread	60.26 ± 16.43	20.35 ± 3.16
5	825	Methyl pyrazine	Nutty, roasted cacao	1.51 ± 0.92	2.02 ± 0.60
6		2-Furan menthol	Roasted cacao, burnt	nd	5.51 ± 0.89
7	877	Isoamyl acetate	Banana, fruity	nd	6.01 ± 1.52
8	885	Heptanal	Green herb, fatty	0.90 ± 0.47	0.70 ± 0.02
9	891	Butrylactone	Creamy, fatty, caramel	nd	1.92 ± 0.18
10	911	2,5-Dimethlypyrazine	Nutty, roasted	6.05 ± 2.52	nd
11	914	Butyl isobutyrate	Pear, pineapple	2.03 ± 0.46	2.04 ± 0.07
12	930	α-Pinene	Herbal, turpentine, woody	0.65 ± 0.18	0.43 ± 0.05
13	942	Isopropyl pentanoate	Fruity, pear	nd	16.20 ± 1.77
14	954	Benzaldehyde	Bitter almond	4.76 ± 1.13	3.14 ± 0.25
15	962	5-Methyl furfural	Sweet, caramel	5.49 ± 2.41	9.52 ± 7.67
16	972	β-Pinene	Woody, green pine	nd	0.54 ± 0.16
17	977	4-Octanone	Unknown	1.20 ± 0.36	1.09 ± 0.06

[a]*RI (Kovat Index) on HP 5 MS column.*
[b]*Aromatic definitions of the volatile compounds are found from the web pages of http://www.thegoodscentscompany.com/index.html# and http://www.flavornet.org/flavornet.html. nd: not detected.*

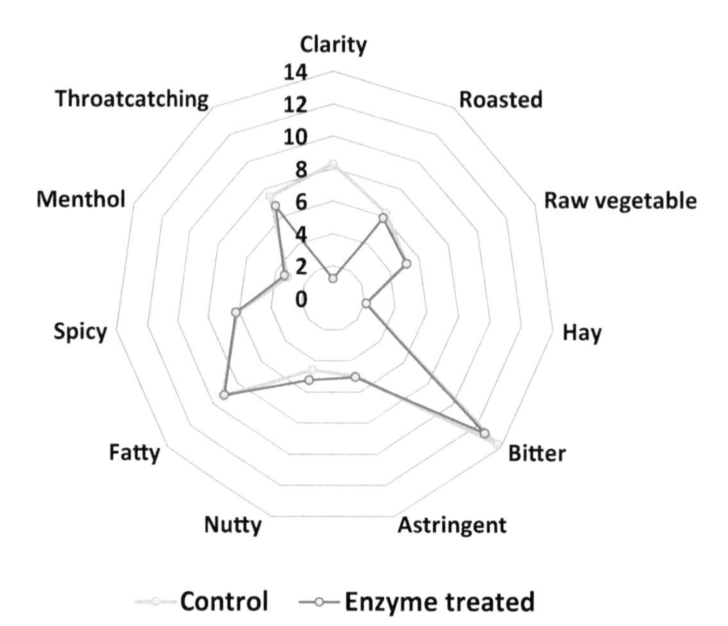

FIG. 1 Sensory descriptors and scores for enzyme-treated and untreated cold pressed grapefruit seed oils (Yilmaz et al., 2018).

Aydeniz-Güneşer and Yilmaz (2018) found that some characteristic flavor and taste descriptors for grapefruit seed oils treated with six different absorbents were bitter, fatty, spicy, raw vegetable, astringent, and throat catching. The researchers indicated that natural sepiolite, natural zeolite, and amberlite XAD7 reduced bitterness and throat-catching scores in grapefruit seed oil by almost half. However, these reductions have not been found adequate enough for people to consume grapefruit seed oil.

6 Health-promoting traits of cold pressed grapefruit oil and oil constituents

Although there is no clear evidence of the adverse effects of essential oils on humans, some potential risks related to inhalation of some essential oils containing high levels of carvone or limonene have been claimed by some researchers. Grapefruit, lemon, caraway, black pepper, fennel, and tarragon oils are some example of these essential oils (Haze et al., 2002; Shen et al., 2005). However, it is important to mention that citrus species and their essential oils contain natural antioxidant, antimicrobial, and anticarcinogenic components (Baratta et al., 1998). Grapefruit essential oils are known for their various functional activities including antimicrobial and antioxidant, and are utilized in industry fields including pharmaceuticals, cosmetics, and food. From a white to red inner color, the unique shape and characteristic flavor of grapefruit are attractive features to citrus-loving consumers. Nutrients, antioxidants, and dietary fibers are contained in grapefruit, making it one of the healthiest and most popular citrus species. Consumption of grapefruit and related products plays a critical role in heart disease, obesity, and diabetes mellitus (Kelebek & Selli, 2011). In an early study (Wattenberg et al., 1985), grapefruit essential oils were found to have chemopreventive and chemotherapeutic activity regarding lung and forestomach cancers. In particular, varying periods (from 3 to 7 min) of inhalation of grapefruit, fennel, black pepper, or tarragon oils have been linked to an ability to increase systolic blood pressure (Haze et al., 2002). Shen et al. (2005) indicated an elevation in blood pressure in rats exposed to inhalation of grapefruit oil for 10 min. Tokoro (1997) suggested grapefruit oil and lemon extracts as antioxidative agents. Grapefruit essential oils were evaluated for different aids and applications in several studies. Grapefruit essential oils could be used in mind-body therapies with the aim of stimulation of immune cells, acceleration of blood circulation, and increasing refreshing/uplifting feelings. Moreover, this essential oil's detox effect on the skin, edema, and chronic fatigue, its diuretic effect, antiseptic and astringent effects, and relaxant effect on muscle and joint pain were also observed (Curtis et al., 2016). Haze et al. (2002) investigated the effects of cold pressed essential oils inhalation on sympathetic activity in normal adults. Tested cold pressed oils (pepper, rose, patchouli, estragon, and fennel oils) have the ability to modulate sympathetic activity and can be used as a regulator in nervous system dysfunctions. In particular, cold pressed grapefruit oils can increase adrenaline and noradrenaline levels and relative sympathetic activity, up to 1.2- and 2.5-fold, respectively. Shen et al. (2005) identified enhanced lipolysis and energy consumption levels and reduced food intake and body weight in rats exposed to grapefruit oil odor and limonene (characteristic odor compound in grapefruit).

Similarly, limonene decreased the activity of the gastric vagal nerve, and appetite suppression, while lipolysis, thermogenesis, and blood pressure were increased (Nagai et al., 2014). Regular consumption of grapefruit and related products rich in vitamin C has been linked with antioxidant defense mechanisms in the immune system and preventive mechanisms against the common cold and flu symptoms (Fashner et al., 2012; Sorice et al., 2014).

Vitamins A and B and also essential trace elements (zinc, iron, and copper) found in grapefruit have crucial effects on human metabolism such as preventing and/or treating inflammation and infectious diseases, promoting the immune response (Mora, Iwata, & von Andrian, 2008; Rink, 2000).

One of the most popular actions of grapefruit has been accepted as weight-reducing effects due to the low calorie (96kcal) and high dietary fiber contents (3.7g) per serving fruit (USDA, 2019). These features have been linked with reducing of blood pressure, total cholesterol, and LDL-cholesterol levels (Dow et al., 2012). In addition, grapefruit consumption plays a role in the control of insulin levels and prevents insulin resistance and type 2 diabetes (Laville & Nazare, 2009).

Grapefruit contains various antioxidants such as lycopene as well as naringin, and hesperidin (flavanones). It has been reported that these antioxidants have a potent antiinflammatory property and ability to protect against some chronic conditions such as heart disease and prostate cancer (Fiedor & Burda, 2014; Gajowik & Dobrzyńska, 2014). Phillips et al. (2012) evaluated the efficacy and adverse effects of citric acid salts, which are the major acid in grapefruit, for the prevention of kidney stones containing calcium. Citrate salts had the ability to inhibit kidney stone formation containing calcium oxalate, and this mechanism was related to the efficacy of citric acid on the inhibition of calcium oxalate crystallization. Curtis et al. (2016) observed that grapefruit essential oil had mild diuretic and beneficial effects on the kidneys, as well as inducing blood circulation in aromatic massage application. Hence organ-specific effects of grapefruit essential oils have been reported by several researchers. Zu et al. (2010) determined the cellular cytotoxicity of grapefruit oils against human prostate and lung cancer, although no cytotoxic effect to breast cancer cells was shown. Hata et al. (2003) drew attention to the fact that limonene in grapefruit essential oils exhibited anticarcinogenic activity and induced apoptosis on tumor cells, especially leukemia cells.

Naringin, hesperidin, neohesperidin, and rutin are the most predominant flavonoids responsible for the bitter taste of grapefruit. Furthermore, clinical and therapeutic effectiveness like antiinflammatory, antioxidant, anticancer, antidiabetic, and protection against cardiovascular and Alzheimer's diseases of these flavonoids has been confirmed by several studies (Giménez-Bastida, González-Sarrías, Vallejo, Espín, & Tomás-Barberán, 2016; Mir & Tiku, 2015).

It was reported that furanocoumarins such as bergamottin and bergapten and flavonoids such as naringin, kaempferol, and quercetin, found in grapefruit, have the ability to inhibit intestinal CYP3A4 and/or P-gp (Bailey, Dresser, & Arnold, 2013; Rodríguez-Fragoso et al., 2011). Moreover, all grapefruit products (juice, concentrated juice, the whole fruit, peel, and essential oil) may be responsible for lower limb edema, gastrointestinal bleeding, arrhythmia, hallucination, and dizziness (Bailey et al., 2013; Seden et al., 2010). Studies have demonstrated the risks and potentially beneficial functions of grapefruit. For many years, the clinical relevance of drug-food or drug-herb interactions has been studied. An interaction of drug-food and drug-herbs is accepted as the consequence of a physical, chemical relationship between a pharmacological drug and food or herbs. Alteration of drug absorption, distribution and excretion via consumed food, herb, and nutrients caused risk factors and serious interactions in human metabolism (Dresser & Bailey, 2002).

6,7-Dihydroxybergamottin is one of the main furanocoumarins and flavonoid components present in grapefruit and responsible for in vivo CYP3A4 inhibition (Ho, Saville, & Wanwimolruk, 2001). Moreover, bergamottin and bergapten have weak inhibition activity for CYP3A4, and their concentrations in grapefruit oil did not cause drug interactions (Tisserand & Young, 2014).

The sesquiterpenoid (+)-nootkatone is accepted as an important chemical compound and plays a remarkable role in the characteristic of the bitter aroma of grapefruit (Iwabuchi et al., 2010). Clinical effects including reduction of body weight, somatic fat ratio, and cholesterol levels (Haze et al., 2002; Uddin et al., 2014), prevention of obesity (Farouk, Mahmoud, El-Sayeh, & Sharaf, 2015), and stimulation of fatty acid oxidation in muscle (Novikova, Ivanov, Garabadzhiu, & Tribulovich, 2017) were reported. Nemmar et al. (2018) reported that nootkatone, naturally occurring in grapefruit essential oil, has antiseptic, antioxidant, antiinflammatory, and antiallergic activities. Another study by Nemmar, Al-Salam, Beegam, Yuvaraju, and Ali (2018) mentioned that nootkatone has the ability to prevent diesel exhaust particles from inducing lung inflammation, and oxidative stress. In particular, nootkatone can provide a protective effect against air pollution-induced respiration dysfunction.

Curtis et al. (2016) drew attention to grapefruit's short shelf life due to the high level of terpenes that can easily oxidize. The maximum recommended dermal use level of grapefruit oil by the International Fragrance Association (IFRA, 2009) was 4% in products excluding bath and body soaps, due to the oxidized oil potentially causing low phototoxic effects and skin sensitization (Tisserand & Young, 2014).

7 Edible and nonedible applications of cold pressed grapefruit oil

Clinical studies have shown that essential oils can be consumed orally and it can be effective in treating mouth and teeth infections against oral pathogens. The efficacy of some citrus-based essential oils including lemons and grapefruit was evaluated against multiresistant bacteria and yeast strains such as *Staphylococcus aureus*, *Staphylococcus epidermidis*, *Streptococcus mutans*, *Streptococcus pyogenes*, *Streptococcus equisimilis*, *Candida albicans*, and *Candida krusei*. It was reported that 20–50mm, 16–43mm and 0–10mm sized inhibition zones were observed for lemongrass, lemon, and grapefruit essential oils, respectively (Warnke et al., 2009). Caccioni et al. (1998) expressed that lemon, grapefruit, and mandarin oils have high antifungal activities against plant pathogenic fungi due to their rich terpene contents. Viuda-Martos, Ruiz-Navajas, Fernández-López, and Perez-Álvarez (2008) examined the relationship between essential oils (0.27, 0.47, 0.71, and 0.94 concentrations) of popular citrus varieties including lemon (*Citrus lemon* L.), mandarin (*Citrus reticulata* L.), grapefruit (*Citrus paradisi* L.), and orange (*Citrus sinensis* L.) and their inhibition efficiency against *Aspergillus* and *Penicillium* spp. using the agar dilution method. The essential oil of grapefruit (0.85 g/mL density and 1.47 refractive indices) obtained by cold pressing the peel did not show an inhibition effect against *Aspergillus niger* and *A. flavus* growth, although it was the best growth reducer to *Penicillium chrysogenum* and *Penicillium verrucosum*. Viuda-Martos, Ruiz-Navajas, Fernandez-Lopez, and Perez-Álvarez (2008) similarly tested the effectiveness of cold pressed essential oils of lemon, mandarin, grapefruit, and orange to inhibit the growth of *Lactobacillus*, *Staphylococcus*, and *Enterobacter* spp. using the disc diffusion and Minimum Inhibition Concentration methods. All citrus cold pressed oils tested exhibited an inhibitory effect against all bacterial strains used in this study. Cold pressed grapefruit oils showed the most antibacterial activity against three bacteria strains by the disc diffusion method. Choi, Sawamura, and Song (2010) determined antioxidative activities of some citrus essential oil samples prepared by cold pressing against a linoleic acid peroxidation model in vitro. When the relative lipid peroxidation rates of citrus oils such as lemon, mandarin, bergamot, lime, orange, yuzu, kumquat, and grapefruit were compared, it was observed that the yuzu and Eureka lemon varieties had the highest antioxidant activity (more than 90%), followed by Italian bergamot, Valencia orange, and Satsuma mandarin (70%–88% activity). Unexpectedly, bergamot (Japanese) and grapefruit cold pressed oils had relatively low antioxidative activities (49%–60%).

8 Conclusion

The grapefruit is characterized by distinct bitterness, sourness, and astringency aroma as well as the high levels of flavonoids, limonoids, carotenoids, minerals, and aromatic content. Global grapefruit production was reported as 9 million metric tonnes for 2017/18, accounting for 6.2% of total citrus production (approximately 146 million metric tonnes). The cold pressing technique, accepted as a green extraction technique, is widely used to extract oil from citrus by-products obtained from citrus processing. Cold pressed grapefruit oil has a remarkable bitter and astringent taste/flavor similar to that of grapefruit, and is rich in nutritional and bioactive compounds. Various techniques such as refining process, debittering enzyme treatment, and adsorbent applications have been advised to remove the undesirable taste. Moreover, clinical studies on protective, functional effects and therapeutic potential of cold pressed grapefruit oils have been carried out. This chapter presented the current scientific knowledge on cold pressing extraction of grapefruit oils, and focused on composition, bioactives, volatile compounds, sensory attributes, and beneficial health effects of cold pressed grapefruit oils.

References

Adeyeye, E. I., & Adesina, A. J. (2015). Citrus seeds oils as sources of quality edible oils. *International Journal of Current Microbiology and Applied Sciences*, *4*(5), 537–554.

Ahmad, M. M., et al. (2006). Genetic variability to essential oil composition in four citrus fruit species. *Pakistan Journal of Botany*, *38*(2), 319–324.

Ajewole, K., & Adeyeye, A. (1993). Characterisation of Nigerian citrus seed oils. *Food Chemistry*, *47*(1), 77–78. https://doi.org/10.1016/0308-8146(93)90306-Z.

Anwar, F., et al. (2008). Physico-chemical characteristics of citrus seeds and seed oils from Pakistan. *Journal of the American Oil Chemists' Society*, *85*(4), 321–330. https://doi.org/10.1007/s11746-008-1204-3.

Aydeniz, B., Güneşer, O., & Yilmaz, E. (2014). Physico-chemical, sensory and aromatic properties of cold press produced safflower oil. *Journal of the American Oil Chemists' Society*. *91*(1). https://doi.org/10.1007/s11746-013-2355-4.

Aydeniz-Güneşer, B. (2016). *Characterization of citrus seed oils produced by cold pressing technique*. Ph.D. thesis Çanakkale, Turkey: Çanakkale Onsekiz Mart University, Department of Food Engineering.

Aydeniz-Güneşer, B., & Yilmaz, E. (2018). Bitterness reduction of cold pressed grapefruit seed oil by adsorbent treatment. *European Journal of Lipid Science and Technology*, *120*(5), 1700308. https://doi.org/10.1002/ejlt.201700308.

Bailey, D. G., Dresser, G., & Arnold, J. M. O. (2013). Grapefruit-medication interactions: Forbidden fruit or avoidable consequences? *Canadian Medical Association Journal*, *185*(4), 309–316. https://doi.org/10.1503/cmaj.120951.

Baratta, M. T., Dorman, H. D., Deans, S. G., Figueiredo, A. C., Barroso, J. G., & Ruberto, G. (1998). Antimicrobial and antioxidant properties of some commercial essential oils. *Flavour and Fragrance Journal*, *13*(4), 235–244.

Barroso, M. A. T., et al. (1972). Grapefruit seed oil sterols. *Journal of the American Oil Chemists' Society*, *49*(1), 85–86. https://doi.org/10.1007/BF02545150.

Brud, W. S. (2015). Industrial uses of essential oils. In K. H. C. Baser, & G. Buchbauer (Eds.), *Handbook of essential oils* (pp. 1015–1025): CRC Press.

Caccioni, D. R. L., et al. (1998). Relationship between volatile components of citrus fruit essential oils and antimicrobial action on Penicillium digitatum and Penicillium italicum. *International Journal of Food Microbiology*, *43*(1), 73–79. https://doi.org/10.1016/S0168-1605(98)00099-3.

Chebrolu, K. K., et al. (2012). Production system and storage temperature influence grapefruit vitamin C, limonoids, and carotenoids. *Journal of Agricultural and Food Chemistry*, *60*(29), 7096–7103. https://doi.org/10.1021/jf301681p.

Choi, H.-S., Sawamura, M., & Song, H.-S. (2010). Functional properties. In *Citrus Essential Oils*: Wiley Online Books. https://doi.org/10.1002/9780470613160.ch6.

Curtis, S., Thomas, P., & Johnson, F. (2016). *Neal's yard remedies essential oils: Restore* rebalance* revitalize* feel the benefits* enhance natural beauty* create blends*. Dorling Kindersley Ltd. https://ndb.nal.usda.gov/ndb/foods/show/09112?Fgcd=&manu=&format=&count=&max=25&offset=&sort=default&order=asc&qlookup=grapefruit&ds=&qt=&qp=&qa=&qn=&q=&ing (Available at 25 March 2019).

Dow, C. A., et al. (2012). The effects of daily consumption of grapefruit on body weight, lipids, and blood pressure in healthy, overweight adults. *Metabolism*, *61*(7), 1026–1035. https://doi.org/10.1016/j.metabol.2011.12.004.

Dresser, G. K., & Bailey, D. G. (2002). A basic conceptual and practical overview of interactions with highly prescribed drugs. *Canadian Journal of Clinical Pharmacology*, *9*(4), 191–198. Available at: http://europepmc.org/abstract/MED/12584577.

Farouk, H., Mahmoud, S. S., El-Sayeh, B. A., & Sharaf, O. A. (2015). Effect of grapefruit juice and sibutramine on body weight loss in obese rats. *African Journal of Pharmacy and Pharmacology*, *9*(8), 265–273.

Fashner, J., Ericson, K., et al. (2012). Treatment of the common cold in children and adults. *American Family Physician*, *86*(6), 153–159. https://doi.org/10.1542/peds.2013-3260.

Fiedor, J., & Burda, K. (2014). Potential role of carotenoids as antioxidants in human health and disease. *Nutrients*. https://doi.org/10.3390/nu6020466.

Gajowik, A., & Dobrzyńska, M. M. (2014). Lycopene-antioxidant with radioprotective and anticancer properties. A review. *Roczniki Państwowego Zakładu Higieny*, *65*(4), 263–271.

Giménez-Bastida, J., González-Sarrías, A., Vallejo, F., Espín, J., & Tomás-Barberán, F. (2016). Hesperetin and its sulfate and glucuronide metabolites inhibit TNF-α induced human aortic endothelial cell migration and decrease plasminogen activator inhibitor-1 (PAI-1) levels. *Food & Function*, *7*, 118–126.

Habib, M. A., et al. (1986). Chemical evaluation of Egyptian citrus seeds as potential sources of vegetable oils. *Journal of the American Oil Chemists' Society*, *63*(9), 1192–1196. https://doi.org/10.1007/BF02663951.

Hata, T., et al. (2003). Induction of apoptosis by Citrus paradisi essential oil in human leukemic (HL-60) cells. *In vivo (Athens, Greece)*, *17*(6), 553–559. Available at: http://europepmc.org/abstract/MED/14758720.

Haze, S., Sakai, K., & Gozu, Y. (2002). Effects of fragrance inhalation on sympathetic activity in normal adults. *The Japanese Journal of Pharmacology*, *90*(3), 247–253. https://doi.org/10.1254/jjp.90.247.

Ho, P. C., Saville, D. J., & Wanwimolruk, S. (2001). Inhibition of human CYP3A4 activity by grapefruit flavonoids, furanocoumarins, and related compounds. *Journal of Pharmacy & Pharmaceutical Sciences*, *4*, 217–227.

IFRA. (2009). *Standards, including amendments as of October 14th, 2009*. Brussels: International Fragrance Association. http://www.İfraorg.org.

Indriyani, L., Rohman, A., & Riyanto, S. (2016). Authentication of avocado oil (*Persea americana* Mill.) using differential scanning calorimetry and multivariate regression. *Asian Journal of Agricultural Research*, *10*, 78–86.

Iwabuchi, H., et al. (2010). Industrial view. In *Citrus essential oils* (pp. 343–380): John Wiley & Sons, Ltd. https://doi.org/10.1002/9780470613160.ch8.

Jian, L., Lee, A. H., & Binns, C. W. (2007). Tea and lycopene protect against prostate cancer. *Asia Pacific Journal of Clinical Nutrition*, *16*(Suppl 1), 453–457. Available at: http://www.ncbi.nlm.nih.gov/pubmed/17392149.

Kelebek, H., & Selli, S. (2011). Determination of volatile, phenolic, organic acid and sugar components in a Turkish cv. Dortyol (*Citrus sinensis* L. Osbeck) orange juice. *Journal of the Science of Food and Agriculture*, *91*(10), 1855–1862. https://doi.org/10.1002/jsfa.4396.

Kesterson, J. W., & Braddock, R. J. (1975). Total pell oil content of the major Florida *Citrus cultivars*. *Journal of Food Science*, *40*(5), 931–933. https://doi.org/10.1111/j.1365-2621.1975.tb02236.x.

Laville, M., & Nazare, J.-A. (2009). Diabetes, insulin resistance and sugars. *Obesity Reviews*, *10*(s1), 24–33. https://doi.org/10.1111/j.1467-789X.2008.00562.x.

Lin, J., & Rouseff, R. L. (2001). Characterization of aroma-impact compounds in cold-pressed grapefruit oil using time-intensity GC-olfactometry and GC-MS. *Flavour and Fragrance Journal*, *16*(6), 457–463. https://doi.org/10.1002/ffj.1041.

Matthaus, B., & Özcan, M. M. (2012). Chemical evaluation of citrus seeds, an agro-industrial waste, as a new potential source of vegetable oils. *Grasas y Aceites*, *63*(3), 313–320. https://doi.org/10.3989/gya.118411.

Mir, I. A., & Tiku, A. B. (2015). Chemopreventive and therapeutic potential of "naringenin," a flavanone present in citrus fruits. *Nutrition and Cancer*, *67*(1), 27–42.

Mora, J. R., Iwata, M., & von Andrian, U. H. (2008). Vitamin effects on the immune system: Vitamins a and D take Centre stage. *Nature Reviews Immunology*, *8*, 685. https://doi.org/10.1038/nri2378.

Nagai, K., et al. (2014). Olfactory stimulatory with grapefruit and lavender oils change autonomic nerve activity and physiological function. *Autonomic Neuroscience*, *185*, 29–35. https://doi.org/10.1016/j.autneu.2014.06.005.

Nemmar, A., Al-Salam, S., Beegam, S., Yuvaraju, P., & Ali, B. H. (2018). Thrombosis and systemic and cardiac oxidative stress and DNA damage induced by pulmonary exposure to diesel exhaust particles and the effect of nootkatone thereon. *American Journal of Physiology—Heart and Circulatory Physiology*, *314*(5), H917–H927. https://doi.org/10.1152/ajpheart.00313.2017.

Nemmar, A., Al-Salam, S., Beegam, S., Yuvaraju, P., Hamadi, N., et al. (2018). *In vivo* protective effects of nootkatone against particles-induced lung injury caused by diesel exhaust is mediated via the NF-κB pathway. *Nutrients*. https://doi.org/10.3390/nu10030263.

Njoroge, S. M., et al. (2005). Volatile constituents of red blush grapefruit (*Citrus paradisi*) and pummelo (*Citrus grandis*) peel essential oils from Kenya. *Journal of Agricultural and Food Chemistry*, *53*(25), 9790–9794. https://doi.org/10.1021/jf051373s.

Novikova, D. S., Ivanov, G. S., Garabadzhiu, A. V., & Tribulovich, V. G. (2017). Compounds of plant origin as AMP-activated protein kinase activators. *Aging*, *4*, 5.

Ortuno, A., et al. (1995). Flavanone and Nootkatone levels in different varieties of grapefruit and pummelo. *Journal of Agricultural and Food Chemistry*, *43*(1), 1–5. https://doi.org/10.1021/jf00049a001.

Phillips, R., et al. (2012). Salts for preventing and treating kidney stones (protocol). *The Cochrane Database of Systematic Reviews*. *9*, CD010057. https://doi.org/10.1002/14651858.CD010057.pub2.www.cochranelibrary.com.

Rink, L. (2000). Zinc and the immune system. *Proceedings of the Nutrition Society*, *59*(4), 541–552. https://doi.org/10.1017/S0029665100000781.

Rodríguez-Fragoso, L., et al. (2011). Potential risks resulting from fruit/vegetable-drug interactions: Effects on drug-metabolizing enzymes and drug transporters. *Journal of Food Science*, *76*(4), R112–R124. https://doi.org/10.1111/j.1750-3841.2011.02155.x.

Saidani, M., Dhifi, W., & Marzouk, B. (2004). Lipid evaluation of some Tunisian citrus seeds. *Journal of Food Lipids*, *11*(3), 242–250. https://doi.org/10.1111/j.1745-4522.2004.01136.x.

Sawamura, M. (2010). Introduction and overview. In *Citrus Essential Oils* (pp. 1–8). John Wiley & Sons, Ltd. https://doi.org/10.1002/9780470613160.ch1.

Seden, K., et al. (2010). Grapefruit-drug interactions. *Drugs*, *70*(18), 2373–2407. https://doi.org/10.2165/11585250-000000000-00000.

Shahidi, F., & Zhong, Y. (2012). Citrus oils and essences. In *Kirk-Othmer encyclopedia of chemical technology* (pp. 1–17). American Cancer Society. https://doi.org/10.1002/0471238961.citrshah.a01.

Shen, J., et al. (2005). Olfactory stimulation with scent of grapefruit oil affects autonomic nerves, lipolysis, and appetite in rats. *Neuroscience Letters*, *380*(3), 289–294. https://doi.org/10.1016/j.neulet.2005.01.058.

Sorice, A., et al. (2014). Ascorbic acid: Its role in immune system and chronic inflammation diseases. *Mini-Reviews in Medicinal Chemistry*, 444–452. https://doi.org/10.2174/1389557514666140428112602.

Sun, D., & Petracek, P. D. (1999). Grapefruit gland oil composition is affected by wax application, storage temperature, and storage time. *Journal of Agricultural and Food Chemistry*, *47*(5), 2067–2069. https://doi.org/10.1021/jf981064k.

Teles, F. F. F., et al. (1972). Triglyceride fatty acids of Arizona grapefruit seed oil. *Journal of Food Science*, *37*(2), 331–332. https://doi.org/10.1111/j.1365-2621.1972.tb05848.x.

Tisserand, R., & Young, R. (2014). 2-Essential oil composition. In R. Tisserand, & R. Young (Eds.), *Essential oil safety* (2nd ed., pp. 5–22). St. Louis: Churchill Livingstone. https://doi.org/10.1016/B978-0-443-06241-4.00002-3.

Tokoro, K. (1997). Antioxidants in grapefruit oil. *Aromatopia*, *24*, 42–44. Available at: http://ci.nii.ac.jp/naid/10012772693/en/ (Accessed 31 May 2019).

Uddin, N., et al. (2014). Assessment of toxic effects of the methanol extract of *Citrus macroptera* Montr. Fruit via biochemical and hematological evaluation in female Sprague-Dawley rats. *PLoS ONE*, *9*(11), e111101. Available at: https://doi.org/10.1371/journal.pone.0111101.

USDA, 2019. Https://apps.fas.usda.gov/psdonline/circulars/citrus.pdf (Available at 07 March 2019)

Vanamala, J., et al. (2005). Bioactive compounds of grapefruit (*Citrus paradisi* Cv. Rio Red) respond differently to postharvest irradiation, storage, and freeze drying. *Journal of Agricultural and Food Chemistry*, *53*(10), 3980–3985. https://doi.org/10.1021/jf048167p.

Viuda-Martos, M., Ruiz-Navajas, Y., Fernández-López, J., & Pérez-Álvarez, J. A. (2009). Chemical composition of mandarin (*C. reticulata* L.), grapefruit (*C. paradisi* L.), lemon (*C. limon* L.) and orange (*C. sinensis* L.) essential oils. *Journal of Essential Oil-Bearing Plants*, *12*(2), 236–243. https://doi.org/10.1080/0972060X.2009.10643716.

Viuda-Martos, M., Ruiz-Navajas, Y., Fernández-López, J., & Perez-Álvarez, J. (2008). Antifungal activity of lemon (*Citrus lemon* L.), mandarin (*Citrus reticulata* L.), grapefruit (*Citrus paradisi* L.) and orange (*Citrus sinensis* L.) essential oils. *Food Control*, *19*(12), 1130–1138. https://doi.org/10.1016/j.foodcont.2007.12.003.

Viuda-Martos, M., Ruiz-Navajas, Y., Fernandez-Lopez, J., & Perez-Álvarez, J. (2008). Antibacterial activity oF lemon (*Citrus lemon* L.), mandarin (*Citrus reticulata* L.), grapefruit (*Citrus paradisi* L.) and orange (*Citrus Sinesis* L.) essential oils. *Journal of Food Safety*, *28*(4), 567–576. https://doi.org/10.1111/j.1745-4565.2008.00131.x.

Waheed, A., Mahmud, S., Saleem, M., & Ahmad, T. (2009). Fatty acid composition of neutral lipid: Classes of citrus seed oil. *Journal of Saudi Chemical Society*, *13*(3), 269–272.

Warnke, P. H., et al. (2009). The battle against multi-resistant strains: Renaissance of antimicrobial essential oils as a promising force to fight hospital-acquired infections. *Journal of Cranio-Maxillofacial Surgery*, *37*(7), 392–397. https://doi.org/10.1016/j.jcms.2009.03.017.

Wattenberg, L. W., et al. (1985). Inhibition of carcinogenesis by some minor dietary constituents. *Princess Takamatsu Symposia*, *16*, 193–203. Available at: https://www.scopus.com/inward/record.uri?eid=2-s2.0-0022322348&partnerID=40&md5=ec31197c33f2a4ac0ad094bed4e8d5bd.

Whitbread, D. (2019). *Top 10 foods highest in lycopene*. https://www.myfooddata.com/articles/high-lycopene-foods.php (Available 03 March 2019).

Wilson, C. W., & Shaw, P. E. (1980). Glass capillary gas chromatography for quantitative determination of volatile constituents in cold-pressed grapefruit oil. *Journal of Agricultural and Food Chemistry*, *28*(5), 919–922. https://doi.org/10.1021/jf60231a024.

Yilmaz, E., Aydeniz-Guneser, B., & Ok, S. (2018). Valorization of grapefruit seeds: Cold press oil production. *Waste and Biomass Valorization*. https://doi.org/10.1007/s12649-018-0286-x.

Zielisnki, E. (2017). The truth about essential oils and cancer. In *A biblically-inspired, evidence-based approach to beating cancer God's way!* Biblical Health Publishing.

Zu, Y., et al. (2010). Activities of ten essential oils towards *Propionibacterium acnes* and PC-3, A-549 and MCF-7 cancer cells. *Molecules*. https://doi.org/10.3390/molecules15053200.

Chapter 46

Cold pressed pecan (*Carya illinoinensis*) oil

Sündüz Sezer Kiralan[a], Mustafa Kiralan[a], and Gulcan Ozkan[b]

[a]*Department of Food Engineering, Faculty of Engineering, Balıkesir University, Balikesir, Turkey,* [b]*Department of Food Engineering, Faculty of Engineering, Suleyman Demirel University, Isparta, Turkey*

Abbreviations

ABTS 2,2′-azino-bis (3-ethylbenzothiazoline-6-sulphonic acid)
DPPH˙ 2,2- diphenyl-1-picrylhydrazyl
GAE gallic acid equivalent
MUFA monounsaturated fatty acid
PUFA polyunsaturated fatty acid
SFE supercritical fluid extraction

1 Introduction

The pecan nut [*Carya illinoinensis* (Wangenh.) K. Koch] belongs to the Juglandaceae family and is a native of the southern United States and northern Mexico. The pecan is one of the most important nut crops worldwide due to its distinctive flavor and desirable nutritional content (Hal, 2000; Ojeda-Barrios, Hernández-Rodríguez, López-Ochoa, & Martínez-Téllez, 2009).

An increasing trend has been observed in global pecan production over the last 10 years, with 92% of pecans being produced in North America. The USA and Mexico accounted for 51% and 41% of the world production, respectively (International Nut and Dried Fruit Council Foundation, 2018). Pecans provide an excellent source of energy due to their high levels of lipids. There are are 691 kcal (kcal) in 100 g of pecan nuts. They are also rich in dietary protein, vitamins, and other bioactive compounds that contribute to health benefits (Amarowicz, Gong, & Pegg, 2017). According to the United States Department of Agriculture National Nutrient Database, pecans, per 100 g, have 2.52 g water, 9.17 g protein, 71.97 g fat, and 13.86 g carbohydrate. Pecans also contain various vitamins and minerals (USDA, 2018).

In addition to basic nutritional components, pecans contain some minor compounds such as phenolics, tocopherols, and sterols, which contribute to human health benefits. Various researches are carried out, especially on phenolics. One of these studies in USA on more than 100 different kinds of foods including fruits, vegetables, nuts, dried fruits, spices, cereals, infant food, and other foods demonstrated that pecan kernels showed the highest total phenolics and antioxidant properties among the nut group, higher than many other fruits and vegetables rich in phenolics (Wu et al., 2004).

Pecans are typically consumed raw, roasted, or salted, as well as being used as an ingredient in a variety of processed foods such as spreads and bakery and confectionery products (Shakerardekani, Karim, Ghazali, & Chin, 2013). The increase in the demand of consumers for novel and healthy oil has focused on oils produced from nuts using the cold pressing method. Pecan oil is generally produced via cold pressing and is used primarily in cooking. These oils are obtained in much smaller amounts than oil extracted using a solvent (Scapinello et al., 2017). In addition to this extraction method, an alternative methodology is the solvent extraction method, in which extracted crude oils must be refined before consumption. The solvent extraction method has some disadvantages, although oil yields are higher than for cold pressed oils. Process temperature in the refining process alters oil composition, and some beneficial compounds for human health decreased during the refining process (Gunstone, 2011).

Triglycerides constitute the major components of pecan, in which monounsaturated fatty acids (MUFAs) and polyunsaturated fatty acids (PUFAs) are present in high levels. Saturated, monounsaturated, and polyunsaturated fatty acids contribute 8.35%, 66.7%, and 24.9% of total fatty acids in pecan oil, respectively. Pecan oils are rich in oleic (C18:1, 66.6%) and linoleic (C18:2, 23.6%) acids (Venkatachalam & Sathe, 2006). Pecan oil also contains other micronutrients, such as tocopherols, sterols, and color pigments, which impart health benefits.

Cold Pressed Oils. https://doi.org/10.1016/B978-0-12-818188-1.00046-3
Copyright © 2020 Elsevier Inc. All rights reserved.

2 Cold press extraction and processing of oils

There are several methods to extract oils from pecans. Mechanical extraction, traditional extraction (solvent extraction), and modern extraction techniques such as supercritical carbon dioxide can be used to obtain the oil. In traditional extraction methods, *n*-hexane is generally used to extract oil. Although the oil yield is high, there is a health risk in this method because of hexane's potential carcinogenic and neurotoxic properties. In addition, the supercritical carbon dioxide technique has been considered a green technology in extraction methods because it involves no pollution. However, a disadvantage of this method is its high cost. Considering public health and cost, cold pressing can be considered the best extraction method. In cold pressing, pecans are processed with a screw press or expeller press (Cockerham, Gorman, Maness, & Lillywhite, 2012).

One of the alternatives and modern methods in extraction of oils from oilseeds is supercritical fluid extraction (SFE) with carbon dioxide (CO_2) as an extraction solvent. Supercritical CO_2 is a safe and efficient solvent for extracting oilseeds, which especially contain a low amount of water. Pecan oil is obtained using solvent extraction and SFE. The extraction time of oils from pecans with SFE (1 h) is shorter than classical organic solvent extraction (4 h). In the SFE method, 98%–99% of the total oil was recovered in 30 min of extraction (Maness, Chrz, Pierce, & Brusewitz, 1995).

In a study about the comparison of extraction methods, 16 phenolic compounds were determined in oil samples using Soxhlet and cold pressing extraction. Catechin and naringenin were major phenolic compounds identified in oils. The catechin and naringenin contents of cold pressed pecan oils were 2.38 and 2.17 μg/100 g, respectively. In the Soxhlet-extracted pecan oils, these compounds were 2.51 and 2.35 μg/100 g, respectively. The amounts of the other phenolics were lower than 1.00 μg/100 g. The total phenolic and antioxidant activity of pecan oils extracted by cold pressing were higher than in oil obtained by Soxhlet extraction. Total phenolics in the oils extracted by Soxhlet and cold pressing methods were determined as 0.78 and 0.45 mg gallic acid equivalent (GAE)/100 g, respectively. Like total phenolics, the antioxidant activity of cold pressed pecan oil was 17.5%, whereas that of oils obtained by solvent extraction was 13.4% (Al Juhaimi, Özcan, Ghafoor, Babiker, & Hussain, 2018).

3 Cold pressed oil recovery, content, uses of oil cake, and economy

The proximate composition of pecan nut expeller cake, a by-product of cold pressing extraction, is given in Table 1. Lipids are a major component in cake. Dietary fiber, proteins, and carbohydrates are valuable components in the composition. Considering the oil content and fatty acid composition of the cake, it could be used to enrich the fatty acid composition of some foods. It could also be used to improve nutritional value of some foods despite its dietary fiber. In addition, this cake could be used as an ingredient for the bakery industry due to its high water and oil absorption capacity (Marchetti et al., 2017; Salvador et al., 2016).

Pecan cake is rich in oil and this residue oil could be extracted using different methods. Salvador et al. (2016) studied different extraction techniques (Soxhlet, ultrasound, and SFE) in terms of yield, fatty acids, and antioxidant properties. Higher extraction yields were obtained by Soxhlet extraction (69.6% with ethanol and 64% with acetone). The higher extraction yield in SFE was obtained at 300 bar and 313.15 K with 58.4%. The major fatty acid was oleic acid, which varied

TABLE 1 Proximate composition of pecan nut cake (%).

Component[a]	
Lipids	36.0–59.4
Dietary fiber	9.4–13.7
Proteins	10.5–19.0
Carbohydrates	11.96–22.00
Water	5.2–5.9
Ash	1.96–3.67

[a]*References: Salvador, Podestá, Block, and Ferreira (2016) and Marchetti, Romero, Andres, and Califano (2017).*

from 60.1% to 73.9%. Phenolic compounds were not detected in extracts obtained by the SFE method, but were determined in samples produced by Soxhlet and ultrasound methods.

The total phenolic content was 100mg GAE/g in the extract obtained by ultrasound with acetone, whereas the higher value among cake samples produced by the Soxhlet method was 71 mg GAE/g (acetone extraction). According to antioxidant experiments such as DPPH·, ABTS, and the β-carotene bleaching test, cake extracts showed antioxidant activity like gallic acid (Salvador et al., 2016).

Pecan cake is rich in total amounts of flavonoids and condensed tannins compared to other nut cakes such as almond, hazelnut, and macadamia. Catechin, epicatechin, and ellagic acid were identified in pecan nut cake. •It was showed that pecan cake had the highest antioxidant activity as well as the highest reducing power among nut cakes (almond, hazelnut, and macadamia) by using antioxidant tests (ABTS and DPPH) (Sarkis et al., 2014).

4 Fatty acid composition and acyl lipids profile of cold pressed oil

The major fatty acids in cold pressed pecan oils are given in Table 2. The content of these fatty acids changes with genotype, maturity, and harvest year. A general decline in linoleic acid was observed with maturation; however, unlike linoleic acid, an increase in oleic acid was observed with maturation (Rudolph, Odell, Hinrichs, Hopfer, & Kays, 1992).

Pecan oil is rich in unsaturated fatty acids especially MUFAs, followed by PUFAs. Total saturated, monounsaturated, and polyunsaturated fatty acids ranges were 7.2%–7.7%, 69.2%–75.9%, and 16.7%–23.1%, respectively (do Prado et al., 2013). The dominant fatty acid occurring in different pecan oils is oleic acid (C18:1), which constitutes about 70%–78% of total fatty acids. Among PUFAs, linoleic acid (C18:2) had high amounts (13.6%–23.1%). Palmitic and stearic acids are major saturated fatty acids with 4.7%–5.59% and 1.49%–2.5%, respectively (Al Juhaimi et al., 2018; do Prado et al., 2013; Scapinello et al., 2017).

Fernandes, Gómez-Coca, Pérez-Camino, Moreda, and Barrera-Arellano (2017) studied the triglyceride profile of oil from Brazil pecans using a laboratory expeller press. While a mixture of PLP, OOO, and PoPP (38.7%) was the major triglyceride in the pecan oil, OOL and LnPP mixture was detected as 23.24% of total triglycerides of the oil. An OOL and LnPP mixture comprised 23.24% of total triglycerides (Table 3).

5 Minor bioactive lipids in cold pressed oil

The tocopherol composition of pecan oils is exhibited in Table 3. The total tocopherols of oils in pecans harvested in the 2009 and 2010 growing seasons from Brazil changed from 24.9 to 39.2mg/100g (do Prado et al., 2013). The major tocopherol in pecan oils is γ-tocopherol, followed by α-tocopherol and a small amount of β-tocopherol and δ-tocopherol. The level of γ-tocopherol varied between 236.4 and 284mg/kg in pecan oil, comprising more than 90% of the total tocopherol content (Table 4).

Major desmethyl-, methyl-, and dimethylsterols, terpenic alcohols, aliphatic alcohols, and squalene of pecan oil are presented in Table 5. The total sterol content of pecan oils from Brazil varied between 0.19% and 0.22% (do Prado et al. (2013). β-Sitosterol, Δ^5-avanasterol, and campesterol are considered as main desmethylsterols, constituting 75.51%, 14.94%, and 4.45% of total desmethylsterols (1791.4mg/kg), respectively. Other minor components were 24-methylene cholesterol, campestanol, stigmasterol, $\Delta^{5,23}$-stigmastadienol, clerosterol, sitostanol, $\Delta^{5,24}$-stigmastadienol, Δ^7-stigamasterol, and Δ7-avenasterol. Citrostadienol, cycloartenol, phytol, and C28-OH are the most abundant constituents

TABLE 2 Major fatty acids in cold pressed pecan oils (%).

Fatty acid	1[a]	2	3
Oleic (C18:1)	69.4	78.0	69.2–75.9
Linoleic (C18:2)	19.3	13.6	16.7–23.1
Palmitic (C16:0)	5.36	5.59	4.7–5.4
Stearic (C18:0)	1.49	2.12	2.3–2.5

[a]*References: 1: Al Juhaimi et al. (2018); 2: Scapinello et al. (2017); 3: do Prado et al. (2013).*

TABLE 3 Triglyceride composition of pecan oil.

Triglyceride[a]	%
LLL[b]	2.63
OLL	9.71
PLL + PoPoO	1.74
OOL + LnPP	23.2
PoOO + SLL + PLO	6.03
PLP + OOO + PoPP	38.7
SOL	2.60
POO	8.12
SOO	4.07

[a]*Reference: Fernandes et al. (2017).*
[b]*P, palmitic acid; Po, palmitoleic acid; S, stearic acid; O, oleic acid; L, linoleic acid; Ln, linolenic acid.*
The content of triglycerides below 1% are not provided.

TABLE 4 Tocopherol profile of cold pressed pecan oils (mg/kg).

Tocopherol	1[a]	2	3
α-Tocopherol	11.85	11–13	1.4
β-Tocopherol	0.27	–	1.9
γ-Tocopherol	278.6	238–381	284
δ-Tocopherol	3.78	–	2.2

[a]*References: 1: Al Juhaimi et al. (2018); 2: do Prado et al. (2013); 3: Castelo-Branco, Santana, Di-Sarli, Freitas, and Torres (2016).*

in methylsterols, dimethylsterols, terpenic alcohols, and aliphatic alcohols, respectively. The squalene amount of oil was 298.8 mg/kg (Fernandes et al., 2017).

Brazil pecan oil has a high phenolics content compared to other nuts and the total phenolic content of the oil varies with the range of 4.00 mg GAE/100 g (Castelo-Branco et al., 2016). Al Juhaimi et al. (2018) identified 16 phenolic compounds in cold pressed pecan oil. Table 6 gives the phenolics with an amount greater than 0.20 μg/100 g (caffeic acid, *p*-coumaric acid, vanillic acid, rutin, resveratrol, kaempferol, quercetin, and pinocembrin). The major phenolics in pecan oil are catechin and naringenin, with amounts of 2.38 and 2.17 μg/100 g, respectively. The amount of other phenolics identified varies between 0.47 and 0.81 μg/100 g. The contents of other minor components in cold pressed pecan oils, β-carotene, flavonoid, carotenoid, and anthocyanin, were 11.89 μg/100 g, 5.27 mg/100 g, 0.53 mg/100 g, and 0.21 mg/100 g, respectively (Al Juhaimi et al., 2018).

6 Composition of cold pressed oil in comparison to other cold pressed oils

Pecan oil can have more than 90% of unsaturated fatty acids (Table 7). Most MUFAs are oleic acid (59.7%). Oleic acid was the predominant fatty acid in all the cold pressed oil samples, comprising more than 70% of the total fatty acids in hazelnut and pecan nut oils. Linoleic acid is the second most abundant fatty acid in pecan oil and also seen in almond, Brazil nut, hazelnut, and pistachio oils. The major saturated fatty acid is palmitic acid (6%). Almond and hazelnut oil contained similar contents of this fatty acid.

γ-Tocopherol is a major constituent among tocopherols in pecan oil, and its amount is 28.4 mg/100 g (Table 7). The amounts of the other isomers identified in pecan oil are very close to each other. Brazil nut oil showed a similarity in terms

TABLE 5 Major desmethyl-, methyl-, and dimethylsterols, terpenic alcohols, aliphatic alcohols, and squalene of pecan oil.[a]

	%
Desmethylsterol	
Campesterol	4.45
β-Sitosterol	75.51
Δ^5-Avenasterol	14.94
Methylsterol	
Obtusifoliol	6.82
Gramisterol	13.88
Citrostadienol	79.31
Dimethylsterol	
Cycloartenol	78.69
24-Methylencycloartanol	9.08
Butyrospermol	6.15
Terpenic alcohol	
Phytol	70.68
Geranylgeraniol	29.32
Aliphatic alcohol	
C28-OH	55.22
C24-OH	21.30
Squalene (mg/kg)	**298.83**

[a]*Reference: Fernandes et al. (2017).*

TABLE 6 Phenolic composition of cold pressed pecan oil.[a]

Phenolic	μg/100g
Gallic	0.47
Protocatechuic	0.53
Catechin	2.38
Ferulic	0.43
Sinapic	0.38
Naringenin	2.17
Chlorogenic	0.53
Luteolin	0.81

[a]*Reference: Al Juhaimi et al. (2018).*

TABLE 7 Fatty acid and tocopherol composition of some cold pressed oils.

	Cold pressed oil[a]						
	PEO[b]	AO	BNO	HO	MO	PO	PUO
Fatty acid (g/100 g)							
C16:0	6.01	6.58	15.2	6.61	8.48	10.95–11.11	11.2–15.5
C18:0	2.40	1.72	12.3	1.84	3.76	1.05–1.11	5.2–6.2
C18:1	72.2	64.6	39.5	79.4	59.7	55.37–56.05	37.1–43.6
C18:2	18.0	26.3	31.6	11.3	1.59	30.46–31.32	37.3–44.5
Tocopherol (mg/100 g)							
α	0.14	64.2	1.89	40.7	0.80	2.44–3.47	2.55–5.20
β	0.19	0.19	0.08	0.68	ND[c]	28.00–30.86	29.92–53.60
γ	28.4	1.70	21.0	0.56	ND		
δ	0.22	0.05	0.20	0.05	ND	2.03–2.55	3.39–10.44

[a]*PEO, pecan oil; AO, almond oil; BNO, Brazil nut oil; HO, hazelnut oil; MO, macadamia oil; PO, pistachio oil; PUO, pumpkin oil.*
[b]*References: PEO, AO, BNO, HO, and MO data from Castelo-Branco et al. (2016); PO data from Ling, Yang, Li, and Wang (2016); PUO data from Rabrenović, Dimić, Novaković, Teševič, and Basić (2014).*
[c]*ND, not detected.*

of γ-tocopherol as a predominant tocopherol in this oil, with 21.0 mg/100 g. In pistachio and pumpkin oils, γ- and β-tocopherols are given in total. The totals of γ- and β-tocopherols of pecan oil were found to be similar for these oils. However, hazelnut and almond oils are rich in α-tocopherols. In addition, the only α-tocopherol isomer is identified in macadamia oil.

The contents of phenolics in cold pressed oils are shown in Table 8. Catechin and naringenin are rich in phenolics, different from some cold pressed oils. Similarly, walnut and pistachio oils contain high levels of catechin and naringenin. Vanillic and caffeic acids are major phenolics in sunflower oil with amounts of 6.9 and 4.9 μg/100 g, respectively. In rapeseed, phenolic acids consist mostly of sinapic acid with 236.0 μg/100 g. The highest luteolin content of oils was found in pistachio oil as 1.17 μg/100 g. The highest vanillic acid was determined in pumpkin oil as 11.4 μg/100 g.

Campesterol, β-sitosterol, and Δ^5-avenasterol are considered the main sterolic components in pecan oil (Table 9). β-Sitosterol is the major phytosterol with content in the range of 87.7% for *Pistacia atlantica* oil, 49.4% for *Nigella sativa* oil, 82.9% for Japanese quince seed oil, 44.1%–47.8% for rapeseed oils, 77.4% for almond oil, and 81.3% for hazelnut oil. Δ^5-avenasterol is considered the second most abundant sterol compound in pecan nut oil after β-sitosterol in pecan oil. Similar to pecan oil, Δ^5-avenasterol is the second most abundant sterol in the cold pressed almond oil. The third sterol compound of pecan nut oil is campesterol. In compared to other nuts, pecan oil has lower amount of campesterol than rapeseed oil (36.7–42.4%) and Nigella sativa oils (13.1%). Brassicasterol is a characteristic sterol for rapeseed oil with a range of 6.8%–13.0%.

7 Edible and nonedible applications of cold pressed oil

Pecans are consumed in a raw or roasted form. The roasting process enhances the aroma of raw pecans (Gong & Pegg, 2015). Magnuson, Kelly, Koppel, and Reid (2016) reported that some sensory properties such as nutty, nutty woody, nutty grain like, nutty buttery, brown, caramelized, roasted, and sweet were higher in roasted pecans versus their raw counterparts. Nut spread is popular and widely consumed due to its delicious flavor and balanced nutritional values. Due to their nutritional properties and unique flavor, pecans could be used to produce pecan spreads (Shakerardekani et al., 2013).

Pecan oil is ideal for cooking and frying at high temperatures due to its high smoke point (243°C) (Ranalli, Andres, & Califano, 2017). In addition to edible applications, pecan oils are also destined for nonedible purposes, especially in cosmetics.

TABLE 8 Phenolics of some cold pressed oils[a] (μg/100 g).

Cold pressed oil	Phenolic										
	Catechin	Naringenin	Chlorogenic	Protocatechuic	*p*-Hydroxybenzoic	Vanillic	Caffeic	*p*-Coumaric	Ferulic	Sinapic	Luteolin
Soybean	ND[b]	ND	ND	ND	0.8	1.1	0.8	1.5	1.2	0.9	ND
Sunflower	ND	ND	ND	ND	1.5	6.9	4.9	1.8	1.3	1.4	ND
Rapeseed	ND	ND	ND	ND	1.6	ND	0.3	13.1	5.6	236.0	ND
Corn	ND	ND	ND	ND	1.7	ND	ND	1.9	5.8	0.6	ND
Grapeseed	ND	ND	ND	ND	ND	0.8	ND	ND	ND	0.2	ND
Hemp	ND	ND	ND	ND	6.0	2.0	ND	2.0	1.0	3.0	ND
Flax	ND	ND	ND	ND	3.1	1.0	ND	ND	1.0	ND	ND
Rice bran	ND	ND	ND	ND	ND	ND	ND	ND	0.4	ND	ND
Pumpkin	ND	ND	ND	3.1	ND	11.4	ND	3.8	3.8	ND	ND
Pecan	2.38	2.17	0.53	0.53	ND	0.11	0.06	0.11	0.43	0.38	0.81
Almond	1.17	1.38	1.41	0.24	ND	0.17	0.43	0.19	0.27	0.73	0.78
Hazelnut	0.89	1.13	0.65	0.47	ND	0.03	0.67	0.17	0.38	0.42	0.97
Peanut	1.34	0.98	0.74	0.32	ND	0.05	021	0.15	0.09	0.11	0.68
Walnut	1.67	3.86	0.61	0.42	ND	0.18	0.18	0.22	0.23	0.41	0.55
Apricot	1.26	0.77	1.07	0.32	ND	0.23	0.15	0.13	0.31	0.68	0.65
Pistachio	3.76	2.75	1.45	0.67	ND	0.34	1.13	0.34	0.13	0.83	1.17
Cashew	0.56	0.54	0.32	0.28	ND	0.16	0.09	0.19	0.16	0.23	0.21

[a] *References: Siger, Nogala-Kalucka, and Lampart-Szczapa (2008) and Al Juhaimi et al. (2018).*
[b] *ND, not detected.*

TABLE 9 Sterol composition of some cold pressed oils (%).

	Cold pressed oils						
Sterols[a]	Pecan oil[b]	*Pistacia patlantica* seed oil[c]	*Nigella sativa* oil[d]	Japanese quince seed oil[e]	Rapeseed oils[f]	Almond oil[b]	Hazelnut oil[b]
Campesterol	4.45	4.35	13.1	5.91	36.7–42.4	2.46	5.21
Stigmasterol	0.90	0.98	17.8	2.37	0.2–0.4	0.91	0.91
β-Sitosterol	75.5	87.7	49.4	82.9	44.1–47.8	77.4	81.3
Δ^5-Avenasterol	14.9	2.28	12.4	–	2.2–6.9	9.89	4.37
Δ^7-Avenasterol	0.69	1.04	2.1	–	–	1.39	0.73
Avenasterol	–	–	–	6.96	–	–	–
Δ^7-Stigmasterol	0.24	–	0.6	–	–	1.94	1.45
Brassicasterol	–	–	–	–	6.8–13.0	–	–

[a]*Some important sterol components are given in table.*
[b]*Fernandes et al. (2017).*
[c]*Saber-Tehrani, Givianrad, Aberoomand-Azar, Waqif-Husain, and Jafari Mohammadi (2012).*
[d]*Gharby et al. (2015).*
[e]*Górnaś, Siger, and Segliņa (2013).*
[f]*Madawala, Kochhar, and Dutta (2012).*

8 Health-promoting traits of cold pressed oil and oil constituents

The fatty acid composition of oils is an important factor for them to become a component in healthy diets. Saturated fatty acids are known as bad fats that increase LDL cholesterol (bad cholesterol); however, MUFAs and PUFAs are considered good fats that reduce LDL cholesterol while increasing HDL cholesterol (good cholesterol). The intake of oils rich in unsaturated fatty acids, especially MUFAs, is associated with some health benefits. In particular, oleic acid among MUFAs exhibited significant beneficial effects on coronary heart disease (Alonso, Ruiz-Gutierrez, & Martínez-González, 2006; Kris-Etherton, Zhao, Binkoski, Coval, & Etherton, 2001). Pecan oil is rich in MUFAs, especially oleic acid. The intake of pecans alters positively serum lipid profiles in healthy men and women (Rajaram, Burke, Connell, Myint, & Sabate, 2001). In another study, consumption of 68 g pecans per day for 8 weeks plus self-selected diets in people with normal lipid levels reduced LDL cholesterol compared with a control group (Morgan & Clayshulte, 2000).

Pecans are rich in tocols; in particular γ-tocopherol isomers are present in higher amounts, which have strong antioxidant activity. These oils are poor in α-tocopherol with weak antioxidant activity. A controlled feeding study showed that the addition of pecans to a healthy diet can lead to improved antioxidant defenses by increasing cholesterol-adjusted γ-tocopherol and decreasing thiobarbituric acid reactive substances (TBARS) (Haddad, Jambazian, Karunia, Tanzman, & Sabaté, 2006).

Pecans are high in fat, and so weight gain is expected to be an associated problem. However, epidemiologic studies and short-term controlled feeding trials demonstrated that nut eaters had lower weight gain compared with nonnut eaters, despite an expected increase in total energy (Albert, Gaziano, Willett, & Manson, 2002; Sabaté, 2003; Soriguer et al., 1995).

More publications have been written about the health benefits of pecans than about pecan oils. As a result of the literature search, a study in the use of pecan oils directly was found. In this work, hexane-extracted pecan oils used and the study demonstrated that pecan oil with a high fat diet in rats reduced the level of triacylglycerols compared with the control group (Domínguez-Avila et al., 2015).

9 Oxidative stability

Pecan oils were stored at room temperature (average 22.5°C) and oxidative stability of oils was determined using some chemical analysis such as peroxide value, anisidin value, and specific extinctions at 232 and 270nm. In addition to chemical analysis, sensory evaluations were carried out. According to sensory analysis, the shelf life of pecan oil at room temperature was 90days. The data from oxidation tests exhibited similarity with sensory analysis during storage (Oro, Bolini, Arellano, & Block, 2009).

Castelo-Branco et al. (2016) evaluated induction periods of refined and cold pressed oils using the Rancimat apparatus at 110°C. Refined commercial oils such as canola (average value 8.63h) and sunflower oils (4.91h) showed lower induction periods than cold pressed pecan oil (9.87h). These data supported the theory that cold pressed oils could be used in cooking like refined vegetable oils. The results showed that pecan oil had a strong activity toward oxidation at room temperature as well as thermal oxidation conditions.

References

Al Juhaimi, F., Özcan, M. M., Ghafoor, K., Babiker, E. E., & Hussain, S. (2018). Comparison of cold-pressing and soxhlet extraction systems for bioactive compounds, antioxidant properties, polyphenols, fatty acids and tocopherols in eight nut oils. *Journal of Food Science and Technology*, *55*(8), 3163–3173.

Albert, C. M., Gaziano, J. M., Willett, W. C., & Manson, J. E. (2002). Nut consumption and decreased risk of sudden cardiac death in the physicians. *Health Study Archives of Internal Medicine*, *162*(12), 1382–1387.

Alonso, A., Ruiz-Gutierrez, V., & Martínez-González, M. Á. (2006). Monounsaturated fatty acids, olive oil and blood pressure: Epidemiological, clinical and experimental evidence. *Public Health Nutrition*, *9*(2), 251–257.

Amarowicz, R., Gong, Y., & Pegg, R. B. (2017). *Recent advances in our knowledge of the biological properties of nuts, wild plants, mushrooms and nuts: Functional food properties and applications.* Chichester, UK: Wiley.

Castelo-Branco, V. N., Santana, I., Di-Sarli, V. O., Freitas, S. P., & Torres, A. G. (2016). Antioxidant capacity is a surrogate measure of the quality and stability of vegetable oils. *European Journal of Lipid Science and Technology*, *118*(2), 224–235.

Cockerham, S. N., Gorman, W., Maness, N., & Lillywhite, J. (2012). *Feasibility assessment of investing in a pecan oil and flour processing facility using new extraction technology.* Ph.D. thesisNew Mexico State University Available from aces.nmsu.edu (5 March 2019).

do Prado, A. C. P., Manion, B. A., Seetharaman, K., Deschamps, F. C., Arellano, D. B., & Block, J. M. (2013). Relationship between antioxidant properties and chemical composition of the oil and the shell of pecan nuts [*Carya illinoinensis* (Wangenh) C. Koch]. *Industrial Crops and Products*, *45*, 64–73.

Domínguez-Avila, J. A., Alvarez-Parrilla, E., López-Díaz, J. A., Maldonado-Mendoza, I. E., del Consuelo Gómez-García, M., & Laura, A. (2015). The pecan nut (*Carya illinoinensis*) and its oil and polyphenolic fractions differentially modulate lipid metabolism and the antioxidant enzyme activities in rats fed high-fat diets. *Food Chemistry*, *168*, 529–537.

Fernandes, G. D., Gómez-Coca, R. B., Pérez-Camino, M. D. C., Moreda, W., & Barrera-Arellano, D. (2017). Chemical characterization of major and minor compounds of nut oils: Almond, hazelnut, and pecan nut. *Journal of Chemistry*, *2017*, 1–11.

Gharby, S., Harhar, H., Guillaume, D., Roudani, A., Boulbaroud, S., Ibrahimi, M., et al. (2015). Chemical investigation of *Nigella sativa* L. seed oil produced in Morocco. *Journal of the Saudi Society of Agricultural Sciences*, *14*(2), 172–177.

Gong, Y., & Pegg, R. B. (2015). *Tree nut oils: Properties and processing for use in food, specialty oils and fats in food and nutrition: Properties, processing and applications.* Cambridge, England: Woodhead Publishing.

Górnaś, P., Siger, A., & Segliņa, D. (2013). Physicochemical characteristics of the cold-pressed Japanese quince seed oil: New promising unconventional bio-oil from by-products for the pharmaceutical and cosmetic industry. *Industrial Crops and Products*, *48*, 178–182.

Gunstone, F. (Ed.). (2011). *Vegetable oils in food technology: Composition, properties and uses.* UK: John Wiley & Sons.

Haddad, E., Jambazian, P., Karunia, M., Tanzman, J., & Sabaté, J. (2006). A pecan-enriched diet increases γ-tocopherol/cholesterol and decreases thiobarbituric acid reactive substances in plasma of adults. *Nutrition Research*, *26*(8), 397–402.

Hal, G. D. (2000). Pecan food potential in prehistoric North America. *Economic Botany*, *54*(1), 103–112.

International Nut and Dried Fruit Council Foundation 2018, Available from https://www.nutfruit.org/files/tech/1523960263_INC_Statistical_Yearbook_2017-2018.pdf (30 December 2018).

Kris-Etherton, P. M., Zhao, G., Binkoski, A. E., Coval, S. M., & Etherton, T. D. (2001). The effects of nuts on coronary heart disease risk. *Nutrition Reviews*, *59*(4), 103–111.

Ling, B., Yang, X., Li, R., & Wang, S. (2016). Physicochemical properties, volatile compounds, and oxidative stability of cold pressed kernel oils from raw and roasted pistachio (*Pistacia vera* L. Var Kerman). *European Journal of Lipid Science and Technology*, *118*(9), 1368–1379.

Madawala, S. R. P., Kochhar, S. P., & Dutta, P. C. (2012). Lipid components and oxidative status of selected specialty oils. *Grasas y Aceites*, *63*(2), 143–151.

Magnuson, S. M., Kelly, B., Koppel, K., & Reid, W. (2016). A comparison of flavor differences between pecan cultivars in raw and roasted forms. *Journal of Food Science*, *8*(5), S1243–S1253.

Maness, N. O., Chrz, D., Pierce, T., & Brusewitz, G. H. (1995). Quantitative extraction of pecan oil from small samples with supercritical carbon dioxide. *Journal of the American Oil Chemists' Society*, *72*(6), 665–669.

Marchetti, L., Romero, L. M., Andres, S. C., & Califano, A. N. (2017). Characterization of pecan nut expeller cake and effect of storage on its microbiological and oxidative quality. *Grasas y Aceites*, *68*(4).

Morgan, W. A., & Clayshulte, B. J. (2000). Pecans lower low density lipoprotein cholesterol in people with normal lipid levels. *Journal of the American Dietetic Association*, *100*(3), 312–318.

Ojeda-Barrios, D. L., Hernández-Rodríguez, O. A., López-Ochoa, G. R., & Martínez-Téllez, J. J. (2009). Evolución de los sistemas de producción de nuez en México. *Tecnociencia Chihuahua*, *3*(3), 115–120.

Oro, T., Bolini, H. M. A., Arellano, D. B., & Block, J. M. (2009). Physicochemical and sensory quality of crude Brazilian pecan nut oil during storage. *Journal of the American Oil Chemists' Society*, *86*(10), 971–976.

Rabrenović, B. B., Dimić, E. B., Novaković, M. M., Tešević, V. V., & Basić, Z. N. (2014). The most important bioactive components of cold pressed oil from different pumpkin (*Cucurbita pepo* L.) seeds. *LWT—Food Science and Technology*, *55*(2), 521–527.

Rajaram, S., Burke, K., Connell, B., Myint, T., & Sabate, J. (2001). A monounsaturated fatty acid-rich pecan-enriched diet favorably alters the serum lipid profile of healthy men and women. *Journal of Nutrition*, *131*(9), 2275–2279.

Ranalli, N., Andres, S. C., & Califano, A. N. (2017). Dulce de leche-like product enriched with emulsified pecan oil: Assessment of physicochemical characteristics, quality attributes, and shelf-life. *European Journal of Lipid Science and Technology*, *119*(7)1600377.

Rudolph, C. J., Odell, G. V., Hinrichs, H. A., Hopfer, D. A., & Kays, S. J. (1992). Genetic, environmental, and maturity effects on pecan kernel lipid, fatty acid, tocopherol, and protein composition. *Journal of Food Quality*, *15*(4), 263–278.

Sabaté, J. (2003). Nut consumption and body weight. *The American Journal of Clinical Nutrition*, *78*(3), 647S–650S.

Saber-Tehrani, M., Givianrad, M. H., Aberoomand-Azar, P., Waqif-Husain, S., & Jafari Mohammadi, S. A. (2012). Chemical composition of Iran's *pistacia atlantica* cold-pressed oil. *Journal of Chemistry*, *2013*, 1–6.

Salvador, A. A., Podestá, R., Block, J. M., & Ferreira, S. R. S. (2016). Increasing the value of pecan nut [*Carya illinoinensis* (Wangenh) C. Koch] cake by means of oil extraction and antioxidant activity evaluation. *Journal of Supercritical Fluids*, *116*, 215–222.

Sarkis, J. R., Côrrea, A. P. F., Michel, I., Brandeli, A., Tessaro, I. C., & Marczak, L. D. (2014). Evaluation of the phenolic content and antioxidant activity of different seed and nut cakes from the edible oil industry. *Journal of the American Oil Chemists' Society*, *91*(10), 1773–1782.

Scapinello, J., Magro, J. D., Block, J. M., Di Luccio, M., Tres, M. V., & Oliveira, J. V. (2017). Fatty acid profile of pecan nut oils obtained from pressurized n-butane and cold pressing compared with commercial oils. *Journal of Food Science and Technology*, *54*(10), 3366–3369.

Shakerardekani, A., Karim, R., Ghazali, H., & Chin, N. (2013). Textural, rheological and sensory properties and oxidative stability of nut spreads-a review. *International Journal of Molecular Sciences*, *14*(2), 4223–4241.

Siger, A., Nogala-Kalucka, M., & Lampart-Szczapa, E. (2008). The content and antioxidant activity of phenolic compounds in cold-pressed plant oils. *Journal of Food Lipids*, *15*(2), 137–149.

Soriguer, F. J., Gonzalez-Romero, S., Esteva, I., García-arnés, J. A., Tinahones, F., De Adana, M. R., et al. (1995). Does the intake of nuts and seeds alter the appearance of menarche? *Acta Obstetricia et Gynecologica Scandinavica*, *74*(6), 455–461.

United States Department of Agriculture Agricultural Research Service 2018, Available from https://ndb.nal.usda.gov/ndb/foods/show/12142 (30 December 2018).

Venkatachalam, M., & Sathe, S. K. (2006). Chemical composition of selected edible nut seeds. *Journal of Agricultural and Food Chemistry*, *54*(13), 4705–4714.

Wu, X., Beecher, G. R., Holden, J. M., Haytowitz, D. B., Gebhardt, S. E., & Prior, R. L. (2004). Lipophilic and hydrophilic antioxidant capacities of common foods in the United States. *Journal of Agricultural and Food Chemistry*, *52*(12), 4026–4037.

Chapter 47

Cold pressed pine (*Pinus koraiensis*) nut oil

Yasemin Incegul[a], Mehmet Aksu[b], Sündüz Sezer Kiralan[c], Mustafa Kiralan[c], and Gulcan Ozkan[a]
[a]*Department of Food Engineering, Faculty of Engineering, Suleyman Demirel University, Isparta, Turkey,* [b]*Agriculture and Rural Development Support Institution, Isparta Provincial Coordination Unit, Isparta, Turkey,* [c]*Department of Food Engineering, Faculty of Engineering, Balıkesir University, Balikesir, Turkey*

Abbreviations

HHP	high hydrostatic pressure
HVED	high voltage electrical discharges
MAE	microwave-assisted extraction
MUFA	monounsaturated fatty acid
PEF	pulsed electric field
PLE	pressurized liquid extraction
PNLA	pinolenic acid
PNO	pine nut oil
PUFA	polyunsaturated fatty acid
SCA	sciadonic acid
SFA	saturated fatty acid
UAE	ultrasound assisted extraction
Δ5-UPIFA	Δ5-unsaturated polymethylene-interrupted fatty acid

1 Introduction

Pine nuts, which are members of the Pinaceae family, are a subbranch of the *Pinus* genus, which consists of approximately 800 species worldwide. Pine nuts are edible seeds obtained from pine tree species grown in temperate and cool temperate forests in East Asia (Piao, Tang, & Swihart, 2011), and their use began in the Paleolithic age (Mutke, González-martínez, Calama, & Montero, 2012). The United Nations Food and Agriculture Organization reported that 29 nut species containing *Pinus koraiensis* (Korean pine) are edible (FAO, 1998). Pine nuts have many areas of use in most parts of the world, especially in the Mediterranean and Asian regions. The increase in demand for pine nuts has increased worldwide production, and the main countries meeting the demand are China, Korea, Russia (Siberia), and Pakistan (FAO, 1998). In the period of 2017/18, China was the top producer with 39% of the world share, followed by North Korea, Pakistan, and Afghanistan with 13% each. According to data from the International Nut and Dried Fruit Council (INC, 2017) statistical yearbook, pine nut production and export percentages are as shown in Figs. 1 and 2.

Pine nuts can be consumed raw or roasted, and are used in the food industry in addition to a variety of traditional dishes such as bread, confectionery, sauces, and cakes, as well as vegetables and meat dishes (Nergiz & Dönmez, 2004; Zadernowski, Naczk, & Czaplicki, 2009). In addition, nuts are widely consumed around the world for their flavors and health benefits due to their unique composition, low sugar content, high unsaturated fatty acids, vitamins and minerals, protein and phytochemicals including phenolics (tannins, ellagic acid, and curcumin), flavonoids (luteolin, quercetin, myricetin, kaempferol, and resveratrol), isoflavones (genistein, and daidzein), terpenes, organosulfuric compounds, and vitamin E (Bravo, 1998; Chang, Alasalvar, & Bolling, 2016; Schlörmann et al., 2015; Venkatachalam & Sathe, 2006). The nuts, with a length of 9–12 mm, contain a high proportion of oil (Zadernowski et al., 2009). Growing area and climate are the two main factors affecting the chemical content of pine nuts. Table 1 presents a comparison of the nutritional values of species having market value worldwide.

Both traditional and green extraction techniques are used to obtain Korean pine nut (*P. koraiensis*) oil. High-quality oils are obtained with different extraction techniques such as cold pressing, Soxhlet extraction, enzymatic extraction, and

Cold Pressed Oils. https://doi.org/10.1016/B978-0-12-818188-1.00047-5
Copyright © 2020 Elsevier Inc. All rights reserved.

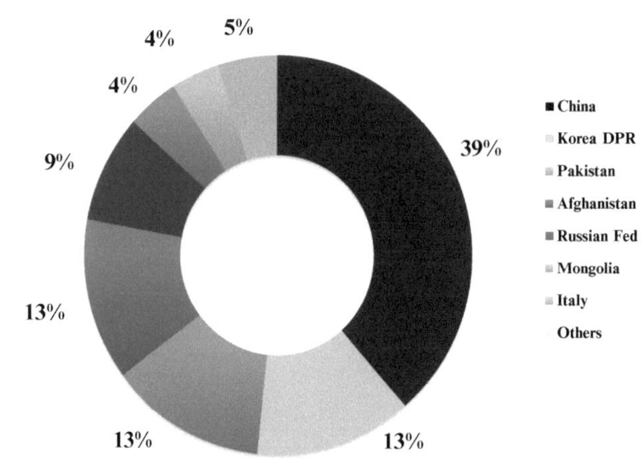

FIG. 1 2017/18 pine nut production (International Nut and Dried Fruit Council (INC), 2017).

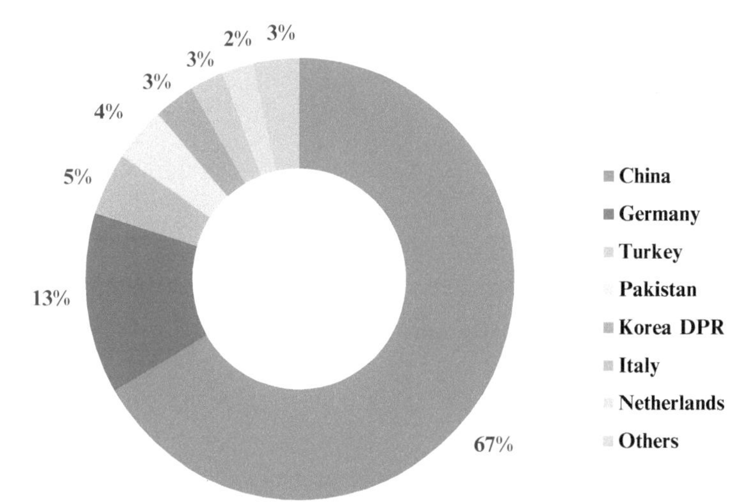

FIG. 2 2016 world pine nut exports (International Nut and Dried Fruit Council (INC), 2017).

TABLE 1 Nutritional values of several species of pine nuts in comparison with other commercially important nuts.[a]

Type of nut	Carbohydrates (%)	Lipids (%)	Protein (%)
P. koraiensis	12	65	18
P. pinea	7	48	34
P. cembroides Zucc.	14	60	19
P. monophylla	54	23	10
P. sabiniana	9	56	28
P. gerardiana	23	51	14
P. edulis	18	62–71	14
P. sibrica	12	51–75	19

[a]*References: Evaristo, Batista, Correia, Paula, and Costa (2010) and Mutke, Pastor, and Picardo (2011).*

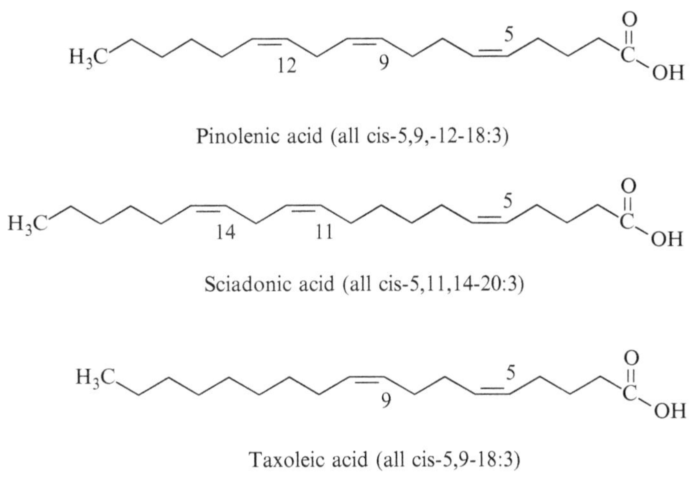

Pinolenic acid (all cis-5,9,-12-18:3)

Sciadonic acid (all cis-5,11,14-20:3)

Taxoleic acid (all cis-5,9-18:3)

FIG. 3 Chemical structures of pinolenic acid, sciadonic acid, and taxoleic acid.

supercritical fluid extraction (Chen, Zhang, Wang, & Zu, 2011; Kornsteiner, Wagner, & Elmadfa, 2006; Li, Jiang, Sui, & Wang, 2011; Salgın, 2013; Wang, Wang, Xiao, & Xu, 2019; Wang, Zhang, et al., 2019).

Pinus koraiensis and *P. sibrica* have the highest oil content with 65%–75% (Nergiz & Dönmez, 2004). Fatty acids were major components of pine nut oils (*P. koraiensis*, *P. sibirica*, *P. pinea*, *P. armandii*, and *P. massoniana*) and these consist of 50% polyunsaturated fatty acids (PUFAs), 40% monounsaturated fatty acids (MUFAs), and 10% saturated fatty acids (SFAs) (Ryan, Galvin, O'Connor, Maguire, & O'Brien, 2006). Linoleic acid (all-*cis*-9-,12-18:2) is the most common fatty acid in pine nut oil (PNO), oleic acid (*cis*-9-18:1) is the second most abundant fatty acid in PNO, and palmitic acid (16:0) and stearic acid (18:0) are the main PUFAs, MUFAs, and SFAs found in PNO, respectively. Pinolenic acid (PNLA; all *cis*-5,-9,-12-18:3), typically constituting 14% to 19% of total fatty acids, is the basic Δ5-unsaturated polymethylene-interrupted fatty acids (Δ5-UPIFA) in pine nuts and their oil (Wolff & Bayard, 1995), while sciadonic acid (SCA; all *cis*-5,-11,-14-20:3) and taxoleic acid (all *cis*-5,-9-18:2) are other Δ5-UPIFAs (Destaillats, Cruz-Hernandez, Giuffrida, & Dionisi, 2010). The chemical structures of widespread fatty acids are given in Fig. 3.

The fatty acid distribution of *P. koraiensis* oil is composed of 14.6%–14.9% PNLA, 45.2%–48.5% linoleic acid, 24.1%–25.5% oleic acid, 1.8%–2.2% stearic acid, and 4.2%–4.9% palmitic acid (Asset et al., 1999; Destaillats et al., 2010; Wolff & Bayard, 1995). Pine nuts are useful for controlling blood lipids and controlling coronary heart disease through unsaturated fatty acids such as linoleic acid and oleic acid (Ryan et al., 2006).

Pine nuts and oils contain phytosterols, tocopherols, and squalene, and these components increase the durability of the oil. In cold pressed pine nut oil (type unspecified), campesterol (17.7 mg/100 g oil) and stigmasterol (3.4 mg/100 g oil) were found in lower amounts, while β-sitosterol was the most common phytosterol present at 120.5 mg/100 g oil (Rabadán, Pardo, Gómez, & Álvarez-ortí, 2018; Ryan et al., 2006). Pine nuts and their oils also contain various minerals such as magnesium, selenium, and potassium (Zadernowski et al., 2009). The total phenolic, flavonoid, and antioxidant contents of Korean pine nut (*P. koraiensis*) were determined as 152.9 mg/100 g, 45.0 mg/100 g, and 14.6 μmol vitamin C equ./g sample, respectively. Regular use of pine nut can reduce the risk of cardiovascular disease. The phytochemicals, especially phenolic substances, are foreseen as bioactive compounds that are essential for health (Yang, Liu, & Halim, 2009). There is a positive correlation between total phenolic content and antioxidant activity (Kanmaz & Saral, 2017; Velioglu, Mazza, Gao, & Oomah, 1998). *P. koraiensis* nut oil has some ideal antioxidant properties (Chen et al., 2011).

Pine nuts are used in edible oil production and health sector due to their high oil content. The oil is used in cosmetics, beauty products, and massage oil as well as in cooking and medical applications. In addition, the by-products of pine nut oil press have been evaluated in granolas, chocolate, and crispy sticks (Miraliakbari & Shahidi, 2008a, 2008b; Sharashkin & Gold, 2004).

2 Cold press extraction and processing of oils

Extraction is the process of separation one or more valuable components from the raw material (Chemat, Vian, & Cravotto, 2012; Lebovka, Vorobiev, & Chemat, 2011). The extraction process is carried out in one of two ways: conventional and modern methods. Soxhlet extraction, maceration, and percolation are the most commonly used conventional methods

(Herrero, Mendiola, Cifuentes, & Ibánez, 2010). The oil from Korean pine nuts (*P. koraiensis*) is generally extracted by traditional methods such as Soxhlet and solvent extraction with *n*-hexane or petroleum ether, or a combination of these methods (Chen et al., 2011; Wang, Wang, et al., 2019; Wang, Zhang, et al., 2019). In addition, different modern extraction techniques such as aqueous enzymatic extraction (Li et al., 2011) and supercritical fluid extraction (Chen et al., 2011; Salgın, 2013) can be used to obtain high-quality oils.

Conventional methods are replaced by modern methods because of the long time requirement, high solvent consumption, and difficulties in raw material preparation (Chemat et al., 2012; Wan & Wong, 1996). Modern techniques or nontraditional techniques are called green or clean techniques thanks to their preference for organic solvents and need for less energy (Chemat et al., 2012; Parker, Adams, Zhou, Harris, & Yu, 2003; Rodríguez-pérez, Quirantes-piné, & Fernández-gutiérrez, 2015). Supercritical fluid (SFE), pressurized liquid (PLE), ultrasound assisted (UAE), microwave-assisted (MAE), pulsed electric field (PEF), high voltage electrical discharges (HVED), and high hydrostatic pressure (HHP) are important green technologies proposed as alternatives to traditional methods (Soquetta, Terra, & Bastos, 2018). These techniques are also known as cold extraction techniques, where the stability of the extracted compound is not affected (Tiwari, 2015). According to the Codex Alimentarius, the cold pressing technique is performed only by mechanical processes without heat application, and the product is produced without destructing the nature of the oil (Matthäus & Spener, 2008). Besides the quality of the raw material, the production conditions and parameters are very important (Rotkiewicz, Konopka, & Żylik, 1999). Therefore, the heat increases to be applied during the process cause a decrease in oil quality (Geçgel, Taşan, & Gürpınar, 2011; Rabadán et al., 2018). At the same time, the oils obtained by the cold pressing technique are low efficiency and may be of inconsistent quality. Cold pressed oils could contain higher amounts of prooxidative compounds, so their shelf life may be shorter compared to that of refined oils (Brühl, 1996; Rotkiewicz et al., 1999). On the other hand, more biologically active substances (natural antioxidants) such as polar phenolic compounds and apolar tocopherols are present in cold pressed oils (Bhatnagar & Krishna, 2014; Prescha, Grajzer, Dedyk, & Grajeta, 2014).

Day by day, demand for natural and safe food, including fats and oils, is increasing. Vegetable oils have been produced using only the pressing method without using a solvent from different oilseeds and fruits, and made usable for consumers. Consumers use food products as an adjunct/supplement to alternative medicine in addition to taste and necessary consumption. As the availability of information increases about the benefits of vegetable oils to human health, the interest for oils produced by cold pressing is increasing (Matthäus & Brühl, 2003; Sloan, 2000).

3 Cold pressed oil recovery, content, uses of oil cake, and economy

Cold pressed oils have been used mainly in the pharmaceuticals and cosmetics industries, but are now starting to appear on our tables (Imer & Taşan, 2018). Oils obtained from some pines are used more often for cooking, while oils from Korean pine nuts are used as nutritional supplements thanks to an unsaturated fatty acid, pinolenic acid, and antioxidants. Pinolenic acid has antiinflammatory properties that can help protect, repair, and strengthen the stomach and stomach lining (Chen et al., 2011; Xie, Miles, & Calder, 2016).

In the pine nut oil industry, the main by-product is pine nut protein, which has good antioxidant activity and immunity (Lin, Yang, Cheng, Wang, & Qin, 2018). After pine nut oil is obtained, some by-products, especially pine nut flakes, bark, and cones, are revealed. Products remaining from cold pressing of pine nut are evaluated in granolas, chocolates, and crunch bars. The pine nut shell can be used as a heat source for the seeds to be processed. After pressing the oil, which constitutes about 30% of the hazelnut, pine nut flour can be obtained from the remaining product as waste and this flour can be used in various kitchen products. It can be used in pastries and pancakes instead of wheat or rye flour to give a rich aroma (NSW Food, 2012).

4 Fatty acids composition and acyl lipids profile of cold pressed oil

Nut species have a low saturated lipid content and they are considered a rich source of healthy lipids, since they contain monounsaturated and polyunsaturated fatty acids as well as vegetable proteins, fiber, phytosterols, phenolics, vitamins, and minerals. MUFAs and PUFAs account for more than 75% of the lipid profiles (Alasalvar & Pelvan, 2011; Alasalvar & Shahidi, 2009). Linoleic acid (C18:2), the typical component of *Pinaceae* nut oil, is the most common fatty acid in the nut oil of *P. koraiensis* (Evaristo, Batista, Correia, Correia, & Costa, 2013). *P. koraiensis* contain other two fatty acids, *cis*-5,9,12 C18:3 (pinolenic acid, 13%–15%) and *cis*-5,11,14-C20:3 (sciadonic acid). The fatty acid composition of *P. koraiensis* is given in Table 2, and major fatty acids of cold pressed pine oils are also mentioned in Table 3.

TABLE 2 The identified fatty acids found in *Pinus koraiensis* oil.[a]

Fatty acid	CN:DBN[b]	Composition range (mol%)
Palmitic acid	16:0	3.89–5.3
Palmitoleic acid	16:1 (Z9)	Traces
14-Methyl-hexadecanoic acid	17:0	Traces
Margaroleic acid	17:1 (Z9)	Traces
Stearic acid	18:0	1.53–2.70
Oleic acid	18:1 (Z9)	19.4–32.1
Asclepic acid (*cis*-vaccenic acid)	18:1 (Z11)	0.40–1.46
Linoleic acid	18:2 (Z9, Z12)	44.1–50.0
Taxoleic acid	18:2 (Z5, 9)	1.8–2.2
Linolenic acid	18:3 (Z9, Z12, Z15)	0.10–0.58
Pinolenic acid	18:3 (Z5, Z9, Z12)	13.30–14.92
Arachidic acid	20:0	0.3–0.5
Gondoic acid	20:1 (Z11)	1.03–1.30
Eicosa dienoic acid	20:2 (Z11, 14)	0.49–0.61
Keteleeronic acid	20:2 (Z5, 11)	0.10
Sciadopinolenic acid	20:3 (Z5, 11, 14)	0.4–1.0
Bis Homopinoleic acid	20:3 (Z7, 11, 14)	0.4
Behenic acid	22:0	Traces

[a]*References: Acheampong, Leveque, Tchapla, and Heron (2011), Evaristo et al. (2013), Imbs, Nevshupova, and Pham (1998), Li et al. (2011), Som No et al. (2015), Xie et al. (2016), and Wolff and Bayard (1995).*
[b]CN, *carbon number;* DBN, *number of double bonds.*

TABLE 3 Major fatty acids in cold pressed pine oils (%).[a]

Fatty acid	%
Palmitic acid (16:0)	4.4
Stearic acid (18:0)	2.3
Oleic acid (9–18:1)	29.1
Taxoleic acid (5,9–18:2)	2.2
Linoleic (9,12–18:2)	43.9
Pinolenic (5,9,12–18:3)	14.2
Gondoic acid (11–20:1)	1.2
Sciadopinolenic acid (5,11,14–20:3)	1.1

[a]*References: Gresti, Mignerot, Bezard, and Wolff (1996).*

As in other nut species, the oil composition of *P. koraiensis* nuts varies according to genetics, harvest season, origin, environmental conditions, soil composition, maturity level, and sowing methods (Kornsteiner et al., 2006). However, when the *P. koraiensis* fatty acid composition is examined, the most common fatty acids are linoleic acid, oleic acid, pinolenic acid, palmitic acid, and stearic acid (Evaristo et al., 2013; Imbs et al., 1998; Li et al., 2011; Wolff & Bayard, 1995).

5 Minor bioactive lipids in cold pressed oil

P. koraiensis contains oil, triglycerides, free fatty acids, phenolic compounds, vitamins, and sugars as well as nearly all essential amino acids and essential fatty acids, so it is a healthy alternative for humans (Chen et al., 2011; Yang et al., 2009).

Tocochromanols are antioxidant substances which are nonsaponifiable in the stability of cold pressed oils. Vitamin E is a fat-soluble compound containing four tocopherols and can be found in foodstuffs in different proportions (Alasalvar & Pelvan, 2011; Ying, Wojciechowska, Siger, Kaczmarek, & Rudzińska, 2018). The total tocochromanols (mg/100 g) content of cold pressed pine nut oil is 18.5 mg/100 g and the basic tocopherols are detected as γ-tocopherol and α-tocopherol (Ying et al., 2018). The distribution and amount of tocopherol derivates in cold pressed pine nut oil are given in Table 4.

Phytosterols, the main component of vegetable oils, are a great source of interest for consumers due to their positive effects on health, such as lowering cholesterol. β-Sitosterol, campesterol, stigmasterol, avenasterol, cycloartenol, and 24-methylene cycloartenol were found in pine nut oil, but the predominant phytosterol is β-sitosterol with a 78% ratio of all phytosterols in cold pressed pine nut oil (Ying et al., 2018). Table 5 summarizes the phytosterols of cold pressed pine nut oil.

In the literature, there is no study investigating the phenolic content of cold pressed Korean pine nut oil. The total phenolic values of pine nut oil extracted using *n*-hexane and chloroform/methanol were determined as 148 and 423 mg gallic acid equivalent/kg oil, respectively (Miraliakbari & Shahidi, 2008a, 2008b). In another study, the total phenolic content of *Pinus koraiensis* seed was determined as 264 mg of gallic acid equivalents/g dry material (Su, Wang, & Liu, 2009).

TABLE 4 Tocopherol profile of cold pressed pine nut oil.[a]

Type of tocopherol	mg/100 g oil
α-Tocopherol	7.4
β-Tocopherol	0.1
γ-Tocopherol	10.3
δ-Tocopherol	0.7

[a]*References: Ying et al. (2018).*

TABLE 5 Phytosterol content of cold pressed pine nut oil.[a]

Type of phytosterol	mg/100 g oil
Campesterol	17.7
Stigmasterol	3.4
β-Sitosterol	120
Avenasterol	3.1
Cycloartenol	3.1
24-Methylene cycloartenol	3.5
Others	2.0

[a]*References: Ying et al. (2018).*

P. koraiensis oil was characterized by a monoterpene content of about 75%, whereas sesquiterpenes were lower with a rate of 9.6%. The main essential oil components of *P. koraiensis* are α-pinene (35.2%), limonene (18.4%), β-pinene (8.7%), β-caryophyllene (3.5%), and myrcene (3.0%) (Yang et al., 2010).

6 Composition of cold pressed pine oil in comparison to other cold pressed oils

Pine nut oil is composed of 90% unsaturated fatty acids, 50% of which are PUFAs, 40% MUFAs, and 10% SFAs (Ryan et al., 2006). Linoleic acid (all-*cis*-9-,12-18:2) is the most abundant fatty acid in pine nut oil (PNO), and oleic acid (*cis*-9-18:1) is the second most abundant fatty acid in PNO. Palmitic acid (16:0) and stearic acid (18:0) are the main SFAs found in PNO (Table 6). While cold pressed pumpkin oil is rich in linoleic acids like pine nut oil, pecan oil, almond oil, Brazil nut oil, hazelnut oil, macadamia oil, and pistachio oil are rich in oleic acid.

γ-Tocopherol is the main tocopherol of pine nut oil and its amount is 10.3 mg/100 g (Table 6). The second main tocopherol is α-tocopherol with an amount of 7.4 mg/100 g. As in pine nut oil, γ-tocopherol is the most abundant tocopherol found in pecan and Brazil nut oils. α- and β-tocopherols are major tocopherols in pistachio and pumpkin oils, and it has been noted that they are higher than total values in pine nut oil. α-Tocopherol is the most common type of tocopherol in hazelnut and almond oils, and the only isomer in macadamia oil.

β-Sitosterol and campesterol are the basic phytosterols of cold pressed pine nut oil (Table 7). The contribution of β-sitosterols to the total sterol content of other oils is the highest in apricot kernel oil (86%), and followed by macadamia nut oil (84%), avocado fruit oil (79%), pine nut oil (77%), blackcurrant seed oil (67%), hemp oil (67%), safflower oil (65%), wheat germ oil (64%), dill seed oil (63%), poppy seed oil (61%), parsley seed oil (60%), black cumin oil, and milk thistle oil (45%–46%). The dominant sterol in borage oil was campesterol (45%), the dominant sterol of watermelon seed oil was stigmasterol (32%), while dominant sterols of argan oil were other sterols (79%).

7 Edible and nonedible applications of cold pressed oil

In the literature, some studies were conducted on *P. koraiensis* seeds and seed oil, but no study was found about cold pressed pine nut oil on edible or nonedible applications. In one study, *P. koraiensis* was treated with a protein film containing green tea extract; at the end of storage, it was reported that the peroxide acid and TBA values had decreased (Lee et al., 2004). In another study, researchers aimed to produce zero-*trans* margarine using pine nut oil and palm stearin in different ratios (Adhikari et al., 2010).

TABLE 6 Fatty acids and tocopherols composition of some cold pressed oils.[a]

	Cold pressed oils[b]								
	PNO[1,2]	PEO[3]	AO[3]	BNO[3]	HO[3]	MO[3]	PO[4]	PUO[5]	CSO[6]
Fatty acid (g/100 g)									
C16:0	4.81	6.01	6.58	15.2	6.61	8.48	10.9–11.1	11.2–15.5	3.71
C18:0	2.77	2.40	1.72	12.3	1.84	3.76	1.05–1.11	5.2–6.2	0.42
C18:1	24.7	72.2	64.6	39.5	79.4	59.7	55.3–56.0	37.1–43.6	82.0
C18:2	46.1	18.0	26.3	31.6	11.3	1.59	30.4–31.3	37.3–44.5	13.1
Tocopherol (mg/100 g)									
α	7.4	0.14	64.2	1.89	40.7	0.80	2.44–3.47	2.55–5.20	–
β	0.1	0.19	0.19	0.08	0.68	ND[c]	28.0–30.8	29.9–53.6	–
γ	10.3	28.4	1.70	21.0	0.56	ND			–
δ	0.7	0.22	0.05	0.20	0.05	ND	2.03–2.55	3.39–10.4	–

[a]*References:* [1] *Gong and Pegg (2015),* [2] *Ying et al. (2018);* [3] *Castelo-Branco, Santana, Di-Sarli, Freitas, and Torres (2016);* [4] *Ling, Yang, Li, and Wang (2016);* [5] *Rabrenović, Dimić, Novaković, Tešević, and Basić (2014);* [6] *Parker et al. (2003).*
[b]*PNO, pine nut oil; PEO, pecan oil; AO, almond oil; BNO, Brazil nut oil; HO, hazelnut oil; MO, macadamia oil; PO, pistachio oil; PUO, pumpkin oil; CSO, carrot seed oil.*
[c]*ND, not detected.*

TABLE 7 Content of sterols (mg/100 g) in cold pressed plant oils.

	Type of phytosterol						
Source of oil	Campesterol	Stigmasterol	β-Sitosterol	Avenasterol	Cycloartenol	24-Methylene cycloartenol	Others
Pine nut oil[a,1]	17.7	3.4	120.5	3.1	3.1	3.5	2.0
Apricot kernel oil[1,2]	5.3	–	56.8	3.8	–	–	–
Argan oil[1,3]	–	–	–	11.8	16.8	5.2	128.9
Avocado fruit oil[1,4,5]	11.6	1.1	107.1	3.6	8.3	2.3	2.1
Black cumin oil[1,2]	6.7	6.1	19.7	2.7	5.7	2.7	–
Blackcurrant seed oil[1,6]	21.7	1.6	170.5	6.6	16.3	18.8	18.2
Borage oil[1]	42.1	–	24.4	14.0	11.7	1.7	–
Dill seed oil[1]	8.5	5.8	39.1	–	3.2	–	5.2
Hemp oil[1,2]	17.4	1.7	63.0	8.5	3.1	–	–
Macadamia seed oil[1,7,8,9]	7.0	1.4	65.6	4.2	–	–	–
Milk thistle seed oil[1,10]	10.3	11.9	51.0	–	32.82	5.3	–
Parsley seed oil[1]	6.3	7.8	41.1	1.9	6.4	–	5.2
Poppy seed oil[1,11]	16.5	10.3	110.0	11.9	22.2	6.5	4.4
Safflower oil[1,12]	26.1	4.4	122.9	11.6	18.7	4.9	–
Watermelon seed oil[1,13,14]	–	27.5	13.8	21.7	21.7	2.5	–
Wheat germ oil[1,15,16,17]	225.3	17.5	545.2	31.6	8.5	–	29.8

[a]References: [1] Ying et al. (2018); [2] Velickovska, Brühl, Miltev, Mirhosseini, and Matthaus (2015); [3] Abbassi, Khalid, Zbakh, and Ahmad (2014); [4] Ansorena, Berasategi, Barriuso, and Astiasaran (2012); [5] Jorge, Polachini, Dias, Jorge, and Telis-romero (2015); [6] Dobson et al. (2012); [7] Navarro and Rodrigues (2016); [8] Maguire et al. (2004); [9] Kaijser, Dutta, and Savage (2000); [10] Dabbour, Al-Ismail, Takruri, and Azzeh (2014); [11] Emir, Aydeniz, and Yılmaz (2015); [12] Aydeniz, Onur, and Yılmaz (2014); [13] Górna and Rudzinska (2016); [14] Wang, Zhang, et al. (2019); [15] Eisenmenger and Dunford (2008); [16] Hassanien et al. (2014); [17] Ghafoor et al. (2017).

Some nut types can be used in the production of various milk-like beverages. In addition to soybean and peanut, *P. koraiensis* can be used as a raw material for making beverages. Lee and Rhee (2003) conducted a study on the production of cereal beverages using *P. koraiensis*. In addition to food and medicinal purposes, pine nut oil is used in cosmetics, beauty products, and massage oil (NSW Food, 2012).

8 Health-promoting traits of cold pressed oil and oil constituents

The intake of high SFA in the body is strongly associated with increased blood cholesterol concentrations and increased hypertension, increased risk of diabetes, and cardiovascular disease. In contrast, intake of foods rich in unsaturated fatty acids found in most vegetable oils is associated with improved lipid profiles, lower atherosclerosis biomarker potential, and fewer cardiovascular diseases (Ros & Mataix, 2006).

The health benefits of pine nut oil are due to a good source of MUFAs (most important: oleic acid, ω-9), PUFAs (most importantly: linoleic acid, ω-6), and pinolenic acid. Unsaturated fatty acids are important nutrients for human health due to their strong lipid and cardioprotective effects, lowering cholesterol and reducing heart disease and cancer risk (No & Kim, 2013).

Pinolenic acid (PLA) is a plant-origin polyunsaturated fatty acid which is designated as all-*cis*-5,9,12-18: 3, and its source is Korean pine nut oil. PLA is an α-linolenic acid (GLA) isomer with ω-6 essential fatty acid. Although PLA is not classified as an essential fatty acid, it plays a similar role. PLA has some positive effects on health, such as alleviating some symptoms of essential fatty acid deficiency, reducing the lipid-lowering effect and appetite (No & Kim, 2013).

Cholecystokinin (CCK-8) is released from the duodenal enteroendocrine cells into the bloodstream and suppresses the appetite (Degen, Matzinger, & Beglinger, 2001). The fatty acids 12C and above support CCK-8 release. Long-chain fatty acids are more effective than middle-chain ones, and PUFAs are more effective than MUFAs (Rovati & Beglinger, 2000). Korean pine nut oil contains more than 92% PUFAs and MUFAs, such as pinolenic acid, linoleic acid, and oleic acid, and reduces appetite by inducing a feeling of satiety (Pasman et al., 2008). In another study that investigated the effect of pressed Korean pine nut oil on appetite and food intake, it was found that pine nut oil increased cholecystokinin secretion, and decreased appetite and food intake (Hughes et al., 2008).

Zhang et al. (2018) investigated the effect of the active constituents of Korean pine nuts on the oxidative stress and low-grade inflammation of D-galactose (D-gal) in rats. The neuroprotective and antioxidative effects of *P. koraiensis* were studied and the active ingredients significantly reduced neuronal apoptosis in mice, improved histopathological findings, and increased superoxide dismutase and catalase activity, but did not increase total antioxidant capacity (Zhang et al., 2018). It has been stated that *P. koraiensis* nut oil obtained by supercritical carbon dioxide extraction improves superoxide dismutase, glutathione peroxidase, and total antioxidant activities, and diminishes the content of malondialdehyde in the serum of the rats; it has also been shown to have excellent antioxidant activity (Chen et al., 2011). The antioxidant activities of polysaccharides extracted from *P. koraiensis*, *P. armandii*, and *P. sylvestris* var. *mongolica* conifers were investigated, and it was concluded that the polysaccharide samples in Korean pine have the highest antioxidant activity (Zhang et al., 2017).

In a study that investigated the effects of a pine nut oil-supplemented diet, the body weight (37.4%), liver weight (13.7%), plasma triglycerides (31.8%), and total cholesterol (28.5%) of mice decreased (Ferramosca, Savy, Einerhand, & Zara, 2008; Rezq & El-Khamisy, 2011).

In rats, *P. koraiensis* oil was found to provide better blood cholesterol concentration compared to safflower oil. Cold pressed pine nut oil contains about 14% of pinolenic acid, which can be consumed for various physiological effects, functions, and prevention or treatment of various degenerative diseases such as hypercholesterolemia, thrombosis, and hypertension (Sugano, Wakamatsu, Ikeda, & Oka, 1994). Korean pine nuts contain vicilins belonging to one of a number of protein families that contain more than 85% of the known nutrient allergens of plant sources, so it may be a food allergen for consumers (Jin et al., 2014; NSW Food, 2012; Zhang, Du, & Fan, 2016).

9 Oxidative stability

Oxidative stability depends on the presence and concentration of antioxidants in the oil, and the composition of fatty acids. Resistance to oxidation or oxidative stability of the oils is measured using the Rancimat tool at 110°C. Ying et al. (2018) evaluated the oxidative stability of some cold pressed edible oils. The pine nut oil was among these oils and the induction periods of the oils were calculated using the Rancimat device. The induction times of the oils (IP) were identified as 0.2 h for apricot kernels, hemp, and wheat germ oil, 14 h for black cumin oil, 11 h for argan oil, 10 h for milk thistle and peanut oil, and 4 h for pine nut oil.

10 Toxicity

Oxygen-related degradation of oils is considered to be a problem in the storage of oils due to undesirable taste, aroma, lowering the nutritional value and releasing toxic compounds. Oxidation of edible oils may be reduced by processing techniques that minimize the loss of tocopherols and other natural antioxidants. Since there is little work on the toxicity of pine nuts, information on this subject is limited (Miraliakbari & Shahidi, 2008a, 2008b). It has been stated that Korean pine nut oil is not toxic to people in quantities up to at least 3 g per person per day (Speijers, Dederen, & Keizer, 2009).

References

Abbassi, A. E., Khalid, N., Zbakh, H., & Ahmad, A. (2014). Physicochemical characteristics, nutritional properties, and health benefits of Argan oil: A reviews. *Food Science and Nutrition*, *54*, 1401–1414.

Acheampong, A., Leveque, N., Tchapla, A., & Heron, S. (2011). Simple complementary liquid chromatography and mass spectrometry approaches for the characterization of triacylglycerols in *Pinus koraiensis* seed oil. *Journal of Chromatography A*, *1218*(31), 5087–5100.

Adhikari, P., Zhu, X., Gautam, A., Shin, J., Hu, J., Lee, J., et al. (2010). Scaled-up production of zero- trans margarine fat using pine nut oil and palm stearin. *Food Chemistry*, *119*(4), 1332–1338.

Alasalvar, C., & Pelvan, E. (2011). Fat-soluble bioactives in nuts. *European Journal of Lipid Science and Technology*, *113*, 943–949.

Alasalvar, C., & Shahidi, F. (2009). Natural antioxidants in tree nuts. *European Journal of Lipid Science and Technology*, *111*, 1056–1062.

Ansorena, D., Berasategi, I., Barriuso, B., & Astiasaran, I. (2012). Stability of avocado oil during heating: Comparative study to olive oil. *Food Chemistry*, *132*, 439–446.

Asset, G., Staels, B., Wolff, R. L., Baugé, E., Madj, Z., Fruchart, J., et al. (1999). Effects of *Pinus pinaster* and *Pinus koraiensis* seed oil supplementation on lipoprotein metabolism in the rat. *Lipids*, *34*(1), 39–44.

Aydeniz, B., Onur, G., & Yılmaz, E. (2014). Physico-chemical, sensory and aromatic properties of cold press produced safflower oil. *Journal of the American Oil Chemists' Society*, *91*(1), 99–110.

Bhatnagar, A. S., & Krishna, G. A. G. (2014). Lipid classes and subclasses of cold-pressed and solvent-extracted oils from commercial Indian Niger (*Guizotia abyssinica* (L.) Cass.) seed. *Journal of the American Oil Chemists' Society*, *91*(7), 1205–1216.

Bravo, L. (1998). Polyphenols: Chemistry, dietary sources, metabolism, and nutritional significance. *Nutrition Reviews*, *56*, 317–333.

Brühl, L. (1996). Trans fatty acids in cold pressed oils and in dried seeds. *Fett/Lipid*, *98*(11), 380–383.

Castelo-Branco, V. N., Santana, I., Di-Sarli, V. O., Freitas, S. P., & Torres, A. G. (2016). Stability of vegetable oils. *European Journal of Lipid Science and Technology*, *118*(2), 224–235.

Chang, K. S., Alasalvar, C., & Bolling, B. W. (2016). Nuts and their co-products: The impact of processing (roasting) on phenolics, bioavailability, and health benefits—A comprehensive review. *Journal of Functional Foods*, *26*, 88–122.

Chemat, F., Vian, M. A., & Cravotto, G. (2012). Green extraction of natural products: Concept and principles. *International Journal of Molecular Sciences*, *13*(7), 8615–8627.

Chen, X., Zhang, Y., Wang, Z., & Zu, Y. (2011). In vivo antioxidant activity of *Pinus koraiensis* nut oil obtained by optimised supercritical carbon dioxide extraction optimised supercritical carbon dioxide extraction. *Natural Product Research*, *25*, 1807–1816.

Dabbour, I. R., Al-Ismail, K. M., Takruri, H. R., & Azzeh, F. S. (2014). Chemical characteristics and antioxidant content properties of cold pressed seed oil of wild milk thistle plant grown in Jordan. *Pakistan Journal of Nutrition*, *13*(2), 67–78.

Degen, L., Matzinger, D., & Beglinger, C. (2001). The effect of cholecystokinin in controlling appetite and food intake in humans. *Peptides*, *22*, 1265–1269.

Destaillats, F., Cruz-Hernandez, C., Giuffrida, F., & Dionisi, F. (2010). Identification of the botanical origin of pine nuts found in food products by gas-liquid chromatography analysis of fatty acid profile. *Journal of Agricultural and Food Chemistry*, *58*(4), 2082–2087.

Dobson, G., Shrestha, M., Hilz, H., Karjalainen, R., Mcdougall, G., & Stewart, D. (2012). Lipophilic components in black currant seed and pomace extracts. *European Journal of Lipid Science and Technology*, *114*, 575–582.

Eisenmenger, M., & Dunford, N. T. (2008). Bioactive components of commercial and supercritical carbon dioxide processed wheat germ oil. *Journal of the American Oil Chemists' Society*, *85*(1), 55–61.

Emir, D. D., Aydeniz, B., & Yılmaz, E. (2015). Effects of roasting and enzyme pretreatments on yield and quality of cold-pressed poppy seed oils. *Turkish Journal of Agriculture and Forestry*, *39*, 260–271.

Evaristo, I., Batista, D., Correia, I., Correia, P., & Costa, R. (2013). Chemical profiling of portuguese *Pinus pinea* L. nuts and comparative analysis with Pinus koraiensis Sieb. & Zucc. commercial kernels. *Options Mediterraneennes*, *105*, 99–104.

Evaristo, I., Batista, D., Correia, I., Paula, C., & Costa, R. (2010). Chemical profiling of Portuguese *Pinus pinea* L. nuts. *Journal of the Science of Food and Agriculture*, *90*, 1041–1049.

Ferramosca, A., Savy, V., Einerhand, A. W. C., & Zara, V. (2008). *Pinus koraiensis* seed oil (PinnoThin™) supplementation reduces body weight gain and lipid concentration in liver and plasma of mice. *Journal of Animal and Feed Sciences*, *17*, 621–630.

Food and Agriculture Organization (FAO) (1998). In W. M. Ciesla (Ed.), *'Seeds, fruits and cones' in non-wood Forest products from conifers* (pp. 72–87). Rome, Italy: FAO.

Geçgel, Ü., Taşan, M., & Gürpınar, Ç. G. (2011). Soğuk presyon tekniği ile üretilen bitkisel yağların özellikleri ve sağlık üzerine etkileri. In: *Proceedings of the 7th Food Engineering Congress, Ankara* (p. 242).

Ghafoor, K., Özcan, M. M., Babıker, E. E., Sarker, Z. I., Ahmed, I. A. M., & Ahmed, M. A. (2017). Nutritional composition, extraction, and utilization of wheat germ oil: A review. *European Journal of Lipid Science and Technology*, *119*, 1–9.

Gong, Y., & Pegg, R. B. (2015). Tree nut oils: Properties and processing for use in food. In G. Talbot (Ed.), *Specialty oils and fats in food and nutrition: Properties, processing and applications* (pp. 65–86). Cambridge, England: Woodhead Publishing.

Górna, P., & Rudzinska, M. (2016). Seeds recovered from industry by-products of nine fruit species with a high potential utility as a source of unconventional oil for biodiesel and cosmetic and pharmaceutical sectors. *Industrial Crops and Products*, *83*, 329–338.

Gresti, J., Mignerot, C., Bezard, J., & Wolff, R. (1996). Distribution of A5-Olefinic acids in the triacylglycerols from *Pinus koraiensis* and *Pinus pinaster* seed oils. *Journal of the American Oil Chemists' Society*, *73*(9), 1539–1547.

Hassanien, M., Abdel-Razek, A., Siger, A., Rudzinska, M., Ratusz, K., & Przybylski, R. (2014). Phytochemical contents and oxidative stability of oils from non-traditional sources. *European Journal of Lipid Science and Technology*, *116*, 1563–1571.

Herrero, M., Mendiola, J. A., Cifuentes, A., & Ibánez, E. (2010). Supercritical fluid extraction: Recent advances and applications. *Journal of Chromatography A*, *1217*, 2495–2511.

Hughes, G. M., Boyland, E. J., Williams, N. J., Mennen, L., Scott, C., Kirkham, T. C., et al. (2008). The effect of Korean pine nut oil (PinnoThin™) on food intake, feeding behaviour and appetite: A double-blind placebo-controlled trial. *Lipids in Health and Disease*, *7*, 1–10.

Imbs, A. B., Nevshupova, N. V., & Pham, L. Q. (1998). Triacylglycerol composition of *Pinus koraiensis* seed oil. *Journal of the American Oil Chemists' Society*, *75*(7), 865–870.

Imer, Y., & Taşan, M. (2018). Determination of some micro and macronutrient elements in various cold press vegetable oils. *Journal of Tekirdag Agricultural Faculty*, *15*(1), 14–25.

International Nut and Dried Fruit Council (INC) (2017). *Nuts and Dried Fruits Statistical Yearbook*. Available from https://www.nutfruit.org/industry/technical-resources?category=statistical-yearbooks. (1 March 2019).

Jin, T., Wang, Y., Chen, Y., Fu, T., Kothary, M. H., Mchugh, T. H., et al. (2014). Crystal structure of Korean pine (*Pinus koraiensis*) 7S seed storage protein with copper ligands. *Journal of Agricultural and Food Chemistry*, *22*, 222–228.

Jorge, T. D. S., Polachini, T. C., Dias, L. S., Jorge, N., & Telis-romero, J. (2015). Physicochemical and rheological characterization of avocado oils. *Ciência e Agrotecnologia*, *39*, 390–400.

Kaijser, A., Dutta, P., & Savage, G. (2000). Oxidative stability and lipid composition of macadamia nuts grown in New Zealand. *Food Chemistry*, *71*, 67–70.

Kanmaz, E. Ö., & Saral, Ö. (2017). The relationship between antioxidant activities and phenolic compounds in subcritical water extracts from orange peel. *Journal of Food*, *42*, 485–493.

Kornsteiner, M., Wagner, K., & Elmadfa, I. (2006). Tocopherols and total phenolics in 10 different nut types. *Food Chemistry*, *98*, 381–387.

Lebovka, N., Vorobiev, E., & Chemat, F. (2011). *Enhancing extraction processes in the food industry*. Cambridge, UK: CRC Press.

Lee, M., Lee, S., Park, S., Bae, D., Ha, S., & Song, K. B. (2004). Changes in quality of pine nuts (*Pinus koraiensis*) and walnuts (*Juglans regia*) coated with protein film containing green tea extract during storage. *Korean Journal of Food Science and Technology*, *36*(5), 842–846.

Lee, S., & Rhee, C. (2003). Processing suitability of a rice and pine nut (*Pinus koraiensis*) beverage. *Food Hydrocolloids*, *17*, 379–385.

Li, Y., Jiang, L., Sui, X., & Wang, S. (2011). Optimization of the aqueous enzymatic extraction of pine kernel oil by response surface methodology. *Procedia Engineering*, *15*, 4641–4652.

Lin, S., Yang, R., Cheng, S., Wang, K., & Qin, L. (2018). Moisture absorption and dynamic flavor changes in hydrolysed and freeze-dried pine nut (*Pinus koraiensis*) by-products during storage. *Food Research International*, *103*, 243–252.

Ling, B., Yang, X., Li, R., & Wang, S. (2016). Physicochemical properties, volatile compounds, and oxidative stability of cold pressed kernel oils from raw and roasted pistachio (*Pistacia vera* L. Var Kerman). *European Journal of Lipid Science and Technology*, *118*(9), 1368–1379.

Maguire, L. S., Sullivan, S. M. O., Galvin, K., Connor, T. P. O., Brien, N. M. O., Sullivan, S. M., et al. (2004). Fatty acid profile, tocopherol, squalene and phytosterol content of walnuts, almonds, peanuts, hazelnuts and the macadamia nut. *International Journal of Food Science and Nutrition*, *55*(3), 171–178.

Matthäus, B., & Brühl, L. (2003). Cold-pressed edible rapeseed oil production in Germany. *Nahrung/Food*, *47*(6), 413–419.

Matthäus, B., & Spener, F. (2008). What we know and what we should know about virgin oils-a general introduction. *European Journal of Lipid Science and Technology*, *110*, 597–601.

Miraliakbari, H., & Shahidi, F. (2008a). Oxidative stability of tree nut oils. *Journal of Agriculture and Food Chemistry*, *56*(12), 4751–4759.

Miraliakbari, H., & Shahidi, F. (2008b). Antioxidant activity of minor components of tree nut oils. *Food Chemistry*, *111*, 421–427.

Mutke, S., González-martínez, S. C., Calama, R., & Montero, G. (2012). Mediterranean stone pine: Botany and Horticulture. In J. Janick (Ed.), *Horticultural reviews,* vol. 39, (pp. 153–201). USA.

Mutke, S., Pastor, A., & Picardo, A. (2011). Toward a traceability of European pine nuts "from forest to fork". In: *Proceedings of the international meeting on Mediterranean stone pine for agroforestry*, (pp. 17–19). Available from: ResearchGate Digital Library [15 February 2019].

Navarro, S. L. B., & Rodrigues, C. E. C. (2016). Macadamia oil extraction methods and uses for the defatted meal by product. *Trends in Food Science & Technology*, *54*, 148–154.

Nergiz, C., & Dönmez, İ. (2004). Chemical composition and nutritive value of *Pinus pinea* L. seeds. *Food Chemistry*, *86*, 365–368.

No, D. S., & Kim, I. (2013). Pinolenic acid as a new source of phyto-polyunsaturated fatty acid. *Lipid Technology*, *25*(6), 135–138.

NSW Food Authority (2012). *Pine nuts and pine mouth*. Available from http://www.foodauthority.nsw.gov.au/aboutus/science/food-risk-studies/pine-nuts-and-pine-mouth (2 March 2019).

Parker, T. D., Adams, D. A., Zhou, K., Harris, M., & Yu, L. (2003). Fatty acid composition and oxidative stability of cold-pressed edible seed oils. *Food Chemistry and Toxicology*, *68*(4), 1240–1243.

Pasman, W. J., Heimerikx, J., Rubingh, C. M., et al. (2008). The effect of Korean pine nut oil on in vitro CCK release, on appetite sensations and on gut hormones in post-menopausal overweight women. *Lipids in Health and Disease*, *7*, 1–10.

Piao, Z., Tang, L., & Swihart, R. K. (2011). Human-wildlife competition for Korean pine seeds: Vertebrate responses and human-wildlife competition for Korean pine seeds: Vertebrate responses and implications for mixed forests on Changbai Mountain, China. *Annals of Forest Science*, *68*, 911–919.

Prescha, A., Grajzer, M., Dedyk, M., & Grajeta, H. (2014). The antioxidant activity and oxidative stability of cold-pressed oils. *Journal of the American Oil Chemists' Society*, *91*, 1291–1301.

Rabadán, A., Pardo, J. E., Gómez, R., & Álvarez-ortí, M. (2018). Influence of temperature in the extraction of nut oils by means of screw pressing. *Food Science and Technology*, *93*, 354–361.

Rabrenović, B. B., Dimić, E. B., Novaković, M. M., Tešević, V. V., & Basić, Z. N. (2014). The most important bioactive components of cold pressed oil from different pumpkin (*Cucurbita pepo* L.) seeds. *Food Science and Technology*, *55*(2), 521–527.

Rezq, A., & El-Khamisy, E. (2011). Hypolipideimic and hypocholestermic effect of pine nuts in rats fed high fat, cholesterol-diet department of nutrition and food science, faculty of specific education. *World Applied Sciences Journal*, *15*(12), 1667–1677.

Rodríguez-pérez, C., Quirantes-piné, R., & Fernández-gutiérrez, A. (2015). Optimization of extraction method to obtain a phenolic compounds-rich extract from *Moringa oleifera* Lam leaves. *Industrial Crops and Products*, *66*, 246–254.

Ros, E., & Mataix, J. (2006). Fatty acid composition of nuts-implications for cardiovascular health. *British Journal of Nutrition*, *96*, 29–35.

Rotkiewicz, D., Konopka, I., & Żylik, S. (1999). State of works on the rapeseed oil processing optimalization' I. Oil obtaining. *Rosliny Oleiste/Oilseed Crops*, 151–168 (in Polish).

Rovati, L., & Beglinger, C. (2000). The role of long chain fatty acids in regulating food intake and cholecystokinin release in humans. *Gut*, *46*, 688–693.

Ryan, E., Galvin, K., O'Connor, T. P., Maguire, A. R., & O'Brien, N. M. (2006). Fatty acid profile, tocopherol, squalene and phytosterol content of Brazil, pecan, pine, pistachio and cashew nuts. *International Journal of Food Sciences and Nutrition*, *57*(3–4), 219–228.

Salgın, S. (2013). Effect of main process parameters on extraction of pine kernel lipid using supercritical green solvents: Solubility models and lipid profiles. *Journal of Supercritical Fluids*, *73*, 18–27.

Schlörmann, W., Birringer, M., Böhm, V., Löber, K., Jahreis, G., Lorkowski, S., et al. (2015). Influence of roasting conditions on health-related compounds in different nuts. *Food Chemistry*, *180*, 77–85.

Sharashkin, L., & Gold, M. (2004). Pine nuts species, products, markets, and potential for U.S. production. In: *Northern nut growers association 95th Annual report, proceedings for the 95th annual meeting, Columbia, Missouri*, pp. 16–19.

Sloan, A. E. (2000). The top ten functional food trends. *Food Technology*, *54*(4), 33–62.

Som No, D., Zhao, T., Kim, Y., Yoon, M., Lee, J., & Kim, I. (2015). Preparation of highly purified pinolenic acid from pine nut oil using a combination of enzymatic esterification and urea complexation. *Food Chemistry*, *170*, 386–393.

Soquetta, M. B., Terra, L. M., & Bastos, C. P. (2018). Green technologies for the extraction of bioactive compounds in fruits and vegetables. *CYTA—Journal of Food*, *16*(1), 400–412.

Speijers, G. J. A., Dederen, L. H. T., & Keizer, H. (2009). A sub-chronic (13 weeks) oral toxicity study in rats and an in vitro genotoxicity study with Korean pine nut oil (PinnoThin TG TM). *Regulatory Toxicology and Pharmacology*, *55*(2), 158–165.

Su, X., Wang, Z., & Liu, J. (2009). In vitro and in vivo antioxidant activity of *Pinus koraiensis* seed extract containing phenolic compounds. *Food Chemistry*, *117*(4), 681–686.

Sugano, B. M., Wakamatsu, K., Ikeda, I., & Oka, T. (1994). Influence of Korean pine (*Pinus koraiensis*) seed oil containing fatty acid metabolism, eicosanoid production and blood pressure of rats. *British Journal of Nutrition*, *72*, 775–783.

Tiwari, B. K. (2015). Ultrasound: A clean, green extraction technology. *Trends in Analytical Chemistry*, *71*, 100–109.

Velickovska, S., Brühl, L., Miltev, S., Mirhosseini, H., & Matthaus, B. (2015). Quality evaluation of cold-pressed edible oils from Macedonia. *European Journal of Lipid Science and Technology*, *117*(12), 1–13.

Velioglu, Y. S., Mazza, G., Gao, L., & Oomah, B. D. (1998). Antioxidant activity and total phenolics in selected fruits, vegetables, and grain products. *Journal of Agricultural and Food Chemistry*, *46*, 4113–4117.

Venkatachalam, M., & Sathe, S. K. (2006). Chemical composition of selected edible nut seeds. *Journal of Agriculture and Food Chemistry*, *54*, 4705–4714.

Wan, H. B., & Wong, M. K. (1996). Minimization of solvent consumption in pesticide residue analysis. *Journal of Chromatography A*, *754*, 43–47.

Wang, W., Wang, H., Xiao, X., & Xu, X. (2019). Chemical composition analysis of seed oil from five wild almond species in China as potential edible oil resource for the future. *South African Journal of Botany*, *121*, 274–281.

Wang, M., Zhang, L., Wu, X., Zhao, Y., Wu, L., & Lu, B. (2019). Quantitative determination of free and esterified phytosterol profile in nuts and seeds commonly consumed in China by SPE/GC-MS. *Food Science and Technology*, *100*, 355–361.

Wolff, R. L., & Bayard, C. C. (1995). Fatty acid composition of some pine seed oils. *Journal of the American Oil Chemists' Society*, *72*(9), 1043–1046.

Xie, K., Miles, E. A., & Calder, P. C. (2016). A review of the potential health benefits of pine nut oil and its characteristic fatty acid pinolenic acid. *Journal of Functional Foods*, *23*, 464–473.

Yang, J., Liu, R. H., & Halim, L. (2009). Antioxidant and antiproliferative activities of common edible nut seeds. *Food Science and Technology*, *42*(1), 1–8.

Yang, X., Zhang, H., Zhang, Y., Zhao, H., Dong, A., Xu, D., et al. (2010). Analysis of the essential oils of pine cones of *Pinus koraiensis* Steb. Et Zucc. and *P. sylvestris* L. from China. *Journal of Essantial Oil Research*, *22*, 446–448.

Ying, Q., Wojciechowska, P., Siger, A., Kaczmarek, A., & Rudzińska, M. (2018). Phytochemical content, oxidative stability, and nutritional properties of unconventional cold-pressed edible oils. *Journal of Food and Nutrition Research*, *6*(7), 476–485.

Zadernowski, R., Naczk, M., & Czaplicki, S. (2009). Chemical composition of *Pinus sibirica* nut oils. *European Journal of Lipid Science and Technology*, *111*, 698–704.

Zhang, Y., Du, W.-X., & Fan, Y. (2016). Prediction and identification of Korean Pine (*Pinus koraiensis*) vicilin as a food. *Journal of Allergy and Clinical Immunology*, *137*(2), 873.

Zhang, J., Lin, W., Wu, R., Liu, Y., Zhu, K., Ren, J., et al. (2018). Mechanisms of the active components from Korean pine nut preventing and treating D-galactose-induced aging rats. *Biomedicine & Pharmacotherapy*, *103*, 680–690.

Zhang, H., Zhao, H., Yao, L., Yang, X., Shen, S., Wang, J., et al. (2017). Isolation, physicochemical properties, and in vitro antioxidant activity of polysaccharides extracted from different parts of *Pinus koraiensis*. *Journal of Wood Chemistry and Technology*, *37*(3), 225–240.

Chapter 48

Cold pressed juniper (*Juniperus communis* L.) oil

Ali Osman
Agricultural Biochemistry Department, Faculty of Agriculture, Zagazig University, Zagazig, Egypt

1 Introduction

Oils constitute a significant origin of bio-components in food such as saturated and unsaturated fatty acids, phospholipids, carotenoids, tocochromanols, phytosterols, squalene, lignans, and phenolic compounds, which promote the health and proper functioning of a organism (Chiou, Kalogeropoulos, Boskou, & Salta, 2012). Cold pressed oils are gaining increasing popularity among consumers due to their higher nutritive value than refined oils. Moreover, they are produced using an ecologically friendly extraction method (Siger, Nogala-Kalucka, & Lampart-Szczapa, 2008). The refining process has an adverse effect on the antioxidant activity of oils, decreasing it by 80%, which is associated with the reduction of carotenoid concentration by 99% and that of polyphenols by 26%–55% (Szydłowska-Czerniak, Trokowski, Karlovits, & Szłyk, 2011). Recently, an increasing number of research projects has focused on finding new sources of oil from unconventional plants. One such promising raw material is juniperus oil. *Juniperus communis* L. (family Cupressaceae), commonly known as "Aaraar" or "haubera" or "abhal" in Hindi, is an evergreen shrub or a small tree (1–3 m high) found in the Himalayas from Kumaon westwards at altitudes of 5000–14,000 ft. in India. *Juniperus communis* is grown wild in many parts of Europe, Asia, North America, and Africa in a temperate or cold climate. Juniper needles are rich in ascorbic acid, and they contain resins, waxes, and esters. The essential oil is shown to possess antimalarial and fungi toxic activity (Milhau et al., 1997).

Juniperus oil is an important natural product known as a source for compounds to be used in fragrance blending. Juniper berries are obtained as fruits used in addition to their essential oil, in liquor, and in the food industry as a spice. Generally, the oil has been used for centuries for its diuretic and antiseptic properties as a rubefacient or in spiritus and especially in alcoholic drinks. The oil also finds use in aromatherapy (Lamparsky & Klimes, 1985). Monoterpenes contribute up to 60% of the juniper oil content. That fraction is probably responsible for the oil's biological properties. The main monoterpenes and their enantiomeric forms have been analyzed in numerous studies (Ochocka et al., 1997; Sybilska et al., 1994; Vernin et al., 1988).

2 Chemical composition

The oil profile of the juniper varies among species, as described in Table 1. The major components of the juniper oil are α-pinene (23%–60%), β-pinene (5.6%), limonene (6.52%), β-phellandrene (2.13%), *p*-cymene (9.96%), bornyl acetate (3.21%), and nerol (2.21%), while myrcene is present at a low concentration (Filipowicz et al., 2003).

A total of 36 components were identified (Angioni et al., 2003). The main constituents in the *J. communis* essential oil were limonene, δ-3-carene, α-pinene, β-pinene, sabinene, myrcene, β-phellandrene, and D-germacrene. Semerdjieva et al. (2019) reported on the essential oil (EO) yield, composition, bioactivity, and leaf morphology of *Juniperus oxycedrus* from Bulgaria and Serbia. The EO content in dried juniper leaves varied from 0.06% (Kopaonik, Serbia) to 0.24% (Markovo, Bulgaria). Fifty-one EO constituents, belonging to monoterpenes, sesquiterpenes, and diterpenes, were detected. Monoterpenes (monoterpene hydrocarbons, and oxygenated monoterpenes) were the predominant compounds, representing 38.6%–65.4% of the total EO, consisting primarily of α-pinene, limonene, sabinene, β-pinene, and β-myrcene. In addition, α-pinene was the major oil constituent in plants from all locations. Sesquiterpenes (sesquiterpene hydrocarbons, and oxygenated sesquiterpenes) were the second-highest class of constituents, which represented 19.3%–33.6% of the total EO. γ-elemene was found only in the EO of *J. oxycedrus* from Bulgaria, while a high concentration of α-curcumene was found only in samples from Serbia (7.5%–7.8%).

Cold Pressed Oils. https://doi.org/10.1016/B978-0-12-818188-1.00048-7
Copyright © 2020 Elsevier Inc. All rights reserved.

TABLE 1 Contents of the main components in juniper essential oil.

Compound	Juniper specie	Reference
α-Pinene	*J. oxycedrus* L. (red juniper) *J. communis* L. (common juniper)	Zheljazkov et al. (2018)
β-Pinene	*J. communis* L. (common juniper)	Filipowicz, Kamiński, Kurlenda, Asztemborska, and Ochocka (2003)
Limonene	*J. communis* L. (common juniper)	Filipowicz et al. (2003)
β-Phellandrene	*J. communis* L. (common juniper)	Filipowicz et al. (2003)
p-Cymene	*J. communis* L. (common juniper)	Filipowicz et al. (2003)
Bornyl acetate	*J. communis* L. (common juniper)	Filipowicz et al. (2003)
Nerol	*J. communis* L. (common juniper)	Filipowicz et al. (2003)
δ-3-Carene	*J. communis* L. (common juniper)	Angioni et al. (2003)
β-Myrcene	*J. oxycedrus* L. (red juniper) *J. communis* L. (common juniper)	Zheljazkov et al. (2018)
γ-Elemene	*J. oxycedrus* (Markovo, Bulgaria)	Semerdjieva et al. (2019)
α-Curcumene	*J. oxycedrus* (Kopaonik, Serbia)	Semerdjieva et al. (2019)
Germacrene D	*J. oxycedrus* L. (red juniper) *J. communis* L. (common juniper)	Zheljazkov et al. (2018)
α-Cedrol	*J. excelsa* M. Bieb. (Gracian juniper)	Zheljazkov et al. (2018)
Sabinene	*J. pygmaea* C. Koch. (Alpine juniper)	Zheljazkov et al. (2018))
Terpinene-4-ol	*J. sabina* L. (Cossack juniper)	Zheljazkov et al. (2018)

Zheljazkov et al. (2018) evaluated the EO composition and antimicrobial and antioxidant activity of the galbuli of the six juniper species naturally distributed in Bulgaria including *Juniperus communis* L. (common juniper), *J. oxycedrus* L. (red juniper), *J. sibirica* Burgsd. (Siberian juniper), *J. sabina* L. (Cossack juniper), *J. pygmaea* C. Koch. (Alpine juniper), and *J. excelsa* M. Bieb. (Gracian juniper). The EO content of the galbuli of the six juniper species varied from 0.47% (in *J. sibirica*) to 1.6% (in *J. sabina*). The oil profile of the galbuli was also different among species. Differences and similarities in the groups of terpenes were established between the six juniper species. The oil constituents with the highest concentration (in descending order) in the galbuli of each of the species were as follows: for *J. oxycedrus*: β-myrcene, α-pinene, germacrene D; for *J. communis*: α-pinene, germacrene D, β-myrcene; for *J. excelsa*: α-pinene, α-cedrol, germacrene D; for *J. sibirica*: α-pinene, β-myrcene, germacrene D; for *J. pygmaea*: α-pinene, sabinene, β-myrcene; and for *J. sabina*: sabinene, α-pinene, terpinene-4-ol.

3 Biological activities

Juniper oil is stated to possess a wide range of pharmacological activities. It has diuretic, antiseptic, carminative, stomachic, and antirheumatic properties, and has been traditionally used for cystitis, flatulence, and colic. Moreover, it has been applied topically for rheumatic pains in joints or muscles. Juniper berries (*Bacchae juniper* and *Juniperi fructus*) are the source of the oils and the subjects of pharmacopeia monographs.

3.1 Antimicrobial activity

Antibacterial and antifungal properties of oil as well as of oil constituents are well documented (Knobloch, Pauli, Iberl, Weigand, & Weis, 1989; Pepeljnjak, Kosalec, Kalodera, & Blazevic, 2005). Volatile oils and some of their constituents have found applications as antimicrobial agents for food preservatives, in clinical microbiology, or in pharmaceutical preparations. Screening for antimicrobial activity has been the subject of many investigations, and oils with very potent antibacterial and antifungal activity could be promising agents for more extensive research and in vivo examination. Among oils are the essential oils from juniper berries (*Juniperus communis* L., Cupressaceae). The plant is widespread in Croatia and it grows in temperate regions of Europe, Asia, and North America. Juniper berries (female cones) are used commercially for the preparation of essential oil, gin, and as a spice (Pepeljnjak et al., 2005).

Juniper volatile oil was tested for the antimicrobial potential against different bacterial species, yeast-like fungi, yeast, and dermatophyte strains. Juniper volatile oil exhibited similar bactericidal activities against Gram-negative and Gram-positive bacteria, with MIC values between 8% and 70% (v/v). Moreover, juniper volatile oil showed a strong fungicidal activity against yeast-like fungi, yeasts, and dermatophytes, with MIC values below 10%. The strongest fungicidal action was found against *Candida* spp. (MIC 0.78%–2.0%, v/v) and dermatophytes (MIC 0.39%–2.0%, v/v).

Filipowicz et al. (2003) recorded that juniper berry oil possesses a wide spectrum of pharmacological activities, and its monographs are included in some national pharmacopeias. The antibacterial and antifungal properties of the oil were reported. The antibacterial and antifungal activity of three different juniper berry oils and their main components were studied. Only one of the oils (labeled A) revealed good antimicrobial properties. None of the single oil components was a stronger antibacterial and antifungal inhibitor than oil A itself. The antimicrobial activity of juniper oil A could be the result of either the specific composition of the oil A (highest concentration of (−)-α-pinene, *p*-cymene, and β-pinene) or the activity of a single nonidentified compound. The presence of an adulterant in the oil was excluded.

Tserennadmid et al. (2011) studied the antiyeast potential of essential oils (EOs) from juniper, clary sage, lemon, and marjoram against isolates from food-related yeasts including *Pichia anomala*, *Geotrichum candidum*, *Saccharomyces cerevisiae*, and *Schizosaccharomyces pombe* in malt extract medium, apple juice, and milk. MIC for the EOs and their major components were recorded, whereas the checkerboard test was used to calculate the fractional inhibitory concentration (FIC) for the combinations of EOs or components. The most sensitive yeast was *S. pombe* (MIC 0.062–0.125 μL/mL), while *G. candidum* was the most insensitive (MIC 0.5–2.0 μL/mL). The lag phases were lengthened by increasing EO levels, while significant reduction of growth rates was recorded at the highest concentration. The antiyeast impact of the EOs was good in the acidic pH, which is optimal for yeast growth. Blends of juniper and clary sage EOs caused additive impacts in the case of *G. candidum* and *S. cerevisiae*. The combination of limonene and α-pinene resulted in synergism, while the combination of α-pinene with linalool caused an additive impact.

3.2 Antioxidants activity

Juniper berry essential oils (JBEOs) have been explored and established for their in vitro antioxidant and antiradical activities, which are mostly dependent on the oil components and on concentrations (Pandey et al., 2018). Reactive oxygen species (ROS) such as H_2O_2, $^{\bullet}O_2^-$, and $OH^{\bullet}$ are produced in organisms during cellular metabolism. At lower concentrations, they participate in cellular physiological reactions (Schopfer, Plachy, & Frahry, 2001). Their overproduction, however, largely determines cell survival. ROS inactivation and removal depends on nonenzymatic and enzymatic protective mechanisms. Research on ROS-induced damage has shown that antioxidant production is genetically controlled in the cells (Kim et al., 2010). The focus on antioxidants naturally contained in essential oils is related to their applications aimed at the prevention of oxidative damage to biological systems by ROS. Low-molecular antioxidants can enhance organism stability under oxidative stress (Martorell et al., 2011). For centuries, juniper berries have been used in folk medicine for the treatment of opportunistic infections, as a spice for meat, and as flavor in the preparation of gin and raki (Höferl et al., 2014). The antioxidant activity of essential oils from different juniper berry species has been established in vitro (Emami, Javadi, & Hassanzadeh, 2007). Antiradical activity depends on the oil components, i.e., their chemical nature and concentration (Wei & Shibamoto, 2007). Regardless of the differences in the composition of juniper berry essential oils, they are dominated by terpene hydrocarbons. In many cases, the essential oil antioxidant activity cannot be attributed to the dominant compounds α- and β-pinene. These monoterpene hydrocarbons in juniper berry essential oil do not contribute to significant inhibition of malondialdehyde formation (Wei & Shibamoto, 2007). The carriers of antioxidant properties in relation to lipid peroxidation in both its stages are α- and γ-terpinenes and, to a lesser extent, their sesquiterpene analogs. This has been established both for juniper essential oils (Misharina & Samusenko, 2008), and for pure terpene hydrocarbons including terpinolene, α-terpinene, and γ-terpinene (Ruberto & Baratta, 2000). Myrcene and α- and β-pinene only inhibit lipid peroxidation in the second stage, while sabinene, limonene, α-pinene, and myrcene demonstrated antiradical activity in relation to DPPH· radicals (Bua-In & Paisooksantivatana, 2009). The scavenging effect of $OH^{\bullet}$ and the protection of deoxyribose against degradation are mainly due to β-pinene and limonene, while the $^{\bullet}O_2^-$ neutralization is determined by germacrene D (Karioti, Hadjipavlou-Litina, Mensah, Fleischer, & Skaltsa, 2004). The 10-membered ring system and the three double bonds acting as electron-rich centers in germacrene D determine its antiradical activity.

References

Angioni, A., Barra, A., Russo, M. T., Coroneo, V., Dessí, S., & Cabras, P. (2003). Chemical composition of the essential oils of *Juniperus* from ripe and unripe berries and leaves and their antimicrobial activity. *Journal of Agricultural and Food Chemistry*, *51*, 3073–3078.

Bua-In, S., & Paisooksantivatana, Y. (2009). Essential oil and antioxidant activity of Cassumunar ginger (Zingiberaceae: *Zingiber montanum* (Koenig) link ex Dietr.) collected from various parts of Thailand. *Kasetsart Journal (Natural Science)*, *43*, 467–475.

Chiou, A., Kalogeropoulos, N., Boskou, G., & Salta, F. N. (2012). Migration of health promoting microconstituents from frying vegetable oils to French fries. *Food Chemistry*, *133*, 1255–1263.

Emami, S., Javadi, B., & Hassanzadeh, M. (2007). Antioxidant activity of the essential oils of different parts of *Juniperus communis*. subsp. hemisphaerica. and *Juniperus oblonga*. *Pharmaceutical Biology*, *45*, 769–776.

Filipowicz, N., Kamiński, M., Kurlenda, J., Asztemborska, M., & Ochocka, J. R. (2003). Antibacterial and antifungal activity of juniper berry oil and its selected components. *Phytotherapy Research*, *17*, 227–231.

Höferl, M., Stoilova, I., Schmidt, E., Wanner, J., Jirovetz, L., Trifonova, D., et al. (2014). Chemical composition and antioxidant properties of Juniper berry (*Juniperus communis* L.) essential oil. Action of the essential oil on the antioxidant protection of *Saccharomyces cerevisiae* model organism. *Antioxidants*, *3*, 81–98.

Karioti, A., Hadjipavlou-Litina, D., Mensah, M. L., Fleischer, T. C., & Skaltsa, H. (2004). Composition and antioxidant activity of the essential oils of *Xylopia aethiopica* (Dun) A. Rich (Annonaceae) leaves, stem bark, root bark, and fresh and dried fruits, growing in Ghana. *Journal of Agricultural and Food Chemistry*, *52*, 8094–8098.

Kim, Y., Kim, H., Kwon, Y., Kang, Y., Lee, I., Jin, B., et al. (2010). Modulation of MnSOD protein in response to different experimental stimulation in *Hyphantria cunea*. *Comparative Biochemistry and Physiology Part B: Biochemistry and Molecular Biology*, *157*, 343–350.

Knobloch, K., Pauli, A., Iberl, B., Weigand, H., & Weis, N. (1989). Antibacterial and antifungal properties of essential oil components. *Journal of Essential Oil Research*, *1*, 119–128.

Lamparsky, D., & Klimes, I. (1985). Neue Ergebnisse der Wacholderbeeröl-Analyse im Hinblick auf terpenoide Inhaltsstoffe. *Parfümerie und Kosmetik*, *66*, 553–560.

Martorell, P., Forment, J. V., De Llanos, R., Montón, F., Llopis, S., González, N., et al. (2011). Use of *Saccharomyces cerevisiae* and *Caenorhabditis elegans* as model organisms to study the effect of cocoa polyphenols in the resistance to oxidative stress. *Journal of Agricultural and Food Chemistry*, *59*, 2077–2085.

Milhau, G., Valentin, A., Benoit, F., Mallié, M., Bastide, J.-M., Pélissier, Y., et al. (1997). In vitro antimalarial activity of eight essential oils. *Journal of Essential Oil Research*, *9*, 329–333.

Misharina, T., & Samusenko, A. (2008). Antioxidant properties of essential oils from lemon, grapefruit, coriander, clove, and their mixtures. *Applied Biochemistry and Microbiology*, *44*, 438–442.

Ochocka, J. R., Asztemborska, M., Zook, D. R., Sybilska, D., Perez, G., & Ossicini, L. (1997). Enantiomers of monoterpenic hydrocarbons in essential oils from *Juniperus communis*. *Phytochemistry*, *44*, 869–873.

Pandey, S., Tiwari, S., Kumar, A., Niranjan, A., Chand, J., Lehri, A., et al. (2018). Antioxidant and anti-aging potential of Juniper berry (*Juniperus communis* L.) essential oil in *Caenorhabditis elegans* model system. *Industrial Crops and Products*, *120*, 113–122.

Pepeljnjak, S., Kosalec, I., Kalodera, Z., & Blazevic, N. (2005). Antimicrobial activity of juniper berry essential oil (*Juniperus communis* L., Cupressaceae). *Acta Pharmaceutica (Zagreb, Croatia)*, *55*, 417.

Ruberto, G., & Baratta, M. T. (2000). Antioxidant activity of selected essential oil components in two lipid model systems. *Food Chemistry*, *69*, 167–174.

Schopfer, P., Plachy, C., & Frahry, G. (2001). Release of reactive oxygen intermediates (superoxide radicals, hydrogen peroxide, and hydroxyl radicals) and peroxidase in germinating radish seeds controlled by light, gibberellin, and abscisic acid. *Plant Physiology*, *125*, 1591–1602.

Semerdjieva, I., Zheljazkov, V. D., Radoukova, T., Radanović, D., Marković, T., Dincheva, I., et al. (2019). Essential oil yield, composition, bioactivity and leaf morphology of *Juniperus oxycedrus* L. from Bulgaria and Serbia. *Biochemical Systematics and Ecology*, *84*, 55–63.

Siger, A., Nogala-Kalucka, M., & Lampart-Szczapa, E. (2008). The content and antioxidant activity of phenolic compounds in cold-pressed plant oils. *Journal of Food Lipids*, *15*, 137–149.

Sybilska, D., Asztemborska, M., Kowalczyk, J., Ochocka, R., Ossicini, L., & Perez, G. (1994). Enantiomeric composition of terpenic hydrocarbons in essential oils from *Juniperus communis* L. *Journal of Chromatography A*, *659*, 389–394.

Szydłowska-Czerniak, A., Trokowski, K., Karlovits, G., & Szłyk, E. (2011). Effect of refining processes on antioxidant capacity, total contents of phenolics and carotenoids in palm oils. *Food Chemistry*, *129*, 1187–1192.

Tserennadmid, R., Takó, M., Galgóczy, L., Papp, T., Pesti, M., Vágvölgyi, C., et al. (2011). Anti yeast activities of some essential oils in growth medium, fruit juices and milk. *International Journal of Food Microbiology*, *144*, 480–486.

Vernin, G., Boniface, C., Metzger, J., Ghiglione, C., Hammoud, A., Suon, K.-N., et al. (1988). GC-MS-SPECMA bank analysis of *Juniperus communis* needles and berries. *Phytochemistry*, *27*, 1061–1064.

Wei, A., & Shibamoto, T. (2007). Antioxidant activities and volatile constituents of various essential oils. *Journal of Agricultural and Food Chemistry*, *55*, 1737–1742.

Zheljazkov, V. D., Kacaniova, M., Dincheva, I., Radoukova, T., Semerdjieva, I. B., Astatkie, T., et al. (2018). Essential oil composition, antioxidant and antimicrobial activity of the galbuli of six juniper species. *Industrial Crops and Products*, *124*, 449–458.

Chapter 49

Cold pressed Japanese quince (*Chaenomeles japonica* (Thunb.) Lindl. ex Spach) seed oil

Ali Osman
Agricultural Biochemistry Department, Faculty of Agriculture, Zagazig University, Zagazig, Egypt

Abbreviations

CAEs caffeic acid equivalents
DHA docosahexaenoic acid
EPA eicosapentaenoic acid
PUFA polyunsaturated fatty acid

1 Introduction

Japanese quince (*Chaenomeles japonica* (Thunb.) Lindl. ex Spach) is an ornamental plant belonging to the Rosaceae family. The Chaenomeles species are native to Asian countries and regions including China, Japan, and Tibet. *C. japonica* is best adapted to the climate of northern Europe (Rumpunen, 2003). Japanese quince (*C. japonica*) fruits are grown mainly in Baltic countries such as Lithuania, Latvia, and Estonia as well as Finland, Poland, and Sweden. Due to their specific physico-chemical and sensory traits, *C. japonica* fruits are processed to produce candied fruits (Seglina, Krasnova, Heidemane, & Ruisa, 2009), juices (Ros et al., 2004), or jams (Wojdyło, Oszmiański, & Bober, 2008). Dried seeds contain about 6%–16% oil. Unsaturated fatty acids (c. 89%), mainly linoleic acid (c. 53%) and oleic acid (c. 44%), are the main fatty acids in Japanese quince seed oil (Rumpunen, 2003), as shown in Fig. 1.

Cold pressing extraction involves neither thermal nor chemical treatments, and it is becoming an interesting substitute for traditional extraction methods because of consumers' desire for safe and natural foodstuffs. The consumption of improved and new products such as cold pressed oils might improve human health and prevent some diseases (Goldberg, 2008). Cold pressed berries and hemp seed oils contain high amounts of α-linolenic acid (18:3n-3), which may be converted to the long-chain n-3 polyunsaturated fatty acid (PUFA), docosahexaenoic acid (DHA, 22:6n-3), and eicosapentaenoic acid (EPA, 20:5n-3) in vivo through elongation and desaturation reaction (Senanayake & Shahidi, 2000). Vegetable and edible oils contain small amounts of bioactive compounds including phenolics, tocols, sterols, phospholipids, stanols, free fatty acids, waxes, squalene, and other hydrocarbons (Lercker & Rodriguez-Estrada, 2000). In several plants, phytochemicals with antioxidative traits have been identified. Such compounds are also present in oilseeds (Kalt, Forney, Martin, & Prior, 1999). Phenolic compounds have a great influence on the sensory, stability, and nutritional properties of the product, and might prevent deterioration through deactivation free radical reactions that are responsible for lipid peroxidation (Koski, Pekkarinen, Hopia, Wähälä, & Heinonen, 2003). Cold pressed oils contain amounts of phenolic compounds, which might have fruitful applications in the promotion of human health and prevention of oxidative damage caused by free radicals. Factors affecting the antioxidant potential of phenolic compounds include the number and position of hydroxyl groups, solubility, polarity, and stability of compounds during processing. The concentration phenolic compounds in cold pressed oil could reach up to 99 ppm as caffeic acid equivalents (CAEs) (Koski et al., 2003).

Cold Pressed Oils. https://doi.org/10.1016/B978-0-12-818188-1.00049-9
Copyright © 2020 Elsevier Inc. All rights reserved.

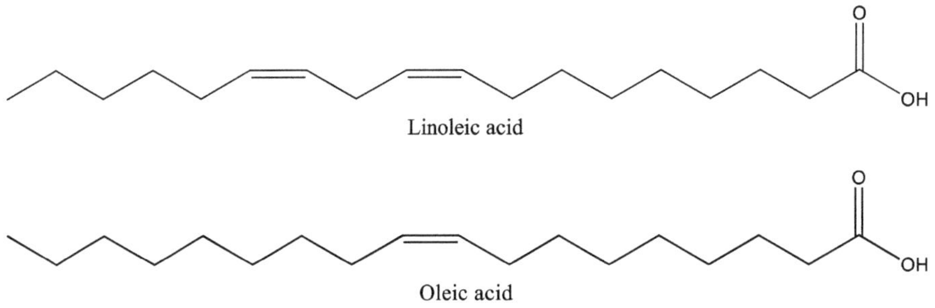

FIG. 1 The major fatty acids in Japanese quince seed oil.

Linoleic acid

Oleic acid

2 Chemical composition of Japanese quince seed oil

Cold pressed oils are an important source of bioactive components including essential unsaturated fatty acids, tocochromanols, phospholipids, carotenoids, sterols, lignans, squalene, and phenolic compounds. The main components in Japanese quince seed cold pressed oil are presented in Table 1.

Japanese quince cold pressed oil recovered from seeds *C. japonica* was characterized and compared with common oils (Górnaś et al., 2014). *C. japonica* seed oil had high amounts of tocopherols (Fig. 2), β-carotene (Fig. 3), and total phenolics (726, 10.7, and 64.0 mg/kg, respectively) and low levels of chlorophyll (0.12 mg/kg) and peroxides (0.59 mEq O_2/kg) compared to sunflower, poppy, sesame, peanut, pumpkin, flaxseed, hazelnut, almond, and walnut oils. A correlation was reported between the levels of tocochromanols, β-carotene, phenolic compounds, and the antiradical action of the oils (0.94, 0.68, and 0.63, respectively). A correlation was also found between the level of chlorophyll and the CIE a^* coordinate (0.80) and the amount of β-carotene and CIE b^* coordinate (0.47). In *C. japonica* seed oil, several fatty acids were detected, wherein linoleic acid (c. 52%), oleic acid (c. 34%), and palmitic acid (c. 10%) were predominating acids. Amygdalin was not detected in Japanese quince seed oil.

TABLE 1 Contents of the main components in Japanese quince seed col-pressed oil.

Compound	Major components	Reference
Fatty acids		Górnaś et al. (2014)
	Linoleic acid	
	Oleic acid	
	Palmitic acid	
Tocopherols		Górnaś et al. (2014)
	α-Tocopherol	
	β-Tocopherol	
	γ-Tocopherol	
β-Carotene		Górnaś et al. (2014)
Phytosterol		Chiou, Kalogeropoulos, Boskou, and Salta (2012)
	β-Sitosterol	
Phenolic compounds		Górnaś, Siger, and Segliņa (2013)
	4-Hydroxybenzoic acid	
	Vanillic acid	
	Vanillin	
	p-Coumaric acid	
	Ferulic acid	
	trans-Cinnamic acid	

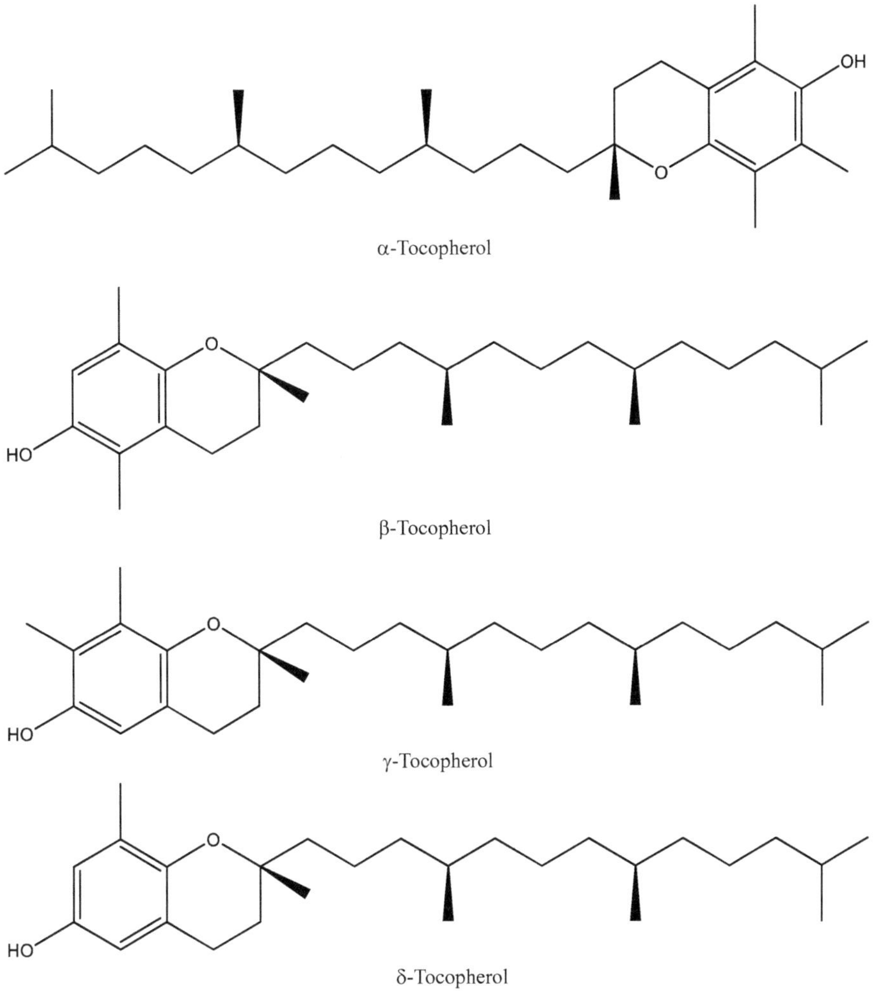

OH

O

α-Tocopherol

O

HO

β-Tocopherol

O

HO

γ-Tocopherol

O

HO

δ-Tocopherol

FIG. 2 The forms of tocopherols found in Japanese quince seed oil.

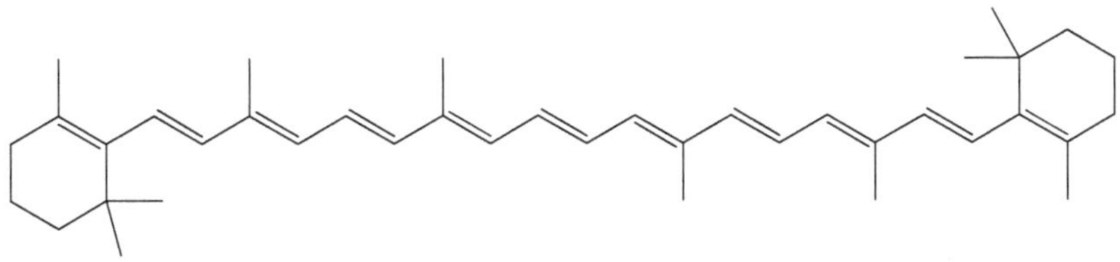

FIG. 3 β-Carotene.

Japanese quince seeds contain up to 20% oil. The chemical composition of *C. japonica* oil is unique because about 90% of fatty acids is formed from two acids, namely linoleic (c. 52%) and oleic (c. 35%) (Mierina et al., 2011). *C. japonica* extracts might be used to improve the oxidative stability of edible oils, wherein the addition of ground seeds (10%) to hemp seed oil and the addition of 5% ground seeds to rapeseed oil could increase the oxidative stability of those oils 2.0 and 1.6 times, respectively. It is worth mentioning that *C. japonica* seeds contain also toxic cianogenic glycoside (amygdalin, Fig. 4). The usage and applications of *C. japonica* seeds are very limited, especially in the case of their hydrophilic extracts due to the presence of amygdalin. The amounts of amygdalin in the ethanolic extracts of *C. japonica* seeds were reported by Mierina et al. (2011).

Górnaś et al. (2013) determined the physicochemical properties of cold pressed *C. japonica* seed oil including: unsaponifiable matter (3.15%), iodine value (96.6 g/100 g), saponification value (196.3 mg KOH/g), acid value (0.88 mg KOH/g), *p*-anisidine value (1.86), peroxide value (0.6 mEq O_2/kg), Totox value (3.06), Rancimat test (7.35 h), carotenoid (10.6 mg/kg), and chlorophyll content (0.11 mg/kg). Linoleic acid ($C_{18:2}$, 52.3%), oleic acid ($C_{18:1}$, 33.8%), and palmitic acid ($C_{16:0}$, 9.46%) are the main fatty acids in *C. japonica* seed oil. The major sterol in *C. japonica* seed oil was β-sitosterol (c. 83%) with the rest being stigmasterol, campesterol, and avenasterol (Fig. 5). The following tocochromanols were identified in *C. japonica*

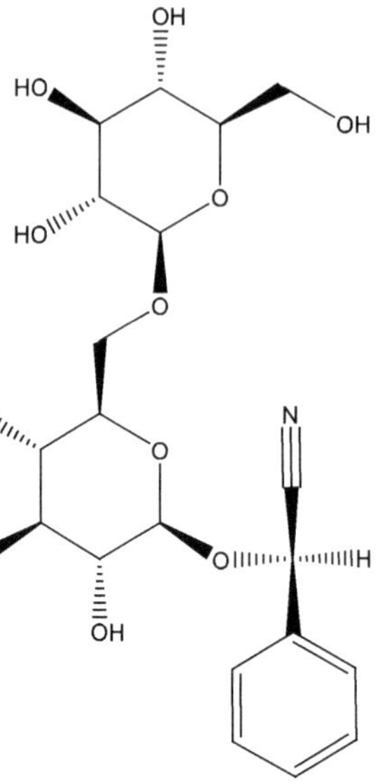

FIG. 4 The formula of amygdalin.

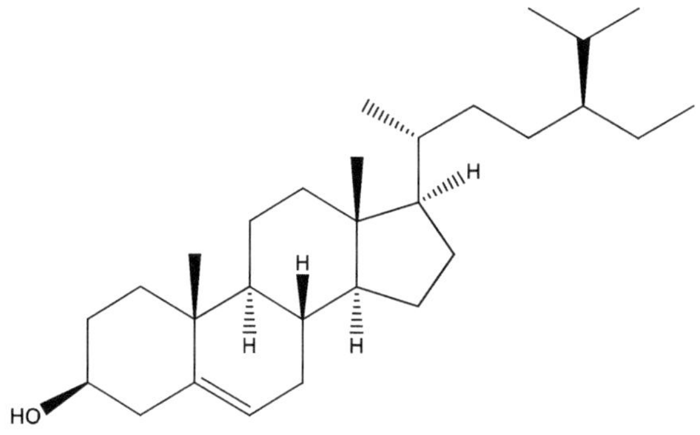

β-Sitosterol

FIG. 5 The major phytosterol in Japanese quince seed oil.

seed oil: α-tocopherol was a main tocochromanol (93.7%), followed by β-tocopherol, γ-tocopherol, and plastochromanol-8. In addition, a large amount of squalene (0.67 mg/g) was detected. Six phenolic compounds (Fig. 6) were detected including *p*-coumaric acid, vanillic acid, 4-hydroxybenzoic acid, vanillin, ferulic acid, and *trans*-cinnamic acid.

Urbanavičiūtė, Rubinskiene, and Viškelis (2019) mentioned that *C. japonica* seeds are discarded as wastes from the fruit processing. The use of *C. japonica* seeds creates added value and opens up the possibility of applying no-waste processing technology. They studied three types of waste: after juicing, after the processing of puree, and syrup. The yield of *C. japonica* seeds left after different processing methods varies from 29.8% to 38.3%. The yield of cold pressed oil ranges from 4.9% to 7.1%, wherein the oil yield obtained from solvent (Soxhlet) extraction varies from 14.6% to 17.3%. Unsaturated fatty acid, especially PUFA, was predominant in *C. japonica* seed oil. The linoleic acid content in *C. japonica* oil was between 47% and 58%. It was concluded that the fatty acid profile of oils from industrial by-products is appropriate in the skincare industry because of the high linoleic acid levels and high omega-6/omega-3 ratio.

Antoniewska, Rutkowska, and Adamska (2017) reported that *C. japonica* fruits contain a considerable amount of phenolic compounds including flavan-3-ols (i.e., epicatechin, catechin, and procyanidin oligomers). The major phenolic compounds in *C. japonica* fruit are the proanthocyanidins (Fig. 7), which showed potential anticancer traits. In addition, the fruit contains high amount of dietary fiber, organic acids, pectins, and vitamin C. *C. japonica* belongs to the group of fruits with a low content of monosaccharides, wherein the proportion of glucose and fructose is good. Owing to its low pH (2.4–2.9) value, the fruits are not suitable for direct consumption. *C. japonica* fruits can be used as a natural acidifying agent. Moreover, the unique sensory quality of *C. japonica* fruit mean that it has a wide scope of food applications, for example, in producing jams, purees, juices, and candied fruits. It is also added to yogurts, lemonades, teas, ice cream, cottage cheese, and confectionery products in order to enhance sensory traits thereof. Fruits are applied as food preserves since they give an attractive original flavor and taste. The seeds of *C. japonica* fruit are a valuable waste material, and could

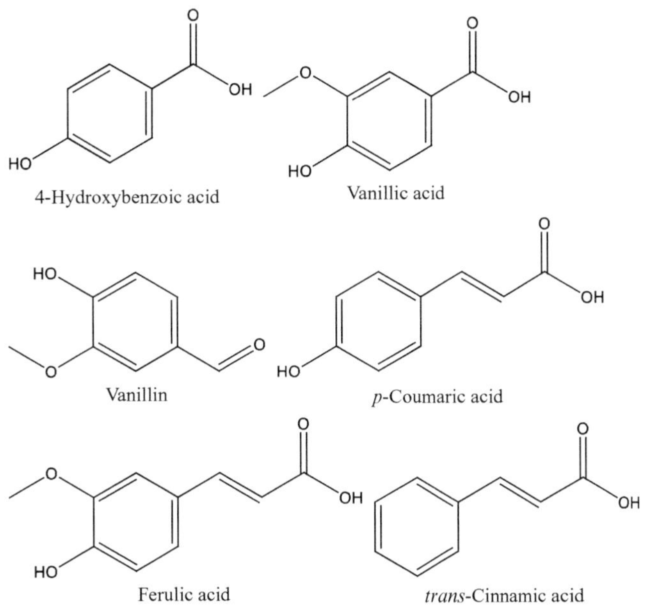

4-Hydroxybenzoic acid

Vanillic acid

Vanillin

p-Coumaric acid

Ferulic acid

trans-Cinnamic acid

FIG. 6 The main phenolic compounds in Japanese quince seed oil.

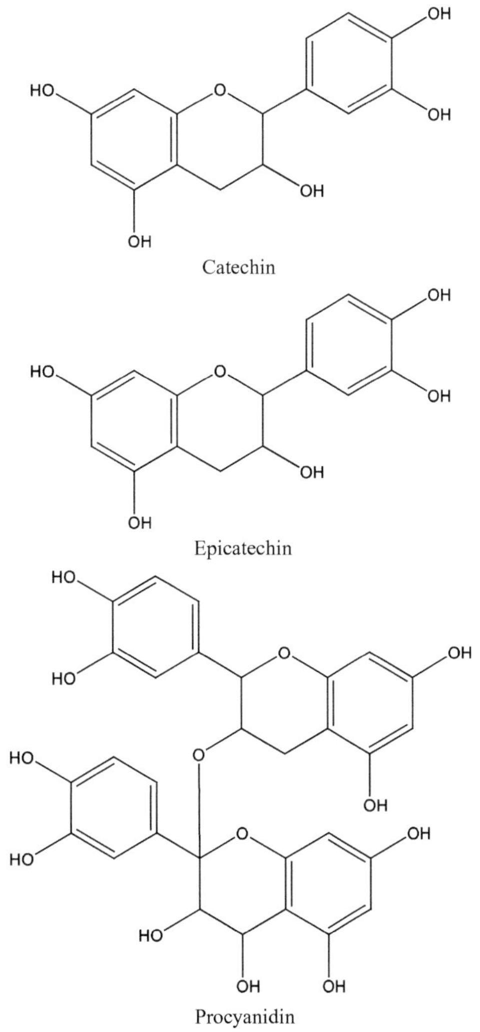

Catechin

Epicatechin

Procyanidin

FIG. 7 The major phenolic compounds in the fruit of Japanese quince.

be utilized to produce a cold pressed oil. *C. japonica* seed oil is characterized by a low content of primary and secondary oxidation products, and its composition is comparable to the commonly used edible oils.

3 Antioxidants activity of Japanese quince seed cold pressed oil

Vegetable oils are an important source of bioactive components including saturated and unsaturated fatty acids, tocochromanols, phospholipids, carotenoids, sterols, lignans, squalene, and phenolic compounds, which promote human health (Chiou et al., 2012). Thus, cold pressed oils are gaining high popularity among consumers because of their better nutritive value than refined oils. In addition, cold pressed oils are produced using an ecologically friendly extraction technique (Siger, Nogala-Kalucka, & Lampart-Szczapa, 2008). Oil refining has an adverse influence on the antioxidant potential of oils, decreasing it by c. 80%, which is linked with the reduction of carotenoid by c. 99% and that of phenolics by c. 26%–55% (Szydłowska-Czerniak, Trokowski, Karlovits, & Szłyk, 2011).

Górnaś et al. (2014) mentioned that *C. japonica* seed oil is characterized by high content of β-carotene, tocopherols, and phenolic compounds, and thereby *C. japonica* seed oil exhibited antioxidant potential in DPPH$^{\bullet}$ assay. Sensory analysis mentioned that 10 tested oils (almond, poppy, sesame, sunflower, pumpkin, hazelnut, peanut, flaxseed, walnut, and Japanese quince) have similar consumer acceptance. The key to attract consumers might be high nutritional value, which meets the properties of *C. japonica* seed oil. *C. japonica* seed oil is a unique, rich, and a promising source of bioactive compounds, which could be successfully applied in several food products. In addition, it is recovered as a by-product of fruit processing; therefore, the use of quince seeds in obtaining healthy products ensures environmental sustainability and more effective management of harvested plant material.

References

Antoniewska, A., Rutkowska, J., & Adamska, A. (2017). Charakterystyka owoców pigwowca japońskiego oraz ich zastosowanie w przemyśle spożywczym. *Żywność: nauka-technologia-jakość*, 5–15.

Chiou, A., Kalogeropoulos, N., Boskou, G., & Salta, F. N. (2012). Migration of health promoting microconstituents from frying vegetable oils to French fries. *Food Chemistry*, *133*, 1255–1263.

Goldberg, G. (2008). *Plants: Diet and health.* John Wiley & Sons.

Górnaś, P., Siger, A., Juhņeviča, K., Lācis, G., Šnē, E., & Segliņa, D. (2014). Cold-pressed Japanese quince (*Chaenomeles japonica* (Thunb.) Lindl. ex Spach) seed oil as a rich source of α-tocopherol, carotenoids and phenolics: A comparison of the composition and antioxidant activity with nine other plant oils. *European Journal of Lipid Science and Technology*, *116*, 563–570.

Górnaś, P., Siger, A., & Segliņa, D. (2013). Physicochemical characteristics of the cold-pressed Japanese quince seed oil: New promising unconventional bio-oil from by-products for the pharmaceutical and cosmetic industry. *Industrial Crops and Products*, *48*, 178–182.

Kalt, W., Forney, C. F., Martin, A., & Prior, R. L. (1999). Antioxidant capacity, vitamin C, phenolics, and anthocyanins after fresh storage of small fruits. *Journal of Agricultural and Food Chemistry*, *47*, 4638–4644.

Koski, A., Pekkarinen, S., Hopia, A., Wähälä, K., & Heinonen, M. (2003). Processing of rapeseed oil: Effects on sinapic acid derivative content and oxidative stability. *European Food Research and Technology*, *217*, 110–114.

Lercker, G., & Rodriguez-Estrada, M. (2000). Chromatographic analysis of unsaponifiable compounds of olive oils and fat-containing foods. *Journal of Chromatography A*, *881*, 105–129.

Mierina, I., Serzane, R., Strele, M., Moskaluka, J., Seglina, D., & Jure, M. (2011). Extracts of Japanese quince seeds-potential source of antioxidants. In: *Conference proceedings of 6th Baltic conference on food science and technology: Innovations for food science and production*, pp. 98–103.

Ros, J., Laencina, J., Hellín, P., Jordan, M., Vila, R., & Rumpunen, K. (2004). Characterization of juice in fruits of different Chaenomeles species. *LWT—Food Science and Technology*, *37*, 301–307.

Rumpunen, K. (2003). *Japanese quince: Potential fruit crop for Northern Europe: Final report FAIR-CT97-3894.* SLU: Department of Crop Science.

Seglina, D., Krasnova, I., Heidemane, G., & Ruisa, S. (2009). Influence of drying technology on the quality of dried candied *Chaenomeles japonica* during storage. *Latvian Journal of Agronomy/Agronomija Vestis*, *12*, 113–118.

Senanayake, S. N., & Shahidi, F. (2000). Concentration of docosahexaenoic acid (DHA) from algal oil via urea complexation. *Journal of Food Lipids*, *7*, 51–61.

Siger, A., Nogala-Kalucka, M., & Lampart-Szczapa, E. (2008). The content and antioxidant activity of phenolic compounds in cold-pressed plant oils. *Journal of Food Lipids*, *15*, 137–149.

Szydłowska-Czerniak, A., Trokowski, K., Karlovits, G., & Szłyk, E. (2011). Effect of refining processes on antioxidant capacity, total contents of phenolics and carotenoids in palm oils. *Food Chemistry*, *129*, 1187–1192.

Urbanavičiūtė, I., Rubinskiene, M., & Viškelis, P. (2019). The fatty acid composition and quality of oils from post-industrial waste of quince *Chaenomeles japonica. Chemistry & Biodiversity*, *16*(8), e1900352.

Wojdyło, A., Oszmiański, J., & Bober, I. (2008). The effect of addition of chokeberry, flowering quince fruits and rhubarb juice to strawberry jams on their polyphenol content, antioxidant activity and colour. *European Food Research and Technology*, *227*, 1043–1051.

Chapter 50

Cold pressed virgin olive oils

Maria Z. Tsimidou[a], Aspasia Mastralexi[a], and Onur Özdikicierler[a,b]
[a]*Laboratory of Food Chemistry and Technology, School of Chemistry, Aristotle University of Thessaloniki, Thessaloniki, Greece,* [b]*Food Engineering Department, Faculty of Engineering, Ege University, Izmir, Turkey*

Abbreviations

EU European Union
IOC international olive council
MUFA monounsaturated fatty acid
OO olive oil
PDO protected denomination of origin
PGI protected geographical indication
PUFA polyunsaturated fatty acid
VOO virgin olive oil

1 Virgin olive oil (VOO) and cold pressed virgin olive oils

According to the Codex Alimentarius's generic definition of what a virgin oil or fat is (Codex Alimentarius Commission, 2017a) and that of the dedicated Codex Standard for olive oils (OO) and olive pomace oil (Codex Alimentarius Commission, 2017b), a virgin olive oil comes from olive fruits that are solely treated by mechanical or other physical means. In cases where thermal conditions are applied, these should not lead to alterations in the oil. The final product should only have undergone washing, decanting, centrifuging, and filtration. Consequently, for cold pressed virgin olive oils, thermal treatment should be avoided. In particular, cold pressing temperature should never exceed 27°C according to the International Olive Council (IOC) glossary (IOC, 2019a). The European Union (EU) olive oil legislation on OO marketing standards (European Union Commission, 2012) is in line with this view and defines further the conditions and types of commercial products that can bear optional indications regarding the "cold" technology applied, as shown in Table 1. It should be noted that the EU legislation permits only two edible commercial categories of VOO: "extra VOO" and "VOO."

At this point, it is important to present briefly the basic types of current OO technology applied at an industrial scale worldwide. Innovation at every stage of olive processing can be introduced avoiding violation of the definition of VOO. Thus, freezing of olives or introduction of ultrasound during malaxation is theoretically in line with the legal definition because they are physical means, unless it is proven that they cause alteration of its composition (Asheri, Sharifani, & Kiani, 2017; Clodoveo, Corbo, & Amirante, 2018). Enzymes are not permitted because they are biochemical means, although their introduction aims at facilitating degradation of hydrocolloids of olive flesh, which is not expected to alter the core compositional characteristics of the oil (Peres, Martins, & Ferreira-Dias, 2017).

2 From olives to virgin olive oil

Technological aspects of OO production, and the impact of different factors have been summarized frequently in different book chapters and review articles (Angerosa, 2002; Aparicio & García-González, 2013; Boselli, Di Lecce, Strabbioli, Pieralisi, & Frega, 2009; Caporaso, 2016; Chiacchierini, Mele, Restuccia, & Vinci, 2007; Clodoveo, 2012; Di Giovacchino, Sestili, & Di Vincenzo, 2002; Kalua et al., 2007; Kiritsakis & Shahidi, 2017; Petrakis, 2006; Tamborrino, 2014; Vitaglione et al., 2015).

As illustrated in the graphical representation of a virgin olive oil production line (Fig. 1), this is a multistep process that may involve heating or nonheating treatment of the raw material and of the intermediate products as well as exposure to oxygen unless the latter is replaced by an inert gas (N_2) stream (Kiritsakis & Shahidi, 2017; Petrakis, 2006).

Cold Pressed Oils. https://doi.org/10.1016/B978-0-12-818188-1.00050-5
Copyright © 2020 Elsevier Inc. All rights reserved.

TABLE 1 Optional indications for VOOs produced without heat treatment according to EU regulation 29/2012, article 5.

Marketing standard	Definition
First cold pressing	May appear only for virgin or extra virgin olive oils obtained at a temperature below 27°C from a first mechanical pressing of the olive paste by a traditional extraction system using hydraulic presses
Cold extraction	May appear only for virgin or extra virgin olive oils obtained at a temperature below 27°C by percolation or centrifugation of the olive paste

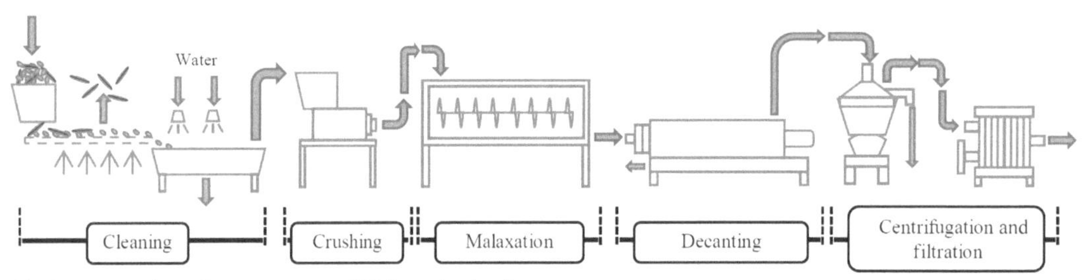

FIG. 1 Graphical representation of a contemporary VOO production line.

2.1 Harvesting and transportation

During the ripening of olive fruit, oil accumulates in the olive flesh under normal climatic conditions and reaches its maximum before the natural fruit drop. Olives should be harvested at the optimal technological maturity stage that guarantees on the one hand, a high oil yield, and on the other hand, that desirable constituents are preserved at high concentrations. Means of olive harvesting have been modernized to avoid damaging or breaking of the olive epidermis, which may trigger enzymatic activities (Castro-Garcia & Ferguson, 2017; IOC, 2011).

Maturity evolution is directly affected by growing conditions of the olive such as climate, olive cultivar, irrigation, and other cultivation practices. Therefore, optimum technological maturity stage even for the same cultivar depends on the location of olive groves, irrigation system, etc., and cannot be categorically predicted (Mele, Islam, Kang, & Giuffrè, 2018). However, it is necessary to accumulate data and make adjustments and introduce good agricultural practices in traditional (for the Mediterranean producing countries) cultivation.

The maturity index (MI) is calculated by a mathematical equation that was developed several decades ago to assess ripeness evolution of the olive fruit by Uceda and Frias (1975). This tool is recommended by the IOC (IOC, 2011) and can be easily used by producers to estimate the optimum harvest time for their production. For calculation of the MI, olives should be harvested from as many trees as possible to ensure representative sampling for the orchard. Two kg of olives can be enough if the oil content is low and the olives are green. Normally, the average sampling should be around 1.5 kg. After harvesting, 100 representative olives are selected for the MI calculation (IOC, 2011). The selected olives are compared and sorted into different categories according to the appearance attributes given in Table 2. In general, an MI value between 3 and 4 is an indication of technological suitability to start harvesting in order to obtain high-quality olive oil (IOC, 2011).

Generally, olives are harvested from the olive tree by hand, by handheld mechanical means, or with the aid of special devices. Hand picking is a slow and expensive practice and labor-intensive. Branch beating with a wooden or plastic pole comes from antiquity and is mainly applicable to mature olives. "Comb" type means help the harvester to detach the stem of the olive fruit from the branches. They are placed on top of a telescopic pole and are operated by vibration. Detached olives can then be collected in nets or cloths laid under the tree. Although the use of this kind of handheld device accelerates harvesting, it may damage, to a certain extent, young branches and cause the fall of many leaves. Branch and trunk shaking are fast and effective methods that are performed using a vibrating arm that grabs a branch to detach the olives. This kind of harvesting is more suitable for relatively smaller trees, which are 2.5–3.5 m high. For higher trees up to 10–15 m high, trunk shaking is more suitable. A special vehicle equipped with a large vibrating arm and a folding flexible net is generally used for trunk shaking (Di Giovacchino et al., 2002). However, this method is not suitable for intensive plantations because the vehicle needs space between trees to maneuver. The best way to transport the olives is using open-mesh plastic crates in order to allow air to circulate to prevent an increase in temperature due to the catabolic activity of enzymes in the olive fruit

TABLE 2 Appearance characteristics and olive categories for the calculation of maturity index (MI).[a]

Appearance	Categories	Number of olives out of 100
Skin color deep green	Category 0	A
Skin color yellow-green	Category 1	B
Skin color green with reddish spots on < half the fruit surface. Start of color change	Category 2	C
Skin color with > half the fruit surface turning reddish or purple. End of color change	Category 3	D
Skin color black with white flesh	Category 4	E
Skin color black with < half the flesh turning purple	Category 5	F
Skin color black with not all the flesh purple to the stone	Category 6	G
Skin color black with all the flesh purple to the stone	Category 7	H

[a] $MI = \frac{A \cdot 0 + B \cdot 1 + C \cdot 2 + D \cdot 3 + E \cdot 4 + F \cdot 5 + G \cdot 6 + H \cdot 7}{100}$

(Angerosa, 2002; Petrakis, 2006). After harvesting, the lipase activity increases and causes hydrolysis of the triacylglycerols to free fatty acids, mono- and di-acylglycerols at high relative humidity levels. Therefore, olives should be processed as soon as possible after harvest to obtain high-quality VOO (Mansour et al., 2015).

2.2 Cleaning

Harvested olives may be contaminated by organic or inorganic impurities, such as leaves or twigs, soil, dust, and stone fragments. These impurities are separated from olives in the very beginning of OO production. In general, olives are cleaned on a vibrating screen or a blower to remove leaves and other light impurities. If necessary, olives can be washed with water to remove solid contamination (e.g., soil mineral or metallic contaminants) that may cause damage to mechanical parts of the machinery or lower the quality of the produced oil (Di Giovacchino et al., 2002).

2.3 Crushing

There are two general types of olive crushing systems: stone mills (traditional system) and metallic crushers (modern systems). Modern metallic crushers can be alternatively equipped with hammers, blades, or disk crushers (Inarejos-García, Fregapane, & Salvador, 2011). During crushing, the oil is liberated from the intracellular oil bodies and the enzymatic reactions affecting the volatile profile and the polar phenolic compound content in the final product are triggered. Moreover, during crushing, due to shear force applied to the paste, a temperature increase inevitably occurs, depending on the crushing method. Studies on the effect of different crushing techniques have shown differences in the impact to olive oil yield, minor component composition, and concentration and sensory properties of VOO (Mele et al., 2018; Polari, Garcí-Aguirre, Olmo-García, Carrasco-Pancorbo, & Wang, 2018a).

2.3.1 Stone mills

The stone mill has been used to extract olive oil for centuries. Stone mills use circular stone disks or wheels, which are rolled rotationally on a slab of granite, and crush the olives, which are placed beneath. Stone mill crushers are only in limited use due to disadvantages such as the low working capacity and yield. On the other hand, olive oils obtained from stone mills are not rich in polar phenolic compounds (Di Giovacchino et al., 2002; Morrone et al., 2017; Preziuso et al., 2010; Veillet, Tomao, Bornard, Ruiz, & Chemat, 2009). When crushing is carried out by stone mills, due to the slow rotation speed of the stone mills, emulsification between the oil phase and the vegetative water does not occur. A partial malaxation takes place because of the slow movement of the olive paste during the crushing process. This type of crushing is not as intense as that using the metallic crushers, so that increase in paste temperature is also limited (4–5°C) (Caponio, Gomes, Summo, & Pasqualone, 2003; Di Giovacchino et al., 2002).

2.3.2 Hammer mills

Hammer crushers have been widely used in OO mills for many years. They consist of a vertically positioned circular crushing chamber. A varying number of hammers rotate in the center and the bottom of the chamber is perforated and allows the crushed paste to exit from the chamber. As the hammers rotate in the chamber, olives are crushed both with intensive force directly applied to olives and shear forces developed between the chamber wall and the tip of each hammer. Like in all metallic crushers, rotation speed impacts oil yield, temperature increase, and chemical composition of the oil (Polari, Garcí-Aguirre, Olmo-García, Carrasco-Pancorbo, & Wang, 2018b). Crushing time extends if mesh size decreases to obtain finer pit sizes. The existence of the mesh in the bottom of the chamber causes greater extent of rupture of the olive cells, which favors oil yield when compared with the effect of other metallic crushers (Morrone et al., 2017). However, extensive cell crush also leads to an increase in enzymatic activity, which causes a higher degree of hydrolysis. In the course of this time-dependent process, a temperature increase of up to 13–15°C in the olive paste is expected, which can lead to various compositional changes in the OO (Caponio & Catalano, 2001; Di Giovacchino et al., 2002; Morrone et al., 2017; Servili, Piacquadio, De Stefano, Taticchi, & Sciancalepore, 2002; Veillet et al., 2009). For example, a greater oxidative degradation of OO is observed when the latter is extracted using hammer mills in comparison with that using metal-toothed crushers. To overcome this disadvantage, the crushing chamber can be equipped with a jacket that allows a coolant to pass and regulate or compensate the temperature increase, which is a crucial factor for cold pressing operation (Caponio et al., 2003).

2.3.3 Blade crushers

This type of crusher is more recent than hammer crushers. The structural architecture of a blade crusher is very similar to the hammer crusher. The hammers are replaced by blades, which tear up the fruit tissue as it rotates. Due to the structure of the blade crusher, the size of olive pit fragments of the paste is larger, hence the OO yield, polar phenolic, and pigment content are generally lower than those obtained with other metallic crushers. The improved sensorial quality of OOs produced using blade crushers compared with that of oils obtained using hammer crushers is due to the higher release of desirable volatiles (Morrone et al., 2017; Servili et al., 2002).

2.3.4 Metal-toothed crushers (disk crushers)

Metal-toothed crushers tear the olives by means of several sharp teeth, which are placed on two disks, one of which is stationary and the other is revolving. An internal rotor directly creates the motion for the revolving ring that has rings of prism-shaped teeth with cutting edges. The stationary ring is attached to the inner surface of the lid and is also assembled with rings of teeth with cutting edges. The paste falls out from the bottom of the chamber. Because the metal-toothed disk crushers do not contain a perforated mesh under the crushing chamber, less intensive crushing can be achieved in comparison to those using hammer crushers. Thus, the increase in the temperature of the paste is lower than that observed using hammer mills. In a study investigating the difference in the impact between metal disk crushers and hammer mills, it was reported that the latter caused a 50% higher temperature increase (Caponio et al., 2003). This piece of information is very important from the point view of the compositional characteristics of OO. In olive oils obtained using metal-toothed crushers, less hydrolysis and oxidative degradation phenomena occur, which can be associated with temperature increase and structural design of the crushers. In addition, studies indicated that the polar phenolic content of OOs obtained using metallic disk crushers is similar to those obtained using hammer crushers (Caponio & Catalano, 2001).

2.4 Malaxation

After crushing, olives turn into a paste. This paste is kneaded in special tanks—malaxers—to prepare the olive paste for separation of the OO. Malaxation conditions affect oil yield and extraction efficiency. They help the merging of small-dispersed droplets into larger ones and facilitate the separation of the latter in the course of mechanical treatment. Breaking up of the emulsions is also promoted (Clodoveo, 2012). In contrary to the positive effects to oil yield, during malaxation, oxidative reactions may prevail. Moreover, removal of polar phenolic compounds from the oil phase into the water phase increases.

Therefore, malaxation time and temperature are critical factors for olive oil quality (Leong, Juliano, & Knoerzer, 2017). In general, semicylindrical tanks that are equipped with stainless steel blades and rotating arms are used for malaxation. Regarding the operational axis of the rotating arm, malaxers are categorized as "vertical" or "horizontal." Horizontal ones are the most popular in the olive oil industry (Clodoveo, 2012). The machinery is usually equipped with a water jacket circulating hot or cold water to regulate malaxation temperature and is hermetically closed to allow regulation of the headspace atmosphere composition (Di Giovacchino et al., 2002). Modern malaxation machines are equipped with different sensors to measure paste temperature and adjust water circulation in the water jacket, paste viscosity, and oxygen concentration of the headspace at the top of the tank. In general, malaxation time and temperature affect olive oil extraction yield positively. These two crucial parameters have different effects on the olive oil yield and quality. Hermetic sealing of the malaxation machine ensures control of the headspace atmosphere to which the olive paste is exposed. Introduction of inert gases to the headspace atmosphere of the malaxer helps to reduce the negative effects caused by exposure of OO to oxygen and it allows the malaxation time to be extended, which increases yield, as summarized by Di Giovacchino et al. (2002).

2.5 Extraction of virgin olive oil

2.5.1 Pressing

In principle, the pressing of olive paste forces the liquid phase to separate from the solid part by the intervention of the drainage effect of the mats (Petrakis, 2006).

2.5.2 Centrifugation (decantation)

Decantation is based on the principle that immiscible liquids of different densities tend to separate by gravity. Horizontal centrifuges (decanters) create centrifugal forces on the olive paste, which are approximately 3000 times higher than gravitational force, as the axis of the decanter rotates. Therefore, during decantation, solid particles (stone fragments and fruit tissue) are pushed against the rotating bowl wall while the liquid phase is gathered closer to the axis of rotation (Petrakis, 2006). The modern decantation systems can be divided into two groups according to the number of outlet phase lines. Usually, they are referred to as three- and two-phase decanters.

Three-phase vs. two-phase decanters

The three-phase decanter is the first type that was used in the olive oil industry. The name "three-phase" comes from the number of outlet phase lines because it allows the separate acquisition of the oil phase, the aqueous phase, and the solid one. In order to operate three-phase decanters efficiently, water addition to paste before decantation is needed. The two-phase decanter is the successor of the previous type in the olive oil industry. The "two-phase" decanter allows the acquisition of two phases, one for the oil and another for vegetative water together with olive pomace. The three-phase system exerts a negative effect on OO quality regarding the polar phenolic content due to its hydrophilicity. An increase in water content of olive paste before decantation results in a greater loss of polar phenolic compounds when compared with two-phase decanters where additional water is not needed to operate. Moreover, two-phase decanters demand lower power consumption compared with the three-phase decanters because of differences in their assembly (Vitaglione et al., 2015). Borja, Raposo, and Rincón (2006) briefly listed advantages and disadvantages of two-phase decanting systems as follows:

- **(i)** architecture and construction of two-phase decanting systems are simpler and reliable and less expensive than the three-phase systems;
- **(ii)** no additional water is needed to operate the two-phase systems;
- **(iii)** OO yield per processing quantity is higher than that using three-phase decanters, hence energy consumption and operation costs in two-phase decanters are lower; and
- **(iv)** VOO that is extracted using two-phase decanting systems has a good oxidative stability and better sensorial characteristics.

The olive paste is obtained from the two-paste decanting system in a diluted form with vegetative water. The moisture percentage of olive pomace obtained from the two-phase systems is about 55%–70%. This is much higher than the moisture percentage (40%–45%) reported for olive pomace obtained from three-phase systems. There are disadvantages in the handling of olive pomace with high moisture content, which is in "semi-solid" form, in terms of storage and transportation. In addition, the moisture content of olive pomace should be reduced prior to pomace oil extraction, hence industrial pomace oil producers consume more energy with higher operational costs for drying of olive pomace obtained from two-phase systems. Therefore, this stage remains controversial in practice.

2.6 Clarification of VOO

Separation of remaining impurities is the last step of olive oil production. In order to separate the remaining impurities, such as solid particles of fruit tissue and trace vegetative water dispersed as microdroplets, a clarifier or a settling tank can be used as a basic option. However, this operation is not compatible with continuous production flows because the time required for settling cannot be controlled and contact of oil with the vegetable water for a long time increases the risk of triacylglycerol hydrolysis (Di Giovacchino et al., 2002; Petrakis, 2006).

In order to perform a fast and efficient separation of solid and liquid impurities from VOO, centrifugation can be performed. Using exogenous water during centrifugation may reduce oil loss from the water outlet of the centrifuge. Moreover, water acts as a coolant in the centrifuge to eliminate a significant amount of heat generated due to frictional forces at the walls of the centrifugal basin (Costagli, 2018). Although veiled (cloudy) olive oils are also desirable commercial products (Lozano-Sánchez, Cerretani, Bendini, Segura-Carretero, & Fernández-Gutiérrez, 2010; Tsimidou, Georgiou, Koidis, & Boskou, 2005), filtration is also applicable as the final step to eliminate suspended trace solids and moisture to increase uniformity of the end product, which positively influences consumer acceptance. It should be noted that the average consumer has been acquainted with the brilliant, transparent, colorless appearance of refined vegetable oils for many decades and is thus expecting something similar for VOOs.

Stainless steel tanks are used for storage of large amounts of OO to block contact with light and air. The bottom of these tanks should be conical in shape and equipped with drainage valves to eliminate the solid particles deposited during storage in order to avoid fermentation reactions. Delay in oxidation evolution, which is promoted at temperatures higher than 20–22°C, is the key point for effective storage of OO (Di Giovacchino, 2013). The temperature should be reduced and maintained preferably around 10°C and an inert gas blanket system should be installed in order to substitute available headspace air with inert gases such as nitrogen on top of the tank during storage of VOO, preferably at ~10°C (Baccioni & Peri, 2014; Kiritsakis & Shahidi, 2017).

3 Olive oil market characteristics and the position of cold pressed olive oil products

Since the seven countries' study, interest in the Mediterranean diet has revealed the importance of olive oil consumption—in particular in its virgin form (not refined)—as a major contributor to the health-promoting properties of this dietary pattern that has been recognized by UNESCO as part of the cultural heritage of humanity (Saulle & La Torre, 2010). The suggestions of scientists worldwide for the consumption of VOO on a regular basis were first focused on its fatty acid composition, which is rich in monounsaturated oleic acid (C18:1) (55%–83%) (Boskou, Blekas, & Tsimidou, 2006) and, soon after, on the presence of rare polar phenolic compounds of secoiridoid type, which are derivatives of oleuropein and ligstroside (Boskou, Tsimidou, & Blekas, 2006). As a consequence, the market for olive oil has been modernized in the last decades and the product can be found at high, medium, or low prices per liter, depending on factors beyond the minimum legal requirements for authenticity and quality criteria set by trade associations and food authorities. Therefore, the edible virgin olive oils are marketed as "extra virgin" (first quality), "virgin" (second quality), and as blends of virgin and refined olive oil. They can also bear optional labels (monovarietal or blends of different varieties, geographical indications, product of mountain regions or organic cultivation, cold pressed technologies, etc.) that add value to the product price. Traditional consumers are now more informed on factors that affect the quality of the product (Tsimidou, 2006) and as a result have changed their habitual way of purchasing OO in bulk and prefer to buy it bottled (5 L and below) depending on expected household use. New consumers from developed economies, but also from emerging ones (India, China, and Brazil), incorporate this fat in their everyday diet, mainly uncooked, as salad oil. Italy is at the forefront of imports of bulk olive oil from Greece or Tunisia and exports to the most demanding markets of northern Europe and the USA. Spain holds the first position in production tonnage despite fluctuations due to climate change. Greece seems to be still blocked by intrinsic weaknesses that are mainly related to the great number of small-sized producers, acting under bureaucratic cooperative Schemes. OO production in California and Australia widened the decision-making scene beyond the activities of the International Olive Council. The EU is still the major world producer, consumer, and exporter of virgin olive oil (67%, 55%, and 67%, respectively, 2012/2013–2016/2017) (IOC, 2019b) (Fig. 2A). Olive oil production seems to have good potential in non-EU countries (Fig. 2B), and this issue should be considered carefully by Spain, Italy, and Greece.

Though no statistical data are available for cold pressed or cold extracted virgin olive oils, this particular technology prevails in products that command a high market price in domestic or international markets. This is exemplified below by two case studies for an EU and a non-EU producing country.

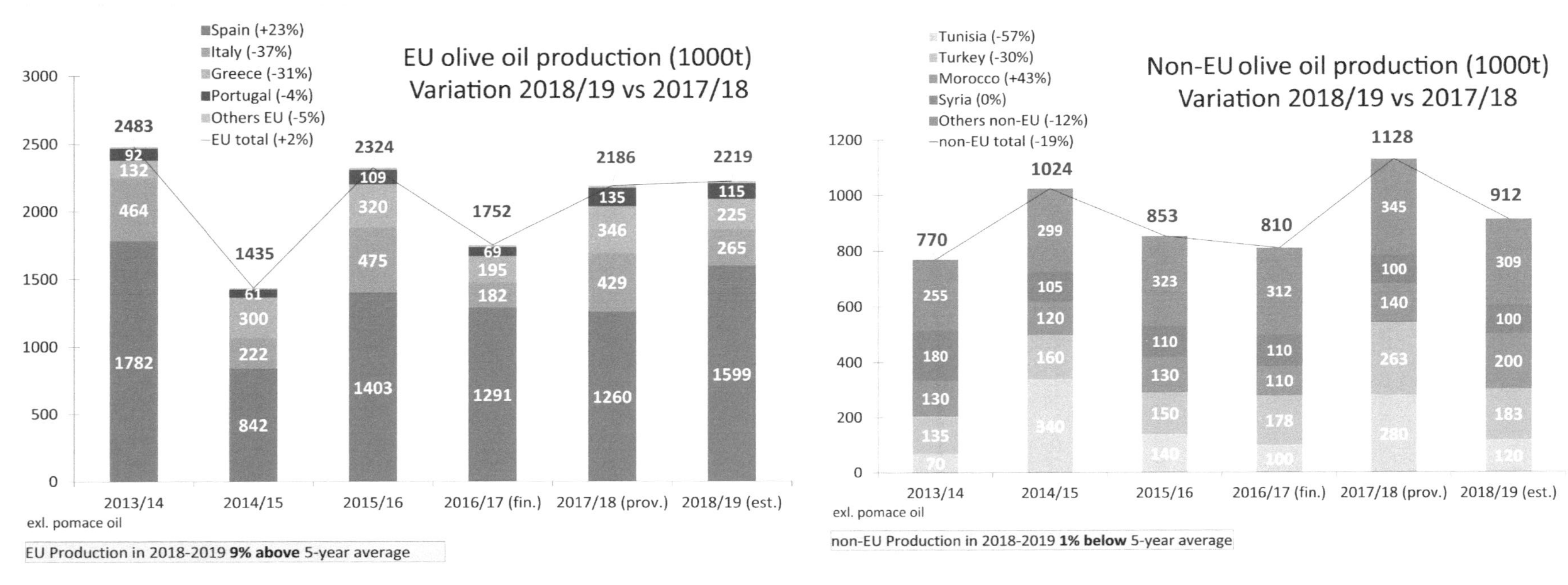

FIG. 2 Olive oil production in EU (A) and non-EU (B) countries (2018/2019 vs. 2017/2018) (IOC, 2019b).

3.1 The case of the Greek olive oil market

In a study of Greek olive oil market structure (Karipidis, 2005) and the identification of differentiation strategies in order for this product to be purchased by consumers at a higher price, it was evidenced that nonthermal processing conditions was one of the most positively influential parameters. A search on the websites of 102 Greek companies that bottle extra VOO oil indicated that only a few of them provide such products to the domestic or international market; this is illustrated in Table 3.

Moreover, a search in the dossiers of Protected Denomination of Origin (PDO) and Protected Geographical Indication (PGI) Greek extra VOOs that command a high value in the international market (Nenadis, Mastralexi, & Tsimidou, 2018) indicated that since the first registrations, this high added-value segment of OO products is obtained under low-temperature treatments, but in most cases because of changes in the EU legislation, some of them cannot take advantage of the optional label "first cold pressing" or "cold extraction" because they refer to temperatures slightly higher than 27°C (Table 4). This observation shows the need for frequent updating of official registrations of PDOs in order for producers and distributors to gain some benefits from the continuous updating of the legislation.

TABLE 3 Companies producing and/or exporting Greek extra virgin olive oil that is bottled, branded, and labeled in Greece with or without the indication "first cold pressing" and/or "cold extraction".

Company No	Indication		Company No	Indication		Company No	Indication	
	First cold pressing	Cold extraction		First cold pressing	Cold extraction		First cold pressing	Cold extraction
1	–	–	35	–	+	69	–	–
2	–	–	36	–	–	70	–	+
3	–	+	37	+	–	71	+	+
4	–	–	38	–	–	72	–	–
5		+	39	–	–	73	–	+
6	–	–	40	–	–	74		+
7	+	+	41	–	–	75	–	–
8	–	–	42	–	–	76	–	–
9		+	43	+	–	77	–	+
10	+	–	44	–	–	78	–	–
11	–	–	45	–	–	79	+	+
12	–	+	46	+	–	80	–	+
13	–	–	47	–	–	81	–	–
14	–	+	48	–	–	82	–	–
15	+	–	49	–	+	83	–	+
16	+	+	50	+	–	84	–	–
17	–	+	51	+	+	85	–	+
18	–	–	52	+	–	86	–	–
19	–	–	53	+	–	87	–	+
20	–	–	54	+	–	88	–	–
21	–	–	55	–	–	89	+	–
22	–	+	56	–	–	90	+	–

TABLE 3 Companies producing and/or exporting Greek extra virgin olive oil that is bottled, branded, and labeled in Greece with or without the indication "first cold pressing" and/or "cold extraction"—cont'd

Company No	Indication		Company No	Indication		Company No	Indication	
	First cold pressing	Cold extraction		First cold pressing	Cold extraction		First cold pressing	Cold extraction
23	–	–	57	+	–	91	–	–
24	–	–	58	–	+	92	–	–
25	–	–	59	+	–	93	–	+
26	–	–	60	–	+	94	+	–
27	–	–	61	–	–	95	–	–
28	–	–	62	–	–	96	–	–
29	+	–	63		+	97	–	–
30	–	+	64	+	–	98	–	+
31	–	+	65	–	–	99	–	–
32	+	–	66	–	+	100	–	–
33	–	+	67	–	–	101	–	–
34	5	+	68	–	+	102	+	–

(Source: Olive Oil Companies, n.d. Website Links).

TABLE 4 Application of heating/nonheating conditions in the production line of Greek PDO/PGI extra virgin olive oil as presented in the registration documents (DOOR, n.d.)

			Processing conditions		
No	Registered name/year of registration	Olive mill	Temperature of malaxation	Time of malaxation	Speed
GREEK PDO EVOOs					
1	"Sitia Lasithiou Kritis"/1998	Traditional or centrifugal	≤27°C	30 min	–
2	"Arxanes Irakliou Kritis"/1996	Traditional or centrifugal	–	–	–
3	"Peza Irakliou Kritis"/1996	Traditional or centrifugal	–	–	–
4	"Exeretiko partheno eleolado Thrapsano"/2002	Centrifugal	≤27°C	–	–
5	"Messara"/2013	Centrifugal	27–33°C	30 min	
6	"Viannos Irakliou Kritis"/1996	Traditional or centrifugal	–	–	–
7	"Vorios Mylopotamos Rethymnis Kritis"/1996	Centrifugal	27–32°C	–	–
8	"Exeretiko Partheno Eleolado Selino Kritis"/2010	Traditional or centrifugal	≤27°C	–	–
9	"Apokoronas Chanion Kritis"/1997	Traditional or centrifugal	≤30°C	30 min	

Continued

TABLE 4 Application of heating/nonheating conditions in the production line of Greek PDO/PGI extra virgin olive oil as presented in the registration documents—cont'd

			Processing conditions		
No	Registered name/year of registration	Olive mill	Temperature of malaxation	Time of malaxation	Speed
10	"Kolymvari Chanion Kritis"/1997	Traditional or centrifugal, or sinolea	≤30°C	30 min	
11	"Lygourio Asklipiou"/1996	Traditional or centrifugal	≤30°C	–	–
12	"Kranidi Argolidas"/1996	Centrifugal	≤32°C	–	–
13	"Finiki Lakonias"/2003	Centrifugal (2- or 3-phase)	≤30°C	30 min	–
14	"Krokees Lakonias"/1996	Traditional or centrifugal	–	–	–
15	"Petrina Lakonias"/1996	Traditional	≤30°C	–	–
16	"Kalamata"/1997	Traditional or centrifugal	≤27°C	20–30 min	–
17	"Exeretiko partheno eleolado "Trizinia"/2007	Traditional or centrifugal	28–29°C	20–30 min	17–19 rev/min
18	"Agoureleo Chalkidikis"/2013	Centrifugal	≤27°C	20–30 min	17–19 rev/min
19	"Galano Metaggitsiou Chalkidikis"/2015	Traditional or centrifugal	≤27°C	~30 min	–
GREEK PGI EVOOs					
1	"Chania Kritis"/1996	Traditional or centrifugal	–	–	–
2	"Lakonia"/1996	Centrifugal	–	–	–
3	"Olympia"/1996	Centrifugal	≤27°C	40 min	–
4	"Thassos"/1996	Traditional or centrifugal	–	–	–
5	"Preveza"/1996	Centrifugal	–	–	–
6	"Lesvos; Mytilini"/2003	Centrifugal	–	–	–
7	"Samos"/1998	Centrifugal	–	–	–
8	"Rodos"/1996	Centrifugal	≤32°C	–	–
9	"Agios Mattheos Kerkyras"/2004	Traditional or centrifugal	≤30°C	30 min	–
10	"Kefalonia"/1996	Traditional or centrifugal or sinolea	≤29°C	30 min	–
11	"Zakynthos"/1998	Traditional or centrifugal or sinolea	≤32°C	–	–

PDOs/PGIs that have been registered before 2002 are gray shaded.

3.2 The case of the Turkish olive oil market

Turkey is an emerging country in olive oil production. In Fig. 2B the fluctuation in OO production between 2013/2014 and 2018/2019 (estimated production) harvesting seasons for Turkey indicates that more has to be done to stabilize its position in world OO production, considering the extended geographical areas where olive trees can thrive (Fig. 3; Veral, 2016) (Erses Yay, Oral, Onay, & Yenigün, 2012). There are approximately 320,000 different small farms and family companies that own olive orchards in Turkey (Koseoglu et al., 2016). Fifty-five percent of the production is dedicated to olive oil,

FIG. 3 Olive tree cultivation regions in Turkey (green (*gray in print version*) highlighted provinces, namely, Tekirdağ, Çanakkale, İzmit, Adapazarı, Bilecik, Bursa, Balıkesir, İzmir, Manisa, Aydın, Denizli, Muğla, Antalya, Karaman, Mersin, Adana, Osmaniye, Kahramanmaraş, Antakya, Kilis, Gaziantep, Adıyaman, Şanlıurfa, Mardin, Trabzon, Artvin). *(Abstracted from Veral, M. G. (2016). Olive gene resources in Turkey.* Olivae, 123, 27–30.*)*

which is obtained from different cultivars (Toplu, Yildiz, Bayazit, & Demirkeser, 2009). Approximately 20% of the olive producers are associated under three olive oil sale cooperatives. According to a market study, the share of small private olive oil brands is very low (approximately 7%); therefore, 70% of the olive oil production is sold in bulk to consumers, cooperatives, other olive oil companies having their own brands, and export brokers (Balkan & Meral, 2017).

Although there are no official statistics available about cold pressing application, the Turkish market follows the general trends observed in the EU market. Among branded virgin olive oils, information on cold pressed or extracted ones were available only for those registered for geographical indications. According to the register of the Republic of Turkey Ministry of Agriculture and Forestry database, by the end of the 2018, among 693 different commercial olive oil brands, some are marketed with a geographical indication (Table 5), together with information about the technology of extraction with focus on temperatures used (Turkish patent and trademark office, n.d.); however, only one qualifies to be designated as obtained by cold technology. It is obvious that the Turkish OO industry should pay more attention to the application of cold conditions in the production line.

4 Impact of "cold technology" on yield, quality parameters, sensory/color attributes, and chemical composition of virgin olive oil

4.1 General

Publications on OO composition, authenticity and quality issues have increased at a tremendous rate since the late 1990s. Introduction of good agricultural practices, good hygiene practices, and the strict legislative framework worldwide for this oil had a positive impact on its quality. This fact does not preclude fraudulent practices, and olive oil is listed among the most frequently adulterated high valued food products, as discussed in a recent article (Tsimidou, Ordoudi, Nenadis, & Mourtzinos, 2015).

A careful search of the titles, abstracts, and keywords among publications revealed that the number of comparative studies for VOOs produced under different heating and nonheating conditions since 2000 are rather limited, but can provide all the necessary information to the reader of the present chapter to make his or her mind up regarding the contribution of this factor to the overall quality, stability, and nutritional value of VOO, which is currently considered as a functional food. The relevant references together with the studied characteristics are presented in Table 6, and the findings are discussed in the following subsections.

TABLE 5 Application of heating/nonheating conditions in the production line of Turkish PDO/PGI extra virgin olive oil as presented in the registration documents (Turkish patent and trademark office, n.d.).

No.	Registered name/Year of Application (App) or Registration (Reg)	Type of Geographic Indicator	Production Method	Temperature of malaxation	Duration of Malaxation
1	"Akhisar Domat Zeytinyaği"/ App:2017	PDO	–	–	–
2	"Akhisar Uslu Zeytinyaği"/ App:2017	PDO	–	–	–
3	"Ayvalik Zeytinyaği"/Reg:2007	PDO	Traditional pressing or centrifugal decantation	–	–
4	"Derik Halhali Zeytinyaği"/ App:2017	PDO	–	–	–
5	"Edremit Zeytinyaği"/Reg:2017	PDO	Traditional pressing, percolation or centrifugal decantation	25–30°C	<1 h
6	"Güney Ege Zeytinyağlari"/ Reg:2004	PDO	Traditional pressing or centrifugal decantation	–	–
7	"Kuzey Ege Zeytinyağlari"/ Reg:2018	PGI	Traditional pressing, percolation or centrifugal decantation	25–35°C	<1 h
8	"Maraş Natürel Sizma Zeytinyaği"/App:2016	PGI	–	–	–
9	"Mut Zeytinyaği"/Reg:2018	PDO	Centrifugal decantation	–	–
10	"Milas Zeytinyaği"/Reg:2016	PDO	Traditional pressing or centrifugal decantation	–	–
11	"Nizip Zeytinyaği"/Reg:2012	PGI	Traditional pressing or centrifugal decantation	20–25°C	Paste of Stone mills: 10–20 min Paste of metallic crushers: 90 min

4.2 Processing yield

Olive oil yield depends on the cultivar, maturity stage of olives, and of course, technology applied. There are three different expressions for olive oil yield of a production line. Hermoso Fernández et al. (1998) used the term "extractability" to express processing yield. The "extractability index" was calculated with reference to that determined by the Soxhlet method (Clodoveo, 2012). Processing yield is also expressed as "extraction performance," which is calculated as the ratio of produced OO divided by the quantity of processed olives (Bordons & Núñez-Reyes, 2008) and is considered as an index of production efficiency. According to the IOC (IOC, 2011), processing yield (%) is calculated according to the following formula:

$$\text{Processing yield} = \frac{\text{kg of oil obtained}}{\text{kg of olives processed}} \times 100$$

In general, processing yield ranges around 15%–20% in the olive oil industry (Aparicio & García-González, 2013). Baccioni and Peri (2014) reported that a value of 15.3% for processing yield is acceptable for industry, whereas 13.5% is a rather poor one. Among all production steps, malaxation temperature is the most critical factor for a high processing yield. Higher malaxation temperatures are considered beneficial for the increase in processing yield, which consequently

TABLE 6 Characteristics of virgin olive oils produced under different heating—nonheating conditions (source: comparative studies since 2000).

		Physicochemical characteristics—oxidative stability—radical scavenging activity						Polar phenolic compounds		Color attributes[a]					Other compounds[b]			
	Yield	Acidity, peroxide values K	Oxidative stability	DPPH	Fatty acid composition	Sensory evaluation	Volatile compounds	Phenolic compounds	Bitterness index	(A)	(B)	(C)	(D)	(E)	Toc	Sq	St	waxes
Angerosa, Mostallino, Basti, and Vito (2001)						+	+	+										
Ranalli, Contento, Schiavone, and Simone (2001)		+	+		+	+	+	+			+	+	+	+	+		+	+
Caponio and Catalano (2001)		+					+					+		+				
García et al. (2001)								+										
Servili, Selvaggini, Taticchi, Esposto, and Montedoro (2003)						+	+	+										
Kalua, Bedgood, Bishop, and Prenzler (2006)	+	+			+		+	+										
Parenti, Spugnoli, Masella, and Calamai (2008)		+						+				+						
Inarejos-García, Gómez-Rico, Salvador, and Fregapane (2009)	+	+	+	+	+			+	+		+				+			
Boselli et al. (2009)		+	+		+	+		+										
Gómez-Rico, Inarejos-García, Salvador, and Fregapane (2009)							+	+										

Continued

TABLE 6 Characteristics of virgin olive oils produced under different heating—nonheating conditions (source: comparative studies since 2000)—cont'd

		Physicochemical characteristics—oxidative stability—radical scavenging activity						Polar phenolic compounds		Color attributes					Other compounds			
	Yield	Acidity, peroxide values K	Oxidative stability	DPPH	Fatty acid composition	Sensory evaluation	Volatile compounds	Phenolic compounds	Bitterness index	(A)	(B)	(C)	(D)	(E)	Toc	Sq	St	waxes
Stefanoudaki, Koutsaftakis, and Harwood (2011)		+	+			+		+				+						
Taticchi et al. (2013)								+										
Reboredo-Rodríguez, González-Barreiro, Cancho-Grande, and Simal-Gándara (2014)		+			+		+										+	+
Selvaggini et al. (2014)								+										
Jolayemi, Tokatli, and Ozen (2016)		+						+		+		+		+				
Veneziani et al. (2017)	+						+	+										
Kula et al. (2018)		+					+	+		+		+					+	

+: parameter is examined in the study.
[a]*Color attributes: (A) carotenoids; (B) β-carotene; (C) chlorophylls; (D) pheophytins; (E) color.*
[b]Toc, *tocopherols;* Sq, *squalene;* St, *sterols.*

increases the profit of oil mill operators. However, this traditional practice is now changing because good manufacturing practices relate higher temperatures with lower quality of the oil in terms of sensory and nutritional attributes (Tsimidou, 2006). Amirante, Cini, Montel, and Pasqualone (2001) have investigated the effect of malaxation conditions such as temperature and time on olive oil quality. Processing yields were found to be 15.1%, 16.9%, and 17.4% when the malaxation conditions were 27°C (60min), 32°C (45min), and 35°C (45min), respectively. From these results, the effect of temperature was reported as more significant than the effect of malaxation time on the yield.

In a study by Inarejos-García et al. (2009), two batches of olives were processed into olive oil with different malaxation conditions. Processing yields were 18.3%, 20.6%, and 20.9% for batch one and 16.6%, 19.0%, and 23.5% for batch two at 20°C, 28°C, and 40°C malaxation temperatures, respectively. The mean increase in the yield was found to be around 13% as the temperature of malaxation increased from 20°C to 28°C, which is associated with the coalescence of the oil droplets, which become easier to extract. As the temperature during malaxation increases, the viscosity decreases and the droplets of olive oil can merge easily. Ranalli et al. (2001) reported that more oil is lost in the pomace and vegetative water outputs of the three-phase extraction system at 20°C and 25°C than at 30°C and 35°C malaxing temperatures. However, the respective loss between 20°C and 25°C was less than 1% in both cases. Malaxation temperature has a great effect on the OO yield. However, an excessive temperature increase induces loss of polar compounds and loss of volatile compounds, which affects sensorial properties and accelerates oxidative degradation of the oil (Clodoveo, 2012; Di Giovacchino et al., 2002; Kalogianni, Georgiou, & Exarhopoulos, 2019). Moreover, if the temperature of the malaxation is higher than 50°C, some constituents such as waxes, aliphatic alcohols, and triterpene di-alcohols solubilize easier in the oil phase and, consequently, their concentration increases.

4.3 Physicochemical characteristics—Oxidative stability—Radical scavenging activity

To study the influence of operative conditions adopted in cold press technology on the quality of the VOO produced, it is important to evaluate first the quality criteria according to the regulation EC No 2568/91 (European Union Commission, 1991). These criteria are related to hydrolytic and oxidative changes occurring in the olive fruit, during oil extraction, and also during storage of the end product. Among them, percentage free acidity, peroxide value (PV), and absorbance in the UV region, expressed as K_{232} and K_{270} values, were evaluated in most of the comparative studies presented in Table 6. Caponio and Catalano (2001) investigated the effect of different crushers and working temperatures using semiindustrial equipment on routine quality characteristics and VOO shelf life. The examined temperature during crushing and the temperature of the output paste caused by the olive-crushing process were 8°C and 14°C, 10°C and 16°C, 15°C and 20°C for the hammer crusher, respectively; 8°C and 12°C, 10°C and 14°C, 15°C and 18°C for the toothed disk crusher, respectively. Hammer crushing caused an average temperature rise of 5.5°C, whereas using disk crushing, the increase was lower by 2°C. Free acidity was not affected so much by crushing means (0.28%–0.33% and 0.22%–0.24% for oils obtained from hammer-crushed and toothed-crushed olives, respectively). Crushing temperature influenced the peroxide values, as there was a slight but constant increase with temperature. Additionally, VOOs produced with hammer crushers were found to be more prone to autooxidation, with a peroxide mean value of 10.9 meqO_2/kg—slightly higher than that produced with the disk crusher, with a mean value of 6.8 meqO_2/kg. Spectrophotometric indices showed the same increasing trend with mean values of 1.37 (hammer crushing) and 1.38 (disk crushing) for K_{232} and 0.18 (hammer crushing) and 0.16 (disk crushing) for K_{270}. Regarding oxidative stability assessed using the Rancimat method (120 °C; 20L/h airflow), a 1.6h (hammer crusher) and 0.2 h (disk crusher) reduction was observed at the highest temperature, considering that at the lowest examined temperature, oils exhibited the same stability (~20.4h) irrespective of the crushing system.

Caponio and Gomes (2001) studied the effect of the temperature at which olives were preserved before crushing in combination with malaxation operation using a pilot hummer crusher. The examined temperatures were 6°C and 18°C. The results obtained showed that oils extracted from olives stored at 6°C presented the lowest values for the legal quality criteria. Increase in storage temperature from 6°C to 18°C affected these values negatively. Thus, for oils obtained without malaxation, the mean values for free acidity, peroxide values, K_{232} and K_{270}, were 0.30% and 0.31%, 3.6 and 3.9 meq O_2/kg, 1.64 and 1.66, 0.177 and 0.173, respectively. Olive paste malaxation at the same temperatures slightly increased the mean values of the parameters considered. Thus, the values recorded were: free acidity, 0.31% and 0.33%; peroxide value, 3.8 and 4.2 meqO_2/kg; K_{232}, 1.64 and 1.70; K_{270}; 0.17 and 0.19, at 6°C and 18°C, respectively. The induction period for the oils obtained from olives stored at 18°C and treated with and without paste malaxation were 14.6h and 14.2h, respectively (Rancimat method, 120°C, 20L/h airflow). These values were on average 20% lower than the corresponding values obtained for oils extracted from the olives preserved at 6°C. These findings were not expected and further examination of this factor is necessary. In another publication, Caponio, Pasqualone, Gomes, and Catalano (2002) examined the effect of 12°C, 16°C, and 20°C crushing temperatures on the quality criteria of the corresponding Cima di Bitonto VOOs extracted in a lab-scale hammer crusher. The operation of the hammer crusher at 12°C caused an 8°C temperature increase during crushing, whereas at room temperature (16°C and 20°C) the increase was gradually less noticeable. At these conditions, percentage free acidity, peroxide values, and K_{232} values did not differ significantly.

Since the "first cold pressing" process seems promising to produce a high-quality final product, a new cooler was designed in the study of Kula et al. (2018) to determine the effect of different working temperatures separately, during crushing and the subsequent malaxation process on the overall quality and chemical composition of VOO obtained from the Edremit yaglik olive cultivar. Oil samples were produced by eight different processes: "first cold pressing" at 13°C, 18°C, and 24°C, crushing and pressing at 30°C, "cold extraction" at 13°C, 18°C, and 24°C, and crushing, malaxation, and pressing at 30°C. The lowest free acidity values were recorded at 30°C crushing temperature for oils produced with or without malaxation of the olive paste. Peroxide values (index of primary oxidation products) were not statistically different. K_{232}, an index of conjugated oxidation products, decreased for oil samples without malaxation, while no difference was observed for VOOs obtained after malaxation. K_{270} value, an index for the evolution of secondary oxidation products, was not found to change under the examined experimental conditions.

The impact of temperature and its management during malaxation has also been investigated in depth since 2000. The results demonstrated that there is a minimal effect with significant changes in the abovementioned quality criteria only under drastic temperature conditions. Ranalli et al. (2001) tested four different malaxation temperatures (20°C, 25°C, 30°C, and 35°C) in the production of VOOs from Caroleo, Leccino, and Dritta olive cultivars. The results for all samples showed that increase in free acidity, peroxide index, and ultraviolet absorption coefficients occurred for temperatures above 25°C. This trend was not reflected in the results from Swift's accelerated oxidation test, which was performed to evaluate the resistance to autoxidation. Indeed, instead of decrease instability at higher temperatures (30°C and 35°C), the opposite was observed, which was related to a lower release of phenolic compounds at lower temperature. This finding is of utmost importance for cold press VOO technology.

In another study, Kalua et al. (2006) evaluated oils extracted from olive pastes using a four-level complete factorial experimental design with four malaxation times (30, 60, 90, and 120 min) and four malaxation temperatures (15°C, 30°C, 45°C, and 60°C) as factors for Frantoio VOO extraction. Their results verified to a certain extent the observations by Ranalli et al. (2001). Moreover, it was found that malaxation time did not significantly alter percentage free acidity, PV value, or the content of volatiles and individual polar phenols at 15°C. Inarejos-García et al. (2009), working with Cornicabra VOO from a two-phase pilot plant, found no differences in the values of the above legal indices up to 28°C; they also confirmed that cold pressed VOOs have lower stability and lower antiradical capacity due to the lower content in polar phenolic antioxidants and not in tocopherols. Boselli et al. (2009) presented similar findings for the initial values of free acidity, peroxide values, and stability expressed as induction period, but their data indicate a better storage behavior of the oils obtained under cold technology conditions in the course of 12-month storage at room temperature in the dark. A positive correlation of oxidative stability with the content of polar phenolic compounds and increase in malaxation temperature is also reported by Stefanoudaki et al. (2011). Similar trends were reported more recently (Jolayemi et al., 2016) on the effect of harvesting and malaxation temperature to olive oil quality from Ayvalik and Memecik olive cultivars in a two-phase continuous olive plant. In conclusion, it seems that application of cold technology does not favor the initial stability of the extracted VOO due to the initial lower content of polar phenols and this is a critical point that has to be optimized per cultivar and also be inspected from crop season to crop season. For example, cultivars that are characterized by a high content of polar phenols may be more suitable for the cold extraction of VOO.

4.4 Fatty acid and acylglycerol composition

Triacylglycerols make up approximately 99% of olive oil composition. Fatty acids generally exist in the esterified form. The limits for the major fatty acids (% methyl esters) adopted by Codex Alimentarius are: palmitic acid (C16:0), 7.5%–20%; palmitoleic acid (C16:1), 0.3%–3.5%; stearic acid (C18:0), 0.5%–5%; oleic acid (C18:1), 75%–83%; linoleic acid (C18:2), 3.5%–21%; and linolenic acid (C18:3), no specific limits are reported. In addition to these fatty acids, the limits of minor fatty acids are myristic acid (C14:0), <0.05%; margaric acid (17:0), <0.3%; margaroleic acid (C17:1), <0.3%; arachidic acid (C20:0), <0.6%; gadoleic acid (C20:1), <0.4%; behenic acid (C22:0), <0.2% and lignoseric acid (C24:0), <0.2% acids; (Codex Alimentarius Commission, 2017b). The major triacylglycerol species in VOO are OOO (40%–59%), POO (12%–20%), OOL (12.5%–20%), POL (5.5%–7%), and SOO (3%–7%). Other species reported are POP, POS, OLnL, LOL, OLnO, PLL, PLnO, and LLL (Boskou, Blekas, & Tsimidou, 2006). The three-letter symbolism represents the triacylglycerols containing the three acyl groups indicated, where P is palmitic, O is oleic, S is stearic, L is linoleic, and Ln is linoleic acid.

Inarejos-García et al. (2009) and Reboredo-Rodríguez et al. (2014) reported no differences in fatty acid composition regarding malaxation temperature. Ranalli et al. (2001) found that monounsaturated/polyunsaturated fatty acid (MUFA/PUFA) ratios of Caroleo, Leccino, and Dritta VOOs produced with malaxation temperatures under 27°C were lower by 1.2-fold on average than the olive oils produced at 35°C, which was associated with the accelerated oxidation of polyunsaturated fatty acids at higher temperatures. Boselli et al. (2009) reported that the oleic/linoleic ratio, which could be related with oxidation stability, did not differ for both Frantoio/Leccino and Coratina VOOs obtained via the cold extraction process, after 12 months of storage. From the findings of the studies mentioned, it can be concluded that although a fluctuation in fatty acid composition was observed, in general, it is not greatly affected by the malaxation temperatures of interest to cold technology (<27°C). With regard to acylglycerols, there is no extensive scientific evidence on the effect of malaxation temperature on triacylglycerol composition. Ranalli et al. (2001) reported that the total triacylglycerol content of VOOs obtained from Caroleo (varying between 95.4% and 96.9%), Leccino (varying between 96.7% and 97.2%), and Dritta (varying between 92.3% and 94.9%) did not differ significantly when the malaxation temperature increased from 20°C to 35°C. However, any increase in malaxation temperature above 27°C favored the isomerization of the 1,2-diacyglycerols present in fresh VOO to the more stable 1,3-forms. This rearrangement resulted in a mean percentage decrease in the 1,2-diacyglycerols/1,3-diacylglycerols ratio. The high 1,2-diacyglycerols/1,3-diacylglycerols ratio is correlated with good manufacturing practices and better organoleptic attributes of VOO (Gertz & Fiebig, 2006). The latter may be possible when cold extraction technology is applied.

4.5 Sensory attributes

Olive oil IOC trade standards have introduced scores for the evaluation of positive attributes such as fruity, pungent, and bitter, which characterize fresh products from "healthy" olives that have been extracted under the most favorable conditions, among which is no heating at any stage of processing. Accordingly, application of temperatures ~27°C and below to healthy olive processing should not be related initially with defects. This is exemplified in Table 7, which contains data from relevant studies published in the 21st century. These comparative studies referred to VOOs from different cultivars, which were produced under cold pressed extraction or extraction at higher temperatures using laboratory or industrial mills. In all cases, the sensory evaluation was conducted by panels accredited by the International Olive Council (IOC, 2018). Regarding positive attributes, "fruity," "bitter," and "pungent" were not uniformly affected by the malaxation temperature increase. In particular, Angerosa et al. (2001) assessed VOOs extracted from two Italian cultivars, Coratina and Frantoio, in a laboratory mill at 25°C and 30°C paste malaxation temperature. A weaker intensity was observed in positive attributes in the samples obtained at higher temperatures.

Boselli et al. (2009) evaluated the sensory profile of VOOs from the same cultivars, subjected not only to similar low malaxation temperatures, but also to one much higher than 45°C in an industrial plant. For the VOO blend from Frantoio and Leccino cultivars (1:1, w/w), the intensity of bitter score decreases slightly, whereas the monovarietal Coratina oil presented an opposite trend. The latter indicated the relationship between bitter taste and the content of polar phenolic compounds, as their content in Coratina oil increased more when the temperature increased from 35°C to 45°C. Pungent score versus malaxation temperature showed a bell-shaped curve for both oils, with maximum scores at 35°C, while fruity was affected negatively by temperature. Green volatiles from the LOX pathway, which are attributed to the fruity sensation, were affected negatively by temperature (Angerosa et al., 2001; Ranalli et al., 2001) due to the enzymatic inactivation above 30°C (Taticchi, Esposto, & Servili, 2014).

Similar results were obtained in the study of Stefanoudaki et al. (2011) where a bitter and pungent taste attribute of oil from Koroneiki cv presented lower values at higher temperatures and, concomitantly, an increase in the total phenol content. From the aforementioned considerations, it is clear that optimization of malaxation temperature regarding sensory characteristics of the final product is cultivar dependent. This dependence is nicely justified by the results obtained by Servili et al. (2003) who, using a response surface model, found that the optimal malaxation temperature for Frantoio and Moraiolo cultivars were not the same (22°C and 26°C, respectively). The presence of negative attributes in VOO is mainly expected when "unhealthy" olives are extracted, due to the development of fermented product off-odors (Taticchi et al., 2014), or when unsuitable malaxation temperature/time combinations are applied (Angerosa et al., 2001). In the studies presented in Table 7, the negative attribute heated/burnt off-flavor was detected only at quite a high temperature (45°C). As evidenced from the data presented, cold pressed conditions ensure high sensory scores, which were maintained for temperatures slightly above 30°C. The malaxation temperature of olive paste has to be adjusted in combination with the olive cultivar characteristics that of course vary from harvest to harvest. Experimental mills that operate using only ~500 g olives may help to achieve optimum sensory attributes. Except for volatiles, nonvolatile compounds,

TABLE 7 **Impact of heating and nonheating conditions on positive attributes of virgin olive oils from different cultivars produced in laboratory or industrial mills (source: comparative studies published since 2000).**

Reference	Olive variety	Country	Processing system	Heating temperature	Positive attributes			
Angerosa et al. (2001)	Frantoio	Italy	Laboratory mill	25°C, 30°C	Fruity; increased with increased temperature Bitter; decreased with increased temperature Pungent; decreased with increased temperature			
	Coratina				Same trend			
Ranalli et al. (2001)	Caroleo	Italy	Two-phase centrifugal	20°C, 25°C, 30°C, 35°C	20°C 7.4	25°C 7.3	30°C 7.1	
	Leccino				20°C 7.9	25°C 7.7	30°C 7.5	
	Dritta				20°C 7.6	25°C 7.5	30°C 7.3	
Boselli et al. (2009)	Frantoio / Leccino 1:1	Italy	Two-way continuous industrial plant	25°C, 35°C, 45°C	Fruity; was not affected Bitter; decreased slightly with increased temperature Pungent; bell-shaped trend with temperature, highest scores at 35°C			
	Coratina				Fruity; decreased with increased temperature Bitter; increased with increased temperature Pungent; bell-shaped trend with temperature, highest scores at 35°C			
Stefanoudaki et al. (2011)	Koroneiki	Greece	Laboratory mill	15°C, 30°C, 37°C, 42°C	15°C Fruity 3.5±1.3 Bitter 3.2±1.0 Pungent 3.2±0.7	30°C Fruity 3.0±0.8 bitter 2.7±0.8 pungent 2.7±1.0	37°C Fruity 3.0±0.8 Bitter 2.0±0.8 Pungent 2.2±0.6	42°C Fruity 2.8±1.0 Bitter 1.0±0.9 Pungent 2.0±0.8

mainly the group of oleuropein and ligstroside derivatives (Vitaglione et al., 2015), influence the overall consumer perception of volatile aroma and taste.

4.6 Volatiles

The molecular weight of volatile components is lower than 300 Da, thus these components can easily vaporize at room temperatures and can dissolve in mucus and bind with olfactory receptors to create the odor sensation. Due to the very low odor thresholds of some volatiles, for example, $<4\,\mu g/kg$, slight changes in volatile compositions may lead to significant sensorial effects (Kalua et al., 2007). Taticchi et al. (2014) refer to the list of 69 volatiles divided into six groups of chemical compounds (aldehydes, ketones, alcohols, esters, acids, and misceleneous) together with odor attributes assigned to each of them first published by Angerosa et al. (2004). Volatile components that contribute to the positive sensorial properties of virgin olive oils are presented in Fig. 4.

Some of the existing volatile components have higher impact to the sensorial characteristics of VOO, such as those derived from hexanal, *trans*-2-hexenal, 1-hexanol, and 3-methylbutan-1-ol, which are the most abundant volatile compounds in VOO (Christophoridou, 2017; Mariotti & Peri, 2014). Some volatiles of VOO are associated with positive sensorial properties of VOO such as for green and fruity perception with *trans*-2-hexenal, hexanal, *cis*-3-hexenal, hexan-1-ol, *cis*-3-hexen-1-ol, hexyl acetate, and *cis*-3-hexenyl acetate, and the hay-like sensory with some alcohols of C_5 and C_6 volatiles (Boskou, Blekas, & Tsimidou, 2006; Taticchi et al., 2014). Others are associated with sensorial defects. Morales et al. (2005) investigated the volatile contents of VOO standards used for the training of sensorial defects and reported that the responsible ones were 1-octen-3-ol (0.25 mg/kg), 2-heptenal (0.34 mg/kg), and 1-octen-3-one (0.13 mg/kg), which were associated with the "musty" defect. In general, acids and esters were associated with the "fusty" defect namely, ethyl butanoate (3.7 mg/kg), propanoic (15.6 mg/kg), and butanoic acids (11.5 mg/kg). For the "vinegary-winey" defect, acetic acid (6.21 mg/kg), 3-methyl butanol (0.71 mg/kg), and ethyl acetate (3.53 mg/kg) were found as the most responsible ones. "Rancid" was associated with 2-octenal (1.10 mg/kg), 2-heptenal (1.18 mg/kg), and 2-decenal (1.54 mg/kg).

The formation pathway of volatiles in olive oil is a chain of numerous enzymatic reactions that starts with the release of linolenic and linoleic acids after hydrolysis of triacylglycerols. Lipoxygenase (LOX) turns linoleic and linolenic acid into 9- and 13-hydroperoxides and hydroperoxide lyase (HPL), *cis*-3-*trans*-2-enal isomerase (CTI), alcohol dehydrogenase (ADH), and alcohol acyltransferase (AAT) are responsible for the cleavage of these hydrocarbons into C_5 and C_6 volatiles through subsequent reactions. Fig. 5 illustrates pathways for the formation of VOO volatiles based on Kalua et al. (2007), Gomes da Silva et al. (2012) and Joaquin et al. (2013).

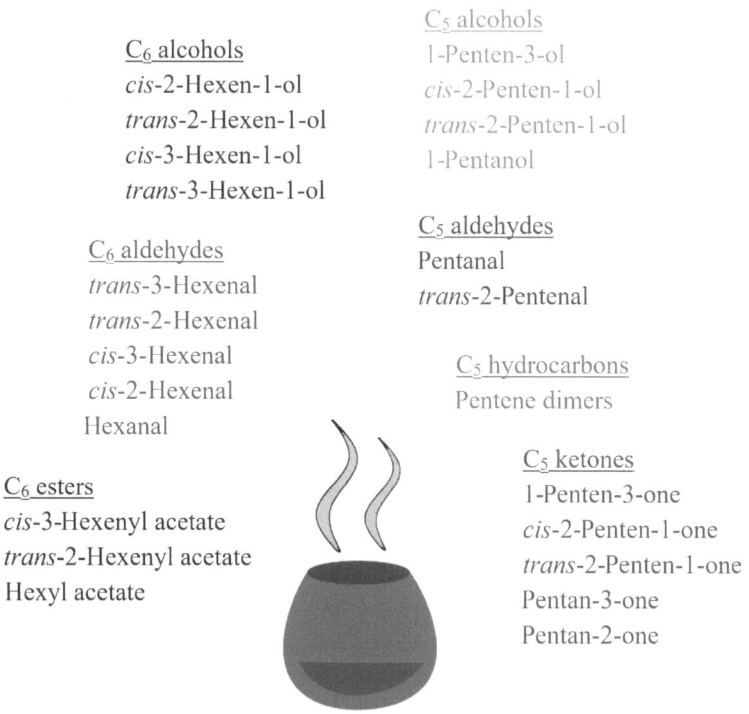

FIG. 4 Major virgin olive oil volatiles contributing to positive sensorial attributes. The list was prepared according to Angerosa (2002), Angerosa et al. (2004), Morales, Luna, and Aparicio (2005), Kalua et al. (2007), and Reboredo-Rodríguez et al. (2014).

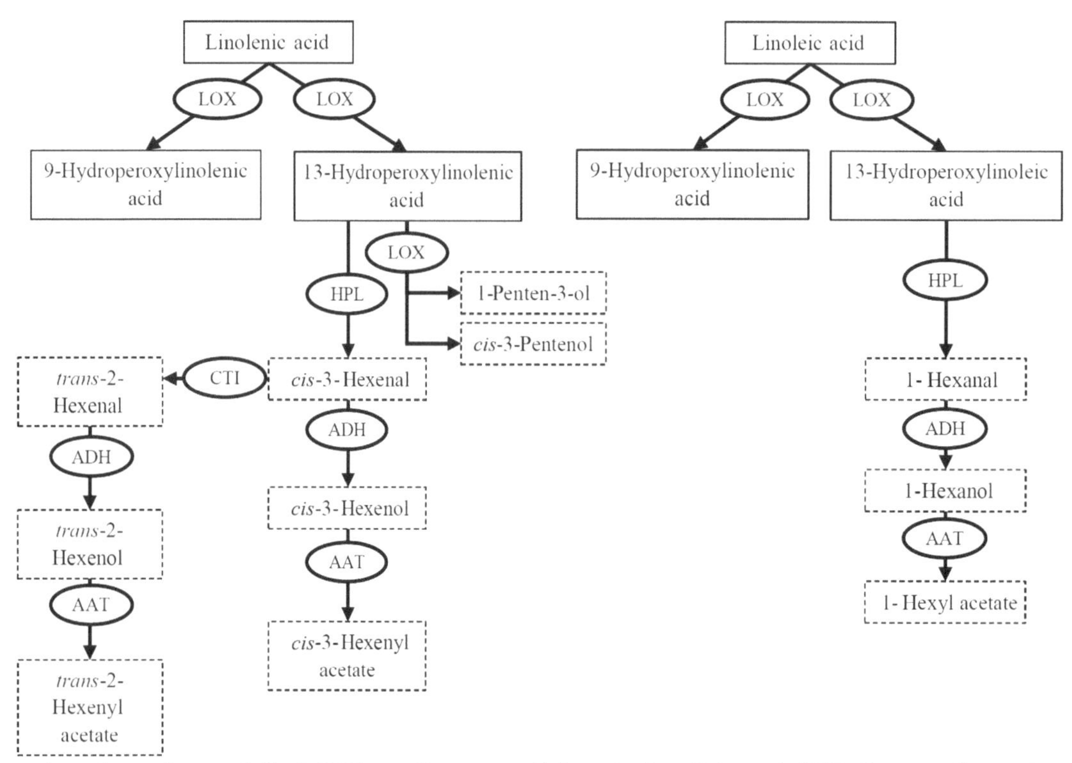

FIG. 5 Formation pathway of some volatiles in VOO according to merged information from Kalua et al. (2007), Gomes da Silva, Freitas, Cabrita, and Garcia (2012), and Joaquín, Harwood, and Martínez-Force (2013). Enzymes given in ellipses and volatiles are represented in dashed rectangles. LOX: lipoxygenase, HPL: hydroperoxide lyase, CTI: *cis*-3:*trans*-2-enal isomerase, ADH: alcohol dehydrogenase, AAT: alcohol acyltransferase.

Volatile compounds in olive oil are mainly formed during the postharvest period and extraction. The changes in volatile composition of VOO determined the final sensory quality (Gomes da Silva et al., 2012). Among the mechanical procedures, crushing and malaxation have an important role in the formation of volatile compounds.

4.6.1 Effect of olive oil extraction steps to volatile composition

Angerosa et al. (2004) stated that as volatile components are mostly liberated during tissue breakage, the crushing method and conditions affect the volatile composition of the end product. In particular, temperature increase during intensive hammer mill crushing leads to a reduction in hydroperoxide lyase activity, which ends up with an OO with low *trans*-2-hexenal, hexanal, and *cis*-3-hexen-1-ol content than the one produced using stone mills (Angerosa et al., 2004). This statement is also supported by findings of Servili et al. (2002), who studied the effect of hammer mill and metal crusher on volatile composition oils of Coratina variety. Crushing temperature is one of the critical factors for volatile composition. In a study by Kula et al. (2018), the effect of crushing temperature on quality of olive oil was investigated. The results of this study revealed that percentages of some volatiles associated with pleasant sensorial attributes, such as *trans*-2-hexenal, decreased as the crushing temperature increased (2.77%, 2.28%, 2.69%, and 1.78%, at 13°C, 18°C, 24°C, and 30°C, respectively). Concomitantly, concentrations of volatile components such as acetic acid and acetate, which are associated with the "vinegary-wine" defect almost doubled as the temperature of crushing increased from 13°C to 30°C.

The major effect of high temperature during malaxation (35°C) on the volatiles is an increase in hexan-1-ol and *trans*-2-hexen-1-ol, and a decrease in C_6 esters (*cis*-3-hexenyl acetate, *trans*-2-hexenyl acetate, hexyl acetate). In parallel, this impact accounts sensorial quality of VOO with a weakening in pleasant sensations such as leaf, freshly cut grass, walnut husk, bitter, and pungent sensory notes (Angerosa, 2002). According to Reboredo-Rodríguez et al. (2014), a high malaxation temperature generally leads to a reduction in total content of volatiles from LOX pathways, which is due to the inevitable inactivation of hydroperoxide lyase. The results of this study showed that an increase in the temperature of the malaxation from 20°C to 30°C decreased the contents of *trans*-2-hexenal (from 1.87 to 1.10 mg/kg), *cis*-3-hexen-1-ol (from 1.10 to 0.89 mg/kg), *cis*-3-hexenyl acetate (from 0.76 to 0.21 mg/kg), hexanal (from 0.41 to 0.28 mg/kg), and 1-hexanol (from 2.36 to 1.84 mg/kg), which are the C_6 volatiles. This decrease in the volatile content also has an impact on the sensorial quality of the olive oil, since these volatiles are more responsible for pleasant odors of cut grass and floral senses (Clodoveo, 2012; Reboredo-Rodríguez et al., 2014). Gómez-Rico et al. (2009) studied the effect of malaxation conditions on the phenolic and volatile composition of different VOOs. According to the findings, *trans*-2-hexenal content decreased from 2.05 to 1.46 mg/kg when the malaxation temperature increased from 20°C to 40°C and duration of malaxation was 50 min. In addition, hexenal content decreased from 0.15 mg/kg (20°C) to 0.03 mg/kg (40°C) with increasing malaxation temperature. These changes are associated with higher activity of lipoxygenase pathway enzymes, especially hydroperoxide lyase at low malaxation temperatures. The findings of the above studies indicate that low temperatures (≤25°C) favor the formation of pleasant volatiles, which are responsible for high sensorial quality of VOO.

4.7 Polar phenolic compounds

Polar phenolic compounds represent a class of secondary plant metabolites, which are transferred from olive fruit to oil during processing. These compounds, which are still often referred to by the inaccurate term "olive oil polyphenols," are contained in the polar fraction of virgin olive oil (Boskou, Tsimidou, & Blekas, 2006). The different classes of phenolic compounds that can be found in this fraction are phenolic alcohols and acids, hydroxy-isochromans, flavonoids, lignans, and secoiridoid derivatives. Among them, secoiridoid derivatives, including the dialdehydic form of decarboxymethyl elenolic acid linked to hydroxytyrosol (3,4-DHPEA-EDA) or tyrosol (p-HPEA-EDA), lignans (+)-1-acetoxypinoresinol and (+)-1-pinoresinol, and the two aglycones of the oleuropein (3,4-DHPEA-EA) and ligstroside (*p*-HPEA-EA), prevail. Other classes like phenolic acids (caffeic, vanillic, *p*-coumaric, *o*-coumaric, ferulic, and cinnamic acid) and hydroxyl-isochromans can be found in lower amounts. The respective alcohols, hydroxytyrosol (3,4-DHPEA) and tyrosol (*p*-HPEA), are present in small amounts in freshly extracted VOOs, whereas during storage their concentration increases (Boskou, 2015). Polar phenolic compounds have been extensively investigated, initially because they contribute to the stability of the oil in the dark, for their organoleptic characteristics such as bitterness, pungency, and astringency (Tsimidou, 2013), and later on for their bioactive properties (Boskou, 2015). Taking into account that the total phenol content is usually high in good-quality oils, there is an imperative need to gain knowledge of the effect of processing conditions on the level of total and individual phenolic compounds.

As the phenolic compounds present in VOO originate from those encountered in the olive drupe, many publications have centered on the agronomic and technological conditions of oil production that may affect their presence in the final product. The most studied agronomic aspects are cultivar, maturity index, pedoclimatic production conditions, and agronomic practices (Boskou, Tsimidou, & Blekas, 2006; Servili et al., 2004).

During crushing, qualitative and quantitative changes of phenolic compounds depend on the type of crusher, operation temperature, and activities of various endogenous enzymes of the olive fruit. Caponio and Catalano (2001) investigated the effect of different crushers and working temperatures, using semiindustrial devices, on the total phenol content of Coratina VOOs. In addition, phenolic concentration did not change substantially with the temperature increase (from 8°C to 15°C) in both types of crushers; from 462 to 453 mg/kg in hammer crusher and from 466 to 481 mg/kg in the toothed disk crusher. In another publication, Caponio et al. (2002) examined the effect of sightly higher crushing temperatures—12°C, 14°C, and 20°C for Cima di Bitonto VOOs—using a laboratory-scale hammer crusher. The results indicated a decrease in phenolic concentration (from 390 to 350 mg/kg) with a temperature increase.

Over the last 20 years, several papers have been published on the effect of malaxation conditions on the level of total or individual phenols in the produced VOO (Table 6). The results regarding this subject are rather conflicting. Ranalli et al. (2001) observed an average increase of ~30% in the concentration of total phenols of VOOs from three different olive cultivars due to change in the malaxation temperature from 20°C to 30°C. This increase was not evident when the paste temperature increased from 30°C to 35°C. A similar increase pattern was observed for *o*-diphenols. A bell-shaped curve of total phenol content with respect to malaxation temperature (21–36°C) was observed by Parenti et al. (2008); a significant increment of phenol concentration was found with a maximum at 27°C. A similar trend was found by Inarejos-García et al. (2009). An increase in temperature in the range 20–35°C during the olive paste malaxing process leads to a greater release of the phenolic compounds from the olive paste into the oil phase (Taticchi et al., 2013). A positive correlation of total phenol content with malaxation temperature was also observed even at higher temperatures (40–45°C) (Boselli et al., 2009; Stefanoudaki et al., 2011). In some investigations, it is reported that an increase in temperature within the range 30–45°C favors the activity of endogenous phenol oxidase and peroxidase, which results in a progressive reduction of their content during malaxation. This effect can be controlled by performing malaxation in "covered malaxers" where operation takes place under a controlled oxygen atmosphere environment (García et al., 2001; Taticchi et al., 2013). From the aforementioned considerations, it is clear that the control of O_2 concentration in the paste during malaxation process can be considered as a technological parameter, in combination with the duration and the temperature of the procedure, to optimize phenolic content in VOO. It should be noted that the optimum operative conditions of malaxation are cultivar dependent (Selvaggini et al., 2014).

With reference to major bound forms of hydroxytyrosol and tyrosol, their content was affected positively up to 30°C, and then decreased sharply. A temperature rise between 21°C and 30°C caused a 4-fold increase in the concentration of 3,4-DHPEA-EDA, which accounted for the 60% of the total secoiridoids in the VOOs analyzed (Parenti et al., 2008). A positive relation of secoiridoid concentration with a malaxation temperature up to 35°C was reported (Taticchi et al., 2013). In particular, three times higher levels of the p-HPEA-EDA compound were found in VOO processed either in an oil mill plant or in an Abencor system (laboratory mill) when the malaxation temperature was increased from 20°C to 40°C (Gómez-Rico et al., 2009). A marked increase in 3,4-DHPEA-EDA content (from 16.0 to 49.9 mg/kg for blended oil and 39.9 to 118 mg/kg for monovarietal oil) and in *p*-HPEA-EDA (from 73.5 to 96.0 mg/kg for blended oil and 101 to 152 mg/kg for monovarietal oil) content was also observed by Boselli et al. (2009), for oils obtained in a two-phase continuous industrial plant. The intensity of pungency attribute is mainly related to the content of *p*-HPEA-EDA, whereas 3,4-DHPEA-EDA and 3,4-DHPEA-EA are associated with the perception of bitterness (Tsimidou, 2013); therefore, it is of particular interest to study the evolution of these compounds in response to malaxation conditions. High amounts of these compounds may produce excessive pungency and bitterness that may affect consumers' acceptance for the final product negatively.

With regard to the content of low molecular weight phenolic compounds (MW < 170), a different trend was presented; hydroxytyrosol and tyrosol the main phenolic alcohols, increased linearly with increasing temperature in the range 30–60°C (Boselli et al., 2009; Kalua et al., 2006; Parenti et al., 2008). The formation of these compounds that results from the hydrolysis of the secoiridoids may be accelerated at elevated temperatures. A positive correlation with temperature was also observed for the content of vanillic acid and vanillin (Parenti et al., 2008).

The major lignans of VOO, (+)-pinoresinol and (+)-1-acetoxypinoresinol, seem not to be significantly affected by the malaxation temperature. Their concentration is only slightly related to the equilibrium between release from cell walls and oxidative degradation catalyzed by oxidoreductases (Parenti et al., 2008; Taticchi et al., 2013).

In conclusion, it is important to note the great effect of crushing and especially malaxation on the fate of phenolic compounds in VOO. Indeed, during malaxation, not only a physical separation takes place but also a complex bioprocess. The abovementioned results indicate that malaxation temperature exerts two contrasting effects: it accelerates the degradation on the one hand, and improves the solubility of the phenolic compounds on the other. The extent to which these phenomena take place determines the concentration of polar phenolic compounds in the final product.

4.8 Other compounds

Other bioactive constituents and nutritionally important compounds of virgin olive oil are tocopherols, squalene, sterols, and waxes.

Regarding tocopherols, α-tocopherol comprises 90% of the total tocopherol content in virgin olive oil. The latter is appreciated for its vitamin E activity and antioxidant properties. Levels of α-tocopherol depending on cultivar, geographical origin, agronomic practices, and processing conditions are expected to fluctuate, as is reported for other constituents of VOO (Tsimidou, 2012). A slight but clear increase in α-tocopherol content by increasing malaxation temperature has been reported by Ranalli et al. (2001); Ranalli, Malfatti, Lucera, Contento, and Sotiriou (2005) and Inarejos-García et al. (2009). It can be said that cold pressed or cold extracted VOOs are expected to contain lower concentrations of tocopherols.

Squalene is the main hydrocarbon present in VOO. It is an aliphatic unsaturated hydrocarbon having 30 carbon atoms and six isolated double bonds. It is regarded as an important precursor in sterol biosynthetic routes and is partially related to the health benefits of VOO and chemopreventive action against certain cancers. Mean squalene concentration in VOO is around 5000 mg/kg. Squalene content seems to be mainly cultivar dependent (Tsimidou, 2010). Extraction technology and malaxation temperature were not found to influence squalene levels (Ranalli et al., 2005).

Sterols are unsaturated alcohols, which are present in fatty tissues of plants and animals. Olive oil contains sterols, in the range of 1000–2300 mg/kg of oil. The most abundant sterol is β-sitosterol (90%–95% of total sterols) followed by Δ5-avenasterol, campesterol, and stigmasterol. Other sterols are present in smaller quantities in VOO. Waxes are another class of the unsaponifiable fraction of VOO. The major olive oil waxes are esters of carbon atoms from C36 to C46.

Total sterol content is affected by the temperature applied during VOO production. Ranalli et al. (2001) studied the effect of malaxation temperature on the sterol composition and content of VOO. The authors indicated that total sterol content of Caroleo, Leccino, and Dritta OO increased steadily with increase in malaxation temperature in the range

20–35°C. In addition, together with the sterol content, wax content of these olive oils increased from 191 to 262 mg/kg, 464 to 562 mg/kg, and 287 to 373 mg/kg for Caroleo, Leccino, and Dritta olive oils, respectively. Inarejos-García et al. (2009) investigated the malaxation temperature on the wax content of Cornicabra virgin olive, which was extracted in an industrial olive oil mill and with an Abencor system. The wax content of VOO produced in an oil mill increased from 76.5 to 85.5 mg/kg, while the wax content of VOOs from an Abencor system increased from 83.5 to 98.5 mg/kg as the temperature of the malaxation increased from 20°C to 40°C. The researchers associated the increase in the content of the waxes with the increasing solubility of the waxes in the oil phase as the temperature rose (Inarejos-García et al., 2009). Reboredo-Rodríguez et al. (2014) and Guillaume and Ravetti (2015) reported similar findings. Kula et al. (2018) investigated the effect of crushing temperature on the sterol composition of VOO, which is obtained with or without a malaxation step. According to the findings of this study, β-sitosterol, Δ5-avenasterol, and campesterol contents of VOOs did not differ significantly in the range 13–30°C. However, the authors indicated that the total sterol content of VOO rose from 83.6% to 86% when malaxation was applied to the olive pastes (Kula et al., 2018).

4.9 Appearance attributes

VOO color, which is one of the main determinants of consumer acceptability, is due to the presence of chlorophylls and carotenoids, which give it the green and yellow hues. Color is considered as a quality index though no specifications have yet been defined in legislation or trade standards. Since 2000, more and more studies have reported that olive oil pigments may also be related to health benefits (Moyano, Heredia, & Meléndez-Martínez, 2010). Their content in virgin olive oil is based on genetic factors, the geographical origin, the maturity index of the fruit, and the conditions of industrial processing and storage (Psomiadou & Tsimidou, 2001). Pigment content in VOO depends heavily on analytical methodology and is found in the magnitude of a few ppm (Gandul-Rojas, Roca, & Gallardo-Guerrero, 2016; Psomiadou & Tsimidou, 2001). Among "green" pigments, pheophytin is the predominant chlorophyll derivate, whereas the main "yellow" pigments are β-carotene and lutein. Changes in the composition and content of the pigments in the final product are mainly related to the stage of paste malaxation during which enzymatic activities take place, due to olive fruit disruption.

Chlorophylls and carotenoids, because of their lipophilic nature, are partially transferred from the crushed olive tissues to the oil. Losses occur either because they remain in the paste or because they degrade to some extent to colorless compounds (Gandul-Rojas et al., 2016). The percentage of losses depends on the conditions applied during OO processing. At the stage of crushing, the amount of chlorophylls depends on the type of crusher, the working temperature, and the temperature reached during crusher operation. Their content is doubled when the temperature rises from 8°C to 10°C and the type of crusher is the disk crusher. Further temperature increase up to 15°C favors pigment extraction but not so intensively. On the other hand, a 7°C temperature rise during hammer crushing led to a decrease of ~26% of the mean chlorophyll content (Caponio & Catalano, 2001). During malaxation, it has been observed that malaxation temperature influences positively the pigment concentration of the final product. Specifically, an increase in an optimal range between 15°C and 30°C could cause an average percentage rise of 45% in the chlorophyll content (Kula et al., 2018; Ranalli et al., 2001; Stefanoudaki et al., 2011) and 55% in the content of β-carotene (Inarejos-García et al., 2009; Ranalli et al., 2001) due to a higher release from the plant tissue. The latter indicates that the effect of the temperature is more evident for the carotenoid content because of their lower retention in the paste, compared to the corresponding one of chlorophylls (Gallardo-Guerrero, Roca, & Mínguez-Mosquera, 2002).

5 Conclusion

Temperatures up to 27°C at the major steps of VOO production have been provisioned in the EU legislation and elsewhere as an option in the processing line of this precious fruit oil. The respective cold technology products command a higher price in the market. They are richer in pleasant odors and more suitable as salad oils. The fatty acid composition is not affected so much by temperature increase. The concentration of volatiles and polar phenolic compounds is drastically influenced by temperature changes, and these changes are cultivar dependent. Extraction of hydrophobic compounds is influenced negatively by cold technology, which is not desirable regarding tocopherol, sterol, and squalene content. Modernization in olive mills is expected to introduce good manufacturing practices that will introduce temperature control throughout processing and storage steps in combination with avoidance of exposure of the olive paste and oil to oxygen and light.

Acknowledgments

M.Z.T. and A.M. acknowledge support of this work by the project "Upgrading the Plant Capital (PlantUp)" (MIS 5002803), which is implemented under the Action "Reinforcement of the Research and Innovation Infrastructure," funded by the Operational Program "Competitiveness, Entrepreneurship, and Innovation" (NSRF 2014–20) and cofinanced by Greece and the European Union (European Regional Development Fund). O.Ö. thanks the Scientific and Technological Research Council of Turkey (TUBITAK) for a scholarship under grant number 1059B191701190 during his stay at Aristotle University (School of Chemistry, Laboratory of Food Chemistry and Technology).

References

Amirante, R., Cini, E., Montel, G. L., & Pasqualone, A. (2001). Influence of mixing and extraction parameters on virgin olive oil quality. *Grasas y Aceites*, *52*, 198–201.

Angerosa, F. (2002). Influence of volatile compounds on virgin olive oil quality evaluated by analytical approaches and sensor panels. *European Journal of Lipid Science and Technology*, *104*, 639–660.

Angerosa, F., Mostallino, R., Basti, C., & Vito, R. (2001). Influence of malaxation temperature and time on the quality of virgin olive oils. *Food Chemistry*, *72*, 19–28.

Angerosa, F., Servili, M., Selvaggini, R., Taticchi, A., Esposto, S., & Montedoro, G. (2004). Volatile compounds in virgin olive oil: Occurrence and their relationship with the quality. *Journal of Chromatography A*, *1054*, 17–31.

Aparicio, R., & García-González, D. L. (2013). Olive oil characterization and traceability. In R. Aparicio, & J. Harwood (Eds.), *Handbook of olive oil, analysis, and properties* (pp. 431–472). Boston, MA: Springer.

Asheri, M., Sharifani, M. M., & Kiani, G. (2017). An examination into the effects of frozen storage of olive fruit on extracted olive oils. *Advances in Horticultural Science*, *31*, 191–198.

Balkan, B., & Meral, S. (2017). Olive oil industry dynamics: The case of Turkey. In: *35th Int. Conf. Syst. Dyn. Soc. 60th Anniv. Syst. Dyn. Celebr*, (pp. 1–26).

Bordons, C., & Núñez-Reyes, A. (2008). Model based predictive control of an olive oil mill. *Journal of Food Engineering*, *84*, 1–11.

Borja, R., Raposo, F., & Rincón, B. (2006). Treatment technologies of liquid and solid wastes from two-phase olive oil mills. *Acta Agriculturae Scandinavica*, *57*, 32–46.

Boselli, E., Di Lecce, G., Strabbioli, R., Pieralisi, G., & Frega, N. G. (2009). Are virgin olive oils obtained below 27°C better than those produced at higher temperatures? *LWT—Food Science and Technology*, *42*, 748–757.

Boskou, D. (2015). Olive fruit, table olives, and olive oil bioactive constituents. In D. Boskou (Ed.), *Olive and olive oil bioactive constituents* (pp. 1–30). Urbana, IL: AOCS Press.

Boskou, D., Blekas, G., & Tsimidou, M. (2006). Olive oil composition. In D. Boskou (Ed.), *Olive oil. Chemistry and technology* (pp. 41–72). Champaign, IL: AOCS Press.

Boskou, D., Tsimidou, M., & Blekas, G. (2006). Polar phenolic compounds. In D. Boskou (Ed.), *Olive oil chemistry and technology* (pp. 73–92). Champaign Illinois: AOCS Press.

Caponio, F., & Catalano, P. (2001). Hammer crushers vs disk crushers : The influence of working temperature on the quality and preservation of virgin olive oil. *European Food Research and Technology*, *213*, 219–224.

Caponio, F., & Gomes, T. (2001). Influence of olive crushing temperature on phenols in olive oils. *European Food Research and Technology*, *212*, 156–159.

Caponio, F., Gomes, T., Summo, C., & Pasqualone, A. (2003). Influence of the type of olive-crusher used on the quality of extra virgin olive oils. *European Journal of Lipid Science and Technology*, *105*, 201–206.

Caponio, F., Pasqualone, A., Gomes, T., & Catalano, P. (2002). Use of HPSEC analysis of polar compounds to assess the influence of crushing temperature on virgin olive oil's quality. *European Food Research and Technology*, *215*, 534–537.

Caporaso, N. (2016). Virgin olive oils: Environmental conditions, agronomical factors and processing technology affecting the chemistry of flavor profile. *Journal of Food Chemistry and Nanotechnology*, *2*, 21–31.

Castro-Garcia, S., & Ferguson, L. (2017). Mechanical harvesting of olives. In A. Kiritsakis, & F. Shahidi (Eds.), *Olives and olive oil as functional foods* (pp. 117–124). New York, USA: John Wiley & Sons Ltd.

Chiacchierini, E., Mele, G., Restuccia, D., & Vinci, G. (2007). Impact evaluation of innovative and sustainable extraction technologies on olive oil quality. *Trends in Food Science and Technology*, *18*, 299–305.

Christophoridou, S. (2017). Authentication of olive oil based on minor components. In A. Kiritsakis, & F. Shahidi (Eds.), *Olives and olive oil as functional foods* (pp. 555–567). New York, USA: John Wiley & Sons Ltd.

Clodoveo, M. L. (2012). Malaxation: Influence on virgin olive oil quality. Past, present, and future-An overview. *Trends in Food Science and Technology*, *25*, 13–23.

Clodoveo, M. L., Corbo, F., & Amirante, R. (2018). *Does the introduction of ultrasound in extra-virgin olive oil extraction process improve the income of the olive millers: The first technology for the simultaneous increment of yield and quality of the product, In: Muzzalupo, I. (Ed.).* IntechOpen, London, UK: Technological innovation in the olive oil production.

Codex Alimentarius Commission, 2017a. Standard for olive oils and olive pomace oils. Codex Stan. 33-1981. Rev. 1989, 2003, 2013; 2017. Amended 2009, 2013.

Codex Alimentarius Commission, 2017b. Standard for edible fats and oils not covered by individual standards. CODEX STAN, 19-1981. Rev. 1987, 1999. Amended 2009, 2013, 2015, 2017.

Costagli, G. (2018). The use of disc stack centrifuge in the virgin olive oil industry. *Journal of Agricultural Engineering*, *49*, 75.

Di Giovacchino, L. (2013). Technological aspects. In R. Aparicio, & J. Harwood (Eds.), *Handbook of olive oil, analysis and properties* (pp. 57–96). Boston, MA: Springer.

Di Giovacchino, L., Sestili, S., & Di Vincenzo, D. (2002). Influence of olive processing on virgin olive oil quality. *European Journal of Lipid Science and Technology*, *104*, 587–601.

DOOR, n.d.: Available from: http://ec.europa.eu/agriculture/quality/door/list.html?locale=el (5 March 2019).

Erses Yay, A. S., Oral, H. V., Onay, T. T., & Yenigün, O. (2012). A study on olive oil mill wastewater management in Turkey: A questionnaire and experimental approach. *Resources, Conservation and Recycling*, *60*, 64–71.

European Union Commission (1991). Commission Regulation (EEC) No 2568/91 on the characteristics of olive oil and olive-residue oil and on the relevant methods of analysis and its amendments. *Official Journal of the European Union*, *L248*, 1–83.

European Union Commission (2012). Implementing Regulation (EU) No 29/2012 on the marketing standards for olive oil. *Official Journal of the European Union*, *L12*, 14–21.

Gallardo-Guerrero, L., Roca, M., & Mínguez-Mosquera, M. I. (2002). Distribution of chlorophylls and carotenoids in ripening olives and between oil and alperujo when processed using a two-phase extraction system. *Journal of the American Oil Chemists' Society*, *79*, 105–109.

Gandul-Rojas, B., Roca, M., & Gallardo-Guerrero, L. (2016). Chlorophylls and carotenoids in food products from olive tree. In D. Boskou, & M. L. Clodoveo (Eds.), *Products from olive tree* (pp. 67–98). Rijeka: InTech.

García, A., Brenes, M., Martínez, F., Alba, J., García, P., & Garrido, A. (2001). High-performance liquid chromatography evaluation of phenols in virgin olive oil during extraction at laboratory and industrial scale. *Journal of the American Oil Chemists' Society*, *78*(6), 625–629.

Gertz, C., & Fiebig, H. J. (2006). Isomeric diacylglycerols—Determination of 1,2- and 1,3-diacylglycerols in virgin olive oil. *European Journal of Lipid Science and Technology*, *108*, 1066–1069.

Gomes da Silva, M. D., Freitas, A. M. C., Cabrita, M. J., & Garcia, R. (2012). *Olive oil composition: Volatile compounds. In D. Boskou (Ed.), Olive oil-constituents, quality, health properties, and bioconversions* (pp. 17–46). Rijeka: InTech-Open Access Company. Available from: (2012). https://www.intechopen.com/books/olive-oil-constituents-quality-health-properties-and-bioconversions/oil-composition-volatiles.

Gómez-Rico, A., Inarejos-García, A. M., Salvador, M. D., & Fregapane, G. (2009). Effect of malaxation conditions on phenol and volatile profiles in olive paste and the corresponding virgin olive oils (*Olea europaea* L. cv. *Cornicabra*). *Journal of Agricultural and Food Chemistry*, *57*, 3587–3595.

Guillaume, C., & Ravetti, L. (2015). Technological and agronomical factors affecting sterols in Australian olive oils. *Rivista Italiana Delle Sostanze Grasse*, *92*, 53–60.

Hermoso Fernández, M., Gonzáles, J., Uceda, M., Garcia-Ortiz, A., Morales, J., Frias, L., et al. (1998). *Elaboración de aceite de oliva de calidad II. Obtención por el sistema de dos fases: (Vol. 61)*. (p. 98). "Venta del Llano" Mengibar (Jaén) Spain: Estación de Olivicultura y Elaiotecnia.

Inarejos-García, A. M., Fregapane, G., & Salvador, M. D. (2011). Effect of crushing on olive paste and virgin olive oil minor components. *European Food Research and Technology*, *232*, 441–451.

Inarejos-García, A. M., Gómez-Rico, A., Salvador, M. D., & Fregapane, G. (2009). Influence of malaxation conditions on virgin olive oil yield, overall quality, and composition. *European Food Research and Technology*, *228*, 671–677.

International Olive Council (IOC) (2011). *Guide for the determination of the characteristics of oil-olives. COI/OH/Doc. No 1*. Available from (2011). www.internationaloliveoil.org/.../5832-co-oh-doc1english (5 March 2019).

International Olive Council (IOC) (2018). *Sensory analysis of olive oil method-organoleptic assessment of virgin olive oil. COI/T.20/Doc. No 15/Rev.10*. Available from *(2018)*. http://www.internationaloliveoil.org/estaticos/view/224-testing-methods (5 March 2019).

International Olive Council (IOC) (2019a). *Glossary*. Available from *(2019a)*. http://www.internationaloliveoil.org/glosario_terminos/index (5 March 2019) .

International Olive Council (IOC) (2019b). *World olive oil figures*. Available from (2019b). http://www.internationaloliveoil.org/estaticos/view/131-world-olive-oil-figures (5 March 2019) .

Joaquín, J. S., Harwood, J. L., & Martínez-Force, E. (2013). Lipid metabolism in olive: Biosynthesis of triacylglycerols and aroma components. In R. Aparicio, & J. Harwood (Eds.), *Handbook of olive oil, analysis and properties* (pp. 97–122). Boston, MA: Springer.

Jolayemi, O. S., Tokatli, F., & Ozen, B. (2016). Effects of malaxation temperature and harvest time on the chemical characteristics of olive oils. *Food Chemistry*, *211*, 776–783.

Kalogianni, E. P., Georgiou, D., & Exarhopoulos, S. (2019). Olive oil droplet coalescence during malaxation. *Journal of Food Engineering*, *240*, 99–104.

Kalua, C. M., Allen, M. S., Bedgood, D. R., Bishop, A. G., Prenzler, P. D., & Robards, K. (2007). Olive oil volatile compounds, flavour development, and quality: A critical review. *Food Chemistry*, *100*, 273–286.

Kalua, C. M., Bedgood, D. R., Bishop, A. G., & Prenzler, P. D. (2006). Changes in volatile and phenolic compounds with malaxation time and temperature during virgin olive oil production. *Journal of Agricultural and Food Chemistry*, *54*, 7641–7651.

Karipidis, P. (2005). The Greek olive oil market structure. *Agricultural Economics Review*, *6*, 64–72.

Kiritsakis, A., & Shahidi, F. (2017). Olive oil quality and its relation to the functional bioactives and their properties. In A. Kiritsakis, & F. Shahidi (Eds.), *Olives and olive oil as functional foods* (pp. 205–216). New York, USA: John Wiley & Sons Ltd.

Koseoglu, O., Ozturk Gungor, F., Altunoglu, Y., Yildirim, A., Irmak, S., & Sevim, D. (2016). The taste panel of olive research institute. *Olivae*, *123*, 20–22.

Kula, Ö., Yıldırım, A., Yorulmaz, A., Duran, M., Mutlu, İ., & Kıvrak, M. (2018). Effect of crushing temperature on virgin olive oil quality and composition. *Grasas y Aceites*, *69*, 1–9.

Leong, T., Juliano, P., & Knoerzer, K. (2017). Advances in ultrasonic and megasonic processing of foods. *Food Engineering Reviews*, *9*, 237–256.

Lozano-Sánchez, J., Cerretani, L., Bendini, A., Segura-Carretero, A., & Fernández-Gutiérrez, A. (2010). Filtration process of extra virgin olive oil: Effect on minor components, oxidative stability and sensorial and physicochemical characteristics. *Trends in Food Science and Technology*, *21*, 201–211.

Mansour, A. B., Flamini, G., Selma, Z. B., Le Dréau, Y., Artaud, J., Abdelhedi, R., et al. (2015). Olive oil quality is strongly affected by cultivar, maturity index and fruit part: chemometrical analysis of volatiles, fatty acids, squalene and quality parameters from whole fruit, pulp and seed oils of two Tunisian olive cultivars. *European Journal of Lipid Science and Technology*, *117*, 976–987.

Mariotti, M., & Peri, C. (2014). The composition and nutritional properties of extra-virgin olive oil. In C. Peri (Ed.), *The extra-virgin olive oil handbook* (pp. 21–34). Chichester, UK: John Wiley & Sons, Ltd.

Mele, M. A., Islam, M. Z., Kang, H. M., & Giuffrè, A. M. (2018). Pre- and post-harvest factors and their impact on oil composition and quality of olive fruit. *Emirates Journal of Food and Agriculture*, *30*, 592–603.

Morales, M. T., Luna, G., & Aparicio, R. (2005). Comparative study of virgin olive oil sensory defects. *Food Chemistry*, *91*, 293–301.

Morrone, L., Pupillo, S., Neri, L., Bertazza, G., Magli, M., & Rotondi, A. (2017). Influence of olive ripening degree and crusher typology on chemical and sensory characteristics of Correggiolo virgin olive oil. *Journal of the Science of Food and Agriculture*, *97*, 1443–1450.

Moyano, M. J., Heredia, F. J., & Meléndez-Martínez, A. J. (2010). The color of olive oils: The pigments and their likely health benefits and visual and instrumental methods of analysis. *Comprehensive Reviews in Food Science and Food Safety*, *9*, 278–291.

Nenadis, N., Mastralexi, A., & Tsimidou, M. Z. (2018). Physicochemical characteristics and antioxidant potential of the Greek PDO and PGI virgin olive oils (VOOs). *European Journal of Lipid Science and Technology*, *1800172*, 1800172.

Olive Oil Companies (). *Website links*. Available from: http://www.greekliquidgold.com/index.php/en/olive-oil-companies/olive-oil-companies-website-links?start=100 (10 January 2019).

Parenti, A., Spugnoli, P., Masella, P., & Calamai, L. (2008). The effect of malaxation temperature on the virgin olive oil phenolic profile under laboratory-scale conditions. *European Journal of Lipid Science and Technology*, *110*, 735–741.

Peres, F., Martins, L. L., & Ferreira-Dias, S. (2017). Influence of enzymes and technology on virgin olive oil composition. *Critical Reviews in Food Science and Nutrition*, *57*, 3104–3126.

Peri, C. (2014). Extra-virgin olive oil storage and handling. In C. Peri (Ed.), *The extra-virgin olive oil handbook* (pp. 165–178). Chichester, UK: John Wiley & Sons, Ltd.

Petrakis, C. (2006). Olive oil extraction. In D. Boskou (Ed.), *Olive oil chemistry and technology* (pp. 191–223). Champaign, IL: AOCS Press.

Polari, J. J., Garcí-Aguirre, D., Olmo-García, L., Carrasco-Pancorbo, A., & Wang, S. C. (2018a). Interactions between hammer mill crushing variables and malaxation time during continuous Clive oil extraction. *European Journal of Lipid Science and Technology*, *1800097*, 1–7.

Polari, J. J., Garcí-Aguirre, D., Olmo-García, L., Carrasco-Pancorbo, A., & Wang, S. C. (2018b). Impact of industrial hammer mill rotor speed on extraction efficiency and quality of extra virgin olive oil. *Food Chemistry*, *242*, 362–368.

Preziuso, S. M., Di Serio, M. G., Biasone, A., Vito, R., Mucciarella, M. R., & Di Giovacchino, L. (2010). Influence of olive crushing methods on the yields and oil characteristics. *European Journal of Lipid Science and Technology*, *112*, 1345–1355.

Psomiadou, E., & Tsimidou, M. (2001). Pigments in Greek virgin olive oils: Occurrence and levels. *Journal of the Science of Food and Agriculture*, *81*, 640–647.

Ranalli, A., Contento, S., Schiavone, C., & Simone, N. (2001). Malaxing temperature affects volatile and phenol composition as well as other analytical features of virgin olive oil. *European Journal of Lipid Science and Technology*, *103*, 228–238.

Ranalli, A., Malfatti, A., Lucera, L., Contento, S., & Sotiriou, E. (2005). Effects of processing techniques on the natural colourings and the other functional constituents in virgin olive oil. *Food Research International*, *38*, 873–878.

Reboredo-Rodríguez, P., González-Barreiro, C., Cancho-Grande, B., & Simal-Gándara, J. (2014). Improvements in the malaxation process to enhance the aroma quality of extra virgin olive oils. *Food Chemistry*, *158*, 534–545.

Saulle, R., & La Torre, G. (2010). The Mediterranean diet, recognized by UNESCO as a cultural heritage of humanity. *Italian Journal of Public Health*, *7*, 414–415.

Selvaggini, R., Esposto, S., Taticchi, A., Urbani, S., Veneziani, G., Di Maio, I., et al. (2014). Optimization of the temperature and oxygen concentration conditions in the malaxation during the oil mechanical extraction process of four Italian olive cultivars. *Journal of Agricultural and Food Chemistry*, *62*, 3813–3822.

Servili, M., Piacquadio, P., De Stefano, G., Taticchi, A., & Sciancalepore, V. (2002). Influence of a new crushing technique on the composition of the volatile compounds and related sensory quality of virgin olive oil. *European Journal of Lipid Science and Technology*, *104*, 483–489.

Servili, M., Selvaggini, R., Esposto, S., Taticchi, A., Montedoro, G. F., & Morozzi, G. (2004). Health and sensory properties of virgin olive oil hydrophilic phenols: Agronomic and technological aspects of production that affect their occurrence in the oil. *Journal of Chromatography A*, *1054*, 113–127.

Servili, M., Selvaggini, R., Taticchi, A., Esposto, S., & Montedoro, G. F. (2003). Volatile compounds and phenolic composition of virgin olive oil: Optimization of temperature and time of exposure of olive pastes to air contact during the mechanical extraction process. *Journal of Agricultural and Food Chemistry*, *51*, 7980–7988.

Stefanoudaki, E., Koutsaftakis, A., & Harwood, J. L. (2011). Influence of malaxation conditions on characteristic qualities of olive oil. *Food Chemistry*, *127*, 1481–1486.

Tamborrino, A. (2014). Olive paste malaxation. In C. Peri (Ed.), *The extra-virgin olive oil handbook* (pp. 127–137). Chichester, UK: John Wiley & Sons, Ltd.

Taticchi, A., Esposto, S., & Servili, M. (2014). The basis of the sensory properties of virgin olive oil. In E. Monteleone, & S. Langstaff (Eds.), *Olive oil sensory science* (pp. 33–54). Chichester, UK: John Wiley & Sons, Ltd.

Taticchi, A., Esposto, S., Veneziani, G., Urbani, S., Selvaggini, R., & Servili, M. (2013). The influence of the malaxation temperature on the activity of polyphenoloxidase and peroxidase and on the phenolic composition of virgin olive oil. *Food Chemistry*, *136*, 975–983.

Toplu, C., Yildiz, E., Bayazit, S., & Demirkeser, T. H. (2009). Assessment of growth behaviour, yield, and quality parameters of some olive (*Olea europaea*) cultivars in Turkey. *New Zealand Journal of Crop and Horticultural Science*, *37*, 61–70.

Tsimidou, M. Z. (2006). Olive oil quality. In D. Boskou (Ed.), *Olive oil, chemistry, and technology* (pp. 93–111). Champaign, IL: AOCS Press.

Tsimidou, M. Z. (2010). Squalene and tocopherols in olive oil: Importance and methods of analysis. In V. R. Preedy, & R. R. Watson (Eds.), *Olives and olive oil in health and disease prevention* (pp. 561–567). London, UK: Academic Press, Elsevier.

Tsimidou, M. Z. (2012). Virgin olive oil (VOO) and other olive tree products as sources of α-tocopherol. Updating and perspective. In A. Catala (Ed.), *Tocopherol: sources, uses, and health benefits* (pp. 1–21). N.Y: Nova Science Publisher.

Tsimidou, M. Z. (2013). Analytical methodologies phenolic compounds related to olive oil taste issues. In R. Aparicio, & J. Harwood (Eds.), *Handbook of olive oil, analysis, and properties* (pp. 311–329). Boston, MA: Springer.

Tsimidou, M. Z., Georgiou, A., Koidis, A., & Boskou, D. (2005). Loss of stability of "veiled" (cloudy) virgin olive oils in storage. *Food Chemistry*, *93*, 377–383.

Tsimidou, M. Z., Ordoudi, S. A., Nenadis, N., & Mourtzinos, I. (2015). Food fraud. In B. Caballero, P. Finglas, & F. Toldrá (Eds.), *Encyclopedia of food and health* (pp. 35–42). London: Academic Press—Elsevier.

Turkish patent and trademark office. n.d.: Available from: http://online.turkpatent.gov.tr/trademark-search/pub/#trademark_result (25 December 2018)

Uceda, M., & Frias, L. (1975). Harvest dates. Evolution of the fruit oil content, oil composition, and oil quality. In: *Proc. Segundo Seminario Oleicola Internacional, Cordoba, Spain 6*, (pp. 125–130).

Veillet, S., Tomao, V., Bornard, I., Ruiz, K., & Chemat, F. (2009). Chemical changes in virgin olive oils as a function of crushing systems: Stone mill and hammer crusher. *Comptes Rendus Chimie*, *12*, 895–904.

Veneziani, G., Esposto, S., Taticchi, A., Urbani, S., Selvaggini, R., Di Maio, I., et al. (2017). Cooling treatment of olive paste during the oil processing: Impact on the yield and extra virgin olive oil quality. *Food Chemistry*, *221*, 107–113.

Veral, M. G. (2016). Olive gene resources in Turkey. *Olivae*, *123*, 27–30.

Vitaglione, P., Savarese, M., Paduano, A., Scalfi, L., Fogliano, V., & Sacchi, R. (2015). Healthy virgin olive oil: A matter of bitterness. *Critical Reviews in Food Science and Nutrition*, *55*, 1808–1818.

Chapter 51

Cold pressed soybean oil

Zeynep Aksoylu Özbek and Pelin Günç Ergönül
Department of Food Engineering, Faculty of Engineering, Manisa Celal Bayar University, Manisa, Turkey

Abbreviations

DPPH 2,2-diphenyl-1-picrylhydrazyl
LDL low-density lipoprotein
PAHs polycyclic aromatic hydrocarbons
TE Trolox equivalent

1 Introduction

Soybean (*Glycine max* L.) is one of the most dominant oilseeds with a total production of 352.6 million tons in the world in 2017. The main producer countries are the United States, Brazil, Argentina, China, India, Paraguay, Canada, Ukraine, the Russian Federation, and Bolivia (Food and Agriculture Organization of the United Nations, 2017). Soybean oil represents the majority of the edible oil consumption (about 80% of the total) in the United States (Liu, 2004b). The cultivation of this annual crop has been considered to have originated from China 4000–5000 years ago; it was then introduced to Europe in 1712 and to the United States in the mid-18th century (Liu, 1997). The advantages of soybean include being an excellent rotational crop due to the ability to fix nitrogen, its high tolerance for various soils and climate conditions, its great edible protein production compared to any other crop, and its wide range of uses (Liu, 2004a).

Seed oil and the meal with high protein content are the major products of soybean in the Western diet, while traditional foods made from soybean occupy an important place in Asia cuisine (Hymowitz, 2008). Generally speaking, soybean contains about 40% protein, and 20% oil with trace amounts of phospholipids, vitamins, minerals, trypsin inhibitors, phytates, oligosaccharides, and isoflavones (Liu, 1997). The numerous uses of this crop may be attributed to its unique amino acid and fatty acid compositions (Asbridge, 1995). The main outputs of cold pressing are crude oil and defatted oil cake (meal). The proximate physicochemical composition of cold pressed soybean oil is summarized in Table 1.

Soybean cake (54%), rapeseed cake (10%), cottonseed cake (10%), groundnut cake, sunflower cake, copra cake, and linseed cake are the dominating oil cakes traded in the world (Sivaramakrishnan & Gangadharan, 2009). The chemical composition of defatted and untreated soybean meal obtained by cold pressing was evaluated by Woyengo, Patterson, and Levesque (2016) and is summarized in Table 2. As this table shows, defatted soybean meal is regarded as an excellent source of crude protein in animal nutrition. The most striking feature of the soybean meal is a significant amount of leucine. Leucine is one of the most potent amino acids in terms of health benefits (Li, Yin, Tan, Kong, & Wu, 2011). Yin et al. (2010) declared that leucine promotes protein synthesis and growth of skeletal muscle, liver, and intestinal tract in weanling pigs.

On the other hand, soybean meal includes several antinutritional compounds. These are trypsin inhibitors, lectins, saponins, antigens (glycinin and β-conglycinin), rachitogenic factors (genistin), and phytic acid. These antinutritional factors may reduce protein digestibility, bind the nutrients and lead to precipitations, cause toxicity, form nonabsorbable chemical structures, and even destroy the development of the gastrointestinal tract in younger animals (Yin, Fatufe, & Blachier, 2011). Some of these antinutritional compounds (trypsin inhibitors and lectins) are removed from oil cake using various high-temperature processes (Banaszkiewicz, 2011; Yasothai, 2016). Extrusion, enzyme treatment (protease, phytase, xylanase, and cellulase), fermentation (*Aspergillus oryzae* GB-107, *Aspergillus usamii*, *Bacillus amyloliquefaciens*, *Bacillus licheniformis*, *Bacillus subtilis*, *Candida utilis*, *Enterococcus faecalis*, *Hansenula anomala*, *Lactobacillus brevis*, *Lactobacillus casei*, *Pediococcus acidilactic*, *Rhodopseudomonas palustris*, *Saccharomyces cerevisae*, and *Streptococcus thermophilus*) are the other elimination techniques used to improve nutritional quality and digestibility of soybean

Cold Pressed Oils. https://doi.org/10.1016/B978-0-12-818188-1.00051-7
Copyright © 2020 Elsevier Inc. All rights reserved.

TABLE 1 Physicochemical properties of cold pressed soybean oil.

Parameter	Amount	Reference
Refractive index (n_D^{25})	1.47287–1.47308	Zeleny and Neustadt (1940)
Peroxide value (meq O_2/kg)	4.21–11.52	Kania, Michalak, Gogolewski, and Hoffmann (2004) and Rudzińska, Kazuś, and Wąsowicz (2001)
p-Anisidine value	2.14	Rudzińska et al. (2001)
Totox value	25.18	
C16:0 (%)	7.6–11.3	Rafalowski, Zegarska, Kuncewicz, and Borejszo (2008) and Tuberoso, Kowalczyk, Sarritzu, and Cabras (2007)
C18:0 (%)	3.2–3.7	
C18:1*n*-9 (%)	24.6–28.5	Rudzińska et al. (2001) and Tuberoso et al. (2007)
C18:2*n*-6 (%)	50.8–55.2	
C18:3*n*-3 (%)	2.3–7.6	Rafalowski et al. (2008) and Tuberoso et al. (2007)
α-Tocopherol (mg/100 g)	9.21–24.16	
β-Tocopherol (mg/100 g)	0–5.98	Kania et al. (2004) and Rafalowski et al. (2008)
γ-Tocopherol (mg/100 g)	33.17–143.23	Rafalowski et al. (2008) and Tuberoso et al. (2007)
δ-Tocopherol (mg/100 g)	14.93–45.46	Kania et al. (2004) and Rafalowski et al. (2008)
β-Carotene (mg/100 g)	0.03–2.73	Rafalowski et al. (2008) and Tuberoso et al. (2007)
Chlorophylls (mg/100 g)	1.2	Tuberoso et al. (2007)
β-Sitosterol (mg/g)	2.16	Rudzińska et al. (2001)
Campesterol (mg/g)	0.49	
Stigmasterol (mg/g)	0.32	
Avenasterol (mg/g)	0.26	
∑ phytosterols (mg/g)	3.23	
Total phenolics content (mg ChAE/kg)[a]	5.38	Kania et al. (2004)
Total phenolics content (mg CAE/kg)[b]	14.8	Siger, Nogala-Kalucka, and Lampart-Szczapa (2008)
p-Coumaric acid (μg/100 g)	1.5	
Ferulic acid (μg/100 g)	1.2	
Vanillic acid (μg/100 g)	1.1	
Sinapic acid (μg/100 g)	0.9	
Caffeic acid (μg/100 g)	0.8	
p-Hydroxybenzoic acid (μg/100 g)	0.8	
DPPH• scavenging activity (%)	17.40–31.95	Kania et al. (2004) and Siger et al. (2008)
DPPH• scavenging activity (mmol TEAC/L)[c]	1.75	Tuberoso et al. (2007)

[a]*Chlorogenic acid equivalent.*
[b]*Caffeic acid equivalent.*
[c]*Trolox equivalent antioxidant capacity.*

TABLE 2 Chemical properties of cold pressed soybean meal (Woyengo et al., 2016).

Parameter	
Moisture (%)	8.43
Crude protein (%)	47.78
Ash (%)	6.05
Crude fiber (%)	5.41
Essential amino acid (%)	
Histidine	1.22
Isoleucine	2.18
Leucine	3.61
Lysine	3.11
Methionine	0.69
Phenylalanine	2.36
Threonine	1.83
Tryptophan	0.67
Valine	2.37
Nonessential amino acid (%)	
Alanine	2.05
Arginine	3.44
Aspartic acid	5.31
Cysteine	0.68
Glutamic acid	8.58
Glycine	2.14
Proline	2.38
Serine	2.22
Tyrosine	1.59

meal. After treatment of soybean meal, it is commonly utilized as a unique protein source in pig, fish, and poultry feeds (Abdul Kader et al., 2012; Barrows, Stone, & Hardy, 2007; Chi & Cho, 2016; Dai et al., 2017; Ding, Zhang, Ye, Du, & Kong, 2015; Feng, Liu, Xu, Wang, & Liu, 2007; Gao, Wang, Zhu, & Qian, 2013; Hirabayashi, Matsui, Yano, & Nakajima, 1998; Hong, Lee, & Kim, 2004; Jacobsen, Samuelsen, Girons, & Kousoulaki, 2018; Kim, van Heugten, Ji, Lee, & Mateo, 2010; Marsman, Gruppen, van der Poel, et al., 1997; Marsman, Gruppen, Mul, & Voragen, 1997; Pongmaneerat & Watanabe, 1993; Refstie, Sahlström, Bråthen, Baeverfjord, & Krogedal, 2005; Romarheim, Aslaksen, Storebakken, Krogdahl, & Skrede, 2005; Seo & Cho, 2016; Sinn, Gibbons, Brown, DeRouchey, & Levesque, 2017; Wang et al., 2014; Yuan et al., 2017; Zhou et al., 2011).

This chapter covers the production of soybean and properties of cold pressed soybean oil as well as its possible health benefits.

2 Extraction and processing of cold pressed oil

Soybean oil is principally obtained by solvent (particularly *n*-hexane) extraction or mechanical expellers. On the other hand, supercritical fluid extraction and extrusion-aided mechanical extraction have emerged as alternative processing techniques in recent years (Ali, 2010). The fundamental problem of cold pressed soybean oils is the lack of deactivation step of lipoxygenase isoenzymes, which impair the stability and quality of the oil. Therefore, Engeseth, Klein, and Warner (1987) suggested to apply tempering procedure to soybeans before oil extraction. However, the cold pressing technique does not allow to use high temperature processes, so that enzyme activity maintains in cold pressed soybean oils. This is the essential limiting factor for use of the cold pressing method in soybean oils.

3 Fatty acids and acyl lipids

Commercial soybean oils are commonly solvent-extracted and refined due to their high phosphatides contents. Hence, few studies exist on characterization of cold pressed soybean oils in the literature. One of these studies revealed that cold pressed soybean oil consists of linoleic (50.8%), oleic (24.6%), palmitic (10.2%), linolenic (7.6%), stearic (3.7%), and vaccenic (1.5%) acids, and trace amounts of myristic, palmitoleic, heptadecanoic, arachidic, eicosenoic, behenic, erucic, and lignoceric acids (Tuberoso et al., 2007). The fatty acid composition of soybean oil is typical of the oleic-linoleic acid group of oils, and these two unsaturated fatty acids form almost 75% of the total fatty acids. This fatty acid profile makes soybean oil one of the healthiest edible oils (Cober, Cianzio, Pantalone, & Rajcan, 2009). However, Brühl (1996) detected the presence of *trans* fatty acids (0.10%–0.15%) in cold pressed soybean oils. The author thus stated that *trans* fatty acids may occur due to high temperature drying of seeds before cold pressing and/or deodorization of cold pressed soybean oils or blending of cold pressed oils with refined ones.

The triacylglycerols (TAGs) profile is used as an indicator of the nutritional quality and authenticity of edible oils (Aparicio & Aparicio-Ruı´z, 2000). Tuberoso et al. (2007) reported that cold pressed soybean oil is characterized by LLP (34.5%), LLO (14.9%), LLL (12.5%), POO (10.4%), LnLL (6.3%), POP (5.1%), LLPn (4.6%), LnLO (3.8%), LOP (3.7%), LOO (2.2%), and LnLnL (1%) with lower levels of PLP (0.5%), OOO (0.4%), and SOO (0.1%). In general, polyunsaturated fatty acids (PUFAs) are present at the *sn-1* and *sn-2* positions, whereas the dominant monounsaturated fatty acid (MUFA), oleic acid, is located at the *sn-2* and *sn-3* positions. This is an expected result because the *sn-2* position is commonly occupied by unsaturated fatty acids in vegetable oils (McClements & Decker, 2017). The high levels of unsaturated fatty acids at the *sn-2* position contribute to the oxidative stability of cold pressed soybean oil (Raghuveer & Hammond, 1967). In addition to oxidative stability, the metabolic fate of dietary fat throughout digestion and its absorption are affected by the stereospecificity and chain length of fatty acids found at the *sn-1*, *sn-2*, and *sn-3* positions in TAGs (Karupaiah & Sundram, 2007).

4 Minor bioactive lipids

Unsaponifiable matters consisting of tocopherols, tocotrienols, phytosterols, carotenoids, phenolics, waxes, squalene, and oryzanols make up about 5% of total lipid content (Aladedunye, Thiyam-Hollander, & Eskin, 2017). The tocopherols and sterols are considered the "fingerprint" of the oil and so, they can be used as important markers in authentication and/or adulteration studies (Eskin, 2002). β-Sitosterol (2.1 mg/g), campesterol (0.5 mg/g), stigmasterol (0.3 mg/g), and avenasterol (0.2 mg/g) were identified in cold pressed soybean oil. It is reported that the total phytosterol content of cold pressed soybean oil (3.2 mg/g) is close to that of cold pressed flaxseed (3.1 mg/g), and sunflower (3.2 mg/g) oils, and less than that of cold pressed corn (9.7 mg/g), and rapeseed (7.7 mg/g) oils (Rudzińska et al., 2001).

Tuberoso et al. (2007) stated that cold pressed soybean oil had the highest total tocopherol content (1797 mg/kg) compared to cold pressed oils of maize (1618 mg/kg), sunflower (634 mg/kg), rapeseed (624 mg/kg), flaxseed (588 mg/kg), pumpkin (508 mg/kg), peanut (398 mg/kg), and grapeseed (142 mg/kg). The researchers agree that the dominant tocopherol isomer of cold pressed soybean oil is γ-tocopherol, ranging in content between 330 and 1432 mg/kg. The oil also contains varying amounts of α-tocopherol (92–246 mg/kg) and δ-tocopherol (149–454 mg/kg) (Kania et al., 2004;Rafalowski et al., 2008 ; Tuberoso et al., 2007). However, the existence of a β-tocopherol isomer (59 mg/kg) was only revealed by Rafalowski et al. (2008) in cold pressed soybean oil. Kania et al. (2004) reported that solvent extraction of soybean oil yielded higher total tocopherols content (1448 mg/kg) than cold pressing (1358 mg/kg), and the refining process led to the loss of α-tocopherol by 37%, β-tocopherol by 56%, γ-tocopherol by 17%, and δ-tocopherol by 34% in solvent-extracted soybean oil. Taking into account all of this, it is obvious that cold pressed oils have higher bioactives content than refined oils. Moreover, despite the lower yields of the cold pressing process, it is considered superior to solvent extraction in terms of reliability due to the lack of chemicals. Evans, Kodali, and Addis (2002) suggested an optimum α-tocopherol concentration of 100 ppm and γ-tocopherol concentration of 300 ppm to inhibit oxidation in soybean oil.

Pigments like carotenoids and chlorophyll influence the appearance and oxidative stability of oil (Shen, Wang, & Das, 2012). β-Carotene is a well-known singlet oxygen quencher, and it therefore acts as an effective photooxidation inhibitor in the oils (Gunstone, 2004). Cold pressed soybean oil is characterized by lower β-carotene content (0.3–27.0 mg/kg) than cold pressed oils of linseed (150 mg/kg), pumpkin seed (5.5–150.0 mg/kg), rapeseed (1.7–80.0 mg/kg), corn (0.9–35.0 mg/kg), and flaxseed (0.7 mg/kg) (Rafalowski et al., 2008; Tuberoso et al., 2007). On the contrary, the activity of other major pigment of seed oils, namely chlorophyll, depends on the environmental conditions. Chlorophyll serves as a weak antioxidant in the absence of light, while it acts as a strong prooxidant in the presence of light (Boskou, 2002). Another critical importance of chlorophylls in soybean oil is their contribution to the formation of 2-pentylfuran and pentenylfuran from linoleic and linolenic acids, which impart an undesirable beany flavor to the oil (Choe & Min, 2006). In order to eliminate these problems, the maximum chlorophyll content should be 50 μg/kg in bleached soybean oil (O'Brien, 2008). Nonetheless, cold pressed oils are not exposed to a refining process, so they contain very high levels of chlorophyll. Tuberoso et al. (2007) reported a chlorophyll content of 12 mg/kg in cold pressed soybean oil. This is the third highest chlorophyll amount, after extra virgin olive oil (33 mg/kg) and cold pressed pumpkin seed oil (30 mg/kg). Cold pressed soybean oil, of which the chlorophyll content exceeds the limit of 50 μg/kg, should be stored in dark conditions to prevent photooxidation.

The phenolics represent an important group of minor constituents in the oil. The total phenolic content in cold pressed soybean oil was reported to be 1.4 mg caffeic acid equivalent (CAE)/100 g. Siger et al. (2008) sorted the cold pressed oils in ascending order of total phenolic content as follows: grapeseed (0.5 mg CAE/100 g) < flaxseed (1.1 mg CAE/100 g) < sunflower (1.2 mg CAE/100 g) < corn (1.2 mg CAE/100 g) < rapeseed (1.3 mg CAE/100 g) < rice bran (1.4 mg CAE/100 g) < soybean (1.4 mg CAE/100 g) < hemp (2.4 mg CAE/100 g) < pumpkin seed (2.4 mg CAE/100 g). Kania et al. (2004) declared that contrary to tocopherols, the cold pressing procedure yielded higher total phenolics content (5.3 mg chlorogenic acid equivalent/kg) than solvent-extraction (2.3 mg chlorogenic acid equivalent/kg). *p*-coumaric (1.5 μg/100 g), ferulic (1.2 μg/100 g), and vanillic (1.1 μg/100 g) acids were the dominant phenolic compounds, accompanied by lower amounts of sinapic (0.9 μg/100 g), caffeic, and *p*-hydroxybenzoic (0.8 μg/100 g) acids in cold pressed soybean oil (Siger et al., 2008).

All the biologically active compounds listed above boost the radical scavenging capacity of cold pressed soybean oil. To date, only a few research studies with conflicting results about antioxidant activity of cold pressed soybean oil have been published. In one of these studies, higher total antioxidant activity (1.7 mmol Trolox equivalent (TE)/L) was reported for cold pressed soybean oil than cold pressed oils of grapeseed (1.4 mmol TE/L), sunflower (1.1 mmol TE/L), flaxseed (1 mmol TE/L), pumpkin seed (0.9 mmol TE/L), rapeseed (0.8 mmol TE/L), peanut (0.4 mmol TE/L), and virgin olive oil (1.2 mmol TE/L) (Tuberoso et al., 2007). On the contrary, cold pressed soybean oil was characterized by remarkably lower 2,2-diphenyl-1-picrylhydrazyl (DPPH·) radical scavenging activity (17.4%) than cold pressed oils of hemp (76.2%), pumpkin seed (65.3%), rapeseed (51.2%), sunflower (23.8%), rice bran (23.7%), and flaxseed (19.3%) (Siger et al., 2008). Differences in results from these two studies may be attributed to various factors such as cultivar of plant, degree of maturation, environmental conditions, pre- and postharvest factors, storage, and polarity of extraction solvent (Kozłowska, Gruczyńska, Ścibisz, & Rudzińska, 2016; Mnari, Harzallah, Amri, Dhaou Aguir, & Hammami, 2016). Thus it seems difficult to classify cold pressed soybean oil with respect to antioxidant activity according to the results of these studies.

In general, the contribution of oil's lipophilic fraction (tocopherols, chlorophyll, carotenoids, and squalene) to antioxidant activity is greater than that of hydrophilic substances (phenolic acids, phenolic alcohols, flavonoids, secoiridoids, and lignans) (El Riachy, Priego-Capote, León, Rallo, & Luque de Castro, 2011; Tuberoso et al., 2007). Kania et al. (2004) stated that cold pressed soybean oil had higher DPPH· scavenging activity than solvent-extracted crude soybean oil due to its higher tocopherol and phenolics content. However, they only found a positive correlation between DPPH· scavenging activity of soybean oils and tocopherol content, a lipophilic antioxidant, as suggested by Tuberoso et al. (2007).

5 Contribution of bioactive compounds in cold pressed oil to organoleptic traits and functions in food or nonfood products

The main problem associated with flavor of soybean oil is known as "reversion flavor." The oil contains substantial amounts of linolenic acid, which is very unstable against oxidation due to its high unsaturation degree. Soybean oil is commonly characterized by a buttery flavor in the early stages of autooxidation. Seals and Hammond (1966) explained the formation of this undesirable flavor with the occurrence of diacetyl. Moreover, they declared that linolenic acid might be the precursor of diacetyl in soybean oil. In addition to diacetyl, the buttery flavor of the oil was linked with the occurrence of 2,3-pentanedione (Seals & Hammond, 1970). Unlike Seals and Hammond (1966), Dutton, Lancaster, Evans, and Cowan (1951) attributed the painty, grassy, fishy, and melony flavor formation in soybean oil to its unique linolenic acid content. On the other side, some researchers believe that reversion compounds (e.g., 2-pentyl furan) originating from linoleic acid rather than linolenic acid are responsible for the beany and grassy flavor of oxidized soybean oil (Chang, Krishnamurthy, & Reddy, 1967; Ho, Smagula, & Chang, 1978; Krishnamurthy, Smouse, Mookherjee, Reddy, & Chang, 1967). Smagula, Ho, and Chang (1979) concluded that the *cis* isomer of 2-pentyl furan provides the beany, grassy, and buttery flavors at 0.50 ppm concentration, whereas the *trans* isomer gives the same flavors at 1 ppm and strong painty and metallic flavors at 4 ppm. Smouse and Chang (1967) identified 71 volatile compounds in reverted soybean oil. The major acidic volatiles were butanoic, pentanoic, hexanoic, heptanoic, and octanoic acid, while nonacidic volatiles were aldehydes (pentanal, hexanal, octanal, nonanal, 2-pentanal, 2-heptanal, 2-octenal, 2-nonenal, 2, *trans*-4, *trans*-heptadienal), ketones (2-heptanone, 2-octanone), alcohols (ethanol, pentanol, 1-penten-3-ol, 1-octen-3-ol), esters (ethyl acetate), hydrocarbons (1-decyne), and aromatic compounds (2-pentyl furan). The flavor intensity of oxidized soybean oil is mainly determined by carbonyl compounds (Dixon & Hammond, 1984).

The autooxidation of soybean oil is controlled by several factors. Metal contamination is considered a fundamental element that accelerates the oxidation rate of the oil. Evans, Schwab, Moser, Hawley, and Melvin (1951) showed that copper is more detrimental than iron for soybean oil quality. Copper may act as a prooxidant at 0.005 ppm and iron has the same effect at a concentration of 0.03 ppm (Flider & Orthoefer, 1981). Evans, Cooney, Moser, and Schwab (1953) suggested that phytic acid addition into soybean oil is a reliable method to inactivate trace metals. However, in the absence of metals, phospholipids such as phosphatidyl choline, phosphatidyl ethanolamine, phosphatidyl inositol, phosphatidyl glycerol, phosphatidic acid, and cardiolipin act as prooxidants and stimulate soybean oil oxidation (Yoon & Min, 1987). The other minor component of soybean oil, chlorophyll, triggers formation of reversion flavor in soybean oil by stimulating singlet oxygen production in the presence of light. Hence removal of chlorophyll during processing may be considered as an alternative way to extend the shelf life of soybean oil (Min, Callison, & Lee, 2003).

6 Health-promoting traits of cold pressed oil and oil constituents

The beneficial effects of soybean oil on human health have been linked with the presence of several bioactive compounds. The most important constituent of these bioactives is fatty acids. Soybean oil is rich in linoleic acid, an essential ω-6 fatty acid. Decreased risk of cardiovascular diseases has been associated with linoleic acid consumption. This effect is attributed to its ability to lower low-density lipoprotein (LDL) level in the blood (Batchu, Chaudhary, Zlobine, Pawa, & Seubert, 2016; Choque, Catheline, Rioux, & Legrand, 2014). It is generally recommended to consume 5%–10% of daily energy from ω-6 fatty acids for the reduction of blood pressure in individuals (Harris, 2008). The other health effects of ω-6 fatty acids are improvement of insulin sensitivity, and protection against DNA damage and apoptosis generated by palmitic acid (Beeharry et al., 2003; Riserus, Willett, & Hu, 2009; Summers et al., 2002). On the other hand, some researchers suggest a diet rich in MUFAs (e.g., oleic acid) is more appropriate than a diet rich in PUFAs (e.g., linoleic acid) for patients with type 2 diabetes (Madigan, Ryan, Owens, Collins, & Tomkin, 2000). Additionally, increased intake of ω-6 fatty acids may stimulate growth of certain cancer cells and have proinflammatory, proadipogenic, and prothrombotic effects on human health (Maggiora et al., 2004; Muhlhausler & Ailhaud, 2013; Park, Allen, & Shultz, 2000; Shultz, Chew, & Seaman, 1992).

Sterols are another beneficial component of soybean oil. Peterson (1951) reported that soybean sterols reduced cholesterol levels in chicks' plasma and liver. The antioxidative effect of β-sitosterol (at lower concentrations), the main phytosterol of soybean oil, was discovered (van Rensburg, Daniels, van Zyl, & Taljaard, 2000). In addition, β-sitosterol is an effective antimicrobial agent against *Staphylococcus aureus*, *Bacillus subtilis*, *Streptococci pyrogene*, *Klebsiella pneumonia*, *Shigella dysentariae*, and *Candida krusei* (Odiba, Musa, Hassan, Yahaya, & Okolo, 2014). By chelating heavy metals, β-sitosterol may form complexes that have antifungal (against *Aspergillus flavus*, *Aspergillus niger*, *Trichophyton rubrum*, *Trichophyton mentagrophytes*, *Trichophyton tonsurans*) and antibacterial (against *Bacillus cereus*, *Bacillus subtilis*, *Bacillus thruingiensis*, *Proteus mirabilis*, *Pseudomonas aeroginosa*, *Pseudomonas aeruginosa* ATCC) activities. It may also be used for the treatment of heavy metal toxicity (Mahmood et al., 2013). However, it should be kept in mind that consumption of plant sterols and stanols may decrease hydrocarbon carotenoid (β-carotene, α-carotene, and lycopene) and oxygenated carotenoid (zeaxanthin and β-cryptoxanthin) concentrations in plasma (Baumgartner, Ras, Trautwein, Mensink, & Plat, 2017). Hence, daily intake recommendations for plant sterols and stanols should be discussed carefully.

The dominant tocopherol isomer of soybean oil, γ-tocopherol, has antiinflammatory and antioxidant characteristics (Dietrich et al., 2006). Several studies have been demonstrated that γ-tocopherol is a very potent chemopreventive agent, in particular for prostate, colon, lung, and mammary cancer types in animal models (Chen et al., 2016, 2017; Guan et al., 2012; Ju et al., 2009, 2010; Lu et al., 2010).

Phenolics are well-known for their superior antioxidative activities. Vegetables, fruits, and cereals are rich sources of phenolic compounds. Soybean oil majorly includes *p*-coumaric acid, which exhibits antiaggregant properties and reduces the risk of cardiovascular diseases as well as having an antiinflammatory effect (Luceri et al., 2007, 2004). Moreover, Yoon et al. (2013) reported that *p*-coumaric acid may inhibit insulin resistance and type 2 diabetes through adjusting lipid and glucose metabolism. Its antibacterial effect against *Staphylococcus aureus*, *Shigella dysenteriae*, *Escherichia coli*,

Salmonella typhimurium, *Streptococcus pneumoniae*, *Bacillus subtilis*, and *Bacillus cereus* has been revealed by many researchers to date (Herald & Davidson, 1983; Lou et al., 2012; Stojković et al., 2013).

All these bioactive components contribute to the health benefits of soybean oil. Nevertheless, research on the direct effects of soybean oil on health is scarce. The outputs of these studies are summarized below:

- Learning skills of young male rats may be improved by dietary soybean oil supplementation (Coscina, Yehuda, Dixon, Kish, & Leprohon-Greenwood, 1986).
- Soybean oil is not as effective as perilla and safflower oils against mammary and colon tumors (Hirose, Masuda, Ito, Kamano, & Okuyama, 1990).
- Soybean oil promotes chemically induced mammary tumors in female rats (Kritchevsky, Weber, & Klurfeld, 1992).
- Soybean oil may serve as a suitable agent for prevention of total parenteral nutrition-induced liver damage (Nishimura, Yamaguchi, Naito, & Yamauchi, 2006).
- Cardioprotective influence of soybean oil is associated with its ability to increase the left ventricular performance in rats (Ribeiro et al., 2010).
- High dietary intake of soybean oil has an adverse effect on bone structure during adultness due to its high linoleic acid content (Da Costa et al., 2015).
- Diets rich in soybean oil increase weight gain, adiposity, diabetes, glucose intolerance, and insulin resistance, and are more harmful to the metabolic health of rats than diets rich in fructose or coconut oil (Deol et al., 2015).

Researchers cannot agree on whether the consumption of soybean oil is healthy or unhealthy. More studies should be carried out to find out the exact health effects of soybean oil.

7 Edible and nonedible applications of cold pressed oil

The characteristic chemical composition of soybean oil makes it a good salad oil, whereas its hydrogenated form serves as margarine or frying oil (Hammond, Johnson, Su, Wang, & White, 2005). Additionally, partially hardened soybean oil is used in semisolid baking, shortenings, mayonnaise, and salad dressings manufacture (Asbridge, 1995; Hammond et al., 2005). Soybean oil has been used as a convenient carrier for some drug substances such as peptides, vitamins, hormones, and insulin by the pharmaceuticals industry (Karasulu et al., 2011).

8 Other issues

Polycyclic aromatic hydrocarbons (PAHs), carcinogenic environmental contaminants, are abundantly found in vegetable oils because of their lipophilic character (Shibamoto & Bjeldanes, 2009). Some research has focused on detection of PAHs in crude soybean oils; however, the production method of soybean oil was not stated clearly in these studies. The existence of phenanthrene (9.8–18.0 μg/kg), fluoranthene (5.7–7.1 μg/kg), pyrene (3.9–5.8 μg/kg), 2-methylphenanthrene (2.5–4.7 μg/kg), benzo[*b*]fluoranthene (2.4–5.2 μg/kg), chrysene/triphenylene (1.7–3.0 μg/kg), 1-methylphenanthren (1.2–3.3 μg/kg), and lower amounts of benzo[*a*]anthracen, indeno[1,2,3∼l]pyrene, benzo[*ghi*]perylene, anthracene, benzo[*b*]fluorene, 4,5-methylphenanthrene, benzo[*a*]pyrene, 2-methylanthracene, perylene, and benzo[*e*]pyrene in crude soybean oils was confirmed by Larsson, Eriksson, and Cervenka (1987). Rojo Camargo, Antoniolli, and Vicente (2012) reported the presence of benz[*a*]anthracene (1.2–42.0 μg/kg), chrysene (2.7–58.0 μg/kg), 5-methylchrysene (0.8– 21.0 μg/kg), benzo[*j*]fluoranthene (up to 16.5 μg/kg), benzo[*b*]fluoranthene (0.3–33.0 μg/kg), benzo[*k*]fluoranthene (0.2–11.3 μg/kg), benzo[*a*]pyrene (up to 58 μg/kg), dibenzo[*al*]pyrene (up to 11 μg/kg), dibenz [*ah*]anthracene (0.9–53.6 μg/kg), indeno[1,2,3-*cd*]pyrene (up to 15.2 μg/kg), dibenzo[*ae*]pyrene (0.4–9.6 μg/kg), dibenzo[*ai*]pyrene (up to 7.8 μg/kg), and dibenzo[*ah*]pyrene (up to 3.8 μg/kg) in crude soybean oils produced in Brazil. The total concentrations of PAHs ranged from 10 μg/kg to 316 μg/kg and they decreased throughout the oil refining process, in particular in the neutralization step. Recently, Hua, Zhao, Wu, and Li (2016) declared the contamination of crude soybean oils with pyrene (7.4 μg/kg), naphthalene (7.3 μg/kg), phenanthrene (6.5 μg/kg), fluoranthene (6.3 μg/kg), fluorene (4 μg/kg), chrysene (3.5 μg/kg), indeno[1,2,3-*c,d*]pyrene (3.3 μg/kg), benzo[*a*]anthracene (3 μg/kg), acenaphthene (2.5 μg/kg), and anthracene (1 μg/kg), and traces of benzo[*b*]fluoranthene, benzo[*k*]fluoranthene, benzo[*a*] pyrene, acenaphthylene, benzo[*ghi*]perylene, and dibenzo[*a,h*]anthracene. Moreover, they detected the residues of oxy-PAHs including 9,10-anthracenedione (13 μg/kg), 7H-benzo[*d,e*]anthracen-7-one (3.5 μg/kg), 9-fluorenone (1.9 μg/kg), benzo[*a*]anthracene-7,12-dione (0.9 μg/kg), and 9,10-dihydrobenzo[*a*]pyren-7(8*H*)-one (0.4 μg/kg) in crude soybean oils. The maximum levels for benzo[*a*]pyrene and sum of benzo[*a*]pyrene, benz[*a*]anthracene, benzo[*b*]fluoranthene, and chrysene are 2 μg/kg and 10 μg/kg in oils and fats (European Commission, 2006).

Compared to other cold pressed oils, the characterization and toxicological studies dealing with soybean oils are very limited. Therefore, further investigations should be carried out to find out the other contaminants and compounds, which may exhibit health-related problems in cold pressed soybean oils.

References

Abdul Kader, M., Koshio, S., Ishikawa, M., Yokoyama, S., Bulbul, M., Nguyen, B. T., et al. (2012). Can fermented soybean meal and squid by-product blend be used as fishmeal replacements for Japanese flounder (*Paralichthys olivaceus*)? *Aquaculture Research*, *43*, 1427–1438.

Aladedunye, F., Thiyam-Hollander, U., & Eskin, N. A. M. (2017). Frying oil chemistry. In C. C. Akoh (Ed.), *Food lipids-chemistry, nutrition, and biotechnology* (pp. 261–294). Boca Raton, FL: CRC Press.

Ali, N. (2010). Soybean processing and utilization. In G. Singh (Ed.), *The soybean: Botany, production, and uses* (pp. 345–374). Oxfordshire, UK: CABI Publishing.

Aparicio, R., & Aparicio-Ruı´z, R. (2000). Authentication of vegetable oils by chromatographic techniques. *Journal of Chromatography A*, *881*, 93–104.

Asbridge, D. D. (1995). Soybeans vs. other vegetable oils as a source of edible oil products. In D. R. Erickson (Ed.), *Practical handbook of soybean processing and utilization* (pp. 1–8). Champaign, Illinois: AOCS Press.

Banaszkiewicz, T. (2011). Nutritional value of soybean meal. In H. El-Shemy (Ed.), *Soybean and nutrition* (pp. 1–20). InTech.

Barrows, F. T., Stone, D. A. J., & Hardy, R. W. (2007). The effects of extrusion conditions on the nutritional value of soybean meal for rainbow trout (*Oncorhynchus mykiss*). *Aquaculture*, *265*, 244–252.

Batchu, S. N., Chaudhary, K., Zlobine, I., Pawa, J., & Seubert, J. M. (2016). Fatty acids and cardiac ischemia-reperfusion injury. In R. R. Watson, & F. De Meester (Eds.), *Handbook of lipids in human function: Fatty acids* (pp. 39–83). San Diego, CA, USA: AOCS Press.

Baumgartner, S., Ras, R. T., Trautwein, E. A., Mensink, R. P., & Plat, J. (2017). Plasma fat-soluble vitamin and carotenoid concentrations after plant sterol and plant stanol consumption: A meta-analysis of randomized controlled trials. *European Journal of Nutrition*, *56*, 909–923.

Beeharry, N., Lowe, J. E., Hernandez, A. R., Chambers, J. A., Fucassi, F., Cragg, P. J., et al. (2003). Linoleic acid and antioxidants protect against DNA damage and apoptosis induced by palmitic acid. *Mutation research Fundamental and Molecular Mechanisms of Mutagenesis*, *530*, 27–33.

Boskou, D. (2002). Olive oil. In F. D. Gunstone (Ed.), *Vegetable oils in food technology: Composition, properties, and uses* (pp. 244–277). Oxford, UK: Blackwell Publishing Ltd.

Brühl, L. (1996). Trans fatty acids in cold pressed oils and in dried seeds. *Fett/Lipid*, *98*, 380–383.

Chang, S. S., Krishnamurthy, R. G., & Reddy, B. R. (1967). The relationship between alkyl furans and the reversion flavor of soybean oil. *Journal of the American Oil Chemists' Society*, *44*, 159.

Chen, J. X., Li, G., Wang, H., Liu, A., Lee, M.-J., Reuhl, K., et al. (2016). Dietary tocopherols inhibit PhIP-induced prostate carcinogenesis in CYP1A-humanized mice. *Cancer Letters*, *371*, 71–78.

Chen, J. X., Liu, A., Lee, M.-J., Wang, H., Yu, S., Chi, E., et al. (2017). δ- and γ-tocopherols inhibit phIP/DSS-induced colon carcinogenesis by protection against early cellular and DNA damages. *Molecular Carcinogenesis*, *56*, 172–183.

Chi, C.-H., & Cho, S.-J. (2016). Improvement of bioactivity of soybean meal by solid-state fermentation with *Bacillus amyloliquefaciens* versus *Lactobacillus* spp. and *Saccharomyces cerevisiae*. *LWT—Food Science and Technology*, *68*, 619–625.

Choe, E., & Min, D. B. (2006). Mechanisms and factors for edible oil oxidation. *Comprehensive Reviews in Food Science and Food Safety*, *5*, 169–186.

Choque, B., Catheline, D., Rioux, V., & Legrand, P. (2014). Linoleic acid: Between doubts and certainties. *Biochimie*, *96*, 14–21.

Cober, E. R., Cianzio, S. R., Pantalone, V. R., & Rajcan, I. (2009). Soybean. In J. Vollmann, & I. Rajcan (Eds.), *Oil crops* (pp. 57–90). New York, NY: Springer New York.

Coscina, D. V., Yehuda, S., Dixon, L. M., Kish, S. J., & Leprohon-Greenwood, C. E. (1986). Learning is improved by a soybean oil diet in rats. *Life Sciences*, *38*, 1789–1794.

Da Costa, C. A. S., Dos De Sousa Santos, A., Carlos, A. S., De Paula Lopes Gonzalez, G., Reis, R. P. G., Carneiro, C., et al. (2015). Impact of a high-fat diet containing canola or soybean oil on body development and bone; parameters in adult male rats. *Nutrición Hospitalaria*, *31*, 2147–2153.

Dai, C., Ma, H., He, R., Huang, L., Zhu, S., Ding, Q., et al. (2017). Improvement of nutritional value and bioactivity of soybean meal by solid-state fermentation with *Bacillus subtilis*. *LWT—Food Science and Technology*, *86*, 1–7.

Deol, P., Evans, J. R., Dhahbi, J., Chellappa, K., Han, D. S., Spindler, S., et al. (2015). Soybean oil is more obesogenic and diabetogenic than coconut oil and fructose in mouse: Potential role for the liver. *PLoS One*, *10*, e0132672.

Dietrich, M., Jacques, P. F., Traber, M. G., Cross, C. E., Hu, Y., & Block, G. (2006). Does γ-Tocopherol play a role in the primary prevention of heart disease and cancer? A review. *Journal of the American College of Nutrition*, *25*, 292–299.

Ding, Z., Zhang, Y., Ye, J., Du, Z., & Kong, Y. (2015). An evaluation of replacing fish meal with fermented soybean meal in the diet of Macrobrachium nipponense: Growth, nonspecific immunity, and resistance to *Aeromonas hydrophila*. *Fish & Shellfish Immunology*, *44*, 295–301.

Dixon, M. D., & Hammond, E. G. (1984). The flavor intensity of some carbonyl compounds important in oxidized fats. *Journal of the American Oil Chemists' Society*, *61*, 1452–1456.

Dutton, H. J., Lancaster, C. R., Evans, C. D., & Cowan, J. C. (1951). The flavor problem of soybean oil. VIII. Linolenic acid. *Journal of the American Oil Chemists' Society*, *28*, 115–118.

El Riachy, M., Priego-Capote, F., León, L., Rallo, L., & Luque de Castro, M. D. (2011). Hydrophilic antioxidants of virgin olive oil. Part 1: Hydrophilic phenols: A key factor for virgin olive oil quality. *European Journal of Lipid Science and Technology*, *113*, 678–691.

Engeseth, N. J., Klein, B. P., & Warner, K. (1987). Lipoxygenase isoenzymes in soybeans: Effects on crude oil quality. *Journal of Food Science*, *52*, 1015–1019.

Eskin, N. A. M. (2002). Authentication of evening primrose, borage and fish oils. In M. Jee (Ed.), *Oils and fats authentication* (pp. 95–114). Oxford, UK: Blackwell Publishing Ltd.

European Commission (2006). *Commission regulation (EC) no 118/2006 of 19 December 2006 setting maximum levels for certain contaminants in foodstuffs*.

Evans, C. D., Cooney, P. M., Moser, H. A., & Schwab, A. W. (1953). The flavor problem of soybean oil. XI. Phytic acid as an inactivating agent for trace metals. *Journal of the American Oil Chemists' Society*, *30*, 143–147.

Evans, J. C., Kodali, D. R., & Addis, P. B. (2002). Optimal tocopherol concentrations to inhibit soybean oil oxidation. *Journal of the American Oil Chemists' Society*, *79*, 47–51.

Evans, C. D., Schwab, A. W., Moser, H. A., Hawley, J. E., & Melvin, E. H. (1951). The flavor problem of soybean oil. VII. Effect of trace metals. *Journal of the American Oil Chemists' Society*, *28*, 68–73.

Feng, J., Liu, X., Xu, Z. R., Wang, Y. Z., & Liu, J. X. (2007). Effects of fermented soybean meal on digestive enzyme activities and intestinal morphology in broilers. *Poultry Science*, *86*, 1149–1154.

Flider, F. J., & Orthoefer, F. T. (1981). Metals in soybean oil. *Journal of the American Oil Chemists' Society*, *58*, 270–272.

Food and Agriculture Organization of the United Nations (2017). *FAOSTAT. [WWW Document](2017)*. http://www.fao.org/faostat/en/#data/QC/visualize (Accessed 6 July 2019).

Gao, Y., Wang, C., Zhu, Q., & Qian, G. (2013). Optimization of solid-state fermentation with *Lactobacillus brevis* and *Aspergillus oryzae* for trypsin inhibitor degradation in soybean meal. *Journal of Integrative Agriculture*, *12*, 869–876.

Guan, F., Li, G., Liu, A. B., Lee, M.-J., Yang, Z., Chen, Y.-K., et al. (2012). δ- and γ-tocopherols, but not α-tocopherol, inhibit colon carcinogenesis in aoxymethane-treated F344 rats. *Cancer Prevention Research*, *5*, 644–654.

Gunstone, F. D. (2004). *The chemistry of oils and fats-sources, composition, properties, and uses*. Oxford, UK: Blackwell Publishing Ltd.

Hammond, E. G., Johnson, L. A., Su, C., Wang, T., & White, P. J. (2005). Soybean oil. In F. Shahidi (Ed.), *Bailey's industrial oil and fat products* (pp. 577–653). Hoboken, NJ, USA: John Wiley & Sons, Inc.

Harris, W. S. (2008). Linoleic acid and coronary heart disease. *Prostaglandins, Leukotrienes & Essential Fatty Acids*, *79*, 169–171.

Herald, P. J., & Davidson, P. M. (1983). Antibacterial activity of selected Hydroxycinnamic acids. *Journal of Food Science*, *48*, 1378–1379.

Hirabayashi, M., Matsui, T., Yano, H., & Nakajima, T. (1998). Fermentation of soybean meal with *Aspergillus usamii* reduces phosphorus excretion in chicks. *Poultry Science*, *77*, 552–556.

Hirose, M., Masuda, A., Ito, N., Kamano, K., & Okuyama, H. (1990). Effects of dietary perilla oil, soybean oil and safflower oil on 7,12-dimethylbenz[a] anthracene (DMBA) and 1,2-dimethylhydrazine (DMH)-induced mammary gland and colon carcinogenesis in female SD rats. *Carcinogenesis*, *11*, 731–735.

Ho, C.-T., Smagula, M. S., & Chang, S. S. (1978). The synthesis of 2-(1-pentenyl) furan and its relationship to the reversion flavor of soybean oil. *Journal of the American Oil Chemists' Society*, *55*, 233–237.

Hong, K.-J., Lee, C.-H., & Kim, S. W. (2004). *Aspergillus oryzae* GB-107 fermentation improves nutritional quality of food soybeans and feed soybean meals. *Journal of Medicinal Food*, *7*, 430–435.

Hua, H., Zhao, X., Wu, S., & Li, G. (2016). Impact of refining on the levels of 4-hydroxy-trans-alkenals, parent and oxygenated polycyclic aromatic hydrocarbons in soybean and rapeseed oils. *Food Control*, *67*, 82–89.

Hymowitz, T. (2008). The history of the soybean. In L. A. Johnson, P. J. White, & R. Galloway (Eds.), *Soybeans: Chemistry, production, processing, and utilization* (pp. 1–31). Urbana, IL: AOCS Press.

Jacobsen, H. J., Samuelsen, T. A., Girons, A., & Kousoulaki, K. (2018). Different enzyme incorporation strategies in Atlantic salmon diet containing soybean meal: Effects on feed quality, fish performance, nutrient digestibility and distal intestinal morphology. *Aquaculture*, *491*, 302–309.

Ju, J., Hao, X., Lee, M.-J., Lambert, J. D., Lu, G., Xiao, H., et al. (2009). A gamma-tocopherol-rich mixture of tocopherols inhibits colon inflammation and carcinogenesis in azoxymethane and dextran sulfate sodium-treated mice. *Cancer Prevention Research*, *2*, 143–152.

Ju, J., Picinich, S. C., Yang, Z., Zhao, Y., Suh, N., Kong, A.-N., et al. (2010). Cancer-preventive activities of tocopherols and tocotrienols. *Carcinogenesis*, *31*, 533–542.

Kania, M., Michalak, M., Gogolewski, M., & Hoffmann, A. (2004). Antioxidative potential of substances contained in cold pressed soybean oil and after each phase of refining process. *Acta Scientiarum Polonorum Technologia Alimentaria*, *3*, 113–121.

Karasulu, H. Y., Karasulu, E., Büyükhelvacıgil, M., Yıldız, M., Ertugrul, A., Büyükhelvacıgil, K., et al. (2011). Soybean oil: Production process, benefits and uses in pharmaceutical dosage form. In H. El-Shemy (Ed.), *Soybean and health* (pp. 283–310). InTech.

Karupaiah, T., & Sundram, K. (2007). Effects of stereospecific positioning of fatty acids in triacylglycerol structures in native and randomized fats: a review of their nutritional implications. *Nutrition & Metabolism (London)*, *4*, 16.

Kim, S. W., van Heugten, E., Ji, F., Lee, C. H., & Mateo, R. D. (2010). Fermented soybean meal as a vegetable protein source for nursery pigs: I. Effects on growth performance of nursery pigs. *Journal of Animal Science*, *88*, 214–224.

Kozłowska, M., Gruczyńska, E., Ścibisz, I., & Rudzińska, M. (2016). Fatty acids and sterols composition, and antioxidant activity of oils extracted from plant seeds. *Food Chemistry*, *213*, 450–456.

Krishnamurthy, R. G., Smouse, T. H., Mookherjee, B. D., Reddy, B. R., & Chang, S. S. (1967). Identification of 2-pentyl furan in fats and oils and its relationship to the reversion flavor of soybean oil. *Journal of Food Science*, *32*, 372–374.

Kritchevsky, D., Weber, M. M., & Klurfeld, D. M. (1992). Influence of different fats (soybean oil, palm olein or hydrogenated soybean oil) on chemically-induced mammary tumors in rats. *Nutrition Research*, *12*, S175–S179.

Larsson, B. K., Eriksson, A. T., & Cervenka, M. (1987). Polycyclic aromatic hydrocarbons in crude and deodorized vegetable oils. *Journal of the American Oil Chemists' Society*, *64*, 365–370.

Li, F., Yin, Y., Tan, B., Kong, X., & Wu, G. (2011). Leucine nutrition in animals and humans: mTOR signaling and beyond. *Amino Acids*, *41*, 1185–1193.

Liu, K. (1997). *Soybeans: Chemistry, technology, and utilization*. Dordrecht: Springer-Science+Business Media.

Liu, K. (2004a). Soybeans as a powerhouse of nutrients and phytochemicals. In K. Liu (Ed.), *Soybeans as functional foods and ingredients* (pp. 1–22). Champaign, Illinois: AOCS Press.

Liu, K. (2004b). Edible soybean products in the current market. In K. Liu (Ed.), *Soybeans as functional foods and ingredients* (pp. 23–51). Champaign, IL: AOCS Press.

Lou, Z., Wang, H., Rao, S., Sun, J., Ma, C., & Li, J. (2012). *p*-Coumaric acid kills bacteria through dual damage mechanisms. *Food Control*, *25*, 550–554.

Lu, G., Xiao, H., Li, G. X., Picinich, S. C., Chen, Y. K., Liu, A., et al. (2010). A gamma-tocopherol-rich mixture of tocopherols inhibits chemically induced lung tumorigenesis in A/J mice and xenograft tumor growth. *Carcinogenesis*, *31*, 687–694.

Luceri, C., Giannini, L., Lodovici, M., Antonucci, E., Abbate, R., Masini, E., et al. (2007). p-Coumaric acid, a common dietary phenol, inhibits platelet activity in vitro and in vivo. *The British Journal of Nutrition*, *97*, 458–463.

Luceri, C., Guglielmi, F., Lodovici, M., Giannini, L., Messerini, L., & Dolara, P. (2004). Plant phenolic 4-coumaric acid protects against intestinal inflammation in rats. *Scandinavian Journal of Gastroenterology*, *39*, 1128–1133.

Madigan, C., Ryan, M., Owens, D., Collins, P., & Tomkin, G. H. (2000). Dietary unsaturated fatty acids in type 2 diabetes: Higher levels of postprandial lipoprotein on a linoleic acid-rich sunflower oil diet compared with an oleic acid-rich olive oil diet. *Diabetes Care*, *23*, 1472–1477.

Maggiora, M., Bologna, M., Cerù, M. P., Possati, L., Angelucci, A., Cimini, A., et al. (2004). An overview of the effect of linoleic and conjugated-linoleic acids on the growth of several human tumor cell lines. *International Journal of Cancer*, *112*, 909–919.

Mahmood, T., Bibi, Y., Ishaq, H., Mahmood, I., Wahab, A., & Sherwani, S. (2013). Complexation and antimicrobial activities of β sitosterol with trace metals (Cu(II), Co(II), and Fe (III)). *European Academic Research*, 677–685.

Marsman, G. J. P., Gruppen, H., Mul, A. J., & Voragen, A. G. J. (1997). In vitro accessibility of untreated, toasted, and extruded soybean meals for proteases and carbohydrases. *Journal of Agricultural and Food Chemistry*, *45*, 4088–4095.

Marsman, G. J. P., Gruppen, H., van der Poel, A. F. B., Kwakkel, R. P., Verstegen, M. W. A., & Voragen, A. G. J. (1997). The effect of thermal processing and enzyme treatments of soybean meal on growth performance, ileal nutrient digestibilities, and chyme characteristics in broiler chicks. *Poultry Science*, *76*, 864–872.

McClements, D. J., & Decker, E. A. (2017). Lipids. In S. Damodaran, & K. L. Parkin (Eds.), *Fennema's food chemistry* (pp. 171–233). Boca Raton, FL: CRC Press.

Min, D. B., Callison, A. L., & Lee, H. O. (2003). Singlet oxygen oxidation for 2-pentylfuran and 2-pentenyfuran formation in soybean oil. *Journal of Food Science*, *68*, 1175–1178.

Mnari, A. B., Harzallah, A., Amri, Z., Dhaou Aguir, S., & Hammami, M. (2016). Phytochemical content, antioxidant properties, and phenolic profile of Tunisian raisin varieties (*Vitis vinifera* L.). *International Journal of Food Properties*, *19*, 578–590.

Muhlhausler, B. S., & Ailhaud, G. P. (2013). Omega-6 polyunsaturated fatty acids and the early origins of obesity. *Current Opinion in Endocrinology, Diabetes, and Obesity*, *20*, 56–61.

Nishimura, M., Yamaguchi, M., Naito, S., & Yamauchi, A. (2006). Soybean oil fat emulsion to prevent TPN-induced liver damage: Possible molecular mechanisms and clinical implications. *Biological & Pharmaceutical Bulletin*, *29*, 855–862.

O'Brien, R. D. (2008). Soybean oil purification. In L. A. Johnson, P. J. White, & R. Galloway (Eds.), *Soybeans: Chemistry, production, processing, and utilization* (pp. 377–408). Urbana, IL: AOCS Press.

Odiba, J., Musa, A., Hassan, H., Yahaya, S., & Okolo, E. (2014). Antimicrobial activity of isolated Stigmast-5-en-3 β-ol (β-Sitosterol) from honeybee. *International Journal of Pharma Sciences and Research*, *5*, 908–918.

Park, Y., Allen, K. G., & Shultz, T. D. (2000). Modulation of MCF-7 breast cancer cell signal transduction by linoleic acid and conjugated linoleic acid in culture. *Anticancer Research*, *20*, 669–676.

Peterson, D. W. (1951). Effect of soybean sterols in the diet on plasma and liver cholesterol in chicks. *Experimental Biology and Medicine*, *78*, 143–147.

Pongmaneerat, J., & Watanabe, T. (1993). Effect of extrusion processing on the utilization of soybean meal diets for rainbow trout. *Nippon Suisan Gakkaishi*, *59*, 1407–1414.

Rafalowski, R., Zegarska, Z., Kuncewicz, A., & Borejszo, Z. (2008). Fatty acid composition, tocopherols and β-carotene content in Polish commercial vegetable oils. *Pakistan Journal of Nutrition*, *7*, 278–282.

Raghuveer, K. G., & Hammond, E. G. (1967). The influence of glyceride structure on the rate of autoxidation. *Journal of the American Oil Chemists' Society*, *44*, 239–243.

Refstie, S., Sahlström, S., Bråthen, E., Baeverfjord, G., & Krogedal, P. (2005). Lactic acid fermentation eliminates indigestible carbohydrates and antinutritional factors in soybean meal for Atlantic salmon (*Salmo salar*). *Aquaculture*, *246*, 331–345.

Ribeiro, R., Fernandes, A., Meira, E., Batista, P., Magalhães Siman, F., Vassallo, D., et al. (2010). Soybean oil increases SERCA2a expression and left ventricular contractility in rats without change in arterial blood pressure. *Lipids in Health and Disease*, *9*, 53.

Riserus, U., Willett, W. C., & Hu, F. B. (2009). Dietary fats and prevention of type 2 diabetes. *Progress in Lipid Research*, *48*, 44–51.

Rojo Camargo, M. C., Antoniolli, P. R., & Vicente, E. (2012). Evaluation of polycyclic aromatic hydrocarbons content in different stages of soybean oils processing. *Food Chemistry*, *135*, 937–942.

Romarheim, O. H., Aslaksen, M. A., Storebakken, T., Krogdahl, Å., & Skrede, A. (2005). Effect of extrusion on trypsin inhibitor activity and nutrient digestibility of diets based on fish meal, soybean meal and white flakes. *Archives of Animal Nutrition*, *59*, 365–375.

Rudzińska, M., Kazuś, T., & Wąsowicz, E. (2001). Sterole i ich utlenione pochodne w olejach roślinnych rafinowanych i tłoczonych na zimno. *Rośliny Oleiste*, 477–494.

Seals, R. G., & Hammond, E. G. (1966). Diacetyl as the buttery flavor component in soybean oil. *Journal of the American Oil Chemists' Society*, *43*, 401–402.

Seals, R. G., & Hammond, E. G. (1970). Some carbonyl flavor compounds of oxidized soybean and linseed oils. *Journal of the American Oil Chemists' Society*, *47*, 278–280.

Seo, S.-H., & Cho, S.-J. (2016). Changes in allergenic and antinutritional protein profiles of soybean meal during solid-state fermentation with *Bacillus subtilis*. *LWT—Food Science and Technology*, *70*, 208–212.

Shen, S., Wang, D., & Das, U. N. (2012). Lipids. In D. Wang, H. Lin, J. Kan, L. Liu, X. Zeng, & S. Shen (Eds.), *Food chemistry* (pp. 107–135). New York: Nova Science Publishers, Inc.

Shibamoto, T., & Bjeldanes, L. (2009). *Introduction to food toxicology* (2nd ed.). Burlington, MA: Academic Press.

Shultz, T. D., Chew, B. P., & Seaman, W. R. (1992). Differential stimulatory and inhibitory responses of human MCF-7 breast cancer cells to linoleic acid and conjugated linoleic acid in culture. *Anticancer Research*, *12*, 2143–2145.

Siger, A., Nogala-Kalucka, M., & Lampart-Szczapa, E. (2008). The content and antioxidant activity of phenolic compounds in cold-pressed plant oils. *Journal of Food Lipids*, *15*, 137–149.

Sinn, S. M., Gibbons, W. R., Brown, M. L., DeRouchey, J. M., & Levesque, C. L. (2017). Evaluation of microbially enhanced soybean meal as an alternative to fishmeal in weaned pig diets. *Animal*, *11*, 784–793.

Sivaramakrishnan, S., & Gangadharan, D. (2009). Edible oil cakes. In P. Singh nee' Nigam, & A. Pandey (Eds.), *Biotechnology for Agro-Industrial Residues Utilisation* (pp. 253–271). Dordrecht: Springer Science+Business Media.

Smagula, M. S., Ho, C.-T., & Chang, S. S. (1979). The synthesis of 2-(2-pentenyl) furans and their relationship to the reversion flavor of soybean oil. *Journal of the American Oil Chemists' Society*, *56*, 516–519.

Smouse, T. H., & Chang, S. S. (1967). A systematic characterization of the reversion flavor of soybean oil. *Journal of the American Oil Chemists' Society*, *44*, 509–514.

Stojković, D., Petrović, J., Soković, M., Glamočlija, J., Kukić-Marković, J., & Petrović, S. (2013). In situ antioxidant and antimicrobial activities of naturally occurring caffeic acid, p-coumaric acid and rutin, using food systems. *Journal of the Science of Food and Agriculture*, *93*, 3205–3208.

Summers, L. K. M., Fielding, B. A., Bradshaw, H. A., Ilic, V., Beysen, C., Clark, M. L., et al. (2002). Substituting dietary saturated fat with polyunsaturated fat changes abdominal fat distribution and improves insulin sensitivity. *Diabetologia*, *45*, 369–377.

Tuberoso, C. I. G., Kowalczyk, A., Sarritzu, E., & Cabras, P. (2007). Determination of antioxidant compounds and antioxidant activity in commercial oilseeds for food use. *Food Chemistry*, *103*, 1494–1501.

van Rensburg, S. J., Daniels, W. M. U., van Zyl, J. M., & Taljaard, J. J. F. (2000). A comparative study of the effects of cholesterol, beta-sitosterol, beta-sitosterol glucoside, dehydro-epiandrosterone sulphate and melatonin on in vitro lipid peroxidation. *Metabolic Brain Disease*, *15*, 257–265.

Wang, Y., Liu, X. T., Wang, H. L., Li, D. F., Piao, X. S., & Lu, W. Q. (2014). Optimization of processing conditions for solid-state fermented soybean meal and its effects on growth performance and nutrient digestibility of weanling pigs. *Livestock Science*, *170*, 91–99.

Woyengo, T. A., Patterson, R., & Levesque, C. L. (2016). Nutritive value of extruded or multi-enzyme supplemented cold-pressed soybean cake for pigs. *Journal of Animal Science*, *94*, 5230–5238.

Yasothai, R. (2016). Antinutritional factors in soybean meal and its deactivation. *International Journal of Science, Environment*, *5*, 3793–3797.

Yin, Y., Fatufe, A. A., & Blachier, F. (2011). Soya bean meal and its extensive use in livestock feeding and nutrition. In H. El-Shemy (Ed.), *Soybean and nutrition* (pp. 369–384). InTech.

Yin, Y., Yao, K., Liu, Z., Gong, M., Ruan, Z., Deng, D., et al. (2010). Supplementing L-leucine to a low-protein diet increases tissue protein synthesis in weanling pigs. *Amino Acids*, *39*, 1477–1486.

Yoon, S.-A., Kang, S.-I., Shin, H.-S., Kang, S.-W., Kim, J.-H., Ko, H.-C., et al. (2013). p-Coumaric acid modulates glucose and lipid metabolism via AMP-activated protein kinase in L6 skeletal muscle cells. *Biochemical and Biophysical Research Communications*, *432*, 553–557.

Yoon, S. H., & Min, D. B. (1987). Roles of phospholipids in flavor stability of soybean oil. *Korean Journal of Food Science and Technology*, *19*, 23–28.

Yuan, L., Chang, J., Yin, Q., Lu, M., Di, Y., Wang, P., et al. (2017). Fermented soybean meal improves the growth performance, nutrient digestibility, and microbial flora in piglets. *Animal Nutrition*, *3*, 19–24.

Zeleny, L., & Neustadt, H. M. (1940). Rapid determination of soybean oil content and of iodine number of soybean oil. *Bulletin of the US Department of Agriculture*, *748*, 1–23.

Zhou, F., Song, W., Shao, Q., Peng, X., Xiao, J., Hua, Y., et al. (2011). Partial replacement of fish meal by fermented soybean meal in diets for Black Sea bream, *Acanthopagrus schlegelii*, Juveniles. *Journal of the World Aquaculture Society*, *42*, 184–197.

Chapter 52

Cold pressed macadamia oil

Chin Xuan Tan[a], Seok Shin Tan[b], and Seok Tyug Tan[c]
[a]Department of Allied Health Sciences, Faculty of Science, Universiti Tunku Abdul Rahman, Kampar, Perak, Malaysia, [b]Department of Nutrition and Dietetics, School of Health Sciences, International Medical University, Kuala Lumpur, Malaysia, [c]Department of Healthcare Professional, Faculty of Health and Life Sciences, Management and Science University, Shah Alam, Selangor, Malaysia

Abbreviations

MUFA monounsaturated fatty acids
PUFA polyunsaturated fatty acids
SFA saturated fatty acids

1 Introduction

The macadamia nut, also known as the "Queensland nut," is the only Australian plant that has been domesticated as a commercial food crop (Wallace & Walton, 2011). The nut belongs to the Proteaceae family, encased in a fibrous green husk and falls to the ground upon maturing. The mature macadamia nut is 2.6cm in length and 2.7cm in width (Koaze, Karanja, Kojima, Baba, & Ishibashi, 2002), consisting of three anatomical regions: husk, shell, and kernel. The edible part of the nut is the kernel, which is rounded in shape and contains a high level of lipids (>60%).

Global macadamia kernel production has increased drastically over the past years and reached approximately 52,000 metric tons in 2017 (Fig. 1). In that year, the largest macadamia kernel producing countries were Australia and South Africa, accounting for more than half of the world's macadamia kernel production (International Nut and Dried Fruit, 2018). Macadamia kernel production in South Africa and China was drastically increased in 2017 compared to the prior season, 26% and 49%, respectively (International Nut and Dried Fruit, 2018).

2 Composition of the macadamia kernel

The proximate composition of the macadamia kernel is affected by the nut maturity, variety, location, and growth conditions (Navanorro & Rodrigues, 2016). Based on the nutrient database of the United States Department of Agriculture (USDA), the macadamia kernel contains 75.8% crude oil, 7.9% crude protein, 8.6% dietary fiber, 1.4% moisture, 1.1% ash, and 13.8% carbohydrate (USDA, 2018). The oil content is the same as the level reported by Moodley, Kindness, and Jonnalagadda (2007). However, Entelmann, Scarpare Filho, Pio, da Silva, and de Souza (2014) reported a lower oil content (65%–68%) of the kernels of macadamia grown in the southwest of São Paulo state, Brazil.

3 Macadamia species targeting oil extraction

Nine macadamia species have been discovered, with seven species from the coastal temperate rainforests of the east coast of Australia and other two species from Sulawesi, Indonesia (Walton, 2005). Some macadamia species contain toxic cyanogen compounds in their kernels and are not suitable for human consumption. Commercial macadamia species for edible oil production and human consumption are *Macadamia integrifolia* and *Macadamia tetraphylla* and their hybrids. Both species and their hybrids are characterized by high oil content in their kernels. *Macadamia integrifolia* is round-shaped with a smooth shell, whereas *Macadamia tetraphylla* is spindle-shaped with a rough shell (Walton, 2005). The mature macadamia nut is characterized by a white or cream kernel enclosed in a dark brown shell. Based on Wall and Gentry (2007), the kernel of a good-quality mature macadamia nut has internal and external L^*(lightness) values of 74.3 and 71.1, respectively. The oil content of a mature macadamia kernel (>60%) is significantly greater than that of an immature one, with a range value of 1%–48% (Koaze et al., 2002). Thus, macadamia oil should be extracted from mature kernels of nontoxic species.

Cold Pressed Oils. https://doi.org/10.1016/B978-0-12-818188-1.00052-9
Copyright © 2020 Elsevier Inc. All rights reserved.

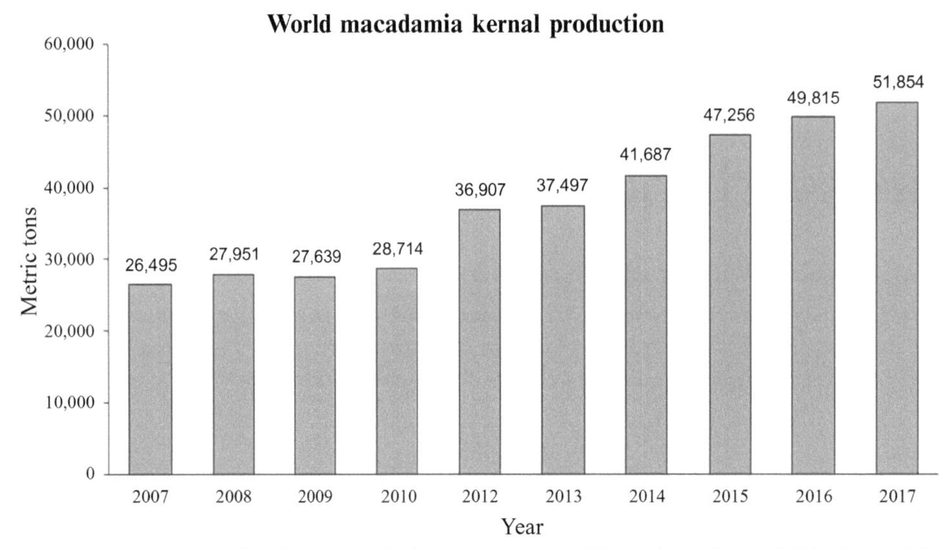

FIG. 1 World macadamia kernel production from 2007 to 2017. *(Source: International Nut and Dried Fruit (2018). Nuts and Dried Fruits Statistical Year Book 2017/2018.)*

4 Extraction and processing of cold pressed macadamia oil

4.1 Preprocessing of macadamia nut

Macadamia oil is extracted from the kernel of the mature macadamia nut. The initial step in macadamia oil production involves the removal of the outer husk of macadamia nut using a de-husking machine and getting the shell. The shell is then dehydrated to a moisture content close to 3.5% dry basis (Navanorro & Rodrigues, 2016). After that, the kernel is separated from the shell using a cracker, vise, or chain breaker.

4.2 Cold pressed extraction

Macadamia kernels contain high oil content and numerous methods have been developed to extract oil from them; these include solvent extraction (Entelmann et al., 2014; Fard, Turner, & Willett, 2003; Moodley et al., 2007; Rodríguez, Silva, & Carrillo, 2011), cold pressing (Albertson et al., 2012; Fard et al., 2003; Rodríguez et al., 2011; Wall, 2010), ultrasound-assisted extraction (Zhu, Lin, Li, Lv, & Huang, 2012), subcritical fluid extraction (Pang et al., 2014), and supercritical fluid extraction (Silva, Mendes, Pessoa, & Queiroz, 2008; Zhu et al., 2012). Among these, the commonly used techniques to extract macadamia oil are cold pressing and solvent extraction.

Cold pressed extraction of macadamia oil is carried out at temperature less than 30°C using mechanical (e.g., hydraulic press or screw press) techniques (Kamal-Eldin & Moreau, 2009; Navanorro & Rodrigues, 2016; Rodríguez et al., 2011). Sometimes, cold pressed macadamia oil is marketed as virgin macadamia oil. To prevent enzymatic hydrolysis of macadamia oil, Rodríguez et al. (2011) recommend that the cold pressed extraction is performed in an environment with minimum light contact. Cold pressing of macadamia kernel produces macadamia oil and pressed macadamia meal.

An early study conducted by Jitngarmkusol, Hongsuwankul, and Tananuwong (2008) showed that the yield of macadamia oil obtained by cold pressing was 75%–79% of the total oil from the kernel. In another study, Rodríguez et al. (2011) reported that the oil yield obtained by cold pressing (35%–40%) was lower than by Soxhlet extraction (40%–60%). Both studies indicated that the pressed macadamia meal still contains oil. Navanorro and Rodrigues (2016) suggest that the pressed macadamia meal could be subjected to a second oil extraction using organic solvents. This is because organic solvents have the ability to extract most of the lipid components such as triacylglycerols, phospholipids, waxes, and pigments from the macadamia kernel. However, solvent-extracted plant oil contains trace amounts of solvent residues and needs to undergo refining, bleaching, and deodorizing (RBD) processes before being edible (Tan, Chong, Hamzah, & Ghazali, 2018a). Cold pressed macadamia oil is a clear golden yellow color with distinctive nutty aroma and flavor (Gustavheess, 2010). This is in contrast to RBD macadamia oil, which is a light yellow color or colorless, with a bland aroma and flavor.

4.3 Utilization of pressed macadamia meal

Pressed macadamia meal, also known as defatted macadamia meal, can be further processed into food products. The chemical and functional characteristics of flour produced from the partially defatted (12%–15% lipid content) and fully defatted (1% lipid content) macadamia meal were reported by Jitngarmkusol et al. (2008). Both partially and fully defatted flours were rich in protein (30.4%–36.5%) and carbohydrate (46.5%–57.1%). However, the water absorption capacity, oil absorption capacity, and foaming capacity of fully defatted flour were greater than those of partially defatted flour, which was probably due to the lower lipid content in fully defatted flour. The authors' study highlights the potential of using high protein defatted flour, either partially or fully defatted, produced from the macadamia meal as raw ingredients to produce an array of bakery products such as biscuits, bread, and cookies.

In another study, the fully defatted macadamia meal was utilized to produce macadamia protein concentrates (Navarro & Rodrigues, 2018). Macadamia protein concentrates contained high levels of protein (70%–72%) and low oil absorption capacity (3.4–3.6 g of oil/g of protein), suggesting their suitability to be used as food emulsifiers.

5 Fatty acids and acyl lipids profile

Cold pressed macadamia oil is a specialty oil with significant production and commercialization. The fatty acids composition of macadamia oil obtained by cold pressing is compared with Soxhlet extraction in Table 1. The relative percentages of saturated fatty acids (SFAs), monounsaturated fatty acids (MUFAs) and polyunsaturated fatty acids (PUFAs) of cold pressed macadamia oil (SFAs: 15.1%–16.4%; MUFAs: 75.1%–81.5% and PUFAs: 2.2%–4.0%) did not vary much from the Soxhlet-extracted macadamia oil (SFAs: 13.2%–21.2%; MUFAs: 72.7%–82.5% and PUFAs: 1.9%–4.7%), indicating that the fatty acids profile of macadamia oil is unaffected by extraction methods.

Cold pressed macadamia oil was a rich source of oleic (54.6%–63.4%), palmitoleic (12.5%–19.3%), and palmitic (8.1%–9.4%) acids (Table 1). More than 50% of the total fatty acids in cold pressed macadamia oil is constituted of oleic acid. The study of Prescha et al. (2014) showed that the level of oleic acid of cold pressed macadamia oil (58.9%) was superior to that of other cold pressed plant (sesame, pumpkin, milk thistle, safflower, flaxseed, walnut, linola, poppy, rose hip, and hempseed) oils, with levels ranging from 12.5% to 39.5%. A high level of oleic acid in cold pressed macadamia oil is desirable, from the perspective of nutrition and oxidative stability (Tan, 2019). As evidenced by Kris-Etherton (1999), the isocaloric substitution of about 5% energy from SFA with oleic acid reduced the risk of cardiovascular disease by 20%–40%. In addition, a high level of palmitoleic acid in cold pressed macadamia oil may be beneficial for diabetes management. It has been demonstrated that the intake of palmitoleic acid was able to reduce muscle insulin resistance and prevent beta-cell apoptosis (Yang, Miyahara, & Hatanaka, 2011).

The polar and neutral lipids of oil extracted from mature macadamia kernels were reported by Koaze et al. (2002). Macadamia oil contained about 3% of polar lipids. These polar lipids were mainly constituted of phospholipids (e.g., phosphatidylcholine and phosphatidylethanolamine; 84%), cerebrosides (7%), acylsteryl-glucosides (6%), and steryl-glucosides (1%). Meanwhile, triacylglycerols were the major fraction of neutral lipids in macadamia oil, accounting for 87% of total neutral lipids. Triolein (OOO: 24.0%), dioleoyl-palmitoleoyl-glycerol (OOPo: 21.6%), and dioleoyl-palmitoyl-glycerol (OOP: 14.7%) were the main triacylglycerol components present in cold pressed macadamia oil (Table 2). A similar trend was observed in commercial macadamia oil.

6 Minor bioactive lipid components

6.1 Tocols

Tocols are constituted by four tocopherols and four tocotrienol isomers, namely, alpha (α), beta (β), gamma (γ), and delta (δ). Both tocopherols and tocotrienols provide positive biological effects by signal transduction, by modulating gene expression, and by regulating cell functions through regulation of protein-membrane interactions (Shahidi & de Camargo, 2016). The amount of tocols in macadamia oil varied considerably due to the methods used to extract the oil, and macadamia oil is generally not a good source of tocols. Castelo-Branco, Santana, Di-Sarli, Freitas, and Torres (2016) and Ying, Wojciechowska, Siger, Kaczmarek, and Rudzińska (2018) reported α-tocopherol (0.8–1.4 mg/100 g) as the only tocopherol component present in cold pressed macadamia oil. Kornsteiner, Wagner, and Elmadfa (2006), however, could not detect the presence of tocopherol components in macadamia oil obtained by solvent (petroleum ether) extraction. Meanwhile, Madawala et al. (2012) reported three tocopherol components

TABLE 1 Fatty acid composition of macadamia oil obtained by cold pressed and Soxhlet extraction.

	Extraction method					
	Cold pressed (%)				Soxhlet (%)	
Fatty acid	Rodrigues, Silva, Marsaioli, and Meirelles (2005)	Madawala, Kochhar, and Dutta (2012)	Prescha, Grajzer, Dedyk, and Grajeta (2014)	Carrillo, Carpio, Morales, Vilcacundo, and Álvarez (2017)	Kaijser, Dutta, and Savage (2000)[a]	Derewiaka, Szwed, and Wołosiak (2014)[b]
C14:0	0.98	0.70	0.80	ND	0.96–1.84	0.90
C16:0	9.38	8.10	8.20	9.11	8.41–11.13	12.10
C16:1	19.28	18.10	18.10	12.48	16.86–33.75	15.00
C17:0	ND	ND	0.10	ND	ND	ND
C18:0	3.40	3.60	3.30	3.93	1.49–3.19	5.20
C18:1	59.76	54.60	58.90	63.36	40.55–59.01	59.90
C18:2	2.03	3.30	3.50	3.22	2.63–4.47	1.90
C18:3	0.14	0.40	0.50	1.69	0.17–0.20	ND
C20:0	2.56	2.70	2.70	3.31	1.23–2.06	3.00
C20:1	2.48	2.40	2.40	2.90	1.36–2.63	2.00
C22:0	ND	ND	ND	ND	0.39–0.61	ND
C24:0	ND	ND	ND	ND	0.11–0.19	ND
SFA	16.32	15.10	15.10	16.35	13.23–17.81	21.20
MUFA	81.52	75.10	79.40	78.74	72.70–82.46	76.90
PUFA	2.17	3.70	4.00	4.91	2.83–4.67	1.90

[a] *The oil was extracted using hexane.*
[b] *The oil was extracted using petroleum ether.*

TABLE 2 The main triacylglycerol components of cold pressed and commercial macadamia oils.

	Cold pressed (%)	Commercial (%)
Triacylglycerol	Rodrigues et al. (2005)	Lee, Jones, Lee, Kim, and Foglia (2003)
OOO	23.97	23.34
OOPo	21.58	20.88
OOP	14.39	13.56
OPPo	8.51	7.89
OPoPo	6.70	8.84
OOA	3.69	3.44
OOL	2.31	1.56
PPoPo	1.78	1.51
POS	1.51	1.25
POA	1.16	0.62

L, linoleic; *Po*, palmitoleic; *O*, oleic; *S*, stearic; *P*, palmitic; *A*, arachidic.

(α-tocopherol (8 µg/g), δ-tocopherol (11 µg/g), and γ-tocopherol (15 µg/g)) and two tocotrienol components (α-tocotrienol (20 µg/g) and γ-tocotrienol (<0.5 µg/g)) in commercial macadamia oil.

6.2 Squalene

Squalene is an unsaturated hydrocarbon steroid precursor with important biological properties. It provides beneficial effects in chemoprevention and reduction of serum cholesterol and triacylglycerol levels (O'brien, 2009). Cicero et al. (2018) reported that squalene content of cold pressed macadamia oil was 23 µg/g, which was lower than that of other cold pressed plant oils (olive, Brazil nut, grapeseed, pequi, and palm oils), with a range value of 88–1046 µg/g. A higher level of squalene content (72–171 µg/g) was reported in macadamia oil obtained by solvent (petroleum ether) extraction (Wall, 2010).

6.3 Phytosterols

Cold pressed plant oil contains a substantial amount of phytosterols and most of the phytosterols in commodity plant oil are degraded during the refining process (Hernandez, 2016; Tan, 2019). The total phytosterol content in cold pressed macadamia oil was 78.2 mg/100 g, with the main phytosterol being β-sitosterol (65.6 mg/100 g), subsequently followed by campesterol (7.0 mg/100 g), avenasterol (4.2 mg/100 g), and stigmasterol (1.4 mg/100 g) (Ying et al., 2018). Cold pressed macadamia oil is rich in β-sitosterol. A study by Ying et al. (2018) reported that the decreasing order of β-sitosterol levels of cold pressed plant oils was macadamia oil (84%) > avocado oil (79%) > pine nut oil (77%) > black currant seed oil (67%) > hemp seed oil (67%) > safflower oil (65%) > wheat germ oil (64%) > dill seed oil (63%) > poppy seed oil (61%) > parsley seed oil (60%) > milk thistle oil (46%) > black cumin seed oil (45%). Compared to other phytosterols, the health benefits of β-sitosterol on disease prevention and management have been well documented. For example, it has been reported to be effective in reducing serum cholesterol and LDL-cholesterol levels (Lu et al., 2010; Tan, Chong, Hamzah, & Ghazali, 2018b).

6.4 Polyphenols

Polyphenols have a great impact on the nutritional and sensory qualities of plant oil. In a study conducted by Cicero et al. (2018), total polyphenol content of selected cold pressed plant oils decreased in the order of palm oil (18.6 µg/g) > canola oil (11.2 µg/g) > coconut oil (10.1 µg/g) > olive oil (7.9 µg/g) > grapeseed oil (4.8 µg/g) > macadamia oil (2.4 µg/g) > pequi oil (2.2 µg/g) > Brazil nut oil (0.6 µg/g). The polyphenol content of cold pressed macadamia oil may increase if shell fragments are crushed together with the kernel during oil production (Walton, 2005). The polyphenols identified in cold pressed macadamia oil were *p*-hydroxybenzoic acid (0.8 µg/g), luteolin (0.7 µg/g), apigenin 7-glucoside (0.5 µg/g), and caffeic acid (0.4 µg/g) (Cicero et al., 2018). The chemical structures of these polyphenols are shown in Fig. 2.

7 Volatile compounds

Seven volatile compounds (hexanal, heptanal, octanal, benzeneacetaldehyde, nonanal, pyridine, and dimethyl disulfide) have been identified in macadamia kernels (Phatanayindee, Borompichaichartkul, Srzednicki, Craske, & Wootton, 2012). Among these, nonanal and heptanal were the main volatiles detected in macadamia kernel, which constituted 42.2% and 19.6%, respectively, of the total volatile composition. Both aldehydes are commonly found in the avocado, olive, and hazelnut oils (Mildner-Szkudlarz & Jeleń, 2008; Tan, 2019). No literature study is available on the volatile composition of cold pressed macadamia oil. However, based on a sensory evaluation performed by Kochhar and Henry (2009), commercial macadamia oil was characterized by a strong nutty flavor.

8 Health-promoting traits of macadamia oil

The unsaturated fatty acids, particularly oleic and palmitoleic acids, and bioactive components are compounds with potential beneficial properties toward human health. Limited data are available regarding the health effects of macadamia oil. The effects of dietary supplementation of commercial macadamia oil (Proteco Gold Pty Ltd., Queensland, Australia) in high-carbohydrate, high-fat diet-induced metabolic syndrome in male Wistar rats were investigated by Poudyal et al. (2013). After 8 weeks' supplementation of 3% macadamia oil, the visceral adiposity index, plasma cholesterol, and plasma liver damage markers (e.g., alkaline phosphate, aspartate aminotransferase, alanine transaminase, and creatine kinase) were reduced when compared to the untreated control rats. In addition, supplementation of macadamia oil normalized the

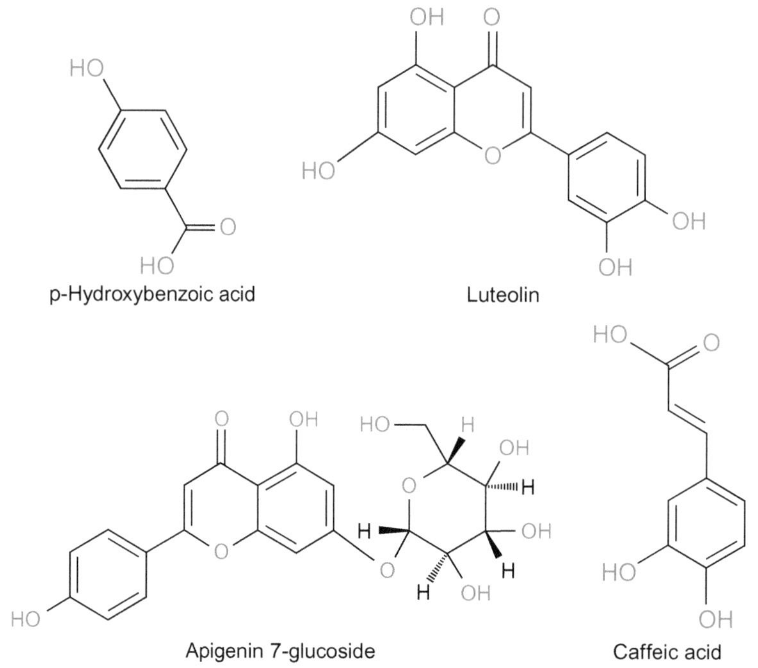

FIG. 2 Chemical structures of polyphenols identified in cold pressed macadamia oil.

systolic blood pressure. In another study, Lima et al. (2014) reported the effects of oral administration (2 g/kg body weight, three times per week) of commercial macadamia oil (Vital Âtman, São Paulo, Brazil) for 12 weeks in high-fat diet induced obesity male C57BL/6 mice. Their study showed that the production of interleukin-1β, a prototypic proinflammatory cytokine that plays key roles in acute and chronic inflammatory and autoimmune disorders, was reduced in peritoneal macrophages and adipose tissue. The increase in adipocytes size, which was associated with disturbances in cellular homeostasis as a result of high-fat diet consumption, was reduced after the ingestion of macadamia oil.

9 Edible and nonedible applications

Cold pressed macadamia oil is often used for salad dressings and for stir-frying because of its pleasant flavor and aroma. The pharmaceuticals industry uses cold pressed macadamia oil to produce an array of dietary supplements and nutraceutical products. It has been reported that macadamia oil is commercially sold as a supplement with heart health benefits (Hernandez, 2016). Owing to high emolliency and rapid skin penetration properties, crude and refined macadamia oils have been extensively used in the production of cosmetics and skin care products such as lipstick, shampoo, and body lotion.

10 Other issues

10.1 Oxidative stability

The resistance of oil to oxidation during processing and storage is known as oxidative stability. The oxidative stability of cold pressed plant oils is usually set for 6 or 12 months and generally limited by the levels of PUFAs and natural antioxidants such as tocopherols and polyphenol compounds (Prescha et al., 2014). Plant oil with high PUFA levels is prone to oxidative degradation (Navanorro & Rodrigues, 2016). Hence, the low PUFA level (<5%) in cold pressed macadamia oil (Table 1) is advantageous in terms of oxidative stability.

The peroxide value measures the degree of primary oxidation products (hydroperoxides). Regardless of the extraction methods, the longer the incubation of macadamia kernels, the greater the peroxide value generated in macadamia oil (Table 3). The peroxide values generated in macadamia oil extracted by cold pressing were greater than by Soxhlet and accelerated solvent extractions through the experimental period. Plant oil with a high peroxide value may be odorless if secondary oxidation has not begun (Miller, 2010). The peroxide value of plant oil decreases as secondary oxidation products (e.g., aldehydes, ketones, carbonyls, and hydroxy compounds) appear (Miller, 2010). Thus, the low peroxide

TABLE 3 Peroxide value of macadamia oil extracted from macadamia kernel with different incubation time.[a]

	Week		
Method	0	12	14
Cold pressed	2.3	2.3	11.6
Soxhlet extraction[b]	1.3	1.4	8.3
Accelerated solvent extraction[c]			
at 40°C	1.3	1.5	9.8
at 100°C	1.3	1.5	10.0

[a] *Data obtained from Albertson et al. (2012).*
[b] *The Soxhlet method with modification was used and the oil was extracted using petroleum ether.*
[c] *The oil was extracted using petroleum ether at different temperatures (40°C and 100°C).*

values in macadamia oil extracted by Soxhlet and accelerated solvent extractions may indicate the occurrence of the secondary stage of oxidation as a result of oil extraction at elevated temperatures (>40°C).

The levels of conjugated dienes and trienes of selected cold pressed plant (macadamia, avocado, sesame, safflower, pumpkin, rose hip, linola, flaxseed, walnut, hempseed, poppy, and milk thistle) oils were investigated by Prescha et al. (2014). In comparison to other cold pressed plant oils, the study showed that the level of conjugated dienes and trienes in cold pressed macadamia oil was the lowest, indicating the high stability of macadamia oil. The researchers suggest that the high stability of cold pressed macadamia oil could be due to the low level of PUFAs and the presence of minor antioxidant compounds.

10.2 Adulteration and authenticity

Raman spectroscopy has been proposed as a simple technique to authenticate the adulteration of virgin plant oil. Cold pressed plant oil, also known as virgin plant oil (Matthäus & Spener, 2008), commands a higher market price than blended plant oil. To enhance the PUFA level, macadamia oil is sometimes blended with plant oils with high PUFA content such as corn and sunflower oils. In a study performed by Carmona, Lafont, Jiménez-Sanchidrián, and Ruiz (2015), increasing the amount (5%–50%) of corn oil or sunflower oil added to virgin macadamia oil increased the degree of unsaturation, thereby increasing the strength of the absorption band.

References

Albertson, P. L., Bursle, J., Turner, A. G., Willett, G. D., McConchie, C. A., & Proschogo, N. W. (2012). Aging effects on macadamia nut oil studied by electrospray ionization Fourier transform ion cyclotron resonance mass spectrometry. *Journal of Agricultural and Food Chemistry*, *60*, 1973–1980.

Carmona, M. A., Lafont, F., Jiménez-Sanchidrián, C., & Ruiz, J. R. (2015). Characterization of macadamia and pecan oils and detection of mixtures with other edible seed oils by Raman spectroscopy. *Grasas y Aceites*, *66*, e094.

Carrillo, W., Carpio, C., Morales, D., Vilcacundo, E., & Álvarez, M. (2017). Fatty acids composition in Macadamia sedes oil (*Macadamia integrifolia*) from Ecuador. *Asian Journal of Pharmaceutical and Clinical Research*, *10*, 303–306.

Castelo-Branco, V. N., Santana, I., Di-Sarli, V. O., Freitas, S. P., & Torres, A. G. (2016). Antioxidant capacity is a surrogate measure of the quality and stability of vegetable oils. *European Journal of Lipid Science and Technology*, *118*, 224–235.

Cicero, N., Albergamo, A., Salvo, A., Bua, G. D., Bartolomeo, G., Mangano, V., et al. (2018). Chemical characterization of a variety of cold-pressed gourmet oils available on the Brazilian market. *Food Research International*, *109*, 517–525.

Derewiaka, D., Szwed, E., & Wołosiak, R. (2014). Physicochemical properties and composition of lipid fraction of selected edible nuts. *Pakistan Journal of Botany*, *46*, 337–343.

Entelmann, F., Scarpare Filho, J., Pio, R., da Silva, S., & de Souza, F. (2014). Production and quality attributes of macadamia cultivars in southwestern São Paulo state. *Revista Brasileira de Fruticultura*, *36*, 192–198.

Fard, A. M., Turner, A. G., & Willett, G. D. (2003). High-resolution electrospray-ionization fourier-transform ion cyclotron resonance and gas chromatography-mass spectrometry of macadamia nut oil. *Australian Journal of Chemistry*, *56*, 499–508.

Gustavheess (2010). *Macadamia nut oil*. (WWW Document)(2010). http://www.gustavheess.de/index.php?option=com_content&view=article&id=219%3Amacadamianussoel&catid=63%3Akosmetik&Itemid=124&lang=en (Accessed 11 January 2018).

Hernandez, E. M. (2016). Specialty oils: Functional and nutraceutical properties. In *Functional dietary lipids* (pp. 69–100). Sawston, UK: Woodhead Publishing.

International Nut and Dried Fruit. (2018). *Nuts and Dried Fruits Statistical Year Book 2017/2018*. .

Jitngarmkusol, S., Hongsuwankul, J., & Tananuwong, K. (2008). Chemical compositions, functional properties, and microstructure of defatted macadamia flours. *Food Chemistry*, *110*, 23–30.

Kaijser, A., Dutta, P., & Savage, G. (2000). Oxidative stability and lipid composition of macadamia nuts grown in New Zealand. *Food Chemistry*, *71*, 67–70.

Kamal-Eldin, A., & Moreau, R. A. (2009). Tree nut oils. In R. A. Moreau, & A. Kamal-Eldin (Eds.), *Gourmet and health-promoting specialty oils* (pp. 127–149). Urbana, IL: AOCS Press.

Koaze, H., Karanja, P. N., Kojima, M., Baba, N., & Ishibashi, K. I. (2002). Lipid accumulation of macadamia nuts during kernel development. *Food Preservation Science*, *28*, 67–73.

Kochhar, S. P., & Henry, C. J. K. (2009). Oxidative stability and shelf-life evaluation of selected culinary oils. *International Journal of Food Sciences and Nutrition*, *60*, 289–296.

Kornsteiner, M., Wagner, K. H., & Elmadfa, I. (2006). Tocopherols and total phenolics in 10 different nut types. *Food Chemistry*, *98*, 381–387.

Kris-Etherton, P. M. (1999). Monounsaturated fatty acids and risk of cardiovascular disease. *Circulation*, *100*, 1253–1258.

Lee, J.-H., Jones, K. C., Lee, K.-T., Kim, M.-R., & Foglia, T. A. (2003). High-performance liquid chromatographic separation of structured lipids produced by interesterification of macadamia oil with tributyrin and tricaprylin. *Chromatographia*, 653–658.

Lima, E. A., Silveira, L. S., Masi, L. N., Crisma, A. R., Davanso, M. R., Souza, G. I. G., et al. (2014). Macadamia oil supplementation attenuates inflammation and adipocyte hypertrophy in obese mice. *Mediators of Inflammation*, *2014*, 1–9.

Lu, B., Xia, D., Huang, W., Wu, X., Zhang, Y., & Yao, Y. (2010). Hypolipidemic effect of bamboo shoot oil (*P. pubescens*) in Sprague-dawley rats. *Journal of Food Science*, *75*, H205–H211.

Madawala, S. R. P., Kochhar, S. P., & Dutta, P. C. (2012). Lipid components and oxidative status of selected specialty oils. *Grasas y Aceites*, *63*, 143–151.

Matthäus, B., & Spener, F. (2008). What we know and what we should know about virgin oils—A general introduction. *European Journal of Lipid Science and Technology*, *110*, 597–601.

Mildner-Szkudlarz, S., & Jeleń, H. H. (2008). The potential of different techniques for volatile compounds analysis coupled with PCA for the detection of the adulteration of olive oil with hazelnut oil. *Food Chemistry*, *110*, 751–761.

Miller, M. (2010). *Oxidation of food grade oils*. (2010). https://www.oilsfats.org.nz/documents/Oxidation%20101.pdf. [Accessed 5 October 2018].

Moodley, R., Kindness, A., & Jonnalagadda, S. B. (2007). Elemental composition and chemical characteristics of five edible nuts (almond, Brazil, pecan, macadamia and walnut) consumed in Southern Africa. *Journal of Environmental Science and Health, Part B, Pesticides, Food Contaminants, and Agricultural Wastes*, *42*, 585–591.

Navanorro, S. L. B., & Rodrigues, C. E. C. (2016). Macadamia oil extraction methods and uses for the defatted meal byproduct. *Trends in Food Science and Technology*, *54*, 148–154.

Navarro, S. L. B., & Rodrigues, C. E. C. (2018). Macadamia oil extraction with alcoholic solvents: Yield and composition of macadamia oil and production of protein concentrates from defatted meal. *European Journal of Lipid Science and Technology*, *120*, 1800092.

O'brien, R. D. (2009). *Fats and oils: Formulating and processing for applications* (3rd ed.). Florida: CRC Press.

Pang, H.L., Xu, L., Chen, M.Y., Kong, L.J., Lu, K.K., Qin, G.Y., Huang, X.S., 2014. Method for extracting macadamia nut oil by adopting subcritical extraction technology. CN103834470A.

Phatanayindee, S., Borompichaichartkul, C., Srzednicki, G., Craske, J., & Wootton, M. (2012). Changes of chemical and physical quality attributes of macadamia nuts during hybrid drying and processing. *Drying Technology*, *30*, 1870–1880.

Poudyal, H., Kumar, S. A., Iyer, A., Waanders, J., Ward, L. C., & Brown, L. (2013). Responses to oleic, linoleic and α-linolenic acids in high-carbohydrate, high-fat diet-induced metabolic syndrome in rats. *Journal of Nutritional Biochemistry*, *24*, 1381–1392.

Prescha, A., Grajzer, M., Dedyk, M., & Grajeta, H. (2014). The antioxidant activity and oxidative stability of cold-pressed oils. *Journal of the American Oil Chemists' Society*, *91*, 1291–1301.

Rodrigues, C. E. C., Silva, F. A., Marsaioli, A., & Meirelles, A. J. A. (2005). Deacidification of Brazil nut and macadamia nut oils by solvent extraction: Liquid-liquid equilibrium data at 298.2 K. *Journal of Chemical & Engineering Data*, *50*, 517–523.

Rodríguez, P. S., Silva, A. S., & Carrillo, M. L. (2011). Physicochemical characterization of Macadamia nut (*Macadamia integrifolia*) oil. *CYTA-Journal of Food*, *9*.

Shahidi, F., & de Camargo, A. C. (2016). Tocopherols and tocotrienols in common and emerging dietary sources: Occurrence, applications, and health benefits. *International Journal of Molecular Sciences*, *17*, 1–29.

Silva, C. F., Mendes, M. F., Pessoa, F. L. P., & Queiroz, E. M. (2008). Supercritical carbon dioxide extraction of macadamia (*Macadamia integrifolia*) nut oil: Experiments and modeling. *Brazilian Journal of Chemical Engineering*, *25*, 175–181.

Tan, C. X. (2019). Virgin avocado oil: An emerging source of functional fruit oil. *Journal of Functional Foods*, *54*, 381–392.

Tan, C. X., Chong, G. H., Hamzah, H., & Ghazali, H. M. (2018a). Characterization of virgin avocado oil obtained via advanced green techniques. *European Journal of Lipid Science and Technology*, *120*, e1800170.

Tan, C. X., Chong, G. H., Hamzah, H., & Ghazali, H. M. (2018b). Hypocholesterolaemic and hepatoprotective effects of virgin avocado oil in diet-induced hypercholesterolaemia rats. *International Journal of Food Science and Technology*, *53*, 2706–2713.

USDA (2018). *National Nutrient Database for standard reference*. (WWW document)(2018). https://ndb.nal.usda.gov/ndb/foods/show/12131 (Accessed 4 September 2018).

Wall, M. M. (2010). Functional lipid characteristics, oxidative stability, and antioxidant activity of macadamia nut (*Macadamia integrifolia*) cultivars. *Food Chemistry*, *121*, 1103–1108.

Wall, M. M., & Gentry, T. S. (2007). Carbohydrate composition and color development during drying and roasting of macadamia nuts (*Macadamia integrifolia*). *LWT—Food Science and Technology*, *40*, 587–593.

Wallace, H. M., & Walton, D. A. (2011). Macadamia (*Macadamia integrifolia, Macadamia tetraphylla* and hybrids). In *Postharvest biology and technology of tropical and subtropical fruits: Cocona to mango* (pp. 450–473). Philadelphia: Woodhead Publishing Limited.

Walton, D. (2005). In University of the Sunshine Coast (Ed.), *Anatomy and handling: Implications for Macadamia nut quality*.

Yang, Z.-H., Miyahara, H., & Hatanaka, A. (2011). Chronic administration of palmitoleic acid reduces insulin resistance and hepatic lipid accumulation in KK-Ay mice with genetic type 2 diabetes. *Lipids in Health and Disease*, *10*, 120.

Ying, Q., Wojciechowska, P., Siger, A., Kaczmarek, A., & Rudzińska, M. (2018). Phytochemical content, oxidative stability, and nutritional properties of unconventional cold-pressed edible oils. *Journal of Food and Nutrition Research*, *6*, 476–485.

Zhu, B. Q., Lin, L. J., Li, J. H., Lv, G. T., & Huang, M. F. (2012). Comparison of four different extraction methods of oil from *macadamia integrifolia*. *Advances in Materials Research*, *610–613*, 3382–3386.

Chapter 53

Cold pressed pomegranate (*Punica granatum*) seed oil

Zinar Pinar Gumus[a], Zeliha Ustun Argon[b,c], and Veysel Umut Celenk[d]

[a]*Central Research Testing and Analysis Laboratory Research and Application Center, EGE-MATAL, Ege University, İzmir, Turkey,* [b]*Department of Biosystems Engineering, Eregli Faculty of Engineering and Natural Sciences, Necmettin Erbakan University, Konya, Turkey,* [c]*Medical and Cosmetic Plants Application and Research Center, Necmettin Erbakan University, Konya, Turkey,* [d]*Drug Research and Pharmacokinetic Development and Applied Center, ARGEFAR, Ege University, İzmir, Turkey*

Abbreviations

CLA	conjugated linoleic acid
CLNA	conjugated linolenic acid
FAME	fatty acid methyl esters
HPLC	high performance liquid chromatography
LC-Q-TOF/MS	liquid chromatography-quadrupole-time of flight-mass spectroscopy
PA	punicic acid
PSO	pomegranate seed oil
PUFA	polyunsaturated fatty acid
TAGs	triacylglycerols

1 Introduction

Pomegranate (*Punica granatum* L.) is one of the oldest edible perennial fruits known to grow in tropical and subtropical regions. Pomegranate is a shrub tree, which is of Iranian origin but can adapt very well to variable climatic conditions. Therefore, it is widely produced in India, America, the Near and the Far East, and European countries (Durante et al., 2017; Tian, Xu, Zheng, & Martin Lo, 2013). Today pomegranates are planted worldwide, but are particularly cultivated and naturalized in the Mediterranean region. There are more than 1000 varieties of *P. granatum* (Alcaraz-Mármol, Nuncio-Jáuregui, García-Sánchez, Martínez-Nicolás, & Hernández, 2017; Levin, 1994). Pomegranate is a very preferred fruit for its bright red color, fleshy and delicious grains. Due to the antioxidants this fruit contains, the commercial use of pomegranate juice products is constantly increasing. Therefore, pomegranate cultivation is correspondingly on the rise (Gil, Tomas-Barberan, Hess-Pierce, Holcroft, & Kader, 2000; Meerts et al., 2009; Schubert, Lansky, & Neeman, 1999; Singh, Murthy, & Jayaprakasha, 2002). The pomegranate (*P. granatum*) is examined in seven anatomical parts: tree kernel, juice, fruit peel, leaf, flower, bark, and root. Each of these components has interesting pharmacological effects. Pomegranate juice, fruit peel, and seed oil have anticancer activity and even prevent tumor cell proliferation, cell cycle, invasion, and angiogenesis. Phytochemical and pharmacological effects of the pomegranate components responsible for cancer prevention and chronic inflammation have been determined by clinical research (Lansky & Newman, 2007).

Today, the pomegranate has gained importance especially due to its antioxidant properties, which provide health benefits against cancer and cardiovascular and hypertensive diseases. Depending on their properties, chemical compositions, and bioactive compounds, the fruits could be designed for fresh consumption, industrial processes, or medical purposes (Nuncio-Jáuregui et al., 2015; Vázquez-Araújo et al., 2014). In parallel with the development of healthy nutrition awareness in the world, there have been significant increases in the numbers of studies on functional foods and functional components of these foods. The pomegranate is included in the functional foods class because it contains essential fatty acids, antioxidants, phenolic substances, and vitamins. These bioactive compounds could prevent cancer and cardiovascular diseases by preventing the formation of free radicals in the body, lower blood pressure in high blood pressure patients,

Cold Pressed Oils. https://doi.org/10.1016/B978-0-12-818188-1.00053-0
Copyright © 2020 Elsevier Inc. All rights reserved.

and have a positive effect on bad cholesterol. The pomegranate is, therefore, an important raw material as a medicinal plant for the pharmaceuticals and cosmetics industries.

The fruit has three main parts: the outer shell, the inner shell (the white septal membranes), and the edible arils which contains fruit juice and pomegranate seeds (Costa, Silva, & Torres, 2019). The fruit could be divided into three parts, including: seeds, accounting for about 3% total weight and containing 20% oil; juice, accounting for approximately 30% total weight and pericarp, including skin; and inner membranous walls, accounting for approximately 67% total weight (Lansky & Newman, 2007).

Seeds of the pomegranate, normally waste products from pomegranate processing, are of great interest, as their oil has a particularly rich composition (Kiralan, Gölükcü, & Tokgöz, 2009). Seeds represent a by-product of a quantitatively large amount of the fruit processing industry. They are usually excreted as pomace with the vascular tissues of the skins and fruits, but can be easily recovered by separation and sieving technologies. Seeds consist of approximately 60% of the total wastes of the pomegranate juice industry (3–6 million tons per year) and 22% of the waste (shelled plus seeds) (Abid et al., 2017; Durante et al., 2017). Seeds represent the part of the fruit that has the highest bioactive molecule concentration, thus representing the cost of disposing of waste for the waste-food industry and the loss of profits for reuse and valuations.

Agricultural and food products produced during processing of fruits and vegetables, including shells, seeds, leaves, buds, stems, and roots, are a major waste disposal problem for the industry (Ezejiofor, Enebaku, & Ogueke, 2014). Pomegranate fruit stores a large amount of substances such as sucrose, starch, polysaccharides, pectin, and inulin. The cell wall of the pomegranate structure is also composed of cellulose, hemicellulose, and pectins. Since pomegranate contains high-value natural compounds such as carbohydrates, proteins, lipids, carotenoids, phenolics, tocophenols, vitamins, and phytosterols, their by-products also have health-enhancing effects (Kohno et al., 2004; Lenucci, Durante, Anna, Dalessandro, & Piro, 2013). Seeds represent a quantitatively large amount of by-product from the fruit processing industry.

The pomegranate (*P. granatum* L.) is a tree species from the Punicace family and commonly found in mountainous regions, Argentina, Mexico, the Middle East, Asia (Iran, India, China), Mediterranean countries (Spain, Turkey, North Africa), and the United States. The pomegranate was seen as a symbol of resurrection, fertility, abundance, unity, and togetherness and was associated with Proserpina (Persephone) in classical mythology, who every spring returns to the world from the underworld. Some historical accounts suggest that the pomegranate was planted together with figs, dates, olives, and grapes as one of the first five garden plants. There are no exact data on pomegranate production in the world, but its annual output is expected to be about 1 million tons. Depending on the pomegranate variety, the total seed weight ranges from 40 to 100 g/kg fruit. The pomegranate seed is surrounded by edible sarcotesta, which is fleshy and very soft. The integument is the woody part of the pomegranate seed, which consists of cotyledons and embryos (Khoddami & Roberts, 2015).

The edible portion of the pomegranate fruit can be made into a fresh or grenadine system and a juice that can be made into a jam. Pomegranate fruit and fruit products contain significant amounts of polysaccharides, organic acids, vitamins, minerals, phenolics, and anthocyanins. Extracts from fruits are used in the food and beverage industry, flavoring and coloring materials, and in handmade products such as health care items, shampoos, creams, and expensive carpets. The woody components of pomegranate seeds are normally left behind from extraction processing but are a rich source of protein and fat that can be processed and used as an additive in the food industry.

2 Cold press extraction and processing of oils

In view of the disadvantages of solvent extraction techniques, environmental damage is considerable and solvents are toxic and expensive. Therefore, alternative extraction methods are important for health and the environment. The desire to obtain environmentally sensitive and high-quality oil is one of the main reasons for using mechanical systems (Richter et al., 1996). The mechanical oil extraction method has many advantages in terms of practicality, speed, low cost, less raw material usage, and application to different oil seeds (Singh & Bargale, 2000). Compared to solvent extraction, the yield of the mechanical press process is lower and 7% of the oil remains in the seed. This method is generally used for oilseeds with oil content below 20% (Zuorro, Lavecchia, Medici, & Piga, 2014).

The mechanical system is generally defined as the solid-liquid separation system, wherein the parameter effective in phase separation is pressure. Mechanical systems include hot and cold pressing, depending on whether or not a high temperature is applied. In the cold process, the seeds can be pretreated by peeling, drying, and enzymatic treatment. Pretreatment parameters such as feed rate, the diameter of restriction paint, temperature, and rotation speed affect the oil yield (Kiritsakis, 2002; Savoire, Lanoisellé, & Vorobiev, 2013).

Since any temperature in cold pressed oils is obtained without the use of solvents, the obtained oil is ready for consumption without the need for refining. Only cold-water washing, filtration, and centrifugation can be used for the purification of cold pressed oils (Matthaus & Brühl, 2003). Since no heat treatment is applied, oxidation reactions do not take place and phenolic compounds with antioxidant properties remain in the oil intact. Therefore, as the oil's unique aroma, color, and taste remain natural, consumers' demand for cold pressed oils is constantly increasing.

3 Cold pressed oil recovery and economy

Different extraction techniques, such as conventional solvent extraction, super-heated hexane, and supercritical CO_2 extraction, have different effects on the chemical properties of pomegranate seed oil (Abbasi, Rezaei, & Rashidi, 2008; Ahangari & Sargolzaei, 2012; Eikani, Golmohammad, & Homami, 2012; Liu, Xu, Hao, & Gao, 2009). There are differences in oil yield between these extraction methods. These techniques also cause differences in the composition profile of the oil. Cold press extraction, a fast mechanical process used to extract oil from a number of matrices, is a pressure-based method. The most important advantage of the cold pressing compared to other methods is that the technique is an environmentally friendly process that does not require organic solvents. In cold extraction, there is no need for the solvent removal process, which reduces the cost (Khoddami, Man, & Roberts, 2014).

4 Fatty acids composition and acyl lipids

Pomegranate seed oil contains between 12% and 20% of the total weight of the pomegranate seeds. The main fatty acids in pomegranate seed oil are punicic acid (C18: 3-9*cis*, 11*trans*, 13*cis*), and α-eleostearic. Punicic acid is synthesized in situ from linoleic acid, an unconjugated octadecadienoic fatty acid present in about 7% of the oil (Lansky & Newman, 2007). Chemical structures of punicic acid (C18: 3-9*cis*, 11*trans*, 13*cis*) and α-eleostearic are shown in Fig. 1.

During the investigation of pomegranate seed oil compositions, a unique fatty acid composition consisting of a high proportion of phytosterol, tocopherol, and mainly punicic acid was observed. Punicic acid, also known as tricosanic acid, is an omega-5 long-chain polyunsaturated fatty acid and a conjugated α-linolenic acid (CLnA) isomer with structural similarities to conjugated linoleic acid (CLA) and α-linolenic acid (LnA) (Viladomiu, Hontecillas, Yuan, Lu, & Bassaganya-Riera, 2013). Due to the potential health benefits of these conjugated fatty acids, they have become increasingly interesting for both scientists and consumers (Carvalho, Melo, & Mancini-Filho, 2010; Grossmann, Mizuno, Schuster, & Cleary, 2010). Other isomers of conjugated linolenic acid (CLnA), such as α-eleostearic acid (C18: 3-9*trans*, 11*trans*, 13*trans*) and catalpic acid (C18: 3-9*trans*, 11*trans*, 13*cis*), are present in lower concentrations of linoleic acid contained in pomegranate seed oil (Melo, Carvalho, & Mancini-Filho, 2014). Agro-climatic parameters affect the fatty acid composition of pomegranate oil. Among these parameters, mainly fruit genotypes, variety, maturity stage, geographic characteristics of cultivated areas, harvest time, and climatic conditions are effective (Kiralan et al., 2009).

The chemical formula of punicic acid (PA), an omega-5 long-chain polyunsaturated fatty acid and a positional and geometric isomer of α-linolenic acid (LnA; C18:3-9c,12c,15c), is C18:3-9c,11t,13c (Yuan, Wahlqvist, Yuan, Wang, & Li, 2009). Punicic acid (PA), also known as trichosanic acid, is a conjugated triene fatty acid naturally found at high concentrations in the seed of *P. granatum* (Sassano et al., 2009).

Fatty acid methyl ester (FAME) analyses showed that punicic acid is the major fatty acid of cold pressed pomegranate seed oil (PSO) with the amount between 55% and 81% (Celenk, Gumus, Ustun Argon, Buyukhelvacıgil, & Karasulu, 2018; Costa et al., 2019). In the works of Costa et al. (2019) with cold pressed pomegranate seed oil, not only punicic acid but also other conjugated linolenic acid isomers were detected, such as α-eleostearic (6.04%), catalpic (4.79%), and β-eleostearic (1.41%). These isomers make the oil rich in conjugated linolenic acid. Among the other fatty acids, high amounts of

FIG. 1 Chemical structures of punicic acid (C18: 3-9*cis*, 11*trans*, 13*cis*) and α-eleostearic acid (http://pubchem.ncbi.nlm.nih.gov).

TABLE 1 Fatty acids composition of cold pressed pomegranate oil.

Fatty acids	%
Punicic acid	55.0–81.0
Palmitic acid (C16:0)	1.9–4.5
Stearic acid (C18:0)	1.6–2.7
Oleic acid (C18:1)	3.6–8.1
Linoleic acid (C18:2)	4.6–8.7
Arachidic acid (C20:0)	0.20–0.55

linoleic, oleic, stearic, and palmitic acid were detected. Table 1 shows the ranges of fatty acids according to the results of different studies (Celenk et al., 2018; Costa et al., 2019).

As shown in Table 1, pomegranate seed oil is rich in unsaturated fatty acids. The high levels of unsaturated fatty acids and low amounts of saturated fatty acids indicate that this oil can be recommended for consumption compared to other vegetable oils. The differences between studies are probably based on pomegranate varieties, climatic factors, harvest times, maturity, and storage conditions.

The fatty acid composition determines the physical properties, stability, and nutritional value of lipids. Most naturally occurring storage lipids are mainly triacylglycerols. These are natural compounds consisting of saturated and unsaturated fatty acids differing in their acyl chain lengths and the number and positions of double bonds: saturated, monoenoic, and polyunsaturated fatty acids (PUFAs) that differ in fatty acid composition. Lansky and Newman (2007) reported significantly high levels of fatty acids (more than 95%) in PSO, practically all in the form of triglycerides (99%). The triglyceride composition of PSO is varied and the most important markers were CLnA-CLnA-CLnA and CLnA-CLnA-P (Elfalleh et al., 2011). The determination of triacylglycerols composition has been shown to have some important implications. Cold pressed PSO contained 99.5% triacylglycerols. Natural oils are known to have a characteristic triacylglycerols profile. Topkafa, Kara, and Sherazi (2015) identified the triacylglycerols with different analytical methods in PSO (Table 2). As can be seen from the table, punicic acid was found in the triglyceride profile of PSO.

5 Minor bioactive lipids

Vegetable oils contain major components such as fatty acids and triacylglycerides as well as minor components. All components determine the content characteristics of the oil and make the oil more beneficial because of the synergistic effect. Therefore, minor components have become important as well as major components. Sterols are important monoacylglycerol constituents of vegetable oil because they relate to the quality of the oil and are widely used to check genuineness, while they can be used to determine properties of PSO. It has been suggested that sterols may be used to classify pomegranate seed oils according to their fruit variety. Sterol composition and content of oils are affected by cultivar, crop year, degree of fruit ripeness, storage time of fruits before oil extraction, and method of oil extraction.

Tocopherols, which belong to the lipophilic group, are derivatives of 2-methyl-6-chromanol with a side chain of three terpene units attached at C2. Their side chains distinguish them. The terpenoid side chain occurs in the saturated form in tocopherols and in the unsaturated form in tocoterienols, with double bonds in positions $3'$, $7'$, and $11'$. Tocopherols are further separated into individual compounds designated by the Greek letter prefixes α, β, γ, and δ depending on the number and position of methyl substitution on the chromanol ring. α-tocopherol is traditionally considered to be the major antioxidant of oils. The amounts of tocopherol isomers vary from oil to oil. Tocopherols are important lipid-soluble antioxidants for the protection of human health because they preserve the oil against lipid oxidation and they are effective against activities causing oxidative stress (Butinar, Bučar-Miklavčič, Mariani, & Raspor, 2011; Guler et al., 2014; Parcerisa, Richardson, Rafecas, Codony, & Boatella, 1998). When a natural tocopherol is compared with synthetic ones, α-tocopherol in particular is superior for radical chain breaking reaction. It also increases the resistance of LDL to oxidation. α-tocopherol and γ-tocopherol have specific biologic activities that potentially protect against chronic diseases, such as inflammation (Saldeen & Saldeen, 2005). Three tocopherols (α, γ, and δ) were identified (Costa et al., 2019) in pomegranate seed oil, while γ-tocopherol was the major isoform in all samples (Table 3).

TABLE 2 Triglyceride composition (%) of pomegranate seed oil.

Triglyceride	Amount (%)
PuPuPu	32.9
PuPuCa	27.7
PuCaCa	10.1
PuCaEl	2.11
PuElEl	0.38
PuPuL + PuElL	4.41
PuLPu + PuCaL	2.74
LPuCa + LPuEl + LLPu	0.88
PuPuO	6.49
PuCaO + PuOPu	2.73
PuPuP + PuPO	2.84
PuOEl + PuPCa	1.32
LPuP + LPuEl + LOCa	0.17
LLL	0.26
OLLn	0.23
SSPu	0.71
OLL	0.70
PLL	2.25
OOL	0.95

TABLE 3 Amounts of tocopherols in cold pressed pomegranate seed oil.

Tocopherol	Amount (mg/100g oil)
α-tocopherol	2.38–4.88
γ-tocopherol	517–1161
δ-tocopherol	4.56–7.23

β-carotene was present in small amounts in commercial pomegranate seed oils, and other carotenoids were undetectable. Low levels of carotenoids in pomegranate seed oils were also found, as reported by Fernandes et al. (2015), who did not find any traces of carotenoids in the seeds of nine pomegranate cultivars grown in Spain.

The levels of total phenolics and individual phenolics in PSO depend on agronomic factors, the maturity of the fruits, processing, packaging, and storage. The phenolic compounds present in PSO are strong antioxidants and radical scavengers. There is a very good correlation between total phenolic content and the stability of the seed oils. It has been demonstrated that phenolic compounds are more effective than tocopherols in enhancing the stability of olive oil toward oxidation. Phenolic compounds are also among the components most responsible for the nutritional and multiple pharmacological effects.

Due to their rich antioxidant properties, phenolics and flavanoids are minor components that increase the health value of fats and oils. They increase the benefits of the oil by creating a synergistic effect with the other bioactive components in the oil. Liquid chromatography methods are generally preferred for determining the phenolic content of cold pressed pomegranate seed oil. Cold pressed pomegranate seeds were qualitatively analyzed by liquid chromatography-quadrupole-time of flight-mass spectroscopy (LC-Q-TOF-MS) to obtain general phenolics and flavonoids profiles. According to the analysis results, sinapyl aldehyde, R-(−)-mandelic acid, pinolenic acid, pinocembrin, petroselinic acid, methylnorlichexanthone, kaempferol, ferulic acid, eupatorin, corosolic acid, 4-hydroxybenzaldehyde, apigenin, naringenin, 2-hydroxycinnamic acid, and coumarin were identified (Ustun-Argon, 2019). 3,4-dihydroxy-phenylacetic, *p*-hydroxy-benzoic acid, vanillic acid, syringic acid, ferulic acid, *m*-coumaric acid, *p*-coumaric acid, trans-cinnamic acid, rosmarinic acid, quercetin, and naringen have been found in some cold pressed oils by using high performance liquid chromatography (HPLC).

Vegetable oils have a characteristic aroma and flavor depending on the profile of alcohol, aldehydes, esters, carboxylic acid, and ketones in the oils. A balanced flavor of green and fruity sensory characteristics of high-quality pomegranate oil has a profile of volatile compounds, mainly aldehydes, esters, alcohols, carboxylic acid, and ketones. The most important hydrocarbons found in pomegranate seed oils are butane, 2-methyl- and cyclohexene, and 1-methyl-4-(1-methylethenyl)-,(*S*)-. Pomegranate seed oil contains: 1-butanol, 3-methyl-2,3-butanediol, phenylethyl alcohol, 1-hexanol, and 1-pentanol as alcohols; benzaldehyde, 2,4-nonadienal, 2-hexenal, 2-heptenal, (*Z*)-, and hexanal as aldehydes; 5-nonanone and 5-decanone as ketones; pentanoic acid and hexanoic acid as carboxylic acids; and butanoic acid ethyl ester, propanoic acid 2-hydroxy-ethyl ester, (*S*)-, acetic acid pentyl ester, 1-butanol 2-methyl-, acetate hexanoic acid ethyl ester, acetic acid hexyl ester, butanedioic acid diethyl ester, and acetic acid 2-phenylethyl ester as esters.

6 Composition of cold pressed oil in comparison to other cold pressed oils

Cold pressed seed oils have a high nutritional value and important chemical properties because the raw material is not subjected to heat treatment, which causes nutrient loss. Unlike conventional methods, in screw pressing no organic solvents are used that cause chemical contamination during the cold pressing process. Therefore, naturally useful phytochemicals and essential fatty acids such as tocols, sterols, carotenoids, phospholipids, and antioxidant phenolic compounds are maintained at higher levels in cold pressed oils (Topkafa, 2016).

When different cold press oils are compared in terms of C16:0, C18:0, C18:1, C18:2, C18:3, C20:0, and C20:1 fatty acids, punicic acid was found only in pomegranate seed oil. In a study conducted by Gecgel et al. (2016), punicic acid was found at 85.8 g/100 g total fatty acids in the pomegranate seed oil. It can also be seen in Table 4 that saturated fatty acids are mostly in argan oil. When evaluated for monounsaturated fatty acids, argan oil has the highest C18:1 level. The most important feature that distinguishes flax seed oil from other oils is that C18:3 was the most abundant. The distribution of fatty acids in grape seed and walnut oils differed from other oils.

Tocopherols are relatively abundant in seed oils. The quantitative compositions of tocopherols from different cold pressed oils are given in Table 5.

When the oils were compared in terms of tocopherol content, γ-tocopherol was found to be the most common tocopherol in PSO in two different studies. δ-tocopherol is mostly found in PSO. In terms of α-tocopherol, wheat germ is the richest oil, followed by safflower and sunflower oils. The richest oil is wheat germ oil in the terms of the β-tocopherol amount.

Phytosterols are found at different levels in vegetable oils and accounted for the largest group of compounds in their unsaponifiable fraction. When the levels of phytosterols in linseed oil and PSO are compared, the total content of phytosterols was greater in PSO than linseed oil. The most abundant phytosterol in both oils was β-sitosterol (De Melo et al., 2016).

7 The contribution of bioactive compounds in cold pressed oil to organoleptic properties and functions in functional foods

Edible oils are one of the most consumed food products worldwide. These oils have beneficial effects on human metabolism. Oils extracted from the seeds of plants also have health-promoting properties. Conjugated fatty acids, having a number of conjugated double bonds, are known to have favorable physiological effects (Nagao & Yanagita, 2005). The main fatty acids included in this category that have been widely researched are conjugated linoleic acid (CLA) and conjugated linolenic acid (CLNA) with two and three double bonds, respectively. These fatty acids have various effects such as antitumor activity and body fat reduction. PSO is characterized by a dominant percentage of punicic acid. This acid is considered an anticancer agent, as demonstrated by its inhibition of human prostate cancer cell invasion, and punicic acid is correlated with the incidence of coronary heart disease and atherosclerosis (Arao, Yotsumoto, Seo-Young, Koji, &

TABLE 4 Fatty acid composition (g/100 g total fatty acids) of the different cold pressed seed oils.

Oil/fatty acid	Pomegranate seed	Black seed	Flax seed	Linseed	Grape seed	Walnut	Safflower	Argan	Date
C16:0	2.77	12.90	5.73	6.06	9.33	6.82	7.30	11.89	11.15
C18:0	2.42	2.56	2.24	5.27	4.54	2.56	2.00	5.05	3.02
C18:1	5.74	22.63	13.44	21.73	20.47	21.83	14.72	53.41	43.50
C18:2	7.29	61.20	17.44	14.67	63.56	57.51	74.51	28.54	7.67
C18:3	0.39	0.26	60.42	52.37	0.24	10.73	0.08	0.07	0.06
C20:0	0.39	0.14	0.13	–	0.19	0.14	0.37	0.35	0.48
C20:1	0.39	0.31	0.14	–	0.95	0.02	0.22	0.42	0.43
Reference	De Melo et al. (2016)	Lutterodt, Slavin, Whent, Turner, and Yu (2011)	Choo, Birch, and Dufour (2007)	De Melo et al. (2016)	Gecgel et al. (2016)	Gecgel et al. (2016)	Gecgel et al. (2016)	Gecgel et al. (2016)	Gecgel et al. (2016)

TABLE 5 Tocopherol composition (mg/100g oil) of the different cold pressed seed oils.

Oil/tocopherols (mg/100g)	α-tocopherol	β-tocopherol	γ-tocopherol	δ-tocopherol	Reference
Pomegranate	7.24	–	262	6.13	Celenk et al. (2018)
Pomegranate	3.81	1.03	153	17.0	De Melo et al. (2016)
Linseed	0.93	18.17	32.9	1.39	De Melo et al. (2016)
Flax seed	0.75	–	43.4	0.43	Celenk et al. (2018)
Sunflower seed	92.0	3.73	0.12	–	Celenk et al. (2018)
Wheat germ	255	106	8.95	–	Celenk et al. (2018)
Black seed	5.81	–	18.5	–	Celenk et al. (2018)
Safflower seed	106	1.84	4.56	–	Celenk et al. (2018)

Teruyoshi, 2004). There is a rising interest in the composition of fatty acids within the oil, namely as triacylglycerol derivatives. Triacylglycerols (TAGs) are the main components of edible oils. They consist of a glycerol backbone attached to three esterified fatty acids (Buchgraber, Ulberth, Emons, & Anklam, 2004). It is suggested that the structure of TAGs, and not only its fatty acid profile, is of special importance regarding its physiological effect (Mu & Porsgaard, 2005). Many studies concerning lipid metabolism and effect within the human body emphasize the importance of the structure-activity relationship (Kaufman & Wiesman, 2007).

Pomegranates are rich in tocopherols. Tocopherols were determined to be α-tocopherol, γ-tocopherol, and δ-tocopherol, and assumed to be the main responsible compounds for the antioxidant activity of the seed oil with phenolic components. Because of its high amount of vitamin E, PSO is considered suitable for use in medicinal and nutritional applications. Flavonoids, tannins, anthocyanins with phenolic components, and tocopherols are the main group of antioxidants due to their free radical scavenging and biological activities (Abbasi, Rezaei, Emamdjomeh, & Mousavi, 2008; Abbasi, Rezaei, & Rashidi, 2008; Negi, Jayaprakasha, & Jena, 2003; Singh et al., 2002). Phenolic compounds and flavonoids are important compounds in various parts of the pomegranate to show pharmaceutical activities (Al-Maiman & Ahmad, 2002). PSO has been reported to prevent skin cancer (Hora, Maydew, Lansky, & Dwivedi, 2003). Flavonoids, often called phytoestrogens, have been consumed at high quantities in places with lower incidence of hormonally dependent cancer (van Elswijk, Schobel, Lansky, Irth, & van der Greef, 2004). This fact suggests a possible role of flavonoids as chemopreventive agents.

Due to the pharmaceutical and nutritional properties of PSO beneficial for human health, the use of seeds, which are by-products of the fruit juice industries, is increasing day by day. In particular, the health effects of phenolic compounds have been effective in increasing the use of by-products.

8 Edible and nonedible applications of cold pressed oil

Pomegranates are known for their ability to enhance and maintain numerous body systems and functions; therefore, for many cultures throughout history, they have been seen as the "fruit of life" for their optimal health-promoting effects. Accordingly, many health benefits of cold pressed pomegranate seed oil have been reported.

PSO is being used topically in skin care and in massages due to its properties such as reinforcing collagen production, increasing skin elasticity, supporting reversal of skin damage and improving scar appearance and preventing inflammation. It also moisturizes skin without causing a greasy residue that clogs pores. It is therefore an ideal ingredient for skin care cosmetics. PSO stimulates cells in the outer layer of the skin. Due to the regenerative property of fatty acid composition, it can be used to heal skin damage from irritation and sunburns. Apart from its medicinal properties, PSO promotes skin with a younger look and feel (Boroushaki, Mollazadeh, & Afshari, 2016).

Pomegranate seed oil also decreases inflammation and mitigates sun damage. These properties open a gateway to use PSO in accelerating the healing of wounds, reconstruction, and alleviating symptoms related to acne, eczema, and psoriasis (Aslam, Lansky, & Varani, 2006).

PSO could be used as a hair supplement since it can be used on all hair types to hydrate dry strands and as a protection against environmental stressors by preventing damage to follicles. A small amount of warmed oil could be used to treat dry scalp dandruff. It also accelerates blood circulation to the scalp, allowing hair cells to flourish, therefore stimulating the growth of stronger, healthier hair (Zaid, Afaq, Syed, Dreher, & Mukhtar, 2007).

Punicic acid in PSO have metabolism-strengthening effects after inflammation. It can be used to relieve muscle pain and to reduce swelling. Moreover, this fatty acid prevents aging on skin by preserving collagens and increasing their production. If applied to the roots of hair, it nourishes follicles, providing healthier-looking hair (Emami et al., 2017).

Oleic acid in PSO has moisturizing properties when applied to the skin. It moisturizes the scalp, eliminates dandruff, and supports growth of thicker, stronger, and longer hair. It softens the texture of hair and skin, providing a more radiant look. Likewise, it has collagen-protecting properties throughout the skin. Oleic acid also has antioxidant properties, supporting the immune system. If applied appropriately, oleic acid can prevent joint inflammation, stiffness, and pain. Linoleic acid has moisture-capturing properties, if applied to hair, it shows moisturizing effects, which will promote hair growth. Linoleic acid also has facilitative properties in wound healing. Apart from the effects already mentioned, linoleic acid has antiinflammatory properties, reduces acne, and prevents future outbreaks due to its fluidity-increasing property when used in oil blends (Lin, Zhong, & Santiago, 2017).

Palmitic acid is one of the most common saturated fatty acids. It has humidifying properties when used on skin, and when used on hair it softens it without leaving any residual material. Stearic acid is known for its dirt, sweat, and excess sebum cleaning properties from hair and scalp. Moreover, it protects hair from pollutants without diminishing its glossiness and fluidity while softening the skin. It has emulsifying properties, binding water and oil together. In addition, it can be used as a preservative (Le Floc'h et al., 2015; Zanzottera, Nobile, Bizzaro, & Michelotti, 2017).

Phytosterols have rejuvenating properties on the skin such as enhancing collagen production and growth of brand-new, firmer skin. Due to this effect, they can be used to allay the effects of sun damage on the skin. They improve the appearance of blemishes and scars. Immunity-boosting abilities of phytosterols are also known (Aruna, Venkataramanamma, Singh, & Singh, 2016; Le Floc'h et al., 2015).

Vitamin E has established remedial properties when applied to burned, scarred, or blemished skin. It prevents water loss from skin and hair and also has antioxidant properties, which slow the aging appearance of skin by boosting circulation. It can be used for deep cleansing of pores and balancing oil production of skin. PSO is generally used as a regenerative, antiinflammatory, antioxidant, and antiaging agent in cosmetics (Keen & Hassan, 2016; Pandel, Poljšak, Godic, & Dahmane, 2013).

9 Health-promoting traits of cold pressed oil and oil constituents

Pomegranate seeds contain high amounts of carbohydrates and protein, while the fat content is low. They also contain vitamin C, which is useful for the immune system and blood clotting. It also contains minerals, which are important for normal regulation of blood pressure, muscles, and the nervous system. The pomegranate is a very rich fruit in terms of antioxidants content (Qamar Abbas, Zara, Rizwan, & Tahir, 2018).

γ-tocopherol inhibits the synthesis of sphingolipids (Jiang, Wong, Fyrst, Saba, & Ames, 2004), COX-2 activity in prostate cells (Jiang, Elson-Schwab, Courtemanche, & Ames, 2000), and protects normal cells, while transporting cancerous cells to apoptosis (Lansky & Newman, 2007). Phytosterols inhibit pro-inflammatory cytokines (Nashed, Yeganeh, HayGlass, & Moghadasian, 2005), release PC-3 apoptosis by releasing ROS changes and prostaglandin, and stops the cell cycle (Awad, Burr, & Fink, 2005; Awad & Fink, 2000; Lansky & Newman, 2007). Punicic acid inhibits PC-3 invasion (Lansky, Harrison, Froom, & Jiang, 2005), B cell function increases in vivo (Yamasaki et al., 2006), and shows cytotoxic effect with lipid peroxidation in leukemia cells (Suzuki et al., 2001).

Phenolics and flavonoids provide fatty acid synthesis activity in human tumor cells (Brusselmans, Schrijver, Heyns, Verhoeven, & Swinnen, 2003). They work synergistically with quercetin to inhibit breast cancer cell proliferation. Focal adhesion kinase activity in tumor cell invasion plays a key regulatory role (Huang et al., 2005). In tumor cells, tumor necrosis provides factor-α expression and interleukin-1β gene expression (Kowalski, Samojedny, Paul, Pietsz, & Wilczok, 2005). The presence of flavones in pomegranate extract has antiinflammatory effects that can reduce collagen-induced arthritis and painful swelling of joints (Bialonska, Kasimsetty, Schrader, & Ferreira, 2009).

No mutagenicity of PSO was observed in the absence and presence of metabolic activation up to precipitating concentrations of 5000lg/plate (Ames test) or 333lg/mL (chromosome aberration test). The results of the mutagenicity studies reveal that PSO is neither mutagenic nor clastogenic, in either the absence or the presence of metabolic activation. In an oral acute toxicity study, no abnormalities were observed at macroscopic postmortem examination of the animals, and no effects on body weight or body weight gain were observed at a concentration of 2000mg/kg body weight (Meerts et al., 2009).

Because of these properties, PSO could have different external or internal applications or supplementary usages for various diseases such as cardiovascular problems and cancer. Customers tend to choose more natural and healthier products, and cold pressed pomegranate seed oil is preferred in different forms such as in foods, supplements, or personal care products.

10 Adulteration and authenticity

The pomegranate is one of the oldest agricultural tree crops worldwide and is an important source of oil with beneficial properties for human health. It is popularly consumed as fresh fruit, beverages, and food products (jams and jellies), and pomegranate extracts are used as botanical ingredients in herbal medicines and dietary supplements. Pomegranate seeds are by-products obtained during pomegranate processing (Verardo et al., 2014). The oil contained in pomegranate seeds rich in punicic acid as conjugated fatty acids and γ-tocopherol.

In comparison to commonly used cold pressed oils, the cost of pomegranates is higher. There is also a possibility of the addition of cheaper rates of pomegranate seed oil to better-grade ones for economic reasons. Food traceability implies the control of the entire chain of food production and marketing, allowing the food to be traced through every step of its production back to its origin. The verification of food traceability is necessary for the prevention of deliberate or accidental mislabeling, which is very important in the assurance of public health. Thus, several regulations provide the basis for the assurance of a high level of protection of human health and consumers' interest in relation to food. Specific analytical methods are used for determination of the oil adulteration. In addition, it has been known that climate, soil, variety of trees (cultivar), and time of harvest may be responsible for the different organoleptic properties of different oils.

Regulation can only go so far in resolving an issue of authenticity. The essential problem in authentication is setting parameters that accurately define the composition of pure fresh oil. Any oil is a complex mixture of components. These include: tri-, di-, and monoglycerides; free fatty acids; saturated, monounsaturated, polyunsaturated, and trans-fatty acids; sterols; aliphatic and other alcohols; flavonoids; and a variety of other organic molecules. For oils extracted from the same species of plant, regional variations in climate and soil conditions may affect the levels of some of these components.

Each type of oil usually has a fatty acid composition that is indicative of its plant source. Therefore, although it may not be possible to confirm the authenticity of fatty acid content alone, the presence of the wrong fatty acids can be a very good indication of adulteration. The ratio of punicic acid and γ-tocopherol in PSO is higher than other fatty acids and tocopherols. Punicic acid and γ-tocopherol are the most critical parameters in the adulteration of pomegranate seeds since the ratio of punicic acid and γ-tocopherol decreases when another oil is mixed with PSO.

References

Abbasi, H., Rezaei, K., Emamdjomeh, Z., & Mousavi, S. M. E. (2008). Effect of various extraction conditions on the phenolic contents of pomegranate seed oil. *European Journal of Lipid Science and Technology*, 435–440.

Abbasi, H., Rezaei, K., & Rashidi, L. (2008). Extraction of essential oils from the seeds of pomegranate using organic solvents and supercritical CO_2. *Journal of the American Oil Chemists' Society*, *85*, 83–89.

Abid, M., Cheikhrouhou, S., Renard, C. M., Bureau, S., Cuvelier, G., Attia, H., et al. (2017). Characterization of pectins extracted from pomegranate peel and their gelling properties. *Food Chemistry*, *215*, 318–325.

Ahangari, B., & Sargolzaei, J. (2012). Extraction of pomegranate seed oil using subcritical propane and supercritical carbon dioxide. *Theoretical Foundations of Chemical Engineering*, *46*, 258–265.

Alcaraz-Mármol, F., Nuncio-Jáuregui, N., García-Sánchez, F., Martínez-Nicolás, J. J., & Hernández, F. (2017). Characterization of twenty pomegranate (*Punica Granatum* L.) cultivars grown in Spain: Aptitudes for fresh consumption and processing. *Scientia Horticulturae*, *219*, 152–160.

Al-Maiman, S. A., & Ahmad, D. (2002). Change in physical and chemical properties during pomegranate (*Punica granatum* L.) fruit maturation. *Food Chemistry*, *76*, 437–441.

Arao, K., Yotsumoto, H., Seo-Young, H., Koji, N., & Teruyoshi, Y. (2004). The 9-cis,11-trans,13-cis isomer of conjugated linolenic acid reduces apolipoprotein B100 secretion and triacylglycerol synthesis in HepG2 cells. *Bioscience, Biotechnology, and Biochemistry*, *68*, 2634–2645.

Aruna, P., Venkataramanamma, D., Singh, A. K., & Singh, R. P. (2016). Health benefits of punicic acid: A review. *Comprehensive Reviews in Food Science and Food Safety*, *15*(1), 16–27.

Aslam, M. N., Lansky, E. P., & Varani, J. (2006). Pomegranate as a cosmeceutical source: Pomegranate fractions promote proliferation and procollagen synthesis and inhibit matrix metalloproteinase-1 production in human skin cells. *Journal of Ethnopharmacology*, *103*(3), 311–318.

Awad, A. B., Burr, A. T., & Fink, C. S. (2005). Effect of resveratrol and beta-sitosterol in combination on reactive oxygen species and prostaglandin release by PC-3 cells. *Prostaglandins, Leukotrienes and Essential Fatty Acids*, *72*, 219–226.

Awad, A. B., & Fink, C. S. (2000). Phytosterols as anticancer dietary components: Evidence and mechanism of action. *Journal of Nutrition*, *130*, 2127–2133.

Bialonska, D., Kasimsetty, S. G., Schrader, K. K., & Ferreira, D. (2009). The effect of pomegranate (*Punica granatum* L.) by-products and ellagitannins on the growth of human gut bacteria. *Journal of Agricultural and Food Chemistry*, *57*(18), 8344–8834.

Boroushaki, M. T., Mollazadeh, H., & Afshari, A. R. (2016). Pomegranate seed oil : A comprehensive review on its therapeutic effects. *International Journal of Pharmaceutical Sciences and Research*, *7*, 430–442.

Brusselmans, K., Schrijver, E. D., Heyns, W., Verhoeven, G., & Swinnen, J. V. (2003). Epigallocatechin-3-gallate is a potent natural inhibitor of fatty acid synthase in intact cells and selectively induces apoptosis in prostate cancer cells. *International Journal of Cancer*, *106*, 856–862.

Buchgraber, M., Ulberth, F., Emons, H., & Anklam, E. (2004). Triacyglycerol profiling by using chromatographic techniques. *European Journal of Lipid Science and Technology*, *106*, 621–648.

Butinar, B., Bučar-Miklavčič, M., Mariani, C., & Raspor, P. (2011). New vitamin E isomers (gamma-tocomonoenol and alpha-tocomonoenol) in seeds, roasted seeds and roasted seed oil from the Slovenian pumpkin variety "Slovenska golica" *Food Chemistry*, *128*, 505–512.

Carvalho, E. B. T., Melo, I. L. P., & Mancini-Filho, J. (2010). Chemical and physiological aspects of isomers of conjugated fatty acids. *Ciência e Tecnologia de Alimentos*, *30*, 295–307.

Celenk, V. U., Gumus, Z. P., Ustun Argon, Z., Buyukhelvacıgil, M., & Karasulu, E. (2018). Analysis of chemical compositions of 15 different cold-pressed oils produced in Turkey: A case study of tocopherol analysis. *Journal of Turkish Chemical Society Section A Chemistry*, *5*(1), 1–18.

Choo, W. S., Birch, J., & Dufour, J. P. (2007). Physicochemical and quality characteristics of cold-pressed flaxseed oils. *Journal of Food Composition and Analysis*, *20*(3–4), 202–211.

Costa, A. M. M., Silva, L. O., & Torres, A. G. (2019). Chemical composition of commercial cold-pressed pomegranate (*Punica Granatum*) seed oil from Turkey and Israel, and the use of bioactive compounds for samples' origin preliminary discrimination. *Journal of Food Composition and Analysis*, *2019*(75), 8–16.

De Melo, I. L. P., De Carvalho, E. B. T., Silva, A. M. D. O. E., Yoshime, L. T., Sattler, J. A. G., Pavan, R. T., et al. (2016). Characterization of constituents, quality and stability of pomegranate seed oil (*Punica Granatum* L.). *Food Science and Technology*, *36*(1), 132–139.

Durante, M., Montefusco, A., Marrese, P. P., Soccio, M., Pastore, D., Piro, G., et al. (2017). Seeds of pomegranate, tomato and grapes: An underestimated source of natural bioactive molecules and antioxidants from agri-food by-products. *Journal of Food Composition and Analysis*, *63*, 65–72.

Eikani, M. H., Golmohammad, F., & Homami, S. S. (2012). Extraction of pomegranate (*Punica granatum* L.) seed oil using superheated hexane. *Food and Bioproducts Processing*, *90*, 32–36.

Elfalleh, W., Ying, M., Nasri, N., Sheng-Hua, H., Guasmi, F., Ferchichi, A., et al. (2011). Fatty acids from Tunisian and Chinese pomegranate (*Punica granatum* L.) seeds. *International Journal of Food Sciences and Nutrition*, *62*, 200–206.

Emami, A., Ganjkhanlou, M., Fathi Nasri, M. H., Zali, A., Rashidi, L., & Sharifi, M. (2017). Antioxidant status of dairy goats fed diets containing pomegranate seed oil or linseed oil. *Small Ruminant Research*, *153*(May), 175–179.

Ezejiofor, T. I. N., Enebaku, U. E., & Ogueke, C. (2014). Waste to wealth- value recovery from agro-food processing wastes using biotechnology: A review. *British Biotechnology Journal*, *4*, 418–481.

Fernandes, L., Pereira, J. A., Lopéz-Cortés, I., Salazar, D. M., Ramalhosa, E., & Casal, S. (2015). Fatty acid, vitamin E and sterols composition of seed oils from nine different pomegranate (*Punica granatum* L.) cultivars grown in Spain. *Journal of Food Composition and Analysis*, *39*, 13–22.

Gecgel, U., Demirci, A. S., Dulger, G. C., Gecgel, U., Tasan, M., Arici, M., et al. (2016). Some physicochemical properties, fatty acid composition and antimicrobial characteristics of different cold-pressed oils. *La Rivista Italiana Delle Sostanze Grasse*, *XCII*, 187–200.

Gil, M. I., Tomas-Barberan, F. A., Hess-Pierce, B., Holcroft, D. M., & Kader, A. A. (2000). Antioxidant activity of pomegranate juice and its relationship with phenolic composition and processing. *Journal of Agricultural and Food Chemistry*, *48*, 4581–4589.

Grossmann, M. E., Mizuno, N. K., Schuster, T., & Cleary, M. P. (2010). Punicic acid is an omega-5 fatty acid capable of inhibiting breast cancer proliferation. *International Journal of Oncology*, *36*, 421–426.

Guler, E., Barlas, F. B., Yavuz, M., Demir, B., Gumus, Z. P., Baspinar, Y., et al. (2014). Bio-active nanoemulsions enriched with gold nanoparticle, marigold extracts and lipoic acid: In vitro investigations. *Colloids and Surfaces B: Biointerfaces*, *121*, 299–306.

Hora, J. J., Maydew, E. R., Lansky, E. P., & Dwivedi, C. (2003). Chemo-preventive effect of pomegranate seed oil on skin tumor development in CD1 mice. *Journal of Medicinal Food*, *6*, 157–161.

Huang, T. H., HPeng, G., Kota, B. P., Li, G. Q., Yamahara, J., Roufogalis, B. D., et al. (2005). Anti-diabetic action of *Punica granatum* flower extract: Activation of PPAR-gamma and identification of an active component. *Toxicology and Applied Pharmacology*, *207*, 160–169.

Jiang, Q., Elson-Schwab, I., Courtemanche, C., & Ames, B. N. (2000). Gamma-tocopherol and its major metabolite, in contrast to alpha-tocopherol, inhibit cyclooxygenase activity in macrophages and epithelial cells. *Proceedings of the National Academy of Sciences of the United States of America*, *97*, 11494–11499.

Jiang, Q., Wong, J., Fyrst, H., Saba, J. D., & Ames, B. N. (2004). Gamma-tocopherol or combinations of vitamin E forms induce cell death in human prostate cancer cells by interrupting sphingolipid synthesis. *Proceedings of the National Academy of Sciences of the United States of America*, *101*, 17825–17830.

Kaufman, M., & Wiesman, Z. (2007). Pomegranate oil analysis with emphasis on MALDI-TOF/MS triacylglycerol fingerprinting. *Journal of Agricultural and Food Chemistry*, *55*, 10405–10413.

Keen, M. A., & Hassan, I. (2016). Vitamin E in dermatology. *Indian Dermatology Online Journal*, *7*(4), 311–315.

Khoddami, A., Man, Y. B. C., & Roberts, T. H. (2014). Physico-chemical properties and fatty acid profile of seed oils from pomegranate (*Punica Granatum* L.) extracted by cold pressing. *European Journal of Lipid Science and Technology*, *116*(5), 553–562.

Khoddami, A., & Roberts, T. H. (2015). Pomegranate oil as a valuable pharmaceutical and nutraceutical. *Lipid Technology*, *27*(2), 40–42.

Kiralan, M., Gölükcü, M., & Tokgöz, H. (2009). Oil and conjugated linolenic acid contents of seeds from important pomegranate cultivars (*Punica granatum* L.) grown in Turkey. *Journal of the American Oil Chemists' Society*, *86*, 985–990.

Kiritsakis, A. K. (2002). Virgin olive oil composition and its effect on human health. *Inform*, *13*, 237–241.
Kohno, H., Suzuki, R., Yasui, Y., Hosokawa, M., Miyashita, K., & Tanaka, T. (2004). Pomegranate seed oil rich in conjugated linolenic acid suppresses chemically induced colon carcinogenesis in rats. *Cancer Science*, *95*, 481–486.
Kowalski, I., Samojedny, A., Paul, M., Pietsz, G., & Wilczok, T. (2005). Effect of kaempferol on the production and gene expression of monocyte chemoattractant protein-1 in J774.2 macrophages. *Pharmacological Reports*, *57*, 107–112.
Lansky, E. P., Harrison, G., Froom, P., & Jiang, W. G. (2005). Pomegranate (*Punica granatum*) pure chemicals show possible synergistic inhibition of human PC-3 prostate cancer cell invasion across MatrigelTM. *Investigational New Drugs*, *23*, 121–122.
Lansky, E. P., & Newman, R. A. (2007). *Punica granatum* (pomegranate) and its potential for prevention and treatment of inflammation and cancer. *Journal of Ethnopharmacology*, *109*, 177–206.
Le Floc'h, C., Cheniti, A., Connétable, S., Piccardi, N., Vincenzi, C., & Tosti, A. (2015). Effect of a nutritional supplement on hair loss in women. *Journal of Cosmetic Dermatology*, *14*(1), 76–82.
Lenucci, M. S., Durante, M., Anna, M., Dalessandro, G., & Piro, G. (2013). Possible use of the carbohydrates present in tomato pomace and in byproducts of the supercritical carbon dioxide lycopene extraction process as biomass for bioethanol production. *Journal of Agricultural and Food Chemistry*, *61*, 3683–3692.
Levin, G. M. (1994). Pomegranate (*Punica granatum*) plant genetic resources in Turkmenistan. *Plant Genetic Resources Newsletter*, *97*, 31–37.
Lin, T. K., Zhong, L., & Santiago, J. L. (2017). Anti-inflammatory and skin barrier repair effects of topical application of some plant oils. *International Journal of Molecular Sciences*, *19*(1), 70.
Liu, G., Xu, X., Hao, Q., & Gao, Y. (2009). Supercritical CO_2 extraction optimization of pomegranate (*Punica granatum* L.) seed oil using response surface methodology. *LWT-Food Science and Technology*, *42*, 1491–1495.
Lutterodt, H., Slavin, M., Whent, M., Turner, E., & Yu, L. (2011). Fatty acid composition, oxidative stability, antioxidant and antiproliferative properties of selected cold-pressed grape seed oils and flours. *Food Chemistry*, *128*(2), 391–399.
Matthaus, B., & Brühl, L. (2003). Quality of cold-pressed edible rapeseed oil in Germany. *Food*, *47*(6), 413–419.
Meerts, I. A. T. M., Verspeek-Rip, C. M., Buskens, C. A. F., Keizer, H. G., Bassaganya-Riera, J., Jouni, Z. E., et al. (2009). Toxicological evaluation of pomegranate seed oil. *Food and Chemical Toxicology*, *47*(6), 1085–1092.
Melo, I. L. P., Carvalho, E. B. T., & Mancini-Filho, J. (2014). Pomegranate seed oil (*Punica Granatum* L.): A source of punicic acid (conjugated α-linolenic acid). *Journal of Human Nutrition & Food Sciences*, *2*(1), 1024.
Mu, H., & Porsgaard, T. (2005). The metabolism of structured triacylglycerols. *Progress in Lipid Research*, *44*, 430–448.
Nagao, K., & Yanagita, T. (2005). Conjugated fatty acid in foods and their health benefits. *Journal of Bioscience and Bioengineering*, *100*, 152–157.
Nashed, B., Yeganeh, B., HayGlass, K. T., & Moghadasian, M. H. (2005). Antiatherogenic effects of dietary plant sterols are associated with inhibition of proinflammatory cytokine production in Apo E-KO mice. *Journal of Nutrition*, *135*, 2438–2444.
Negi, P. S., Jayaprakasha, G. K., & Jena, B. S. (2003). Antioxidant and antimutagenic activities of pomegranate peel extracts. *Food Chemistry*, *80*, 393–397.
Nuncio-Jáuregui, N., Munera-Picazo, S., Calín-Sánchez, A., Wojdyło, A., Hernández, F., & Carbonell-Barrachina, A. A. (2015). Potential of pomegranate fruits removed during thinning as source of bioactive compounds. *Journal of Food Composition and Analysis*, *37*, 11–19.
Pandel, R., Poljšak, B., Godic, A., & Dahmane, R. (2013). Skin photoaging and the role of antioxidants in its prevention. *ISRN Dermatology*, *2013*, 930164 (11 pp.).
Parcerisa, J., Richardson, D. G., Rafecas, M., Codony, R., & Boatella, J. (1998). Fatty acid, tocopherol and sterol content of some hazelnut varieties (*Corylus avellana* L.) harvested in Oregon (USA). *Journal of Chromatography A*, *805*, 259–268.
Qamar Abbas, S., Zara, B., Rizwan, S., & Tahir, Z. (2018). Nutritional and therapeutic properties of pomegranate. *Scholarly Journal of Food and Nutrition*, *1*(4), 115–120.
Richter, B. E., Jones, B. A., Ezzell, J. L., Porter, N. L., Avdalovic, N., & Pohl, C. (1996). Accelerated solvent extraction: A technique for sample preparation. *Analytical Chemistry*, *68*, 1033.
Saldeen, K., & Saldeen, T. (2005). Importance of tocopherols beyond a-tocopherol: Evidence from animal and human studies. *Nutrition Research*, *25*, 877–889.
Sassano, G., et al. (2009). Analysis of pomegranate seed oil for the presence of jacaric acid. *Journal of the Science of Food and Agriculture*, *89*(6), 1046–1052.
Savoire, R., Lanoisellé, J. L., & Vorobiev, E. (2013). Mechanical continuous oil expression from oilseeds: A review. *Food and Bioprocess Technology*, *6*(1), 1–16.
Schubert, S. Y., Lansky, E. P., & Neeman, I. (1999). Antioxidant and eicosanoid enzyme inhibition properties of pomegranate seed oil and fermented juice flavonoids. *Journal of Ethnopharmacology*, *66*, 11–17.
Singh, J., & Bargale, P. C. (2000). Development of a small capacity double stage compression cold press for oil expression. *Journal of Food Engineering*, *43*(2), 75–82.
Singh, R. P., Murthy, K. N. C., & Jayaprakasha, G. K. (2002). Studies on the antioxidant activity of pomegranate (*Punica granatum*) peel and seed extracts using in vitro models. *Journal of Agricultural and Food Chemistry*, *50*, 81–86.
Suzuki, R., Noguchi, R., Ota, T., Abe, M., Miyashita, K., & Kawada, T. (2001). Cytotoxic effect of conjugated trienoic fatty acids on mouse tumor and human monocytic leukemia cells. *Lipids*, *36*, 477–482.
Tian, Y., Xu, Z., Zheng, B., & Martin Lo, Y. (2013). Optimization of ultrasonic-assisted extraction of pomegranate (*Punica Granatum* L.) seed oil. *Ultrasonics Sonochemistry*, *20*(1), 202–208.

Topkafa, M. (2016). Evaluation of chemical properties of cold pressed onion, okra, rosehip, safflower and carrot seed oils: Triglyceride, fatty acid and tocol compositions. *Analytical Methods*, *8*(21), 4220–4225.

Topkafa, M., Kara, H., & Sherazi, S. T. H. (2015). Evaluation of the triglyceride composition of pomegranate seed oil by RP-HPLC followed by GC-MS. *Journal of the American Oil Chemists' Society*, *92*(6), 791–800.

Ustun-Argon, Z. (2019). Phenolic compounds, fatty acid compositions and antioxidant activity of commercial coldpressed pomegranate (*Punica granatum*) seed oils from Turkey. *International Journal of Scientific & Engineering Research*, *10*(4), 1347–1352.

van Elswijk, D. A., Schobel, U. P., Lansky, E. P., Irth, H., & van der Greef, J. (2004). Rapid dereplication of estrogenic compounds in pomegranate (*Punica granatum*) using on-line biochemical detection coupled to mass spectrometry. *Phytochemistry*, *65*, 233–241.

Vázquez-Araújo, L., Nuncio-Jáuregui, P. N., Cherdchu, P., Hernández, F., Chambers, E., IV, & Carbonell-Barrachina, A. A. (2014). Use of descriptive sensory analysis to find the best market option for Spanish pomegranates. *International Journal of Food Science and Technology*, *49*, 2259–2265.

Verardo, V., Garcia-Salas, P., Baldi, E. B., Segura-Carretero, A., Fernandez-Gutierrez, A., & Cabonia, M. F. (2014). Pomegranate seeds as a source of nutraceutical oil naturally rich in bioactive lipids. *Food Research International*, *65*, 445–452.

Viladomiu, M., Hontecillas, R., Yuan, L., Lu, P., & Bassaganya-Riera, J. (2013). Nutritional protective mechanisms against gut inflammation. *The Journal of Nutritional Biochemistry*, *24*, 929–939.

Yamasaki, M., Kitagawa, T., Koyanagi, N., Chujo, H., Maeda, H., Kohno-Murase, J., et al. (2006). Dietary effect of pomegranate seed oil on immune function and lipid metabolism in mice. *Nutrition*, *22*, 54–59.

Yuan, G. F., Wahlqvist, M. L., Yuan, J. Q., Wang, Q. M., & Li, D. (2009). Effect of punicic acid naturally occurring in food on lipid peroxidation in healthy young humans. *Journal of Science of Food and Agriculture*, *89*, 2331–2335.

Zaid, M. A., Afaq, F., Syed, D. N., Dreher, M., & Mukhtar, H. (2007). Inhibition of UVB-mediated oxidative stress and markers of photoaging in immortalized HaCaT keratinocytes by pomegranate polyphenol extractPOMx. *Photochemistry and Photobiology*, *83*, 882–888.

Zanzottera, F., Nobile, B., Bizzaro, G., & Michelotti, A. (2017). Efficacy of a nutritional supplement, standardized in fatty acids and phytosterols, on hair loss and hair health in both women and men. *Journal of Cosmetology & Trichology*, *3*, 2.

Zuorro, A., Lavecchia, R., Medici, F., & Piga, L. (2014). Use of cell wall degrading enzymes for the production of high-quality functional products from tomato processing waste. *Chemical Engineering Transactions*, *38*, 355–360.

Chapter 54

Cold pressed *Cucumis melo* L. seed oil

Leila Rezig[a,b], Moncef Chouaibi[c], Kamel Msaada[d], and Salem Hamdi[e]

[a]*High Institute of Food Industries, El Khadra City, Tunis, Tunisia,* [b]*University of Carthage, National Institute of Applied Sciences and Technology, LR11ES26, 'Laboratory of Protein Engineering and Bioactive Molecules', Tunis, Tunisia,* [c]*Research Unit: 'Bio-Preservation and Valorization of Agricultural Products UR13-AGR 02', University of Carthage, High Institute of Food Industries of Tunisia, Tunis, Tunisia,* [d]*Laboratory of Aromatic and Medicinal Plants, Biotechnology Center in Borj Cedria Technopole, Hammam-Lif, Tunisia,* [e]*Food Conservation and Valorization Laboratory, High Institute of Food Industries of Tunisia, Tunis, Tunisia*

1 Introduction

Melon fruits are consumed *in natura* or in their processed forms as purees, jams, juices, nectars, fruit cocktails, and alcoholic beverages (Veronezi & Jorge, 2018; Wani, Sogi, Singh, et al., 2011). In the process of fruit manufacturing, large amounts of waste, such as peels and seeds, are discarded (Mallek-Ayadi, Bahloul, & Kechaou, 2018; Veronezi & Jorge, 2015, 2018). Moreover, the high costs of drying, storing, and transporting these by-products are economically unprofitable. In recent years, fruit seeds have received growing interest due to the important nutritional and medicinal properties of their bioactive components. Obviously, fruit seeds can be used for plant oil extraction. Such oils contain a large number of valuable biocomponents and natural antioxidants (Mallek-Ayadi et al., 2018).

Grown extensively in temperate, tropical, and subtropical regions of the world, melon (*Cucumis melo* L.) belongs to the Cucurbitaceae family—also referred to as cucurbits—which comprises 825 species, including cucumber, pumpkin, watermelon, and squash (ITIS, 2017; Vishwakarma, Gupta, & Upadhyay, 2017). Thanks to the great diversity in the morphology of its fruits and their features like size, shape, color, texture, and flavor, the melon is considered to be one of the most diversified species belonging to the genus Cucumis (Maynard & Maynard, 2000).

Melon seeds are extensively consumed in Tunisia and in African countries as snacks. In some other countries, melon seed kernels are used as a dressing for bread, cakes, confectionery, and foods instead of almonds and pistachios (Karakaya, Kavas, Nehir El, et al., 1995). In addition, melon seed oil is frequently used for cooking in some sub-Saharan African countries and in the Middle East (Hemavatahy, 1992; Maran & Priya, 2015; Maynard & Maynard, 2000). Melon seeds also have favorable medicinal properties and are used to treat chronic or acute eczema (Rezig, Chouaibi, Meddeb, et al., 2019). Such health benefits require effective extraction techniques in order to guarantee high-quality melon seed oil (MSO).

It is known that organic solvents are frequently used to extract oil from the seeds. *n*-hexane is a typical solvent that is commercially used in oil extraction from different plant materials. However, its usage is now under close examination due to growing governmental restrictions and concerns about consumer safety, especially those regarding organic solvents' use in food-processing industries. Consequently, environmental rules and regulations plus an increasing awareness of health risks pushed manufacturers to adopt novel alternatives in the use of organic solvents for oil extraction (Mitra, Ramaswamy, & Chang, 2009). In this regard, several studies have already been conducted with the sole aim of developing new extraction methods as viable alternatives to organic solvents for the sake of meeting the growing consumer demand for more natural and safe food products. Such alternatives include supercritical CO_2 extraction (Nyam, Tan, Karim, et al., 2010) or cold pressing extraction (Rezig et al., 2019; Veronezi & Jorge, 2018). It is worth noting that the cold pressing procedure involves neither heat nor chemical treatments. It is essentially a mechanical technique in which the extraction is established using a small capacity screw presses (6–40 kg/h). In this case, cold pressed seed oil is extracted by pressing raw, dried, and mainly hull-less seeds, on a continuous screw press (Rabrenović, Dimić, Novaković, et al., 2014).

According to Dimić (2005), with the intention of producing cold pressed oils, the temperature of the oil leaving the press during the process of squeezing oil seed should not exceed 50°C. This method would enable the preservation of bioactive components, such as vitamins, provitamins, phytosterols, phospholipids, and squalene, which together with some fatty acids possess significant nutritional value factors.

As far as we know, the physicochemical properties of melon seed oil remain fairly unexplored. Accordingly, the determination of both the physicochemical properties and the oxidative stability would significantly contribute to the potential

Cold Pressed Oils. https://doi.org/10.1016/B978-0-12-818188-1.00054-2
Copyright © 2020 Elsevier Inc. All rights reserved.

valorization of melon seed oil in the cosmetic, pharmaceutical, and food industries. This chapter aims first at identifying the bioactive compounds in melon seed oil from a Tunisian variety (Ananas) of melon seed (*Cucumis melo* L.) extracted by a cold pressing technique. A comparative study will next be undertaken to compare the natural bioactive components with those obtained from different varieties originating from different geographical areas all over the world. This might lead to the innovative utilization of melon seed oil as an alternative for industrial uses.

2 Oil content in melon seeds

Melon (*Cucumis melo* L.) seeds are a rich source of oil. Table 1 summarizes the origins, the extraction procedures, and the oil contents of different melon seed varieties cultivated all over the world. High percentages of oil make these seeds eminently suitable for many oil industry applications. According to Nyam, Tan, Lai, et al. (2009), variation in oil yield may be caused by differences in plant variety, cultivation climate, ripening stage, harvesting time of seeds, and extraction method used. The findings showed that melon seeds might have the potential to be used as high oil sources in some food formulations.

3 Extraction and processing of cold pressed oil

Studies dealing with cold pressed oil extraction were undertaken to confirm that such an extraction procedure is an interesting substitute for conventional practices, satisfying consumers' demand for safe food products. For this reason, the consumption of new and improved products such as cold pressed oils may improve human health and prevent certain diseases, especially those related to the cardiovascular system (Siger, Nogala-Kalucka, & Lampert-Szczapa, 2008). Over the last few years, there has been an increasing interest in cold pressed plant oils as they have better nutritive properties than those obtained after being refined.

Fig. 1 provides a general schema of cold pressed seed oils that can perfectly be applied for all plant seeds, while Fig. 2 shows a vegetable screw press hole cylinder (type CA 59, IBGMonforts Oekotec GmbH & Co. KGMönchengladbach, Germany). Cleaning and pressing of oilseeds are the main steps of processing any plant seed oil. The seeds should be first

TABLE 1 Oil content of melon seeds.

Sample	Origin	Extraction procedure	Oil content (%)	Reference
Cucumis melo var. "Ananas"	Tunisia	Cold pressing	28.4	Rezig et al. (2019)
Cucumis melo var. "Maazoun"	Tunisia	*n*-Hexane	30.6	Mallek-Ayadi et al. (2018)
Honeydew melon (*Cucumis melo* L.)	Spain	*n*-Hexane	27.6	Górnaś and Rudzińska (2016)
Mashhadi melon (*Cucumis melo* var. Iranians cv. Mashhadi)	Iran	*n*-Hexane	14.4	Hashemi, Khaneghag, Koubaa, et al. (2017)
Cucumis melo var. "inodorus"	Brazil	*n*-Hexane	15.3	Veronezi and Jorge (2018)
Cucumis melo var. "honeydew"	Bulgaria	*n*-Hexane	41.6	Petkova and Antova (2015)
Cucumis melo Var. "Dessert 5"			41.6	
Cucumis melo var. "Hybrid 1"			44.5	
Cucumis melo var. "tibish"	Sudan	*n*-Hexane	31.1	Azhari, Xu, Jiang, et al. (2014)
Cucumis melo var. "inodorus Naudin"	Brazil	Chloroform/ methanol/ water	30.6	da Silva and Jorge (2014)
Cucumis melo var. "agrestis"	Sudan	Petroleum ether	23.3	Mariod, Ahmed, Matthäus, et al. (2009)
Cucumis melo var. "flexuosus"			22.3	
Cucumis melo var. "inodorus"	Malaysia	Petroleum ether	25.0	Yanty, Lai, Osman, et al. (2008)

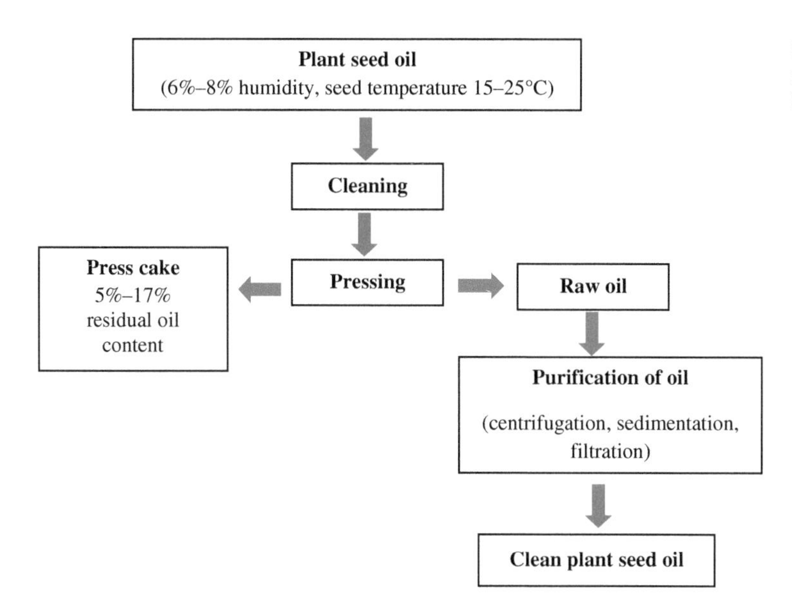

FIG. 1 The schema of cold pressed plant seed oil (Kulaitienė, Černiauskienė, Jarienė, et al., 2018; Salgin & Korkmaz, 2011).

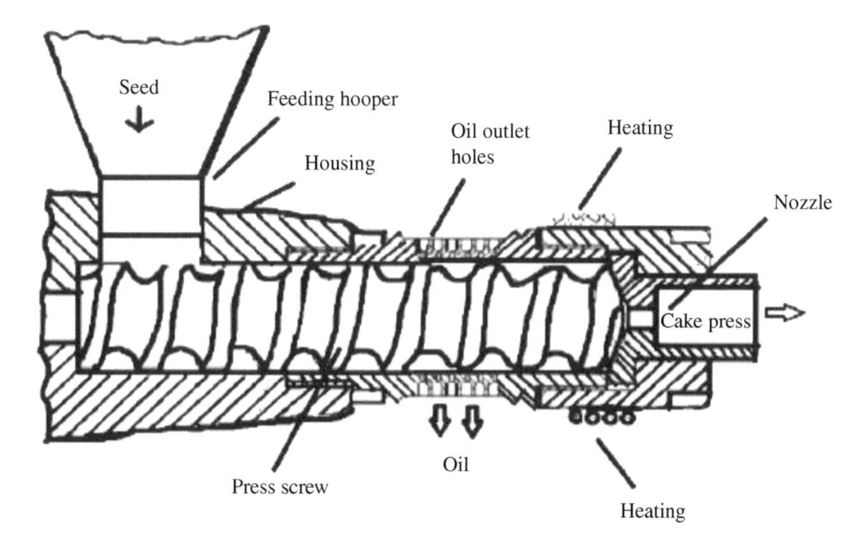

FIG. 2 Vegetable screw press hole cylinder type CA 59 (IBG Monforts Oekotec GmbH & Co. KG, Mönchengladbach, Germany) (Kulaitienė et al., 2018).

cleaned of extraneous impurities such as debris, plant parts, and damaged seeds. This would inevitably provide an unchanging and more reliable oil. Generally, the seeds are prewarmed to about 25°C by a special unit or by a heat exchanger that makes use of the heat from the warm press cake. Note that preheating the seeds to over 25°C has no favorable outcome (Kulaitienė et al., 2018).

It is well known that cold pressing is much simpler, ecological preferable, and does not require much energy. However, this process includes some drawbacks such as low productivity and difficulties in obtaining a product of consistent quality (Rotkiewicz, Konopka, & Zylik, 1999). In the same way, Salgin and Korkmaz (2011) and Kulaitienė et al. (2018) reported that this extraction procedure has a major drawback: the level of residual oil left in the seeds is high and varies from 5 to 17 wt%. This finding was confirmed by Rezig, Chouaibi, and Fregapane (2018) in the "Béjaoui" *Cucurbita maxima* variety seed flours in which the levels of residual oil in the seed flour were 8.22% and 14.0%, respectively, after having used a German screw press (Komet DD 85G, IBG Monforts Oekotec GmbH & Co. KG, Mönchengladbach, Germany) and a Moroccan one (MUV2 65-PSO2, Smir Technotour, Agadir, Morocco).

4 Fatty acid composition of cold pressed melon seed oil

To the best of our knowledge, scientific literature in relation to the fatty acid composition of cold pressed melon seed oils is scarce. Table 2 illustrates the fatty acid composition of the "Ananas" Tunisian melon seed oil variety obtained by cold

TABLE 2 Fatty acid (%) composition of melon (*Cucumis melo* var. "Ananas") cold pressed seed oil (Rezig et al., 2019).

Fatty acid	%
Myristic (C14:0)	tr.
Palmitic (C16:0)	14.4
Palmitoleic (C16:1)	tr.
Stearic (C18:0)	5.81
Oleic (C18:1)	23.5
Linoleic (C18:2)	59.2
Linolenic (C18:3)	0.22
Arachidic (C20:0)	0.24
Eicosanoic (C20:1)	tr.
Behenic (C22:0)	tr.
Lignoceric (C24:0)	tr.
SAFAs	20.24
MUFAs	23.52
PUFAs	59.48

MUFAs, monounsaturated fatty acids; *PUFAs*, polyunsaturated fatty acids; *SAFAs*, saturated fatty acids; *tr.*, trace amounts (<0.2%).

pressing using a Komet DD 85 G vegetable oil screw press (IBG Monforts Oekotec GmbH & Co. KG, Mönchengladbach, Germany). Eleven fatty acids were identified in the *Cucumis melo* lipid fraction. Concerning the saturated fatty acids, most of them were palmitic (14.4%) and stearic (5.81%). For the unsaturated fatty acids, linoleic (59.2%) and oleic (23.5%) stand out. These values are close to those obtained by Górnaś and Rudzińska (2016), and Hashemi et al. (2017) for honeydew melon (*Cucumis melo* L.) and Mashhadi melon (*Cucumis melo* var. Iranians cv. Mashhadi) seed oils in terms of stearic, oleic, and linoleic fatty acid contents. However, the fatty acid composition slightly differs when comparing the results with those obtained by Mallek-Ayadi et al. (2018) and Yanty et al. (2008). The "Maazoun" *Cucumis melo* L. and "inodorus" melon seed oils both witnessed lesser amounts of oleic acid and a higher content of linoleic acid. In the *Cucumis melo* var. "agrestis" and var. "flexuosus" seed oils, originating from Sudan, Mariod et al. (2009) reported a similar fatty acid composition to that of Rezig et al. (2019). In fact, linoleic and oleic acids were the main unsaturated fatty acids and together represented 80.8% and 78.9% of the total fatty acids for the "agrestis" and "flexuosus" varieties, respectively. As for the saturated fatty acids, palmitic and stearic acids were the most dominant detected ones for both Sudanese varieties and accounted for 10.8% and 12.8%, and 9.4% and 6%, respectively in var. "agrestis" and var. "flexuosus."

The fatty acid composition of seed oils heavily depends on both the climatic conditions and the cultivar of the crop or fruit itself (de Melo, Narain, & Bora, 2000; Fadavi, Barzegar, & Azizi, 2006; Tangolar, Özogul, Tangolar, et al., 2009). From a nutritional point of view, the presence of essential fatty acids is vital to human metabolism. Take, for instance, one of the most important polyunsaturated fatty acids (PUFAs): linoleic acid. This acid is not produced by the human body and is essential for the formation of cell membranes, vitamin D, and several hormones (Fruhwirth & Hermetter, 2007). In the same line of thought, the oleic and linoleic acids together represented 82.7% of the total fatty acids present in the oil, hence the oil is considered as linoleic-oleic oil. As the examined cold pressed *Cucumis melo* seed oil sample is rich in both oleic and linoleic acids, it may thus be used as a cooking oil, a salad oil, or as an ingredient for manufacturing margarine.

5 Triacylglycerol composition

The determination of the fatty acid composition is undoubtedly decisive for evaluating both the stability and the nutritional quality of vegetable oils. However, the determination of the type and the amounts of triacylglycerol (TAG) species within the *Cucumis melo* seed oils is much more important for understanding their physical and functional properties. As far as we

TABLE 3 Triglyceride profile (%) of the cold pressed "Ananas" *Cucumis melo* seed oil (%) (Rezig et al., 2019).

TAG molecule	%
LLnLn	18.3
LLLn	0.55
LLL	23.3
OOLn	17.8
PLLn	11.2
OLL	15.0
LnPO	1.08
OOL	2.92
POL	6.26
PPL	1.27
OOO	1.36
SOL	0.53
PLS	0.24

L, linoleic acid; *Ln*, linolenic acid; *O*, oleic acid; *P*, palmitic acid; *S*, stearic acid.

know, scientific literature dealing with the TAG composition of cold pressed *Cucumis melo* seed oil obtained by a vegetable screw press is still scarce. Meanwhile, we believe that a study conducted by Rezig et al. (2019) was the first to report about TAG composition in cold pressed Cucurbitaceae seed oils obtained by a vegetable screw press.

The distribution of TAG is illustrated in Table 3, and it highlights the fact that the *Cucumis melo* var. "Ananas" seed oil contained 13 TAG. LLL, LLnLn, OOLn, OLL, and PLLn were the five major TAG isomers identified in the cold pressed "Ananas" *Cucumis melo* seed oil with a composition ranging between 11.2% and 23.3%. In honeydew melon seed oil, Yanty et al. (2008) reported that the LLL isomer was the predominant one (24.9%) followed in descending order by OLL (21.5%), PLL (15.9%), POL (12.4%), and OOL (6.0%). The contents of the three main types of triacylglycerol (LLL, LLO, and LLP) in the melon seed oil were close to the data obtained by Petkova and Antova (2015) in the "honeydew" *Cucumis melo* seed oil variety. According to Reske, Siebrecht, and Hazebroek (1997), the amounts of each TAG type depend on the proportions of the individual fatty acids, the fat or oil source, and the product's processing history. Otherwise, Neff, Mounts, Rinsch, et al. (1993, 1994) demonstrated that TAG types such as OLL, OOL, POL, SOL, and POO increased the oils' oxidative stability.

6 Minor bioactive lipids in the cold pressed melon seed oil

Minor components include a number of chemical compounds like tocopherols, hydrocarbons, sterols, alcohols, phospholipids, and others. These minor lipid compounds have always been of great interest to food analysts because their composition and ratios can provide a "fingerprint" for edible oils. Several classes of these compounds can be present in the oil and act as anti- or prooxidants, and they can be helpful in preventing diseases and boosting health (Nederal Nakić, Rade, Škevin, et al., 2006).

6.1 Sterols

Sterols are found in vegetable oils and can be free either or esterified with fatty acids. Depending on the position of the double bond in the ring, sterols may be subdivided into Δ5- and Δ7-sterols. Most plant species contain dominant Δ5-sterols, while Δ7-sterols are characteristic of only a small number of plant families, Cucurbitaceae being one of them (Akihisa, Kokke, & Tamura, 1992).

Sterols are used in order to determine the identity of any specific oil and to detect fraud when some people mix cheap oils with more expensive ones (Nederal Nakić et al., 2006). The scientific literature, however, does not offer enough information about the sterol composition of *Cucumis melo* seed oils and especially those obtained with cold pressing techniques. Rezig et al. (2019) reported that in the Tunisian "Ananas" *Cucumis melo* L. variety, sterols accounted for 5162 mg/kg oil. This content available in the cold pressed seed oil was higher than that found by Mallek-Ayadi et al. (2018) and Górnaś and Rudzińska (2016), respectively, in the "Maazoun" *Cucumis melo* L. and the honeydew melon (*Cucumis melo*) seed oils, and lower than that observed by Veronezi and Jorge (2018) in the *Cucumis melo* var. "inodorus" seed oil obtained by *n*-hexane extraction. In the cold pressed "Ananas" *Cucumis melo* seed oil, 15 phytosterols were identified: cholesterol, brassicasterol, 24-methylenecholesterol, campesterol, campestanol, stigmasterol, Δ5-avenasterol, Δ5,23-stigmastadienol, Δ5,24-stigmastadienol, Δ7-campesterol, Δ7-stigmastenol, Δ7-avenasterol, clerosterol, sitosterol, and sitostanol. β-sitosterol, which accounted for 3248.4 mg/kg oil, was the predominant sterol in melon seed oil, followed by the Δ5-avenasterol, which accounted for 1533.1 mg/kg. Thus, β-sitosterol and Δ5-avenasterol were the major sterols and together they made up 92.6% of total sterols in melon seed oil. β-sitosterol was also the sterol marker in the *Cucurbita maxima* var. "Béjaoui," *Cucurbita pepo* L., Kalahari melon (*Citrullus lanatus*), Bittermelon (*Momordica charantia* L.), *Cucumis melo* L., and watermelon (*Citrullus lanatus* (Thunb.) Matsum. & Nakai) seed oils, all of which belong to the same botanical family, namely *Cucurbitaceae* (Górnaś & Rudzińska, 2016; Nyam et al., 2009; Rezig, Chouaibi, Msaada, et al., 2012; Veronezi & Jorge, 2018).

The level of cholesterol in the "Ananas" cold pressed seed oil was 3.61 mg/kg oil. Such a level was lower than those reported by Mallek-Ayadi et al. (2018) and Azhari et al. (2014) for the "Maazoun" (0.88 mg/100 g) and "tibish" melon seed oils, which accounted for 0.88 mg/100 g and 1.30 mg/100 g, respectively. According to Hirsinger (1989), the concentration of phytosterols was only mildly affected by environmental factors and/or by the cultivation of new breeding lines.

It is known that sterols found in plant oils help prevent a number of cardiovascular diseases. In fact, phytosterols have a positive impact on the human body as they inhibit cholesterol absorption from the intestine and decrease blood levels of low-density lipoprotein cholesterol fraction (LDLc) (Jones, Racini-Sarjaz, Ntanios, et al., 2000; Tapiero, Townsend, & Tew, 2003). △5-avenasterol was reported by White and Armstrong (1986) and Savage, Mcneil, and Dutta (1997) to act as an antioxidant and as an antipolymerization agent in frying oils. As reported by Plat and Mensink (2005), β-sitosterol has some positive effects on the prevention of apoptosis disease and certain cancers. Similarly, sitosterol contributes to lower blood LDL cholesterol by 10%–15% as part of a healthy diet (Ntanios, 2001). Therefore, the oil extracted from melon seeds may potentially be used as a new therapeutic agent for the treatment of hypercholesterolemia.

6.2 Tocopherols and tocotrienols (tocochromanols)

Tocopherols (a saturated side chain) and tocotrienols (three unsaturated bonds in the side chain) (Fig. 3) are lipid-soluble compounds commonly known as vitamin E. α-T among all forms of vitamin E plays a vital role in human nutrition and

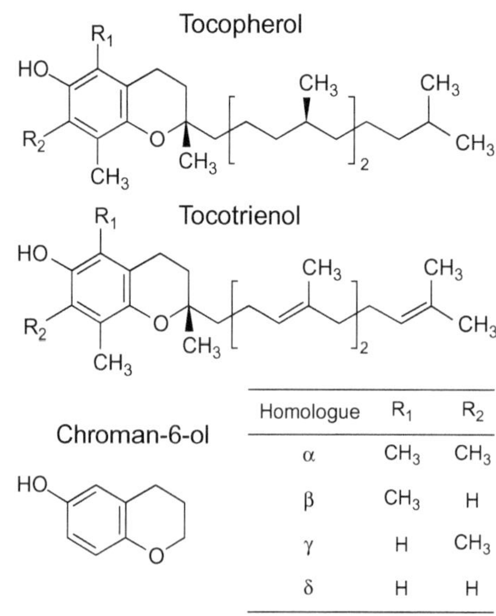

Homologue	R1	R2
α	CH3	CH3
β	CH3	H
γ	H	CH3
δ	H	H

FIG. 3 Chemical structure of tocochromanols (Górnaś, Soliven, & Seglina, 2015).

health. It protects unsaturated fatty acids from oxidation and ensures the stability of lipid membranes (Dwiecki, Górnaś, Jackowiak, et al., 2007; Dwiecki, Górnaś, Wilk, et al., 2007). Tocochromanols, especially tocotrienols, could possibly be utilized as a potential natural cure for cancer and some other chronic diseases (Aggarwal, Sundaram, Prasad, et al., 2010). There has recently been an increasing interest in recovering oils from fruit seeds. The majority of previous investigations on fruit seed oils have only focused on fatty acid composition. Likewise, there is no adequate information in relation to the profile of tocochromanols, especially those found in cold pressed *Cucumis melo* seed oils.

Table 4 illustrates the tocopherol and tocotrienol composition of *Cucumis melo* seed oils available in the literature, obtained by solvent extraction methods. As shown in the table, γ-tocopherol was the predominant tocopherol in all samples, representing 78.2% and 100%, respectively, of the total tocopherols in the *Cucumis melo* var. "agrestis" originating from Gezira provinces and that collected from a local market in Sudan (Mariod & Matthäus, 2008; Mariod et al., 2009). Petkova and Antova (2019) confirmed that γ-tocopherol accounted for 87.8% of the total tocopherols. According to Fatnassi, Nehdi, and Zarrouk (2009), α-tocopherol is recommended for human and animal consumption because it has a higher biological activity than other tocopherols, but the γ-tocopherol shows a higher antioxidant capacity when compared to the α-tocopherol. The high γ-tocopherol content in melon seed oils might help prepare antioxidant formulations for commercial purposes.

TABLE 4 Tocopherol and tocotrienol composition of *Cucumis melo* L. seed oils

Sample	Origin	α-T	β-T	(β+γ)-T	(β+δ)-T	γ-T	δ-T	α-T3	β-T3	γ-T3	δ-T3	Reference
Cucumis melo var. "Maazoun"	Tunisia	2.85	ND	18.1	ND	6.09	ND	ND	ND	ND	ND	Mallek-Ayadi et al. (2018)
Cucumis melo var. "flexuosus"	Sudan	0.4	ND	ND	ND	33.5	0.8	ND	ND	ND	ND	Mariod et al. (2009)
Cucumis melo var. "agrestis"		ND	ND	ND	ND	29.1	ND					
Cucumis melo var. "inodorus"	Brazil	3.98	ND	ND	ND	46.7	ND	ND	ND	ND	ND	Veronezi and Jorge (2018)
Canary melon (*Cucumis melo* L.)	Spain	6.88	ND	ND	ND	63.0	0.77	0.47	ND	0.9	ND	Górnaś, Soliven, et al. (2015)
Yellow melon (*Cucumis melo* var. "inodorus Naudin")	Brazil	2.05	ND	ND	ND	24.9	ND	ND	ND	ND	ND	da Silva and Jorge (2014)
Cucumis melo var. "tibish"	Sudan	2.7	ND	ND	ND	13.1	27.4	ND	ND	ND	ND	Azhari et al. (2014)
Cucumis melo var. "agrestis"	Sudan (Ghibaish)	7.21	ND	ND	ND	32.2	0.17	ND	0.33	ND	ND	Mariod and Matthäus (2008)
Cucumis melo var. "agrestis"	Sudan (Gezira)	8.21	ND	ND	ND	30.2	0.2	ND	0.3	ND	ND	Mariod and Matthäus (2008)

ND, not determined.

6.3 Phenolic compounds

Phenolic compounds represent a part of the unsaponifiable matter that determines some of the oils' characteristics such as flavor, oxidative stability, and shelf life (Mariod & Matthäus, 2008). To the best of our knowledge, there are very limited data in relation to the identification and quantification of phenolic compounds in the cold pressed *Cucumis melo* seed oils. Rezig et al. (2019) reported five phenolic compounds. There were three phenolic acids (caffeic, vanillic, and ferulic), one secoiridoid (oleuropein), and one lignan (pinoresinol). The main identified phenolic compounds were phenolic acids (139.1 μg/g) followed by lignans (10.8 μg/g) and secoiridoids (2.31 μg/g). These phenolic compounds would certainly determine some of the oil's features and can affect its astringency. They also bear some biological traits that account for their antioxidant and antiradical scavenging activities (Ahmad, Hashim, Noor, et al., 2011). Recent studies have shown that polyphenols are being added to foodstuffs, as natural antioxidants, in order to prevent off-flavor development and to achieve stabilization (Mallek-Ayadi et al., 2018).

Nyam et al. (2009) recognized other phenolic acids such as gallic acid, protocatechuic acid, *p*-hydroxybenzoic acid, syringic acid, and *p*-coumaric acid in Bittermelon and Kalahari melon seed oils. In fact, vanillic acid was the predominant phenolic acid, accounting for 0.74 mg/100 g and 0.55 mg/100 g, respectively, in the Bittermelon and Kalahari melon seed oils. Apart from caffeic acid, Mallek-Ayadi et al. (2018) identified three other phenolic acids—gallic, protocatechuic, and rosmarinic—in "Maazoun" melon seed oil. Nevertheless, "Ananas" melon seed oil presented a lower amount of caffeic acid when compared to those found by Mallek-Ayadi et al. (2018) and Nyam et al. (2009). It also presented a higher amount of ferulic acid in comparison to that reported by Nyam et al. (2009). According to Martínez-Cruz and Paredes-López (2014), several biological activities have been described for rosmarinic and caffeic acids. It has been stated that rosmarinic acid acts as an antimutagenic and an antiinflammatory agent.

Mariod and Matthäus (2008) identified seven phenolic compounds in the *Cucumis melo* var. "agrestis" seed oils originating from two different provinces in Sudan (Ghibish and Gezira). Among these seven phenolic compounds, three were unknown and the other four were callistephin, sinapic acid, vanillic acid, and catechin. Callistephin was the major phenolic compound followed by sinapic acid, which accounted for 30.6% and 27.5%, respectively, and 6.6% and 6.3% in Ghibaish and Gezira *Cucumis melo* L. var. "agrestis" samples. The "Ananas" melon seed oil also contained oleuropein, which is the major phenolic compound found in olive oil and leaves, and has a high antioxidant capacity (Rodríguez-Morató, Xicota, Fitó, et al., 2015). Oleuropein was identically detected in the "Maazoun" melon seed oil with an amount (1.65 μg/g) lower than that found by Rezig et al. (2019).

6.4 Total phenolics and flavonoids contents

Phenolic constituents seem to be of unmatched importance for the quality assessment of edible oils as far as shelf life and flavor stability are concerned (Tsimidou, Lytridou, Boskou, et al., 1996). To the best of our knowledge, there is very limited scientific evidence when it comes to total phenolic and flavonoid contents in cold pressed Cucurbitaceae seed oils and especially in the *Cucumis melo* L. variety. Rezig et al. (2019) reported a total phenolic content of cold pressed "Ananas" *Cucumismelo* variety of 24.04 mg gallic acid equivalents per 100 g (GAE/100 g). In Mashhadi melon (*Cucumis melo* var. Iranians cv. Mashhadi) and yellow melon *Cucumis melo* var. "inodorus Naudin" seed oils, the total phenolic content accounted for 125 mg GAE/kg oil and 130.7 mg GAE/kg oil, respectively (da Silva & Jorge, 2014; Hashemi et al., 2017). These total phenolic content were lower than those reported for the Sudanese "agrestis" *Cucumis melo* L. seed oils variety originating from Ghibaish and Gezira provinces (33 mg/g oil vs 31.9 mg/g), melon seed oil (1007.8 mg GAE/kg), *Cucumis melo* var. "Maazoun" seeds (304.1 mg/100 g), cantaloupe seeds (285 mg/100 g), melon (*Cucumis melo*) seeds (99.69 mg/100 g), and honeydew seeds (80 mg/100 g) (Ismail, Chan, Mariod, et al., 2010; Jorge, da Silva, & Malacrida, 2015; Mallek-Ayadi et al., 2018; Mariod & Matthäus, 2008; Morais, Rotta, Sargi, et al., 2015; Zeb, 2016). Thanks to their diverse chemical structures, phenolics have the potential to protect cellular components against oxidation (Bahloul, Kechaou, & Mihoubi, 2014). Similarly, polyphenols have been known for their potential effects to prevent cancer and cardiovascular diseases (Liara, Débora, & Neuza, 2013). It has also been reported that naturally synthesized polyphenols contribute to the prevention of neurodegenerative disorders such as brain aging (Sarubbo, Esteban, Miralles, et al., 2018).

Flavonoids (anthocyanidins, catechins, tannins, and flavones) are the most common and widely distributed group of plant phenolic compounds, and are usually considered as very effective antioxidants (Ismail et al., 2010). Moreover, these bioactive compounds were also reported to reduce and delay the development of some chronic diseases (Arvanitoyannis & Van Houwelingen-Koukaliaroglou, 2005). The total flavonoid content of cold pressed "Ananas" *Cucumis melo* L. variety seed oil was about 6.01 mg rutin eq/100 g oil. Mallek-Ayadi et al. (2018), Morais et al. (2015), and Ismail et al. (2010) all reported a total flavonoid content of 87.5 mg quercetin equivalent (QE) per 100 g, 3.61 mg QE per 100 g, and 1.62 μg rutin equivalent (RE) per g in the "Maazoun," melon, and cantaloupe "*Cucumis melo*" varieties seed extracts, respectively.

6.5 Carotenoids

Carotenoids are lipophilic substances present in fruit seed oils. Known as nonnutritive nutrients, these bioactive compounds possess antioxidant properties since they are efficient singlet oxygen quenchers and function as chain-breaking antioxidants, thus protecting cells and other body components from any free radical attack (Murkovic, Mülleder, & Neunteufl, 2002). Carotenoids were also reported to have excellent benefits for human health such as preventing cardiovascular diseases, prostate cancer, and macular degeneration (Silva, Albuquerque, Alves, et al., 2018; Stahl & Sies, 2005). da Silva and Jorge (2014) also reported a carotenoid content of 6.3 μg/g in the yellow melon (*Cucumis melo* var. inodorus Naudin) seed oil obtained by cold extraction with chloroform, methanol, and water. This content was higher than those found in Mashhadi melon (*Cucumis melo* var. Iranians cv. Mashhadi) and *Cucumis melo* var. "Maazoun" seed oils (3.4 vs 2.43 mg/kg) (Hashemi et al., 2017; Mallek-Ayadi, Bahloul, & Kechaou, 2019) and lower than that reported by Szydlowska-Czerniak, Trokowski, Karlovits, et al. (2011) in palm oil, which accounted for 38.5 μg/g. It should also be noted that high carotenoid concentration oils are very useful from a technological point of view, as this pigment is rich in vitamin A and acts as an effective antioxidant. On the other hand, Tuberoso, Kowalczyk, Sarritzu, et al. (2007) quantified β-carotene in linseed, grape, corn, peanut, canola, soybean, and sunflower seed oils, and in olive oil as well. Quantities above 1.0 μg/g were found only in olive oil (6.9 μg/g), pumpkin (5.7 μg/g), and canola (1.7 μg/g) seed oils.

7 Oxidative stability

The determination of the oxidative stability of vegetable oil would certainly serve as a revealing factor and could hence provide a good estimation of the oil's resistance during oxidation. As far as we know, there is little information about the oxidative stability of cold pressed *Cucumis melo* seed oil. Rezig et al. (2019) reported oxidative stability of 3.8 h in cold pressed *Cucumis melo* L. var. "Ananas" seed oil obtained by a vegetable screw press (IBG Monforts Oekotec GmbH & Co. KG, Mönchengladbach, Germany). This value was lower than those found in other varieties and obtained by cold solvent extraction. In fact, Mallek-Ayadi et al. (2019) and Petkova and Antova (2015) reported oxidative stability of 7.2 h in "Maazoun," "honeydew," and "Dessert 5" *Cucumis melo* L. seed oil varieties. In other respects, Veronezi and Jorge (2018), showed that in the "inodorus" *Cucumis melo* seed oil variety, the induction time was 13.7 h. Besbes, Blecker, Deroanne, et al. (2004) reported that high stability may be explained by a relatively low content of polyunsaturated fatty acids and a high content of natural antioxidants, such as phenolic compounds. This statement was largely in agreement with a previous study conducted by Kaijser, Dutta, and Savage (2000). In fact, the authors showed that the stability of the oil was much influenced by factors like the positions of the individual fatty acids within the triacylglycerol molecule, along with the presence of tocopherols, carotenoids, and sterols.

Comparably, Aparicio, Roda, Albi, et al. (1999) found that a linear regression based on the oleic/linoleic ratio and the contents of phenols and tocopherols in virgin olive oil showed a good correlation with the oxidative stability measured by Rancimat. The contribution of phenolic and orthophenolic compounds in the oxidation stability of olive oil was about 51%, 24% for fatty acids and, to lesser percentages, α-tocopherols, carotenoids, and chlorophylls.

8 Antioxidant properties

The antioxidant ability of melon by-products is essentially caused by the presence of specific compounds, such as phenolics and flavonoids (Silva et al., 2018). Other studies have suggested that some fatty acids and trace components such as linoleic acid were important in relation to the antioxidant activity of oils (Ramaprasad, Srinivasan, Baskaran, et al., 2006). There is no adequate scientific literature dealing with the antioxidant capacity of cold pressed *Cucumis melo* seed oil using a vegetable screw press. The antiradical activity by the hydrogen-donating antioxidants model is widely used to evaluate the antioxidant properties in a relatively short time. To assess this ability, the DPPH· free radical scavenging assay is generally used. In this case, IC_{50} corresponds to the concentration of the sample required to scavenge 50% of DPPH· radicals. Thus, low IC_{50} is equivalent to high scavenging capacity, and is calculated by plotting percentage inhibition against different concentrations of oil (Guergouri, Sobhi, & Benboubetra, 2017).

A recent study investigated cold pressed Cucurbitaceae seed oils (Rezig et al., 2019). It showed that "Ananas" cold pressed *Cucumis melo* seed oil exhibited an IC_{50} of 52.5 μg/g. Moreover, Azhari et al. (2014) reported antioxidant activity in Seinat (*Cucumis melo* var. "tibish") seed oil higher than that found by Rezig et al. (2019), with an IC_{50} of 25.2 mg/mL. For comparative reasons, *Cucumis melo* seed oil showed the highest antioxidant activity when compared to cold pressed "Essahli" *Cucurbita pepo* L., and "Crimson" *Citrullus lanatus* seed oils, all of which were extracted in the same conditions by a vegetable screw press (IBG Monforts Oekotec GmbH & Co. KG, Mönchengladbach, Germany). Rezig et al. (2019)

observed a linear correlation between the antioxidant activity of Cucurbitaceae seed oils and oxidative stability. In fact, the *Cucumis melo* seed oil, exhibiting the highest antioxidant activity, was the most stable against oxidation, followed by *Cucurbita pepo* and *Citrullus lanatus* seed oils. Hashemi et al. (2017) reported the highest antioxidant activity for pumpkin seed oil (~90%), followed by those of Mashhadi melon (between 80% and 90%) and Iranian watermelon seed oils (~80%). According to Górnaś, Pugajeva, et al. (2014); Górnaś, Soliven, et al. (2015), watermelon (*Citrullus lanatus* (Thunb.) Matsum. & Nakai) and canary melon (*Cucumis melo* L.) seeds and seed oils are rich sources of tocochromanols (tocopherols and tocotrienols), amounting to 33.4 mg/100 g and 20.4 mg/100 g for seeds, and 117.89 mg/100 g oil and 72.1 mg/100 g for seed oils, respectively. The DPPH· radical scavenging activity mainly depended on the concentration of tocochromanols in these seed oils. According to Górnaś, Soliven, et al. (2015), a significant correlation ($r = 0.994$) was found between the total content of tocochromanols in the seed oils and the DPPH· radical scavenging activity. Such a result demonstrates that different tocopherol and tocotrienol homologs quench DPPH· free radicals with similar intensity.

Comparably, in the cantaloupe (*Cucumis melo*) methanolic extract, Ismail et al. (2010) reported an IC_{50} of 25.4 mg/mL. Such antioxidant activity was the lowest when compared to flesh, stem, leaf, and skin cantaloupe methanolic extracts. The DPPH· scavenging activity of cantaloupe extracts showed a similar trend with the result of total phenolic content ($r = 0.9228$) and total flavonoid content ($r = 0.8478$), indicating that DPPH· radical scavenging activity of the cantaloupe extracts is highly related to the amount of phenolic compounds, especially those flavonoids present in the extracts. According to Chang, Wu, Wang, et al. (2001), this is probably due to their redox properties, which allow them to act as reducing agents, hydrogen donors, and singlet oxygen quenchers.

9 CIE L*, a*, and b* coordinates

The objective measurement of color is of great importance for food producers due to the existing relationships between such attributes and the acceptability of foods standard by consumers (Moyano, Meléndez-Martínez, Alba, et al., 2008). CieLab coordinates values (L*, a*, b*) of cold pressed *Cucumis melo* seed oils are very scarce in the literature. Rezig et al. (2019) reported in "Ananas" *Cucumis melo* seed oil obtained by a vegetable screw press (IBG Monforts Oekotec GmbH & Co. KG, Mönchengladbach, Germany) CieLab coordinate values (L*, a*, b*) of 80.57, −2.05, and 55.5. *Cucumis melo* var. "Ananas" exhibited a higher L* value than those observed by Nyam et al. (2009) in bittermelon and Kalahari melon seed oils extracted by hexane. This means that the "Ananas" *Cucumis melo* seed oil variety was lighter in color than the compared samples. According to Hsu and Yu (2002), the CieLab (L*, a*, b*) values of other vegetable oils, such as palm, soybean, sunflower, olive, and corn, ranged from 63.4 to 69.5, 3.8 to 4.4, and 9.2 to 10.4, respectively. Hence, the melon (*Cucumis melo*) seed oil's b* value was higher than those of other vegetable oils. Such a result suggests that melon seed oil was also yellower than the vegetable oils studied by Hsu and Yu (2002). In addition, we should mention that Nyam et al. (2009) found a b* value in the same range as that discovered by Rezig et al. (2019) in the bittermelon seed oil. This suggests the presence of yellow pigments such as carotenoids. Melon seed oil was also characterized by another color specification: an a* negative value which was markedly lower than the a* value of common vegetable oils.

10 Conclusion

It seems fair to claim that the cold pressed "Ananas" *Cucumis melo* seed oil originating from Tunisia is a promising source of biologically active substances and can undoubtedly be used as a functional food component. The high amounts of polyunsaturated fatty acids might reasonably possess very beneficial physiological effects that would greatly contribute to the prevention of coronary heart disease and cancer. The high levels of sterols and phenolics in the melon seed oil could help prevent oxidation. Such a feature will inevitably improve the stability of the melon seed oil, thus extending its shelf life. The melon seed oil is also rich in bioactive compounds, which again confirms its high nutritional value and its potential use in many industrial applications. Similarly, the appropriate utilization of melon seeds may set a good example for the valorization of fruit by-products and produce an alternative oil that is no lower in quality than other common vegetable oils. However, since data about cold pressed melon seed oil are very scarce, future studies should be carried out in order to highlight the nutraceutical side of cold pressed seed oils that are obtained by a vegetable screw press. The use of melon seed oil for industrial applications might require its exposure to high thermal treatments that would eventually lead to changes in its quality. Therefore, a study of thermo-oxidation effects on the physicochemical properties of cold pressed melon seed oil must be undertaken. Nevertheless, we must retain that *Cucumis melo* seed oil production will provide new resources for industrial processes, and open fields for promoting agricultural products.

References

Aggarwal, B. B., Sundaram, C., Prasad, S., et al. (2010). Tocotrienols, the vitamin E of the 21st century: Its potential against cancer and other chronic diseases. *Biochemical Pharmacology*, *80*, 1613–1631.

Ahmad, R., Hashim, H. M., Noor, Z. M., et al. (2011). Antioxidant and antidiabetic potential of Malaysian *uncaria*. *Research Journal of Medicinal Plants*, *5*(5), 587–595.

Akihisa, T., Kokke, W. C. M. C., & Tamura, T. (1992). In G. W. Patterson & W. D. Nes (Eds.), *Naturally occurring sterols in physiology and biochemistry of sterols* (pp. 172–178). Champaign, IL: American Oil Chemists' Society Press.

Aparicio, R., Roda, L., Albi, M. A., et al. (1999). Effect of various compounds on virgin olive oil stability measured by Rancimat. *Journal of Agricultural and Food Chemistry*, *47*, 4150–4155.

Arvanitoyannis, I. S., & Van Houwelingen-Koukaliaroglou, M. (2005). Functional foods: A survey of health claims, pros and cons, and current legislation. *Critical Reviews in Food Science and Nutrition*, *45*(5), 385–404.

Azhari, S., Xu, Y. S., Jiang, Q. X., et al. (2014). Physicochemical properties and chemical composition of seinat (*Cucumis melo* var. tibish) seed oil and its antioxidant activity. *Grasas y Aceites*, *65*, 1–8.

Bahloul, N., Kechaou, N., & Mihoubi, N. B. (2014). Comparative investigation of minerals, chlorophylls contents, fatty acid composition and thermal profiles of olive leaves (*Olea europeae* L.) as by-product. *Grasas Aceites*, *65*(3), 3–35.

Besbes, S., Blecker, C., Deroanne, C., et al. (2004). Date seeds: Chemical composition and characteristic profiles of the lipid fraction. *Food Chemistry*, *84*, 577–584.

Chang, S. T., Wu, J. H., Wang, S. Y., et al. (2001). Antioxidant activity of extracts from Acacia confusa bark and heartwood. *Journal of Agricultural and Food Chemistry*, *49*(7), 3420–3424.

da Silva, A. C., & Jorge, N. (2014). Bioactive compounds of the lipid fractions of agro-industrial waste. *Food Research International*, *66*, 493–500.

de Melo, M. L. S., Narain, N., & Bora, P. S. (2000). Characterisation of some nutritional constituents of melon (*Cucumis melo* hybrid AF-522) seeds. *Food Chemistry*, *68*, 411–414.

Dimić, E. (2005). Cold pressed oils. In *Monography* (pp. 170–172). Novi Sad: The University of Novi Sad, Faculty of Technology.

Dwiecki, K., Górnaś, P., Jackowiak, H., et al. (2007). The effect of D-alpha-tocopherol on the solubilisation of dipalmitoyl phosphatidylcholine membrane by anionic detergent sodium dodecyl sulphate. *Journal of Food Lipids*, *14*(1), 50–61.

Dwiecki, K., Górnaś, P., Wilk, A., et al. (2007). Spectroscopic studies of a D-α-tocopherol concentration-induced transformation in egg phosphatidylcholine vesicles. *Cellular & Molecular Biology Letters*, *12*, 51–69.

Fadavi, A., Barzegar, M., & Azizi, M. H. (2006). Determination of fatty acids and total lipid content in oilseed of 25 pomegranates varieties grown in Iran. *Journal of Food Composition and Analysis*, *19*, 676–680.

Fatnassi, S., Nehdi, I., & Zarrouk, H. (2009). Chemical composition and profile characteristics of Osage orange *Maclura pomifera* (Rafin.) Schneider seed and seed oil. *Industrial Crops and Products*, *29*, 1–8.

Fruhwirth, G. O., & Hermetter, A. (2007). Seeds and oil of the Styrian oil pumpkin: Components and biological activities. *European Journal of Lipid Science and Technology*, *109*(11), 1128–1140.

Górnaś, P., Pugajeva, I., & Seglina, D. (2014). Seeds recovered from by-products of selected fruit processing as a rich source of tocochromanols: RP-HPLC/FLD and RP-UPLC-ESI/MSn study. *European Food Research and Technology*, *239*, 519–524.

Górnaś, P., & Rudzińska, M. (2016). Seeds recovered from industry by-products of nine fruit species with a high potential utility as a source of unconventional oil for biodiesel and cosmetic and pharmaceutical sectors. *Industrial Crops and Products*, *83*, 329–338.

Górnaś, P., Soliven, A., & Seglina, D. (2015). Seed oils recovered from industrial fruit byproducts are a rich source of tocopherols and tocotrienols: Rapid separation of α/β/γ/δhomologues by RP-HPLC/FLD. *European Journal of Lipid Science and Technology*, *117*, 773–777.

Guergouri, F. Z., Sobhi, W., & Benboubetra, M. (2017). Antioxidant activity of Algerian *Nigella Sativa* total oil and its unsaponifiable fraction. *Journal of Phytopharmacology*, *6*(4), 234–238.

Hashemi, S. M. B., Khaneghag, A. M., Koubaa, M., et al. (2017). Novel edible oil sources: Microwave heating and chemical properties. *Food Research International*, *92*, 147–153.

Hemavatahy, J. (1992). Lipid composition of melon (*Cucumis melo*) kernel. *Journal of Food Composition and Analysis*, *5*, 90–95.

Hirsinger, F. (1989). New annual oil crops. In G. Roebbelen, R. K. Downey, & A. Ashri (Eds.), *Oil crops of the world* (pp. 518–532). New York: McGraw Hill.

Hsu, S. Y., & Yu, S. H. (2002). Comparisons on 11 plant oil fat substitutes for low-fat kung-wans. *Journal of Food Engineering*, *51*, 215–220.

Ismail, H. I., Chan, K. W., Mariod, A. A., et al. (2010). Phenolic content and antioxidant activity of cantaloupe (*Cucumis melo*) methanolic extracts. *Food Chemistry*, *119*, 643–647.

ITIS, Integrated taxonomic information system. (2017). *Integrated taxonomic information system report*. Retrieved from *(2017).* https://www.itis.gov/servlet/SingleRpt/SingleRptsearch_topic=TSN&search_value=22362#null.

Jones, P., Racini-Sarjaz, M., Ntanios, et al. (2000). Modulation of plasma lipid levels and cholesterol kinetics by phytosterol versus phytostanol ester. *Journal of Lipid Research*, *41*, 697–705.

Jorge, N., da Silva, A. C., & Malacrida, C. R. (2015). Physicochemical characterisation and radical-scavenging activity of Cucurbitaceae seed oils. *Natural Product Research*. https://doi.org/10.1080/14786419.14782015.11007135.

Kaijser, A., Dutta, P., & Savage, G. (2000). Oxidative stability and lipid composition of macadamia nuts grown in New Zealand. *Food Chemistry*, *71*, 67–70.

Karakaya, S., Kavas, A., Nehir El, S., et al. (1995). Nutritive value of a melon seed beverage. *Food Chemistry*, *52*, 139–141.

Kulaitienė, J., Černiauskienė, J., Jarienė, E., et al. (2018). Antioxidant activity and other quality parameters of cold pressing pumpkin seed oil. *Notulae Botanicae Horti Agrobotanici Cluj-Napoca*, *46*(1), 161–166.

Liara, S. D., Débora, M. M. L., & Neuza, J. (2013). Physicochemical and bioactive properties of *Hymenaea courbaril* L. pulp and seed lipid fraction. *Industrial Crops and Products*, *49*, 610–618.

Mallek-Ayadi, S., Bahloul, N., & Kechaou, N. (2018). Chemical composition and bioactive compounds of *Cucumis melo* L. seeds: Potential source for new trends of plant oils. *Process Safety and Environment Protection*, *113*, 68–77.

Mallek-Ayadi, S., Bahloul, N., & Kechaou, N. (2019). *Cucumis melo* L. seeds as a promising source of oil naturally rich in biologically active substances: Compositional characteristics, phenolic compounds, and thermal properties. *Grasas Aceitas*, *70*(1), 1–10.

Maran, J. P., & Priya, B. (2015). Supercritical fluid extraction of oil from muskmelon (*Cucumis melo*) seeds. *Journal of the Taiwan Institute of Chemical Engineers*, *47*, 71–78.

Mariod, A. A., Ahmed, Y. M., Matthäus, B., et al. (2009). A comparative study of the properties of six Sudanese cucurbit seeds and seed oils. *Journal of the American Oil Chemists' Society*, *86*, 1181–1188.

Mariod, A., & Matthäus, B. (2008). Fatty acids, tocopherols, sterols, phenolic profiles and oxidative stability of *Cucumis melo* Var. 'agrestis' oil. *Journal of Food Lipids*, *15*(1), 56–67.

Martínez-Cruz, O., & Paredes-López, O. (2014). Phytochemical profile and nutraceutical potential of chia seeds (*Salvia hispanica* L.) by ultra high performance liquid chromatography. *Journal of Chromatography A*, *1346*, 43–48.

Maynard, D., & Maynard, D. (2000). Cucumbers, melons, and watermelons. In K. F. Kiple (Ed.), *The cambridge world history of food*: Cambridge University Press.

Mitra, P., Ramaswamy, H. S., & Chang, K. S. (2009). Pumpkin (*Cucurbita maxima*) seed oil extraction using supercritical carbon dioxide and physicochemical properties of the oil. *Journal of Food Engineering*, *95*, 208–213.

Morais, D. R., Rotta, E. M., Sargi, S. C., et al. (2015). Antioxidant activity, phenolics, and UPLC—ESI-MS of different tropical fruits parts and processed peels. *Food Research International*. https://doi.org/10.1016/j.foodres.2015.08.036.

Moyano, M. J., Meléndez-Martínez, A. J., Alba, J., et al. (2008). A comprehensive study on the colour of virgin olive oils and its relationship with their chlorophylls and carotenoids indexes (II): CIELUV and CIELAB uniform colour spaces. *Food Research International*, *41*, 513–521.

Murkovic, M., Mülleder, U., & Neunteufl, H. (2002). Carotenoid content in different varieties of pumpkins. *Journal of Food Composition and Analysis*, *15*, 633–638.

Nederal Nakić, S., Rade, D., Škevin, D., et al. (2006). Chemical characteristics of oils from naked and husk seeds of *Cucurbita pepo* L. *European Journal of Lipid Science and Technology*, *108*, 936–943.

Neff, W. E., Mounts, T. L., Rinsch, W. M., et al. (1993). Photooxidation of soybean oils as affected by triacylglycerol composition and structure. *Journal of the American Oil Chemists' Society*, *71*, 163–168.

Neff, W. E., Mounts, T. L., Rinsch, W. M., et al. (1994). Triacylglycerols with altered fatty acid compositions as affected by triacylglycerol composition and structure. *Journal of the American Oil Chemists' Society*, *71*, 1101–1109.

Ntanios, F. (2001). Plant sterol-ester-enriched spreads as an example of a new functional food. *European Journal of Lipid Science and Technology*, *103*, 102–106.

Nyam, K. L., Tan, C. P., Karim, R., et al. (2010). Extraction of tocopherol-enriched oils from Kalahari melon and roselle seeds by supercritical fluid extraction (SFE-CO_2). *Food Chemistry*, *119*, 1278–1283.

Nyam, K. L., Tan, C. P., Lai, O. M., et al. (2009). Physicochemical properties and bioactive compounds of selected seed oils. *Food Science and Technology*, *42*, 1396–1403.

Petkova, Z., & Antova, G. (2015). Proximate composition of seeds and seed oils from melon (*Cucumis melo* L.) cultivated in Bulgaria. *Cogent Food and Agriculture*, *1*(1), 1–15.

Petkova, Z., & Antova, G. (2019). A comparative study on quality parameters of pumpkin, melon and sunflower oils during thermal treatment. *Oilseeds and Fats Crops and Lipids*, *26*, 32.

Plat, J., & Mensink, R. P. (2005). Plant stanol and sterol esters in the control of blood cholesterol levels: Mechanism and safety aspects. *The American Journal of Cardiology*, *96*, 15–22.

Rabrenović, B. B., Dimić, E. B., Novaković, M. M., et al. (2014). The most important bioactive components of cold pressed oil from different pumpkin (*Cucurbita pepo* L.) seeds. *Food Science and Technology*, *55*(2), 521–527.

Ramaprasad, T. R., Srinivasan, K., Baskaran, V., et al. (2006). Spray-dried milk supplemented with a-linoleic acid and eicosapentaenoic acid and docosahexaenoic acid decrease HMG Co A reductase activity and increases biliary secretion of lipids of rats. *Steroids*, *71*, 409–415.

Reske, J., Siebrecht, J., & Hazebroek, J. (1997). Triacylglycerol composition and structure in genetically modified sunflower and soybean oils. *Journal of the American Oil Chemists' Society*, *74*, 989–998.

Rezig, L., Chouaibi, M., & Fregapane, G. (2018). *Profile characterisation of pumpkin (*Cucurbita maxima*) seed oils* (p. 53). Lambert Academic Publishing. ISBN: 978-613-9-87846-8.

Rezig, L., Chouaibi, M., Meddeb, W., et al. (2019). Chemical composition and bioactive compounds of Cucurbitaceae seeds: Potential sources for new trends of plant oils. *Process Safety and Environment Protection*. https://doi.org/10.1016/j.psep.2019.05.005.

Rezig, L., Chouaibi, M., Msaada, K., et al. (2012). Chemical composition and profile characterization of pumpkin (*Cucurbita maxima*) seed oil. *Industrial Crops and Products*, *37*, 82–87.

Rodríguez-Morató, J., Xicota, L., Fitó, M., et al. (2015). Potential role of olive oil phenolic compounds in the prevention of neurodegenerative diseases. *Molecules*, *20*, 4655–4680.

Rotkiewicz, D., Konopka, I., & Zylik, S. (1999). State of works on the rapeseed oil processing optimalization. I. *Oil obtaining Rośliny Oleiste/Oilseed Crops XX*, 151–168. (in Polish).

Salgin, U., & Korkmaz, H. (2011). A green separation process for recovery of healthy oil from pumpkin seed. *Journal of Supercritical Fluids*, *58*, 239–248.

Sarubbo, F., Esteban, S., Miralles, A., et al. (2018). Effects of resveratrol and other polyphenols on Sirt 1: Relevance to brain function during aging. *Current Neuropharmacology*, *16*(2), 126–136.

Savage, G. P., Mcneil, D. L., & Dutta, P. C. (1997). Lipid composition and oxidative stability of oils in hazelnuts (*Corylus avellana* L.) grown in New Zealand. *Journal of the American Oil Chemists' Society*, *74*, 755–759.

Siger, A., Nogala-Kalucka, M., & Lampert-Szczapa, E. (2008). The content and antioxidant activity of phenolic compounds in cold-pressed plant oils. *Journal of Food Lipids*, *15*, 137–149.

Silva, M. A., Albuquerque, T. G., Alves, R. C., et al. (2018). Melon (*Cucumis melo* L.) by-products: Potential food ingredients for novel foods. *Trends in Food Science and Technology*. https://doi.org/10.1016/j.tifs.2018.07.005.

Stahl, W., & Sies, H. (2005). Bioactivity and protective effects of natural carotenoids. *Biochimica et Biophysica Acta*, *1740*(2), 101–107.

Szydlowska-Czerniak, A., Trokowski, K., Karlovits, G., et al. (2011). Effect of refining process on antioxidant capacity, total content of phenolics and carotenoids in palm oils. *Food Chemistry*, *129*, 1187–1192.

Tangolar, S. G., Özogul, Y., Tangolar, S., et al. (2009). Evaluation of fatty acid profiles and mineral content of grape seed oil of some grape genotypes. *International Journal of Food Sciences and Nutrition*, *60*, 32–39.

Tapiero, H., Townsend, D. M., & Tew, K. D. (2003). Phytosterols in the prevention of human pathologies. *Biomedicine & Pharmacotherapy*, *57*, 321–325.

Tsimidou, M., Lytridou, M., Boskou, D., et al. (1996). On the determination of minor phenolic acids of virgin olive oil by RP-HPLC. *Grasas y Aceites*, *47*, 151–157.

Tuberoso, C. I. G., Kowalczyk, A., Sarritzu, E., et al. (2007). Determination of antioxidant compounds and antioxidant activity in commercial oilseeds for food use. *Food Chemistry*, *103*, 1494–1501.

Veronezi, C. M., & Jorge, N. (2015). Chemical characterization of the lipid fractions of pumpkin seeds. *Nutrition & Food Sciences*, *45*(1), 164–173.

Veronezi, C. M., & Jorge, N. (2018). Effect of *Carica papaya* and *Cucumis melo* seed oils on the soybean oil stability. *Food Science and Biotechnology*, *27*(4), 1031–1040.

Vishwakarma, V. K., Gupta, J. K., & Upadhyay, P. K. (2017). Pharmacological importance of *Cucumis melo* L.: An overview. *Asian Journal of Pharmaceutical and Clinical Research*, *10*(3), 8–12.

Wani, A. A., Sogi, D. S., Singh, P., et al. (2011). Characterization and functional properties of watermelon (*Citrullus lanatus*) seed proteins. *Journal of the Science of Food and Agriculture*, *91*, 113–121.

White, P. J., & Armstrong, L. S. A. (1986). Effect of selected oat sterols on the determination of heated soybean oil. *Journal of the American Oil Chemists' Society*, *63*, 525–529.

Yanty, N. A. M., Lai, O. M., Osman, A., et al. (2008). Physicochemical properties of *Cucumis melo* var. inodorus (honeydew melon) seed and seed oil. *Journal of Food Lipids*, *15*, 42–55.

Zeb, A. (2016). Phenolic profile and antioxidant activity of melon (*Cucumis melo* L.) seeds from Pakistan. *Foods*, *5*, 67–74.

Chapter 55

Cold pressed *Citrullus lanatus* seed oil

Leila Rezig[a,b], Moncef Chouaibi[c], Kamel Msaada[d], and Salem Hamdi[e]
[a]*High Institute of Food Industries, El Khadra City, Tunis, Tunisia,* [b]*University of Carthage, National Institute of Applied Sciences and Technology, LR11ES26, 'Laboratory of Protein Engineering and Bioactive Molecules', Tunis, Tunisia,* [c]*Research Unit: 'Bio-Preservation and Valorization of Agricultural Products UR13-AGR 02', University of Carthage, High Institute of Food Industries of Tunisia, Tunis, Tunisia,* [d]*Laboratory of Aromatic and Medicinal Plants, Biotechnology Center in Borj Cedria Technopole, Hammam-Lif, Tunisia,* [e]*Food Conservation and Valorization Laboratory, High Institute of Food Industries of Tunisia, Tunis, Tunisia*

1 Introduction

Cucurbits are one of the highest-yielding vegetable crop in the world that grow in temperate and tropical regions. Watermelon is a particularly important such horticultural crop. It is appreciated most for its sweet and juicy flesh. Taxonomically, *Citrullus lanatus* belongs to the Cucurbitaceae family, which is also referred to as cucurbits. It is an annual species that includes both wild and semidomesticated forms, and is mainly distributed in tropical and subtropical areas (Acar, Özcan, Kanbur, & Dursun, 2012). Watermelon has been cultivated globally for its fruit and has particularly been grown in Africa for the last 4000 years. Its genesis has been attributed to two different theories, one of which claims that it was derived from the perennial relative *Citrullus colocynthis* present in wild archeological sites, while the other theory states that it was domesticated from wild forms of *Citrullus lanatus*. Originating in West Africa, watermelon was subsequently introduced to India in about 800 CE and to China in 900 CE, then further extended to Southeast Asia, Japan, Europe, and America in the late 1500s (Erhirhie & Ekene, 2013). Sultana and Ashraf (2019) reported about 1200 varieties of watermelon in various sizes ranging from less than a pound to more than 200 pounds. These varieties are generally known as melons (Mabaleha, Mitei, & Yeboah, 2007).

Thanks to its sweet and red juicy flesh, recognized as a rich source of carotenoids, watermelon is utilized for the production of juices, nectars, and fruit cocktails (Ahmed, 1996; Hour, Ahmed, & Carter, 1980). Moreover, as a major by-product, the rind is utilized for products like pickle, preserve, and pectin (Godawa & Jalali, 1995; Hasan, 1993). Watermelon seeds have also been reported to be a valuable and rich source of proteins (25%–37%) (El-Adawy & Taha, 2001; Sultana & Ashraf, 2019; Wani, Sogi, Singh, et al., 2011a, 2011b), and lipids (37.8%–45.4%) (El-Adawy & Taha, 2001; Lazos, 1986; Rezig, Chouaibi, Meddeb, et al., 2019; Sultana & Ashraf, 2019). The seeds have equally been recognized as an excellent source of dietary oil (Wani et al., 2011a). The watermelon seeds are first dried, then used in various forms, such as snacks in India, while in Arabian and African countries, seeds are used as a masticatory article, as a food additive in various dishes, as an ornament for cakes, and as a stuffing for the indigenous pudding "kheer" (Mariod, Ahmed, Matthäus, et al., 2009; Wani et al., 2011b). Even the residues are used as a fuel source for cooking (Ziyada & Elhussien, 2008). On an industrial scale, however, watermelon seeds generally remain underexplored in both African and Arabian countries, specifically in Tunisia. It is fair to say, nevertheless, that the scientific literature dealing with the bioactive compounds present in watermelon seed oil remains scarce. It is also fair to say that the presence of such bioactive compounds as essential fatty acids, phytosterols, tocopherols, phenolic compounds, and carotenoids makes the extracted seed oils an ideal alternative for a number of functional edible oils.

In the same line of thought, the present chapter aims at evaluating the chemical properties of the oils obtained by extracting the watermelon seed oil of a Tunisian variety "Crimson" through a vegetable screw press (Komet DD 85 G, IBG Monforts Oekotec GmbH & Co. KG Mönchengladbach, Germany). The investigation first involves the study of peroxide, acid, iodine, and saponification values, specific gravity, refractive index, unsaponifiable matter, and specific absorptivity values K_{232} and K_{270}. The study of the bioactive compounds, such as fatty acids, sterols, phenolic compounds, and total flavonoid and polyphenol contents, will be undertaken in comparison with other varieties originating from different countries. It is also important to investigate herein, along with the nutritional aspect of the cold pressed "Crimson" *Citrullus lanatus* seed oil variety, the oxidative stability and the antioxidant properties for better valorization at an industrial level.

Cold Pressed Oils. https://doi.org/10.1016/B978-0-12-818188-1.00055-4
Copyright © 2020 Elsevier Inc. All rights reserved.

2 Oil content in watermelon seeds

Numerous studies have already been conducted on the oil yield of watermelon seed varieties originating from countries all over the world. Hashemi, Khaneghag, Koubaa, et al. (2017) reported a 13.1% of oil content in the Iranian watermelon (*Citrullus lanatus* cv Fire Fon) seeds extracted by *n*-hexane cold extraction. This content was lower than those reported by other researchers. In fact, Al-Khalifa (1996) showed that the oil content in Iranian, Chinese, and Egyptian *Citrullus lanatus* seeds, following a chloroform/methanol extraction, ranged between 20% and 24%. Such results are close to those found by Morais, Rotta, Sargi, et al. (2017) and Nehdi, Sbihi, Tan, et al. (2013) in watermelon (*Citrullus lanatus*) and in *Citrullus colocynthis* L. "Schrad" seeds, accounting for 24.1% and 23.1%, respectively, but lower than those found by Górnaś and Rudzińska (2016) and Mariod et al. (2009) in *Citrullus lanatus* and *Citrullus lanatus* var. "colocynthoide" seeds, exhibiting oil yields of 27.1% and 28.5%, respectively. Acar et al. (2012) reported a higher oil yield of 52.3% in *Citrullus lanatus* var. "citroides" seeds following a petroleum ether Soxhlet extraction. It is important to note that the above-cited oil yields were obtained either by solvent extraction or by a Soxhlet apparatus. None of the previously conducted studies reported a cold extraction procedure using a vegetable screw press. As far as we are concerned, de Conto, Gragnani, Maus, et al. (2011) were the first to report the oil content in cold pressed watermelon seeds (*Citrullus lanatus* (Thunb.) Matsum. & Nakai) (21.6%) using a mechanical expeller (Komet, model CA 59G, Mönchengladbach, Germany). Otherwise, Rezig et al. (2019) reported in cold pressed "Crimson" watermelon seeds a lipid content of 19.2% using a German screw press (Komet DD 85 G, IBG Monforts Oekotec GmbH & Co. KG Mönchengladbach). The oil content was lower than those found in "Essahli" *Cucurbita pepo* (35.5%) and "Ananas" *Cucumis melo* (28.4%) seed oils, obtained under the same extraction conditions, and both belonging to the same botanical family (Cucurbitaceae).

The high concentrations of oil make the watermelon seeds ideal for many oil industry applications. Nyam, Tan, Lai, et al. (2009) reported that the variation in the oil yield was due to the differences in plant variety, cultivation climate, ripening stage, harvesting time of seeds, and extraction method used.

3 Extracting and processing of watermelon seed oil

Organic solvents are commonly used to extract lipids from oilseeds, but solvent extraction entails some drawbacks such as the risk of the thermal degradation of the unsaturated fatty acids and functional compounds, depending on the used extracting conditions and the need to eliminate the organic solvent's residues from the oil (Bozan & Temelli, 2002). The mechanical cold extraction of oils using an expeller is mainly used for fiber-rich sources with an oil content above 20%, aiming at a preextraction or the preservation of compounds that might be degraded in the solvent extraction process. However, the expeller usually results in a lower extraction yield in comparison to that obtained using solvent extraction (10%–18% of the oil) (Carr, 1989). It is noteworthy that in both cases, plant seed moisture should generally be lowered by sun-drying in order to facilitate the extraction procedures.

In the *Citrullus* genus, the commonly used varieties for extracting oils are *Citrullus lanatus* and *Citrullus vulgaris*, although the wild *Citrullus colocynthis* variety remains underexploited (Sultana & Ashraf, 2019). To the best of our knowledge, the most common way of extracting oil from watermelon seeds is the Soxhlet apparatus. In such an extraction technique, nonpolar solvents like *n*-hexane and chloroform are often more used than petroleum ether (Al-Khalifa, 1996; Mariod et al., 2009; Nehdi et al., 2013). Oil yield was reported to be primarily affected by the solvent/kernel ratio, rather than by time and temperature (Sultana & Ashraf, 2019).

Some researchers have resorted to solvent extraction for obtaining watermelon seed oil. In this case, the most commonly used solvent is *n*-hexane (Górnaś & Rudzińska, 2016), which is reported to be a typical commercial solvent used for oil extraction from several plant materials. Nevertheless, solvent usage is now under close inspection due to growing governmental concerns about consumer safety, especially regarding organic solvents' use in food industries. Consequently, environmental rules and regulations plus an increasing awareness of health risks pushed manufacturers to look for novel alternatives in the use of organic solvents for oil extraction (Mitra, Ramaswamy, & Chang, 2009). In this regard, a number of researches have already been conducted. Their sole purpose was to develop new extracting methods as viable alternatives to organic solvents and to meet the consumers' growing demand for more natural and safer food products.

Cold pressed extraction is among such alternative extraction processes (de Conto et al., 2011; Rezig et al., 2019). It is also worth noting that the cold press procedure involves neither heat nor chemical treatments. It is essentially a mechanical technique in which the extraction is applied using small capacity screw presses (6–40kg/h). For this purpose, the cold pressed seed oil is obtained by pressing the raw, dried, mainly hull-less seeds, on a continuous screw press (Rabrenović, Dimić, Novaković, et al., 2014). According to Dimić (2005), the oil temperature leaving the press during the process of pressing seeds, with the aim of producing cold pressed oils, should not exceed 50°C. This would enable the preservation of bioactive components, such as vitamins, pro-vitamins, phytosterols, phospholipids, and squalene, which represent, along with some fatty acids, highly nutritional seed oils.

4 Fatty acid composition of watermelon seed oil

We duly believe that the scientific literature is scarce in relation to the fatty acid composition of cold pressed watermelon seed oils. Table 1 presents the fatty acid composition of "Crimson" Tunisian watermelon seed oil variety obtained by cold pressing using a Komet DD 85 G vegetable oil screw press (IBG Monforts Oekotec GmbH & Co. KG, Mönchengladbach, Germany). Stearic, palmitic, oleic, and linoleic acids are the main fatty acids present in the oil, making up more than 90% of the triacylglycerol (TAG) portion, with a predominance of linoleic acid (68.07%). These results are very similar to those obtained by Al-Khalifa (1996), de Conto et al. (2011) and Górnaś and Rudzińska (2016) in terms of palmitic and stearic contents, respectively, in Iranian and Chinese *Citrullus lanatus* as well as Brazilian and Bulgarian *Citrullus lanatus* (Thunb.) Matsum. & Nakai seed oils. But above all, the Tunisian "Crimson" *Citrullus lanatus* seed oil variety exhibited lower amounts of oleic and linoleic acids when compared to those found by Al-Khalifa (1996), de Conto et al. (2011), and Górnaś and Rudzińska (2016). In fact, oleic acid accounted for 14.8%, 16.0%, 15.4%, and 17.8%, respectively, in Bulgarian and Brazilian *Citrullus lanatus* (Thunb.) Matsum. & Nakai and in Egyptian and Iranian *Citrullus lanatus* seed oils. Yet in terms of linoleic acid content, the Tunisian "Crimson" *Citrullus lanatus* variety exhibited the highest amount (68.0%) when compared to those reported by de Conto et al. (2011) and Górnaś and Rudzińska (2016) (65.6% vs 66.4%). It is worth noting that even Egyptian and Iranian *Citrullus lanatus* seed oils exhibited lower contents of linoleic acid when compared to those found by Rezig et al. (2019) (62.4% vs 63.7%). For comparative reasons, the *Citrullus colocynthis* seed oil exhibited the highest contents in terms of palmitic and stearic acids (12% vs 11.6%) and the lowest content in terms of oleic acid (10.7%) (Al-Khalifa, 1996).

These differences can probably be explained by the difference in the used watermelon varieties in each experiment. de Melo, Narain, and Bora (2000), Fadavi, Barzegar, and Azizi (2006), and Tangolar, Özogul, Tangolar, et al. (2009) had already confirmed this hypothesis as they stipulated that the fatty acid composition of seed oils greatly depends on both climatic conditions and the cultivar of the fruit itself. According to Jorge, da Silva, and Malacrida (2015), fatty acid profiles

TABLE 1 Fatty acid (%) composition of watermelon (*Citrullus lanatus* var. "Crimson") cold pressed seed oil (Rezig et al., 2019).

Fatty acid	%
Myristic (C14:0)	tr.
Palmitic (C16:0)	9.88
Palmitoleic (C16:1)	tr.
Stearic (C18:0)	6.96
Oleic (C18:1)	14.2
Linoleic (C18:2)	68.0
Linolenic (C18:3)	tr.
Arachidic (C20:0)	0.26
Eicosanoic (C20:1)	tr.
Behenic (C22:0)	tr.
Lignoceric (C24:0)	tr.
SAFAs	17.10
MUFAs	14.25
PUFAs	68.07

MUFAs, monounsaturated fatty acids; *PUFAs*, polyunsaturated fatty acids; *SAFAs*, saturated fatty acids; *tr.*, trace amounts (<0.2%).

are ideal for edible oils since they present a higher percentage of unsaturated fatty acids and a lower one in terms of saturated fatty acids. In consequence, watermelon seed oil could be regarded as appropriate to culinary utilization, to the manufacture of margarine, and especially to the home use as a salad dressing.

5 TAG composition of watermelon seed oil

The determination of the type and amounts of TAG species within watermelon seed oils is fundamental to understand their physical and functional properties. To the best of our knowledge, there is no adequate scientific literature covering the distribution of TAG in watermelon seed oils obtained by a mechanical expeller. Table 2 illustrates the TAG composition of the Tunisian "Crimson" Citrullus lanatus seed oil variety that was reported by Rezig et al. (2019) and extracted using a German vegetable screw press (Komet DD 85 G, IBG Monforts Oekotec GmbH & Co. KG Mönchengladbach). In the light of the corresponding results, the Citrullus lanatus var. "Crimson" seed oil contained 12 TAGs, among which glycerol-palmitate-linoleate-stearate was not detected. The main TAGs were glycerol-linoleate-dilinoleate (LLnLn, 29.1%), followed by glycerol-trilinoleate (LLL, 19.4%), glycerol-dioleate-linolenate (OOLn, 18.0%), and glycerol-oleate-dilinoleate (OLL, 16.3%). de Conto et al. (2011) reported a theoretical triacylglycerol composition in watermelon (Citrullus lanatus (Thunb.) Matsum. & Nakai) seed oil assumingly obtained by a random distribution of a computer program based on the fatty acid composition. The authors found that the predominant TAGs were made up of LLL, OLL, SLL, and PLL, which indicate a low melting point in regards to the watermelon seed oil. For comparative purposes, no data concerning the TAG composition of watermelon seed oil have been found.

6 Physicochemical properties of crude watermelon seed oil

Table 3 exhibits the physicochemical properties of crude watermelon seed oil obtained by a mechanical expeller of the "Crimson" Citrullus lanatus and Citrullus lanatus (Thunb.) Matsum. & Nakai varieties. Cold extraction was established using German Komet vegetable screw presses (Komet model DD 85 G and Komet model CA 59G, respectively). The specific gravitational value of the "Crimson" Citrullus lanatus seed oil did not exceed 0.9. Such a result is in perfect agreement with the values reported by El-Adawy and Taha (2001) and Nehdi et al. (2013) for the Citrullus vulgaris and Citrullus colocynthis (L.) Schrad seed oils. However, the specific gravity value was lower than those reported by Al-Khalifa (1996)

TABLE 2 Triglyceride profile (%) of the cold pressed "Crimson" *Citrullus lanatus* seed oil (%) (Rezig et al., 2019).

TAG molecule	%
LLnLn	29.1
LLLn	0.69
LLL	19.4
OOLn	18.0
PLLn	5.53
OLL	16.3
LnPO	1.64
OOL	1.04
POL	4.44
PPL	1.96
OOO	0.65
SOL	1.10
PLS	nd

L, linoleic acid; *Ln*, linolenic acid; *O*, oleic acid; *P*, palmitic acid; *S*, stearic acid.

TABLE 3 Physicochemical characterization of cold pressed watermelon seed oils.

Parameter	(*Citrullus lanatus* (Thunb.) Matsum. & Nakai) (de Conto et al., 2011)	*"Crimson" Citrullus lanatus* (Rezig et al., 2019)
Refractive index (40°C)	1.47	1.46
Specific gravity (25°C)	–	0.90
Acid value (mg KOH/g oil)	0.19	9.50
Saponification value (mg KOH/g oil)	–	174.2
Iodine value (gI_2/100g)		135.3
Peroxide value (meq O_2/kg oil)	0.27	8.16
Unsaponifiable matter (%)	–	1.04

for the Egyptian, Iranian, and Chinese Citrullus lanatus varieties, ranging between 0.9235 and 0.9283, and those reported by Acar et al. (2012) for the Citrullus lanatus *vercitroides* and Citrullus lanatus (Thumb) *Matsumura & Nakai* var. *Citroides* (Balley) *Mansf.*

The refraction index is mainly related to the saturation degree and the double bond *cis*/*trans* fatty acid ratio. Refractive indexes of 1.46 for the Citrullus lanatus var. "Crimson" and 1.4696 for the Citrullus vulgaris seed oils were obtained (El-Adawy & Taha, 2001; Rezig et al., 2019). These values were lower than those achieved by de Conto et al. (2011) and Nehdi et al. (2013) for the Citrullus lanatus (Thunb.) Matsum. & Nakai (1.4744) and the Citrullus colocynthis (L.) Schrad seed oils (1.4748).

Just like the refraction index, the iodine is a test that indicates the degree of oil unsaturation. The Tunisian cold pressed "Crimson" Citrullus lanatus seed oil variety exhibited an iodine value of 135.3 gI_2/100 g oil (Rezig et al., 2019). This value was higher than those reported by Al-Khalifa (1996) and Mokwala and Shai (2017) for the Egyptian, Iranian, and Chinese Citrullus lanatus and sweet, bitter, and weedy Citrullus lanatus Thumb (Matsum and Nakai) seed oil varieties. According to O'Brien (2009), a high iodine value indicates that watermelon seed oil is rich in double bonds. This indicates that the Tunisian watermelon seed oil is convenient for edible and/or drying oil purposes (Meddeb, Rezig, Abderrabba, et al., 2017).

The saponification value of Citrullus lanatus var. "Crimson" was slightly higher than that reported by Nyam et al. (2009) on Kalahari melon (Citrullus lanatus) seed oil. High saponification values indicate the presence of large contents of high triacylglycerol molecular weight, which are useful for soap production (Meddeb et al., 2017).

The acid and peroxide values of the "Crimson" Citrullus lanatus seed oil were low. These values were even lower than those determined by the Codex Alimentarius (2009), which stipulated a permitted maximum acid value of no more than 10 meq of peroxide oxygen/kg oil for vegetable oils and an average maximum acid value of 10 mg KOH/mg oil. These results are in perfect accordance with those reported by de Conto et al. (2011), Nehdi et al. (2013), Nyam et al. (2009), and El-Adawy and Taha (2001). The lower acidity of the Tunisian watermelon seed oil indicates that it is edible and might have a longer shelf life. In fact, the higher acidity index is related to the presence of free fatty acids (FFAs), which are much more susceptible to oxidation than the fatty acids that are present in the triacylglycerol molecules.

The amount of unsaponifiable matter in the analyzed "Crimson" Citrullus lanatus lipid fraction was about 1.04%. Such an amount is higher than those observed by Al-Khalifa (1996), de Conto et al. (2011), and El-Adawy and Taha (2001) for the Egyptian, Iranian, and Chinese Citrullus lanatus, Citrullus *l*anatus (Thunb.) Matsum. & Nakai, and Citrullus vulgaris seed oils. Considering the fact that the unsaponifiable matter corresponding to the compounds present in oils after saponification with alkalis are insoluble in an aqueous solution, including naturally occurring substances such as sterols, tocopherols, pigments, and hydrocarbons, "Crimson" Citrullus lanatus seed oil must contain high amounts of these substances. According to the Codex Alimentarius Commission (2009), the maximum amount of unsaponifiable matter for soybean and cotton oil is 1.5%, for corn oil 2.8%, and for rice oil 6.5%. Based on these values, Tunisian cold pressed watermelon seed oil is in the range of edible vegetable oils.

7 Minor bioactive lipids in cold pressed watermelon seed oil

Minor lipid components include numerous chemical compounds such as tocopherols, hydrocarbons, sterols, alcohols, phospholipids, and a few others. These minor compounds are of significant interest to food analysts mainly because both their components and ratios can fairly provide a fingerprint for edible oils. In fact, many classes of these compounds can be found in watermelon seed oil and act as anti- or pro-oxidants. They may help considerably in preventing diseases and stimulating the human physiological system (Shahidi & Shukla, 1996).

7.1 Sterols

Phytosterols are of great interest owing to their cholesterol-lowering properties and positive impact on health. The determination of the levels of phytosterols found in vegetable oils has significantly contributed to the distinction and determination of the different qualities of oils and their derivatives (De-Blas & Del-Valle, 1996). Furthermore, the phytosterol concentration has reportedly been attributed to some environmental factors and/or to the cultivation of new breeding lines (Hirsinger, 1989). Among the different plant sterols, β-sitosterol has intensively been investigated with respect to its physiological effects on human health. Many experiments have thus demonstrated the favorable properties of sitosterol (Yang, Karlsson, Oksman, et al., 2001). This phytosterol is now available in the market and has scientifically been proven to lower LDL cholesterol by 10%–15% as part of a healthy diet (Ntanios, 2001).

As far as we can tell, the scientific literature about the phytosterols found in cold pressed watermelon seed oils is very scarce. We can say that de Conto et al. (2011) were pioneering researchers with regard to the separation and quantification of sterols from the Brazilian Citrullus lanatus (Thunb.) Matsum. & Nakai watermelon seed oil. The experiment consisted of compressing the seed material using an expeller (Komet, model CA 59G, Mönchengladbach, Germany) and by chemically using *n*-hexane as a solvent. The researchers found that from among the nine identified phytosterols, stigmasterol was the prevailing one, followed by β-sitosterol. In fact, stigmasterol accounted for 114.6 mg/100 g and 96.95 mg/100 g in the watermelon seed oil extracted by *n*-hexane and cold pressing procedure, respectively. Hence, this phytosterol represents 47% and 30% of the total sterols. For comparative reasons, the watermelon seed oil contains higher amounts of stigmasterol than both soybean (56 mg/100 g) and corn seed oils (66 mg/100 g) (Normén, Johnsson, Andersson, et al., 1999). Stigmasterol is used for the synthesis of the steroid hormones, which are subsequently used for the treatment of certain human deficiencies. However, its presence in soybean oil steroid mixtures makes stigmasterol one of the abundant resources for the synthesis of steroid hormones (Fieser & Fieser, 1959). Furthermore, Górnaś and Rudzińska (2016) reported eight sterols in Citrullus lanatus seed oil extracted by *n*-hexane totaling an amount of 172 mg/100 g. β-sitosterol was the predominant compound (49 mg/100 g), followed by Δ 7-stigmasterol (42 mg/100 g), campesterol (33 mg/100 g), and stigmasterol (22 mg/100 g). Rezig et al. (2019) reported in the Tunisian "Crimson" Citrullus lanatus seed oil, mechanically extracted by a vegetable screw press, a total sterol content of 4208 mg/kg (Table 4). Such content is obviously higher than that reported by Górnaś and Rudzińska (2016). Among the 15 identified sterols, β-sitosterol was the main element followed by the Δ5-avenasterol. Thus, β-sitosterol and Δ5-avenasterol were the major two sterols and together made up 85.9%. On the grounds of the higher content of sitosterol reported in the "Crimson" watermelon seed oil variety, it is now fair to say that such oil can be applied as a new therapeutic agent for the treatment of hypercholesterolemia.

7.2 Tocopherols and tocotrienols (tocochromanols)

Tocochromanols, a group of fat-soluble molecules of a unique biological activity, include four different homologues (α, β, γ, and δ) of tocopherol (T) and tocotrienol (T3), and belong to the group of vitamin E. The highest biological activity of vitamin E among all tocochromanol homologues is reported for α-T (100%) (Eitenmiller & Lee, 2004). α-T plays a key role in promoting the physiological processes of many organisms. α-T can also enhance membrane stability and prevent unsaturated fatty acids oxidation (Dwiecki, Górnas, Jackowiak, et al., 2007; Dwiecki, Górnas, Wilk, et al., 2007). Table 5 presents the tocopherol and tocotrienol composition of Citrullus lanatus seed oils available in the literature and obtained by both solvent and compressing extraction techniques.

As reported by Górnaś et al. (2014), in the *Citrullus lanatus* (Thunb.) Matsum. & Nakai seed oil, the γ-T dominated, at a concentration 25 times higher than that recorded for α-T. Low amounts of β-T, δ-T, α-T3, and γ-T3 were detected in the watermelon seeds. Similarly, in the seeds of four different watermelon varieties grown in Pakistan, only a high level of α-T and a low amount of δ-T were detected (Raziq et al., 2012). However, a similar tocochromanol composition, as found by Górnaś et al. (2014), was recorded in the Kalahari melon seeds from the northern part of Namibia and in the *Citrullus lanatus* (Thunb.) Matsum. & Nakai originating from Brazil (de Conto et al., 2011; Nyam et al., 2009). The high content of γ-tocopherol is in close agreement with that of Mariod et al. (2009), Jorge et al. (2015) and Angelova-Romova, Simeonova, Petkova, et al. (2019). They all asserted that γ-tocopherol predominated. According to Rossel (1991), the original concentration of tocopherols may vary from a few mg/kg or even more depending on the oil type and the oil's fatty acid composition. It should be noted that very limited data are available about the tocopherol contents of the cold

TABLE 4 Sterol composition (mg/kg oil) of "Crimson" *Citrullus lanatus* seed oil.

Sterol	mg/kg oil
Cholesterol	5.05
Brassicasterol	38.7
24-Methylenecholesterol	0.84
Campesterol	12.6
Campestanol	2.10
Stigmasterol	77.4
Δ5-Avenasterol	1319.2
Δ5,23-Stigmastadienol	4.21
Δ5,24-Stigmastadienol	158.2
Δ7-Campesterol	24.8
Δ7-Stigmastenol	21.8
Δ7-Avenasterol	83.3
Clerosterol	21.4
Sitosterol	2298.8
Sitostanol	139.2
Total	4208

TABLE 5 Tocopherol and tocotrienol composition of *Citrullus lanatus* seed oils (mg/kg).

Sample	Origin	Extraction procedure	α-T	β-T	γ-T	δ-T	α-T3	β-T3	γ-T3	δ-T3	References
Citrullus lanatus (Thunb.) Matsum. & Nakai	Brazil	*n*-Hexane	1.68	ND	62.8	0.69	ND	ND	ND	ND	de Conto et al. (2011)
		Expeller	1.43	ND	71.0	0.69	ND	ND	ND	ND	
Citrullus lanatus (Thunb.) Matsum. & Nakai	Latvia	*n*-Hexane-ethyl acetate (9:1; v/v)	12.9	0.6	313	7.2	0.4	ND	0.5	ND	Górnaś, Pugajeva, and Seglina (2014)
Kalahari melon (*Citrullus lanatus*)	Namibia	Soxhlet extractor (petroleum ether)	259.4	32.7	705.6	93.3	ND	ND	ND	ND	Nyam et al. (2009)
Citrullus lanatus var. Sugar Baby	Pakistan	Soxhlet extractor (*n*-hexane)	195.6	ND	ND	12.3	ND	ND	ND	ND	Raziq, Anwar, Mahmoud, et al. (2012)
Citrullus lanatus var. Q-F-12			164.3	ND	ND	58.3	ND	ND	ND	ND	
Citrullus lanatus var. D-W-H-21			122.0	ND	ND	9.1	ND	ND	ND	ND	
Citrullus lanatus var. Red Circle—1885			120.6	ND	ND	20	ND	ND	ND	ND	
Citrullus colocynthoides	Sudan	Soxhlet extractor (petroleum ether)	ND	ND	359	10	ND	ND	ND	ND	Mariod et al. (2009)

ND, not determined; *T*, tocopherol; *T3*, tocotrienol.

pressed watermelon seeds. As far as we know, de Conto et al. (2011) were the first researchers who compared the tocopherol contents in watermelon seeds chemically extracted by solvent and mechanically by an expeller. The authors also reported that the extraction procedure did not significantly affect the preservation of the functional compounds such as tocopherols.

7.3 Phenolic compounds

In recent years, there has been a growing interest in studying phenolic compounds from oilseeds, their skins, hulls, and oil cake meals. This interest is ascribed to the fact that these compounds are potentially health-stimulating substances and have many industrial applications (Peschel, Sanchez-Rabaneda, Diekmann, et al., 2006; Wang, Yaun, Jin, et al., 2007). Still, there is very scarce literature about the presence of phenolic compounds in the watermelon *Citrullus lanatus* seed oils, especially those mechanically extracted using a vegetable screw press. Rezig et al. (2019) identified in the "Crimson" watermelon seed oil variety extracted by a screw press. Two classes of phenolic compounds including phenolic acids and lignans were identified by HPLC, which are represented by caffeic acid (1.33 μg/g) and pinoresinol (1.02 μg/g). Caffeic acid was reported to possess several biological activities (Martínez-Cruz & Paredes-López, 2014). Its content was lower than that reported by Nyam et al. (2009) in Kalahari melon (*Citrullus lanatus*) seed oil (0.41 mg/100 g). Apart from caffeic acid, vanillic acid was also determined (0.55 mg/100 g), followed in descending order by gallic acid (0.23 mg/100 g), *p*-hydroxybenzoic acid (0.21 mg/100 g), *p*-coumaric acid (0.18 mg/100 g), ferulic acid (0.17 mg/100 g), and protocatechuic acid (0.05 mg/100 g). Syringic acid was also detected in trace amounts. The pinoresinol, however, detected in "Crimson" seed oil variety was present in "Ananas" melon seed and "Essahli" pumpkin seed oils, totaling amounts of 10.8 mg/100 g and 7.92 mg/100 g, respectively (Rezig et al., 2019). According to Chtourou, Gargouri, Jaber, et al. (2013), the contents present in Cucurbitaceae seed oils were higher than those reported in virgin olive oil (Chemlali cultivar) with an amount totaling 1.7 μg/g. Furthermore, the oleuropein, a potently high antioxidant phenolic compound present in olive oil and leaves (Rodríguez-Morató, Xicota, Fitó, et al., 2015), was found in the Tunisian "Ananas" (2.31 μg/g) and "Maazoun" (1.65 μg/g) melon seed oil varieties (Mallek-Ayadi, Bahloul, & Kechaou, 2018; Rezig et al., 2019). These discrepancies probably arise from the sample differences (seeds vs whole plants) as well as the different extracting methods.

According to Nyam et al. (2009), the objectionable flavor of some seed oils is due to the presence of phenolics that possess sour, bitter, astringent, and/or phenolic-like flavor characteristics. It is known that phenolic acids possess high antioxidant, antimicrobial, antiproliferative, and preservative properties. They account for 30% of total dietary plants and play a key role in enhancing plant resistance toward diseases, in maintaining human health, and in preventing food deterioration (Balasundram, Sundram, & Samman, 2006; Peschel et al., 2006).

7.4 Total phenolics and flavonoids contents

Phenolics are bioactive secondary plant metabolites that are widely present in plant-based foods. There are three main types of phenolics, namely flavonoids, phenolic acids, and tannins, and these act as powerful antioxidants in vitro. These compounds are thought to have many potential health benefits like preventing diabetes, osteoporosis, and a number of cardiovascular, carcinogenic, and neurodegenerative diseases. When present in food, polyphenols may contribute to certain features like bitterness, astringency, color, flavor, and oxidative stability (Han, Shen, & Lou, 2007; Scalbert, Manach, Morand, et al., 2005). There is very scarce scientific literature about total phenolic and flavonoid contents in Cucurbitaceae seed oils. Hashemi et al. (2017) reported a total phenolic content (TPC) of 111 mg gallic acid equivalent (GAE) per kg of oil in the Iranian watermelon *Citrullus lanatus* cv. Fire Fon seed oil extracted by *n*-hexane. This content was lower than that reported in the *Citrullus lanatus* seed oil (1428.9 mg GAE/kg), in the *Cucumis melo* "Maazoun" variety (22.6 mg GAE/100 g), and in the yellow melon (*Cucumis melo* var. inodorus Naudin) (130.7 mg GAE/kg) (da Silva & Jorge, 2014; Jorge et al., 2015; Mallek-Ayadi, Bahloul, & Kechaou, 2019). The findings of Hashemi et al. (2017) were in accordance with the findings of Parry, Su, Luther, Zhou, et al. (2005) and Parry, Hao, Luther, et al. (2006) for other vegetable oils. Differences in the TPC in the literature findings may be attributed to differences in the varieties, growing conditions, and the extracting solvents (Rahman, Manjulak, Anoosha, et al., 2013).

Flavonoids are, on the other hand, the most common and widely distributed group of plant phenolic compounds. They play a fundamental role in plant growth and defense against infection and injury. These plant secondary metabolites have been shown to have a wide range of antiallergic, antiinflammatory, antimicrobial, and anticancer activities thanks to their significant antioxidant and chelating properties (Heim, Tagliaferro, & Bobliya, 2002; Khatiwora, Adsul, Kulkarni, et al., 2010). Morais, Rotta, Sargi, et al. (2015) reported a total flavonoid content of 24.7 mg quercetin equivalent (QE) and 3.61 mg QE per 100 g, respectively in watermelon (*Citrullus lanatnus*) and melon (*Cucumis melo*) dry weight seeds.

The results are obviously lower than those reported by Mallek-Ayadi et al. (2018) in melon (*Cucumis melo* var. "Maazoun") seeds (87.52 mg QE/100 g), which belong to the same botanical family "Cucurbitaceae." Another study revealed that the total flavonoid content totaled an amount of 3.06 μg catechin per mg in watermelon seeds (Mehra, Pasricha, & Gupta, 2015).

7.5 Carotenoids

Carotenoids play a key role in the oxidative stability of edible oils thanks to their antioxidant activity (Criado, Romero, Casanovas, et al., 2008). From a nutritional viewpoint and apart from their antioxidant properties, these pigments are reported to be rich in vitamin A (Mallek-Ayadi et al., 2019). de Conto et al. (2011) were the first researchers to report the carotenoid content in watermelon (*Citrullus lanatus* (Thunb.) Matsum. & Nakai) seed oil extracted by a mechanical expeller. They found that the watermelon seed oil exhibited a carotenoid content of 39.1 mg/kg. Such content was almost similar to that found in palm oil (38.5 μg/g) (Szydlowska-Czerniak, Trokowski, Karlovits, et al., 2011), but higher than that found in the watermelon seed oil (30.5 mg/kg) extracted by *n*-hexane. The findings of de Conto et al. (2011) are not in agreement with those of da Silva and Jorge (2014), as the latter found that the total carotenoid content, expressed as μg β-carotene per g of oil, amounted to 7.5 μg/g and 6.3 μg/g, respectively, in both pumpkin (*Cucurbita moschata*) and yellow melon (*Cucumis melo* var. inodorus Naudin) seed oils. High carotenoid concentrations in oils have great nutritional and technological features thanks to their high potential antioxidant activity.

8 Oxidative stability

Oxidative stability is known as the determination of the oil's oxidation induction time and hence its resistance to oxidation phenomena. The oxidative stability index is undoubtedly a significant criterion to evaluate edible oils' quality. To the best of our knowledge, Rezig et al. (2019) were the first researchers to investigate the stability of cold pressed watermelon seed oil by means of a mechanical expeller. The Tunisian watermelon "Crimson" variety exhibited an induction time of 1.72 h using the Rancimat at 100°C with an airflow of 10 L/h. Statistical differences were recorded for "Essahli" *Cucurbita pepo* and "Ananas" *Cucumis melo* in the induction time when compared to that found in *Citrullus lanatus* seed oil, both of which belong to the same botanical family Cucurbitaceae. In fact, Rezig et al. (2019) showed that within the same experimental conditions, the oxidative stability was 3.74 and 3.8 h in melon (*Cucumis melo*) and pumpkin (*Cucurbita pepo*) seed oils, respectively. These values were lower than those reported for Mashahadi melon (*Cucumis melo* var. Iranians cv. Mashhadi), Iranian watermelon (*Citrullus lanatus* cv. Fire Fon), and pumpkin (*Cucurbita pepo* subsp. *pepo* var. Stryriaca) seed oils exhibiting an oxidative stability between 6 and 9 h using the Rancimat at 120°C with an airflow of 15 L/h (Hashemi et al., 2017). Da Silva and Jorge (2014) observed a negative correlation (0.72) between the oxidative stability and the percentage of unsaturated fatty acids in yellow melon (*Cucumis melo* var. inodorus Naudin), and pumpkin (*Cucurbita moschata*) seed oils cold extracted with chloroform, methanol, and water 2:1:0.8 (*v*/*v*/v), respectively. Such a finding reveals that the higher the number is of unsaturation in the fatty acids (85.6% vs 75.3%), the lower the oil induction periods are (29.5 vs 65.3 h).

9 Antioxidant properties

The antioxidant capacity of the seed oils could be measured by DPPH· free radical scavenging assay. This technique is a powerful tool to investigate the antioxidant potential of numerous molecules (Koubaa, Barba, Mhemdi, et al., 2015). As far as we know, Rezig et al. (2019) were the first researchers to report the antioxidant properties of Cucurbitaceae seed oils extracted using a vegetable screw press through the DPPH· assay. Such a technique allows calculating the IC_{50} corresponding to the concentration of the sample required to scavenge 50% of DPPH· radicals. Hence, low IC_{50} is equivalent to high scavenging capacity. According to Guergouri, Sobhi, and Benboubetra (2017), IC_{50} is calculated by plotting percentage inhibition against different concentrations of oil. Citrullus lanatus var. "Crimson" exhibited an IC_{50} higher than those observed in melon and pumpkin seed oils extracted in the same experimental conditions (159.6 μg/g) (Rezig et al., 2019). Such results are in close agreement with those of Hashemi et al. (2017). They found that the antioxidant capacity of Iranian watermelon seed oil (~80%) was the lowest when compared to those of pumpkin and Mashhadi melon seed oils' antioxidant capacities. However, Jorge et al. (2015) reported that DPPH$^{\bullet}$ percentages, remaining after 30 min of reaction between the oils and DPPH, were 36.4% and 30.6% in melon and watermelon seed oils, respectively. This indicates that these oils possess free radical scavenging activity, and they contain compounds that react directly with DPPH·. Watermelon seed oil showed an antioxidant capacity higher than the melon seed oil. Such a result was consistent with the higher content of tocopherols and phenolic compounds in Citrullus lanatus seed oil.

10 CIE L*, a*, and b* coordinates

CieLab coordinates values (L*, a*, b*) of cold pressed Citrullus lanatus seed oils are scarce in the literature. Rezig et al. (2019) reported CieLab coordinates values (L*, a*, b*) of 90.9, 0.12, and 0.56 in the "Crimson" Citrullus lanatus seed oil obtained by a vegetable screw press. The higher L* value proves that Citrullus lanatus seed oil is light in color. According to Hsu and Yu (2002), the CieLab (L*, a*, b*) values of other vegetable oils such as palm, soybean, sunflower, olive, and corn ranged from 63.4 to 69.5, 3.8 to 4.4, and 9.2 to 10.4, respectively. Therefore, the watermelon (Citrullus lanatus) seed oil a* and b* values were lower than those of other vegetable oils. Such a result suggests that the watermelon seed oil was less yellow than the vegetable oils studied by Hsu and Yu (2002) and contains low amounts of yellow pigments such as carotenoids.

11 Conclusion

Keeping in view the results reported in the present chapter, it is reasonable to say that cold pressed Tunisian watermelon seed oil of the "Crimson" variety extracted by using a mechanical expeller is a potential source of nutrients, especially essential fatty acids (oleic and linoleic acids) and antioxidant compounds (total phenolics, tocopherols, total flavonoids, and phytosterols). Hence extracting watermelon seed oil mechanically offers a novel alternative in avoiding chemical solvents and thus ensuring consumer safety. According to most studies, the values of the physicochemical characteristics of cold pressed watermelon seed oil is within the recommended limits. This suggests that watermelon seed oil might be used as a seasoning or frying oil, mayonnaise and table margarine, and in commercial natural antioxidant formulations. After investigating all its nutritional and physiological benefits, watermelon seed oil ought to be commercially exploited and used as nutraceuticals, functional food commodities, and potential antidotes for fighting a number of diseases. Further research is still needed in order to explore the presence of other bioactive phytoconstituents responsible for the positive health benefits of cold pressed watermelon seed oil. Moreover, since this oil is susceptible to be valorized on an industrial scale, supplementary studies are needed to explore its resistance toward thermo-oxidation. Similarly, a thorough assessment of the evolution of the physicochemical parameters of such edible oil is also needed in future studies.

References

Acar, R., Özcan, M. M., Kanbur, G., & Dursun, N. (2012). Some physic-chemical properties of edible and forage watermelon seeds. *Iranian journal of chemistry and chemical engineering*, *31*(4), 41–47.

Ahmed, J. (1996). Studies on watermelon products. *Indian Food Packer*, *50*, 15–20.

Al-Khalifa, A. S. (1996). Physicochemical characteristics, fatty acid composition, and lipoxygenase activity of crude pumpkin and melon seed oils. *Journal of Agricultural and Food Chemistry*, *44*, 964–966.

Angelova-Romova, M. Y., Simeonova, Z. B., Petkova, Z., et al. (2019). Lipid composition of watermelon seed oil. *Bulgarian Chemical Communications*, *51*, 268–272.

Balasundram, N., Sundram, K., & Samman, S. (2006). Phenolic compounds in plants and agri-industrial byproducts: Antioxidant activity, occurrence, and potential uses. *Food Chemistry*, *99*, 191–203.

Bozan, B., & Temelli, F. (2002). Supercritical CO_2 extraction of flaxseed. *Journal of the American Oil Chemists' Society*, *79*, 231–235.

Carr, R. A. (1989). Processing of oilseed crops. In G. Robbelon, R. K. Downey, & A. Ashiri (Eds.), *Oil crops of the world* (pp. 226–259). New York: McGraw-Hill.

Chtourou, M., Gargouri, B., Jaber, H., et al. (2013). Comparative study of olive oil quality from *Chemlali* Sfax versus *Arbequina* cultivated in Tunisia. *European Journal of Lipid Science and Technology*, *115*, 631–640.

Codex Alimentarius Commission (2009). *Codex-Stan 210: codex standard for named vegetable oils, Rome.*

Criado, M. N., Romero, M. P., Casanovas, M., et al. (2008). Pigment profile and colour of monovarietal virgin olive oils from *Arbequina* cultivar obtained during two consecutive crop seasons. *Food Chemistry*, *110*(4), 873–880.

da Silva, A. C., & Jorge, N. (2014). Bioactive compounds of the lipid fractions of agro-industrial waste. *Food Research International*, *66*, 493–500.

de Conto, L. C., Gragnani, M. A. L., Maus, D., et al. (2011). Characterization of crude watermelon seed oil by two different extractions methods. *Journal of the American Oil Chemists' Society*, *88*, 1709–1714.

de Melo, M. L. S., Narain, N., & Bora, P. S. (2000). Characterisation of some nutritional constituents of melon (*Cucumismelo* hybrid AF-522) seeds. *Food Chemistry*, *68*, 411–414.

De-Blas, O. J., & Del-Valle, G. A. (1996). Determination of sterols by capillary column gas chromatography. Differentiation among different types of olive oil: Virgin, refined and solvent extracted. *Journal of the American Oil Chemists' Society*, *73*, 1685–1689.

Dimić, E. (2005). Cold pressed oils. In *Monography* (pp. 170–172). Novi Sad: The University of Novi Sad, Faculty of Technology.

Dwiecki, K., Górnas, P., Jackowiak, H., et al. (2007). The effect of D-alpha-tocopherol on the solubilisation of dipalmitoyl phosphatidylcholine membrane by anionic detergent sodium dodecyl sulphate. *Journal of Food Lipids*, *14*(1), 50–61.

Dwiecki, K., Górnas, P., Wilk, A., et al. (2007). Spectroscopic studies of a D-α-tocopherol concentration-induced transformation in egg phosphatidylcholine vesicles. *Cellular & Molecular Biology Letters*, *12*, 51–69.

Eitenmiller, R., & Lee, J. (2004). *Vitamin E: Food chemistry, composition, and analysis.* New York: Marcel Dekker Inc.

El-Adawy, T. A., & Taha, K. M. (2001). Characteristics and composition of different seed oils and flours. *Food Chemistry*, *74*, 47–54.

Erhirhie, E., & Ekene, N. (2013). Medicinal values on *Citrullus lanatus* (watermelon): Pharmacological review. *International Journal of Research in Pharmaceutical and Biomedical Sciences*, *4*, 1305–1312.

Fadavi, A., Barzegar, M., & Azizi, M. H. (2006). Determination of fatty acids and total lipid content in oilseed of 25 pomegranates varieties grown in Iran. *Journal of Food Composition and Analysis*, *19*, 676–680.

Fieser, L. F., & Fieser, M. (1959). *Steroids* (pp. 341–363). New York: Reinhold Publishing Corporation.

Godawa, I. N. D., & Jalali, M. (1995). Studies on juice making from watermelon fruits. *Indian Food Packer*, *49*(3), 33–41.

Górnaś, P., Pugajeva, I., & Seglina, D. (2014). Seeds recovered from by-products of selected fruit processing as a rich source of tocochromanols: RP-HPLC/FLD and RP-UPLC-ESI/MSN study. *European Food Research and Technology*, *239*, 519–524.

Górnaś, P., & Rudzińska, M. (2016). Seeds recovered from industry by-products of nine fruit species with a high potential utility as a source of unconventional oil for biodiesel and cosmetic and pharmaceutical sectors. *Industrial Crops and Products*, *83*, 329–338.

Guergouri, F. Z., Sobhi, W., & Benboubetra, M. (2017). Antioxidant activity of Algerian *Nigella sativa* total oil and its unsaponifiable fraction. *The Journal of Phytopharmacology*, *6*(4), 234–238.

Han, X., Shen, T., & Lou, H. (2007). Dietary polyphenols and their biological significance. *International Journal of Molecular Sciences*, *8*, 950–988.

Hasan, F. (1993). Research on the possibility of using watermelon juice in the fruit juice cocktails. *Gida*, *18*(6), 369–371.

Hashemi, S. M. B., Khaneghag, A. M., Koubaa, M., et al. (2017). Novel edible oil sources: Microwave heating and chemical properties. *Food Research International*, *92*, 147–153.

Heim, K., Tagliaferro, A., & Bobliya, D. (2002). Flavonoid antioxidants: Chemistry, metabolism and structure-activity relationship. *The Journal of Nutritional Biochemistry*, *13*(10), 572–584.

Hirsinger, F. (1989). New annual oil crops. In G. Roebbelen, R. K. Downey, & A. Ashri (Eds.), *Oil crops of the world* (pp. 518–532). New York: McGraw Hill.

Hour, S. S., Ahmed, E. M., & Carter, R. D. (1980). Concentration of watermelon juice. *Journal of Food Science*, *45*, 718–719.

Hsu, S. Y., & Yu, S. H. (2002). Comparisons on 11 plant oil fat substitutes for low-fat kung-wans. *Journal of Food Engineering*, *51*, 215–220.

Jorge, N., da Silva, A. C., & Malacrida, C. R. (2015). Physicochemical characterisation and radical-scavenging activity of Cucurbitaceae seed oils. *Natural Product Research*. https://doi.org/10.1080/14786419.14782015.11007135.

Khatiwora, E., Adsul, V., Kulkarni, M., et al. (2010). Spectroscopic determination of total phenol and flavonoid contents of *Ipomoera Carnea*. *International Journal of ChemTech Research*, *2*, 1698–1701.

Koubaa, M., Barba, F. J., Mhemdi, H., et al. (2015). Gas assisted mechanical expression (GAME) as a promising technology for oil and phenolic compound recovery from tiger nuts. *Innovative Food Science & Emerging Technologies*, *32*, 172–180.

Lazos, E. (1986). Nutritional, fatty acid and oil characteristics of pumpkin and melon seeds. *Journal of Food Science*, *51*, 1381–1383.

Mabaleha, M. B., Mitei, Y. C., & Yeboah, S. O. (2007). A comparative study of the properties of selected melon seed oils as potential candidates for development into commercial edible vegetable oil. *Journal of the American Oil Chemists' Society*, *84*, 31–36.

Mallek-Ayadi, S., Bahloul, N., & Kechaou, N. (2018). Chemical composition and bioactive compounds of *Cucumis melo* L. seeds: Potential source for new trends of plant oils. *Process Safety and Environment Protection*, *113*, 68–77.

Mallek-Ayadi, S., Bahloul, N., & Kechaou, N. (2019). *Cucumis melo* L. seeds as a promising source of oil naturally rich in biologically active substances: Compositional characteristics, phenolic compounds, and thermal properties. *Grasas y Aceites*, *70*(1), 1–10.

Mariod, A. A., Ahmed, Y. M., Matthäus, B., et al. (2009). A comparative study of the properties of six Sudanese cucurbit seeds and seed oils. *Journal of the American Oil Chemists' Society*, *86*, 1181–1188.

Martínez-Cruz, O., & Paredes-López, O. (2014). Phytochemical profile and nutraceutical potential of chia seeds (*Salvia hispanica* L.) by ultra high performance liquid chromatography. *Journal of Chromatography A*, *1346*, 43–48.

Meddeb, W., Rezig, L., Abderrabba, M., et al. (2017). Tunisian milk thistle: An investigation of the chemical composition and the characterization of its cold-pressed seed oils. *International Journal of Molecular Sciences*, *18*, 2582–2595.

Mehra, M., Pasricha, V., & Gupta, R. K. (2015). Estimation of nutritional, phytochemical and antioxidant activity of seeds of musk melon (*Cucumis melo*) and watermelon (*Citrullus lanatus*) and nutritional analysis of their respective oils. *Journal of Pharmacognosy and Phytochemistry*, *3*(6), 98–102.

Mitra, P., Ramaswamy, H. S., & Chang, K. S. (2009). Pumpkin (*Cucurbita maxima*) seed oil extraction using supercritical carbon dioxide and physicochemical properties of the oil. *Journal of Food Engineering*, *95*, 208–213.

Mokwala, P. W., & Shai, T. (2017). Physicochemical analysis of seed oil from indigenous watermelons. *South African Journal of Botany*. https://doi.org/10.1016/j.sajb.2017.01.120.

Morais, D. R., Rotta, E. M., Sargi, S. C., et al. (2015). Antioxidant activity, phenolics and UPLC- ESI (-) - MS of extracts from different tropical fruits parts and processed peels. *Food Research International*. https://doi.org/10.1016/j.foodres.2015.08.036.

Morais, D. R., Rotta, E. M., Sargi, S. C., et al. (2017). Proximate composition, mineral contents and fatty acid composition of the different parts and dried peels of tropical fruits cultivated in Brazil. *Journal of the Brazilian Chemical Society*, *28*(2), 308–318.

Nehdi, I. A., Sbihi, H., Tan, C. P., et al. (2013). Evaluation and characterisation of *Citrullus colocynthis* (L.) Schrad seed oil: Comparison with *Helianthus annuus* (sunflower) seed oil. *Food Chemistry*, *136*, 348–353.

Normén, L., Johnsson, M., Andersson, H., et al. (1999). Plant sterols in vegetables and fruits commonly consumed in Sweden. *European Journal of Nutrition*, *38*, 84–89.

Ntanios, F. (2001). Plant sterol-ester-enriched spreads as an example of a new functional food. *European Journal of Lipid Science and Technology*, *103*, 102–106.

Nyam, K. L., Tan, C. P., Lai, O. M., et al. (2009). Physicochemical properties and bioactive compounds of selected seed oils. *Food Science and Technology*, *42*, 1396–1403.

O'Brien, R. D. (2009). *Fats and oils: Formulations and processing for applications* (2nd ed.). Boca Raton, FL: CRC Press.

Parry, J., Hao, Z., Luther, M., et al. (2006). Characterization of cold-pressed onion, parsley, cardamom, mullein, roasted pumpkin, and milk thistle seed oils. *Journal of the American Oil Chemists' Society*, *83*, 847–854.

Parry, J., Su, L., Luther, M., Zhou, K., et al. (2005). Fatty acid composition and antioxidant properties of cold-pressed marionberry, boysenberry, red raspberry, and blueberry seed oils. *Journal of Agricultural and Food Chemistry*, *53*(566–573), 2005.

Peschel, W., Sanchez-Rabaneda, F., Diekmann, W., et al. (2006). An industrial approach in the search of natural antioxidants from vegetable and fruit wastes. *Food Chemistry*, *97*, 137–150.

Rabrenović, B. B., Dimić, E. B., Novaković, M. M., et al. (2014). The most important bioactive components of cold pressed oil from different pumpkin (Cucurbita pepo L.) seeds. *Food Science and Technology*, *55*(2), 521–527.

Rahman, H., Manjulak, K., Anoosha, T., et al. (2013). In vitro antioxidant activity of *Citrullus lanatus* seed extracts. *Asian Journal of Pharmaceutical and Clinical Research*, *6*(3), 152–157.

Raziq, B. S., Anwar, F., Mahmoud, Z., et al. (2012). Characterization of seed oils from different varieties of watermelon [*Citrullus lanatus* (Thunb)] from Pakistan. *Grasas y Aceites*, *63*(4), 365–372.

Rezig, L., Chouaibi, M., Meddeb, W., et al. (2019). Chemical composition and bioactive compounds of Cucurbitaceae seeds: Potential sources for new trends of plant oils. *Process Safety and Environment Protection.* https://doi.org/10.1016/j.psep.2019.05.005.

Rodríguez-Morató, J., Xicota, L., Fitó, M., et al. (2015). Potential role of olive oil phenolic compounds in the prevention of neurodegenerative diseases. *Molecules*, *20*, 4655–4680.

Rossel, J. B. (1991). Vegetable oil and fats. In J. B. Rossel, & J. L. R. Pritchard (Eds.), *Analysis of oilseeds, fats and fatty foods* (pp. 261–328). New York: Elsevier Applied Science.

Scalbert, A., Manach, C., Morand, C., et al. (2005). Dietary polyphenols and the prevention of diseases. *Critical Reviews in Food Science and Nutrition*, *45*, 287–306.

Shahidi, F., & Shukla, V. K. S. (1996). Non triacylglycerol constituents of fats, oils. *Inform*, *7*, 1227–1232.

Sultana, B., & Ashraf, R. (2019). Watermelon (*Citrullus lanatus*) oil. In M. Ramadan (Ed.), *Fruit oils: Chemistry and functionality* (pp. 741–756). Cham: Springer. https://doi.org/10.1007/978-3-030-12473-1_39.

Szydlowska-Czerniak, A., Trokowski, K., Karlovits, G., et al. (2011). Effect of refining process on antioxidant capacity, total content of phenolics and carotenoids in palm oils. *Food Chemistry*, *129*, 1187–1192.

Tangolar, S. G., Özogul, Y., Tangolar, S., et al. (2009). Evaluation of fatty acid profiles and mineral content of grape seed oil of some grape genotypes. *International Journal of Food Sciences and Nutrition*, *60*, 32–39.

Wang, J., Yaun, X., Jin, Z., et al. (2007). Free radical and reactive oxygen species scavenging activities of peanut skins extracts. *Food Chemistry*, *104*, 242–250.

Wani, A. A., Sogi, D. S., Singh, P., et al. (2011a). Characterization and functional properties of watermelon (*Citrullus lanatus)* seed protein isolates and salt assisted protein concentrates. *Food Science and Biotechnology*, *20*(4), 877–887.

Wani, A. A., Sogi, D. S., Singh, P., et al. (2011b). Characterization and functional properties of watermelon (*Citrullus lanatus*) seed proteins. *Journal of the Science of Food and Agriculture*, *91*, 113–121.

Yang, B., Karlsson, R. M., Oksman, P. H., et al. (2001). Phytosterols in sea buckthorn (*Hippophaë rhamnoides* L.) berries: Identification and effects of different origins and harvesting times. *Journal of Agricultural and Food Chemistry*, *49*, 5620–5629.

Ziyada, A. K., & Elhussien, S. A. (2008). Physical and chemical characteristics of *Citrullus lanatus* var. colocynthoide seed oil. *Journal of Physical Science*, *19*, 69–75.

Chapter 56

Cold pressed oils from genus *Prunus*

Maja Natić[a], Dragana Dabić Zagorac[b], Ivanka Ćirić[b], Mekjell Meland[c], Biljana Rabrenović[d], and Milica Fotirić Akšić[d]

[a]*University of Belgrade—Faculty of Chemistry, Belgrade, Serbia,* [b]*Innovation Centre of Faculty of Chemistry Ltd, Belgrade, Serbia,* [c]*Norwegian Institute of Bioeconomy Research, Ås, Norway,* [d]*University of Belgrade—Faculty of Agriculture, Belgrade-Zemun, Serbia*

Abbreviations

E	α-eleostearic
FA	fatty acids
HCN	hydrocyanic acid
L	linoleic
O	oleic
OOO	triolein
P	palmitic
PC	phosphatidylcholine
PE	phosphatidylethanolamine
PI	phosphatidylinositol
PUFA	polyunsaturated fatty acid
S	stearic
SFA	saturated fatty acid
SFE	supercritical fluid extraction
TAG	triacylglycerol
UAE	ultrasound-assisted extraction
UFA	unsaturated fatty acid

1 Introduction

The genus *Prunus* originates from the Asian continent and belongs to the Rosaceae family, Amygdaloideae subfamily (Potter et al., 2007). It includes about 430 species of deciduous or evergreen trees and shrubs naturally widespread throughout temperate regions. The genus *Prunus* is divided into six subgenera: (1) *Amygdalus* (almonds and peaches); (2) *Prunus* (plums and apricots); (3) *Cerasus* (cherries); (4) *Lithocerasus* (one of the representatives is *Prunus pumila*—sand cherry); (5) *Padus* (*Prunus padus*—bird cherry); and (6) *Laurocerasus* (*Prunus laurocerasus*—cherry laurel).

Prunus species cultivation is common in temperate regions throughout the world (encompassing much of Europe, North Africa, the Near and Far East, South Australia and New Zealand, and temperate zones of the American continent) (Mariette et al., 2010). Among stone fruits, the highest production is of peaches/nectarines with ~25 million tons, followed by plums with ~11.5 million tons. Apricot production is in third place with 4.2 million tons, then come almonds (2.4 million tons), sweet cherries (2.2 million tons), and sour cherries (1.1 million tons). By far the highest peach/nectarine production is organized in China (14 million tons or 56% of world production), followed by Italy and Spain. Most plum production is in China (59% world production), Romania, and Serbia. Almond production is mostly done in the United States, Spain, and Morocco. Apricot has the highest annual production in Turkey, Uzbekistan, and Italy, sweet cherries in Turkey, the United States, and Iran, and sour cherry in Russia, Ukraine, and the United States (FAOStat, 2017).

The majority of species within this genus are cultivated for their fruit, but some are used for decoration or timber production. The fruit is a fleshy drupe ("prune"), but unlike almonds, where the seed is consumed, the edible part in most stone fruits includes the mesocarp and/or the exocarp. All *Prunus* species are very popular temperate fruits, being highly

Cold Pressed Oils. https://doi.org/10.1016/B978-0-12-818188-1.00056-6
Copyright © 2020 Elsevier Inc. All rights reserved.

appreciated by consumers and studied due to their taste, color, and sweetness, but also for their nutritional and bioactive properties that are beneficial to human health (Alrgei et al., 2016; Colaric, Veberic, Stampar, & Hudina, 2005; Čolić et al., 2017; Kim, Chun, Kim, Moon, & Lee, 2003; Mratinić, Popovski, Milošević, & Popovska, 2011; Usenik, Fabcic, & Stampar, 2008). These properties have been linked to preventing different diseases and disorders, such as reducing cardiovascular diseases, cancer, diabetes, and other age-related declines (Cevallos-Casals, Byrne, Okie, & Cisneros-Zevallos, 2006; Liu, 2013).

The mesocarp surrounds a single, relatively large shell of hardened endocarp ("stone" or "pit") with a seed inside. The seed contains all necessary genetic material and nutrients, which supports growth until the developing seedling becomes autotrophic and establishes the next generation (Baud, 2018). Although proteins and carbohydrates are the main constituents of a stone fruit kernel, recently, great attention has been given to oil and fatty acids. Those compounds are important as metabolites in all living organisms, like food supplements, and ingredients in the cosmetics industry (Wu et al., 2011), showing antioxidant and antimicrobial activities (Tian, Zhang, Zhan, & Tian, 2011). Although stone fruit kernels contain high amounts of lipids, the lipid fraction does not contribute to cholesterol formation in humans, due to high level of unsaturated fatty acids, mostly oleic, linoleic, and palmitic acids. The relative percentages of kernel and seed together with the oil percentage in seed are given in Table 1. The kernel weight based on fruit weight ranged from 2.82% for plum to 28% for nectarine. The weight of seed based on kernel weight is the highest for apricot (38%) but only 9.1% and

TABLE 1 Percentage kernel weight based on fruit weight, percentage of seed weight based on kernel weight, percentage of oil in seed in different stone fruits.

Species	Kernel weight (%)	References	Seed weight (%)	References	Oil in seed (%)	References
Almond (*Prunus amygdalus*)	19.8–44	Godini (1984)	17–25	Čolić, Rakonjac, Zec, Nikolić, and Fotirić Akšić (2012)	36.3–62.8	Čolić et al. (2017)
	17.32–66.1	Comas, Socias, Company, and Segura (2019)				
Apricot (*Prunus armeniaca*)	5.5–11.34	Mratinić et al. (2011)	18.8–38	Gezer and Dikilitas (2002)	27.7–66.7	Femenia, Rosello, Mulet, and Canellas (1995)
Cherry plum (*Prunus cerasifera*)	6–10	Cosmulescu, Ionica, and Mutu (2018)	14.3–23.7	–	22.6–53.1	Górnaś et al. (2016)
Peach/ nectarine (*Prunus persicae*)	18–28	Chaurasiya and Mishra (2017)	10–15	Malcolm, Holford, Mc-Glasson, and Newman (2003)	48	Lazos (1991)
					50	Sánchez-Vicente, Cabañas, Renuncio, and Pando (2009)
Plum (*Prunus domestica*)	2.82–4.91	Dimkova, Ivanova, Todorova, and Marinova (2017)	9.1	Kamel and Kakuda (1992)	24.2–46.9	Górnaś, Rudzińska, Raczyk, Mišina, Soliven, et al. (2016)
Sweet cherry (*Prunus avium*)	3.7–8.4	Blažková (1988)	26.6	Kamel and Kakuda (1992)	25–30	Siano et al. (2016)
					30.3–40.3	Górnaś, Rudzińska, Raczyk, Mišina, and Segliņa (2016)

TABLE 1 Percentage kernel weight based on fruit weight, percentage of seed weight based on kernel weight, percentage of oil in seed in different stone fruits—cont'd

Species	Kernel weight (%)	References	Seed weight (%)	References	Oil in seed (%)	References
Sour cherry (*Prunus cerasus*)	10	Rakonjac, Fotirić Akšić, Nikolić, Milatović, and Čolić (2010)	22.7	Lin, Chen, and Grethlein (1990)	18–31	Górnaś et al. (2016)
					31.8	Lin et al. (1990)
Dwarf almond (*Prunus tenella*)	–		–		46.15	Matthäus and Özcan (2014)
Blackthorn (*Prunus spinose*)	12–22	Călușaru, Claudia, Ionica, and Cosmulescu (2017)	–		53.5	Matthäus and Özcan (2009)
Cherry laurel (*Prunus laurocerasus*)	10	Islam (2002)	–		38.1	Erciyes, Tüter-Erim, Kabasakal, and Dandik (1995)
Mahaleb cherry (*Prunus mahaleb*)	–		–		31.26	Sbihi, Imededdine, and Al-Resayes (2014)

10% for peach/nectarine and plum, respectively. Oil content is very high, going up to 66.7% in apricot, up to 62.8% in almond, and up to 53.1% in cherry plum. From the wild *Prunus* species, the highest level of seed oil (53.5%) was detected in *Prunus spinosa* (Matthäus & Özcan, 2009).

The composition and concentration of biochemical components depend on the location of cultivation, climate factors, varieties, and fruit ripeness (Campbell & Padilla-Zakour, 2013; Stanojević et al., 2015). In addition, stone fruit seeds are rich in cyanogenic glycosides, mostly amygdalin, which gives hydrogen cyanide (characteristic taste of a "bitter almond") during hydrolysis. In stone fruit, it can be found in the range between 0.1 and 17.5 mg/g (Bolarinwa, Orfila, & Morgan, 2014). In plant metabolism, the role of cyanogenic glycosides is to act in defense against herbivores, to store nitrogen required for seedling growth, or either to promote or inhibit seed germination (Ganjewala, Kumar, Asha, & Ambika, 2010).

Currently, large amounts of fruit seeds, especially fruit stones, are discarded yearly in juice, marmalades, pie fillings, preserves, or other conserve-producing industries. The residues can account for up to 60% of the fruit mass and often discarded at the expense of the manufacturer (Pande & Akoh, 2010). At the moment, there is no organized collection and utilization of stone fruit kernels, and therefore potentially beneficial products are lost (Koprivica et al., 2018; Stanojević et al., 2015). Although small amounts of pits are utilized in heating and cooling packs, as a heating fuel, biodiesel, or added as filler in cement manufacturing (Segers, 2002; Wang & Yu, 2012), fruit seeds, which were previously considered as food wastes, can also be potential ingredients in food formulations or raw matter useful for the extraction of bioactive compounds, especially essential oils.

2 Extraction and processing of cold pressed oil

In ancient times, people used oils and fats primarily as food. Evidence has been found in many places, from Pharaonic Egypt, ancient Rome, Pompeii, and Herculaneum (and the whole Mediterranean region including Tunis and Northern Africa (Bockisch, 1998). The revolution in oil pressing started in 1902 with the mechanical screw press so-called "expeller." In parallel, an oil extraction technology with organic solvents and oil refining process has been developed. The current processing technologies and increased oilseed production developed in the 20th century have been very successful in the industrial-scale production of commercially available plant oils. The extraction of vegetable oil by solvent leads the seed material into contact with the solvent and affects the environment due to pollution with solvent vapors, which is undesirable today. Therefore, consumers want to buy "green" food products that require, in addition to minimal processing, less energy for production and transport, and, therefore, are considered ecologically responsible. This could be achieved by alternative green technologies applied in different stages of processing oilseeds and vegetable oils, which primarily includes oil extraction technologies such as mechanical screw presses, green extraction processes, including enzyme-assisted aqueous extractions, supercritical methods, and ultrasound-assisted extractions.

2.1 Mechanical screw presses

The major objectives of green processing are optimization of the usage of energy to give minimal or no impact on the environment. For that purpose, a mechanical screw press (expeller) would be a proven piece of equipment to use. In addition, the great advantage of this equipment is producing vegetable oil free of contact with any chemicals, with high quality and ready for consumption. Mechanical screw presses are worm conveyors that move the material, compress it under very high pressure and at the same time, eliminate the oil and produce the cake. The universal press for all types of oilseeds does not exist and most of them are intended for pressing 6–8 types of oilseeds.

The process begins by putting the oilseeds inside the feed hopper. The screw press has a horizontal main shaft that carries the worm, which is formed integrally with the shaft. The rotation of the worm occurs inside the slotted-wall cage, which forms a drained barrel around the worm assembly and allows the oil to be expelled while retaining the solid material within it. The pressure releases the oil, most of which flows out through the slotted-wall cage. A movable cone or choke control is installed at the discharged end. This device has the function of regulating pressure by changing the width of the annular space through which cake must pass (Khan & Hanna, 1983).

It is important that the temperature of the outlet oil during cold pressing is not higher than 50°C, in order to preserve all nutritionally important components. In order to achieve this, presses of special construction are required, or the pressing must be done under milder conditions—that is, at lower pressures. In this case, the content of the residual oil in the cake is usually higher, i.e., the oil yield is lower. Modern worm assembly design reduces the energy cost per ton of material processed by a press, enabling temperatures to be better controlled.

Screw presses are used to produce cold pressed oils, whose variety in the market today is large. By-products from the fruit processing industry, such as apricot, plum, peach, sour cherries, and sweet cherries, represent raw materials with a significant content of quality oil with a high concentration of various bioactive compounds, but the literature does not yet have enough data on this subject. Gupta and Sharma (2009) examined the effect of different oil extraction methods, including cold pressing, on the quality characteristics of apricot kernel oil. Data presented by Uluata (2016) showed that cold pressed apricot oil had better oxidative stability and higher total tocopherols content than oil extracted with solvent. The same research showed that the method of oil extraction did not significantly affect the fatty acid composition, color, and total phenolics content. A similar study on chemical composition and quality was done by Kostadinovic-Velickovska, Bruhl, Mitrev, Mirhosseini, and Matthaus (2015) on eight cold pressed oils, among which they also studied apricot kernel oil. Medium roasting of apricot kernels prior to cold pressing can increase oxidative stability and antioxidant capacity, although the content of tocopherols decreases. Similarly, Durmaz, Karabulut, Topçu, Asiltürk, and Kutlu (2010) concluded that Maillard reaction products, which appeared during the heat pretreatment, are responsible for better oxidative stability.

In our recent research (unpublished data), cold pressed plum kernel oils from various types of waste material in brandy production were examined, and the difference in the composition of fatty acids of cold pressed oil from the kernels obtained after fermentation and distillation.

2.2 Enzyme-assisted aqueous extraction

In order to improve oil yield by aqueous extraction, specific food-grade enzymes have been introduced that were able to hydrolyze some or all components of seeds. Enzymes hydrolyze different constituents of cell walls (cellulose, hemicellulose, pectins, proteins, etc.) and are able to enhance the release of the oil and significantly increase the yields of extraction. The advantages of the method reside in the fact that processing occurs at relatively low temperatures with water as a solvent, ensuring superior product quality, and making the method safe and environmentally friendly (Ricochon & Muniglia, 2010).

Aqueous-enzymatic extraction of plum kernel oil, chemical composition, and oxidative stability of obtained oil were investigated by Picuric-Jovanovic, Vrbaski, and Milovanovic (1997, 1999). The authors concluded that this innovative method for oil extraction did not affect the composition of fatty acids and tocopherols, iodine value, and saponification value, while the oxidative stability of oil was lower when compared with conventional solvent extraction. Bisht et al.

(2011) studied the improvement of the efficiency of the extraction of oil from apricot kernels using enzymes, and found that enzymes increased the yield of oil from kernels and that this extraction process does not affect nutritional characteristics of obtained oil.

2.3 Supercritical fluid extraction

Supercritical fluid extraction (SFE) has been widely used in recent years in the chemical, food, and pharmaceuticals industries. So far, this method has been used for oil extraction from some fruit kernels of sweet cherry, sour cherry, and apricot (Bernardo-Gil, Oneto, Antunes, Rodrigues, & Empis, 2001; Straccia, Siano, Coppola, La Cara, & Volpe, 2012). Bernardo-Gil et al. (2001) found no significant differences in fatty acid and sterol composition, oxidative stability, and phenolics content in oils from cherry kernels extracted by SFE and solvent extraction. A similar study was done by Straccia et al. (2012), but the results on the fatty acids composition of cherry seed oils differed significantly between the two extractions. Cherry seed oil obtained with SFE had a higher content of polyunsaturated fatty acids (PUFAs) and also a higher content of phytosterols than the oil obtained with solvent extraction, i.e., the oil obtained with SFE was of better quality.

Pavlović et al. (2018) used two green technologies, cold pressing and SFE, to extract oil from apricot kernels. The total oil yield was higher with SFE. In addition, obtained oils were compared for fatty acids composition and tocopherol content and composition. The tocopherols content was higher in the oil sample obtained with SFE, while no significant difference was obtained for the fatty acids composition.

2.4 Ultrasound-assisted extraction

Ultrasound-assisted extraction (UAE) is considered a promising technique for the extraction of apricot and sour cherry kernel oil. Gayas, Kaur, and Gul (2016) studied the influence of various parameters, such as temperature, time, and solvent content on UAE of apricot kernel oil. They proved that the temperature is the parameter that has a major influence on UAE. In order to obtain a higher yield of oil, Górnaś, Rudzińska, Raczyk, Mišina, and Segliņa (2016), Górnaś, Rudzińska, Raczyk, Mišina, Soliven, and Seglina (2016), and Górnaś, Rudzińska, Raczyk, Mišina, Soliven, et al. (2016) used an ultrasonic technique to extract oil from sour cherries kernels. Ultrasound indicates the use of high intensity and frequency sounds and their interaction with materials. Efficient cell disruption and effective mass transfer are two key factors leading to the enhancement of extraction with ultrasonic power. The main advantages of using ultrasound are reduction of the extraction time and the consumption of solvents with a very good extraction yield as well as more efficient mixing, faster energy transfer, reduced extraction and degradation temperature, selective extraction, reduced equipment size, and eliminated process steps (Lloyd & van Wyk, 2012).

3 Fatty acids and acyl lipids profile of cold pressed oil

3.1 Fatty acids

Although seed oils usually contain the same profile of 10–20 fatty acids, the composition varies significantly among different plant species (Ozcan, Unvera, Erkanb, & Arslana, 2011). The fatty acids profile and the presence of specific not common fatty acids could be used as a marker for the differentiation of a particular plant family (Aitzetmüller, 1993; Gur, 1980). In addition, composition and the ratio of unsaturated fatty acids (UFAs) and saturated fatty acids (SFAs) provides useful information about possibilities of the nutritional or technical application of the oils (Matthäus & Özcan, 2009). The contents of most abundant fatty acids in kernel oils recovered from the genus *Prunus* are presented in Table 2. Due to limited literature data on the FA composition of cold pressed kernel oil from *Prunus* species, Table 2 also contains information of FA contents in oils extracted from the kernels using organic solvents, such as *n*-hexane, petroleum ether, chloroform, and on oils obtained by supercritical extraction with carbon dioxide (SC-CO_2).

As reported in the literature, only four FAs were quantified in the cold pressed oil recovered from sour cherry (*Prunus cerasus*) kernels (Uluata & Özdemir, 2017). The highest number of fatty acids in sour cherry oils was reported by Lazos (1991), where petroleum ether was used for extraction. Górnaś, Rudzińska, Raczyk, Mišina, Soliven, and Seglina (2016), in a study of six cultivars of sour cherry, quantified nine fatty acids. This investigation showed that the composition of quantified fatty acids is largely affected by the cultivar. A study by Yılmaz and Gökmen (2013) showed that the extraction technique had no significant effect on FA profiles of sour cherry kernel oil. Generally, the sour cherry oils are characterized by high contents of monounsaturated oleic acid and polyunsaturated linoleic acid (the sum is higher than 80%). The content of two major saturated fatty acids (SFAs) quantified in sour cherry oil, palmitic acid, and stearic acid is significantly lower

TABLE 2 The content of the most abundant fatty acids in *Prunus* kernel oils.

	% of total fatty acids								
Prunus species	Oleic 18:1	Linoleic 18:2	Palmitic 16:0	Stearic 18:0	Arachidic 20:0	α-eleostearic 18:3	Origin	Extraction method	References
Sour cherry (*Prunus cerasus*)	45.8	41.8	6.9	2.6			Turkey	Laboratory type oil press	Uluata and Özdemir (2017)
	42.9	38.2	11	6.4	0.9		Romania	Petroleum ether	Popa et al. (2011)
	63.9	27.0	5.3	1.5	0.1		Turkey	Petroleum ether	Matthäus and Özcan (2009)
	25.3–45.3	35.5–46.1	5.1–7.4	2.2–3.4	1.0–1.4	7.4–15.8	Latvia	*n*-Hexane	Górnaś, Rudzińska, Raczyk, Mišina, Soliven, and Seglina (2016)
	52.9	35.0	7.6	2.3	1.4		Canada	Chloroform: methanol	Kamel and Kakuda (1992)
	46.8	40.6	6.2	1.3	5.1		Turkey	*n*-Hexane	Yılmaz and Gökmen (2013)
	47.9	41.2	5.9	1.1	3.8			*n*-Hexane with 3% ethanol	
	45.0	41.8	7.2	1.3	4.6			SC-CO_2	
	45.54	42.26	6.14	1.2	4.9			SC-CO_2 with 3% ethanol	
	46.9	41.7	9.4	2.0			Bulgaria	*n*-Hexane	Zlatanov and Janakieva (1998)
	46.0	41.6	8.6	2.9			Greece	Petroleum ether	Lazos (1991)
	41.1	40.8	5.5	4.4			Russia	Acetone	Deineka, Gabruk, Deineka, and Manokhina (2002)
Sweet cherry (*Prunus avium*)	42.9	40.7	9.7	3.5			Turkey	Laboratory type oil press	Uluata and Özdemir (2017)
	36.9	45.7	15.4	1.0			Bulgaria	*n*-Hexane	Zlatanov and Janakieva (1998)
	32.0	41.0	6.0	2.4	1.7	10.5	Portugal	*n*-Hexane	Bernardo-Gil et al. (2001)
	32.6	40.8	5.3	2.2	1.5	10.1		SC-CO_2	
	23.9–37.5	40.0–48.9	6.8–9.4	1.6–2.6		9.9–13.2	France	*n*-Hexane	Comes, Farines, Aumelas, and Soulier (1992)

	41.1	40.8	6.5	4.4		Russia	Acetone	Deineka et al. (2002)
	35.1	41.5	9.1	3.0		Italy	Diethyl ether	Siano et al. (2016)
Plum (*Prunus domestica*)	63.5–66.6	24.4–26.4	5.2–5.8	2.4–3.6		Pakistan	*n*-Hexane	Anwar, Manzoor, Bukhari, and Aladedunye (2014)
	59.5	27.1	7.5	1.5	0.1	Serbia	Petroleum ether	Veličković et al. (2016)
	63.9–78.5	9.7–26.9	5.4–7.3	1.3–1.4	0.1–0.2	Turkey	Petroleum ether	Matthäus and Özcan (2009)
	75.4	15.9	5.8	1.4	0.1	Turkey	Cold pressed	Kiralan, Kayahan, Kiralan, and Ramadan (2018)
	62.0	29.6	6.3	1.4	0.3	Canada	Chloroform: methanol	Kamel and Kakuda (1992)
	70.5	15.7	9.2	3.9		Bulgaria	*n*-Hexane	Zlatanov and Janakieva (1998)
	63.6	29.0	3.2	4.1		Russia	Acetone	Deineka et al. (2002)
Blackthorn (*Prunus spinosa*)	43.9	37.0	5.2	2.1	0.2	Turkey	Petroleum ether	Matthäus and Özcan (2009)
Cherry plum (*Prunus cerasifera*)	67.3	23.4	6.1	2.1		Turkey	Laboratory type oil press	Uluata and Özdemir (2017)
Peach (*Prunus persica*)	69.3–73.6	16.0–20.5	5.6–6.1	1.9		Tunisia	Petroleum ether	Chamli et al. (2017)
	65.8	26.0	5.6	2.0	0.1	Canada	Petroleum ether	Wu et al. (2011)
	65.7	26.0	5.6	2.0	0.1		Chloroform	
	65.8	26.0	5.7	3.0	0.1		Ethyl ether	
	61.9	29.1	6.4	1.9	0.1		*n*-Hexane	
	59.8–64.6	28.0–32.8	4.1–5.8	1.2–1.4		Pakistan	*n*-Hexane	Anwar et al. (2014)
	70.3	19.5	5.9	1.6	0.1	Turkey	Petroleum ether	Matthäus and Özcan (2009)

Continued

TABLE 2 The content of the most abundant fatty acids in *Prunus* kernel oils—cont'd

Prunus species	% of total fatty acids Oleic 18:1	Linoleic 18:2	Palmitic 16:0	Stearic 18:0	Arachidic 20:0	α-eleostearic 18:3	Origin	Extraction method	References
	58.5	32.8	8.1		0.3		Canada	Chloroform: methanol	Kamel and Kakuda (1992)
	67.6	21.3	8.5	2.0			Bulgaria	*n*-Hexane	Zlatanov and Janakieva (1998)
	63.8	15.4	13.4	6.41			Egypt	*n*-Hexane	Rahma (1988)
	69.0	22.0	6.3	1.6	0.1		Greece	Petroleum ether	Lazos (1991)
	41.1	48.4	8.4	1.2	0.2		Brazil	–	Pelentir, Block, Fritz, Reginatto, and Amante (2011)
	32.5	59.8	6.1	1.8			Russia	Acetone	Deineka et al. (2002)
Nectarine (*Prunus persica*)	67.7–75	15.7–22.1	5.7–6.3	2.0			Tunisia	Petroleum ether	Chamli et al. (2017)
	66.3	26.8	6.1		0.3		Canada	Chloroform: methanol	Kamel and Kakuda (1992)
	38.6	50.6	6.1	4.5			Russia	Acetone	Deineka et al. (2002)

in comparison with amounts of unsaturated FAs. Some FAs, such as C12:0, C14:0, C15:0, C16:0, and C20:0, in sour cherry kernel oil were found in amounts less than 1%. α-eleostearic acid was detected in an insignificant amount (7.43%–15.7%) in sour cherry kernel oils in only one study (Górnaś, Rudzińska, Raczyk, Mišina, Soliven, & Seglina, 2016). The presence of α-eleostearic acid in each cultivar in this study was explained by the impact of plant material different genetic basis, although misidentification in other studies also cannot be excluded.

FA profiles of sweet cherry (*Prunus avium*) kernel oil are similar to the composition of sour cherry oils FA (Table 2). The main difference between the sour and sweet cherry oils is in the higher content of linoleic acid compared to oleic acid, which was found in the majority of studied sweet cherry oils. On the contrary, in two reports, the cold pressed oil recovered from sweet cherry kernels (Uluata & Özdemir, 2017) and sweet cherry kernel oil extracted with acetone (Deineka et al., 2002) were characterized with higher content of oleic acid when compared with linoleic acid. No significant differences between FA profiles were found when the sweet cherry oils were extracted by hexane or by SC-CO_2 (Bernardo-Gil et al., 2001). Along with oleic and linoleic acid as the most abundant, sweet cherry oils contain significant amounts of palmitic acid and α-eleostearic acid, the last one being specific for this kind of cherry seed oil (Farines, Soulier, & Comes, 1986). The results of Siano et al. (2016) showed that sweet cherry seed oil contained *cis*-5,8,11,14,17-eicosapentaenoic and nervonic acids, among other characteristic fatty acids given in Table 2.

The major fatty acids in kernel oil of plum (*Prunus domestica*) are oleic, linoleic, and palmitic acids. The content of oleic acid (59.5%–78.5%) is significantly higher than the content of linoleic acid (9.7%–29.6%). The oleic acid/linoleic acid ratio is considered to be an important criterion of oil stability and quality, with high ratios being preferable (Kodad et al., 2014). A total of 10 FAs were quantified in cold pressed plum kernel oil (Kiralan et al., 2018), and five of them—C12:0, C14:0, C20:0, C18:3, and C21:0 (n-11)—were detected in very small amounts (<1%). The findings of Uluata and Özdemir (2017), who reported the FA compositions of cherry plum (*Prunus cerasifera*) kernel oil, are in line with the FA profiles of *Prunus domestica* kernel oil.

The seed oil from the blackthorn (*Prunus spinose*) is characterized by the high contents of unsaturated oleic (43.9%) and linoleic acid (37.0%), while the sum of quantified saturated FA is less than 10% (Matthäus & Özcan, 2009). According to Mead, El-Shafiey, and Sabry (2016) mahaleb (*P. mahaleb*) seed oil has 92.8% of unsaturated fatty acids (timnodonic, oleic, linoleic, and erucic acids) and some saturated fatty acids (palmitic acid and stearic acid, 2.74%, and 1.73%, respectively).

The FA profiles of peach and nectarine (*Prunus persica*) kernel oils are similar to the FA profiles of *Prunus domestica* kernel oil. The most abundant fatty acids in peach and nectarine kernel oils are oleic, linoleic, palmitic, and stearic acids. *Prunus persica* kernel oils have a similar O/L ratio to *Prunus domestica* and a higher value of the O/L compared with *Prunus cerasus* and *Prunus avium*. Wu et al. (2011) studied peach kernel oil originating from Canada and quantified 14 FAs. Some of these, like C10:0, C20:1 (n-9), and C24:0, were found only in *Prunus persica* kernel oil. In addition, no significant differences among FA profiles of the peach kernel oils extracted with various solvents were found.

The general conclusion is that *Prunus* oils contain high amounts of the unsaturated fatty acids that primarily are oleic and linoleic acids, and relatively low contents of saturated fatty acid. From the FA compositions of the oils recovered from kernels of *Prunus* species, it can be concluded that these oils are comparable to mid- to high-oleic oil such as canola, corn, grape seed, olive, peanut, sesame, soybean, and walnut (Fasina, Craig-Schmidt, Colley, & Hallman, 2008; USDA, 2006).

3.2 Neutral and polar lipids

The oils are mainly composed of very diverse triacylglycerol (TAG) species, consisting of glycerol molecule esterified with three fatty acids (palmitic = P, oleic = O, linoleic = L, α-eleostearic = E, stearic = S). Hassanein (1999) determined 11 triacylglycerol molecular species in kernel oils from plum and peach, at various amounts. It should be noted that examined oils were extracted with chloroform:methanol (2:1, *v*/v). Triolein (OOO) was the most abundant component, with 35.6% and 42.9% in peach and plum kernel oils, respectively. LOO and LLO contributed with 28.5 and 16.0% in peach kernel oil, respectively, whereas in plum kernel oil their quantities were 23.9% and 9.6%, respectively. LLL in peach kernel oil was present in a higher percentage than in plum kernel oil. Both oils (peach and plum kernel oils) were free of linolenate; therefore, no TAGs including this acid were detected. Chamli et al. (2017) detected 14 TAG species in oils from two *Prunus persica* varieties, peach and nectarine, grown in Tunisia. Oils were extracted with petroleum ether as a solvent. The predominant triacylglycerol species were those containing oleic, linoleic, and palmitic acids. Trioleylglycerol species (OOO) were found in the range 38.4%–50.5%, followed by OOL in range 18.2%–23.2%, while POO and OLL were found in smaller amounts, 8.3%–9.7% and 6.3%–10.1%, respectively.

Comes et al. (1992) analyzed cherry seed oil and revealed that LLO, LOO, OOO, and EEL species of TAGs were the most abundant, while other TAGs (EEL, LLL, ELO, ELP, LLP/EOO, EOP, LOP, LPP, and OOP) were present in proportions under 10%. In the peach kernel oil extracted with hexane, the total neutral lipid content was 98.1% of total lipids

(Rahma, 1988). Deineka et al. (2002) analyzed sweet cherry, sour cherry, plum, peach, and nectarine kernel oils extracted with acetone. Major triglyceride components in all examined oils were LLO, LOO, and OOO. LLO content was in the range 16.2%–16.6% in plum, sweet cherry, and sour cherry kernel oil, while in peach and nectarine kernel oil, it was 30.9% and 24.5%, respectively. Content of LOO was in the range 14.5%–16.2% in sweet cherry, sour cherry, and nectarine kernel oil, while plum kernel oil had the highest content at 26.9%. The most abundant TAG in plum kernel oil was OOO with 33.3%, whereas in other examined oils it was in the range 12.2%–17.1%. It was found that sweet cherry kernel oil contained significant amounts (13.0%) of triglyceride ELL. Peach and nectarine kernel oils also contained LLL triglyceride, at about 16%, and to a lesser extent LLP and LLS species. Zlatanov and Janakieva (1998) determined neutral lipids content in peach, plum, and sour and sweet cherry kernel oils extracted by hexane in the range 97.2%–98.7%. The only report about polar lipids (phospholipids) was from the same authors, Zlatanov and Janakieva (1998). Polar lipids contributed to the content of examined oils from 0.4% to 1.1%. Among the identified classes of phospholipids, phosphatidylcholine (PC), phosphatidylethanolamine (PE), and phosphatidylinositol (PI) were the major components. Plum oil phospholipid fraction had the highest amount of PC, at 59.8%, which makes it a valuable source for obtaining corresponding concentrates. Morello cherry had significant amounts of PE (19.5%), and sweet cherry phospholipids had 30.0% of PI.

4 Minor bioactive compounds in cold pressed oil

4.1 Tocopherols and tocotrienols

The term tocochromanols applies to homologues of tocopherol and tocotrienol, named α-, β-, γ-, and δ-, differing in the methylation of the chroman ring. They are known as vitamin E active compounds, synthesized only in photosynthetic organisms and acting as vitamin E in human nutrition. Kernel oils from genus *Prunus* are natural sources of tocopherols, therefore unsaturated fatty acids in oils are protected against oxidative damage (Güçlü-Üstündağ & Temelli, 2004). The predominant tocopherol in most kernel oils included in this work was γ-tocopherol, whose content in the oils was in the range 46.53–302.1 mg/100 g. Results obtained from a literature survey are summarized in Table 3. Only in the *Prunus persica* kernel oils from Turkey (Matthäus & Özcan, 2009) and Pakistan (Anwar et al., 2014) was the dominating tocopherol α-tocopherol, with 37.3 mg/100 g and 175.4–187.5 mg/100 g, respectively. Another exception was found in *Prunus domestica* kernel oils from Pakistan (Anwar et al., 2014), where the main component among tocopherols was δ-tocopherol in quantities 204.7–221.8 mg/100 g. Minor levels (up to 4.0 mg/100 g) were detected for β-tocopherol (Table 3).

Two tocotrienols, α- and γ-, were detected in low concentrations, with the exception of oils from *Prunus cerasus*, *Prunus domestica*, *Prunus spinosa*, and *Prunus persica* from Turkey (Matthäus & Özcan, 2009) in which amounts of α-tocotrienol were in the range from 212.5 to 41.6 mg/100 g. Plastochromanol-8, another member of vitamin E active compounds, was detected only in minor concentration (1.1–1.2 mg/100 g) in *Prunus cerasus* and *Prunus domestica* from Turkey (Matthäus & Özcan, 2009). Finally, not only oxidative stability but also nutritional quality of kernel oils from genus *Prunus* could be attributed to high concentrations of tocopherols (Anwar, Przybylski, Rudzinska, Gruczynska, & Bain, 2008).

4.2 Phytosterols

Phytosterols, which include plant stanols and sterols, have high importance for human health, such as ability to decrease the level of cholesterol in blood serum (Chen, Jiao, & Ma, 2008), anticarcinogenic effects (Ostlund Jr, 2002), and improving immune functions of the body (Anwar et al., 2008). Kernel oils recovered from *Prunus* species are a valuable source of minor lipophilic compounds such as phytosterols (Caligiani, Bonzanini, Palla, Cirlini, & Bruni, 2010; Hassanien et al., 2014). Nine sterols (campesterol, Δ-sitosterol, Δ5-avenasterol, 24-methylene-cycloartanol, cholesterol, gramisterol, Δ7-stigmasterol, Δ7-avenasterol, and citrostadienol) were quantified in a study of kernel oils of six sour cherry cultivars (Górnaś, Rudzińska, Raczyk, Mišina, Soliven, & Seglina, 2016). The highest levels were determined for β-sitosterol (77%–82% of the total phytosterols), which is the most common plant sterol. Significant concentrations were found for campesterol, Δ5-avenasterol, and 24-methylenecycloartanol, while the contents of other quantified sterols were considerably lower. The findings of Górnaś, Rudzińska, Raczyk, Mišina, Soliven, and Seglina (2016) showed that the amount of phytosterols in sour cherry kernel oils is largely affected by the cultivar. A study of peach and plum kernel oils (Anwar et al., 2014) showed that the main phytosterol was β-sitosterol (78.8%–80.0% and 82.9%–84.9% of total phytosterols, respectively), followed by Δ5-avenasterol (8.9%–12.2% and 5.0%–7.8%, respectively). A considerable content of campesterol and Δ7-avenasterol was also detected in the tested peach and plum kernel oils. A total of 10 phytosterols and cholesterol were identified in kernel oils recovered from 21 *Prunus domestica* and seven *Prunus cerasifera* cultivars

TABLE 3 The content (mg/100 g oil) of tocochromanols in *Prunus* kernel oils.

Prunus species	Tocopherol				Tocotrienol		Origin	Extraction method	References
	α-	β-	γ-	δ-	α-	γ-			
Sour cherry (*Prunus cerasus*)	4.7	0.4	197.2	15.1	21.5		Turkey	Petroleum ether	Matthäus and Özcan (2009)
	9.2–38.5	0.5–2.5	89.1–133.3	9.5–18.2	0.5–2.2	0.1–0.4	Latvia	*n*-Hexane	Górnaś, Rudzińska, Raczyk, Mišina, Soliven, and Seglina (2016)
	7.47		57.99	0.87			Turkey	Laboratory type oil press	Uluata and Özdemir (2017)
Sweet cherry (*Prunus avium*)	10.05	1.67	46.53	3.65			Turkey	Laboratory type oil press	Uluata and Özdemir (2017)
Plum (*Prunus domestica*)	24.1–27.1	2.3–4.0	133.1–302.1	11.4–18.9	31.4–41.6		Turkey	Petroleum ether	Matthäus and Özcan (2009)
	53.4–62.8		150.9–172.9	204.7–221.8			Pakistan	*n*-Hexane	Anwar et al. (2014)
	5.7–19.9	0.1–0.7	76.2–182.0	2.9–11.6	0.1–2.9	0.1–0.8	Latvia	*n*-Hexane	Górnaś, Rudzińska, Raczyk, Mišina, Soliven, et al. (2016)
	7.8		60.7	2.5			Egypt	Chloroform: methanol	Hassanein (1999)
Blackthorn (*Prunus spinosa*)	8.0		278.6	12.0	23.1		Turkey	Petroleum ether	Matthäus and Özcan (2009)
Cherry plum (*Prunus cerasifera*)	4.3–17.2	0.1–0.3	60.5–170.6	1.7–6.9	0.1–1.4	0.1–0.3	Latvia	*n*-Hexane	Górnaś, Rudzińska, Raczyk, Mišina, Soliven, et al. (2016)
	9.21	0.71	61.45	2.44			Turkey	Laboratory type oil press	Uluata and Özdemir (2017)
Peach (*Prunus persica*)	37.3		1.6		24.0		Turkey	Petroleum ether	Matthäus and Özcan (2009)
	175.4–187.5		110.2–126.7	74.5–85.9			Pakistan	*n*-Hexane	Anwar et al. (2014)
			50.8	1.2			Egypt	Chloroform: methanol	Hassanein (1999)

(Górnaś, Rudzińska, Raczyk, Mišina, Soliven, et al., 2016). As expected, β-sitosterol was the most abundant phytosterols (64%–80% of the total content of phytosterols) in all investigated *Prunus* varieties. Other phytosterols with significant amounts in plum kernel oils were Δ5-avenasterol and campesterol (4%–18% and 3%–6%, respectively). Findings of Górnaś, Rudzińska, Raczyk, Mišina, Soliven, et al. (2016) showed that *Prunus domestica* had a higher average percentage of β-sitosterol and lower average percentage of Δ5-avenasterol compared to *Prunus cerasifera*.

4.3 Other minor bioactive compounds

In sour cherry kernel oils from Latvia (Górnaś, Rudzińska, Raczyk, Mišina, Soliven, & Seglina, 2016), only minor content of carotenoids were determined, with an average value of 0.94 mg/100 g oil and significant amounts of squalene (65.8–102.8 mg/100 g oil). The findings of Yılmaz and Gökmen (2013) proved that extraction method had an impact on β-carotene content in sour cherry kernel oil. Hexane extracted oil had significantly higher levels of β-carotene compared to oil extracted with SC-CO_2. Górnaś, Rudzińska, Raczyk, Mišina, Soliven, et al. (2016) reported lower concentrations of carotenoids in *Prunus cerasifera* species (0.41–0.91 mg/100 g oil) compared to *Prunus domestica* (0.91–3.07 mg/100 g oil). In the same study, the content of squalene was determined to be in the range 25.7–80.4 mg/100 g oil, and it was dependent on the variety.

Generally, total phenolic content in oils extracted with different organic solvents from *Prunus* kernels is low (Wu et al., 2011; Yılmaz & Gökmen, 2013) due to the limited transfer of phenolic compounds to oils. Reports found in the literature show the existence of a relationship between total phenolics content and antioxidant capacity, as individual phenolics are considered to contribute the most to scavenging ability (Natić et al., 2015; Pantelić et al., 2016; Pavlovic et al., 2016). Wu et al. (2011) reported the highest antioxidant activity (16.93 mg trolox/100 g oil) in peach kernel oil extracted with hexane. In cold pressed oils from *Prunus cerasus*, *Prunus avium*, and *Prunus cerasifera*, DPPH· scavenging capacities were found to be 57.4, 60.5, and 63.3 mg trolox/100 g oil, respectively. In plant material, phenolic compounds are analyzed by using hyphenated chromatographic techniques proven to be useful for estimating phenolic profiles based on both quantitative and semiquantitative data (Pantelić et al., 2014; Pavlović et al., 2013). Phenolic profiles of different parts of *Prunus* fruits and leaves have been investigated in many studies so far (Alrgei et al., 2016; Colaric et al., 2005; Davidović et al., 2013; Fotirić Akšić et al., 2015; Kim et al., 2003; Mratinić et al., 2011; Usenik et al., 2008), but data on polyphenols from kernel oils are limited. To the best of our knowledge, only Wu et al. (2011) have presented results of phenolic profile of *Prunus persica* kernel oil. Among 15 quantified polyphenols, dithiothereitol, rutin, and caffeic acid were the most abundant in peach kernel oil. The study showed that amounts of individual phenolic compounds were dependent on the type of solvents used for obtaining the oils. In the oil extracted with hexane, the most abundant phenolic was rutin, at 2.737 mg/g oil.

The literature survey showed only one result concerning volatile compounds in cold pressed oil. A total of 19 volatile compounds, including six aldehydes, four terpenes, three hydrocarbons, two alcohols, two acids, one ester, and one furan, were found in *Prunus domestica* kernel oil. Among identified volatile compounds in plum kernel oil, benzaldehyde was the major one (Kiralan et al., 2018).

5 Contribution of bioactive compounds in cold pressed oil to organoleptic traits and functions in food or nonfood products

Cold pressed oils are produced under gentle pressure conditions, where the temperature of the outlet oil must not be higher than 50°C. Therefore, this procedure allows all bioactive compounds to be preserved in the oil, and above all, the oil keeps the flavor and taste of the raw material from which it is obtained.

Due to the presence of amygdalin, the kernels and obtained cold pressed oil have a typical "almond" flavor. If the content of amygdalin is higher than 500 ppm, it may be toxic, because hydrolysis of amygdalin by enzyme β-glucosidase can result in the formation of hydrocyanic acid (HCN), benzaldehyde, and glucose. Pavlović et al. (2018) found a very low concentration of amygdalin in cold pressed apricot kernel oil (0.4 mg/g oil), while amygdalin content in apricot kernels was 5.0 mg/g. On the other hand, Ghiulai, Socaciu, Jianu, Ranga, and Fetea (2006) found no amygdalin in oils from peach, plum, and apricot kernels, but kernels from apricots and plums had amygdalin content between 3 and 24 mg/kg. They concluded that the absence of amygdalin from oil samples indicates that this oil is very suitable for use in the food industry.

In wild apricot kernels, HCN was detected and associated with bitter taste and toxicity of these kernels, reported by Gupta and Sharma (2009). The content of HCN in bitter apricot kernels ranged from 148 to 173 mg/100 g, but it is not registered in kernels of sweet apricot. The authors examined an effective method that would allow the removal of HCN from the seed, since a lower amount was detected in cold pressed oil. The most effective method was dipping of kernels in 10% sodium thiosulfate solution.

Color, as one of the sensory characteristics of food, is also very important when it comes to cold pressed oils. It can be a key factor that determines the use of cold pressed oil in food or cosmetics. The color of the oil is usually determined by measuring CIEL* (lightness), a* (redness), and b* (yellowness) values with a colorimeter. The oil extraction process usually does not affect the color of the oil, as confirmed by Uluata (2016), when comparing the cold pressing and solvent extraction of the apricot kernels. However, cold pressing of roasted and unroasted apricot kernels produces oil with different organoleptic traits (Durmaz et al., 2010) and the color difference was proved in all color parameters (L*, a*, b*). The authors assume that Maillard reactions during roasting are responsible for this when brown compounds and less colored intermediate compounds are formed, contributing to the specific color of food, which is, in this case, the oil. Studies have shown that these components can have antioxidant properties at a certain level (Morales & Perez, 2001).

6 Health-promoting traits of cold pressed oil and oil constituents

Secondary metabolites are chemical constituents of seeds that are mainly responsible for the specific therapeutic properties and which contribute to the medicinal properties. Seeds can serve as herbal drugs and as a valuable source of medicinally active compounds, such as various vitamins, tocopherols, tocotrienols, carotenoids, phenolic compounds, and some minerals, which are important health and nutritional constituents. Therefore, oils could be used as a source of products in both traditional and contemporary medicine. This section deals with the functionality of the cold pressed oils constituents, which are characteristic for seed kernels belonging to the genus *Prunus*. Some of the pharmacologically active chemical constituents are presented below, together with the potential medicinal actions and therapeutic activities, according to the available literature. A literature overview on kernel oil bioactive components and potential pharmacological (or therapeutic) activities exhibited by the main bioactive ingredients found in the genus *Prunus* kernel oils is resumed in Table 4.

As has already been stated, kernel oils from the genus *Prunus* are abundant in MUFAs, while being low in saturated acids, such as palmitic and stearic acids. In addition, oils are considered to be an excellent source of bioelements. Elemental

TABLE 4 Potential pharmacological (or therapeutic) activities of some of the main bioactive ingredients found in the kernel oils from genus *Prunus*.

Compound/group of compounds		Potential medicinal properties	References
Unsaturated fatty acids	Oleic acid Linoleic acid	Reduces the prospect of cardiovascular diseases by reducing the low density lipoprotein cholesterol levels in the blood, reduce triglycerides, total cholesterol and glycemic index	Eduardo (2010)
	α-Eleostearic acid	Preventing cardiovascular diseases; role as a suppressor of various tumors	Harris, Kris-Etherton, and Harris (2008), Yasui, Hosokawa, Kohno, Tanaka, and Miyashita (2006), and Tsuzuki, Tokuyama, Igarashi, and Miyazawa (2004)
	Nervonic acid	A neurotrophic factor treating neuromuscular diseases	Lauritzen et al. (2000)
	γ-Linolenic acid	Modulator of inflammation within humans	Wathes, Abayasekara, and Aitken (2007) and Kapoor and Huang (2006)
		A beneficial effect toward alleviating symptoms of atopic dermatitis	Finch, Munhutu, and Whitaker-Worth (2010)
		GLA exerting tumoricidal activity against numerous types of cancer	Mainou-Fowler, Proctor, and Dickinson (2001), Das (2004), and Watkins, Martin, Bryce, Mansel, and Jiang (2005)
Triacylglycerols	Trioleoylglycerol, linoleoyl-dioleoyl glycerol, dilinoleoyloleoyl glycerol	An important source of energy for the cardiac muscle	Perona and Ruiz-Gutierrez (2005)

Continued

TABLE 4 Potential pharmacological (or therapeutic) activities of some of the main bioactive ingredients found in the kernel oils from genus *Prunus*—cont'd

Compound/group of compounds		Potential medicinal properties	References
Tocopherols		Antioxidative action	Traber and Atkinson (2007)
		Improvements to macular degeneration and glaucoma, a reduced risk of Alzheimer's disease, reduction of tumor formation, lowered occurrence of CHD, and decreased risk of Parkinson's disease	Engin, Engin, Kucuksahin, Oncu, and Guvener (2007), and Morris et al. (2005)
Phytosterols/ phytostanols	Stigmasterol, campesterol, β-sitosterol, phytostanols	Effective agents to lower blood levels of low-density lipoprotein (LDL) lowering blood cholesterol levels in humans	Ostlund Jr (2004), Katan et al. (2003), Kritchevsky and Chen (2005), Chen et al. (2008), and Law (2000)
		Decreased risks of heart disease by supplementation	Matvienko et al. (2002) and Jones and AbuMweis (2009)
		Reduce oxidative stress by promoting the activity of antioxidative enzymes, reduce risk of cancer	Vivancos and Moreno (2005), Woyengo, Ramprasath, and Jones (2009), De Stefani et al. (2000), Rudkowska (2010), and Kritchevsky and Chen (2005)
Squalene		Acts as a potent scavenger of singlet oxygen—inhibiting oxidative damage induced by UV radiation and presenting possible protection against the formation of cancerous tumors in the colon, breast, and prostate	Ghanbari, Anwar, Alkharfy, Gilani, and Saari (2012)
		Protective effect against the damage of UV radiation linked to a-photoprotection of the skin	Owen et al. (2004)
		Immune-stimulating properties that may inhibit tumor formation	Reddy and Couvreur (2009)
		Reduce side effects associated with chemotherapy	
		Increases effectiveness of the vaccine	
Carotenoids		Decreased risk of skin damage, cataract, cardiovascular diseases, and cancer	Aust, Sies, Stahl, and Polidori (2001)
Enzymes	Heme oxygenase-1	Protects the organism against a diverse variety of stressors, protect against ischemic damage to both retinal and cardiac tissue	Bak et al. (2006)

composition is characterized by the high quantity of K, P, Mg, and Ca, and some essential elements such as Fe, Mn, Cu, and Zn. Although kernel oils are characterized by the presence of some biologically active substances, such as tocopherols, phenolic compounds, β-carotene, and various phytosterols, health-promoting traits of cold pressed oil from the genus *Prunus* have not been studied much. Most papers on the functionality in the literature are related to apricot and almonds, as these were studied largely in comparison to other species of the genus *Prunus*. Functional properties of apricot kernels are related to phenolics and carotenoids, exhibiting liver and heart protective, antioxidant, and antiinflammatory effects (Fratianni, Ombra, d'Acierno, Cipriano, & Nazzaro, 2018). As reported by Kutlu, Durmaz, Ateş, and Erdoğan (2009), supplementation with apricot seed oil caused a significant increase in the activity of antioxidant enzymes (glutathione peroxidase and catalase) in hypercholesteremic rats. Similarly, Zhang et al. (2011) reported on the cardioprotective effects of apricot seed oil.

Although cyanogenic glycosides have been used as nutritional supplements, one should be aware of detrimental effects of natural amygdalin, a chemical compound incorrectly referred to as vitamin B17, and synthetic compound laetrile, as revealed in clinical trials (Milazzo, Lejeune, & Ernst, 2006). As already mentioned, Rosaceous plants, such as almonds, apricots, peaches, and cherries, contain toxic cyanogenic compounds in their seeds. Consumption in greater amounts can lead to severe symptoms of cyanide poisoning or even death. Amygdalin was reported to show antioxidative and antimicrobial activity, while in small amounts it can have a therapeutic effect, stimulating respiration and improving digestion; it may also be used for the treatment of cancer (Li, 2009; Savić et al., 2016). In addition, toxicological evaluation of the oral application of sour cherry seed extracts did not result in any adverse effects, as evidenced by studies of liver and kidney enzymatic functions and immunohistochemistry (Bak et al., 2011).

7 Edible and nonedible applications of cold pressed oils

The results of studies presented in literature regarding utilization of kernels provide useful information on their industrial potential as a biowaste, especially in the food industry. Due to favorable fatty acid composition and biologically active compounds, kernel oil has received much attention as a nutritionally important source for edible and other industrial applications, such as formulation of new functional foods and pharmaceutical applications (Turan, Topcu, Karabulut, Vural, & Hayaloglu, 2007; Zhou, Wang, Kang, Zhong, & Prenzler, 2016). Seeds from *Prunus* species are valued for vitamin-E-active compounds, high content of oleic acid, moderate contents of linoleic acid, and low amounts of saturated fatty acids, and therefore are considered good substitutes for olive oil and sunflower oil (Matthäus & Özcan, 2009).

In comparison to traditional oils, such as sunflower and rapeseed oils, peach and nectarine oils are considered valuable oils for industrial use (Chamli et al., 2017). Peach kernel oil was proposed as a substitute for olive oil and grapevine seed oil in products such as salad oil, salad dressing, dips, and sauces, and it can be used as a cooking oil. Moreover, stronger antioxidant activities of peach kernel meals in comparison to traditional oils justify investigation of its application as ingredients of functionally enriched foods. Apricot kernel oil was shown to be useful as a substitute for corn oil, and it was used in preparing some food products, such as biscuits, cakes, and macaroon paste, and for preparing fixed oil (Femenia et al., 1995). In addition, oxidative stability of apricot kernel oil is considered higher than that of rapeseed oil (Matthaus, Özcan, & Al Juhaimi, 2016; Matthäus & Özcan, 2009). Similarly, plum kernel oil was proposed as a substitute for almond oil and olive oil (Matthäus & Özcan, 2009). *Prunus* stones are used for oil production in the cosmetics industry, especially peach kernel oil, which is an ingredient in bath products, lotions, and creams (Wu et al., 2011), and apricot kernel oil (Alpaslan & Hayta, 2006).

Numerous reports exist in the literature on the oils recovered from kernel by-products as potential biodiesel feedstock (Górnás & Rudzínska, 2016; Gumus & Kasifoglu, 2010; Kostić, Veličković, Joković, Stamenković, & Veljković, 2016). Plum stones were examined as biofuel (Kostić et al., 2016). Recently, Górnaś, Rudzińska, and Soliven (2017) examined the potential of application of two species, *Prunus domestica* L. and *Prunus cerasifera* Ehrh., in the biodiesel industry, while Nowicki, Wachowska, and Pietrzak (2010) reported the application of plum stones in chemical preparation of active carbons and carbonaceous adsorbents.

7.1 Plant protection

Since the application of synthetic insecticides can often have disadvantages such as mammalian toxicity, insect resistance, and ecological hazards, integrated pest management (IPM) and organic production are trying to find alternatives for plant protection. Plants' insecticidal allelochemicals gave good results because many of these compounds are secondary plant substances, including alkaloids, quinones, and oils. The volatility, ephemeral nature, and biodegradability of essential oils make them especially advantageous in exploitation as pesticides (Tripathi, Dubey, & Shukla, 2008).

Prunus oils are proved to be nontoxic to animals and noninjurious to treated food commodities, and in some cases they have shown enhancement in the shelf life of commodities (Tripathi, 2016). Perumalsamy, Jang, Kim, Kadarkarai, and Ahn (2015) showed that some unsaturated and saturated fatty acids have ovicidal and larvicidal activity. *Prunus mahaleb* kernel oil shows strong antimicrobial, antifungal, and antioxidant activities due to its specific content of fatty acids (Mead et al., 2016), and therefore can be used for pharmaceutical and industrial purposes (Özçelik, Koca, Kaya, & Şekeroğlu, 2012). Mead et al. (2016) showed that mahaleb seed oil has ovicidal activity against cotton leafworm, *Spodoptera littoralis* (Boisd.), and it can be used as a bioretional pesticide in insect control. Choi et al. (2006) tested almond and apricot kernel oil, which showed some toxicity against *Lycoriella mali* (primary insect pests on mushrooms). Almond oil prevents the penetration of pest insects, including *Tribolium castaneum*, *Sitophilus granarius*, *Stegobium paniceum*, and *Rhyzopertha dominica*, to packaged cereals (Allahvaisi, 2010). Nollet and Rathore (2017) showed toxicity and repellent effect of *Prunus amygdalus* var. *amara* on the adult stage of some stored product insects, while Moawad and Sadek (2018) proved insecticidal and antifeedant activities against the eggs and larval stages of the cotton leaf worm insect. Peach oil is active after 48 months, being fungi toxic in nature at different temperatures between 5°C and 50°C, and exhibits a broad fungitoxicity to the mycelial growth of many fruit rotting fungi such as *Alternaria citri*, *Aspergillus niger*, *Aspergillus flavus*, *Botryodiplodia theobromae*, *Botrytis cinerea*, *Ceratocystis paradoxa*, *Colletotrichum gloeosporioides*, *Fusarium roseum*, *Geotrichum candidum*, *Monilinia fructicola*, *Mucor pyriformis*, *Penicillium italicum*, *Penicillium expansum*, *Phomopsis citri*, and *Rhizopus stolonifer* (Tripathi et al., 2008). According to Tworkoski (2002), plum seed oil has some herbicidal effect on dandelions (*Taraxacum officinale*).

7.2 Waste and residues application

Generally, some waste materials are in usage as animal feeds or fertilizers. Bearing in mind the fact that seeds from genus *Prunus* are high in protein content, it should be noted that residues of edible oils (after removal of amygdalin) are an important source of protein-rich livestock feed.

After the removal of the oil fraction, the kernel residues rich in sugars, minerals, fibers, vitamins (B1, B3, B5, and B6), and antioxidants (Stanojević et al., 2015; Wu et al., 2011; Yılmaz & Gökmen, 2013) also show high potential health benefits. The apricot kernels are rich in crude proteins, which, after oil extraction, leave behind a residue (press cake) that contains about 50% of crude protein (Hallabo, Wakeil-El, & Morsi, 1977). Similarly, plum stones contain a seed with high content in proteins and lipids (González-García, Luisa Marina, & Concepción García, 2014). Protein-rich by-products of peach, plum, apricot, and cherry were reported as a source of bioactive peptides with potential bioactivities, such as antioxidant activity, contributing to the prevention of oxidative stress and antihypertensive activity by inhibiting the in vitro activity of ACE inhibitors, and antiinflammatory properties (García, Endermann, González-García, & Marina, 2015; González-García, Puchalska, Marina, & García, 2015; Vásquez-Villanueva, Marina, & García, 2015; Zhu, Qiu, & Yi, 2010).

Defatted stone fruit kernel extracts are rich in phenolic compounds with strong cardioprotective effect (Bak et al., 2010). A study on the content of the defatted sour cherry seed extract showed abundance in some phenolic acids, flavonoids, and stilbenes, active compounds that were shown in some preclinical studies to be responsible for a strong cardioprotective role (Bak et al., 2006; Varga et al., 2017).

Some of the seeds can be ground to gluten-free flour (almond flour), which can be consumed by gluten-sensitive people having grain allergies and celiac disease (Ballhorn, 2011). Apricot kernels are also added to bakery products either whole or ground and are consumed as an appetizer, while peach kernels have been increasingly used in the production of persipan, a substitute for marzipan (Bruening, Haase, Matissek, & Fischer, 2011; Hayta & Alpaslan, 2011).

8 Future perspectives

Seeds from the fruit processing industry can be processed to obtain oils highly valued for their taste and composition, suitable for both edible and nonedible applications. Oils from the genus *Prunus* are considered good alternatives to other commonly used edible oils, especially as nontraditional sources in the food processing industry. Nevertheless, the usage of the oils obtained from *Prunus* species kernels in supplementation, the cosmetics industry, plant protection, as animal feeds, and fertilizers is not so wide and it is not well documented as it could be. Exploring the composition of the oil, which is obviously complex and rich in biochemical compounds, can affect its wider usage. For its more extensive utilization, not only identifying the bioactive compounds is required, but also determining the health benefits of the oil as a whole is needed. As was reviewed in this chapter, the individual chemical components have already been demonstrated to possess certain medicinal properties. As synergy of acting of bioactive components is possible, it would be particularly useful to examine the health benefits of each oil.

Therefore, further examination should be oriented to investigation of the chemical composition of different both major and minor *Prunus* species, while attention should be given to differences that could arise among cultivars. This is especially of interest as specific traits of oils could lead to new aspects for their usage along with minimizing agricultural wastes.

Acknowledgments

The authors are grateful to the Ministry of Science of the Republic of Serbia (Grant No. 172017) for financial support and to Miss Anita Smailagić for technical assistance.

References

Aitzetmüller, K. (1993). Capillary GLC fatty acid fingerprints of seed lipids—A tool in plant chemotaxonomy. *Journal of High Resolution Chromatography*, *16*, 488–490.

Allahvaisi, S. (2010). Reducing insects contaminations through stored foodstuffs by use of packaging and repellency essential oils. *Notulae Botanicae Horti Agrobotanici Cluj-Napoca*, *38*(3), 21–24.

Alpaslan, M., & Hayta, M. (2006). Apricot kernel: Physical and chemical properties. *Journal of the American Oil Chemists' Society*, *83*, 469–471.

Alrgei, H. O., Dabić, D., Natić, M., Rakonjac, V., Milojković-Opsenica, D., Tešić, Ž., et al. (2016). Chemical profile of major taste- and health-related compounds of (Oblačinska) sour cherry. *Journal of Agricultural and Food Chemistry*, *63*, 7476–7486.

Anwar, F., Manzoor, M., Bukhari, I. H., & Aladedunye, F. (2014). Physico-chemical attributes of fruit seed oils from different varieties of peach and plum. *Journal of Advances in Biology*, *4*, 384–392.

Anwar, F., Przybylski, R., Rudzinska, M., Gruczynska, E., & Bain, J. (2008). Fatty acid, tocopherol and sterol compositions of *Canadian prairie* fruit seed lipids. *Journal of the American Oil Chemists' Society*, *85*, 953–959.

Aust, O., Sies, H., Stahl, W., & Polidori, M. C. (2001). Analysis of lipophilic antioxidants in human serum and tissues: Tocopherols and carotenoids. *Journal of Chromatography A*, *936*, 83–93.

Bak, I., Czompa, A., Csepanyi, E., Juhasz, B., Kalantari, H., Najm, K., et al. (2011). Evaluation of systemic and dermal toxicity and dermal photoprotection by sour cherry kernels. *Phytotherapy Research*, *25*, 1714–1720.

Bak, I., Lekli, I., Juhasz, B., Nagy, N., Varga, E., Varadi, J., et al. (2006). Cardioprotective mechanisms of *Prunus cerasus* (sour cherry) seed extract against ischemia-reperfusion-induced damage in isolated rat hearts. *American Journal of Physiology Heart and Circulatory Physiology*, *291*, H1329–H1336.

Bak, I., Lekli, I., Juhasz, B., Varga, E., Varga, B., Gesztelyi, R., et al. (2010). Isolation and analysis of bioactive constituents of sour cherry (*Prunus cerasus*) seed kernel: An emerging functional food. *Journal of Medicinal Food*, *13*, 905–910.

Ballhorn, D. J. (2011). Cyanogenic glycosides in nuts and seeds. In V. R. Preedy (Ed.), *Nuts & seeds in health and disease prevention* (pp. 129–136). London, Burlington: RR Watson & VB Patel, Elsevier.

Baud, S. (2018). Seeds as oil factories. *Plant Reproduction*, *31*, 213–235.

Bernardo-Gil, G., Oneto, C., Antunes, P., Rodrigues, M. F., & Empis, J. M. (2001). Extraction of lipids from cherry seed oil using supercritical carbon dioxide. *European Food Research and Technology*, *212*, 170–174.

Bisht, T. S., Sharma, S. K., Sati, R. C., Rao, V. K., Yadav, V. K., Dixit, A. K., et al. (2011). Coronary heart disease prevention: Nutrients, foods, and dietary patterns. *Clinica Chimica Acta*, *412*, 1493–1514.

Blažková, J. (1988). The evaluation of stone character for the identification of sweet cherry cultivars. *Acta Horticulturae*, *224*, 285–294.

Bockisch, M. (1998). *Fats and oils handbook*: (pp. 2–4). Champaign, IL: AOCS Press.

Bolarinwa, I. F., Orfila, C., & Morgan, M. R. A. (2014). Amygdalin content of seeds, kernels, and food products commercially—Available in the UK. *Food Chemistry*, *152*, 133–139.

Bruening, P., Haase, I., Matissek, R., & Fischer, M. (2011). Marzipan: Polymerase chain reaction-driven methods for authenticity control. *Journal of Agricultural and Food Chemistry*, *59*, 11910–11917.

Caligiani, A., Bonzanini, F., Palla, G., Cirlini, M., & Bruni, R. (2010). Characterization of a potential nutraceutical ingredient: Pomegranate (*Punica granatum* L.) seed oil unsaponifiable fraction. *Plant Foods for Human Nutrition*, *65*, 277–283.

Călușaru, G., Claudia, F., Ionica, M. E., & Cosmulescu, S. N. (2017). Some fruit characteristics of blackthorn (*Prunus spinosa* L.). *Annals of the University of Craiova*, *22*, 129–136.

Campbell, O. E., & Padilla-Zakour, O. I. (2013). Phenolic and carotenoid composition of canned peaches (*Prunus persica*) and apricots (*Prunus armeniaca*) as affected by variety and peeling. *Food Research International*, *54*, 448–455.

Cevallos-Casals, B. A., Byrne, D., Okie, W. R., & Cisneros-Zevallos, L. (2006). Selecting new peach and plum genotypes rich in phenolic compounds and enhanced functional properties. *Food Chemistry*, *96*, 273–280.

Chamli, D., Bootello, M. A., Bouali, I., Jouhri, S., Boukhchina, S., & Martínez-Force, E. (2017). Chemical characterization and thermal properties of kernel oils from Tunisian peach and nectarine varieties of *Prunus persica'*. *Grasas y Aceites*, *68*(3), e211.

Chaurasiya, P. C., & Mishra, R. K. (2017). Varietal performance of Peach [*Prunus persica* (L.) Batsch] under northern hill zone of Chhattisgarh. *International Journal of Chemistry*, *5*, 37–40.

Chen, Z.-Y., Jiao, R., & Ma, K. Y. (2008). Cholesterol-lowering nutraceuticals and functional foods. *Journal of Agricultural and Food Chemistry*, *56*, 8761–8773.

Choi, W.-S., Park, B.-S., Lee, Y.-H., Jang, D. Y., Yoon, H. Y., & Lee, S.-E. (2006). Fumigant toxicities of essential oils and monoterpenes against Lycoriella mali adults. *Crop Protection*, *25*, 398–401.

Colaric, M., Veberic, R., Stampar, F., & Hudina, M. (2005). Evaluation of peach and nectarine fruit quality and correlations between sensory and chemical attributes. *Journal of the Science of Food and Agriculture*, *85*, 2611–2616.

Čolić, S. D., Fotirić Akšić, M. M., Lazarević, K. B., Zec, G. N., Gašić, U. M., Dabić Zagorac, D. Č., et al. (2017). Fatty acid and phenolic profiles of almond grown in Serbia. *Food Chemistry*, *234*, 455–463.

Čolić, S., Rakonjac, V., Zec, G., Nikolić, D., & Fotirić Akšić, M. (2012). Morphological and biochemical evaluation of selected almond [*Prunus dulcis* (Mill.) D.A.Webb] genotypes in northern Serbia. *Turkish Journal of Agriculture and Forestry*, *36*, 429–438.

Comas, J. F., Socias, I., Company, R., & Segura, J. M. A. (2019). Shell hardness in almond: Cracking load and kernel percentage. *Scientia Horticulturae*, *245*, 7–11.

Comes, F., Farines, M., Aumelas, A., & Soulier, J. (1992). Fatty acids and triacylglycerols of cherry seed oil. *Journal of the American Oil Chemists' Society*, *69*, 1224–1227.

Cosmulescu, S. N., Ionica, M. E., & Mutu, N. (2018). Evaluation on genetic diversity of phenotypictraits in Myrobalan plum (*Prunus cerasifera* Ehrh.). *South-Western Journal of Horticulture, Biology, and Environment*, *9*, 25–34.

Das, U. N. (2004). From bench to the clinic: γ-linolenic acid therapy of human gliomas. *Prostaglandins, Leukotrienes, and Essential Fatty Acids*, *70*, 539–552.

Davidović, S., Veljović, M., Pantelić, M., Baošić, R., Natić, M., Dabić, D., et al. (2013). Physicochemical, antioxidant and sensory properties of peach wine made from Redhaven cultivar. *Journal of Agricultural and Food Chemistry*, *61*(6), 1357–1363.

De Stefani, E., Boffetta, P., Ronco, A. L., Brennan, P., Deneo-Pellegrini, H., Carzoglio, J. C., et al. (2000). Plant sterols and risk of stomach cancer: A case-control study in Uruguay. *Nutrition and Cancer*, *37*, 140–144.

Deineka, V. I., Gabruk, N. G., Deineka, L. A., & Manokhina, L. A. (2002). Tryglyceride composition of oil from stones of nine Rosaceae plants. *Chemistry of Natural Compounds*, *38*, 410–412.

Dimkova, S., Ivanova, D., Todorova, S., & Marinova, N. (2017). Biometrical indicators of fresh fruits of bulgarian and introduced plum cultivars of *Prunus domestica* L. *Bulgarian Journal of Agricultural Science*, *23*, 947–950.

Durmaz, G., Karabulut, I., Topçu, A., Asiltürk, M., & Kutlu, T. (2010). Roasting-related changes in oxidative stability and antioxidant capacity of apricot kernel oil. *Journal of the American Oil Chemist's Society*, *87*, 401–409.

Eduardo, L. H. (2010). Health effects of oleic acid and long chain omega-3 fatty acids (EPA and DHA) enriched milks. A review of intervention studies. *Pharmacological Research*, *61*, 200–207.

Engin, K. N., Engin, G., Kucuksahin, H., Oncu, M., & Guvener, B. (2007). Clinical evaluation of the neuroprotective effect of alpha-tocopherol against glaucomatous damage. *European Journal of Ophthalmology*, *17*, 528–533.

Erciyes, A. T., Tüter-Erim, M., Kabasakal, O. S., & Dandik, L. (1995). Seed oil characteristics of *Onopordum tauricum* Willd. and *Prunus laurocerasus* L. *European Journal of Lipid Science and Technology*, *97*, 387–388.

FAOStat (2017). Available from: (2017). http://www.fao.org/faostat/en/#data/QC [February 2019].

Farines, M., Soulier, J., & Comes, F. (1986). Study of the glyceridic fraction of lipids of seeds from rosaceae prunoids. *Revue Francaise des Corps Gras*, *33*, 115–118.

Fasina, O. O., Craig-Schmidt, M., Colley, Z., & Hallman, H. (2008). Predicting melting characteristics of vegetable oils from fatty acid composition. *Journal of Food Science and Technology*, *41*, 1501–1505.

Femenia, A., Rosello, C., Mulet, A., & Canellas, J. (1995). Chemical composition of bitter and sweet apricot kernels. *Journal of Agricultural and Food Chemistry*, *43*, 356–361.

Finch, J., Munhutu, M. N., & Whitaker-Worth, D. L. (2010). Atopic dermatitis and nutrition. *Clinics in Dermatology*, *28*, 605–614.

Fotirić Akšić, M., Dabić, D., Gašić, U., Zec, G., Vulić, T., Tešić, Ž., et al. (2015). Polyphenolic profile of pear leaves with different resistance to pear psylla (*Cacopsylla pyri*). *Journal of Agricultural and Food Chemistry*, *63*(34), 7476–7486.

Fratianni, F., Ombra, M. N., d'Acierno, A., Cipriano, L., & Nazzaro, F. (2018). Apricots: Biochemistry and functional properties. *Current Opinion in Food Science*, *19*, 23–29.

Ganjewala, D., Kumar, S., Asha, D. S., & Ambika, K. (2010). Advances in cyanogenic glycosides biosynthesis and analysis in plants: A review. *Acta Biologica Szegediensis*, *54*, 1–14.

García, M. C., Endermann, J., González-García, E., & Marina, M. L. (2015). Identification of antioxidant and antihypertensive peptides in cherry (*Prunus cerasus* L.) kernels by HPLC-Q-TOF-MS. *Journal of Agricultural and Food Chemistry*, *63*, 1514–1520.

Gayas, B., Kaur, G., & Gul, K. (2016). Ultrasound-assisted extraction of apricot kernel oil: Effects on functional and rheological properties. *Journal of Food Process Engineering*, *40*, e12439.

Gezer, I., & Dikilitas, S. (2002). The study of work process and determination of some working parameters in an apricot pit processing plant in Turkey. *Journal of Food Engineering*, *53*, 111–114.

Ghanbari, R., Anwar, F., Alkharfy, K. M., Gilani, A.-H., & Saari, N. (2012). Valuable nutrients and functional bioactives in different parts of olive (*Olea europaea* L.)—A review. *International Journal of Molecular Sciences*, *13*, 3291–3340.

Ghiulai, V. M., Socaciu, C., Jianu, I., Ranga, F., & Fetea, F. (2006). Identification and quantitative evaluation of amygdalin from apricot, plum and peach oils, and kernels. *Bulletin of University of Agricultural Sciences and Veterinary Medicine Cluj-Napoca Agriculture*, *62*, 246–253.

Godini, A. (1984). Hull, shell, and kernel relationships in almond fresh fruits. *Options Méditerranéennes*, *84*, 53–56.

González-García, E., Luisa Marina, M., & Concepción García, M. (2014). Plum (*Prunus Domestica* L.) by-product as a new and cheap source of bioactive peptides: Extraction method and peptides characterization. *Journal of Functional Foods*, *11*, 428–437.

González-García, E., Puchalska, P., Marina, M. L., & García, M. C. (2015). Fractionation and identification of antioxidant and angiotensin-converting enzyme-inhibitory peptides obtained from plum (*Prunus Domestica* L.) stones. *Journal of Functional Foods*, *19*, 376–384.

Górnás, P., & Rudzínska, M. (2016). Seeds recovered from industry by-products of nine fruit species with a high potential utility as a source of unconventional oil for biodiesel and cosmetic and pharmaceutical sectors. *Industrial Crops and Products*, *83*, 329–338.

Górnaś, P., Rudzińska, M., Raczyk, M., Mišina, I., & Segliņa, D. (2016). Impact of cultivar on profile and concentration of lipophilic bioactive compounds in kernel oils recovered from sweet cherry (*Prunus avium* L.) by-product. *Plant Foods for Human Nutrition*, *71*, 158–164.

Górnaś, P., Rudzińska, M., Raczyk, M., Mišina, I., Soliven, A., Lacis, G., et al. (2016). Impact of species and variety on concentrations of minor lipophilic bioactive compounds in oils recovered from plum kernels. *Journal of Agricultural and Food Chemistry*, *64*, 898–905.

Górnaś, P., Rudzińska, M., Raczyk, M., Mišina, I., Soliven, A., & Seglina, D. (2016). Composition of bioactive compounds in kernel oils recovered from sour cherry (*Prunus cerasus* L.) by-products: Impact of the cultivar on potential applications. *Industrial Crops and Products*, *82*, 44–50.

Górnaś, P., Rudzińska, M., & Soliven, A. (2017). Industrial by-products of plum *Prunus domestica* L. and *Prunus cerasifera* Ehrh. as potential biodiesel feedstock: Impact of variety. *Industrial Crops and Products*, *100*, 77–84.

Güçlü-Üstündağ, Ö., & Temelli, F. (2004). Correlating the solubility behavior of minor lipid components in supercritical carbon dioxide. *Journal of Supercritical Fluids*, *31*, 235–253.

Gumus, M., & Kasifoglu, S. (2010). Performance and emission evaluation of a compression ignition engine using a biodiesel (apricot seed kernel oil methyl ester) and its blends with diesel fuel. *Biomass & Bioenergy*, *34*, 134–139.

Gupta, A., & Sharma, P. C. (2009). Standardization of technology for extraction of wild apricot kernel oil at semi-pilot scale. *Biological Forum-an International Journal*, *1*(1), 51–64.

Gur, M. I. (1980). The biosynthesis of triacylglycerols. P. K. Stumpf (Ed.), *The biochemistry of plants: Vol. 4* (pp.205–248). New York, NY: Academic Press.

Hallabo, S. A. S., Wakeil-El, F. A., & Morsi, M. K. S. (1977). Chemical and physical properties of apricot kernels, apricot kernel oil, and almond kernel oil. *Egyptian Journal of Food Science*, *3*, 1–6.

Harris, W. S., Kris-Etherton, P. M., & Harris, K. A. (2008). Intakes of long-chain omega-3 fatty acid associated with reduced risk for death from coronary heart disease in healthy adults. *Current Atheroscerosis Reports*, *10*, 503–509.

Hassanein, M. M. M. (1999). Studies on non-traditional oils: I. Detailed studies on different lipid profiles of some Rosaceae kernel oils. *Grasas y Aceites*, *50*, 379–384.

Hassanien, M. M. M., Abdel-Razek, A. G., Rudzińska, M., Siger, A., Ratusz, K., & Przybylski, R. (2014). Phytochemical contents and oxidative stability of oils from non-traditional sources. *European Journal of Lipid Science and Technology*, *116*, 1563–1571.

Hayta, M., & Alpaslan, M. (2011). Apricot kernel flour and its use in maintaining health. In V. R. Preedy (Ed.), *Flour and breads and their fortification in health and disease prevention* (pp. 213–221). London: RR Watson & VB Patel, Academic Press.

Islam, A. (2002). 'Kiraz'-cherry laurel (*Prunus laurocerasus*). *New Zealand Journal of Crop and Horticultural Science*, *30*, 301–302.

Jones, P. J. H., & AbuMweis, S. S. (2009). Phytosterols as functional food ingredients: Linkages to cardiovascular disease and cancer. *Current Opinion in Clinical Nutrition and Metabolic Care*, *12*, 147–151.

Kamel, B. S., & Kakuda, Y. (1992). 'Characterization of the seed oil and meal from apricot, cherry, nectarine, peach and plum. *Journal of the American Oil Chemists' Society*, *69*, 492–494.

Kapoor, R., & Huang, Y.-S. (2006). Gamma linolenic acid: An antiinflammatory omega-6 fatty acid. *Current Pharmaceutical Biotechnology*, *7*, 531–534.

Katan, M. B., Grundy, S. M., Jones, P., Law, M., Miettinen, T., & Paoletti, R. (2003). Efficacy and safety of plant stanols and sterols in the management of blood cholesterol levels. *Mayo Clinic Proceedings*, *78*, 965–978.

Khan, L. M., & Hanna, M. A. (1983). Expression of oil from oilseeds—A review. *Journal of Agricultural Engineering Research*, *28*, 495–503.

Kim, D.-O., Chun, O. K., Kim, Y. J., Moon, H.-Y., & Lee, C. Y. (2003). Quantification of polyphenolics and their antioxidant capacity in fresh plums. *Journal of Agricultural and Food Chemistry*, *51*, 6509–6515.

Kiralan, M., Kayahan, M., Kiralan, S. S., & Ramadan, M. F. (2018). Effect of thermal and photo oxidation on the stability of cold-pressed plum and apricot kernel oils. *European Food Research and Technology*, *244*, 31–42.

Kodad, O., Estopañán Muñoz, G., Juan Esteban, T., Alonso Segura, J. M., Espiau Ramírez, M. T., Socias, I., et al. (2014). Kernel quality of local Spanish almond cultivars: Provenance variability and end uses. *FAO-CIHEAM—Nucis-Newsletter*, *16*, 16–19.

Koprivica, M. R., Trifković, J. Đ., Dramićanin, A. M., Gašić, U. M., Fotirić Akšić, M., & Milojković-Opsenica, D. M. (2018). Determination of the phenolic profile of peach (*Prunus persica* L.) kernels using UHPLC–LTQ OrbiTrap MS/MS technique. *European Food Research and Technology*, *244*, 2051–2064.

Kostadinovic-Velickovska, S., Bruhl, L., Mitrev, S., Mirhosseini, H., & Matthaus, M. (2015). Quality evaluation of cold-pressed edible oils from Macedonia. *European Journal of Lipid Science and Technology*, *117*, 2023–2035.

Kostić, M. D., Veličković, A. V., Joković, N. M., Stamenković, O. S., & Veljković, V. B. (2016). Optimization and kinetic modeling of esterification of the oil obtained from waste plum stones as a pretreatment step in biodiesel production. *Waste Management*, *48*, 619–629.

Kritchevsky, D., & Chen, S. C. (2005). Phytosterols-health benefits and potential concerns: A review. *Nutrition Research*, *25*, 413–428.

Kutlu, T., Durmaz, G., Ateş, B., & Erdoğan, A. (2009). Protective effect of dietary apricot kernel oil supplementation on cholesterol levels and antioxidant status of liver in hypercholesteremic rats. *Journal of Food, Agriculture and Environment*, *7*, 61–65.

Lauritzen, I., Blondeau, N., Heurteaux, C., Widmann, C., Romey, G., & Lazdunski, M. (2000). Polyunsaturated fatty acids are potent neuroprotectors. *EMBO Journal*, *19*(8), 1784–1793.

Law, M. R. (2000). Plant sterol and stanol margarines and health. *Best Practice*, *173*(1), 43–47.

Lazos, E. S. (1991). Composition and oil characteristics of apricot, peach and cherry kernel. *Grasas y Aceites*, *42*, 127–131.

Li, T. S. C. (2009). *Vegetables and fruits: Nutritional and therapeutic values*. Boca Raton, FL: CRC Press, Taylor & Francis Group.

Lin, J. E., Chen, A. H. C., & Grethlein, H. (1990). *Feasibility study of using cherry pits as a natural benzaldehyde source*. Report 1 to cherry marketing institute; Michigan Biotechnology Institute, MI, USA.

Liu, R. H. (2013). Dietary bioactive compounds and their health implications. *Journal of Food Science*, *78*, A18–A25.

Lloyd, P. J., & van Wyk, J. (2012). Introduction to extraction in food processing. In N. Lebovka, E. Vorobiev, & F. Chemat (Eds.), *Enhancing extraction processes in the food industry* (pp. 1–25). CRC Press, Taylor & Francis Group, BR.

Mainou-Fowler, T., Proctor, S. J., & Dickinson, A. M. (2001). γ-Linolenic acid induces apoptosis in B-chronic lymphocytic leukemia cells in vitro. *Leukemia & Lymphoma*, *40*, 393–403.

Malcolm, P. J., Holford, P., Mc-Glasson, W. B., & Newman, S. (2003). Temperature and seed weight affect the germination of Peach rootstock seeds and the growth of root stock seedlings. *Scientia Horticulturae*, *98*, 247–256.

Mariette, S., Tavaud, M., Arunyawat, U., Capdeville, G., Millan, M., & Salin, F. (2010). Population structure and genetic bottleneck in sweet cherry estimated with SSRs and the gametophytic self-incompatibility locus. *BMC Genetics*, *11*, 77–80.

Matthäus, B., & Özcan, M. M. (2009). Fatty acids and tocopherol contents of some *Prunus* spp. kernel oils. *Journal of Food Lipids*, *16*, 187–199.

Matthäus, B., & Özcan, M. M. (2014). Fatty acid, tocopherol and squalene contents of Rosaceae seed oils. *Botanical Studies*, *55*, 48.

Matthaus, B., Özcan, M. M., & Al Juhaimi, F. (2016). Fatty acid composition and tocopherol content of the kernel oil from apricot varieties (Hasanbey, Hacihaliloglu, Kabaasi and Soganci) collected at different harvest times. *European Food Research and Technology*, *242*, 221–226.

Matvienko, O. A., Lewis, D. S., Swanson, M., Arndt, B., Rainwater, D. L., Stewart, J., et al. (2002). A single daily dose of soybean phytosterols in ground beef decreases serum total cholesterol and LDL cholesterol in young, mildly hypercholesterolemic men. *American Journal of Clinical Nutrition*, *76*, 57–64.

Mead, H. M., El-Shafiey, S. N., & Sabry, H. M. (2016). Chemical constituents and ovicidal effects of mahlab, *Prunus mahaleb* L. kernels oil on cotton leaf worm, *Spodoptera littoralis* (Boisd.) eggs. *Journal of Plant Protection Research*, *56*, 279–290.

Milazzo, S., Lejeune, S., & Ernst, E. (2006). Laetrile for cancer: A systematic review of the clinical evidence. *Supportive Care in Cancer*, *15*, 583–595.

Moawad, S. S., & Sadek, H. E. (2018). Evaluation of two eco friendly botanical oils on cotton leaf worm, *Spodoptera littoralis* (Boisd) (Lepidoptera/ Noctuidae). *Annals of Agricultural Science*, *63*, 141–144.

Morales, F. J., & Perez, S. J. (2001). Free radical scavenging capacity of Maillard reaction products as related to colour and fluorescence. *Food Chemistry*, *72*, 119–125.

Morris, M. C., Evans, D. A., Tangney, C. C., Bienias, J. L., Wilson, R. S., Aggarwal, N. T., et al. (2005). Relation of the tocopherol forms to incident Alzheimer disease and to cognitive change. *American Journal of Clinical Nutrition*, *81*, 508–514.

Mratinić, E., Popovski, B., Milošević, T., & Popovska, M. (2011). Evaluation of apricot fruit quality and correlations between physical and chemical attributes. *Czech Journal of Food Sciences*, *29*, 161–170.

Natić, M., Dabić, D., Papetti, A., Fotirić Akšić, M., Ognjanov, V., Ljubojević, M., et al. (2015). Analysis and characterisation of phytochemicals in Mulberry (*Morus alba* L.) fruits grown in Vojvodina, North Serbia. *Food Chemistry*, *171*, 128–136.

Nollet, L. M. L., & Rathore, H. S. (2017). *Green pesticides handbook: Essential oils for pest control.* CRC Press.

Nowicki, P., Wachowska, H., & Pietrzak, R. (2010). Active carbons prepared by chemical activation of plum stones and their application in removal of NO_2. *Journal of Hazardous Materials*, *181*, 1088–1094.

Ostlund, R. E., Jr. (2002). Phytosterols in human nutrition. *Annual Review of Nutrition*, *22*, 533–549.

Ostlund, R. E., Jr. (2004). Phytosterols and cholesterol metabolism. *Current Opinion in Lipidology*, *15*, 37–41.

Owen, R. W., Haubner, R., Würtele, G., Hull, W. E., Spiegelhalder, B., & Bartsch, H. (2004). Olives and olive oil in cancer prevention. *European Journal of Cancer Prevention*, *13*, 319–326.

Ozcan, M. M., Unvera, A., Erkanb, E., & Arslana, D. (2011). Characterization of some almond kernel and oils. *Food Chemistry*, *127*, 330–333.

Özçelik, B., Koca, U., Kaya, D. A., & Şekeroğlu, N. (2012). Evaluation of the in vitro bioactivities of mahaleb cherry (*Prunus mahaleb* L.). *Romanian Biotechnological Letters*, *17*, 7863–7872.

Pande, G., & Akoh, C. C. (2010). Organic acids, antioxidant capacity, phenolic content and lipid characterisation of Georgia-grown underutilized fruit crops. *Food Chemistry*, *120*, 1067–1075.

Pantelić, M., Dabić, D., Matijašević, S., Davidović, S., Dojčinović, B., Milojković-Opsenica, D., et al. (2014). Chemical characterization of fruit wine made from Oblačinska sour cherry. *The Scientific World Journal*, *2014*, 9.

Pantelić, M., Dabić Zagorac, D., Davidović, S., Todić, S., Bešlić, Z., Gašić, U., et al. (2016). Identification and quantification of phenolic compounds in berry skin, pulp, and seeds in 13 grapevine varieties grown in Serbia. *Food Chemistry*, *211*, 243–252.

Pavlović, A., Dabić, D., Momirović, N., Dojčinović, B., Milojković-Opsenica, D., Tešić, Ž., et al. (2013). Chemical composition of two different extracts of berries harvested in Serbia. *Journal of Agricultural and Food Chemistry*, *61*(17), 4188–4194.

Pavlovic, A., Papetti, A., Dabic Zagorac, D., Gasic, U., Misic, D., Tesic, Z., et al. (2016). Phenolics composition of leaf extracts of raspberry and blackberry cultivars grown in Serbia. *Industrial Crops and Products*, *87*, 304–314.

Pavlović, N., Vidović, S., Vladić, J., Popović, L., Moslavac, T., Jakobović, S., et al. (2018). Recovery of tocopherols, amygdalin, and fatty acids from apricot kernel oil: Cold pressing versus supercritical carbon dioxide. *European Journal of Lipid Science and Technology*, *120*(11), 1800043.

Pelentir, N., Block, J. M., Fritz, A. R. M., Reginatto, V., & Amante, E. R. (2011). Production and chemical characterization of peach (*Prunus persica*) kernel flour. *Journal of Food Process Engineering*, *34*, 1253–1265.

Perona, J. S., & Ruiz-Gutierrez, V. (2005). Triacylglycerol molecular species are depleted to different extents in the myocardium of spontaneously hypertensive rats fed two oleic acid-rich oils. *American Journal of Hypertension*, *18*, 72–80.

Perumalsamy, H., Jang, M. J., Kim, J. R., Kadarkarai, M., & Ahn, Y. J. (2015). Larvicidal activity and possible mode of action of four flavonoids and two fatty acids identified in *Millettia pinnata* seed toward three mosquito species. *Parasites & Vectors*, *8*, 237.

Picuric-Jovanovic, K., Vrbaski, Z., & Milovanovic, M. (1997). Aqueous-enzymatic extraction of plum kernel oil. *Fett-Lipid*, *99*, 433–435.

Picuric-Jovanovic, K., Vrbaski, Z., & Milovanovic, M. (1999). Influence of the aqueous-enzymatic method on the oxidative stability of plum kernel oil. *Fett-Lipid*, *101*, 109–112.

Popa, V. M., Misca, C., Bordean, D., Raba, D. N., Stef, D., & Dumbrava, D. (2011). Characterization of sour cherries (*Prunus cerasus*) kernel oil cultivars from Banat. *Journal of Agroalimentary Processes and Technologies*, *17*, 398–401.

Potter, D., Eriksson, T., Evans, R. C., Oh, S., Smedmark, J. E. E., Morgan, D. R., et al. (2007). Phylogeny and classification of Rosaceae. *Plant Systematics and Evolution*, *266*, 5–43.

Rahma, E. H. (1988). Chemical characterization of peach kernel oil and protein: Functional properties, in vitro digestibility and amino acids profile of the flour. *Food Chemistry*, *28*, 31–43.

Rakonjac, V., Fotirić Akšić, M., Nikolić, D., Milatović, D., & Čolić, S. (2010). Morphological characterization of 'Oblačinska' sour cherry by multivariate analysis. *Scientia Horticulturae*, *125*, 679–684.

Reddy, L. H., & Couvreur, P. (2009). Squalene: A natural triterpene for use in disease management and therapy. *Advanced Drug Delivery Reviews*, *61*, 1412–1426.

Ricochon, G., & Muniglia, L. (2010). Influence of enzymes on the oil extraction processes in aqueous media. *Oléagineux, Corps Gras, Lipides*, *17*(6), 356–359.

Rudkowska, I. (2010). Plant sterols and stanols for healthy ageing. *Maturitas*, *66*, 158–162.

Sánchez-Vicente, Y., Cabañas, A., Renuncio, J. A. R., & Pando, C. (2009). Supercritical fluid extraction of peach (*Prunus persica*) seed oil using carbon dioxide and ethanol. *Journal of Supercritical Fluids*, *49*, 167–173.

Savić, I. M., Nikolić, V. D., Savić-Gajić, I. M., Kundaković, T. D., Stanojković, T. P., & Najman, S. J. (2016). Chemical composition and biological activity of the plum seed extract. *Advanced Technologies*, *5*, 38–45.

Sbihi, H. M., Imededdine, A. N., & Al-Resayes, S. I. (2014). Characterization of white mahlab (*Prunus mahaleb* L.) seed oil: A rich source of α-eleostearic acid. *Journal of Food Science*, *79*, C795–C801.

Segers, F. (2002). *Therapeutic products containing cherry pits.* US patent 6383053 B1.

Siano, F., Straccia, M. C., Paolucci, M., Fasulo, G., Boscaino, F., & Volope, M. G. (2016). Physico-chemical properties and fatty acid composition of pomegranate, cherry and pumpkin seed oils. *Journal of the Science of Food and Agriculture*, *96*, 1730–1735.

Stanojević, M., Trifković, J., Fotirić Akšić, M., Rakonjac, V., Nikolić, D., Šegan, S., et al. (2015). Sugar profile of kernels as a marker of origin and ripening time of peach (*Prunus persicae* L.). *Plant Foods for Human Nutrition*, *70*, 433–440.

Straccia, M. C., Siano, F., Coppola, R., La Cara, F., & Volpe, M. G. (2012). Extraction and characterization of vegetable oils from cherry seed by different extraction processes. *Chemical Engineering Transactions*, *27*, 391–396.

Tian, H., Zhang, H., Zhan, P., & Tian, F. (2011). Composition and antioxidant and antimicrobial activities of white apricot almond (*Amygdalus communis* L.) oil. *European Journal of Lipid Science and Technology*, *113*, 1138–1144.

Traber, M. G., & Atkinson, J. (2007). Vitamin E, antioxidant and nothing more. *Free Radical Biology and Medicine*, *43*, 4–15.

Tripathi, P. (2016). Pharmacological evaluation of essential oils of *Ocimum sanctum*, *Prunus persica* and *Zingiber officinale*. *Journal of Innovations in Pharmaceutical and Biological Sciences*, *3*, 7–11.

Tripathi, P., Dubey, N. K., & Shukla, A. K. (2008). Use of some essential oils as post-harvest botanical fungicides in the management of grey mould of grapes caused by Botrytis cinerea. *World Journal of Microbiology and Biotechnology*, *24*, 39–46.

Tsuzuki, T., Tokuyama, Y., Igarashi, M., & Miyazawa, T. (2004). Tumor growth suppression by alpha-eleostearic acid, a linolenic acid isomer with a conjugated triene system, via lipid peroxidation. *Carcinogenesis*, *25*, 1417–1425.

Turan, S., Topcu, A., Karabulut, I., Vural, H., & Hayaloglu, A. A. (2007). Fatty acid, triacylglycerol, phytosterol, and tocopherol variations in kernel oil of Malatya apricots from Turkey. *Journal of Agricultural and Food Chemistry*, *55*, 10787–10794.

Tworkoski, T. (2002). Herbicide effects of essential oils. *Weed Science*, *50*, 425–431.

Uluata, S. (2016). Effect of extraction method on biochemical properties and oxidative stability of apricot seed oil. *Akademik Gıda*, *14*, 333–340.

Uluata, S., & Özdemir, N. (2017). Evaluation of chemical characterization, antioxidant activity and oxidative stability of some waste seed oil. *Turkish Journal of Agriculture: Food Science and Technology*, *5*, 48–53.

USDA (2006). *What's in the foods you eat search tool.* Available from: http://www.ars.usda.gov/ba/bhnrc/fsrg [February 2006].

Usenik, V., Fabcic, J., & Stampar, F. (2008). Sugar, organic acids, phenolic composition and antioxidant activity of sweet cherry (*Prunus avium* L.). *Food Chemistry*, *107*, 185–192.

Varga, B., Priksz, D., Lampé, N., Bombicz, M., Kurucz, A., Szabó, A. M., et al. (2017). Protective effect of *Prunus cerasus* (sour cherry) seed extract on the recovery of ischemia/reperfusion-induced retinal damage in Zucker diabetic fatty rat. *Molecules*, *22*(10), 1782.

Vásquez-Villanueva, R., Marina, M. L., & García, M. C. (2015). Revalorization of a peach (*Prunus Persica* L. Batsch) byproducts: Extraction and characterizatiion of ACE-inhibitory peptides from peach stones. *Journal of Functional Foods*, *18*, 137–146.

Veličković, D., Ristić, M., Karabegović, I., Stojičević, S., Nikolić, N., & Lazić, M. (2016). Plum (*Prunus domestica*) and walnut (*Juglans regia*): Volatiles and fatty oils. *Advanced Technologies*, *5*, 10–16.

Vivancos, M., & Moreno, J. J. (2005). β-Sitosterol modulates antioxidant enzyme response in RAW 264.7 macrophages. *Free Radical Biology and Medicine*, *39*, 91–97.

Wang, L., & Yu, H. (2012). Biodiesel from Siberian apricot (*Prunus sibirica* L.) seed kernel oil. *Bioresource Technology*, *112*, 355–358.

Wathes, D. C., Abayasekara, D. R. E., & Aitken, R. J. (2007). Polyunsaturated fatty acids in male and female reproduction. *Biology of Reproduction*, *77*, 190–201.

Watkins, G., Martin, T. A., Bryce, R., Mansel, R. E., & Jiang, W. G. (2005). γ-Linolenic acid regulates the expression and secretion of SPARC in human cancer cells. *Prostaglandins, Leukotrienes and Essential Fatty Acids*, *72*, 273–278.

Woyengo, T. A., Ramprasath, V. R., & Jones, P. J. H. (2009). Anticancer effects of phytosterols. *European Journal of Clinical Nutrition*, *63*, 813–820.

Wu, H., Shi, J., Xue, S. J., Kakuda, Y., Wang, D., Jiang, Y. M., et al. (2011). Essential oil extracted from peach (*Prunus persica*) kernel and its physicochemical and antioxidant properties. *LWT - Food Science and Technology*, *44*, 2032–2039.

Yasui, Y., Hosokawa, M., Kohno, H., Tanaka, T., & Miyashita, K. (2006). Growth inhibition and apoptosis induction by all-trans-conjugated linolenic acids on human colon cancer cells. *Anticancer Research*, *26*, 1855–1860.

Yılmaz, C., & Gökmen, V. (2013). Compositional characteristics of sour cherry kernel and its oil as influenced by different extraction and roasting conditions. *Industrial Crops and Products*, *49*, 130–135.

Zhang, J., Gu, H.-D., Zhang, L., Tian, Z. J., Zhang, Z. Q., Shi, X. C., et al. (2011). Protective effects of apricot kernel oil on myocardium against ischemia-reperfusion injury in rats. *Food and Chemical Toxicology*, *49*, 3136–3141.

Zhou, B., Wang, Y., Kang, J., Zhong, J., & Prenzler, P. D. (2016). The quality and volatile-profile changes of Longwangmo apricot (*Prunus armeniaca* L.) kernel oil prepared by different oil-producing processes. *European Journal of Lipid Science and Technology*, *118*, 236–243.

Zhu, Z., Qiu, N., & Yi, J. (2010). Production and characterization of angiotensin converting enzyme (ACE) inhibitory peptides from apricot (*Prunus armeniaca* L.) kernel protein hydrolysate. *European Food Research and Technology*, *231*, 13–19.

Zlatanov, M., & Janakieva, I. (1998). Phospholipid composition of some fruit-stone oils of Rosaceae species. *Fett-Lipid*, *100*, 312–315.

Chapter 57

Cold pressed *Eucommia ulmoides* oliver oil

Monika Choudhary
Punjab Agricultural University, Ludhiana, India

1 Introduction

Eucommia ulmoides (EU) oliver being native to China is commonly known as Du Zhong and belongs to the Eucommiaceae family (Zhu, Wen, Zhu, Su, & Sun, 2016). This plant is widely popular in local dwellers owing to its oil, which possesses proven medicinal properties and it is widely distributed from subtropical to temperate areas. About 112 compounds, namely iridoids, phenolics, and steroids, have been isolated from *Eucommia ulmoides* (Deyama, Nishibe, & Nakazawa, 2001). These compounds have been investigated for numerous pharmacological functions, such as antihypertensive, antioxidative, and antiinflammatory effects, immunity enhancement, and blood lipids regulation (Berry & Hirsch, 1986; Deyama et al., 2001; Ren, Yang, Zheng, & Hu, 2009; Xiang & Li, 2011). Discussing the major by-products of EU, researchers were primarily focused on the isolation of monomeric compounds, such as polysaccharide, chlorogenic acid, lignans, flavonoids, and other trace elements from bark and leaves, but seeds of this medicinal fruit have received less attention (He et al., 2014). Nevertheless, recent studies have recognized seed oil of this woody species as the main by-product of EU. The oil content of EU seeds is approximately 10%–15%, but after peeling oil content may increase to 35%, of which 63% accounts for unsaturated fatty acids, predominantly essential omega-3 (linolenic acid) fatty acid (Zhang, Liu, & Che, 2018a). The oil content of the EU is higher than that of soybean oil (18%–22%), which is one of the most important oilseeds in the world (Medic, Atkinson, & Hurburgh, 2014). In recent times, EU oil has shown excellent market potential owing to its health-promoting traits, and is being used as an edible oil and as a concentrated source of linolenic acid in the form of soft gel capsules.

2 Extraction and processing cold pressed oil

In the conventional method, extraction is carried out following the drying of seeds of *Eucommia* fruits (Du et al., 2018). Firstly, fruits are picked and stored in a cool, dry place until the pericarp is brown. The seed kernel is separated by peeling off the pericarp by hand and then oven-dried at 45°C. The seed oil is extracted using Soxhlet extraction method. For this, 10 g of seed sample is ground and packed with filter paper tube and put into a container, then petroleum ether (b.p. 60–90°C) is added and the seeds are subjected to water bath reflux extraction for 10 h at 65°C. After extraction, residual petroleum ether is removed using a rotary evaporator apparatus and dried at 105°C for 0.5 h.

In the conventional method, a nonpolar solvent, usually *n*-hexane, is used for extraction in the edible oil industry to obtain crude oil (Johnson, 1997). However, in 2002 *n*-hexane was listed among 187 hazardous air pollutants by National-Scale Air Toxics Assessments, because of its toxic nature (United States Environmental Protection Agency, 2011). Compared to the conventional method, extraction of oils from seeds using cold pressing technology has been evolved as a better alternative due to the usage of carbon dioxide (CO_2) as a solvent, which has many advantages. CO_2 has emerged as a substitute for *n*-hexane as it is easy to separate from the extracted oil and leaves no residual effect on the final product (Eller, Cermak, & Taylor, 2011). CO_2 also acts as a better solvent for the extraction of other nonpolar solutes in plants and other biological substrates due to its nontoxic, nonflammable properties with zero solvent disposal cost (Sarmento, Ferreira, & Hense, 2006; Silva, Mendes, Pessoa, & Queiroz, 2008; Zacchi, Daghero, Jaeger, & Eggers, 2006).

Cold Pressed Oils. https://doi.org/10.1016/B978-0-12-818188-1.00057-8
Copyright © 2020 Elsevier Inc. All rights reserved.

3 Extraction of *Eucommia ulmoides* seed oil using cold pressing technology

Cold pressing technology provides the means for extracting and preparing *Eucommia* seed oil at a low temperature and enables the high yield of α-linolenic acid (Uquiche, Romero, Ortíz, & Del Valle, 2012). This invention consists of various steps of primary and secondary processing such as the careful selection of fruit seed followed by husking, drying, metering charging, pressing, filtration, purification, examination, and sterilization packaging processes. The whole process is performed at a standard temperature, which ensures the stability of essential substances such as α-linolenic acid in *Eucommia* seed oil. Following this processing method, extracted oil does not adopt any chemical substances during pressing, but rather avoids the pollution-induced by a solvent on *Eucommia* seed oil. The nano-filtration purification technology used during processing facilitates the α-linolenic acid content in the final product. The method for extracting *Eucommia* seed has a low production cost and is quite simple to adopt; it also consumes low quantities of energy at the same time. This trait may provide good raw material for the development of foods, health care foods, or medical products. Moreover, nutrient content present in oil cakes which are obtained after the pressing of *Eucommia* seeds is not destroyed, which imposes impactful results for further development of *Eucommia* products (Du et al., 2018).

Optimum extraction parameters used in this technology are identified as pressure, temperature, and time, which further affect the qualities of the extracted oil. The qualities of the extracted oil are evaluated by physicochemical properties, fatty acid composition, and vitamin E composition. The optimum conditions and significant parameters in this technology are obtained using response surface methodology (RSM). It has been proved that the most optimal extraction parameters are at pressure of 37 MPa, the temperature of 40°C, extraction time of 125 min, and CO_2 flow rate of 2.6 SL/min. The importance of evaluated parameters is decreased in the order of pressure > extraction time > temperature > CO_2 flow rate. In one study, investigators standardized the method of extraction using supercritical carbon dioxide (SC-CO_2) in which they took 200 g of EU seed powder into a 1000 mL extraction vessel under controlled temperature. In this method, the flow rate of CO_2 was manually controlled using a micro-metering valve. The volume of CO_2 was measured by a digital mass flow meter in standard liters per minute (SL/min) at a standard state ($P_{CO_2} = 100$ kPa, $T_{CO_2} = 20°$C, $\rho_{CO_2} = 0.0018$ g/mL). The extracts were collected in glass bottles. The yield was calculated using the equation: Yield of oil (g/100 g seed powder) = weight of extracted oil (g)/weight of EU seed powder (g) × 100 (Zhang, Liu, & Che, 2018b).

4 Fatty acid composition

The fatty acid analysis of EU oil extracted by the conventional method has revealed that linolenic acid (56.6%–60.7%), oleic acid (16.3%–17.8%), and linoleic acid (11.0%–13.3%) are the major components. Palmitic acid, which is the fourth most abundant fatty acid, ranged between 6.16% and 7.03%, whereas stearic acid varied between 2.07% and 2.77%. EU seed oils possess high concentrations of unsaturated fatty acids (UFAs, 88.5%) which are further bifurcated into 71.34% polyunsaturated fatty acid (PUFA), and 17.17% monounsaturated fatty acid (MUFA). The most characteristic possession of EU seed oil is α-linolenic acid (59.3%), which occupies the major proportion of PUFA, as this essential fatty acid can further be metabolized into docosahexaenoic acid (DHA) and eicosapentaenoic acid (EPA) (Jiao, Liu, & Zhang, 2015; Wu et al., 2013). A diet rich in PUFA is important for the structure and function of many membrane proteins, and MUFA has been shown to lower blood cholesterol levels (Hussain et al., 2016). These findings suggest that EU seed oils can be better material for the food and health care industries. Moreover, the higher ratio of UFA and saturated fatty acid (SFA) in daily diets have proven to have beneficial effects on human health. This statement is in concordance with the UFA/SFA ratio of EU seed oil, which ranges from 8.98 to 10.6 (Shu, Li, Wang, Yang, & Yuan, 2015; Zhang et al., 2018a).

However, the fatty acid composition of cold pressed EU seed oil consists of 61% of linolenic acid, which is slightly higher than the oil extracted by a conventional method. Other major fatty acids present in cold pressed EU oils are oleic acid (~17%), linoleic (~12%), palmitic (~6.5%), and stearic (~2%) (Zhang et al., 2018b). These results are in line with those reported in the previous studies (Zhang et al., 2010). The fatty acid composition of EU seed oil has been found to be similar to flaxseed oil and perilla oil (Table 1), which enables this oil to be used as an alternative to flaxseed and perilla oil. Likewise, this edible oil may also perform efficaciously in the prevention of noncommunicable diseases.

5 Minor bioactive components

Among minor bioactive components, vitamin E (tocols) content has been placed on the topmost position in EU seed oil the same as that among other edible oils. Vitamin E is well known for its antioxidant properties, which help in the stabilization of vegetable oils. The total vitamin E content in EU seed oil has been estimated as 190.7 mg/100 g (Zhang et al., 2018b). This amount is greater than that of other edible oils, viz. sunflower oil (154.7 mg/100 g), soybean oil (118.8 mg/100 g), and

TABLE 1 Fatty acid composition of *Eucommia ulmoides* seed oil, flaxseed oil, and perilla oil.

	SFAs							MUFAs				PUFAs			
Oil	Octanoic acid (C 8:0)	Palmitic acid (C 16:0)	Stearic acid (C 18:0)	Arachidic acid (C 20:0)	Behenic acid (C 22:0)	Lignoceric acid (C 24:0)	Total SFAs	Palmitoleic acid (C 16:1)	Oleic acid (C 18:1)	Gondoic acid (C 20:1)	Total MUFAs	Linoleic acid (C 18:2)	Linolenic acid (C 18:3)	Total PUFAs	PUFAs/ SFAs
EU seed oil	–	6.68	2.40	0.49	0.30	0.19	10.0	0.09	17.3	0.06	17.4	12.0	60.5	72.5	7.21
Flaxseed oil	–	5.22	4.12	0.20	0.15	0.13	9.82	0.00	18.3	0.00	18.3	15.6	56.2	71.8	7.32
Perilla oil	0.68	4.11	1.87	0.18	0.08	0.07	6.99	0.11	16.6	–	16.7	14.3	61.9	76.2	10.9

MUFA, monounsaturated fatty acids; *PUFA*, polyunsaturated fatty acids; *SFAs*, saturated fatty acids.
The data in "Flaxseed oil" row are adapted from Khattab, R. Y., & Zeitoun, M. A., (2013). Quality evaluation of flaxseed oil obtained by different extraction techniques. *LWT- Food Science and Technology*, *53*, 338–345. The data in "Perilla oil" row are adapted from Li, H. Z., Zhang, Z. J., Hou, T. Y., Li, X. J., & Chen, T., (2015). Optimization of ultrasound-assisted hexane extraction of perilla oil using response surface methodology. *Industrial Crops and Products*, *76*, 18–24.

rapeseed oil (79.7 mg/100 g) (Nagy, Kerrihardb, Beggio, Craft, & Pegg, 2016; Veličkovska, Brühl, Mitrev, Mirhosseini, & Matthäus, 2015). Vitamin E can be divided into two groups of tocopherols (TP) and tocotrienols (TT), each of which have four isomers, named as α-, β-, γ-, and δ-, which differ in number and position of methyl groups in the chromanol ring (Zhang et al., 2018b). The major isomers of vitamin E in EU seed oil are γ- and δ-tocopherol, accounting for 70.87% and 24.8%, respectively. γ- and δ-tocopherol have been shown to be beneficial for the human body as anticancer, antiinflammatory, cytotoxic, and antimutagenic agents (Zarogoulidis et al., 2013). Several studies have witnessed the protective health effect of vitamin E in many noncommunicable diseases such as cardiovascular diseases, atherosclerosis, and cancer (Deyama et al., 2001). These qualities make EU seed oil a suitable ingredient in functional foods. In addition, one scientific study estimated 68 and 73 compounds from the essential oils of the EU barks and leaves, respectively (Zhou, Zhang, Chen, & Liang, 2009). The total components made up 92.9%–97.7% of total essential oils from the two samples of EU, wherein the contents of the rutin, quercetin, and kaempferol in the barks and leaves of EU were 0.0169, 0.0036, 0.0021, and 0.0644, 0.0302, 0.0100 mg/g, respectively (Zhou, Zhang, et al., 2009).

6 Contribution of bioactive compounds in cold pressed oil to organoleptic traits and food/nonfood products

EU leaves have been used as a functional food and drink in China. However, the data depicting the use of EU oil as such in cooking or any food preparation are scanty. No doubt minor vegetable oils and cold pressed oils are becoming popular in the food industry due to their specific characteristics, but there is still a massive need for extensive research in this particular area. Cold pressed EU oil is considered as one of the important minor vegetable oils due to the bioactive components present in it. Owing to its biological activities, EU oil can be used as a cooking oil or dietary supplements by consumers. An upper limit for its daily intake has been kept as 3 mL/day to keep consumers safe and healthy (Yang et al., 2012). Usually, cold pressed oils have a distinct aroma and taste. They are used both as an additive to fresh food and as an ingredient enhancing various products with specific bioactive nutrients. Oxidative stability is a crucial quality marker of cold pressed oils containing natural antioxidants as well as undesirable substances, with oxidative properties (e.g., chlorophylls, metals) being removed in the refining process (Obiedzińska & Waszkiewicz-Robak, 2012). Therefore, there are great possibilities for the investigation of EU oil applications in the nutraceuticals and pharmaceuticals industries.

7 Health-promoting traits of cold pressed oil and oil constituents

Studies on health-promoting traits of cold pressed EU oil as such have not been documented in the literature, but many studies have been conducted in relation to the health outcomes owing to the active components present in different parts of the EU plant. For instance, the extract of the EU bark is used as an active component to formulate antihypertensive preparations. In many human and animal studies, EU extract is documented as a vasorelaxant. For example, supplementation of lignan from EU (300 mg/kg for 16 weeks) showed an improved vascular remodeling and reduced mean arterial blood pressure (Jin et al., 2010; Lang, Liu, Taylor, & Baker, 2005). Likewise, feeding trails of leaf extract of EU (0.175 g/100 g for 10 weeks) in a hyperlipidemic animal model resulted in an improved lipid profile (Choi et al., 2008). Similarly, 1 or 5 mg intraduodenal injection of EU leaf extract also showed a significant reduction in plasma triglyceride levels (Horii et al., 2010).

Several studies have also proven the antioxidant properties of EU under in vivo as well as in vitro conditions (Lin et al., 2011; Park et al., 2006). This particular trait of EU extract gets held in the body by enhancing the immunity system through increasing the actions of natural antioxidants like superoxide dismutase, catalase, and glutathione peroxidase, and reducing the concentration of hydrogen peroxide and lipid peroxide in erythrocytes, liver, and kidneys (Park et al., 2006). Furthermore, *Eucommia* also boosts levels of antioxidant enzymes in the blood to counteract free radicals (Park et al., 2006). In this context, a study conducted on the antioxidant properties of EU leaves by calculating the radical scavenging activity of 2,2-diphenyl-1-picrylhydrazyl (DPPH·) indicated a maximum DPPH· radical scavenging activity of EU leaf extract with an inhibition rate of 81.4%, followed by butylated hydroxytoluene (BHT) (76.6%) and the roasted cortex extract (16.7%) (Xu et al., 2010).

Both extract and powder from *Eucommia* leaves have also shown a noticeable suppression in body weight and white adipose tissue in the rat model. This suppression effect may be due to the presence of various active components such as geniposidic acid, asperuloside, and chlorogenic acid isolated from the extract (Hirata et al., 2011). Furthermore, the stem bark extract of the EU exhibited acetylcholinesterase inhibition properties and the application of extract (2.5 μg/mL) also inhibited 30%–70% of cytotoxicity (Zhou et al., 2009). Moreover, *Eucommia* cortex extract can be used in the control of osteoporosis as bioactive components such as glycosides initiate osteoblast, enhance osteogenesis, decrease osteoclast, and thus prevent osteolysis (Li et al., 2011; Zhang et al., 2009). Thus, the fatty acid composition, bioactive components, and health-promoting properties have attracted the immense interest of investigators toward extraction and functional development of vital constituents of EU oil. Yet further studies are required to explore the impact of supplementation of EU oil on human health and its inclusion in functional foods as an active ingredient.

References

Berry, E. M., & Hirsch, J. (1986). Does dietary linolenic acid influence blood pressure? *American Journal of Clinical Nutrition*, *44*, 336.

Choi, M.-S., Jung, U. J., Kim, H.-J., Do, G.-M., Jeon, S.-M., Kim, M.-J., et al. (2008). Du-Zhong (*Eucommia ulmoides* Oliver) leaf extract mediates hypolipidemic action in hamsters fed a high-fat diet. *The American Journal of Chinese Medicine*, *36*(1), 81–93.

Deyama, T., Nishibe, S., & Nakazawa, Y. (2001). Constituents and pharmacological effects of Eucommia and Siberian ginseng. *Acta Pharmacologica Sinica*, *22*(12), 1057–1070.

Du, Q., Wang, L., Liu, P., Qing, J., Sun, C., Sun, Z., et al. (2018). Fatty acids variation in seed of *Eucommia ulmoides* populations collected from different regions in China. *Forests*, *9*, 505.

Eller, F. J., Cermak, S. C., & Taylor, S. L. (2011). Supercritical carbon dioxide extraction of cuphea seed oil. *Industrial Crops and Products*, *33*, 554–557.

He, X. R., Wang, J. H., Li, M. X., Hao, D. J., Yang, Y., Zhang, C. L., et al. (2014). *Eucommia ulmoides* Oliv.: Ethnopharmacology, phytochemistry, and pharmacology of an important traditional Chinese medicine. *Journal of Ethnopharmacology*, *151*, 78–92.

Hirata, T., Kobayashi, T., Wada, A., Ueda, T., Fujikawa, T., Miyashita, H., et al. (2011). Anti-obesity compounds in green leaves of *Eucommia ulmoides*. *Bioorganic and Medicinal Chemistry Letters*, *21*(6), 1786–1791.

Horii, Y., Tanida, M., Shen, J., Hirata, T., Kawamura, N., Wada, A., et al. (2010). Effects of *Eucommia* leaf extracts on autonomic nerves, body temperature, lipolysis, food intake, and body weight. *Neuroscience Letters*, *479*(3), 181–186.

Hussain, T., Tan, B. 'e., Liu, G., Oladele, O. A., Rahu, N., Tossou, M. C., et al. (2016). Health-promoting properties of *Eucommia ulmoides*: A review. *Evidence-based Complementary and Alternative Medicine*, *2016*, 1–9.

Jiao, H. L., Liu, Y. L., & Zhang, D. D. (2015). Qualities of *Eucommia ulmoides* seeds oils from different areas and oil extraction techniques. *Cereals Oils*, *28*, 53–57.

Jin, X., Amitani, K., Zamami, Y., Takatori, S., Hobara, N., Kawamura, N., et al. (2010). Ameliorative effect of *Eucommia ulmoides* Oliv. Leaves extract (ELE) on insulin resistance and abnormal perivascular innervation in fructose drinking rats. *Journal of Ethnopharmacology*, *128*(3), 672–678.

Johnson, L. A. (1997). Theoretical, comparative, and historical analysis of alternative technologies for oilseeds extraction. In P. J. Wan, & P. J. Wakelin (Eds.), *Technology and solvents for extracting oilseeds and non petroleum oils* (p. 4). Champaign, IL: AOCS Press.

Lang, C., Liu, Z., Taylor, H. W., & Baker, D. G. (2005). Effect of *Eucommia ulmoides* on systolic blood pressure in the spontaneous hypertensive rat. *American Journal of Chinese Medicine*, *33*(2), 215–230.

Li, Y., Wang, M.-J., Li, S., Zhang, Y. M., Zhao, Y., Xie, R. M., et al. (2011). Effect of total glycosides from *Eucommia ulmoides* seed on bone microarchitecture in rats. *Phytotherapy Research*, *25*(12), 1895–1897.

Lin, J., Fan, Y. J., Mehl, C., Zhu, J.-J., Chen, H., Jin, L.-Y., et al. (2011). *Eucommia ulmoides* Oliv. Antagonizes H_2O_2-induced rat osteoblastic MC3T3-E1 apoptosis by inhibiting expressions of caspases 3, 6, 7, and 9. *Journal of Zhejiang University Science B*, *12*(1), 47–54.

Medic, J., Atkinson, C., & Hurburgh, C. R. (2014). Current knowledge in soybean composition. *Journal of the American Oil Chemists' Society*, *91*, 363–384.

Nagy, K., Kerrihardb, A. L., Beggio, M., Craft, B. D., & Pegg, R. B. (2016). Modeling the impact of residual fat-soluble vitamin (FSV) contents on the oxidative stability of commercially refined vegetable oils. *Food Research International*, *84*, 26–32.

Obiedzińska, A., & Waszkiewicz-Robak, B. (2012). Cold pressed oils as functional food. *Zywnosc Nauka Technologia Jakosc*, *1*(80), 27–44.

Park, S. A., Choi, M.-S., Jung, U. J., Kim, M. J., Kim, D. J., Park, H. M., et al. (2006). *Eucommia ulmoides* oliver leaf extract increases endogenous antioxidant activity in type 2 diabetic mice. *Journal of Medicinal Food*, *9*(4), 474–479.

Ren, J., Yang, Z. H., Zheng, S. X., & Hu, K. (2009). Studies on in vitro anti-inflammatory mechanism of α-linolenic acid. *Journal of Yunnan Minzu University*, *31*, 419–426.

Sarmento, C. M. P., Ferreira, S. R. S., & Hense, H. (2006). Supercritical fluid extraction (SFE) of rice bran oil to obtain fractions enriched with tocopherols and tocotrienols. *Brazilian Journal of Chemical Engineering*, *23*, 243.

Shu, X. M., Li, J. X., Wang, X. Y., Yang, J. J., & Yuan, Q. H. (2015). Optimization of subcritical extraction of *Eucommia ulmoides* seed oil and its fatty acid composition analysis. *China Oils Fats*, *40*, 15–18.

Silva, C. F., Mendes, M. F., Pessoa, F. L. P., & Queiroz, E. M. (2008). Supercritical carbon dioxide extraction of macadamia (*Macadamia integrifolia*) nut oil: Experiments and modeling. *Brazilian Journal of Chemical Engineering*, *25*, 175.

United States Environmental Protection Agency (2011). Available from: http://www.epa.gov/ttn/atw/nata2002/tables.htm#table1. [January 2011].

Uquiche, E., Romero, V., Ortíz, J., & Del Valle, J. M. (2012). Extraction of oil and minor lipids from cold-press rapeseed cake with supercritical CO_2. *Brazilian Journal of Chemical Engineering*, *29*(03), 585–597.

Veličkovska, S. K., Brühl, L., Mitrev, S., Mirhosseini, H., & Matthäus, B. (2015). Quality evaluation of cold-pressed edible oils from Macedonia. *European Journal of Lipid Science and Technology*, *117*(12), 2023–2035.

Wu, L. Y., Yang, W. G., Huang, Q., Yu, J., Chen, G. X., & Ma, C. J. (2013). Separation of unsaturated fatty acid from Eucommia seed oil and the analysis of its α-linolenic acid content. *Journal of Food Safety and Food Quality*, *4*, 1393–1400.

Xiang, Z. G., & Li, X. H. (2011). Antioxidant activity in vitro of seed oil from *Eucommia ulmoides Oliver* key fruits. *Food Science*, *32*, 133–136.

Xu, Z., Tang, M., Li, Y., Liu, F., Li, X., & Dai, R. (2010). Antioxidant properties of du-Zhong (*Eucommia ulmoides* Oliv.) extracts and their effects on color stability and lipid oxidation of raw pork patties. *Journal of Agricultural and Food Chemistry*, *58*(12), 7289–7296.

Yang, R., Zhang, L., Li, P., Yu, L., Mao, J., Wang, X., et al. (2012). A review of chemical composition and nutritional properties of minor vegetable oils in China. *Trends in Food Science and Technology*, *74*, 26–32.

Zacchi, P., Daghero, J., Jaeger, P., & Eggers, R. (2006). Extraction/fractionation and deacidification of wheat germ oil using supercritical carbon dioxide. *Brazilian Journal of Chemical Engineering*, *23*, 105.

Zarogoulidis, P., Cheva, A., Zarampouka, K., Huang, H., Li, C., Huang, Y., et al. (2013). Tocopherols and tocotrienols as anticancer treatment for lung cancer: Future nutrition. *Journal of Thoracic Disease*, *5*(3), 349–352.

Zhang, L. X., Ji, X. Y., Tan, B. B., Liang, Y. Z., Liang, N. N., Wang, X. L., et al. (2010). Identification of the composition of fatty acids in *Eucommia ulmoides* seed oil by fraction chain length and mass spectrometry. *Food Chemistry*, *121*, 815–819.

Zhang, Z.-S., Liu, Y.-L., & Che, L.-M. (2018a). Characterization of a new α-Linolenic acid-rich oil: *Eucommia ulmoides* seed oil. *Journal of Food Science*, *83*(3), 617–623.

Zhang, Z.-S., Liu, Y.-L., & Che, L.-M. (2018b). Optimization of supercritical carbon dioxide extraction of *Eucommia ulmoides* seed oil and quality evaluation of the oil. *Journal of Oleo Science*, *67*(3), 255–263.

Zhang, R., Liu, Z. G., Li, C., Hu, S. J., Liu, L., Wang, J. P., et al. (2009). Du-Zhong (*Eucommia ulmoides* Oliv.) cortex extract prevent OVX-induced osteoporosis in rats. *Bone*, *45*(3), 553–559.

Zhou, Y., Liang, M., Li, W., Li, K., Li, P., Hu, Y., et al. (2009). Protective effects of *Eucommia ulmoides* Oliv. Bark and leaf on amyloid β-induced cytotoxicity. *Environmental Toxicology and Pharmacology*, *28*(3), 342–349.

Zhou, J.-F., Zhang, T.-M., Chen, W.-A., & Liang, Y.-Z. (2009). Comparative analysis of chemical components between barks and leaves of *Eucommia ulmoides* Oliver. *Journal of Central South University of Technology*, *16*(3), 371–379.

Zhu, M. Q., Wen, J. L., Zhu, Y. H., Su, Y. Q., & Sun, R. C. (2016). Isolation and analysis of four constituents from barks and leaves of *Eucommia ulmoides* Oliver by a multi-step process. *Industrial Crops and Products*, *83*, 124–132.

Chapter 58

Cold pressed oils from the *Capsicum* genus

Alan-Javier Hernández-Álvarez[a], **Martin Mondor**[b], **and Guiomar Melgar-Lalanne**[c]

[a]*School of Food Science & Nutrition, University of Leeds, Leeds, United Kingdom,* [b]*Saint-Hyacinthe Research and Development Centre, Agriculture and Agri-Food Canada, Saint-Hyacinthe, QC, Canada,* [c]*Institute for Basic Sciences, Veracruzana University, Veracruz, Mexico*

Abbreviations

ASTA American Spice Trade Association
SC-CO_2 supercritical fluid extraction technology with CO_2 as fluid
SHU Scoville heat unit

1 Introduction

The genus Capsicum belongs to the Solanaceae family and currently includes 38 species, of which only five are domesticated: *C. annuum*, *C. chinense*, *C. frutensce*, *C. pubescens*, and *C. baccatum* (Moscone et al., 2007; Silvar & García-González, 2016). Fruits display a broad variety of sizes, shapes, and colors. However, these physical characteristics do not dictate the sensorial attributes in chili varieties, which are mostly determined by pungency (Patel et al., 2016). Hence, chilies are commercially divided into two main groups: sweet (nonpungent) and hot (pungent). Color (green, yellow, orange, and red) can be used to characterize the fruit, mostly in fresh form.

The final quality of the fruit is determined by the variety, the maturity stage, and the growing conditions (Padilha et al., 2015). Generally, pungency and color are the main sensorial attributes used to determine global chili quality. However, some secondary sensorial characteristics like flavor (acid, bell pepper, sweet, and tomato) and aroma (apple, citrus, fruity, herbal, oregano, and passion fruit) should be considered to determine the bouquet (Patel et al., 2016).

Their sensorial properties make *Capsicum* spp. a unique ingredient in many ethnic countries around the world. Moreover, there are many medicinal (Salehi et al., 2018), agronomic (Silvar & García-González, 2016), and even shelf-defense (Krishnatreyya, Hazarika, Saha, & Chattopadhyay, 2018) uses reported. All these applications make chilies one of the most important vegetables and spices in the worldwide trade (Silvar & García-González, 2016), which is currently still dominated by the production of fresh peppers, with around 34.5 million tons produced, followed by dry chili, with 3.9 million tons (Baenas, Belović, Ilic, Moreno, & García-Viguera, 2019).

The intensity of the red color and the degree of pungency are the most important quality parameters in the *Capsicum* spp. This is quantified through Scoville Heat Units (SHU), a sensorial test in which ethanolic serial dilutions of pepper extracts are translated into SHU. SHU are then calculated as parts of capsaicin per million parts of pepper fruit dry mass multiplied by 16 (Rezende-Naves et al., 2019). However, the scale is subjective and does not consider changes in population sensitivity or the presence of other capsaicinoids that are also responsible for chili pungency (Melgar-Lalanne, Hernández-Álvarez, Jiménez-Fernández, & Azuara, 2016). *Capsicum* pungency level depends on several factors, from the fruit variety to environmental conditions (mainly soil nutrients, moisture, and temperature) (Rezende-Naves et al., 2019), although it remains constant during the drying process (Melgar-Lalanne et al., 2016). Capsaicinoids are synthesized and accumulated by vesicles, small structures within the fruit placenta. In fact, seeds do not produce capsaicinoids, but they can retain them from the surrounding tissue containing the vesicles (Guzmán & Bosland, 2017). Capsaicinoids are mostly soluble in polar aprotic solvents like acetonitrile and acetone, as well as in water (Amruthraj, Raj, & Lebel, 2013). However, nonpolar solvents and comestible oils are not good solvents (Amruthraj, Raj, & Lebel, 2014).

Carotenoids are the main lipophilic compounds localized in the pods of mature fruits. They impart the red color and vary from 0.1 to 3.2 g/100 g dry weight. They are susceptible to oxidization by light and oxygen during the drying and storage processes, which negatively affects their quality (Arimboor, Natarajan, Menon, Chandrasekhar, & Moorkoth, 2015). Carotenoid biosynthesis is associated with the transformation of chloroplast to chromoplast in fruits (De Masi et al., 2007).

Cold Pressed Oils. https://doi.org/10.1016/B978-0-12-818188-1.00058-X
Crown Copyright © 2020 Published by Elsevier Inc. All rights reserved.

Capsicum spp. are also rich in nutrients and secondary metabolites with biological activity and have a low caloric content, which make these fruits an ideal ingredient to flavor foods without caloric addition. The fruit is rich in vitamins and antioxidants (ascorbic acid, carotenoids, B complex, and phenolic compounds). Red peppers are one of the agricultural products with the highest carotenoid and ascorbate content. One hundred grams of *Capsicum* fruit provides approximately 25% of the recommended daily amount of vitamin A (Guil-Guerrero, Martínez-Guirado, del Mar Rebolloso-Fuentes, & Carrique-Pérez, 2006; Palma et al., 2015).

The inherent quality limitations in the homogeneity of secondary metabolites in dry chilies make it necessary to develop new products that are more stable in both color and pungency for industrial purposes. Oleoresins are natural products that consist of a complex mixture of lipophilic molecules rich in bioactive metabolites that can be used as natural additives. Oleoresins are used in the pharmaceuticals, cosmetics, agricultural, and food industries. However, traditionally *Capsicum* oleoresins have been extracted by organic solvents (mainly *n*-hexane and ethanol) that are both potentially risky for the environment and human health, requiring efforts to be made in the development of cost-efficient green technologies to extract components of interest in *Capsicum* (Melgar-Lalanne et al., 2016).

When peppers are processed, the seeds, stems, and placenta are usually removed as waste products, and many solid wastes are produced. These wastes are still rich in oils, proteins, and antioxidant compounds, such as polyphenols, carotenoids, and terpenes. *C. annuum* seeds have high levels of crude protein (18.3 g/100 g dry matter), crude oil (11.0 g/100 g dry matter), and dietary fiber (60.9 g/100 g dry matter). Moreover, the lipophilic fraction of these wastes showed higher radical scavenging activity compared to their hydrophilic fraction, and their fatty acid profile was characterized by a high polyunsaturated content (84.2%), with linoleic acid being the major polyunsaturated fatty acid (70.9%). The main monounsaturated fatty acid was oleic acid (12.1%), while the main saturated fatty acid was palmitic acid (11.9%) (Azabou, Taheur, Jridi, Bouaziz, & Nasri, 2017). Hence, sweet pepper oil could be used as salad or cooking oil, UV protectants, and nutraceuticals (Ma, Wu, Zhao, Wang, & Liao, 2019). In hot peppers, many capsaicinoids are also extracted depending on the extraction technology used. The capsaicin recovered from the seeds depends on the fruit and the extraction method used. In hot Tunisian *C. annuum*, amounts varied from 2.81 mg capsaicin/kg oil with cold pressed extraction to 18.14 mg/kg oil in microwave-assisted extraction (Chouaibi, Rezig, Hamdi, & Ferrari, 2019). These could be used in a wide range of health applications, such as weight management, treating neurogenic and dermatological pains, and as a pungent additive in the food industry (Chouaibi et al., 2019).

The most common solvent-free technique for extracting these compounds is the mechanical pressing of whole fruits or fruit wastes, which preserves most of the volatile and nonvolatile compounds responsible for providing aroma and natural antioxidants that prevent autooxidation. The main limitation of these technologies is the elevated temperature produced during the pressing procedure, which negatively affects the compounds of interest and boosts oxidation processes. To avoid this heating, diverse techniques have been developed, mainly involving cold extraction of essential oils without the use of solvents as infusion in oils, supercritical fluid extraction, and cold pressing. In cold pressing, the oilseeds and/or the paste is not precooked to assist extraction. The result is high-quality oil that retains the original flavor, aroma, and nutritional value (Grosshagauer, Steinschaden, & Pignitter, 2019).

The objective of this chapter is to explore the wide variety of solvent-free technologies, with emphasis on cold pressing to extract compounds of interest in hot chili peppers.

2 Cold pressed *Capsicum* oil

Extraction of oil from distinct species, varieties, and even cultivars of *Capsicum* spp. has been widely explored. However, it is important to note that research has been focused on the technological aspects of extraction and not on the biological and agronomical aspects of the *Capsicum* used. This lack of information becomes a great limitation when the same extraction method is compared because many of the physicochemical differences obtained may be from the cultivar or variety of *Capsicum* used and even from the environmental conditions in the crop and not from the technology used.

In *Capsicum* oil extraction, most research has focused on extraction with different organic solvents, and only a few studies have been carried out using green technologies (Melgar-Lalanne et al., 2016). Among green technologies, supercritical carbon dioxide extraction (SC-CO_2) is probably the most studied for its absence of toxicity and explosion risks, low operational temperature, and the renewable character of CO_2 (Koubaa, Mhemdi, & Vorobiev, 2016). However, the investment cost is high, and the extraction conditions are difficult to obtain, meaning that highly trained staff are required (Cvjetko Bubalo, Vidović, Radojčić Redovniković, & Jokić, 2018). Some of the cost limitations have recently been overcome using homemade SC-CO_2 equipment (Castro-Vargas, Rodríguez-Varela, & Parada-Alfonso, 2011; Moslavac et al., 2014). Simultaneous extraction from avocado and red bell pepper using SC-CO_2 as solvent has also been tested (Barros, Coutinho, Grimaldi, Godoy, & Cabral, 2016). In another approach, maceration of the fruit in different oils has been tested by Cerecedo-Cruz, Azuara-Nieto, Hernández-Álvarez, González-González, and Melgar-Lalanne (2018), using avocado oil as a vehicle. The effect of different vegetable oils in the extraction of capsaicinoids from Bhut jolokia peppers was studied by Amruthraj et al. (2013).

Cold pressed oils are preferred to refined oils as they contain more antioxidants and bioactive substances like carotenoids, tocopherols, phytosterols, and phenolics. Moreover, cold pressed oils are more resistant to lipid oxidation over time, although initial oxidation is higher since the compounds that can start the oxidation process are not removed (Grosshagauer et al., 2019). Cold pressing is mostly used by small and medium-sized industries located close to the source of agricultural production, with capacities lower than 25 tons per day. The advantages of this technology at the industrial level include the low investment cost and low energy consumption (six times lower than that of solvent extraction) (Huang, Li, Huang, Nju, & Wan, 2007). This extraction does not use chemical solvents or thermal conditioning of the seeds and does not generate wastewater. It ensures a safe working environment for employees, has a lower environmental impact in comparison with solvent extraction, and shows higher flexibility because processing diverse types of seeds is easy and fast (Wong, Eyres, & Ravetti, 2014).

Despite its nutritional, environmental, and economic advantages, cold pressing has been scarcely explored for extraction of the lipid fraction in wastes and seeds of *Capsicum* spp. *Capsicum annuum* L. seeds are shown in Fig. 1.

The authors found only two articles focusing on cold pressing oil from *capsicum* seeds. Yılmaz, Sevgi Arsunar, Aydeniz, and Güneşer (2015) used a laboratory-scale cold press machine with 2 kg seed/h capacity, a single head, 2 hp., and 1.5 kW of power to extract Capia seed oil. The maximum extraction temperature used was 40°C, and the final oil was centrifuged to eliminate impurities. Chouaibi et al. (2019) extracted pepper seeds at room temperature and centrifuged them at 2800 g for 3 min after pressing. Unfortunately, the cold pressing conditions were not described.

Yılmaz et al. (2015) extracted oil from Capia (*C. annuum* L.) seeds harvested for canned products with an oil yield of 5.12% (*w*/w), when no pretreatment was performed. Chouaibi et al. (2019) compared different extraction methods for red pepper seed oils with an oil yield of 14.6% (w/w) in the case of cold pressed oil, which was significantly lower than the yield obtained with Soxhlet extraction (18.3%) and SC-CO_2 (25.2%). None of the authors explained the complete taxonomy of

FIG. 1 *Capsicum annuum* L. seeds.

the *Capsicum* seeds used. Yılmaz et al. (2015) explained that they used Capia seeds (*C. annuum L.*), and Chouaibi et al. (2019) only specified the use of pepper seeds (no species). Although both authors extracted hot varieties, capsaicinoids were not quantified and/or classified. Moreover, other minority compounds with bioactive properties, such as polyphenols, were analyzed.

Seeds were cleaned by air blowing before use to reduce impurities after they were sun-dried (Yılmaz et al., 2015) or air-dried (Chouaibi et al., 2019). The cleaned and dried seeds were then stored under low moisture and temperature until use (Melgar-Lalanne et al., 2016). Extraction can be carried out with different equipment and under different conditions, resulting in different oil yields (Melgar-Lalanne et al., 2016). Moreover, heat treatment can increase the oil yield of screw pressing.

Different pretreatments can be performed to increase oil extraction yield. The treatments used most are roasting and enzymatic. Roasting affects the sensorial properties of seeds and enhances flavor, color, and texture, mostly due to the presence of volatiles (Zhang, Liu, Ma, & Wang, 2019). Moreover, it inactivates degradative enzymes and facilitates oil extraction by destroying cell walls though microstructure changes. Maillard reactions are responsible for flavor enrichment and may increase the antioxidant activity of the oil (Ji, Liu, Shi, Wang, & Wang, 2019). Maillard reactions may also bring some heat-induced contaminants like acrylamide, furans, and furfurals. However, roasting does not change the fatty acid composition of seeds, but if polyunsaturated fatty acids are abundant and/or the temperature is too high, the composition could be affected by oxidative reactions and triglyceride degradation (Durmaz & Gökmen, 2010).

In Capia seeds, a roasting pretreatment was carried out using an air oven at 150°C for 25 min, and an enzymatic pretreatment was applied by incubating the seeds with 100 U/100 g of hemicellulase and 0.25 U/g seeds of protease (Yılmaz et al., 2015). However, the oil yield was lower (6.65 g oil/100 g seeds for roasted seeds, and 5.41 g/100 g seeds for enzyme-treated seeds) than those obtained by Chouaibi et al. (2019) without pretreatment (14.6 g/100 g seeds) in red pepper seeds.

Enzyme-assisted extraction is used as a complementary pretreatment to increase oil extraction yield. Enzymes degrade the protein and complex polysaccharides present in the cell walls and facilitate the release of oil. In general, mixtures of enzymes are employed to degrade the fibrous material more effectively (usually cellulose linked with hemicellulose in a pectin matrix) (Candido & Maziero, 2017). Enzymes used in pepper seeds include cellulase, α-amylase, β-glucosidase, xylanase, β-gluconate, and pectinase, as well as industrial enzyme preparations such as Viscozyme L (cellulolytic enzyme complex), Pectinex, and Lallzyme BetaTM (Marathe, Jadhav, Bankar, Kumari Dubey, & Singhal, 2019). Many factors, such as temperature, pH, quantity, and kind of enzymes, time reaction, agitation, and water content, influence the performance of the treatment (Peng, Ye, Liu, Liu, & Meng, 2019). The enzymatic pretreatment also increases the unsaturation, the oxidative stability of oils, and the amount of anthocyanins present in the final product (Malacrida, Moraes, de Rosso, Rodrigues, & de Souza, 2018; Soto, Acosta, Vaillant, & Pérez, 2016). To improve the final quality of cold pressed extracted oils, some technological strategies have been developed, including limitation of light, oxygen, and elevated temperatures.

3 Physicochemical and biochemical properties of *Capsicum* seeds oil

The main physicochemical properties studied in cold pressed *Capsicum* oleoresins are listed in Table 1.

Oil density, which is defined as the ratio of the densities of a given oil and of water when both are at identical temperatures, is an important quality parameter in oils (Zellner, Dugo, Dugo, & Mondello, 2010). The attained value is characteristic for each oil (Ceriani, Paiva, Gonçalves, Batista, & Meirelles, 2008). In cold pressed *capsicum* oils, nonsignificant differences were found when different *capsicum* varieties and extraction methods were used (Chouaibi et al., 2019; Yılmaz et al., 2015). The refractive index of oils depends on their molecular weight, fatty acid chain length, degree of unsaturation, and degree of conjugation. In *Capsicum* seed oils, nonsignificant differences were found among different samples using different treatments, as shown in Table 1.

Viscosity is affected by elevated temperatures, air, and increased numbers of frying cycles, thus enhancing the formation of oxidative and polymeric compounds (Nayak, Dash, Rayaguru, & Krishnan, 2016). High viscosity is related to long-chain structures (Moncef et al., 2012). When oil from red pepper seeds was extracted using subcritical butane, viscosity was 52.094 cP (Gu et al., 2017). In the case of *Capsicum annuum* cv *California wonder* and *C. baccatum* cv. *Aji´ Colorado*, the viscosity at 40°C was 25.1 and 25.6 cP, respectively, and it reduced to 6.6 and 6.7 cP at 100°C (Jarret, Levy, Potter, & Cermak, 2013).

Cold pressed oils show more turbidity than those extracted using solvents, probably due to the winterization and refinement processes after solvent extraction, which removes certain particles (Yilmaz & Güneşer, 2017). Turbidity can also depend on the pretreatment given to seeds, because pretreatments may modify the microstructure and the presence of low-chain fatty acids in seeds. In pretreated Capia seeds, turbidity was significantly lower than for nonpretreated cold pressed oils (Güneşer, Yılmaz, & Ok, 2017; Yılmaz et al., 2015).

TABLE 1 Physicochemical properties of cold pressed oils.

	Capia seeds			
	No pretreatment	Preroasted	Preenzymatic assistance	Red pepper seeds
Density (20°C)	0.9222±0.001	0.9234±0.001	0.9203±0.001	0.918
Refractive index (20°C)	1.4743±0.001	1.4736±0.001	1.4740±0.001	1.46590
Viscosity (20°C) cP	45.57±0.28	44.96±0.28	43.23±0.11	64.35±2.15
Turbidity (20°C, cP)	4.75±0.75	1.50±0.29	1.75±0.48	Nd
Color L	24.04±1.72	24.38±0.14	24.81±0.83	38.25±0.05
Color a*	13.06±2.27	14.95±0.22	15.05±0.52	30.10±0.06
Color b*	13.90±4.38	16.90±0.44	15.20±0.41	40.95±0.12
Peroxide value (meq O_2/kg oil)	5.81±0.95	4.58±0.22	5.23±0.08	5.37±0.07
Free fatty acids (% FFA)	4.97±0.07	3.69±0.04	8.82±1.29	0.60±0.02
Iodine number (g/100 g)	141.8±2.4	145.4±1.7	149.3±0.1	142.52±2.28

Capia seeds: Adapted from Yılmaz, E., Sevgi Arsunar, E., Aydeniz, B., & Güneşer, O. (2015). Cold pressed capia pepperseed (Capsicum annuum L.) oils: Composition, aroma, and sensory properties. *European Journal of Lipid Science and Technology, 117*(7), 1016–1026. doi:10.1002/ejlt.201400276 and *Red pepper seeds:* Adapted from Chouaibi, M., Rezig, L., Hamdi, S., & Ferrari, G. (2019). Chemical characteristics and compositions of red pepper seed oils extracted by different methods. *Industrial Crops and Products, 128*, 363–370. doi:10.1016/j.indcrop.2018.11.030.

Color of foods is considered one of the most important quality attributes. In red pepper oleoresins, color is dictated by the presence of natural pigments, mostly carotenoids (Minguez-Mosquera, Jaren-Galan, & Garrido-Fernandez, 1992). The fruit's red color is imparted mostly by carotenoids, which contain more than 50 identified compounds, with a total content varying from 0.1 to 3.2 g/100 g dry weight with differences in composition between varieties tested and depending on the extraction method. There are several standard methods for rapidly obtaining quantifiable data on oleoresin color, such as the ASTA measurement, which consists of recording the absorbance of the sample at 460 nm then applying a dilution factor that gives a color value in ASTA units (ASTA, 1986). Color determination both for fresh *Capsicum* and oleoresins using colorimeters is gaining importance (Güneşer et al., 2017; Wall & Bosland, 1998). Differences in color are produced mostly by: (1) fruit maturity stage; (2) fruit variety; (3) extraction method used; and (4) storage conditions and life of the final products. Thus, when oleoresins are stored, they may suffer oxidation by light and oxygen when conditions are not optimal (Cerecedo-Cruz et al., 2018). Results obtained with different extraction technologies and identical *capsicum* varieties support the hypothesis that oleoresin color is a function more of the chili variety than of the extraction technology used (Melgar-Lalanne et al., 2016) (Table 1).

There is sufficient evidence that several components in oil, especially polyunsaturated fatty acids (PUFAs), have positive health effects on the human body. However, PUFAs are susceptible to lipid oxidation, forming free radicals that may promote oxidation and the generation of further radicals (Frankel, 1980). Formation of free radicals could be reduced by limiting oil exposure to oxidation-inducing factors, such as light, heat, or metals (Grosshagauer et al., 2019). The degree of oxidation is given by the peroxide index, which is one of the main causes of deterioration, reduced stability, and formation of off-flavors that negatively affect the quality and storage life of food products (Hornero-Méndez, Pérez-Gálvez, & Mínguez-Mosquera, 2001). Peroxides are primary oxidation products that can further undergo degradation to form low molecular weight aldehydes and ketones (Cerecedo-Cruz et al., 2018). The Codex Standard 19-1981 for virgin oils allows up to 15 meq O_2/kg oil and 10 meq O_2/kg oil for other fats and oils. All results obtained for cold pressed *Capsicum* seeds complied with the Codex Standard and are regarded as edible oils (Table 1). It has been well established that cold pressing increases the presence of natural antioxidants, such as tocopherols and phenols (McDowell, Elliott, & Koidis, 2017). Moreover, a macerated *capsicum* oleoresin in avocado oil showed that the presence of *Capsicum* clearly increased the shelf life of the product by a month by reducing the Totox index, probably due to the increased presence of carotenoids and phenolics (Cerecedo-Cruz et al., 2018).

The free fatty acid value is expressed as a percentage of free oleic acid. This value is considered a good indicator of oil quality and measures the degree of hydrolysis of an oil. Oils with a higher free fatty acid content are considered to be of lower quality than oils with lower free fatty acid content and could even be considered as not edible when the percentage of fatty acids is higher than 0.4 mg/kg in cold pressed oils (Codex Standard, 1999; Konuskan, Arslan, & Oksuz, 2019).

The iodine value of an oil is an indicator of unsaturation degree. Cold pressed *Capsicum* seeds showed a high iodine value, indicating that these could be used for soap production and human diets (Konuskan et al., 2019). This seems to be independent of the extraction method (Chouaibi et al., 2019) (Table 1).

4 Potential health benefits of cold pressed *Capsicum* seed oil

There has been no specific research about health benefits of cold pressed *Capsicum* seeds oils. However, the composition of these kinds of extracted oils is similar to that of others, especially to those not extracted using organic solvents, and so it is possible to infer that cold pressed oils have potentially the same health benefits as all other extracted *capsicum* without the health risk implied with the presence of organic solvents.

There are many compositional differences between species, varieties, and maturation stages of fruits, as well as in the seeds. As previously said, *Capsicum* seeds are the most abundant, cheap, and readily available by-product obtained with the industrialization of *capsicum* fruits (Azabou et al., 2017), and their composition is like that of the fruit. Seeds are rich in proteins, oils, and antioxidant compounds, such as phenolics, carotenoids, and terpenes (Silva, Azevedo, Pereira, Valentão, & Andrade, 2013). Moreover, seeds from hot *capsicum* fruits are rich in capsaicinoids. Seeds are also rich in minerals, especially in potassium and sodium, and have a fair content of fiber (Azabou et al., 2017). The oil content in seeds varies from 10.8 g/100 g dry matter to 35.9 g/100 g dry matter, and the content of linoleic acid, which is the main fatty acid present, varies from 73.9% to 77.9% (Jarret et al., 2013). More than 125 volatile compounds have been identified in fruits, although their sensorial and health significance have not yet been well established (El-Ghorab, Javed, Anjum, Hamed, & Shaaban, 2013). Whole *Capsicum* fruits are rich in antioxidants such as carotenoids, phenolics, and ascorbic acid (Álvarez-Parrilla, De La Rosa, Amarowicz, & Shahidi, 2011), which protect the lipidic fraction from oxidation (Álvarez-Parrilla, de la Rosa, Amarowicz, & Shahidi, 2012).

There are more than 50 identified capsaicinoids, with capsaicin, dihydrocapsaicin, nordihydrocapsaicin, homodihydrocapsaicin, and homocapsaicin being the most abundant. Although capsaicinoids are related with the hotness sensation of the chili, there are some nonpungent capsaicinoids, such as capsinoids (Kawabata et al., 2009) and capsiate (Macho et al., 2003). Capsaicinoids are condensed from fatty acids and vanillylamine in the placenta of the fruit and they are stable to polar and nonpolar solvents (Luo, Peng, & Li, 2011). Consumed worldwide, capsaicin has a long story of controversy regarding its safety for human consumption. The U.S. Food and Drug Administration considers chili (as fruit) to be safe, but not the pure chemical capsaicin. Moreover, the consumption of great quantities of hot fruits may be considered as a risk factor for gastric cancer (Bode & Dong, 2011). However, when capsaicinoids are consumed moderately, they can stimulate the secretion of saliva and bile salts. The increase in saliva production boosts the presence of enzymes (mainly amylases and proteases); the increase in the bile salt secretion improves the digestion of fat (Maji Amal & Pratim, 2016). The presence of capsaicin also increases intestinal motility and thus diminishes functional dyspepsia (Bortolotti, Coccia, Grossi, & Miglioli, 2002). The protective role of capsaicin on the gastric mucosa is still controversial because it inhibits the acid secretion, prevents ulcers, and even inhibits the growth of *Helicobacter pylori*, but has been also associated with cases of stomach cancer (López-Carrillo et al., 2003).

Capsaicin is effective for some topical pain relief and can be used for postherpetic neuralgia pain, the most common complication of shingles caused by reactivation of the herpes zoster virus (Mankowski et al., 2017). However, the side effects found with the use of topical capsaicin are severe and include redness, dryness, irritation, and burning. As such, topical capsaicin cannot be proposed as a first-line treatment (Yong et al., 2017). The positive influence of capsaicin on lipid metabolism has been proven though animal models, especially its effect as an antihypercholesterolemic and antilithogenic. Moreover, capsaicin may be hypotriglyceridemic and prevent the accumulation of fat in the liver by enhancing triglyceride transport out of the liver, which has been proven in animal models (Singletary, 2011). In addition, capsaicin's capacity to reduce adiposity in rats has been reported. Capsaicin can induce apoptosis in different lines of cancer cells, such as pancreatic, colonic, prostatic, liver, esophageal, bladder, skin, leukemia, lung, and endothelial cells, while normal cells are unharmed (Clark & Lee, 2016).

Increased energy expenditure and reduced appetite have been tested in some human studies with volunteers (Whiting, Derbyshire, & Tiwari, 2012). In short-term studies, an improvement of diet-induced thermogenesis has been found as well as a decrease in the respiratory quotient after the intake of food supplemented with capsaicin (Yoshioka, Matsuo, Lim, Tremblay, & Suzuki, 2000). Thus, capsaicinoids may increase energy expenditure and reduce body fat via stimulation of brown adipose tissue through the specific receptor TRPV1, increasing stamina consumption and decreasing body fat modestly but consistently (Saito & Yoneshiro, 2013). However, in long-term studies (3 months) the effect of capsaicin supplementation on weight loss was limited to weight maintenance (Lejeune, Kovacs, & Westerterp-Plantenga, 2003).

The effect on lipid oxidation of capsaicinoids has also been reported as being higher in people with a high body mass index (Inoue, Matsunaga, Satoh, & Takahashi, 2007).

Capsinoids show thermogenic activity and reduction of visceral fat with the advantage of their low pungency, which makes them a promising treatment in weight management (Galgani, Ryan, & Ravussin, 2010). The health benefits of carotenoids from *Capsicum* spp. have been less explored. Total carotenoids are responsible for the yellow-orange-red color in mature fruits, and are composed of more than 50 compounds, such as β-carotene, α-carotene, capsanthin, capsorubin, cryptocapsin, α-cryptoxanthin, β-cryptoxanthin, lutein, antheraxanthin, violaxanthin, and zeaxanthin (Giuffrida et al., 2013). Carotenoids have a strong antioxidant activity, mainly in the form of reactive oxygen scavenging, due to the bearing of a k-ring as end group, which gives them a protective character against some chronic disease and cancer (Arimboor et al., 2015).

The presence of phenolic compounds is scarce in capsicum oils because of their hydrophilic nature (Cerecedo-Cruz et al., 2018). However, the antioxidant activity of these compounds may reduce the risk of cardiovascular diseases and certain types of cancer. Olive oil aromatized with *C. frutescens* showed an increase in its content of all vitamin E isoforms, also increasing the nutritional value, but the phenolic content decreased (Baiano, Gambacorta, & La Notte, 2010; Sousa et al., 2015) due to the hydrophilic nature of the phenolics.

Caporaso, Paduano, Nicoletti, and Sacchi (2013) found that olive oil infused with hot *C. annuum* had higher antioxidant activity than noninfused olive oil and was able to reduce the autooxidation rate in the oil while increasing the shelf life of the oil. Similar results were found by Gambacorta et al. (2007), and Cerecedo-Cruz et al. (2018) in other infused oils.

The compounds found in extracts of *capsicum* seeds and fruits depend on the nature of the organic solvent used. Thus, in Soxhlet extraction with *n*-hexane, high quantities of capsaicinoids and carotenoids are found; however, in ethanol extracts, higher levels of phenolic compounds and ascorbate are found (Bae, Jayaprakasha, Jifon, & Patil, 2012; Castro-Concha, Tuyub-Che, Moo-Mukul, Vazquez-Flota, & Miranda-Ham, 2014; Rahman et al., 2014). Moreover, methanol extracts from *C. annuum* L. seeds showed strong antiproliferative activity against MCF7, MKn45, and HCT116 tumor cells at a concentration of 500 μg/mL, due to an increase in apoptosis for the phenolic compounds (Jeon et al., 2012). The ethanolic and butanolic extracts from *C. baccatum* contained potential antioxidant and antiinflammatory compounds, which were tested against oxidative and inflammation-related pathological processes (Zimmer et al., 2012).

Capsicum fruits and seeds show an interesting α-glycosidase inhibitory capacity, which is strongly related to the absorption of glucose (Kwon, Apostolidis, & Shetty, 2007). This antidiabetic potential is higher in nonpungent *Capsicum* varieties and so seems to be related with the carotenoid profile. The pungent *C. chinense* (habanero pepper) showed an inhibitory effect against α-amylase, which is also related to antidiabetic potential (Tundis et al., 2011). The combination through a diet of different chilies seems to increase this inhibitory activity against enzymes related to carbohydrate degradation (Oboh, Puntel, & Rocha, 2007). Moreover, regular consumption of chilies might attenuate postprandial hyperinsulinemia (Ahuja, Robertson, Geraghty, & Ball, 2006). In the largest population-based study to describe the association between hot *capsicum* food habits and insulin resistance, carried out in China, an association between chili foods consumption habits and insulin resistance showing a protective effect was observed. Regular supplementation of capsaicin in the diet improved postprandial hyperglycemia and hyperinsulinemia as well as fasting lipid metabolic disorder in women with gestational diabetes mellitus (Yuan et al., 2016).

Capsicum also shows some antimicrobial activity related to its antioxidant activity. Dorantes et al. (2000) found that sweet red pepper (*C. annuum* L.) showed more antimicrobial activity against common food pathogens than hot peppers such as Jalapeño. Moreover, water extracts of different *Capsicum* spp. have exhibited different degrees of inhibition against *Bacillus cereus*, *Bacillus subtilis*, *Clostridium sporogenes*, *Clostridium tetani*, and *Streptococcus pyogenes* (Cichewicz & Thorpe, 1996). Oil extracts of *Capsicum* spp. have also shown some antimicrobial activity. Hexane and chloroform oleoresins of *C. frutescens* L. demonstrated inhibitory activity against *Pseudomonas aeruginosa*, *Klebsiella pneumonia*, *Staphylococcus aureus*, and *Candida albicans* (Gurnani, Gupta, Mehta, & Mehta, 2016), and supercritical CO_2 oils and ethanol extracts from *C. annuum* and *C. frutescens* showed antibacterial effects against *Streptococcus sobrinus* and *Streptococcus salivarius* (Pilna et al., 2015).

5 Conclusion

Cold pressed oils are simple to obtain, economical, and environmentally friendly. The final oil does not require refinement. In general, cold pressing requires no specialized workforce and represents a good alternative for small companies close to the production fields. The global quality of the oil is higher than for those extracted with organic solvents since more antioxidants are recovered. The use of some pretreatments can increase extraction yield. However, only roasting and enzymatic-assisted treatments have been tested, and more studies are required. The obtained *Capsicum* oil is rich in carotenoids, capsaicin, and linoleic acid, but not in phenolic compounds due to its hydrophilic nature. The presence of carotenoids and capsaicinoids may provide some interesting health properties to the oil, such as antiinflammatory, antidiabetic and antiobesity properties. However, specific and in-depth research on the health benefits of cold pressed *Capsicum* seed oils should be carried out. Hence, *Capsicum* cold pressed oils may be utilized in functional foods as ingredients, mostly the less pungent varieties. However, further studies about its biochemical effects and applications in product formulations are needed.

References

Ahuja, K. D. K., Robertson, I. K., Geraghty, D. P., & Ball, M. J. (2006). Effects of chili consumption on postprandial glucose, insulin, and energy metabolism. *The American Journal of Clinical Nutrition*, *84*(1), 63–69. https://doi.org/10.1093/ajcn/84.1.63.

Álvarez-Parrilla, E., De La Rosa, L. A., Amarowicz, R., & Shahidi, F. (2011). Antioxidant activity of fresh and processed Jalapeño and serrano peppers. *Journal of Agricultural and Food Chemistry*, *59*(1), 163–173. https://doi.org/10.1021/jf103434u.

Álvarez-Parrilla, E., de la Rosa, L. A., Amarowicz, R., & Shahidi, F. (2012). Protective effect of fresh and processed Jalapeño and serrano peppers against food lipid and human LDL cholesterol oxidation. *Food Chemistry*, *133*(3), 827–834. https://doi.org/10.1016/j.foodchem.2012.01.100.

Amruthraj, N. J., Raj, J. P. P., & Lebel, L. A. (2013). Polar aprotic extraction of Capsaicinoids from *Capsicum chinense* Bhut jolokia fruit for antimicrobial activity. *International Journal of Biological & Pharmaceutical Research*, *4*(12), 959–964.

Amruthraj, N. J., Raj, J. P. P., & Lebel, L. A. (2014). Effect of vegetable oil in the solubility of capsaicinoids extracted from *Capsicum chinense* Bhut Jolokia. *Asian Journal of Pharmaceutical and Clinical Research*, *7*(1), 48–51.

Arimboor, R., Natarajan, R. B., Menon, K. R., Chandrasekhar, L. P., & Moorkoth, V. (2015). Red pepper (*Capsicum annuum*) carotenoids as a source of natural food colors: Analysis and stability-a review. *Journal of Food Science and Technology*, *52*(3), 1258–1271. https://doi.org/10.1007/s13197-014-1260-7.

ASTA (Ed.), (1986). *Official analytical method of the American spice trade association* (2nd ed.): ASTA analytical method 20.1.

Azabou, S., Taheur, F. B., Jridi, M., Bouaziz, M., & Nasri, M. (2017). Discarded seeds from red pepper (*Capsicum annum*) processing industry as a sustainable source of high added-value compounds and edible oil. *Environmental Science and Pollution Research*, *24*(28), 22196–22203. https://doi.org/10.1007/s11356-017-9857-9.

Bae, H., Jayaprakasha, G. K., Jifon, J., & Patil, B. S. (2012). Variation of antioxidant activity and the levels of bioactive compounds in lipophilic and hydrophilic extracts from hot pepper (*Capsicum* spp.) cultivars. *Food Chemistry*, *134*(4), 1912–1918. https://doi.org/10.1016/j.foodchem.2012.03.108.

Baenas, N., Belović, M., Ilic, N., Moreno, D. A., & García-Viguera, C. (2019). Industrial use of pepper (*Capsicum annum* L.) derived products: Technological benefits and biological advantages. *Food Chemistry*, *274*, 872–885. https://doi.org/10.1016/j.foodchem.2018.09.047.

Baiano, A., Gambacorta, G., & La Notte, E. (2010). Aromatization of olive oil. *Transworld Research Network*, *661*, 1–29.

Barros, H. D. F. Q., Coutinho, J. P., Grimaldi, R., Godoy, H. T., & Cabral, F. A. (2016). Simultaneous extraction of edible oil from avocado and capsanthin from red bell pepper using supercritical carbon dioxide as solvent. *The Journal of Supercritical Fluids*, *107*, 315–320. https://doi.org/10.1016/j.supflu.2015.09.025.

Bode, A. M., & Dong, Z. (2011). The two faces of capsaicin. *Cancer Research*, *71*(8), 2809–2814. https://doi.org/10.1158/0008-5472.CAN-10-3756.

Bortolotti, M., Coccia, G., Grossi, G., & Miglioli, M. (2002). The treatment of functional dyspepsia with red pepper. *Alimentary Pharmacology & Therapeutics*, *16*(6), 1075–1082. https://doi.org/10.1046/j.1365-2036.2002.01280.x.

Candido, R. G., & Maziero, P. (2017). Enzymatic pretreatment for edible oils extraction. In S. Chemat (Ed.), *Edible oils: Extraction, processing, and applications*: CRC Press.

Caporaso, N., Paduano, A., Nicoletti, G., & Sacchi, R. (2013). Capsaicinoids, antioxidant activity, and volatile compounds in olive oil flavored with dried chili pepper (*Capsicum annuum*). *European Journal of Lipid Science and Technology*, *115*(12), 1434–1442. https://doi.org/10.1002/ejlt.201300158.

Castro-Concha, L. A., Tuyub-Che, J., Moo-Mukul, A., Vazquez-Flota, F. A., & Miranda-Ham, M. L. (2014). Antioxidant capacity and total phenolic content in fruit tissues from accessions of *Capsicum chinense* Jacq. (habanero pepper) at different stages of ripening. *The Scientific World Journal*, *2014*, 5. https://doi.org/10.1155/2014/809073.

Castro-Vargas, H. I., Rodríguez-Varela, L. I., & Parada-Alfonso, F. (2011). Guava (Psidium guajava L.) seed oil obtained with a homemade supercritical fluid extraction system using supercritical CO_2 and CO-solvent. *The Journal of Supercritical Fluids*, *56*(3), 238–242. https://doi.org/10.1016/j.supflu.2010.10.040.

Cerecedo-Cruz, L., Azuara-Nieto, E., Hernández-Álvarez, A. J., González-González, C. R., & Melgar-Lalanne, G. (2018). Evaluation of the oxidative stability of chipotle chili (*Capsicum annuum* L.) oleoresins in avocado oil. *Grasas y Aceites*, *69*(1), 1–4. https://doi.org/10.3989/gya.0884171.

Ceriani, R., Paiva, F. R., Gonçalves, C. B., Batista, E. A. C., & Meirelles, A. J. A. (2008). Densities and viscosities of vegetable oils of nutritional value. *Journal of Chemical & Engineering Data*, *53*(8), 1846–1853. https://doi.org/10.1021/je800177e.

Chouaibi, M., Rezig, L., Hamdi, S., & Ferrari, G. (2019). Chemical characteristics and compositions of red pepper seed oils extracted by different methods. *Industrial Crops and Products*, *128*, 363–370. https://doi.org/10.1016/j.indcrop.2018.11.030.

Cichewicz, R. H., & Thorpe, P. A. (1996). The antimicrobial properties of chile peppers (*Capsicum species*) and their uses in Mayan medicine. *Journal of Ethnopharmacology*, *52*(2), 61–70. https://doi.org/10.1016/0378-8741(96)01384-0.

Clark, R., & Lee, S.-H. (2016). Anticancer properties of capsaicin against human cancer. *Anticancer Research*, *36*(3), 837–843.

Codex Standard 19-1981, F. C. S. (1999). *Norma del Codex para Grasas y Aceites Comestibles No Regulados por Normas Individuales. Codex Alimentarius Official Standards.*

Cvjetko Bubalo, M., Vidović, S., Radojčić Redovniković, I., & Jokić, S. (2018). New perspective in extraction of plant biologically active compounds by green solvents. *Food and Bioproducts Processing*, *109*(Part C), 52–73.

De Masi, L., Siviero, P., Castaldo, D., Cautela, D., Esposito, C., & Laratta, B. (2007). Agronomic, chemical and genetic profiles of hot peppers (*Capsicum annuum* ssp.). *Molecular Nutrition & Food Research*, *51*(8), 1053–1062. https://doi.org/10.1002/mnfr.200600233.

Dorantes, L., Colmenero, R., Hernandez, H., Mota, L., Jaramillo, M. E., Fernandez, E., et al. (2000). Inhibition of growth of some foodborne pathogenic bacteria by Capsicum annum extracts. *International Journal of Food Microbiology*, *57*(1), 125–128. https://doi.org/10.1016/S0168-1605(00)00216-6.

Durmaz, G., & Gökmen, V. (2010). Impacts of roasting oily seeds and nuts on their extracted oils. *Lipid Technology*, *22*(8), 179–182. https://doi.org/10.1002/lite.201000042.

El-Ghorab, A. H., Javed, Q., Anjum, F. M., Hamed, S. F., & Shaaban, H. A. (2013). Pakistani bell pepper (*Capsicum annum* L.): Chemical compositions and its antioxidant activity. *International Journal of Food Properties*, *16*(1), 18–32. https://doi.org/10.1080/10942912.2010.513616.

Frankel, E. N. (1980). Lipid oxidation. *Progress in Lipid Research*, *19*(1), 1–22. https://doi.org/10.1016/0163-7827(80)90006-5.

Galgani, J. E., Ryan, D. H., & Ravussin, E. (2010). Effect of capsinoids on energy metabolism in human subjects. *British Journal of Nutrition*, *103*(1), 38–42. https://doi.org/10.1017/S0007114509991358.

Gambacorta, G., Faccia, M., Pati, S., Lamacchia, C., Baiano, A., & Notte, E. L. A. (2007). Changes in the chemical and sensorial profile of spices during storage. *Journal of Food Lipids*, *14*(2007), 202–215.

Giuffrida, D., Dugo, P., Torre, G., Bignardi, C., Cavazza, A., Corradini, C., et al. (2013). Characterization of 12 *Capsicum* varieties by evaluation of their carotenoid profile and pungency determination. *Food Chemistry*, *140*(4), 794–802. https://doi.org/10.1016/j.foodchem.2012.09.060.

Grosshagauer, S., Steinschaden, R., & Pignitter, M. (2019). Strategies to increase the oxidative stability of cold pressed oils. *LWT*, *106*, 72–77. https://doi.org/10.1016/j.lwt.2019.02.046.

Gu, L.-B., Pang, H.-L., Lu, K.-K., Liu, H.-M., Wang, X.-D., & Qin, G.-Y. (2017). Process optimization and characterization of fragrant oil from red pepper (*Capsicum annuum* L.) seed extracted by subcritical butane extraction. *Journal of the Science of Food and Agriculture*, *97*(6), 1894–1903. https://doi.org/10.1002/jsfa.7992.

Guil-Guerrero, J. L., Martínez-Guirado, C., del Mar Rebolloso-Fuentes, M., & Carrique-Pérez, A. (2006). Nutrient composition and antioxidant activity of 10 pepper (*Capsicum annuun*) varieties. *European Food Research and Technology*, *224*(1), 1–9. https://doi.org/10.1007/s00217-006-0281-5.

Güneşer, B., Yılmaz, E., & Ok, S. (2017). Cold pressed versus refined winterized corn oils: Quality, composition and aroma. *Grasas y Aceites*, *68*(2). https://doi.org/10.3989/gya.1168162.

Gurnani, N., Gupta, M., Mehta, D., & Mehta, B. K. (2016). Chemical composition, total phenolic and flavonoid contents, and in vitro antimicrobial and antioxidant activities of crude extracts from red chilli seeds (*Capsicum frutescens* L.). *Journal of Taibah University for Science*, *10*(4), 462–470. https://doi.org/10.1016/j.jtusci.2015.06.011.

Guzmán, I., & Bosland, P. W. (2017). Sensory properties of Chile pepper heat—And its importance to food quality and cultural preference. *Appetite*, *117*, 186–190. https://doi.org/10.1016/j.appet.2017.06.026.

Hornero-Méndez, D., Pérez-Gálvez, A., & Mínguez-Mosquera, M. I. (2001). A rapid spectrophotometric method for the determination of peroxide value in food lipids with high carotenoid content. *Journal of the American Oil Chemists' Society*, *78*(11), 1151–1155. https://doi.org/10.1007/s11746-001-0404-y.

Huang, F., Li, W., Huang, Q., Nju, Y., & Wan, C. (2007). New process of dehulling-cold pressing-expansion for double rapeseed. *Proc.* In *Vol. 5. Proc. 12th international rapeseed congress, Wuhan, China* (pp. 126–130).

Inoue, N., Matsunaga, Y., Satoh, H., & Takahashi, M. (2007). Enhanced energy expenditure and fat oxidation in humans with high BMI scores by the ingestion of novel and non-pungent capsaicin analogues (Capsinoids). *Bioscience, Biotechnology, and Biochemistry*, *71*(2), 380–389. https://doi.org/10.1271/bbb.60341.

Jarret, R. L., Levy, I. J., Potter, T. L., & Cermak, S. C. (2013). Seed oil and fatty acid composition in *Capsicum* spp. *Journal of Food Composition and Analysis*, *30*(2), 102–108. https://doi.org/10.1016/j.jfca.2013.02.005.

Jeon, G., Choi, Y., Lee, S.-M., Kim, Y., Oh, M., Jeong, H.-S., et al. (2012). Antioxidant and antiproliferative properties of hot pepper (*Capsicum annuum* L.) seeds. *Journal of Food Biochemistry*, *36*(5), 595–603. https://doi.org/10.1111/j.1745-4514.2011.00571.x.

Ji, J., Liu, Y., Shi, L., Wang, N., & Wang, X. (2019). Effect of roasting treatment on the chemical composition of sesame oil. *LWT*, *101*, 191–200. https://doi.org/10.1016/j.lwt.2018.11.008.

Kawabata, F., Inoue, N., Masamoto, Y., Matsumura, S., Kimura, W., Kadowaki, M., et al. (2009). Non-pungent capsaicin analogs (Capsinoids) increase metabolic rate and enhance thermogenesis via gastrointestinal TRPV1 in mice. *Bioscience, Biotechnology, and Biochemistry*, *73*(12), 2690–2697. https://doi.org/10.1271/bbb.90555.

Konuskan, D. B., Arslan, M., & Oksuz, A. (2019). Physicochemical properties of cold pressed sunflower, peanut, rapeseed, mustard and olive oils grown in the eastern Mediterranean region. *Saudi Journal of Biological Sciences*, *26*(2), 340–344. https://doi.org/10.1016/j.sjbs.2018.04.005.

Koubaa, M., Mhemdi, H., & Vorobiev, E. (2016). Influence of canola seed dehulling on the oil recovery by cold pressing and supercritical CO_2 extraction. *Journal of Food Engineering*, *182*, 18–25. https://doi.org/10.1016/j.jfoodeng.2016.02.021.

Krishnatreyya, H., Hazarika, H., Saha, A., & Chattopadhyay, P. (2018). Fundamental pharmacological expressions on ocular exposure to capsaicin, the principal constituent in pepper sprays. *Scientific Reports*, *8*(1), 12153. https://doi.org/10.1038/s41598-018-30542-2.

Kwon, Y.-I., Apostolidis, E., & Shetty, K. (2007). Evaluation of pepper (*Capsicum annuum*) for management of diabetes and hypertension. *Journal of Food Biochemistry*, *31*(3), 370–385. https://doi.org/10.1111/j.1745-4514.2007.00120.x.

Lejeune, M. P. G. M., Kovacs, E. M. R., & Westerterp-Plantenga, M. S. (2003). Effect of capsaicin on substrate oxidation and weight maintenance after modest body-weight loss in human subjects. *British Journal of Nutrition*, *90*(3), 651–659. https://doi.org/10.1079/BJN2003938.

López-Carrillo, L., López-Cervantes, M., Robles-Díaz, G., Ramírez-Espitia, A., Mohar-Betancourt, A., Meneses-García, A., et al. (2003). Capsaicin consumption, helicobacter pylori positivity and gastric cancer in Mexico. *International Journal of Cancer*, *106*(2), 277–282. https://doi.org/10.1002/ijc.11195.

Luo, X.-J., Peng, J., & Li, Y.-J. (2011). Recent advances in the study on capsaicinoids and capsinoids. *European Journal of Pharmacology*, *650*(1), 1–7. https://doi.org/10.1016/j.ejphar.2010.09.074.

Ma, Y., Wu, X., Zhao, L., Wang, Y., & Liao, X. (2019). Comparison of the compounds and characteristics of pepper seed oil by pressure-assisted, ultrasound-assisted and conventional solvent extraction. *Innovative Food Science & Emerging Technologies*, *54*, 78–86. https://doi.org/10.1016/j.ifset.2019.03.011.

Macho, A., Lucena, C., Sancho, R., Daddario, N., Minassi, A., Muñoz, E., et al. (2003). Non-pungent capsaicinoids from sweet pepper. *European Journal of Nutrition*, *42*(1), 2–9. https://doi.org/10.1007/s00394-003-0394-6.

Maji Amal, K., & Pratim, B. (2016). Phytochemistry and gastrointestinal benefits of the medicinal spice, *Capsicum annuum* L. (chilli): A review. *Journal of Complementary and Integrative Medicine*, *13*, 97. https://doi.org/10.1515/jcim-2015-0037.

Malacrida, C. R., Moraes, I. C. F., de Rosso, V. V., Rodrigues, C. E. d. C., & de Souza, A. C. (2018). Effect of the application of an enzymatic pretreatment on bioactive compounds of *Caryocar brasiliense* Camb pulp oil. *Journal of Food Processing and Preservation*, *42*(12), e13828. https://doi.org/10.1111/jfpp.13828.

Mankowski, C., Poole, C. D., Ernault, E., Thomas, R., Berni, E., Currie, C. J., et al. (2017). Effectiveness of the capsaicin patch in the management of peripheral neuropathic pain in European clinical practice: The ASCEND study. *BMC Neurology*, *17*(1), 80. https://doi.org/10.1186/s12883-017-0836-z.

Marathe, S. J., Jadhav, S. B., Bankar, S. B., Kumari Dubey, K., & Singhal, R. S. (2019). Improvements in the extraction of bioactive compounds by enzymes. *Current Opinion in Food Science*, *25*, 62–72. https://doi.org/10.1016/j.cofs.2019.02.009.

McDowell, D., Elliott, C. T., & Koidis, A. (2017). Pre-processing effects on cold pressed rapeseed oil quality indicators and phenolic compounds. *European Journal of Lipid Science and Technology*, *119*(9), 1600357. https://doi.org/10.1002/ejlt.201600357.

Melgar-Lalanne, G., Hernández-Álvarez, A. J. A. J., Jiménez-Fernández, M., & Azuara, E. (2016). Oleoresins from *Capsicum* spp.: Extraction methods and bioactivity. *Food and Bioprocess Technology*, *10*(1), 1–26. https://doi.org/10.1007/s11947-016-1793-z.

Minguez-Mosquera, M. I., Jaren-Galan, M., & Garrido-Fernandez, J. (1992). Color quality in paprika. *Journal of Agricultural and Food Chemistry*, *40*(12), 2384–2388. https://doi.org/10.1021/jf00024a012.

Moncef, C., Nesrine, M., Leila, R., Francesco, D., Giovanna, F., & Salem, H. (2012). A comparative study on physicochemical, rheological and surface tension properties of Tunisian jujube (*Zizyphus lotus* L.) seed and vegetable oils. *International Journal of Food Engineering*, *8*(2), 1556–3758. https://doi.org/10.1515/1556-3758.2759.

Moscone, E. A., Scaldaferro, M. A., Grabiele, M., Cecchini, N. M., Sánchez García, Y., Jarret, R., et al. (2007). The evolution of chili peppers (*Capsicum-Solanaceae*): A cytogenetic perspective. *Acta Horticulturae*, *745*, 137–170. https://doi.org/10.17660/ActaHortic.2007.745.5.

Moslavac, T., Jokić, S., Šubarić, D., Aladić, K., Vukoja, J., & Prce, N. (2014). Pressing and supercritical CO_2 extraction of Camelina sativa oil. *Industrial Crops and Products*, *54*, 122–129. https://doi.org/10.1016/j.indcrop.2014.01.019.

Nayak, P. K., Dash, U., Rayaguru, K., & Krishnan, K. R. (2016). Physio-chemical changes during repeated frying of cooked oil: A review. *Journal of Food Biochemistry*, *40*(3), 371–390. https://doi.org/10.1111/jfbc.12215.

Oboh, G., Puntel, R. L., & Rocha, J. B. T. (2007). Hot pepper (*Capsicum annuum*, Tepin and *Capsicum chinese*, habanero) prevents Fe^{2+}–induced lipid peroxidation in brain—In vitro. *Food Chemistry*, *102*(1), 178–185. https://doi.org/10.1016/j.foodchem.2006.05.048.

Padilha, H. K. M., Pereira, E., Dos, S., Munhoz, P. C., Vizzotto, M., Valgas, R. A., et al. (2015). Genetic variability for synthesis of bioactive compounds in peppers (*Capsicum annuum*) from Brazil. *Food Science and Technology*, *35*, 516–523.

Palma, J. M., Sevilla, F., Jiménez, A., del Río, L. A., Corpas, F. J., Álvarez de Morales, P., et al. (2015). Physiology of pepper fruit and the metabolism of antioxidants: Chloroplasts, mitochondria and peroxisomes. *Annals of Botany*, *116*(4), 627–636. https://doi.org/10.1093/aob/mcv121.

Patel, K., Calderon, R., Asencios, E., Vilchez, D., Marcelo, M., & Rojas, R. (2016). Agro-morphological characteristics and sensory evaluation of native Peruvian chili peppers. Journal of Agricultural Science and Technology *B*, *6*(2016), 180–187. https://doi.org/10.17265/2161-6264/2016.03.005.

Peng, L., Ye, Q., Liu, X., Liu, S., & Meng, X. (2019). Optimization of aqueous enzymatic method for *Camellia sinensis* oil extraction and reuse of enzymes in the process. *Journal of Bioscience and Bioengineering*. https://doi.org/10.1016/j.jbiosc.2019.05.010.

Pilna, J., Vlkova, E., Krofta, K., Nesvadba, V., Rada, V., & Kokoska, L. (2015). *In vitro* growth-inhibitory effect of ethanol GRAS plant and supercritical CO_2 hop extracts on planktonic cultures of oral pathogenic microorganisms. *Fitoterapia*, *105*, 260–268. https://doi.org/10.1016/j.fitote.2015.07.016.

Rahman, M., Habib, M., Hasan, M., Al Amin, M., Saha, A., & Mannan, A. (2014). Comparative assessment on in vitro antioxidant activities of ethanol extracts of *Averrhoa bilimbi, Gymnema sylvestre* and *Capsicum frutescens. Pharmacognosy research*, *6*(1), 36–41. https://doi.org/10.4103/0974-8490.122915.

Rezende-Naves, E., de Ávila Silva, L., Sulpice, R., Araújo, W. L., Nunes-Nesi, A., Peres, L. E. P., et al. (2019). Capsaicinoids: Pungency beyond capsicum. *Trends in Plant Science*, *24*(2), 109–120. https://doi.org/10.1016/j.tplants.2018.11.001.

Saito, M., & Yoneshiro, T. (2013). Capsinoids and related food ingredients activating brown fat thermogenesis and reducing body fat in humans. *Current Opinion in Lipidology*, *24*(1), 71–77. https://doi.org/10.1097/MOL.0b013e32835a4f40.

Salehi, B., Hernández-Álvarez, A. J., del Mar Contreras, M., Martorell, M., Ramírez-Alarcón, K., Melgar-Lalanne, G., et al. (2018). Potential phytopharmacy and food applications of capsicum spp.: A comprehensive review. *Natural Product Communications*. *13*(11) 1934578X1801301133https://doi.org/10.1177/1934578X1801301133.

Silva, L. R., Azevedo, J., Pereira, M. J., Valentão, P., & Andrade, P. B. (2013). Chemical assessment and antioxidant capacity of pepper (*Capsicum annuum* L.) seeds. *Food and* Chemical Toxicology, *53*, 240–248. https://doi.org/10.1016/j.fct.2012.11.036.

Silvar, C., & García-González, C. A. (2016). Deciphering genetic diversity in the origins of pepper (*Capsicum* spp.) and comparison with worldwide variability. *Crop Science*, *56*, 3100–3111. https://doi.org/10.2135/cropsci2016.02.0128.

Singletary, K. (2011). Red pepper: Overview of potential health benefits. *Nutrition Today*, *46*(1), 33–47. https://doi.org/10.1097/NT.0b013e3182076ff2.

Soto, M., Acosta, O., Vaillant, F., & Pérez, A. (2016). Effects of mechanical and enzymatic pretreatments on extraction of polyphenols from blackberry fruits. *Journal of Food Process Engineering*, *39*(5), 492–500. https://doi.org/10.1111/jfpe.12240.

Sousa, A., Casal, S., Malheiro, R., Lamas, H., Bento, A., & Pereira, J. A. (2015). Aromatized olive oils: Influence of flavouring in quality;. *LWT—Food Science and Technology*, *60*(1), 22–28. https://doi.org/10.1016/j.lwt.2014.08.026.

Tundis, R., Loizzo, M. R., Menichini, F. F., Bonesi, M., Conforti, F., Statti, G., et al. (2011). Comparative study on the chemical composition, antioxidant properties and hypoglycaemic activities of two *Capsicum annuum* L. cultivars (Acuminatum small and Cerasiferum). *Plant Foods for Human Nutrition*, *66*(3), 261–269. https://doi.org/10.1007/s11130-011-0248-.

Wall, M. M., & Bosland, P. W. (1998). Analytical methods for color and pungency of chiles (capsicums). In D. L. B. Wetzel, & G. B. T.-D. in F. S. Charalambous (Eds.), *Vol. 39. Instrumental methods in food and beverage analysis* (pp. 347–373). https://doi.org/10.1016/S0167-4501(98)80014-9.

Whiting, S., Derbyshire, E., & Tiwari, B. K. (2012). Capsaicinoids and capsinoids. A potential role for weight management? A systematic review of the evidence. *Appetite*, *59*(2), 341–348. https://doi.org/10.1016/j.appet.2012.05.015.

Wong, M., Eyres, L., & Ravetti, L. (2014). In W. E. Farr, & A. Proctor (Eds.), *Modern aqueous oil extraction-centrifugation systems for olive and avocado oils* (pp. 19–51). AOCS Press. https://doi.org/10.1016/B978-0-9888565-3-0.50005-4.

Yilmaz, E., & Güneşer, B. A. (2017). Cold pressed versus solvent extracted lemon (*Citrus limon* L.) seed oils: Yield and properties. *Journal of Food Science and Technology*, *54*(7), 1891–1900. https://doi.org/10.1007/s13197-017-2622-8.

Yılmaz, E., Sevgi Arsunar, E., Aydeniz, B., & Güneşer, O. (2015). Cold pressed capia pepperseed (*Capsicum annuum* L.) oils: Composition, aroma, and sensory properties. *European Journal of Lipid Science and Technology*, *117*(7), 1016–1026. https://doi.org/10.1002/ejlt.201400276.

Yong, Y. L., Tan, L. T.-H., Ming, L. C., Chan, K.-G., Lee, L.-H., Goh, B.-H., et al. (2017). The effectiveness and safety of topical capsaicin in Postherpetic neuralgia: A systematic review and meta-analysis. *Frontiers in Pharmacology*, *7*, 538. https://doi.org/10.3389/fphar.2016.00538.

Yoshioka, M., Matsuo, T., Lim, K., Tremblay, A., & Suzuki, M. (2000). Effects of capsaicin on abdominal fat and serum free-fatty acids in exercise-trained rats. *Nutrition Research*, *20*(7), 1041–1045. https://doi.org/10.1016/S0271-5317(00)00180-9.

Yuan, L.-J., Qin, Y., Wang, L., Zeng, Y., Chang, H., Wang, J., et al. (2016). Capsaicin-containing chili improved postprandial hyperglycemia, hyperinsulinemia, and fasting lipid disorders in women with gestational diabetes mellitus and lowered the incidence of large-for-gestational-age newborns. *Clinical Nutrition*, *35*(2), 388–393. https://doi.org/10.1016/j.clnu.2015.02.011.

Zellner, B. d'Acampora, Dugo, P., Dugo, G., & Mondello, L. (2010). Analysis of essential oils. In *Handbook of essential oils: Science, technology and applications*. London, UK: CRC Press, Taylor and Francis Group.

Zhang, R.-Y., Liu, H.-M., Ma, Y.-X., & Wang, X.-D. (2019). Effects of roasting on composition of chili seed and storage stability of chili seed oil. *Food Science and Biotechnology*. https://doi.org/10.1007/s10068-019-00578-9.

Zimmer, A. R., Leonardi, B., Miron, D., Schapoval, E., de Oliveira, J. R., & Gosmann, G. (2012). Antioxidant and anti-inflammatory properties of *Capsicum baccatum*: From traditional use to scientific approach. *Journal of Ethnopharmacology*, *139*(1), 228–233. https://doi.org/10.1016/j.jep.2011.11.005.

Chapter 59

Cold pressed ginger (*Zingiber officinale*) oil

Mohamed Fawzy Ramadan
Agricultural Biochemistry Department, Faculty of Agriculture, Zagazig University, Zagazig, Egypt

Abbreviations

CAT	catalase
DPPH•	2,2-diphenyl-1-picrylhydrazyl
FSH	follicle-stimulating hormone
GO	cold pressed ginger oil
LH	luteinizing hormone
RSA	radical scavenging activity
SC-CO_2	supercritical CO_2
SOD	superoxide dismutase

1 Introduction

Ginger, the rhizome of *Zingiber officinale* Roscoe (family Zingiberaceae), is a plant used as a dietary supplement with health-promoting benefits (An et al., 2016; El Makawy, Ibrahim, Mabrouk, Ahmed, & Ramadan, 2019). *Z. officinale* is believed to have originated in southeast Asia (Bailey, 1949; Munda, Dutta, Haldar, & Lal, 2018; Parry, 1969) and to have spread to China, Latin America, Jamaica, Austria, Japan, and Africa (Demin & Yingying, 2010; Hasan, Rasheed Raauf, Abd Razik, & Rasool Hassan, 2012; Sasidharan & Menon, 2010).

The interest in *Z. officinale* is endorsed to its bioactive phytochemicals including α-zingiberene, gingerol, curcumin, gingerdiol, shogaols, gingerdione, and β-sesquiphellandrene (Zhao et al., 2011). *Z. officinale* is used as a condiment worldwide and used in the treatment of ailments, colds, gastrointestinal disorders, arthritis, hypertension, and migraines (El Makawy et al., 2019; Hosseini & Mirazi, 2015; Shanmugam, Mallikarjuna, Kesireddy, & Sathyavelu, 2011; White, 2007). Ginger was reported to exhibit antiinflammatory and antimicrobial activities (Dugasani et al., 2010; El-Ghorab, Nauman, Anjum, Hussain, & Nadeem, 2010; Noori, Zeynali, & Almasi, 2018). *Z. officinale* essential oil is widely used in therapeutic and flavoring applications worldwide. Maghbooli, Golipour, Esfandabadi, and Yousefi (2014) reported the impact of *Z. officinale* powder in the treatment of migraine attacks and its similarity to antiepileptic drugs. Studies highlighted the antioxidant activity of *Z. officinale* due to its ability to scavenge free radicals (El-Ghorab et al., 2010; Mesomo, Scheer, Perez, Ndiaye, & Corazza, 2012; Oboh, Akinyemi, & Ademiluyi, 2012), which were associated with the prevention of many diseases (Gupta & Sharma, 2014; Liu, Zhou, Zhao, Chen, & Li, 2014; Przygodzka, Zielinska, Ciesarová, Kukurová, & Zielinski, 2014).

Current interest in green technologies has increased the demand for natural products on the international market. Cold pressing provides high levels of bioactive compounds extracted by a simple technique that requires no thermal or chemical treatments (Assiri, Elbanna, Abulreesh, & Ramadan, 2016; El-Beeh, Aljabri, Orabi, Qari, & Ramadan, 2019; El-Hadary, Elsanhoty, & Ramadan, 2019; El-Hadary & Ramadan, 2016; Ramadan, 2013). It is hard to find literature about cold pressed ginger oil (GO) in international sources. Only one recent study (El Makawy et al., 2019) reported the effect of GO and an antiepileptic drug on testicular gene expression, and sex hormones and in mice. This chapter summarizes the chemical composition and applications of GO.

2 Extraction and processing of *Z. officinale* oils

To extract Z. *officinale* oil from Z. *officinale*, authentication of the plant is required. Fresh Z. *officinale* rhizome is dried at 50°C for 24 h and ground before essential oil extraction. The essential oil recovery is affected by the plant origin, its variety,

Cold Pressed Oils. https://doi.org/10.1016/B978-0-12-818188-1.00059-1
Copyright © 2020 Elsevier Inc. All rights reserved.

humidity during harvesting, cultivation process, extraction process, and the age of the plant (Munda et al., 2018; Onyenekwe & Hashimoto, 1999). Chemical analyses showed that *Z. officinale* essential oil contains high levels of sesquiterpene hydrocarbons and relatively low monoterpene hydrocarbons. There are differences in the chemical composition of fresh and dry *Z. officinale* essential oil. Fresh *Z. officinale* oil contains higher oxygenated compounds (29%) compared to dry *Z. officinale* oil (14%). Studies have reported that α-zingiberene, ar-curcumene, geranial, and camphenes are the main compounds (Munda et al., 2018).

Jaiswal and Naik (2018) applied Soxhlet (with ethanol) and green supercritical CO_2 (SC-CO_2), to compare yields of 6-gingerol rich extracts. The yield of 6-gingerol was greater in the SC-CO_2 extract than the Soxhlet extract. A 6-gingerol rich extract from SC-CO_2 had comparable flavonoid, antioxidant, and stability potential to those found in the *Z. officinale* extract from the Soxhlet method. Shukla, Naik, Goud, and Das (2019) used a single-step SC-CO_2 extraction coupled with fractionation of dry ginger to obtain gingerols-rich oleoresin and essential oil. *Z. officinale* oleoresin and volatile oil find a wide range of applications in the beverage, food, fragrance, and pharmaceuticals industries. SC-CO_2 fractionation at 175 bar/40°C resulted in a 5.95% oleoresin (96.1% pure) yield.

3 Fatty acids and tocols composition of GO

El Makawy et al. (2019) reported the fatty acids and tocols composition of GO found in the market. Table 1 summarizes the fatty acids profile (relative percentages) of GO. Linoleic and oleic acids were the main fatty acids, accounting for about 81% of total detected acids, whereas palmitic and stearic acids were the main saturated fatty acids (SFAs), accounting for about 14% of total measured acids. C16:0, C18:0, C18:1*n*-9, and C18:2 accounted for 8.81%, 5.40%, 39.5%, and 42.6%, respectively. GO contained high levels of polyunsaturated fatty acids (PUFAs, 43.1% of total acids), and monounsaturated fatty acids (MUFAs, 41.0% of total acids) which was considered the striking feature of GO.

GO contained high amounts of unsaponifiable matter (3.2 g/kg total lipids) and Fig. 1 presents the percentages of tocols in the cold pressed oil. The amounts of γ- and δ-tocopherols were 500 and 13.7 mg/kg total lipids, respectively. The amounts of α-, β-, and δ-tocotrienols were 167, 9.11, and 190 mg/kg oil, respectively. Approximately 56% of tocols was found as γ-tocopherol, followed by δ-tocotrienol. β-tocopherol was not detected in the cold pressed oil, while other tocols were detected in traces.

TABLE 1 Fatty acids profile (%) of GO.

Fatty acid	%
C 10:0	1.01
C 14:0	0.64
C 16:0	8.81
C 16:1	0.96
C 18:0	5.40
C 18:1*n*-9	39.5
C 18:2	42.6
C 20:1	0.53
C 18:3	0.45
Total SFA	15.9
Total MUFA	41.0
Total PUFA	43.1

Data from El Makawy, A. I., Ibrahim, F. M., Mabrouk, D. M., Ahmed, K. A., & Ramadan, M. (2019). Effect of antiepileptic drug (Topiramate) and cold pressed ginger oil on testicular genes expression, sexual hormones and histopathological alterations in mice. *Biomedicine & Pharmacotherapy, 110*, 409–419.

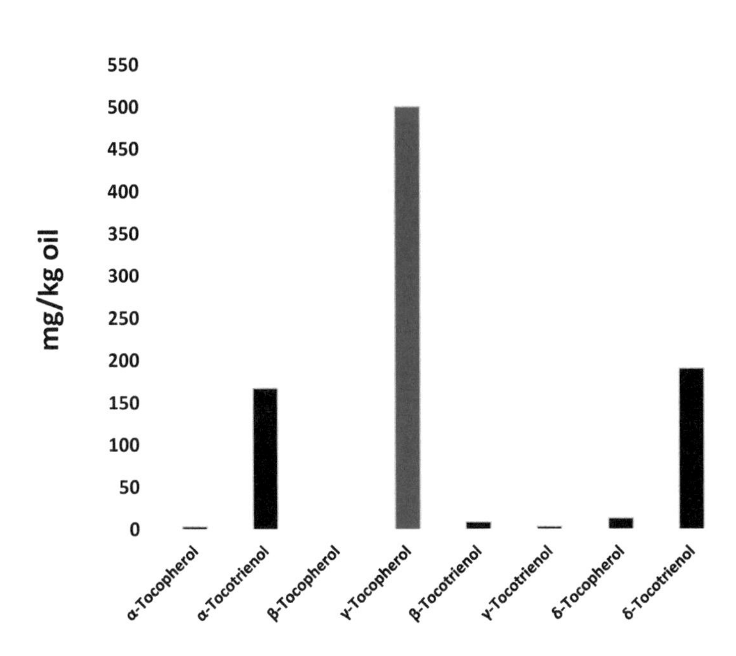

FIG. 1 Tocols composition of GO. *(Adapted from El Makawy, A. I., Ibrahim, F. M., Mabrouk, D. M., Ahmed, K. A., & Ramadan, M. (2019). Effect of antiepileptic drug (Topiramate) and cold pressed ginger oil on testicular genes expression, sexual hormones and histopathological alterations in mice.* Biomedicine & Pharmacotherapy, 110, *409–419.)*

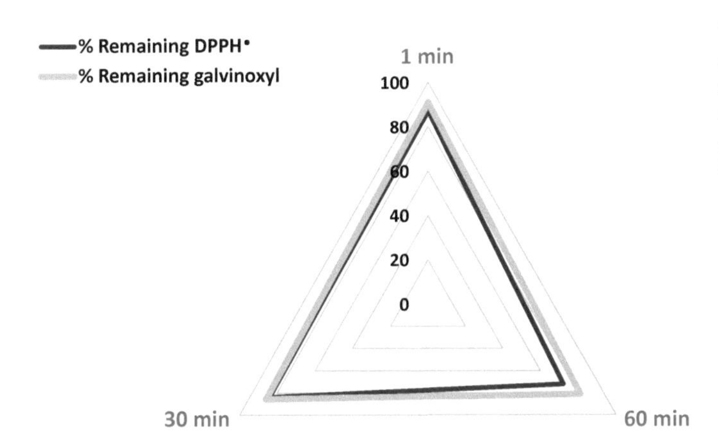

FIG. 2 Radical scavenging activity of GO against DPPH• and galvinoxyl free radicals. *(Reproduced from El Makawy, A. I., Ibrahim, F. M., Mabrouk, D. M., Ahmed, K. A., & Ramadan, M. (2019). Effect of antiepileptic drug (Topiramate) and cold pressed ginger oil on testicular genes expression, sexual hormones and histopathological alterations in mice.* Biomedicine & Pharmacotherapy, 110, *409–419.)*

4 Health-promoting properties of GO and oil constituents

The radical scavenging activities of GO were compared by applying DPPH• and galvinoxyl stable free radicals (El Makawy et al., 2019). GO exhibited high antiradical action against DPPH• (Fig. 2). After incubation with DPPH• radicals for 1, 30, and 60 min, 13%, 17%, and 28% of free radicals were quenched by GO, respectively. An electron spin resonance test exhibited the same pattern. After incubation with galvinoxyl radicals for 1, 30, and 60 min, 9%, 14%, and 19% of galvinoxyl free radicals were quenched by GO, respectively (Fig. 2).

Z. officinale essential oil had dominative protection against H_2O_2 induced DNA damage. The oil acts as an oxygen radicals scavenger and could be applied as an antioxidant (Ma, Jin, Yang, & Liu, 2004; Munda et al., 2018). The essential oil exhibited antioxidant activity, which was revealed by DPPH• radical scavenging potential and β-carotene bleaching tests when compared to BHT, BHA, and α-tocopherol (El-Baroty, Abd El-Baky, Farag, & Saleh, 2010). In addition, the antioxidant potential of *Z. officinale* was measured by the temperature changes of the oleic acid oxidation, which changed from 119.9°C to 125.3°C after the addition of *Z. officinale* essential oil (Ali & Mehri, 2017; Munda et al., 2018). Jayanudin, Fahrurrozi, Wirawan, and Rochmadi (2019) studied the antioxidant potential and controlled release of red ginger (*Z. officinale* var. *rubrum*) oleoresin encapsulated in chitosan. The study investigated particle size, encapsulation efficiency, and oleoresin microcapsule such as morphology and controlled release. The antioxidant effect was between 38.2% and 61.9%.

Munda et al. (2018) reviewed the biological and pharmacological traits of *Z. officinale* essential oil. The dry oil is used against *Pseudomonas aeruginosa*, *Penicillium* spp., and *Candida albicans*, while the oil extracted from fresh plants could be used against *Aspergillus niger* and *C. albicans* (Munda et al., 2018; Sasidharan & Menon, 2010). *Z. officinale* oil changed the morphological traits of bacterial cells, proving to be an effective antibacterial agent (El-Shouny, Ali, Sun, Samy, & Ali, 2018; Munda et al., 2018). β-sesquiphellandrene, zingiberene, and caryophyllene possess antimicrobial action against Gram negative bacteria (*Klebsiella pneumoniae* and *Serratia marcescens*), Gram positive bacteria (*Bacillus subtilis*, *B. cereus*, *Staphylococcus aureus*, and *Micrococcus luteus*), and fungal (*A. niger*, *Penicillium notatum*, *Mucor hiemalis*, and *Fusarium oxysporum*) strains (El-Baroty et al., 2010; Munda et al., 2018). Souza et al. (2019) studied the in vitro bioactivity of chitosan nanocomposites incorporated with *Zingiber officinale* Rosc. essential oil. Films were tested for their antimicrobial action against foodborne pathogenic bacteria (*Enterococcus faecalis*, *Listeria monocytogenes*, *S. aureus*, *Escherichia coli*, and *Pseudomonas aeruginosa*). Ginger oil improved the antimicrobial action of chitosan toward *Enterococcus faecalis* and *S. aureus*. Cutrim, Teles, Mouchrek, Mouchrek Filho, and Everton (2019) evaluated antimicrobial activity of *Z. officinale* essential oil and extract (alcohol-based product). The oil exhibited high antimicrobial action against *E. coli* and *S. aureus* with a minimum inhibitory concentration of 200 μg/mL. Baldin et al. (2019) determined the composition of *Z. officinale* Rosc. essential oil obtained by hydrodistillation and fractionation against *Mycobacterium* spp., as well as their cytotoxicity. The essential oil contained a mixture of monoterpenes and sesquiterpenes with high antimycobacterial action against *M. chelonae*, and *M. abscessus* sub. *massiliense*.

The inhibitory impact of ginger essential oil was screened in vitro against herpes simplex virus type 2 (HSV-2) on RC-37 cells. The oil was considered as a virucidal agent that changed HSV-2 before adsorption by interacting with the viral envelope (Koch, Reichling, Schneele, & Schnitzler, 2008). The essential oil had direct and systemic impacts on leukocyte migration. The oil reduced leukocyte adherence after 120 min of carrageenan injection into the scrotal chamber. Moreover, leukocyte migration to the perivascular tissue was decreased after 4 h (Munda et al., 2018; Nogueira de Melo et al., 2011).

Acute and chronic inflammation induced by formalin and carrageenan were reduced after application of *Z. officinale* essential oil. The essential oil possessed antiinflammatory potential and lipoxygenase inhibition of 51% (at 0.4 mg/mL), which was comparable to quercetin that exhibited 52.3% inhibition (at 0.1 mg/mL) as well as 100% inhibition at 8 mg/mL (Bayala et al., 2014; Munda et al., 2018).

Propranolol (a β-adrenergic receptor antagonist), found in *Z. officinale* essential oil, had the trait to reverse the bronchodilatory impacts, suggesting a pharmacological effect for the *Z. officinale* rhizomes to be a bronchodilator. The essential oil was reported to recover humoral immune response in immune-suppressed animals (Carrasco et al., 2009). Monoterpenes in the essential oil possessed antitumor potential and could serve as a cancer chemopreventive agent (Crowell, 1999; Kelloff et al., 1996; Munda et al., 2018). Feng et al. (2018) reported the skin-protective impact of *Z. officinale* Rosc. essential oil. Treatment with oil ameliorated ultraviolet B-induced skin inflammation and inhibited interleukin-1β and tumor necrosis factor-α expression in the skin tissue. Thus, the oil could be used to protect the skin from UVB irradiation and photoaging.

El Makawy et al. (2019) studied the impact of Topiramate (antiepileptic drug) and GO on sexual hormones, testicular genes expression, and histopathological alteration in the experimental animals. In a dose-dependent manner, GO significantly elevated SOD and CAT levels. Coadministration of Topiramate and GO attenuated the superoxide dismutase (SOD) and catalase (CAT) approximately to control. Coadministration of Topiramate and GO raised the serum luteinizing hormone (LH), and follicle-stimulating hormone (FSH) over the control. In addition, GO upregulated the Vegfa gene expression and Sycp3 gene expression in a dose-dependent pattern. GO decreased the mRNA expression compared to Topiramate.

5 Conclusion

The current chapter reported on the extraction, chemical composition, and applications of GO. The oil contains high levels of tocols and essential fatty acids as well as some bioactive phytochemicals including zingiberene, β-bisabolene, ar-curcumene, citral, and geranial. The ginger essential oil has several medicinal values and therapeutic uses. The oil demonstrated anticancer, analgesic, antiarthritic, antiinflammatory, antitussive, and antimicrobial traits. The oil might be used as preservatives for food products to prevent spoilage. The oil could be utilized as topical medication for bacterial and fungal diseases. GO administration could protect from sexual dysfunction that might harm epilepsy patients due to the consuming of antiepileptic drugs. In addition, it is recommended that epileptic men could use GO to minimize the effect of this medication on their fertility.

References

Ali, S., & Mehri, A. (2017). Separation identification and antioxidant evaluation of *Zingiber officinale* essential oil. *Iranian Chemical Communication*, *5*, 278–285.

An, K., Zhao, D., Wang, Z., Wu, J., Xu, Y., & Xiao, G. (2016). Comparison of different drying methods on Chinese ginger (*Zingiber officinale* roscoe): Changes in volatiles, chemical profile, antioxidant properties, and microstructure. *Food Chemistry*, *197*, 1292–1300.

Assiri, A. M. A., Elbanna, K., Abulreesh, H. H., & Ramadan, M. F. (2016). Bioactive compounds of cold-pressed thyme (*Thymus vulgaris*) oil with antioxidant and antimicrobial properties. *Journal of Oleo Science*, *65*, 629–640.

Bailey, L. B. (1949). *Manual of cultivated plants* (2nd ed.). New York: Macmillan.

Baldin, V. P., de Lima Scodro, R. B., Fernandez, C. M. M., Ieque, A. L., Caleffi-Ferracioli, K. R., Siqueira, V. L. D., et al. (2019). Ginger essential oil and fractions against *Mycobacterium* spp. *Journal of Ethnopharmacology*, *244*, 112095.

Bayala, B., Bassole, I. H. N., Gnoula, C., Nebie, R., Yonli, A., Morel, L., et al. (2014). Chemical composition, antioxidant, anti-inflammatory and antiproliferative activities of essential oils of plants from Burkina Faso. *PLoS One*, *9*(3), e92122.

Carrasco, F. R., Schmidt, G., Romero, A. L., Sartoretto, J. L., Caparroz-Assef, S. M., Amado, C. A. B., et al. (2009). Immunomodulatory activity of *Zingiber officinale* roscoe, *Salvia officinalis* L. and *Syzygium aromaticum* L. essential oils: Evidence for humor-and cell-mediated responses. *The Journal of Pharmacy and Pharmacology*, *61*, 961–967.

Crowell, P. L. (1999). Prevention and therapy of cancer by dietary monoterpenes. *The Journal of Nutrition*, *129*, 775S–778S.

Cutrim, E. S. M., Teles, A. M., Mouchrek, A. N., Mouchrek Filho, V. E., & Everton, G. O. (2019). Evaluation of antimicrobial and antioxidant activity of essential oils and hydroalcoholic extracts of *Zingiber officinale* (ginger) and *Rosmarinus officinalis* (rosemary). *Revista Virtual de Quimica*, *11*(1), 60–81.

Demin, G., & Yingying, Z. (2010). Comparative antibacterial activities of crude polysaccharides and flavonoids from *Zingiber officinale* and their extraction. *American Journal of Tropical Medicine*, *5*, 235–238.

Dugasani, S., Pichikac, M. R., Nadarajahc, V. D., Balijepalli, M. K., Tandra, S., & Korlakunta, J. N. (2010). Comparative antioxidant and anti-inflammatory effects of [6]-gingerol, [8]-gingerol, [10]-gingerol and [6]-shogaol. *Journal of Ethnopharmacology*, *127*, 515–520.

El Makawy, A. I., Ibrahim, F. M., Mabrouk, D. M., Ahmed, K. A., & Ramadan, M. (2019). Effect of antiepileptic drug (Topiramate) and cold pressed ginger oil on testicular genes expression, sexual hormones and histopathological alterations in mice. *Biomedicine & Pharmacotherapy*, *110*, 409–419.

El-Baroty, G. S., Abd El-Baky, H. H., Farag, R. S., & Saleh, M. A. (2010). Characterization of antioxidant and antimicrobial compounds of cinnamon and ginger essential oils. *African Journal of Biochemistry Research*, *4*(6), 167–174.

El-Beeh, M. E., Aljabri, M., Orabi, H. F., Qari, S. H., & Ramadan, M. F. (2019). Ameliorative impact of cold-pressed *Rosmarinus officinalis* oil against liver toxicity and genotoxic effects in streptozotocin-induced diabetic rats and their offspring. *Journal of Food Biochemistry*. *43*, e12905. https://doi.org/10.1111/jfbc.12905.

El-Ghorab, A. H., Nauman, M., Anjum, F. M., Hussain, S., & Nadeem, M. (2010). A comparative study on chemical composition and antioxidant activity of ginger (*Zingiber officinale*) and cumin (*Cuminum cyminum*). *Journal of Agricultural and Food Chemistry*, *58*, 8231–8237.

El-Hadary, A. E., Elsanhoty, R. M., & Ramadan, M. F. (2019). *In vivo* protective effect of Rosmarinus officinalis oil against carbon tetrachloride (CCl_4)-induced hepatotoxicity in rats. *PharmaNutrition*, *9*, 100151.

El-Hadary, A. E., & Ramadan, M. F. (2016). Hepatoprotective effect of cold-pressed *Syzygium aromaticum* oil against carbon tetrachloride (CCl_4)-induced hepatotoxicity in rats. *Pharmaceutical Biology*, *54*, 1364–1372. https://doi.org/10.3109/13880209.2015.1078381.

El-Shouny, W. A., Ali, S. S., Sun, J., Samy, S. M., & Ali, A. (2018). Drug resistance profile and molecular characterization of extended spectrum beta-lactamase (esâl)-producing *Pseudomonas aeruginosa* isolated from burn wound infections. Essential oils and their potential for utilization. *Microbial Pathogenesis*, *116*, 301–312.

Feng, J., Du, Z., Zhang, L., Luo, W., Zheng, Y., Chen, D., et al. (2018). Chemical composition and skin protective effects of essential oil obtained from ginger (*Zingiber officinale* roscoe). *Journal of Essential Oil-Bearing Plants*, *21*(6), 1542–1549. https://doi.org/10.1080/0972060X.2018.1533436.

Gupta, S. K., & Sharma, A. (2014). Medicinal properties of *Zingiber officinale* roscoe-A review. *Int J Pharm Bio Sci*, *9*, 124–129.

Hasan, H. A., Rasheed Raauf, A. M., Abd Razik, B. M., & Rasool Hassan, B. A. (2012). Chemical composition and antimicrobial activity of the crude extracts isolated from *Zingiber officinale* by different solvents. *Pharmaceut Anal Acta*, *3*, 9.

Hosseini, A., & Mirazi, N. (2015). Alteration of pentylenetetrazole-induced seizure threshold by chronic administration of ginger (*Zingiber officinale*) extract in male mice. *Pharmaceutical Biology*, *53*, 752–757.

Jaiswal, S., & Naik, S. (2018). Contribution of agricultural produce spice *Zingiber officinale* to a sustainable food system: Green extraction and stability study of antioxidant compounds. *Open Agriculture*, *3*(1), 326–338. https://doi.org/10.1515/opag-2018-0036.

Jayanudin, M., Fahrurrozi, S., Wirawan, K., & Rochmadi (2019). Antioxidant activity and controlled release analysis of red ginger oleoresin (*Zingiber officinale* var *rubrum*) encapsulated in chitosan cross-linked by glutaraldehyde saturated toluene. *Sustainable Chemistry and Pharmacy*, *12*, 100132.

Kelloff, G. J., Boone, C. W., Crowell, J. A., Steele, V. E., Lubet, R. A., Doody, L. A., et al. (1996). New agents for cancer chemoprevention. *Journal of Cellular Biochemistry*, *26*, 1–28.

Koch, C., Reichling, J., Schneele, J., & Schnitzler, P. (2008). Inhibitory effect of essential oils against herpes simplex virus type 2. *Phytomedicine*, *15*, 71–78.

Liu, W., Zhou, C. L., Zhao, J., Chen, D., & Li, Q. H. (2014). Optimized microwave-assisted extraction of 6-gingerol from *Zingiber officinale* roscoe and evaluation of antioxidant activity in vitro. *Acta Scientiarum Polonorum Technologia Alimentaria*, *13*, 155–168.

Ma, J., Jin, X., Yang, L., & Liu, Z. L. (2004). Diarylheptanoids from the rhizomes of *Zingiber officinale*. *Phytochemistry*, *65*(8), 1137–1143.

Maghbooli, M., Golipour, F., Esfandabadi, M. A., & Yousefi, M. (2014). Comparison between the efficacy of ginger and sumatriptan in the ablative treatment of the common migraine. *Phytotherapy Research*, *28*, 412–415.

Mesomo, M. C., Scheer, A. P., Perez, E., Ndiaye, P. M., & Corazza, M. L. (2012). Ginger (*Zingiber officinale* R.) extracts obtained using supercritical CO_2 and compressed propane: Kinetics and antioxidant activity evaluation. *Journal of Supercritical Fluids*, *71*, 102–109.

Munda, S., Dutta, S., Haldar, S., & Lal, M. (2018). Chemical analysis and therapeutic uses of ginger (*Zingiber officinale* Rosc.) essential oil: A review. *Journal of Essential Oil-Bearing Plants*, *21*(4), 994–1002. https://doi.org/10.1080/0972060X.2018.1524794.

Nogueira de Melo, G. A., Grespan, R., Fonseca, J. P., Farinha, T. O., Leite da Silva, E., Romero, A. L., et al. (2011). Inhibitory effects of ginger (*Zingiber officinale* Roscoe) essential oil on leukocyte migration in vivo and in vitro. *Journal of Natural Medicines*, *65*, 241.

Noori, S., Zeynali, F., & Almasi, H. (2018). Antimicrobial and antioxidant efficiency of nanoemulsion-based edible coating containing ginger (*Zingiber officinale*) essential oil and its effect on safety and quality attributes of chicken breast fillets. *Food Control*, *84*, 312–320.

Oboh, G., Akinyemi, A. J., & Ademiluyi, A. O. (2012). Antioxidant and inhibitory effect of red ginger (*Zingiber officinale* var. *rubra*) and white ginger (*Zingiber officinale* roscoe) on Fe^{2+} induced lipid peroxidation in rat brain in vitro. *Experimental and Toxicologic Pathology*, *64*, 31–36.

Onyenekwe, P. C., & Hashimoto, S. (1999). The composition of the essential oil of dried Nigerian ginger (*Zingiber officinale* roscoe). *European Food Research and Technology*, *209*, 407–410.

Parry, J. W. (1969). *Ginger morphology, history, chemistry*: (p. 49). (Vol. 2). London Food Trade Press Ltd.

Przygodzka, M., Zielinska, D., Ciesarová, Z., Kukurová, K., & Zielinski, H. (2014). Comparison of methods for evaluation of the antioxidant capacity and phenolic compounds in common spices. *LWT- Food Science and Technology*, *58*, 321–326.

Ramadan, M. F. (2013). Healthy blends of high linoleic sunflower oil with selected cold pressed oils: Functionality, stability and antioxidative characteristics. *Industrial Crops and Products*, *43*, 65–72. https://doi.org/10.1016/j.indcrop.2012.07.013.

Sasidharan, I., & Menon, A. N. (2010). Comparative chemical composition and antimicrobial activity fresh and dry ginger oils (*Zingiber officinale* Roscoe). *International Journal of Current Pharmaceutical Research*, *2*(4), 40–43.

Shanmugam, K. R., Mallikarjuna, K., Kesireddy, N., & Sathyavelu, R. K. (2011). Neuroprotective effect of ginger on antioxidant enzymes in streptozotocin-induced diabetic rats. *Food and Chemical Toxicology*, *49*, 893–897.

Shukla, A., Naik, S. N., Goud, V. V., & Das, C. (2019). Supercritical CO_2 extraction and online fractionation of dry ginger for production of high-quality volatile oil and gingerols enriched oleoresin. *Industrial Crops and Products*, *130*, 352–362.

Souza, V. G. L., Rodrigues, C., Ferreira, L., Pires, J. R. A., Duarte, M. P., Coelhoso, I., et al. (2019). In vitro bioactivity of novel chitosan bionanocomposites incorporated with different essential oils. *Industrial Crops and Products*, *140*, 111563.

White, B. (2007). Ginger: An overview. *American Family Physician*, *75*, 1689–1691.

Zhao, X., Zingiber, B., Yang, W. R., Yang, Y., Wang, S., Jiang, Z., et al. (2011). Effects of ginger root (*Zingiber officinale*) on laying performance and antioxidant status of laying hens and on dietary oxidation stability. *Poultry Science*, *90*, 1720–1727.

Chapter 60

Cold pressed rosemary (*Rosmarinus officinalis*) oil

Mohamed Fawzy Ramadan[a,b], Manal Khider[c], Hussein H. Abulreesh[d], Adel M.A. Assiri[e], Rafaat M. Elsanhoty[f], Abdelrahaman Assaeedi[d], and Khaled Elbanna[d,g]

[a]*Agricultural Biochemistry Department, Faculty of Agriculture, Zagazig University, Zagazig, Egypt,* [b]*Deanship of Scientific Research, Umm Al-Qura University, Makkah, Saudi Arabia,* [c]*Department of Dairy Science, Faculty of Agriculture, Fayoum University, Fayoum, Egypt,* [d]*Department of Biology, Faculty of Applied Science, Umm Al-Qura University, Makkah, Saudi Arabia,* [e]*Biochemistry Department, Faculty of Medicine, Umm Al-Qura University, Makkah, Saudi Arabia,* [f]*Department of Industrial Biotechnology, Institute of Genetic Engineering and Biotechnology, University of Sadat City, Sadat City, Egypt,* [g]*Department of Agricultural Microbiology, Faculty of Agriculture, Fayoum University, Fayoum, Egypt*

Abbreviations

CERs cerebrosides
CK creatine phosphokinase
CPRO cold pressed rosemary (*Rosmarinus officinalis*) oil
DG diacylglycerol
DGDs digalactosyldiglycerides
DM diabetes mellitus
DPPH· 2,2-diphenyl-1-picrylhydrazyl
ESGs esterified sterylglycosides
FBG fasting blood glucose
FFA free fatty acid
GC-FID gas chromatography-flame ionization detector
HDL-C high dentistry lipoprotein-cholesterol
HPLC high-performance liquid chromatography
LDL-C low dentistry lipoprotein-cholesterol
MG monoacylglycerol
MGDs monogalactosyldiglycerides
MIC minimum inhibitory concentration
REO rosemary essential oil
RSA radical scavenging activity
SCF supercritical fluid
SFA saturated fatty acid
SQD sulphoquinovosyldiacylglycerol
STEs esterified sterols
STZ streptozotocin
T1DM DM type 1
T2DM DM type 2
TAG triacylglycerol
TC total cholesterol
TGs total triacylglycerols
TLs total lipids

Cold Pressed Oils. https://doi.org/10.1016/B978-0-12-818188-1.00060-8
Copyright © 2020 Elsevier Inc. All rights reserved.

TPCs total phenolic compounds
VLDL-C very low dentistry lipoprotein-cholesterol
WHO World Health Organization

1 Introduction

Medicinal and aromatic plants are rich in bioactive phytochemicals, which are used in nutraceuticals, functional foods, and pharmaceuticals with biological properties to promote health (Al-Kalaldeh, Abu-Dahab, & Afifi, 2010; Albano & Miguel, 2011; Elbanna et al., 2014; El-Hadary & Ramadan, 2019; Ramadan, Amer, & Awad, 2008; Ramadan & Elsanhoty, 2012; Ramadan, Sharanabasappa, Seetharam, Seshagiri, & Moersel, 2006). Rosemary (*Rosmarinus officinalis*, family Lamiaceae) is a perennial herb and native Mediterranean shrub (El-Naggara, Abdel-Farid, Germoush, Elgebaly, & Alm-Eldeen, 2016; Elbanna et al., 2018). Andrade et al. (2018) showed an interest in *R.* officinalis, with about 120 studies every year since 2010. Sadeh et al. (2019) highlighted the impact of production process and the genetic variation on *R. officinalis* oil composition. They reflected the importance of studying the impact of genetic and environmental factors and processing on oil composition for industrial breeding and oil production. Sarmoum et al. (2019) investigated the impact of water stresses and salinity on the constituents of *R. officinalis* essential oil (REO). *R. officinalis* plants were subjected to tap water, salt water, and without irrigation. Nonirrigated plants contained the highest oil yield. Differences in the oil constituents were highlighted in relation to water stress.

R. officinalis extracts and REO are in edible applications as preservatives and for treating some diseases (Elbanna et al., 2018; Olmedo, Nepote, & Grosso, 2013; Wollinger et al., 2016). *R. officinalis* extracts are natural sources of antioxidants (Commission Regulation (EU) no. 1130/2011, 2011; Ojeda-Sana, Baren, Elechosa, Juárez, & Moreno, 2013; Yang et al., 2016), wherein the antioxidant potential has been attributed to phenolic diterpenes (Gallego, Gordon, Segovia, Skowyra, & Almajano, 2013). Bioactive constituents of REO are 1,8-cineole, camphor, carnosic acid, and rosmarinic acid (Borges, Ortiz, Pereira, Keita, & Carvalho, 2019; Terpinc, Bezjak, & Abramovic, 2009). REO exhibited antibacterial (Ojeda-Sana et al., 2013), antifungal (Soylu, Kurt, & Soylu, 2010), and hepatoprotective (Amin & Hamza, 2005) properties. *R. officinalis* extracts showed hepatoprotective effects against hepatotoxic agents such as t-BHP (Joyeux, Roland, Fleurentin, Mortier, & Dorfman, 1990), CCl_4 (Fahim, Esmat, Fadel, & Hassan, 1999), and cyclophosphamide (Fahim et al., 1999). In addition, *R.* officinalis showed a protective impact against Azathioprine-induced liver damage in animals and blocked serum high levels of alanine aminotransferase and aspartate aminotransferase (Amin & Hamza, 2005). Extract from *R.* officinalis leaves mitigated cyclophosphamide-induced (El-Naggara et al., 2016) and creosote-induced (El-Demerdash, Abbady, & Baghdadi, 2016) hepatotoxicity in rats. *R. officinalis* oleoresin was used to develop stabile vegetable oil blends used for frying (Upadhyay, Sehwag, & Mishra, 2017; Yang et al., 2016).

Crude extracted oils are rich sources of bioactive compounds, such as phenolics, tocols, phytosterols, and fatty acids with health-promoting and functional properties (Assiri, Elbanna, Abulreesh, & Ramadan, 2016; Kiralan et al., 2017; Ramadan, Asker, & Tadros, 2012). Current interest in environmentally friendly technologies has resulted in a huge international market of natural products (Ibrahim, Attia, Maklad, Ahmed, & Ramadan, 2017). Cold pressing is popular due to the high levels of bioactives in the recovered oil. Cold pressing is an environmentally safe and simple technique that requires no chemical or thermal treatments (Assiri, Elbanna, Al-Thubiani, & Ramadan, 2016; El-Hadary & Ramadan, 2016a; Ramadan, 2013). Some cold pressed oils showed a protective impact against CCl_4-induced hepatotoxicity in rats (El-Hadary & Ramadan, 2016a, 2016b).

This chapter reviews the lipids profile, phenolics content, and antioxidant, antimicrobial, antidiabetic, and hepatoprotective properties of cold pressed rosemary (*R. officinalis*) oil (CPRO).

2 Extraction and processing of cold pressed *R. officinalis* oil

Ali, Chua, and Chow (2019) reviewed the history, chemical profile, and analysis of *R. officinalis* extraction technologies. The execution of extraction methods is endless because they stretch from conventional (maceration, Soxhlet, microwave distillation) to developed technologies (supercritical fluid extraction, ultrasound-assisted, pressurized liquid). Carnosol, carnosic acid, and rosmarinic acid, the major markers of bioassays with the highest activities in drugs, were the main compounds investigated. To produce fractions rich in those bioactive compounds, pressurized liquid and SCF extraction followed by supercritical antisolvent fractionation are among the most tested methods. With the development of novel techniques, extracting plant bioactive phytochemicals according to the desired applications is possible. The major bioactive compounds in the *R. officinalis* volatiles are 1α-pinene, 8-cineole, and camphor, while bioactive compounds in the non-volatile extract are carnosol, carnosic acid, and rosmarinic acid.

3 Acyl lipids and fatty acid profile of cold pressed *R. officinalis* oil

3.1 Lipid classes

Neutral lipids (NL), phospholipids (PL), and glycolipids (GL) represent the main lipid classes in the most of crude vegetable oils. The proportions of lipid classes in CPRO are shown in Table 1 (Elbanna et al., 2018). In the CPRO, NL fraction was the highest, followed by PL and GL. Triacylglycerol (TAG), diacylglycerol (DG), free fatty acids (FFAs), monoacylglycerol (MG), and esterified sterols (STEs) were the main NL classes. Classes of GL were sterylglycosides (SGs), sulphoquinovosyldiacylglycerol (SQD), digalactosyldiglycerides (DGDs), monogalactosyldiglycerides (MGDs), cerebrosides (CERs), and esterified sterylglycosides (ESGs). The main PL subclasses were phosphatidylcholine, followed by phosphatidylethanolamine. Polar lipids (GL and PL) in crude vegetable oils exhibited antiradical and antioxidant traits (Ramadan, 2008, 2012; Ramadan and Asker, 2009).

3.2 Fatty acids

Fatty acid profiles of the R. officinalis total lipids (TL) and lipid classes in CPRO are given in Table 2 (Elbanna et al., 2018). Lipid classes have a similar fatty acid composition, whereas linoleic (C18:2) and oleic (C18:1) acids were the major acids. Palmitic (C16:0) and stearic (C18:0) acids were the major identified saturated fatty acids (SFAs). The amounts of polyunsaturated fatty acids (PUFAs), monounsaturated fatty acids (MUFAs), and SFAs were 42.30%, 41.70%, and 15.80%, respectively. The U/S ratio of CPRO was 5.3, and CPRO might be included into oleic/linoleic oils group. The CPRO fatty acids profile was similar to that of pumpkin, sunflower, and maize oils (Tuberoso, Kowalczyk, Sarritzu, & Cabras, 2007). MUFA levels were comparable to cranberry, blueberry, hemp, and onion cold pressed oils (Parker, Adams, Zhou, Harris, & Yu, 2003; Parry et al., 2006). High levels of PUFAs and MUFAs make CPRO a valuable oil in human nutrition (Elbanna et al., 2018).

4 Minor bioactive lipids in cold pressed *R. officinalis* oil

Tocols are inhibitors of free radical chain reactions and delay lipids oxidation (Hassanien et al., 2014Parry et al., 2006). CPRO contained high amounts (25 g/kg oil) of unsaponifiables. α-, β-, γ-, and δ-tocopherols in CPRO accounted for 291, 22, 1145, and 41 mg/100 g CPRO, respectively. Levels of α-, β-, γ-, and δ-tocotrienols accounted for 18, 12, 29, and 158 mg/100 g CPRO, respectively (Elbanna et al., 2018). The major tocopherol homologue was γ-tocopherol (more than 66% of the tocols), followed by δ-tocopherol. According to the levels of tocochromanols (Hassanien et al., 2014), oils could be divided to oils with high α-tocopherol amounts (sunflower, almond, olive oil, hazelnut, and wheat germ) and oils with high γ-tocopherol amounts (pumpkin, black cumin, flaxseed, poppy, apricot kernel, and sesame). The most efficient antioxidant

TABLE 1 Neutral lipids (NL), glycolipids (GL), and phospholipids (PL) classes (g/kg total lipids) of CPRO.

NL class	Total lipids (g/kg)	GL class	Total lipids (g/kg)	PL class	Total lipids (g/kg)
MG	2.55	SQD	0.70	PS	1.20
DG	7.59	DGD	0.55	PI	2.10
FFAs	10.9	CER	3.26	PC	3.44
TG	833	SG	2.45	PE	2.50
STEs	6.88	MGD	0.33		
		ESG	1.55		

MAGs, monoacylglycerols; *DAGs*, diacylglycerols; *TAGs*, triacylglycerols; *FFAs*, free fatty acids; *STEs*, sterol esters; *SQD*, sulphoquinovosyldiacylglycerol; *DGD*, digalactosyldiacylglycerol; *CERs*, cerebrosides; *SG*, steryl glucoside; *MGD*, monogalactosyldiacylglycerol; *ESG*, esterified steryl glucoside; *PS*, phosphatidylserine; *PI*, phosphatidylinositol; *PC*, phosphatidylcholine; *PE*, phosphatidylethanolamine.
(Adapted from Elbanna, K., Assiri, A. M. A., Tadros, M., Khider, M., Assaeedi, A., Mohdaly, A. A. A., & Ramadan, M. F. (2018). Rosemary (*Rosmarinus officinalis*) oil: Composition and functionality of the cold-pressed extract. *Journal of Food Measurement and Characterization, 12*, 1601–1609.)

TABLE 2 Fatty acid profile of CPRO and lipid classes.

	CPRO	NL	GL	PL
C 10:0	0.07	0.07	0.06	0.07
C 12:0	0.04	0.03	0.02	0.02
C 14:0	0.09	0.05	0.03	0.03
C 16:0	9.10	8.95	8.40	8.90
C 16:1	0.54	0.53	0.55	0.54
C 18:0	6.55	6.41	5.50	5.96
C 18:1*n*-9	41.0	41.1	41.6	41.2
C 18:2*n*-6	41.1	41.4	41.9	41.7
C 20:1	0.25	0.26	0.38	0.28
C 18:3	1.21	1.20	1.50	1.30
ΣSFA	15.85	15.51	14.01	14.98
ΣMUFA	41.79	41.89	42.59	42.02
ΣPUFA	42.36	42.60	43.40	43.00
U/S	5.30	5.44	6.09	5.67
n-6/*n*-3	34.01	34.50	27.93	32.08

(Adapted from Elbanna, K., Assiri, A. M. A., Tadros, M., Khider, M., Assaeedi, A., Mohdaly, A. A. A., & Ramadan, M. F. (2018). Rosemary (*Rosmarinus officinalis*) oil: Composition and functionality of the cold-pressed extract. *Journal of Food Measurement and Characterization, 12,* 1601–1609.)

of tocols is α-tocopherol, while γ-tocopherol has 35% of the antioxidant impact of γ-tocopherol (Ramadan, 2013). The amounts of tocochromanols in CPRO suggest that it might effectively resist oxidation in vivo and in vitro.

5 Health-promoting properties of cold pressed *R. officinalis* oil and oil constituents

5.1 Antioxidant activity

Radical scavenging activity (RSA) of CPRO and extra virgin olive oil (EVOO) was evaluated against galvinoxyl and DPPH• free radicals (Elbanna et al., 2018). CPRO exhibited more RSA than EVOO (Fig. 1). After 60 min of incubation with DPPH• free radicals, 67% of free radicals were quenched by CPRO. In the course of 60 min, CPRO quenched 55% of galvinoxyl free radicals. CPRO contained higher total phenolics (7.2 mg GAE/g) than EVOO (3.6 mg GAE/g). The same study (Elbanna et al., 2018) reported that the induction period (IP) of CPRO and sunflower oil blend (1:9, *v*/*v*) was 390 min, while the IP of CPRO blended with sunflower oil (2:8, *v*/*v*) was 540 min. The antioxidant effect of CPRO was likely to be due to the high amounts of tocochromanols and phenolic compounds in the cold pressed oil.

5.2 Antimicrobial activity

Nowadays, there is an interest in use of natural preservatives in the production of foodstuffs. Rosemary extracts showed biological activities such as insecticide, antioxidant, hepatoprotective, antifungal, and antibacterial. The antimicrobial activity (AA) of CPRO was tested against dermatophyte fungi (*Trichophyton mentagrophytes*, and T*richophyton* rubrum), and food pathogens (*Salmonella enteritidis*, L*isteria* monocytogenes, and E*scherichia* coli). CPRO exhibited a broad AA spectrum (Table 3) against food-borne pathogens and dermatophyte fungi (Elbanna et al., 2018). The inhibition effect of CPRO against selected microorganisms is shown in Fig. 2. AA, measured as a clear zone diameter (CZD), was 29 mm, 17 mm, and 14 mm for E. coli, *S. enteritidis*, and L. monocytogenes, respectively. CPRO had activity against the pathogenic microorganisms with minimum inhibitory concentration (MIC) ranging from 160 to 320 μg/mL. CPRO exhibited a high AA compared to different antibiotics (Chloramphenicol, Flucoral, Augmentin, and Mycosat) as given in Table 3. On the other hand, CPRO did not inhibit *Staphylococcus* aureus growth. Dermatophytic fungi such as *T. rubrum* are anthropophilic

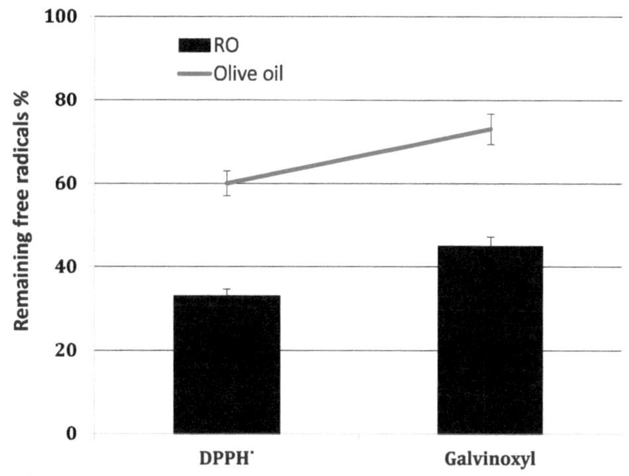

FIG. 1 Antiradical impact after 60 min of incubation for CPRO and EVOO on DPPH· and galvinoxyl free radicals. *(Adapted from Elbanna, K., Assiri, A. M. A., Tadros, M., Khider, M., Assaeedi, A., Mohdaly, A. A. A., & Ramadan, M. F. (2018). Rosemary (*Rosmarinus officinalis*) oil: Composition and functionality of the cold-pressed extract.* Journal of Food Measurement and Characterization, 12, *1601–1609.)*

and cause acute inflammatory tinea corporis, which is dermatophytosis of the arms and legs (Burmester et al., 2011). CPRO inhibited *T. mentagrophytes* and T. rubrum with CZDs of 25 mm and 23 mm, respectively, while the MLC for both fungi was 160 μg/mL. It was worth noting that the AA of CPRO against tested microorganisms kept the same CZD for long time (more than 10 days). The AA of CPRO might be due to the presence of tococls and phenolics in CPRO (Elbanna et al., 2018). The proposed mode of action of CPRO is that the cell walls and membranes of the pathogen microorganisms were damaged with the loss of cytoplasmic materials. A similar observation was reported by Carson, Mee, & Riley (2002), who showed the electron micrographs captured for the cell wall and membrane of *S. aureus* MRSA by the cold Valencia orange. Cell walls were damaged with the loss of cytoplasmic materials.

Burt (2004) and Sirocchi et al. (2013) found that the rosemary oil exhibited potential AA against both Gram-negative and Gram-positive bacteria including *Bacillus cereus*, *S. aureus*, *Clostridium perfringens*, *E. coli*, *Aeromonas hydrophila*, and *Salmonella choleraesuis*. Fung, Taylor, and Kahan (1977) mentioned that the inhibitory effect of the rosemary essential oil refers to the action of carnosic acid, rosmanol, carnosol, epirosmanol, rosmarinic acid, rosmaridiphenol, and isorosmanol on the cell membrane, causing changes in nutrients and genetic material and nutrients, leakage of cellular components, altering the transport of electrons, and changing the fatty acids. Furthermore, it produces an interaction with the membrane of proteins that causes the loss of membrane structure and functionality. Khezri, Farahpour, and Mounesi Rad (2019) evaluated the efficiency of rosemary essential oil loaded in the nanostructured lipid carrier (REO-NLC) on in vitro AA and in vivo infected wound healing in experimental animals. REO-NLC showed AA against *E. coli*, *Staphylococcus epidermidis*, *S. aureus*, *L. monocytogenes*, and *Psudomonas aeruginosa*. REO-NLCs reduced the rate of tissue bacterial colonization and wound size. In a recent study, Risaliti et al. (2019) reported that *R. officinalis* oil-loaded liposomes and exhibited significant AA, antiinflammatory, and antioxidant activities. These findings might lead to effective applications of CPRO as a natural antimicrobial agent to control food-borne and food spoilage pathogens.

5.3 Antidiabetic activity

Diabetes mellitus (DM) is the fast-growing chronic disease and one of the main disorders threatening human health (Assiri, El-Beeh, Amin, & Ramadan et al., 2017; Esteves et al., 2008; Hebi, Farid, Ajebli, & Eddouks, 2017; Wang et al., 2018). The WHO reported that DM patients by 2025 will reach 300 million (Jeszka-Skowron et al., 2014; Rahimi-Madiseh, Heidarian, Kheiri, & Rafieian-Kopaei, 2017; Sarfraz, Khaliq, Khan, & Aslam, 2017). DM is classified as type 1 (T1DM) and type 2 (T2DM), with T2DM accounting for about 95% of cases (Taslimi et al., 2018). Pathogenesis of T2DM involved resistance to insulin activities, an abnormality in glucose, and inadequate insulin secretion from β-cells (Achenbach, Bonifacio, Koczwara, & Ziegler, 2005; Goldstein, 2007).

El-Beeh et al. (2019) studied the ameliorative effect of CPRO against liver injury and genotoxic impacts in streptozotocin (STZ)-induced diabetic rats and offspring. Treatment with CPRO reduced the harmful impact of diabetes on the weight loss and caused an increase in the animal's body weight except of the animals in the control group. According to Wang et al. (2018), body weight reduction is one of the major markers of diabetic. On the other hand, CPRO supplementation decreased the impact of diabetes on the liver weight.

TABLE 3 Antimicrobial effects of CPRO recorded by clear zone diameter (CZD, mm) and minimal lethal concentration (MLC, μg/mL).[a]

	Food-borne pathogen bacteria								Dermatophytic fungi			
	S. enteritidis		*L. monocytogenes*		*E. coli*		*S. aureus*		*T. mentagrophytes*		*T. rubrum*	
	CZD	**MLC**	**CZD**	**MLC**	**CZD**	**MLC**	**CZD**	**MLC**	**CZD**	**MLC**	**CZD**	**MLC**
CPRO (100 μL)	17	320	14	160	29	320	nd	nd	25	160	23	160
Augmentin (30 μg)	28	Nd[b]	28	nd	30	nd	40	nd	nd	nd	nd	nd
Chloramphenicol (30 μg)	20	nd	22	nd	27	nd	25	nd	nd	nd	nd	nd
Flucoral (100 μg/mL)	nd	nd	nd	nd	nd	nd	nd	nd	35	nd	34	nd
Mycosat (100 μg/mL)	nd	nd	nd	nd	nd	nd	nd	nd	40	nd	38	nd

[a] *The diameter of the inhibition zone was measured as the clear area centered on the agar well containing the sample.*
[b] *nd: not determined.*
(Adapted from Elbanna, K., Assiri, A. M. A., Tadros, M., Khider, M., Assaeedi, A., Mohdaly, A. A. A., & Ramadan, M. F. (2018). Rosemary (*Rosmarinus officinalis*) oil: Composition and functionality of the cold-pressed extract. *Journal of Food Measurement and Characterization, 12,* 1601–1609.)

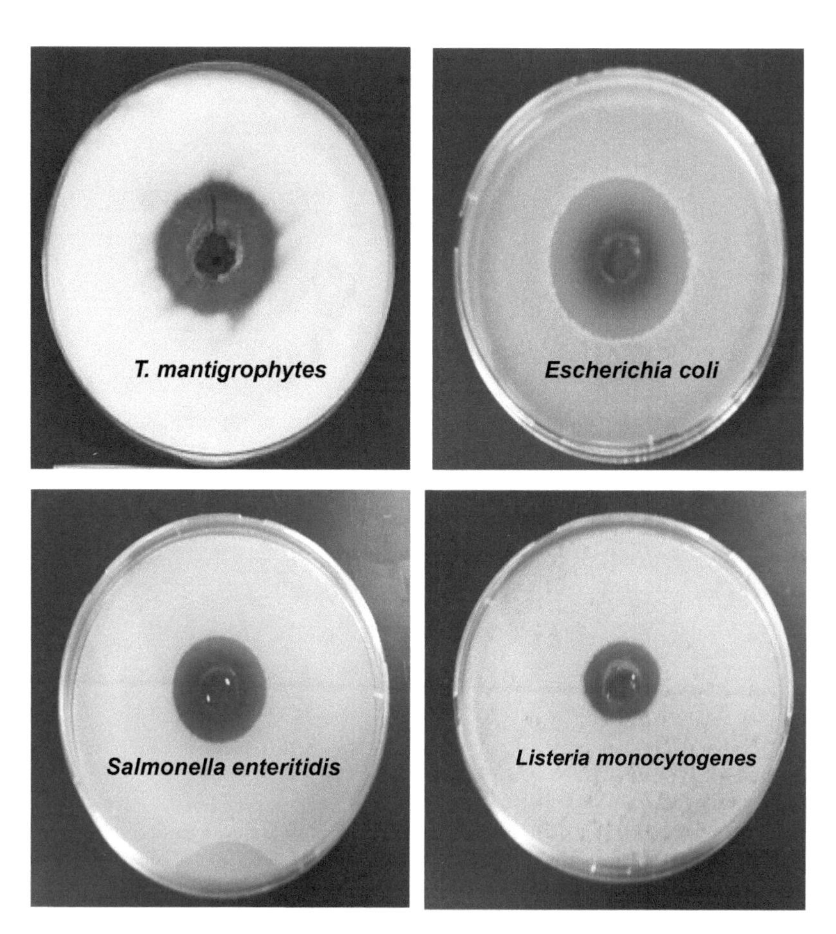

FIG. 2 Antimicrobial effect of CPRO against food-borne pathogens and dermatophyte fungi (Elbanna et al., 2018).

Liver histological graphs of maternal and offspring are shown in Fig. 3. DM rats showed degeneration of hepatic cells, congestion of blood vessels with an edematous central vein, as well as fatty degeneration. In addition, livers of offspring from DM animals showed degeneration of hepatic cords and vascularization of blood vessels, while DM animals treated with CPRO showed good recovery of histological characteristics (El-Beeh et al., 2019).

The serum biochemical parameters are given in Table 4. Group D contained high TC, TG, LDL-C, FBG, and CK as well as low HDL-C levels. Supplementation with CPRO reduced TC, TG, LDL-C, FBG, and CK levels. FBG reached the highest value (630 mg/dL) in DM animals, while supplementation with CPRO decreased the serum FBG. Table 5 represents the biochemical parameters of antioxidant enzymes, liver function, lipid oxidation, tumor markers, and free radicals of DM animals and offspring. DM increased alanine and aspartate transaminase. Liver bilirubin and albumen contents, as well as the activities of α-L-fucosidases and arginase, increased in DM animals. Antioxidant enzymes (i.e., superoxide dismutase, catalase, glutathione *S*-transferase, peroxidase, and reductase) showed a reduction in DM mothers and offspring. Fig. 4 shows photomicrographs from Comet assay for the livers. DM animals and offspring showed increased stretching of apoptotic cells, while in the CPRO-treated group, normal cell content was shown. DNA fragments were detected in the livers of DM animals and offspring, but there was some amelioration upon CPRO supplementation (El-Beeh et al., 2019).

Tocols are in vivo lipid radical scavengers and play a key role in the maintenance of cell membrane integrity (Gugliandolo, Bramanti, & Mazzon, 2017). Tocols maintain the membrane structure, limit lipid oxidation, and prevent inflammation in the neuroglia and hippocampal neurons (Galli et al., 2017). CPRO contained high levels of tocols and phenolics, which have unique antioxidative effects and play a role as antioxidant agents in treating DM and cardiovascular diseases (El-Beeh et al., 2019; Wang et al., 2018).

5.4 Hepatoprotective activity

El-Hadary, Elsanhoty, and Ramadan (2019) studied the hepatoprotective impact of CPRO against carbon tetrachloride (CCl_4)-induced toxicity in rats. Using concentrations of CPRO (100, 200, and 400 mg/kg, p.o.), only 400 mg/kg decreased the body weight. Based on a 24-h toxicity investigation, the LD_{50} of CPRO was 5780 mg/kg. The defensive aptitude of

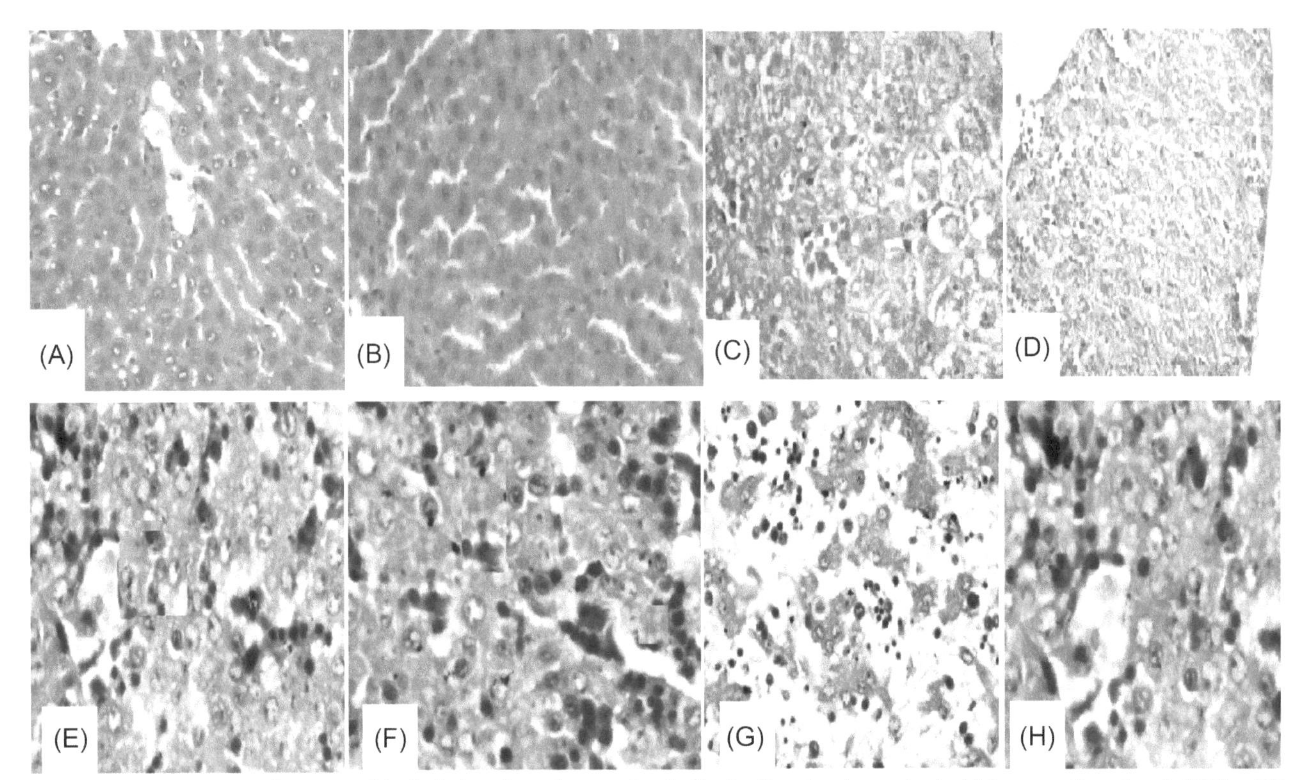

FIG. 3 Photomicrographs of transverse histological sections of maternal and offspring liver (sections stained with hematoxylin and eosin H&E × 400) (El-Beeh, Aljabri, Orabi, Qari, Ramadan, 2019). (A) Control group showing a normal arrangement of hepatic cells in cords around the central vein. (B) CPRO-treated group showing a normal arrangement of hepatocytes. (C) Experimental diabetic group (group D) showing degeneration of hepatic cells. (D) Experimental D-CPRO group. (E) Control pups liver showing normal hepatocytes. (F) Liver of pups maternally CPRO-treated group showing a normal arrangement of hepatocytes. (G) Liver of offspring group (D) showing degeneration of hepatic cells. (H) Liver of offspring of diabetic mothers treated with CPRO (D-CPRO) showing some degree of recovery.

TABLE 4 Biochemical profiles in the serum of nondiabetic and diabetic animals treated with CPRO (El-Beeh et al., 2019).

	FBG (mg/dL)	LDL-C (mg/dL)	HDL-C (mg/dL)	TG (mg/dL)	TC (mg/dL)	CK (U/L)
Control pregnant (C)	238	52	41	33	30	87
CPRO-treated group (CPRO)	245	57	40	32	34	88
Diabetic group (D)	630	110	39.2	52	45	348
Combined diabetic and CPRO-treated group (D-CPRO)	367	65	57.3	47	42	153

CPRO was validated by determining serum ALT, AST, and ALP. Serum markers levels were reduced upon CPRO supplementation at low dose. In addition, CPRO attenuated the increased levels of AST, ALT, and ALP enzymes, and caused a recovery toward normalization. Table 6 represents the CPRO impact on the protein profiles (albumin (A), globulin (G), A/G ratio, and total protein) of CCl_4-adminstrated animals. The impact of CPRO on the kidney function markers is also shown in Table 6. CCl_4 resulted in an increase in creatinine, urea, and uric acid levels, while CPRO administration decreased urea, creatinine, and uric acid amounts (El-Hadary et al., 2019). CPRO administration prevented the harmful impacts of CCl_4, indicating that CPRO might attenuate lipid per-oxidation resulted from CCl_4. Bioactive phytoconstituents might be the cause of CPRO protective potential. The mechanism of CPRO hepatoprotection might be due to its antioxidant activities. CPRO is rich in essential fatty acids. MUFAs were reported to reduce "bad" LDL-C and retain "good" HDL-C. CPRO also

TABLE 5 Antioxidant markers and enzymes levels in the livers of mothers and offspring in nondiabetic and diabetic rats treated with CPRO (El-Beeh et al., 2019).

	Mother				Offspring			
	C	CPRO	D	D-CPRO	C	CPRO	D	D-CPRO
MDA (nmol/g tissue)	10.47	10.27	19.24	12.37	11.37	10.37	22.4	12.87
H_2O_2 (mmol/g tissue)	166.09	156.19	380.03	199.65	164.0	154.0	231.6	195.6
SOD (U/g tissue)	19.25	19.2	10.48	17.37	25.35	25.35	12.88	15.3
CAT (U/g tissue)	362.08	362.01	323.20	330.24	379.1	379.7	333.87	360.86
GSH (mg/g tissue)	1.85	1.82	0.58	0.99	2.91	2.99	1.60	2.68
GST (U/g tissue)	87.51	86.51	144.8	113.6	81.99	80.09	368.33	220.93
GSPase (U/g tissue)	41.05	42.05	25.29	32.57	45.1	45.91	24.9	25.7
ALT (U/mL)	1.25	1.24	2.11	1.11	1.05	1.14	2.01	1.01
AST (U/mL)	29.7	30.5	45.85	33.99	28.79	31.5	44.85	35.99
Albumin (mg/dL)	12.43	13.3	23.56	15.29	13.43	13.9	25.56	14.29
Bilirubin (mg/dL)	1.02	1.05	1.4	1.31	1.52	1.55	1.04	1.30
Arginase (U/L)	4.08	4.0	50.6	10.75	4.8	3.9	45.6	10.75
α-L-fucosidase (U/L)	21.81	20.5	45.09	25.5	22.8	22.5	46.19	25.9

MDA, malondialdhyde concentration; *H_2O_2*, hydrogen peroxide; *SOD*, superoxide dismutase activity; *CAT*, catalase activity; *GSH*, glutathione reduced concentration; *GST*, glutathione-*S*-transferase activity; *GSPase*, glutathione peroxidase activity; *ALT*, alanine transaminase; *AST*, asparate transaminase.

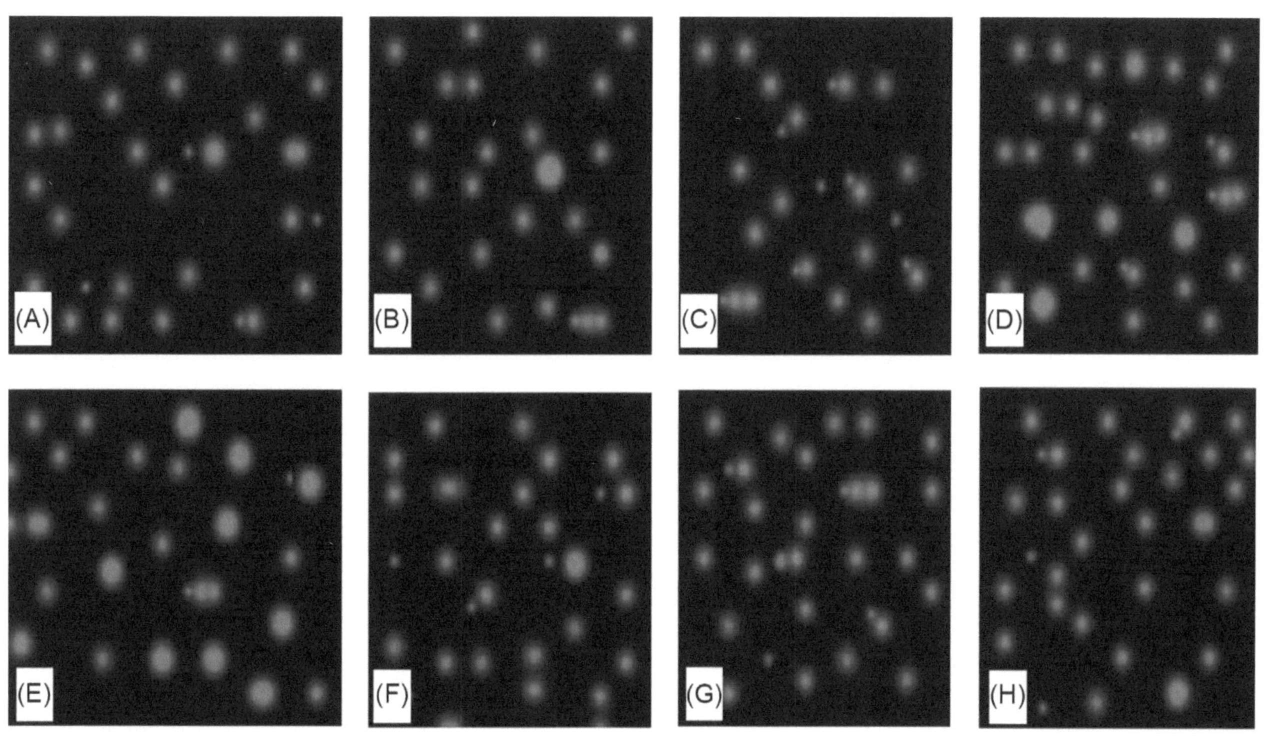

FIG. 4 Photomicrograph of Comet cells of the mothers and their offspring liver of rats treated with CPRO (El-Beeh et al., 2019). (A) Control group showing a normal structure of the liver cells. (B) RO-treated rats showing a normal structure of the liver. (C) Experimental diabetic group (D) showing increased stretching of apoptic cells. (D) Experimental D-CPRO group showing some amelioration. (E) Control offspring group showing a normal structure of the liver cells. (F) CPRO-maternally treated offspring group showing a normal structure of the liver. (G) Experimental maternally diabetic offspring (D) showing increased stretching of apoptic cells. (H) Experimental D-CPRO treated offspring group showing amelioration traits.

TABLE 6 Effect of CPRO on kidney function indicators and protein profile in CCl_4-induced injury in rats.

Treatment	Creatinine (mg/dL)	Urea (mg/dL)	Uric acid (mg/dL)	T-protein (g/dL)	Albumin (g/dL)	Globulin (g/dL)	A/G ratio
Negative control (normal)	0.716	25.3	3.63	6.50	3.78	2.72	1.40
Positive control (CCl_4)	1.85	56.3	7.93	5.40	3.01	2.38	1.31
CPRO (100mg/kg) + CCl_4	0.69	28.3	3.30	6.60	3.78	2.81	1.34
CPRO (200mg/kg) + CCl_4	0.66	28.3	3.42	6.66	3.91	2.74	1.43

contains high amounts of unsaponifiables (tocols, sterols, and phenolics). Healthy oils rich in phenolic compounds play an important role in preventing several diseases (Ramadan, 2013; Ramadan, Kinni, Seshagiri, & Mörsel, 2010).

6 Edible and nonedible applications of cold pressed *R. officinalis* oil

CPRO is a promising healthy oil rich in phytonutrients (i.e., PUFAs, MUFAs, tocols, and phenolic compounds), and exhibits unique antiradical, antioxidant, and antimicrobial traits. CPRO could be used in novel pharmaceutical, cosmetic, and edible applications. Recently, Mahgoub et al. (2019) studied the effects of CPRO on growth performance, biostimulating health, and intestinal bacterial populations in Japanese quail. The addition of CPRO increased the body weight and the body weight gain of birds. CPRO administration showed an increase in metabolic hormones levels and serum total protein, while it reduced serum TC, LDL-C, 8-hydroxy-2′-deoxyguanosine, and protein carbonyl levels. In addition, CPRO increased antioxidant enzymes and reduced lipids oxidation in quail liver. Supplementation with CPRO also reduced the populations of total cultural bacterial count, *E. coli*, coliforms, and *Salmonella* spp.

References

Achenbach, P., Bonifacio, E., Koczwara, K., & Ziegler, A. G. (2005). Natural history of type1 diabetes. *Diabetes*, *54*, S25–S31.

Albano, S. M., & Miguel, M. G. (2011). Biological activities of extracts of plants grown in Portugal. *Industrial Crops and Products*, *33*, 338–343.

Ali, A., Chua, B. L., & Chow, Y. H. (2019). An insight into the extraction and fractionation technologies of the essential oils and bioactive compounds in *Rosmarinus officinalis* L.: Past, present and future. *TrAC Trends in Analytical Chemistry*, *118*, 338–351.

Al-Kalaldeh, J. Z., Abu-Dahab, R., & Afifi, F. U. (2010). Volatile oil composition and antiproliferative activity of *Laurus nobilis*, *Origanum syriacum*, *Origanum vulgare*, and *Salvia triloba* against human breast adenocarcinoma cells. *Nutrition Research*, *30*, 271–278.

Amin, A., & Hamza, A. A. (2005). Hepatoprotective effects of Hibiscus, Rosmarinus and Salvia on azathioprine-induced toxicity in rats. *Life Science*, *77*, 266–278.

Andrade, J. M., Faustino, C., Garcia, C., Ladeiras, D., Reis, C. P., & Rijo, P. (2018). *Rosmarinus officinalis* L.: An update review of its phytochemistry and biological activity. *Future Science OA*, *4*, FSO283. https://doi.org/10.4155/fsoa-2017-0124.

Assiri, A. M. A., Elbanna, K., Abulreesh, H. H., & Ramadan, M. F. (2016). Bioactive compounds of cold-pressed thyme (*Thymus vulgaris*) oil with antioxidant and antimicrobial properties. *Journal of Oleo Science*, *65*, 629–640.

Assiri, A. M. A., Elbanna, K., Al-Thubiani, A., & Ramadan, M. F. (2016). Cold pressed oregano (*Origanum vulgare*) oil: A rich source of bioactive lipids with novel antioxidant and antimicrobial properties. *European Food Research and Technology*, *242*, 1013–1023.

Assiri, A. M. A., El-Beeh, M. E., Amin, A. H., & Ramadan, M. F. (2017). Ameliorative impact of *Morus alba* leaves' aqueous extract against embryonic ophthalmic tissue malformation in streptozotocin-induced diabetic rats. *Biomedicine & Pharmacotherapy*, *95*, 1072–1081.

Borges, R. S., Ortiz, B. L. S., Pereira, A. C. M., Keita, H., & Carvalho, J. C. T. (2019). *Rosmarinus officinalis* essential oil: A review of its phytochemistry, anti-inflammatory activity, and mechanisms of action involved. *Journal of Ethnopharmacology*, *229*, 29–45.

Burmester, A., Shelest, E., Glöckner, G., Heddergott, C., Schindler, S., Staib, P., et al. (2011). Comparative and functional genomics provide insights into the pathogenicity of dermatophytic fungi. *Genome Biology*, *12*, R7.

Burt, S. (2004). Essential oils: Their antibacterial properties and potential applications in foods—A review. *International Journal of Food Microbiology*, *94*, 223–253.

Carson, C. F., Mee, B. J., & Riley, T. V. (2002). Mechanism of action of *Melaleuca alternifolia* (tea tree) oil on *Staphylococcus aureus* determined by time-kill, lysis, leakage, and salt tolerance assays and electron microscopy. *Antimicrobial Agents and Chemotherapy*, *46*, 1914–1920.

Commission Regulation (EU) no. 1130/2011, 11 November 2011.

Elbanna, K., Assiri, A. M. A., Tadros, M., Khider, M., Assaeedi, A., Mohdaly, A. A. A., et al. (2018). Rosemary (*Rosmarinus officinalis*) oil: Composition and functionality of the cold-pressed extract. *Journal of Food Measurement and Characterization*, *12*, 1601–1609.

Elbanna, K., Attalla, K., Elbadry, M., Abdeltawab, A., Gamal-Eldin, H., & Ramadan, M. F. (2014). Impact of floral sources and processing on the antimicrobial activities of different unifloral honeys. *Asian Pacific Journal of Tropical Disease*, *4*, 194–200.

El-Beeh, M. E., Aljabri, M., Orabi, H. F., Qari, S. H., & Ramadan, M. F. (2019). Ameliorative impact of cold-pressed *Rosmarinus officinalis* oil against liver toxicity and genotoxic effects in streptozotocin-induced diabetic rats and their offspring. *Journal of Food Biochemistry*, *43*, e12905. https://doi.org/10.1111/jfbc.12905.

El-Demerdash, F. M., Abbady, E. A., & Baghdadi, H. H. (2016). Oxidative stress modulation by *Rosmarinus officinalis* in creosote-induced hepatotoxicity. *Environmental Toxicology*, *31*, 85–92. https://doi.org/10.1002/tox.22024.

El-Hadary, A. E., Elsanhoty, R. M., & Ramadan, M. F. (2019). *In vivo* protective effect of *Rosmarinus officinalis* oil against carbon tetrachloride (CCl_4)-induced hepatotoxicity in rats. *PharmaNutrition*, *9*, 100151.

El-Hadary, A. E., & Ramadan, M. F. (2016a). Potential protective effect of cold-pressed *Coriandrum sativum* oil against carbon tetrachloride-induced hepatotoxicity in rats. *Journal of Food Biochemistry*, *40*, 190–200.

El-Hadary, A. E., & Ramadan, M. F. (2016b). Hepatoprotective effect of cold-pressed *Syzygium aromaticum* oil against carbon tetrachloride (CCl_4)-induced hepatotoxicity in rats. *Pharmaceutical Biology*, *54*, 1364–1372. https://doi.org/10.3109/13880209.2015.1078381.

El-Hadary, A. E., & Ramadan, M. F. (2019). Antioxidant traits and protective impact of *Moringa oleifera* leaf extract against diclofenac sodium-induced liver toxicity in rats. *Journal of Food Biochemistry*, *43*, e12704. https://doi.org/10.1111/jfbc.12704.

El-Naggara, S. A., Abdel-Farid, I. B., Germoush, M. O., Elgebaly, H. A., & Alm-Eldeen, A. A. (2016). Efficacy of *Rosmarinus officinalis* leaves extract against cyclophosphamide induced hepatotoxicity. *Pharmaceutical Biology*, *54*, 2007–2016. https://doi.org/10.3109/13880209.2015.1137954.

Esteves, J. F., Dal Pizzol, M. M., Sccoco, C. A., Roggia, M. F., Milano, S. B., Guarienti, J. A., et al. (2008). Cataract and type 1 diabetes mellitus. *Diabetes Research and Clinical Practice*, *82*, 324–328.

Fahim, F., Esmat, A., Fadel, H., & Hassan, K. (1999). Allied studies on the effect of *Rosemarinus officinalis* L. on experimental hepatotoxicity and mutagenesis. *International Journal of Food Sciences and Nutrition*, *50*, 413–427.

Fung, D. Y. C., Taylor, S., & Kahan, J. (1977). Effect of butylated hydroxyanisole (BHA) and buthylated hydroxytoluebe (BHT) on growth and aflatoxin production of *Aspergillus flavus*. *Journal of Food Safety*, *1*, 39–51.

Gallego, M. G., Gordon, M., Segovia, F., Skowyra, M., & Almajano, M. (2013). Antioxidant properties of three aromatic herbs (rosemary, thyme and lavender) in oil-in-water emulsions. *Journal of the American Oil Chemists' Society*, *90*, 1559–1568. https://doi.org/10.1007/s11746-013-2303-3.

Galli, F., Azzi, A., Birringer, M., Cook-Mills, J. M., Eggersdorfer, M., Frank, J., et al. (2017). Vitamin E: Emerging aspects and new directions. *Free Radical Biology & Medicine*, *102*, 16–36.

Goldstein, B. J. (2007). Clinical translation of "a diabetes outcome progression trial": ADOPT appropriate combination oral therapies in type 2 diabetes. *Journal of Clinical Endocrinology & Metabolism*, *92*, 1226–1228.

Gugliandolo, A., Bramanti, P., & Mazzon, E. (2017). Role of vitamin E in the treatment of Alzheimer's disease: Evidence from animal models. *International Journal of Molecular Sciences*, *18*. https://doi.org/10.3390/ijms18122504.

Hassanien, M. M. M., Abdel-Razek, A. G. M., Rudzińska, M., Siger, A., Ratusz, K., & Przybylski, R. (2014). Phytochemical contents and oxidative stability of oils from non-traditional sources. *European Journal of Lipid Science and Technology*, *116*, 1563–1571.

Hebi, M., Farid, O., Ajebli, M., & Eddouks, M. (2017). Potent antihyperglycemic and hypoglycemic effect of *Tamarix articulate* Vahl. in normal and streptozotocin-induced diabetic rats. *Biomedicine & Pharmacotherapy*, *87*, 230–239.

Ibrahim, F. M., Attia, H. N., Maklad, Y. A. A., Ahmed, A. A., & Ramadan, M. F. (2017). Biochemical characterization, anti-inflammatory properties and ulcerogenic traits of some cold-pressed oils in experimental animals. *Pharmaceutical Biology*, *55*, 740–748. https://doi.org/10.1080/13880209.2016.1275705.

Jeszka-Skowron, M., Flaczyk, E., Jeszka, J., Krejpcio, Z., Król, E., & Buchowski, M. S. (2014). Mulberry leaf extract intake reduces hyperglycaemia in streptozotocin (STZ)-induced diabetic rats fed high-fat diet. *Journal of Functional Foods*, *8*, 9–17.

Joyeux, M., Roland, A., Fleurentin, J., Mortier, F., & Dorfman, P. (1990). Tert-butyl hydroperoxide induced injury isolated rat hepatocytes: A model for studying anti-hepatotoxoxic crude drugs. *Planta Medica*, *56*, 171–174. https://doi.org/10.1055/s-2006-960918.

Khezri, K., Farahpour, M. R., & Mounesi Rad, S. (2019). Accelerated infected wound healing by topical application of encapsulated Rosemary essential oil into nanostructured lipid carriers. *Artificial Cells, Nanomedicine and Biotechnology*, *47*(1), 980–988.

Kiralan, M., Ulaş, M., Özaydin, A. G., Özdemir, N., Özkan, G., Bayrak, A., et al. (2017). Blends of cold pressed black cumin oil and sunflower oil with improved stability: A study based on changes in the levels of volatiles, tocopherols and thymoquinone during accelerated oxidation conditions. *Journal of Food Biochemistry*, *41*, e12272.

Mahgoub, S. A. M., El-Hack, M. E. A., Saadeldin, I. M., Hussein, M. A., Swelum, A. A., & Alagawany, M. (2019). Impact of *Rosmarinus officinalis* cold-pressed oil on health, growth performance, intestinal bacterial populations, and immunocompetence of Japanese quail. *Poultry Science*, *98*(5), 2139–2149.

Ojeda-Sana, A. M., Baren, C. M. V., Elechosa, M. A., Juárez, M. A., & Moreno, S. (2013). New insights into antibacterial and antioxidant activities of rosemary essential oils and their main components. *Food Control*, *31*, 189–195.

Olmedo, R. H., Nepote, V., & Grosso, N. R. (2013). Preservation of sensory and chemical properties in flavoured cheese prepared with cream cheese base using oregano and rosemary essential oils. *LWT—Food Science and Technology*, *53*, 409–417.

Parker, T. D., Adams, D. A., Zhou, K., Harris, M., & Yu, L. (2003). Fatty acid composition and oxidative stability of cold-pressed edible seed oils. *Journal of Food Science*, *68*, 1240–1243.

Parry, J., Su, L., Moore, J., Cheng, Z., Luther, M., Rao, J. N., et al. (2006). Chemical compositions, antioxidant capacities, and antiproliferative activities of selected fruit seed flours. *Journal of Agricultural and Food Chemistry*, *54*, 3773–3778.

Rahimi-Madiseh, M., Heidarian, E., Kheiri, S., & Rafieian-Kopaei, M. (2017). Effect of hydroalcoholic *Allium ampeloprasum* extract on oxidative stress, diabetes mellitus and dyslipidemia in alloxan-induced diabetic rats. *Biomedicine & Pharmacotherapy*, *86*, 363–367.

Ramadan, M. F. (2008). Quercetin increases antioxidant activity of soy lecithin in a triolein model system. *LWT—Food Science and Technology*, *41*, 581–587.

Ramadan, M. F. (2012). Antioxidant characteristics of phenolipids (quercetin-enriched lecithin) in lipid matrices. *Industrial Crops and Products*, *36*, 363–369.

Ramadan, M. F. (2013). Healthy blends of high linoleic sunflower oil with selected cold pressed oils: Functionality, stability and antioxidative characteristics. *Industrial Crops and Products*, *43*, 65–72. https://doi.org/10.1016/j.indcrop.2012.07.013.

Ramadan, M. F., Amer, M. M. A., & Awad, A. (2008). Coriander (*Coriandrum sativum* L.) seed oil improves plasma lipid profile in rats fed diet containing cholesterol. *European Food Research and Technology*, *227*, 1173–1182. https://doi.org/10.1007/s00217-008-0833-y.

Ramadan, M. F., & Asker, M. M. S. (2009). Antimicrobical and antivirial impact of novel quercetin-enriched lecithin. *Journal of Food Biochemistry*, *33*, 557–571.

Ramadan, M. F., Asker, M. M. S., & Tadros, M. (2012). Antiradical and antimicrobial properties of cold-pressed black cumin and cumin oils. *European Food Research and Technology*, *234*, 833–844.

Ramadan, M. F., & Elsanhoty, R. M. (2012). Lipid classes, fatty acids and bioactive lipids of genetically modified potato Spunta with Cry V gene. *Food Chemistry*, *133*, 1169–1176.

Ramadan, M. F., Kinni, S. G., Seshagiri, M., & Mörsel, J.-T. (2010). Fat-soluble bioactives, fatty acid profile and radical scavenging activity of *Semecarpus anacardium* seed oil. *Journal of the American Oil Chemists' Society*, *87*, 885–894. https://doi.org/10.1007/s11746-010-1567-0.

Ramadan, M. F., Sharanabasappa, G., Seetharam, Y. N., Seshagiri, M., & Moersel, J.-T. (2006). Characterisation of fatty acids and bioactive compounds of kachnar (*Bauhinia purpurea* L.) seed oil. *Food Chemistry*, *98*, 359–365.

Risaliti, L., Kehagia, A., Daoultzi, E., Lazari, D., Bergonzi, M. C., Vergkizi-Nikolakaki, S., et al. (2019). Liposomes loaded with Salvia triloba and Rosmarinus officinalis essential oils: In vitro assessment of antioxidant, antiinflammatory and antibacterial activities. *Journal of Drug Delivery Science and Technology*, *51*, 493–498. https://doi.org/10.1016/j.jddst.2019.03.034.

Sadeh, D., Nitzan, N., Chaimovitsh, D., Shachter, A., Ghanim, M., & Dudai, N. (2019). Interactive effects of genotype, seasonality and extraction method on chemical compositions and yield of essential oil from rosemary (*Rosmarinus officinalis* L.). *Industrial Crops and Products*, *138*, 111419.

Sarfraz, M., Khaliq, T., Khan, J. A., & Aslam, B. (2017). Effect of aqueous extract of black pepper and ajwa seed on liver enzymes in alloxan-induced diabetic Wister albino rats. *Saudi Pharmaceutical Journal*, *25*, 449–452.

Sarmoum, R., Haid, S., Biche, M., Djazouli, Z., Zebib, B., & Merah, O. (2019). Effect of salinity and water stress on the essential oil components of rosemary (*Rosmarinus officinalis* L.). *Agronomy*, *9*, 214.

Sirocchi, V., Caprioli, G., Cecchini, C., Coman, M. M., Cresci, A., Maggi, F., et al. (2013). Biogenic amines as freshness index of meat wrapped in a new active packaging system formulated with essential oils of *Rosmarinus officinalis*. *International Journal of Food Sciences and Nutrition*, *64*, 921–928.

Soylu, E. M., Kurt, S., & Soylu, S. (2010). In vitro and in vivo antifungal activities of the essential oils of various plants against tomato grey mould disease agent *Botrytis cinerea*. *International Journal of Food Microbiology*, *143*, 183–189.

Taslimi, P., Aslan, H. E., Demir, Y., Oztaskin, N., Maraş, A., Gulçin, İ., et al. (2018). Diarylmethanon, bromophenol and diarylmethane compounds: Discovery of potent aldose reductase, α-amylase and α-glycosidase inhibitors as new therapeutic approach in diabetes and functional hyperglycemia. *International Journal of Biological Macromolecules*, *119*, 857–863.

Terpinc, P., Bezjak, M., & Abramovic, H. (2009). A kinetic model for evaluation of the antioxidant activity of several rosemary extracts. *Food Chemistry*, *115*, 740–744.

Tuberoso, C. I. G., Kowalczyk, A., Sarritzu, E., & Cabras, P. (2007). Determination of antioxidant compounds and antioxidant activity in commercial oilseeds for food use. *Food Chemistry*, *103*, 1494–1501.

Upadhyay, R., Sehwag, S., & Mishra, H. (2017). Chemometric approach to develop frying stable sunflower oil blends stabilized with oleoresin rosemary and ascorbyl palmitate. *Food Chemistry*, *218*, 496–504.

Wang, F., Li, H., Zhao, H., Zhang, Y., Qiu, P., Li, J., et al. (2018). Antidiabetic activity and chemical composition of Sanbai melon seed oil. *Evidence-Based Complementary and Alternative Medicine*, 5434156. https://doi.org/10.1155/2018/5434156.

Wollinger, A., Perrin, É., Chahboun, J., Jeannot, V., Touraud, D., & Kunz, W. (2016). Antioxidant activity of hydro distillation water residues from *Rosmarinus officinalis* L. leaves determined by DPPH assays. *Comptes Rendus Chimie*, *19*, 754–765.

Yang, Y., Song, X., Sui, X., Qi, B., Wang, Z., Li, Y., et al. (2016). Rosemary extract can be used as a synthetic antioxidant to improve vegetable oil oxidative stability. *Industrial Crops and Products*, *80*, 141–147.

Chapter 61

Cold pressed cumin (*Cuminum cyminum*) oil

Mohamed Fawzy Ramadan[a,b]

[a]*Agricultural Biochemistry Department, Faculty of Agriculture, Zagazig University, Zagazig, Egypt,* [b]*Deanship of Scientific Research, Umm Al-Qura University, Makkah, Saudi Arabia*

Abbreviations

AgNPs silver nanoparticles
CEO cumin essential oil
CPCSO cold pressed cumin seed oil
CSO cumin seed oil
DPPH· 2,2-diphenyl-1-picrylhydrazyl
ESGs esterified sterylglycosides
ESR electron spin resonance
EVOO extra virgin olive oil
HDL-C high dentistry lipoprotein-cholesterol
LDL-C low dentistry lipoprotein-cholesterol
MAE microwave-assisted extraction
MDA malondialdehyde
MHG microwave hydrodiffusion and gravity
MIC minimum inhibitory concentration
NADES natural deep eutectic solvents
RSA radical scavenging activity
SEM scanning electron microscopy
SFA saturated fatty acid
TC total cholesterol
TPCs total phenolic compounds

1 Introduction

Cumin (*Cuminum cyminum* L., family Apiaceae) is an annual herb native of Syria and Egypt. Cumin is commercially marketed as a dried seed (ripe fruit) of *C. cyminum* with a bitter taste and an aromatic smell. *C. cyminum* seed contains c.4% essential oil and c.15% crude fixed oil. *C. cyminum* seeds are widely used as a condiment in dishes, as a flavor in meat, sausage, confectionery, and bread, and as a preservative in foodstuffs. Cumin also has excellent medicinal value in the treatment of flatulence, digestive disorders, diarrhea, and wounds. Its seeds are a valuable remedy for diarrhea, dyspepsia, hoarseness, indigestion, and colic (Al-Yahya, 1986; El-Ghorab, Nauman, Anjum, Hussin, & Nadeem, 2010; Shahnaz, Hifza, Bushra, & Khan, 2004; Singh, Gangadharappa, & Mruthunjaya, 2017). *C. cyminum* seeds are also used in Ayurvedic medicine as a carminative, stimulant, and astringent (Allahghdri et al., 2010). Cumin is a potent antioxidant that is able to deactivate free radicals such as peroxyl, hydroxyl, and DPPH, and hence inhibits radical-mediated lipids oxidation (Thippeswamy & Naidu, 2005). Phytochemical studies were conducted to investigate the composition of cumin essential oil (CEO). The main component of CEO is cuminaldehyde (Ani, Varadaraj, & Naidu, 2006; Bettaieb et al., 2010; El-Ghorab et al., 2010; Sowbhagya, Sathyendra Rao, & Krishnamurthy, 2008).

Cold Pressed Oils. https://doi.org/10.1016/B978-0-12-818188-1.00061-X
Copyright © 2020 Elsevier Inc. All rights reserved.

Cold pressed oils have become a focus of interest as they have health-promoting and nutritive properties. The cold pressing procedure is a safe traditional practice that has gained popularity because of consumers' interest in safe and natural products (Lutterodt et al., 2010; Parry et al., 2006). The technology involves no thermal or chemical treatment. In addition, cold pressing technology involves no refining process; therefore, cold pressed oils contain high levels of hydrophilic bioactive compounds (Lutterodt et al., 2010).

Few reports have been published on the chemical composition and functional traits of cold pressed cumin seed oil (CPCSO) and oil blends (Ramadan, 2013; Ramadan, Asker, & Tadros, 2012). This chapter summarizes the health-beneficial components of CPCSO including fatty acids, phytosterols, tocopherols, and phenolics. In addition, radical scavenging activity, antimicrobial properties, health-promoting impacts, and industrial applications of CPCSO are highlighted.

2 Extraction and processing of CPCSO

Recently, natural deep eutectic solvents (NADES), as a promising green medium, have attracted more attention. A method based on microwave-assisted NADES pretreatment linked with microwave hydro-distillation was applied to extract *C. cyminum* seed oil. NADES were used as pretreatment solvents for adsorbing microwaves and dissolving cellulose. The highest oil yield obtained with this microwave extraction technology was 2.22% (*w*/*w*) when the NADES were composed of choline chloride and L-lactic acid with 40% of water. NADES had an impact on oil extraction with higher yield and premium quality, especially when combined with microwave technology (Zhao et al., 2019).

Microwave hydro-diffusion and gravity (MHG) pushed microwave-assisted extraction (MAE) further toward being an efficient, innovative, quick, and eco-friendly extraction method. The MHG process for the extraction of CEO was optimized (Benmoussa et al., 2018). The effects of different parameters on the extraction yield were tested using the central composite design and response surface methodology. The optimum conditions were 203.3 W of irradiation power, 16 min of microwave irradiation, and 44.67% of moisture.

Guo, An, Jia, and Xu (2018) investigated the effect of different drying techniques (infrared drying, air drying, and hot-air oven drying) on the composition and activity of CEO. Different drying techniques resulted in differences in the yield, chemical composition, antioxidative and antibacterial activities, as well as the DNA damage protective impact of CEO. However, no differences in main compositions were detected among different CEO samples.

3 Acyl lipids and fatty acid profile of CPCSO

3.1 Fatty acid profile of CPCSO

Chromatographic analysis of CPCSO revealed that petroselinic acid (C18:1*n*-12) is the highest (c.41%), while linoleic acid (c.34%) is the second major unsaturated acid (Table 1). Knothe and Steidley (2019) studied hexane-extracted *C. cyminum* seed oil. The main fatty acid was petroselinic (6(*Z*)-octadecenoic) acid. The identification of petroselinic acid was performed with their 2-methoxyethyl ester to resolve the overlap of methyl esters. The mass spectra of 2-methoxyethyl ester was recorded, with a salient feature being the base peak in the spectrum of 2-methoxyethyl petroselinate at *m*/*z* 84. Minor levels of C16:1 isomers ($\Delta 4$, $\Delta 6$, and $\Delta 9$) were also detected in the oil.

Low levels of saturated fatty acids, except palmitic acid (c.8.70%), were also reported (Ramadan et al., 2012). Earlier reports (Ramadan, Kroh, & Moersel, 2003; Shahnaz et al., 2004) showed also that petroselinic acid was the main fatty acid in several species of Apiaceae family. CPCSO contains high levels of monounsaturated fatty acids (MUFAs, 47.5% fatty acids) that are comparable to those of blueberry, cranberry, hemp, onion, and milk thistle cold pressed oils, but less than those of parsley and carrot cold pressed oils (Parker, Adams, Zhou, Harris, & Yu, 2003; Parry et al., 2005). PUFAs in CPCSO accounted for 58.7% (Table 1), which is comparable to those in cranberry (67.6%), milk thistle (61%), onion (64%–65%), and blueberry (69%) cold pressed oils, but less than those in hemp, boysenberry, marionberry, red raspberry, and mullein cold pressed oils (Parker et al., 2003; Parry et al., 2006; Ramadan et al., 2012). MUFAs have been shown to lower LDL-C and retain HDL-C. This is the main beneficial effect of olive oil over the highly polyunsaturated seed oils, wherein PUFAs reduce LDL-C and retain HDL-C levels (Ramadan, Kinni, Seshagiri, & Mörsel, 2010; Ramadan, Sharanabasappa, Seetharam, Seshagiri, & Moersel, 2006). Interest in PUFAs as health-promoting compounds has expanded and several published articles illustrated the health-promoting effects of PUFAs in alleviating inflammatory problems, cardiovascular diseases, autoimmune disorders, heart diseases, atherosclerosis, and diabetes.

TABLE 1 Fatty acids, unsaponifiables, and total phenolic compounds of CPCSO.[a]

Compound	%	Compound	g/kg
C14:0	0.5	α-Tocopherol	0.08
C16:0	8.70	β-Tocopherol	0.77
C18:0	5.70	γ-Tocopherol	0.15
C18:1*n*-12	41.3	δ-Tocopherol	0.35
C18:1*n*-9	6.2	Ergosterol	0.19
C18:2*n*-6	34.2	Campesterol	0.76
C18:3*n*	3.40	Stigmasterol	1.53
Σ SFAs[b]	14.9	Lanosterol	0.15
Σ MUFAs[c]	47.5	β-Sitosterol	1.57
Σ PUFAs[d]	37.6	Δ5-Avenasterol	1.49
		Δ7-Avenasterol	0.38
		Total unsaponifiables	22.8
		Total phenolics (g gallic acid/kg oil)	4.3

[a]*Adapted from Ramadan et al. (2012).*
[b]*Total saturated fatty acids.*
[c]*Total monounsaturated fatty acids.*
[d]*Total polyunsaturated fatty acids.*

4 Minor bioactive lipids in CPCSO

4.1 Sterols

CPCSO contained high amounts of unsaponifiable materials (22.8 g/kg). In CPCSO, β-sitosterol (c.26% total phytosterols), stigmasterol (c.26% total phytosterols), and Δ5-avenasterol were the main components (Table 1). Other constituents, e.g., campesterol and lanosterol, were detected in lower levels (Ramadan et al., 2012). Δ5, 24-Stigmastadienol, and brassicasterol were not detected in CPCSO. β-Sitosterol has been the most studied phytosterol with respect to its beneficial health-promoting and physiological impact (Ramadan, 2013; Yang, Karlsson, Oksman, & Kallio, 2001).

4.2 Tocols

The quantitative and qualitative analyses of tocochromanols in CPCSO are given in Table 1. All tocopherol isomers were present in CPCSO, wherein β-tocopherol was the major isomer in the oil (c.40% of total tocopherols). γ- and δ-tocopherols were detected in lower levels (Ramadan et al., 2012). α- and γ-Tocopherols reported to be the major tocopherols in edible oils. γ-Tocopherol was found in high levels in linseed, rapeseed, camelina, and corn oils (Schwartz, Ollilainen, Piironen, & Lampi, 2008). The most efficient antioxidant of all tocopherol isomers is α-tocopherol, while β-tocopherol has c.50% of the antioxidant action of α-tocopherol, and the γ-isomer c.35% (Kallio, Yang, Peippo, Tahvonen, & Pan, 2002). Amounts of tocopherols measured in CPCSO might contribute to the oxidative stability of the oil.

4.3 Phenolic compounds

CPCSO was characterized by a high amount of TPC, which accounted for 4.3 g/kg oil (Ramadan et al., 2012). TPC values of CPCSO were comparable to those of pumpkin and marionberry cold pressed oils, but were higher than those of red raspberry, blueberry, and boysenberry cold pressed oils, and that of 2–3.5 mg GAE/g oil for milk thistle, mullein, cardamom, parsley, and onion cold pressed oils (Parry et al., 2005, 2006). Ultraviolet (UV) absorption was scanned between 200 and 400 nm to assist in characterizing phenolics in CPCSO. The UV spectra (Fig. 1) showed absorption maxima at 280 nm,

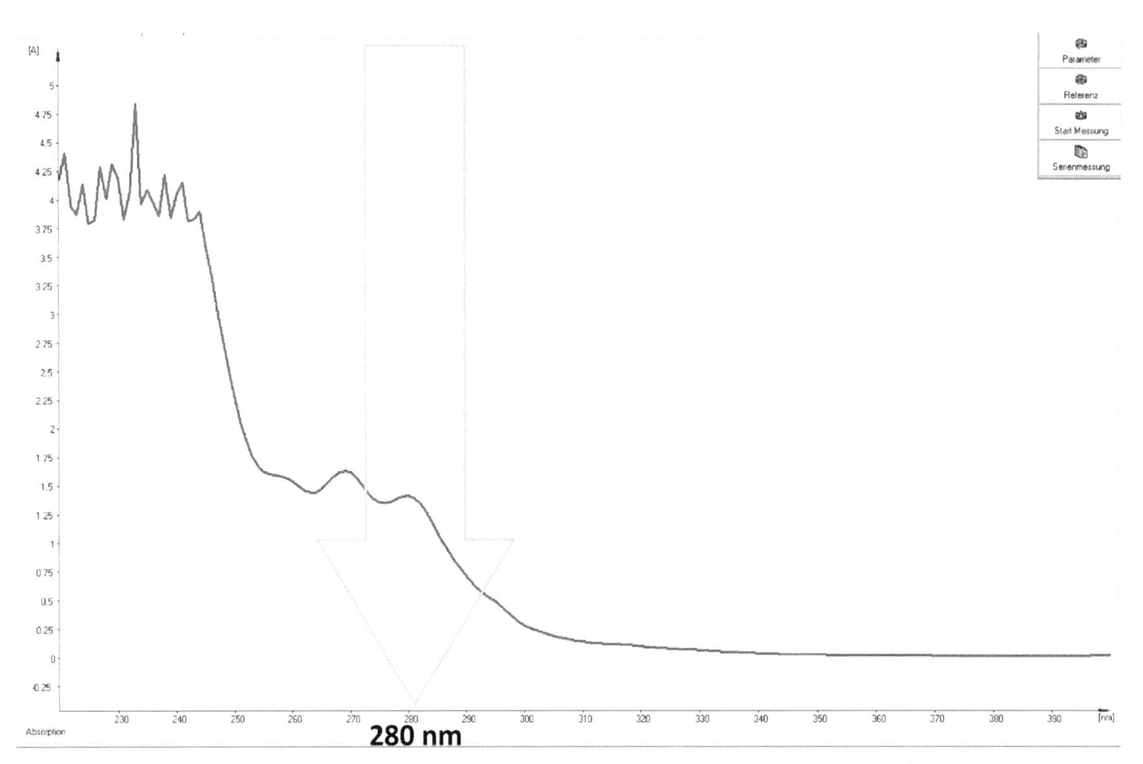

FIG. 1 Ultraviolet scan (200–400 nm) of CPCSO. *(Reproduced from Ramadan, M. F., Kroh, L. W., & Moersel, J.-T. (2013). Radical scavenging activity of black cumin (Nigella sativa L.), coriander (Coriandrum sativum L.) and Niger (Guizotia abyssinica Cass.) crude seed oils and oil fractions. Journal of Agricultural and Food Chemistry, 51, 6961–6969).*

which may be due to the presence of flavone/flavonol derivatives and *p*-hydroxybenzoic acid (Ramadan, 2013; Ramadan et al., 2003). CPCSO could be considered as a rich source of phenolics that may act as natural antioxidants for disease prevention and health promotion.

5 Health-promoting properties of CPCSO and oil constituents

Radical scavenging activity (RSA) of the CPCSO and extra virgin olive oil (EVOO) were compared using stable DPPH·and galvinoxyl radicals. A simple and rapid test used toluene to dissolve the free radicals, and oil samples were developed (Ramadan et al., 2003). This test allowed comparison and characterization of antiradical action of oil samples under the same conditions. CPCSO had higher RSA than EVOO (Ramadan et al., 2012). After 1 h of incubation of CPCSO with DPPH·free radicals, c.60% of DPPH·free radicals were deactivated by CPCSO, while EVOO quenched 45%. An electron spin resonance (ESR) test showed the same results, whereas CPCSO deactivated c.50% of galvinoxyl free radicals and EVOO deactivated c.38% after 60 min of reaction. Scavenging DPPH· by phenolics recovered from rapeseeds was affected by the number of hydroxyl groups in the aromatic ring of phenolic compounds (Ramadan et al., 2012; Sroka & Cisowski, 2003). Keerthiga, Anitha, Rajeshkumar, and Lakshmi (2019) reported good antioxidant potential of cumin oil-mediated silver nanoparticles.

Different bacteria (Gram-negative and Gram-positive) and molds (fungi and yeasts) were tested to screen CPCSO antimicrobial activity. An agar diffusion test revealed that CPCSO inhibited the growth of tested microorganisms. CPCSO led to high inhibition zones of molds (between 10 and 13 mm). The MICs of CPCSO for Gram-positive bacteria, Gram-negative bacteria, fungi, and yeasts were reported (Ramadan et al., 2012).

The effects of different levels of CPCSO on the biosynthesis of total lipids, acid-soluble phosphorus, protein, and nucleic acids (DNA and RNA) in *Bacillus subtilis* cells were studied (Fig. 2). CPCSO had a drastic impact on the biosynthesis of total lipids and protein in *B. subtilis* cells. This effect increased with increasing the incubation period and the concentration (1/8–1/2 MIC). CPCSO showed a slight effect on the biosynthesis of acid-soluble phosphorus compounds, while the oil showed a weak effect on the biosynthesis of nucleic acids (Ramadan et al., 2012). CPCSO strongly influenced the biosynthesis of protein by inhibiting some steps in the translation process (Shuichi et al., 2000). It was concluded that CPCSO could have useful applications as a natural antimicrobial agent and food preservative (Ramadan et al., 2012).

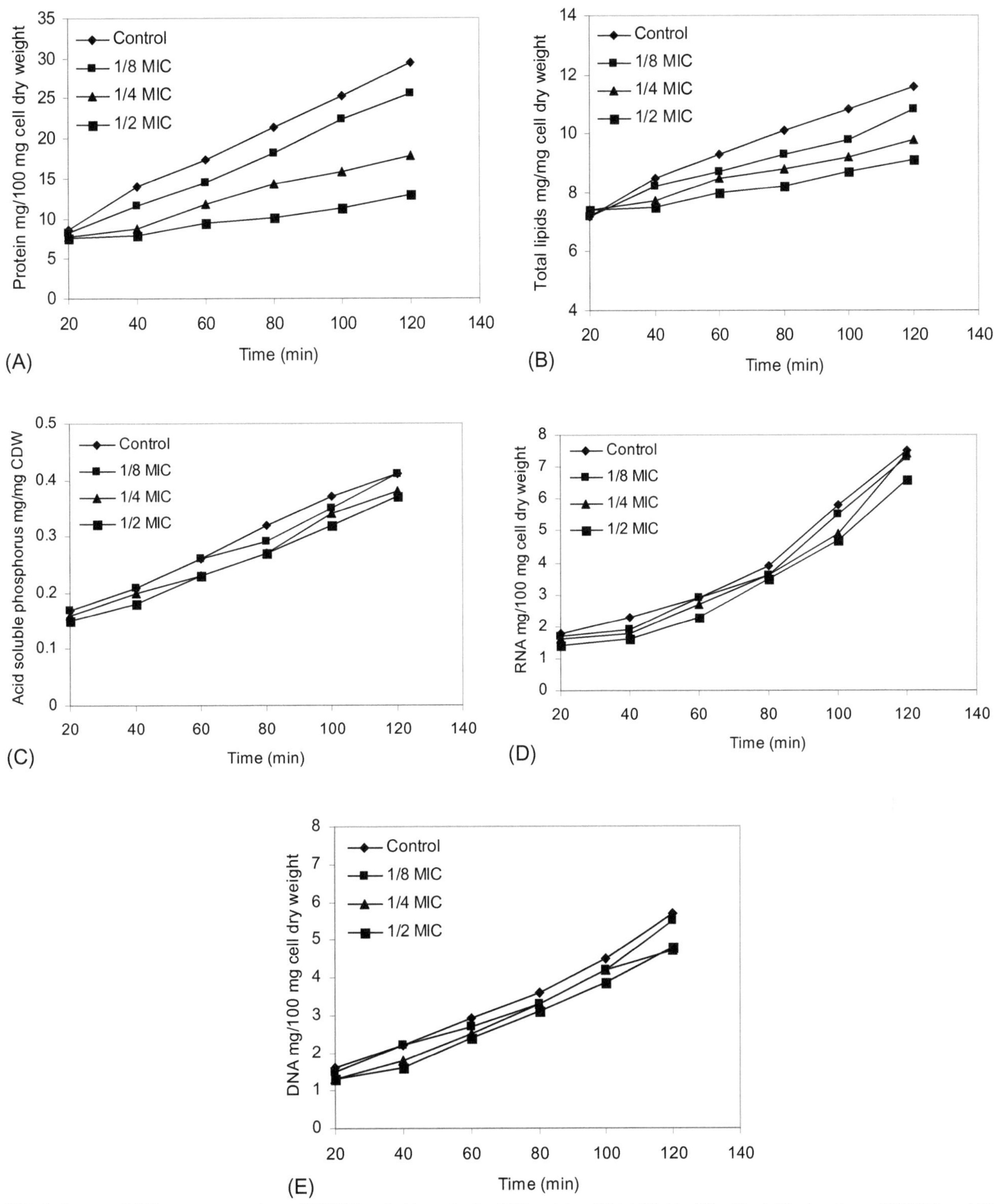

FIG. 2 Impact of different concentrations of CPCSO on the biosynthesis of (A) proteins, (B) total lipids, (C) acid-soluble phosphorus, (D) RNA, and (E) DNA in the *B. subtilis* (NRRL B-94) cells. *(From Ramadan, M. F., Asker, M. M. S., & Tadros, M. (2012). Antiradical and antimicrobial properties of cold-pressed black cumin and cumin oils. European Food Research and Technology, 234(5), 833–844).*

Amrutha, Sundar, and Shetty (2017) used nanoemulsions of CEO in controlling the microbial load and preventing outbreaks. CEO nanoemulsions were formulated by ultrasonic emulsification. They showed antiquorum sensing activity with 42% inhibition in violacein production by *Chromobacterium violaceum* CV026. CEO nanoemulsion reduced quorum-regulated phenotypes in *Escherichia coli* and *Salmonella enterica* such as bacterial swarming, swimming, and biofilm formation. Cumin oil nanoemulsion inhibited biofilm formation to 42.5% and 38.9% in *E. coli* and *S. enterica*, respectively. CEO nanoemulsions might be used as an inhibitor against pathogens in fresh vegetables and fruits by disrupting interbacterial communication. Alizadeh Behbahani, Noshad, and Falah (2019) investigated the antimicrobial impacts of CEO and its mechanism of action through SEM against *E. coli* and *Listeria innocua*. GC-MS analysis of CEO revealed that cuminal was the major compound and the oil contained TPC (89.4 mg GAE/g). The RSA of CEO (IC_{50}) was equal to 9.10 μg/mL and the oil showed a strong inhibitory potential (c.63%) on β-carotene bleaching. The electron micrographs showed that CEO increased the permeabilization and disrupted the membrane integrity of the cells.

Jafari et al. (2018) assessed the effect of the CEO on serum leptin and metabolic markers in prediabetic subjects. Serum leptin and glycemic indices did not improve in the intervention group. Markers of lipid profile, except serum total cholesterol, as well as anthropometric indexes, were enhanced in the intervention group. It was concluded that CEO could be utilized in prediabetics as an adjuvant therapy to ameliorate metabolic status. Goyal, Vincent, Pandey, and Nirogi (2019) evaluated *C. cyminum* oil for its clastogenic and anticlastogenic effects using a micronucleus assay in CHO-K1 cells. No increase in micronuclei was recorded, indicating that CSO did not have clastogenic potential. In addition, CSO did not show any anticlastogenic action against tested clastogens.

6 Edible and nonedible applications of CPCSO

Ramadan (2013) formulated blends of CPCSO with high linoleic sunflower oil. Changes were noted in the contents of C18:1 and C18:2 of blended oils, due to blending with CPCSO. By increasing the ratio of CPCSO in sunflower oil, the linoleic acid amount decreased, while tocols increased. The stability of oil blend was better than sunflower oil, probably as a consequence of changes in tocols' profile, fatty acids, and minor active lipids in CPCSO.

Karimirad, Behnamian, and Dezhsetan (2019) investigated the impact of CEO-loaded chitosan nanoparticles (CEO-CSNP) on the shelf life of mushrooms during cold storage. CEO and CEO-CSNP were spotted onto the Whatman filter paper inside the containers. CEO-CSNP was effective in maintaining firmness, color, and overall acceptability of mushrooms and inhibited the tested bacteria, mold, and yeast growth.

Saleh, Kirrella, Dawood, and Ebeid (2019) studied the effect of CSO on the ovarian follicular development, laying performance, and immune response in laying hens subjected to high-temperature conditions. By feeding CSO, yolk weight, yolk width, yolk color, shell weight, and shell thickness were increased. Egg production was not affected, while CSO had a positive impact on egg weight. Yolk cholesterol level was decreased, while the liver malondialdehyde (MDA) level was slightly decreased. Moreover, CSO improved the antibody titers against avian influenza H9N1 and Newcastle disease.

Benelli et al. (2018) evaluated the effect of the CEO on pests (i.e., the peach-potato aphid *Myzus persicae* and the tobacco cutworm *Spodoptera littoralis*), on insect vectors (i.e., the housefly *Musca domestica* and the *Lymphatic filariasis*), and the Zika virus vector *Culex quinquefasciatus*. Moreover, CEO safety on some beneficial organisms such as the earthworm *Eisenia fetida* and the aphid predator *Harmonia axyridis* was assessed. CEO was active on adults of *M. persicae* and *M. domestica*.

7 Conclusion

The industrial, nutritional, and medical applications of oils are dependent on their composition and functional properties. CPCSO is a rich source of fatty acids and lipid-soluble bioactive compounds. It contains significant levels of petroselinic acid and natural antioxidants. Tocols and phytosterols in CPCSO are of nutritional importance as natural antioxidants. CPCSO is considered a nontraditional source for pharmaceutical applications and edible products that provide health-promoting traits.

References

Alizadeh Behbahani, B., Noshad, M., & Falah, F. (2019). Cumin essential oil: Phytochemical analysis, antimicrobial activity and investigation of its mechanism of action through scanning electron microscopy. *Microbial Pathogenesis*, *136*, 103716.

Allahghdri, T., Rasooli, I., Owlia, P., Nadooshan, M. J., Ghazanfari, T., Taghizadeh, M., et al. (2010). Antimicrobial property, antioxidant capacity, and cytotoxicity of essential oil from cumin produced in Iran. *Journal of Food Science*, *75*, H54–H61.

Al-Yahya, M. (1986). Phyto chemical studies of plant used in traditional medicine in Saudi Arabia. *Fitoterapia*, *57*, 179–182.

Amrutha, B., Sundar, K., & Shetty, P. H. (2017). Spice oil nanoemulsions: Potential natural inhibitors against pathogenic *E. coli* and *Salmonella* spp. from fresh fruits and vegetables. *LWT—Food Science and Technology*, *79*, 152–159.

Ani, V., Varadaraj, M. C., & Naidu, K. A. (2006). Antioxidant and antibacterial activities of polyphenolic compounds from bitter cumin (*Cuminum nigrum* L.). *European Food Research and Technology*, *224*, 109–115.

Benelli, G., Pavela, R., Petrelli, R., Cappellacci, L., Canale, A., Senthil-Nathan, S., et al. (2018). Not just popular spices! Essential oils from *Cuminum cyminum* and *Pimpinella anisum* are toxic to insect pests and vectors without affecting non-target invertebrates. *Industrial Crops and Products*, *124*, 236–243.

Benmoussa, H., Elfalleh, W., He, S., Romdhane, M., Benhamou, A., & Chawech, R. (2018). Microwave hydrodiffusion and gravity for rapid extraction of essential oil from Tunisian cumin (*Cuminum cyminum* L.) seeds: Optimization by response surface methodology. *Industrial Crops and Products*, *124*, 633–642.

Bettaieb, I., Bourgou, S., Wannes, W. A., Hamrouni, I., Limam, F., & Marzouk, B. (2010). Essential oils, phenolics, and antioxidant activities of different parts of cumin (*Cuminum cyminum* L.). *Journal of Agricultural and Food Chemistry*, *58*, 10410–10418.

El-Ghorab, A. H., Nauman, M., Anjum, F. M., Hussin, S., & Nadeem, M. (2010). A comparative study on chemical composition and antioxidant activity of ginger (*Zingiber officinale*) and cumin (*Cuminum cyminum*). *Journal of Agricultural and Food Chemistry*, *58*, 8231–8237.

Goyal, V. K., Vincent, S., Pandey, S. K., & Nirogi, R. (2019). Determination of Clastogenic and Anticlastogenic potential of *Cuminum Cyminum* seed oil using *in vitro* micronucleus assay in CHO-K1 cells. *Journal of Herbs, Spices & Medicinal Plants*, *25*(3), 259–270. https://doi.org/10.1080/10496475.2019.1602574.

Guo, Y.-R., An, Y.-M., Jia, Y. X., & Xu, J.-G. (2018). Effect of drying methods on chemical composition and biological activity of essential oil from cumin (*Cuminum cyminum* L.). *Journal of Essential Oil Bearing Plants*, *21*(5), 1295–1302. https://doi.org/10.1080/0972060X.2018.1538818.

Jafari, T., Mahmoodnia, L., Tahmasebi, P., Memarzadeh, M. R., Sedehi, M., Beigi, M., et al. (2018). Effect of cumin (*Cuminum cyminum*) essential oil supplementation on metabolic profile and serum leptin in pre-diabetic subjects: A randomized double-blind placebo-controlled clinical trial. *Journal of Functional Foods*, *47*, 416–422.

Kallio, H., Yang, B., Peippo, P., Tahvonen, R., & Pan, R. (2002). Triacylglycerols, glycerophospholipids, tocopherols and tocotrienols in berries and seeds of two subspecies (ssp. *sinensis* and *mongolica*) of sea buckthorn (*Hippophaë rhamnoides*). *Journal of Agricultural and Food Chemistry*, *50*, 3004–3009.

Karimirad, R., Behnamian, M., & Dezhsetan, S. (2019). Application of chitosan nanoparticles containing *Cuminum cyminum* oil as a delivery system for shelf life extension of *Agaricus bisporus*. *LWT*, *106*, 218–228.

Keerthiga, N., Anitha, R., Rajeshkumar, S., & Lakshmi, T. (2019). Antioxidant activity of cumin oil mediated silver nanoparticles. *Pharmacognosy Journal*, *11*(4), 787–789.

Knothe, G., & Steidley, K. R. (2019). Composition of some Apiaceae seed oils includes phytochemicals, and mass spectrometry of fatty acid 2-Methoxyethyl esters. *European Journal of Lipid Science and Technology*, *121*, 1800386. https://doi.org/10.1002/ejlt.201800386.

Lutterodt, H., Luther, M., Slavin, M., Yin, J.-J., Parry, J., Gao, J.-M., et al. (2010). Fatty acid profile, thymoquinone content, oxidative stability, and antioxidant properties of cold-pressed black cumin seed oils. *LWT—Food Science and Technology*, *43*, 1409–1413.

Parker, T. D., Adams, D. A., Zhou, K., Harris, M., & Yu, L. (2003). Fatty acid composition and oxidative stability of cold-pressed edible seed oils. *Journal of Food Science*, *68*, 1240–1243.

Parry, J., Su, L., Luther, M., Zhou, K., Yurawecz, M. P., Whittaker, P., et al. (2005). Fatty acid composition and antioxidant properties of cold-pressed marionberry, boysenberry, red raspberry, and blueberry seed oils. *Journal of Agricultural and Food Chemistry*, *53*, 566–573.

Parry, J., Su, L., Moore, J., Cheng, Z., Luther, M., Rao, J. N., et al. (2006). Chemical compositions, antioxidant capacities, and antiproliferative activities of selected fruit seed flours. *Journal of Agricultural and Food Chemistry*, *54*, 3773–3778.

Ramadan, M. F. (2013). Healthy blends of high linoleic sunflower oil with selected cold pressed oils: Functionality, stability and antioxidative characteristics. *Industrial Crops and Products*, *43*, 65–72.

Ramadan, M. F., Asker, M. M. S., & Tadros, M. (2012). Antiradical and antimicrobial properties of cold-pressed black cumin and cumin oils. *European Food Research and Technology*, *234*(5), 833–844.

Ramadan, M. F., Kinni, S. G., Seshagiri, M., & Mörsel, J.-T. (2010). Fat-soluble bioactives, fatty acid profile and radical scavenging activity of *Semecarpus anacardium* seed oil. *Journal of the American Oil Chemists' Society*, *87*, 885–894.

Ramadan, M. F., Kroh, L. W., & Moersel, J.-T. (2003). Radical scavenging activity of black cumin (*Nigella sativa* L.), coriander (*Coriandrum sativum* L.) and Niger (*Guizotia abyssinica* Cass.) crude seed oils and oil fractions. *Journal of Agricultural and Food Chemistry*, *51*, 6961–6969.

Ramadan, M. F., Sharanabasappa, G., Seetharam, Y. N., Seshagiri, M., & Moersel, J.-T. (2006). Profile and levels of fatty acids and bioactive constituents in mahua butter from fruit-seeds of buttercup tree [*Madhuca longifolia* (Koenig)]. *European Food Research and Technology*, *222*, 710–718.

Saleh, A. A., Kirrella, A. A., Dawood, M. A. O., & Ebeid, T. A. (2019). Effect of dietary inclusion of cumin seed oil on the performance, egg quality, immune response and ovarian development in laying hens under high ambient temperature. *Journal of Animal Physiology and Animal Nutrition (Berlin)*, *103*(6), 1810–1817.

Schwartz, H., Ollilainen, V., Piironen, V., & Lampi, A.-M. (2008). Tocopherol, tocotrienol and plant sterol contents of vegetable oils and industrial fats. *Journal of Food Composition and Analysis*, *21*, 152–161.

Shahnaz, H., Hifza, A., Bushra, K., & Khan, J. I. (2004). Lipid studies of *Cuminum Cyminum* fixed oil. *Pakistan Journal of Botany*, *36*, 395–401.

Shuichi, A., Hidenori, N., Sumito, I., Hiroaki, T., Hiroshi, S., Shuichi, K., et al. (2000). Interleukin-8 gene repression by clarithormyecin is mediated by the activator protein-1 binding site in human bronchial epithelial cells. *American Journal of Respiratory Cell and Molecular Biology*, *22*, 51–60.

Singh, R. P., Gangadharappa, H. V., & Mruthunjaya, K. (2017). *Cuminum cyminum*—A popular spice: An updated review. *Pharmacognosy Journal*, *9*(3), 292–301.

Sowbhagya, H. B., Sathyendra Rao, B. V., & Krishnamurthy, N. (2008). Evaluation of size reduction and expansion on yield and quality of cumin (*Cuminum cyminum*) seed oil. *Journal of Food Engineering*, *84*, 595–600.

Sroka, Z., & Cisowski, W. (2003). Hydrogen peroxide scavenging, antioxidant and anti-radical activity of some phenolic acids. *Food Chemistry and Toxicology*, *41*, 753–758.

Thippeswamy, N. B., & Naidu, K. A. (2005). Antioxidant potency of cumin varieties; cumin, black cumin and bitter cumin on antioxidant systems. *European Food Research and Technology*, *220*, 472–476.

Yang, B., Karlsson, R. M., Oksman, P. H., & Kallio, H. P. (2001). Phytosterols in sea buckthorn (*Hippophaë rhamnoides* L.) berries: Identification and effects of different origins and harvesting times. *Journal of Agricultural and Food Chemistry*, *49*, 5620–5629.

Zhao, Y., Wang, P., Zheng, W., Yu, G., Li, Z., She, Y., et al. (2019). Three-stage microwave extraction of cumin (*Cuminum cyminum* L.) Seed essential oil with natural deep eutectic solvents. *Industrial Crops and Products*, *140*, 111660.

Chapter 62

Cold pressed green coffee oil

Ahmed A. Hussein
Chemistry Department, Faculty of Applied Sciences, Cape Peninsula University of Technology, Bellville, South Africa

Abbreviations

AFB(1)	aflatoxin B1
CP-GCO	cold press green coffee oil
GAE/g	gallic acid equivalent/g
GC-MS	gas chromatography coupled with mass spectrometry
GCO	green coffee oil
GSTs	glutathione S-transferases
HaCat	keratinocyte cell line from human skin
HDL	high-density lipoprotein
HPLC	high performance liquid chromatography
IL-23	interleukin-23 (a proinflammatory cytokine)
IL-6	interleukin-6 (acts as both a proinflammatory cytokine and an antiinflammatory myokine)
LDL	low-density lipoprotein
mRNA	single-stranded RNA molecule
NMR	nuclear magnetic resonance
P450s	cytochromes P450
SPF	sun protection factor
Starch Hi-Cap 100	octenyl succinic anhydride derivative of starch (has high retention and antioxidant capability)
UVA	long UV light (315–400 nm)
UVB	intermediate UV light (280–315 nm)

1 Introduction

The importance of coffee to the world economy cannot be overstated. World coffee production for 2018/19 was forecast at 174.5 million bags (60 kg each, ~10.5 million tons). It is one of the most valuable primary products in world trade. Historically, it originated in the Ethiopian province of Kaffa, was cultivated later in Yemen (during the 15th century), then quickly spread throughout the Arab world, and was taken by the Dutch to India, Indonesia, then Latin America by the 17th century. The coffee tree or shrub belongs to the family Rubiaceae, wherein more than 70 species belonging to the genus *Coffea* L. were reported. However, only two of these species are commercially explored worldwide: *Coffea arabica* (Arabica) and *Coffea canephora* (Robusta). Arabica was recognized as the most common cultivated species among others, which developed at a very early stage from the African region and provides 75% of world production, while Robusta provides the other 25%. Coffee beans contain several classes of health-related chemicals such as alkaloids (caffeine, trigonelline, theophylline, and theobromine), phenolic acids (caffeic, chlorogenic, caffeoylquinic, dicaffeoylquinic, feruloylquinic, *p*-coumaroylquinic acids, and mixed diesters of caffeic and ferulic acids with quinic acid), diterpenes (cafestol and kahweol derivatives), tannins, carbohydrates, lipids, and vitamin precursors (Ashihara, 2006; Farah & Donangelo, 2006; Mehari et al., 2016; Speer & Kölling-Speer, 2006).

The oil extracted from green coffee beans (*C. arabica* L./*C. canephora* var. Robusta) is called green coffee oil (GCO). It is a natural source of valuable bioactive compounds, such as diterpene esters, fatty acids, and unsaponifiable matter, with biological properties that include antioxidant and cancer chemoprevention activities. Some cosmetic products containing GCO and/or its isolated compounds/fractions have been used to improve skin-aging conditions (Affonso et al., 2016).

Cold Pressed Oils. https://doi.org/10.1016/B978-0-12-818188-1.00062-1
Copyright © 2020 Elsevier Inc. All rights reserved.

2 Phytochemistry of green coffee oil

The green coffee bean contains ~18% lipids. Cold pressed GCO is obtained traditionally by pressing of unroasted "green" coffee beans under low-temperature conditions. The oil is a rich source of unique bioactive compounds not found in other seed oils, such as diterpene esters. In general, and as indicated in Fig. 1, the coffee oil consists mainly of saponifiable matters such as triglycerides, esters of diterpenes, and sterols. On the other hand, the unsaponifiable matter of the free forms of fatty acids, diterpenes, sterols, tocopherols, ceramides, phosphatides, and alkaloids are represented in low percentages, dependent on the bean type or variety. In the current literature, there is a clear gap in the analysis of CP-GCO; only a few articles have mentioned the chemical structure of the major constituents like triglycerides, fatty acids composition, and major diterpenes, while the other constituents are not covered properly, such as phenolic compounds (Kurzrock & Speer, 2001).

The chemical studies on lipid constituents are usually limited to the GC-MS and related techniques, where the sample after solvent extraction is subjected to hydrolysis and methylation. However, other methods have been introduced such as HPLC (for caffeine, triglycerides, and diterpenes), NMR (lipid constituents), and cyclic voltammetry (caffeine) (Nuhu, 2014).

NMR was introduced recently as a new, easy, and fast tool for quantification of main components. Using ^{1}H and ^{13}C NMR spectroscopy, the quantitative analysis of glycerides and their fatty acid components (oleic, linoleic, linolenic) and minor constituents (caffeine, cafestol, kahweol, 16-*O*-methylcafestol) was achieved in a short time with a small sample quantity (D'Amelio, De Angelis, Navarini, Schievano, & Mammi, 2013).

2.1 Fatty acids and acyl lipids

Glycerides (also known as acylglycerols) are the major chemical constituents of vegetable oils. They consist of different fatty acids esterified with glycerol. The physicochemical properties of vegetable oil are largely dependent on the nature of their constituent fatty acids, which further determine the field of application(s). Triglycerides from GCO can be separated from other esters by gel permeation chromatography using dichloromethane (Kurzrock & Speer, 2001). After hydrolysis, the fatty acids are methylated and injected into GC-MS for identification and determination of their relative percentage. According to Bitencourt, Ferreira, Oliveira, Cabral, and Meirelles (2018) and others (Table 1), fatty acid composition of GCO is dominated by linoleic and palmitic acids.

The comparison of the fatty acids content of CP-GCO (Table 2) with common edible oils (Rueda et al., 2014) showed unique chemical structure, where the palmitic acid content is higher than in other oils.

2.2 Terpenoides

In addition to the common chemical constituents, the oil contains unique and rare furanic *ent*-kaurane-type diterpenes (as esters), which are only identified from the *Coffea* genus, namely cafestol and kahweol. The only difference between the two compounds is the presence of the C_1-C_2 double bond in kahweol; such similarity makes the purification process difficult.

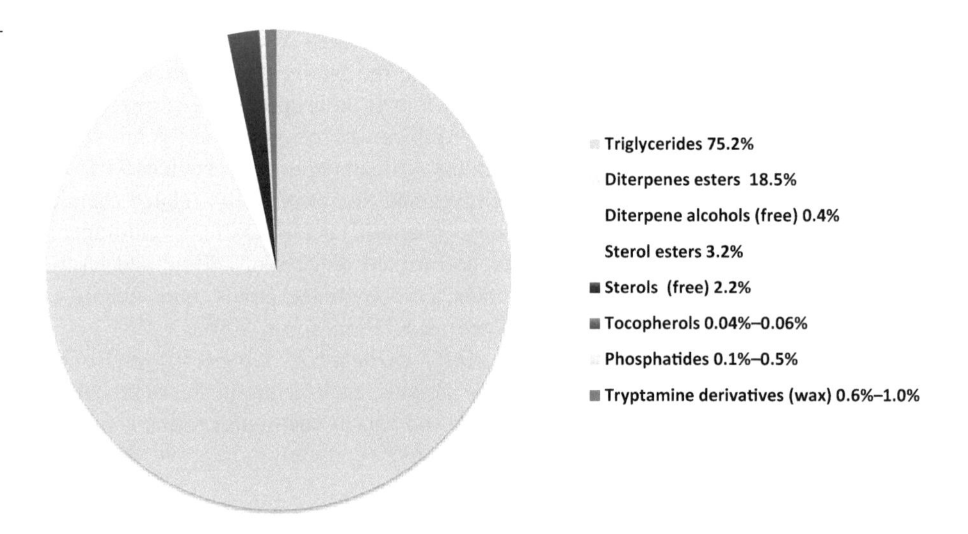

FIG. 1 The average percentage of different compounds in coffee lipid.

TABLE 1 Relative percentage of major fatty acids in GCO.

Fatty acids	1	2	3	4
Palmitic (C16:0)	33.78	35.2–38.6	30.7–35.3	35.2–36.7
Stearic (C18:0)	7.8	6.6–8.35	6.6–9.0	7.2–9.7
Oleic (C18:1)	9.31	7.55–10.9	7.6–10.1	9.5–11.9
Linoleic (C18:2)	44.26	38.4–43.0	43.2–45.9	41.2–42.6
Linolenic (C18:3)	2.62		1.1–1.7	1.3–2.7
Arachidic (C20:0)	1.39	4.05–4.75	2.7–3.3	0.3–1.5
Gondoic (C20:1)	0.25			
Behenic (C22:0)	33.78	0.65–2.6	0.3–0.5	

1, Bitencourt et al. (2018); 2, Calzorali and Cerma (1963); 3, Kroplien (1963); 4, Chassevent, Dalger, Gerwig, and Vincent (1974).

TABLE 2 Relative percentage of major fatty acids in common edible oils.

Oil type	Palmitic C16:0	Palmitoleic C16:1	Stearic C18:0	Oleic C18:1	Linoleic C18:2	Arachidic C20:0	Linolenic C18:3	Gadoleic C20:1	Behenic C22:0
Sunflower	4.98	0.10	3.24	53.11	37.8	0.10	0.28	0.04	0.03
Sesame	9.82	0.11	5.96	41.12	41.8	0.30	0.51	0.12	0.1
Grape seed	7.66	0.03[a]	4.53	17.65	69.3	0.16	0.33[a]	0.02	0.04
Soybean	9.17	0.08	3.81	24.44	54.0	0.05[a]	8.03	ND	0.05
Olive	13.6	0.50	3.28	75.40[f]	5.50	0.55	0.78	0.42	ND
CP-GCO	33.78	ND	7.8	9.31	44.26	1.39	2.62	ND	0.6

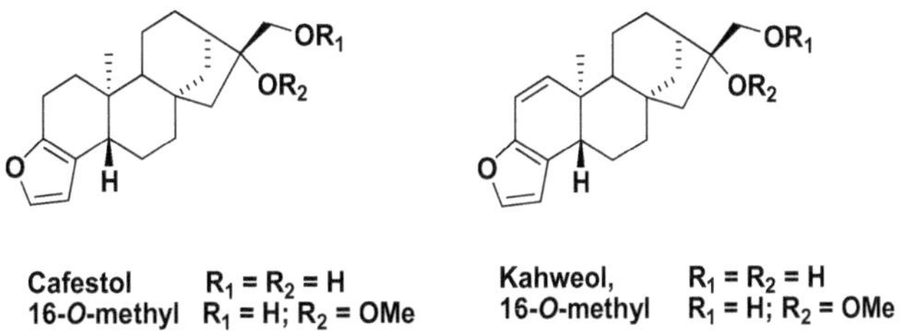

Cafestol $R_1 = R_2 = H$
16-*O*-methyl $R_1 = H$; $R_2 = OMe$
Kahweol, $R_1 = R_2 = H$
16-*O*-methyl $R_1 = H$; $R_2 = OMe$
R_1 = Linoleic or eicosanoic or behenic acids

FIG. 2 Structures of common *ent*-kaurane diterpenes and their esters in green coffee oil.

The ester bonds were found to be connected to C_{17}. 16-*O*-methyl derivatives of cafestol and kahweol were also identified (Fig. 2). The higher percentage of these diterpene esters renders GCO unsuitable as edible oil. Cafestol is found in both *C. arabica* and *C. canephora*, while kahweol is more specific to *C. arabica* (Speer & Kölling-Speer, 2006).

The quantitative analysis of two samples of CP-GCO (2 years difference: 2010, 2012) showed two major compounds with quantities of 3.5 and 7.5 g/kg GCO for cafestol and 8.6 and 12.8 g/kg GCO for kahweol (the stability of the two compounds was largely affected by storage time). The diterpenes were found to be esterified mainly with meristic, linoleic, eicosanoic (arachidic acid), and behenic acids (Nikolova-Damyanova, Velikova, & Jham, 1998). The sterols composition of CP-GCO has not been studied yet in the literature.

2.3 Phenolics

Phenolic compounds in GCO are important metabolites due to their wide spectrum of biological activities. Hydroxycinnamic acid derivatives are one of the main classes of phenolic compounds in green coffee beans, for example, caffeic, ferulic, and *p*-coumaric acids, were identified as free and esters with quinic acid (de Oliveira et al., 2014). No chemical studies have so far reported on the phenolic contents of CP-GCO and/or coffee oil in general. However, the determination of total phenolic contents in two samples of CP-GCO (2 years difference: 2010, 2012) were found to be 2.9 and 18.31 g GAE/g (Speer & Kölling-Speer, 2006).

3 Extraction and processing of cold pressed oil

The vegetable oils, in addition to their nutrition values, showed different industrial applications including biodiesel, pharmaceuticals, and cosmetics. The extraction of plant oils commenced at the beginning of human civilization, and various traditional methods have been applied. Conventional extraction methods such as mechanical expression are well known and widely practiced on an industrial scale for extraction of GCO. There are two types of mechanical press methods: cold press and hot press. The cold press method is a high-pressure single-step extraction and usually carried out at low temperature (<50°C) to avoid adverse side effects and to retain the purity and chemical composition of seed oils (Bhatol, 2013).

The current literature does not cover the cold press extraction of GCO yield percentage from different coffee varieties; however, a single report has mentioned that the yield of cold pressed oil from coffee bean (CP-GCO) ranges from 7% to 11% (Lopez, 2007). Variations of the oil yield and chemical constituents profile are expected and depend on many physical factors such as the nature of the bean (species type, cultivation conditions, processing methods, and moisture content), heat and pressure applied, and duration of extraction. On the other hand, the yield is lower than that of the supercritical CO_2 and solvent extraction methods. The quality of CP-GCO is remarkable when compared with other methods, especially in terms of antioxidant activity.

The oil obtained is aromatic and shows high chlorophyll content and dark green color. The cosmetic grade oils are required to be colorless and odorless, and possess favorable oxidation, chemical, and light stabilities. To reach such a grade, an invention was developed to reduce the color using clarifying clay (Casagrande, 2016a). Another invention was developed to remove the odor of the oil using activated charcoal (Casagrande, 2016b).

The high percentage (18%, *w/w*) of diterpene fatty acid esters in GCO renders it unsuitable for use as an edible vegetable oil. The diterpene fatty acid esters and free fatty acids enriched fraction could be separated from the triglycerides through the fractionation of GCO using the molecular distillation method. According to their chemical constituents, different fractions from GCO could be used in nutritional, cosmetic, and pharmaceutical applications (Rincon, Maciel, & Maciel, 2011a, 2011b).

4 Contribution of CP-GCO bioactive compounds in the cosmetic field

Due to the unique composition of CP-GCO and the high percentage of diterpene esters, the oil is not suitable for the food industry and direct human consumption. The reported emollient and antioxidant properties make the oil more suitable for cosmetic applications.

4.1 Cosmetic and pharmaceutical applications of CP-GCO

UV radiation has the ability to penetrate and generate free radicals in the biological system of the skin, and causes direct damage to the skin such as sunburn, aging, DNA damage, and cancer. Different sunscreen products have been developed and used as a protective shield from harmful solar radiation. These contain a biological factor able to interact with the radiation and prevent it from reaching the skin. Natural products play important roles in skin protection and are highly acceptable ingredients in sunscreen products.

Different interesting biological effects have been associated with natural products such as antioxidant, photoprotection, antiinflammatory, antibacterial, wound healing, and many other premature skin-aging signs, which make them very appropriate for the cosmetics industry. On the other hand, products termed "natural" are expected to meet with customer needs and to improve the quality of human life. Green coffee beans are rich in phenolic content, especially chlorogenic acid derivatives, which are reported to have remarkable biological activities such as antioxidant, and support the cosmetic applications of coffee bean by-products such as CP-GCO. The cosmetic quality of CP-GCO is evident, especially in terms of sun protection factor, as reflected in recent publications (Chiari et al., 2014; Marto, Ascenso, et al., 2016; Marto, Gouveia, Chiari, et al., 2016; Marto, Gouveia, Goncalves, et al., 2016; Wagemaker et al., 2016).

Comparing the sun protection factor (SPF) with that of other natural oils (e.g., olive (7.5); coconut (7.1); castor oil (5.7); lemon oil (2.81); tea tree oil (1.70); rose oil (0.248)), CP-GCO showed (Kaur & Saraf, 2010) a range of SPF values (2.0–5.0), which suggested its use as a potential natural product for improving SPF in sunscreen formulations (Chiari et al., 2014; Wagemaker et al., 2016). More in-depth study was performed to determine the active fractions in GCO that could be responsible for desirable cosmetic properties such as UVB absorption activity, and showed that the unsaponifiable matter extracted from GCO has SPF 10 times higher than in GCO and is considered as a novel potential UV-B absorbent (Wagemaker et al., 2016).

The enhancement of SPF of CP-GCO was achieved through combination with other ingredients such as melatonin. The combination with melatonin gave advantage for multiple pharmacological actions in addition to enhancing SPF. The formula has high SPF (50+) against UVA and UVB and could be considered as a new potential formulation in the personal care industry (Marto, Ascenso, et al., 2016). In addition, the combination with ethylhexylmethoxycinnamate showed a synergistic effect in SPF value, leading to an increase of 20% (Chiari et al., 2014). The addition of modified starch product (aluminum starch octenylsuccinate) effectively enhanced the SPF and showed an excellent SPF with a value of ~82.0 (Marto, Gouveia, Chiari, et al., 2016; Marto, Gouveia, Goncalves, et al., 2016).

The stability of GCO against heat, light, and oxygen was improved by microencapsulation of GCO using gum arabic as a carrier agent (Nosari, Lima, Serra, & Freitas, 2015). In addition, modified starch Hi-Cap 100, in combination with corn syrup using lecithin-chitosan and lecithin emulsion stabilizers, showed good results (Carvalho, Silva, & Hubinger, 2014).

The cosmetic quality of CP-GCO was reported in vitro and in vivo and introduced the oil as a high standard natural ingredient for protection of skin from harmful radiation. It showed enhanced photoprotection effect either alone or in combination with other ingredients. The addition of starch derivative, melatonin, cinnamic acid derivative, and propolis (Ribeiro, Paulino, & Marques dos Santos, 2014) enhanced not only the SPF but also other cosmetic-related factors.

The in vitro cytotoxicity against human keratinocytes of GCO showed no cytotoxicity effects up to 100 μg mL^{-1} (Wagemaker et al., 2015). This finding was supported by Chiari et al.'s (2014) report, which showed no cytotoxicity of GCO against skin and liver cells in vitro, even at high concentrations. Furthermore, the safety and efficiency of GCO for topical applications and skin compatibility were evaluated in vivo, and the GCO (intact) and formulations containing 2.5%–15% were evaluated in humans. The 15% formulation of GCO was applied in the volar forearm of human volunteers over 3 days, and was also probed for skin tolerance through a patch test. The results showed a slight water loss reduction when the formulation was applied. Stratum corneum water content and erythema index did not show significant differences. None of the volunteers displayed any reaction after using an occlusive patch (Wagemaker et al., 2015).

The topical application of GCO in vivo showed fast wound healing action. Rats treated with GCO showed higher expression of mRNA IL-6 and IL-23 in the early stages. The nonlinear imaging analysis showed a higher collagen density (Lania et al., 2017). The use of GCO formula with propolis was able to maintain the physical characteristics of the skin and skin hydration and showed antiinflammatory, antioxidant, and antiaging properties, while reducing the roughness of the skin and helping to reduce and prevent stretch marks (Ribeiro et al., 2014).

The combination of melatonin and GCO achieved a stable sunscreen. The results showed that the formulation was beneficial and presented powerful protection against UVB-induced damage in HaCat cells, including inhibition of apoptosis (Marto, Ascenso, et al., 2016).

4.2 Pharmacological activity of GCO and its chemical constituents

Coffee oil and/or its chemical constituents showed different pharmacological effects in vivo. Cholesterol and liver enzymes were greatly affected by coffee oil constituents, as indicated by different human studies. A randomized double-blind parallel study among 36 subjects on the effect of coffee lipid from Arabica and Robusta on serum cholesterol showed elevation of serum total cholesterol and plasma triglycerides by Arabica oil, while the effect of Robusta oil was not statistically significant (van Rooij et al., 1995). Another study indicated the serum lipid level was elevated by both Arabica and Robusta oils (Mensink et al., 1995). The responses of total and LDL cholesterol to coffee oil were poorly reproducible among 32 healthy volunteers administered coffee oil (69 mg cafestol/day), while the responses of HDL and triglycerides were found to be highly reproducible (Boekschoten, Engberink, Katan, & Schouten, 2003). The diterpenoids fraction was reported to be responsible for such effects on liver enzymes and cholesterol. Boekschoten, Schouten, and Katan (2004) showed that a mixture of cafestol and kahweol raised liver enzyme levels more potently than pure cafestol, whereas the effect on serum cholesterol levels was similar. In a randomized, double-blind crossover study, 10 healthy male volunteers were given either pure cafestol (61–64 mg/day) or a mixture of cafestol (60 mg/day) and kahweol (48–54 mg/day) for 28 days. The mixture of cafestol plus kahweol increased total cholesterol LDL cholesterol triacylglycerols and alanine aminotransferase (Urgert & Katan, 1997).

On the other hand, GCO was found to stimulate liver regeneration in hepatectomized rats fed diets supplemented with up to 7.0%. The activity was proposed to be attributed to the diterpene fraction containing cafestol and kahweol (Gershbein & Baburao, 1980).

The chemopreventive effects of cafestol and kahweol have been reported in several animal models. These effects were correlated with the compounds' ability to induce glutathione S-transferases (GSTs) and suggested a potential role for GSTs in detoxifying carcinogenic compounds (Schilter, Perrin, Cavin, & Huggett, 1996). Kahweol palmitate and cafestol palmitate were isolated and found to induce this enzyme activity; however, kahweol palmitate was more potent (Lam, Sparnins, & Wattenberg, 1982).

The anticarcinogenic effects of coffee oil constituents (kahweol and cafestol) were further confirmed by Cavin et al. (2002). The treatment of human liver epithelial cell lines, transfected to express AFB(1)-activating P450s, with kahweol and cafestol resulted in a reduction of AFB(1)-DNA binding. This protection was correlated with an induction of GST-mu, an enzyme known to be involved in AFB(1) detoxification. In addition, kahweol and cafestol mixtures were found to inhibit P450 2B6, one of the human enzymes responsible for AFB(1) activation (Cavin et al., 2002).

5 Authenticity of cold pressed oil

The chemistry of coffee oil showed a unique chemical composition. The oil contains a higher percentage of palmitic acid than other edible vegetable oils (Bitencourt et al., 2018). Additionally, the presence of diterpene kahweol and cafestol and their derivatives make the oil very characteristic and easy to distinguish from other oils (Speer & Kölling-Speer, 2006). Different lipid components, such as sterols, triacylglycerols, tocopherols, and diterpenic alcohols have been reported to distinguish between Arabica and Robusta species (Cossignani, Montesano, Simonetti, & Blasi, 2016). On the other hand, the relative percentage of diterpenes vary among different coffee cultivars and are controlled by many factors, which make this chemical marker very difficult to distinguish between different coffee oil cultivars.

6 Conclusion

Green chemistry offers solutions to overcome the current industrial shortcomings and improve the sustainable usage of natural resources. The food industry is one of the major industries that results in huge amounts of biowaste; these by-products need scientific efforts to optimize their usage for human benefits. Coffee is a major food product and produces more than 95% waste material from its starting raw materials. Different biowaste products produced include the oil. Green technology for producing oils is highly appreciated where the intact natural properties of the produced oils are preserved. Green coffee oil produced by cold pressing has been introduced to the market as a valuable product with different applications excluding direct human consumption. The cosmetic quality of the final products has been proved; however, the toxicity of the oil against the liver, as mentioned above, needs to be studied further and different aspects like biodistribution and accumulation of the diterpenes in the body as well as half-life and monitoring toxicity against the liver still need to be determined.

References

Affonso, R. C., Voytena, A. P., Fanan, S., Pitz, H., Coelho, D. S., Horstmann, A. L., et al. (2016). Phytochemical composition, antioxidant activity, and the effect of the aqueous extract of coffee (*Coffea arabica* L.) bean residual press cake on the skin wound healing. *Oxidative Medicine and Cellular Longevity*, *2016*, 1923754. https://doi.org/10.1155/2016/1923754.

Ashihara, H. (2006). Metabolism of alkaloids in coffee plants. *Brazilian Journal of Plant Physiology*, *18*(1), 1–8.

Bhatol, K. (2013). *Castor oil obtained by cold press method*. Banaskantha, Gujarat: Shri Bhgwati Oil Mill (SBOM) Manufacturer's Info.

Bitencourt, R. G., Ferreira, N. J., Oliveira, A. L., Cabral, F. A., & Meirelles, A. J. A. (2018). High pressure phase equilibrium of the crude green coffee oil - CO_2- ethanol system and the oil bioactive compounds. *Journal of Supercritical Fluids*, *133*, 49–57. https://doi.org/10.1016/j.supflu.2017.09.017.

Boekschoten, M. V., Engberink, M. F., Katan, M. B., & Schouten, E. G. (2003). Reproducibility of the serum lipid response to coffee oil in healthy volunteers. *Nutrition Journal*, *2*, 8. https://doi.org/10.1186/1475-2891-2-8.

Boekschoten, M. V., Schouten, E. G., & Katan, M. B. (2004). Coffee bean extracts rich and poor in kahweol both give rise to elevation of liver enzymes in healthy volunteers. *Nutrition Journal*, *3*, 7. https://doi.org/10.1186/1475-2891-3-7.

Calzorali, C., & Cerma, E. (1963). Sulle sostanze grasse del caffé. *Rivista Italiana Delle Sostanze Grasse*, *40*, 176–180.

Carvalho, A. G. S., Silva, V. M., & Hubinger, M. D. (2014). Microencapsulation by spray drying of emulsified green coffee oil with two-layered membranes. *Food Research International*, *61*, 236–245. https://doi.org/10.1016/S0963-9969(99)00016-2.

Casagrande, M. (2016a). *Productive process for clarifying green coffee oil.* Brazilian Patent, PI BR 102014026048 A2 20160524.

Casagrande, M. (2016b). *Production method for deodorizing green coffee oil.* Brazilian Patent PI, BR 102014026042 A2 20160524.

Cavin, C., Holzhaeuser, D., Scharf, G., Constable, A., Huber, W. W., & Schilter, B. (2002). Cafestol and kahweol, two coffee specific diterpenes with anticarcinogenic activity. *Food and Chemical Toxicology*, *40*, 1155–1163. https://doi.org/10.1016/S0278-6915(02)00029-7.

Chassevent, F., Dalger, G., Gerwig, S., & Vincent, J. C. (1974). Contribution à l'étude des Mascarocoffea. *Café, Cacao, Thé*, *18*, 49–56.

Chiari, B. G., Trovattib, E., Pecorarob, E., Corrêaa, M. A., Cicarellia, R. M., Ribeiro, S., et al. (2014). Synergistic effect of green coffee oil and synthetic sunscreen for health care application. *Industrial Crops and Products*, *52*, 389–393.

Cossignani, L., Montesano, D., Simonetti, M. S., & Blasi, F. (2016). Authentication of Coffea arabica according to triacylglycerol stereospecific composition. *Journal of Analytical Methods in Chemistry*. *7482620*, https://doi.org/10.1155/2016/7482620.

D'Amelio, N., De Angelis, E., Navarini, L., Schievano, E., & Mammi, S. (2013). Green coffee oil analysis by high-resolution nuclear magnetic resonance spectroscopy. *Talanta*, *110*, 118–127. https://doi.org/10.1016/j.talanta.2013.02.024.

de Oliveira, P. M. A., de Almeida, R. H., de Oliveira, N. A., Bostyn, S., Goncalves, C. B., & de Oliveiraa, A. L. (2014). Enrichment of diterpenes in green coffee oil using supercritical fluid extraction-characterization and comparison with green coffee oil from pressing. *Journal of Supercritical Fluids*, *95*, 137–145.

Farah, A., & Donangelo, C. M. (2006). Phenolic compounds in coffee. *Brazilian Journal of Plant Physiology*, *18*(1), 23–36. https://doi.org/10.1590/S1677-04202006000100003.

Gershbein, L. L., & Baburao, K. (1980). Effect of feeding coffee and its lipids on regenerating and intact liver. *Research Communications in Chemical Pathology and Pharmacology*, *28*(3), 457–472.

Kaur, C. D., & Saraf, S. (2010). In vitro sun protection factor determination of herbal oils used in cosmetics. *Pharmacognosy Research*, *2*(1), 22–25. https://doi.org/10.4103/0974-8490.60586.

Kroplien, U. (1963). *Green and roasted coffee tests.* Hamburg: Gordian-Max Rieck.

Kurzrock, T., & Speer, K. (2001). Diterpenes and diterpene esters in coffee. *Food Reviews International*, *17*(4), 433–450. https://doi.org/10.1081/FRI-100108532.

Lam, L. K. T., Sparnins, V. L., & Wattenberg, L. W. (1982). Isolation and identification of kahweol palmitate and cafestol palmitate as active constituents of green coffee beans that enhance glutathione S-transferase activity in the mouse. *Cancer Research*, *42*, 1193–1198.

Lania, B. G., Morari, J., Souza, A. L., Silva, M. N. D., de Almeida, A. R., Veira-Damiani, G., et al. (2017). Topical use and systemic action of green and roasted coffee oils and ground oils in a cutaneous incision model in rats (*Rattus norvegicus albinus*). *PLoS One*, *12*(12), e0188779/1–e0188779/17. https://doi.org/10.1372/journal.pone.0188779.

Lopez, E. M. (2007). Extracción de aceite de café. *Revista Ingenieria e Investigation*, *27*(1), 25–31.

Marto, J., Ascenso, A., Goncalves, L., Gouveia, L. F., Manteigas, P., Pinto, P., et al. (2016). Melatonin-based pickering emulsion for skin's photoprotection. *Drug Delivery*, *23*(5), 1594–1607. https://doi.org/10.3109/10727544.2015.1128496.

Marto, J., Gouveia, L. F., Chiari, B. G., Paiva, A., Isaac, V., Pinto, P., et al. (2016). The green generation of sunscreens: Using coffee industrial sub-products. *Industrial Crops and Products*, *80*, 93–100. https://doi.org/10.1016/j.indcrop.2015.11.033.

Marto, J., Gouveia, L. F., Goncalves, L., Chiari-Andreo, B. G., Isaac, V., Pinto, P., et al. (2016). Design of novel starch-based Pickering emulsions as platforms for skin photoprotection. *Journal of Photochemistry and Photobiology, B: Biology*, *162*, 56–64. https://doi.org/10.1016/j.jphotobiol.2016.06.026.

Mehari, B., Redi-Abshiro, M., Chandravanshi, B. S., Atlabachew, M., Combrinck, S., & McCrindle, R. (2016). Simultaneous determination of alkaloids in green coffee beans from Ethiopia: Chemometric evaluation of geographical origin. *Food Analytical Methods*, *9*, 1627. https://doi.org/10.1007/s12161-015-0340-2.

Mensink, R. P., Lebrink, W. J., Lobbezoo, I. E., Weusten-Van der Wouw, M. P., Zock, P. L., & Katan, M. B. (1995). Diterpene composition of oils from Arabica and Robusta coffee beans and their effects on serum lipids in man. *Journal of International Medicine*, *237*, 543–550.

Nikolova-Damyanova, B., Velikova, R., & Jham, G. N. (1998). Lipid classes, fatty acid composition and triacylglycerol molecular species in crude coffee beans harvested in Brazil. *Food Research International*, *31*(6–7), 479–486. https://doi.org/10.1016/S0963-9969(99)00016-2.

Nosari, A. B. F. L., Lima, J. F., Serra, O. A., & Freitas, L. A. P. (2015). Improved green coffee oil antioxidant activity for cosmetical purpose by spray drying microencapsulation. *Revista Brasileira de Farmacognosia*, *25*(3), 307–311. https://doi.org/10.1016/j.bjp.2015.04.006.

Nuhu, A. A. (2014). Bioactive micronutrients in coffee: Recent analytical approaches for characterization and quantification. *ISRN Nutrition*, *2014*, 384230. https://doi.org/10.1155/2014/384230.

Ribeiro, M. C. M., Paulino, N., & Marques dos Santos, A. (2014). *Cosmetic and pharmaceutical formulations based on coffee oil and propolis.* Brazilian Patent, PI, BR 102012019950 A2 20140909.

Rincon, M. A. D., Maciel, F. R., & Maciel, M. R. W. (2011a). Fractionation of green coffee oil by molecular distillation: Modeling and simulation. *Journal of Materials Science and Engineering A*, *1*(2), 264–272.

Rincon, M. A. D., Maciel, F. R., & Maciel, M. R. W. (2011b). Modeling of molecular distillation parameters: Case study of green coffee oil (*Coffea arabica*). *Journal of Chemistry and Chemical Engineering*, *5*(8), 706–720.

Rueda, A., Seiquer, I., Olalla, M., Giménez, R., Lara, L., & Cabrera-Vique, C. (2014). Characterization of fatty acid profile of argan oil and other edible vegetable oils by gas chromatography and discriminant analysis. *Journal of Chemistry*. *2014*, https://doi.org/10.1155/2014/843908.

Schilter, B., Perrin, I., Cavin, C., & Huggett, A. C. (1996). Placental glutathione *S*-transferase (GST-P) induction as a potential mechanism for the anti-carcinogenic effect of the coffee-specific components cafestol and kahweol. *Carcinogenesis*, *17*, 2377–2384.

Speer, K., & Kölling-Speer, K. (2006). The lipid fraction of the coffee bean. *Brazilian Journal of Plant Physiology*, *18*(1), 201–216.

Urgert, R., & Katan, M. B. (1997). The cholesterol-raising factor from coffee beans. *Annual Review of Nutrition*, *17*(1), 305–324.

van Rooij, J., van der Stegen, G. H. D., Shoemaker, R. C., Kroon, C., Burggraaf, J., Hollaar, L., et al. (1995). A placebo-controlled parallel study of the effect of two types of coffee oil on serum lipids and transaminases: Identification of chemical substances involved in the cholesterol-raising effect of coffee. *American Journal of Clinical Nutrition*, *61*, 1277–1283.

Wagemaker, T. A. L., Campos, P. M., Fernandes, A. S., Rijo, P., Nicolai, M., Roberto, A., et al. (2016). Unsaponifiable matter from oil of green coffee beans: Cosmetic properties and safety evaluation. *Drug Development and Industrial Pharmacy*, *42*(10), 1695–1699. https://doi.org/10.3109/03639045.2016.1165692.

Wagemaker, T. A. L., Rijo, P., Rodrigues, L. M., Maia Campos, P. M. B. G., Fernandes, A. S., & Rosado, C. (2015). Integrated approach in the assessment of skin compatibility of cosmetic formulations with green coffee oil. *International Journal of Cosmetic Science*, *37*(5), 506–510. https://doi.org/10.1111/ics.12225.

Chapter 63

Cold pressed yuzu (*Citrus junos* Sieb. ex Tanaka) oil

Sayed A. El-Toumy[a] and Ahmed A. Hussein[b]

[a]*Chemistry of Tannins Department, National Research Centre, Cairo, Egypt,* [b]*Chemistry Department, Faculty of Applied Sciences, Cape Peninsula University of Technology, Bellville, South Africa*

Abbreviations

BHT	butylated hydroxytoluene
CgA	the salivary chromogranin A
CPO	cold pressed oil
FD factor	flavor dilution factor
GC	gas chromatography
NDMA	*N*-nitrosodimethylamine
SOS	singlet oxygen species

1 Introduction

The *Citrus* genus is widely distributed in temperate and tropical zones of the southern hemisphere and considered one of the most important crops in the world. The fruit products have different food as well as perfumery and medical-related applications. Essential oil (~1%/wt of the fruit) is an important product of citrus; it accumulates in the oil glands of the peel and has a unique and pleasant aroma. The oil is very popular and has different applications such as food flavorings, toiletry products, cosmetics, and perfumes (Sawamura, 2005). It is estimated that there are ~10,000 different *Citrus* species in the world. *Citrus junos* Tanaka (yuzu) is among the most important citrus crops and is very popular in Japan and Korea, where many interesting applications have been reported. The origin of yuzu is not certain and it is considered to be a hybrid of sour mandarin (*C. reticulata* var. *Austera*) and Ichang papeda (*C. ichangensis*) (Rahman, Nito, & Isshiki, 2001; Taninaka, Otoi, & Morimoto, 1981). Japan and Korea are the main users/producers of Yuzu fruits; in 2013, Japan produced ~23,000 tons (Kuraya, Touyama, Nakada, Higa, & Itoh, 2017).

2 Phytochemistry of yuzu oil

Like other *Citrus* species, yuzu oil is obtained mainly from the fruit rind (flavedo), using different methods including, cold pressing, solvent extraction, and steam distillation. The citrus essential oils consist mainly of volatile compounds, especially when steam distillation and similar methods are used; the cold press method allows some nonvolatile constituents to be extracted. Volatile and semivolatile compounds represent 85%–99% of the entire oil fraction. Yuzu oil, in general, has the same chemical profile as other citrus oils and shows a high content of monoterpenes including limonene, α-phellandrene, β-phellandrene, α-pinene, β-pinene, and myrcene (Fig. 1). Volatile compounds such mono- and sesquiterpene hydrocarbons usually represent the highest concentration, followed by oxygenated (mainly alcohol) compounds, and then carbonyl compounds (ketones and aldehydes). Fatty acids, sulfur compounds, and coumarin derivatives exist in very small amounts, not exceeding 0.02%.

Limonene, as the major constituent of the citrus fruits family, has important industrial applications as a degreasing solvent and as a flavor and fragrance additive in food, household cleaning products, and perfumes. Limonene containing chiral carbon (C-4) and *d* (orange flavor) and *l* (lemon flavor) isomers exists in nature; however, the *d* isomer is more common and can be isolated up to 98% purity.

Cold Pressed Oils. https://doi.org/10.1016/B978-0-12-818188-1.00063-3
Copyright © 2020 Elsevier Inc. All rights reserved.

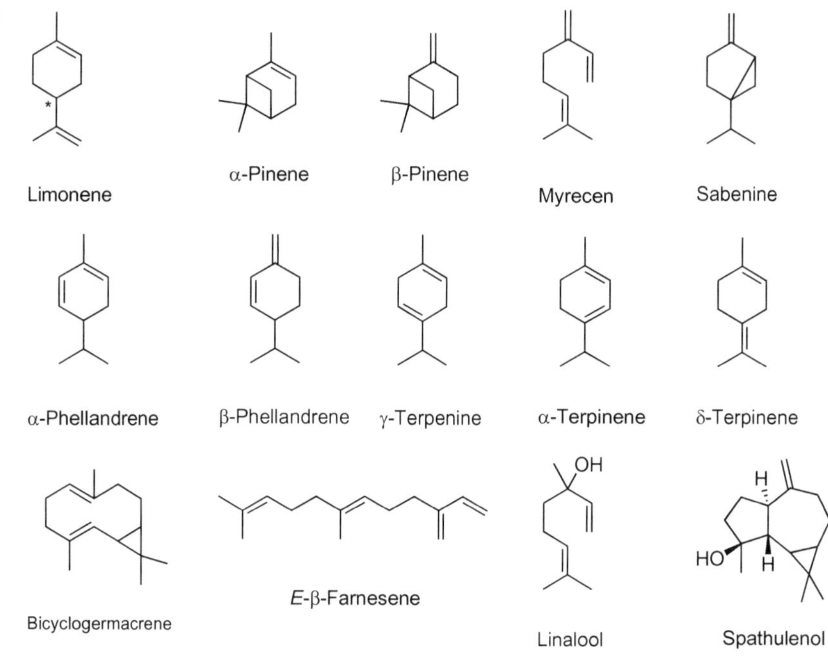

FIG. 1 Major mono- and sesquiterpene identified from yuzu oil.

An earlier report (Shinoda, Shiga, & Nishimura, 1970) on the analysis of yuzu cold pressed oil, after fractionation, showed a similar profile to other studies (Table 1). However, the fractionation process led to some chemical transformation and/or isomerization for unstable compounds; the authors also reported the presence of some fatty acids such as capric, lauric, myristic, palmitic, and stearic acids, in addition to two coumarins: phellopterin and aurapten (Fig. 2). The authors proposed that the presence of some compounds contributed to the characteristic flavor of yuzu in very small amounts, but could not be isolated or identified.

Furanocoumarins and coumarins are either undetectable or exist in small amounts in commercial yuzu oils (after refinement). On the other hand, bergapten (10.3 mg/kg) and xanthotoxin (195 mg/kg) were detected in unrefined yuzu seed oil (Sawamura et al., 2018). Lan-Phi and Sawamura (2008) compared 15 samples of cold pressed oil (CPO) collected from 5 different locations (Table 1). Limonene was the dominant monoterpene hydrocarbon, ranging between 55.8% and 64.7%. Major components were γ-terpinene, β-phellandrene, myrcene, and α-pinene. The major sesquiterpenes were bicyclogermacrene and *E*-β-farnesene. In addition, the yuzu oil showed a relatively higher content of sesquiterpenes to that of monoterpenes when compared to other citrus oils (Lan-Phi & Sawamura, 2008). The same group studied the variation of the volatile content among six cultivars, and the results are listed in Table 1. The relative concentrations of the volatile contents have a range like previous studies. The authors applied GC-olfactometry and aroma extraction dilution analysis technique in threefold stepwise dilution of the neat oil for all samples and the results indicated that limonene, α-pinene, α- and β-phellandrene, myrcene, γ-terpinene, (*E*)-β-farnesene, and linalool have the highest flavor dilution (FD) values (Lan-Phi, Shimamura, Ukeda, & Sawamura, 2009).

Other reports by the same group contradicted the above the results and attributed the unique flavor of yuzu oil to minor (sometimes undetectable) sulfur compounds in addition to alcohols and aldehydes. Song, Sawamura, Ito, Kawashimo, and Ukeda (2000) used the FD factor to determine the volatile compounds contributing to the characteristic flavors of yuzu, and they concluded that compounds like 6-methyl-5-hepten-2-ol and methyl trisulfide have the highest contribution with an FD factor of 2^{18}, followed by borneol, *n*-octanol, *trans*-undec-2-enal, (+)-*p*-mentha-1-en-9-ol, eugenol, and carvacrol. On the other hand, the major compounds like limonene do not have the characteristic flavor or a yuzu-like odor according to the sniffing test in spite of their high concentrations. The sum of characteristic flavor components of yuzu amounted to only 0.02% weight. These results suggest that minor components (Fig. 3) often contribute significantly to the characteristic flavor (Njoroge, Ukeda, Kusunose, & Sawamura, 1994; Song et al., 2000). Others supported the major contribution of the minor sulfur-related monoterpenes to yuzu flavor, and two other compounds were detected and identified: 1-*p*-menthene-8-thiol and 4-methyl-4-mercapto-pentan-2-one (Yukawa, Osaki, & Iwabuchi, 1995).

Miyazawa et al. (2009) conducted a sensory evaluation of yuzu oil to clarify the odor-active volatiles that differentiates yuzu from other citrus fruits. Seven polar compounds were identified as the source of the characteristic

TABLE 1 List of yuzu volatile oil constituents using different methods of extraction.

Compound	Cold pressed oil				Steam distillation		Supercritical-CO_2	
	A	B	C	D	E	F	G	H
α-Thujone					0.35–1.71			tr
α-Pinene	1.4	2.52–3.17	2.3–2.7	2.03	0.88–3.97	2.29	0.7	0.43
Camphene			tr	1.09		tr		1.87
β-Pinene	0.5	1.15–1.45	0.9–1.1	0.01	1.64–3.96	0.92	0.8	0.34
Sabinene		0.48–0.63	0.4–0.5	0.81	0–2.21	0.29	0.2	tr
Myrcene	2.2	2.22–3.09	3.0–3.2	0.38	0.55–1.55	2.22	3.1	0.83
α-Phellandrene		0.90–1.19	0.7–0.9	0.94	0.04–0.3	0.64	0.6	0.33
β-Phellandrene		3.18–4.11	4.6–5.4	2.16		3.90		
3-Carene							10.2	
4-Carene					0.54–1.23			tr
Limonene	79.4	55.86–64.79	63.1–68.1	77.44	35.53–60.60	72.65	74.3	60.32
α-Terpinene		0.37–0.51	0.3	0.24	0.88–3.97	0.34		
γ-Terpinene	9.5	9.69–13.92	11.4–12.5	9.43	21.71–46.74	11.20		11.81
o-Cymene					0.38–4.96			
p-Cymene	0.2	0.49–0.90	0.4–0.6	0.33	0–1.23	0.67		
Terpinolene	tr	0.67–0.89	0.6–0.7	0.01		0.59		0.16
Elixene		0.04–0.06						0.5
α-Copaene	0.3	0.11–0.17	tr–0.1	0.02		0.03		0.18
β-Cubebene		0.06–0.09	tr–0.1	0.01		0.02		0.in
Linalool	0.87	2.12–3.50	2.1–2.9	1.56	0.73–7.95	1.32	5.5	11.59
Linalyl acetate				0.01			0.5	
β-Elemene	0.4	0.07–0.13	0.1–0.2	tr		0.05		
γ-Elemene	0.51					0.01		
δ-Elemene	0.30					0.09		
β-Caryophyllene	0.30	0.36–0.58	0.3	0.13		0.18	0.7	0.76
(*E*)-Decenal			tr	0.01		0.02	0.2	
α-Humulene	tr		0.1	0.02		0.03		
(*E*)-β-Farnesene	0.86		0.9–1.3	0.47	0–0.15	0.50	0.6	
α-Terpineol	0.12	0.25–0.39	0.2–0.3	0.07	0.3–0.85		0.3	
Terpinen-4-ol	0.02				0.21–1.02	0.09		
α-Muurolene	tr		tr	0.01		0.02		
Piperitone			tr	0.02				
Bicyclogermacrene		1.73–2.38	1.5–2.0	0.66		0.87		
α-Sesquiphellandrene		0.03–0.05	0.1	0.04		0.04		
β-Sesquiphellandrene		0.07–0.13	0.1					
δ-Cadinene		0.13–0.17				0.10		0.32
Germacrene B		0.18–0.36	0.1–0.2	0.01		0.02		
Germacrene D		0.38–0.62	0.3–0.4	0.12	0–0.55	0.02	0.8	
Germacrene D 4-ol		0.35–0.60	0.3–0.4					
Thymol	0.15	0.16–0.48	0.2–0.3	0.09		0.12	0.3	1.18

(A) Shinoda et al. (1970); (B) Lan-Phi and Sawamura (2008); (C) Lan-Phi et al. (2009); (D) Song et al. (2000); (E) Liu et al. (2014); (F) Kashiwagi et al. (2010); (G) Roy et al. (2007); (H) Hoshino, Tanaka, Sasaki, and Goto (2006).

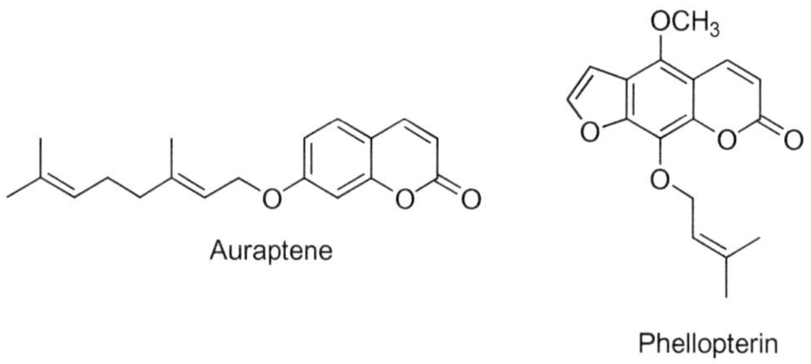

FIG. 2 Auraptene and phellopterin identified in yuzu oil.

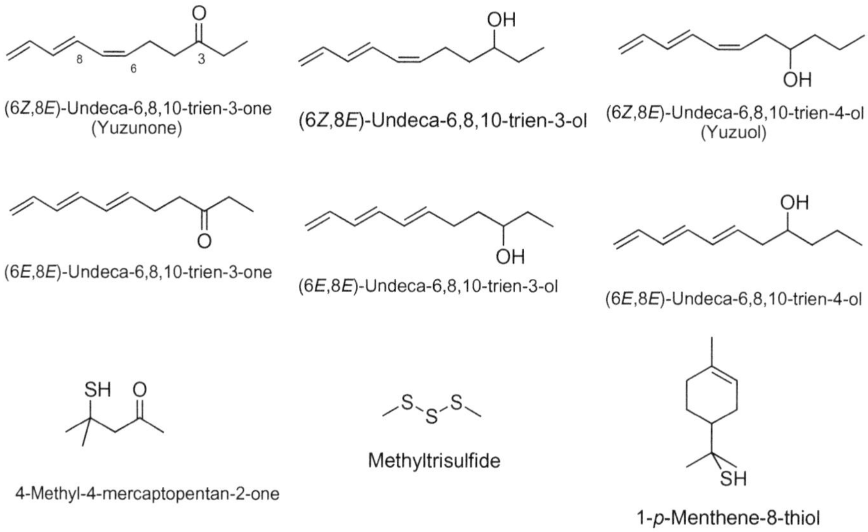

FIG. 3 Minor constituents strongly contribute to yuzu oil aroma.

aroma: oct-1-en-3-one, (*E*)-non-4-enal, (*E*)-dec-4-enal, 4-methyl-4-mercaptopentan-2-one, (*E*)-non-6-enal, (6*Z*,8*E*)-undeca-6,8,10-trien-3-one (yuzunone), and (6*Z*,8*E*)-undeca-6,8,10-trien-4-ol (yuzuol).

3 Extraction and processing of yuzu oil

Different methods have been used to extract yuzu essential oils; however, the cold press method is considered the oldest and most widely used one. Other methods like supercritical CO_2, steam distillation, and solvent extraction have been reported. Low boiling point solvents like hexane and methylene chloride are employed for oil extraction, where other conventional methods cannot be used and traces amount of oils exist; however, this method is not specific to the oil fraction, and other compounds can be extracted according to the polarity of the solvent used and will be excluded from this section.

3.1 Cold press method

In addition to its simplicity, the cold press method keeps the oil intact without change. The isomerization, chemical transformation, and oxidation are very limited throughout the extraction process. Volatile constituents of yuzu consist of mainly mono- and sesquiterpene hydrocarbons, and this type of compound is easily affected by oxygen and moisture even at room temperature. In order to minimize these chemical changes, careful and temperature-controlled cold press processes are usually designed.

The peel of *Citrus* fruit consists of flavedo and albedo. The flavedo is the outer peel and contains the oil glands. Sawamura and Kuriyama (1988) described small-scale extraction of the oil, where the fruit is sliced into six to eight pieces lengthwise, then the flavedo is separated and pressed by hand to express the oil, collected in a cold brine solution. After centrifugation at $4000 \times g$ for 15 min at 5°C, the oils are collected carefully, then dehydrated over anhydrous sodium sulfate. The oil is stored at −21°C until analysis. On the commercial scale, the citrus fruit is generally squeezed for commercial juice products, and the oil produced as a by-product. Because of the nature of yuzu fruits, the yield of the oil can reach ~1.0% of the fruit weight.

Different studies have been conducted on the extraction and chemical evaluation of the oil constituents using cold press methods and the results tabulated in columns A–D of Table 1 (Lan-Phi & Sawamura, 2008; Lan-Phi et al., 2009; Shinoda et al., 1970; Song et al., 2000). From the values (relative percentage) of the chemical individuals, limonene with α-pinene, β-pinene, myrcene β-phellandrene, and γ-terpinene are the major chemical constituents. The quantitative variation is expected, and it is interesting to note that the extraction performed in 1970 did not show β-phellandrene, while the same compound was found with values >3% in columns B and C. In addition, the reports did not specify the limonene isomer isolated (whether *d* or *l*). On the other hand, the majority of safety reports are dealing with the *d* isomer rather than the *l* (Dosoky & Setzer, 2018), while the olfactometry test proposed the *l* isomer (lemon-like) (Song et al., 2000).

3.2 Steam distillation

Steam distillation is a widely utilized method for extraction of volatile constituents, especially in the case of using leaves and flowers as starting materials. Two studies are available in the literature and report the use of steam distillation method to extract yuzu oil. The results of the oil are presented in Table 1 (columns E and F) and show a similar profile to the cold pressed oil extraction method (Kashiwagi, Lan-Phi, & Sawamura, 2010; Liu et al., 2014). In column E, the values show a decrease of limonene concentration and an increase of γ-terpinene from the average.

3.3 Supercritical CO_2 extraction

The low critical temperature of the CO_2 extraction method prevents thermal degradation in addition to the residue-free, odorless, colorless, and harmless (nontoxic solvent) condition of the resulting oil. Two studies reported the average percentage of yuzu chemical constituents (Hoshino et al., 2006; Roy, Hoshino, Ueno, Sasaki, & Goto, 2007). The first method obtained a 1.27% yield at a pressure of 20 MPa; the quality of the extracted oil is similar to other methods, where the major compounds were identified as limonene, γ-terpinene, linalool, sabinene, β-myrcene, α-pinene, β-farnesene, α-terpineol, and terpinolene. Variation also exists: column H (Table 1) shows a higher concentration of linalool than other methods, and column G (Table 1) shows another monoterpene, 3-carene (10.2%), not reported by other methods.

4 Oil stability

The stability of cold pressed peel oil was studied during storage under different temperatures (−21.0°C, 5°C, and 20°C). Major qualitative and quantitative changes were found after 12 months at 20°C. Monoterpene hydrocarbons were largely affected, and showed a decrease from 93.3% to 37.7%, especially limonene, γ-terpinene, myrcene, and α-pinene. On the other hand, *p*-cymene and the monoterpene alcohols increased from 2.1% to 13.2%. The main sesquiterpene hydrocarbon, bicyclogermacrene, was depleted. Sesquiterpene alcohols were formed as dominant artifacts (36.1%). (−)-Spathulenol was the main artifact detected. The sensitivity of the oil at 20°C is reflected from the instability of the oil individuals, especially limonene, and this can be directly attributed to the destructive oxidation process of the oil. The sesquiterpene spathulenol as a major artifact product in yuzu could be used to monitor the quality and freshness of yuzu flavor during storage (Njoroge, Ukeda, & Sawamura, 1996).

Another study, published in 2010, supported the instability of the oil under storage conditions. The authors studied the effect of common antioxidants such as α-tocopherol, β-carotene, and BHT, which showed protective effects and decreased the transformation/degradation of the volatile oil constituents. The authors proposed the addition of an antioxidant to maintain the high quality of the oil during storage (Kashiwagi et al., 2010).

5 Applications/uses

Yuzu is a typical sour citrus fruit; it has a pleasant fragrance with a special overtone, and is widely used in Japan and Korea. The fruit is rarely consumed directly as food, due to its strong, tart flavor. However, yuzu juice is part of many food recipes such as salads. In China, yuzu has been reported as a stomachic and sweating medicine. A hot yuzu bath is recommended to improve blood circulation and prevent colds. Yuzu showed antiinflammatory (Hirota et al., 2010), antioxidant (Yoo, Lee, Park, Lee, & Hwang, 2004), antiaging (Adhikari, Panthi, Pangeni, Kim, & Park, 2017; Shim, Chae, & Cho, 2019), anticancer (Kim et al., 2014), antidiabetic (Kim et al., 2013), and antiobesity (Zang, Maddison, & Chen, 2018) activities.

5.1 Pharmacological properties

The precursor of the carcinogens *N*-nitroso compounds such as nitrates and secondary amines occur in different food products such as fish, meat, and vegetables. Consuming excess of such products such as vegetables may lead to stomach cancer. Strawberries, kale juice, and garlic juice are effective inhibitors of *N*-nitrosodimethylamine (NDMA) (Chung, Lee, & Sung, 2002). Yuzu cold pressed oil showed inhibitory activity against the formation of NDMA in a given reaction medium even in the presence of vegetables and saliva. The individual common terpene hydrocarbons terpinolene, α-terpinene, myrcene, and limonene were effectively inhibiting NDMA formation >80%. Only limonene showed a low inhibitory effect, as much as 38% (Sawamura, Wu, Fujiwara, & Urushibata, 2005). Limonene from yuzu oil showed antioxidant activity to human eosinophilic cells, and it may prevent the damage from diesel exhaust particles in the lungs. Furthermore, limonene prevented eosinophil migration and could be useful in treating bronchial asthma (Hirota et al., 2010).

Lavender is well-known for the treatment of anxiety symptoms; yuzu showed biological and chemical similarities with lavender. It contains linalool and pinene similar to lavender. Additionally, limonene, the major constituent of the oil, is reported to inhibit monoamine-induced elevation of psychological stress (Lee, Wu, Tsang, Leung, & Cheung, 2011). A controlled clinical trial on the effect of cold pressed yuzu oil on the maternal anxiety of 60 mothers showed a decrease of maternal anxiety after inhalation of the oil (Ueki et al., 2014).

The improving of women's anxiety due to premenstrual symptoms using yuzu essential oil was also reported by Matsumoto, Kimura, and Hayashi (2016a, 2016b). This clinical study was conducted on 17 women in their 20s with premenstrual symptoms. The oil significantly decreased premenstrual symptoms (tension-anxiety, anger-hostility, and fatigue). These premenstrual psychoneurophysiologic effects of yuzu fragrance are similar to those of lavender oil (Matsumoto et al., 2016a).

5.2 Cosmetic-related studies

Yuzu cold pressed oil showed strong inhibitory effects against linoleic acid oxidation. Tocopherols α- and γ- were detected in the oil; however, the authors attributed the antioxidant activity to the terpenoidal fraction of the oil. For example, α-pinene, β-pinene, and γ-terpinene limonene, the major constituents of yuzu oils in addition to octanal, along with decanal showed more potent antioxidant activity than tocopherols (Song, Ukeda, & Sawamura, 2001).

Singlet oxygen species (SOS) are one of the key problems causing acceleration of the aging process among humans. Antioxidants are believed to be an effective inhibitor of SOS. Aromatherapy is focused on well-being and skin conditioning, among other aspects. Yuzu oil was found to enhance the production of SOS at low concentrations (0.1%–0.6%), and scavenge radicals at higher concentrations (0.8%–2.0%). Furthermore, limonene, the major constituent of citrus oils, enhanced SOS production. On the other hand, yuzu oil showed cytotoxicity against normal human dermal fibroblasts; the authors attributed such an effect to the high content of limonene and proposed blending the oil with others before using it (Ao, Satoh, Shibano, Kawahito, & Shioda, 2008).

5.3 Aromatherapy uses

The oil has been traditionally used to promote mind and body health in Japan. The hot yuzu winter bath (*yuzu-yu*) is an old custom and is intended to relieve stress and create a feeling of well-being; it is also thought to fight against colds, and treat arthritis, rheumatism, and roughness of the skin (Sawamura, Kashiwagi, & Tanabe, 2012). The oil has been used to promote psychosomatic health in Japan. The effects of the fragrance of yuzu essential oil on autonomic nervous system activity, which plays a crucial role in the integrity of the mind-body connection, have been investigated. Twenty-one women in their 20s participated in a single-blind randomized controlled crossover study. The results showed the potential of the oil to alleviate negative emotional stress, which would contribute to the improvement of parasympathetic nervous system activity (Matsumoto et al., 2016b).

Another study supported the psychological effects of yuzu oil. The authors measured the salivary chromogranin A (CgA) level study as a noninvasive biomarker to evaluate psychophysiological changes or improvement resulted from health care modalities, including aromatherapy. The oil could alleviate negative emotional stress, which (at least in part) would help suppress sympathetic nervous system activity (Matsumoto, Asakura, & Hayashi, 2014). The effect of yuzu inhalation on anesthetized rats showed stimulation of olfactory sense and decreased activities of sympathetic nerves innervating brown and white adipose tissues (Kumagai et al., 2009).

A postoperational clinical study was conducted by Sawamura and colleagues on the effects of yuzu oil. Members of the inpatient group who inhaled the oil fell asleep more easily on the night before an operation, and woke up feeling better, compared with those in the control group (Sawamura et al., 2009). On the other hand, a clinical study on the psychological effects of yuzu essential oil on presymptomatic depression patients (4 men and 4 women; mean age 22.3 years) as an olfactory stimulus was reported by Konno (2009). Inhalation of yuzu oil for 7 min significantly reduced fatigue and feelings of anger, while vigor recovered, but the changes are not statistically significant.

6 Safety

The majority of *Citrus* essential oils have good safety margins. They are not hazardous in pregnancy, and do not alter maternal reproductive outcomes. However, there are no in vivo cytotoxicity studies available so far on yuzu oil. It is currently used safely in aromatherapy, perfumes, and the food industry. According to Dosoky and Setzer (2018), the coumarins and/or furanocoumarins are undetectable and the oil is irritant-free. Additionally, the major constituents like (*d*) limonene (acute dermal LD_{50} >5 g/kg in rabbits, and acute oral LD_{50} >5 g/kg in rats), linalool (5.61 g/kg and 2.79 g/kg), α-pinene (>3 g/kg and 4.3 g/kg), and γ-terpinene (>5 g/kg and 3.65 g/kg) are safe as well.

7 Conclusion

Yuzu fruits have a unique flavor and interesting oil structure, where many minor sulfur-containing compounds contribute to the oil's amazing aroma. However, limited bioevaluation studies have been reported, especially in terms of biosafety. The available data indicate a high safety margin of the oil and its individual constituents. Quantitative (and sometimes qualitative) variations are present among the oil collections, and continuous assessment is highly recommended. The oil currently has a market limited to Japan and Korea, but is expected to expand to a global level due to the high demand of the perfume industry.

References

Adhikari, D., Panthi, V. K., Pangeni, R., Kim, H. J., & Park, J. W. (2017). Preparation, characterization, and biological activities of topical anti-aging ingredients in a *Citrus junos* callus extract. *Molecules*, *22*(12), 2198. https://doi.org/10.3390/molecules22122198.

Ao, Y., Satoh, K., Shibano, K., Kawahito, Y., & Shioda, S. (2008). Singlet oxygen scavenging activity and cytotoxicity of essential oils from rutaceae. *Journal of Clinical Biochemistry and Nutrition*, *43*(1), 6–12. https://doi.org/10.3164/jcbn.2008037.

Chung, M. J., Lee, S. H., & Sung, N. J. (2002). Inhibitory effect of whole strawberries, garlic juice or kale juice on endogenous formation of *N*-nitrosodimethylamine in humans. *Cancer Letters*, *182*, 1–10.

Dosoky, N. S., & Setzer, W. N. (2018). Biological activities and safety of *Citrus* spp. essential oils. *International Journal of Molecular Sciences*, *19*(7), 1966. https://doi.org/10.3390/ijms19071966.

Hirota, R., Roger, N. N., Nakamura, H., Song, H. S., Sawamura, M., & Suganuma, N. (2010). Anti-inflammatory effects of limonene from yuzu (*Citrus junos* Tanaka) essential oil on eosinophils. *Journal of Food Science*, *75*, H87–H92. https://doi.org/10.1111/j.1750-3841.2010.01541.x.

Hoshino, M., Tanaka, M., Sasaki, M., & Goto, M. (2006). https://www.researchgate.net/publication/260169857_Extraction_of_Essential_Oil_from_Citrus_junos_Peel_using_Supercritical_Carbon_Dioxide (Accessed September 2019).

Kashiwagi, T., Lan-Phi, N. T., & Sawamura, M. (2010). Compositional changes in yuzu (*Citrus junos*) steam-distilled oil and effects of antioxidants on oil quality during storage. *Food Science and Technology Research*, *16*(1), 51–58.

Kim, S. H., Hur, H. J., Yang, H. J., Kim, H. J., Kim, M. J., Park, J. H., et al. (2013). *Citrus junos* Tanaka peel extract exerts antidiabetic effects via AMPK and PPAR-gamma both in vitro and in vivo in mice fed a high-fat diet. *Evidence-Based Complementary and Alternative Medicine*, *2013*, 921012.

Kim, S. H., Shin, E. J., Hur, H. J., Park, J. H., Sung, M. J., Kwon, D. Y., et al. (2014). *Citrus junos* Tanaka peel extract attenuates experimental colitis and inhibits tumour growth in a mouse xenograft model. *Journal of Functional Foods*, *8*, 301–308.

Konno, N. (2009). Mental caring for pre symptomatic depression by aroma-effects of lemon and yuzu essential oil. *Aroma Research*, *10*(3), 260–263.

Kumagai, C., Horii, Y., Shen, J., Tanida, M., Niijima, A., Okada, Y., et al. (2009). Effects of olfactory stimulation with yuzu (*Citrus junos*) peel oil on autonomic nerves, lipolysis and body temperature. *Aroma Research*, *38*, 156–161.

Kuraya, E., Touyama, A., Nakada, S., Higa, O., & Itoh, S. (2017). Underwater shockwave pretreatment process to improve the scent of extracted *Citrus junos* Tanaka (yuzu) juice. *International Journal of Food Science*. *2017*, 2375181. https://doi.org/10.1155/2017/2375181.

Lan-Phi, N. T., & Sawamura, M. (2008). Characteristic aroma composition profile of mature stage *Citrus junos* (yuzu) peel oil from different origins. *Food Science and Technology Research*, *14*(4), 359–366.

Lan-Phi, N. T., Shimamura, T., Ukeda, H., & Sawamura, M. (2009). Chemical and aroma profiles of yuzu (*Citrus junos*) peel oils of different cultivars. *Food Chemistry*, *115*, 1042–1047.

Lee, Y. L., Wu, Y., Tsang, H. W. H., Leung, A. Y., & Cheung, W. M. (2011). A systematic review on the anxiolytic effects of aromatherapy in people with anxiety symptoms. *The Journal of Alternative and Complementary Medicine*, *17*(2), 101–108.

Liu, Y., Liu, Z., Wang, C., Zha, Q., Lu, C., Song, Z., et al. (2014). Study on essential oils from four species of Zhishi with gas chromatography-mass spectrometry. *Chemistry Central Journal*, *8*(1), 22. https://doi.org/10.1186/1752-153X-8-22.

Matsumoto, T., Asakura, H., & Hayashi, T. (2014). Effects of olfactory stimulation from the fragrance of the Japanese citrus fruit yuzu (*Citrus junos* Sieb. ex Tanaka) on mood states and salivary chromogranin A as an endocrinologic stress marker. *Journal of Alternative and Complementary Medicine*, *20*(6), 500–506. https://doi.org/10.1089/acm.2013.0425.

Matsumoto, T., Kimura, T., & Hayashi, T. (2016a). Does Japanese *Citrus* fruit yuzu (*Citrus junos* Sieb. ex Tanaka) fragrance have lavender-like therapeutic effects that alleviate premenstrual emotional symptoms? A single-blind randomized crossover study. *The Journal of Alternative and Complementary Medicine*, *23*, 6. https://doi.org/10.1089/acm.2016.0328.

Matsumoto, T., Kimura, T., & Hayashi, T. (2016b). Aromatic effects of a Japanese citrus fruit-yuzu (*Citrus junos* Sieb. ex Tanaka)-on psychoemotional states and autonomic nervous system activity during the menstrual cycle: A single-blind randomized controlled crossover study. *Biopsychosocial Medicine*, *10*, 11. https://doi.org/10.1186/s13030-016-0063-7.

Miyazawa, N., Tomita, N., Kurobayashi, Y., Nakanishi, A., Ohkubo, Y., Maeda, T., et al. (2009). Novel character impact compounds in yuzu (*Citrus junos* Sieb. ex Tanaka) peel oil. *Journal of Agricultural and Food Chemistry*, *57*(5), 1990–1996. https://doi.org/10.1021/jf803257x.

Njoroge, S. M., Ukeda, H., Kusunose, H., & Sawamura, M. (1994). Volatile components of Japanese yuzu and lemon oils. *Flavour and Fragrance Journal*, *9*, 159–166.

Njoroge, S. M., Ukeda, H., & Sawamura, M. (1996). Changes in the volatile composition of yuzu (*Citrus junos* Tanaka) cold-pressed oil during storage. *Journal of Agricultural and Food Chemistry*, *44*, 550–556.

Rahman, M. M., Nito, N., & Isshiki, S. (2001). Cultivar identification of 'yuzu' (*Citrus junos* Sieb. ex Tanaka) and related acid citrus by leaf isozymes. *Scientia Horticulturae*, *87*(3), 191–198.

Roy, B. C., Hoshino, M., Ueno, H., Sasaki, M., & Goto, M. (2007). Supercritical carbon dioxide extraction of the volatiles from the peel of Japanese citrus fruits. *Journal of Essential Oil Research*, *19*(1), 78–84. https://doi.org/10.1080/10412905.2007.9699234.

Sawamura, M. (2005). *Citrus Junos* Sieb. Ex Tanaka (yuzu) fruit. In *Fruits, growth, nutrition and quality* (pp. 1–24). Helsinki, Finland: World Food Ltd.

Sawamura, M., Fukata, J., Kumagai, C., Lan-Phi, N., Mizushima, N., Hori, N., et al. (2009). Functional activities of Japanese yuzu essential oil. *Japanese Journal of Aromatherapy*, *9*, 55–65.

Sawamura, M., Kashiwagi, T., & Tanabe, K. (2012). Eco-conscious technology for essential oil extraction from post-squeezing citrus waste-effective utilization of Japanese yuzu fruits. *Journal of Japan Association on Odor Environment*, *43*, 102–111. https://doi.org/10.2171/jao.43.102.

Sawamura, M., & Kuriyama, T. (1988). Quantitative determination of volatile constituents in the pummelo (*Citrus grandis* Osbeck forma Tosa-buntan). *Journal of Agriculture and Food Chemistry*, *36*(3), 567–569. https://doi.org/10.1021/jf00081a040.

Sawamura, M., Suzuki, S., Asano, K., Sato, M., Kinoshita-Kitagawa, A., Sasaki, K., et al. (2018). Determination of furocoumarins concentrations in several essential oils and yuzu seed oils. *Japan Journal of Aromatherapy*, *19*(1), 2018.

Sawamura, M., Wu, Y., Fujiwara, C., & Urushibata, M. (2005). Inhibitory effect of yuzu essential oil on the formation of *N*-nitrosodimethylamine in vegetables. *Journal of Agriculture and Food Chemistry*, *53*, 4281–4287.

Shim, J. H., Chae, J. I., & Cho, S. S. (2019). Identification and extraction optimization of active constituents in *Citrus junos* Seib ex TANAKA peel and its biological evaluation. *Molecules*, *24*(4), 680.

Shinoda, N., Shiga, M., & Nishimura, K. (1970). Constituents of yuzu (*Citrus junos*) oil. *Agricultural and Biological Chemistry*, *34*, 234–242.

Song, H. S., Sawamura, M., Ito, T., Kawashimo, K., & Ukeda, H. (2000). Quantitative determination and characteristic flavour of *Citrus junos* (yuzu) peel oil. *Flavour and Fragrance Journal*, *15*, 245–250.

Song, H.-S., Ukeda, H., & Sawamura, M. (2001). Antioxidative activities of citrus peel essential oils and their components against linoleic acid oxidation. *Food Science and Technology Research*, *7*(1), 50–56.

Taninaka, T., Otoi, N., & Morimoto, J. (1981). Acid citrus cultivars related to the yuzu (*Citrus junos* Sieb. ex Tanaka) in Japan. *Proceedings—International Society of Citriculture, 73–76*.

Ueki, S., Niinomi, K., Takashima, Y., Kimura, R., Komai, K., Murakami, K., et al. (2014). Effectiveness of aromatherapy in decreasing maternal anxiety for a sick child undergoing infusion in a paediatric clinic. *Complementary Therapies in Medicine*, *22*(6), 1019–1026. https://doi.org/10.1016/j.ctim.2014.09.004.

Yoo, K. M., Lee, K. W., Park, J. B., Lee, H. J., & Hwang, I. K. (2004). Variation in major antioxidants and total antioxidant activity of yuzu (*Citrus junos* Sieb ex Tanaka) during maturation and between cultivars. *Journal of Agricultural and Food Chemistry*, *52*, 5907–5913.

Yukawa, C., Osaki, K., & Iwabuchi, H. (1995). Study on the volatile components of yuzu (*Citrus junos* Sieb ex Tanaka). *Japanese Journal of Food Chemistry*, *1*(1), 46–49. https://doi.org/10.18891/jjfcs.1.1_46.

Zang, L., Maddison, L. A., & Chen, W. (2018). Zebrafish as a model for obesity and diabetes. *Frontiers in Cell and Developmental Biology*, *6*, 91. https://doi.org/10.3389/fcell.2018.00091.

Chapter 64

Cold pressed thyme (*Thymus vulgaris*) oil

Mohamed Fawzy Ramadan
Agricultural Biochemistry Department, Faculty of Agriculture, Zagazig University, Zagazig, Egypt

List of abbreviations

CERs cerebrosides
CPTO cold pressed thyme oil
DG diacylglycerol
DGDs digalactosyldiglycerides
DPPH• 2,2-diphenyl-1-picrylhydrazyl
ESG esterified sterylglycosides
FFA free fatty acid
HDL-C high density lipoprotein-cholesterol
IT induction time
LDL-C low density lipoprotein-cholesterol
MG monoacylglycerol
MGDs monogalactosyldiglycerides
MIC minimum inhibitory concentration
RSA radical scavenging activity
SFA saturated fatty acid
SQD sulfoquinovosyldiacylglycerol
STEs esterified sterol
TAG triacylglycerol
TC total cholesterol
TEO thyme essential oil
TPCs total phenolic compounds

1 Introduction

The genus *Thymus*, distributed over Asia, North Africa, Europe, and the Canary Islands, is an important and taxonomically complex genera within the Lamiaceae family (Casiglia, Bruno, Scandolera, Senatore, & Senatore, 2015; Golmakani & Rezaei, 2008). The genus *Thymus* includes aromatic plants, which produce high amounts of bioactive compounds (Pavel, Ristić, & Stević, 2010). The plants have been utilized from the ancient time due to their health-promoting traits, which are linked with their bioactive phytochemicals. Among those species, *T. vulgaris* (common name thyme) is a culinary herb with a history of applications in several food and nonfood products (Salehi et al., 2019, 2018).

Thyme is commonly used in foodstuffs as a spice. *Thymus vulgaris* essential oil extracted from leaves could be used as a spice aroma additive in foodstuffs, cosmetics, and pharmaceuticals (Assiri, Elbanna, Abulreesh, & Ramadan, 2016; Chizzola, Michitsch, & Franz, 2008; Lee, Umano, Shibamoto, & Lee, 2005; Zarzuelo & Crespo, 2002). Numerous bioactive compounds were reported in thyme including essential oil constituents, phenolics, and flavonoids (Miura, Kikuzaki, & Nakatani, 2002; Tomaino et al., 2005; Vila, 2002; Viuda-Martos et al., 2010, 2011). Thymol, α-terpinolene, and *p*-cymene were the major compounds found in *T. vulgaris* essential oil (Viuda-Martos et al., 2011).

Cold pressed oils rich in natural bioactive are nationally and internationally marketed. However, information on their composition and biological actions has not been fully studied. Cold press extraction is considered an important oil extraction method due to consumers' desire for safe and natural products (El Makawya, Ibrahimb, Mabrouka, Ahmedc, & Ramadan, 2019; El-Hadary & Ramadan, 2016; Ibrahim, Attia, Maklad, Ahmed, & Ramadan, 2017; Kiralan, Çalik, Kiralan, & Ramadan, 2018; Kiralan & Ramadan, 2016; Mahgoub, Ramadan, & El-Zahar, 2013; Ramadan, 2013;

Cold Pressed Oils. https://doi.org/10.1016/B978-0-12-818188-1.00064-5
Copyright © 2020 Elsevier Inc. All rights reserved.

Ramadan, Asker, & Tadros, 2012). This chapter covers the lipids and phenolics profile of cold pressed thyme (*Thymus vulgaris*) oil (CPTO), and describes its functional properties as a healthy agent as well as the recent food and nonfood uses of CPTO.

2 Extraction and processing of CPTO

Several reports were published on the extraction of *T. vulgaris* essential oils. In a recent article, Gedikoglu, Sokmen, and Civit (2019) obtained *T. vulgaris* oils by hydrodistillation and microwave-assisted extraction (MAE) techniques. The authors determined the impact of the extraction technique on the oil yield (%), chemical composition, ferric reducing antioxidant power (FRAP), radical scavenging activity (IC_{50}), and antimicrobial traits of oils. In addition, they determined the impact of methanol and ethanol on the extraction by means of the total phenolics, phenolic acids profile, total flavonoids, IC_{50}, and FRAP. The amount of thymol and *p*-cymene was found to be highest in thyme, when using hydrodistillation and MAE, respectively. Methanol extract of thyme had high phenolic composition in comparison with ethanol extract. In addition, the essential oil of thyme showed antimicrobial action against *St. aureus*.

3 Acyl lipids and fatty acid profile of CPTO

3.1 Fatty acid profile of CPTO

Fatty acid profiles of CPTO were studied by Assiri et al. (2016). Nine fatty acid methyl esters (FAMEs) were identified in CPTO, and linoleic and oleic acids were the major FAMEs. C10:0, C12:0, C14:0, and C16:1 were detected in traces. CPTO contains high amounts of monounsaturated fatty acids (MUFAs, 40% FAMEs) which is comparable to those in blueberry, hemp, onion, cranberry, and milk thistle cold pressed oils, but less than the 81%–82% in parsley and carrot cold pressed oils (Parker, Adams, Zhou, Harris, & Yu, 2003; Parry et al., 2005). CPTO is rich in polyunsaturated fatty acids (PUFAs) which accounted for 43% of the FAMEs. PUFA content is less than that in cold pressed oils from onion (c.65%), milk thistle (c.61%), cranberry (67.6%), and blueberry (69%). In CPTO, palmitic and stearic acids are the main saturated fatty acids (SFAs). CPTO contains c.16% SFAs, which is less than in cardamom cold pressed oil (30.8%) and comparable to that of 14% and 16% found in milk thistle and pumpkin cold pressed oils, respectively (Parry et al., 2006). SFA levels were higher than those of 7.5%–9.7% in hemp, mullein, parsley, onion, and cranberry cold pressed oils (Assiri et al., 2016; Parker et al., 2003).

3.2 Lipid classes of CPTO

In the CPTO, the amount of neutral lipids (NLs) was the highest (90%), followed by glycolipids (GLs, 0.78%) and phospholipids (PLs, 0.53%). NLs contained, in decreasing order, triacylglycerol (TAG, c.95%), free fatty acids (FFAs, c.2%), diacylglycerol (DG), esterified sterols (STEs), and monoacylglycerol (MG) (Assiri et al., 2016). DG and STEs were detected in lesser amounts. GL classes in CPTO were SQD, DGD, CER, SG, MGD, and ESG. The major PL classes in CPTO were PC (c.50%) followed by PE (c.25%), PI, and PS. Fatty acids of different lipid classes (NLs, GLs, and PLs) did not differ from each other, whereas linoleic and oleic acids were the major fatty acids. The ratio of unsaturated fatty acids to SFAs was not higher in NLs than in GLs and PLs. CPTO contains high levels of MUFAs and PUFAs, which are reported to reduce LDL-cholesterol and retain HDL-cholesterol (Assiri et al., 2016; Ramadan, Kinni, Seshagiri, & Mörsel, 2010). Fatty acid composition of CPTO reveals that CPTO is a rich source of essential fatty acids.

4 Minor bioactive lipids in CPTO

Fatty acid profiles of CPTO were studied by Assiri et al. (2016). CPTO contained high amount of unsaponifiables (c.27 g/kg oil). Amounts (mg/kg oil) of γ-, α-, β-, and δ-tocopherols in CPTO were 4800, 718, 23, and 26, respectively. On the other side, amounts (mg/kg oil) of α-, β-, γ-, and δ-tocotrienols were 2146, 63, 62, and 134, respectively. α- and γ-Tocopherols were considered to be the main tocols in edible oils. γ-Tocopherol was detected in high amounts in rapeseed, linseed, camelina, and corn cold pressed oils (Schwartz, Ollilainen, Piironen, & Lampi, 2008). Amounts of tocols reported in CPTO might be associated with the stability of CPTO against oxidation. Moreover, the total phenolic content of CPTO was greater than that of 1.7–2.0 mg GAE/g oil for blueberry, boysenberry, and red raspberry cold pressed oils, and that of 1.7–3.5 mg GAE/g oil for parsley, cardamom, onion, and mullein cold pressed oils (Assiri et al., 2016; Parry et al., 2005, 2006). On the other side, tocols levels in the oils might have a great effect on their antiradical and antioxidant potential. Increasing ring methyl substitution led to an increase of antiradical action against the DPPH˙ radicals, and to a decrease in oxygen radical absorbance effect (Assiri et al., 2016; Müller, Theile, & Böhm, 2010).

5 Health-promoting properties of CPTO and oil constituents

5.1 Antioxidant potential and RSA of CPTO

Antiradical properties of the CPTO against DPPH˙ and galvinoxyl free radicals were compared with extra virgin olive oil. CPTO exhibited stronger RSA than extra virgin olive oil. After incubation for 30 min with DPPH˙ radicals, CPTO deactivated c.65% of DPPH˙ free radicals, while extra virgin olive oil quenched only 45% (Fig. 1). An electron spin resonance test exhibited the same pattern, whereas CPTO deactivated c.55% galvinoxyl radical and extra virgin olive oil quenched 38%. The strong RSA of CPTO might be due to the variations in the profiles of unsaponifiables, the differences in the structural traits of bioactive compounds, the synergism of bioactive compounds, and the different kinetics of antioxidants.

5.2 Antimicrobial action of CPTO

Microbial and fungal spoilage is an important factor influencing food and nonfood products. Antimicrobial action of CPTO was tested against yeast and bacterial strains selected upon their relevance as food spoilage or human pathogenic agents. CPTO inhibited the growth of all tested microorganisms. CPTO showed different degrees of antimicrobial action against different pathogens bacteria and fungi, whereas the highest antimicrobial activity was measured against the dermatophytic fungi and yeasts including *T. rubrum*, *T. mentagrophytes*, and *C. albicans* followed by the spoilage fungi *A. flavus* (MLC between 80 and 320 μg/mL). CPTO had a strong antifungal activity against dermatophytic fungi (*T. rubrum* and *T. mentagrophytes*), while CPTO showed very low MLC (between 80 and 160 μg/mL). CPTO showed activity against *A. flavus*,

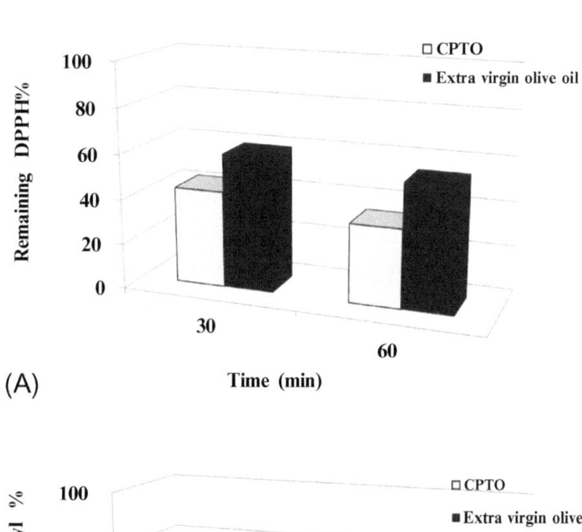

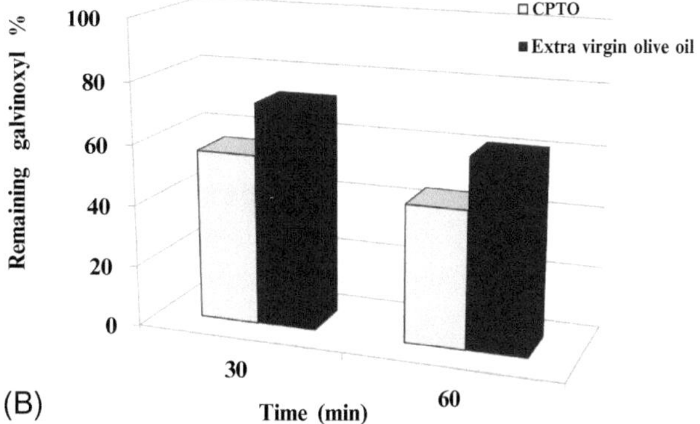

FIG. 1 Antiradical action of CPTO and extra virgin olive oil on (A) DPPH˙ and on (B) galvinoxyl radicals.

E. coli, *S. aureus*, and *L. monocytogenes*. The lowest antimicrobial activity was reported for *S. enteritidis* (Assiri et al., 2016). In a study carried out by Ghabraie, Vu, Tata, Salmieri, and Lacroix (2016), red thyme exhibited substantial antimicrobial action against *E. coli*, *S. aureus*, and *L. monocytogenes*. In addition, Ivanovic, Misic, Zizovic, and Ristic (2012) mentioned a strong antimicrobial activity of thyme against food-associated bacteria (i.e., *Salmonella*, *E. coli*, and *Bacillus*).

6 Edible and nonedible applications of CPTO

Encapsulated thyme oil was tested as an antifungal agent in cake (Gonçalves et al., 2017). Enrichment with thyme oil extended the shelf life of cakes by about 30days. Encapsulation could protect the oil against volatilization or oxidation. It could also minimize the loss of bioactive compounds and guarantee their effect in the food matrix. Therefore, encapsulation of oil could control the decay of processed vegetables and extend the shelf life of foodstuffs.

Application of *Thymus* derivatives in the meat products as preservative have been reported. Abed (2011) reported the antimicrobial action of *T. capitata* essential oil on *L. monocytogenes* inoculated minced beef. Beef treated with 1% of TEO reduced the bacteria population. Jayari et al. (2018) tested the antibacterial action of *T. capitatus* and *T. algeriensis* essential oils against *S. aureus*, *E. coli*, *S. typhimurium*, and *P. aeruginosa* in minced beef. *P. aeruginosa* was the most resistant pathogen on the minced beef. Boskovic et al. (2019) studied the antioxidant stability of minced pork treated with TEO. Minced pork containing TEO packaged under vacuum was evaluated during 15days of cold storage. TEO increased the stability of minced pork with respect to lipid oxidation, wherein the antioxidative effect was concentration-dependent.

Thyme oil is a potential fumigant agent for controlling postharvest diseases in fresh vegetables and fruits. The oil was applied in table grapes to test their quality properties during storage (Geransayeh, Mostofi, Abdossi, & Nejatian, 2012). A difference was found in the pH, decay percentage, vitamin C, weight loss, and the storage quality of the treated samples. The chroma reduced during the storage, but a decrease in the chroma rate was noted. In addition, the effect of thyme oil on anthracnose during the postharvest of avocados was observed (Bill, Korsten, Remize, Glowacz, & Sivakumar, 2017; Sellamuthu, Sivakumar, Soundy, & Korsten, 2013). The antifungal action of thyme oil was revealed against the anthracnose caused by *C. musae* in bananas (Vilaplana, Pazmiño, & Valencia-Chamorro, 2018). Feng, Chen, Zheng, and Liu (2011) evaluated the fumigant effect of thyme oil on artificially *Alternaria alternate* inoculated tomatoes. Thyme oil reduced *A. alternate* population on tomato, wherein vapor phase application was more effective than the direct contact application. In addition, Viacava, Ayala-Zavala, González-Aguilar, and Ansorena (2018) studied the impact of thyme oil (free and encapsulated) on the quality characteristics of processed Romaine lettuce during the cold storage. The functional and organoleptic traits of Romaine lettuce treated with encapsulated oil were much better than those used in free oil and the control sample.

Enrichment of vegetable and edible oil with natural phytochemicals could be an important way to improve the stability of oils against oxidation (Ramadan, 2008, 2012). In this context, Karoui, Dhifi, Ben Jemia, and Marzouk (2011) investigated the heat stability of corn oil enriched with thyme oil. Thyme oil enhanced the thermal stability of corn oil after heating. Assiri et al. (2016) studied the effect of blending CPTO with sunflower oil on the oxidative stability of oil blends. The induction time (IT) of refined sunflower oil was c.3h. IT for the CPTO:sunflower oil blend (1:9, w/w) was 6.5h, wherein a 2:8 (w/w) blend recorded a higher value (9h). Dagdemir, Cakmakci, and Gundogdu (2009) studied the antioxidant potential of *T. haussknechtii* essential oil on the stability of milk butter. TEO exhibited a high antioxidant effect when applied in milk butter at 0.2%. The high antioxidant effect of TEO was attributed to the high thymol content. In another study, Boroujeni and Hojjatoleslamy (2018) fried potato chips by sunflower oil enriched with *T. carmanicus* essential oil. The addition of TEO to sunflower oil improved the physicochemical traits of chips, decreased the peroxide and acid values of chips, and inhibited the increase of thiobarbituric acid value of chips.

Attia, Bakhashwain, and Bertu (2017) utilized *T. vulgaris* oil, in comparison to mannanoligosaccharides, as a growth enhancer for broilers raised under a hot climate. The oil groups had high plasma protein, and the oil increased plasma globulin, but decreased plasma albumin/globulin ratio. *T. vulgaris* oil groups exhibited high antibody titer to infectious bursa disease. The authors concluded that *T. vulgaris* oil (1 g/kg diet) might be utilized as a growth promoter for broilers in hot regions.

7 Conclusion

Few studies have been reported on CPTO. The oil is a rich source of essential fatty acids, phenolics, and tocols. CPTO exhibits strong antimicrobial potential and could be used in cosmetics, foodstuffs, and pharmaceuticals to prevent microbial growth. CPTO, as a rich source of natural antioxidants, might be the basis of the formulation of new pharmacological active ingredients to be used in functional foods and nutraceuticals as a healthy agents.

References

Abed, R. M. (2011). Cytotoxic, cytogenetics and immunomodulatory effects of thymol from *Thymus vulgaris* on cancer and normal cell lines in vitro and in vivo. *Al-Mustansiriyah Journal of Science*, *22*, 41–53.

Assiri, A. M., Elbanna, K., Abulreesh, H. H., & Ramadan, M. F. (2016). Bioactive compounds of cold-pressed thyme (*Thymus vulgaris*) oil with antioxidant and antimicrobial properties. *Journal of Oleo Science*, *65*(8), 629–640.

Attia, Y. A., Bakhashwain, A. A., & Bertu, N. K. (2017). Thyme oil (*Thyme vulgaris* L.) as a natural growth promoter for broiler chickens reared under hot climate. *Italian Journal of Animal Science*, *16*(2), 275–282. https://doi.org/10.1080/1828051X.2016.1245594.

Bill, M., Korsten, L., Remize, F., Glowacz, M., & Sivakumar, D. (2017). Effect of thyme oil vapours exposure on phenylalanine ammonia-lyase (PAL) and lipoxygenase (LOX) genes expression, and control of anthracnose in 'Hass' and 'Ryan' avocado fruit. *Scientia Horticulturae*, *224*, 232–237.

Boroujeni, L. S., & Hojjatoleslamy, M. (2018). Using *Thymus carmanicus* and *Myrtus communis* essential oils to enhance the physicochemical properties of potato chips. *Food Science and Nutrition*, *6*, 1006–1014.

Boskovic, M., Glisic, M., Djordjevic, J., Starcevic, M., Glamoclija, N., Djordjevic, V., et al. (2019). Antioxidative activity of thyme (*Thymus vulgaris*) and oregano (*Origanum vulgare*) essential oils and their effect on oxidative stability of minced pork packaged under vacuum and modified atmosphere. *Journal of Food Science*, *84*(9), 2467–2474.

Casiglia, S., Bruno, M., Scandolera, E., Senatore, F., & Senatore, F. (2015). Influence of harvesting time on composition of the essential oil of *Thymus capitatus* (L.) from *Thymus vulgaris* on cancer and normal cell lines in vitro and in vivo. *Al-Mustansiriyah Journal of Science*, *22*(5), 41–53.

Chizzola, R., Michitsch, H., & Franz, C. (2008). Antioxidative properties of *Thymus vulgaris* leaves: Comparison of different extracts and essential oil chemotypes. *Journal of Agricultural and Food Chemistry*, *56*, 6897–6904.

Dagdemir, E., Cakmakci, S., & Gundogdu, E. (2009). Effect of *Thymus haussknechtii* and *Origanum acutidens* essential oils on the stability of cow milk butter. *European Journal of Lipid Science and Technology*, *111*(11), 1118–1123.

El Makawya, A. I., Ibrahimb, F. M., Mabrouka, D. M., Ahmedc, K. A., & Ramadan, M. F. (2019). Effect of antiepileptic drug (Topiramate) and cold pressed ginger oil on testicular genes expression, sexual hormones and histopathological alterations in mice. *Biomedicine & Pharmacotherapy*, *110*, 409–419. https://doi.org/10.1016/j.biopha.2018.11.146.

El-Hadary, A. E., & Ramadan, M. F. (2016). Potential protective effect of cold-pressed *Coriandrum sativum* oil against carbon tetrachloride-induced hepatotoxicity in rats. *Journal of Food Biochemistry*, *40*, 190–200.

Feng, W., Chen, J., Zheng, X., & Liu, Q. (2011). Thyme oil to control *Alternaria alternata* in vitro and in vivo as fumigant and contact treatments. *Food Control*, *22*(1), 78–81.

Gedikoglu, A., Sokmen, M., & Civit, A. (2019). Evaluation of *Thymus vulgaris* and *Thymbra spicata* essential oils and plant extracts for chemical composition, antioxidant, and antimicrobial properties. *Food Science & Nutrition*, *7*(5), 1704–1714.

Geransayeh, M., Mostofi, Y., Abdossi, V., & Nejatian, M. A. (2012). Use of *Thymus vulgaris* essential oil to improve the safety and shelf-life of Iranian table grape. *Journal of Essential Oil-Bearing Plants*, *15*(1), 164–173.

Ghabraie, M., Vu, K. D., Tata, L., Salmieri, S., & Lacroix, M. (2016). Antimicrobial effect of essential oils in combinations against five bacteria and their effect on sensorial quality of ground meat. *LWT: Food Science and Technology*, *66*, 332–339.

Golmakani, M.-T., & Rezaei, K. (2008). Comparison of microwave-assisted hydrodistillation with the traditional hydrodistillation method in the extraction of essential oils from *Thymus vulgaris* L. *Food Chemistry*, *109*(4), 925–930.

Gonçalves, N. D., de L. Pena, F., Sartoratto, A., Derlamelina, C., Duarte, M. C. T., Antunes, A. E. C., et al. (2017). Encapsulated thyme (*Thymus vulgaris*) essential oil used as a natural preservative in bakery product. *Food Research International*, *96*, 154–160.

Ibrahim, F. M., Attia, H. N., Maklad, Y. A. A., Ahmed, K. A., & Ramadan, M. F. (2017). Biochemical characterization, anti-inflammatory properties and ulcerogenic traits of some cold pressed oils in experimental animals. *Pharmaceutical Biology*, *55*, 740–748. https://doi.org/10.1080/13880209.2016.1275705.

Ivanovic, J., Misic, D., Zizovic, I., & Ristic, M. (2012). In vitro control of multiplication of some food-associated bacteria by thyme, rosemary and sage isolates. *Food Control*, *25*, 110–116.

Jayari, A., El Abed, N., Jouini, A., Mohammed Saed Abdul-Wahab, O., Maaroufi, A., & Ben Hadj Ahmed, S. (2018). Antibacterial activity of *Thymus capitatus* and *Thymus algeriensis* essential oils against four food-borne pathogens inoculated in minced beef meat. *Journal of Food Safety*, *38*(1), 1–10.

Karoui, I. J., Dhifi, W., Ben Jemia, M., & Marzouk, B. (2011). Thermal stability of corn oil flavoured with *Thymus capitatus* under heating and deep-frying conditions. *Journal of the Science of Food and Agriculture*, *91*(5), 927–933.

Kiralan, M., Çalik, G., Kiralan, S., & Ramadan, M. F. (2018). Monitoring stability and volatile oxidation compounds of cold pressed flax seed, grape seed and black cumin seed oils upon photo-oxidation. *Journal of Food Measurement and Characterization*, *12*, 616–621.

Kiralan, M., & Ramadan, M. F. (2016). Volatile oxidation compounds and stability of safflower, sesame and canola cold-pressed oils as affected by thermal and microwave treatments. *Journal of Oleo Science*, *65*, 825–833.

Lee, S.-J., Umano, K., Shibamoto, T., & Lee, K.-G. (2005). Identification of volatile components in basil (*Ocimum basilicum* L.) and thyme leaves (*Thymus vulgaris* L.) and their antioxidant properties. *Food Chemistry*, *91*, 131–137.

Mahgoub, S. A., Ramadan, M. F., & El-Zahar, K. M. (2013). Cold pressed *Nigella sativa* oil inhibits the growth of forborne pathogens and improves the quality of Domiati cheese. *Journal of Food Safety*, *33*, 470–480.

Miura, K., Kikuzaki, H., & Nakatani, N. (2002). Antioxidant activity of chemical compounds of sage (*Salvia officinalis*) and thyme (*Thymus vulgaris*) measured by the oil stability index method. *Journal of Agricultural and Food Chemistry*, *50*, 1850–1855.

Müller, L., Theile, K., & Böhm, V. (2010). In vitro antioxidant activity of tocopherols and tocotrienols and comparison of vitamin E concentration and lipophilic antioxidant capacity in human plasma. *Molecular Nutrition & Food Research*, *54*, 731–742.

Parker, T. D., Adams, D. A., Zhou, K., Harris, M., & Yu, L. (2003). Fatty acid composition and oxidative stability of cold-pressed edible seed oils. *Journal of Food Science*, *68*, 1240–1243.

Parry, J., Su, L., Luther, M., Zhou, K., Yurawecz, M. P., Whittaker, P., et al. (2005). Fatty acid composition and antioxidant properties of cold-pressed marionberry, boysenberry, red raspberry, and blueberry seed oils. *Journal of Agricultural and Food Chemistry*, *53*, 566–573.

Parry, J., Su, L., Moore, J., Cheng, Z., Luther, M., Rao, J. N., et al. (2006). Chemical compositions, antioxidant capacities, and antiproliferative activities of selected fruit seed flours. *Journal of Agricultural and Food Chemistry*, *54*, 3773–3778.

Pavel, M., Ristić, M., & Stević, T. (2010). Essential oils of *Thymus pulegioides* and *Thymus glabrescens* from Romania: Chemical composition and antimicrobial activity. *Journal of the Serbian Chemical Society*, *75*(1), 27–34.

Ramadan, M. F. (2008). Quercetin increases antioxidant activity of soy lecithin in a triolein model system. *LWT: Food Science and Technology*, *41*, 581–587.

Ramadan, M. F. (2012). Antioxidant characteristics of phenolipids (quercetin-enriched lecithin) in lipid matrices. *Industrial Crops and Products*, *36*, 363–369.

Ramadan, M. F. (2013). Healthy blends of high linoleic sunflower oil with selected cold pressed oils: Functionality, stability and antioxidative characteristics. *Industrial Crops and Products*, *43*, 65–72.

Ramadan, M. F., Asker, M. M. S., & Tadros, M. (2012). Antiradical and antimicrobial properties of cold-pressed black cumin and cumin oils. *European Food Research and Technology*, *234*, 833–844.

Ramadan, M. F., Kinni, S. G., Seshagiri, M., & Mörsel, J.-T. (2010). Fat-soluble bioactives, fatty acid profile and radical scavenging activity of *Semecarpus anacardium* seed oil. *Journal of the American Oil Chemists' Society*, *87*, 885–894.

Salehi, B., Abu-Darwish, M. S., Tarawneh, A. H., Cabral, C., Gadetskaya, A. V., Salgueiro, L., et al. (2019). *Thymus* spp. plants-food applications and phytopharmacy properties. *Trends in Food Science & Technology*, *85*, 287–306.

Salehi, B., Mishra, A. P., Shukla, I., Sharifi-Rad, M., Contreras, M. D. M., Segura-Carretero, A., et al. (2018). Thymol, thyme, and other plant sources: Health and potential uses. *Phytotherapy Research*, *32*(9), 1688–1706.

Schwartz, H., Ollilainen, V., Piironen, V., & Lampi, A.-M. (2008). Tocopherol, tocotrienol and plant sterol contents of vegetable oils and industrial fats. *Journal of Food Composition and Analysis*, *21*, 152–161.

Sellamuthu, P. S., Sivakumar, D., Soundy, P., & Korsten, L. (2013). Essential oil vapours suppress the development of anthracnose and enhance defence related and antioxidant enzyme activities in avocado fruit. *Postharvest Biology and Technology*, *81*, 66–72.

Tomaino, A., Cimino, F., Zimbalatti, V., Venuti, V., Sulfaro, V., De Pasquale, A., et al. (2005). Influence of heating on antioxidant activity and the chemical composition of some spice essential oils. *Food Chemistry*, *89*, 549–554.

Viacava, G. E., Ayala-Zavala, J. F., González-Aguilar, G. A., & Ansorena, M. R. (2018). Effect of free and microencapsulated thyme essential oil on quality attributes of minimally processed lettuce. *Postharvest Biology and Technology*, *145*, 125–133.

Vila, R. (2002). Flavonoids and further polyphenols in the genus Thymus. In: E. Stahl-Biskup, & F. Saez (Eds.), *Thyme. The genus Thymus. Medicinal and aromatic plants industrial profiles: Vol. 24*. New York: Taylor and Francis.

Vilaplana, R., Pazmiño, L., & Valencia-Chamorro, S. (2018). Control of anthracnose, caused by *Colletotrichum musae*, on postharvest organic banana by thyme oil. *Postharvest Biology and Technology*, *138*, 56–63.

Viuda-Martos, M., El Gendy, N. G. S., Sendra, E., Fernandez-Lopez, J., ElRazik, K. A. A., El-Sayed, A., et al. (2010). Chemical composition and antioxidant and anti-listeria activities of essential oils obtained from some Egyptian plants. *Journal of Agricultural and Food Chemistry*, *58*, 9063–9070.

Viuda-Martos, M., Mohamady, M. A., Fernández-López, J., Abd ElRazik, K. A., Omer, E. A., Pérez-Alvarez, J. A., et al. (2011). In vitro antioxidant and antibacterial activities of essentials oils obtained from Egyptian aromatic plants. *Food Control*, *22*, 1715–1722.

Zarzuelo, A., & Crespo, E. (2002). The medicinal and non-medicinal uses of thyme. In: E. Stahl-Biskup, & F. Saez (Eds.), *Thyme. The genus Thymus. Medicinal and aromatic plants industrial profiles: Vol. 24*. New York: Taylor and Francis.

Chapter 65

Cold pressed apricot (*Prunus armeniaca* L.) kernel oil

Muhammad Iqbal Bhanger[a], Farooq Anwar[b], Najma Memon[c], and Rahman Qadir[b]
[a]*H.E.J. Research Institute of Chemistry, International Centre for Chemical and Biological Sciences, University of Karachi, Karachi, Pakistan,* [b]*Department of Chemistry, University of Sargodha, Sargodha, Pakistan,* [c]*National Centre of Excellence in Analytical Chemistry, University of Sindh, Jamshoro, Pakistan*

List of abbreviations

AKO apricot kernel oil
FA fatty acid

1 Introduction

Apricot (*Prunus armeniaca*), a member of the Rosaceae family, has been widely cultivated in Mediterranean countries as well as in Russia, Pakistan, the United States, and Iran (Hussain, Gulzar, & Shakir, 2011). There has been a worldwide increase in apricots production due to new plantings in Asia. Apricot dried fruit (Fig. 1) is available throughout the year, while fresh fruit is usually available from May to September (Faqir, Saeed, & Maqam, 2004; Gezer et al., 2011). Apricots can be used as fresh fruit or in dried form. Due to their unique aroma, these can also be employed for the preparation of jams, juice, wine, and sauce (Han, 2001). Most importantly, apricot has been used as a folk medicine for the treatment of different ailments (Otsuka et al., 2005; Panda, 2004). The apricot kernel is a byproduct of apricot fruit and can be eaten as an appetizer (either raw or roasted) (Asma, Kan, & Birhanlı, 2007). Nevertheless, the kernel is especially important for the oleo-chemical industry due to its valuable oil. Apricot kernels can be used for the production of oils, benzaldehyde, and cosmetics. The seed kernels possess antioxidant, antiasthmatic, and antispasmodic abilities (Alpaslan & Hayta, 2006; Durmaz, Karabulut, Topçu, Asiltürk, & Kutlu, 2009). Chinese folk medicine used apricot kernels in the treatment of various diseases (Li, 1997). Crushed shells of the apricot kernel can also be used in water filtration (Aksogan, Basturk, Yuksel, & Akgiray, 2003).

Recently, there has been increased interest from nutritionists in exploring new sources of edible oils enriched with bioactives with nutra-pharmaceutical attributes (Cerchiara et al., 2010; Nehdi, 2011). Agro-wastes such as fruit seeds are discarded by the food processing industry every year in bulk that can be explored as a potential source of oil and valuable minor bioactives (El-Adawy & Taha, 2001). Likewise, apricot kernels, mostly discarded in large quantities, can also be utilized both for the production of oil, so-called apricot kernel oil (AKO) and other valuable byproducts such as oil cake (Liangli, Parry, & Zhou, 2005; Matthaus & Ozcan, 2009). Apricot oilseed cake can be employed in animal or poultry feed. However, it is reported that oilseed cake obtained from the bitter kernel by the mechanical method is toxic to animals and should not be used as animal feed.

Currently, the cosmetic industry is the main utilizer of AKO, but it can also be utilized for edible purposes pertaining to its fatty acids profile and high nutritive values (Durmaz et al., 2009; Turan, Topcu, Karabulut, Vural, & Hayaloglu, 2007). The AKO oil contains around 13% saturated and 86% unsaturated fatty acids (Femenia, Rossello, Mulet, & Cañellas, 1995) including those of 60%–70.9% oleic acid, 20%–30% linoleic acid, 0.08%–0.13% linolenic acid, 4%–4.5% palmitic acid, 1%–1.24% stearic acid, and 0.10%–0.12% arachidic acid (Ramadan, Zayed, Abozid, & Asker, 2011). AKO has the potential for medicinal and pharmaceutical applications due to antimicrobial, antifungal, antioxidant, antiseptic, antiaging, antibacterial, and emollient activities/characteristics. The superior oxidative stability of AKO can be attributed to its higher content of oleic acid (C18:1), antioxidant phenolics, and tocopherols, which contribute to a longer shelf life (Alpaslan & Hayta, 2006; Ramadan et al., 2011).

Cold Pressed Oils. https://doi.org/10.1016/B978-0-12-818188-1.00065-7
Copyright © 2020 Elsevier Inc. All rights reserved.

FIG. 1 Dried apricot fruit (right) and its kernel (left).

2 Extraction and processing of AKO

Vegetable oils from oilseeds can be extracted by different extraction techniques such as cold pressing, solvent (*n*-hexane) extraction, supercritical fluid extraction, enzyme-assisted extraction, or a combination of these methods. On a commercial scale, vegetable oils are usually produced by organic solvent extraction as this process recovered high yield. However, cold pressing is now becoming more popular as this process does not need any organic solvents or heat and enables the retention of appreciable contents of minor bioactive components phenolics, phytosterols, free radical scavengers, and tocopherols in the recovered oils (Teh & Birch, 2013). In terms of oil yield, the efficiency of this technique is lower than that of solvent extraction (Siger, Nogala-Kalucka, & Lampart-Szczapae, 2007). AKO extracted using cold pressing technique consists of 92%–98% lipids as major constituent as well as phytosterols (β-sitosterol), β-carotene, and tocopherols as a minor fraction. The oil yield from apricot kernels ranged from 28% to 56% (Femenia et al., 1995; Matthaus, Ozcan, & Juhaimi, 2016). Due to the high amount of unsaturated FAs (91%–92%) and low amount of saturated FAs (7%–8%) in cold pressed AKO, it can be utilized as an edible oil (Uluata, 2016). According to Gezer et al. (2011), the oil contents recovered from different cultivars of apricot ranged from 28% to 42%. Ozcan, Ozalp, Unver, Arslan, and Dursun (2010) reported the oil yield from 42.2% to 57.2% for different varieties of apricot kernels using a simple conventional extraction technique.

3 Fatty acids profile of cold pressed AKO

FAs are vital for human health, especially the essential FAs, which play a key role in growth, maintenance, and physiological processes. Overall, nine fatty acids have been identified so far in AKO including minor quantities of arachidic and eicosanoic acids. AKO prepared using different techniques such as baking (80°C), sun-drying, cold pressing, and pressing at 40°C and 120°C contained oleic acid (51.0%–83.3%), linoleic acid (9.6%–45.9%), palmitic acid (3.2%–19%), and tocopherols (300–900 mg/kg of oil) as major components (Femenia et al., 1995). Both unsaturated FAs and natural antioxidants like tocopherols have positive effects on human health.

The main FAs reported by Ozcan et al. (2010) in AKO were oleic (C18:1), linoleic (C18:2), and palmitic (C16:0) acids. The level of oleic acid in AKO varied from 53.06% to 70.90%, while the amount of C18:2 was detected in the range of 21.43%–35.67%. According to Gandhi, Mulky, Mukerji, Iyer, and Cherian (1997), AKO is rich in unsaturated FAs (94.4%), especially C18:1 (66.2%) and C18:2 (28.2%). Fatty acid compositions (%) of AKO reported by different researchers using different extraction techniques are given in Table 1.

4 Minor bioactive lipids in cold pressed AKO

In addition to containing an impressive amount of unsaturated FAs (C18:1 and C18:2), AKO is also a good source of minor bioactive components including those of tocols, phenolics, and carotenoids with antioxidant potential (Bozan & Temelli, 2008; Waraho, McClements, & Decker, 2011). It is well accepted that phenolics natural antioxidants effectively contribute toward improved oxidative stability and sensory and nutraceutical characteristics of oils (Decker, 2002; Qadir, Anwar, Batool, Mushtaq, & Jabbar, 2019; Qadir, Anwar, Gilani, et al., 2019). Rudzinska, Górnaś, Raczyk, and Soliven (2017) demonstrated in their study that AKO is a good source of various bioactive such as squalene (12–44 mg/100 g), phytosterols

TABLE 1 Fatty acid composition (%) of AKO extracted using different extraction techniques.

C16:0 (palmitic acid)	C16:1 (palmitoleic acid)	C18:0 (stearic acid)	C18:1 (oleic acid)	C18:2 (linoleic acid)	C18:3 (linolenic acid)	C20:1 (11-eicosanoic)	Extraction technique	Reference
19	1.50	3.50	83.3	45.6	2.0	0.18	Solvent extraction	Ozcan et al. (2010)
6.1	0.60	1.20	70.4	21.7	0.14	–	Cold pressing	Uluata (2016)
5.0	–	0.30	71.8	27.7	–	–	Solvent extraction	Filsoof, Mehran, and Farrohi (1976)
4.6	–	3.60	66.7	21.8	21.8	–	Cold pressing	Velickovska et al. (2018)
5.93	0.71	1.68	80.9	30.3	–	–	Solvent extraction	Manzoor, Anwar, Ashraf, and Alkharfy (2012)
6.0	–	1.20	65.7	28.5	1.00	0.20	Solvent extraction	Burnett et al. (2017)
8.8	1.20	1.40	56.5	31.7	0.20	–	Solvent extraction	Ogihara, Itah, and Tsuyuki (1982)
7.31	1.95	2.81	51.4	32.3	0.14	0.10	Solid-state fermentation	Dulf, Vodnar, Dulf, and Pintea (2017)
4.37	0.12	0.46	66.2	28.6	0.12	–	Solvent extraction	Abd El-Aal, Khalil, and Rahma (1986)

(215–970 mg/100 g), β-sitosterol (76%–86% of total sterols), tocochromanols (0.15–0.53 mg/100 g), γ-tocopherol (91%–94%), carotenoids (78.8–258.5 mg/100 g), β-cryptoxanthin, β-carotene, lutein, and zeaxanthin (76–94 mg/100 g). In particular, β-sitosterol was found to be the major phytosterol in different varieties of AKO. Ramadan et al. (2011) identified minute quantities (<35 mg/kg) of stigmasterol, Δ5, 24-stigmastanol, and Δ7-stigmastanol in cold pressed AKO along with other minor bioactives. In another study by Velickovska, Brühl, Mitrev, Mirhosseini, and Matthaus (2015), similar sterols were quantified in AKO, wherein the total sterol content was determined to be 4243.9 mg/kg of total AKO. Out of different sterols detected, β-sitosterol was found to be the major sterol, followed by Δ5-avenasterol (336.1 mg/kg) and campesterol (199.17 mg/kg), whereas Δ7-campesterol was recorded at the lowest amount (4.20 mg/kg).

Tocopherols have a vital role in human health with regard to reducing oxidative stress and in protecting lipids and lipid-containing foodstuffs from oxidation damage during storage, thus prolonging their shelf life. Tocopherol is basically a lipid-soluble antioxidant. In AKO, γ-tocopherol was found to be the major tocopherol isomer followed by α-tocopherol, while δ- and β-tocopherols were determined in lower quantities (Pavlovicc et al., 2018). Interestingly, tocopherols were found to be more in cold pressed oil than in supercritical fluid extracted oil. Tocopherols contents (mg/kg oil) of AKO extracted using different extraction techniques are presented in Table 2.

Apricot kernel oil contains antioxidants such as flavonoids and phenolic acids. Four different methods (DPPH·, ABTS, ORAC, and TPC assays) were used for assessment of the antioxidant potential of AKO (Yi, Yi, & Mavi, 2009). In another study, the total phenolic substances (total phenolics and flavonoids) in AKO were established to be 37.95 mg/kg of oil (gallic acid equivalent) and 19.17 mg/kg of oil (luteolin equivalent), respectively (Velickovska et al., 2015). The fresh and dried apricot fruits of apricot have been characterized by the presence of notable amounts of phenolics such as chlorogenic acid proanthocyanidin (dimers and trimers), kaempferol glycosides, quercetin, and cyanidin 3-glucoside, among others. Meanwhile, the total carotenoids concentration in AKO of different cultivars ranged from 0.15 to 0.53 mg/100 g

TABLE 2 Tocols content (mg/kg oil) of AKO extracted using different extraction techniques.

α-Tocopherol	β-Tocopherol	γ-Tocopherol	δ-Tocopherol	β-Tocotrienol	Extraction technique	Reference
43.5	10	550	60.2	7.5	Cold pressing	Pavlovicc et al. (2018)
27.4	11.3	318.9	17.2	–	Cold pressing	Uluata (2016)
11.6	0.1	237	8.4	0.1	Solvent extraction	Gornas et al. (2015)
40.4	–	520.8	60.20	–	Solvent extraction	Manzoor et al. (2012)
18.1	0.32	563.4	18.94	–	Solvent extraction	Turan et al. (2007)
2.8	–	67.3	2.2	–	Solvent extraction	Matthaus et al. (2016)

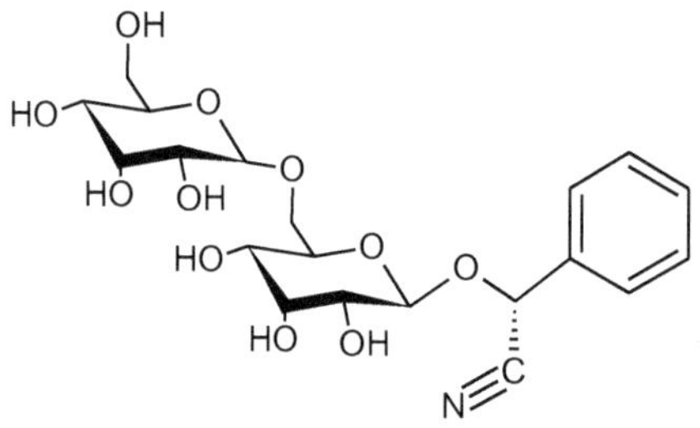

OH
HO O
O
HO OH
HO O
O
HO OH
N≡C

FIG. 2 Chemical structure of amygdalin.

of oil. Overall, β-carotene, lutein, β-cryptoxanthin, and zeaxanthin comprised 76%–94% of the total carotenoids (Gornas et al., 2015).

As a member of the Rosaceae family, the apricot comprises a significant quantity of amygdalin in seeds. The amygdalin holds a cyanide substituent (R-CN) that can release toxic HCN and beta-glycoside when in contact with water as a plant defense system (Fig. 2). Apricot seeds contain amygdalin ranging from 3% to 12% (Al-Bakri, 2010; Femenia et al., 1995). Amygdalin below 5 mg/kg in AKO is considered safe, while in refined AKO, it ranges from 0.17 to 2.15 mg/kg oil.

5 Organoleptic traits and cold pressed AKO as an ingredient in food/nonfood products

Consumers acknowledge the superior characteristic taste and aroma of cold pressed oils compared to solvent-extracted oil. Likewise, a considerable amount of monounsaturated oleic acid (>60%) and a low percentage of polyunsaturated linoleic acid (<30%) make AKO suitable for cooking as well as deep frying purposes. As far as amygdalin-related toxicity is concerned, toxicology tests showed no toxic effect of AKO (Ramadan et al., 2011). In view of its physicochemical attributes, AKO is not only valuable for edible purposes, but can also be employed as an ingredient of lubricants and cosmetics, as well as surfactants. AKO can be employed as the major constituent in cosmetic products and is suitable for all skin types including old, dried, and irritated skin. It is equally useful in food and cosmetology, and equally valuable for high-quality toilet soaps, lip balms, and creams (Zhang, Wang, Yuan, Yang, & Liu, 2016).

6 Health-promoting traits of cold pressed AKO constituents

The medicinal potential of AKO can be due partly to the high proportion of unsaturated FAs, squalene, and phytosterols; special physiological benefits can also be linked to carotenoids. AKO has been valued for pharmaceutical preparations and has notable antimicrobial, antifungal, antioxidant, antiseptic, antioxidant, antiaging, and emollient characteristics (Yi et al.,

2009). In particular, unsaturated FAs and antioxidant bioactives in AKO are valuable in terms of imparting medicinal benefits. Such biologically active substances in AKO are important to prevent cardiovascular diseases and reduce plasma cholesterol levels. Apricot oil has been used in both oriental and European traditional medicine. From the perspective of nutra-cosmeceuticals, estrogens are detected in AKO, which has an antiaging effect on the skin (Stenson et al., 2014). Beta-estradiol (E2), present in AKO, enhances collagen in sun-protected skin (Elsner & Maibach, 2000; Rittie, Kang, Voorhees, & Fisher, 2008). In the light of an impressive FAs and bioactives profile, AKO can be explored as an important ingredient for nutra-pharmaceutical and cosmeceutical applications.

7 Other issues

In addition to the fatty acids profile and degree of unsaturation, the concentration of natural antioxidants, postharvest processing, and storage conditions affect the oxidation state of vegetable oils. Generally, the oil has an elevated amount of polyunsaturated fatty acids that have a higher susceptibility to oxidative deterioration. Data show that AKO contains a high amount of unsaturated FAs with oleic acid (with relatively better oxidative stability) as a major ingredient. Moreover, antioxidant compounds such as phenolics present in AKO make an effective contribution toward stability, sensory, and nutritional values, and prevent lipid oxidation (Uluata, 2016). It has been reported that cold pressed sweet and bitter AKO possesses good oxidative stability and shelf life with induction periods of 15.6 and 11.8 h, respectively. The relatively high oxidative stability of AKO can be attributed partly to the high content of oleic acid and antioxidant compounds (tocols and phenolics) that can effectively contribute to oxidative stability of this oil (Matthaus & Ozcan, 2009; Sies & Murphy, 1991). Conclusively, studies reveal that AKO is nutritionally beneficial for human health due to its favorable fatty acid composition and biologically active components. It has also been proven that cold pressing in addition to medicinal potential may lead to an enhanced flavor which is of commercial value.

References

Abd El-Aal, M. H., Khalil, M. K. M., & Rahma, E. H. (1986). Apricot kernel oil: Characterization, chemical composition and utilization in some baked products. *Food Chemistry*, *19*, 287–298.

Aksogan, S., Basturk, A., Yuksel, E., & Akgiray, O. (2003). On the use of crushed shells of apricot as the upper layer in dual media filters. *Water Science and Technology*, *48*, 497–503.

Al-Bakri, S. A. (2010). Antibacterial activity of apricot kernel extract containing amygdalin. *Iraqi Journal of Science*, *51*(4), 571–576.

Alpaslan, M., & Hayta, M. (2006). Apricot kernel: Physical and chemical properties. *Journal of the American Oil Chemists Society*, *83*, 469.

Asma, B. M., Kan, T., & Birhanlı, O. (2007). Characterization of promising apricot (*Prunus armeniaca* L.) genetic resources in Malatya, Turkey. *Genetic Resources and Crop Evolution*, *54*, 205–212.

Bozan, B., & Temelli, F. (2008). Chemical composition and oxidative stability of flax, safflower, poppy seed and seed oils. *Bioresource Technology*, *99* (14), 6354–6359.

Burnett, C. L., Fiume, M. M., Bergfeld, W. F., Belsito, D. V., Hill, R. A., Klaassen, C. D., et al. (2017). Safety assessment of plant-derived fatty acid oils. *International Journal of Toxicology*, *36*(3), 51–129.

Cerchiara, T., Chidichimo, G., Ragusa, M. I., Belsito, E. L., Liguori, A., & Arioli, A. (2010). Characterization and utilization of Spanish broom (*Spartium junceum* L.) seed oil. *Industrial Crops and Production*, *31*, 423–426.

Decker, E. A. (2002). Antioxidant mechanism. In C. C. Akoh, & D. B. Min (Eds.), *Food lipids: Chemistry, nutrition and biotechnology* (pp. 475–492p). New York: Marcel Dekker.

Dulf, V. F., Vodnar, D. C., Dulf, E. H., & Pintea, A. (2017). Phenolic compounds, flavonoids, lipids and antioxidant potential of apricot (*Prunus armeniaca* L.) pomace fermented by two filamentous fungal strains in solid state system. *Chemistry Central Journal*, *11*(92), 1–10.

Durmaz, G., Karabulut, İ., Topçu, A., Asiltürk, M., & Kutlu, T. (2009). Roasting-related changes in oxidative stability and antioxidant capacity of apricot kernel oil. *Journal of the American Oil Chemists Society*, *87*(4), 401–409.

El-Adawy, T. A., & Taha, K. M. (2001). Characterization and composition of different seed oils and flours. *Food Chemistry*, *74*, 47–54.

Elsner, P., & Maibach, H. I. (2000). Cosmeceuticals: Drugs vs. cosmetics. *Cosmetics science and technology series* (pp. 1–338). Vol. 23(pp. 1–338). New York, NY: Marcel Dekker, Inc.

Faqir, M. A., Saeed, A., & Maqam, D. (2004). Storage effect on physicochemical and sensory characteristics of dried apricot jam. *Pakistan Journal of Food Science*, *14*, 43–47.

Femenia, A., Rossello, C., Mulet, A., & Cañellas, J. (1995). Chemical composition of bitter and sweet apricot kernels. *Journal of Agriculture and Food Chemistry*, *43*, 356.

Filsoof, M., Mehran, M., & Farrohi, F. (1976). Determination and comparison of oil characteristics in Iranian almond, apricot and peach nuts. *Fette, Seifen, Anstrichmittel*, *78*, 150–151.

Gandhi, V. M., Mulky, M. J., Mukerji, B., Iyer, V. J., & Cherian, K. M. (1997). Safety evaluation of wild apricot oil. *Food Chemistry*, *35*, 583–587.

Gezer, I., Haciseferogullari, H., Ozcan, M. M., Arslan, D., Asma, B. M., & Unver, A. (2011). Physico-chemical properties of apricot kernel. South Western South Western. *Journal of Horticulture, Biology, and Environment*, *2*, 1–13.

Gornas, P., Misina, I., Grāvīte, I., Soliven, A., Kaufmane, E., & Segliņa, D. (2015). Tocochromanols composition in kernels recovered from different apricot varieties: RP-HPLC/FLD and RP-UPLC-ESI/MSn study. *Natural Products Research*, *29*(13), 1222–1227.

Han, Z. (2001). *Fruit wine continuous production*. Patent Num CN1172851-A.

Hussain, I., Gulzar, S., & Shakir, I. (2011). Physico-chemical properties of bitter and sweet apricot kernel flour and oil from north of Pakistan. *International Journal of Food Safety*, *13*, 11–15.

Li, S. (1997). *Feiyangling medicine for curing infantile virus pneumonia*. Patent Num CN1105570-A and CN1048882-C.

Liangli, Y., Parry, J. W., & Zhou, K. (2005). Oils from herbs, spices, and fruit seeds. F. Shahidi, & A. E. Bailey (Eds.), *Bailey's industrial oil and fat product* (pp. 233–258). Vol. 3(pp. 233–258). Hoboken, NJ: John Wiley & Sons, Inc.

Manzoor, M., Anwar, F., Ashraf, M., & Alkharfy, K. M. (2012). Physico-chemical characteristics of seed oils extracted from different apricot (*Prunus armeniaca* L.) varieties from Pakistan. *Grasas y Aceites*, *63*(2), 193–201.

Matthaus, B., & Ozcan, M. M. (2009). Fatty acids and tocopherol contents of some *Prunus* spp. kernel oil. *Journal of Food Lipids*, *16*, 187–199.

Matthaus, B., Ozcan, M. M., & Juhaimi, F. (2016). Fatty acid composition and tocopherol content of the kernel oil from apricot varieties (Hasanbey, Hacihaliloglu, Kabaasi, and Soganci) collected at different harvest times. *European Food Research and Technology*, *1*(242), 221.

Nehdi, I. A. (2011). Characteristics and composition of *Washingtonia filifera* (Linden ex André) H. Wendl. seed and seed oil. *Food Chemistry*, *126*, 197–202.

Ogihara, H., Itah, S., & Tsuyuki, H. (1982). Studies on lipids of ume apricot. *Journal of the Japanese Society for Food Science and Technology*, *29*, 221–227.

Otsuka, T., Tsukamoto, T., Tanaka, H., Inada, K., Utsunomiya, H., et al. (2005). Suppressive effects of fruit-juice concentrate of *Prunus mume* Sieb. et Zucc. (Japanese apricot, Ume) on *Helicobacter pylori*-induced glandular stomach lesions in Mongolian gerbils. *Asian Pacific Journal of Cancer Prevention*, *6*, 337–341.

Ozcan, M. M., Ozalp, C., Unver, A., Arslan, D., & Dursun, N. (2010). Properties of apricot kernel and oils as fruit juice processing waste. *Food and Nutrition Sciences*, *1*, 31–37.

Panda, H. (2004). *Herbal food and its medicinal value*: (p. 182). Kamla Nagar, Delhi, India: National Institute of Industrial Research.

Pavlovicc, N., Vidović, S., Vladić, J., Popović, L., Moslavac, T., Jaković, S., et al. (2018). Recovery of tocopherols, amygdalin and fatty acids from apricot kernel oil: Cold pressing vs. supercritical carbon dioxide. *European Journal of Lipid Science and Technology*, 1–29.

Qadir, R., Anwar, F., Batool, F., Mushtaq, M., & Jabbar, A. (2019). Enzyme-assisted extraction of *Momordica balsamina* L. fruit phenolics: Process optimized by response surface methodology. *Journal of Food Measurement and Characterization*, *13*, 697–706.

Qadir, R., Anwar, F., Gilani, M. A., Zahoor, S., Rehman, M., & Mustaqeem, M. (2019). RSM/ANN based optimized recovery of phenolics from mulberry leaves by enzyme-assisted extraction. *Czech Journal of Food Sciences*, *37*(2), 99–105.

Ramadan, M. F., Zayed, R., Abozid, M., & Asker, M. M. S. (2011). Apricot and pumpkin oils reduce plasma cholesterol and triacylglycerol concentrations in rats fed a high-fat diet. *Grasas* y *Aceites*, *62*(4), 443–452.

Rittie, L., Kang, S., Voorhees, J. J., & Fisher, G. J. (2008). Induction of collagen by estradiol: Difference between sun-protected and photodamaged human skin *in vivo*. *Archives of Dermatological Research*, *144*(9), 1129–1140.

Rudzinska, M., Górnaś, P., Raczyk, M., & Soliven, A. (2017). Sterols and squalene in apricot (*Prunus armeniaca* L.) kernel oils: The variety as a key factor. *Natural Products Research*, *31*(1), 1–12.

Sies, H., & Murphy, M. E. (1991). Role of tocopherols in protection of biological systems against oxidative damage. *Journal of Photochemistry and Photobiology, (B)*, *8*, 211–224.

Siger, A., Nogala-Kalucka, M., & Lampart-Szczapae, E. (2007). The content and antioxidant activity of phenolic compounds in cold-pressed plant oils. *Journal of Lipids*, *15*, 137–149.

Stenson, P. D., Mort, M., Ball, E. V., Shaw, K., Phillips, A., & Cooper, D. N. (2014). The human gene mutation database: Building a comprehensive mutation repository for clinical and molecular genetics, diagnostic testing and personalized genomic medicine. *Journal of Human Genetics*, *133*, 1–9.

Teh, S. S., & Birch, J. (2013). Physicochemical and quality characteristics of cold-pressed hemp, flax and canola seed oils. *Journal of Food Composition Analysis*, *3*, 20–26.

Turan, S., Topcu, A., Karabulut, I., Vural, H., & Hayaloglu, A. A. (2007). Fatty acid, triacylglycerol, phytosterol, and tocopherol variations in kernel oil of Malatya apricots from Turkey. *Journal of Agriculture and Food Chemistry*, *55*, 10787–10794.

Uluata, S. (2016). Effect of extraction method on biochemical properties and oxidative stability of apricot seed oil. *Akademik Gıda*, *14*(4), 333–340.

Velickovska, S. K., Brühl, L., Mitrev, S., Mirhosseini, H., & Matthaus, B. (2015). Quality evaluation of cold-pressed edible oils from Macedonia. *European Journal of Lipid Science and Technology*, *117*(12), 2023–2025.

Velickovska, S. K., Mot, A. C., Mitrev, S., Gulaboski, R., Bruhl, L., Mirhosseini, H., et al. (2018). Bioactive compounds and ''in vitro'' antioxidant activity of some traditional and non-traditional cold-pressed edible oils from Macedonia. *Journal of Food Science and Technology*, *55*(5), 1614–1623.

Waraho, T., McClements, D. J., & Decker, E. A. (2011). Mechanisms of lipid oxidation in food dispersions. *Trends in Food Science and Technology*, *22*(1), 3–13.

Yi, D., Yi, N., & Mavi, A. (2009). Antioxidant and antimicrobial activities of bitter and sweet apricot (*Prunus armeniaca* L.) kernels. *Brazilian Journal of Medical and Biological Research*, *42*(4), 346–352.

Zhang, W. C., Wang, R., Yuan, Y. H., Yang, T. K., & Liu, S. Q. (2016). Changes in volatiles of palm kernel oil before and after kernel roasting. *LWT: Food Science and Technology*, *73*, 432–441.

Index

Note: Page numbers followed by *f* indicate figures and *t* indicate tables.

B

C

D

E

N

O

P

T

U

V

W